ICC

INTERNATIONAL
CODE COUNCIL®

INTERNATIONAL
BUILDING
CODE®

2003

2003 International Building Code®

First Printing: December 2002
Second Printing: June 2003
Third Printing: November 2003

ISBN # 1-892395-56-8 (soft)
ISBN # 1-892395-55-X (loose-leaf)
ISBN # 1-892395-79-7 (e-document)

COPYRIGHT © 2002
by
INTERNATIONAL CODE COUNCIL, INC.

PREFACE

Introduction

Internationally, code officials recognize the need for a modern, up-to-date building code addressing the design and installation of building systems through requirements emphasizing performance. The *International Building Code®*, in this 2003 edition, is designed to meet these needs through model code regulations that safeguard the public health and safety in all communities, large and small.

This comprehensive building code establishes minimum regulations for building systems using prescriptive and performance-related provisions. It is founded on broad-based principles that make possible the use of new materials and new building designs. This 2003 edition is fully compatible with all the International Codes ("I-Codes") published by the International Code Council (ICC), including the ICC *Electrical Code, International Energy Conservation Code, International Existing Building Code, International Fire Code, International Fuel Gas Code, International Mechanical Code,* ICC *Performance Code, International Plumbing Code, International Private Sewage Disposal Code, International Property Maintenance Code, International Residential Code, International Urban-Wildland Interface Code* and *International Zoning Code.*

The *International Building Code* provisions provide many benefits, among which is the model code development process that offers an international forum for building professionals to discuss performance and prescriptive code requirements. This forum provides an excellent arena to debate proposed revisions. This model code also encourages international consistency in the application of provisions.

Development

The first edition of the *International Building Code* (2000) was the culmination of an effort initiated in 1997 by the ICC. This included five drafting subcommittees appointed by ICC and consisting of representatives of the three statutory members of the International Code Council: Building Officials and Code Administrators International, Inc. (BOCA), International Conference of Building Officials (ICBO) and Southern Building Code Congress International (SBCCI). The intent was to draft a comprehensive set of regulations for building systems consistent with and inclusive of the scope of the existing model codes. Technical content of the latest model codes promulgated by BOCA, ICBO and SBCCI was utilized as the basis for the development, followed by public hearings in 1997, 1998 and 1999 to consider proposed changes. This 2003 edition presents the code as originally issued, with changes approved through the ICC Code Development Process through 2002. A new edition such as this is promulgated every three years.

With the development and publication of the family of International Codes in 2000, the continued development and maintenance of the model codes individually promulgated by BOCA ("BOCA National Codes"), ICBO ("Uniform Codes") and SBCCI ("Standard Codes") was discontinued. This 2003 *International Building Code,* as well as its predecessor—the 2000 edition, is intended to be the successor building code to those codes previously developed by BOCA, ICBO and SBCCI.

The development of a single set of comprehensive and coordinated family of International Codes was a significant milestone in the development of regulations for the built environment. The timing of this publication mirrors a milestone in the change in structure of the model codes, namely, the pending Consolidation of BOCA, ICBO and SBCCI into the ICC. The activities and services previously provided by the individual model code organizations will be the responsibility of the Consolidated ICC.

This code is founded on principles intended to establish provisions consistent with the scope of a building code that adequately protects public health, safety and welfare; provisions that do not unnecessarily increase construction costs; provisions that do not restrict the use of new materials, products or methods of construction; and provisions that do not give preferential treatment to particular types or classes of materials, products or methods of construction.

Adoption

The *International Building Code* is available for adoption and use by jurisdictions internationally. Its use within a governmental jurisdiction is intended to be accomplished through adoption by reference in accordance with proceedings establishing the jurisdiction's laws. At the time of adoption, jurisdictions should insert the appropriate information in provisions requiring specific local information, such as the name of the adopting jurisdiction. These locations are shown in bracketed words in small capital letters in the code and in the sample ordinance. The sample adoption ordinance on page v addresses several key elements of a code adoption ordinance, including the information required for insertion into the code text.

Maintenance

The *International Building Code* is kept up to date through the review of proposed changes submitted by code enforcing officials, industry representatives, design professionals and other interested parties. Proposed changes are carefully considered through an open code development process in which all interested and affected parties may participate.

The contents of this work are subject to change both through the Code Development Cycles and the governmental body that enacts the code into law. For more information regarding the code development process, contact the Code and Standard Development Department of the International Code Council.

While the development procedure of the *International Building Code* assures the highest degree of care, ICC and the founding members of ICC—BOCA, ICBO, SBCCI—their members and those participating in the development of this code do not accept any liability resulting from compliance or noncompliance with the provisions because ICC and its founding members do not have the power or authority to police or enforce compliance with the contents of this code. Only the governmental body that enacts the code into law has such authority.

Letter Designations in Front of Section Numbers

In each code development cycle, proposed changes to this code are considered at the Code Development Hearing by the International Building Code Development Committee, whose action constitutes a recommendation to the voting membership for final action on the proposed change. Proposed changes to a code section whose number begins with a letter in brackets are considered by a different code development committee. For instance, proposed changes to code sections which have the letter [F] in front (e.g., [F] 1001.3), are considered by the International Fire Code Development Committee at the Code Development Hearing. Where this designation is applicable to the entire content of a main section of the code, the designation appears at the main section number and title and is not repeated at every subsection in that section.

The content of sections in this code which begin with a letter designation is maintained by another code development committee in accordance with the following: [E] = International Energy Conservation Code Development Committee; [EB] = International Existing Building Code Development Committee; [EL] = ICC Electrical Code Development Committee; [F] = International Fire Code Development Committee; [M] = International Mechanical Code Development Committee; [P] = International Plumbing Code Development Committee; [PC] = ICC Performance Code Development Committee; [PM] = International Property Maintenance Code Development Committee; [RBE] = International Residential Code Building and Energy Development Committee; [RMP] = International Residential Code Mechanical/Plumbing Development Committee; [UW] = International Urban-Wildland Interface Code Development Committee; and [Z] = International Zoning Code Development Committee.

Marginal Markings

Solid vertical lines in the margins within the body of the code indicate a technical change from the requirements of the 2000 edition. Deletion indicators (➡) are provided in the margin where a paragraph or item has been deleted.

Chapter 10 user note: Chapter 10 of the code has been reorganized from the 2000 edition as a result of an approved code change proposal. This resulted in a renumbering of the chapter from nine sections to 25. The presentation of text predominantly follows that of the 2000 edition; however, the section numbers have been revised. Marginal markings are included at each section number but have not been included to reflect the subsection renumbering. A comprehensive 2000/2003 Chapter 10 section number cross index is posted on the ICC website at www.intlcode.org.

ORDINANCE

The *International Codes* are designed and promulgated to be adopted by reference by ordinance. Jurisdictions wishing to adopt the 2003 *International Building Code* as an enforceable regulation governing structures and premises should ensure that certain factual information is included in the adopting ordinance at the time adoption is being considered by the appropriate governmental body. The following sample adoption ordinance addresses several key elements of a code adoption ordinance, including the information required for insertion into the code text.

SAMPLE ORDINANCE FOR ADOPTION OF THE INTERNATIONAL BUILDING CODE ORDINANCE NO._____

An ordinance of the **[JURISDICTION]** adopting the 2003 edition of the *International Building Code*, regulating and governing the conditions and maintenance of all property, buildings and structures; by providing the standards for supplied utilities and facilities and other physical things and conditions essential to ensure that structures are safe, sanitary and fit for occupation and use; and the condemnation of buildings and structures unfit for human occupancy and use and the demolition of such structures in the **[JURISDICTION]**; providing for the issuance of permits and collection of fees therefor; repealing Ordinance No. _____ of the **[JURISDICTION]** and all other ordinances and parts of the ordinances in conflict therewith.

The **[GOVERNING BODY]** of the **[JURISDICTION]** does ordain as follows:

Section 1. That a certain document, three (3) copies of which are on file in the office of the **[TITLE OF JURISDICTION'S KEEPER OF RECORDS]** of **[NAME OF JURISDICTION]**, being marked and designated as the *International Building Code*, 2003 edition, including Appendix Chapters **[FILL IN THE APPENDIX CHAPTERS BEING ADOPTED]** (see *International Building Code* Section 101.2.1, 2003 edition), as published by the International Code Council, be and is hereby adopted as the Building Code of the **[JURISDICTION]**, in the State of **[STATE NAME]** for regulating and governing the conditions and maintenance of all property, buildings and structures; by providing the standards for supplied utilities and facilities and other physical things and conditions essential to ensure that structures are safe, sanitary and fit for occupation and use; and the condemnation of buildings and structures unfit for human occupancy and use and the demolition of such structures as herein provided; providing for the issuance of permits and collection of fees therefor; and each and all of the regulations, provisions, penalties, conditions and terms of said Building Code on file in the office of the **[JURISDICTION]** are hereby referred to, adopted, and made a part hereof, as if fully set out in this ordinance, with the additions, insertions, deletions and changes, if any, prescribed in Section 2 of this ordinance.

Section 2. The following sections are hereby revised:

Section 101.1. Insert: **[NAME OF JURISDICTION]**

Section 1612.3. Insert: **[NAME OF JURISDICTION]**

Section 1612.3. Insert: **[DATE OF ISSUANCE]**

Section 3410.2. Insert: **[DATE IN ONE LOCATION]**

Section 3. That Ordinance No. _____ of **[JURISDICTION]** entitled **[FILL IN HERE THE COMPLETE TITLE OF THE ORDINANCE OR ORDINANCES IN EFFECT AT THE PRESENT TIME SO THAT THEY WILL BE REPEALED BY DEFINITE MENTION]** and all other ordinances or parts of ordinances in conflict herewith are hereby repealed.

Section 4. That if any section, subsection, sentence, clause or phrase of this ordinance is, for any reason, held to be unconstitutional, such decision shall not affect the validity of the remaining portions of this ordinance. The **[GOVERNING BODY]** hereby declares that it would have passed this ordinance, and each section, subsection, clause or phrase thereof, irrespective of the fact that any one or more sections, subsections, sentences, clauses and phrases be declared unconstitutional.

Section 5. That nothing in this ordinance or in the Building Code hereby adopted shall be construed to affect any suit or proceeding impending in any court, or any rights acquired, or liability incurred, or any cause or causes of action acquired or existing, under any act or ordinance hereby repealed as cited in Section 2 of this ordinance; nor shall any just or legal right or remedy of any character be lost, impaired or affected by this ordinance.

Section 6. That the **[JURISDICTION'S KEEPER OF RECORDS]** is hereby ordered and directed to cause this ordinance to be published. (An additional provision may be required to direct the number of times the ordinance is to be published and to specify that it is to be in a newspaper in general circulation. Posting may also be required.)

Section 7. That this ordinance and the rules, regulations, provisions, requirements, orders and matters established and adopted hereby shall take effect and be in full force and effect **[TIME PERIOD]** from and after the date of its final passage and adoption.

TABLE OF CONTENTS

CHAPTER 1
ADMINISTRATION

SECTION 101
GENERAL

101.1 Title. These regulations shall be known as the *Building Code* of [NAME OF JURISDICTION], hereinafter referred to as "this code."

101.2 Scope. The provisions of this code shall apply to the construction, alteration, movement, enlargement, replacement, repair, equipment, use and occupancy, location, maintenance, removal and demolition of every building or structure or any appurtenances connected or attached to such buildings or structures.

Exceptions:

1. Detached one- and two-family dwellings and multiple single-family dwellings (town houses) not more than three stories above grade plane in height with a separate means of egress and their accessory structures shall comply with the *International Residential Code.*

2. Existing buildings undergoing repair, alterations or additions and change of occupancy shall be permitted to comply with the *International Existing Building Code.*

101.2.1 Appendices. Provisions in the appendices shall not apply unless specifically adopted.

101.3 Intent. The purpose of this code is to establish the minimum requirements to safeguard the public health, safety and general welfare through structural strength, means of egress facilities, stability, sanitation, adequate light and ventilation, energy conservation, and safety to life and property from fire and other hazards attributed to the built environment and to provide safety to fire fighters and emergency responders during emergency operations.

101.4 Referenced codes. The other codes listed in Sections 101.4.1 through 101.4.7 and referenced elsewhere in this code shall be considered part of the requirements of this code to the prescribed extent of each such reference.

101.4.1 Electrical. The provisions of the ICC *Electrical Code* shall apply to the installation of electrical systems, including alterations, repairs, replacement, equipment, appliances, fixtures, fittings and appurtenances thereto.

101.4.2 Gas. The provisions of the *International Fuel Gas Code* shall apply to the installation of gas piping from the point of delivery, gas appliances and related accessories as covered in this code. These requirements apply to gas piping systems extending from the point of delivery to the inlet connections of appliances and the installation and operation of residential and commercial gas appliances and related accessories.

101.4.3 Mechanical. The provisions of the *International Mechanical Code* shall apply to the installation, alterations, repairs and replacement of mechanical systems, including equipment, appliances, fixtures, fittings and/or appurte-nances, including ventilating, heating, cooling, air-conditioning and refrigeration systems, incinerators and other energy-related systems.

101.4.4 Plumbing. The provisions of the *International Plumbing Code* shall apply to the installation, alteration, repair and replacement of plumbing systems, including equipment, appliances, fixtures, fittings and appurtenances, and where connected to a water or sewage system and all aspects of a medical gas system. The provisions of the *International Private Sewage Disposal Code* shall apply to private sewage disposal systems.

101.4.5 Property maintenance. The provisions of the *International Property Maintenance Code* shall apply to existing structures and premises; equipment and facilities; light, ventilation, space heating, sanitation, life and fire safety hazards; responsibilities of owners, operators and occupants; and occupancy of existing premises and structures.

101.4.6 Fire prevention. The provisions of the *International Fire Code* shall apply to matters affecting or relating to structures, processes and premises from the hazard of fire and explosion arising from the storage, handling or use of structures, materials or devices; from conditions hazardous to life, property or public welfare in the occupancy of structures or premises; and from the construction, extension, repair, alteration or removal of fire suppression and alarm systems or fire hazards in the structure or on the premises from occupancy or operation.

101.4.7 Energy. The provisions of the *International Energy Conservation Code* shall apply to all matters governing the design and construction of buildings for energy efficiency.

SECTION 102
APPLICABILITY

102.1 General. Where, in any specific case, different sections of this code specify different materials, methods of construction or other requirements, the most restrictive shall govern. Where there is a conflict between a general requirement and a specific requirement, the specific requirement shall be applicable.

102.2 Other laws. The provisions of this code shall not be deemed to nullify any provisions of local, state or federal law.

102.3 Application of references. References to chapter or section numbers, or to provisions not specifically identified by number, shall be construed to refer to such chapter, section or provision of this code.

102.4 Referenced codes and standards. The codes and standards referenced in this code shall be considered part of the requirements of this code to the prescribed extent of each such reference. Where differences occur between provisions of this code and referenced codes and standards, the provisions of this code shall apply.

102.5 Partial invalidity. In the event that any part or provision of this code is held to be illegal or void, this shall not have the effect of making void or illegal any of the other parts or provisions.

102.6 Existing structures. The legal occupancy of any structure existing on the date of adoption of this code shall be permitted to continue without change, except as is specifically covered in this code, the *International Property Maintenance Code* or the *International Fire Code*, or as is deemed necessary by the building official for the general safety and welfare of the occupants and the public.

SECTION 103
DEPARTMENT OF BUILDING SAFETY

103.1 Creation of enforcement agency. The Department of Building Safety is hereby created and the official in charge thereof shall be known as the building official.

103.2 Appointment. The building official shall be appointed by the chief appointing authority of the jurisdiction.

103.3 Deputies. In accordance with the prescribed procedures of this jurisdiction and with the concurrence of the appointing authority, the building official shall have the authority to appoint a deputy building official, the related technical officers, inspectors, plan examiners and other employees. Such employees shall have powers as delegated by the building official. For the maintenance of existing properties, see the *International Property Maintenance Code.*

SECTION 104
DUTIES AND POWERS OF BUILDING OFFICIAL

104.1 General. The building official is hereby authorized and directed to enforce the provisions of this code. The building official shall have the authority to render interpretations of this code and to adopt policies and procedures in order to clarify the application of its provisions. Such interpretations, policies and procedures shall be in compliance with the intent and purpose of this code. Such policies and procedures shall not have the effect of waiving requirements specifically provided for in this code.

104.2 Applications and permits. The building official shall receive applications, review construction documents and issue permits for the erection, and alteration, demolition and moving of buildings and structures, inspect the premises for which such permits have been issued and enforce compliance with the provisions of this code.

104.3 Notices and orders. The building official shall issue all necessary notices or orders to ensure compliance with this code.

104.4 Inspections. The building official shall make all of the required inspections, or the building official shall have the authority to accept reports of inspection by approved agencies or individuals. Reports of such inspections shall be in writing and be certified by a responsible officer of such approved agency or by the responsible individual. The building official is authorized to engage such expert opinion as deemed necessary to report upon unusual technical issues that arise, subject to the approval of the appointing authority.

104.5 Identification. The building official shall carry proper identification when inspecting structures or premises in the performance of duties under this code.

104.6 Right of entry. Where it is necessary to make an inspection to enforce the provisions of this code, or where the building official has reasonable cause to believe that there exists in a structure or upon a premises a condition which is contrary to or in violation of this code which makes the structure or premises unsafe, dangerous or hazardous, the building official is authorized to enter the structure or premises at reasonable times to inspect or to perform the duties imposed by this code, provided that if such structure or premises be occupied that credentials be presented to the occupant and entry requested. If such structure or premises is unoccupied, the building official shall first make a reasonable effort to locate the owner or other person having charge or control of the structure or premises and request entry. If entry is refused, the building official shall have recourse to the remedies provided by law to secure entry.

104.7 Department records. The building official shall keep official records of applications received, permits and certificates issued, fees collected, reports of inspections, and notices and orders issued. Such records shall be retained in the official records for the period required for retention of public records.

104.8 Liability. The building official, member of the board of appeals or employee charged with the enforcement of this code, while acting for the jurisdiction in good faith and without malice in the discharge of the duties required by this code or other pertinent law or ordinance, shall not thereby be rendered liable personally and is hereby relieved from personal liability for any damage accruing to persons or property as a result of any act or by reason of an act or omission in the discharge of official duties. Any suit instituted against an officer or employee because of an act performed by that officer or employee in the lawful discharge of duties and under the provisions of this code shall be defended by legal representative of the jurisdiction until the final termination of the proceedings. The building official or any subordinate shall not be liable for cost in any action, suit or proceeding that is instituted in pursuance of the provisions of this code.

104.9 Approved materials and equipment. Materials, equipment and devices approved by the building official shall be constructed and installed in accordance with such approval.

> **104.9.1 Used materials and equipment.** The use of used materials which meet the requirements of this code for new materials is permitted. Used equipment and devices shall not be reused unless approved by the building official.

104.10 Modifications. Wherever there are practical difficulties involved in carrying out the provisions of this code, the building official shall have the authority to grant modifications for individual cases, upon application of the owner or owner's representative, provided the building official shall first find that special individual reason makes the strict letter of this code impractical and the modification is in compliance with the intent and purpose of this code and that such modification does not lessen health, accessibility, life and fire safety, or structural requirements. The details of action granting modifications shall be recorded and entered in the files of the department of building safety.

104.11 Alternative materials, design and methods of construction and equipment. The provisions of this code are not intended to prevent the installation of any material or to prohibit any design or method of construction not specifically prescribed by this code, provided that any such alternative has been approved. An alternative material, design or method of construction shall be approved where the building official finds that the proposed design is satisfactory and complies with the intent of the provisions of this code, and that the material, method or work offered is, for the purpose intended, at least the equivalent of that prescribed in this code in quality, strength, effectiveness, fire resistance, durability and safety.

104.11.1 Research reports. Supporting data, where necessary to assist in the approval of materials or assemblies not specifically provided for in this code, shall consist of valid research reports from approved sources.

104.11.2 Tests. Whenever there is insufficient evidence of compliance with the provisions of this code, or evidence that a material or method does not conform to the requirements of this code, or in order to substantiate claims for alternative materials or methods, the building official shall have the authority to require tests as evidence of compliance to be made at no expense to the jurisdiction. Test methods shall be as specified in this code or by other recognized test standards. In the absence of recognized and accepted test methods, the building official shall approve the testing procedures. Tests shall be performed by an approved agency. Reports of such tests shall be retained by the building official for the period required for retention of public records.

SECTION 105
PERMITS

105.1 Required. Any owner or authorized agent who intends to construct, enlarge, alter, repair, move, demolish, or change the occupancy of a building or structure, or to erect, install, enlarge, alter, repair, remove, convert or replace any electrical, gas, mechanical or plumbing system, the installation of which is regulated by this code, or to cause any such work to be done, shall first make application to the building official and obtain the required permit.

105.1.1 Annual permit. In lieu of an individual permit for each alteration to an already approved electrical, gas, mechanical or plumbing installation, the building official is authorized to issue an annual permit upon application therefor to any person, firm or corporation regularly employing one or more qualified tradepersons in the building, structure or on the premises owned or operated by the applicant for the permit.

105.1.2 Annual permit records. The person to whom an annual permit is issued shall keep a detailed record of alterations made under such annual permit. The building official shall have access to such records at all times or such records shall be filed with the building official as designated.

105.2 Work exempt from permit. Exemptions from permit requirements of this code shall not be deemed to grant authori-

zation for any work to be done in any manner in violation of the provisions of this code or any other laws or ordinances of this jurisdiction. Permits shall not be required for the following:

Building:

1. One-story detached accessory structures used as tool and storage sheds, playhouses and similar uses, provided the floor area does not exceed 120 square feet (11.15 m²).

2. Fences not over 6 feet (1829 mm) high.

3. Oil derricks.

4. Retaining walls which are not over 4 feet (1219 mm) in height measured from the bottom of the footing to the top of the wall, unless supporting a surcharge or impounding Class I, II or III-A liquids.

5. Water tanks supported directly on grade if the capacity does not exceed 5,000 gallons (18 925 L) and the ratio of height to diameter or width does not exceed 2 to 1.

6. Sidewalks and driveways not more than 30 inches (762 mm) above grade and not over any basement or story below and which are not part of an accessible route.

7. Painting, papering, tiling, carpeting, cabinets, counter tops and similar finish work.

8. Temporary motion picture, television and theater stage sets and scenery.

9. Prefabricated swimming pools accessory to a Group R-3 occupancy, as applicable in Section 101.2, which are less than 24 inches (610 mm) deep, do not exceed 5,000 gallons (18 925 L) and are installed entirely above ground.

10. Shade cloth structures constructed for nursery or agricultural purposes and not including service systems.

11. Swings and other playground equipment accessory to detached one- and two-family dwellings.

12. Window awnings supported by an exterior wall which do not project more than 54 inches (1372 mm) from the exterior wall and do not require additional support of Group R-3, as applicable in Section 101.2, and Group U occupancies.

13. Movable cases, counters and partitions not over 5 feet 9 inches (1753 mm) in height.

Electrical:

Repairs and maintenance: Minor repair work, including the replacement of lamps or the connection of approved portable electrical equipment to approved permanently installed receptacles.

Radio and television transmitting stations: The provisions of this code shall not apply to electrical equipment used for radio and television transmissions, but do apply to equipment and wiring for power supply, the installations of towers and antennas.

Temporary testing systems: A permit shall not be required for the installation of any temporary system required for the testing or servicing of electrical equipment or apparatus.

Gas:

1. Portable heating appliance.
2. Replacement of any minor part that does not alter approval of equipment or make such equipment unsafe.

Mechanical:

1. Portable heating appliance.
2. Portable ventilation equipment.
3. Portable cooling unit.
4. Steam, hot or chilled water piping within any heating or cooling equipment regulated by this code.
5. Replacement of any part which does not alter its approval or make it unsafe.
6. Portable evaporative cooler.
7. Self-contained refrigeration system containing 10 pounds (4.54 kg) or less of refrigerant and actuated by motors of 1 horsepower (746 W) or less.

Plumbing:

1. The stopping of leaks in drains, water, soil, waste or vent pipe provided, however, that if any concealed trap, drain pipe, water, soil, waste or vent pipe becomes defective and it becomes necessary to remove and replace the same with new material, such work shall be considered as new work and a permit shall be obtained and inspection made as provided in this code.
2. The clearing of stoppages or the repairing of leaks in pipes, valves or fixtures, and the removal and reinstallation of water closets, provided such repairs do not involve or require the replacement or rearrangement of valves, pipes or fixtures.

105.2.1 Emergency repairs. Where equipment replacements and repairs must be performed in an emergency situation, the permit application shall be submitted within the next working business day to the building official.

105.2.2 Repairs. Application or notice to the building official is not required for ordinary repairs to structures, replacement of lamps or the connection of approved portable electrical equipment to approved permanently installed receptacles. Such repairs shall not include the cutting away of any wall, partition or portion thereof, the removal or cutting of any structural beam or load-bearing support, or the removal or change of any required means of egress, or rearrangement of parts of a structure affecting the egress requirements; nor shall ordinary repairs include addition to, alteration of, replacement or relocation of any standpipe, water supply, sewer, drainage, drain leader, gas, soil, waste, vent or similar piping, electric wiring or mechanical or other work affecting public health or general safety.

105.2.3 Public service agencies. A permit shall not be required for the installation, alteration or repair of generation, transmission, distribution or metering or other related equipment that is under the ownership and control of public service agencies by established right.

105.3 Application for permit. To obtain a permit, the applicant shall first file an application therefor in writing on a form furnished by the department of building safety for that purpose. Such application shall:

1. Identify and describe the work to be covered by the permit for which application is made.
2. Describe the land on which the proposed work is to be done by legal description, street address or similar description that will readily identify and definitely locate the proposed building or work.
3. Indicate the use and occupancy for which the proposed work is intended.
4. Be accompanied by construction documents and other information as required in Section 106.3.
5. State the valuation of the proposed work.
6. Be signed by the applicant, or the applicant's authorized agent.
7. Give such other data and information as required by the building official.

105.3.1 Action on application. The building official shall examine or cause to be examined applications for permits and amendments thereto within a reasonable time after filing. If the application or the construction documents do not conform to the requirements of pertinent laws, the building official shall reject such application in writing, stating the reasons therefor. If the building official is satisfied that the proposed work conforms to the requirements of this code and laws and ordinances applicable thereto, the building official shall issue a permit therefor as soon as practicable.

105.3.2 Time limitation of application. An application for a permit for any proposed work shall be deemed to have been abandoned 180 days after the date of filing, unless such application has been pursued in good faith or a permit has been issued; except that the building official is authorized to grant one or more extensions of time for additional periods not exceeding 90 days each. The extension shall be requested in writing and justifiable cause demonstrated.

105.4 Validity of permit. The issuance or granting of a permit shall not be construed to be a permit for, or an approval of, any violation of any of the provisions of this code or of any other ordinance of the jurisdiction. Permits presuming to give authority to violate or cancel the provisions of this code or other ordinances of the jurisdiction shall not be valid. The issuance of a permit based on construction documents and other data shall not prevent the building official from requiring the correction of errors in the construction documents and other data. The building official is also authorized to prevent occupancy or use of a structure where in violation of this code or of any other ordinances of this jurisdiction.

105.5 Expiration. Every permit issued shall become invalid unless the work on the site authorized by such permit is commenced within 180 days after its issuance, or if the work authorized on the site by such permit is suspended or abandoned for a period of 180 days after the time the work is commenced. The building official is authorized to grant, in writing, one or more extensions of time, for periods not more than 180 days each.

The extension shall be requested in writing and justifiable cause demonstrated.

105.6 Suspension or revocation. The building official is authorized to suspend or revoke a permit issued under the provisions of this code wherever the permit is issued in error or on the basis of incorrect, inaccurate or incomplete information, or in violation of any ordinance or regulation or any of the provisions of this code.

105.7 Placement of permit. The building permit or copy shall be kept on the site of the work until the completion of the project.

SECTION 106
CONSTRUCTION DOCUMENTS

106.1 Submittal documents. Construction documents, special inspection and structural observation programs, and other data shall be submitted in one or more sets with each application for a permit. The construction documents shall be prepared by a registered design professional where required by the statutes of the jurisdiction in which the project is to be constructed. Where special conditions exist, the building official is authorized to require additional construction documents to be prepared by a registered design professional.

> **Exception:** The building official is authorized to waive the submission of construction documents and other data not required to be prepared by a registered design professional if it is found that the nature of the work applied for is such that review of construction documents is not necessary to obtain compliance with this code.

106.1.1 Information on construction documents. Construction documents shall be dimensioned and drawn upon suitable material. Electronic media documents are permitted to be submitted when approved by the building official. Construction documents shall be of sufficient clarity to indicate the location, nature and extent of the work proposed and show in detail that it will conform to the provisions of this code and relevant laws, ordinances, rules and regulations, as determined by the building official.

106.1.1.1 Fire protection system shop drawings. Shop drawings for the fire protection system(s) shall be submitted to indicate conformance with this code and the construction documents and shall be approved prior to the start of system installation. Shop drawings shall contain all information as required by the referenced installation standards in Chapter 9.

106.1.2 Means of egress. The construction documents shall show in sufficient detail the location, construction, size and character of all portions of the means of egress in compliance with the provisions of this code. In other than occupancies in Groups R-2, R-3, as applicable in Section 101.2, and I-1, the construction documents shall designate the number of occupants to be accommodated on every floor, and in all rooms and spaces.

106.1.3 Exterior wall envelope. Construction documents for all buildings shall describe the exterior wall envelope in sufficient detail to determine compliance with this code. The construction documents shall provide details of the ex-

terior wall envelope as required, including flashing, intersections with dissimilar materials, corners, end details, control joints, intersections at roof, eaves or parapets, means of drainage, water-resistive membrane and details around openings.

The construction documents shall include manufacturer's installation instructions that provide supporting documentation that the proposed penetration and opening details described in the construction documents maintain the weather resistance of the exterior wall envelope. The supporting documentation shall fully describe the exterior wall system which was tested, where applicable, as well as the test procedure used.

106.2 Site plan. The construction documents submitted with the application for permit shall be accompanied by a site plan showing to scale the size and location of new construction and existing structures on the site, distances from lot lines, the established street grades and the proposed finished grades and, as applicable, flood hazard areas, floodways, and design flood elevations; and it shall be drawn in accordance with an accurate boundary line survey. In the case of demolition, the site plan shall show construction to be demolished and the location and size of existing structures and construction that are to remain on the site or plot. The building official is authorized to waive or modify the requirement for a site plan when the application for permit is for alteration or repair or when otherwise warranted.

106.3 Examination of documents. The building official shall examine or cause to be examined the accompanying construction documents and shall ascertain by such examinations whether the construction indicated and described is in accordance with the requirements of this code and other pertinent laws or ordinances.

106.3.1 Approval of construction documents. When the building official issues a permit, the construction documents shall be approved, in writing or by stamp, as "Reviewed for Code Compliance." One set of construction documents so reviewed shall be retained by the building official. The other set shall be returned to the applicant, shall be kept at the site of work and shall be open to inspection by the building official or a duly authorized representative.

106.3.2 Previous approvals. This code shall not require changes in the construction documents, construction or designated occupancy of a structure for which a lawful permit has been heretofore issued or otherwise lawfully authorized, and the construction of which has been pursued in good faith within 180 days after the effective date of this code and has not been abandoned.

106.3.3 Phased approval. The building official is authorized to issue a permit for the construction of foundations or any other part of a building or structure before the construction documents for the whole building or structure have been submitted, provided that adequate information and detailed statements have been filed complying with pertinent requirements of this code. The holder of such permit for the foundation or other parts of a building or structure shall proceed at the holder's own risk with the building operation and without assurance that a permit for the entire structure will be granted.

106.3.4 Design professional in responsible charge.

106.3.4.1 General. When it is required that documents be prepared by a registered design professional, the building official shall be authorized to require the owner to engage and designate on the building permit application a registered design professional who shall act as the registered design professional in responsible charge. If the circumstances require, the owner shall designate a substitute registered design professional in responsible charge who shall perform the duties required of the original registered design professional in responsible charge. The building official shall be notified in writing by the owner if the registered design professional in responsible charge is changed or is unable to continue to perform the duties.

The registered design professional in responsible charge shall be responsible for reviewing and coordinating submittal documents prepared by others, including phased and deferred submittal items, for compatibility with the design of the building.

Where structural observation is required by Section 1709, the inspection program shall name the individual or firms who are to perform structural observation and describe the stages of construction at which structural observation is to occur (see also duties specified in Section 1704).

106.3.4.2 Deferred submittals. For the purposes of this section, deferred submittals are defined as those portions of the design that are not submitted at the time of the application and that are to be submitted to the building official within a specified period.

Deferral of any submittal items shall have the prior approval of the building official. The registered design professional in responsible charge shall list the deferred submittals on the construction documents for review by the building official.

Documents for deferred submittal items shall be submitted to the registered design professional in responsible charge who shall review them and forward them to the building official with a notation indicating that the deferred submittal documents have been reviewed and been found to be in general conformance to the design of the building. The deferred submittal items shall not be installed until the design and submittal documents have been approved by the building official.

106.4 Amended construction documents. Work shall be installed in accordance with the approved construction documents, and any changes made during construction that are not in compliance with the approved construction documents shall be resubmitted for approval as an amended set of construction documents.

106.5 Retention of construction documents. One set of approved construction documents shall be retained by the building official for a period of not less than 180 days from date of completion of the permitted work, or as required by state or local laws.

SECTION 107
TEMPORARY STRUCTURES AND USES

107.1 General. The building official is authorized to issue a permit for temporary structures and temporary uses. Such permits shall be limited as to time of service, but shall not be permitted for more than 180 days. The building official is authorized to grant extensions for demonstrated cause.

107.2 Conformance. Temporary structures and uses shall conform to the structural strength, fire safety, means of egress, accessibility, light, ventilation and sanitary requirements of this code as necessary to ensure the public health, safety and general welfare.

107.3 Temporary power. The building official is authorized to give permission to temporarily supply and use power in part of an electric installation before such installation has been fully completed and the final certificate of completion has been issued. The part covered by the temporary certificate shall comply with the requirements specified for temporary lighting, heat or power in the ICC *Electrical Code*.

107.4 Termination of approval. The building official is authorized to terminate such permit for a temporary structure or use and to order the temporary structure or use to be discontinued.

SECTION 108
FEES

108.1 Payment of fees. A permit shall not be valid until the fees prescribed by law have been paid, nor shall an amendment to a permit be released until the additional fee, if any, has been paid.

108.2 Schedule of permit fees. On buildings, structures, electrical, gas, mechanical, and plumbing systems or alterations requiring a permit, a fee for each permit shall be paid as required, in accordance with the schedule as established by the applicable governing authority.

108.3 Building permit valuations. The applicant for a permit shall provide an estimated permit value at time of application. Permit valuations shall include total value of work, including materials and labor, for which the permit is being issued, such as electrical, gas, mechanical, plumbing equipment and permanent systems. If, in the opinion of the building official, the valuation is underestimated on the application, the permit shall be denied, unless the applicant can show detailed estimates to meet the approval of the building official. Final building permit valuation shall be set by the building official.

108.4 Work commencing before permit issuance. Any person who commences any work on a building, structure, electrical, gas, mechanical or plumbing system before obtaining the necessary permits shall be subject to a fee established by the building official that shall be in addition to the required permit fees.

108.5 Related fees. The payment of the fee for the construction, alteration, removal or demolition for work done in connection to or concurrently with the work authorized by a building permit shall not relieve the applicant or holder of the permit from the payment of other fees that are prescribed by law.

108.6 Refunds. The building official is authorized to establish a refund policy.

SECTION 109
INSPECTIONS

109.1 General. Construction or work for which a permit is required shall be subject to inspection by the building official and such construction or work shall remain accessible and exposed for inspection purposes until approved. Approval as a result of an inspection shall not be construed to be an approval of a violation of the provisions of this code or of other ordinances of the jurisdiction. Inspections presuming to give authority to violate or cancel the provisions of this code or of other ordinances of the jurisdiction shall not be valid. It shall be the duty of the permit applicant to cause the work to remain accessible and exposed for inspection purposes. Neither the building official nor the jurisdiction shall be liable for expense entailed in the removal or replacement of any material required to allow inspection.

109.2 Preliminary inspection. Before issuing a permit, the building official is authorized to examine or cause to be examined buildings, structures and sites for which an application has been filed.

109.3 Required inspections. The building official, upon notification, shall make the inspections set forth in Sections 109.3.1 through 109.3.10.

109.3.1 Footing and foundation inspection. Footing and foundation inspections shall be made after excavations for footings are complete and any required reinforcing steel is in place. For concrete foundations, any required forms shall be in place prior to inspection. Materials for the foundation shall be on the job, except where concrete is ready mixed in accordance with ASTM C 94, the concrete need not be on the job.

109.3.2 Concrete slab and under-floor inspection. Concrete slab and under-floor inspections shall be made after in-slab or under-floor reinforcing steel and building service equipment, conduit, piping accessories and other ancillary equipment items are in place, but before any concrete is placed or floor sheathing installed, including the subfloor.

109.3.3 Lowest floor elevation. In flood hazard areas, upon placement of the lowest floor, including the basement, and prior to further vertical construction, the elevation certification required in Section 1612.5 shall be submitted to the building official.

109.3.4 Frame inspection. Framing inspections shall be made after the roof deck or sheathing, all framing, fireblocking and bracing are in place and pipes, chimneys and vents to be concealed are complete and the rough electrical, plumbing, heating wires, pipes and ducts are approved.

109.3.5 Lath and gypsum board inspection. Lath and gypsum board inspections shall be made after lathing and gypsum board, interior and exterior, is in place, but before any plastering is applied or gypsum board joints and fasteners are taped and finished.

> **Exception:** Gypsum board that is not part of a fire-resistance-rated assembly or a shear assembly.

109.3.6 Fire-resistant penetrations. Protection of joints and penetrations in fire-resistance-rated assemblies shall not be concealed from view until inspected and approved.

109.3.7 Energy efficiency inspections. Inspections shall be made to determine compliance with Chapter 13 and shall include, but not be limited to, inspections for: envelope insulation R and U values, fenestration U value, duct system R value, and HVAC and water-heating equipment efficiency.

109.3.8 Other inspections. In addition to the inspections specified above, the building official is authorized to make or require other inspections of any construction work to ascertain compliance with the provisions of this code and other laws that are enforced by the department of building safety.

109.3.9 Special inspections. For special inspections, see Section 1704.

109.3.10 Final inspection. The final inspection shall be made after all work required by the building permit is completed.

109.4 Inspection agencies. The building official is authorized to accept reports of approved inspection agencies, provided such agencies satisfy the requirements as to qualifications and reliability.

109.5 Inspection requests. It shall be the duty of the holder of the building permit or their duly authorized agent to notify the building official when work is ready for inspection. It shall be the duty of the permit holder to provide access to and means for inspections of such work that are required by this code.

109.6 Approval required. Work shall not be done beyond the point indicated in each successive inspection without first obtaining the approval of the building official. The building official, upon notification, shall make the requested inspections and shall either indicate the portion of the construction that is satisfactory as completed, or notify the permit holder or his or her agent wherein the same fails to comply with this code. Any portions that do not comply shall be corrected and such portion shall not be covered or concealed until authorized by the building official.

SECTION 110
CERTIFICATE OF OCCUPANCY

110.1 Use and occupancy. No building or structure shall be used or occupied, and no change in the existing occupancy classification of a building or structure or portion thereof shall be made until the building official has issued a certificate of occupancy therefor as provided herein. Issuance of a certificate of occupancy shall not be construed as an approval of a violation of the provisions of this code or of other ordinances of the jurisdiction.

110.2 Certificate issued. After the building official inspects the building or structure and finds no violations of the provisions of this code or other laws that are enforced by the department of building safety, the building official shall issue a certificate of occupancy that contains the following:

1. The building permit number.
2. The address of the structure.
3. The name and address of the owner.
4. A description of that portion of the structure for which the certificate is issued.
5. A statement that the described portion of the structure has been inspected for compliance with the requirements of this code for the occupancy and division of occupancy and the use for which the proposed occupancy is classified.
6. The name of the building official.
7. The edition of the code under which the permit was issued.
8. The use and occupancy, in accordance with the provisions of Chapter 3.
9. The type of construction as defined in Chapter 6.
10. The design occupant load.
11. If an automatic sprinkler system is provided, whether the sprinkler system is required.
12. Any special stipulations and conditions of the building permit.

110.3 Temporary occupancy. The building official is authorized to issue a temporary certificate of occupancy before the completion of the entire work covered by the permit, provided that such portion or portions shall be occupied safely. The building official shall set a time period during which the temporary certificate of occupancy is valid.

110.4 Revocation. The building official is authorized to, in writing, suspend or revoke a certificate of occupancy or completion issued under the provisions of this code wherever the certificate is issued in error, or on the basis of incorrect information supplied, or where it is determined that the building or structure or portion thereof is in violation of any ordinance or regulation or any of the provisions of this code.

SECTION 111
SERVICE UTILITIES

111.1 Connection of service utilities. No person shall make connections from a utility, source of energy, fuel or power to any building or system that is regulated by this code for which a permit is required, until released by the building official.

111.2 Temporary connection. The building official shall have the authority to authorize the temporary connection of the building or system to the utility source of energy, fuel or power.

111.3 Authority to disconnect service utilities. The building official shall have the authority to authorize disconnection of utility service to the building, structure or system regulated by this code and the codes referenced in case of emergency where necessary to eliminate an immediate hazard to life or property. The building official shall notify the serving utility, and wherever possible the owner and occupant of the building, structure

or service system of the decision to disconnect prior to taking such action. If not notified prior to disconnecting, the owner or occupant of the building, structure or service system shall be notified in writing, as soon as practical thereafter.

SECTION 112
BOARD OF APPEALS

112.1 General. In order to hear and decide appeals of orders, decisions or determinations made by the building official relative to the application and interpretation of this code, there shall be and is hereby created a board of appeals. The board of appeals shall be appointed by the governing body and shall hold office at its pleasure. The board shall adopt rules of procedure for conducting its business.

112.2 Limitations on authority. An application for appeal shall be based on a claim that the true intent of this code or the rules legally adopted thereunder have been incorrectly interpreted, the provisions of this code do not fully apply or an equally good or better form of construction is proposed. The board shall have no authority to waive requirements of this code.

112.3 Qualifications. The board of appeals shall consist of members who are qualified by experience and training to pass on matters pertaining to building construction and are not employees of the jurisdiction.

SECTION 113
VIOLATIONS

113.1 Unlawful acts. It shall be unlawful for any person, firm or corporation to erect, construct, alter, extend, repair, move, remove, demolish or occupy any building, structure or equipment regulated by this code, or cause same to be done, in conflict with or in violation of any of the provisions of this code.

113.2 Notice of violation. The building official is authorized to serve a notice of violation or order on the person responsible for the erection, construction, alteration, extension, repair, moving, removal, demolition or occupancy of a building or structure in violation of the provisions of this code, or in violation of a permit or certificate issued under the provisions of this code. Such order shall direct the discontinuance of the illegal action or condition and the abatement of the violation.

113.3 Prosecution of violation. If the notice of violation is not complied with promptly, the building official is authorized to request the legal counsel of the jurisdiction to institute the appropriate proceeding at law or in equity to restrain, correct or abate such violation, or to require the removal or termination of the unlawful occupancy of the building or structure in violation of the provisions of this code or of the order or direction made pursuant thereto.

113.4 Violation penalties. Any person who violates a provision of this code or fails to comply with any of the requirements thereof or who erects, constructs, alters or repairs a building or structure in violation of the approved construction documents or directive of the building official, or of a permit or certificate issued under the provisions of this code, shall be subject to penalties as prescribed by law.

SECTION 114
STOP WORK ORDER

114.1 Authority. Whenever the building official finds any work regulated by this code being performed in a manner either contrary to the provisions of this code or dangerous or unsafe, the building official is authorized to issue a stop work order.

114.2 Issuance. The stop work order shall be in writing and shall be given to the owner of the property involved, or to the owner's agent, or to the person doing the work. Upon issuance of a stop work order, the cited work shall immediately cease. The stop work order shall state the reason for the order, and the conditions under which the cited work will be permitted to resume.

114.3 Unlawful continuance. Any person who shall continue any work after having been served with a stop work order, except such work as that person is directed to perform to remove a violation or unsafe condition, shall be subject to penalties as prescribed by law.

SECTION 115
UNSAFE STRUCTURES AND EQUIPMENT

115.1 Conditions. Structures or existing equipment that are or hereafter become unsafe, insanitary or deficient because of inadequate means of egress facilities, inadequate light and ventilation, or which constitute a fire hazard, or are otherwise dangerous to human life or the public welfare, or that involve illegal or improper occupancy or inadequate maintenance, shall be deemed an unsafe condition. Unsafe structures shall be taken down and removed or made safe, as the building official deems necessary and as provided for in this section. A vacant structure that is not secured against entry shall be deemed unsafe.

115.2 Record. The building official shall cause a report to be filed on an unsafe condition. The report shall state the occupancy of the structure and the nature of the unsafe condition.

115.3 Notice. If an unsafe condition is found, the building official shall serve on the owner, agent or person in control of the structure, a written notice that describes the condition deemed unsafe and specifies the required repairs or improvements to be made to abate the unsafe condition, or that requires the unsafe structure to be demolished within a stipulated time. Such notice shall require the person thus notified to declare immediately to the building official acceptance or rejection of the terms of the order.

115.4 Method of service. Such notice shall be deemed properly served if a copy thereof is (a) delivered to the owner personally; (b) sent by certified or registered mail addressed to the owner at the last known address with the return receipt requested; or (c) delivered in any other manner as prescribed by local law. If the certified or registered letter is returned showing that the letter was not delivered, a copy thereof shall be posted in a conspicuous place in or about the structure affected by such notice. Service of such notice in the foregoing manner upon the owner's agent or upon the person responsible for the structure shall constitute service of notice upon the owner.

115.5 Restoration. The structure or equipment determined to be unsafe by the building official is permitted to be restored to a safe condition. To the extent that repairs, alterations or additions are made or a change of occupancy occurs during the restoration of the structure, such repairs, alterations, additions or change of occupancy shall comply with the requirements of Section 105.2.2 and Chapter 34.

CHAPTER 2
DEFINITIONS

SECTION 201
GENERAL

201.1 Scope. Unless otherwise expressly stated, the following words and terms shall, for the purposes of this code, have the meanings shown in this chapter.

201.2 Interchangeability. Words used in the present tense include the future; words stated in the masculine gender include the feminine and neuter; the singular number includes the plural and the plural, the singular.

201.3 Terms defined in other codes. Where terms are not defined in this code and are defined in the *International Fuel Gas Code, International Fire Code, International Mechanical Code* or *International Plumbing Code,* such terms shall have the meanings ascribed to them as in those codes.

201.4 Terms not defined. Where terms are not defined through the methods authorized by this section, such terms shall have ordinarily accepted meanings such as the context implies.

SECTION 202
DEFINITIONS

ACCESSIBLE. See Section 1102.1.

ACCESSIBLE MEANS OF EGRESS. See Section 1002.1.

ACCESSIBLE ROUTE. See Section 1102.1.

ACCESSIBLE UNIT. See Section 1102.

ACCREDITATION BODY. See Section 2302.1.

ACTIVE FAULT/ACTIVE FAULT TRACE. See Section 1613.1.

ADDITION. An extension or increase in floor area or height of a building or structure.

ADHERED MASONRY VENEER. See Section 1402.1.

ADJUSTED SHEAR RESISTANCE. (Steel Construction). See Section 2202.1.

ADJUSTED SHEAR RESISTANCE. (Wood Construction). See Section 2302.1.

ADMIXTURE. See Section 1902.1.

ADOBE CONSTRUCTION. See Section 2102.1.

    Stabilized adobe. See Section 2102.1.
    Unstabilized adobe. See Section 2102.1.

[F] AEROSOL. See Section 307.2.

    Level 1 aerosol products. See Section 307.2.
    Level 2 aerosol products. See Section 307.2.
    Level 3 aerosol products. See Section 307.2.

[F] AEROSOL CONTAINER. See Section 307.2.

AGGREGATE. See Section 1902.1.

AGGREGATE, LIGHTWEIGHT. See Section 1902.1.

AGRICULTURAL, BUILDING. A structure designed and constructed to house farm implements, hay, grain, poultry, livestock or other horticultural products. This structure shall not be a place of human habitation or a place of employment where agricultural products are processed, treated or packaged, nor shall it be a place used by the public.

AIR-INFLATED STRUCTURE. See Section 3102.2.

AIR-SUPPORTED STRUCTURE. See Section 3102.2.

    Double skin. See Section 3102.2.
    Single skin. See Section 3102.2.

AISLE ACCESSWAY. See Section 1002.1.

[F] ALARM NOTIFICATION APPLIANCE. See Section 902.1.

[F] ALARM SIGNAL. See Section 902.1.

[F] ALARM VERIFICATION FEATURE. See Section 902.1.

ALLEY. See "Public way."

ALLOWABLE STRESS DESIGN. See Section 1602.1.

ALTERATION. Any construction or renovation to an existing structure other than repair or addition.

ALTERNATING TREAD DEVICE. See Section 1002.1.

ANCHOR. See Sections 1913.2.2 and 2102.1.

ANCHOR BUILDING. See Section 402.2.

ANCHORED MASONRY VENEER. See Section 1402.1.

ANNULAR SPACE. See Section 702.1.

[F] ANNUNCIATOR. See Section 902.1.

APPROVED. Acceptable to the building official.

APPROVED AGENCY. See Section 1702.1.

APPROVED FABRICATOR. See Section 1702.1.

APPROVED SOURCE. An independent person, firm or corporation, approved by the building official, who is competent and experienced in the application of engineering principles to materials, methods or systems analyses.

ARCHITECTURAL TERRA COTTA. See Section 2102.1.

AREA. See Section 2102.1.

    Bedded. See Section 2102.1.
    Gross cross-sectional. See Section 2102.1.
    Net cross-sectional. See Section 2102.1.

AREA, BUILDING. See Section 502.1.

AREA OF REFUGE. See Section 1002.1.

AREAWAY. A subsurface space adjacent to a building open at the top or protected at the top by a grating or guard.

ATRIUM. See Section 404.1.1.

ATTACHMENT. See Section 1913.2.2.

ATTACHMENTS, SEISMIC. See Section 1613.1.

ATTIC. The space between the ceiling beams of the top story and the roof rafters.

[F] AUDIBLE ALARM NOTIFICATION APPLIANCE. See Section 902.1.

[F] AUTOMATIC. See Section 902.1.

[F] AUTOMATIC FIRE-EXTINGUISHING SYSTEM. See Section 902.1.

[F] AUTOMATIC SPRINKLER SYSTEM. See Section 902.1.

[F] AVERAGE AMBIENT SOUND LEVEL. See Section 902.1

AWNING. An architectural projection that provides weather protection, identity or decoration and is wholly supported by the building to which it is attached. An awning is comprised of a lightweight, rigid skeleton structure over which a covering is attached.

BACKING. See Section 1402.1.

BALCONY, EXTERIOR. See Section 1602.1.

[F] BARRICADE. See Section 307.2.

   Artificial barricade. See Section 307.2.

   Natural barricade. See Section 307.2.

BASE. See Section 1613.1.

BASE FLOOD. See Section 1612.2.

BASE FLOOD ELEVATION. See Section 1612.2.

BASE SHEAR. See Section 1602.1.

BASIC SEISMIC-FORCE-RESISTING SYSTEMS. See Section 1602.1.

   Bearing wall system. See Section 1602.1.
   Building frame system. See Section 1602.1.
   Dual system. See Section 1602.1.
   Inverted pendulum system. See Section 1602.1.
   Moment-resisting frame system. See Section 1602.1.
   Shear wall-frame interactive system. See Section 1602.1.

BASEMENT. That portion of a building that is partly or completely below grade (see "Story above grade plane" and Sections 502.1 and 1612.2).

BED JOINT. See Section 2102.1.

BLEACHERS. See Section 1002.1.

BOARDING HOUSE. See Section 310.2.

[F] BOILING POINT. See Section 307.2.

BOND BEAM. See Section 2102.1.

BOND REINFORCING. See Section 2102.1.

BOUNDARY ELEMENT. See Sections 1602.1 and 1613.1.

BOUNDARY MEMBERS. See Section 1602.1.

BRACED WALL LINE. See Section 2302.1.

BRACED WALL PANEL. See Section 2302.1.

BRICK. See Section 2102.1.

   Calcium silicate (sand lime brick). See Section 2102.1.

   Clay or shale. See Section 2102.1.
   Concrete. See Section 2102.1.

BRITTLE. See Section 1613.1.

BRITTLE STEEL ELEMENT. See Section 1913.2.2.

BUILDING. Any structure used or intended for supporting or sheltering any use or occupancy.

BUILDING, ENCLOSED. See Section 1609.2.

BUILDING LINE. The line established by law, beyond which a building shall not extend, except as specifically provided by law.

BUILDING, LOW-RISE. See Section 1609.2.

BUILDING OFFICIAL. The officer or other designated authority charged with the administration and enforcement of this code, or a duly authorized representative.

BUILDING, OPEN. See Section 1609.2.

BUILDING, PARTIALLY ENCLOSED. See Section 1609.2.

BUILDING, SIMPLE DIAPHRAGM. See Section 1609.2.

BUILT-UP ROOF COVERING. See Section 1502.1.

BUTTRESS. See Section 2102.1.

CABLE-RESTRAINED, AIR-SUPPORTED STRUCTURE. See Section 3102.2.

CANOPY. An architectural projection that provides weather protection, identity or decoration and is supported by the building to which it is attached and at the outer end by not less than one stanchion. A canopy is comprised of a rigid structure over which a covering is attached.

CANTILEVERED COLUMN SYSTEM. See Section 1602.1.

[F] CARBON DIOXIDE EXTINGUISHING SYSTEMS. See Section 902.1.

CAST STONE. See Section 2102.1.

[F] CEILING LIMIT. See Section 902.1.

CEILING RADIATION DAMPER. See Section 702.1.

CELL. See Section 2102.1.

CEMENT PLASTER. See Section 2502.1.

CEMENTITIOUS MATERIALS. See Section 1902.1.

CERAMIC FIBER BLANKET. See Section 721.1.1.

CERTIFICATE OF COMPLIANCE. See Section 1702.1.

CHIMNEY. See Section 2102.1.

CHIMNEY TYPES. See Section 2102.1.

   High-heat appliance type. See Section 2102.1.
   Low-heat appliance type. See Section 2102.1.
   Masonry type. See Section 2102.1.
   Medium-heat appliance type. See Section 2102.1.

CIRCULATION PATH. See Section 1102.1.

CLADDING. See "Components and cladding."

[F] CLEAN AGENT. See Section 902.1.

CLEANOUT. See Section 2102.1.

[F] CLOSED SYSTEM. See Section 307.2.

COLLAR JOINT. See Section 2102.1.

COLLECTOR. See Sections 1613.1 and 2302.1.

COLLECTOR ELEMENTS. See Section 1602.1.

COLUMN. See Section 1902.1.

COLUMN, MASONRY. See Section 2102.1.

COMBINATION FIRE/SMOKE DAMPER. See Section 702.1.

[F] COMBUSTIBLE DUST. See Section 307.2.

[F] COMBUSTIBLE FIBERS. See Section 307.2.

[F] COMBUSTIBLE LIQUID. See Section 307.2.

   Class II. See Section 307.2.
   Class IIIA. See Section 307.2.
   Class IIIB. See Section 307.2.

COMMON PATH OF EGRESS TRAVEL. See Section 1002.1.

COMPONENT. See Section 1613.1.

   Component equipment. See Section 1613.1.
   Component, flexible. See Section 1613.1.
   Component, rigid. See Section 1613.1.

COMPONENTS AND CLADDING. See Section 1609.2.

COMPOSITE MASONRY. See Section 2102.1.

[F] COMPRESSED GAS. See Section 307.2.

COMPRESSIVE STRENGTH OF MASONRY. See Section 2102.1.

CONCRETE. See Section 1902.1.

CONCRETE BREAKOUT STRENGTH. See Section 1913.2.2.

CONCRETE CARBONATE AGGREGATE. See Section 721.1.1.

CONCRETE, CELLULAR. See Section 721.1.1.

CONCRETE, LIGHTWEIGHT AGGREGATE. See Section 721.1.1.

CONCRETE, PERLITE. See Section 721.1.1.

CONCRETE PRYOUT STRENGTH. See Section 1913.2.2.

CONCRETE, SAND-LIGHTWEIGHT. See Section 721.1.1.

CONCRETE, SILICEOUS AGGREGATE. See Section 721.1.1.

CONCRETE (F'_c), SPECIFIED COMPRESSIVE STRENGTH OF. See Section 1902.1.

CONCRETE, VERMICULITE. See Section 721.1.1.

CONFINED REGION. See Section 1602.1.

CONNECTOR. See Section 2102.1.

[F] CONSTANTLY ATTENDED LOCATION. See Section 902.1.

CONSTRUCTION DOCUMENTS. Written, graphic and pictorial documents prepared or assembled for describing the design, location and physical characteristics of the elements of a project necessary for obtaining a building permit.

CONSTRUCTION TYPES. See Section 602.

   Type I. See Section 602.2.
   Type II. See Section 602.2.
   Type III. See Section 602.3.
   Type IV. See Section 602.4.
   Type V. See Section 602.5.

[F] CONTINUOUS GAS-DETECTION SYSTEM. See Section 415.2.

CONTRACTION JOINT. See Section 1902.1.

[F] CONTROL AREA. See Section 307.2.

CONTROLLED LOW-STRENGTH MATERIAL. A self-compacted, cementitious material used primarily as a backfill in place of compacted fill.

CONVENTIONAL LIGHT-FRAME WOOD CONSTRUCTION. See Section 2302.1.

CORRIDOR. See Section 1002.1.

CORROSION RESISTANCE. The ability of a material to withstand deterioration of its surface or its properties when exposed to its environment.

CORROSION RESISTANT. See Section 1502.1.

[F] CORROSIVE. See Section 307.2.

COURT. An open, uncovered space, unobstructed to the sky, bounded on three or more sides by exterior building walls or other enclosing devices.

COVER. See Section 2102.1.

COVERED MALL BUILDING. See Section 402.2.

CRIPPLE WALL. See Section 2302.1.

CRYOGENIC FLUID. See Section 307.2.

DALLE GLASS. See Section 2402.1.

DAMPER. See Section 702.1.

DEAD LOADS. See Section 1602.1.

DECK. See Section 1602.1.

DECORATIVE GLASS. See Section 2402.1.

[F] DEFLAGRATION. See Section 307.2.

DEFORMABILITY. See Section 1602.1.

   High deformability element. See Section 1602.1.
   Limited deformability element. See Section 1602.1.
   Low deformability element. See Section 1602.1.

DEFORMATION. See Section 1602.1.

   Limited deformation. See Section 1602.1.
   Ultimate deformation. See Section 1602.1.

DEFORMED REINFORCEMENT. See Section 1902.1.

[F] DELUGE SYSTEM. See Section 902.1.

DESIGN EARTHQUAKE. See Section 1613.1.

DESIGN FLOOD. See Section 1612.2.

DESIGN FLOOD ELEVATION. See Section 1612.2.

DESIGN STRENGTH. See Section 1602.1.

DESIGNATED SEISMIC SYSTEM. See Section 1613.1.

[F] DETACHED STORAGE BUILDING. See Section 307.2.

DETECTABLE WARNING. See Section 1102.1.

[F] DETECTOR, HEAT. See Section 902.1.

[F] DETONATION. See Section 307.2.

DIAPHRAGM. See Sections 1602.1 and 2102.1.

   Diaphragm, blocked. See Sections 1602.1 and 2102.1.

   Diaphragm, boundary. See Section 1602.1.

   Diaphragm, chord. See Section 1602.1.

   Diaphragm, flexible. See Section 1602.1.

   Diaphragm, rigid. See Section 1602.1.

DIAPHRAGM, UNBLOCKED. See Section 2302.1.

DIMENSIONS. See Section 2102.1.

   Actual. See Section 2102.1.

   Nominal. See Section 2102.1.

   Specified. See Section 2102.1.

DISPENSING. See Section 307.2.

DISPLACEMENT. See Section 1613.1.

   Design displacement. See Section 1613.1.

   Total design displacement. See Section 1613.1.

   Total maximum displacement. See Section 1613.1.

DISPLACEMENT RESTRAINT SYSTEM. See Section 1613.1

DOOR, BALANCED. See Section 1002.1.

DORMITORY. See Section 310.2.

DRAFTSTOP. See Section 702.1.

DRAG STRUT. See Section 2302.1.

[F] DRY-CHEMICAL EXTINGUISHING AGENT. See Section 902.1.

DRY FLOODPROOFING. See Section 1612.2.

DUCTILE STEEL ELEMENT. See Section 1913.2.2.

DURATION OF LOAD. See Section 1602.1.

DWELLING. A building that contains one or two dwelling units used, intended or designed to be used, rented, leased, let or hired out to be occupied for living purposes.

DWELLING UNIT. See Section 310.2.

DWELLING UNIT OR SLEEPING UNIT, MULTI-STORY. See Section 1102.

DWELLING UNIT OR SLEEPING UNIT, TYPE A. See Section 1102.

DWELLING UNIT OR SLEEPING UNIT, TYPE B. See Section 1102.

EDGE DISTANCE. See Section 1913.2.2.

EFFECTIVE DAMPING. See Section 1613.1.

EFFECTIVE DEPTH OF SECTION (*d*). See Section 1902.1.

EFFECTIVE EMBEDMENT DEPTH. See Section 1913.2.2.

EFFECTIVE HEIGHT. See Section 2102.1.

EFFECTIVE STIFFNESS. See Section 1613.1.

EFFECTIVE WIND AREA. See Section 1609.2.

EGRESS COURT. See Section 1002.1.

ELEMENT. See Section 1602.1.

   Ductile element. See Section 1602.1.

   Limited ductile element. See Section 1602.1.

   Nonductile element. See Section 1602.1.

[F] EMERGENCY ALARM SYSTEM. See Section 902.1.

[F] EMERGENCY CONTROL STATION. See Section 415.2.

EMERGENCY ESCAPE AND RESCUE OPENING. See Section 1002.1.

[F] EMERGENCY VOICE/ALARM COMMUNICATIONS. See Section 902.1.

EMPLOYEE WORK AREA. See Section 1102.1.

EQUIPMENT SUPPORT. See Section 1602.1.

ESSENTIAL FACILITIES. See Section 1602.1.

[F] EXHAUSTED ENCLOSURE. See Section 415.2.

EXISTING CONSTRUCTION. See Section 1612.2.

EXISTING STRUCTURE. A structure erected prior to the date of adoption of the appropriate code, or one for which a legal building permit has been issued.

EXIT. See Section 1002.1.

EXIT ACCESS. See Section 1002.1.

EXIT DISCHARGE. See Section 1002.1.

EXIT DISCHARGE, LEVEL OF. See Section 1002.1.

EXIT ENCLOSURE. See Section 1002.1.

EXIT PASSAGEWAY. See Section 1002.1.

EXPANDED VINYL WALL COVERING. See Section 802.1.

[F] EXPLOSION. See Section 902.1.

[F] EXPLOSIVE. See Section 307.2.

   High explosive. See Section 307.2.

   Low explosive. See Section 307.2.

   Mass detonating explosives. See Section 307.2.

   UN/DOTn Class 1 Explosives. See Section 307.2.

      Division 1.1. See Section 307.2.

      Division 1.2. See Section 307.2.

      Division 1.3. See Section 307.2.

      Division 1.4. See Section 307.2.

      Division 1.5. See Section 307.2.

      Division 1.6. See Section 307.2.

EXTERIOR SURFACES. See Section 2502.1.

EXTERIOR WALL. See Section 1402.1.

EXTERIOR WALL COVERING. See Section 1402.1.

EXTERIOR WALL ENVELOPE. See Section 1402.1.

F RATING. See Section 702.1.

FABRICATED ITEM. See Section 1702.1.

[F] FABRICATION AREA. See Section 415.2.

FACILITY. See Section 1102.1.

FACTORED LOAD. See Section 1602.1.

FIBERBOARD. See Section 2302.1.

[F] FIRE ALARM CONTROL UNIT. See Section 902.1.

[F] FIRE ALARM SIGNAL. See Section 902.1.

[F] FIRE ALARM SYSTEM. See Section 902.1.

FIRE AREA. See Section 702.1.

FIRE BARRIER. See Section 702.1.

[F] FIRE COMMAND CENTER. See Section 902.1.

FIRE DAMPER. See Section 702.1.

[F] FIRE DETECTOR, AUTOMATIC. See Section 902.1.

FIRE DOOR. See Section 702.1.

FIRE DOOR ASSEMBLY. See Section 702.1.

FIRE EXIT HARDWARE. See Section 1002.1.

FIRE PARTITION. See Section 702.1.

FIRE PROTECTION RATING. See Section 702.1.

[F] FIRE PROTECTION SYSTEM. See Section 902.1.

FIRE RESISTANCE. See Section 702.1.

FIRE-RESISTANCE RATING. See Section 702.1.

FIRE-RESISTANT JOINT SYSTEM. See Section 702.1.

[F] FIRE SAFETY FUNCTIONS. See Section 902.1.

FIRE SEPARATION DISTANCE. See Section 702.1.

FIRE WALL. See Section 702.1.

FIRE WINDOW ASSEMBLY. See Section 702.1.

FIREBLOCKING. See Section 702.1.

FIREPLACE. See Section 2102.1.

FIREPLACE THROAT. See Section 2102.1.

FIREWORKS. See Section 307.2.

FIREWORKS, 1.3G. See Section 307.2.

FIREWORKS, 1.4G. See Section 307.2.

5-PERCENT FRACTILE. See Section 1913.2.2.

FLAME RESISTANCE. See Section 802.1.

FLAME SPREAD. See Section 802.1.

FLAME SPREAD INDEX. See Section 802.1.

[F] FLAMMABLE GAS. See Section 307.2.

[F] FLAMMABLE LIQUEFIED GAS. See Section 307.2.

[F] FLAMMABLE LIQUID. See Section 307.2.

    Class IA. See Section 307.2.

    Class IB. See Section 307.2.

    Class IC. See Section 307.2.

[F] FLAMMABLE MATERIAL. See Section 307.2.

[F] FLAMMABLE SOLID. See Section 307.2.

[F] FLAMMABLE VAPORS OR FUMES. See Section 415.2.

[F] FLASH POINT. See Section 307.2.

FLEXIBLE BUILDINGS AND OTHER STRUCTURES. See Section 1609.2.

FLEXIBLE EQUIPMENT CONNECTIONS. See Section 1602.1.

FLEXURAL LENGTH. See Section 1808.1.

FLOOD OR FLOODING. See Section 1612.2.

FLOOD DAMAGE-RESISTANT MATERIALS. See Section 1612.2.

FLOOD HAZARD AREA. See Section 1612.2.

FLOOD HAZARD AREA SUBJECT TO HIGH VELOCITY WAVE ACTION. See Section 1612.2.

FLOOD INSURANCE RATE MAP (FIRM). See Section 1612.2.

FLOOD INSURANCE STUDY. See Section 1612.2.

FLOODWAY. See Section 1612.2.

FLOOR AREA, GROSS. See Section 1002.1.

FLOOR AREA, NET. See Section 1002.1.

FLOOR FIRE DOOR ASSEMBLY. See Section 702.1.

FLY GALLERY. See Section 410.2.

[F] FOAM-EXTINGUISHING SYSTEMS. See Section 902.1.

FOAM PLASTIC INSULATION. See Section 2602.1.

FOLDING AND TELESCOPIC SEATING. See Section 1002.1.

FOOD COURT. See Section 402.2.

FRAME. See Section 1602.1.

    Braced frame. See Section 1602.1.

    Concentrically braced frame (CBF). See Section 1602.1.

    Eccentrically braced frame (EBF). See Section 1602.1.

    Ordinary concentrically braced frame (OCBF). See Section 1602.1.

    Special concentrically braced frame (SCBF). See Section 1602.1.

    Moment frame. See Section 1602.1.

[F] GAS CABINET. See Section 415.2.

[F] GAS ROOM. See Section 415.2.

GLASS FIBERBOARD. See Section 721.1.1.

GLUED BUILT-UP MEMBER. See Section 2302.1.

GRADE FLOOR OPENING. A window or other opening located such that the sill height of the opening is not more than 44 inches (1118 mm) above or below the finished ground level adjacent to the opening.

GRADE (LUMBER). See Section 2302.1.

GRADE PLANE. See Section 502.1.

GRANDSTAND. See Section 1002.1.

GRAVITY LOAD. See Section 1613.1.

GRIDIRON. See Section 410.2.

GROSS LEASABLE AREA. See Section 402.2.

GROUTED MASONRY. See Section 2102.1.

    Grouted hollow-unit masonry. See Section 2102.1.

    Grouted multiwythe masonry. See Section 2102.1.

GUARD. See Section 1002.1.

GYPSUM BOARD. See Section 2502.1.

GYPSUM PLASTER. See Section 2502.1.

GYPSUM VENEER PLASTER. See Section 2502.1.

HABITABLE SPACE. A space in a building for living, sleeping, eating or cooking. Bathrooms, toilet rooms, closets, halls, storage or utility spaces and similar areas are not considered habitable spaces.

[F] HALOGENATED EXTINGUISHING SYSTEMS. See Section 902.1.

[F] HANDLING. See Section 307.2.

HANDRAIL. See Section 1002.1.

HARDBOARD. See Section 2302.1.

HAZARDOUS CONTENTS. See Section 1613.1.

[F] HAZARDOUS MATERIALS. See Section 307.2.

[F] HAZARDOUS PRODUCTION MATERIAL (HPM). See Section 415.2.

HEAD JOINT. See Section 2102.1.

HEADER (Bonder). See Section 2102.1.

[F] HEALTH HAZARD. See Section 307.2.

HEIGHT, BUILDING. See Section 502.1.

HEIGHT, STORY. See Section 502.1.

HEIGHT, WALLS. See Section 2102.1.

HELIPORT. See Section 412.5.2.

HELISTOP. See Section 412.5.2.

[F] HIGHLY TOXIC. See Section 307.2.

HISTORIC BUILDINGS. Buildings that are listed in or eligible for listing in the National Register of Historic Places, or designated as historic under an appropriate state or local law (see Section 3406).

HOOKED BOLT. See Section 1913.2.2.

HORIZONTAL EXIT. See Section 1002.1.

[F] HPM FLAMMABLE LIQUID. See Section 415.2.

[F] HPM ROOM. See Section 415.2.

HURRICANE-PRONE REGIONS. See Section 1609.2.

IMMEDIATELY DANGEROUS TO LIFE AND HEALTH (IDLH). See Section 415.2.

IMPACT LOAD. See Section 1602.1.

IMPORTANCE FACTOR, *I*. See Section 1609.2.

INCOMPATIBLE MATERIALS. See Section 307.2.

INDUSTRIAL EQUIPMENT PLATFORM. See Section 502.1.

[F] INITIATING DEVICE. See Section 902.1.

INSPECTION CERTIFICATE. See Section 1702.1.

INTENDED TO BE OCCUPIED AS A RESIDENCE. See Section 1102.

INTERIOR FINISH. See Section 802.1.

INTERIOR FLOOR FINISH. See Section 802.1.

INTERIOR SURFACES. See Section 2502.1.

INTERIOR WALL AND CEILING FINISH. See Section 802.1.

INTERLAYMENT. See Section 1502.1.

INVERTED PENDULUM-TYPE STRUCTURES. See Section 1613.1.

ISOLATION INTERFACE. See Section 1613.1.

ISOLATION JOINT. See Section 1902.1.

ISOLATION SYSTEM. See Section 1613.1.

ISOLATOR UNIT. See Section 1613.1.

JOINT. See Sections 702.1 and 1602.1.

JURISDICTION. The governmental unit that has adopted this code under due legislative authority.

LABEL. See Section 1702.1.

LIGHT-DIFFUSING SYSTEM. See Section 2602.1.

LIGHT-FRAME CONSTRUCTION. A type of construction whose vertical and horizontal structural elements are primarily formed by a system of repetitive wood or light gage steel framing members.

LIGHT-TRANSMITTING PLASTIC ROOF PANELS. See Section 2602.1.

LIGHT-TRANSMITTING PLASTIC WALL PANELS. See Section 2602.1.

LIMIT STATE. See Section 1602.1.

[F] LIQUID. See Section 415.2.

[F] LIQUID STORAGE ROOM. See Section 415.2.

[F] LIQUID USE, DISPENSING AND MIXING ROOMS. See Section 415.2.

LISTED. See Section 902.1.

LIVE LOADS. See Section 1602.1.

LIVE LOADS (ROOF). See Section 1602.1.

LOAD. See Section 1613.1.

    Gravity load (*W*). See Section 1613.1.

LOAD AND RESISTANCE FACTOR DESIGN (LRFD). See Section 1602.1.

LOAD FACTOR. See Section 1602.1.

LOADS. See Section 1602.1.

LOADS EFFECTS. See Section 1602.1.

LOT. A portion or parcel of land considered as a unit.

LOT LINE. A line dividing one lot from another, or from a street or any public place.

[F] LOWER FLAMMABLE LIMIT (LFL). See Section 415.2.

LOWEST FLOOR. See Section 1612.2.

MAIN WINDFORCE-RESISTING SYSTEM. See Section 1609.2.

MALL. See Section 402.2.

[F] MANUAL FIRE ALARM BOX. See Section 902.1.

MANUFACTURER'S DESIGNATION. See Section 1702.1.

MARK. See Section 1702.1.

MARQUEE. A permanent roofed structure attached to and supported by the building and that projects into the public right-of-way.

MASONRY. See Section 2102.1.

    Ashlar masonry. See Section 2102.1.
    Coursed ashlar. See Section 2102.1.
    Glass unit masonry. See Section 2102.1.
    Plain masonry. See Section 2102.1.
    Random ashlar. See Section 2102.1.
    Reinforced masonry. See Section 2102.1.
    Solid masonry. See Section 2102.1.

MASONRY UNIT. See Section 2102.1.

    Clay. See Section 2102.1.
    Concrete. See Section 2102.1.
    Hollow. See Section 2102.1.
    Solid. See Section 2102.1.

MAXIMUM CONSIDERED EARTHQUAKE. See Section 1613.1.

MEAN DAILY TEMPERATURE. See Section 2102.1.

MEAN ROOF HEIGHT. See Section 1609.2.

MEANS OF EGRESS. See Section 1002.1.

MECHANICAL-ACCESS OPEN PARKING GARAGES. See Section 406.3.2.

MECHANICAL EQUIPMENT SCREEN. See Section 1502.1.

MEMBRANE-COVERED CABLE STRUCTURE. See Section 3102.2.

MEMBRANE-COVERED FRAME STRUCTURE. See Section 3102.2.

MEMBRANE PENETRATION. See Section 702.1.

MEMBRANE-PENETRATION FIRESTOP. See Section 702.1.

METAL COMPOSITE MATERIAL (MCM). See Section 1402.

METAL COMPOSITE MATERIAL SYSTEM. See Section 1402.

METAL ROOF PANEL. See Section 1502.1.

METAL ROOF SHINGLE. See Section 1502.1.

MEZZANINE. See Section 502.1.

MINERAL BOARD. See Section 721.1.1.

MODIFIED BITUMEN ROOF COVERING. See Section 1502.1.

MORTAR. See Section 2102.1.

MORTAR, SURFACE-BONDING. See Section 2102.1.

[F] MULTIPLE-STATION ALARM DEVICE. See Section 902.1.

[F] MULTIPLE-STATION SMOKE ALARM. See Section 902.1.

NAILING, BOUNDARY. See Section 2302.1.

NAILING, EDGE. See Section 2302.1.

NAILING, FIELD. See Section 2302.1.

NATURALLY DURABLE WOOD. See Section 2302.1.

    Decay resistant. See Section 2302.1.
    Termite resistant. See Section 2302.1.

NOMINAL LOADS. See Section 1602.1.

NOMINAL SIZE (LUMBER). See Section 2302.1.

NONBUILDING STRUCTURE. See Section 1613.1.

NONCOMBUSTIBLE MEMBRANE STRUCTURE. See Section 3102.2.

[F] NORMAL TEMPERATURE AND PRESSURE (NTP). See Section 415.2.

NOSING. See Section 1002.1.

[F] NUISANCE ALARM. See Section 902.1.

OCCUPANCY IMPORTANCE FACTOR. See Section 1613.1.

OCCUPANT LOAD. See Section 1002.1.

OCCUPIABLE SPACE. A room or enclosed space designed for human occupancy in which individuals congregate for amusement, educational or similar purposes or in which occupants are engaged at labor, and which is equipped with means of egress and light and ventilation facilities meeting the requirements of this code.

OPEN PARKING GARAGE. See Section 406.3.2.

[F] OPEN SYSTEM. See Section 307.2.

OPERATING BUILDING. See Section 307.2.

[F] ORGANIC PEROXIDE. See Section 307.2.

    Class I. See Section 307.2.
    Class II. See Section 307.2.
    Class III. See Section 307.2.
    Class IV. See Section 307.2.
    Class V. See Section 307.2.
    Unclassified detonable. See Section 307.2.

OTHER STRUCTURES. See Section 1602.1.

OWNER. Any person, agent, firm or corporation having a legal or equitable interest in the property.

[F] OXIDIZER. See Section 307.2.

    Class 4. See Section 307.2.
    Class 3. See Section 307.2.
    Class 2. See Section 307.2.
    Class 1. See Section 307.2.

[F] OXIDIZING GAS. See Section 307.2.

P-DELTA EFFECT. See Section 1602.1.

PANEL (PART OF A STRUCTURE). See Section 1602.1.

PANIC HARDWARE. See Section 1002.1.

PARTICLEBOARD. See Section 2302.1.

PEDESTAL. See Section 1902.1.

PENETRATION FIRESTOP. See Section 702.1.

PENTHOUSE. See Section 1502.1.

PERMIT. An official document or certificate issued by the authority having jurisdiction which authorizes performance of a specified activity.

PERSON. An individual, heirs, executors, administrators or assigns, and also includes a firm, partnership or corporation, its or their successors or assigns, or the agent of any of the aforesaid.

PERSONAL CARE SERVICE. See Section 310.2.

[F] PHYSICAL HAZARD. See Section 307.2.

PIER FOUNDATIONS. See Section 1808.1.

  Belled piers. See Section 1808.1.

PILE FOUNDATIONS. See Section 1808.1.

  Auger uncased piles. See Section 1808.1.
  Caisson piles. See Section 1808.1.
  Concrete-filled steel pipe and tube piles. See Section 1808.1.
  Driven uncased piles. See Section 1808.1.
  Enlarged base piles. See Section 1808.1.
  Piles. See Section 1808.1.
  Steel-cased piles. See Section 1808.1.

PINRAIL. See Section 410.2.

PLAIN CONCRETE. See Section 1902.1.

PLAIN REINFORCEMENT. See Section 1902.1.

PLASTIC, APPROVED. See Section 2602.1.

PLASTIC GLAZING. See Section 2602.1.

PLASTIC HINGE. See Section 2102.1.

PLATFORM. See Section 410.2.

POSITIVE ROOF DRAINAGE. See Section 1502.1.

PRECAST CONCRETE. See Section 1902.1.

PRESERVATIVE-TREATED WOOD. See Section 2302.1.

PRESTRESSED CONCRETE. See Section 1902.1.

PRESTRESSED MASONRY. See Section 2102.1.

  Prestressed masonry shear wall. See Section 2102.1.
  Ordinary plain prestressed masonry shear wall. See Section 2102.1.
  Special prestressed masonry shear wall. See Section 2102.1.
  Special reinforced masonry shear wall. See Section 2102.1.

PRISM. See Section 2102.1.

PROJECTED AREA. See Section 1913.2.2.

PROSCENIUM WALL. See Section 410.2.

PUBLIC ENTRANCE. See Section 1102.1.

PUBLIC-USE AREAS. See Section 1102.1.

PUBLIC WAY. See Section 1002.1.

[F] PYROPHORIC. See Section 307.2.

[F] PYROTECHNIC COMPOSITION. See Section 307.2.

QUALITY ASSURANCE PLAN. A written procedure complying with the requirements of Section 1705.

RAMP. See Section 1002.1.

RAMP-ACCESS OPEN PARKING GARAGES. See Section 406.3.2.

[F] RECORD DRAWINGS. See Section 902.1.

REFERENCE RESISTANCE (D). See Section 2302.1.

REGISTERED DESIGN PROFESSIONAL. An individual who is registered or licensed to practice their respective design profession as defined by the statutory requirements of the professional registration laws of the state or jurisdiction in which the project is to be constructed.

REINFORCED CONCRETE. See Section 1902.1.

REINFORCED PLASTIC, GLASS FIBER. See Section 2602.1.

REINFORCEMENT. See Section 1902.1.

REPAIR. The reconstruction or renewal of any part of an existing building for the purpose of its maintenance.

REQUIRED STRENGTH. See Sections 1602.1 and 2102.1.

REROOFING. See Section 1502.1.

RESHORES. See Section 1902.1.

RESIDENTIAL AIRCRAFT HANGAR. See Section 412.3.1.

RESIDENTIAL CARE/ASSISTED LIVING FACILITIES. See Section 310.2.

RESISTANCE FACTOR. See Section 1602.1.

RESTRICTED ENTRANCE. See Section 1102.1.

RETRACTABLE AWNING. See Section 3105.2.

ROOF ASSEMBLY. See Section 1502.1.

ROOF COVERING. See Section 1502.1.

ROOF COVERING SYSTEM. See Section 1502.1.

ROOF DECK. See Section 1502.1.

ROOF RECOVER. See Section 1502.1.

ROOF REPAIR. See Section 1502.1.

ROOF REPLACEMENT. See Section 1502.1.

ROOF VENTILATION. See Section 1502.1.

ROOFTOP STRUCTURE. See Section 1502.1.

RUBBLE MASONRY. See Section 2102.1.

  Coursed rubble. See Section 2102.1.
  Random rubble. See Section 2102.1.
  Rough or ordinary rubble. See Section 2102.1.

RUNNING BOND. See Section 2102.1.

SCISSOR STAIR See Section 1002.1.

SCUPPER. See Section 1502.1.

SEISMIC DESIGN CATEGORY. See Section 1613.1.

SEISMIC-FORCE-RESISTING SYSTEM. See Section 1613.1.

SEISMIC FORCES. See Section 1613.1.

SEISMIC RESPONSE COEFFICIENT. See Section 1613.1.

SEISMIC USE GROUP. See Section 1613.1.

SELF-CLOSING. See Section 702.1.

SELF-SERVICE STORAGE FACILITY. See Section 1102.1.

[F] SERVICE CORRIDOR. See Section 415.2.

SERVICE ENTRANCE. See Section 1102.1.

SHAFT. See Section 702.1.

SHAFT ENCLOSURE. See Section 702.1.

SHALLOW ANCHORS. See Section 1602.1.

SHEAR PANEL. See Section 1602.1.

SHEAR WALL. See Sections 1602.1, 1613.1 and 2102.1.

    Detailed plain masonry shear wall. See Section 2102.1.

    Intermediate reinforced masonry shear wall. See Section 2102.1.

    Ordinary plain masonry shear wall. See Section 2102.1.

    Ordinary reinforced masonry shear wall. See Section 2102.1.

    Perforated shear wall. See Section 2302.1.

    Perforated shear wall segment. See Section 2302.1.

    Special reinforced masonry shear wall. See Section 2102.1.

    Type I shear wall. See Section 2202.1.

    Type II shear wall. See Section 2202.1.

    Type II shear wall segment. See Section 2202.1.

SHEAR WALL-FRAME INTERACTIVE SYSTEM. See Section 1613.1.

SHELL. See Section 2102.1.

SHORES. See Section 1902.1.

SHOTCRETE. See Section 1914.1.

SIDE-FACE BLOWOUT STRENGTH. See Section 1913.2.2.

SINGLE-PLY MEMBRANE. See Section 1502.1.

[F] SINGLE-STATION SMOKE ALARM. See Section 902.1.

SITE. See Section 1102.1.

SITE CLASS. See Section 1613.1.

SITE COEFFICIENTS. See Section 1613.1.

SKYLIGHT, UNIT. A factory-assembled, glazed fenestration unit, containing one panel of glazing material that allows for natural lighting through an opening in the roof assembly while preserving the weather-resistant barrier of the roof.

SKYLIGHTS AND SLOPED GLAZING. Glass or other transparent or translucent glazing material installed at a slope of 15 degrees (0.26 rad) or more from vertical. Glazing material in skylights, including unit skylights, solariums, sunrooms, roofs and sloped walls, are included in this definition.

SLEEPING UNIT. A room or space in which people sleep, which can also include permanent provisions for living, eating, and either sanitation or kitchen facilities but not both. Such rooms and spaces that are also part of a dwelling unit are not sleeping units.

[F] SMOKE ALARM. See Section 902.1.

SMOKE BARRIER. See Section 702.1.

SMOKE COMPARTMENT. See Section 702.1.

SMOKE DAMPER. See Section 702.1.

[F] SMOKE DETECTOR. See Section 902.1.

SMOKE-DEVELOPED INDEX. See Section 802.1.

SMOKE-PROTECTED ASSEMBLY SEATING. See Section 1002.1.

SMOKEPROOF ENCLOSURE. See Section 902.1.

[F] SOLID. See Section 415.2.

SPACE FRAME. See Section 1602.1.

SPECIAL AMUSEMENT BUILDING. See Section 411.2.

SPECIAL INSPECTION. See Section 1702.1.

    Special continuous inspection. See Section 1702.1.

    Special periodic inspection. See Section 1702.1.

SPECIAL FLOOD HAZARD AREA. See Section 1612.2.

SPECIAL TRANSVERSE REINFORCEMENT. See Section 1602.1.

SPECIFIED. See Section 2102.1.

SPECIFIED COMPRESSIVE STRENGTH OF MASONRY (f'_m). See Section 2102.1.

SPIRAL REINFORCEMENT. See Section 1902.1.

SPLICE. See Section 702.1.

SPRAYED FIRE-RESISTANT MATERIALS. See Section 1702.1.

STACK BOND. See Section 2102.1.

STAGE. See Section 410.2.

STAIR. See Section 1002.1.

STAIRWAY. See Section 1002.1.

STAIRWAY, EXTERIOR. See Section 1002.1.

STAIRWAY, INTERIOR. See Section 1002.1.

STAIRWAY, SPIRAL. See Section 1002.1.

[F] STANDPIPE SYSTEM, CLASSES OF. See Section 902.1.

    Class I system. See Section 902.1.

    Class II system. See Section 902.1.

    Class III system. See Section 902.1.

[F] STANDPIPE, TYPES OF. See Section 902.1.

    Automatic dry. See Section 902.1.
    Automatic wet. See Section 902.1.
    Manual dry. See Section 902.1.
    Manual wet. See Section 902.1.
    Semiautomatic dry. See Section 902.1.

START OF CONSTRUCTION. See Section 1612.2.

STEEL CONSTRUCTION, COLD-FORMED. See Section 2202.1.

STEEL JOIST. See Section 2202.1.

STEEL MEMBER, STRUCTURAL. See Section 2202.1.

STEEP SLOPE. A roof slope greater than two units vertical in 12 units horizontal (17-percent slope).

STONE MASONRY. See Section 2102.1.

    Ashlar stone masonry. See Section 2102.1.
    Rubble stone masonry. See Section 2102.1.

[F] STORAGE, HAZARDOUS MATERIALS. See Section 415.2.

STORY. That portion of a building included between the upper surface of a floor and the upper surface of the floor or roof next above (also see "Basement," "Mezzanine" and Section 502.1). It is measured as the vertical distance from top to top of two successive tiers of beams or finished floor surfaces and, for the topmost story, from the top of the floor finish to the top of the ceiling joists or, where there is not a ceiling, to the top of the roof rafters.

STORY ABOVE GRADE PLANE. Any story having its finished floor surface entirely above grade plane, except that a basement shall be considered as a story above grade plane where the finished surface of the floor above the basement is:

1. More than 6 feet (1829 mm) above grade plane;
2. More than 6 feet (1829 mm) above the finished ground level for more than 50 percent of the total building perimeter; or
3. More than 12 feet (3658 mm) above the finished ground level at any point.

STORY DRIFT RATIO. See Section 1613.1.

STRENGTH. See Section 2102.1.

    Design strength. See Section 2102.1.
    Nominal strength. See Sections 1602.1 and 2102.1.

STRENGTH DESIGN. See Section 1602.1.

STRUCTURAL CONCRETE. See Section 1902.1.

STRUCTURAL GLUED-LAMINATED TIMBER. See Section 2302.1.

STRUCTURAL OBSERVATION. See Section 1702.1.

STRUCTURE. That which is built or constructed.

SUBDIAPHRAGM. See Section 2302.1.

SUBSTANTIAL DAMAGE. See Section 1612.2.

SUBSTANTIAL IMPROVEMENT. See Section 1612.2.

[F] SUPERVISING STATION. See Section 902.1.

[F] SUPERVISORY SERVICE. See Section 902.1.

[F] SUPERVISORY SIGNAL. See Section 902.1.

[F] SUPERVISORY SIGNAL-INITIATING DEVICE. See Section 902.1.

SWIMMING POOLS. See Section 3109.2.

T RATING. See Section 702.1.

TECHNICALLY INFEASIBLE. See Section 3402.

TENDON. See Section 1902.1.

TENT. Any structure, enclosure or shelter which is constructed of canvas or pliable material supported in any manner except by air or the contents it protects.

THERMOPLASTIC MATERIAL. See Section 2602.1.

THERMOSETTING MATERIAL. See Section 2602.1.

THROUGH PENETRATION. See Section 702.1.

THROUGH-PENETRATION FIRESTOP SYSTEM. See Section 702.1.

TIE-DOWN (HOLD-DOWN). See Section 2302.1.

TIE, LATERAL. See Section 2102.1.

TIE, WALL. See Section 2102.1.

TILE. See Section 2102.1.

TILE, STRUCTURAL CLAY. See Section 2102.1.

[F] TIRES, BULK STORAGE OF. See Section 902.1.

TORSIONAL FORCE DISTRIBUTION. See Section 1613.1.

TOUGHNESS. See Section 1613.1.

[F] TOXIC. See Section 307.2.

TREATED WOOD. See Section 2302.1.

TRIM. See Section 802.1.

[F] TROUBLE SIGNAL. See Section 902.1.

UNADJUSTED SHEAR RESISTANCE. See Section 2202.1.

UNDERLAYMENT. See Section 1502.1.

[F] UNSTABLE (REACTIVE) MATERIAL. See Section 307.2.

    Class 4. See Section 307.2.
    Class 3. See Section 307.2.
    Class 2. See Section 307.2.
    Class 1. See Section 307.2.

[F] USE (MATERIAL). See Section 415.2.

VAPOR-PERMEABLE MEMBRANE. A material or covering having a permeance rating of 5 perms (52.9×10^{-10} kg/Pa · s · m²) or greater, when tested in accordance with the dessicant method using Procedure A of ASTM E 96. A vapor-permeable material permits the passage of moisture vapor.

VAPOR RETARDER. A vapor-resistant material, membrane or covering such as foil, plastic sheeting or insulation facing having a permeance rating of 1 perm (5.7×10^{-11} kg/Pa · s · m²) or less, when tested in accordance with the dessicant method using Procedure A of ASTM E 96. Vapor retarders limit the amount of moisture vapor that passes through a material or wall assembly.

VENEER. See Section 1402.1.

VENTILATION. The natural or mechanical process of supplying conditioned or unconditioned air to, or removing such air from, any space.

[F] VISIBLE ALARM NOTIFICATION APPLIANCE. See Section 902.1.

WALKWAY, PEDESTRIAN. A walkway used exclusively as a pedestrian trafficway.

WALL. See Section 2102.1.

   Cavity wall. See Section 2102.1.

   Composite wall. See Section 2102.1.

   Dry-stacked, surface-bonded wall. See Section 2102.1.

   Masonry-bonded hollow wall. See Section 2102.1.

   Parapet wall. See Section 2102.1.

WALL, LOAD-BEARING. See Section 1602.1.

WALL, NONLOAD-BEARING. See Section 1602.1.

[F] WATER-REACTIVE MATERIAL. See Section 307.2.

   Class 3. See Section 307.2.

   Class 2. See Section 307.2.

   Class 1. See Section 307.2.

WEATHER-EXPOSED SURFACES. See Section 2502.1.

WEB. See Section 2102.1.

[F] WET-CHEMICAL EXTINGUISHING SYSTEM. See Section 902.1.

WHEELCHAIR SPACE. See Section 1102.1.

WHEELCHAIR SPACE CLUSTER. See Section 1102.1.

WIND-BORNE DEBRIS REGION. See Section 1609.2.

WIND-RESTRAINT SEISMIC SYSTEM. See Section 1613.

WIRE BACKING. See Section 2502.1.

[F] WIRELESS PROTECTION SYSTEM. See Section 902.1.

WOOD SHEAR PANEL. See Section 2302.1.

WOOD STRUCTURAL PANEL. See Section 2302.1.

   Composite panels. See Section 2302.1.

   Oriented strand board (OSB). See Section 2302.1.

   Plywood. See Section 2302.1.

[F] WORKSTATION. See Section 415.2.

WYTHE. See Section 2102.1.

YARD. An open space, other than a court, unobstructed from the ground to the sky, except where specifically provided by this code, on the lot on which a building is situated.

[F] ZONE. See Section 902.1.

CHAPTER 3

USE AND OCCUPANCY CLASSIFICATION

SECTION 301
GENERAL

301.1 Scope. The provisions of this chapter shall control the classification of all buildings and structures as to use and occupancy.

SECTION 302
CLASSIFICATION

302.1 General. Structures or portions of structures shall be classified with respect to occupancy in one or more of the groups listed below. Structures with multiple uses shall be classified according to Section 302.3. Where a structure is proposed for a purpose which is not specifically provided for in this code, such structure shall be classified in the group which the occupancy most nearly resembles, according to the fire safety and relative hazard involved.

1. Assembly (see Section 303): Groups A-1, A-2, A-3, A-4 and A-5

2. Business (see Section 304): Group B

3. Educational (see Section 305): Group E

4. Factory and Industrial (see Section 306): Groups F-1 and F-2

5. High Hazard (see Section 307): Groups H-1, H-2, H-3, H-4 and H-5

6. Institutional (see Section 308): Groups I-1, I-2, I-3 and I-4

7. Mercantile (see Section 309): Group M

8. Residential (see Section 310): Groups R-1, R-2, R-3 as applicable in Section 101.2, and R-4

9. Storage (see Section 311): Groups S-1 and S-2

10. Utility and Miscellaneous (see Section 312): Group U

302.1.1 Incidental use areas. Spaces which are incidental to the main occupancy shall be separated or protected, or both, in accordance with Table 302.1.1 or the building shall be classified as a mixed occupancy and comply with Section 302.3. Areas that are incidental to the main occupancy shall be classified in accordance with the main occupancy of the portion of the building in which the incidental use area is located.

> **Exception:** Incidental use areas within and serving a dwelling unit are not required to comply with this section.

302.1.1.1 Separation. Where Table 302.1.1 requires a fire-resistance-rated separation, the incidental use area shall be separated from the remainder of the building with a fire barrier. Where Table 302.1.1 permits an automatic fire-extinguishing system without a fire barrier, the incidental use area shall be separated by construction capable of resisting the passage of smoke. The partitions shall extend from the floor to the underside of the fire-resistance-rated floor/ceiling assembly or fire-resistance-rated roof/ceiling assembly or to the underside of the floor or roof deck above. Doors shall be self-closing or automatic-closing upon detection of smoke. Doors shall not have air transfer openings and shall not be undercut in excess of the clearance permitted in accordance with NFPA 80.

TABLE 302.1.1
INCIDENTAL USE AREAS

ROOM OR AREA	SEPARATION[a]
Furnace room where any piece of equipment is over 400,000 Btu per hour input	1 hour or provide automatic fire-extinguishing system
Rooms with any boiler over 15 psi and 10 horsepower	1 hour or provide automatic fire-extinguishing system
Refrigerant machinery rooms	1 hour or provide automatic sprinkler system
Parking garage (Section 406.2)	2 hours; or 1 hour and provide automatic fire-extinguishing system
Hydrogen cut-off rooms	1-hour fire barriers and floor/ceiling assemblies in Group B, F, H, M, S and U occupancies. 2-hour fire barriers and floor/ceiling assemblies in Group A, E, I and R occupancies.
Incinerator rooms	2 hours and automatic sprinkler system
Paint shops, not classified as Group H, located in occupancies other than Group F	2 hours; or 1 hour and provide automatic fire-extinguishing system
Laboratories and vocational shops, not classified as Group H, located in Group E or I-2 occupancies	1 hour or provide automatic fire-extinguishing system
Laundry rooms over 100 square feet	1 hour or provide automatic fire-extinguishing system
Storage rooms over 100 square feet	1 hour or provide automatic fire-extinguishing system
Group I-3 cells equipped with padded surfaces	1 hour
Group I-2 waste and linen collection rooms	1 hour
Waste and linen collection rooms over 100 square feet	1 hour or provide automatic fire-extinguishing system
Stationary lead-acid battery systems having a liquid capacity of more than 100 gallons used for facility standby power, emergency power or uninterrupted power supplies	1-hour fire barriers and floor/ceiling assemblies in Group B, F, H, M, S and U occupancies. 2-hour fire barriers and floor/ceiling assemblies in Group A, E, I and R occupancies

For SI: 1 square foot = 0.0929 m^2, 1 pound per square inch = 6.9 kPa,
1 British thermal unit = 0.293 watts, 1 horsepower = 746 watts,
1 gallon = 3.785 L.

a. Where an automatic fire-extinguishing system is provided, it need only be provided in the incidental use room or area.

302.2 Accessory use areas. A fire barrier shall be required to separate accessory use areas classified as Group H in accordance with Section 302.3.2, and incidental use areas in accordance with Section 302.1.1. Any other accessory use area shall not be required to be separated by a fire barrier provided the accessory use area occupies an area not more than 10 percent of the area of the story in which it is located and does not exceed the tabular values in Table 503 for the allowable height or area for such use.

302.2.1 Assembly areas. Accessory assembly areas are not considered separate occupancies if the floor area is equal to or less than 750 square feet (69.7 m²). Assembly areas that are accessory to Group E are not considered separate occupancies. Accessory religious educational rooms and religious auditoriums with occupant loads of less than 100 are not considered separate occupancies.

302.3 Mixed occupancies. Where a building is occupied by two or more uses not included in the same occupancy classification, the building or portion thereof shall comply with Section 302.3.1 or 302.3.2 or a combination of these sections.

Exceptions:

1. Occupancies separated in accordance with Section 508.

2. Areas of Group H-2, H-3, H-4 or H-5 occupancies shall be separated from any other occupancy in accordance with Section 302.3.2.

3. Where required by Table 415.3.2, areas of Group H-1, H-2 or H-3 occupancy shall be located in a separate and detached building or structure.

4. Accessory use areas in accordance with Section 302.2.

5. Incidental use areas in accordance with Section 302.1.1.

302.3.1 Nonseparated uses. Each portion of the building shall be individually classified as to use. The required type of construction for the building shall be determined by applying the height and area limitations for each of the applicable occupancies to the entire building. The most restrictive type of construction, so determined, shall apply to the entire building. All other code requirements shall apply to each portion of the building based on the use of that space except that the most restrictive applicable provisions of Section 403 and Chapter 9 shall apply to these nonseparated uses. Fire separations are not required between uses, except as required by other provisions.

302.3.2 Separated uses. Each portion of the building shall be individually classified as to use and shall be completely separated from adjacent areas by fire barrier walls or horizontal assemblies or both having a fire-resistance rating determined in accordance with Table 302.3.2 for uses being separated. Each fire area shall comply with this code based on the use of that space. Each fire area shall comply with the height limitations based on the use of that space and the type of construction classification. In each story, the building area shall be such that the sum of the ratios of the floor area of each use divided by the allowable area for each use shall not exceed one.

Exception: Except for Group H and I-2 areas, where the building is equipped throughout with an automatic sprinkler system, installed in accordance with Section 903.3.1.1, the fire-resistance ratings in Table 302.3.2 shall be reduced by 1 hour but to not less than 1 hour and to not less than that required for floor construction according to the type of construction.

302.4 Spaces used for different purposes. A room or space that is intended to be occupied at different times for different purposes shall comply with all the requirements that are applicable to each of the purposes for which the room or space will be occupied.

SECTION 303
ASSEMBLY GROUP A

303.1 Assembly Group A. Assembly Group A occupancy includes, among others, the use of a building or structure, or a portion thereof, for the gathering together of persons for purposes such as civic, social or religious functions, recreation, food or drink consumption or awaiting transportation. A room or space used for assembly purposes by less than 50 persons and accessory to another occupancy shall be included as a part of that occupancy. Assembly areas with less than 750 square feet (69.7 m²) and which are accessory to another occupancy according to Section 302.2.1 are not assembly occupancies. Assembly occupancies which are accessory to Group E in accordance with Section 302.2 are not considered assembly occupancies. Religious educational rooms and religious auditoriums which are accessory to churches in accordance with Section 302.2 and which have occupant loads of less than 100 shall be classified as A-3.

Assembly occupancies shall include the following:

A-1 Assembly uses, usually with fixed seating, intended for the production and viewing of the performing arts or motion pictures including, but not limited to:
   Motion picture theaters
   Symphony and concert halls
   Television and radio studios admitting an audience
   Theaters

A-2 Assembly uses intended for food and/or drink consumption including, but not limited to:
   Banquet halls
   Night clubs
   Restaurants
   Taverns and bars

A-3 Assembly uses intended for worship, recreation or amusement and other assembly uses not classified elsewhere in Group A including, but not limited to:
   Amusement arcades
   Art galleries
   Bowling alleys
   Churches
   Community halls
   Courtrooms
   Dance halls (not including food or drink consumption)
   Exhibition halls

TABLE 302.3.2
REQUIRED SEPARATION OF OCCUPANCIES (HOURS)[a]

USE	A-1	A-2	A-3	A-4	A-5	B[b]	E	F-1	F-2	H-1	H-2	H-3	H-4	H-5	I-1	I-2	I-3	I-4	M[b]	R-1	R-2	R-3, R-4	S-1	S-2[c]	U
A-1	—	2	2	2	2	2	2	3	2	NP	4	3	2	4	2	2	2	2	2	2	2	2	3	2	1
A-2[e]	—	—	2	2	2	—	2	3	2	NP	4	3	2	4	2	2	2	2	2	2	2	2	3	2	1
A-3	—	—	—	—	—	—	2	3	2	NP	4	3	2	4	2	2	2	2	2	2	2	2	3	2	1
A-4	—	—	—	—	2	—	2	3	2	NP	4	3	2	4	2	2	2	2	2	2	2	2	3	2	1
A-5	—	—	—	—	—	2	2	3	2	NP	4	3	2	4	2	2	2	2	2	2	2	2	3	2	1
B[b]	—	—	—	—	—	—	2	3	2	NP	2	1	1	1	2	2	2	2	—	2	2	2	3	2	1
E	—	—	—	—	—	—	—	3	2	NP	4	3	2	3	2	2	2	2	2	2	2	2	3	2	1
F-1	—	—	—	—	—	—	—	—	3	NP	2	1	1	1	3	3	3	3	3	3	3	3	3	3	3
F-2	—	—	—	—	—	—	—	—	—	NP	2	1	1	1	2	2	2	2	2	2	2	2	3	2	1
H-1	—	—	—	—	—	—	—	—	—	—	NP	NP	NP	NP	NP	NP	NP	NP	NP	NP	NP	NP	NP	NP	NP
H-2	—	—	—	—	—	—	—	—	—	—	—	1	2	2	4	4	4	4	2	4	4	4	2	2	1
H-3	—	—	—	—	—	—	—	—	—	—	—	—	1	1	4	3	4	3	2	3	3	3	1	1	1
H-4	—	—	—	—	—	—	—	—	—	—	—	—	—	1	4	4	4	4	2	4	4	4	1	1	1
H-5	—	—	—	—	—	—	—	—	—	—	—	—	—	—	4	4	4	3	1	4	4	4	1	1	3
I-1	—	—	—	—	—	—	—	—	—	—	—	—	—	—	—	2	2	2	2	2	2	2	4	3	2
I-2	—	—	—	—	—	—	—	—	—	—	—	—	—	—	—	—	2	2	2	2	2	2	3	2	1
I-3	—	—	—	—	—	—	—	—	—	—	—	—	—	—	—	—	—	2	2	2	2	2	3	2	1
I-4	—	—	—	—	—	—	—	—	—	—	—	—	—	—	—	—	—	—	2	2	2	2	3	2	1
M[b]	—	—	—	—	—	—	—	—	—	—	—	—	—	—	—	—	—	—	—	2	2	2	3	2	1
R-1	—	—	—	—	—	—	—	—	—	—	—	—	—	—	—	—	—	—	—	—	2	2	3	2	1
R-2	—	—	—	—	—	—	—	—	—	—	—	—	—	—	—	—	—	—	—	—	—	2	3	2	1
R-3, R-4	—	—	—	—	—	—	—	—	—	—	—	—	—	—	—	—	—	—	—	—	—	—	3	2[d]	1[d]
S-1	—	—	—	—	—	—	—	—	—	—	—	—	—	—	—	—	—	—	—	—	—	—	—	3	3
S-2[c]	—	—	—	—	—	—	—	—	—	—	—	—	—	—	—	—	—	—	—	—	—	—	—	—	1
U	—	—	—	—	—	—	—	—	—	—	—	—	—	—	—	—	—	—	—	—	—	—	—	—	—

For SI: 1 square foot = 0.0929 m².

NP = Not permitted.

a. See exception to Section 302.3.2 for reductions permitted.

b. Occupancy separation need not be provided for storage areas within Groups B and M if the:
1. Area is less than 10 percent of the floor area;
2. Area is provided with an automatic fire-extinguishing system and is less than 3,000 square feet; or
3. Area is less than 1,000 square feet.

c. Areas used only for private or pleasure vehicles shall be allowed to reduce separation by 1 hour.

d. See Section 406.1.4.

e. Commercial kitchens need not be separated from the restaurant seating areas that they serve.

Funeral parlors
Gymnasiums (without spectator seating)
Indoor swimming pools (without spectator seating)
Indoor tennis courts (without spectator seating)
Lecture halls
Libraries
Museums
Waiting areas in transportation terminals
Pool and billiard parlors

A-4 Assembly uses intended for viewing of indoor sporting events and activities with spectator seating including, but not limited to:
Arenas
Skating rinks
Swimming pools
Tennis courts

A-5 Assembly uses intended for participation in or viewing outdoor activities including, but not limited to:
Amusement park structures
Bleachers
Grandstands
Stadiums

303.1.1 Nonaccessory assembly use. A building or tenant space used for assembly purposes by less than 50 persons shall be considered a Group B occupancy.

SECTION 304
BUSINESS GROUP B

304.1 Business Group B. Business Group B occupancy includes, among others, the use of a building or structure, or a portion thereof, for office, professional or service-type transactions, including storage of records and accounts. Business occupancies shall include, but not be limited to, the following:

Airport traffic control towers
Animal hospitals, kennels and pounds
Banks
Barber and beauty shops
Car wash
Civic administration
Clinic—outpatient
Dry cleaning and laundries; pick-up and delivery stations and self-service
Educational occupancies above the 12th grade
Electronic data processing
Laboratories; testing and research
Motor vehicle showrooms
Post offices
Print shops
Professional services (architects, attorneys, dentists, physicians, engineers, etc.)
Radio and television stations
Telephone exchanges

SECTION 305
EDUCATIONAL GROUP E

305.1 Educational Group E. Educational Group E occupancy includes, among others, the use of a building or structure, or a portion thereof, by six or more persons at any one time for educational purposes through the 12th grade. Religious educational rooms and religious auditoriums, which are accessory to churches in accordance with Section 302.2 and have occupant loads of less than 100, shall be classified as A-3 occupancies.

305.2 Day care. The use of a building or structure, or portion thereof, for educational, supervision or personal care services for more than five children older than 2$^1/_2$ years of age, shall be classified as a Group E occupancy.

SECTION 306
FACTORY GROUP F

306.1 Factory Industrial Group F. Factory Industrial Group F occupancy includes, among others, the use of a building or structure, or a portion thereof, for assembling, disassembling, fabricating, finishing, manufacturing, packaging, repair or processing operations that are not classified as a Group H hazardous or Group S storage occupancy.

306.2 Factory Industrial F-1 Moderate-Hazard Occupancy. Factory industrial uses which are not classified as Factory Industrial F-2 Low Hazard shall be classified as F-1 Moderate Hazard and shall include, but not be limited to, the following:

Aircraft
Appliances
Athletic equipment
Automobiles and other motor vehicles
Bakeries
Beverages; over 12-percent alcohol content
Bicycles
Boats
Brooms or brushes
Business machines
Cameras and photo equipment
Canvas or similar fabric
Carpets and rugs (includes cleaning)
Clothing
Construction and agricultural machinery
Disinfectants
Dry cleaning and dyeing
Electric generation plants
Electronics
Engines (including rebuilding)
Food processing
Furniture
Hemp products
Jute products
Laundries
Leather products
Machinery
Metals
Millwork (sash & door)
Motion pictures and television filming (without spectators)
Musical instruments
Optical goods
Paper mills or products
Photographic film
Plastic products

Printing or publishing
Recreational vehicles
Refuse incineration
Shoes
Soaps and detergents
Textiles
Tobacco
Trailers
Upholstering
Wood; distillation
Woodworking (cabinet)

306.3 Factory Industrial F-2 Low-Hazard Occupancy. Factory industrial uses that involve the fabrication or manufacturing of noncombustible materials which during finishing, packing or processing do not involve a significant fire hazard shall be classified as F-2 occupancies and shall include, but not be limited to, the following:

Beverages; up to and including 12-percent alcohol content
Brick and masonry
Ceramic products
Foundries
Glass products
Gypsum
Ice
Metal products (fabrication and assembly)

SECTION 307
HIGH-HAZARD GROUP H

[F] 307.1 High-Hazard Group H. High-Hazard Group H occupancy includes, among others, the use of a building or structure, or a portion thereof, that involves the manufacturing, processing, generation or storage of materials that constitute a physical or health hazard in quantities in excess of those found in Tables 307.7(1) and 307.7(2) (see also definition of "Control area").

[F] 307.2 Definitions. The following words and terms shall, for the purposes of this section and as used elsewhere in this code, have the meanings shown herein.

AEROSOL. A product that is dispensed from an aerosol container by a propellant.

Aerosol products shall be classified by means of the calculation of their chemical heats of combustion and shall be designated Level 1, 2 or 3.

Level 1 aerosol products. Those with a total chemical heat of combustion that is less than or equal to 8,600 British thermal units per pound (Btu/lb) (20 kJ/g).

Level 2 aerosol products. Those with a total chemical heat of combustion that is greater than 8,600 Btu/lb (20 kJ/g), but less than or equal to 13,000 Btu/lb (30 kJ/g).

Level 3 aerosol products. Those with a total chemical heat combustion that is greater than 13,000 Btu/lb (30 kJ/g).

AEROSOL CONTAINER. A metal can or a glass or plastic bottle designed to dispense an aerosol. Metal cans shall be limited to a maximum size of 33.8 fluid ounces (1,000 ml). Glass or plastic bottles shall be limited to a maximum size of 4 fluid ounces (118 ml).

BARRICADE. A structure that consists of a combination of walls, floor and roof, which is designed to withstand the rapid release of energy in an explosion and which is fully confined, partially vented or fully vented; or other effective method of shielding from explosive materials by a natural or artificial barrier.

Artificial barricade. An artificial mound or revetment a minimum thickness of 3 feet (914 mm).

Natural barricade. Natural features of the ground, such as hills, or timber of sufficient density that the surrounding exposures that require protection cannot be seen from the magazine or building containing explosives when the trees are bare of leaves.

BOILING POINT. The temperature at which the vapor pressure of a liquid equals the atmospheric pressure of 14.7 pounds per square inch (psi) (101 kPa) gage or 760 mm of mercury. Where an accurate boiling point is unavailable for the material in question, or for mixtures which do not have a constant boiling point, for the purposes of this classification, the 20-percent evaporated point of a distillation performed in accordance with ASTM D 86 shall be used as the boiling point of the liquid.

CLOSED SYSTEM. The use of a solid or liquid hazardous material involving a closed vessel or system that remains closed during normal operations where vapors emitted by the product are not liberated outside of the vessel or system and the product is not exposed to the atmosphere during normal operations; and all uses of compressed gases. Examples of closed systems for solids and liquids include product conveyed through a piping system into a closed vessel, system or piece of equipment.

COMBUSTIBLE DUST. Finely divided solid material that is 420 microns or less in diameter and which, when dispersed in air in the proper proportions, could be ignited by a flame, spark or other source of ignition. Combustible dust will pass through a U.S. No. 40 standard sieve.

COMBUSTIBLE FIBERS. Readily ignitable and free-burning fibers, such as cocoa fiber, cloth, cotton, excelsior, hay, hemp, henequen, istle, jute, kapok, oakum, rags, sisal, Spanish moss, straw, tow, wastepaper, certain synthetic fibers or other like materials.

COMBUSTIBLE LIQUID. A liquid having a closed cup flash point at or above 100°F (38°C). Combustible liquids shall be subdivided as follows:

Class II. Liquids having a closed cup flash point at or above 100°F (38°C) and below 140°F (60°C).

Class IIIA. Liquids having a closed cup flash point at or above 140°F (60°C) and below 200°F (93°C).

Class IIIB. Liquids having a closed cup flash point at or above 200°F (93°C).

The category of combustible liquids does not include compressed gases or cryogenic fluids.

[F] TABLE 307.7(1)
MAXIMUM ALLOWABLE QUANTITY PER CONTROL AREA OF HAZARDOUS MATERIALS POSING A PHYSICAL HAZARD[a, j, m, n]

MATERIAL	CLASS	GROUP WHEN THE MAXIMUM ALLOWABLE QUANTITY IS EXCEEDED	STORAGE[b] Solid pounds (cubic feet)	STORAGE[b] Liquid gallons (pounds)	STORAGE[b] Gas (cubic feet at NTP)	USE-CLOSED SYSTEMS[b] Solid pounds (cubic feet)	USE-CLOSED SYSTEMS[b] Liquid gallons (pounds)	USE-CLOSED SYSTEMS[b] Gas (cubic feet at NTP)	USE-OPEN SYSTEMS[a, j, m, n] Solid pounds (cubic feet)	USE-OPEN SYSTEMS[a, j, m, n] Liquid gallons (pounds)
Combustible liquid[c, i]	II	H-2 or H-3	N/A	120[d, e]	N/A	N/A	120[d]	N/A	N/A	30[d]
	IIIA	H-2 or H-3	N/A	330[d, e]	N/A	N/A	330[d]	N/A	N/A	80[d]
	IIIB	N/A	N/A	13,200[e, f]	N/A	N/A	13,200[e, f]	N/A	N/A	3,300[f]
Combustible fiber	Loose	H-3	(100)	N/A	N/A	(100)	N/A	N/A	(20)	N/A
	Baled	H-3	(1,000)	N/A	N/A	(1,000)	N/A	N/A	(200)	N/A
Consumer fireworks (Class C, Common)	1.4G	H-3	125[d, e, l]	N/A	N/A	N/A	N/A	N/A	N/A	N/A
Cryogenics flammable	N/A	H-2	N/A	45[d]	N/A	N/A	45[d]	N/A	N/A	10[d]
Cryogenics, oxidizing	N/A	H-3	N/A	45[d]	N/A	N/A	45[d]	N/A	N/A	10[d]
Explosives	Division 1.1	H-1	1[e, g]	(1)[e, g]	N/A	0.25[g]	(0.25)[g]	N/A	0.25[g]	(0.25)[g]
	Division 1.2	H-1	1[e, g]	(1)[e, g]	N/A	0.25[g]	(0.25)[g]	N/A	0.25[g]	(0.25)[g]
	Division 1.3	H-1 or 2	5[e, g]	(5)[e, g]	N/A	1[g]	(1)[g]	N/A	1[g]	(1)[g]
	Division 1.4	H-3	50[e, g]	(50)[e, g]	N/A	50[g]	(50)[g]	N/A	N/A	N/A
	Division 1.4G	H-3	125[d, e, l]	N/A	N/A	N/A	N/A	N/A	N/A	N/A
	Division 1.5	H-1	1[e, g]	(1)[e, g]	N/A	0.25[g]	(0.25)[g]	N/A	0.25[g]	(0.25)[g]
	Division 1.6	H-1	1[d, e, g]	N/A	N/A	N/A	N/A	N/A	N/A	N/A
Flammable gas	Gaseous	H-2	N/A	N/A	1,000[d, e]	N/A	N/A	1,000[d, e]	N/A	N/A
	liquefied		N/A	30[d, e]	N/A	N/A	30[d, e]	N/A	N/A	N/A
Flammable liquid[c]	1A	H-2	N/A	30[d, e]	N/A	N/A	30[d]	N/A	N/A	10[d]
	1B and 1C	H-2 or H-3	N/A	120[d, e]	N/A	N/A	120[d]	N/A	N/A	30[d]
Combination flammable liquid (1A, 1B, 1C)	N/A	H-2 or H-3	N/A	120[d, e, h]	N/A	N/A	120[d, h]	N/A	N/A	30[d, h]
Flammable solid	N/A	H-3	125[d, e]	N/A	N/A	125[d, e]	N/A	N/A	25[d]	N/A
Organic peroxide	UD	H-1	1[e, g]	(1)[e, g]	N/A	0.25[g]	(0.25)[g]	N/A	0.25[g]	(0.25)[g]
	I	H-2	5[d, e]	(5)[d, e]	N/A	1[d]	(1)	N/A	1[d]	(1)[d]
	II	H-3	50[d, e]	(50)[d, e]	N/A	50[d]	(50)[d]	N/A	10[d]	(10)[d]
	III	H-3	125[d, e]	(125)[d, e]	N/A	125[d]	(125)[d]	N/A	25[d]	(25)[d]
	IV	N/A	NL	NL	N/A	NL	NL	N/A	NL	NL
	V	N/A	NL	NL	N/A	NL	NL	N/A	NL	NL
Oxidizer	4	H-1	1[e, g]	(1)[e, g]	N/A	0.25[g]	(0.25)[g]	N/A	0.25[g]	(0.25)[g]
	3[k]	H-2	10[d, e]	(10)[d, e]	N/A	2[d]	(2)[d]	N/A	2[d]	(2)[d]
	2	H-3	250[d, e]	(250)[d, e]	N/A	250[d]	(250)[d]	N/A	50[d]	(50)[d]
	1	H-3	4,000[e, f]	(4,000)[e, f]	N/A	4,000[f]	(4,000)[f]	N/A	1,000[f]	(1,000)[f]
Oxidizing gas	Gaseous	H-3	N/A	N/A	1,500[d, e]	N/A	N/A	1,500[d, e]	N/A	N/A
	liquefied		N/A	15[d, e]	N/A	N/A	15[d, e]	N/A	N/A	N/A

(continued)

[F] TABLE 307.7(1)—continued
MAXIMUM ALLOWABLE QUANTITY PER CONTROL AREA OF HAZARDOUS MATERIALS POSING A PHYSICAL HAZARD[a, j, m, n]

MATERIAL	CLASS	GROUP WHEN THE MAXIMUM ALLOWABLE QUANTITY IS EXCEEDED	STORAGE[b]			USE-CLOSED SYSTEMS[b]			USE-OPEN SYSTEMS[b]	
			Solid pounds (cubic feet)	Liquid gallons (pounds)	Gas (cubic feet at NTP)	Solid pounds (cubic feet)	Liquid gallons (pounds)	Gas (cubic feet at NTP)	Solid pounds (cubic feet)	Liquid gallons (pounds)
Pyrophoric material	N/A	H-2	4[e, g]	(4)[e, g]	50[e, g]	1[g]	(1)[g]	10[e, g]	0	0
Unstable (reactive)	4	H-1	1[e, g]	(1)[e, g]	10[d, g]	0.25[g]	(0.25)[g]	2[e, g]	0.25[g]	(0.25)[g]
	3	H-1 or H-2	5[d, e]	(5)[d, e]	50[d, e]	1[d]	(1)	10[d, e]	1[d]	(1)[d]
	2	H-3	50[d, e]	(50)[d, e]	250[d, e]	50[d]	(50)[d]	250[d, e]	10[d]	(10)[d]
	1	N/A	NL	NL	NL	NL	N/L	NL	NL	NL
Water reactive	3	H-2	5[d, e]	(5)[d, e]	N/A	5[d]	(5)[d]	N/A	1[d]	(1)[d]
	2	H-3	50[d, e]	(50)[d, e]	N/A	50[d]	(50)[d]	N/A	10[d]	(10)[d]
	1	N/A	NL	NL	N/A	NL	NL	N/A	NL	NL

For SI: 1 cubic foot = 0.023 m^3, 1 pound = 0.454 kg, 1 gallon = 3.785 L.

NL = Not Limited; N/A = Not Applicable; UD = Unclassified Detonable

a. For use of control areas, see Section 414.2.

b. The aggregate quantity in use and storage shall not exceed the quantity listed for storage.

c. The quantities of alcoholic beverages in retail and wholesale sales occupancies shall not be limited providing the liquids are packaged in individual containers not exceeding 1.3 gallons. In retail and wholesale sales occupancies, the quantities of medicines, foodstuffs, consumer or industrial products, and cosmetics containing not more than 50 percent by volume of water-miscible liquids with the remainder of the solutions not being flammable, shall not be limited, provided that such materials are packaged in individual containers not exceeding 1.3 gallons.

d. Maximum allowable quantities shall be increased 100 percent in buildings equipped throughout with an automatic sprinkler system in accordance with Section 903.3.1.1. Where Note e also applies, the increase for both notes shall be applied accumulatively.

e. Maximum allowable quantities shall be increased 100 percent when stored in approved storage cabinets, gas cabinets, exhausted enclosures or safety cans as specified in the *International Fire Code*. Where Noted d also applies, the increase for both notes shall be applied accumulatively.

f. The permitted quantities shall not be limited in a building equipped throughout with an automatic sprinkler system in accordance with Section 903.3.1.1.

g. Permitted only in buildings equipped throughout with an automatic sprinkler system in accordance with Section 903.3.1.1.

h. Containing not more than the maximum allowable quantity per control area of Class IA, IB or IC flammable liquids.

i. Inside a building, the maximum capacity of a combustible liquid storage system that is connected to a fuel-oil piping system shall be 660 gallons provided such system conforms to the *International Fire Code*.

j. Quantities in parenthesis indicate quantity units in parenthesis at the head of each column.

k. A maximum quantity of 200 pounds of solid or 20 gallons of liquid Class 3 oxidizers is allowed when such materials are necessary for maintenance purposes, operation or sanitation of equipment. Storage containers and the manner of storage shall be approved.

l. Net weight of the pyrotechnic composition of the fireworks. Where the net weight of the pyrotechnic composition of the fireworks is not known, 25 percent of the gross weight of the fireworks, including packaging, shall be used.

m. For gallons of liquids, divide in pounds by 10 in accordance with Section 2703.1.2 of the *International Fire Code*.

n. For storage and display quantities in Group M and storage quantities in Group S occupancies complying with Section 414.2.4, see Table 414.2.4.

[F] TABLE 307.7(2)
MAXIMUM ALLOWABLE QUANTITY PER CONTROL AREA OF HAZARDOUS MATERIAL POSING A HEALTH HAZARD[a, b, c]

MATERIAL	STORAGE[d]			USE-CLOSED SYSTEMS[d]			USE-OPEN SYSTEMS[d]	
	Solid pounds[e, f]	Liquid gallons (pounds)[e, f]	Gas (cubic feet at NTP)[e]	Solid pounds[e]	Liquid gallons (pounds)[e]	Gas (cubic feet at NTP)[e]	Solid pounds[e]	Liquid gallons (pounds)[e]
Corrosive	5,000	500	810[f, g]	5,000	500	810[f, g]	1,000	100
Highly toxic	10	(10)[i]	20[h]	10	(10)[i]	20[h]	3	(3)[i]
Toxic	500	(500)[i]	810[f]	500	(500)[i]	810[f]	125	(125)[i]

For SI: 1 cubic foot = 0.028 m³, 1 pound = 0.454 kg, 1 gallon = 3.785 L.

a. For use of control areas, see Section 414.2.

b. In retail and wholesale sales occupancies, the quantities of medicines, foodstuffs, consumer or industrial products, and cosmetics, containing not more than 50 percent by volume of water-miscible liquids and with the remainder of the solutions not being flammable, shall not be limited, provided that such materials are packaged in individual containers not exceeding 1.3 gallons.

c. For storage and display quantities in Group M and storage quantities in Group S occupancies complying with Section 414.2.4, see Table 414.2.4.

d. The aggregate quantity in use and storage shall not exceed the quantity listed for storage.

e. Quantities shall be increased 100 percent in buildings equipped throughout with an approved automatic sprinkler system in accordance with Section 903.3.1.1. Where Note f also applies, the increase for both notes shall be applied accumulatively.

f. Quantities shall be increased 100 percent when stored in approved storage cabinets, gas cabinets or exhausted enclosures as specified in the *International Fire Code*. Where Note e also applies, the increase for both notes shall be applied accumulatively.

g. A single cylinder containing 150 pounds or less of anhydrous ammonia in a single control area in a nonsprinklered building shall be considered a maximum allowable quantity. Two cylinders, each containing 150 pounds or less in a single control area, shall be considered a maximum allowable quantity provided the building is equipped throughout with an automatic sprinkler system in accordance with Section 903.3.1.1.

h. Allowed only when stored in approved exhausted gas cabinets or exhausted enclosures as specified in the *International Fire Code*.

i. Quantities in parenthesis indicate quantity units in parenthesis at the head of each column.

COMPRESSED GAS. A material, or mixture of materials which:

1. Is a gas at 68°F (20°C) or less at 14.7 pounds per square inch atmosphere (psia) (101 kPa) of pressure; and

2. Has a boiling point of 68°F (20°C) or less at 14.7 psia (101 kPa) which is either liquefied, nonliquefied or in solution, except those gases which have no other health- or physical-hazard properties are not considered to be compressed until the pressure in the packaging exceeds 41 psia (282 kPa) at 68°F (20°C).

The states of a compressed gas are categorized as follows:

1. Nonliquefied compressed gases are gases, other than those in solution, which are in a packaging under the charged pressure and are entirely gaseous at a temperature of 68°F (20°C).

2. Liquefied compressed gases are gases that, in a packaging under the charged pressure, are partially liquid at a temperature of 68°F (20°C).

3. Compressed gases in solution are nonliquefied gases that are dissolved in a solvent.

4. Compressed gas mixtures consist of a mixture of two or more compressed gases contained in a packaging, the hazard properties of which are represented by the properties of the mixture as a whole.

CONTROL AREA. Spaces within a building that are enclosed and bounded by exterior walls, fire walls, fire barriers and roofs, or a combination thereof, where quantities of hazardous materials not exceeding the maximum allowable quantities per control area are stored, dispensed, used or handled.

CORROSIVE. A chemical that causes visible destruction of, or irreversible alterations in, living tissue by chemical action at the point of contact. A chemical shall be considered corrosive if, when tested on the intact skin of albino rabbits by the method described in DOTn 49 CFR, Part 173.137, such a chemical destroys or changes irreversibly the structure of the tissue at the point of contact following an exposure period of 4 hours. This term does not refer to action on inanimate surfaces.

CRYOGENIC FLUID. A liquid having a boiling point lower than -150°F (-101°C) at 14.7 pounds per square inch atmosphere (psia) (an absolute pressure of 101 kPa).

DEFLAGRATION. An exothermic reaction, such as the extremely rapid oxidation of a flammable dust or vapor in air, in which the reaction progresses through the unburned material at a rate less than the velocity of sound. A deflagration can have an explosive effect.

DETACHED BUILDING. A separate single-story building, without a basement or crawl space, used for the storage or use of hazardous materials and located an approved distance from all structures.

DETONATION. An exothermic reaction characterized by the presence of a shock wave in the material which establishes and maintains the reaction. The reaction zone progresses through the material at a rate greater than the velocity of sound. The principal heating mechanism is one of shock compression. Detonations have an explosive effect.

DISPENSING. The pouring or transferring of any material from a container, tank or similar vessel, whereby vapors, dusts, fumes, mists or gases are liberated to the atmosphere.

EXPLOSIVE. Any chemical compound, mixture or device, the primary or common purpose of which is to function by explosion. The term includes, but is not limited to, dynamite, black powder, pellet powder, initiating explosives, detonators, safety fuses, squibs, detonating cord, igniter cord, igniters and display fireworks, 1.3G (Class B, Special).

The term "explosive" includes any material determined to be within the scope of USC Title 18: Chapter 40 and also includes any material classified as an explosive other than consumer fireworks, 1.4G (Class C, Common) by the hazardous materials regulations of DOTn 49 CFR.

High explosive. Explosive material, such as dynamite, which can be caused to detonate by means of a No. 8 test blasting cap when unconfined.

Low explosive. Explosive material that will burn or deflagrate when ignited. It is characterized by a rate of reaction that is less than the speed of sound. Examples of low explosives include, but are not limited to, black powder; safety fuse; igniters; igniter cord; fuse lighters; fireworks, 1.3G (Class B, Special) and propellants, 1.3C.

Mass-detonating explosives. Division 1.1, 1.2 and 1.5 explosives alone or in combination, or loaded into various types of ammunition or containers, most of which can be expected to explode virtually instantaneously when a small portion is subjected to fire, severe concussion, impact, the impulse of an initiating agent or the effect of a considerable discharge of energy from without. Materials that react in this manner represent a mass explosion hazard. Such an explosive will normally cause severe structural damage to adjacent objects. Explosive propagation could occur immediately to other items of ammunition and explosives stored sufficiently close to and not adequately protected from the initially exploding pile with a time interval short enough so that two or more quantities must be considered as one for quantity-distance purposes.

UN/DOTn Class 1 explosives. The former classification system used by DOTn included the terms "high" and "low" explosives as defined herein. The following terms further define explosives under the current system applied by DOTn for all explosive materials defined as hazard Class 1 materials. Compatibility group letters are used in concert with the division to specify further limitations on each division noted (i.e., the letter G identifies the material as a pyrotechnic substance or article containing a pyrotechnic substance and similar materials).

Division 1.1. Explosives that have a mass explosion hazard. A mass explosion is one which affects almost the entire load instantaneously.

Division 1.2. Explosives that have a projection hazard but not a mass explosion hazard.

Division 1.3. Explosives that have a fire hazard and either a minor blast hazard or a minor projection hazard or both, but not a mass explosion hazard.

Division 1.4. Explosives that pose a minor explosion hazard. The explosive effects are largely confined to the package and no projection of fragments of appreciable size or range is to be expected. An external fire must not cause virtually instantaneous explosion of almost the entire contents of the package.

Division 1.5. Very insensitive explosives. This division is comprised of substances that have a mass explosion hazard, but that are so insensitive there is very little probability of initiation or of transition from burning to detonation under normal conditions of transport.

Division 1.6. Extremely insensitive articles which do not have a mass explosion hazard. This division is comprised of articles that contain only extremely insensitive detonating substances and which demonstrate a negligible probability of accidental initiation or propagation.

FIREWORKS. Any composition or device for the purpose of producing a visible or audible effect for entertainment purposes by combustion, deflagration or detonation that meets the definition of 1.4G fireworks or 1.3G fireworks as set forth herein.

FIREWORKS, 1.3G. (Formerly Class B, Special Fireworks.) Large fireworks devices, which are explosive materials, intended for use in fireworks displays and designed to produce audible or visible effects by combustion, deflagration or detonation. Such 1.3G fireworks include, but are not limited to, firecrackers containing more than 130 milligrams (2 grains) of explosive composition, aerial shells containing more than 40 grams of pyrotechnic composition, and other display pieces which exceed the limits for classification as 1.4G fireworks. Such 1.3G fireworks are also described as fireworks, 49 CFR (172) by the DOTn.

FIREWORKS, 1.4G. (Formerly Class C, Common Fireworks.) Small fireworks devices containing restricted amounts of pyrotechnic composition designed primarily to produce visible or audible effects by combustion. Such 1.4G fireworks which comply with the construction, chemical composition and labeling regulations of the DOTn for fireworks, 49 CFR (172), and the U.S. Consumer Product Safety Commission (CPSC) as set forth in CPSC 16 CFR: Parts 1500 and 1507, are not explosive materials for the purpose of this code.

FLAMMABLE GAS. A material that is a gas at 68°F (20°C) or less at 14.7 pounds per square inch atmosphere (psia) (101 kPa) of pressure [a material that has a boiling point of 68°F (20°C) or less at 14.7 psia (101 kPa)] which:

1. Is ignitable at 14.7 psia (101 kPa) when in a mixture of 13 percent or less by volume with air; or
2. Has a flammable range at 14.7 psia (101 kPa) with air of at least 12 percent, regardless of the lower limit.

The limits specified shall be determined at 14.7 psi (101 kPa) of pressure and a temperature of 68°F (20°C) in accordance with ASTM E 681.

FLAMMABLE LIQUEFIED GAS. A liquefied compressed gas which, under a charged pressure, is partially liquid at a temperature of 68°F (20°C) and which is flammable.

FLAMMABLE LIQUID. A liquid having a closed cup flash point below 100°F (38°C). Flammable liquids are further categorized into a group known as Class I liquids. The Class I category is subdivided as follows:

Class IA. Liquids having a flash point below 73°F (23°C) and a boiling point below 100°F (38°C).

Class IB. Liquids having a flash point below 73°F (23°C) and a boiling point at or above 100°F (38°C).

Class IC. Liquids having a flash point at or above 73°F (23°C) and below 100°F (38°C).

The category of flammable liquids does not include compressed gases or cryogenic fluids.

FLAMMABLE MATERIAL. A material capable of being readily ignited from common sources of heat or at a temperature of 600°F (316°C) or less.

FLAMMABLE SOLID. A solid, other than a blasting agent or explosive, that is capable of causing fire through friction, absorption or moisture, spontaneous chemical change, or retained heat from manufacturing or processing, or which has an ignition temperature below 212°F (100°C) or which burns so vigorously and persistently when ignited as to create a serious hazard. A chemical shall be considered a flammable solid as determined in accordance with the test method of CPSC 16 CFR; Part 1500.44, if it ignites and burns with a self-sustained flame at a rate greater than 0.1 inch (2.5 mm) per second along its major axis.

FLASH POINT. The minimum temperature in degrees Fahrenheit at which a liquid will give off sufficient vapors to form an ignitable mixture with air near the surface or in the container, but will not sustain combustion. The flash point of a liquid shall be determined by appropriate test procedure and apparatus as specified in ASTM D 56, ASTM D 93 or ASTM D 3278.

HANDLING. The deliberate transport by any means to a point of storage or use.

HAZARDOUS MATERIALS. Those chemicals or substances that are physical hazards or health hazards as defined and classified in this section and the *International Fire Code*, whether the materials are in usable or waste condition.

HEALTH HAZARD. A classification of a chemical for which there is statistically significant evidence that acute or chronic health effects are capable of occurring in exposed persons. The term "health hazard" includes chemicals that are toxic or highly toxic, and corrosive.

HIGHLY TOXIC. A material which produces a lethal dose or lethal concentration that falls within any of the following categories:

1. A chemical that has a median lethal dose (LD50) of 50 milligrams or less per kilogram of body weight when administered orally to albino rats weighing between 200 and 300 grams each.

2. A chemical that has a median lethal dose (LD50) of 200 milligrams or less per kilogram of body weight when administered by continuous contact for 24 hours (or less if death occurs within 24 hours) with the bare skin of albino rabbits weighing between 2 and 3 kilograms each.

3. A chemical that has a median lethal concentration (LC50) in air of 200 parts per million by volume or less of gas or vapor, or 2 milligrams per liter or less of mist, fume or dust, when administered by continuous inhalation for 1 hour (or less if death occurs within 1 hour) to albino rats weighing between 200 and 300 grams each.

Mixtures of these materials with ordinary materials, such as water, might not warrant classification as highly toxic. While this system is basically simple in application, any hazard evaluation that is required for the precise categorization of this type of material shall be performed by experienced, technically competent persons.

INCOMPATIBLE MATERIALS. Materials that, when mixed, have the potential to react in a manner that generates heat, fumes, gases or byproducts which are hazardous to life or property.

OPEN SYSTEM. The use of a solid or liquid hazardous material involving a vessel or system that is continuously open to the atmosphere during normal operations and where vapors are liberated, or the product is exposed to the atmosphere during normal operations. Examples of open systems for solids and liquids include dispensing from or into open beakers or containers, dip tank and plating tank operations.

OPERATING BUILDING. A building occupied in conjunction with the manufacture, transportation or use of explosive materials. Operating buildings are separated from one another with the use of intraplant or intraline distances.

ORGANIC PEROXIDE. An organic compound that contains the bivalent -O-O- structure and which may be considered to be a structural derivative of hydrogen peroxide where one or both of the hydrogen atoms have been replaced by an organic radical. Organic peroxides can pose an explosion hazard (detonation or deflagration) or they can be shock sensitive. They can also decompose into various unstable compounds over an extended period of time.

Class I. Those formulations that are capable of deflagration but not detonation.

Class II. Those formulations that burn very rapidly and that pose a moderate reactivity hazard.

Class III. Those formulations that burn rapidly and that pose a moderate reactivity hazard.

Class IV. Those formulations that burn in the same manner as ordinary combustibles and that pose a minimal reactivity hazard.

Class V. Those formulations that burn with less intensity than ordinary combustibles or do not sustain combustion and that pose no reactivity hazard.

Unclassified detonable. Organic peroxides that are capable of detonation. These peroxides pose an extremely high explosion hazard through rapid explosive decomposition.

OXIDIZER. A material that readily yields oxygen or other oxidizing gas, or that readily reacts to promote or initiate combustion of combustible materials. Examples of other oxidizing gases include bromine, chlorine and fluorine.

Class 4. An oxidizer that can undergo an explosive reaction due to contamination or exposure to thermal or physical shock. Additionally, the oxidizer will enhance the burning rate and can cause spontaneous ignition of combustibles.

Class 3. An oxidizer that will cause a severe increase in the burning rate of combustible materials with which it comes in contact or that will undergo vigorous self-sustained decomposition due to contamination or exposure to heat.

Class 2. An oxidizer that will cause a moderate increase in the burning rate or that causes spontaneous ignition of combustible materials with which it comes in contact.

Class 1. An oxidizer whose primary hazard is that it slightly increases the burning rate but which does not cause spontaneous ignition when it comes in contact with combustible materials.

OXIDIZING GAS. A gas that can support and accelerate combustion of other materials.

PHYSICAL HAZARD. A chemical for which there is evidence that it is a combustible liquid, compressed gas, cryogenic, explosive, flammable gas, flammable liquid, flammable solid, organic peroxide, oxidizer, pyrophoric or unstable (reactive) or water-reactive material.

PYROPHORIC. A chemical with an autoignition temperature in air, at or below a temperature of 130°F (54.4°C).

PYROTECHNIC COMPOSITION. A chemical mixture that produces visible light displays or sounds through a self-propagating, heat-releasing chemical reaction which is initiated by ignition.

TOXIC. A chemical falling within any of the following categories:

1. A chemical that has a median lethal dose (LD50) of more than 50 milligrams per kilogram, but not more than 500 milligrams per kilogram of body weight when administered orally to albino rats weighing between 200 and 300 grams each.

2. A chemical that has a median lethal dose (LD50) of more than 200 milligrams per kilogram but not more than 1,000 milligrams per kilogram of body weight when administered by continuous contact for 24 hours (or less if death occurs within 24 hours) with the bare skin of albino rabbits weighing between 2 and 3 kilograms each.

3. A chemical that has a median lethal concentration (LC50) in air of more than 200 parts per million but not more than 2,000 parts per million by volume of gas or vapor, or more than 2 milligrams per liter but not more than 20 milligrams per liter of mist, fume or dust, when administered by continuous inhalation for 1 hour (or less if death occurs within 1 hour) to albino rats weighing between 200 and 300 grams each.

UNSTABLE (REACTIVE) MATERIAL. A material, other than an explosive, which in the pure state or as commercially produced, will vigorously polymerize, decompose, condense or become self-reactive and undergo other violent chemical changes, including explosion, when exposed to heat, friction or shock, or in the absence of an inhibitor, or in the presence of contaminants, or in contact with incompatible materials. Unstable (reactive) materials are subdivided as follows:

Class 4. Materials that in themselves are readily capable of detonation or explosive decomposition or explosive reaction at normal temperatures and pressures. This class includes materials that are sensitive to mechanical or localized thermal shock at normal temperatures and pressures.

Class 3. Materials that in themselves are capable of detonation or of explosive decomposition or explosive reaction but which require a strong initiating source or which must be heated under confinement before initiation. This class includes materials that are sensitive to thermal or mechanical shock at elevated temperatures and pressures.

Class 2. Materials that in themselves are normally unstable and readily undergo violent chemical change but do not detonate. This class includes materials that can undergo chemical change with rapid release of energy at normal temperatures and pressures, and that can undergo violent chemical change at elevated temperatures and pressures.

Class 1. Materials that in themselves are normally stable but which can become unstable at elevated temperatures and pressure.

WATER-REACTIVE MATERIAL. A material that explodes; violently reacts; produces flammable, toxic or other hazardous gases; or evolves enough heat to cause self-ignition or ignition of nearby combustibles upon exposure to water or moisture. Water-reactive materials are subdivided as follows:

Class 3. Materials that react explosively with water without requiring heat or confinement.

Class 2. Materials that may form potentially explosive mixtures with water.

Class 1. Materials that may react with water with some release of energy, but not violently.

[F] 307.3 High-Hazard Group H-1. Buildings and structures containing materials that pose a detonation hazard shall be classified as Group H-1. Such materials shall include, but not be limited to, the following:

Explosives:

Division 1.1
Division 1.2
Division 1.3

> **Exception:** Materials that are used and maintained in a form where either confinement or configuration will not elevate the hazard from a mass fire to mass explosion hazard shall be allowed in H-2 occupancies.

Division 1.4

> **Exception:** Articles, including articles packaged for shipment, that are not regulated as an explosive under Bureau of Alcohol, Tobacco and Firearms regulations, or unpackaged articles used in process operations that do not propagate a detonation or deflagration between articles shall be allowed in H-3 occupancies.

Division 1.5
Division 1.6

Organic peroxides, unclassified detonable
Oxidizers, Class 4
Unstable (reactive) materials, Class 3 detonable and Class 4
Detonable pyrophoric materials

[F] 307.4 High-Hazard Group H-2. Buildings and structures containing materials that pose a deflagration hazard or a hazard from accelerated burning shall be classified as Group H-2. Such materials shall include, but not be limited to, the following:

Class I, II or IIIA flammable or combustible liquids which are used or stored in normally open containers or systems, or in closed containers or systems pressurized at more than 15 psi (103.4 kPa) gage.

> Combustible dusts
> Cryogenic fluids, flammable
> Flammable gases
> Organic peroxides, Class I
> Oxidizers, Class 3, that are used or stored in normally open containers or systems, or in closed containers or systems pressurized at more than 15 psi (103 kPa) gage
> Pyrophoric liquids, solids and gases, nondetonable
> Unstable (reactive) materials, Class 3, nondetonable
> Water-reactive materials, Class 3

[F] 307.5 High-Hazard Group H-3. Buildings and structures containing materials that readily support combustion or that pose a physical hazard shall be classified as Group H-3. Such materials shall include, but not be limited to, the following:

> Class I, II or IIIA flammable or combustible liquids which are used or stored in normally closed containers or systems pressurized at less than 15 psi (103.4 kPa) gage.
> Combustible fibers
> Consumer fireworks, 1.4G (Class C Common)
> Cryogenic fluids, oxidizing
> Flammable solids
> Organic peroxides, Classes II and III
> Oxidizers, Class 2
> Oxidizers, Class 3, that are used or stored in normally closed containers or systems pressurized at less than 15 pounds per square inch (103 kPa) gauge
> Oxidizing gases
> Unstable (reactive) materials, Class 2
> Water-reactive materials, Class 2

[F] 307.6 High-Hazard Group H-4. Buildings and structures which contain materials that are health hazards shall be classified as Group H-4. Such materials shall include, but not be limited to, the following:

> Corrosives
> Highly toxic materials
> Toxic materials

[F] 307.7 Group H-5 structures. Semiconductor fabrication facilities and comparable research and development areas in which hazardous production materials (HPM) are used and the aggregate quantity of materials is in excess of those listed in Tables 307.7(1) and 307.7(2). Such facilities and areas shall be designed and constructed in accordance with Section 415.9.

[F] 307.8 Multiple hazards. Buildings and structures containing a material or materials representing hazards that are classified in one or more of Groups H-1, H-2, H-3 and H-4 shall conform to the code requirements for each of the occupancies so classified.

[F] 307.9 Exceptions: The following shall not be classified in Group H, but shall be classified in the occupancy that they most nearly resemble. Hazardous materials in any quantity shall conform to the requirements of this code, including Section 414, and the *International Fire Code.*

1. Buildings and structures that contain not more than the maximum allowable quantities per control area of hazardous materials as shown in Tables 307.7(1) and 307.7(2) provided that such buildings are maintained in accordance with the *International Fire Code.*

2. Buildings utilizing control areas in accordance with Section 414.2 that contain not more than the maximum allowable quantities per control area of hazardous materials as shown in Tables 307.7(1) and 307.7(2).

3. Buildings and structures occupied for the application of flammable finishes, provided that such buildings or areas conform to the requirements of Section 416 and the *International Fire Code.*

4. Wholesale and retail sales and storage of flammable and combustible liquids in mercantile occupancies conforming to the *International Fire Code.*

5. Closed systems housing flammable or combustible liquids or gases utilized for the operation of machinery or equipment.

6. Cleaning establishments that utilize combustible liquid solvents having a flash point of 140°F (60°C) or higher in closed systems employing equipment listed by an approved testing agency, provided that this occupancy is separated from all other areas of the building by 1-hour fire-resistance-rated fire barrier walls or horizontal assemblies or both.

7. Cleaning establishments which utilize a liquid solvent having a flash point at or above 200°F (93°C).

8. Liquor stores and distributors without bulk storage.

9. Refrigeration systems.

10. The storage or utilization of materials for agricultural purposes on the premises.

11. Stationary batteries utilized for facility emergency power, uninterrupted power supply or telecommunication facilities provided that the batteries are provided with safety venting caps and ventilation is provided in accordance with the *International Mechanical Code.*

12. Corrosives shall not include personal or household products in their original packaging used in retail display or commonly used building materials.

13. Buildings and structures occupied for aerosol storage shall be classified as Group S-1, provided that such buildings conform to the requirements of the *International Fire Code.*

14. Display and storage of nonflammable solid and nonflammable or noncombustible liquid hazardous materials in quantities not exceeding the maximum allowable quantity per control area in Group M or S occupancies complying with Section 414.2.4.

15. The storage of black powder, smokeless propellant and small arms primers in Groups M and R-3 and special industrial explosive devices in Groups B, F, M and S, provided such storage conforms to the quantity limits and requirements prescribed in the *International Fire Code.*

SECTION 308
INSTITUTIONAL GROUP I

308.1 Institutional Group I. Institutional Group I occupancy includes, among others, the use of a building or structure, or a portion thereof, in which people are cared for or live in a supervised environment, having physical limitations because of health or age are harbored for medical treatment or other care or treatment, or in which people are detained for penal or correctional purposes or in which the liberty of the occupants is restricted. Institutional occupancies shall be classified as Group I-1, I-2, I-3 or I-4.

308.2 Group I-1. This occupancy shall include buildings, structures or parts thereof housing more than 16 persons, on a 24-hour basis, who because of age, mental disability or other reasons, live in a supervised residential environment that provides personal care services. The occupants are capable of responding to an emergency situation without physical assistance from staff. This group shall include, but not be limited to, the following:

> Residential board and care facilities
> Assisted living facilities
> Halfway houses
> Group homes
> Congregate care facilities
> Social rehabilitation facilities
> Alcohol and drug centers
> Convalescent facilities

A facility such as the above with five or fewer persons shall be classified as a Group R-3 or shall comply with the *International Residential Code* in accordance with Section 101.2. A facility such as above, housing at least six and not more than 16 persons, shall be classified as Group R-4.

308.3 Group I-2. This occupancy shall include buildings and structures used for medical, surgical, psychiatric, nursing or custodial care on a 24-hour basis of more than five persons who are not capable of self-preservation. This group shall include, but not be limited to, the following:

> Hospitals
> Nursing homes (both intermediate-care facilities and skilled nursing facilities)
> Mental hospitals
> Detoxification facilities

A facility such as the above with five or fewer persons shall be classified as Group R-3 or shall comply with the *International Residential Code* in accordance with Section 101.2.

308.3.1 Child care facility. A child care facility that provides care on a 24-hour basis to more than five children $2^1/_2$ years of age or less shall be classified as Group I-2.

308.4 Group I-3. This occupancy shall include buildings and structures that are inhabited by more than five persons who are under restraint or security. An I-3 facility is occupied by persons who are generally incapable of self-preservation due to security measures not under the occupants' control. This group shall include, but not be limited to, the following:

> Prisons
> Jails
> Reformatories
> Detention centers
> Correctional centers
> Prerelease centers

Buildings of Group I-3 shall be classified as one of the occupancy conditions indicated in Sections 308.4.1 through 308.4.5 (see Section 408.1).

308.4.1 Condition 1. This occupancy condition shall include buildings in which free movement is allowed from sleeping areas, and other spaces where access or occupancy is permitted, to the exterior via means of egress without restraint. A Condition 1 facility is permitted to be constructed as Group R.

308.4.2 Condition 2. This occupancy condition shall include buildings in which free movement is allowed from sleeping areas and any other occupied smoke compartment to one or more other smoke compartments. Egress to the exterior is impeded by locked exits.

308.4.3 Condition 3. This occupancy condition shall include buildings in which free movement is allowed within individual smoke compartments, such as within a residential unit comprised of individual sleeping units and group activity spaces, where egress is impeded by remote-controlled release of means of egress from such a smoke compartment to another smoke compartment.

308.4.4 Condition 4. This occupancy condition shall include buildings in which free movement is restricted from an occupied space. Remote-controlled release is provided to permit movement from sleeping units, activity spaces and other occupied areas within the smoke compartment to other smoke compartments.

308.4.5 Condition 5. This occupancy condition shall include buildings in which free movement is restricted from an occupied space. Staff-controlled manual release is provided to permit movement from sleeping units, activity spaces and other occupied areas within the smoke compartment to other smoke compartments.

308.5 Group I-4, day care facilities. This group shall include buildings and structures occupied by persons of any age who receive custodial care for less than 24 hours by individuals other than parents or guardians, relatives by blood, marriage or adoption, and in a place other than the home of the person cared for. A facility such as the above with five or fewer persons shall be classified as a Group R-3 or shall comply with the *International Residential Code* in accordance with Section 101.2. Places of worship during religious functions are not included.

308.5.1 Adult care facility. A facility that provides accommodations for less than 24 hours for more than five unrelated adults and provides supervision and personal care services shall be classified as Group I-4.

Exception: A facility where occupants are capable of responding to an emergency situation without physical assistance from the staff shall be classified as Group A-3.

308.5.2 Child care facility. A facility that provides supervision and personal care on less than a 24-hour basis for more than five children $2^1/_2$ years of age or less shall be classified as Group I-4.

Exception: A child day care facility that provides care for more than five but no more than 100 children $2^1/_2$ years or less of age, when the rooms where such children are cared for are located on the level of exit discharge and each of these child care rooms has an exit door directly to the exterior, shall be classified as Group E.

SECTION 309
MERCANTILE GROUP M

309.1 Mercantile Group M. Mercantile Group M occupancy includes, among others, buildings and structures or a portion thereof, for the display and sale of merchandise, and involves stocks of goods, wares or merchandise incidental to such purposes and accessible to the public. Mercantile occupancies shall include, but not be limited to, the following:

Department stores
Drug stores
Markets
Motor fuel-dispensing facilities
Retail or wholesale stores
Sales rooms

309.2 Quantity of hazardous materials. The aggregate quantity of nonflammable solid and nonflammable or noncombustible liquid hazardous materials stored or displayed in a single control area of a Group M occupancy shall not exceed the quantities in Table 414.2.4.

SECTION 310
RESIDENTIAL GROUP R

310.1 Residential Group R. Residential Group R includes, among others, the use of a building or structure, or a portion thereof, for sleeping purposes when not classified as an Institutional Group I. Residential occupancies shall include the following:

R-1 Residential occupancies where the occupants are primarily transient in nature, including:

Boarding houses (transient)
Hotels (transient)
Motels (transient)

R-2 Residential occupancies containing sleeping units or more than two dwelling units where the occupants are primarily permanent in nature, including:

Apartment houses
Boarding houses (not transient)
Convents
Dormitories
Fraternities and sororities
Monasteries
Vacation timeshare properties
Hotels (nontransient)
Motels (nontransient)

R-3 Residential occupancies where the occupants are primarily permanent in nature and not classified as R-1, R-2, R-4 or I and where buildings do not contain more than two dwelling units as applicable in Section 101.2, or adult and child care facilities that provide accommodations for five or fewer persons of any age for less than 24 hours. Adult and child care facilities that are within a single-family home are permitted to comply with the *International Residential Code* in accordance with Section 101.2.

R-4 Residential occupancies shall include buildings arranged for occupancy as residential care/assisted living facilities including more than five but not more than 16 occupants, excluding staff.

Group R-4 occupancies shall meet the requirements for construction as defined for Group R-3 except as otherwise provided for in this code or shall comply with the *International Residential Code* in accordance with Section 101.2.

310.2 Definitions. The following words and terms shall, for the purposes of this section and as used elsewhere in this code, have the meanings shown herein.

BOARDING HOUSE. A building arranged or used for lodging for compensation, with or without meals, and not occupied as a single-family unit.

DORMITORY. A space in a building where group sleeping accommodations are provided in one room, or in a series of closely associated rooms, for persons not members of the same family group, under joint occupancy and single management, as in college dormitories or fraternity houses.

DWELLING UNIT. A single unit providing complete, independent living facilities for one or more persons, including permanent provisions for living, sleeping, eating, cooking and sanitation.

PERSONAL CARE SERVICE. The care of residents who do not require chronic or convalescent medical or nursing care. Personal care involves responsibility for the safety of the resident while inside the building.

RESIDENTIAL CARE/ASSISTED LIVING FACILITIES. A building or part thereof housing persons, on a 24-hour basis, who because of age, mental disability or other reasons, live in a supervised residential environment which provides personal care services. The occupants are capable of responding to an emergency situation without physical assistance from staff. This classification shall include, but not be limited to, the following: residential board and care facilities, assisted living facilities, halfway houses, group homes, congregate care facilities, social rehabilitation facilities, alcohol and drug abuse centers and convalescent facilities.

SECTION 311
STORAGE GROUP S

311.1 Storage Group S. Storage Group S occupancy includes, among others, the use of a building or structure, or a portion thereof, for storage that is not classified as a hazardous occupancy.

311.2 Moderate-hazard storage, Group S-1. Buildings occupied for storage uses which are not classified as Group S-2 including, but not limited to, storage of the following:

Aerosols, Levels 2 and 3
Aircraft repair hangar
Bags; cloth, burlap and paper
Bamboos and rattan
Baskets

Belting; canvas and leather
Books and paper in rolls or packs
Boots and shoes
Buttons, including cloth covered, pearl or bone
Cardboard and cardboard boxes
Clothing, woolen wearing apparel
Cordage
Furniture
Furs
Glues, mucilage, pastes and size
Grains
Horns and combs, other than celluloid
Leather
Linoleum
Lumber
Motor vehicle repair garages complying with the maximum allowable quantities of hazardous materials listed in Table 307.7(1) (see Section 406.6)
Photo engravings
Resilient flooring
Silks
Soaps
Sugar
Tires, bulk storage of
Tobacco, cigars, cigarettes and snuff
Upholstery and mattresses
Wax candles

311.3 Low-hazard storage, Group S-2. Includes, among others, buildings used for the storage of noncombustible materials such as products on wood pallets or in paper cartons with or without single thickness divisions; or in paper wrappings. Such products are permitted to have a negligible amount of plastic trim, such as knobs, handles or film wrapping. Storage uses shall include, but not be limited to, storage of the following:

Aircraft hangar
Asbestos
Beverages up to and including 12-percent alcohol in metal, glass or ceramic containers
Cement in bags
Chalk and crayons
Dairy products in nonwaxed coated paper containers
Dry cell batteries
Electrical coils
Electrical motors
Empty cans
Food products
Foods in noncombustible containers
Fresh fruits and vegetables in nonplastic trays or containers
Frozen foods
Glass
Glass bottles, empty or filled with noncombustible liquids
Gypsum board
Inert pigments
Ivory
Meats
Metal cabinets
Metal desks with plastic tops and trim
Metal parts
Metals
Mirrors

Oil-filled and other types of distribution transformers
Parking garages, open or enclosed
Porcelain and pottery
Stoves
Talc and soapstones
Washers and dryers

SECTION 312
UTILITY AND MISCELLANEOUS GROUP U

312.1 General. Buildings and structures of an accessory character and miscellaneous structures not classified in any specific occupancy shall be constructed, equipped and maintained to conform to the requirements of this code commensurate with the fire and life hazard incidental to their occupancy. Group U shall include, but not be limited to, the following:

Agricultural buildings
Aircraft hangars, accessory to a one- or two-family residence (see Section 412.3)
Barns
Carports
Fences more than 6 feet (1829 mm) high
Grain silos, accessory to a residential occupancy
Greenhouses
Livestock shelters
Private garages
Retaining walls
Sheds
Stables
Tanks
Towers

CHAPTER 4

SPECIAL DETAILED REQUIREMENTS BASED ON USE AND OCCUPANCY

SECTION 401
SCOPE

401.1 Detailed use and occupancy requirements. In addition to the occupancy and construction requirements in this code, the provisions of this chapter apply to the special uses and occupancies described herein.

SECTION 402
COVERED MALL BUILDINGS

402.1 Scope. The provisions of this section shall apply to buildings or structures defined herein as covered mall buildings not exceeding three floor levels at any point nor more than three stories above grade. Except as specifically required by this section, covered mall buildings shall meet applicable provisions of this code.

Exceptions:

1. Foyers and lobbies of Groups B, R-1 and R-2 are not required to comply with this section.

2. Buildings need not comply with the provisions of this section where they totally comply with other applicable provisions of this code.

402.2 Definitions. The following words and terms shall, for the purposes of this chapter and as used elsewhere in this code, have the meanings shown herein.

ANCHOR BUILDING. An exterior perimeter building of a group other than H having direct access to a covered mall building but having required means of egress independent of the mall.

COVERED MALL BUILDING. A single building enclosing a number of tenants and occupants such as retail stores, drinking and dining establishments, entertainment and amusement facilities, passenger transportation terminals, offices, and other similar uses wherein two or more tenants have a main entrance into one or more malls. For the purpose of this chapter, anchor buildings shall not be considered as a part of the covered mall building.

FOOD COURT. A public seating area located in the mall that serves adjacent food preparation tenant spaces.

GROSS LEASABLE AREA. The total floor area designed for tenant occupancy and exclusive use. The area of tenant occupancy is measured from the centerlines of joint partitions to the outside of the tenant walls. All tenant areas, including areas used for storage, shall be included in calculating gross leasable area.

MALL. A roofed or covered common pedestrian area within a covered mall building that serves as access for two or more tenants and not to exceed three levels that are open to each other.

402.3 Lease plan. Each covered mall building owner shall provide both the building and fire departments with a lease plan showing the location of each occupancy and its exits after the certificate of occupancy has been issued. No modifications or changes in occupancy or use shall be made from that shown on the lease plan without prior approval of the building official.

402.4 Means of egress. Each tenant space and the covered mall building shall be provided with means of egress as required by this section and this code. Where there is a conflict between the requirements of this code and the requirements of this section, the requirements of this section shall apply.

402.4.1 Determination of occupant load. The occupant load permitted in any individual tenant space in a covered mall building shall be determined as required by this code. Means of egress requirements for individual tenant spaces shall be based on the occupant load thus determined.

402.4.1.1 Occupant formula. In determining required means of egress of the mall, the number of occupants for whom means of egress are to be provided shall be based on gross leasable area of the covered mall building (excluding anchor buildings) and the occupant load factor as determined by the following equation.

$$OLF = (0.00007)(GLA) + 25 \qquad \textbf{(Equation 4-1)}$$

where:

OLF = The occupant load factor (square feet per person).

GLA = The gross leasable area (square feet).

402.4.1.2 OLF range. The occupant load factor (OLF) is not required to be less than 30 and shall not exceed 50.

402.4.1.3 Anchor buildings. The occupant load of anchor buildings opening into the mall shall not be included in computing the total number of occupants for the mall.

402.4.1.4 Food courts. The occupant load of a food court shall be determined in accordance with Section 1004. For the purposes of determining the means of egress requirements for the mall, the food court occupant load shall be added to the occupant load of the covered mall building as calculated above.

402.4.2 Number of means of egress. Wherever the distance of travel to the mall from any location within a tenant space used by persons other than employees exceeds 75 feet (22 860 mm) or the tenant space exceeds an occupant load of 50, not less than two means of egress shall be provided.

402.4.3 Arrangements of means of egress. Assembly occupancies with an occupant load of 500 or more shall be so located in the covered mall building that their entrance will be immediately adjacent to a principal entrance to the mall and shall have not less than one-half of their required means of egress opening directly to the exterior of the covered mall building.

402.4.3.1 Anchor building means of egress. Required means of egress for anchor buildings shall be provided independently from the mall means of egress system. The occupant load of anchor buildings opening into the mall shall not be included in determining means of egress requirements for the mall. The path of egress travel of malls shall not exit through anchor buildings. Malls terminating at an anchor building where no other means of egress has been provided shall be considered as a dead-end mall.

402.4.4 Distance to exits. Within each individual tenant space in a covered mall building, the maximum distance of travel from any point to an exit or entrance to the mall shall not exceed 200 feet (60 960 mm).

The maximum distance of travel from any point within a mall to an exit shall not exceed 200 feet (60 960 mm).

402.4.5 Access to exits. Where more than one exit is required, they shall be so arranged that it is possible to travel in either direction from any point in a mall to separate exits. The minimum width of an exit passageway or corridor from a mall shall be 66 inches (1676 mm).

Exception: Dead ends not exceeding a length equal to twice the width of the mall measured at the narrowest location within the dead-end portion of the mall.

402.4.5.1 Exit passageway enclosures. Where exit passageway enclosures provide a secondary means of egress from a tenant space, doors to the exit passageway enclosures shall be 1-hour fire doors. Such doors shall be self-closing and be so maintained or shall be automatic-closing by smoke detection.

402.4.6 Service areas fronting on exit passageways. Mechanical rooms, electrical rooms, building service areas and service elevators are permitted to open directly into exit passageways provided that the exit passageway is separated from such rooms with 1-hour fire-resistance-rated walls and 1-hour opening protectives.

402.5 Mall width. For the purpose of providing required egress, malls are permitted to be considered as corridors but need not comply with the requirements of Section 1005.1 of this code where the width of the mall is as specified in this section.

402.5.1 Minimum width. The minimum width of the mall shall be 20 feet (6096 mm). The mall width shall be sufficient to accommodate the occupant load served. There shall be a minimum of 10 feet (3048 mm) clear exit width to a height of 8 feet (2438 mm) between any projection of a tenant space bordering the mall and the nearest kiosk, vending machine, bench, display opening, food court or other obstruction to means of egress travel.

402.6 Types of construction. The area of any covered mall building, including anchor buildings, of Type I, II, III and IV construction, shall not be limited provided the covered mall building and attached anchor buildings and parking garages are surrounded on all sides by a permanent open space of not less than 60 feet (18 288 mm) and the anchor buildings do not exceed three stories in height. The allowable height and area of anchor buildings greater than three stores in height shall comply with Section 503, as modified by Sections 504 and 506. The construction type of open parking garages and enclosed parking garages shall comply with Sections 406.3 and 406.4, respectively.

402.7 Fire-resistance-rated separation. Fire-resistance-rated separation is not required between tenant spaces and the mall. Fire-resistance-rated separation is not required between a food court and adjacent tenant spaces or the mall.

402.7.1 Attached garage. An attached garage for the storage of passenger vehicles having a capacity of not more than nine persons and open parking garages shall be considered as a separate building where it is separated from the covered mall building by a fire barrier having a fire-resistance rating of at least 2 hours.

Exception: Where an open parking garage or enclosed parking garage is separated from the covered mall building or anchor building a distance greater than 10 feet (3048 mm), the provisions of Table 602 shall apply. Pedestrian walkways and tunnels which attach the open parking garage or enclosed parking garage to the covered mall building or anchor building shall be constructed in accordance with Section 3104.

402.7.2 Tenant separations. Each tenant space shall be separated from other tenant spaces by a fire partition complying with Section 708. A tenant separation wall is not required between any tenant space and the mall.

402.7.3 Anchor building separation. An anchor building shall be separated from the covered mall building by fire walls complying with Section 705.

Exception: Anchor buildings of not more than three stories above grade which have an occupancy classification of the same uses permitted as tenants of the covered mall building shall be separated by 2-hour fire-barriers complying with Section 705.

402.7.3.1 Openings between anchor building and mall. Except for the separation between Group R-1 sleeping units and the mall, openings between anchor buildings of Type IA, IB, IIA and IIB construction and the mall need not be protected.

[F] 402.8 Automatic sprinkler system. The covered mall building and buildings connected shall be equipped throughout with an automatic sprinkler system in accordance with Section 903.3.1.1, which shall comply with the following:

1. The automatic sprinkler system shall be complete and operative throughout occupied space in the covered mall building prior to occupancy of any of the tenant spaces. Unoccupied tenant spaces shall be similarly protected unless provided with approved alternate protection.

2. Sprinkler protection for the mall shall be independent from that provided for tenant spaces or anchors. Where tenant spaces are supplied by the same system, they shall be independently controlled.

Exception: An automatic sprinkler system shall not be required in spaces or areas of open parking garages constructed in accordance with Section 406.2.

402.8.1 Standpipe system. The covered mall building shall be equipped throughout with a standpipe system as required by Section 905.3.3.

402.9 Smoke control. A smoke control system shall be provided where required for atriums in Section 404.

402.10 Kiosks. Kiosks and similar structures (temporary or permanent) shall meet the following requirements:

1. Combustible kiosks or other structures shall not be located within the mall unless constructed of any of the following materials:

    1.1. Fire-retardant-treated wood complying with Section 2303.2.

    1.2. Foam plastics having a maximum heat release rate not greater than 100kW (105 Btu/h) when tested in accordance with the exhibit booth protocol in UL 1975.

    1.3. Aluminum composite material (ACM) having a flame spread index of not more than 25 and a smoke-developed index of not more than 450 when tested as an assembly in the maximum thickness intended for use in accordance with ASTM E 84.

2. Kiosks or similar structures located within the mall shall be provided with approved fire suppression and detection devices.

3. The minimum horizontal separation between kiosks or groupings thereof and other structures within the mall shall be 20 feet (6096 mm).

4. Each kiosk or similar structure or groupings thereof shall have a maximum area of 300 square feet (28 m²).

402.11 Security grilles and doors. Horizontal sliding or vertical security grilles or doors that are a part of a required means of egress shall conform to the following:

1. They shall remain in the full open position during the period of occupancy by the general public.

2. Doors or grilles shall not be brought to the closed position when there are more than 10 persons occupying spaces served by a single exit or 50 persons occupying spaces served by more than one exit.

3. The doors or grilles shall be openable from within without the use of any special knowledge or effort where the space is occupied.

4. Where two or more exits are required, not more than one-half of the exits shall be permitted to include either a horizontal sliding or vertical rolling grille or doors.

402.12 Standby power. Covered mall buildings exceeding 50,000 square feet (4645 m²) shall be provided with standby power systems that are capable of operating the emergency voice/alarm communication system.

[F] 402.13 Emergency voice/alarm communication system. Covered mall buildings exceeding 50,000 square feet (4645 m²) in total floor area shall be provided with an emergency voice/alarm communication system. Emergency voice/alarm communication systems serving a mall, required or otherwise,

shall be accessible to the fire department. The system shall be provided in accordance with Section 907.2.12.2.

402.14 Plastic signs. Within every store or level and from sidewall to sidewall of each tenant space facing the mall, plastic signs shall be limited as specified in Sections 402.14.1 through 402.14.5.

402.14.1 Area. Plastic signs shall not exceed 20 percent of the wall area facing the mall.

402.14.2 Height and width. Plastic signs shall not exceed a height of 36 inches (914 mm), except if the sign is vertical, the height shall not exceed 96 inches (2438 mm) and the width shall not exceed 36 inches (914 mm).

402.14.3 Location. Plastic signs shall be located a minimum distance of 18 inches (457 mm) from adjacent tenants.

402.14.4 Plastics other than foam plastics. Plastics other than foam plastics used in signs shall be light-transmitting plastics complying with Section 2606.4 or shall have a self-ignition temperature of 650°F (343°C) or greater when tested in accordance with ASTM D 1929, and a flame spread index not greater than 75 and smoke-developed index not greater than 450 when tested in the manner intended for use in accordance with ASTM E 84 or meet the acceptance criteria of Section 803.2.1 when tested in accordance with NFPA 286.

402.14.4.1 Encasement. Edges and backs of plastic signs in the mall shall be fully encased in metal.

402.14.5 Foam plastics. Foam plastics used in signs shall have flame-retardant characteristics such that the sign has a maximum heat-release rate of 150 kilowatts when tested in accordance with UL 1975 and the foam plastics shall have the physical characteristics specified in this section. Foam plastics used in signs installed in accordance with Section 402.14 shall not be required to comply with the flame spread and smoke-developed indexes specified in Section 2603.3.

402.14.5.1 Density. The minimum density of foam plastics used in signs shall be not less than 20 pounds per cubic foot (pcf) (320 kg/M³).

402.14.5.2 Thickness. The thickness of foam plastic signs shall not be greater than ¹/₂-inch (12.7 mm).

402.15 Fire department access to equipment. Rooms or areas containing controls for air-conditioning systems, automatic fire-extinguishing systems or other detection, suppression or control elements shall be identified for use by the fire department.

SECTION 403
HIGH-RISE BUILDINGS

403.1 Applicability. The provisions of this section shall apply to buildings having occupied floors located more than 75 feet (22 860 mm) above the lowest level of fire department vehicle access.

Exception: The provisions of this section shall not apply to the following buildings and structures:

1. Airport traffic control towers in accordance with Section 412.

2. Open parking garages in accordance with Section 406.3.

3. Buildings with an occupancy in Group A-5 in accordance with Section 303.1.

4. Low-hazard special industrial occupancies in accordance with Section 503.1.2.

5. Buildings with an occupancy in Group H-1, H-2 or H-3 in accordance with Section 415.

[F] 403.2 Automatic sprinkler system. Buildings and structures shall be equipped throughout with an automatic sprinkler system in accordance with Section 903.3.1.1 and a secondary water supply where required by Section 903.3.5.2.

Exception: An automatic sprinkler system shall not be required in spaces or areas of:

1. Open parking garages in accordance with Section 406.3.

2. Telecommunications equipment buildings used exclusively for telecommunications equipment, associated electrical power distribution equipment, batteries and standby engines, provided that those spaces or areas are equipped throughout with an automatic fire detection system in accordance with Section 907.2 and are separated from the remainder of the building with fire barriers consisting of 1-hour fire-resistance-rated walls and 2-hour fire-resistance-rated floor/ceiling assemblies.

403.3 Reduction in fire-resistance rating. The fire-resistance-rating reductions listed in Sections 403.3.1 and 403.3.2 shall be allowed in buildings that have sprinkler control valves equipped with supervisory initiating devices and water-flow initiating devices for each floor.

403.3.1 Type of construction. The following reductions in the minimum construction type allowed in Table 601 shall be allowed as provided in Section 403.3:

1. Type IA construction shall be allowed to be reduced to Type IB.

2. In other than Groups F-1, M and S-1, Type IB construction shall be allowed to be reduced to Type IIA.

3. The height and area limitations of the reduced construction type shall be allowed to be the same as for the original construction type.

403.3.2 Shaft enclosures. The required fire-resistance rating of the fire barrier walls enclosing vertical shafts, other than exit enclosures and elevator hoistway enclosures, shall be reduced to 1 hour where automatic sprinklers are installed within the shafts at the top and at alternate floor levels.

403.4 Emergency escape and rescue. Emergency escape and rescue openings required by Section 1025 are not required.

[F] 403.5 Automatic fire detection. Smoke detection shall be provided in accordance with Section 907.2.12.1.

[F] 403.6 Emergency voice/alarm communication systems. An emergency voice/alarm communication system shall be provided in accordance with Section 907.2.12.2.

[F] 403.7 Fire department communications system. A two-way fire department communications system shall be provided for fire department use in accordance with Section 907.2.12.3.

[F] 403.8 Fire command. A fire command center complying with Section 911 shall be provided in a location approved by the fire department.

403.9 Elevators. Elevator operation and installation shall be in accordance with Chapter 30.

403.10 Standby power. A standby power system complying with Section 2702 shall be provided for standby power loads specified in Section 403.10.2.

403.10.1 Special requirements for standby power systems. If the standby system is a generator set inside a building, the system shall be located in a separate room enclosed with 2-hour fire-resistance-rated fire barrier assemblies. System supervision with manual start and transfer features shall be provided at the fire command center.

403.10.2 Standby power loads. The following are classified as standby power loads:

1. Power and lighting for the fire command center required by Section 403.8;

2. Electrically powered fire pumps;

3. Ventilation and automatic fire detection equipment for smokeproof enclosures.

Standby power shall be provided for elevators in accordance with Section 3003.

403.11 Emergency power systems. An emergency power system complying with Section 2702 shall be provided for emergency power loads specified in Section 403.11.1.

403.11.1 Emergency power loads. The following are classified as emergency power loads:

1. Exit signs and means of egress illumination required by Chapter 10;

2. Elevator car lighting;

3. Emergency voice/alarm communications systems;

4. Automatic fire detection systems; and

5. Fire alarm systems.

403.12 Stairway door operation. Stairway doors other than the exit discharge doors shall be permitted to be locked from stairway side. Stairway doors that are locked from the stairway side shall be capable of being unlocked simultaneously without unlatching upon a signal from the fire command center.

403.12.1 Stairway communications system. A telephone or other two-way communications system connected to an approved constantly attended station shall be provided at not less than every fifth floor in each required stairway where the doors to the stairway are locked.

403.13 Smokeproof exit enclosures. Every required stairway serving floors more than 75 feet (22 860 mm) above the lowest level of fire department vehicle access shall comply with Sections 909.20 and 1019.1.8.

403.14 Seismic considerations. For seismic considerations, see Chapter 16.

SECTION 404
ATRIUMS

404.1 General. Vertical openings meeting the requirements of this section are not required to be enclosed in other than Group H occupancies.

404.1.1 Definition. The following word and term shall, for the purposes of this chapter and as used elsewhere in this code, have the meaning shown herein.

ATRIUM. An opening connecting two or more stories other than enclosed stairways, elevators, hoistways, escalators, plumbing, electrical, air-conditioning or other equipment, which is closed at the top and not defined as a mall. Stories, as used in this definition, do not include balconies within assembly groups or mezzanines that comply with Section 505.

404.2 Use. The floor of the atrium shall not be used for other than low fire hazard uses and only approved materials and decorations in accordance with the *International Fire Code* shall be used in the atrium space.

Exception: The atrium floor area is permitted to be used for any approved use where the individual space is provided with an automatic sprinkler system in accordance with Section 903.3.1.1.

[F] 404.3 Automatic sprinkler protection. An approved automatic sprinkler system shall be installed throughout the entire building.

Exceptions:

1. That area of a building adjacent to or above the atrium need not be sprinklered provided that portion of the building is separated from the atrium portion by a 2-hour fire-resistance-rated fire barrier wall or horizontal assembly or both.

2. Where the ceiling of the atrium is more than 55 feet (16 764 mm) above the floor, sprinkler protection at the ceiling of the atrium is not required.

404.4 Smoke control. A smoke control system shall be installed in accordance with Section 909.

Exceptions:

1. Smoke control is not required for floor openings meeting the requirements of Section 707.2, Exception 2,7, 8 or 9.

2. Smoke control is not required for floor openings meeting the requirements of Section 1019.1, Exception 8 or 9.

404.5 Enclosure of atriums. Atrium spaces shall be separated from adjacent spaces by a 1-hour fire barrier wall.

Exceptions:

1. A glass wall forming a smoke partition where automatic sprinklers are spaced 6 feet (1829 mm) or less along both sides of the separation wall, or on the room side only if there is not a walkway on the atrium side, and between 4 inches and 12 inches (102 mm and 305 mm) away from the glass and so designed that the entire surface of the glass is wet upon activation of the sprinkler system. The glass shall be installed in a gasketed frame so that the framing system deflects without breaking (loading) the glass before the sprinkler system operates.

2. A glass-block wall assembly in accordance with Section 2110 and having a $^3/_4$-hour fire protection rating.

3. The adjacent spaces of any three floors of the atrium shall not be required to be separated from the atrium where such spaces are included in computing the atrium volume for the design of the smoke control system.

404.6 Standby power. Equipment required to provide smoke control shall be connected to a standby power system in accordance with Section 909.11.

404.7 Interior finish. The interior finish of walls and ceilings of the atrium shall not be less than Class B with no reduction in class for sprinkler protection.

404.8 Travel distance. In other than the lowest level of the atrium, where the required means of egress is through the atrium space, the portion of exit access travel distance within the atrium space shall not exceed 200 feet (60 960 mm).

SECTION 405
UNDERGROUND BUILDINGS

405.1 General. The provisions of this section apply to building spaces having a floor level used for human occupancy more than 30 feet (9144 mm) below the lowest level of exit discharge.

Exceptions:

1. One- and two-family dwellings, sprinklered in accordance with Section 903.3.1.3.

2. Parking garages with automatic fire suppression systems in compliance with Section 405.3.

3. Fixed guideway transit systems.

4. Grandstands, bleachers, stadiums, arenas and similar facilities.

5. Where the lowest story is the only story that would qualify the building as an underground building and has an area not exceeding 1,500 square feet (139 m²) and has an occupant load less than 10.

405.2 Construction requirements. The underground portion of the building shall be of Type I construction.

[F] 405.3 Automatic sprinkler system. The highest level of exit discharge serving the underground portions of the building and all levels below shall be equipped with an automatic sprinkler system installed in accordance with Section 903.3.1.1. Water-flow switches and control valves shall be supervised in accordance with Section 903.4.

405.4 Compartmentation. Compartmentation shall be in accordance with Sections 405.4.1 through 405.4.3.

405.4.1 Number of compartments. A building having a floor level more than 60 feet (18 288 mm) below the lowest level of exit discharge shall be divided into a minimum of two compartments of approximately equal size. Such compartmentation shall extend through the highest level of exit discharge serving the underground portions of the building and all levels below.

Exception: The lowest story need not be compartmented where the area does not exceed 1,500 square feet (139 m²) and has an occupant load of less than 10.

405.4.2 Smoke barrier penetration. The separation between the two compartments shall be of minimum 1-hour fire barrier wall construction that shall extend from floor slab to floor deck above. Openings between the two compartments shall be limited to plumbing and electrical piping and conduit penetrations firestopped in accordance with Section 712. Doorways shall be protected by fire door assemblies that are automatic-closing by smoke detection in accordance with Section 715.3 and shall be provided with gasketing and a drop sill to minimize smoke leakage. Where provided, each compartment shall have an air supply and an exhaust system independent of the other compartments.

405.4.3 Elevators. Where elevators are provided, each compartment shall have direct access to an elevator. Where an elevator serves more than one compartment, an elevator lobby shall be provided and shall be separated from each compartment by a 1-hour fire barrier wall. Doors shall be gasketed, have a drop sill, and be automatic-closing by smoke detection installed in accordance with Section 907.10.

405.5 Smoke control system. A smoke control system shall be provided in accordance with Sections 405.5.1 and 405.5.2.

405.5.1 Control system. A smoke control system is required to control the migration of products of combustion in accordance with Section 909 and the provisions of this section. Smoke control shall restrict movement of smoke to the general area of fire origin and maintain means of egress in a usable condition.

405.5.2 Smoke exhaust system. Where compartmentation is required, each compartment shall have an independent smoke control system. The system shall be automatically activated and capable of manual operation in accordance with Section 907.2.18.

[F] 405.6 Fire alarm systems. A fire alarm system shall be provided where required by Section 907.2.19.

[F] 405.7 Public address. A public address system shall be provided where required by Section 907.2.19.1.

405.8 Means of egress. Means of egress shall be in accordance with Sections 405.8.1 and 405.8.2.

405.8.1 Number of exits. Each floor level shall be provided with a minimum of two exits. Where compartmentation is required by Section 405.4, each compartment shall have a minimum of one exit and shall also have an exit access doorway into the adjoining compartment.

405.8.2 Smokeproof enclosure. Every required stairway serving floor levels more than 30 feet (9144 mm) below its level of exit discharge shall comply with the requirements for a smokeproof enclosure as provided in Section 1019.1.8.

[F] 405.9 Standby power. A standby power system complying with Section 2702 shall be provided standby power loads specified in Section 405.9.1.

405.9.1 Standby power loads. The following loads are classified as standby power loads.

1. Smoke control system.
2. Ventilation and automatic fire detection equipment for smokeproof enclosures.
3. Fire pumps.

Standby power shall be provided for elevators in accordance with Section 3003.

405.9.2 Pick-up time. The standby power system shall pick up its connected loads within 60 seconds of failure of the normal power supply.

[F] 405.10 Emergency power. An emergency power system complying with Section 2702 shall be provided for emergency power loads specified in Section 405.10.1.

405.10.1 Emergency power loads. The following loads are classified as emergency power loads:

1. Emergency voice/alarm communications systems.
2. Fire alarm systems.
3. Automatic fire detection systems.
4. Elevator car lighting.
5. Means of egress and exit sign illumination as required by Chapter 10.

405.11 Standpipe system. The underground building shall be equipped throughout with a standpipe system in accordance with Section 905.

SECTION 406
MOTOR-VEHICLE-RELATED OCCUPANCIES

406.1 Private garages and carports.

406.1.1 Classification. Buildings or parts of buildings classified as Group U occupancies because of the use or character of the occupancy shall not exceed 1,000 square feet (93 m²) in area or one story in height except as provided in Section 406.1.2. Any building or portion thereof that exceeds the limitations specified in this section shall be classified in the occupancy group other than Group U that it most nearly resembles.

406.1.2 Area increase. Group U occupancies used for the storage of private or pleasure-type motor vehicles where no repair work is done or fuel dispensed are permitted to be 3,000 square feet (279 m²), when the following provisions are met:

1. For a mixed occupancy building, the exterior wall and opening protection for the Group U portion of the building shall be as required for the major occupancy of the building. For such mixed occupancy building, the allowable floor area of the building shall be as permitted for the major occupancy contained therein.

2. For a building containing only a Group U occupancy, the exterior wall and opening protection shall be as required for a Group R-1 or R-2 occupancy.

More than one 3,000-square-foot (279 m²) Group U occupancy shall be permitted to be in the same building, provided each 3,000-square-foot (279 m²) area is separated by fire walls complying with Section 705.

406.1.3 Garages and carports. Carports shall be open on at least two sides. Carport floor surfaces shall be of approved noncombustible material. Carports not open on at least two sides shall be considered a garage and shall comply with the provisions of this section for garages.

Exception: Asphalt surfaces shall be permitted at ground level in carports.

The area of floor used for parking of automobiles or other vehicles shall be sloped to facilitate the movement of liquids to a drain or toward the main vehicle entry doorway.

406.1.4 Separation. Separations shall comply with the following:

1. The private garage shall be separated from the dwelling unit and its attic area by means of a minimum ¹/₂-inch (12.7 mm) gypsum board applied to the garage side. Garages beneath habitable rooms shall be separated from all habitable rooms above by not less than ⁵/₈-inch Type X gypsum board or equivalent. Door openings between a private garage and the dwelling unit shall be equipped with either solid wood doors, or solid or honeycomb core steel doors not less than 1³/₈ inches (34.9 mm) thick, or doors in compliance with Section 715.3.3. Openings from a private garage directly into a room used for sleeping purposes shall not be permitted.

2. Ducts in a private garage and ducts penetrating the walls or ceilings separating the dwelling unit from the garage shall be constructed of a minimum 0.019-inch (0.48 mm) sheet steel and shall have no openings into the garage.

3. A separation is not required between a Group R-3 and U carport provided the carport is entirely open on two or more sides and there are not enclosed areas above.

406.2 Parking garages.

406.2.1 Classification. Parking garages shall be classified as either open, as defined in Section 406.3, or enclosed and shall meet the appropriate criteria in Section 406.4. Also see Section 508 for special provisions for parking garages.

406.2.2 Clear height. The clear height of each floor level in vehicle and pedestrian traffic areas shall not be less than 7 feet (2134 mm). Vehicle and pedestrian areas accommodating van-accessible parking required by Section 1106.5 shall conform to ICC A117.1.

406.2.3 Guards. Guards shall be provided in accordance with Section 1012 at exterior and interior vertical openings on floor and roof areas where vehicles are parked or moved and where the vertical distance to the ground or surface directly below exceeds 30 inches (762 mm).

406.2.4 Vehicle barriers. Parking areas shall be provided with exterior or interior walls or vehicle barriers, except at pedestrian or vehicular accesses, designed in accordance with Section 1607.7. Vehicle barriers not less than 2 feet (607 mm) high shall be placed at the ends of drive lanes, and

at the end of parking spaces where the difference in adjacent floor elevation is greater than 1 foot (305 mm).

406.2.5 Ramps. Vehicle ramps shall not serve as an exit element.

406.2.6 Floor surface. Parking surfaces shall be of concrete or similar noncombustible and nonabsorbent materials.

Exception: Asphalt parking surfaces are permitted at ground level.

The area of floor used for parking of automobiles or other vehicles shall be sloped to facilitate the movement of liquids to a drain or toward the main vehicle entry doorway.

406.2.7 Mixed separation. Parking garages shall be separated from other occupancies in accordance with Section 302.3.2.

406.2.8 Special hazards. Connection of a parking garage with any room in which there is a fuel-fired appliance shall be by means of a vestibule providing a two-doorway separation.

Exception: A single door shall be allowed provided the sources of ignition in the appliance are at least 18 inches (457 mm) above the floor.

406.2.9 Attached to rooms. Openings from a parking garage directly into a room used for sleeping purposes shall not be permitted.

406.3 Open parking garages.

406.3.1 Scope. Except where specific provisions are made in the following subsections, other requirements of this code shall apply.

406.3.2 Definitions. The following words and terms shall, for the purposes of this chapter and as used elsewhere in this code, have the meanings shown herein.

MECHANICAL-ACCESS OPEN PARKING GARAGES. Open parking garages employing parking machines, lifts, elevators or other mechanical devices for vehicles moving from and to street level and in which public occupancy is prohibited above the street level.

OPEN PARKING GARAGE. A structure or portion of a structure with the openings as described in Section 406.3.3.1 on two or more sides that is used for the parking or storage of private motor vehicles as described in Section 406.3.4.

RAMP-ACCESS OPEN PARKING GARAGES. Open parking garages employing a series of continuously rising floors or a series of interconnecting ramps between floors permitting the movement of vehicles under their own power from and to the street level.

406.3.3 Construction. Open parking garages shall be of Type I, II or IV construction. Open parking garages shall meet the design requirements of Chapter 16. For vehicle barriers, see Section 406.2.4.

406.3.3.1 Openings. For natural ventilation purposes, the exterior side of the structure shall have uniformly distributed openings on two or more sides. The area of such openings in exterior walls on a tier must be at least 20 percent of the total perimeter wall area of each tier. The aggregate length of the openings considered to be providing natural ventilation shall constitute a minimum of 40 percent of the perimeter of the tier. Interior walls shall be at least 20 percent open with uniformly distributed openings.

Exception: Openings are not required to be distributed over 40 percent of the building perimeter where the required openings are uniformly distributed over two opposing sides of the building.

406.3.4 Uses. Mixed uses shall be allowed in the same building as an open parking garage subject to the provisions of Sections 302.3, 402.7.1, 406.3.13, 508.3, 508.4 and 508.7.

406.3.5 Area and height. Area and height of open parking garages shall be limited as set forth in Chapter 5 for Group S-2 occupancies and as further provided for in Section 302.3.

406.3.5.1 Single use. When the open parking garage is used exclusively for the parking or storage of private motor vehicles, with no other uses in the building, the area and height shall be permitted to comply with Table 406.3.5, along with increases allowed by Section 406.3.6.

Exception: The grade-level tier is permitted to contain an office, waiting and toilet rooms having a total combined area of not more than 1,000 square feet (93 m²). Such area need not be separated from the open parking garage.

In open parking garages having a spiral or sloping floor, the horizontal projection of the structure at any cross section shall not exceed the allowable area per parking tier. In the case of an open parking garage having a continuous spiral floor, each 9 feet 6 inches (2896 mm) of height, or portion thereof, shall be considered a tier.

The clear height of a parking tier shall not be less than 7 feet (2134 mm), except that a lower clear height is permitted in mechanical-access open parking garages where approved by the building official.

406.3.6 Area and height increases. The allowable area and height of open parking garages shall be increased in accordance with the provisions of this section. Garages with sides open on three-fourths of the building perimeter are permitted to be increased by 25 percent in area and one tier in height. Garages with sides open around the entire building perimeter are permitted to be increased 50 percent in area and one tier in height. For a side to be considered open under the above provisions, the total area of openings along the side shall not be less than 50 percent of the interior area of the side at each tier, and such openings shall be equally distributed along the length of the tier.

Allowable tier areas in Table 406.3.5 shall be increased for open parking garages constructed to heights less than the table maximum. The gross tier area of the garage shall not exceed that permitted for the higher structure. At least three sides of each such larger tier shall have continuous horizontal openings not less than 30 inches (762 mm) in clear height extending for at least 80 percent of the length of the sides, and no part of such larger tier shall be more than 200 feet (60 960 mm) horizontally from such an opening. In addition, each such opening shall face a street or yard accessible to a street with a width of at least 30 feet (9144 mm) for the full length of the opening, and standpipes shall be provided in each such tier.

Open parking garages of Type IB and II construction, with all sides open, shall be unlimited in allowable area where the height does not exceed 75 feet (22 860 mm). For a side to be considered open, the total area of openings along the side shall not be less than 50 percent of the interior area of the side at each tier, and such openings shall be equally distributed along the length of the tier. All portions of tiers shall be within 200 feet (60 960 mm) horizontally from such openings.

406.3.7 Location on property. Exterior walls and openings in exterior walls shall comply with Tables 601 and 602. The distance from an adjacent lot line shall be determined in accordance with Table 602 and Section 704.

406.3.8 Means of egress. Where persons other than parking attendants are permitted, open parking garages shall meet the means of egress requirements of Chapter 10. Where no persons other than parking attendants are permitted, there shall not be less than two 36-inch-wide (914 mm) exit stairways. Lifts shall be permitted to be installed for use of em-

TABLE 406.3.5
OPEN PARKING GARAGES AREA AND HEIGHT

TYPE OF CONSTRUCTION	AREA PER TIER (square feet)	HEIGHT (in tiers)		
		Ramp access	Mechanical access	
			Automatic sprinkler system	
			No	Yes
IA	Unlimited	Unlimited	Unlimited	Unlimited
IB	Unlimited	12 tiers	12 tiers	18 tiers
IIA	50,000	10 tiers	10 tiers	15 tiers
IIB	50,000	8 tiers	8 tiers	12 tiers
IV	50,000	4 tiers	4 tiers	4 tiers

For SI: 1 square foot = 0.0929 m².

ployees only, provided they are completely enclosed by noncombustible materials.

406.3.9 Standpipes. Standpipes shall be installed where required by the provisions of Chapter 9.

406.3.10 Sprinkler systems. Where required by other provisions or this code, automatic sprinkler systems and standpipes shall be installed in accordance with the provisions of Chapter 9.

406.3.11 Enclosure of vertical openings. Enclosure shall not be required for vertical openings except as specified in Section 406.3.8.

406.3.12 Ventilation. Ventilation, other than the percentage of openings specified in Section 406.3.3.1, shall not be required.

406.3.13 Prohibitions. The following uses and alterations are not permitted:

1. Vehicle repair work.

2. Parking of buses, trucks and similar vehicles.

3. Partial or complete closing of required openings in exterior walls by tarpaulins or any other means.

4. Dispensing of fuel.

406.4 Enclosed parking garages.

406.4.1 Heights and areas. Enclosed vehicle parking garages and portions thereof that do not meet the definition of open parking garages shall be limited to the allowable heights and areas specified in Table 503. Roof parking is permitted.

406.4.2 Ventilation. A mechanical ventilation system shall be provided in accordance with the *International Mechanical Code.*

406.5 Motor fuel-dispensing facilities.

406.5.1 Construction. Motor fuel-dispensing facilities shall be constructed in accordance with the *International Fire Code* and this section.

406.5.2 Canopies. Canopies under which fuels are dispensed shall have a clear, unobstructed height of not less than 13 feet 6 inches (4115 mm) to the lowest projecting element in the vehicle drive-through area. Canopies and their supports over pumps shall be of noncombustible materials, fire-retardant-treated wood complying with Chapter 23, wood of Type IV sizes or of construction providing 1-hour fire resistance. Combustible materials used in or on a canopy shall comply with one of the following:

1. Shielded from the pumps by a noncombustible element of the canopy, or wood of Type IV sizes;

2. Plastics covered by aluminum facing having a minimum thickness of 0.010 inch (0.30 mm) or corrosion-resistant steel having a minimum base metal thickness of 0.016 inch (0.41 mm). The plastic shall have a flame spread index of 25 or less and a smoke-developed index of 450 or less when tested in the form intended for use in accordance with ASTM E 84 and a self-ignition temperature of 650°F (343°C) or greater when tested in accordance with ASTM D 1929; or

3. Panels constructed of light-transmitting plastic materials shall be permitted to be installed in canopies erected over motor vehicle fuel-dispensing station fuel dispensers, provided the panels are located at least 10 feet (3048 mm) from any building on the same property and face yards or streets not less than 40 feet (12 192 mm) in width on the other sides. The aggregate areas of plastics shall not exceed 1,000 square feet (93 m²). The maximum area of any individual panel shall not exceed 100 square feet (9.3 m²).

406.6 Repair garages.

406.6.1 General. Repair garages shall be constructed in accordance with the *International Fire Code* and this section. This occupancy shall not include motor fuel-dispensing facilities, as regulated in Section 406.5.

406.6.2 Mixed uses. Mixed uses shall be allowed in the same building as a repair garage subject to the provisions of Section 302.3.

406.6.3 Ventilation. Repair garages shall be mechanically ventilated in accordance with the *International Mechanical Code*. The ventilation system shall be controlled at the entrance to the garage.

406.6.4 Floor surface. Repair garage floors shall be of concrete or similar noncombustible and nonabsorbent materials.

> **Exception:** Slip-resistant, nonabsorbent, interior floor finishes having a critical radiant flux not more than 0.45 W/cm2, as determined by NFPA 253, shall be permitted.

406.6.5 Heating equipment. Heating equipment shall be installed in accordance with the *International Mechanical Code*.

[F] 406.6.6 Gas detection system. Repair garages used for repair of vehicles fueled by nonodorized gases, such as hydrogen and nonodorized LNG, shall be provided with an approved flammable gas-detection system.

[F] 406.6.6.1 System design. The flammable gas-detection system shall be calibrated to the types of fuels or gases used by vehicles to be repaired. The gas detection system shall be designed to activate when the level of flammable gas exceeds 25 percent of the lower explosive limit. Gas detection shall also be provided in lubrication or chassis repair pits of garages used for repairing nonodorized LNG-fueled vehicles.

[F] 406.6.6.2 Operation. Activation of the gas detection system shall result in all of the following:

1. Initiation of distinct audible and visual alarm signals in the repair garage.

2. Deactivation of all heating systems located in the repair garage.

3. Activation of the mechanical ventilation system, where the system is interlocked with gas detection.

[F] 406.6.6.3 Failure of the gas detection system. Failure of the gas detection system shall result in the deactivation of the heating system, activation of the mechanical ventilation system when the system is interlocked with the gas detection system and cause a trouble signal to sound in an approved location.

SECTION 407
GROUP I-2

407.1 General. Occupancies in Group I-2 shall comply with the provisions of this section and other applicable provisions of this code.

407.2 Corridors. Corridors in occupancies in Group I-2 shall be continuous to the exits and separated from other areas in accordance with Section 407.3 except spaces conforming to Sections 407.2.1 through 407.2.4.

407.2.1 Spaces of unlimited area. Waiting areas and similar spaces constructed as required for corridors shall be permitted to be open to a corridor, only where all of the following criteria are met:

1. The spaces are not occupied for patient sleeping units, treatment rooms, hazardous or incidental use areas as defined in Section 302.1.1.

2. The open space is protected by an automatic fire detection system installed in accordance with Section 907.

3. The corridors onto which the spaces open, in the same smoke compartment, are protected by an automatic fire detection system installed in accordance with Section 907, or the smoke compartment in which the spaces are located is equipped throughout with quick-response sprinklers in accordance with Section 903.3.2.

4. The space is arranged so as not to obstruct access to the required exits.

407.2.2 Nurses' stations. Spaces for doctors' and nurses' charting, communications and related clerical areas shall be permitted to be open to the corridor, when such spaces are constructed as required for corridors.

407.2.3 Mental health treatment areas. Areas wherein mental health patients who are not capable of self-preservation are housed, or group meeting or multipurpose therapeutic spaces other than incidental use areas as defined in Section 302.1.1, under continuous supervision by facility staff, shall be permitted to be open to the corridor, where the following criteria are met:

1. Each area does not exceed 1,500 square feet (140 m^2).

2. The area is located to permit supervision by the facility staff.

3. The area is arranged so as not to obstruct any access to the required exits.

4. The area is equipped with an automatic fire detection system installed in accordance with Section 907.2.

5. Not more than one such space is permitted in any one smoke compartment.

6. The walls and ceilings of the space are constructed as required for corridors.

407.2.4 Gift shops. Gift shops less than 500 square feet (46.5 m^2) in area shall be permitted to be open to the corridor provided the gift shop and storage areas are fully sprinklered and storage areas are protected in accordance with Section 302.1.1.

407.3 Corridor walls. Corridor walls shall be constructed as smoke partitions.

407.3.1 Corridor doors. Corridor doors, other than those in a wall required to be rated by Section 302.1.1 or for the enclosure of a vertical opening or an exit, shall not have a required fire protection rating and shall not be required to be equipped with self-closing or automatic-closing devices, but shall provide an effective barrier to limit the transfer of smoke and shall be equipped with positive latching. Roller latches are not permitted. Other doors shall conform to Section 715.3.

407.3.2 Locking devices. Locking devices that restrict access to the patient room from the corridor, and that are operable only by staff from the corridor side, shall not restrict the means of egress from the patient room except for patient rooms in mental health facilities.

407.4 Smoke barriers. Smoke barriers shall be provided to subdivide every story used by patients for sleeping or treatment and to divide other stories with an occupant load of 50 or more persons, into at least two smoke compartments. Such stories shall be divided into smoke compartments with an area of not more than 22,500 square feet (2092 m^2) and the travel distance from any point in a smoke compartment to a smoke barrier door shall not exceed 200 feet (60 960 mm). The smoke barrier shall be in accordance with Section 709.

407.4.1 Refuge area. At least 30 net square feet (2.8 m^2) per patient shall be provided within the aggregate area of corridors, patient rooms, treatment rooms, lounge or dining areas and other low-hazard areas on each side of each smoke barrier. On floors not housing patients confined to a bed or litter, at least 6 net square feet (0.56 m^2) per occupant shall be provided on each side of each smoke barrier for the total number of occupants in adjoining smoke compartments.

407.4.2 Independent egress. A means of egress shall be provided from each smoke compartment created by smoke barriers without having to return through the smoke compartment from which means of egress originated.

[F] 407.5 Automatic sprinkler system. Smoke compartments containing patient sleeping units shall be equipped throughout with an automatic fire sprinkler system in accordance with Section 903.3.1.1. The smoke compartments shall be equipped with approved quick-response or residential sprinklers in accordance with Section 903.3.2.

[F] 407.6 Automatic fire detection. Corridors in nursing homes (both intermediate-care and skilled nursing facilities), detoxification facilities and spaces permitted to be open to corridors by Section 407.2 shall be protected by an automatic fire detection system installed in accordance with Section 907.

Exceptions:

1. Corridor smoke detection is not required where patient sleeping units are provided with smoke detectors that comply with UL 268. Such detectors shall pro-

vide a visual display on the corridor side of each patient sleeping unit and an audible and visual alarm at the nursing station attending each unit.

2. Corridor smoke detection is not required where patient sleeping unit doors are equipped with automatic door-closing devices with integral smoke detectors on the unit sides installed in accordance with their listing, provided that the integral detectors perform the required alerting function.

407.7 Secured yards. Grounds are permitted to be fenced and gates therein are permitted to be equipped with locks, provided that safe dispersal areas having 30 net square feet (2.8 m²) for bed and litter patients and 6 net square feet (0.56 m²) for ambulatory patients and other occupants are located between the building and the fence. Such provided safe dispersal areas shall not be located less than 50 feet (15 240 mm) from the building they serve.

SECTION 408
GROUP I-3

408.1 General. Occupancies in Group I-3 shall comply with the provisions of this section and other applicable provisions of this code (see Section 308.4).

408.2 Mixed occupancies. Portions of buildings with an occupancy in Group I-3 that are classified as a different occupancy shall meet the applicable requirements of this code for such occupancies. Where security operations necessitate the locking of required means of egress, provisions shall be made for the release of occupants at all times.

Means of egress from detention and correctional occupancies that traverse other use areas shall, as a minimum, conform to requirements for detention and correctional occupancies.

Exception: It is permissible to exit through a horizontal exit into other contiguous occupancies that do not conform to detention and correctional occupancy egress provisions but that do comply with requirements set forth in the appropriate occupancy, as long as the occupancy is not a high-hazard use.

408.3 Means of egress. Except as modified or as provided for in this section, the provisions of Chapter 10 shall apply.

408.3.1 Door width. Doors to resident sleeping units shall have a clear width of not less than 28 inches (711 mm).

408.3.2 Sliding doors. Where doors in a means of egress are of the horizontal-sliding type, the force to slide the door to its fully open position shall not exceed 50 pounds (220 N) with a perpendicular force against the door of 50 pounds (220 N).

408.3.3 Spiral stairs. Spiral stairs that conform to the requirements of Section 1009.9 are permitted for access to and between staff locations.

408.3.4 Exit discharge. Exits are permitted to discharge into a fenced or walled courtyard. Enclosed yards or courts shall be of a size to accommodate all occupants, a minimum of 50 feet (15 240 mm) from the building with a net area of 15 square feet (1.4 m²) per person.

408.3.5 Sallyports. A sallyport shall be permitted in a means of egress where there are provisions for continuous

and unobstructed passage through the sallyport during an emergency egress condition.

408.3.6 Vertical exit enclosures. One of the required vertical exit enclosures in each building shall be permitted to have glazing installed in doors and interior walls at each landing level providing access to the enclosure, provided that the following conditions are met:

1. The vertical exit enclosure shall not serve more than four floor levels.

2. Vertical exit enclosure doors shall not be less than ³/₄-hour fire doors complying with Section 715.3.

3. The total area of glazing at each floor level shall not exceed 5,000 square inches (3.23 m²) and individual panels of glazing shall not exceed 1,296 square inches (0.84 m²).

4. The glazing shall be protected on both sides by an automatic fire sprinkler system. The sprinkler system shall be designed to wet completely the entire surface of any glazing affected by fire when actuated.

5. The glazing shall be in a gasketed frame and installed in such a manner that the framing system will deflect without breaking (loading) the glass before the sprinkler system operates.

6. Obstructions, such as curtain rods, drapery traverse rods, curtains, drapes or similar materials shall not be installed between the automatic sprinklers and the glazing.

408.4 Locks. Egress doors are permitted to be locked in accordance with the applicable use condition. Doors from an area of refuge to the exterior are permitted to be locked with a key in lieu of locking methods described in Section 408.4.1. The keys to unlock the exterior doors shall be available at all times and the locks shall be operable from both sides of the door.

408.4.1 Remote release. Remote release of locks on doors in a means of egress shall be provided with reliable means of operation, remote from the resident living areas, to release locks on all required doors. In Occupancy Conditions 3 or 4, the arrangement, accessibility and security of the release mechanism(s) required for egress shall be such that with the minimum available staff at any time, the lock mechanisms are capable of being released within 2 minutes.

Exception: Provisions for remote locking and unlocking of occupied rooms in Occupancy Condition 4 are not required provided that not more than 10 locks are necessary to be unlocked in order to move occupants from one smoke compartment to a refuge area within 3 minutes. The opening of necessary locks shall be accomplished with not more than two separate keys.

[F] 408.4.2 Power-operated doors and locks. Power-operated sliding doors or power-operated locks for swinging doors shall be operable by a manual release mechanism at the door, and either emergency power or a remote mechanical operating release shall be provided.

Exception: Emergency power is not required in facilities with 10 locks or less complying with the exception to Section 408.4.1.

408.4.3 Redundant operation. Remote release, mechanically operated sliding doors or remote release, mechanically operated locks shall be provided with a mechanically operated release mechanism at each door, or shall be provided with a redundant remote release control.

408.4.4 Relock capability. Doors remotely unlocked under emergency conditions shall not automatically relock when closed unless specific action is taken at the remote location to enable doors to relock.

408.5 Vertical openings. Vertical openings shall be enclosed in accordance with Section 707.

Exception: A floor opening between floor levels of residential housing areas is permitted without enclosure protection between the levels, provided that both of the following conditions are met:

1. The entire normally occupied areas so interconnected are open and unobstructed so as to enable observation of the areas by supervisory personnel.

2. Means of egress capacity is sufficient to provide simultaneous egress for all occupants from all interconnected levels and areas.

The height difference between the highest and lowest finished floor levels shall not exceed 23 feet (7010 mm). Each story, considered separately, has at least one-half of its individual required means of egress capacity provided by exits leading directly out of that story without traversing another story within the interconnected area.

408.6 Smoke barrier. Occupancies in Group I-3 shall have smoke barriers complying with Section 709 to divide every story occupied by residents for sleeping, or any other story having an occupant load of 50 or more persons, into at least two smoke compartments.

Exception: Spaces having direct exit to one of the following, provided that the locking arrangement of the doors involved complies with the requirements for doors at the compartment barrier for the use condition involved:

1. A public way.

2. A building separated from the resident housing area by a 2-hour fire-resistance-rated assembly or 50 feet (15 240 mm) of open space.

3. A secured yard or court having a holding space 50 feet (15 240 mm) from the housing area that provides 6 square feet (0.56 m²) or more of refuge area per occupant, including residents, staff and visitors.

408.6.1 Smoke compartments. The maximum number of residents in any smoke compartment shall be 200. The travel distance to a door in a smoke barrier from any room door required as exit access shall not exceed 150 feet (45 720 mm). The travel distance to a door in a smoke barrier from any point in a room shall not exceed 200 feet (60 960 mm).

408.6.2 Refuge area. At least 6 net square feet (0.56 m²) per occupant shall be provided on each side of each smoke barrier for the total number of occupants in adjoining smoke compartments. This space shall be readily available wherever the occupants are moved across the smoke barrier in a fire emergency.

408.6.3 Independent egress. A means of egress shall be provided from each smoke compartment created by smoke barriers without having to return through the smoke compartment from which means of egress originates.

408.7 Subdivision of resident housing areas. Sleeping areas and any contiguous day room, group activity space or other common spaces where residents are housed shall be separated from other spaces in accordance with Sections 408.7.1 through 408.7.4.

408.7.1 Occupancy Conditions 3 and 4. Each sleeping area in Occupancy Conditions 3 and 4 shall be separated from the adjacent common spaces by a smoke-tight partition where the travel distance from the sleeping area through the common space to the exit access corridor exceeds 50 feet (15 240 mm).

408.7.2 Occupancy Condition 5. Each sleeping area in Occupancy Condition 5 shall be separated from adjacent sleeping areas, corridors and common spaces by a smoke-tight partition. Additionally, common spaces shall be separated from the exit access corridor by a smoke-tight partition.

408.7.3 Openings in room face. The aggregate area of openings in a solid sleeping room face in Occupancy Conditions 2, 3, 4 and 5 shall not exceed 120 square inches (77 419 mm²). The aggregate area shall include all openings including door undercuts, food passes and grilles. Openings shall be not more than 36 inches (914 mm) above the floor. In Occupancy Condition 5, the openings shall be closeable from the room side.

408.7.4 Smoke-tight doors. Doors in openings in partitions required to be smoke tight by Section 408.7 shall be substantial doors, of construction that will resist the passage of smoke. Latches and door closures are not required on cell doors.

408.8 Windowless buildings. For the purposes of this section, a windowless building or portion of a building is one with nonopenable windows, windows not readily breakable or without windows. Windowless buildings shall be provided with an engineered smoke control system to provide ventilation (mechanical or natural) in accordance with Section 909 for each windowless smoke compartment.

SECTION 409
MOTION PICTURE PROJECTION ROOMS

409.1 General. The provisions of this section shall apply to rooms in which ribbon-type cellulose acetate or other safety film is utilized in conjunction with electric arc, xenon or other light-source projection equipment that develops hazardous gases, dust or radiation. Where cellulose nitrate film is utilized or stored, such rooms shall comply with NFPA 40.

409.1.1 Projection room required. Every motion picture machine projecting film as mentioned within the scope of this section shall be enclosed in a projection room. Appurtenant electrical equipment, such as rheostats, transformers and generators, shall be within the projection room or in an adjacent room of equivalent construction.

409.2 Construction of projection rooms. Every projection room shall be of permanent construction consistent with the construction requirements for the type of building in which the projection room is located. Openings are not required to be protected.

The room shall have a floor area of not less than 80 square feet (7.44 m²) for a single machine and at least 40 square feet (3.7 m²) for each additional machine. Each motion picture projector, floodlight, spotlight or similar piece of equipment shall have a clear working space of not less than 30 inches by 30 inches (762 mm by 762 mm) on each side and at the rear thereof, but only one such space shall be required between two adjacent projectors. The projection room and the rooms appurtenant thereto shall have a ceiling height of not less than 7 feet 6 inches (2286 mm). The aggregate of openings for projection equipment shall not exceed 25 percent of the area of the wall between the projection room and the auditorium. Openings shall be provided with glass or other approved material, so as to close completely the opening.

409.3 Projection room and equipment ventilation. Ventilation shall be provided in accordance with the *International Mechanical Code*.

409.3.1 Projection room.

409.3.1.1 Supply air. Each projection room shall be provided with adequate air supply inlets so arranged as to provide well-distributed air throughout the room. Air inlet ducts shall provide an amount of air equivalent to the amount of air being exhausted by projection equipment. Air is permitted to be taken from the outside, from adjacent spaces within the building, provided the volume and infiltration rate is sufficient; or from the building air-conditioning system, provided it is so arranged as to provide sufficient air when other systems are not in operation.

409.3.1.2 Exhaust air. Projection rooms are permitted to be exhausted through the lamp exhaust system. The lamp exhaust system shall be positively interconnected with the lamp so that the lamp will not operate unless there is the required airflow. Exhaust air ducts shall terminate at the exterior of the building in such a location that the exhaust air cannot be readily recirculated into any air supply system. The projection room ventilation system is permitted to also serve appurtenant rooms, such as the generator and rewind rooms.

Each projection machine shall be provided with an exhaust duct that will draw air from each lamp and exhaust it directly to the outside of the building. The lamp exhaust is permitted to serve to exhaust air from the projection room to provide room air circulation. Such ducts shall be of rigid materials, except for a flexible connector approved for the purpose. The projection lamp or projection room exhaust system, or both, is permitted to be combined but shall not be interconnected with any other exhaust or return system, or both, within the building.

409.4 Lighting control. Provisions shall be made for control of the auditorium lighting and the means of egress lighting systems of theaters from inside the projection room and from at least one other convenient point in the building.

409.5 Miscellaneous equipment. Each projection room shall be provided with rewind and film storage facilities.

SECTION 410
STAGES AND PLATFORMS

410.1 Applicability. The provisions of this section shall apply to all parts of buildings and structures that contain stages or platforms and similar appurtenances as herein defined.

410.2 Definitions. The following words and terms shall, for the purposes of this section and as used elsewhere in this code, have the meanings shown herein.

FLY GALLERY. A raised floor area above a stage from which the movement of scenery and operation of other stage effects are controlled.

GRIDIRON. The structural framing over a stage supporting equipment for hanging or flying scenery and other stage effects.

PINRAIL. A rail on or above a stage through which belaying pins are inserted and to which lines are fastened.

PLATFORM. A raised area within a building used for worship, the presentation of music, plays or other entertainment; the head table for special guests; the raised area for lecturers and speakers; boxing and wrestling rings; theater-in-the-round stages; and similar purposes wherein there are no overhead hanging curtains, drops, scenery or stage effects other than lighting and sound. A temporary platform is one installed for not more than 30 days.

PROSCENIUM WALL. The wall that separates the stage from the auditorium or assembly seating area.

STAGE. A space within a building utilized for entertainment or presentations, which includes overhead hanging curtains, drops, scenery or stage effects other than lighting and sound. Stage area shall be measured to include the entire performance area and adjacent backstage and support areas not separated from the performance area by fire-resistance-rated construction. Stage height shall be measured from the lowest point on the stage floor to the highest point of the roof or floor deck above the stage.

410.3 Stages. Stage construction shall comply with Sections 410.3.1 through 410.3.7.

410.3.1 Stage construction. Stages shall be constructed of materials as required for floors for the type of construction of the building in which such stages are located.

Exceptions:

1. Stages of Type IIB or IV construction with a nominal 2-inch (51 mm) wood deck, provided that the stage is separated from other areas in accordance with Section 410.3.4.

2. In buildings of Type IIA, IIIA and VA construction, a fire-resistance-rated floor is not required, provided the space below the stage is equipped with an automatic fire-extinguishing system in accordance with Section 903 or 904.

3. In all types of construction, the finished floor shall be constructed of wood or approved noncombustible

materials. Openings through stage floors shall be equipped with tight-fitting, solid wood trap doors with approved safety locks.

410.3.1.1 Stage height and area. Stage areas shall be measured to include the entire performance area and adjacent backstage and support areas not separated from the performance area by fire-resistance-rated construction. Stage height shall be measured from the lowest point on the stage floor to the highest point of the roof or floor deck above the stage.

410.3.2 Galleries, gridirons, catwalks and pinrails. Beams designed only for the attachment of portable or fixed theater equipment, gridirons, galleries and catwalks shall be constructed of approved materials consistent with the requirements for the type of construction of the building; and a fire-resistance rating shall not be required. These areas shall not be considered to be floors, stories, mezzanines or levels in applying this code.

Exception: Floors of fly galleries and catwalks shall be constructed of any approved material.

410.3.3 Exterior stage doors. Where protection of openings is required, exterior exit doors shall be protected with fire doors that comply with Section 715. Exterior openings that are located on the stage for means of egress or loading and unloading purposes, and that are likely to be open during occupancy of the theater, shall be constructed with vestibules to prevent air drafts into the auditorium.

410.3.4 Proscenium wall. Where the stage height is greater than 50 feet (15 240 mm), all portions of the stage shall be completely separated from the seating area by a proscenium wall with not less than a 2-hour fire-resistance rating extending continuously from the foundation to the roof.

410.3.5 Proscenium curtain. The proscenium opening of every stage with a height greater than 50 feet (15 240 mm) shall be provided with a curtain of approved material or an approved water curtain complying with Section 903.3.1.1. The curtain shall be designed and installed to intercept hot gases, flames and smoke, and to prevent a glow from a severe fire on the stage from showing on the auditorium side for a period of 20 minutes. The closing of the curtain from the full open position shall be effected in less than 30 seconds, but the last 8 feet (2438 mm) of travel shall require not less than 5 seconds.

410.3.5.1 Activation. The curtain shall be activated by rate-of-rise heat detection installed in accordance with Section 907.10 operating at a rate of temperature rise of 15 to 20°F per minute (8 to 11°C per minute), and by an auxiliary manual control.

410.3.5.2 Fire test. A sample curtain with a minimum of two vertical seams shall be subjected to the standard fire test specified in ASTM E 119 for a period of 30 minutes. The curtain shall overlap the furnace edges by an amount that is appropriate to seal the top and sides. The curtain shall have a bottom pocket containing a minimum of 4 pounds per linear foot (58 N/m) of batten. The exposed surface of the curtain shall not glow, and flame or smoke shall not penetrate the curtain during the test period. Un-

exposed surface temperature and hose stream test requirements are not applicable to the proscenium fire safety curtain test.

410.3.5.3 Smoke test. Curtain fabrics shall have a smoke-developed rating of 25 or less when tested in accordance with ASTM E 84.

410.3.5.4 Tests. The completed proscenium curtain shall be subjected to operating tests prior to the issuance of a certificate of occupancy.

410.3.6 Scenery. Combustible materials used in sets and scenery shall be rendered flame resistant in accordance with Section 805 and the *International Fire Code*. Foam plastics and materials containing foam plastics shall comply with Section 2603 and the *International Fire Code*.

410.3.7 Stage ventilation. Emergency ventilation shall be provided for stages larger than 1,000 square feet (93 m²) in floor area, or with a stage height greater than 50 feet (15 240 mm). Such ventilation shall comply with Section 410.3.7.1 or 410.3.7.2.

410.3.7.1 Roof vents. Two or more vents constructed to open automatically by approved heat-activated devices and with an aggregate clear opening area of not less than 5 percent of the area of the stage shall be located near the center and above the highest part of the stage area. Supplemental means shall be provided for manual operation of the ventilator. Curbs shall be provided as required for skylights in Section 2610.2. Vents shall be labeled.

410.3.7.2 Smoke control. Smoke control in accordance with Section 909 shall be provided to maintain the smoke layer interface not less than 6 feet (1829 mm) above the highest level of the assembly seating or above the top of the proscenium opening where a proscenium wall is provided in compliance with Section 410.3.4.

410.4 Platform construction. Permanent platforms shall be constructed of materials as required for the type of construction of the building in which the permanent platform is located. Permanent platforms are permitted to be constructed of fire-retardant-treated wood for Type I, II, and IV construction where the platforms are not more than 30 inches (762 mm) above the main floor, and not more than one-third of the room floor area and not more than 3,000 square feet (279 m²) in area. Where the space beneath the permanent platform is used for storage or any other purpose other than equipment, wiring or plumbing, the floor construction shall not be less than 1-hour fire-resistance-rated construction. Where the space beneath the permanent platform is used only for equipment, wiring or plumbing, the underside of the permanent platform need not be protected.

410.4.1 Temporary platforms. Platforms installed for a period of not more than 30 days are permitted to be constructed of any materials permitted by the code. The space between the floor and the platform above shall only be used for plumbing and electrical wiring to platform equipment.

410.5 Dressing and appurtenant rooms. Dressing and appurtenant rooms shall comply with Sections 410.5.1 through 410.5.4.

410.5.1 Separation from stage. Where the stage height is greater than 50 feet (15 240 mm), the stage shall be separated from dressing rooms, scene docks, property rooms, workshops, storerooms and compartments appurtenant to the stage and other parts of the building by a fire barrier wall and horizontal assemblies, or both, with not less than a 2-hour fire-resistance rating with approved opening protectives. For stage heights of 50 feet (15 240 mm) or less, the required stage separation shall be a fire barrier wall and horizontal assemblies, or both, with not less a 1-hour fire-resistance rating with approved opening protectives.

410.5.2 Separation from each other. Dressing rooms, scene docks, property rooms, workshops, storerooms and compartments appurtenant to the stage shall be separated from each other by fire barrier wall and horizontal assemblies, or both, with not less than a 1-hour fire-resistance rating with approved opening protectives.

410.5.3 Opening protectives. Openings other than to trunk rooms and the necessary doorways at stage level shall not connect such rooms with the stage, and such openings shall be protected with fire door assemblies that comply with Section 715.

410.5.4 Stage exits. At least one approved means of egress shall be provided from each side of the stage; and from each side of the space under the stage. At least one means of escape shall be provided from each fly gallery and from the gridiron. A steel ladder, alternating tread stairway or spiral stairway is permitted to be provided from the gridiron to a scuttle in the stage roof.

[F] 410.6 Automatic sprinkler system. Stages shall be equipped with an automatic fire-extinguishing system in accordance with Chapter 9. The system shall be installed under the roof and gridiron, in the tie and fly galleries, in places behind the proscenium wall of the stage, and in dressing rooms, lounges, workshops and storerooms accessory to such stages.

Exceptions:

1. Sprinklers are not required under stage areas less than 4 feet (1219 mm) in clear height utilized exclusively for storage of tables and chairs, provided the concealed space is separated from the adjacent spaces by not less than $^5/_8$-inch (15.9 mm) Type X gypsum board.

2. Sprinklers are not required for stages 1,000 square feet (93 m^2) or less in area and 50 feet (15 240 mm) or less in height where curtains, scenery or other combustible hangings are not retractable vertically. Combustible hangings shall be limited to a single main curtain, borders, legs and a single backdrop.

[F] 410.7 Standpipes. Standpipe systems shall be provided in accordance with Section 905.

SECTION 411
SPECIAL AMUSEMENT BUILDINGS

411.1 General. Special amusement buildings having an occupant load of 50 or more shall comply with the requirements for the appropriate Group A occupancy and this section. Amusement buildings having an occupant load of less than 50 shall comply with the requirements for a Group B occupancy and this section.

Exception: Amusement buildings or portions thereof that are without walls or a roof and constructed to prevent the accumulation of smoke.

For flammable decorative materials, see the *International Fire Code.*

411.2 Special amusement building. A special amusement building is any temporary or permanent building or portion thereof that is occupied for amusement, entertainment or educational purposes and that contains a device or system that conveys passengers or provides a walkway along, around or over a course in any direction so arranged that the means of egress path is not readily apparent due to visual or audio distractions or is intentionally confounded or is not readily available because of the nature of the attraction or mode of conveyance through the building or structure.

[F] 411.3 Automatic fire detection. Special amusement buildings shall be equipped with an automatic fire detection system in accordance with Section 907.

[F] 411.4 Automatic sprinkler system. Special amusement buildings shall be equipped throughout with an automatic sprinkler system in accordance with Section 903.3.1.1. Where the special amusement building is temporary, the sprinkler water supply shall be of an approved temporary means.

Exception: Automatic fire sprinklers are not required where the total floor area of a temporary special amusement building is less than 1,000 square feet (93 m^2) and the travel distance from any point to an exit is less than 50 feet (15 240 mm).

[F] 411.5 Alarm. Actuation of a single smoke detector, the automatic sprinkler system or other automatic fire detection device shall immediately sound an alarm at the building at a constantly attended location from which emergency action can be initiated including the capability of manual initiation of requirements in Section 907.2.11.2.

[F] 411.6 Emergency voice/alarm communications system. An emergency voice/alarm communications system shall be provided in accordance with Sections 907.2.11 and 907.2.12.2, which is also permitted to serve as a public address system and shall be audible throughout the entire special amusement building.

411.7 Exit marking. Exit signs shall be installed at the required exit or exit access doorways of amusement buildings. Approved directional exit markings shall also be provided. Where mirrors, mazes or other designs are utilized that disguise the path of egress travel such that they are not apparent, approved low-level exit signs and directional path markings shall be provided and located not more than 8 inches (203 mm) above the walking surface and on or near the path of egress travel. Such markings shall become visible in an emergency. The directional exit marking shall be activated by the automatic fire detection system and the automatic sprinkler system in accordance with Section 907.2.11.2.

411.8 Interior finish. The interior finish shall be Class A in accordance with Section 803.1.

SECTION 412
AIRCRAFT-RELATED OCCUPANCIES

412.1 Airport traffic control towers.

412.1.1 General. The provisions of this section shall apply to airport traffic control towers not exceeding 1,500 square feet (140 m²) per floor occupied only for the following uses:

1. Airport traffic control cab.

2. Electrical and mechanical equipment rooms.

3. Airport terminal radar and electronics rooms.

4. Office spaces incidental to the tower operation.

5. Lounges for employees, including sanitary facilities.

412.1.2 Type of construction. Airport traffic control towers shall be constructed to conform to the height and area limitations of Table 412.1.2.

TABLE 412.1.2
HEIGHT AND AREA LIMITATIONS FOR AIRPORT TRAFFIC CONTROL TOWERS

TYPE OF CONSTRUCTION	HEIGHT[a] (feet)	MAXIMUM AREA (square feet)
IA	Unlimited	1,500
IB	240	1,500
IIA	100	1,500
IIB	85	1,500
IIIA	65	1,500

For SI: 1 foot = 304.8 mm, 1 square foot = 0.093 m².

a. Height to be measured from grade to cab floor.

412.1.3 Egress. A minimum of one exit stairway shall be permitted for airport traffic control towers of any height provided that the occupant load per floor does not exceed 15. The stairway shall conform to the requirements of Section 1009. The stairway shall be separated from elevators by a minimum distance of one-half of the diagonal of the area served measured in a straight line. The exit stairway and elevator hoistway are permitted to be located in the same shaft enclosure, provided they are separated from each other by a 4-hour separation having no openings. Such stairway shall be pressurized to a minimum of 0.15 inch of water column (43 Pa) and a maximum of 0.35 inch of water column (101 Pa) in the shaft relative to the building with stairway doors closed. Stairways need not extend to the roof as specified in Section 1009.12. The provisions of Section 403 do not apply.

Exception: Smokeproof enclosures as set forth in Section 1019.1.8 are not required where required stairways are pressurized.

[F] 412.1.4 Automatic fire detection systems. Airport traffic control towers shall be provided with an automatic fire detection system installed in accordance with Section 907.2.

[F] 412.1.5 Standby power. A standby power system that conforms to Section 2702 shall be provided in airport traffic control towers more than 65 feet (19 812 mm) in height. Power shall be provided to the following equipment:

1. Pressurization equipment, mechanical equipment and lighting.

2. Elevator operating equipment.

3. Fire alarm and smoke detection systems.

412.1.6 Accessibility. Airport traffic control towers need not be accessible as specified in the provisions of Chapter 11.

412.2 Aircraft hangar.

412.2.1 Exterior walls. Exterior walls located less than 30 feet (9 144 mm) from property lines, lot lines or a public way shall have a fire-resistance rating not less than 2 hours.

412.2.2 Basements. Where hangars have basements, the floor over the basement shall be of Type IA construction and shall be made tight against seepage of water, oil or vapors. There shall be no opening or communication between the basement and the hangar. Access to the basement shall be from outside only.

412.2.3 Floor surface. Floors shall be graded and drained to prevent water or fuel from remaining on the floor. Floor drains shall discharge through an oil separator to the sewer or to an outside vented sump.

412.2.4 Heating equipment. Heating equipment shall be placed in another room separated by 2-hour fire-resistance-rated construction. Entrance shall be from the outside or by means of a vestibule providing a two-doorway separation.

Exceptions:

1. Unit heaters suspended at least 10 feet (3048 mm) above the upper surface of wings or engine enclosures of the highest aircraft that are permitted to be housed in the hangar and at least 8 feet (2438 mm) above the floor in shops, offices and other sections of the hangar communicating with storage or service areas.

2. A single interior door shall be allowed, provided the sources of ignition in the appliances are at least 18 inches (457 mm) above the floor.

412.2.5 Finishing. The process of "doping," involving use of a volatile flammable solvent, or of painting, shall be carried on in a separate detached building equipped with automatic fire-extinguishing equipment in accordance with Section 903.

[F] 412.2.6 Fire suppression. Aircraft hangars shall be provided with fire suppression as required in NFPA 409.

Exception: Group II hangars as defined in NFPA 409 storing private aircraft without major maintenance or overhaul are exempt from foam suppression requirements.

412.3 Residential aircraft hangars. Residential aircraft hangars as defined in Section 412.3.1 shall comply with Sections 412.3.2 through 412.3.6.

412.3.1 Definition. The following word and term shall, for the purposes of this chapter and as used elsewhere in this code, have the meaning shown herein.

RESIDENTIAL AIRCRAFT HANGAR. An accessory building less than 2,000 square feet (186 m²) and 20 feet (6096 mm) in height, constructed on a one- or two-family residential property where aircraft are stored. Such use will be considered as a residential accessory use incidental to the dwelling.

412.3.2 Fire separation. A hangar shall not be attached to a dwelling unless separated by walls having a fire-resistance rating of not less than 1 hour. Such separation shall be continuous from the foundation to the underside of the roof and unpierced except for doors leading to the dwelling unit. Doors into the dwelling unit must be equipped with self-closing devices and conform to the requirements of Section 715 with at least a 4-inch (102 mm) noncombustible raised sill. Openings from a hanger directly into a room used for sleeping purposes shall not be permitted.

412.3.3 Egress. A hangar shall provide two means of egress. One of the doors into the dwelling shall be considered as meeting only one of the two means of egress.

[F] 412.3.4 Smoke detection. Smoke alarms shall be provided within the hangar in accordance with Section 907.2.21.

412.3.5 Independent systems. Mechanical and plumbing drain, waste and vent (DWV) systems installed within the hangar shall be independent of the systems installed within the dwelling. Building sewer lines may connect outside the structures.

> **Exception:** Smoke detector wiring and feed for electrical subpanels in the hangar.

412.3.6 Height and area limits. Residential aircraft hangars shall not exceed 2,000 square feet (186 m²) in area and 20 feet (6096 mm) in height.

412.4 Aircraft paint hangars. Aircraft painting operations where flammable liquids are used in excess of the maximum allowable quantities per control area listed in Table 307.7(1) shall be conducted in an aircraft paint hangar that complies with the provisions of Section 412.4.

412.4.1 Occupancy group. Aircraft paint hangars shall be classified as Group H-2. Aircraft paint hangars shall comply with the applicable requirements of this code and the *International Fire Code* for such occupancy.

412.4.2 Construction. The aircraft paint hangar shall be of Type I or II construction.

412.4.3 Operations. Only those flammable liquids necessary for painting operations shall be permitted in quantities less than the maximum allowable quantities per control area in Table 307.7(1). Spray equipment cleaning operations shall be conducted in a liquid use, dispensing and mixing room.

412.4.4 Storage. Storage of flammable liquids shall be in a liquid storage room.

412.4.5 Fire suppression. Aircraft paint hangars shall be provided with fire suppression as required in NFPA 409.

412.4.6 Ventilation. Aircraft paint hangars shall be provided with ventilation as required in the *International Mechanical Code.*

412.5 Heliports and helistops.

412.5.1 General. Heliports and helistops may be erected on buildings or other locations where they are constructed in accordance with this section.

412.5.2 Definitions. The following words and terms shall, for the purposes of this chapter and as used elsewhere in this code, have the meanings shown herein.

HELIPORT. An area of land or water or a structural surface that is used, or intended for use, for the landing and taking off of helicopters, and any appurtenant areas that are used, or intended for use, for heliport buildings and other heliport facilities.

HELISTOP. The same as a "Heliport," except that no fueling, defueling, maintenance, repairs or storage of helicopters is permitted.

412.5.3 Size. The touchdown or landing area for helicopters of less than 3,500 pounds (1588 kg) shall be a minimum of 20 feet (6096 mm) in length and width. The touchdown area shall be surrounded on all sides by a clear area having a minimum average width at roof level of 15 feet (4572 mm) but with no width less than 5 feet (1524 mm).

412.5.4 Design. Helicopter landing areas and the supports thereof on the roof of a building shall be noncombustible construction. Landing areas shall be designed to confine any flammable liquid spillage to the landing area itself and provisions shall be made to drain such spillage away from any exit or stairway serving the helicopter landing area or from a structure housing such exit or stairway. For structural design requirements, see Section 1605.5.

412.5.5 Means of egress. The means of egress from heliports and helistops shall comply with the provisions of Chapter 10. Landing areas located on buildings or structures shall have two or more means of egress. For landing platforms or roof areas less than 60 feet (18 288 mm) in length, or less than 2,000 square feet (187 m²) in area, the second means of egress may be a fire escape or ladder leading to the floor below.

412.5.6 Rooftop heliports and helistops. Rooftop heliports and helistops shall comply with NFPA 418.

SECTION 413
COMBUSTIBLE STORAGE

413.1 General. High-piled stock or rack storage in any occupancy group shall comply with the *International Fire Code*.

413.2 Attic, under-floor and concealed spaces. Attic, under-floor and concealed spaces used for storage of combustible materials shall be protected on the storage side as required for 1-hour fire-resistance-rated construction. Openings shall be protected by assemblies that are self-closing and are of noncombustible construction or solid wood core not less than $1^3/_4$ inch (45 mm) in thickness.

Exceptions:

1. Areas protected by approved automatic sprinkler systems.

2. Group R-3 and U occupancies.

SECTION 414
HAZARDOUS MATERIALS

[F] 414.1 General. The provisions of this section shall apply to buildings and structures occupied for the manufacturing, processing, dispensing, use or storage of hazardous materials.

[F] 414.1.1 Other provisions. Buildings and structures with an occupancy in Group H shall also comply with the applicable provisions of Section 415 and the *International Fire Code.*

[F] 414.1.2 Materials. The safe design of hazardous material occupancies is material dependent. Individual material requirements are also found in Sections 307 and 415, and in the *International Mechanical Code* and the *International Fire Code.*

[F] 414.1.2.1 Aerosols. Level 2 and 3 aerosol products shall be stored and displayed in accordance with the *International Fire Code.* See Section 311.2 and the *International Fire Code* for occupancy group requirements.

[F] 414.1.3 Information required. Separate floor plans shall be submitted for buildings and structures with an occupancy in Group H, identifying the locations of anticipated contents and processes so as to reflect the nature of each occupied portion of every building and structure. A report identifying hazardous materials including, but not limited to, materials representing hazards that are classified in Group H to be stored or used, shall be submitted and the methods of protection from such hazards shall be indicated on the construction documents. The opinion and report shall be prepared by a qualified person, firm or corporation approved by the building official and shall be provided without charge to the enforcing agency.

[F] 414.2 Control areas. Control areas shall be those spaces within a building where quantities of hazardous materials not exceeding the maximum quantities allowed by this code are stored, dispensed, used or handled.

[F] 414.2.1 Construction requirements. Control areas shall be separated from each other by not less than a 1-hour fire barrier constructed in accordance with Chapter 7.

[F] 414.2.2 Number. The maximum number of control areas within a building shall be in accordance with Table 414.2.2.

[F] 414.2.3 Separation. The required fire-resistance rating for fire barrier assemblies shall be in accordance with Table 414.2.2. The floor construction of the control area, and the construction supporting the floor of the control area, shall have a minimum 2-hour fire-resistance rating.

[F] 414.2.4 Hazardous material in Group M display and storage areas and in Group S storage areas. The aggregate quantity of nonflammable solid and nonflammable or noncombustible liquid hazardous materials permitted within a single control area of a Group M or S occupancy or an outdoor control area is permitted to exceed the maximum allowable quantities per control area specified in Tables 307.7(1) and 307.7(2) without classifying the building or use as a Group H occupancy, provided that the materials are displayed and stored in accordance with the *International*

Fire Code and quantities do not exceed the maximum allowable specified in Table 414.2.4.

[F] 414.3 Ventilation. Rooms, areas or spaces of Group H in which explosive, corrosive, combustible, flammable or highly toxic dusts, mists, fumes, vapors or gases are or may be emitted due to the processing, use, handling or storage of materials shall be mechanically ventilated as required by the *International Fire Code* and the *International Mechanical Code.*

Ducts conveying explosives or flammable vapors, fumes or dusts shall extend directly to the exterior of the building without entering other spaces. Exhaust ducts shall not extend into or through ducts and plenums.

Exception: Ducts conveying vapor or fumes having flammable constituents less than 25 percent of their lower flammable limit (LFL) are permitted to pass through other spaces.

Emissions generated at workstations shall be confined to the area in which they are generated as specified in the *International Fire Code* and the *International Mechanical Code.*

The location of supply and exhaust openings shall be in accordance with the *International Mechanical Code.* Exhaust air contaminated by highly toxic material shall be treated in accordance with the *International Fire Code.*

A manual shutoff control for ventilation equipment required by this section shall be provided outside the room adjacent to the principal access door to the room. The switch shall be of the break-glass type and shall be labeled: VENTILATION SYSTEM EMERGENCY SHUTOFF.

[F] 414.4 Hazardous material systems. Systems involving hazardous materials shall be suitable for the intended application. Controls shall be designed to prevent materials from entering or leaving process or reaction systems at other than the intended time, rate or path. Automatic controls, where provided, shall be designed to be fail safe.

[F] 414.5 Inside storage, dispensing and use. The inside storage, dispensing and use of hazardous materials in excess of the maximum allowable quantities per control area of Tables 307.7(1) and 307.7(2) shall be in accordance with Sections 414.5.1 through 414.5.5 of this code and the *International Fire Code.*

[F] 414.5.1 Explosion control. Explosion control shall be provided in accordance with the *International Fire Code* as required by Table 414.5.1 where quantities of hazardous materials specified in that table exceed the maximum allowable quantities in Table 307.7(1) or where a structure, room or space is occupied for purposes involving explosion hazards as required by Section 415 or the *International Fire Code.*

[F] 414.5.2 Monitor control equipment. Monitor control equipment shall be provided where required by the *International Fire Code.*

[F] 414.5.3 Automatic fire detection systems. Group H occupancies shall be provided with an automatic fire detection system in accordance with Section 907.2.

[F] TABLE 414.2.2
DESIGN AND NUMBER OF CONTROL AREAS

FLOOR LEVEL		PERCENTAGE OF THE MAXIMUM ALLOWABLE QUANTITY PER CONTROL AREA[a]	NUMBER OF CONTROL AREAS PER FLOOR[b]	FIRE-RESISTANCE RATING FOR FIRE BARRIERS IN HOURS[c]
Above grade	Higher than 9	5	1	2
	7-9	5	2	2
	6	12.5	2	2
	5	12.5	2	2
	4	12.5	2	2
	3	50	2	1
	2	75	3	1
	1	100	4	1
Below grade	1	75	3	1
	2	50	2	1
	Lower than 2	Not Allowed	Not Allowed	Not Allowed

a. Percentages shall be of the maximum allowable quantity per control area shown in Tables 307.7(1) and 307.7(2), with all increases allowed in the notes to those tables.

b. There shall be a maximum of two control areas per floor in Group M occupancies and in buildings or portions of buildings having Group S occupancies with storage conditions and quantities in accordance with Section 414.2.4.

c. Fire barriers shall include walls and floors as necessary to provide separation from other portions of the building.

[F] TABLE 414.2.4
MAXIMUM ALLOWABLE QUANTITY PER INDOOR AND OUTDOOR CONTROL AREA IN GROUP M AND S OCCUPANCIES
NONFLAMMABLE SOLIDS AND NONFLAMMABLE AND NONCOMBUSTIBLE LIQUIDS [d,e,f]

CONDITION		MAXIMUM ALLOWABLE QUANTITY PER CONTROL AREA	
Material[a]	Class	Solids pounds	Liquids gallons
A. Health-hazard materials—nonflammable and noncombustible solids and liquids			
1. Corrosives[b,c]	Not Applicable	9,750	975
2. Highly toxics	Not Applicable	20[b,c]	2[b,c]
3. Toxics[b,c]	Not Applicable	1,000	100
B. Physical-hazard materials—nonflammable and noncombustible solids and liquids			
1. Oxidizers[b,c]	4	Not Allowed	Not Allowed
	3	1,150[g]	115
	2	2,250[h]	225
	1	18,000[i,j]	1,800[i,j]
2. Unstable (reactives)[b,c]	4	Not Allowed	Not Allowed
	3	550	55
	2	1,150	115
	1	Not Limited	Not Limited
3. Water (reactives)	3[b,c]	550	55
	2[b,c]	1,150	115
	1	Not Limited	Not Limited

For SI: 1 pound = 0.454 kg, 1 gallon = 3.785 L.

a. Hazard categories are as specified in the *International Fire Code*.

b. Maximum allowable quantities shall be increased 100 percent in buildings that are sprinklered in accordance with Section 903.3.1.1. When Note c also applies, the increase for both notes shall be applied accumulatively.

c. Maximum allowable quantities shall be increased 100 percent when stored in approved storage cabinets, in accordance with the *International Fire Code*. When Note b also applies, the increase for both notes shall be applied accumulatively.

d. See Table 414.2.2 for design and number of control areas.

e. Allowable quantities for other hazardous material categories shall be in accordance with Section 307.

f. Maximum quantities shall be increased 100 percent in outdoor control areas.

g. Maximum amounts are permitted to be increased to 2,250 pounds when individual packages are in the original sealed containers from the manufacturer or packager and do not exceed 10 pounds each.

h. Maximum amounts are permitted to be increased to 4,500 pounds when individual packages are in the original sealed containers from the manufacturer or packager and do not exceed 10 pounds each.

i. The permitted quantities shall not be limited in a building equipped throughout with an automatic sprinkler system in accordance with Section 903.3.1.1.

j. Quantities are unlimited in an outdoor control area.

[F] TABLE 414.5.1
EXPLOSION CONTROL REQUIREMENTS[a]

MATERIAL	CLASS	EXPLOSION CONTROL METHODS	
		Barricade construction	Explosion (deflagration) venting or explosion (deflagration) prevention systems[b]
HAZARD CATEGORY			
Combustible dusts[c]	—	Not Required	Required
Cryogenic flammables	—	Not Required	Required
Explosives	Division 1.1	Required	Not Required
	Division 1.2	Required	Not Required
	Division 1.3	Not Required	Required
	Division 1.4	Not Required	Required
	Division 1.5	Required	Not Required
	Division 1.6	Required	Not Required
Flammable gas	Gaseous	Not Required	Required
	Liquefied	Not Required	Required
Flammable liquid	IA[d]	Not Required	Required
	IB[e]	Not Required	Required
Organic peroxides	U	Required	Not Permitted
	I	Required	Not Permitted
Oxidizer liquids and solids	4	Required	Not Permitted
Pyrophoric gas	—	Not Required	Required
Unstable (reactive)	4	Required	Not Permitted
	3 Detonable	Required	Not Permitted
	3 Nondetonable	Not Required	Required
Water-reactive liquids and solids	3	Not Required	Required
	2[g]	Not Required	Required
SPECIAL USES		•	
Acetylene generator rooms	—	Not Required	Required
Grain processing	—	Not Required	Required
Liquefied petroleum gas-distribution facilities	—	Not Required	Required
Where explosion hazards exist[f]	Detonation	Required	Not Permitted
	Deflagration	Not Required	Required

a. See Section 414.1.3.
b. See the *International Fire Code*.
c. As generated during manufacturing or processing. See definition of "Combustible dust" in Chapter 3.
d. Storage or use.
e. In open use or dispensing.
f. Rooms containing dispensing and use of hazardous materials when an explosive environment can occur because of the characteristics or nature of the hazardous materials or as a result of the dispensing or use process.
g. A method of explosion control shall be provided when Class 2 water-reactive materials can form potentially explosive mixtures.

[F] 414.5.4 Standby or emergency power. Where mechanical ventilation, treatment systems, temperature control, alarm, detection or other electrically operated systems are required, such systems shall be provided with an emergency or standby power system in accordance with the ICC *Electrical Code*.

Exceptions:

1. Storage areas for Class I and II oxidizers.

2. Storage areas for Class III, IV and V organic peroxides.

3. Storage, use and handling areas for highly toxic or toxic materials as provided for in the *International Fire Code*.

4. Standby power for mechanical ventilation, treatment systems and temperature control systems shall not be required where an approved fail-safe engineered system is installed.

[F] 414.5.5 Spill control, drainage and containment. Rooms, buildings or areas occupied for the storage of solid and liquid hazardous materials shall be provided with a means to control spillage and to contain or drain off spillage and fire protection water discharged in the storage area where required in the *International Fire Code*. The methods of spill control shall be in accordance with the *International Fire Code*.

[F] 414.6 Outdoor storage, dispensing and use. The outdoor storage, dispensing and use of hazardous materials shall be in accordance with the *International Fire Code*.

[F] 414.6.1 Weather protection. Where weather protection is provided for sheltering outdoor hazardous material storage or use areas, such storage or use shall be considered outdoor storage or use, provided that all of the following conditions are met:

1. Structure supports and walls shall not obstruct more than one side nor more than 25 percent of the perimeter of the storage or use area.

2. The distance from the structure and the structure supports to buildings, lot lines, public ways or means of egress to a public way shall not be less than the distance required for an outside hazardous material storage or use area without weather protection.

3. The overhead structure shall be of approved noncombustible construction with a maximum area of 1,500 square feet (140 m^2).

Exception: The increases permitted by Section 506 apply.

[F] 414.7 Emergency alarms. Emergency alarms for the detection and notification of an emergency condition in Group H occupancies shall be provided as set forth herein.

[F] 414.7.1 Storage. An approved manual emergency alarm system shall be provided in buildings, rooms or areas used for storage of hazardous materials. Emergency alarm-initiating devices shall be installed outside of each interior exit or exit access door of storage buildings, rooms or areas. Activation of an emergency alarm-initiating device shall sound a local alarm to alert occupants of an emergency situation involving hazardous materials.

[F] 414.7.2 Dispensing, use and handling. Where hazardous materials having a hazard ranking of 3 or 4 in accordance with NFPA 704 are transported through corridors or exit enclosures, there shall be an emergency telephone system, a local manual alarm station or an approved alarm-initiating device at not more than 150-foot (45 720 mm) intervals and at each exit and exit access doorway throughout the transport route. The signal shall be relayed to an approved central, proprietary or remote station service or constantly attended on-site location and shall also initiate a local audible alarm.

[F] 414.7.3 Supervision. Emergency alarm systems shall be supervised by an approved central, proprietary or remote station service or shall initiate an audible and visual signal at a constantly attended on-site location.

[F] SECTION 415
GROUPS H-1, H-2, H-3, H-4 AND H-5

415.1 Scope. The provisions of this section shall apply to the storage and use of hazardous materials in excess of the maximum allowable quantities per control area listed in Section 307.9. Buildings and structures with an occupancy in Group H shall also comply with the applicable provisions of Section 414 and the *International Fire Code*.

415.2 Definitions. The following words and terms shall, for the purposes of this chapter and as used elsewhere in the code, have the meanings shown herein.

CONTINUOUS GAS-DETECTION SYSTEM. A gas detection system where the analytical instrument is maintained in continuous operation and sampling is performed without interruption. Analysis is allowed to be performed on a cyclical basis at intervals not to exceed 30 minutes.

EMERGENCY CONTROL STATION. An approved location on the premises where signals from emergency equipment are received and which is staffed by trained personnel.

EXHAUSTED ENCLOSURE. An appliance or piece of equipment that consists of a top, a back and two sides providing a means of local exhaust for capturing gases, fumes, vapors and mists. Such enclosures include laboratory hoods, exhaust fume hoods and similar appliances and equipment used to locally retain and exhaust the gases, fumes, vapors and mists that could be released. Rooms or areas provided with general ventilation, in themselves, are not exhausted enclosures.

FABRICATION AREA. An area within a semiconductor fabrication facility and related research and development areas in which there are processes using hazardous production materials. Such areas are allowed to include ancillary rooms or areas such as dressing rooms and offices that are directly related to the fabrication area processes.

FLAMMABLE VAPORS OR FUMES. The concentration of flammable constituents in air that exceed 10 percent of their lower flammable limit (LFL).

GAS CABINET. A fully enclosed, noncombustible enclosure used to provide an isolated environment for compressed gas cylinders in storage or use. Doors and access ports for exchanging cylinders and accessing pressure-regulating controls are allowed to be included.

GAS ROOM. A separately ventilated, fully enclosed room in which only compressed gases and associated equipment and supplies are stored or used.

HAZARDOUS PRODUCTION MATERIAL (HPM). A solid, liquid or gas associated with semiconductor manufacturing that has a degree-of-hazard rating in health, flammability or reactivity of Class 3 or 4 as ranked by NFPA 704 and which is used directly in research, laboratory or production processes that have as their end product materials that are not hazardous.

HPM FLAMMABLE LIQUID. An HPM liquid that is defined as either a Class I flammable liquid or a Class II or Class IIIA combustible liquid.

HPM ROOM. A room used in conjunction with or serving a Group H-5 occupancy, where HPM is stored or used and which is classified as a Group H-2, H-3 or H-4 occupancy.

IMMEDIATELY DANGEROUS TO LIFE AND HEALTH (IDLH). The concentration of air-borne contaminants which poses a threat of death, immediate or delayed permanent adverse health effects, or effects that could prevent escape from such an environment. This contaminant concentration level is established by the National Institute of Occupational Safety and Health (NIOSH) based on both toxicity and flammability. It generally is expressed in parts per million by volume (ppm v/v) or milligrams per cubic meter (mg/m³). If adequate data do not exist for precise establishment of IDLH concentrations, an independent certified industrial hygienist, industrial toxicologist, appropriate regulatory agency or other source approved by the code official shall make such determination.

LIQUID. A material that has a melting point that is equal to or less than 68°F (20°C) and a boiling point that is greater than 68°F (20°C) at 14.7 pounds per square inch absolute (psia) (101 kPa). When not otherwise identified, the term "liquid" includes both flammable and combustible liquids.

LIQUID STORAGE ROOM. A room classified as a Group H-3 occupancy used for the storage of flammable or combustible liquids in a closed condition.

LIQUID USE, DISPENSING AND MIXING ROOMS. Rooms in which Class I, II and IIIA flammable or combustible liquids are used, dispensed or mixed in open containers.

LOWER FLAMMABLE LIMIT (LFL). The minimum concentration of vapor in air at which propagation of flame will occur in the presence of an ignition source. The LFL is sometimes referred to as "LEL" or "lower explosive limit."

NORMAL TEMPERATURE AND PRESSURE (NTP). A temperature of 70°F (21°C) and a pressure of 1 atmosphere [14.7 psia (101 kPa)].

SERVICE CORRIDOR. A fully enclosed passage used for transporting HPM and purposes other than required means of egress.

SOLID. A material that has a melting point, decomposes or sublimes at a temperature greater than 68°F (20°C).

STORAGE, HAZARDOUS MATERIALS.

1. The keeping, retention or leaving of hazardous materials in closed containers, tanks, cylinders or similar vessels, or

2. Vessels supplying operations through closed connections to the vessel.

USE (MATERIAL). Placing a material into action, including solids, liquids and gases.

WORKSTATION. A defined space or an independent principal piece of equipment using HPM within a fabrication area where a specific function, laboratory procedure or research activity occurs. Approved or listed hazardous materials storage cabinets, flammable liquid storage cabinets or gas cabinets serving a workstation are included as part of the workstation. A workstation is allowed to contain ventilation equipment, fire protection devices, detection devices, electrical devices and other processing and scientific equipment.

415.3 Location on property. Group H shall be located on property in accordance with the other provisions of this chapter. In Group H-2 or H-3, not less than 25 percent of the perimeter wall of the occupancy shall be an exterior wall.

Exceptions:

1. Liquid use, dispensing and mixing rooms having a floor area of not more than 500 square feet (46.5 m2) need not be located on the outer perimeter of the building where they are in accordance with the *International Fire Code* and NFPA 30.

2. Liquid storage rooms having a floor area of not more than 1,000 square feet (93 m2) need not be located on the outer perimeter where they are in accordance with the *International Fire Code* and NFPA 30.

3. Spray paint booths that comply with the *International Fire Code* need not be located on the outer perimeter.

415.3.1 Group H minimum distance to lot lines. Regardless of any other provisions, buildings containing Group H occupancies shall be set back a minimum distance from lot lines as set forth in Items 1 through 4 below. Distances shall be measured from the walls enclosing the occupancy to lot lines, including those on a public way. Distances to assumed property lines drawn for the purposes of determination of exterior wall and opening protection are not to be used to establish the minimum distance for separation of buildings on sites where explosives are manufactured or used when separation is provided in accordance with the quantity distance tables specified for explosive materials in the *International Fire Code*.

1. Group H-1. Not less than 75 feet (22 860 mm) and not less than required by the *International Fire Code*.

    Exceptions:

    1. Fireworks manufacturing buildings separated in accordance with NFPA 1124.

2. Buildings containing the following materials when separated in accordance with Table 415.3.1:
    2.1. Organic peroxides, unclassified detonable.
    2.2. Unstable reactive materials Class 4.
    2.3. Unstable reactive materials, Class 3 detonable.
    2.4. Detonable pyrophoric materials.

2. Group H-2. Not less than 30 feet (9144 mm) where the area of the occupancy exceeds 1,000 square feet (93 m^2) and it is not required to be located in a detached building.

3. Groups H-2 and H-3. Not less than 50 feet (15 240 mm) where a detached building is required (see Table 415.3.2).

4. Groups H-2 and H-3. Occupancies containing materials with explosive characteristics shall be separated as required by the *International Fire Code*. Where separations are not specified, the distances required shall not be less than the distances required by Table 415.3.1.

415.3.2 Group H-1 and H-2 or H-3 detached buildings. Where a detached building is required by Table 415.3.2, there are no requirements for wall and opening protection based on location on property.

415.4 Special provisions for Group H-1 occupancies. Group H-1 occupancies shall be in buildings used for no other purpose, shall not exceed one story in height and be without basement, crawl spaces or other under-floor spaces. Roofs shall be of lightweight construction with suitable thermal insulation to prevent sensitive material from reaching its decomposition temperature.

Group H-1 occupancies containing materials which are in themselves both physical and health hazards in quantities exceeding the maximum allowable quantities per control area in Table 307.7.(2) shall comply with requirements for both Group H-1 and H-4 occupancies.

415.4.1 Floors in storage rooms. Floors in storage areas for organic peroxides, pyrophoric materials and unstable (reactive) materials shall be of liquid-tight, noncombustible construction.

415.5 Special provisions for Group H-2 and H-3 occupancies. Group H-2 and H-3 occupancies containing quantities of hazardous materials in excess of those set forth in Table 415.3.2 shall be in buildings used for no other purpose, shall not exceed one story in height and shall be without basements, crawl spaces or other under-floor spaces.

Group H-2 and H-3 occupancies containing water-reactive materials shall be resistant to water penetration. Piping for conveying liquids shall not be over or through areas containing water reactives, unless isolated by approved liquid-tight construction.

Exception: Fire protection piping.

415.5.1 Floors in storage rooms. Floors in storage areas for organic peroxides, oxidizers, pyrophoric materials, unstable (reactive) materials and water-reactive solids and liquids shall be of liquid-tight, noncombustible construction.

415.5.2 Waterproof room. Rooms or areas used for the storage of water-reactive solids and liquids shall be constructed in a manner that resists the penetration of water through the use of waterproof materials. Piping carrying water for other than approved automatic fire sprinkler systems shall not be within such rooms or areas.

415.6 Smoke and heat venting. Smoke and heat vents complying with Section 910 shall be installed in the following locations:

1. In occupancies classified as Group H-2 or H-3, any of which are over 15,000 square feet (1394 m^2) in single floor area.

    Exception: Buildings of noncombustible construction containing only noncombustible materials.

2. In areas of buildings in Group H used for storing Class 2, 3 and 4 liquid and solid oxidizers, Class 1 and unclassified detonable organic peroxides, Class 3 and 4 unstable (reactive) materials, or Class 2 or 3 water-reactive materials as required for a Class V hazard classification.

    Exception: Buildings of noncombustible construction containing only noncombustible materials.

415.7 Group H-2. Occupancies in Group H-2 shall be constructed in accordance with Sections 415.7.1 through 415.7.4 and the *International Fire Code*.

415.7.1 Combustible dusts, grain processing and storage. The provisions of Sections 415.7.1.1 through 415.7.1.5 shall apply to buildings in which materials that produce combustible dusts are stored or handled. Buildings that store or handle combustible dusts shall comply with the applicable provisions of NFPA 61, NFPA 120, NFPA 651, NFPA 654, NFPA 655, NFPA 664 and NFPA 85, and the *International Fire Code*.

415.7.1.1 Type of construction and height exceptions. Buildings shall be constructed in compliance with the height and area limitations of Table 503 for Group H-2; except that where erected of Type I or II construction, the heights and areas of grain elevators and similar structures shall be unlimited, and where of Type IV construction, the maximum height shall be 65 feet (19 812 mm) and except further that, in isolated areas, the maximum height of Type IV structures shall be increased to 85 feet (25 908 mm).

415.7.1.2 Grinding rooms. Every room or space occupied for grinding or other operations that produce combustible dusts shall be enclosed with fire barriers and horizontal assemblies or both that have not less than a 2-hour fire-resistance rating where the area is not more than 3,000 square feet (279 m^2), and not less than a 4-hour fire-resistance rating where the area is greater than 3,000 square feet (279 m^2).

TABLE 415.3.1
MINIMUM SEPARATION DISTANCES FOR BUILDINGS CONTAINING EXPLOSIVE MATERIALS

QUANTITY OF EXPLOSIVE MATERIAL[a]		MINIMUM DISTANCE (feet)		
		Lot lines[b] and inhabited buildings[c]		
Pounds over	Pounds not over	Barricaded[d]	Unbarricaded	Separation of magazines[d, e, f]
2	5	70	140	12
5	10	90	180	16
10	20	110	220	20
20	30	125	250	22
30	40	140	280	24
40	50	150	300	28
50	75	170	340	30
75	100	190	380	32
100	125	200	400	36
125	150	215	430	38
150	200	235	470	42
200	250	255	510	46
250	300	270	540	48
300	400	295	590	54
400	500	320	640	58
500	600	340	680	62
600	700	355	710	64
700	800	375	750	66
800	900	390	780	70
900	1,000	400	800	72
1,000	1,200	425	850	78
1,200	1,400	450	900	82
1,400	1,600	470	940	86
1,600	1,800	490	980	88
1,800	2,000	505	1,010	90
2,000	2,500	545	1,090	98
2,500	3,000	580	1,160	104
3,000	4,000	635	1,270	116
4,000	5,000	685	1,370	122
5,000	6,000	730	1,460	130
6,000	7,000	770	1,540	136
7,000	8,000	800	1,600	144
8,000	9,000	835	1,670	150
9,000	10,000	865	1,730	156
10,000	12,000	875	1,750	164
12,000	14,000	885	1,770	174
14,000	16,000	900	1,800	180
16,000	18,000	940	1,880	188
18,000	20,000	975	1,950	196

(continued)

TABLE 415.3.1—continued
MINIMUM SEPARATION DISTANCES FOR BUILDINGS CONTAINING EXPLOSIVE MATERIALS

QUANTITY OF EXPLOSIVE MATERIAL[a]		MINIMUM DISTANCE (feet)		
		Lot lines[b] and inhabited buildings[c]		
Pounds over	Pounds not over	Barricaded[d]	Unbarricaded	Separation of magazines[d, e, f]
20,000	25,000	1,055	2,000	210
25,000	30,000	1,130	2,000	224
30,000	35,000	1,205	2,000	238
35,000	40,000	1,275	2,000	248
40,000	45,000	1,340	2,000	258
45,000	50,000	1,400	2,000	270
50,000	55,000	1,460	2,000	280
55,000	60,000	1,515	2,000	290
60,000	65,000	1,565	2,000	300
65,000	70,000	1,610	2,000	310
70,000	75,000	1,655	2,000	320
75,000	80,000	1,695	2,000	330
80,000	85,000	1,730	2,000	340
85,000	90,000	1,760	2,000	350
90,000	95,000	1,790	2,000	360
95,000	100,000	1,815	2,000	370
100,000	110,000	1,835	2,000	390
110,000	120,000	1,855	2,000	410
120,000	130,000	1,875	2,000	430
130,000	140,000	1,890	2,000	450
140,000	150,000	1,900	2,000	470
150,000	160,000	1,935	2,000	490
160,000	170,000	1,965	2,000	510
170,000	180,000	1,990	2,000	530
180,000	190,000	2,010	2,010	550
190,000	200,000	2,030	2,030	570
200,000	210,000	2,055	2,055	590
210,000	230,000	2,100	2,100	630
230,000	250,000	2,155	2,155	670
250,000	275,000	2,215	2,215	720
275,000	300,000	2,275	2,275	770

For SI: 1 pound = 0.454 kg, 1 foot = 304.8 mm.

a. The number of pounds of explosives listed is the number of pounds of trinitrotoluene (TNT) or the equivalent pounds of other explosive.

b. The distance listed is the distance to lot line, including lot lines at public ways.

c. For the purpose of this table, an inhabited building is any building on the same property that is regularly occupied by people. Where two or more buildings containing explosives or magazines are located on the same property, each building or magazine shall comply with the minimum distances specified from inhabited buildings and, in addition, they shall be separated from each other by not less than the distance shown for "Separation of magazines," except that the quantity of explosive materials contained in detonator buildings or magazines shall govern in regard to the spacing of said buildings or magazines from buildings or magazines, as a group, shall be considered as one building or magazine, and the total quantity of explosive materials stored in such group shall be treated as if the explosive were in a single building or magazine located on the site of any building or magazine of the group, and shall comply with the minimum distance specified from other magazines or inhabited buildings.

d. Barricades shall effectively screen the building containing explosives from other buildings, public ways or magazines. Where mounds or revetted walls of earth are used for barricades, they shall not be less than 3 feet in thickness. A straight line from the top of any side wall of the building containing explosive materials to the eave line of any other building, magazine or a point 12 feet above the centerline of a public way shall pass through the barricades.

e. Magazine is a building or structure, other than an operating building, approved for storage of explosive materials. Portable or mobile magazines not exceeding 120 square feet (11 m^2) in area need not comply with the requirements of this code, however, all magazines shall comply with the *International Fire Code*.

f. The distance listed is permitted be reduced by 50 percent where approved natural or artificial barriers are provided in accordance with the requirements in Note d.

TABLE 415.3.2
REQUIRED DETACHED STORAGE

DETACHED STORAGE IS REQUIRED WHEN THE QUANTITY OF MATERIAL EXCEEDS THAT LISTED HEREIN			
Material	Class	Solids and Liquids (tons)[a,b]	Gases (cubic feet)[a,b]
Explosives	Division 1.1 Division 1.2 Division 1.3 Division 1.4 Division 1.4[c] Division 1.5 Division 1.6	Maximum Allowable Quantity Maximum Allowable Quantity Maximum Allowable Quantity Maximum Allowable Quantity 1 Maximum Allowable Quantity Maximum Allowable Quantity	Not Applicable
Oxidizers	Class 4	Maximum Allowable Quantity	Maximum Allowable Quantity
Unstable (reactives) detonable	Class 3 or 4	Maximum Allowable Quantity	Maximum Allowable Quantity
Oxidizer, liquids and solids	Class 3 Class 2	1,200 2,000	Not Applicable Not Applicable
Organic peroxides	Detonable Class I Class II Class III	Maximum Allowable Quantity Maximum Allowable Quantity 25 50	Not Applicable Not Applicable Not Applicable Not Applicable
Unstable (reactives) nondetonable	Class 3 Class 2	1 25	2,000 10,000
Water reactives	Class 3 Class 2	1 25	Not Applicable Not Applicable
Pyrophoric gases	Not Applicable	Not Applicable	2,000

For SI: 1 ton = 906 kg, 1 cubic foot = 0.02832 M^3.

a. For materials that are detonable, the distance to other buildings or lot lines shall be as specified in Table 415.3.1 based on trinitrotoluene (TNT) equivalence of the material. For materials classified as explosives, see Chapter 33 the *International Fire Code*. For all other materials, the distance shall be as indicated in Section 415.3.1.

b. "Maximum Allowable Quantity" means the maximum allowable quantity per control area set forth in Table 307.7(1).

c. Limited to Division 1.4 materials and articles, including articles packaged for shipment, that are not regulated as an explosive under Bureau of Alcohol, Tobacco and Firearms (BATF) regulations or unpackaged articles used in process operations that do not propagate a detonation or deflagration between articles, providing the net explosive weight of individual articles does not exceed 1 pound.

415.7.1.3 Conveyors. Conveyors, chutes, piping and similar equipment passing through the enclosures of rooms or spaces shall be constructed dirt tight and vapor tight, and be of approved noncombustible materials complying with Chapter 30.

415.7.1.4 Explosion control. Explosion control shall be provided as specified in the *International Fire Code*, or spaces shall be equipped with the equivalent mechanical ventilation complying with the *International Mechanical Code*.

415.7.1.5 Grain elevators. Grain elevators, malt houses and buildings for similar occupancies shall not be located within 30 feet (9144 mm) of interior lot lines or structures on the same lot, except where erected along a railroad right-of-way.

415.7.1.6 Coal pockets. Coal pockets located less than 30 feet (9144 mm) from interior lot lines or from structures on the same lot shall be constructed of not less than Type IB construction. Where more than 30 feet (9144 mm) from interior lot lines, or where erected along a railroad right-of-way, the minimum type of construction of such structures not more than 65 feet (19 812 mm) in height shall be Type IV.

415.7.2 Flammable and combustible liquids. The storage, handling, processing and transporting of flammable and combustible liquids shall be in accordance with the *International Mechanical Code* and the *International Fire Code*.

415.7.2.1 Mixed occupancies. Where the storage tank area is located in a building of two or more occupancies, and the quantity of liquid exceeds the maximum allowable quantity for one control area, the use shall be completely separated from adjacent fire areas in accordance with the requirements of Section 302.3.2.

415.7.2.1.1 Height exception. Where storage tanks are located within only a single-story building, the height limitation of Section 503 shall not apply for Group H.

415.7.2.2 Tank protection. Storage tanks shall be noncombustible and protected from physical damage. A fire barrier wall or horizontal assemblies or both around the storage tank(s) shall be permitted as the method of protection from physical damage.

415.7.2.3 Tanks. Storage tanks shall be approved tanks conforming to the requirements of the *International Fire Code*.

415.7.2.4 Suppression. Group H shall be equipped throughout with an approved automatic sprinkler system, installed in accordance with Section 903.

415.7.2.5 Leakage containment. A liquid-tight containment area compatible with the stored liquid shall be provided. The method of spill control, drainage control and secondary containment shall be in accordance with the *International Fire Code.*

> **Exception:** Rooms where only double-wall storage tanks conforming to Section 415.7.2.3 are used to store Class I, II and IIIA flammable and combustible liquids shall not be required to have a leakage containment area.

415.7.2.6 Leakage alarm. An approved automatic alarm shall be provided to indicate a leak in a storage tank and room. The alarm shall sound an audible signal, 15 dBa above the ambient sound level, at every point of entry into the room in which the leaking storage tank is located. An approved sign shall be posted on every entry door to the tank storage room indicating the potential hazard of the interior room environment, or the sign shall state: WARNING, WHEN ALARM SOUNDS, THE ENVIRONMENT WITHIN THE ROOM MAY BE HAZARDOUS. The leakage alarm shall also be supervised in accordance with Chapter 9 to transmit a trouble signal.

415.7.2.7 Tank vent. Storage tank vents for Class I, II or IIIA liquids shall terminate to the outdoor air in accordance with the *International Fire Code.*

415.7.2.8 Room ventilation. Storage tank areas storing Class I, II or IIIA liquids shall be provided with mechanical ventilation. The mechanical ventilation system shall be in accordance with the *International Mechanical Code* and the *International Fire Code.*

415.7.2.9 Explosion venting. Where Class I liquids are being stored, explosion venting shall be provided in accordance with the *International Fire Code.*

415.7.2.10 Tank openings other than vents. Tank openings other than vents from tanks inside buildings shall be designed to ensure that liquids or vapor concentrations are not released inside the building.

415.7.3 Liquefied petroleum gas-distribution facilities. The design and construction of propane, butane, propylene, butylene and other liquefied petroleum gas-distribution facilities shall conform to the applicable provisions of Sections 415.7.3.1 through 415.7.3.5.2. The storage and handling of liquefied petroleum gas systems shall conform to the *International Fire Code.* The design and installation of piping, equipment and systems that utilize liquefied petroleum gas shall be in accordance with the *International Fuel Gas Code.* Liquefied petroleum gas-distribution facilities shall be ventilated in accordance with the *International Mechanical Code* and Section 415.7.3.1.

415.7.3.1 Air movement. Liquefied petroleum gas-distribution facilities shall be provided with air inlets and outlets arranged so that air movement across the floor of the facility will be uniform. The total area of both inlet and outlet openings shall be at least 1 square inch (645

mm²) for each 1 square foot (0.093 m²) of floor area. The bottom of such openings shall not be more than 6 inches (152 mm) above the floor.

415.7.3.2 Construction. Liquefied petroleum gas-distribution facilities shall be constructed in accordance with Section 415.7.3.3 for separate buildings, Section 415.7.3.4 for attached buildings or Section 415.7.3.5 for rooms within buildings.

415.7.3.3 Separate buildings. Where located in separate buildings, liquefied petroleum gas-distribution facilities shall be occupied exclusively for that purpose or for other purposes having similar hazards. Such buildings shall be limited to one story in height and shall conform to Sections 415.7.3.3.1 through 415.7.3.3.3.

415.7.3.3.1 Floors. The floor shall not be located below ground level and any spaces beneath the floor shall be solidly filled or shall be unenclosed.

415.7.3.3.2 Materials. Walls, floors, ceilings, columns and roofs shall be constructed of noncombustible materials.

415.7.3.3.3 Explosion venting. Explosion venting shall be provided in accordance with the *International Fire Code.*

415.7.3.4 Attached buildings. Where liquefied petroleum gas-distribution facilities are located in an attached structure, the attached perimeter shall not exceed 50 percent of the perimeter of the space enclosed and the facility shall comply with Sections 415.7.3.3 and 415.7.3.4.1. Where the attached perimeter exceeds 50 percent, such facilities shall comply with Section 415.7.3.5.

415.7.3.4.1 Fire separation assemblies. Separation of the attached structures shall be provided by fire barrier walls and horizontal assemblies, or both, having a fire-resistance rating of not less than 1 hour and shall not have openings. Fire barrier walls and horizontal assemblies, or both, between attached structures occupied only for the storage of LP-gas are permitted to have fire doors that comply with Section 715. Such fire barrier walls and horizontal assemblies, or both, shall be designed to withstand a static pressure of at least 100 pounds per square foot (psf) (4788 Pa), except where the building to which the structure is attached is occupied by operations or processes having a similar hazard.

415.7.3.5 Rooms within buildings. Where liquefied petroleum gas-distribution facilities are located in rooms within buildings, such rooms shall be located in the first story above grade plane and shall have at least one exterior wall with sufficient exposed area to provide explosion venting as required in the *International Fire Code.* The building in which the room is located shall not have a basement or unventilated crawl space and the room shall comply with Sections 415.7.3.5.1 and 415.7.3.5.2.

415.7.3.5.1 Materials. Walls, floors, ceilings and roofs of such rooms shall be constructed of approved noncombustible materials.

415.7.3.5.2 Common construction. Walls and floor/ceiling assemblies common to the room and to the building within which the room is located shall have a fire barrier wall and horizontal assembly or both of not less than a 1-hour fire-resistance rating and without openings. Common walls for rooms occupied only for storage of LP-gas are permitted to have opening protectives complying with Section 715. Such walls and ceiling shall be designed to withstand a static pressure of at least 100 psf (4788 Pa).

> **Exception:** Where the building, within which the room is located, is occupied by operations or processes having a similar hazard.

415.7.4 Dry cleaning plants. The construction and installation of dry cleaning plants shall be in accordance with the requirements of this code, the *International Mechanical Code*, the *International Plumbing Code* and NFPA 32. Dry cleaning solvents and systems shall be classified in accordance with the *International Fire Code*.

415.8 Groups H-3 and H-4. Groups H-3 and H-4 shall be constructed in accordance with the applicable provisions of this code and the *International Fire Code*.

415.8.1 Gas rooms. When gas rooms are provided, such rooms shall be separated from other areas by not less than a 1-hour fire barrier.

415.8.2 Floors in storage rooms. Floors in storage areas for corrosive liquids and highly toxic or toxic materials shall be of liquid-tight, noncombustible construction.

415.8.3 Separation—highly toxic solids and liquids. Highly toxic solids and liquids not stored in approved hazardous materials storage cabinets shall be isolated from other hazardous materials storage by construction having a 1-hour fire-resistance rating.

415.9 Group H-5.

415.9.1 General. In addition to the requirements set forth elsewhere in this code, Group H-5 shall comply with the provisions of Section 415.9 and the *International Fire Code*.

415.9.2 Fabrication areas.

415.9.2.1 Hazardous materials in fabrication areas.

415.9.2.1.1 Aggregate quantities. The aggregate quantities of hazardous materials stored and used in a single fabrication area shall not exceed the quantities set forth in Table 415.9.2.1.1.

> **Exception:** The quantity limitations for any hazard category in Table 415.9.2.1.1 shall not apply where the fabrication area contains quantities of hazardous materials not exceeding the maximum allowable quantities per control area established by Tables 307.7(1) and 307.7(2).

415.9.2.1.2 Hazardous production materials. The maximum quantities of hazardous production materials stored in a single fabrication area shall not exceed the maximum allowable quantities per control area established by Tables 307.7(1) and 307.7(2).

415.9.2.2 Separation. Fabrication areas, whose sizes are limited by the quantity of hazardous materials allowed by Table 415.9.2.1.1, shall be separated from each other, from exit access corridors, and from other parts of the building by not less than 1-hour fire barriers.

> **Exceptions:**
> 1. Doors within such fire barrier walls, including doors to corridors, shall be only self-closing fire assemblies having a fire-protection rating of not less than $^3/_4$ hour.
> 2. Windows between fabrication areas and exit access corridors are permitted to be fixed glazing listed and labeled for a fire protection rating of at least $^3/_4$ hour in accordance with Section 715.

415.9.2.3 Location of occupied levels. Occupied levels of fabrication areas shall be located at or above the first story above grade plane.

415.9.2.4 Floors. Except for surfacing, floors within fabrication areas shall be of noncombustible construction.

Openings through floors of fabrication areas are permitted to be unprotected where the interconnected levels are used solely for mechanical equipment directly related to such fabrication areas (see also Section 415.9.2.5).

Floors forming a part of an occupancy separation shall be liquid tight.

415.9.2.5 Shafts and openings through floors. Elevator shafts, vent shafts and other openings through floors shall be enclosed when required by Section 707. Mechanical, duct and piping penetrations within a fabrication area shall not extend through more than two floors. The annular space around penetrations for cables, cable trays, tubing, piping, conduit or ducts shall be sealed at the floor level to restrict the movement of air. The fabrication area, including the areas through which the ductwork and piping extend, shall be considered a single conditioned environment.

415.9.2.6 Ventilation. Mechanical exhaust ventilation shall be provided throughout the fabrication area at the rate of not less than 1 cubic foot per minute per square foot (0.044 L/S/m²) of floor area. The exhaust air duct system of one fabrication area shall not connect to another duct system outside that fabrication area within the building.

A ventilation system shall be provided to capture and exhaust fumes and vapors at workstations.

Two or more operations at a workstation shall not be connected to the same exhaust system where either one or the combination of the substances removed could constitute a fire, explosion or hazardous chemical reaction within the exhaust duct system.

Exhaust ducts penetrating occupancy separations shall be contained in a shaft of equivalent fire-resistance construction. Exhaust ducts shall not penetrate fire walls.

Fire dampers shall not be installed in exhaust ducts.

TABLE 415.9.2.1.1
QUANTITY LIMITS FOR HAZARDOUS MATERIALS IN A SINGLE FABRICATION AREA IN GROUP H-5[a]

HAZARD CATEGORY		SOLIDS (pounds per square feet)	LIQUIDS (gallons per square feet)	GAS (feet^3 @ NTP/square feet)
PHYSICAL-HAZARD MATERIALS				
Combustible dust		Note b	Not Applicable	Not Applicable
Combustible fiber	Loose	Note b	Not Applicable	Not Applicable
	Baled	Note b		
Combustible liquid	II		0.01	
	IIIA	Not Applicable	0.02	Not Applicable
	IIIB		Not Limited	
Combination Class I, II and IIIA			0.04	
Cryogenic gas	Flammable	Not Applicable	Not Applicable	Note c
	Oxidizing			1.25
Explosives		Note b	Note b	Note b
Flammable gas	Gaseous	Not Applicable	Not Applicable	Note b
	Liquefied			Note c
Flammable liquid	IA		0.0025	
	IB		0.025	
	IC	Not Applicable	0.025	Not Applicable
Combination Class IA, IB and IC			0.025	
Combination Class I, II and IIIA			0.04	
Flammable solid		0.001	Not Applicable	Not Applicable
Organic peroxide				
Unclassified detonable		Note b		
Class I		Note b		
Class II		0.025	Not Applicable	Not Applicable
Class III		0.1		
Class IV		Not Limited		
Class V		Not limited		
Oxidizing gas	Gaseous			1.25
	Liquefied	Not Applicable	Not Applicable	1.25
Combination of gaseous and liquefied				1.25
Oxidizer	Class 4	Note b	Note b	
	Class 3	0.003	0.003	
	Class 2	0.003	0.003	Not Applicable
	Class 1	0.003	0.003	
Combination	Class 1, 2, 3	0.003	0.003	
Pyrophoric material		Note b	0.00125	Notes c and d
Unstable reactive	Class 4	Note b	Note b	Note b
	Class 3	0.025	0.0025	Note b
	Class 2	0.1	0.01	Note b
	Class 1	Not Limited	Not Limited	Not Limited
Water reactive	Class 3	Note b	0.00125	
	Class 2	0.25	0.025	Not Applicable
	Class 1	Not Limited	Not Limited	
HEALTH-HAZARD MATERIALS				
Corrosives		Not Limited	Not Limited	Not Limited
Highly toxic		Not Limited	Not Limited	Note c
Toxics		Not Limited	Not Limited	Note c

For SI: 1 pound per square foot = 4.882 kg/m^2, 1 gallon per square foot = 0.025 L/m^2, 1 cubic foot @ NTP/square foot = 0.305 M^3 @ NTP/m^2, 1 cubic foot = 0.02832 M^3.

a. Hazardous materials within piping shall not be included in the calculated quantities.
b. Quantity of hazardous materials in a single fabrication shall not exceed the maximum allowable quantities per control area in Tables 307.7(1) and 307.7(2).
c. The aggregate quantity of flammable, pyrophoric, toxic and highly toxic gases shall not exceed 9,000 cubic feet at NTP.
d. The aggregate quantity of pyrophoric gases in the building shall not exceed the amounts set forth in Table 415.3.2.

415.9.2.7 Transporting hazardous production materials to fabrication areas. Hazardous production materials shall be transported to fabrication areas through enclosed piping or tubing systems that comply with Section 415.9.6.1, through service corridors complying with Section 415.9.4, or in exit access corridors as permitted in the exception to Section 415.9.3. The handling or transporting of hazardous production materials within service corridors shall comply with the *International Fire Code*.

415.9.2.8 Electrical.

415.9.2.8.1 General. Electrical equipment and devices within the fabrication area shall comply with the ICC *Electrical Code*. The requirements for hazardous locations need not be applied where the average air change is at least four times that set forth in Section 415.9.2.6 and where the number of air changes at any location is not less than three times that required by Section 415.9.2.6. The use of recirculated air shall be permitted.

415.9.2.8.2 Workstations. Workstations shall not be energized without adequate exhaust ventilation. See Section 415.9.2.6 for workstation exhaust ventilation requirements.

415.9.3 Exit access corridors. Exit access corridors shall comply with Chapter 10 and shall be separated from fabrication areas as specified in Section 415.9.2.2. Exit access corridors shall not contain HPM and shall not be used for transporting such materials, except through closed piping systems as provided in Section 415.9.6.3.

Exception: Where existing fabrication areas are altered or modified, HPM is allowed to be transported in existing exit access corridors, subject to the following conditions:

1. Corridors. Exit access corridors adjacent to the fabrication area where the alteration work is to be done shall comply with Section 1016 for a length determined as follows:

    1.1. The length of the common wall of the corridor and the fabrication area; and

    1.2. For the distance along the exit access corridor to the point of entry of HPM into the exit access corridor serving that fabrication area.

2. Emergency alarm system. There shall be an emergency telephone system, a local manual alarm station or other approved alarm-initiating device within exit access corridors at not more than 150-foot (45 720 mm) intervals and at each exit and exit access doorway. The signal shall be relayed to an approved central, proprietary or remote station service or the emergency control station and shall also initiate a local audible alarm.

3. Pass-throughs. Self-closing doors having a fire-protection rating of not less than 1 hour shall separate pass-throughs from existing exit access corridors. Pass-throughs shall be constructed as required for the exit access corridors, and protected

by an approved automatic fire-extinguishing system.

415.9.4 Service corridors.

415.9.4.1 Occupancy. Service corridors shall be classified as Group H-5.

415.9.4.2 Use conditions. Service corridors shall be separated from exit access corridors as required by Section 415.9.2.2. Service corridors shall not be used as a required exit access corridor.

415.9.4.3 Mechanical ventilation. Service corridors shall be mechanically ventilated as required by Section 415.9.2.6 or at not less than six air changes per hour, whichever is greater.

415.9.4.4 Means of egress. The maximum distance of travel from any point in a service corridor to an exit, exit access corridor or door into a fabrication area shall not exceed 75 feet (22 860 mm). Dead ends shall not exceed 4 feet (1219 mm) in length. There shall be not less than two exits, and not more than one-half of the required means of egress shall require travel into a fabrication area. Doors from service corridors shall swing in the direction of egress travel and shall be self-closing.

415.9.4.5 Minimum width. The minimum clear width of a service corridor shall be 5 feet (1524 mm), or 33 inches (838 mm) wider than the widest cart or truck used in the corridor, whichever is greater.

415.9.4.6 Emergency alarm system. Emergency alarm systems shall be provided in accordance with this section and Sections 414.7.1 and 414.7.2. The maximum allowable quantity per control area provisions shall not apply to emergency alarm systems required for HPM.

415.9.4.6.1 Service corridors. An emergency alarm system shall be provided in service corridors, with at least one alarm device in each service corridor.

415.9.4.6.2 Exit access corridors and exit enclosures. Emergency alarms for exit access corridors and exit enclosures shall comply with Section 414.7.2.

415.9.4.6.3 Liquid storage rooms, HPM rooms and gas rooms. Emergency alarms for liquid storage rooms, HPM rooms and gas rooms shall comply with Section 414.7.1.

415.9.4.6.4 Alarm-initiating devices. An approved emergency telephone system, local alarm manual pull stations, or other approved alarm-initiating devices are allowed to be used as emergency alarm-initiating devices.

415.9.4.6.5 Alarm signals. Activation of the emergency alarm system shall sound a local alarm and transmit a signal to the emergency control station.

415.9.5 Storage of hazardous production materials.

415.9.5.1 General. Storage of HPM in fabrication areas shall be within approved or listed storage cabinets or gas cabinets, or within a workstation. The storage of hazardous production materials in quantities greater than those listed in Tables 307.7(1) or 307.7(2) shall be in liquid storage rooms, HPM rooms or gas rooms as appropriate for the materials stored. The storage of other hazardous materials shall be in accordance with other applicable provisions of this code and the *International Fire Code*.

415.9.5.2 Construction.

415.9.5.2.1 HPM rooms and gas rooms. HPM rooms and gas rooms shall be separated from other areas by not less than a 2-hour fire barrier where the area is 300 square feet (27.9 m²) or more and not less than a 1-hour fire barrier where the area is less than 300 square feet (27.9 m²).

415.9.5.2.2 Liquid storage rooms. Liquid storage rooms shall be constructed in accordance with the following requirements:

1. Rooms in excess of 500 square feet (46.5 m²) shall have at least one exterior door approved for fire department access.

2. Rooms shall be separated from other areas by fire barriers having a fire-resistance rating of not less than 1-hour for rooms up to 150 square feet (13.9 m²) in area and not less than 2 hours where the room is more than 150 square feet (13.9 m²) in area.

3. Shelving, racks and wainscoting in such areas shall be of noncombustible construction or wood of not less than 1inch (25 mm) nominal thickness.

4. Rooms used for the storage of Class I flammable liquids shall not be located in a basement.

415.9.5.2.3 Floors. Except for surfacing, floors of HPM rooms and liquid storage rooms shall be of noncombustible liquid-tight construction. Raised grating over floors shall be of noncombustible materials.

415.9.5.3 Location. Where HPM rooms, liquid storage rooms and gas rooms are provided, they shall have at least one exterior wall and such wall shall be not less than 30 feet (9144 mm) from property lines, including property lines adjacent to public ways.

415.9.5.4 Explosion control. Explosion control shall be provided where required by Section 414.5.1.

415.9.5.5 Exits. Where two exits are required from HPM rooms, liquid storage rooms and gas rooms, one shall be directly to the outside of the building.

415.9.5.6 Doors. Doors in a fire barrier wall, including doors to corridors, shall be self-closing fire assemblies having a fire-protection rating of not less than ³/₄ hour.

415.9.5.7 Ventilation. Mechanical exhaust ventilation shall be provided in liquid storage rooms, HPM rooms and gas rooms at the rate of not less than 1 cubic foot per minute per square foot (0.044 L/S/m2) of floor area or six air changes per hour, whichever is greater, for categories of material.

Exhaust ventilation for gas rooms shall be designed to operate at a negative pressure in relation to the surrounding areas and direct the exhaust ventilation to an exhaust system.

415.9.5.8 Emergency alarm system. An approved emergency alarm system shall be provided for HPM rooms, liquid storage rooms and gas rooms.

Emergency alarm-initiating devices shall be installed outside of each interior exit door of such rooms.

Activation of an emergency alarm-initiating device shall sound a local alarm and transmit a signal to the emergency control station.

An approved emergency telephone system, local alarm manual pull stations or other approved alarm-initiating devices are allowed to be used as emergency alarm-initiating devices.

415.9.6 Piping and tubing.

415.9.6.1 General. Hazardous production materials piping and tubing shall comply with this section and ANSI B31.3.

415.9.6.2 Supply piping and tubing.

415.9.6.2.1 HPM having a health-hazard ranking of 3 or 4. Systems supplying HPM liquids or gases having a health-hazard ranking of 3 or 4 shall be welded throughout, except for connections, to the systems that are within a ventilated enclosure if the material is a gas, or an approved method of drainage or containment is provided for the connections if the material is a liquid.

415.9.6.2.2 Location in service corridors. Hazardous production materials supply piping or tubing in service corridors shall be exposed to view.

415.9.6.2.3 Excess flow control. Where HPM gases or liquids are carried in pressurized piping above 15 pounds per square inch gauge (psig) (103.4 kPa), excess flow control shall be provided. Where the piping originates from within a liquid storage room, HPM room or gas room, the excess flow control shall be located within the liquid storage room, HPM room or gas room. Where the piping originates from a bulk source, the excess flow control shall be located as close to the bulk source as practical.

415.9.6.3 Installations in exit access corridors and above other occupancies. The installation of hazardous production material piping and tubing within the space defined by the walls of exit access corridors and the floor or roof above or in concealed spaces above other occupancies shall be in accordance with Section 415.9.6.2 and the following conditions:

1. Automatic sprinklers shall be installed within the space unless the space is less than 6 inches (152 mm) in the least dimension.

2. Ventilation not less than six air changes per hour shall be provided. The space shall not be used to convey air from any other area.

3. Where the piping or tubing is used to transport HPM liquids, a receptor shall be installed below

such piping or tubing. The receptor shall be designed to collect any discharge or leakage and drain it to an approved location. The 1-hour enclosure shall not be used as part of the receptor.

4. HPM supply piping and tubing and HPM nonmetallic waste lines shall be separated from the exit access corridor and from occupancies other than Group H-5 by construction as required for walls or partitions that have a fire protection rating of not less than 1 hour. Where gypsum wallboard is used, joints on the piping side of the enclosure are not required to be taped, provided the joints occur over framing members. Access openings into the enclosure shall be protected by approved fire-resistance-rated assemblies.

5. Readily accessible manual or automatic remotely activated fail-safe emergency shutoff valves shall be installed on piping and tubing other than waste lines at the following locations:

    5.1. At branch connections into the fabrication area.

    5.2. At entries into exit access corridors.

Exception: Transverse crossings of the corridors by supply piping that is enclosed within a ferrous pipe or tube for the width of corridor need not comply with Items 1 through 5.

415.9.6.4 Identification. Piping, tubing and HPM waste lines shall be identified in accordance with ANSI A13.1 to indicate the material being transported.

415.9.7 Continuous gas-detection systems. A continuous gas-detection system shall be provided for HPM gases when the physiological warning properties of the gas are at a higher level than the accepted permissible exposure limit (PEL) for the gas and for flammable gases in accordance with this section.

415.9.7.1 Where required. A continuous gas-detection system shall be provided in the areas identified in Sections 415.9.7.1.1 through 415.9.7.1.4.

415.9.7.1.1 Fabrication areas. A continuous gas-detection system shall be provided in fabrication areas when gas is used in the fabrication area.

415.9.7.1.2 HPM rooms. A continuous gas-detection system shall be provided in HPM rooms when gas is used in the room.

415.9.7.1.3 Gas cabinets, exhausted enclosures and gas rooms. A continuous gas-detection system shall be provided in gas cabinets and exhausted enclosures. A continuous gas-detection system shall be provided in gas rooms when gases are not located in gas cabinets or exhausted enclosures.

415.9.7.1.4 Exit access corridors. When gases are transported in piping placed within the space defined by the walls of an exit access corridor, and the floor or roof above the exit access corridor, a continuous gas-detection system shall be provided where piping is located and in the exit access corridor.

Exception: A continuous gas-detection system is not required for occasional transverse crossings of the corridors by supply piping that is enclosed in a ferrous pipe or tube for the width of the corridor.

415.9.7.2 Gas-detection system operation. The continuous gas-detection system shall be capable of monitoring the room, area or equipment in which the gas is located at or below the PEL or ceiling limit of the gas for which detection is provided. For flammable gases, the monitoring detection threshold level shall be vapor concentrations in excess of 20 percent of the lower explosive limit (LFL). Monitoring for highly toxic and toxic gases shall also comply with the requirements for such material in the *International Fire Code*.

415.9.7.2.1 Alarms. The gas detection system shall initiate a local alarm and transmit a signal to the emergency control station when a short-term hazard condition is detected. The alarm shall be both visual and audible and shall provide warning both inside and outside the area where the gas is detected. The audible alarm shall be distinct from all other alarms.

415.9.7.2.2 Shutoff of gas supply. The gas detection system shall automatically close the shutoff valve at the source on gas supply piping and tubing related to the system being monitored for which gas is detected when a short-term hazard condition is detected. Automatic closure of shutoff valves shall comply with the following:

1. Where the gas-detection sampling point initiating the gas detection system alarm is within a gas cabinet or exhausted enclosure, the shutoff valve in the gas cabinet or exhausted enclosure for the specific gas detected shall automatically close.

2. Where the gas-detection sampling point initiating the gas detection system alarm is within a room and compressed gas containers are not in gas cabinets or an exhausted enclosure, the shutoff valves on all gas lines for the specific gas detected shall automatically close.

3. Where the gas-detection sampling point initiating the gas detection system alarm is within a piping distribution manifold enclosure, the shutoff valve supplying the manifold for the compressed gas container of the specific gas detected shall automatically close.

Exception: Where the gas-detection sampling point initiating the gas detection system alarm is at the use location or within a gas valve enclosure of a branch line downstream of a piping distribution manifold, the shutoff valve for the branch line located in the piping distribution manifold enclosure shall automatically close.

415.9.8 Manual fire alarm system. An approved manual fire alarm system shall be provided throughout buildings containing Group H-5. Activation of the alarm system shall initiate a local alarm and transmit a signal to the emergency control station. The fire alarm system shall be designed and installed in accordance with Section 907.

415.9.9 Emergency control station. An emergency control station shall be provided on the premises at an approved location, outside of the fabrication area and shall be continuously staffed by trained personnel. The emergency control station shall receive signals from emergency equipment and alarm and detection systems. Such emergency equipment and alarm and detection systems shall include, but not necessarily be limited to, the following where such equipment or systems are required to be provided either in Section 415.9 or elsewhere in this code:

1. Automatic fire sprinkler system alarm and monitoring systems.
2. Manual fire alarm systems.
3. Emergency alarm systems.
4. Continuous gas-detection systems.
5. Smoke detection systems.
6. Emergency power system.

415.9.10 Emergency power system. An emergency power system shall be provided in Group H-5 occupancies where required in Section 415.9.10.1. The emergency power system shall be designed to supply power automatically to required electrical systems when the normal electrical supply system is interrupted.

415.9.10.1 Where required. Emergency power shall be provided for electrically operated equipment and connected control circuits for the following systems:

1. HPM exhaust ventilation systems.
2. HPM gas cabinet ventilation systems.
3. HPM exhausted enclosure ventilation systems.
4. HPM gas room ventilation systems.
5. HPM gas detection systems.
6. Emergency alarm systems.
7. Manual fire alarm systems.
8. Automatic sprinkler system monitoring and alarm systems.
9. Electrically operated systems required elsewhere in this code applicable to the use, storage or handling of HPM.

415.9.10.2 Exhaust ventilation systems. Exhaust ventilation systems are allowed to be designed to operate at not less than one-half the normal fan speed on the emergency power system where it is demonstrated that the level of exhaust will maintain a safe atmosphere.

415.9.11 Fire sprinkler system protection in exhaust ducts for HPM.

415.9.11.1 General. Automatic fire sprinkler system protection shall be provided in exhaust ducts conveying vapors, fumes, mists or dusts generated from HPM in ac-

cordance with this section and the *International Mechanical Code.*

415.9.11.2 Metallic and noncombustible, nonmetallic exhaust ducts. Automatic fire sprinkler system protection shall be provided in metallic and noncombustible, nonmetallic exhaust ducts where all of the following conditions apply:

1. Where the largest cross-sectional diameter is equal to or greater than 10 inches (254 mm).
2. The ducts are within the building.
3. The ducts are conveying flammable vapors or fumes.

415.9.11.3 Combustible nonmetallic exhaust ducts. Automatic fire sprinkler system protection shall be provided in combustible nonmetallic exhaust ducts where the largest cross-sectional diameter of the duct is equal to or greater than 10 inches (254 mm).

Exceptions:

1. Ducts listed or approved for applications without automatic fire sprinkler system protection.
2. Ducts not more than 12 feet (3658 mm) in length installed below ceiling level.

415.9.11.4 Automatic sprinkler locations. Sprinkler systems shall be installed at 12-foot (3658 mm) intervals in horizontal ducts and at changes in direction. In vertical ducts, sprinklers shall be installed at the top and at alternate floor levels.

[F] SECTION 416
APPLICATION OF FLAMMABLE FINISHES

416.1 General. The provisions of this section shall apply to the construction, installation and use of buildings and structures, or parts thereof, for the spraying of flammable paints, varnishes and lacquers or other flammable materials or mixtures or compounds used for painting, varnishing, staining or similar purposes. Such construction and equipment shall comply with the *International Fire Code.*

416.2 Spray rooms. Spray rooms shall be enclosed with fire barrier walls and horizontal assemblies or both with not less than a 1-hour fire-resistance rating. Floors shall be waterproofed and drained in an approved manner.

416.2.1 Surfaces. The interior surfaces of spray rooms shall be smooth and shall be so constructed to permit the free passage of exhaust air from all parts of the interior and to facilitate washing and cleaning, and shall be so designed to confine residues within the room. Aluminum shall not be used.

416.3 Spraying spaces. Spraying spaces shall be ventilated with an exhaust system to prevent the accumulation of flammable mist or vapors in accordance with the *International Mechanical Code.* Where such spaces are not separately enclosed, noncombustible spray curtains shall be provided to restrict the spread of flammable vapors.

416.3.1 Surfaces. The interior surfaces of spraying spaces shall be smooth and continuous without edges, and shall be so constructed to permit the free passage of exhaust air from all parts of the interior and to facilitate washing and cleaning, and shall be so designed to confine residues within the spraying space. Aluminum shall not be used.

416.4 Fire protection. An automatic fire-extinguishing system shall be provided in all spray, dip and immersing spaces and storage rooms, and shall be installed in accordance with Chapter 9.

[F] SECTION 417
DRYING ROOMS

417.1 General. A drying room or dry kiln installed within a building shall be constructed entirely of approved noncombustible materials or assemblies of such materials regulated by the approved rules or as required in the general and specific sections of Chapter 4 for special occupancies and where applicable to the general requirements of Chapter 28.

417.2 Piping clearance. Overhead heating pipes shall have a clearance of not less than 2 inches (51 mm) from combustible contents in the dryer.

417.3 Insulation. Where the operating temperature of the dryer is 175°F (79°C) or more, metal enclosures shall be insulated from adjacent combustible materials by not less than 12 inches (305 mm) of airspace, or the metal walls shall be lined with $1/4$-inch (6.35 mm) insulating mill board or other approved equivalent insulation.

417.4 Fire protection. Drying rooms designed for high-hazard materials and processes, including special occupancies as provided for in Chapter 4, shall be protected by an approved automatic fire-extinguishing system conforming to the provisions of Chapter 9.

[F] SECTION 418
ORGANIC COATINGS

418.1 Building features. Manufacturing of organic coatings shall be done only in buildings that do not have pits or basements.

418.2 Location. Organic coating manufacturing operations and operations incidental to or connected therewith shall not be located in buildings having other occupancies.

418.3 Process mills. Mills operating with close clearances and that process flammable and heat-sensitive materials, such as nitrocellulose, shall be located in a detached building or noncombustible structure.

418.4 Tank storage. Storage areas for flammable and combustible liquid tanks inside of structures shall be located at or above grade and shall be separated from the processing area by not less than 2-hour fire-resistance-rated fire barriers.

418.5 Nitrocellulose storage. Nitrocellulose storage shall be located on a detached pad or in a separate structure or a room enclosed with no less than 2-hour fire-resistance-rated fire barriers.

418.6 Finished products. Storage rooms for finished products that are flammable or combustible liquids shall be separated from the processing area by fire barriers having a fire-resistance rating of at least 2 hours, and openings in the walls shall be protected with approved opening protectives.

CHAPTER 5
GENERAL BUILDING HEIGHTS AND AREAS

SECTION 501
GENERAL

501.1 Scope. The provisions of this chapter control the height and area of structures hereafter erected and additions to existing structures.

[F] 501.2 Premises identification. Approved numbers or addresses shall be provided for new buildings in such a position as to be clearly visible and legible from the street or roadway fronting the property. Letters or numbers shall be a minimum 3 inches (76 mm) in height and stroke of minimum 0.5 inch (12.7 mm) of a contrasting color to the background itself.

SECTIONS 502
DEFINITIONS

502.1 Definitions. The following words and terms shall, for the purposes of this chapter and as used elsewhere in this code, have the meanings shown herein.

AREA, BUILDING. The area included within surrounding exterior walls (or exterior walls and fire walls) exclusive of vent shafts and courts. Areas of the building not provided with surrounding walls shall be included in the building area if such areas are included within the horizontal projection of the roof or floor above.

BASEMENT. That portion of a building that is partly or completely below grade plane (See "Story above grade plane" in Section 202). A basement shall be considered as a story above grade plane where the finished surface of the floor above the basement is:

1. More than 6 feet (1829 mm) above grade plane;

2. More than 6 feet (1829 mm) above the finished ground level for more than 50 percent of the total building perimeter; or

3. More than 12 feet (3658 mm) above the finished ground level at any point.

GRADE PLANE. A reference plane representing the average of finished ground level adjoining the building at exterior walls. Where the finished ground level slopes away from the exterior walls, the reference plane shall be established by the lowest points within the area between the building and the lot line or, where the lot line is more than 6 feet (1829 mm) from the building, between the building and a point 6 feet (1829 mm) from the building.

HEIGHT, BUILDING. The vertical distance from grade plane to the average height of the highest roof surface.

HEIGHT, STORY. The vertical distance from top to top of two successive finished floor surfaces; and, for the topmost story, from the top of the floor finish to the top of the ceiling joists or, where there is not a ceiling, to the top of the roof rafters.

INDUSTRIAL EQUIPMENT PLATFORM. An unoccupied, elevated platform in an industrial occupancy used exclusively for mechanical systems or industrial process equipment, including the associated elevated walkways, stairs and ladders necessary to access the platform (see Section 505.5).

MEZZANINE. An intermediate level or levels between the floor and ceiling of any story with an aggregate floor area of not more than one-third of the area of the room or space in which the level or levels are located (see Section 505).

STORY. That portion of a building included between the upper surface of a floor and the upper surface of the floor or roof next above (also see "Basement" and "Mezzanine").

SECTION 503
GENERAL HEIGHT AND AREA LIMITATIONS

503.1 General. The height and area for buildings of different construction types shall be governed by the intended use of the building and shall not exceed the limits in Table 503 except as modified hereafter. Each part of a building included within the exterior walls or the exterior walls and fire walls where provided shall be permitted to be a separate building.

503.1.1 Basements. Basements need not be included in the total allowable area provided they do not exceed the area permitted for a one-story building.

503.1.2 Special industrial occupancies. Buildings and structures designed to house low-hazard industrial processes that require large areas and unusual heights to accommodate craneways or special machinery and equipment including, among others, rolling mills; structural metal fabrication shops and foundries; or the production and distribution of electric, gas or steam power, shall be exempt from the height and area limitations of Table 503.

503.1.3 Buildings on same lot. Two or more buildings on the same lot shall be regulated as separate buildings or shall be considered as portions of one building if the height of each building and the aggregate area of buildings are within the limitations of Table 503 as modified by Sections 504 and 506. The provisions of this code applicable to the aggregate building shall be applicable to each building.

503.1.4 Type I construction. Buildings of Type I construction permitted to be of unlimited tabular heights and areas are not subject to the special requirements that allow unlimited area buildings in Section 507 or unlimited height in Sections 503.1.2 and 504.3 or increased height and areas for other types of construction.

503.2 Party walls. Any wall located on a lot line between adjacent buildings, which is used or adapted for joint service between the two buildings, shall be constructed as a fire wall in accordance with Section 705, without openings and shall create separate buildings.

TABLE 503
ALLOWABLE HEIGHT AND BUILDING AREAS
Height limitations shown as stories and feet above grade plane.
Area limitations as determined by the definition of "Area, building," per floor.

GROUP	Hgt(feet) Hgt(S)	TYPE I A UL	TYPE I B 160	TYPE II A 65	TYPE II B 55	TYPE III A 65	TYPE III B 55	TYPE IV HT 65	TYPE V A 50	TYPE V B 40
A-1	S	UL	5	3	2	3	2	3	2	1
	A	UL	UL	15,500	8,500	14,000	8,500	15,000	11,500	5,500
A-2	S	UL	11	3	2	3	2	3	2	1
	A	UL	UL	15,500	9,500	14,000	9,500	15,000	11,500	6,000
A-3	S	UL	11	3	2	3	2	3	2	1
	A	UL	UL	15,500	9,500	14,000	9,500	15,000	11,500	6,000
A-4	S	UL	11	3	2	3	2	3	2	1
	A	UL	UL	15,500	9,500	14,000	9,500	15,000	11,500	6,000
A-5	S	UL	UL	UL	UL	UL	UL	UL	UL	UL
	A	UL	UL	UL	UL	UL	UL	UL	UL	UL
B	S	UL	11	5	4	5	4	5	3	2
	A	UL	UL	37,500	23,000	28,500	19,000	36,000	18,000	9,000
E	S	UL	5	3	2	3	2	3	1	1
	A	UL	UL	26,500	14,500	23,500	14,500	25,500	18,500	9,500
F-1	S	UL	11	4	2	3	2	4	2	1
	A	UL	UL	25,000	15,500	19,000	12,000	33,500	14,000	8,500
F-2	S	UL	11	5	3	4	3	5	3	2
	A	UL	UL	37,500	23,000	28,500	18,000	50,500	21,000	13,000
H-1	S	1	1	1	1	1	1	1	1	NP
	A	21,000	16,500	11,000	7,000	9,500	7,000	10,500	7,500	NP
H-2	S	UL	3	2	1	2	1	2	1	1
	A	21,000	16,500	11,000	7,000	9,500	7,000	10,500	7,500	3,000
H-3	S	UL	6	4	2	4	2	4	2	1
	A	UL	60,000	26,500	14,000	17,500	13,000	25,500	10,000	5,000
H-4	S	UL	7	5	3	5	3	5	3	2
	A	UL	UL	37,500	17,500	28,500	17,500	36,000	18,000	6,500
H-5	S	3	3	3	3	3	3	3	3	2
	A	UL	UL	37,500	23,000	28,500	19,000	36,000	18,000	9,000
I-1	S	UL	9	4	3	4	3	4	3	2
	A	UL	55,000	19,000	10,000	16,500	10,000	18,000	10,500	4,500
I-2	S	UL	4	2	1	1	NP	1	1	NP
	A	UL	UL	15,000	11,000	12,000	NP	12,000	9,500	NP
I-3	S	UL	4	2	1	2	1	2	2	1
	A	UL	UL	15,000	10,000	10,500	7,500	12,000	7,500	5,000
I-4	S	UL	5	3	2	3	2	3	1	1
	A	UL	60,500	26,500	13,000	23,500	13,000	25,500	18,500	9,000
M	S	UL	11	4	4	4	4	4	3	1
	A	UL	UL	21,500	12,500	18,500	12,500	20,500	14,000	9,000
R-1	S	UL	11	4	4	4	4	4	3	2
	A	UL	UL	24,000	16,000	24,000	16,000	20,500	12,000	7,000
R-2[a]	S	UL	11	4	4	4	4	4	3	2
	A	UL	UL	24,000	16,000	24,000	16,000	20,500	12,000	7,000
R-3[a]	S	UL	11	4	4	4	4	4	3	3
	A	UL	UL	UL	UL	UL	UL	UL	UL	UL
R-4	S	UL	11	4	4	4	4	4	3	2
	A	UL	UL	24,000	16,000	24,000	16,000	20,500	12,000	7,000
S-1	S	UL	11	4	3	3	3	4	3	1
	A	UL	48,000	26,000	17,500	26,000	17,500	25,500	14,000	9,000
S-2[b, c]	S	UL	11	5	4	4	4	5	4	2
	A	UL	79,000	39,000	26,000	39,000	26,000	38,500	21,000	13,500
U[c]	S	UL	5	4	2	3	2	4	2	1
	A	UL	35,500	19,000	8,500	14,000	8,500	18,000	9,000	5,500

For SI: 1 foot = 304.8 mm, 1 square foot = 0.0929 m².

UL = Unlimited, NP = Not permitted.

a. As applicable in Section 101.2.

b. For open parking structures, see Section 406.3.

c. For private garages, see Section 406.1.

SECTION 504
HEIGHT MODIFICATIONS

504.1 General. The heights permitted by Table 503 shall only be increased in accordance with this section.

Exception: The height of one-story aircraft hangars, aircraft paint hangars and buildings used for the manufacturing of aircraft shall not be limited if the building is provided with an automatic fire-extinguishing system in accordance with Chapter 9 and is entirely surrounded by public ways or yards not less in width than one and one-half times the height of the building.

504.2 Automatic sprinkler system increase. Where a building is equipped throughout with an approved automatic sprinkler system in accordance with Section 903.3.1.1, the value specified in Table 503 for maximum height is increased by 20 feet (6096 mm) and the maximum number of stories is increased by one story. These increases are permitted in addition to the area increase in accordance with Sections 506.2 and 506.3. For Group R buildings equipped throughout with an approved automatic sprinkler system in accordance with Section 903.3.1.2, the value specified in Table 503 for maximum height is increased by 20 feet (6096 mm) and the maximum number of stories is increased by one story, but shall not exceed four stories or 60 feet (18 288 mm), respectively.

Exceptions:

1. Group I-2 of Type IIB, III, IV or V construction.
2. Group H-1, H-2, H-3 or H-5.
3. Fire-resistance rating substitution in accordance with Table 601, Note d.

504.3 Roof structures. Towers, spires, steeples and other roof structures shall be constructed of materials consistent with the required type of construction of the building except where other construction is permitted by Section 1509.2.1. Such structures shall not be used for habitation or storage. The structures shall be unlimited in height if of noncombustible materials and shall not extend more than 20 feet (6096 mm) above the allowable height if of combustible materials (see Chapter 15 for additional requirements).

SECTION 505
MEZZANINES

505.1 General. A mezzanine or mezzanines in compliance with this section shall be considered a portion of the floor below. Such mezzanines shall not contribute to either the building area or number of stories as regulated by Section 503.1. The area of the mezzanine shall be included in determining the fire area defined in Section 702. The clear height above and below the mezzanine floor construction shall not be less than 7 feet (2134 mm).

505.2 Area limitation. The aggregate area of a mezzanine or mezzanines within a room shall not exceed one-third of the area of that room or space in which they are located. The enclosed portions of rooms shall not be included in a determination of the size of the room in which the mezzanine is located. In determining the allowable mezzanine area, the area of the mezzanine shall not be included in the area of the room.

Exception: The aggregate area of mezzanines in buildings and structures of Type I or II construction for special industrial occupancies in accordance with Section 503.1.2 shall not exceed two-thirds of the area of the room.

505.3 Egress. Each occupant of a mezzanine shall have access to at least two independent means of egress where the common path of egress travel exceeds the limitations of Section 1013.3. Where a stairway provides a means of exit access from a mezzanine, the maximum travel distance includes the distance traveled on the stairway measured in the plane of the tread nosing.

Exceptions:

1. A single means of egress shall be permitted in accordance with Section 1014.1.
2. Accessible means of egress shall be provided in accordance with Section 1007.

505.4 Openness. A mezzanine shall be open and unobstructed to the room in which such mezzanine is located except for walls not more than 42 inches (1067 mm) high, columns and posts.

Exceptions:

1. Mezzanines or portions thereof are not required to be open to the room in which the mezzanines are located, provided that the occupant load of the aggregate area of the enclosed space does not exceed 10.
2. A mezzanine having two or more means of egress is not required to be open to the room in which the mezzanine is located, if at least one of the means of egress provides direct access to an exit from the mezzanine level.
3. Mezzanines or portions thereof are not required to be open to the room in which the mezzanines are located, provided that the aggregate floor area of the enclosed space does not exceed 10 percent of the mezzanine area.
4. In industrial facilities, mezzanines used for control equipment are permitted to be glazed on all sides.
5. In Group F occupancies of unlimited area, meeting the requirements of Section 507.2 or 507.3, mezzanines or portions thereof are not required to be open to the room in which the mezzanines are located, provided that an approved fire alarm system is installed throughout the entire building or structure and notification appliances are installed throughout the mezzanines in accordance with the provisions of NFPA 72. In addition, the fire alarm system shall be initiated by automatic sprinkler water flow.

505.5 Industrial equipment platforms. Industrial equipment platforms in buildings shall not be considered as a portion of the floor below. Such equipment platforms shall not contribute to either the building area or the number of stories as regulated by Section 503.1. The area of the industrial equipment platform shall not be included in determining the fire area. Industrial equipment platforms shall not be a part of any mezzanine, and such platforms and the walkways, stairs and ladders providing access to an equipment platform shall not serve as a part of the means of egress from the building.

505.5.1 Area limitations. The aggregate area of all industrial equipment platforms within a room shall not exceed two-thirds of the area of the room in which they occur. Where an equipment platform is located in the same room as a mezzanine, the area of the mezzanine shall be determined by Section 505.2, and the combined aggregate area of the equipment platforms and mezzanines shall not exceed two-thirds of the room in which they occur.

505.5.2 Fire suppression. Where located in a building that is required to be protected by an automatic sprinkler system, industrial equipment platforms shall be fully protected by sprinklers above and below the platform, where required by the standards referenced in Section 903.3.

505.5.3 Guards. Equipment platforms shall have guards where required by Section 1012.1.

SECTION 506
AREA MODIFICATIONS

506.1 General. The areas limited by Table 503 shall be permitted to be increased due to frontage (I_f) and automatic sprinkler system protection (I_s) in accordance with the following:

$$A_a = A_t + \left[\frac{A_t I_f}{100}\right] + \left[\frac{A_t I_s}{100}\right]$$

(Equation 5-1)

where:

A_a = Allowable area per floor (square feet).

A_t = Tabular area per floor in accordance with Table 503 (square feet).

I_f = Area increase due to frontage (percent) as calculated in accordance with Section 506.2.

I_s = Area increase due to sprinkler protection (percent) as calculated in accordance with Section 506.3.

506.1.1 Basements. A single basement need not be included in the total allowable area provided such basement does not exceed the area permitted for a one-story building.

506.2 Frontage increase. Every building shall adjoin or have access to a public way to receive an area increase for frontage. Where a building has more than 25 percent of its perimeter on a public way or open space having a minimum width of 20 feet (6096 mm), the frontage increase shall be determined in accordance with the following:

$$I_f = 100\left[\frac{F}{P} - 0.25\right]\frac{W}{30}$$

(Equation 5-2)

where:

I_f = Area increase due to frontage.

F = Building perimeter which fronts on a public way or open space having 20 feet (6096 mm) open minimum width (feet).

P = Perimeter of entire building (feet).

W = Width of public way or open space (feet) in accordance with Section 506.2.1.

506.2.1 Width limits. W must be at least 20 feet (6096 mm) and the quantity W divided by 30 shall not exceed 1.0. Where the value of W varies along the perimeter of the building, the calculation performed in accordance with Equation 5-2 shall be based on the weighted average of each portion of exterior wall and open space where the value of W is between 20 and 30 feet (6096 and 9144 mm).

> **Exception:** The quantity W divided by 30 shall be permitted to not exceed 2.0 when all of the following conditions exist:
>
> 1. The building is permitted to be unlimited in area by Section 507; and
>
> 2. The only provision preventing unlimited area is compliance with the 60-foot (18 288 mm) public way or yard requirement, as applicable.

506.2.2 Open space limits. Such open space shall be either on the same lot or dedicated for public use and shall be accessed from a street or approved fire lane.

506.3 Automatic sprinkler system increase. Where a building is equipped throughout with an approved automatic sprinkler system in accordance with Section 903.3.1.1, the area limitation in Table 503 is permitted to be increased by an additional 200 percent (I_s = 200 percent) for multistory buildings and an additional 300 percent (I_s = 300 percent) for single-story buildings. These increases are permitted in addition to the height and story increases in accordance with Section 504.2.

Exceptions:

1. Buildings with an occupancy in Group H-1, H-2 or H-3.

2. Fire-resistance rating substitution in accordance with Table 601, Note d.

506.4 Area determination. The maximum area of a building with more than one story shall be determined by multiplying the allowable area of the first floor (A_a), as determined in Section 506.1, by the number of stories as listed below.

1. For two-story buildings, multiply by 2;

2. For three-story or higher buildings, multiply by 3; and,

3. No story shall exceed the allowable area per floor (A_a), as determined in Section 506.1 for the occupancies on that floor.

Exceptions:

1. Unlimited area buildings in accordance with Section 507.

2. The maximum area of a building equipped throughout with an automatic sprinkler system in accordance with Section 903.3.1.2 shall be determined by multiplying the allowable area per floor (A_a), as determined in Section 506.1 by the number of stories.

SECTION 507
UNLIMITED AREA BUILDINGS

507.1 Nonsprinklered, one story. The area of a one-story, Group F-2 or S-2 building shall not be limited when the building is surrounded and adjoined by public ways or yards not less than 60 feet (18 288 mm) in width.

507.2 Sprinklered, one story. The area of a one-story, Group B, F, M or S building or a one-story Group A-4 building of other than Type V construction shall not be limited when the building is provided with an automatic sprinkler system throughout in accordance with Section 903.3.1.1, and is surrounded and adjoined by public ways or yards not less than 60 feet (18 288 mm) in width.

Exceptions:

1. Buildings and structures of Type I and II construction for rack storage facilities which do not have access by the public shall not be limited in height provided that such buildings conform to the requirements of Section 507.1 and NFPA 231C.

2. The automatic sprinkler system shall not be required in areas occupied for indoor participant sports, such as tennis, skating, swimming and equestrian activities, in occupancies in Group A-4, provided that:

   2.1. Exit doors directly to the outside are provided for occupants of the participant sports areas, and

   2.2. The building is equipped with a fire alarm system with manual fire alarm boxes installed in accordance with Section 907.

507.3 Two story. The area of a two-story, Group B, F, M or S building shall not be limited when the building is provided with an automatic sprinkler system in accordance with Section 903.3.1.1 throughout, and is surrounded and adjoined by public ways or yards not less than 60 feet (18 288 mm) in width.

507.4 Reduced open space. The permanent open space of 60 feet (18 288 mm) required in Sections 507.1, 507.2 and 507.3 shall be permitted to be reduced to not less than 40 feet (12 192 mm) provided the following requirements are met:

1. The reduced open space shall not be allowed for more than 75 percent of the perimeter of the building.

2. The exterior wall facing the reduced open space shall have a minimum fire-resistance rating of 3 hours.

3. Openings in the exterior wall, facing the reduced open space, shall have opening protectives with a fire-resistance rating of 3 hours.

507.5 Group A-3 buildings. The area of a one-story, Group A-3 building used as a church, community hall, dance hall, exhibition hall, gymnasium, lecture hall, indoor swimming pool or tennis court of Type I or II construction shall not be limited when all of the following criteria are met:

1. The building shall not have a stage other than a platform.

2. The building shall be equipped throughout with an automatic sprinkler system in accordance with Section 903.3.1.1.

3. The assembly floor shall be located at or within 21 inches (533 mm) of street or grade level and all exits are provided with ramps complying with Section 1010.1 to the street or grade level.

4. The building shall be surrounded and adjoined by public ways or yards not less than 60 feet (18 288 mm) in width.

507.6 High-hazard use groups. Group H-2, H-3 and H-4 fire areas shall be permitted in unlimited area buildings having occupancies in Groups F and S, in accordance with the limitations of this section. Fire areas located at the perimeter of the unlimited area building shall not exceed 10 percent of the area of the building nor the area limitations specified in Table 503 as modified by Section 506.2, based upon the percentage of the perimeter of the fire area that fronts on a street or other unoccupied space. Other fire areas shall not exceed 25 percent of the area limitations specified in Table 503. Fire-resistance-rating requirements of fire barrier assemblies shall be in accordance with Table 302.3.2.

507.7 Aircraft paint hangar. The area of a one-story, Group H-2 aircraft paint hangar shall not be limited where such aircraft paint hangar complies with the provisions of Section 412.4 and is entirely surrounded by public ways or yards not less in width than one and one-half times the height of the building.

507.8 Group E buildings. The area of a one-story Group E building of Type II, IIIA or IV construction shall not be limited when the following criteria are met:

1. Each classroom shall have not less than two means of egress, with one of the means of egress being a direct exit to the outside of the building complying with Section 1017.

2. The building is equipped throughout with an automatic sprinkler system in accordance with Section 903.3.1.1.

3. The building is surrounded and adjoined by public ways or yards not less than 60 feet (18 288 mm) in width.

507.9 Motion picture theaters. In buildings of Type I or II construction, the area of one-story motion picture theaters shall not be limited when the building is provided with an automatic sprinkler system throughout in accordance with Section 903.3.1.1 and is surrounded and adjoined by public ways or yards not less than 60 feet (18 288 mm) in width.

SECTION 508
SPECIAL PROVISIONS

508.1 General. The provisions in this section shall permit the use of special conditions that are exempt from, or modify, the specific requirements of this chapter regarding the allowable heights and areas of buildings based on the occupancy classification and type of construction, provided the special condition complies with the provisions specified in this section for such condition and other applicable requirements of this code.

508.2 Group S-2 enclosed parking garage with Group A, B, M or R above. A basement and/or the first story above grade plane of a building shall be considered as a separate and distinct building for the purpose of determining area limitations, continuity of fire walls, limitation of number of stories and type of construction, when all of the following conditions are met:

1. The basement and/or the first story above grade plane is of Type IA construction and is separated from the build-

ing above with a horizontal assembly having a minimum 3-hour fire-resistance rating.

2. Shaft, stairway, ramp or escalator enclosures through the horizontal assembly shall have not less than a 2-hour fire-resistance rating with opening protectives in accordance with Table 715.3.

Exception: Where the enclosure walls below the horizontal assembly have not less than a 3-hour fire-resistance rating with opening protectives in accordance with Table 715.3, the enclosure walls extending above the horizontal assembly shall be permitted to have a 1-hour fire-resistance rating provided:

1. The building above the horizontal assembly is not required to be of Type I construction;

2. The enclosure connects less than four stories, and

3. The enclosure opening protectives above the horizontal assembly have a minimum 1-hour fire protection rating.

3. The building above the horizontal assembly contains only Group A having an assembly room with an occupant load of less than 300, or Group B, M or R.

4. The building below the horizontal assembly is a Group S-2 enclosed parking garage, used for the parking and storage of private motor vehicles.

Exceptions:

1. Entry lobbies, mechanical rooms and similar uses incidental to the operation of the building shall be permitted.

2. Group A having an assembly room with an occupant load of less than 300, or Group B or M shall be permitted in addition to those uses incidental to the operation of the building (including storage areas), provided that the entire structure below the horizontal assembly is protected throughout by an approved automatic sprinkler system.

5. The maximum building height in feet shall not exceed the limits set forth in Table 503 for the least restrictive type of construction involved.

508.3 Group S-2 enclosed parking garage with Group S-2 open parking garage above. A Group S-2 enclosed parking garage located in the basement or first story below a Group S-2 open parking garage shall be classified as a separate and distinct building for the purpose of determining the type of construction when the following conditions are met:

1. The allowable area of the structure shall be such that the sum of the ratios of the actual area divided by the allowable area for each separate occupancy shall not exceed 1.0.

2. The Group S-2 enclosed parking garage is of Type I or II construction and is at least equal to the fire-resistance requirements of the Group S-2 open parking garage.

3. The height and the number of the floors above the basement shall be limited as specified in Table 406.3.5.

4. The floor assembly separating the Group S-2 enclosed parking garage and Group S-2 open parking garage shall be protected as required for the floor assembly of the Group S-2 enclosed parking garage. Openings between the Group S-2 enclosed parking garage and Group S-2 open parking garage, except exit openings, shall not be required to be protected.

5. The Group S-2 enclosed parking garage is used exclusively for the parking or storage of private motor vehicles, but shall be permitted to contain an office, waiting room and toilet room having a total area of not more than 1,000 square feet (93 m^2), and mechanical equipment rooms incidental to the operation of the building.

508.4 Parking beneath Group R. Where a maximum one-story above grade plane Group S-2 parking garage, enclosed or open, or combination thereof, of Type I construction or open of Type IV construction, with grade entrance, is provided under a building of Group R, the number of stories to be used in determining the minimum type of construction shall be measured from the floor above such a parking area. The floor assembly between the parking garage and the Group R above shall comply with the type of construction required for the parking garage and shall also provide a fire-resistance rating not less than the mixed occupancy separation required in Section 302.3.2.

508.5 Group R-2 buildings of Type IIIA construction. The height limitation for buildings of Type IIIA construction in Group R-2 shall be increased to six stories and 75 feet (22 860 mm) where the first-floor construction above the basement has a fire-resistance rating of not less than 3 hours and the floor area is subdivided by 2-hour fire-resistance-rated fire walls into areas of not more than 3,000 square feet (279 m^2).

508.6 Group R-2 buildings of Type IIA construction. The height limitation for buildings of Type IIA construction in Group R-2 shall be increased to nine stories and 100 feet (30 480 mm) where the building is separated by not less than 50 feet (15 240 mm) from any other building on the lot and from property lines, the exits are segregated in an area enclosed by a 2-hour fire-resistance-rated fire wall and the first-floor construction has a fire-resistance rating of not less than 1$^1/_2$ hours.

508.7 Open parking garage beneath Groups A, I, B, M and R. Open parking garages constructed under Groups A, I, B, M and R shall not exceed the height and area limitations permitted under Section 406.3. The height and area of the portion of the building above the open parking garage shall not exceed the limitations in Section 503 for the upper occupancy. The height, in both feet and stories, of the portion of the building above the open parking garage shall be measured from grade plane and shall include both the open parking garage and the portion of the building above the parking garage.

508.7.1 Fire separation. Fire separation assemblies between the parking occupancy and the upper occupancy shall correspond to the required fire-resistance rating prescribed in Table 302.3.2 for the uses involved. The type of construction shall apply to each occupancy individually, except that structural members, including main bracing within the open parking structure, which is necessary to support the upper occupancy, shall be protected with the more restrictive fire-resistance-rated assemblies of the groups involved as shown in Table 601. Means of egress for the upper occupancy shall conform to Chapter 10 and shall be separated from the parking occupancy by fire barriers having at least a 2-hour fire-resistance rating as required by Section 706, with self-closing doors complying with Section 715. Means of egress from the open parking garage shall comply with Section 406.3.

CHAPTER 6

TYPES OF CONSTRUCTION

SECTION 601
GENERAL

601.1 Scope. The provisions of this chapter shall control the classification of buildings as to type of construction.

SECTION 602
CONSTRUCTION CLASSIFICATION

602.1 General. Buildings and structures erected or to be erected, altered or extended in height or area shall be classified in one of the five construction types defined in Sections 602.2 through 602.5. The building elements shall have a fire-resistance rating not less than that specified in Table 601 and exterior walls shall have a fire-resistance rating not less than that specified in Table 602.

602.1.1 Minimum requirements. A building or portion thereof shall not be required to conform to the details of a type of construction higher than that type, which meets the minimum requirements based on occupancy even though certain features of such a building actually conform to a higher type of construction.

602.2 Types I and II. Type I and II construction are those types of construction in which the building elements listed in Table 601 are of noncombustible materials.

602.3 Type III. Type III construction is that type of construction in which the exterior walls are of noncombustible materials and the interior building elements are of any material permitted by this code. Fire-retardant-treated wood framing complying with Section 2303.2 shall be permitted within exterior wall assemblies of a 2-hour rating or less.

602.4 Type IV. Type IV construction (Heavy Timber, HT) is that type of construction in which the exterior walls are of noncombustible materials and the interior building elements are of solid or laminated wood without concealed spaces. The details of Type IV construction shall comply with the provisions of this section. Fire-retardant-treated wood framing complying with Section 2303.2 shall be permitted within exterior wall assemblies with a 2-hour rating or less.

602.4.1 Columns. Wood columns shall be sawn or glued laminated and shall not be less than 8 inches (203 mm), nominal, in any dimension where supporting floor loads and not less than 6 inches (152 mm) nominal in width and not less than 8 inches (203 mm) nominal in depth where supporting roof and ceiling loads only. Columns shall be continuous or superimposed and connected in an approved manner.

602.4.2 Floor framing. Wood beams and girders shall be of sawn or glued-laminated timber and shall be not less than 6 inches (152 mm) nominal in width and not less than 10 inches (254 mm) nominal in depth. Framed sawn or glued-laminated timber arches, which spring from the floor line and support floor loads, shall be not less than 8 inches (203 mm) nominal in any dimension. Framed timber trusses supporting floor loads shall have members of not less than 8 inches (203 mm) nominal in any dimension.

602.4.3 Roof framing. Wood-frame or glued-laminated arches for roof construction, which spring from the floor line or from grade and do not support floor loads, shall have members not less than 6 inches (152 mm) nominal in width and have less than 8 inches (203 mm) nominal in depth for the lower half of the height and not less than 6 inches (152 mm) nominal in depth for the upper half. Framed or glued-laminated arches for roof construction that spring from the top of walls or wall abutments, framed timber trusses and other roof framing, which do not support floor loads, shall have members not less than 4 inches (102 mm) nominal in width and not less than 6 inches (152 mm) nominal in depth. Spaced members shall be permitted to be composed of two or more pieces not less than 3 inches (76 mm) nominal in thickness where blocked solidly throughout their intervening spaces or where spaces are tightly closed by a continuous wood cover plate of not less than 2 inches (51 mm) nominal in thickness secured to the underside of the members. Splice plates shall be not less than 3 inches (76 mm) nominal in thickness. Where protected by approved automatic sprinklers under the roof deck, framing members shall be not less than 3 inches (76 mm) nominal in width.

602.4.4 Floors. Floors shall be without concealed spaces. Wood floors shall be of sawn or glued-laminated planks, splined or tongue-and-groove, of not less than 3 inches (76 mm) nominal in thickness covered with 1-inch (25 mm) nominal dimension tongue-and-groove flooring, laid crosswise or diagonally, or 0.5-inch (12.7 mm) particleboard or planks not less than 4 inches (102 mm) nominal in width set on edge close together and well spiked and covered with 1-inch (25 mm) nominal dimension flooring or $^{15}/_{32}$-inch (12 mm) wood structural panel or 0.5-inch (12.7 mm) particleboard. The lumber shall be laid so that no continuous line of joints will occur except at points of support. Floors shall not extend closer than 0.5 inch (12.7 mm) to walls. Such 0.5-inch (12.7 mm) space shall be covered by a molding fastened to the wall and so arranged that it will not obstruct the swelling or shrinkage movements of the floor. Corbeling of masonry walls under the floor shall be permitted to be used in place of molding.

602.4.5 Roofs. Roofs shall be without concealed spaces and wood roof decks shall be sawn or glued laminated, splined or tongue-and-groove plank, not less than 2 inches (51 mm) thick, $1^1/_8$-inch-thick (32 mm) wood structural panel (exterior glue), or of planks not less than 3 inches (76 mm) nominal in width, set on edge close together and laid as required for floors. Other types of decking shall be permitted to be used if providing equivalent fire resistance and structural properties.

602.4.6 Partitions. Partitions shall be of solid wood construction formed by not less than two layers of 1-inch (25 mm) matched boards or laminated construction 4 inches (102 mm) thick, or of 1-hour fire-resistance-rated construction.

602.4.7 Exterior structural members. Where a horizontal separation of 20 feet (6096 mm) or more is provided, wood columns and arches conforming to heavy timber sizes shall be permitted to be used externally.

602.5 Type V. Type V construction is that type of construction in which the structural elements, exterior walls and interior walls are of any materials permitted by this code.

SECTION 603
COMBUSTIBLE MATERIAL IN TYPE I AND II CONSTRUCTION

603.1 Allowable materials. Combustible materials shall be permitted in buildings of Type I or II construction in the following applications and in accordance with Sections 603.1.1 through 603.1.3:

1. Fire-retardant-treated wood shall be permitted in:

    1.1. Nonbearing partitions where the required fire-resistance rating is 2 hours or less.

    1.2. Nonbearing exterior walls where no fire rating is required.

    1.3. Roof construction as permitted in Table 601, Note c, Item 3.

2. Thermal and acoustical insulation, other than foam plastics, having a flame spread index of not more than 25.

    Exceptions:

    1. Insulation placed between two layers of noncombustible materials without an intervening airspace shall be allowed to have a flame spread index of not more than 100.

    2. Insulation installed between a finished floor and solid decking without intervening airspace shall be allowed to have a flame spread index of not more than 200.

3. Foam plastics in accordance with Chapter 26.

4. Roof coverings that have an A, B or C classification.

5. Interior floor finish and interior finish, trim and millwork such as doors, door frames, window sashes and frames.

6. Where not installed over 15 feet (4572 mm) above grade, show windows, nailing or furring strips, wooden bulkheads below show windows, their frames, aprons and show cases.

7. Finished flooring applied directly to the floor slab or to wood sleepers that are firestopped in accordance with Section 717.2.7.

8. Partitions dividing portions of stores, offices or similar places occupied by one tenant only and which do not establish a corridor serving an occupant load of 30 or more shall be permitted to be constructed of fire-retardant-treated wood, 1-hour fire-resistance-rated construction or of wood panels or similar light construction up to 6 feet (1829 mm) in height.

9. Platforms as permitted in Section 410.

10. Combustible exterior wall coverings, balconies, bay or oriel windows, or similar appendages in accordance with Chapter 14.

11. Blocking such as for handrails, millwork, cabinets, and window and door frames.

12. Light-transmitting plastics as permitted by Chapter 26.

13. Mastics and caulking materials applied to provide flexible seals between components of exterior wall construction.

14. Exterior plastic veneer installed in accordance with Section 2605.2.

15. Nailing or furring strips as permitted by Section 803.4.

16. Heavy timber as permitted by Note c, Item 2, to Table 601 and Sections 602.4.7 and 1406.3.

17. Aggregates, component materials and admixtures as permitted by Section 703.2.2.

18. Sprayed cementitious and mineral fiber fire-resistance-rated materials installed to comply with Section 1704.11.

19. Materials used to protect penetrations in fire-resistance-rated assemblies in accordance with Section 712.

20. Materials used to protect joints in fire-resistance-rated assemblies in accordance with Section 713.

21. Materials allowed in the concealed spaces of buildings of Type I and II construction in accordance with Section 717.5.

22. Materials exposed within plenums complying with Section 602 of the *International Mechanical Code.*

603.1.1 Ducts. The use of nonmetallic ducts shall be permitted when installed in accordance with the limitations of the *International Mechanical Code.*

603.1.2 Piping. The use of combustible piping materials shall be permitted when installed in accordance with the limitations of the *International Mechanical Code* and the *International Plumbing Code.*

603.1.3 Electrical. The use of electrical wiring methods with combustible insulation, tubing, raceways and related components shall be permitted when installed in accordance with the limitations of the ICC *Electrical Code.*

TABLE 601
FIRE-RESISTANCE RATING REQUIREMENTS FOR BUILDING ELEMENTS (hours)

BUILDING ELEMENT	TYPE I		TYPE II		TYPE III		TYPE IV	TYPE V	
	A	B	A[d]	B	A[d]	B	HT	A[d]	B
Structural frame[a] Including columns, girders, trusses	3[b]	2[b]	1	0	1	0	HT	1	0
Bearing walls Exterior[f] Interior	3 3[b]	2 2[b]	1 1	0 0	2 1	2 0	2 1/HT	1 1	0 0
Nonbearing walls and partitions Exterior	See Table 602								
Nonbearing walls and partitions Interior[e]	0	0	0	0	0	0	See Section 602.4.6	0	0
Floor construction Including supporting beams and joists	2	2	1	0	1	0	HT	1	0
Roof construction Including supporting beams and joists	$1^1/_2$[c]	1[c]	1[c]	0	1[c]	0	HT	1[c]	0

For SI: 1 foot = 304.8 mm.

a. The structural frame shall be considered to be the columns and the girders, beams, trusses and spandrels having direct connections to the columns and bracing members designed to carry gravity loads. The members of floor or roof panels which have no connection to the columns shall be considered secondary members and not a part of the structural frame.

b. Roof supports: Fire-resistance ratings of structural frame and bearing walls are permitted to be reduced by 1 hour where supporting a roof only.

c. 1. Except in Factory-Industrial (F-1), Hazardous (H), Mercantile (M) and Moderate-Hazard Storage (S-1) occupancies, fire protection of structural members shall not be required, including protection of roof framing and decking where every part of the roof construction is 20 feet or more above any floor immediately below. Fire-retardant-treated wood members shall be allowed to be used for such unprotected members.

   2. In all occupancies, heavy timber shall be allowed where a 1-hour or less fire-resistance rating is required.

   3. In Type I and II construction, fire-retardant-treated wood shall be allowed in buildings including girders and trusses as part of the roof construction when the building is:
     i. Two stories or less in height;
     ii. Type II construction over two stories; or
     iii. Type I construction over two stories and the vertical distance from the upper floor to the roof is 20 feet or more.

d. An approved automatic sprinkler system in accordance with Section 903.3.1.1 shall be allowed to be substituted for 1-hour fire-resistance-rated construction, provided such system is not otherwise required by other provisions of the code or used for an allowable area increase in accordance with Section 506.3 or an allowable height increase in accordance with Section 504.2. The 1-hour substitution for the fire resistance of exterior walls shall not be permitted.

e. Not less than the fire-resistance rating required by other sections of this code.

f. Not less than the fire-resistance rating based on fire separation distance (see Table 602).

TABLE 602
FIRE-RESISTANCE RATING REQUIREMENTS FOR EXTERIOR WALLS BASED ON FIRE SEPARATION DISTANCE[a]

FIRE SEPARATION DISTANCE (feet)	TYPE OF CONSTRUCTION	GROUP H	GROUP F-1, M, S-1	GROUP A, B, E, F-2, I, R[b], S-2, U
< 5[c]	All	3	2	1
≥ 5 < 10	IA Others	3 2	2 1	1 1
≥ 10 < 30	IA, IB IIB, VB Others	2 1 1	1 0 1	1 0 1
≥ 30	All	0	0	0

For SI: 1 foot = 304.8 mm.

a. Load-bearing exterior walls shall also comply with the fire-resistance rating requirements of Table 601.

b. Group R-3 and Group U when used as accessory to Group R-3, as applicable in Section 101.2 shall not be required to have a fire-resistance rating where the fire separation distance is 3 feet or more.

c. See Section 503.2 for party walls.

CHAPTER 7

FIRE-RESISTANCE-RATED CONSTRUCTION

SECTION 701
GENERAL

701.1 Scope. The provisions of this chapter shall govern the materials and assemblies used for structural fire resistance and fire-resistance-rated construction separation of adjacent spaces to safeguard against the spread of fire and smoke within a building and the spread of fire to or from buildings.

SECTION 702
DEFINITIONS

702.1 Definitions. The following words and terms shall, for the purposes of this chapter, and as used elsewhere in this code, have the meanings shown herein.

ANNULAR SPACE. The opening around the penetrating item.

CEILING RADIATION DAMPER. A listed device installed in a ceiling membrane of a fire-resistance-rated floor/ceiling or roof/ceiling assembly to limit automatically the radiative heat transfer through an air inlet/outlet opening.

COMBINATION FIRE/SMOKE DAMPER. A listed device installed in ducts and air transfer openings designed to close automatically upon the detection of heat and to also resist the passage of air and smoke. The device is installed to operate automatically, controlled by a smoke detection system, and where required, is capable of being positioned from a remote command station.

DAMPER. See "Ceiling radiation damper," "Combination fire/smoke damper," "Fire damper" and "Smoke damper."

DRAFTSTOP. A material, device or construction installed to restrict the movement of air within open spaces of concealed areas of building components such as crawl spaces, floor/ceiling assemblies, roof/ceiling assemblies and attics.

F RATING. The time period that the through-penetration firestop system limits the spread of fire through the penetration when tested in accordance with ASTM E 814.

FIRE AREA. The aggregate floor area enclosed and bounded by fire walls, fire barriers, exterior walls or fire-resistance-rated horizontal assemblies of a building.

FIRE BARRIER. A fire-resistance-rated vertical or horizontal assembly of materials designed to restrict the spread of fire in which openings are protected.

FIRE DAMPER. A listed device, installed in ducts and air transfer openings of an air distribution system or smoke control system, designed to close automatically upon detection of heat, to interrupt migratory airflow, and to restrict the passage of flame. Fire dampers are classified for use in either static systems that will automatically shut down in the event of a fire, or in dynamic systems that continue to operate during a fire. A dynamic fire damper is tested and rated for closure under airflow.

FIRE DOOR. The door component of a fire door assembly.

FIRE DOOR ASSEMBLY. Any combination of a fire door, frame, hardware, and other accessories that together provide a specific degree of fire protection to the opening.

FIRE PARTITION. A vertical assembly of materials designed to restrict the spread of fire in which openings are protected.

FIRE PROTECTION RATING. The period of time that an opening protective assembly will maintain the ability to confine a fire as determined by tests prescribed in Section 715. Ratings are stated in hours or minutes.

FIRE RESISTANCE. That property of materials or their assemblies that prevents or retards the passage of excessive heat, hot gases or flames under conditions of use.

FIRE-RESISTANCE RATING. The period of time a building element, component or assembly maintains the ability to confine a fire, continues to perform a given structural function, or both, as determined by the tests, or the methods based on tests, prescribed in Section 703.

FIRE-RESISTANT JOINT SYSTEM. An assemblage of specific materials or products that are designed, tested, and fire-resistance rated in accordance with either ASTM E 1966 or UL 2079 to resist for a prescribed period of time the passage of fire through joints made in or between fire-resistance-rated assemblies.

FIRE SEPARATION DISTANCE. The distance measured from the building face to the closest interior lot line, to the centerline of a street, alley or public way, or to an imaginary line between two buildings on the lot. The distance shall be measured at right angles from the face of the wall.

FIRE WALL. A fire-resistance-rated wall having protected openings, which restricts the spread of fire and extends continuously from the foundation to or through the roof, with sufficient structural stability under fire conditions to allow collapse of construction on either side without collapse of the wall.

FIRE WINDOW ASSEMBLY. A window constructed and glazed to give protection against the passage of fire.

FIREBLOCKING. Building materials installed to resist the free passage of flame to other areas of the building through concealed spaces.

FLOOR FIRE DOOR ASSEMBLY. A combination of a fire door, a frame, hardware and other accessories installed in a horizontal plane, which together provide a specific degree of fire protection to a through opening in a fire-resistance-rated floor (see Section 712.4.6).

JOINT. The linear opening in or between adjacent fire-resistance-rated assemblies that is designed to allow independent movement of the building in any plane caused by thermal, seismic, wind or any other loading.

MEMBRANE PENETRATION. An opening made through one side (wall, floor or ceiling membrane) of an assembly.

MEMBRANE-PENETRATION FIRESTOP. A material, device or construction installed to resist for a prescribed time period the passage of flame and heat through openings in a protective membrane in order to accommodate cables, cable trays, conduit, tubing, pipes or similar items.

PENETRATION FIRESTOP. A through-penetration firestop or a membrane-penetration firestop.

SELF-CLOSING. As applied to a fire door or other opening, means equipped with an approved device that will ensure closing after having been opened.

SHAFT. An enclosed space extending through one or more stories of a building, connecting vertical openings in successive floors, or floors and roof.

SHAFT ENCLOSURE. The walls or construction forming the boundaries of a shaft.

SMOKE BARRIER. A continuous membrane, either vertical or horizontal, such as a wall, floor, or ceiling assembly, that is designed and constructed to restrict the movement of smoke.

SMOKE COMPARTMENT. A space within a building enclosed by smoke barriers on all sides, including the top and bottom.

SMOKE DAMPER. A listed device installed in ducts and air transfer openings that is designed to resist the passage of air and smoke. The device is installed to operate automatically, controlled by a smoke detection system, and where required, is capable of being positioned from a remote command station.

SPLICE. The result of a factory and/or field method of joining or connecting two or more lengths of a fire-resistant joint system into a continuous entity.

T RATING. The time period that the penetration firestop system, including the penetrating item, limits the maximum temperature rise to 325°F (163°C) above its initial temperature through the penetration on the nonfire side when tested in accordance with ASTM E 814.

THROUGH PENETRATION. An opening that passes through an entire assembly.

THROUGH-PENETRATION FIRESTOP SYSTEM. An assemblage of specific materials or products that are designed, tested and fire-resistance rated to resist for a prescribed period of time the spread of fire through penetrations. The F and T rating criteria for penetration firestop systems shall be in accordance with ASTM E 814. See definitions of "F rating" and "T rating."

SECTION 703
FIRE-RESISTANCE RATINGS AND FIRE TESTS

703.1 Scope. Materials prescribed herein for fire resistance shall conform to the requirements of this chapter.

703.2 Fire-resistance ratings. The fire-resistance rating of building elements shall be determined in accordance with the test procedures set forth in ASTM E 119 or in accordance with Section 703.3. Where materials, systems or devices that have not been tested as part of a fire-resistance-rated assembly are incorporated into the assembly, sufficient data shall be made available to the building official to show that the required fire-resistance rating is not reduced. Materials and methods of construction used to protect joints and penetrations in fire-resistance-rated building elements shall not reduce the required fire-resistance rating.

Exception: In determining the fire-resistance rating of exterior bearing walls, compliance with the ASTM E 119 criteria for unexposed surface temperature rise and ignition of cotton waste due to passage of flame or gases is required only for a period of time corresponding to the required fire-resistance rating of an exterior nonbearing wall with the same fire separation distance, and in a building of the same group. When the fire-resistance rating determined in accordance with this exception exceeds the fire-resistance rating determined in accordance with ASTM E 119, the fire exposure time period, water pressure, and application duration criteria for the hose stream test of ASTM E 119 shall be based upon the fire-resistance rating determined in accordance with this exception.

703.2.1 Nonsymmetrical wall construction. Interior walls and partitions of nonsymmetrical construction shall be tested with both faces exposed to the furnace, and the assigned fire-resistance rating shall be the shortest duration obtained from the two tests conducted in compliance with ASTM E 119. When evidence is furnished to show that the wall was tested with the least fire-resistant side exposed to the furnace, subject to acceptance of the building official, the wall need not be subjected to tests from the opposite side (see Section 704.5 for exterior walls).

703.2.2 Combustible components. Combustible aggregates are permitted in gypsum and portland cement concrete mixtures approved for fire-resistance-rated construction. Any approved component material or admixture is permitted in assemblies if the resulting tested assembly meets the fire-resistance test requirements of this code.

703.2.3 Restrained classification. Fire-resistance-rated assemblies tested under ASTM E 119 shall not be considered to be restrained unless evidence satisfactory to the building official is furnished by the registered design professional showing that the construction qualifies for a restrained classification in accordance with ASTM E 119. Restrained construction shall be identified on the plans.

703.3 Alternative methods for determining fire resistance. The application of any of the alternative methods listed in this section shall be based on the fire exposure and acceptance criteria specified in ASTM E 119. The required fire resistance of a building element shall be permitted to be established by any of the following methods or procedures:

1. Fire-resistance designs documented in approved sources.
2. Prescriptive designs of fire-resistance-rated building elements as prescribed in Section 720.
3. Calculations in accordance with Section 721.

4. Engineering analysis based on a comparison of building element designs having fire-resistance ratings as determined by the test procedures set forth in ASTM E 119.

5. Alternative protection methods as allowed by Section 104.11.

703.4 Noncombustibility tests. The tests indicated in Sections 703.4.1 and 703.4.2 shall serve as criteria for acceptance of building materials as set forth in Sections 602.2, 602.3 and 602.4 in Type I, II, III and IV construction. The term "noncombustible" does not apply to the flame spread characteristics of interior finish or trim materials. A material shall not be classified as a noncombustible building construction material if it is subject to an increase in combustibility or flame spread beyond the limitations herein established through the effects of age, moisture or other atmospheric conditions.

703.4.1 Elementary materials. Materials required to be noncombustible shall be tested in accordance with ASTM E 136.

703.4.2 Composite materials. Materials having a structural base of noncombustible material as determined in accordance with Section 703.4.1 with a surfacing not more than 0.125 inch (3.18 mm) thick that has a flame spread index not greater than 50 when tested in accordance with ASTM E 84 shall be acceptable as noncombustible materials.

SECTION 704
EXTERIOR WALLS

704.1 General. Exterior walls shall be fire-resistance rated and have opening protection as required by this section.

704.2 Projections. Cornices, eave overhangs, exterior balconies and similar architectural appendages extending beyond the floor area shall conform to the requirements of this section and Section 1406. Exterior egress balconies and exterior exit stairways shall also comply with Sections 1013.5 and 1022.1. Projections shall not extend beyond the distance determined by the following two methods, whichever results in the lesser projection:

1. A point one-third the distance to the lot line from an assumed vertical plane located where protected openings are required in accordance with Section 704.8.

2. More than 12 inches (305 mm) into areas where openings are prohibited.

704.2.1 Type I and II construction. Projections from walls of Type I or II construction shall be of noncombustible materials or combustible materials as allowed by Sections 1406.3 and 1406.4.

704.2.2 Type III, IV or V construction. Projections from walls of Type III, IV or V construction shall be of any approved material.

704.2.3 Combustible projections. Combustible projections located where openings are not permitted or where protection of openings is required shall be of at least 1-hour fire-resistance-rated construction, Type IV construction or as required by Section 1406.3.

Exception: Type V construction shall be allowed for R-3 occupancies, as applicable in Section 101.2.

704.3 Buildings on the same lot. For the purposes of determining the required wall and opening protection and roof-covering requirements, buildings on the same lot shall be assumed to have an imaginary line between them.

Where a new building is to be erected on the same lot as an existing building, the location of the assumed imaginary line with relation to the existing building shall be such that the exterior wall and opening protection of the existing building meet the criteria as set forth in Sections 704.5 and 704.8.

Exception: Two or more buildings on the same lot shall either be regulated as separate buildings or shall be considered as portions of one building if the aggregate area of such buildings is within the limits specified in Chapter 5 for a single building. Where the buildings contain different occupancy groups or are of different types of construction, the area shall be that allowed for the most restrictive occupancy or construction.

704.4 Materials. Exterior walls shall be of materials permitted by the building type of construction.

704.5 Fire-resistance ratings. Exterior walls shall be fire-resistance rated in accordance with Tables 601 and 602. The fire-resistance rating of exterior walls with a fire separation distance of greater than 5 feet (1524 mm) shall be rated for exposure to fire from the inside. The fire-resistance rating of exterior walls with a fire separation distance of 5 feet (1524 mm) or less shall be rated for exposure to fire from both sides.

704.6 Structural stability. The wall shall extend to the height required by Section 704.11 and shall have sufficient structural stability such that it will remain in place for the duration of time indicated by the required fire-resistance rating.

704.7 Unexposed surface temperature. Where protected openings are not limited by Section 704.8, the limitation on the rise of temperature on the unexposed surface of exterior walls as required by ASTM E 119 shall not apply. Where protected openings are limited by Section 704.8, the limitation on the rise of temperature on the unexposed surface of exterior walls as required by ASTM E 119 shall not apply provided that a correction is made for radiation from the unexposed exterior wall surface in accordance with the following formula:

$$A_e = A + (A_f \times F_{eo}) \qquad \text{(Equation 7-1)}$$

where:

A_e = Equivalent area of protected openings.

A = Actual area of protected openings.

A_f = Area of exterior wall surface in the story under consideration exclusive of openings, on which the temperature limitations of ASTM E 119 for walls are exceeded.

F_{eo} = An "equivalent opening factor" derived from Figure 704.7 based on the average temperature of the unexposed wall surface and the fire-resistance rating of the wall.

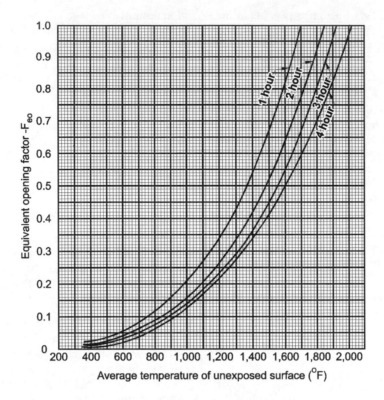

For SI: °C = [(°F) - 32] / 1.8.

FIGURE 704.7
EQUIVALENT OPENING FACTOR

704.8 Allowable area of openings. The maximum area of unprotected or protected openings permitted in an exterior wall in any story shall not exceed the values set forth in Table 704.8. Where both unprotected and protected openings are located in the exterior wall in any story, the total area of the openings shall comply with the following formula:

$$\frac{A}{a} + \frac{A_u}{a_u} \leq 1.0 \qquad \textbf{(Equation 7-2)}$$

where:

A = Actual area of protected openings, or the equivalent area of protected openings, A_e (see Section 704.7).

a = Allowable area of protected openings.

A_u = Actual area of unprotected openings.

a_u = Allowable area of unprotected openings.

704.8.1 Automatic sprinkler system. In buildings equipped throughout with an automatic sprinkler system in accordance with Section 903.3.1.1, the maximum allowable area of unprotected openings in occupancies other than Groups H-1, H-2 and H-3 shall be the same as the tabulated limitations for protected openings.

704.8.2 First story. In occupancies other than Group H, unlimited unprotected openings are permitted in the first story of exterior walls facing a street that have a fire separation distance of greater than 15 feet (4572 mm), or facing an unoccupied space. The unoccupied space shall be on the same lot or dedicated for public use, shall not be less than 30 feet (9144 mm) in width, and shall have access from a street by a posted fire lane in accordance with the *International Fire Code.*

704.9 Vertical separation of openings. Openings in exterior walls in adjacent stories shall be separated vertically to protect against fire spread on the exterior of the buildings where the openings are within 5 feet (1524 mm) of each other horizontally and the opening in the lower story is not a protected opening in accordance with Section 715.4.8. Such openings shall be separated vertically at least 3 feet (914 mm) by spandrel girders, exterior walls or other similar assemblies that have a fire-resistance rating of at least 1 hour or by flame barriers that extend horizontally at least 30 inches (762 mm) beyond the exterior wall. Flame barriers shall also have a fire-resistance rating of at least 1 hour. The unexposed surface temperature limitations specified in ASTM E 119 shall not apply to the flame barriers or vertical separation unless otherwise required by the provisions of this code.

Exceptions:

1. This section shall not apply to buildings that are three stories or less in height.

2. This section shall not apply to buildings equipped throughout with an automatic sprinkler system in accordance with Section 903.3.1.1 or 903.3.1.2.

3. Open parking garages.

TABLE 704.8
MAXIMUM AREA OF EXTERIOR WALL OPENINGS[a]

CLASSIFICATION OF OPENING	FIRE SEPARATION DISTANCE (feet)							
	0 to 3[e,h]	Greater than 3 to 5[b]	Greater than 5 to 10[d,f]	Greater than 10 to 15[c,d,f]	Greater than 15 to 20[c,f]	Greater than 20 to 25[c,f]	Greater than 25 to 30[c,f]	Greater than 30
Unprotected	Not Permitted[g]	Not Permitted[b,g]	10%[g]	15%[g]	25%[g]	45%[g]	70%[g]	No Limit
Protected	Not Permitted	15%	25%	45%	75%	No Limit	No Limit	No Limit

For SI: 1 foot = 304.8 mm.

a. Values given are percentage of the area of the exterior wall.

b. For occupancies in Group R-3, as applicable in Section 101.2, the maximum percentage of unprotected and protected exterior wall openings shall be 25 percent.

c. The area of openings in an open parking structure with a fire separation distance of greater than 10 feet shall not be limited.

d. For occupancies in Group H-2 or H-3, unprotected openings shall not be permitted for openings with a fire separation distance of 15 feet or less.

e. For requirements for fire walls for buildings with differing roof heights, see Section 705.6.1.

f. The area of unprotected and protected openings is not limited for occupancies in Group R-3, as applicable in Section 101.2, with a fire separation distance greater than 5 feet.

g. Buildings whose exterior bearing wall, exterior nonbearing wall and exterior structural frame are not required to be fire-resistance rated shall be permitted to have unlimited unprotected openings.

h. Includes accessory buildings to Group R-3 as applicable in Section 101.2.

704.10 Vertical exposure. For buildings on the same lot, approved protectives shall be provided in every opening that is less than 15 feet (4572 mm) vertically above the roof of an adjoining building or adjacent structure that is within a horizontal fire separation distance of 15 feet (4572 mm) of the wall in which the opening is located.

Exception: Opening protectives are not required where the roof construction has a fire-resistance rating of not less than 1 hour for a minimum distance of 10 feet (3048 mm) from the adjoining building and the entire length and span of the supporting elements for the fire-resistance-rated roof assembly has a fire-resistance rating of not less than 1 hour.

704.11 Parapets. Parapets shall be provided on exterior walls of buildings.

Exceptions: A parapet need not be provided on an exterior wall where any of the following conditions exist:

1. The wall is not required to be fire-resistance rated in accordance with Table 602 because of fire separation distance.

2. The building has an area of not more than 1,000 square feet (93 m²) on any floor.

3. Walls that terminate at roofs of not less than 2-hour fire-resistance-rated construction or where the roof, including the deck and supporting construction, is constructed entirely of noncombustible materials.

4. One-hour fire-resistance-rated exterior walls that terminate at the underside of the roof sheathing, deck or slab, provided:

    4.1. Where the roof/ceiling framing elements are parallel to the walls, such framing and elements supporting such framing shall not be of less than 1-hour fire-resistance-rated construction for a width of 4 feet (1220 mm) measured from the interior side of the wall for Groups R and U and 10 feet (3048 mm) for other occupancies.

    4.2. Where roof/ceiling framing elements are not parallel to the wall, the entire span of such framing and elements supporting such framing shall not be of less than 1-hour fire-resistance-rated construction.

    4.3. Openings in the roof shall not be located within 5 feet (1524 mm) of the 1-hour fire-resistance-rated exterior wall for Groups R and U and 10 feet (3048 mm) for other occupancies.

    4.4. The entire building shall be provided with not less than a Class B roof covering.

5. In occupancies of Groups R-2 and R-3 as applicable in Section 101.2, both provided with a Class C roof covering, the exterior wall shall be permitted to terminate at the roof sheathing or deck in Type III, IV and V construction provided:

    5.1. The roof sheathing or deck is constructed of approved noncombustible materials or of fire-retardant-treated wood, for a distance of 4 feet (1220 mm); or

    5.2. The roof is protected with 0.625-inch (15.88 mm) Type X gypsum board directly beneath the underside of the roof sheathing or deck, supported by a minimum of nominal 2-inch (51 mm) ledgers attached to the sides of the roof framing members, for a minimum distance of 4 feet (1220 mm).

6. Where the wall is permitted to have at least 25 percent of the exterior wall areas containing unprotected openings based on fire separation distance as determined in accordance with Section 704.8.

704.11.1 Parapet construction. Parapets shall have the same fire-resistance rating as that required for the supporting wall, and on any side adjacent to a roof surface, shall have noncombustible faces for the uppermost 18 inches (457 mm), including counterflashing and coping materials.

The height of the parapet shall not be less than 30 inches (762 mm) above the point where the roof surface and the wall intersect. Where the roof slopes toward a parapet at a slope greater than two units vertical in 12 units horizontal (16.7-percent slope), the parapet shall extend to the same height as any portion of the roof within a fire separation distance where protection of wall openings is required, but in no case shall the height be less than 30 inches (762 mm).

704.12 Opening protection. Windows required to be protected in accordance with Section 704.8, 704.9, or 704.10 shall comply with Section 715.4.8. Other openings required to be protected with fire doors or shutters in accordance with Sections 704.8, 704.9 and 704.10 shall comply with Section 715.3.

> **Exception:** Fire protective assemblies are not required where the building is protected throughout by an automatic sprinkler system and the exterior openings are protected by an approved water curtain using automatic sprinklers approved for that use. The sprinklers and the water curtain shall be installed in accordance with NFPA 13.

704.12.1 Unprotected openings. Where protected openings are not required by Section 704, windows and doors shall be constructed of any approved materials. Glazing shall conform to the requirements of Chapters 24 and 26.

704.13 Joints. Joints made in or between exterior walls required by this section to have a fire-resistance rating shall comply with Section 713.

> **Exception:** Joints in exterior walls that are permitted to have unprotected openings.

704.13.1 Voids. The void created at the intersection of a floor/ceiling assembly and an exterior curtain wall assembly shall be protected in accordance with Section 713.4.

704.14 Ducts and air transfer openings. Penetrations by air ducts and air transfer openings in fire-resistance-rated exterior walls required to have protected openings shall comply with Section 716.

> **Exception:** Foundation vents installed in accordance with this code are permitted.

SECTION 705
FIRE WALLS

705.1 General. Each portion of a building separated by one or more fire walls that comply with the provisions of this section shall be considered a separate building. The extent and location of such fire walls shall provide a complete separation. Where a fire wall also separates groups that are required to be separated by a fire barrier wall, the most restrictive requirements of each separation shall apply. Fire walls located on lot lines shall also comply with Section 503.2. Such fire walls (party walls) shall be constructed without openings.

705.2 Structural stability. Fire walls shall have sufficient structural stability under fire conditions to allow collapse of construction on either side without collapse of the wall for the duration of time indicated by the required fire-resistance rating.

705.3 Materials. Fire walls shall be of any approved noncombustible materials.

> **Exception:** Buildings of Type V construction.

705.4 Fire-resistance rating. Fire walls shall have a fire-resistance rating of not less than that required by Table 705.4.

TABLE 705.4
FIRE WALL FIRE-RESISTANCE RATINGS

GROUP	FIRE-RESISTANCE RATING (hours)
A, B, E, H-4, I, R-1, R-2, U	3[a]
F-1, H-3[b], H-5, M, S-1	3
H-1, H-2	4[b]
F-2, S-2, R-3, R-4	2

a. Walls shall be not less than 2-hour fire-resistance rated where separating buildings of Type II or V construction.
b. For Group H-1, H-2 or H-3 buildings, also see Sections 415.4 and 415.5.

705.5 Horizontal continuity. Fire walls shall be continuous from exterior wall to exterior wall and shall extend at least 18 inches (457 mm) beyond the exterior surface of exterior walls.

> **Exceptions:**
>
> 1. Fire walls shall be permitted to terminate at the interior surface of combustible exterior sheathing or siding provided the exterior wall has a fire-resistance rating of at least 1 hour for a horizontal distance of at least 4 feet (1220 mm) on both sides of the fire wall. Openings within such exterior walls shall be protected by fire assemblies having a fire protection rating of not less than $^3/_4$ hour.
>
> 2. Fire walls shall be permitted to terminate at the interior surface of noncombustible exterior sheathing, exterior siding or other noncombustible exterior finishes provided the sheathing, siding, or other exterior noncombustible finish extends a horizontal distance of at least 4 feet (1220 mm) on both sides of the fire wall.
>
> 3. Fire walls shall be permitted to terminate at the interior surface of noncombustible exterior sheathing where the building on each side of the fire wall is protected by an automatic sprinkler system installed in accordance with Section 903.3.1.1 or 903.3.1.2.

705.5.1 Exterior walls. Where the fire wall intersects the exterior walls, the fire-resistance rating for the exterior walls on both sides of the fire wall shall have a 1-hour fire-resistance rating with $^3/_4$-hour opening protection where opening protection is required. The fire-resistance rating of the exterior wall shall extend a minimum of 4 feet (1220 mm) on each side of the intersection of the fire wall to exterior wall. Exterior wall intersections at fire walls that form an angle equal to or greater than 180 degrees (3.14 rad) do not need exterior wall protection.

705.5.2 Horizontal projecting elements. Fire walls shall extend to the outer edge of horizontal projecting elements such as balconies, roof overhangs, canopies, marquees and architectural projections that are within 4 feet (1220 mm) of the fire wall.

Exceptions:

1. Horizontal projecting elements without concealed spaces provided the exterior wall behind and below the projecting element has not less than 1-hour fire-resistance-rated construction for a distance not less than the depth of the projecting element on both sides of the fire wall. Openings within such exterior walls shall be protected by fire assemblies having a fire protection rating of not less than $^3/_4$ hour.

2. Noncombustible horizontal projecting elements with concealed spaces, provided a minimum 1-hour fire-resistance-rated wall extends through the concealed space. The projecting element shall be separated from the building by a minimum of 1-hour fire-resistance-rated construction for a distance on each side of the fire wall equal to the depth of the projecting element. The wall is not required to extend under the projecting element where the building exterior wall is not less than 1-hour fire-resistance rated for a distance on each side of the fire wall equal to the depth of the projecting element. Openings within such exterior walls shall be protected by fire assemblies having a fire protection rating of not less than $^3/_4$ hour.

3. For combustible horizontal projecting elements with concealed spaces, the fire wall need only extend through the concealed space to the outer edges of the projecting elements. The exterior wall behind and below the projecting element shall be of not less than 1-hour fire-resistance-rated construction for a distance not less than the depth of the projecting elements on both sides of the fire wall. Openings within such exterior walls shall be protected by fire assemblies having a fire-protection rating of not less than $^3/_4$ hour.

705.6 Vertical continuity. Fire walls shall extend from the foundation to a termination point at least 30 inches (762 mm) above both adjacent roofs.

Exceptions:

1. Stepped buildings in accordance with Section 705.6.1.

2. Two-hour fire-resistance-rated walls shall be permitted to terminate at the underside of the roof sheathing, deck or slab provided:

    2.1. The lower roof assembly within 4 feet (1220 mm) of the wall has not less than a 1-hour fire-resistance rating and the entire length and span of supporting elements for the rated roof assembly has a fire-resistance rating of not less than 1 hour.

    2.2. Openings in the roof shall not be located within 4 feet (1220 mm) of the fire wall.

    2.3. Each building shall be provided with not less than a Class B roof covering.

3. Walls shall be permitted to terminate at the underside of noncombustible roof sheathing, deck, or slabs where both buildings are provided with not less than a Class B roof covering. Openings in the roof shall not be located within 4 feet (1220 mm) of the fire wall.

4. In buildings of Type III, IV and V construction, walls shall be permitted to terminate at the underside of combustible roof sheathing or decks provided:

    4.1. There are no openings in the roof within 4 feet (1220 mm) of the fire wall,

    4.2. The roof is covered with a minimum Class B roof covering, and

    4.3. The roof sheathing or deck is constructed of fire-retardant-treated wood for a distance of 4 feet (1220 mm) on both sides of the wall or the roof is protected with $^5/_8$ inch (15.9 mm) Type X gypsum board directly beneath the underside of the roof sheathing or deck, supported by a minimum of 2-inch (51 mm) nominal ledgers attached to the sides of the roof framing members for a minimum distance of 4 feet (1220 mm) on both sides of the fire wall.

5. Buildings located above a parking garage designed in accordance with Section 508.2 shall be permitted to have the fire walls for the buildings located above the parking garage extend from the horizontal separation between the parking garage and the buildings.

705.6.1 Stepped buildings. Where a fire wall serves as an exterior wall for a building and separates buildings having different roof levels, such wall shall terminate at a point not less than 30 inches (762 mm) above the lower roof level, provided the exterior wall for a height of 15 feet (4572 mm) above the lower roof is not less than 1-hour fire-resistance-rated construction from both sides with openings protected by assemblies having a $^3/_4$-hour fire protection rating.

Exception: Where the fire wall terminates at the underside of the roof sheathing, deck or slab of the lower roof, provided:

1. The lower roof assembly within 10 feet (3048 mm) of the wall has not less than a 1-hour fire-resistance rating and the entire length and span of supporting elements for the rated roof assembly has a fire-resistance rating of not less than 1 hour.

2. Openings in the lower roof shall not be located within 10 feet (3048 mm) of the fire wall.

705.7 Combustible framing in fire walls. Adjacent combustible members entering into a concrete or masonry fire wall from opposite sides shall not have less than a 4-inch (102 mm) distance between embedded ends. Where combustible members frame into hollow walls or walls of hollow units, hollow spaces shall be solidly filled for the full thickness of the wall and for a distance not less than 4 inches (102 mm) above, below and between the structural members, with noncombustible materials approved for fireblocking.

705.8 Openings. Each opening through a fire wall shall be protected in accordance with Section 715.3 and shall not exceed 120 square feet (11 m²). The aggregate width of openings at any floor level shall not exceed 25 percent of the length of the wall.

Exceptions:

1. Openings are not permitted in party walls constructed in accordance with Section 503.2.

2. Openings shall not be limited to 120 square feet (11 m²) where both buildings are equipped throughout with an automatic sprinkler system installed in accordance with Section 903.3.1.1.

705.9 Penetrations. Penetrations through fire walls shall comply with Section 712.

705.10 Joints. Joints made in or between fire walls shall comply with Section 713.

705.11 Ducts and air transfer openings. Ducts and air transfer openings shall not penetrate fire walls.

Exception: Penetrations by ducts and air transfer openings of fire walls that are not on a lot line shall be allowed provided the penetrations comply with Sections 712 and 716. The size and aggregate width of all openings shall not exceed the limitations of Section 705.8.

SECTION 706
FIRE BARRIERS

706.1 General. Fire barriers used for separation of shafts, exits, exit passageways, horizontal exits or incidental use areas, to separate different occupancies, to separate a single occupancy into different fire areas, or to separate other areas where a fire barrier is required elsewhere in this code or the *International Fire Code*, shall comply with this section.

706.2 Materials. The walls and floor assemblies shall be of materials permitted by the building type of construction.

706.3 Fire-resistance rating. The fire-resistance rating of the walls and floor assemblies shall comply with this section.

706.3.1 Shaft enclosures. The fire-resistance rating of the fire barrier separating building areas from a shaft shall comply with Section 707.4.

706.3.2 Exit enclosures. The fire-resistance rating of the fire barrier separating building areas from an exit shall comply with Section 1019.1.

706.3.3 Exit passageway. The fire-resistance rating of the separation between building areas and an exit passageway shall comply with Section 1020.1.

706.3.4 Horizontal exit. The fire-resistance rating of the separation between building areas connected by a horizontal exit shall comply with Section 1021.1.

706.3.5 Incidental use areas. The fire barrier separating incidental use areas shall have a fire-resistance rating of not less than that indicated in Table 302.1.1.

706.3.6 Separation of mixed occupancies. Where the provisions of Section 302.3.2 are applicable, the fire barrier separating mixed occupancies shall have a fire-resistance rating of not less than that indicated in Section 302.3.2 based on the occupancies being separated.

706.3.7 Single-occupancy fire areas. The fire barrier separating a single occupancy into different fire areas shall have a fire-resistance rating of not less than that indicated in Table 706.3.7.

TABLE 706.3.7
FIRE-RESISTANCE RATING REQUIREMENTS FOR FIRE BARRIER ASSEMBLIES BETWEEN FIRE AREAS

OCCUPANCY GROUP	FIRE-RESISTANCE RATING (hours)
H-1, H-2	4
F-1, H-3, S-1	3
A, B, E, F-2, H-4, H-5, I, M, R, S-2	2
U	1

706.4 Continuity of fire barrier walls. Fire barrier walls shall extend from the top of the floor/ceiling assembly below to the underside of the floor or roof slab or deck above and shall be securely attached thereto. These walls shall be continuous through concealed spaces such as the space above a suspended ceiling. The supporting construction for fire barrier walls shall be protected to afford the required fire-resistance rating of the fire barrier supported except for 1-hour fire-resistance-rated incidental use area separations as required by Table 302.1.1 in buildings of Type IIB, IIIB and VB construction. Hollow vertical spaces within the fire barrier wall shall be firestopped at every floor level.

Exceptions:

1. The maximum required fire-resistance rating for assemblies supporting fire barriers separating tank storage as provided for in Section 415.7.2.1 shall be 2 hours, but not less than required by Table 601 for the building construction type.

2. Shaft enclosure shall be permitted to terminate at a top enclosure complying with Section 707.12.

706.5 Horizontal fire barriers. Horizontal fire barriers shall be constructed in accordance with Section 711.

706.6 Exterior walls. Where exterior walls serve as a part of a required fire-resistance-rated enclosure, such walls shall comply with the requirements of Section 704 for exterior walls and the fire-resistance-rated enclosure requirements shall not apply.

Exception: Exterior walls required to be fire-resistance rated in accordance with Section 1022.6.

706.7 Openings. Openings in a fire barrier wall shall be protected in accordance with Section 715. Openings shall be limited to a maximum aggregate width of 25 percent of the length of the wall, and the maximum area of any single opening shall not exceed 120 square feet (11 m²). Openings in exit enclosures shall also comply with Section 1019.1.1.

Exceptions:

1. Openings shall not be limited to 120 square feet (11 m²) where adjoining fire areas are equipped throughout with an automatic sprinkler system in accordance with Section 903.3.1.1.

2. Fire doors serving an exit enclosure.

3. Openings shall not be limited to 120 square feet (11 m^2) or an aggregate width of 25 percent of the length of the wall where the opening protective assembly has been tested in accordance with ASTM E 119 and has a minimum fire-resistance rating not less than the fire-resistance rating of the wall.

706.8 Penetrations. Penetrations through fire barriers shall comply with Section 712.

706.8.1 Prohibited penetrations. Penetrations into an exit enclosure shall only be allowed when permitted by Section 1019.1.2.

706.9 Joints. Joints made in or between fire barriers shall comply with Section 713.

706.10 Ducts and air transfer openings. Penetrations by ducts and air transfer openings shall comply with Sections 712 and 716.

SECTION 707
SHAFT ENCLOSURES

707.1 General. The provisions of this section shall apply to vertical shafts where such shafts are required to protect openings and penetrations through floor/ceiling and roof/ceiling assemblies.

707.2 Shaft enclosure required. Openings through a floor/ceiling assembly shall be protected by a shaft enclosure complying with this section.

Exceptions:

1. A shaft enclosure is not required for openings totally within an individual residential dwelling unit and connecting four stories or less.

2. A shaft enclosure is not required in a building equipped throughout with an automatic sprinkler system in accordance with Section 903.3.1.1 for an escalator opening or stairway which is not a portion of the means of egress protected according to Item 2.1 or 2.2:

    2.1. Where the area of the floor opening between stories does not exceed twice the horizontal projected area of the escalator or stairway and the opening is protected by a draft curtain and closely spaced sprinklers in accordance with NFPA 13. In other than Groups B and M, this application is limited to openings that do not connect more than four stories.

    2.2. Where the opening is protected by approved power-operated automatic shutters at every floor penetrated. The shutters shall be of noncombustible construction and have a fire-resistance rating of not less than 1.5 hours. The shutter shall be so constructed as to close immediately upon the actuation of a smoke detector installed in accordance with Section 907.10 and shall completely shut off the well opening. Escalators shall cease oper-

ation when the shutter begins to close. The shutter shall operate at a speed of not more than 30 feet per minute (152.4 mm/s) and shall be equipped with a sensitive leading edge to arrest its progress where in contact with any obstacle, and to continue its progress on release therefrom.

3. A shaft enclosure is not required for penetrations by pipe, tube, conduit, wire, cable, and vents protected in accordance with Section 712.4.

4. A shaft enclosure is not required for penetrations by ducts protected in accordance with Section 712.4. Grease ducts shall be protected in accordance with the *International Mechanical Code*.

5. A shaft enclosure is not required for floor openings complying with the provisions for covered malls or atriums.

6. A shaft enclosure is not required for approved masonry chimneys, where annular space protection is provided at each floor level in accordance with Section 717.2.5.

7. In other than Groups I-2 and I-3, a shaft enclosure is not required for a floor opening that complies with the following:

    7.1. Does not connect more than two stories.

    7.2. Is not part of the required means of egress system except as permitted in Section 1019.1.

    7.3. Is not concealed within the building construction.

    7.4. Is not open to a corridor in Group I and R occupancies.

    7.5. Is not open to a corridor on nonsprinklered floors in any occupancy.

    7.6. Is separated from floor openings serving other floors by construction conforming to required shaft enclosures.

8. A shaft enclosure is not required for automobile ramps in open parking garages and enclosed parking garages constructed in accordance with Sections 406.3 and 406.4, respectively.

9. A shaft enclosure is not required for floor openings between a mezzanine and the floor below.

10. A shaft enclosure is not required for joints protected by a fire-resistant joint system in accordance with Section 713.

11. Where permitted by other sections of this code.

707.3 Materials. The shaft enclosure shall be of materials permitted by the building type of construction.

707.4 Fire-resistance rating. Shaft enclosures shall have a fire-resistance rating of not less than 2 hours where connecting four stories or more and not less than 1 hour where connecting less than four stories. The number of stories connected by the shaft enclosure shall include any basements but not any mezzanines. Shaft enclosures shall be constructed as fire barriers in accordance with Section 706. Shaft enclosures shall have a fire-resistance rating not less than the floor assembly penetrated, but need not exceed 2 hours.

707.5 Continuity. Shaft enclosure walls shall extend from the top of the floor/ceiling assembly below to the underside of the floor or roof slab or deck above and shall be securely attached thereto. These walls shall be continuous through concealed spaces such as the space above a suspended ceiling. The supporting construction shall be protected to afford the required fire-resistance rating of the element supported. Hollow vertical spaces within the shaft enclosure construction wall shall be firestopped at every floor level.

707.6 Exterior walls. Where exterior walls serve as a part of a required shaft enclosure, such walls shall comply with the requirements of Section 704 for exterior walls and the fire-resistance-rated enclosure requirements shall not apply.

Exception: Exterior walls required to be fire-resistance rated in accordance with Section 1022.6.

707.7 Openings. Openings in a shaft enclosure shall be protected in accordance with Section 715 as required for fire barriers. Such openings shall be self-closing or automatic-closing by smoke detection.

707.7.1 Prohibited openings. Openings other than those necessary for the purpose of the shaft shall not be permitted in shaft enclosures.

707.8 Penetrations. Penetrations in a shaft enclosure shall be protected in accordance with Section 712 as required for fire barriers.

707.8.1 Prohibited penetrations. Penetrations other than those necessary for the purpose of the shaft shall not be permitted in shaft enclosures. Ducts shall not penetrate exit shaft enclosures.

Exception: Duct penetrations as permitted in Section 1019.1.2.

707.9 Joints. Joints in a shaft enclosure shall comply with Section 713.

707.10 Ducts and air transfer openings. Penetrations of a shaft enclosure by ducts and air transfer openings shall comply with Sections 712 and 716.

707.11 Enclosure at the bottom. Shafts that do not extend to the bottom of the building or structure shall:

1. Be enclosed at the lowest level with construction of the same fire-resistance rating as the lowest floor through which the shaft passes, but not less than the rating required for the shaft enclosure;

2. Terminate in a room having a use related to the purpose of the shaft. The room shall be separated from the remainder of the building by construction having a fire-resistance rating and opening protectives at least equal to the protection required for the shaft enclosure; or

3. Be protected by approved fire dampers installed in accordance with their listing at the lowest floor level within the shaft enclosure.

Exceptions:

1. The fire-resistance-rated room separation is not required provided there are no openings in or penetra-

tions of the shaft enclosure to the interior of the building except at the bottom. The bottom of the shaft shall be closed off around the penetrating items with materials permitted by Section 717.3.1 for draftstopping, or the room shall be provided with an approved automatic fire suppression system.

2. A shaft enclosure containing a refuse chute or laundry chute shall not be used for any other purpose and shall terminate in a room protected in accordance with Section 707.13.4.

3. The fire-resistance-rated room separation and the protection at the bottom of the shaft are not required provided there are no combustibles in the shaft and there are no openings or other penetrations through the shaft enclosure to the interior of the building.

707.12 Enclosure at the top. A shaft enclosure that does not extend to the underside of the roof deck of the building shall be enclosed at the top with construction of the same fire-resistance rating as the topmost floor penetrated by the shaft, but not less than the fire-resistance rating required for the shaft enclosure.

707.13 Refuse and laundry chutes. Refuse and laundry chutes, access and termination rooms and incinerator rooms shall meet the requirements of Sections 707.13.1 through 707.13.6.

Exception: Chutes serving and contained within a single dwelling unit.

707.13.1 Refuse and laundry chute enclosures. A shaft enclosure containing a refuse or laundry chute shall not be used for any other purpose and shall be enclosed in accordance with Section 707.4. Openings into the shaft, including those from access rooms and termination rooms, shall be protected in accordance with this section and Section 715. Openings into chutes shall not be located in exit access corridors. Opening protectives shall be self-closing or automatic-closing upon the actuation of a smoke detector installed in accordance with Section 907.10, except that heat-activated closing devices shall be permitted between the shaft and the termination room.

707.13.2 Materials. A shaft enclosure containing a refuse or laundry chute shall be constructed of materials as permitted by the building type of construction.

707.13.3 Refuse and laundry chute access rooms. Access openings for refuse and laundry chutes shall be located in rooms or compartments completely enclosed by construction that has a fire-resistance rating of not less than 1 hour and openings into the access rooms shall be protected by opening protectives having a fire protection rating of not less than $3/4$ hour and shall be self-closing or automatic-closing upon the detection of smoke.

707.13.4 Termination room. Refuse and laundry chutes shall discharge into an enclosed room completely separated from the remainder of the building by construction that has a fire-resistance rating of not less than 1 hour and openings into the termination room shall be protected by opening protectives having a fire protection rating of not less than $3/4$ hour and shall be self-closing or automatic-closing upon the detection of smoke. Refuse chutes shall not terminate in an incinerator room. Refuse and laundry rooms that are not provided with chutes need only comply with Table 302.1.1.

707.13.5 Incinerator room. Incinerator rooms shall comply with Table 302.1.1.

707.13.6 Automatic fire sprinkler system. An approved automatic fire sprinkler system shall be installed in accordance with Section 903.2.10.2.

707.14 Elevator and dumbwaiter shafts. Elevator hoistway and dumbwaiter enclosures shall be constructed in accordance with Section 707.4 and Chapter 30.

707.14.1 Elevator lobby. Elevators opening into a fire-resistance-rated corridor as required by Section 1016.1 shall be provided with an elevator lobby at each floor containing such a corridor. The lobby shall separate the elevators from the corridor by fire partitions and the required opening protection. Elevator lobbies shall have at least one means of egress complying with Chapter 10 and other provisions within this code.

Exceptions:

1. In office buildings, separations are not required from a street-floor elevator lobby provided the entire street floor is equipped with an automatic sprinkler system in accordance with Section 903.3.1.1.

2. Elevators not required to be located in a shaft in accordance with Section 707.2.

3. Where additional doors are provided in accordance with Section 3002.6. Such doors shall be tested in accordance with UL 1784 without an artificial bottom seal.

4. In other than Group I-3, and buildings more than four stories above the lowest level of fire department vehicle access, lobby separation is not required where the building, including the lobby and corridors leading to the lobby, is protected by an automatic sprinkler system installed throughout in accordance with Section 903.3.1.1 or 903.3.1.2.

SECTION 708
FIRE PARTITIONS

708.1 General. The following wall assemblies shall comply with this section.

1. Walls separating dwelling units in the same building.

2. Walls separating sleeping units in occupancies in Group R-1, hotel occupancies, R-2 and I-1.

3. Walls separating tenant spaces in covered mall buildings as required by Section 402.7.2.

4. Corridor walls as required by Section 1016.1.

5. Elevator lobby separation as required by Section 707.14.1.

708.2 Materials. The walls shall be of materials permitted by the building type of construction.

708.3 Fire-resistance rating. The fire-resistance rating of the walls shall be at least 1 hour.

Exceptions:

1. Corridor walls as permitted by Table 1016.1.

2. Dwelling unit and sleeping unit separations in buildings of Type IIB, IIIB and VB construction shall have fire-resistance ratings of not less than $1/2$ hour in buildings equipped throughout with an automatic sprinkler system in accordance with Section 903.3.1.1.

708.4 Continuity. Fire partitions shall extend from the top of the floor assembly below to the underside of the floor or roof slab or deck above or to the fire-resistance-rated floor/ceiling or roof/ceiling assembly above, and shall be securely attached thereto. If the partitions are not continuous to the deck, and where constructed of combustible construction, the space between the ceiling and the deck above shall be fireblocked or draftstopped in accordance with Sections 717.2.1 and 717.3.1 at the partition line. The supporting construction shall be protected to afford the required fire-resistance rating of the wall supported, except for tenant and sleeping unit separation walls and exit access corridor walls in buildings of Type IIB, IIIB and VB construction.

Exceptions:

1. The wall need not be extended into the crawl space below where the floor above the crawl space has a minimum 1-hour fire-resistance rating.

2. Where the room-side fire-resistance-rated membrane of the corridor is carried through to the underside of a fire-resistance-rated floor or roof above, the ceiling of the corridor shall be permitted to be protected by the use of ceiling materials as required for a 1-hour fire-resistance-rated floor or roof system.

3. Where the corridor ceiling is constructed as required for the corridor walls, the walls shall be permitted to terminate at the upper membrane of such ceiling assembly.

4. The fire partition separating tenant spaces in a mall, complying with Section 402.7.2, is not required to extend beyond the underside of a ceiling that is not part of a fire-resistance-rated assembly. A wall is not required in attic or ceiling spaces above tenant separation walls.

5. Fireblocking or draftstopping is not required at the partition line in Group R-2 buildings that do not exceed four stories in height provided the attic space is subdivided by draftstopping into areas not exceeding 3,000 square feet (279 m²) or above every two dwelling units, whichever is smaller.

6. Fireblocking or draftstopping is not required at the partition line in buildings equipped with an automatic sprinkler system installed throughout in accordance with Section 903.3.1.1 or 903.3.1.2 provided that automatic sprinklers are installed in combustible floor/ceiling and roof/ceiling spaces.

708.5 Exterior walls. Where exterior walls serve as a part of a required fire-resistance-rated enclosure, such walls shall comply with the requirements of Section 704 for exterior walls and the fire-resistance-rated enclosure requirements shall not apply.

708.6 Openings. Openings in a fire partition shall be protected in accordance with Section 715.

708.7 Penetrations. Penetrations through fire partitions shall comply with Section 712.

708.8 Joints. Joints made in or between fire partitions shall comply with Section 713.

708.9 Ducts and air transfer openings. Penetrations by ducts and air transfer openings shall comply with Sections 712 and 716.

SECTION 709
SMOKE BARRIERS

709.1 General. Smoke barriers shall comply with this section.

709.2 Materials. Smoke barriers shall be of materials permitted by the building type of construction.

709.3 Fire-resistance rating. A 1-hour fire-resistance rating is required for smoke barriers.

> **Exception:** Smoke barriers constructed of minimum 0.10-inch-thick (2.5 mm) steel in Group I-3 buildings.

709.4 Continuity. Smoke barriers shall form an effective membrane continuous from outside wall to outside wall and from floor slab to floor or roof deck above, including continuity through concealed spaces, such as those found above suspended ceilings, and interstitial structural and mechanical spaces. The supporting construction shall be protected to afford the required fire-resistance rating of the wall or floor supported in buildings of other than Type IIB, IIIB or VB construction.

> **Exception:** Smoke barrier walls are not required in interstitial spaces where such spaces are designed and constructed with ceilings that provide resistance to the passage of fire and smoke equivalent to that provided by the smoke barrier walls.

709.5 Openings. Openings in a smoke barrier shall be protected in accordance with Section 715.

> **Exception:** In Group I-2, where such doors are installed across corridors, a pair of opposite-swinging doors without a center mullion shall be installed having vision panels with approved fire-resistance-rated glazing materials in approved fire-resistance-rated frames, the area of which shall not exceed that tested. The doors shall be close fitting within operational tolerances, and shall not have undercuts, louvers or grilles. The doors shall have head and jamb stops, astragals or rabbets at meeting edges and automatic-closing devices. Positive-latching devices are not required.

709.6 Penetrations. Penetrations through smoke barriers shall comply with Section 712.

709.7 Joints. Joints made in or between smoke barriers shall comply with Section 713.

709.8 Ducts and air transfer openings. Penetrations by ducts and air transfer openings shall comply with Sections 712 and 716.

SECTION 710
SMOKE PARTITIONS

710.1 General. Smoke partitions installed as required elsewhere in the code shall comply with this section.

710.2 Materials. The walls shall be of materials permitted by the building type of construction.

710.3 Fire-resistance rating. Unless required elsewhere in the code, smoke partitions are not required to have a fire-resistance rating.

710.4 Continuity. Smoke partitions shall extend from the floor to the underside of the floor or roof deck above or to the underside of the ceiling above where the ceiling membrane is constructed to limit the transfer of smoke.

710.5 Openings. Windows shall be sealed to resist the free passage of smoke or be automatic-closing upon detection of smoke. Doors in smoke partitions shall comply with this section.

> **710.5.1 Louvers.** Doors in smoke partitions shall not include louvers.

> **710.5.2 Smoke and draft-control doors.** Where required elsewhere in the code, doors in smoke partitions shall be tested in accordance with UL 1784 with an artificial bottom seal installed across the full width of the bottom of the door assembly. The air leakage rate of the door assembly shall not exceed 3.0 cubic feet per minute per square foot [ft^3/(min ft^2)] (0.015424 m^3/sm^2) of door opening at 0.10 inch (24.9 Pa) of water for both the ambient temperature test and the elevated temperature exposure test.

> **710.5.3 Self-closing or automatic-closing doors.** Where required elsewhere in the code, doors in smoke partitions shall be self-closing or automatic-closing in accordance with Section 715.3.7.3.

710.6 Penetrations and joints. The space around penetrating items and in joints shall be filled with an approved material to limit the free passage of smoke.

710.7 Ducts and air transfer openings. Air transfer openings in smoke partitions shall be provided with a smoke damper complying with Section 716.3.2.

> **Exception:** Where the installation of a smoke damper will interfere with the operation of a required smoke control system in accordance with Section 909, approved alternative protection shall be utilized.

SECTION 711
HORIZONTAL ASSEMBLIES

711.1 General. Floor and roof assemblies required to have a fire-resistance rating shall comply with this section.

711.2 Materials. The floor and roof assemblies shall be of materials permitted by the building type of construction.

711.3 Fire-resistance rating. The fire-resistance rating of floor and roof assemblies shall not be less than that required by the building type of construction. Where the floor assembly separates mixed occupancies, the assembly shall have a fire-resistance rating of not less than that required by Section 302.3.2 based on the occupancies being separated. Where the floor assembly separates a single occupancy into different fire areas, the assembly shall have a fire-resistance rating of not less than that required by Section 706.3.7. Floor assemblies separating dwelling units in the same building or sleeping units in occupancies in Group R-1, hotel occupancies, R-2 and I-1 shall be a minimum of 1-hour fire-resistance-rated construction.

Exception: Dwelling unit and sleeping unit separations in buildings of Type IIB, IIIB, and VB construction shall have fire-resistance ratings of not less than $^1/_2$ hour in buildings equipped throughout with an automatic sprinkler system in accordance with Section 903.3.1.1.

711.3.1 Ceiling panels. Where the weight of lay-in ceiling panels, used as part of fire-resistance-rated floor/ceiling or roof/ceiling assemblies, is not adequate to resist an upward force of 1 lb/ft.^{2} (48 Pa), wire or other approved devices shall be installed above the panels to prevent vertical displacement under such upward force.

711.3.2 Access doors. Access doors shall be permitted in ceilings of fire-resistance-rated floor/ceiling and roof/ceiling assemblies provided such doors are tested in accordance with ASTM E 119 as horizontal assemblies and labeled by an approved agency for such purpose.

711.3.3 Unusable space. In 1-hour fire-resistance-rated floor construction, the ceiling membrane is not required to be installed over unusable crawl spaces. In 1-hour fire-resistance-rated roof construction, the floor membrane is not required to be installed where unusable attic space occurs above.

711.4 Continuity. Assemblies shall be continuous without openings, penetrations or joints except as permitted by this section and Sections 707.2, 712.4 and 713. Skylights and other penetrations through a fire-resistance-rated roof deck are permitted to be unprotected, provided that the structural integrity of the fire-resistance-rated roof construction is maintained. Unprotected skylights shall not be permitted in roof construction required to be fire-resistance rated in accordance with Section 704.10. The supporting construction shall be protected to afford the required fire-resistance rating of the horizontal assembly supported.

711.5 Penetrations. Penetrations through fire-resistance-rated horizontal assemblies shall comply with Section 712.

711.6 Joints. Joints made in or between fire-resistance-rated horizontal assemblies shall comply with Section 713. The void created at the intersection of a floor/ceiling assembly and an exterior curtain wall assembly shall be protected in accordance with Section 713.4.

711.7 Ducts and air transfer openings. Penetrations by ducts and air transfer openings shall comply with Sections 712 and 716.

SECTION 712
PENETRATIONS

712.1 Scope. The provisions of this section shall govern the materials and methods of construction used to protect through penetrations and membrane penetrations.

712.2 Installation details. Where sleeves are used, they shall be securely fastened to the assembly penetrated. The space between the item contained in the sleeve and the sleeve itself and any space between the sleeve and the assembly penetrated shall be protected in accordance with this section. Insulation and coverings on or in the penetrating item shall not penetrate the assembly unless the specific material used has been tested as part of the assembly in accordance with this section.

712.3 Fire-resistance-rated walls. Penetrations into or through fire walls, fire barriers, smoke barrier walls, and fire partitions shall comply with this section.

712.3.1 Through penetrations. Through penetrations of fire-resistance-rated walls shall comply with Section 712.3.1.1 or 712.3.1.2.

Exception: Where the penetrating items are steel, ferrous or copper pipes or steel conduits, the annular space between the penetrating item and the fire-resistance-rated wall shall be permitted to be protected as follows:

1. In concrete or masonry walls where the penetrating item is a maximum 6-inch (152 mm) nominal diameter and the opening is a maximum 144 square inches (0.0929 m^2), concrete, grout or mortar shall be permitted where installed the full thickness of the wall or the thickness required to maintain the fire-resistance rating; or

2. The material used to fill the annular space shall prevent the passage of flame and hot gases sufficient to ignite cotton waste when subjected to ASTM E 119 time-temperature fire conditions under a minimum positive pressure differential of 0.01 inch (2.49 Pa) of water at the location of the penetration for the time period equivalent to the fire-resistance rating of the construction penetrated.

712.3.1.1 Fire-resistance-rated assemblies. Penetrations shall be installed as tested in an approved fire-resistance-rated assembly.

712.3.1.2 Through-penetration firestop system. Through penetrations shall be protected by an approved penetration firestop system installed as tested in accordance with ASTM E 814 or UL 1479, with a minimum positive pressure differential of 0.01 inch (2.49 Pa) of water and shall have an F rating of not less than the required fire-resistance rating of the wall penetrated.

712.3.2 Membrane penetrations. Membrane penetrations shall comply with Section 712.3.1. Where walls and partitions are required to have a minimum 1-hour fire-resistance rating, recessed fixtures shall be installed such that the required fire resistance will not be reduced.

Exceptions:

1. Steel electrical boxes that do not exceed 16 square inches (0.0103 m^2) in area provided the total area of such openings does not exceed 100 square inches (0.0645 m^2) for any 100 square feet (9.29 m^2) of wall area. Outlet boxes on opposite sides of the wall shall be separated as shown:

1.1. By a horizontal distance of not less than 24 inches (610 mm);

1.2. By a horizontal distance of not less than the depth of the wall cavity where the wall cavity is filled with cellulose loose fill, rockwool or slag mineral wool insulation;

1.3. By solid fireblocking in accordance with Section 717.2.1;

1.4. By protecting both outlet boxes with listed putty pads; or

1.5. By other listed materials and methods.

2. Membrane penetrations for listed electrical outlet boxes of any material are permitted provided such boxes have been tested for use in fire-resistance-rated assemblies and are installed in accordance with the instructions included in the listing. Outlet boxes on opposite sides of the wall shall be separated as follows:

2.1. By a horizontal distance of not less than 24 inches (610 mm);

2.2. By solid fireblocking in accordance with Section 717.2.1;

2.3. By protecting both outlet boxes with listed putty pads; or

2.4. By other listed materials and methods.

3. The annular space created by the penetration of a fire sprinkler provided it is covered by a metal escutcheon plate.

712.3.3 Ducts and air transfer openings. Penetrations of fire-resistance-rated walls by ducts and air transfer openings that are not protected with fire dampers shall comply with this section.

712.3.4 Dissimilar materials. Noncombustible penetrating items shall not connect to combustible items beyond the point of firestopping unless it can be demonstrated that the fire-resistance integrity of the wall is maintained.

712.4 Horizontal assemblies. Penetrations of a floor, floor/ceiling assembly or the ceiling membrane of a roof/ceiling assembly shall be protected in accordance with Section 707. Penetrations permitted by Exceptions 3 and 4 of Section 707.2 shall comply with Sections 712.4.1 through 712.4.4.

Exception: Penetrations located within the same room or undivided area as floor openings are not required to have a shaft enclosure in accordance with Exception 1, 2, 5, 7, 8 or 9 in Section 707.2.

712.4.1 Through penetrations. Through penetrations of fire-resistance-rated horizontal assemblies shall comply with Section 712.4.1.1 or 712.4.1.2.

Exceptions:

1. Penetrations by steel, ferrous or copper conduits, pipes, tubes, vents, concrete, or masonry through a single fire-resistance-rated floor assembly where the annular space is protected with materials that prevent the passage of flame and hot gases sufficient to ignite cotton waste when subjected to

ASTM E 119 time-temperature fire conditions under a minimum positive pressure differential of 0.01 inch (2.49 Pa) of water at the location of the penetration for the time period equivalent to the fire-resistance rating of the construction penetrated. Penetrating items with a maximum 6-inch (152 mm) nominal diameter shall not be limited to the penetration of a single fire-resistance-rated floor assembly provided that the area of the penetration does not exceed 144 square inches (92 900 mm^2) in any 100 square feet (9.3 m^2) of floor area.

2. Penetrations in a single concrete floor by steel, ferrous or copper conduits, pipes, tubes and vents with a maximum 6-inch (152 mm) nominal diameter provided concrete, grout or mortar is installed the full thickness of the floor or the thickness required to maintain the fire-resistance rating. The penetrating items with a maximum 6-inch (152 mm) nominal diameter shall not be limited to the penetration of a single concrete floor provided that the area of the penetration does not exceed 144 square inches (0.0929 m^2).

3. Electrical outlet boxes of any material are permitted provided that such boxes are tested for use in fire-resistance-rated assemblies and installed in accordance with the tested assembly.

712.4.1.1 Fire-resistance-rated assemblies. Penetrations shall be installed as tested in an approved fire-resistance-rated assembly.

712.4.1.2 Through-penetration firestop system. Through penetrations shall be protected by an approved through-penetration firestop system installed and tested in accordance with ASTM E 814 or UL 1479, with a minimum positive pressure differential of 0.01 inch (2.49 Pa) of water. The system shall have an F rating and a T rating of not less than 1 hour but not less than the required rating of the floor penetrated.

Exception: Floor penetrations contained and located within the cavity of a wall do not require a T rating.

712.4.2 Membrane penetrations. Penetrations of membranes that are part of a fire-resistance-rated horizontal assembly shall comply with Section 712.4.1.1 or 712.4.1.2. Where floor/ceiling assemblies are required to have a minimum 1-hour fire-resistance rating, recessed fixtures shall be installed such that the required fire resistance will not be reduced.

Exceptions:

1. Membrane penetrations by steel, ferrous or copper conduits, electrical outlet boxes, pipes, tubes, vents, concrete, or masonry-penetrating items where the annular space is protected either in accordance with Section 712.4.1 or to prevent the free passage of flame and the products of combustion. Such penetrations shall not exceed an aggregate area of 100 square inches (64 500 mm^2) in any 100 square feet (9.3 m^2) of ceiling area in assemblies tested without penetrations.

2. Membrane penetrations by listed electrical outlet boxes of any material are permitted provided such boxes have been tested for use in fire-resistance-rated assemblies and are installed in accordance with the instructions included in the listing.

3. The annular space created by the penetration of a fire sprinkler provided it is covered by a metal escutcheon plate.

712.4.3 Nonfire-resistance-rated assemblies. Penetrations of horizontal assemblies without a required fire-resistance rating shall meet the requirements of Section 707 or shall comply with Sections 712.4.3.1 through 712.4.3.2.

712.4.3.1 Noncombustible penetrating items. Noncombustible penetrating items that connect not more than three stories are permitted provided that the annular space is filled with an approved noncombustible material to resist the free passage of flame and the products of combustion.

712.4.3.2 Penetrating items. Penetrating items that connect not more than two stories are permitted provided that the annular space is filled with an approved material to resist the free passage of flame and the products of combustion.

712.4.4 Ducts and air transfer openings. Penetrations of horizontal assemblies by ducts and air transfer openings that are not required to have dampers shall comply with this section. Ducts and air transfer openings that are protected with dampers shall comply with Section 716.

712.4.5 Dissimilar materials. Noncombustible penetrating items shall not connect to combustible materials beyond the point of firestopping unless it can be demonstrated that the fire-resistance integrity of the horizontal assembly is maintained.

712.4.6 Floor fire doors. Floor fire doors used to protect openings in fire-resistance-rated floors shall be tested in the horizontal position in accordance with ASTM E 119, and shall achieve a fire-resistance rating not less than the assembly being penetrated. Floor fire doors shall be labeled by an approved agency.

SECTION 713
FIRE-RESISTANT JOINT SYSTEMS

713.1 General. Joints installed in or between fire-resistance-rated walls, floor or floor/ceiling assemblies and roofs or roof/ceiling assemblies shall be protected by an approved fire-resistant joint system designed to resist the passage of fire for a time period not less than the required fire-resistance rating of the wall, floor or roof in or between which it is installed. Fire-resistant joint systems shall be tested in accordance with Section 713.3. The void created at the intersection of a floor/ceiling assembly and an exterior curtain wall assembly shall be protected in accordance with Section 713.4.

Exception: Fire-resistant joint systems shall not be required for joints in all of the following locations:

1. Floors within a single dwelling unit.

2. Floors where the joint is protected by a shaft enclosure in accordance with Section 707.

3. Floors within atriums where the space adjacent to the atrium is included in the volume of the atrium for smoke control purposes.

4. Floors within malls.

5. Floors within open parking structures.

6. Mezzanine floors.

7. Walls that are permitted to have unprotected openings.

8. Roofs where openings are permitted.

9. Control joints not exceeding a maximum width of 0.625 inch (15.9 mm) and tested in accordance with ASTM E 119.

713.2 Installation. Fire-resistant joint systems shall be securely installed in or on the joint for its entire length so as not to dislodge, loosen or otherwise impair its ability to accommodate expected building movements and to resist the passage of fire and hot gases.

713.3 Fire test criteria. Fire-resistant joint systems shall be tested in accordance with the requirements of either ASTM E 1966 or UL 2079. Nonsymmetrical wall joint systems shall be tested with both faces exposed to the furnace, and the assigned fire-resistance rating shall be the shortest duration obtained from the two tests. When evidence is furnished to show that the wall was tested with the least fire-resistant side exposed to the furnace, subject to acceptance of the building official, the wall need not be subjected to tests from the opposite side.

Exception: For exterior walls with a horizontal fire separation distance greater than 5 feet (1524 mm), the joint system shall be required to be tested for interior fire exposure only.

713.4 Exterior curtain wall/floor intersection. Where fire-resistance-rated floor or floor/ceiling assemblies are required, voids created at the intersection of the exterior curtain wall assemblies and such floor assemblies shall be sealed with an approved material or system to prevent the interior spread of fire. Such material or systems shall be securely installed and capable of preventing the passage of flame and hot gases sufficient to ignite cotton waste where subjected to ASTM E 119 time-temperature fire conditions under a minimum positive pressure differential of 0.01 inch of water (2.49 Pa) for the time period at least equal to the fire-resistance rating of the floor assembly. Height and fire-resistance requirements for curtain wall spandrels shall comply with Section 704.9.

SECTION 714
FIRE-RESISTANCE RATING OF STRUCTURAL MEMBERS

714.1 Requirements. The fire-resistance rating of structural members and assemblies shall comply with the requirements for the type of construction and shall not be less than the rating required for the fire-resistance-rated assemblies supported.

Exception: Fire barriers and fire partitions as provided in Sections 706.4 and 708.4, respectively.

714.2 Protection of structural members. Protection of columns, girders, trusses, beams, lintels or other structural members that are required to have a fire-resistance rating shall comply with this section.

714.2.1 Individual protection. Columns, girders, trusses, beams, lintels or other structural members that are required to have a fire-resistance rating and that support more than two floors or one floor and roof, or support a load-bearing wall or a nonload-bearing wall more than two stories high, shall be individually protected on all sides for the full length with materials having the required fire-resistance rating. Other structural members required to have a fire-resistance rating shall be protected by individual encasement, by a membrane or ceiling protection as specified in Section 711, or by a combination of both. Columns shall also comply with Section 714.2.2.

714.2.2 Column protection above ceilings. Where columns require a fire-resistance rating, the entire column, including its connections to beams or girders, shall be protected. Where the column extends through a ceiling, fire resistance of the column shall be continuous from the top of the floor through the ceiling space to the top of the column.

714.2.3 Truss protection. The required thickness and construction of fire-resistance-rated assemblies enclosing trusses shall be based on the results of full-scale tests or combinations of tests on truss components or on approved calculations based on such tests that satisfactorily demonstrate that the assembly has the required fire resistance.

714.2.4 Attachments to structural members. The edges of lugs, brackets, rivets and bolt heads attached to structural members shall be permitted to extend to within 1 inch (25 mm) of the surface of the fire protection.

714.2.5 Reinforcing. Thickness of protection for concrete or masonry reinforcement shall be measured to the outside of the reinforcement except that stirrups and spiral reinforcement ties are permitted to project not more than 0.5-inch (12.7 mm) into the protection.

714.3 Embedments and enclosures. Pipes, wires, conduits, ducts or other service facilities shall not be embedded in the required fire protective covering of a structural member that is required to be individually encased.

714.4 Impact protection. Where the fire protective covering of a structural member is subject to impact damage from moving vehicles, the handling of merchandise or other activity, the fire protective covering shall be protected by corner guards or by a substantial jacket of metal or other noncombustible material to a height adequate to provide full protection, but not less than 5 feet (1524 mm) from the finished floor.

714.5 Exterior structural members. Load-bearing structural members located within the exterior walls or on the outside of a building or structure shall be provided with the highest fire-resistance rating as determined in accordance with the following:

1. As required by Table 601 for the type of building element based on the type of construction of the building;

2. As required by Table 601 for exterior bearing walls based on the type of construction; and

3. As required by Table 602 for exterior walls based on the fire separation distance.

714.6 Bottom flange protection. Fire protection is not required at the bottom flange of lintels, shelf angles and plates, spanning not more than 6 feet (1829 mm) whether part of the structural frame or not, and from the bottom flange of lintels, shelf angles and plates not part of the structural frame, regardless of span.

714.7 Seismic isolation systems. Fire-resistance ratings for the isolation system shall meet the fire-resistance rating required for the columns, walls, or other structural elements in which the isolation system is installed in accordance with Table 601.

Isolation systems required to have a fire-resistance rating shall be protected with approved materials or construction assemblies designed to provide the same degree of fire resistance as the structural element in which it is installed when tested in accordance with ASTM E 119 (see Section 703.2).

Such isolation system protection applied to isolator units shall be capable of retarding the transfer of heat to the isolator unit in such a manner that the required gravity load-carrying capacity of the isolator unit will not be impaired after exposure to the standard time-temperature curve fire test prescribed in ASTM E 119 for a duration not less than that required for the fire resistance rating of the structure element in which it is installed.

Such isolation system protection applied to isolator units shall be suitably designed and securely installed so as not to dislodge, loosen, sustain damage, or otherwise impair its ability to accommodate the seismic movements for which the isolator unit is designed and to maintain its integrity for the purpose of providing the required fire-resistance protection.

SECTION 715
OPENING PROTECTIVES

715.1 General. Opening protectives required by other sections of this code shall comply with the provisions of this section.

715.2 Fire-resistance-rated glazing. Labeled fire-resistance-rated glazing tested as part of a fire-resistance-rated wall assembly in accordance with ASTM E 119 shall not be required to comply with this section.

715.3 Fire door and shutter assemblies. Approved fire door and fire shutter assemblies shall be constructed of any material or assembly of component materials that conforms to the test requirements of Section 715.3.1, 715.3.2 or 715.3.3 and the fire protection rating indicated in Table 715.3. Fire door assemblies and shutters shall be installed in accordance with the provisions of this section and NFPA 80.

Exceptions:

1. Labeled protective assemblies that conform to the requirements of this section or UL 10A, UL 14B and UL 14C for tin-clad fire door assemblies.

2. Floor fire doors in accordance with Section 712.4.6.

TABLE 715.3
FIRE DOOR AND FIRE SHUTTER FIRE PROTECTION RATINGS

TYPE OF ASSEMBLY	REQUIRED ASSEMBLY RATING (hours)	MINIMUM FIRE DOOR AND FIRE SHUTTER ASSEMBLY RATING (hours)
Fire walls and fire barriers having a required fire-resistance rating greater than 1 hour	4 3 2 $1^1/_2$	3 3^a $1^1/_2$ $1^1/_2$
Fire barriers having a required fire-resistance rating of 1 hour: 　Shaft, exit enclosure and exit passageway walls 　Other fire barriers	 1 1	 1 $^3/_4$
Fire partitions: 　Corridor walls 　Other fire partitions	 1 0.5 1	 $^1/_3{}^b$ $^1/_3{}^b$ $^3/_4$
Exterior walls	3 2 1	$1^1/_2$ $1^1/_2$ $^3/_4$

a. Two doors, each with a fire protection rating of $1^1/_2$ hours, installed on opposite sides of the same opening in a fire wall, shall be deemed equivalent in fire protection rating to one 3-hour fire door.

b. For testing requirements, see Section 715.3.3.

715.3.1 Side-hinged or pivoted swinging doors. Side-hinged and pivoted swinging doors shall be tested in accordance with NFPA 252 or UL 10C. After 5 minutes into the NFPA 252 test, the neutral pressure level in the furnace shall be established at 40 inches (1016 mm) or less above the sill.

715.3.2 Other types of doors. Other types of doors, including swinging elevator doors, shall be tested in accordance with NFPA 252 or UL 10B. The pressure in the furnace shall be maintained as nearly equal to the atmospheric pressure as possible. Once established, the pressure shall be maintained during the entire test period.

715.3.3 Door assemblies in corridors and smoke barriers. Fire door assemblies required to have a minimum fire protection rating of 20 minutes where located in corridor walls or smoke barrier walls having a fire-resistance rating in accordance with Table 715.3 shall be tested in accordance with NFPA 252 or UL 10C without the hose stream test. If a 20-minute fire door assembly contains glazing material, the glazing material in the door itself shall have a minimum fire protection rating of 20 minutes and be exempt from the hose stream test. Glazing material in any other part of the door assembly, including transom lites and sidelites, shall be tested in accordance with NFPA 257, including the hose stream test, in accordance with Section 715.4. Fire door assemblies shall also meet the requirements for a smoke- and draft-control door assembly tested in accordance with UL 1784 with an artificial bottom seal installed across the full width of the bottom of the door assembly. The air leakage rate of the door assembly shall not exceed 3.0 cfm per square foot (0.01524 m³/s·m²) of door opening at 0.10 inch (24.9 Pa) of water for

both the ambient temperature and elevated temperature tests. Louvers shall be prohibited.

Exceptions:

1. Viewports that require a hole not larger than 1 inch (25 mm) in diameter through the door, have at least an 0.25-inch-thick (6.4 mm) glass disc and the holder is of metal that will not melt out where subject to temperatures of 1,700°F (927°C).

2. Corridor door assemblies in occupancies of Group I-2 shall be in accordance with Section 407.3.1.

3. Unprotected openings shall be permitted for corridors in multitheater complexes where each motion picture auditorium has at least one-half of its required exit or exit access doorways opening directly to the exterior or into an exit passageway.

715.3.4 Doors in vertical exit enclosures and exit passageways. Fire door assemblies in vertical exit enclosures and exit passageways shall have a maximum transmitted temperature end point of not more than 450°F (232°C) above ambient at the end of 30 minutes of standard fire test exposure.

Exception: The maximum transmitted temperature end point is not required in buildings equipped throughout with an automatic sprinkler system installed in accordance with Section 903.3.1.1 or 903.3.1.2.

715.3.4.1 Glazing in doors. Fire-protection-rated glazing in excess of 100 square inches (0.065 m²) shall be permitted in fire door assemblies when tested in accordance with NFPA 252 as components of the door assemblies and not as glass lights, and shall have a maximum transmitted temperature end point of 450°F (232°C) in accordance with Section 715.3.4.

Exception: The maximum transmitted temperature end point is not required in buildings equipped throughout with an automatic sprinkler system installed in accordance with Section 903.3.1.1 or 903.3.1.2.

715.3.5 Labeled protective assemblies. Fire door assemblies shall be labeled by an approved agency. The labels shall comply with NFPA 80, and shall be permanently affixed to the door or frame.

715.3.5.1 Fire door labeling requirements. Fire doors shall be labeled showing the name of the manufacturer, the name of the third-party inspection agency, the fire protection rating and, where required for fire doors in exit enclosures by Section 715.3.4, the maximum transmitted temperature end point. Smoke and draft control doors complying with UL 1784 shall be labeled as such. Labels shall be approved and permanently affixed. The label shall be applied at the factory or location where fabrication and assembly are performed.

715.3.5.2 Oversized doors. Oversized fire doors shall bear an oversized fire door label by an approved agency or shall be provided with a certificate of inspection furnished by an approved testing agency. When a certificate of inspection is furnished by an approved testing agency, the certificate shall state that the door conforms to the requirements of design, materials and construction, but has not been subjected to the fire test.

715.3.5.3 Smoke and draft control door labeling requirements. Smoke and draft control doors complying with UL 1784 shall be labeled in accordance with Section 715.3.5.1 and shall show the letter "S" on the fire rating label of the door. This marking shall indicate that the door and frame assembly are in compliance when listed or labeled gasketing is also installed.

715.3.5.4 Fire door frame labeling requirements. Fire door frames shall be labeled showing the names of the manufacturer and the third-party inspection agency.

715.3.6 Glazing material. Fire-protection-rated glazing conforming to the opening protection requirements in Section 715.3 shall be permitted in fire door assemblies.

715.3.6.1 Size limitations. Wired glass used in fire doors shall comply with Table 715.4.3. Other fire-protection-rated glazing shall comply with the size limitations of NFPA 80.

Exceptions:

1. Fire-protection-rated glazing in fire doors located in fire walls shall be prohibited except that where serving as a horizontal exit, a self-closing swinging door shall be permitted to have a vision panel of not more than 100 square inches (0.065 m²) without a dimension exceeding 10 inches (254 mm).

2. Fire-protection-rated glazing shall not be installed in fire doors having a $1^1/_2$-hour fire protection rating intended for installation in fire barriers, unless the glazing is not more than 100 square inches (0.065 m²) in area.

715.3.6.2 Exit and elevator protectives. Approved fire-protection-rated glazing used in fire doors in elevator and stairway shaft enclosures shall be so located as to furnish clear vision of the passageway or approach to the elevator or stairway.

715.3.6.3 Labeling. Fire-protection-rated glazing shall bear a label or other identification showing the name of the manufacturer, the test standard and the fire protection rating. Such label or other identification shall be issued by an approved agency and shall be permanently affixed.

715.3.6.4 Safety glazing. Fire-protection-rated glazing installed in fire doors or fire window assemblies in areas subject to human impact in hazardous locations shall comply with Chapter 24.

715.3.7 Door closing. Fire doors shall be self-closing or automatic-closing in accordance with this section.

Exception: Fire doors located in common walls separating sleeping units in Group R-1 shall be permitted without automatic-closing or self-closing devices.

715.3.7.1 Latch required. Unless otherwise specifically permitted, single fire doors and both leaves of pairs of side-hinged swinging fire doors shall be provided with an active latch bolt that will secure the door when it is closed.

715.3.7.2 Automatic-closing fire door assemblies. Automatic-closing fire door assemblies shall be self-closing in accordance with NFPA 80.

715.3.7.3 Smoke-activated doors. Automatic-closing fire doors installed in the following locations shall be automatic-closing by the actuation of smoke detectors installed in accordance with Section 907.10 or by loss of power to the smoke detector or hold-open device. Fire doors that are automatic-closing by smoke detection shall not have more than a 10-second delay before the door starts to close after the smoke detector is actuated.

1. Doors installed across a corridor.
2. Doors that protect openings in horizontal exits, exits or exit access corridors required to be of fire-resistance-rated construction.
3. Doors that protect openings in walls required to be fire-resistance rated by Table 302.1.1.
4. Doors installed in smoke barriers in accordance with Section 709.5.
5. Doors installed in fire partitions in accordance with Section 708.6.
6. Doors installed in a fire wall in accordance with Section 705.8.

715.3.7.4 Doors in pedestrian ways. Vertical sliding or vertical rolling steel fire doors in openings through which pedestrians travel shall be heat activated or activated by smoke detectors with alarm verification.

715.3.8 Swinging fire shutters. Where fire shutters of the swinging type are installed in exterior openings, not less than one row in every three vertical rows shall be arranged to be readily opened from the outside, and shall be identified by distinguishing marks or letters not less than 6 inches (152 mm) high.

715.3.9 Rolling fire shutters. Where fire shutters of the rolling type are installed, such shutters shall include approved automatic-closing devices.

715.4 Fire-protection rated glazing. Glazing in fire window assemblies shall be fire protection rated in accordance with this section and Table 715.4. Glazing in fire doors shall comply with Section 715.3.6. Fire-protection-rated glazing installed as an opening protective in fire partitions, smoke barriers and fire barriers shall be tested in accordance with and shall meet the acceptance criteria of NFPA 257 for a fire protection rating of 45 minutes. Fire-protection-rated glazing shall also comply with NFPA 80. Fire-protection-rated glazing required in accordance with Section 704.12 for exterior wall opening protection shall be tested in accordance with and shall meet the acceptance criteria of NFPA 257 for a fire protection rating as required in Section 715.4.8.

Exceptions:

1. Wired glass in accordance with Section 715.4.3.
2. Fire-protection-rated glazing in 0.5-hour fire-resistance-rated partitions is permitted to have an 0.33-hour fire protection rating.

TABLE 715.4
FIRE WINDOW ASSEMBLY FIRE PROTECTION RATINGS

TYPE OF ASSEMBLY	REQUIRED ASSEMBLY RATING (hours)	MINIMUM FIRE WINDOW ASSEMBLY RATING (hours)
Interior walls:		
Fire walls	All	NP[a]
Fire barriers and fire partitions	>1	NP[a]
	1	$^3/_4$
Smoke barriers	1	$^3/_4$
Exterior walls	>1	$1^1/_2$
	1	$^3/_4$
Party walls	All	NP[a]

a. Not permitted except as specified in Section 715.2.

715.4.1 Testing under positive pressure. NFPA 257 shall evaluate fire-protection-rated glazing under positive pressure. Within the first 10 minutes of a test, the pressure in the furnace shall be adjusted so at least two-thirds of the test specimen is above the neutral pressure plane, and the neutral pressure plane shall be maintained at that height for the balance of the test.

715.4.2 Nonsymmetrical glazing systems. Nonsymmetrical fire-protection-rated glazing systems in fire partitions, fire barriers or in exterior walls with a fire separation of 5 feet (1524 mm) or less pursuant to Section 704 shall be tested with both faces exposed to the furnace, and the assigned fire protection rating shall be the shortest duration obtained from the two tests conducted in compliance with NFPA 257.

715.4.3 Wired glass. Steel window frame assemblies of 0.125-inch (3.2 mm) minimum solid section or of not less than nominal 0.048-inch-thick (1.2 mm) formed sheet steel members fabricated by pressing, mitering, riveting, interlocking or welding and having provision for glazing with $^1/_4$-inch (6.4 mm) wired glass where securely installed in the building construction and glazed with $^1/_4$-inch (6.4 mm) labeled wired glass shall be deemed to meet the requirements for a $^3/_4$-hour fire window assembly. Wired glass panels shall conform to the size limitations set forth in Table 715.4.3.

TABLE 715.4.3
LIMITING SIZES OF WIRED GLASS PANELS

OPENING FIRE PROTECTION RATING	MAXIMUM AREA (square inches)	MAXIMUM HEIGHT (inches)	MAXIMUM WIDTH (inches)
3 hours	0	0	0
$1^1/_2$-hour doors in exterior walls	0	0	0
1 and $1^1/_2$ hours	100	33	10
$^3/_4$ hour	1,296	54	54
20 minutes	Not Limited	Not Limited	Not Limited
Fire window assemblies	1,296	54	54

For SI: 1 inch = 25.4 mm, 1 square inch = 645.2 mm².

715.4.4 Nonwired glass. Glazing other than wired glass in fire window assemblies shall be fire-protection-rated glazing installed in accordance with and complying with the size limitations set forth in NFPA 80.

715.4.5 Installation. Fire-protection-rated glazing shall be in the fixed position or be automatic-closing and shall be installed in approved frames.

715.4.6 Window mullions. Metal mullions that exceed a nominal height of 12 feet (3658 mm) shall be protected with materials to afford the same fire-resistance rating as required for the wall construction in which the protective is located.

715.4.7 Interior fire window assemblies. Fire-protection-rated glazing used in fire window assemblies located in fire partitions and fire barriers shall be limited to use in assemblies with a maximum fire-resistance rating of 1 hour in accordance with this section.

715.4.7.1 Where permitted. Fire-protection-rated glazing shall be limited to fire partitions designed in accordance with Section 708 and fire barriers utilized in the applications set forth in Sections 706.3.5 and 706.3.6 where the fire-resistance rating does not exceed 1 hour.

715.4.7.2 Size limitations. The total area of windows shall not exceed 25 percent of the area of a common wall with any room.

715.4.8 Exterior fire window assemblies. Exterior openings, other than doors, required to be protected by Section 704.12, where located in a wall required by Table 602 to have a fire-resistance rating of greater than 1 hour, shall be protected with an assembly having a fire protection rating of not less than $1^1/_2$ hours. Exterior openings required to be protected by Section 704.8, where located in a wall required by Table 602 to have a fire-resistance rating of 1 hour, shall be protected with an assembly having a fire protection rating of not less than $^3/_4$ hour. Exterior openings required to be protected by Section 704.9 or 704.10 shall be protected with an assembly having a fire protection rating of not less than $^3/_4$ hour. Openings in nonfire-resistance-rated exterior wall assemblies that require protection in accordance with Section 704.8, 704.9 or 704.10 shall have a fire protection rating of not less than $^3/_4$ hour.

715.4.9 Labeling requirements. Fire-protection-rated glazing shall bear a label or other identification showing the name of the manufacturer, the test standard, and the fire protection rating. Such label or identification shall be issued by an approved agency and shall be permanently affixed.

SECTION 716
DUCTS AND AIR TRANSFER OPENINGS

716.1 General. The provisions of this section shall govern the protection of ducts and air transfer openings in fire-resistance-rated assemblies.

716.1.1 Ducts and air transfer openings without dampers. Ducts and air transfer openings that penetrate fire-resistance-rated assemblies and are not required by this section to have dampers shall comply with the requirements of Section 712.

716.2 Installation. Fire dampers, smoke dampers, combination fire/smoke dampers and ceiling dampers located within air distribution and smoke control systems shall be installed in accordance with the requirements of this section, the manufacturer's installation instructions and listing.

716.2.1 Smoke control system. Where the installation of a fire damper will interfere with the operation of a required smoke control system in accordance with Section 909, approved alternative protection shall be utilized.

716.2.2 Hazardous exhaust ducts. Fire dampers for hazardous exhaust duct systems shall comply with the *International Mechanical Code*.

716.3 Damper testing and ratings. Dampers shall be listed and bear the label of an approved testing agency indicating compliance with the standards in this section. Fire dampers shall comply with the requirements of UL 555. Only fire dampers labeled for use in dynamic systems shall be installed in heating, ventilation and air-conditioning systems designed to operate with fans on during a fire. Smoke dampers shall comply with the requirements of UL 555S. Combination fire/smoke dampers shall comply with the requirements of both UL 555 and UL 555S. Ceiling radiation dampers shall comply with the requirements of UL 555C.

716.3.1 Fire protection rating. Fire dampers shall have the minimum fire protection rating specified in Table 716.3.1 for the type of penetration.

TABLE 716.3.1
FIRE DAMPER RATING

TYPE OF PENETRATION	MINIMUM DAMPER RATING (hours)
Less than 3-hour fire-resistance-rated assemblies	1.5
3-hour or greater fire-resistance-rated assemblies	3

716.3.1.1 Fire damper actuating device. The fire damper actuating device shall meet one of the following requirements:

1. The operating temperature shall be approximately 50°F (10°C) above the normal temperature within the duct system, but not less than 160°F (71°C).

2. The operating temperature shall be not more than 286°F (141°C) where located in a smoke control system complying with Section 909.

3. Where a combination fire/smoke damper is located in a smoke control system complying with Section 909, the operating temperature rating shall be approximately 50°F (10°C) above the maximum smoke control system designed operating temperature, or a maximum temperature of 350°F (177°C). The temperature shall not exceed the UL 555S degradation test temperature rating for a combination fire/smoke damper.

716.3.2 Smoke damper ratings. Smoke damper leakage ratings shall not be less than Class II. Elevated temperature ratings shall not be less than 250°F (121°C).

716.3.2.1 Smoke damper actuation methods. The smoke damper shall close upon actuation of a listed smoke detector or detectors installed in accordance with

Section 907.10 and one of the following methods, as applicable:

1. Where a damper is installed within a duct, a smoke detector shall be installed in the duct within 5 feet (1524 mm) of the damper with no air outlets or inlets between the detector and the damper. The detector shall be listed for the air velocity, temperature and humidity anticipated at the point where it is installed. Other than in mechanical smoke control systems, dampers shall be closed upon fan shutdown where local smoke detectors require a minimum velocity to operate.

2. Where a damper is installed above smoke barrier doors in a smoke barrier, a spot-type detector listed for releasing service shall be installed on either side of the smoke barrier door opening.

3. Where a damper is installed within an unducted opening in a wall, a spot-type detector listed for releasing service shall be installed within 5 feet (1524 mm) horizontally of the damper.

4. Where a damper is installed in a corridor wall, the damper shall be permitted to be controlled by a smoke detection system installed in the corridor.

5. Where a total-coverage smoke detector system is provided within areas served by a heating, ventilation and air-conditioning (HVAC) system, dampers shall be permitted to be controlled by the smoke detection system.

716.4 Access and identification. Fire and smoke dampers shall be provided with an approved means of access, large enough to permit inspection and maintenance of the damper and its operating parts. The access shall not affect the integrity of fire-resistance-rated assemblies. The access openings shall not reduce the fire-resistance rating of the assembly. Access points shall be permanently identified on the exterior by a label having letters not less than 0.5 inch (12.7 mm) in height reading: SMOKE DAMPER or FIRE DAMPER. Access doors in ducts shall be tight fitting and suitable for the required duct construction.

716.5 Where required. Fire dampers, smoke dampers, combination fire/smoke dampers and ceiling radiation dampers shall be provided at the locations prescribed in this section. Where an assembly is required to have both fire dampers and smoke dampers, combination fire/smoke dampers or a fire damper and a smoke damper shall be required.

716.5.1 Fire walls. Ducts and air transfer openings permitted in fire walls in accordance with Section 705.11 shall be protected with approved fire dampers installed in accordance with their listing.

716.5.2 Fire barriers. Ducts and air transfer openings in fire barriers shall be protected with approved fire dampers installed in accordance with their listing.

Exception: Fire dampers are not required at penetrations of fire barriers where any of the following apply:

1. Penetrations are tested in accordance with ASTM E 119 as part of the fire-resistance-rated assembly.

2. Ducts are used as part of an approved smoke control system in accordance with Section 909.

3. Such walls are penetrated by ducted HVAC systems, have a required fire-resistance rating of 1 hour or less, are in areas of other than Group H and are in buildings equipped throughout with an automatic sprinkler system in accordance with Section 903.3.1.1 or 903.3.1.2. For the purposes of this exception, a ducted HVAC system shall be a duct system for conveying supply, return or exhaust air as part of the structure's HVAC system. Such a duct system shall be constructed of sheet steel not less than 26 gage thickness and shall be continuous from the air-handling appliance or equipment to the air outlet and inlet terminals.

716.5.3 Shaft enclosures. Ducts and air transfer openings shall not penetrate a shaft serving as an exit enclosure except as permitted by Section 1019.1.2.

716.5.3.1 Penetrations of shaft enclosures. Shaft enclosures that are permitted to be penetrated by ducts and air transfer openings shall be protected with approved fire and smoke dampers installed in accordance with their listing.

Exceptions:

1. Fire dampers are not required at penetrations of shafts where:

   1.1. Steel exhaust subducts are extended at least 22 inches (559 mm) vertically in exhaust shafts, provided there is a continuous airflow upward to the outside;

   1.2. Penetrations are tested in accordance with ASTM E 119 as part of the rated assembly;

   1.3. Ducts are used as part of an approved smoke control system designed and installed in accordance with Section 909, and where the fire damper will interfere with the operation of the smoke control system; or

   1.4. The penetrations are in parking garage exhaust or supply shafts that are separated from other building shafts by not less than 2-hour fire-resistance-rated construction.

2. In Group B occupancies, equipped throughout with an automatic sprinkler system in accordance with Section 903.3.1.1, smoke dampers are not required at penetrations of shafts where:

   2.1. Bathroom and toilet room exhaust openings with steel exhaust subducts, having a wall thickness of at least 0.019 inches (0.48 mm) that extend at least 22 inches (559 mm) vertically and the exhaust fan at the upper terminus, pow-

ered continuously in accordance with the provisions of Section 909.11, maintains airflow upward to the outside; or

   2.2. Ducts are used as part of an approved smoke control system, designed and installed in accordance with Section 909, and where the smoke damper will interfere with the operation of the smoke control system.

3. Smoke dampers are not required at penetration of exhaust or supply shafts in parking garages that are separated from other building shafts by not less than 2-hour fire-resistance-rated construction.

716.5.4 Fire partitions. Duct penetrations in fire partitions shall be protected with approved fire dampers installed in accordance with their listing.

Exceptions: In occupancies other than Group H, fire dampers are not required where any of the following apply:

1. The partitions are tenant separation and corridor walls in buildings equipped throughout with an automatic sprinkler system in accordance with Section 903.3.1.1 or 903.3.1.2 and the duct is protected as a through penetration in accordance with Section 712.

2. The duct system is constructed of approved materials in accordance with the *International Mechanical Code* and the duct penetrating the wall meets all of the following minimum requirements:

   2.1. The duct shall not exceed 100 square inches (0.06 m²).

   2.2. The duct shall be constructed of steel a minimum of 0.0217 inch (0.55 mm) in thickness.

   2.3. The duct shall not have openings that communicate the corridor with adjacent spaces or rooms.

   2.4. The duct shall be installed above a ceiling.

   2.5. The duct shall not terminate at a wall register in the fire-resistance-rated wall.

   2.6. A minimum 12-inch-long (0.30 m) by 0.060-inch-thick (1.52 mm) steel sleeve shall be centered in each duct opening. The sleeve shall be secured to both sides of the wall and all four sides of the sleeve with minimum 1¹/₂-inch by 1¹/₂-inch by 0.060-inch (0.038 m by 0.038 m by 1.52 mm) steel retaining angles. The retaining angles shall be secured to the sleeve and the wall with No. 10 (M5) screws. The annular space between the steel sleeve and wall opening shall be filled with rock (mineral) wool batting on all sides.

716.5.4.1 Corridors. A listed smoke damper designed to resist the passage of smoke shall be provided at each point a duct or air transfer opening penetrates a corridor enclosure required to have smoke and draft control doors in accordance with Section 715.3.3.

Exceptions:

1. Smoke dampers are not required where the building is equipped throughout with an approved smoke control system in accordance with Section 909, and smoke dampers are not necessary for the operation and control of the system.

2. Smoke dampers are not required in corridor penetrations where the duct is constructed of steel not less than 0.019-inch (0.48 mm) in thickness and there are no openings serving the corridor.

716.5.5 Smoke barriers. A listed smoke damper designed to resist the passage of smoke shall be provided at each point a duct or air transfer opening penetrates a smoke barrier. Smoke dampers and smoke damper actuation methods shall comply with Section 716.3.2.1.

Exception: Smoke dampers are not required where the openings in ducts are limited to a single smoke compartment and the ducts are constructed of steel.

716.6 Horizontal assemblies. Penetrations by ducts and air transfer openings of a floor, floor/ceiling assembly or the ceiling membrane of a roof/ceiling assembly shall be protected by a shaft enclosure that complies with Section 707 or shall comply with this section.

716.6.1 Through penetrations. In occupancies other than Groups I-2 and I-3, a duct and air transfer opening system constructed of approved materials in accordance with the *International Mechanical Code* that penetrates a fire-resistance-rated floor/ceiling assembly that connects not more than two stories is permitted without shaft enclosure protection provided a fire damper is installed at the floor line.

Exception: A duct is permitted to penetrate three floors or less without a fire damper at each floor provided it meets all of the following requirements.

1. The duct shall be contained and located within the cavity of a wall and shall be constructed of steel not less than 0.019 inch (0.48 mm) (26 gage) in thickness.

2. The duct shall open into only one dwelling unit or sleeping unit and the duct system shall be continuous from the unit to the exterior of the building.

3. The duct shall not exceed 4-inch (102 mm) nominal diameter and the total area of such ducts shall not exceed 100 square inches (0.065 m²) in any 100 square feet (9.3 m²) of floor area.

4. The annular space around the duct is protected with materials that prevent the passage of flame and hot gases sufficient to ignite cotton waste where subjected to ASTM E 119 time-temperature

conditions under a minimum positive pressure differential of 0.01 inch (2.49 Pa) of water at the location of the penetration for the time period equivalent to the fire-resistance rating of the construction penetrated

5. Grille openings located in a ceiling of a fire-resistance-rated floor/ceiling or roof/ceiling assembly shall be protected with a ceiling radiation damper in accordance with Section 716.6.2.

716.6.2 Membrane penetrations. Where duct systems constructed of approved materials in accordance with the *International Mechanical Code* penetrate a ceiling of a fire-resistance-rated floor/ceiling or roof/ceiling assembly, shaft enclosure protection is not required provided an approved ceiling radiation damper is installed at the ceiling line. Where a duct is not attached to a diffuser that penetrates a ceiling of a fire-resistance-rated floor/ceiling or roof/ceiling assembly, shaft enclosure protection is not required provided an approved ceiling radiation damper is installed at the ceiling line. Ceiling radiation dampers shall be tested in accordance with UL 555C and constructed in accordance with the details listed in a fire-resistance-rated assembly or shall be labeled to function as a heat barrier for air-handling outlet/inlet penetrations in the ceiling of a fire-resistance-rated assembly. Ceiling radiation dampers shall not be required where ASTM E 119 fire tests have shown that ceiling radiation dampers are not necessary in order to maintain the fire-resistance rating of the assembly. Ceiling radiation dampers shall not be required where exhaust duct penetrations are protected in accordance with Section 712.4.2, where exhaust ducts are located within the cavity of a wall, and where exhaust ducts do not pass through another dwelling unit or tenant space.

716.6.3 Nonfire-resistance-rated assemblies. Duct systems constructed of approved materials in accordance with the *International Mechanical Code* that penetrate nonfire-resistance-rated floor assemblies and that connect not more than two stories are permitted without shaft enclosure protection provided that the annular space between the assembly and the penetrating duct is filled with an approved noncombustible material to resist the free passage of flame and the products of combustion. Duct systems constructed of approved materials in accordance with the *International Mechanical Code* that penetrate nonfire-resistance-rated floor assemblies and that connect not more than three stories are permitted without shaft enclosure protection provided that the annular space between the assembly and the penetrating duct is filled with an approved noncombustible material to resist the free passage of flame and the products of combustion, and a fire damper is installed at each floor line.

Exception: Fire dampers are not required in ducts within individual residential dwelling units.

716.7 Flexible ducts and air connectors. Flexible ducts and air connectors shall not pass through any fire-resistance-rated assembly. Flexible air connectors shall not pass through any wall, floor or ceiling.

SECTION 717
CONCEALED SPACES

717.1 General. Fireblocking and draftstopping shall be installed in combustible concealed locations in accordance with this section. Fireblocking shall comply with Section 717.2. Draftstopping in floor/ceiling spaces and attic spaces shall comply with Sections 717.3 and 717.4, respectively. The permitted use of combustible materials in concealed spaces of noncombustible buildings shall be limited to the applications indicated in Section 717.5.

717.2 Fireblocking. In combustible construction, fireblocking shall be installed to cut off concealed draft openings (both vertical and horizontal) and shall form an effective barrier between floors, between a top story and a roof or attic space. Fireblocking shall be installed in the locations specified in Sections 717.2.2 through 717.2.7.

717.2.1 Fireblocking materials. Fireblocking shall consist of 2-inch (51 mm) nominal lumber or two thicknesses of 1-inch (25 mm) nominal lumber with broken lap joints or one thickness of 0.719-inch (18.3 mm) wood structural panel with joints backed by 0.719-inch (18.3 mm) wood structural panel or one thickness of 0.75-inch (19 mm) particleboard with joints backed by 0.75-inch (19 mm) particleboard. Gypsum board, cement fiber board, batts or blankets of mineral wool or glass fiber or other approved materials installed in such a manner as to be securely retained in place shall be permitted as an acceptable fireblock. Batts or blankets of mineral or glass fiber or other approved nonrigid materials shall be permitted for compliance with the 10-foot (3048 mm) horizontal fireblocking in walls constructed using parallel rows of studs or staggered studs. Loose-fill insulation material shall not be used as a fireblock unless specifically tested in the form and manner intended for use to demonstrate its ability to remain in place and to retard the spread of fire and hot gases. The integrity of fireblocks shall be maintained.

717.2.1.1 Double stud walls. Batts or blankets of mineral or glass fiber or other approved nonrigid materials shall be allowed as fireblocking in walls constructed using parallel rows of studs or staggered studs.

717.2.2 Concealed wall spaces. Fireblocking shall be provided in concealed spaces of stud walls and partitions, including furred spaces, and parallel rows of studs or staggered studs, as follows:

a. Vertically at the ceiling and floor levels.

b. Horizontally at intervals not exceeding 10 feet (3048 mm).

717.2.3 Connections between horizontal and vertical spaces. Fireblocking shall be provided at interconnections between concealed vertical stud wall or partition spaces and concealed horizontal spaces created by an assembly of floor joists or trusses, and between concealed vertical and horizontal spaces such as occur at soffits, drop ceilings, cove ceilings and similar locations.

717.2.4 Stairways. Fireblocking shall be provided in concealed spaces between stair stringers at the top and bottom of the run. Enclosed spaces under stairs shall also comply with Section 1019.1.5.

717.2.5 Ceiling and floor openings. Where annular space protection is provided in accordance with Exception 6 of Section 707.2, Exception 1 of Section 712.4.2, or Section 712.4.3, fireblocking shall be installed at openings around vents, pipes, ducts, chimneys and fireplaces at ceiling and floor levels, with an approved material to resist the free passage of flame and the products of combustion. Factory-built chimneys and fireplaces shall be fireblocked in accordance with UL 103 and UL 127.

717.2.6 Architectural trim. Fireblocking shall be installed within concealed spaces of exterior wall finish and other exterior architectural elements where permitted to be of combustible construction as specified in Section 1406 or where erected with combustible frames, at maximum intervals of 20 feet (6096 mm). If noncontinuous, such elements shall have closed ends, with at least 4 inches (102 mm) of separation between sections.

Exceptions:

1. Fireblocking of cornices is not required in single-family dwellings, as applicable in Section 101.2. Fireblocking of cornices of a two-family dwelling as applicable in Section 101.2 is required only at the line of dwelling unit separation.

2. Fireblocking shall not be required where installed on noncombustible framing and the face of the exterior wall finish exposed to the concealed space is covered by one of the following materials:

   2.1. Aluminum having a minimum thickness of 0.019 inch (0.5 mm).

   2.2. Corrosion-resistant steel having a base metal thickness not less than 0.016 inch (0.4 mm) at any point.

   2.3. Other approved noncombustible materials.

717.2.7 Concealed sleeper spaces. Where wood sleepers are used for laying wood flooring on masonry or concrete fire-resistance-rated floors, the space between the floor slab and the underside of the wood flooring shall be filled with an approved material to resist the free passage of flame and products of combustion or fireblocked in such a manner that there will be no open spaces under the flooring that will exceed 100 square feet (9.3 m²) in area and such space shall be filled solidly under permanent partitions so that there is no communication under the flooring between adjoining rooms.

Exceptions:

1. Fireblocking is not required for slab-on-grade floors in gymnasiums.

2. Fireblocking is required only at the juncture of each alternate lane and at the ends of each lane in a bowling facility.

717.3 Draftstopping in floors. In combustible construction, draftstopping shall be installed to subdivide floor/ceiling assemblies in the locations prescribed in Sections 717.3.2 through 717.3.3.

717.3.1 Draftstopping materials. Draftstopping materials shall not be less than 0.5-inch (12.7 mm) gypsum board, 0.375-inch (9.5 mm) wood structural panel, 0.375-inch (9.5 mm) particleboard or other approved materials adequately supported. The integrity of draftstops shall be maintained.

717.3.2 Groups R-1, R-2, R-3 and R-4. Draftstopping shall be provided in floor/ceiling spaces in Group R-1 buildings, in Group R-2 buildings as applicable in Section 101.2 with three or more dwelling units, in Group R-3 buildings as applicable in Section 101.2 with two dwelling units and in Group R-4 buildings. Draftstopping shall be located above and in line with the dwelling unit and sleeping unit separations.

Exceptions:

1. Draftstopping is not required in buildings equipped throughout with an automatic sprinkler system in accordance with Section 903.3.1.1.

2. Draftstopping is not required in buildings equipped throughout with an automatic sprinkler system in accordance with Section 903.3.1.2, provided that automatic sprinklers are also installed in the combustible concealed spaces.

717.3.3 Other groups. In other groups, draftstopping shall be installed so that horizontal floor areas do not exceed 1,000 square feet (93 m^2).

Exception: Draftstopping is not required in buildings equipped throughout with an automatic sprinkler system in accordance with Section 903.3.1.1.

717.4 Draftstopping in attics. In combustible construction, draftstopping shall be installed to subdivide attic spaces and concealed roof spaces in the locations prescribed in Sections 717.4.2 and 717.4.3. Ventilation of concealed roof spaces shall be maintained in accordance with Section 1203.2.

717.4.1 Draftstopping materials. Materials utilized for draftstopping of attic spaces shall comply with Section 717.3.1.

717.4.1.1 Openings. Openings in the partitions shall be protected by self-closing doors with automatic latches constructed as required for the partitions.

717.4.2 Groups R-1 and R-2. Draftstopping shall be provided in attics, mansards, overhangs or other concealed roof spaces of Group R-2 buildings with three or more dwelling units and in all Group R-1 buildings. Draftstopping shall be installed above, and in line with, sleeping unit and dwelling unit separation walls that do not extend to the underside of the roof sheathing above.

Exceptions:

1. Where corridor walls provide a sleeping unit or dwelling unit separation, draftstopping shall only be required above one of the corridor walls.

2. Draftstopping is not required in buildings equipped throughout with an automatic sprinkler system in accordance with Section 903.3.1.1.

3. In occupancies in Group R-2 that do not exceed four stories in height, the attic space shall be subdivided by draftstops into areas not exceeding 3,000 square feet (279 m^2) or above every two dwelling units, whichever is smaller.

4. Draftstopping is not required in buildings equipped throughout with an automatic sprinkler system in accordance with Section 903.3.1.2, provided that automatic sprinklers are also installed in the combustible concealed spaces.

717.4.3 Other groups. Draftstopping shall be installed in attics and concealed roof spaces, such that any horizontal area does not exceed 3,000 square feet (279 m^2).

Exception: Draftstopping is not required in buildings equipped throughout with an automatic sprinkler system in accordance with Section 903.3.1.1.

717.5 Combustibles in concealed spaces in Type I or II construction. Combustibles shall not be permitted in concealed spaces of buildings of Type I or II construction.

Exceptions:

1. Combustible materials in accordance with Section 603.

2. Combustible materials complying with Section 602 of the *International Mechanical Code*.

3. Class A interior finish materials.

4. Combustible piping within partitions or enclosed shafts installed in accordance with the provisions of this code. Combustible piping shall be permitted within concealed ceiling spaces where installed in accordance with the *International Mechanical Code* and the *International Plumbing Code*.

SECTION 718
FIRE-RESISTANCE REQUIREMENTS
FOR PLASTER

718.1 Thickness of plaster. The minimum thickness of gypsum plaster or portland cement plaster used in a fire-resistance-rated system shall be determined by the prescribed fire tests. The plaster thickness shall be measured from the face of the lath where applied to gypsum lath or metal lath.

718.2 Plaster equivalents. For fire-resistance purposes, 0.5 inch (12.7 mm) of unsanded gypsum plaster shall be deemed equivalent to 0.75 inch (19.1 mm) of one-to-three gypsum sand plaster or 1 inch (25 mm) of portland cement sand plaster.

718.3 Noncombustible furring. In buildings of Type I and II construction, plaster shall be applied directly on concrete or masonry or on approved noncombustible plastering base and furring.

718.4 Double reinforcement. Plaster protection more than 1 inch (25 mm) in thickness shall be reinforced with an additional layer of approved lath embedded at least 0.75 inch (19.1 mm) from the outer surface and fixed securely in place.

> **Exception:** Solid plaster partitions or where otherwise determined by fire tests.

718.5 Plaster alternatives for concrete. In reinforced concrete construction, gypsum plaster or portland cement plaster is permitted to be substituted for 0.5 inch (12.7 mm) of the required poured concrete protection, except that a minimum thickness of 0.375 inch (9.5 mm) of poured concrete shall be provided in reinforced concrete floors and 1 inch (25 mm) in reinforced concrete columns in addition to the plaster finish. The concrete base shall be prepared in accordance with Section 2510.7.

SECTION 719
THERMAL- AND SOUND-INSULATING MATERIALS

719.1 General. Insulating materials, including facings such as vapor retarders and vapor-permeable membranes, similar coverings, and all layers of single and multilayer reflective foil insulations, shall comply with the requirements of this section. Where a flame spread index or a smoke-developed index is specified in this section, such index shall be determined in accordance with ASTM E 84. Any material that is subject to an increase in flame spread index or smoke-developed index beyond the limits herein established through the effects of age, moisture, or other atmospheric conditions shall not be permitted.

> **Exceptions:**
> 1. Fiberboard insulation shall comply with Chapter 23.
> 2. Foam plastic insulation shall comply with Chapter 26.
> 3. Duct and pipe insulation and duct and pipe coverings and linings in plenums shall comply with the *International Mechanical Code*.

719.2 Concealed installation. Insulating materials, where concealed as installed in buildings of any type of construction, shall have a flame spread index of not more than 25 and a smoke-developed index of not more than 450.

> **Exception:** Cellulose loose-fill insulation that is not spray applied, complying with the requirements of Section 719.6, shall only be required to meet the smoke-developed index of not more than 450.

719.2.1 Facings. Where such materials are installed in concealed spaces in buildings of Type III, IV or V construction, the flame spread and smoke-developed limitations do not apply to facings, coverings, and layers of reflective foil insulation that are installed behind and in substantial contact with the unexposed surface of the ceiling, wall or floor finish.

719.3 Exposed installation. Insulating materials, where exposed as installed in buildings of any type of construction, shall

have a flame spread index of not more than 25 and a smoke-developed index of not more than 450.

> **Exception:** Cellulose loose-fill insulation that is not spray applied complying with the requirements of Section 719.6 shall only be required to meet the smoke-developed index of not more than 450.

719.3.1 Attic floors. Exposed insulation materials installed on attic floors shall have a critical radiant flux of not less than 0.12 watt per square centimeter when tested in accordance with ASTM E 970.

719.4 Loose-fill insulation. Loose-fill insulation materials that cannot be mounted in the ASTM E 84 apparatus without a screen or artificial supports shall comply with the flame spread and smoke-developed limits of Sections 719.2 and 719.3 when tested in accordance with CAN/ULC S102.2.

> **Exception:** Cellulose loose-fill insulation shall not be required to comply with this test method, provided such insulation complies with the requirements of Section 719.6.

719.5 Roof insulation. The use of combustible roof insulation not complying with Sections 719.2 and 719.3 shall be permitted in any type of construction provided it is covered with approved roof coverings directly applied thereto.

719.6 Cellulose loose-fill insulation. Cellulose loose-fill insulation shall comply with CPSC 16 CFR, Part 1209 and CPSC 16 CFR, Part 1404. Each package of such insulating material shall be clearly labeled in accordance with CPSC 16 CFR, Part 1209 and CPSC 16 CFR, Part 1404.

719.7 Insulation and covering on pipe and tubing. Insulation and covering on pipe and tubing shall have a flame spread index of not more than 25 and a smoke-developed index of not more than 450.

SECTION 720
PRESCRIPTIVE FIRE RESISTANCE

720.1 General. The provisions of this section contain prescriptive details of fire-resistance-rated building elements. The materials of construction listed in Tables 720.1(1), 720.1(2), and 720.1(3) shall be assumed to have the fire-resistance ratings prescribed therein. Where materials that change the capacity for heat dissipation are incorporated into a fire-resistance-rated assembly, fire test results or other substantiating data shall be made available to the building official to show that the required fire-resistance-rating time period is not reduced.

720.1.1 Thickness of protective coverings. The thickness of fire-resistant materials required for protection of structural members shall be not less than set forth in Table 720.1(1), except as modified in this section. The figures shown shall be the net thickness of the protecting materials and shall not include any hollow space in back of the protection.

720.1.2 Unit masonry protection. Where required, metal ties shall be embedded in transverse joints of unit masonry for protection of steel columns. Such ties shall be as set forth in Table 720.1(1) or be equivalent thereto.

720.1.3 Reinforcement for cast-in-place concrete column protection. Cast-in-place concrete protection for steel columns shall be reinforced at the edges of such members with wire ties of not less than 0.18 inch (4.6 mm) in diameter wound spirally around the columns on a pitch of not more than 8 inches (203 mm) or by equivalent reinforcement.

720.1.4 Plaster application. The finish coat is not required for plaster protective coatings where they comply with the design mix and thickness requirements of Tables 720.1(1), 720.1(2) and 720.1(3).

720.1.5 Bonded prestressed concrete tendons. For members having a single tendon or more than one tendon installed with equal concrete cover measured from the nearest surface, the cover shall not be less than that set forth in Table 720.1(1). For members having multiple tendons installed with variable concrete cover, the average tendon cover shall not be less than that set forth in Table 720.1(1), provided:

1. The clearance from each tendon to the nearest exposed surface is used to determine the average cover.

2. In no case can the clear cover for individual tendons be less than one-half of that set forth in Table 720.1(1). A minimum cover of 0.75 inch (19.1 mm) for slabs and 1 inch (25 mm) for beams is required for any aggregate concrete.

3. For the purpose of establishing a fire-resistance rating, tendons having a clear covering less than that set forth in Table 720.1(1) shall not contribute more than 50 percent of the required ultimate moment capacity for members less than 350 square inches (0.226 m^2) in cross-sectional area and 65 percent for larger members. For structural design purposes, however, tendons having a reduced cover are assumed to be fully effective.

SECTION 721
CALCULATED FIRE RESISTANCE

721.1 General. The provisions of this section contain procedures by which the fire resistance of specific materials or combinations of materials is established by calculations. These procedures apply only to the information contained in this section and shall not be otherwise used. The calculated fire resistance of concrete, concrete masonry, and clay masonry assemblies shall be permitted in accordance with ACI 216.1/TMS 0216.1. The calculated fire resistance of steel assemblies shall be permitted in accordance with Chapter 5 of ASCE/SFPE 29.

721.1.1 Definitions. The following words and terms shall, for the purposes of this chapter and as used elsewhere in this code, have the meanings shown herein.

CERAMIC FIBER BLANKET. A mineral wool insulation material made of alumina-silica fibers and weighing 4 to 10 pounds per cubic foot (pcf) (64 to 160 kg/m^3).

CONCRETE, CARBONATE AGGREGATE. Concrete made with aggregates consisting mainly of calcium or magnesium carbonate, such as limestone or dolomite, and containing 40 percent or less quartz, chert, or flint.

CONCRETE, CELLULAR. A lightweight insulating concrete made by mixing a preformed foam with portland cement slurry and having a dry unit weight of approximately 30 pcf (480 kg/m^3).

CONCRETE, LIGHTWEIGHT AGGREGATE. Concrete made with aggregates of expanded clay, shale, slag or slate or sintered fly ash or any natural lightweight aggregate meeting ASTM C 330 and possessing equivalent fire-resistance properties and weighing 85 to 115 pcf (1360 to 1840 kg/m^3).

CONCRETE, PERLITE. A lightweight insulating concrete having a dry unit weight of approximately 30 pcf (480 kg/m^3) made with perlite concrete aggregate. Perlite aggregate is produced from a volcanic rock which, when heated, expands to form a glass-like material of cellular structure.

CONCRETE, SAND-LIGHTWEIGHT. Concrete made with a combination of expanded clay, shale, slag, slate, sintered fly ash, or any natural lightweight aggregate meeting ASTM C 330 and possessing equivalent fire-resistance properties and natural sand. Its unit weight is generally between 105 and 120 pcf (1680 and 1920 kg/m^3).

CONCRETE, SILICEOUS AGGREGATE. Concrete made with normal-weight aggregates consisting mainly of silica or compounds other than calcium or magnesium carbonate, which contains more than 40-percent quartz, chert, or flint.

CONCRETE, VERMICULITE. A lightweight insulating concrete made with vermiculite concrete aggregate which is laminated micaceous material produced by expanding the ore at high temperatures. When added to a portland cement slurry the resulting concrete has a dry unit weight of approximately 30 pcf (480 kg/m^3).

GLASS FIBERBOARD. Fibrous glass roof insulation consisting of inorganic glass fibers formed into rigid boards using a binder. The board has a top surface faced with asphalt and kraft reinforced with glass fiber.

MINERAL BOARD. A rigid felted thermal insulation board consisting of either felted mineral fiber or cellular beads of expanded aggregate formed into flat rectangular units.

TABLE 720.1(1)
MINIMUM PROTECTION OF STRUCTURAL PARTS BASED ON TIME PERIODS
FOR VARIOUS NONCOMBUSTIBLE INSULATING MATERIALS^m

STRUCTURAL PARTS TO BE PROTECTED	ITEM NUMBER	INSULATING MATERIAL USED	MINIMUM THICKNESS OF INSULATING MATERIAL FOR THE FOLLOWING FIRE-RESISTANCE PERIODS (inches)			
			4 hour	3 hour	2 hour	1 hour
1. Steel columns and all of primary trusses	1-1.1	Carbonate, lightweight and sand-lightweight aggregate concrete, members 6″ × 6″ or greater (not including sandstone, granite and siliceous gravel).^a	$2^1/_2$	2	$1^1/_2$	1
	1-1.2	Carbonate, lightweight and sand-lightweight aggregate concrete, members 8″ × 8″ or greater (not including sandstone, granite and siliceous gravel).^a	2	$1^1/_2$	1	1
	1-1.3	Carbonate, lightweight and sand-lightweight aggregate concrete, members 12″ × 12″or greater (not including sandstone, granite and siliceous gravel).^a	$1^1/_2$	1	1	1
	1-1.4	Siliceous aggregate concrete and concrete excluded in Item 1-1.1, members 6″ × 6″ or greater.^a	3	2	$1^1/_2$	1
	1-1.5	Siliceous aggregate concrete and concrete excluded in Item 1-1.1, members 8″ × 8″ or greater.^a	$2^1/_2$	2	1	1
	1-1.6	Siliceous aggregate concrete and concrete excluded in Item 1-1.1, members 12″ × 12″ or greater.^a	2	1	1	1
	1-.2.1	Clay or shale brick with brick and mortar fill.^a	$3^3/_4$	—	—	$2^1/_4$
	1-3.1	4″ hollow clay tile in two 2″ layers; $^1/_2$″ mortar between tile and column; $^3/_8$″ metal mesh 0.046″ wire diameter in horizontal joints; tile fill.^a	4	—	—	—
	1-3.2	2″ hollow clay tile; $^3/_4$″ mortar between tile and column; $^3/_8$″metal mesh 0.046″ wire diameter in horizontal joints; limestone concrete fill;^a plastered with $^3/_4$″ gypsum plaster.	3	—	—	—
	1-3.3	2″ hollow clay tile with outside wire ties 0.08″ diameter at each course of tile or $^3/_8$″ metal mesh 0.046″ diameter wire in horizontal joints; limestone or trap-rock concrete fill^1 extending 1″ outside column on all sides	—	—	3	—
	1-3.4	2″ hollow clay tile with outside wire ties 0.08″ diameter at each course of tile with or without concrete fill; $^3/_4$″ mortar between tile and column.	—	—	—	2
	1-4.1	Cement plaster over metal lath wire tied to $^3/_4$″ cold-rolled vertical channels with 0.049″ (No. 18 B.W. gage) wire ties spaced 3″ to 6″ on center. Plaster mixed 1:2 $^1/_2$ by volume, cement to sand.	—	—	$2^1/_2$^b	$^7/_8$
	1-5.1	Vermiculite concrete, 1:4 mix by volume over paperbacked wire fabric lath wrapped directly around column with additional 2″ × 2″ 0.065″/0.065″ (No. 16/16 B.W. gage) wire fabric placed $^3/_4$″ from outer concrete surface. Wire fabric tied with 0.049″ (No. 18 B.W. gage) wire spaced 6″ on center for inner layer and 2″ on center for outer layer.	2	—	—	—
	1-6.1	Perlite or vermiculite gypsum plaster over metal lath wrapped around column and furred $1^1/_4$″ from column flanges. Sheets lapped at ends and tied at 6″ intervals with 0.049″ (No. 18 B.W. gage) tie wire. Plaster pushed through to flanges.	$1^1/_2$	1	—	—
	1-6.2	Perlite or vermiculite gypsum plaster over self-furring metal lath wrapped directly around column, lapped 1″ and tied at 6″ intervals with 0.049″ (No. 18 B.W. gage) wire.	$1^3/_4$	$1^3/_8$	1	—
	1-6.3	Perlite or vermiculite gypsum plaster on metal lath applied to $^3/_4$″ cold-rolled channels spaced 24″ apart vertically and wrapped flatwise around column.	$1^1/_2$	—	—	—
	1-6.4	Perlite or vermiculite gypsum plaster over two layers of $^1/_2$″ plain full-length gypsum lath applied tight to column flanges. Lath wrapped with 1″ hexagonal mesh of No. 20 gage wire and tied with doubled 0.035″ diameter (No. 18 B.W. gage) wire ties spaced 23″ on center. For three-coat work, the plaster mix for the second coat shall not exceed 100 pounds of gypsum to $2^1/_2$ cubic feet of aggregate for the 3-hour system.	$2^1/_2$	2	—	—

(continued)

TABLE 720.1(1)—continued
MINIMUM PROTECTION OF STRUCTURAL PARTS BASED ON TIME PERIODS
FOR VARIOUS NONCOMBUSTIBLE INSULATING MATERIALS[m]

STRUCTURAL PARTS TO BE PROTECTED	ITEM NUMBER	INSULATING MATERIAL USED	MINIMUM THICKNESS OF INSULATING MATERIAL FOR THE FOLLOWING FIRE-RESISTANCE PERIODS (inches)			
			4 hour	3 hour	2 hour	1 hour
1. Steel columns and all of primary trusses (continued)	1-6.5	Perlite or vermiculate gypsum plaster over one layer of $^1/_2''$ plain full-length gypsum lath applied tight to column flanges. Lath tied with doubled 0.049" (No. 18 B.W. gage) wire ties spaced 23" on center and scratch coat wrapped with 1" hexagonal mesh 0.035" (No. 20 B.W. gage) wire fabric. For three-coat work, the plaster mix for the second coat shall not exceed 100 pounds of gypsum to $2^1/_2$ cubic feet of aggregate.	—	2	—	—
	1-7.1	Multiple layers of $^1/_2''$ gypsum wallboard[c] adhesively[d] secured to column flanges and successive layers. Wallboard applied without horizontal joints. Corner edges of each layer staggered. Wallboard layer below outer layer secured to column with doubled 0.049" (No. 18 B.W. gage) steel wire ties spaced 15" on center. Exposed corners taped and treated.	—	—	2	1
	1-7.2	Three layers of $^5/_8''$ Type X gypsum wallboard.[c] First and second layer held in place by $^1/_8''$ diameter by $1^3/_8''$ long ring shank nails with $^5/_{16}''$ diameter heads spaced 24" on center at corners. Middle layer also secured with metal straps at mid-height and 18" from each end, and by metal corner bead at each corner held by the metal straps. Third layer attached to corner bead with 1" long gypsum wallboard screws spaced 12" on center.	—	—	$1^7/_8$	—
	1-7.3	Three layers of $^5/_8''$ Type X gypsum wallboard,[c] each layer screw attached to $1^5/_8''$ steel studs 0.018" thick (No. 25 carbon sheet steel gage) at each corner of column. Middle layer also secured with 0.049" (No. 18 B.W. gage) double-strand steel wire ties, 24" on center. Screws are No. 6 by 1" spaced 24" on center for inner layer, No. 6 by $1^5/_8''$ spaced 12" on center for middle layer and No. 8 by $2^1/_4''$ spaced 12" on center for outer layer.	—	$1^7/_8$	—	—
	1-8.1	Wood-fibered gypsum plaster mixed 1:1 by weight gypsum-to-sand aggregate applied over metal lath. Lath lapped 1" and tied 6" on center at all end, edges and spacers with 0.049" (No. 18 B.W. gage) steel tie wires. Lath applied over $^1/_2''$ spacers made of $^3/_4''$ furring channel with 2" legs bent around each corner. Spacers located 1" from top and bottom of member and a maximum of 40" on center and wire tied with a single strand of 0.049" (No. 18 B.W. gage) steel tie wires. Corner bead tied to the lath at 6" on center along each corner to provide plaster thickness.	—	—	$1^5/_8$	—
2. Webs or flanges of steel beams and girders	2-1.1	Carbonate, lightweight and sand-lightweight aggregate concrete (not including sandstone, granite and siliceous gravel) with 3" or finer metal mesh placed 1" from the finished surface anchored to the top flange and providing not less than 0.025 square inch of steel area per foot in each direction.	2	$1^1/_2$	1	1
	2-1.2	Siliceous aggregate concrete and concrete excluded in Item 2-1.1 with 3" or finer metal mesh placed 1" from the finished surface anchored to the top flange and providing not less than 0.025 square inch of steel area per foot in each direction.	$2^1/_2$	2	$1^1/_2$	1
	2-2.1	Cement plaster on metal lath attached to $^3/_4''$ cold-rolled channels with 0.049" (No. 18 B.W. gage) wire ties spaced 3" to 6" on center. Plaster mixed $1:2^1/_2$ by volume, cement to sand.	—	—	$2^1/_2$[b]	$^7/_8$
	2-3.1	Vermiculite gypsum plaster on a metal lath cage, wire tied to 0.165" diameter (No. 8 B.W. gage) steel wire hangers wrapped around beam and spaced 16" on center. Metal lath ties spaced approximately 5" on center at cage sides and bottom.	—	$^7/_8$	—	—

(continued)

TABLE 720.1(1)—continued
**MINIMUM PROTECTION OF STRUCTURAL PARTS BASED ON TIME PERIODS
FOR VARIOUS NONCOMBUSTIBLE INSULATING MATERIALS[m]**

STRUCTURAL PARTS TO BE PROTECTED	ITEM NUMBER	INSULATING MATERIAL USED	MINIMUM THICKNESS OF INSULATING MATERIAL FOR THE FOLLOWING FIRE-RESISTANCE PERIODS (inches)			
			4 hour	3 hour	2 hour	1 hour
2. Webs or flanges of steel beams and girders (continued)	2-4.1	Two layers of ⁵⁄₈″ Type X gypsum wallboard[c] are attached to U-shaped brackets spaced 24″ on center. 0.018″ thick (No. 25 carbon sheet steel gage) 1⁵⁄₈″ deep by 1″ galvanized steel runner channels are first installed parallel to and on each side of the top beam flange to provide a ¹⁄₂″ clearance to the flange. The channel runners are attached to steel deck or concrete floor construction with approved fasteners spaced 12″ on center. U-shaped brackets are formed from members identical to the channel runners. At the bent portion of the U-shaped bracket, the flanges of the channel are cut out so that 1⁵⁄₈″ deep corner channels can be inserted without attachment parallel to each side of the lower flange. As an alternate, 0.021″ thick (No. 24 carbon sheet steel gage) 1″ × 2″ runner and corner angles may be used in lieu of channels, and the web cutouts in the U-shaped brackets may be omitted. Each angle is attached to the bracket with ¹⁄₂″-long No. 8 self-drilling screws. The vertical legs of the U-shaped bracket are attached to the runners with one ¹⁄₂″ long No. 8 self-drilling screw. The completed steel framing provides a 2¹⁄₈″ and 1¹⁄₂″ space between the inner layer of wallboard and the sides and bottom of the steel beam, respectively. The inner layer of wallboard is attached to the top runners and bottom corner channels or corner angles with 1¹⁄₄″-long No. 6 self-drilling screws spaced 16″ on center. The outer layer of wallboard is applied with 1³⁄₄″-long No. 6 self-drilling screws spaced 8″ on center. The bottom corners are reinforced with metal corner beads.	—	—	1¹⁄₄	—
	2-4.2	Three layers of ⁵⁄₈″ Type X gypsum wallboard[c] attached to a steel suspension system as described immediately above utilizing the 0.018″ thick (No. 25 carbon sheet steel gage) 1″ × 2″ lower corner angles. The framing is located so that a 2¹⁄₈″ and 2″ space is provided between the inner layer of wallboard and the sides and bottom of the beam, respectively. The first two layers of wallboard are attached as described immediately above. A layer of 0.035″ thick (No. 20 B.W. gage) 1″ hexagonal galvanized wire mesh is applied under the soffit of the middle layer and up the sides approximately 2″. The mesh is held in position with the No. 6 1⁵⁄₈″-long screws installed in the vertical leg of the bottom corner angles. The outer layer of wallboard is attached with No. 6 2¹⁄₄″-long screws spaced 8″ on center. One screw is also installed at the mid-depth of the bracket in each layer. Bottom corners are finished as described above.	—	1⁷⁄₈	—	—

3. Bonded pretensioned reinforcement in prestressed concrete[e]	3-1.1	Carbonate, lightweight, sand-lightweight and siliceous[f] aggregate concrete
		        Beams or girders
		
		        Solid slabs[h]
			4[g]
			
				3[g]
				
				2
					2¹⁄₂
					
					1¹⁄₂
						1¹⁄₂
						
						1

4. Bonded or unbonded post-tensioned tendons in prestressed concrete[e, i]	4-1.1	Carbonate, lightweight, sand-lightweight and siliceous[f] aggregate concrete
		Unrestrained members:
		    Solid slabs[h]
		    Beams and girders[j]
		        8″ wide
		        greater than 12″ wide
			—
			
			3
				2
				
				4¹⁄₂
				2¹⁄₂
					1¹⁄₂
					
					2¹⁄₂
					2
						—
						
						1³⁄₄
						1¹⁄₂

	4-1.2	Carbonate, lightweight, sand-lightweight and siliceous aggregate
		Restrained members:[k]
		    Solid slabs[h]
		    Beams and girders[j]
		        8″ wide
		        greater than 12″ wide
			1¹⁄₄
			
			2¹⁄₂
			2
				1
				
				2
				1³⁄₄
					³⁄₄
					
					1³⁄₄
					1¹⁄₂
						—
						
						—
						—

(continued)

TABLE 720.1(1)—continued
MINIMUM PROTECTION OF STRUCTURAL PARTS BASED ON TIME PERIODS
FOR VARIOUS NONCOMBUSTIBLE INSULATING MATERIALS[m]

STRUCTURAL PARTS TO BE PROTECTED	ITEM NUMBER	INSULATING MATERIAL USED	MINIMUM THICKNESS OF INSULATING MATERIAL FOR THE FOLLOWING FIRE-RESISTANCE PERIODS (inches)			
			4 hour	3 hour	2 hour	1 hour
5. Reinforcing steel in reinforced concrete columns, beams girders and trusses	5-1.1	Carbonate, lightweight and sand-lightweight aggregate concrete, members 12″ or larger, square or round. (Size limit does not apply to beams and girders monolithic with floors.)	$1^1/_2$	$1^1/_2$	$1^1/_2$	$1^1/_2$
		Siliceous aggregate concrete, members 12″ or larger, square or round. (Size limit does not apply to beams and girders monolithic with floors.)	2	$1^1/_2$	$1^1/_2$	$1^1/_2$
6. Reinforcing steel in reinforced concrete joists[l]	6-1.1	Carbonate, lightweight and sand-lightweight aggregate concrete.	$1^1/_4$	$1^1/_4$	1	$^3/_4$
	6-1.2	Siliceous aggregate concrete.	$1^3/_4$	$1^1/_2$	1	$^3/_4$
7. Reinforcing and tie rods in floor and roof slabs[l]	7-1.1	Carbonate, lightweight and sand-lightweight aggregate concrete.	1	1	$^3/_4$	$^3/_4$
	7-1.2	Siliceous aggregate concrete.	$1^1/_4$	1	1	$^3/_4$

For SI: 1 inch = 25.4 mm, 1 square inch = 645.2 mm^2, 1 cubic foot = 0.0283 m^3.

a. Reentrant parts of protected members to be filled solidly.

b. Two layers of equal thickness with a $^3/_4$-inch airspace between.

c. For all of the construction with gypsum wallboard described in Table 720.1(1), gypsum base for veneer plaster of the same size, thickness and core type shall be permitted to be substituted for gypsum wallboard, provided attachment is identical to that specified for the wallboard and the joints on the face layer are reinforced, and the entire surface is covered with a minimum of $^1/_{16}$-inch gypsum veneer plaster.

d. An approved adhesive qualified under ASTM E 119.

e. Where lightweight or sand-lightweight concrete having an oven-dry weight of 110 pounds per cubic foot or less is used, the tabulated minimum cover shall be permitted to be reduced 25 percent, except that in no case shall the cover be less than $^3/_4$ inch in slabs or $1^1/_2$ inches in beams or girders.

f. For solid slabs of siliceous aggregate concrete, increase tendon cover 20 percent.

g. Adequate provisions against spalling shall be provided by U-shaped or hooped stirrups spaced not to exceed the depth of the member with a clear cover of 1 inch.

h. Prestressed slabs shall have a thickness not less than that required in Table 720.1(3) for the respective fire resistance time period.

i. Fire coverage and end anchorages shall be as follows: Cover to the prestressing steel at the anchor shall be $^1/_2$ inch greater than that required away from the anchor. Minimum cover to steel-bearing plate shall be 1 inch in beams and $^3/_4$ inch in slabs.

j. For beam widths between 8 inches and 12 inches, cover thickness shall be permitted to be determined by interpolation.

k. Interior spans of continuous slabs, beams and girders shall be permitted to be considered restrained.

l. For use with concrete slabs having a comparable fire endurance where members are framed into the structure in such a manner as to provide equivalent performance to that of monolithic concrete construction.

m. Generic fire-resistance ratings (those not designated as PROPRIETARY* in the listing) in GA 600 shall be accepted as if herein listed.

TABLE 720.1(2)
RATED FIRE-RESISTANCE PERIODS FOR VARIOUS WALLS AND PARTITIONS [a,o,p]

MATERIAL	ITEM NUMBER	CONSTRUCTION	MINIMUM FINISHED THICKNESS FACE-TO-FACE[b] (inches)			
			4 hour	3 hour	2 hour	1 hour
1. Brick of clay or shale	1-1.1	Solid brick of clay or shale[c]	6	4.9	3.8	2.7
	1-1.2	Hollow brick, not filled.	5.0	4.3	3.4	2.3
	1-1.3	Hollow brick unit wall, grout or filled with perlite vermiculite or expanded shale aggregate.	6.6	5.5	4.4	3.0
	1-2.1	4″ nominal thick units at least 75 percent solid backed with a hat-shaped metal furring channel $^3/_4$″ thick formed from 0.021″ sheet metal attached to the brick wall on 24″ centers with approved fasteners, and $^1/_2$″ Type X gypsum wallboard attached to the metal furring strips with 1″-long Type S screws spaced 8″ on center.	—	—	5[d]	—
2. Combination of clay brick and load-bearing hollow clay tile	2-1.1	4″ solid brick and 4″ tile (at least 40 percent solid).	—	8	—	—
	2-1.2	4″ solid brick and 8″ tile (at least 40 percent solid).	12	—	—	—
3. Concrete masonry units	3-1.1[f, g]	Expanded slag or pumice.	4.7	4.0	3.2	2.1
	3-1.2[f, g]	Expanded clay, shale or slate.	5.1	4.4	3.6	2.6
	3-1.3[f]	Limestone, cinders or air-cooled slag.	5.9	5.0	4.0	2.7
	3-1.4[f, g]	Calcareous or siliceous gravel.	6.2	5.3	4.2	2.8
4. Solid concrete[h, i]	4-1.1	Siliceous aggregate concrete.	7.0	6.2	5.0	3.5
		Carbonate aggregate concrete.	6.6	5.7	4.6	3.2
		Sand-lightweight concrete.	5.4	4.6	3.8	2.7
		Lightweight concrete.	5.1	4.4	3.6	2.5
5. Glazed or unglazed facing tile, nonload-bearing	5-1.1	One 2″ unit cored 15 percent maximum and one 4″ unit cored 25 percent maximum with $^3/_4$″ mortar-filled collar joint. Unit positions reversed in alternate courses.	—	$6^3/_8$	—	—
	5-1.2	One 2″ unit cored 15 percent maximum and one 4″ unit cored 40 percent maximum with $^3/_4$″mortar-filled collar joint. Unit positions side with $^3/_4$″ gypsum plaster. Two wythes tied together every fourth course with No. 22 gage corrugated metal ties.	—	$6^3/_4$	—	—
	5-1.3	One unit with three cells in wall thickness, cored 29 percent maximum.	—	—	6	—
	5-1.4	One 2″ unit cored 22 percent maximum and one 4″ unit cored 41 percent maximum with $^1/_4$″mortar-filled collar joint. Two wythes tied together every third course with 0.030″ (No. 22 galvanized sheet steel gage) corrugated metal ties.	—	—	6	—
	5-1.5	One 4″ unit cored 25 percent maximum with $^3/_4$″ gypsum plaster on one side.	—	—	$4^3/_4$	—
	5-1.6	One 4″ unit with two cells in wall thickness, cored 22 percent maximum.	—	—	—	4
	5-1.7	One 4″ unit cored 30 percent maximum with $^3/_4$″ vermiculite gypsum plaster on one side.	—	—	$4^1/_2$	—
	5-1.8	One 4″ unit cored 39 percent maximum with $^3/_4$″ gypsum plaster on one side.	—	—	—	$4^1/_2$

(continued)

TABLE 720.1(2)—continued
RATED FIRE-RESISTANCE PERIODS FOR VARIOUS WALLS AND PARTITIONS [a,o,p]

MATERIAL	ITEM NUMBER	CONSTRUCTION	MINIMUM FINISHED THICKNESS FACE-TO-FACE[b] (inches)			
			4 hour	3 hour	2 hour	1 hour
6. Solid gypsum plaster	6-1.1	$^3/_4''$ by 0.055″ (No. 16 carbon sheet steel gage) vertical cold-rolled channels, 16″ on center with 2.6-pound flat metal lath applied to one face and tied with 0.049″ (No. 18 B.W. Gage) wire at 6″ spacing. Gypsum plaster each side mixed 1:2 by weight, gypsum to sand aggregate.	—	—	—	2^d
	6-1.2	$^3/_4''$ by 0.055″ (No. 16 carbon sheet steel gage) cold-rolled channels 16″ on center with metal lath applied to one face and tied with 0.049″ (No. 18 B.W. gage) wire at 6″ spacing. Perlite or vermiculite gypsum plaster each side. For three-coat work, the plaster mix for the second coat shall not exceed 100 pounds of gypsum to $2^1/_2$ cubic feet of aggregate for the 1-hour system.	—	—	$2^1/_2{}^d$	2^d
	6-1.3	$^3/_4''$ by 0.055″ (No. 16 carbon sheet steel gage) vertical cold-rolled channels, 16″ on center with $^3/_8''$ gypsum lath applied to one face and attached with sheet metal clips. Gypsum plaster each side mixed 1:2 by weight, gypsum to sand aggregate.	—	—	—	2^d
	6-2.1	Studless with $^1/_2''$ full-length plain gypsum lath and gypsum plaster each side. Plaster mixed 1:1 for scratch coat and 1:2 for brown coat, by weight, gypsum to sand aggregate.	—	—	—	2^d
	6-2.2	Studless with $^1/_2''$ full-length plain gypsum lath and perlite or vermiculite gypsum plaster each side.	—	—	$2^1/_2{}^d$	2^d
	6-2.3	Studless partition with $^3/_8''$ rib metal lath installed vertically adjacent edges tied 6″ on center with No. 18 gage wire ties, gypsum plaster each side mixed 1:2 by weight, gypsum to sand aggregate.	—	—	—	2^d
7. Solid perlite and portland cement	7-1.1	Perlite mixed in the ratio of 3 cubic feet to 100 pounds of portland cement and machine applied to stud side of $1^1/_2''$ mesh by 0.058-inch (No. 17 B.W. gage) paper-backed woven wire fabric lath wire-tied to 4″-deep steel trussed wire[j] studs 16″ on center. Wire ties of 0.049″ (No. 18 B.W. gage) galvanized steel wire 6″ on center vertically.	—	—	$3^1/_8{}^d$	—
8. Solid neat wood fibered gypsum plaster	8-1.1	$^3/_4''$ by 0.055-inch (No. 16 carbon sheet steel gage) cold-rolled channels, 12″ on center with 2.5-pound flat metal lath applied to one face and tied with 0.049″ (No. 18 B.W. gage) wire at 6″ spacing. Neat gypsum plaster applied each side.	—	—	2^d	—
9. Solid wallboard partition	9-1.1	One full-length layer $^1/_2''$ Type X gypsum wallboard[e] laminated to each side of 1″ full-length V-edge gypsum coreboard with approved laminating compound. Vertical joints of face layer and coreboard staggered at least 3″.	—	—	2^d	—
10. Hollow (studless) gypsum wallboard partition	10-1.1	One full-length layer of $^5/_8''$ Type X gypsum wallboard[e] attached to both sides of wood or metal top and bottom runners laminated to each side of 1″ × 6″ full-length gypsum coreboard ribs spaced 24″ on center with approved laminating compound. Ribs centered at vertical joints of face plies and joints staggered 24″ in opposing faces. Ribs may be recessed 6″ from the top and bottom.	—	—	—	$2^1/_4{}^d$
	10-1.2	1″ regular gypsum V-edge full-length backing board attached to both sides of wood or metal top and bottom runners with nails or $1^5/_8''$ drywall screws at 24″ on center. Minimum width of rumors $1^5/_8''$. Face layer of $^1/_2''$ regular full-length gypsum wallboard laminated to outer faces of backing board with approved laminating compound.	—	—	$4^5/_8{}^d$	—

(continued)

TABLE 720.1(2)—continued
RATED FIRE-RESISTANCE PERIODS FOR VARIOUS WALLS AND PARTITIONS [a,o,p]

MATERIAL	ITEM NUMBER	CONSTRUCTION	MINIMUM FINISHED THICKNESS FACE-TO-FACE[b] (inches)			
			4 hour	3 hour	2 hour	1 hour
11. Noncombustible studs—interior partition with plaster each side	11-1.1	$3^1/_4'' \times 0.044''$ (No. 18 carbon sheet steel gage) steel studs spaced 24″ on center. $^5/_8''$ gypsum plaster on metal lath each side mixed 1:2 by weight, gypsum to sand aggregate.	—	—	—	$4^3/_4^d$
	11-1.2	$3^3/_8'' \times 0.055''$ (No. 16 carbon sheet steel gage) approved nailable[k] studs spaced 24" on center. $^5/_8''$ neat gypsum wood-fibered plaster each side over $^3/_8''$ rib metal lath nailed to studs with 6d common nails, 8″ on center. Nails driven $1^1/_4''$ and bent over.	—	—	$5^5/_8$	—
	11-1.3	$4'' \times 0.044''$ (No. 18 carbon sheet steel gage) channel-shaped steel studs at 16″ on center. On each side approved resilient clips pressed onto stud flange at 16″ vertical spacing, $^1/_4''$ pencil rods snapped into or wire tied onto outer loop of clips, metal lath wire-tied to pencil rods at 6″ intervals, 1″ perlite gypsum plaster, each side.	—	$7^5/_8^d$	—	—
	11-1.4	$2^1/_2'' \times 0.044''$ (No. 18 carbon sheet steel gage) steel studs spaced 16″ on center. Wood fibered gypsum plaster mixed 1:1 by weight gypsum to sand aggregate applied on $^3/_4$-pound metal lath wire tied to studs, each side. $^3/_4''$ plaster applied over each face, including finish coat.	—	—	$4^1/_4^d$	—
12. Wood studs interior partition with plaster each side	12-1.1[l,m]	$2'' \times 4''$ wood studs 16″ on center with $^5/_8''$ gypsum plaster on metal lath. Lath attached by 4d common nails bent over or No. 14 gage by $1^1/_4''$ by $^3/_4''$ crown width staples spaced 6″ on center. Plaster mixed $1:1^1/_2$ for scratch coat and 1:3 for brown coat, by weight, gypsum to sand aggregate.	—	—	—	$5^1/_8$
	12-1.2[l]	$2'' \times 4''$ wood studs 16″ on center with metal lath and $^7/_8''$ neat wood-fibered gypsum plaster each side. Lath attached by 6d common nails, 7″ on center. Nails driven $1^1/_4''$ and bent over.	—	—	$5^1/_2^d$	—
	12-1.3[l]	$2'' \times 4''$ wood studs 16″ on center with $^3/_8''$ perforated or plain gypsum lath and $^1/_2''$ gypsum plaster each side. Lath nailed with $1^1/_8''$ by No. 13 gage by $^{19}/_{64}''$ head plasterboard blued nails, 4″ on center. Plaster mixed 1:2 by weight, gypsum to sand aggregate.	—	—	—	$5^1/_4$
	12-1.4[l]	$2'' \times 4''$ wood studs 16″ on center with $^3/_8''$ Type X gypsum lath and $^1/_2''$ gypsum plaster each side. Lath nailed with $1^1/_8''$ by No. 13 gage by $^{19}/_{64}''$ head plasterboard blued nails, 5″ on center. Plaster mixed 1:2 by weight, gypsum to sand aggregate.	—	—	—	$5^1/_4$
13.Noncumbustible studs—interior partition with gypsum wallboard each side	13-1.1	0.018″ (No. 25 carbon sheet steel gage) channel-shaped studs 24″ on center with one full-length layer of $^5/_8''$ Type X gypsum wallboard[e] applied vertically attached with 1″ long No. 6 drywall screws to each stud. Screws are 8″ on center around the perimeter and 12″ on center on the intermediate stud. The wallboard may be applied horizontally when attached to $3^5/_8''$ studs and the horizontal joints are staggered with those on the opposite side. Screws for the horizontal application shall be 8″ on center at vertical edges and 12″ on center at intermediate studs.	—	—	—	$2^7/_8^d$
	13-1.2	0.018″ (No. 25 carbon sheet steel gage) channel-shaped studs 25″ on center with two full-length layers of $^1/_2''$ Type X gypsum wallboard[e] applied vertically each side. First layer attached with 1″-long, No. 6 drywall screws, 8″ on center around the perimeter and 12″ on center on the intermediate stud. Second layer applied with vertical joints offset one stud space from first layer using $1^5/_8''$ long, No. 6 drywall screws spaced 9″ on center along vertical joints, 12″ on center at intermediate studs and 24″ on center along top and bottom runners.	—	—	$3^5/_8^d$	—
	13-1.3	0.055″ (No. 16 carbon sheet steel gage) approved nailable metal studs[e] 24″ on center with full-length $^5/_8''$ Type X gypsum wallboard[e] applied vertically and nailed 7″ on center with 6d cement-coated common nails. Approved metal fastener grips used with nails at vertical butt joints along studs.	—	—	—	$4^7/_8$

(continued)

TABLE 720.1(2)—continued
RATED FIRE-RESISTANCE PERIODS FOR VARIOUS WALLS AND PARTITIONS [a,o,p]

MATERIAL	ITEM NUMBER	CONSTRUCTION	MINIMUM FINISHED THICKNESS FACE-TO-FACE[b] (inches)			
			4 hour	3 hour	2 hour	1 hour
14. Wood studs—interior partition with gypsum wallboard each side	14-1.1[h, m]	$2'' \times 4''$ wood studs $16''$ on center with two layers of $3/8''$ regular gypsum wallboard[e] each side, 4d cooler[n] or wallboard[n] nails at $8''$ on center first layer, 5d cooler[n] or wallboard[n] nails at $8''$ on center second layer with laminating compound between layers, joints staggered. First layer applied full length vertically, second layer applied horizontally or vertically	—	—	—	5
	14-1.2[l, m]	$2'' \times 4''$ wood studs $16''$ on center with two layers $1/2''$ regular gypsum wallboard[e] applied vertically or horizontally each side[k], joints staggered. Nail base layer with 5d cooler[n] or wallboard[n] nails at $8''$ on center face layer with 8d cooler[n] or wallboard[n] nails at $8''$ on center.	—	—	—	$5^{1}/_{2}$
	14-1.3[l, m]	$2'' \times 4''$ wood studs $24''$ on center with $5/8''$ Type X gypsum wallboard[e] applied vertically or horizontally nailed with 6d cooler[n] or wallboard[n] nails at $7''$ on center with end joints on nailing members. Stagger joints each side.	—	—	—	$4^{3}/_{4}$
	14-1.4[l]	$2'' \times 4''$ fire-retardant-treated wood studs spaced $24''$ on center with one layer of $5/8''$ Type X gypsum wallboard[e] applied with face paper grain (long dimension) parallel to studs. Wallboard attached with 6d cooler[n] or wallboard[n] nails at $7''$ on center.	—	—	—	$4^{3}/_{4}$[d]
	14-1.5[l, m]	$2'' \times 4''$ wood studs $16''$ on center with two layers $5/8''$ Type X gypsum wallboard[e] each side. Base layers applied vertically and nailed with 6d cooler[n] or wallboard[n] nails at $9''$ on center. Face layer applied vertically or horizontally and nailed with 8d cooler[n] or wallboard[n] nails at $7''$ on center. For nail-adhesive application, base layers are nailed $6''$ on center. Face layers applied with coating of approved wallboard adhesive and nailed $12''$ on center.	—	—	6	—
	14-1.6[l]	$2'' \times 3''$ fire-retardant-treated wood studs spaced $24''$ on center with one layer of $5/8''$ Type X gypsum wallboard[e] applied with face paper grain (long dimension) at right angles to studs. Wallboard attached with 6d cement-coated box nails spaced $7''$ on center.	—	—	—	$3^{5}/_{8}$[d]
15. Exterior or interior walls	15-1.1[l, m]	Exterior surface with $3/4''$ drop siding over $1/2''$ gypsum sheathing on $2'' \times 4''$ wood studs at $16''$ on center, interior surface treatment as required for 1-hour-rated exterior or interior $2'' \times 4''$ wood stud partitions. Gypsum sheathing nailed with $1^{3}/_{4}''$ by No. 11 gage by $7/16''$ head galvanized nails at $8''$ on center. Siding nailed with 7d galvanized smooth box nails.	—	—	—	Varies
	15-1.2[l, m]	$2'' \times 4''$ wood studs $16''$ on center with metal lath and $3/4''$ cement plaster on each side. Lath attached with 6d common nails $7''$ on center driven to $1''$ minimum penetration and bent over. Plaster mix 1:4 for scratch coat and 1:5 for brown coat, by volume, cement to sand.	—	—	—	$5^{3}/_{8}$
	15-1.3[l, m]	$2'' \times 4''$ wood studs $16''$ on center with $7/8''$ cement plaster (measured from the face of studs) on the exterior surface with interior surface treatment as required for interior wood stud partitions in this table. Plaster mix 1:4 for scratch coat and 1:5 for brown coat, by volume, cement to sand.	—	—	—	Varies
	15-1.4	$3^{5}/_{8}''$ No. 16 gage noncombustible studs $16''$ on center with $7/8''$ cement plaster (measured from the face of the studs) on the exterior surface with interior surface treatment as required for interior, nonbearing, noncombustible stud partitions in this table. Plaster mix 1:4 for scratch coat and 1:5 for brown coat, by volume, cement to sand.	—	—	—	Varies[d]

(continued)

TABLE 720.1(2)—continued
RATED FIRE-RESISTANCE PERIODS FOR VARIOUS WALLS AND PARTITIONS [a,o,p]

MATERIAL	ITEM NUMBER	CONSTRUCTION	MINIMUM FINISHED THICKNESS FACE-TO-FACE[b] (inches)			
			4 hour	3 hour	2 hour	1 hour
15. Exterior or interior walls (continued)	15-1.5[m]	$2^1/_4'' \times 3^3/_4''$ clay face brick with cored holes over $^1/_2''$ gypsum sheathing on exterior surface of $2'' \times 4''$ wood studs at 16″ on center and two layers $^5/_8''$ Type X gypsum wallboard[e] on interior surface. Sheathing placed horizontally or vertically with vertical joints over studs nailed 6″ on center with $1^3/_4'' \times$ No. 11 gage by $^7/_{16}''$ head galvanized nails. Inner layer of wallboard placed horizontally or vertically and nailed 8″ on center with 6d cooler[n] or wallboard[n] nails. Outer layer of wallboard placed horizontally or vertically and nailed 8″ on center with 8d cooler[n] or wallboard[n] nails. All joints staggered with vertical joints over studs. Outer layer joints taped and finished with compound. Nail heads covered with joint compound. 0.035 inch (No. 20 galvanized sheet gage) corrugated galvanized steel wall ties $^3/_4''$ by $6^5/_8''$ attached to each stud with two 8d cooler[n] or wallboard[n] nails every sixth course of bricks.	—	—	10	—
	15-1.6[l, m]	$2'' \times 6''$ fire-retardant-treated wood studs 16″ on center. Interior face has two layers of $^5/_8''$ Type X gypsum with the base layer placed vertically and attached with 6d box nails 12″ on center. The face layer is placed horizontally and attached with 8d box nails 8″ on center at joints and 12″ on center elsewhere. The exterior face has a base layer of $^5/_8''$ Type X gypsum sheathing placed vertically with 6d box nails 8″ on center at joints and 12″ on center elsewhere. An approved building paper is next applied, followed by self-furred exterior lath attached with $2^1/_2''$, No. 12 gage galvanized roofing nails with a $^3/_8''$ diameter head and spaced 6″ on center along each stud. Cement plaster consisting of a $^1/_2''$ brown coat is then applied. The scratch coat is mixed in the proportion of 1:3 by weight, cement to sand with 10 pounds of hydrated lime and 3 pounds of approved additives or admixtures per sack of cement. The brown coat is mixed in the proportion of 1:4 by weight, cement to sand with the same amounts of hydrated lime and approved additives or admixtures used in the scratch coat.	—	—	$8^1/_4$	—
	15-1.7[l, m]	$2'' \times 6''$ wood studs 16″ on center. The exterior face has a layer of $^5/_8''$ Type X gypsum sheathing placed vertically with 6d box nails 8″ on center at joints and 12″ on center elsewhere. An approved building paper is next applied, followed by 1″ by No. 18 gage self-furred exterior lath attached with 8d by $2^1/_2''$ long galvanized roofing nails spaced 6″ on center along each stud. Cement plaster consisting of a $^1/_2''$ scratch coat, a bonding agent and a $^1/_2''$ brown coat and a finish coat is then applied. The scratch coat is mixed in the proportion of 1:3 by weight, cement to sand with 10 pounds of hydrated lime and 3 pounds of approved additives or admixtures per sack of cement. The brown coat is mixed in the proportion of 1:4 by weight, cement to sand with the same amounts of hydrated lime and approved additives or admixtures used in the scratch coat. The interior is covered with $^3/_8''$ gypsum lath with 1″ hexagonal mesh of 0.035 inch (No. 20 B.W. gage) woven wire lath furred out $^5/_{16}''$ and 1″ perlite or vermiculite gypsum plaster. Lath nailed with $1^1/_8''$ by No. 13 gage by $^{19}/_{64}''$ head plasterboard glued nails spaced 5″ on center. Mesh attached by $1^3/_4''$ by No. 12 gage by $^3/_8''$ head nails with $^3/_8''$ furrings, spaced 8″ on center. The plaster mix shall not exceed 100 pounds of gypsum to $2^1/_2$ cubic feet of aggregate.	—	—	$8^3/_8$	—
	15-1.8[l, m]	$2'' \times 6''$ wood studs 16″ on center. The exterior face has a layer of $^5/_8''$ Type X gypsum sheathing placed vertically with 6d box nails 8″ on center at joints and 12″ on center elsewhere. An approved building paper is next applied, followed by $1^1/_2''$ by No. 17 gage self-furred exterior lath attached with 8d by $2^1/_2''$ long galvanized roofing nails spaced 6″ on center along each stud. Cement plaster consisting of a $^1/_2''$ scratch coat, and a $^1/_2''$ brown coat is then applied. The plaster may be placed by machine. The scratch coat is mixed in the proportion of 1:4 by weight, plastic cement to sand. The brown coat is mixed in the proportion of 1:5 by weight, plastic cement to sand. The interior is covered with $^3/_8''$ gypsum lath with 1″ hexagonal mesh of No. 20 gage woven wire lath furred out $^5/_{16}''$ and 1″ perlite or vermiculite gypsum plaster. Lath nailed with $1^1/_8''$ by No. 13 gage by $^{19}/_{64}''$ head plasterboard glued nails spaced 5″ on center. Mesh attached by $1^3/_4''$ by No. 12 gage by $^3/_8''$ head nails with $^3/_8''$ furrings, spaced 8″ on center. The plaster mix shall not exceed 100 pounds of gypsum to $2^1/_2$ cubic feet of aggregate.	—	—	$8^3/_8$	—

(continued)

TABLE 720.1(2)—continued
RATED FIRE-RESISTANCE PERIODS FOR VARIOUS WALLS AND PARTITIONS [a,o,p]

MATERIAL	ITEM NUMBER	CONSTRUCTION	MINIMUM FINISHED THICKNESS FACE-TO-FACE[b] (inches)			
			4 hour	3 hour	2 hour	1 hour
15. Exterior or interior walls (continued)	15-1.9	4″ No. 18 gage, nonload-bearing metal studs, 16″ on center, with 1″ portland cement lime plaster [measured from the back side of the $^3/_4$-pound expanded metal lath] on the exterior surface. Interior surface to be covered with 1″ of gypsum plaster on $^3/_4$-pound expanded metal lath proportioned by weight—1:2 for scratch coat, 1:3 for brown, gypsum to sand. Lath on one side of the partition fastened to $^1/_4$″ diameter pencil rods supported by No. 20 gage metal clips, located 16″ on center vertically, on each stud. 3″ thick mineral fiber insulating batts friction fitted between the studs.	—	—	$6^1/_2$[d]	—
	15-1.10	Steel studs 0.060″ thick, 4″ deep or 6″ at 16″ or 24″ centers, with $^1/_2$″ Glass Fiber Reinforced Concrete (GFRC) on the exterior surface. GFRC is attached with flex anchors at 24″ on center, with 5″ leg welded to studs with two $^1/_2$″-long flare-bevel welds, and 4″ foot attached to the GFRC skin with $^5/_8$″ thick GFRC bonding pads that extend $2^1/_2$″ beyond the flex anchor foot on both sides. Interior surface to have two layers of $^1/_2$″ Type X gypsum wallboard.[e] The first layer of wallboard to be attached with 1″-long Type S buglehead screws spaced 24″ on center and the second layer is attached with $1^5/_8$″-long Type S screws spaced at 12″ on center. Cavity is to be filled with 5″ of 4 pcf (nominal) mineral fiber batts. GFRC has $1^1/_2$″ returns packed with mineral fiber and caulked on the exterior.	—	—	$6^1/_2$	—
	15-1.11	Steel studs 0.060″ thick, 4″ deep or 6″ at 16″ or 24″ centers, respectively, with $^1/_2$″ Glass Fiber Reinforced Concrete (GFRC) on the exterior surface. GFRC is attached with flex anchors at 24″ on center, with 5″ leg welded to studs with two $^1/_2$″-long flare-bevel welds, and 4″ foot attached to the GFRC skin with $^5/_8$″-thick GFRC bonding pads that extend $2^1/_2$″ beyond the flex anchor foot on both sides. Interior surface to have one layer of $^5/_8$″ Type X gypsum wallboard[e], attached with $1^1/_4$″-long Type S buglehead screws spaced 12″ on center. Cavity is to be filled with 5″ of 4 pcf (nominal) mineral fiber batts. GFRC has $1^1/_2$″ returns packed with mineral fiber and caulked on the exterior.	—	—	—	$6^1/_8$
	15-1.12[q]	2″ × 6″ wood studs at 16″ with double top plates, single bottom plate; interior and exterior sides covered with $^5/_8$″ Type X gypsum wallboard, 4′ wide, applied horizontally or vertically with vertical joints over studs, and fastened with $2^1/_4$″ Type S drywall screws, spaced 12″ on center. Cavity filled with $5^1/_2$″ mineral wool insulation.	—	—	—	$6^3/_4$
	15-1.13[q]	2″ × 6″ wood studs at 16″ with double top plates, single bottom plate; interior and exterior sides covered with $^5/_8$″ Type X gypsum wallboard, 4′ wide, applied horizontally or vertically with vertical joints over studs, and fastened with $2^1/_4$″ Type S drywall screws, spaced 7″ on center. Cavity to be filled with $5^1/_2$″ mineral wool insulation minimum 2.58 pcf (nominal).	—	—	—	$6^3/_4$
	15-1.14[q]	2″ × 4″ wood studs at 16″ with double top plates, single bottom plate; interior and exterior sides covered with $^5/_8$″ Type X gypsum wallboard and sheathing, respectively, 4′ wide, applied horizontally or vertically with vertical joints over studs, and fastened with $2^1/_4$″ Type S drywall screws, spaced 12″ on center. Cavity to be filled with $3^1/_2$″ mineral wool insulation.	—	—	—	$4^3/_4$
	15-1.15[q]	2″ × 4″ wood studs at 16″ with double top plates, single bottom plate; interior sides covered with $^5/_8$″ Type X gypsum wallboard, 4′ wide, applied horizontally unblocked, and fastened with $2^1/_4$″ Type S drywall screws, spaced 12″ on center, wallboard joints covered with paper tape and joint compound, fastener heads covered with joint compound. Exterior covered with $^3/_8$″ wood structural panels (oriented strand board), applied vertically, horizontal joints blocked and fastened with 6d common nails (bright)—12″ on center in the field, 6″ on center panel edges. Cavity to be filled with $3^1/_2$″ mineral wool insulation. Rating established for exposure from interior side only.	—	—	—	$4^1/_2$

(continued)

TABLE 720.1(2)—continued
RATED FIRE-RESISTANCE PERIODS FOR VARIOUS WALLS AND PARTITIONS [a,o,p]

MATERIAL	ITEM NUMBER	CONSTRUCTION	MINIMUM FINISHED THICKNESS FACE-TO-FACE[b] (inches)			
			4 hour	3 hour	2 hour	1 hour
15. Exterior or interior walls (continued)	15-1.16[q]	$2'' \times 6''$ (51mm x 152 mm) wood studs at 16″ centers with double top plates, single bottom plate; interior side covered with $5/8''$ Type X gypsum wallboard, 4′ wide, applied horizontally or vertically with vertical joints over studs and fastened with $2^{1}/_{4}''$ Type S drywall screws, spaced 12″ on center, wallboard joints covered with paper tape and joint compound, fastener heads covered with joint compound, exterior side covered with $7/_{16}''$; wood structural panels (oriented strand board) fastened with 6d common nails (bright) spaced 12″ on center in the field and 6″ on center along the panel edges. Cavity to be filled with $5^{1}/_{2}''$ mineral wool insulation. Rating established from the gypsum-covered side only.	—	—	—	$6^{9}/_{16}$
	15-1.17[q]	$2'' \times 6''$ wood studs at 24″ centers with double top plates, single bottom plate; interior and exterior side covered with two layers of $5/8''$ Type X gypsum wallboard, 4′ wide, applied horizontally with vertical joints over studs. Base layer fastened with $2^{1}/_{4}''$ Type S drywall screws, spaced 24″ on center, and face layer fastened with Type S drywall screws, spaced 8″ on center, wallboard joints covered with paper tape and joint compound, fastened heads covered with joint compound. Cavity to be filled with $5^{1}/_{2}''$ mineral wool insulation.	—	—	$7^{3}/_{4}$	—
16. Exterior walls rated for fire resistance from the inside only in accordance with Section 704.5.	16-1.1[q]	$2'' \times 4''$ wood studs at 16″ centers with double top plates, single bottom plate; interior side covered with $5/8''$ Type X gypsum wallboard, 4′ wide, applied horizontally unblocked, and fastened with $2^{1}/_{4}''$ Type S drywall screws, spaced 12″ on center, wallboard joints covered with paper tape and joint compound, fastener heads covered with joint compound. Exterior covered with $3/8''$ wood structural panels (oriented strand board), applied vertically, horizontal joints blocked and fastened with 6d common nails (bright) — 12″ on center in the field, and 6″ on center panel edges. Cavity to be filled with $3^{1}/_{2}''$ mineral wool insulation. Rating established for exposure from interior side only.	—	—	—	$4^{1}/_{2}$

For SI: 1 inch = 25.4 mm, 1 square inch = 645.2 mm^2, 1 cubic foot = 0.0283 m^3.

a. Staples with equivalent holding power and penetration shall be permitted to be used as alternate fasteners to nails for attachment to wood framing.

b. Thickness shown for brick and clay tile are nominal thicknesses unless plastered, in which case thicknesses are net. Thickness shown for concrete masonry and clay masonry is equivalent thickness defined in Section 721.3.1 for concrete masonry and Section 721.4.1.1 for clay masonry. Where all cells are solid grouted or filled with silicone-treated perlite loose-fill insulation; vermiculite loose-fill insulation; or expanded clay, shale or slate lightweight aggregate, the equivalent thickness shall be the thickness of the block or brick using specified dimensions as defined in Chapter 21. Equivalent thickness may also include the thickness of applied plaster and lath or gypsum wallboard, where specified.

c. For units in which the net cross-sectional area of cored brick in any plane parallel to the surface containing the cores is at least 75 percent of the gross cross-sectional area measured in the same plane.

d. Shall be used for nonbearing purposes only.

e. For all of the construction with gypsum wallboard described in this table, gypsum base for veneer plaster of the same size, thickness and core type shall be permitted to be substituted for gypsum wallboard, provided attachment is identical to that specified for the wallboard, and the joints on the face layer are reinforced and the entire surface is covered with a minimum of $1/_{16}$-inch gypsum veneer plaster.

f. The fire-resistance time period for concrete masonry units meeting the equivalent thicknesses required for a 2-hour fire-resistance rating in Item 3, and having a thickness of not less than $7^{5}/_{8}$ inches is 4 hours when cores which are not grouted are filled with silicone-treated perlite loose-fill insulation; vermiculite loose-fill insulation; or expanded clay, shale or slate lightweight aggregate, sand or slag having a maximum particle size of $3/_{8}$ inch.

g. The fire-resistance rating of concrete masonry units composed of a combination of aggregate types or where plaster is applied directly to the concrete masonry shall be determined in accordance with ACI 216.1/TMS 216. Lightweight aggregates shall have a maximum combined density of 65 pounds per cubic foot.

h. See also Note b. The equivalent thickness shall be permitted to include the thickness of cement plaster or 1.5 times the thickness of gypsum plaster applied in accordance with the requirements of Chapter 25.

i. Concrete walls shall be reinforced with horizontal and vertical temperature reinforcement as required by Chapter 19.

j. Studs are welded truss wire studs with 0.18 inch (No. 7 B.W. gage) flange wire and 0.18 inch (No. 7 B.W. gage) truss wires.

k. Nailable metal studs consist of two channel studs spot welded back to back with a crimped web forming a nailing groove.

l. Wood structural panels shall be permitted to be installed between the fire protection and the wood studs on either the interior or exterior side of the wood frame assemblies in this table, provided the length of the fasteners used to attach the fire protection are increased by an amount at least equal to the thickness of the wood structural panel.

m. The design stress of studs shall be reduced to 78 percent of allowable F'_c with the maximum not greater than 78 percent of the calculated stress with studs having a slenderness ratio l_e/d of 33.

n. For properties of cooler or wallboard nails, see ASTM C 514, ASTM C 547 or ASTM F 1667.

o. Generic fire-resistance ratings (those not designated as PROPRIETARY* in the listing) in the GA 600 shall be accepted as if herein listed.

p. NCMA TEK 5-8, shall be permitted for the design of fire walls.

q. The design stress of studs shall be equal to a maximum of 100 percent of the allowable F'_c calculated in accordance with Section 2306.

TABLE 720.1(3)
MINIMUM PROTECTION FOR FLOOR AND ROOF SYSTEMS[a,q]

FLOOR OR ROOF CONSTRUCTION	ITEM NUMBER	CEILING CONSTRUCTION	THICKNESS OF FLOOR OR ROOF SLAB (inches)				MINIMUM THICKNESS OF CEILING (inches)			
			4 hour	3 hour	2 hour	1 hour	4 hour	3 hour	2 hour	1 hour
1. Siliceous aggregate concrete	1-1.1	Slab (no ceiling required). Minimum cover over nonprestressed reinforcement shall not be less than $^3/_4$ inch.[b]	7.0	6.2	5.0	3.5	—	—	—	—
2. Carbonate aggregate concrete	2-1.1		6.6	5.7	4.6	3.2	—	—	—	—
3. Sand-lightweight concrete	3-1.1		5.4	4.6	3.8	2.7	—	—	—	—
4. Lightweight concrete	4-1.1		5.1	4.4	3.6	2.5	—	—	—	—
5. Reinforced concrete	5-1.1	Slab with suspended ceiling of vermiculite gypsum plaster over metal lath attached to $^3/_4$″ cold-rolled channels spaced 12″ on center. Ceiling located 6″ minimum below joists.	3	2	—	—	1	$^3/_4$	—	—
	5-2.1	$^3/_8$″ Type X gypsum wallboard[c] attached to 0.018 inch (No. 25 carbon sheet steel gage) by $^7/_8$″ deep by $2^5/_8$″ hat-shaped galvanized steel channels with 1″-long No. 6 screws. The channels are spaced 24″ on center, span 35″ and are supported along their length at 35″ intervals by 0.033-inch (No. 21 galvanized sheet gage) galvanized steel flat strap hangers having formed edges that engage the lips of the channel. The strap hangers are attached to the side of the concrete joists with $^5/_{32}$″ by $1^1/_4$″ long power-driven fasteners. The wallboard is installed with the long dimension perpendicular to the channels. All end joints occur on channels and supplementary channels are installed parallel to the main channels, 12″ each side, at end joint occurrences. The finished ceiling is located approximately 12″ below the soffit of the floor slab.	—	—	$2^1/_2$	—	—	—	$^5/_8$	—
6. Steel joists constructed with a poured reinforced concrete slab on metal lath forms or steel form units[d, e]	6-1.1	Gypsum plaster on metal lath attached to the bottom cord with single No. 16 gage or doubled No. 18 gage wire ties spaced 6″ on center. Plaster mixed 1:2 for scratch coat, 1:3 for brown coat, by weight, gypsum-to-sand aggregate for 2-hour system. For 3-hour system plaster is neat.	—	—	$2^1/_2$	$2^1/_4$	—	—	$^3/_4$	$^5/_8$
	6-2.1	Vermiculite gypsum plaster on metal lath attached to the bottom chord with single No.16 gage or doubled 0.049-inch (No. 18 B.W. gage) wire ties 6″ on center.	—	2	—	—	—	$^5/_8$	—	—
	6-3.1	Cement plaster over metal lath attached to the bottom chord of joists with single No. 16 gage or doubled 0.049-inch (No. 18 B.W. gage) wire ties spaced 6″ on center. Plaster mixed 1:2 for scratch coat, 1:3 for brown coat for 1-hour system and 1:1 for scratch coat, $1:1 \, ^1/_2$ for brown coat for 2-hour system, by weight, cement to sand.	—	—	—	2	—	—	—	$^5/_8$[f]
	6-4.1	Ceiling of $^5/_8$″ Type X wallboard[c] attached to $^7/_8$″ deep by $2^5/_8$″ by 0.021 inch (No. 25 carbon sheet steel gage) hat-shaped furring channels 12″ on center with 1″ long No. 6 wallboard screws at 8″ on center. Channels wire tied to bottom chord of joists with doubled 0.049 inch (No. 18 B.W. gage) wire or suspended below joists on wire hangers.[g]	—	—	$2^1/_2$	—	—	—	$^5/_8$	—
	6-5.1	Wood-fibered gypsum plaster mixed 1:1 by weight gypsum to sand aggregate applied over metal lath. Lath tied 6″ on center to $^3/_4$″ channels spaced $13^1/_2$″ on center. Channels secured to joists at each intersection with two strands of 0.049 inch (No. 18 B.W. gage) galvanized wire.	—	—	$2^1/_2$	—	—	—	$^3/_4$	—

(continued)

TABLE 720.1(3)—continued
MINIMUM PROTECTION FOR FLOOR AND ROOF SYSTEMS[a,q]

FLOOR OR ROOF CONSTRUCTION	ITEM NUMBER	CEILING CONSTRUCTION	THICKNESS OF FLOOR OR ROOF SLAB (inches)				MINIMUM THICKNESS OF CEILING (inches)			
			4 hour	3 hour	2 hour	1 hour	4 hour	3 hour	2 hour	1 hour
7. Reinforced concrete slabs and joists with hollow clay tile fillers laid end to end in rows $2^1/_2$″ or more apart; reinforcement placed between rows and concrete cast around and over tile.	7-1.1	$^5/_8$″ gypsum plaster on bottom of floor or roof construction.	—	—	8[h]	—	—	—	$^5/_8$	—
	7-1.2	None	—	—	—	$5^1/_2$[i]	—	—	—	—
8. Steel joists constructed with a reinforced concrete slab on top poured on a $^1/_2$″ deep steel deck.[e]	8-1.1	Vermiculite gypsum plaster on metal lath attached to $^3/_4$″ cold-rolled channels with 0.049″ (No. 18 B.W. gage) wire ties spaced 6″ on center.	$2^1/_2$[j]	—	—	—	$^3/_4$	—	—	—
9. 3″ deep cellular steel deck with concrete slab on top. Slab thickness measured to top.	9-1.1	Suspended ceiling of vermiculite gypsum plaster base coat and vermiculite acoustical plaster on metal lath attached at 6″ intervals to $^3/_4$″ cold-rolled channels spaced 12″ on center and secured to $1^1/_2$″ cold-rolled channels spaced 36″ on center with 0.065″ (No. 16 B.W. gage) wire. $1^1/_2$″ channels supported by No. 8 gage wire hangers at 36″ on center. Beams within envelope and with a $2^1/_2$″airspace between beam soffit and lath have a 4-hour rating.	$2^1/_2$	—	—	—	$1^1/_8$[k]	—	—	—
10. $1^1/_2$″-deep steel roof deck on steel framing. Insulation board, 30 pcf density, composed of wood fibers with cement binders of thickness shown bonded to deck with unified asphalt adhesive. Covered with a Class A or B roof covering.	10-1.1	Ceiling of gypsum plaster on metal lath. Lath attached to $^3/_4$″ furring channels with 0.049″ (No. 18 B.W. gage) wire ties spaced 6″ on center. $^3/_4$″ channel saddle tied to 2″ channels with doubled 0.065″ (No. 16 B.W. gage) wire ties. 2″ channels spaced 36″ on center suspended 2″ below steel framing and saddle-tied with 0.165″ (No. 8 B.W. gage) wire. Plaster mixed 1:2 by weight, gypsum-to-sand aggregate.	—	—	$1^7/_8$	1	—	—	$^3/_4$[l]	$^3/_4$[l]
11. $1^1/_2$″-deep steel roof deck on steel-framing wood fiber insulation board, 17.5 pcf density on top applied over a 15-lb asphalt-saturated felt. Class A or B roof covering.	11-1.1	Ceiling of gypsum plaster on metal lath. Lath attached to $^3/_4$″ furring channels with 0.049″ (No. 18 B.W. gage) wire ties spaced 6″ on center. $^3/_4$″ channels saddle tied to 2″ channels with doubled 0.065″ (No. 16 B.W. gage) wire ties. 2″ channels spaced 36″ on center suspended 2″ below steel framing and saddle tied with 0.165″ (No. 8 B.W. gage) wire. Plaster mixed 1:2 for scratch coat and 1:3 for brown coat, by weight, gypsum-to-sand aggregate for 1-hour system. For 2-hour system, plaster mix is 1:2 by weight, gypsum-to-sand aggregate.	—	—	$1^1/_2$	1	—	—	$^7/_8$[g]	$^3/_4$[l]

(continued)

TABLE 720.1(3)—continued
MINIMUM PROTECTION FOR FLOOR AND ROOF SYSTEMS[a,q]

FLOOR OR ROOF CONSTRUCTION	ITEM NUMBER	CEILING CONSTRUCTION	THICKNESS OF FLOOR OR ROOF SLAB (inches)				MINIMUM THICKNESS OF CEILING (inches)			
			4 hour	3 hour	2 hour	1 hour	4 hour	3 hour	2 hour	1 hour
12. $1^1/_2$″ deep steel roof deck on steel-framing insulation of rigid board consisting of expanded perlite and fibers impregnated with integral asphalt waterproofing; density 9 to 12 pcf secured to metal roof deck by $^1/_2$″ wide ribbons of waterproof, cold-process liquid adhesive spaced 6″ apart. Steel joist or light steel construction with metal roof deck, insulation, and Class A or B built-up roof covering.[e]	12-1.1	Gypsum-vermiculite plaster on metal lath wire tied at 6″ intervals to $^3/_4$″ furring channels spaced 12″ on center and wire tied to 2″ runner channels spaced 32″ on center. Runners wire tied to bottom chord of steel joists.	—	—	1	—	—	—	$^7/_8$	—
13. Double wood floor over wood joists spaced 16″ on center.[m,n]	13-1.1	Gypsum plaster over $^3/_8$″ Type X gypsum lath. Lath initially applied with not less than four $1^1/_8$″ by No. 13 gage by $^{19}/_{64}$″ head plasterboard blued nails per bearing. Continuous stripping over lath along all joist lines. Stripping consists of 3″ wide strips of metal lath attached by $1^1/_2$″ by No. 11 gage by $^1/_2$″ head roofing nails spaced 6″ on center. Alternate stripping consists of 3″ wide 0.049″ diameter wire stripping weighing 1 pound per square yard and attached by No.16 gage by $1^1/_2$″ by $^3/_4$″ crown width staples, spaced 4″ on center. Where alternate stripping is used, the lath nailing may consist of two nails at each end and one nail at each intermediate bearing. Plaster mixed 1:2 by weight, gypsum-to-sand aggregate.	—	—	—	—	—	—	—	$^7/_8$
	13-1.2	Cement or gypsum plaster on metal lath. Lath fastened with $1^1/_2$″ by No. 11 gage by $^7/_{16}$″ head barbed shank roofing nails spaced 5″ on center. Plaster mixed 1:2 for scratch coat and 1:3 for brown coat, by weight, cement to sand aggregate.	—	—	—	—	—	—	—	$^5/_8$
	13-1.3	Perlite or vermiculite gypsum plaster on metal lath secured to joists with $1^1/_2$″ by No. 11 gage by $^7/_{16}$″ head barbed shank roofing nails spaced 5″ on center.	—	—	—	—	—	—	—	$^5/_8$
	13-1.4	$^1/_2$″ Type X gypsum wallboard[c] nailed to joists with 5d cooler[o] or wallboard[o] nails at 6″ on center. End joints of wallboard centered on joists.	—	—	—	—	—	—	—	$^1/_2$
14. Plywood stressed skin panels consisting of $^5/_8$″-thick interior C-D (exterior glue) top stressed skin on 2″ × 6″nominal (minimum) stringers. Adjacent panel edges joined with 8d common wire nails spaced 6″ on center. Stringers spaced 12″ maximum on center.	14-1.1	$^1/_2$″-thick wood fiberboard weighing 15 to 18 pounds per cubic foot installed with long dimension parallel to stringers or $^3/_8$″ C-D (exterior glue) plywood glued and/or nailed to stringers. Nailing to be with 5d cooler[o] or wallboard[o] nails at 12″ on center. Second layer of $^1/_2$″ Type X gypsum wallboard[c] applied with long dimension perpendicular to joists and attached with 8d cooler[o] or wallboard[o] nails at 6″ on center at end joints and 8″ on center elsewhere. Wallboard joints staggered with respect to fiberboard joints.	—	—	—	—	—	—	—	1

(continued)

TABLE 720.1(3)—continued
MINIMUM PROTECTION FOR FLOOR AND ROOF SYSTEMS[a,q]

FLOOR OR ROOF CONSTRUCTION	ITEM NUMBER	CEILING CONSTRUCTION	THICKNESS OF FLOOR OR ROOF SLAB (inches)				MINIMUM THICKNESS OF CEILING (inches)			
			4 hour	3 hour	2 hour	1 hour	4 hour	3 hour	2 hour	1 hour
15. Vermiculite concrete slab proportioned 1:4 (portland cement to vermiculite aggregate) on a $1^1/_2$″-deep steel deck supported on individually protected steel framing. Maximum span of deck 6′-10″ where deck is less than 0.019 inch (No. 26 carbon steel sheet gage) or greater. Slab reinforced with 4″ × 8″ 0.109/0.083″ (No. $^{12}/_{14}$ B.W. gage) welded wire mesh.	15-1.1	None	—	—	—	3^j	—	—	—	—
16. Perlite concrete slab proportioned 1:6 (portland cement to perlite aggregate) on a $1^1/_4$″-deep steel deck supported on individually protected steel framing. Slab reinforced with 4″ × 8″ 0.109/0.083″ (No. $^{12}/_{14}$ B.W. gage) welded wire mesh.	16-1.1	None	—	—	—	$3^1/_2{}^j$	—	—	—	—
17. Perlite concrete slab proportioned 1:6 (portland cement to perlite aggregate) on a $^9/_{16}$″-deep steel deck supported by steel joists 4′ on center. Class A or B roof covering on top.	17-1.1	Perlite gypsum plaster on metal lath wire tied to $^3/_4$″ furring channels attached with 0.065-inch (No. 16 B.W. gage) wire ties to lower chord of joists.	—	2^p	2^p	—	—	$^7/_8$	$^3/_4$	—
18. Perlite concrete slab proportioned 1:6 (portland cement to perlite aggregate) on $1^1/_4$″-deep steel deck supported on individually protected steel framing. Maximum span of deck 6′-10″ where deck is less than 0.019″ (No. 26 carbon sheet steel gage) and 8′-0″ where deck is 0.019″ (No. 26 carbon sheet steel gage) or greater. Slab reinforced with 0.042″ (No. 19 B.W. gage) hexagonal wire mesh. Class A or B roof covering on top.	18-1.1	None	—	$2^1/_4{}^p$	$2^1/_4{}^p$	—	—	—	—	—

(continued)

TABLE 720.1(3)—continued
MINIMUM PROTECTION FOR FLOOR AND ROOF SYSTEMS[a,q]

FLOOR OR ROOF CONSTRUCTION	ITEM NUMBER	CEILING CONSTRUCTION	THICKNESS OF FLOOR OR ROOF SLAB (inches)				MINIMUM THICKNESS OF CEILING (inches)			
			4 hour	3 hour	2 hour	1 hour	4 hour	3 hour	2 hour	1 hour
19. Floor and beam construction consisting of 3″-deep cellular steel floor unit mounted on steel members with 1:4 (proportion of portland cement to perlite aggregate) perlite-concrete floor slab on top.	19-1.1	Suspended envelope ceiling of perlite gypsum plaster on metal lath attached to $^3/_4$″ cold-rolled channels, secured to $1^1/_2$″ cold-rolled channels spaced 42″ on center supported by 0.203 inch (No. 6 B.W. gage) wire 36″ on center. Beams in envelope with 3″ minimum airspace between beam soffit and lath have a 4-hour rating.	2[p]	—	—	—	1[l]	—	—	—
20. Perlite concrete proportioned 1:6 (portland cement to perlite aggregate) poured to $^1/_8$-inch thickness above top of corrugations of $1^5/_{16}$″-deep galvanized steel deck maximum span 8′-0″ for 0.024-inch (No. 24 galvanized sheet gage) or 6′ 0″ for 0.019-inch (No. 26 galvanized sheet gage) with deck supported by individually protected steel framing. Approved polystyrene foam plastic insulation board having a flame spread not exceeding 75 (1″ to 4″ thickness) with vent holes that approximate 3 percent of the board surface area placed on top of perlite slurry. A 2′ by 4′ insulation board contains six $2^3/_4$″ diameter holes. Board covered with $2^1/_4$″ minimum perlite concrete slab.	20-1.1	None	—	—	Varies	—	—	—	—	—

(continued)

FLOOR OR ROOF CONSTRUCTION	ITEM NUMBER	CEILING CONSTRUCTION	THICKNESS OF FLOOR OR ROOF SLAB (inches)				MINIMUM THICKNESS OF CEILING (inches)			
			4 hour	3 hour	2 hour	1 hour	4 hour	3 hour	2 hour	1 hour
(continued) 20. Slab reinforced with mesh consisting of 0.042 inch (No. 19 B.W. gage) galvanized steel wire twisted together to form 2″ hexagons with straight 0.065 inch (No. 16 B.W. gage) galvanized steel wire woven into mesh and spaced 3″. Alternate slab reinforcement shall be permitted to consist of 4″ × 8″, 0.109/0.238-inch (No. 12/4 B.W. gage), or 2″ × 2″, 0.083/0.083-inch (No. 14/14 B.W. gage) welded wire fabric. Class A or B roof covering on top.	20-1.1	None	—	—	Varies	—	—	—	—	—
21. Wood joists, floor trusses and flat or pitched roof trusses spaced a maximum 24″ o.c. with $^1/_2$″ wood structural panels with exterior glue applied at right angles to top of joist or top chord of trusses with 8d nails. The wood structural panel thickness shall not be less than nominal $^1/_2$″ less than required by Chapter 23.	21-1.1	Base layer $^5/_8$″ Type X gypsum wallboard applied at right angles to joist or truss 24″ o.c. with $1^1/_4$″ Type S or Type W drywall screws 24″ o.c. Face layer $^5/_8$″ Type X gypsum wallboard or veneer base applied at right angles to joist or truss through base layer with $1^7/_8$″ Type S or Type W drywall screws 12″ o.c. at joints and intermediate joist or truss. Face layer Type G drywall screws placed 2″ back on either side of face layer end joints, 12″ o.c.	—	—	—	Varies	—	—	—	$1^1/_4$
22. Steel joists, floor trusses and flat or pitched roof trusses spaced a maximum 24″ o.c. with $^1/_2$″ wood structural panels with exterior glue applied at right angles to top of joist or top chord of trusses with No. 8 screws. The wood structural panel thickness shall not be less than nominal $^1/_2$″ nor less than required by Chapter 22.	22-1.1	Base layer $^5/_8$″ Type X gypsum board applied at right angles to steel framing 24″ on center with 1″ Type S drywall screws spaced 24″ on center. Face layer $^5/_8$″ Type X gypsum board applied at right angles to steel framing attached through base layer with $1^5/_8$″ Type S drywall screws 12″ on center at end joints and intermediate joints and $1^1/_2$″ Type G drywall screws 12 inches on center placed 2″ back on either side of face layer end joints. Joints of the face layer are offset 24″ from the joints of the base layer.	—	—	—	Varies	—	—	—	$1^1/_4$
23. Wood I-joist (minimum joist depth $9^1/_4$″ with a minimum flange depth of $1^5/_{16}$″ and a minimum flange cross-sectional area of 2.3 square inches) at 24″ o.c. spacing with 1 × 4 (nominal) wood furring strip spacer applied parallel to and covering the bottom of the bottom flange of each member, tacked in place. 2″ mineral fiber insulation, 3.5 pcf (nominal) installed adjacent to the bottom flange of the I-joist and supported by the 1 × 4 furring strip spacer.	23-1.1	$^1/_2$″ deep single leg resilient channel 16″ on center (channels doubled at wallboard end joints), placed perpendicular to the furring strip and joist and attached to each joist by $1^7/_8$″ Type S drywall screws. $^5/_8$″ Type C gypsum wallboard applied perpendicular to the channel with end joints staggered at least 4′ and fastened with $1^1/_8$″ Type S drywall screws spaced 7″ on center. Wallboard joints to be taped and covered with joint compound.	—	—	—	Varies	—	—	—	—

Table 720.1(3) Notes.

For SI: 1 inch = 25.4 mm, 1 foot = 304.8 mm, 1 pound = 0.454 kg, 1 cubic foot = 0.0283 m^3,
  1 pound per square inch = 6.895 kPa = 1 pound per lineal foot = 1.4882 kg/m.

a. Staples with equivalent holding power and penetration shall be permitted to be used as alternate fasteners to nails for attachment to wood framing.

b. When the slab is in an unrestrained condition, minimum reinforcement cover shall not be less than $1^5/_8$ inches for 4-hour (siliceous aggregate only); $1^1/_4$ inches for 4- and 3-hour; 1 inch for 2-hour (siliceous aggregate only); and $^3/_4$ inch for all other restrained and unrestrained conditions.

c. For all of the construction with gypsum wallboard described in this table, gypsum base for veneer plaster of the same size, thickness and core type shall be permitted to be substituted for gypsum wallboard, provided attachment is identical to that specified for the wallboard, and the joints on the face layer are reinforced and the entire surface is covered with a minimum of $^1/_{16}$-inch gypsum veneer plaster.

d. Slab thickness over steel joists measured at the joists for metal lath form and at the top of the form for steel form units.

e. (a) The maximum allowable stress level for H-Series joists shall not exceed 22,000 psi.
  (b) The allowable stress for K-Series joists shall not exceed 26,000 psi, the nominal depth of such joist shall not be less than 10 inches and the nominal joist weight shall not be less than 5 pounds per lineal foot.

f. Cement plaster with 15 pounds of hydrated lime and 3 pounds of approved additives or admixtures per bag of cement.

g. Gypsum wallboard ceilings attached to steel framing shall be permitted to be suspended with $1^1/_2$-inch cold-formed carrying channels spaced 48 inches on center, which are suspended with No. 8 SWG galvanized wire hangers spaced 48 inches on center. Cross-furring channels are tied to the carrying channels with No. 18 SWG galvanized wire hangers spaced 48 inches on center. Cross-furring channels are tied to the carrying channels with No. 18 SWG galvanized wire (double strand) and spaced as required for direct attachment to the framing. This alternative is also applicable to those steel framing assemblies recognized under Note q.

h. Six-inch hollow clay tile with 2-inch concrete slab above.

i. Four-inch hollow clay tile with $1^1/_2$-inch concrete slab above.

j. Thickness measured to bottom of steel form units.

k. Five-eighths inch of vermiculite gypsum plaster plus $^1/_2$ inch of approved vermiculite acoustical plastic.

l. Furring channels spaced 12 inches on center.

m. Double wood floor shall be permitted to be either of the following:
  (a) Subfloor of 1-inch nominal boarding, a layer of asbestos paper weighing not less than 14 pounds per 100 square feet and a layer of 1-inch nominal tongue-and-groove finished flooring; or
  (b) Subfloor of 1-inch nominal tongue-and-groove boarding or $^{15}/_{32}$-inch wood structural panels with exterior glue and a layer of 1-inch nominal tongue-and-groove finished flooring or $^{19}/_{32}$-inch wood structural panel finish flooring or a layer of Type I Grade M-1 particleboard not less than $^5/_8$-inch thick.

n. The ceiling shall be permitted to be omitted over unusable space, and flooring shall be permitted to be omitted where unusable space occurs above.

o. For properties of cooler or wallboard nails, see ASTM C 514, ASTM C 547 or ASTM F 1667.

p. Thickness measured on top of steel deck unit.

q. Generic fire-resistance ratings (those not designated as PROPRIETARY* in the listing) in the GA 600 shall be accepted as if herein listed.

721.2 Concrete assemblies. The provisions of this section contain procedures by which the fire-resistance ratings of concrete assemblies are established by calculations.

721.2.1 Concrete walls. Cast-in-place and precast concrete walls shall comply with Section 721.2.1.1. Multiwythe concrete walls shall comply with Section 721.2.1.2. Joints between precast panels shall comply with Section 721.2.1.3. Concrete walls with gypsum wallboard or plaster finish shall comply with Section 721.2.1.4.

721.2.1.1 Cast-in-place or precast walls. The minimum equivalent thicknesses of cast-in-place or precast concrete walls for fire-resistance ratings of 1 hour to 4 hours are shown in Table 721.2.1.1. For solid walls with flat vertical surfaces, the equivalent thickness is the same as the actual thickness. The values in Table 721.2.1.1 apply to plain, reinforced or prestressed concrete walls.

TABLE 721.2.1.1
MINIMUM EQUIVALENT THICKNESS OF CAST-IN-PLACE OR PRECAST CONCRETE WALLS, LOAD-BEARING OR NONLOAD-BEARING

CONCRETE TYPE	MINIMUM SLAB THICKNESS (inches) FOR FIRE-RESISTANCE RATING OF				
	1-hour	$1^1/_2$-hour	2-hour	3-hour	4-hour
Siliceous	3.5	4.3	5.0	6.2	7.0
Carbonate	3.2	4.0	4.6	5.7	6.6
Sand-Lightweight	2.7	3.3	3.8	4.6	5.4
Lightweight	2.5	3.1	3.6	4.4	5.1

For SI: 1 inch = 25.4 mm.

721.2.1.1.1 Hollow-core precast wall panels. For hollow-core precast concrete wall panels in which the cores are of constant cross section throughout the length, calculation of the equivalent thickness by dividing the net cross-sectional area (the gross cross section minus the area of the cores) of the panel by its width shall be permitted.

721.2.1.1.2 Core spaces filled. Where all of the core spaces of hollow-core wall panels are filled with loose-fill material, such as expanded shale, clay, or slag, or vermiculite or perlite, the fire-resistance rating of the wall is the same as that of a solid wall of the same concrete type and of the same overall thickness.

721.2.1.1.3 Tapered cross sections. The thickness of panels with tapered cross sections shall be that determined at a distance $2t$ or 6 inches (152 mm), whichever is less, from the point of minimum thickness, where t is the minimum thickness.

721.2.1.1.4 Ribbed or undulating surfaces. The equivalent thickness of panels with ribbed or undulating surfaces shall be determined by one of the following expressions:

For $s \geq 4t$, the thickness to be used shall be t

For $s \leq 2t$, the thickness to be used shall be t_e

For $4t > s > 2t$, the thickness to be used shall be

$$t + \left(\frac{4t}{s} - 1 \right) \left(t_e - t \right) \qquad \text{(Equation 7-3)}$$

where:

s = Spacing of ribs or undulations.

t = Minimum thickness.

t_e = Equivalent thickness of the panel calculated as the net cross-sectional area of the panel divided by the width, in which the maximum thickness used in the calculation shall not exceed $2t$.

721.2.1.2 Multiwythe walls. For walls that consist of two wythes of different types of concrete, the fire-resistance ratings shall be permitted to be determined from Figure 721.2.1.2.

721.2.1.2.1 Two or more wythes. The fire-resistance rating for wall panels consisting of two or more wythes shall be permitted to be determined by the formula:

$$R = (R_1^{0.59} + R_2^{0.59} + ... + R_n^{0.59})^{1.7}$$ **(Equation 7-4)**

where:

R = The fire endurance of the assembly, minutes.

R_1, R_2, and R_n = The fire endurances of the individual wythes, minutes. Values of $R_n^{0.59}$ for use in Equation 7-4 are given in Table 721.2.1.2(1). Calculated fire-resistance ratings are shown in Table 721.2.1.2(2).

721.2.1.2.2 Foam plastic insulation. The fire-resistance ratings of precast concrete wall panels consisting of a layer of foam plastic insulation sandwiched between two wythes of concrete shall be permitted to be determined by use of Equation 7-4. Foam plastic insulation with a total thickness of less than 1 inch (25 mm) shall be disregarded. The R_n value for thickness of foam plastic insulation of 1 inch (25 mm) or greater, for use in the calculation, is 5 minutes; therefore $R_n^{0.59} = 2.5$.

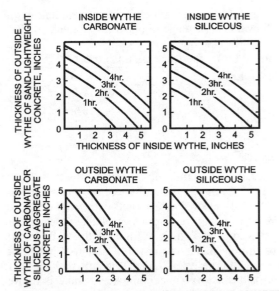

For SI: 1 inch = 25.4 mm.

**FIGURE 721.2.1.2
FIRE-RESISTANCE RATINGS OF
TWO-WYTHE CONCRETE WALLS**

721.2.1.3 Joints between precast wall panels. Joints between precast concrete wall panels which are not insulated as required by this section shall be considered as openings in walls. Uninsulated joints shall be included in determining the percentage of openings permitted by Table 704.8. Where openings are not permitted or are required by this code to be protected, the provisions of this section shall be used to determine the amount of joint insulation required. Insulated joints shall not be considered

**TABLE 721.2.1.2(1)
VALUES OF $R_n^{0.59}$ FOR USE IN EQUATION 7-4**

TYPE OF MATERIAL	THICKNESS OF MATERIAL (inches)											
	1½	2	2½	3	3½	4	4½	5	5½	6	6½	7
Siliceous aggregate concrete	5.3	6.5	8.1	9.5	11.3	13.0	14.9	16.9	18.8	20.7	22.8	25.1
Carbonate aggregate concrete	5.5	7.1	8.9	10.4	12.0	14.0	16.2	18.1	20.3	21.9	24.7	27.2[c]
Sand-lightweight concrete	6.5	8.2	10.5	12.8	15.5	18.1	20.7	23.3	26.0[c]	Note c	Note c	Note c
Lightweight concrete	6.6	8.8	11.2	13.7	16.5	19.1	21.9	24.7	27.8[c]	Note c	Note c	Note c
Insulating concrete[a]	9.3	13.3	16.6	18.3	23.1	26.5[c]	Note c	Note c	Note c	Note c	Note c	Note c
Airspace[b]	—	—	—	—	—	—	—	—	—	—	—	—

For SI: 1 inch = 25.4 mm, 1 pound per cubic foot = 16.02 kg/m³.

a. Dry unit weight of 35 pcf or less and consisting of cellular, perlite or vermiculite concrete.
b. The $R_n^{0.59}$ value for one ½″ to 3½″ airspace is 3.3. The $R_n^{0.59}$ value for two ½″ to 3½″ airspaces is 6.7.
c. The fire-resistance rating for this thickness exceeds 4 hours.

openings for purposes of determining compliance with the allowable percentage of openings in Table 704.8.

TABLE 721.2.1.2(2)
FIRE-RESISTANCE RATINGS BASED ON $R^{0.59}$

R [a], MINUTES	$R^{0.59}$
60	11.20
120	16.85
180	21.41
240	25.37

a. Based on Equation 7-4.

721.2.1.3.1 Ceramic fiber joint protection. Figure 721.2.1.3.1 shows thicknesses of ceramic fiber blankets to be used to insulate joints between precast concrete wall panels for various panel thicknesses and for joint widths of $^3/_8$ inch (9.5 mm) and 1 inch (25 mm) for fire-resistance ratings of 1 hour to 4 hours. For joint widths between $^3/_8$ inch (9.5 mm) and 1 inch (25 mm), the thickness of ceramic fiber blanket is allowed to be determined by direct interpolation. Other tested and labeled materials are acceptable in place of ceramic fiber blankets.

721.2.1.4 Walls with gypsum wallboard or plaster finishes. The fire-resistance rating of cast-in-place or precast concrete walls with finishes of gypsum wallboard or plaster applied to one or both sides shall be permitted to be calculated in accordance with the provisions of this section.

721.2.1.4.1 Nonfire-exposed side. Where the finish of gypsum wallboard or plaster is applied to the side of the wall not exposed to fire, the contribution of the finish to the total fire-resistance rating shall be determined as follows: The thickness of the finish shall first be corrected by multiplying the actual thickness of the finish by the applicable factor determined from Table 721.2.1.4(1) based on the type of aggregate in the concrete. The corrected thickness of finish shall then be added to the actual or equivalent thickness of concrete and fire-resistance rating of the concrete and finish determined from Table 721.2.1.1, Figure 721.2.1.2 or Table 721.2.1.2(1).

721.2.1.4.2 Fire-exposed side. Where gypsum wallboard or plaster is applied to the fire-exposed side of the wall, the contribution of the finish to the total fire-resistance rating shall be determined as follows: The time assigned to the finish as established by Table 721.2.1.4(2) shall be added to the fire-resistance rating determined from Table 721.2.1.1 or Figure 721.2.1.2, or Table 721.2.1.2(1) for the concrete alone, or to the rating determined in Section 721.2.1.4.1 for the concrete and finish on the nonfire-exposed side.

721.2.1.4.3 Nonsymmetrical assemblies. For a wall having no finish on one side or different types or thicknesses of finish on each side, the calculation procedures of Sections 721.2.1.4.1 and 721.2.1.4.2 shall be performed twice, assuming either side of the wall to be the fire-exposed side. The fire-restance rating of the wall shall not exceed the lower of the two values.

Exception: For an exterior wall with more than 5 feet (1524 mm) of horizontal separation, the fire shall be assumed to occur on the interior side only.

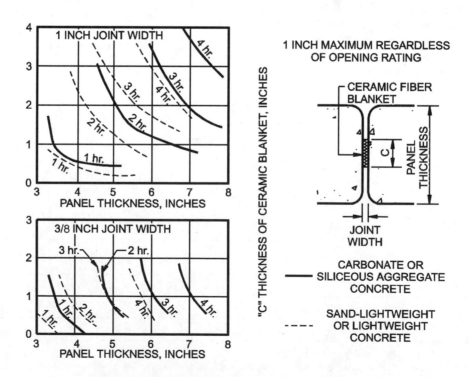

For SI: 1 inch = 25.4 mm.

FIGURE 721.2.1.3.1
CERAMIC FIBER JOINT PROTECTION

TABLE 721.2.1.4(1)
MULTIPLYING FACTOR FOR FINISHES ON NONFIRE-EXPOSED SIDE OF WALL

TYPE OF FINISH APPLIED TO MASONRY WALL	TYPE OF AGGREGATE USED IN CONCRETE OR CONCRETE MASONRY			
	Concrete: siliceous or carbonate Masonry: siliceous or calcareous gravel	Concrete: sand lightweight concrete Masonry: limestone, cinders or unexpected slag	Concrete: lightweight concrete Masonry: expanded shale, clay or slate	Concrete: pumice, or expanded slag
Portland cement-sand plaster	1.00	0.75[a]	0.75[a]	0.50[a]
Gypsum-sand plaster or gypsum wallboard	1.25	1.00	1.00	1.00
Gypsum-vermiculite or perlite plaster	1.75	1.50	1.50	1.25

For SI: 1 inch = 25.4 mm.

a. For portland cement-sand plaster $^5/_8$ inch or less in thickness and applied directly to the masonry on the nonfire-exposed side of the wall, the multiplying factor shall be 1.00.

721.2.1.4.4 Minimum concrete fire-resistance rating. Where finishes applied to one or both sides of a concrete wall contribute to the fire-resistance rating, the concrete alone shall provide not less than one-half of the total required fire-resistance rating. Additionally, the contribution to the fire resistance of the finish on the nonfire-exposed side of a load-bearing wall shall not exceed one-half the contribution of the concrete alone.

TABLE 721.2.1.4(2)
TIME ASSIGNED TO FINISH MATERIALS ON FIRE-EXPOSED SIDE OF WALL

FINISH DESCRIPTION	TIME (minute)
Gypsum wallboard	
$^3/_8$ inch	10
$^1/_2$ inch	15
$^5/_8$ inch	20
2 layers of $^3/_8$ inch	25
1 layer $^3/_8$ inch, 1 layer $^1/_2$ inch	35
2 layers $^1/_2$ inch	40
Type X gypsum wallboard	
$^1/_2$ inch	25
$^5/_8$ inch	40
Portland cement-sand plaster applied directly to concrete masonry	See Note a
Portland cement-sand plaster on metal lath	
$^3/_4$ inch	20
$^7/_8$ inch	25
1 inch	30
Gypsum sand plaster on $^3/_8$-inch gypsum lath	
$^1/_2$ inch	35
$^5/_8$ inch	40
$^3/_4$ inch	50
Gypsum sand plaster on metal lath	
$^3/_4$ inch	50
$^7/_8$ inch	60
1 inch	80

For SI: 1 inch = 25.4 mm.

a. The actual thickness of portland cement-sand plaster, provided it is $^5/_8$ inch or less in thickness, shall be permitted to be included in determining the equivalent thickness of the masonry for use in Table 721.3.2.

721.2.1.4.5 Concrete finishes. Finishes on concrete walls that are assumed to contribute to the total fire-resistance rating of the wall shall comply with the installation requirements of Section 721.3.2.5.

721.2.2 Concrete floor and roof slabs. Reinforced and prestressed floors and roofs shall comply with Section 721.2.2.1. Multicourse floors and roofs shall comply with Sections 721.2.2.2 and 721.2.2.3, respectively.

721.2.2.1 Reinforced and prestressed floors and roofs. The minimum thicknesses of reinforced and prestressed concrete floor or roof slabs for fire-resistance ratings of 1 hour to 4 hours are shown in Table 721.2.2.1.

TABLE 721.2.2.1
MINIMUM SLAB THICKNESS (inches)

CONCRETE TYPE	FIRE-RESISTANCE RATING (hour)				
	1	$1^1/_2$	2	3	4
Siliceous	3.5	4.3	5.0	6.2	7.0
Carbonate	3.2	4.0	4.6	5.7	6.6
Sand-lightweight	2.7	3.3	3.8	4.6	5.4
Lightweight	2.5	3.1	3.6	4.4	5.1

For SI: 1 inch = 25.4 mm.

721.2.2.1.1 Hollow-core prestressed slabs. For hollow-core prestressed concrete slabs in which the cores are of constant cross section throughout the length, the equivalent thickness shall be permitted to be obtained by dividing the net cross-sectional area of the slab including grout in the joints, by its width.

721.2.2.1.2 Slabs with sloping soffits. The thickness of slabs with sloping soffits (see Figure 721.2.2.1.2) shall be determined at a distance $2t$ or 6 inches (152 mm), whichever is less, from the point of minimum thickness, where t is the minimum thickness.

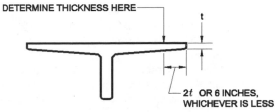

For SI: 1 inch = 25.4 mm.

FIGURE 721.2.2.1.2
DETERMINATION OF SLAB THICKNESS
FOR SLOPING SOFFITS

721.2.2.1.3 Slabs with ribbed soffits. The thickness of slabs with ribbed or undulating soffits (see Figure 721.2.2.1.3) shall be determined by one of the following expressions, whichever is applicable:

For $s \geq 4t$, the thickness to be used shall be t

For $s \leq 2t$, the thickness to be used shall be t_e

For $4t > s > 2t$, the thickness to be used shall be

$$t + \left(\frac{4t}{s} - 1\right)\left(t_e - t\right) \qquad \textbf{(Equation 7-5)}$$

where:

s = Spacing of ribs or undulations.

t = Minimum thickness.

t_e = Equivalent thickness of the slab calculated as the net area of the slab divided by the width, in which the maximum thickness used in the calculation shall not exceed $2t$.

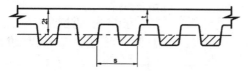

NEGLECT SHADED AREA IN CALCULATION OF EQUIVALENT THICKNESS

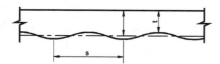

For SI: 1 inch = 25.4 mm.

FIGURE 721.2.2.1.3
SLABS WITH RIBBED OR UNDULATING SOFFITS

721.2.2.2 Multicourse floors. The fire-resistance ratings of floors that consist of a base slab of concrete with a topping (overlay) of a different type of concrete shall comply with Figure 721.2.2.2.

721.2.2.3 Multicourse roofs. The fire-resistance ratings of roofs which consist of a base slab of concrete with a topping (overlay) of an insulating concrete or with an insulating board and built-up roofing shall comply with Figures 721.2.2.3(1) and 721.2.2.3(2).

721.2.2.3.1 Heat transfer. For the transfer of heat, three-ply built-up roofing contributes 10 minutes to the fire-resistance rating. The fire-resistance rating for concrete assemblies such as those shown in Figure 721.2.2.3(1) shall be increased by 10 minutes. This increase is not applicable to those shown in Figure 721.2.2.3(2).

721.2.2.4 Joints in precast slabs. Joints between adjacent precast concrete slabs need not be considered in calculating the slab thickness provided that a concrete topping at least 1 inch (25 mm) thick is used. Where no

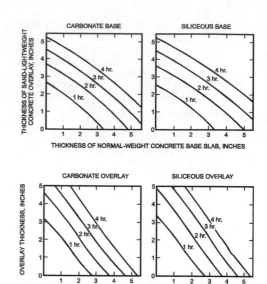

For SI: 1 inch = 25.4 mm.

FIGURE 721.2.2.2
FIRE-RESISTANCE RATINGS FOR TWO-COURSE
CONCRETE FLOORS

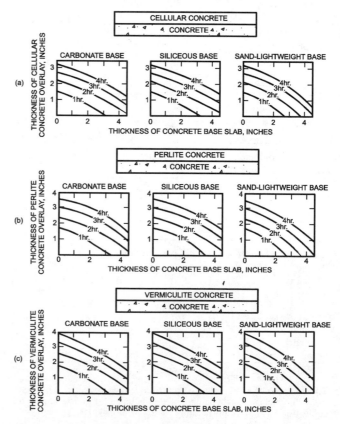

For SI: 1 inch = 25.4 mm.

FIGURE 721.2.2.3(1)
FIRE-RESISTANCE RATINGS FOR CONCRETE
ROOF ASSEMBLIES

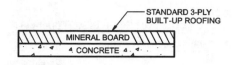

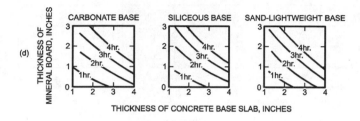

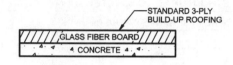

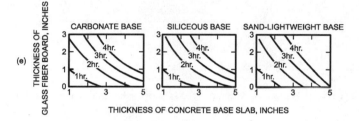

For SI: 1 inch = 25.4 mm.

FIGURE 721.2.2.3(2)
FIRE-RESISTANCE RATINGS FOR CONCRETE
ROOF ASSEMBLIES

concrete topping is used, joints must be grouted to a depth of at least one-third the slab thickness at the joint, but not less than 1 inch (25 mm), or the joints must be made fire resistant by other approved methods.

721.2.3 Concrete cover over reinforcement. The minimum thickness of concrete cover over reinforcement in concrete slabs, reinforced beams and prestressed beams shall comply with this section.

721.2.3.1 Slab cover. The minimum thickness of concrete cover to the positive moment reinforcement shall

comply with Table 721.2.3(1) for reinforced concrete and Table 721.2.3(2) for prestressed concrete. These tables are applicable for solid or hollow-core one-way or two-way slabs with flat undersurfaces. These tables are applicable to slabs that are either cast in place or precast. For precast prestressed concrete not covered elsewhere, the procedures contained in PCI MNL 124 shall be acceptable.

721.2.3.2 Reinforced beam cover. The minimum thickness of concrete cover to the positive moment reinforcement (bottom steel) for reinforced concrete beams is shown in Table 721.2.3(3) for fire-resistance ratings of 1 hour to 4 hours.

721.2.3.3 Prestressed beam cover. The minimum thickness of concrete cover to the positive moment prestressing tendons (bottom steel) for restrained and unrestrained prestressed concrete beams and stemmed units shall comply with the values shown in Tables 721.2.3(4) and 721.2.3(5) for fire-resistance ratings of 1 hour to 4 hours. Values in Table 721.2.3(4) apply to beams 8 inches (203 mm) or greater in width. Values in Table 721.2.3(5) apply to beams or stems of any width, provided the cross-section area is not less than 40 square inches (25 806 mm²). In case of differences between the values determined from Table 721.2.3(4) or 721.2.3(5), it is permitted to use the smaller value. The concrete cover shall be calculated in accordance with Section 721.2.3.3.1. The minimum concrete cover for nonprestressed reinforcement in prestressed concrete beams shall comply with Section 721.2.3.2.

721.2.3.3.1 Calculating concrete cover. The concrete cover for an individual tendon is the minimum thickness of concrete between the surface of the tendon and the fire-exposed surface of the beam, except that for ungrouped ducts, the assumed cover thickness is the minimum thickness of concrete between the surface of the duct and the fire-exposed surface of the beam. For beams in which two or more tendons are used, the cover is assumed to be the average of the minimum cover of the individual tendons. For corner tendons (tendons equal distance from the bottom and side), the minimum cover used in the calculation shall be one-half the actual value. For stemmed members with two or more prestressing

TABLE 721.2.3(1)
COVER THICKNESS FOR REINFORCED CONCRETE FLOOR OR ROOF SLABS (inches)

| CONCRETE AGGREGATE TYPE | FIRE-RESISTANCE RATING (hours) | | | | | | | | | |
| | Restrained | | | | | Unrestrained | | | | |
	1	1¹/₂	2	3	4	1	1¹/₂	2	3	4
Siliceous	³/₄	³/₄	³/₄	³/₄	³/₄	³/₄	³/₄	1	1¹/₄	1⁵/₈
Carbonate	³/₄	³/₄	³/₄	³/₄	³/₄	³/₄	³/₄	³/₄	1¹/₄	1¹/₄
Sand-lightweight or lightweight	³/₄	³/₄	³/₄	³/₄	³/₄	³/₄	³/₄	³/₄	1¹/₄	1¹/₄

For SI: 1 inch = 25.4 mm.

TABLE 721.2.3(2)
COVER THICKNESS FOR PRESTRESSED CONCRETE FLOOR OR ROOF SLABS (inches)

CONCRETE AGGREGATE TYPE	FIRE-RESISTANCE RATING (hours)									
	Restrained					Unrestrained				
	1	$1^1/_2$	2	3	4	1	$1^1/_2$	2	3	4
Siliceous	$^3/_4$	$^3/_4$	$^3/_4$	$^3/_4$	$^3/_4$	$1^1/_8$	$1^1/_2$	$1^3/_4$	$2^3/_8$	$2^3/_4$
Carbonate	$^3/_4$	$^3/_4$	$^3/_4$	$^3/_4$	$^3/_4$	1	$1^3/_8$	$1^5/_8$	$2^1/_8$	$2^1/_4$
Sand-lightweight or lightweight	$^3/_4$	$^3/_4$	$^3/_4$	$^3/_4$	$^3/_4$	1	$1^3/_8$	$1^1/_2$	2	$2^1/_4$

For SI: 1 inch = 25.4 mm.

TABLE 721.2.3(3)
MINIMUM COVER FOR MAIN REINFORCING BARS OF REINFORCED CONCRETE BEAMS[c]
(APPLICABLE TO ALL TYPES OF STRUCTURAL CONCRETE)

RESTRAINED OR UNRESTRAINED[a]	BEAM WIDTH[b] (inches)	FIRE-RESISTANCE RATING (hours)				
		1	$1^1/_2$	2	3	4
Restrained	5	$^3/_4$	$^3/_4$	$^3/_4$	1^a	$1^1/_4{}^a$
	7	$^3/_4$	$^3/_4$	$^3/_4$	$^3/_4$	$^3/_4$
	≥ 10	$^3/_4$	$^3/_4$	$^3/_4$	$^3/_4$	$^3/_4$
Unrestrained	5	$^3/_4$	1	$1^1/_4$	—	—
	7	$^3/_4$	$^3/_4$	$^3/_4$	$1^3/_4$	3
	≥ 10	$^3/_4$	$^3/_4$	$^3/_4$	1	$1^3/_4$

For SI: 1 inch = 25.4 mm, 1 foot = 304.8 mm.

a. Tabulated values for restrained assemblies apply to beams spaced more than 4 feet on center. For restrained beams spaced 4 feet or less on center, minimum cover of $^3/_4$ inch is adequate for ratings of 4 hours or less.

b. For beam widths between the tabulated values, the minimum cover thickness can be determined by direct interpolation.

c. The cover for an individual reinforcing bar is the minimum thickness of concrete between the surface of the bar and the fire-exposed surface of the beam. For beams in which several bars are used, the cover for corner bars used in the calculation shall be reduced to one-half of the actual value. The cover for an individual bar must be not less than one-half of the value given in Table 721.2.3(3) nor less than $^3/_4$ inch.

TABLE 721.2.3(4)
MINIMUM COVER FOR PRESTRESSED CONCRETE BEAMS 8 INCHES OR GREATER IN WIDTH

RESTRAINED OR UNRESTRAINED[a]	CONCRETE AGGREGATE TYPE	BEAM WIDTH[b] (inches)	FIRE-RESISTANCE RATING (hours)				
			1	$1^1/_2$	2	3	4
Restrained	Carbonate or siliceous	8	$1^1/_2$	$1^1/_2$	$1^1/_2$	$1^3/_4{}^a$	$2^1/_2{}^a$
	Carbonate or siliceous	≥ 12	$1^1/_2$	$1^1/_2$	$1^1/_2$	$1^1/_2$	$1^7/_8{}^a$
	Sand lightweight	8	$1^1/_2$	$1^1/_2$	$1^1/_2$	$1^1/_2$	2^a
	Sand lightweight	≥ 12	$1^1/_2$	$1^1/_2$	$1^1/_2$	$1^1/_2$	$1^5/_8{}^a$
Unrestrained	Carbonate or siliceous	8	$1^1/_2$	$1^3/_4$	$2^1/_2$	5^c	—
	Carbonate or siliceous	≥ 12	$1^1/_2$	$1^1/_2$	$1^7/_8{}^a$	$2^1/_2$	3
	Sand lightweight	8	$1^1/_2$	$1^1/_2$	2	$3^1/_4$	—
	Sand lightweight	≥ 12	$1^1/_2$	$1^1/_2$	$1^5/_8$	2	$2^1/_2$

For SI: 1 inch = 25.4 mm, 1 foot = 304.8 mm.

a. Tabulated values for restrained assemblies apply to beams spaced more than 4 feet on center. For restrained beams spaced 4 feet or less on center, minimum cover of $^3/_4$ inch is adequate for 4-hour ratings or less.

b. For beam widths between 8 inches and 12 inches, minimum cover thickness can be determined by direct interpolation.

c. Not practical for 8-inch-wide beam but shown for purposes of interpolation.

TABLE 721.2.3(5)
MINIMUM COVER FOR PRESTRESSED CONCRETE BEAMS OF ALL WIDTHS

RESTRAINED OR UNRESTRAINED[a]	CONCRETE AGGREGATE TYPE	BEAM AREA[b] A (square inches)	FIRE-RESISTANCE RATING (hours)				
			1	$1^1/_2$	2	3	4
Restrained	All	$40 \leq A \leq 150$	$1^1/_2$	$1^1/_2$	2	$2^1/_2$	—
	Carbonate or siliceous	$150 < A \leq 300$	$1^1/_2$	$1^1/_2$	$1^1/_2$	$1^3/_4$	$2^1/_2$
		$300 < A$	$1^1/_2$	$1^1/_2$	$1^1/_2$	$1^1/_2$	2
	Sand lightweight	$150 < A$	$1^1/_2$	$1^1/_2$	$1^1/_2$	$1^1/_2$	2
Unrestrained	All	$40 \leq A \leq 150$	2	$2^1/_2$	—	—	—
	Carbonate or siliceous	$150 < A \leq 300$	$1^1/_2$	$1^3/_4$	$2^1/_2$	—	—
		$300 < A$	$1^1/_2$	$1^1/_2$	2	3^c	4^c
	Sand lightweight	$150 < A$	$1^1/_2$	$1^1/_2$	2	3^c	4^c

For SI: 1 inch = 25.4 mm, 1 foot = 304.8 mm.

a. Tabulated values for restrained assemblies apply to beams spaced more than 4 feet on center. For restrained beams spaced 4 feet or less on center, minimum cover of $^3/_4$ inch is adequate for 4-hour ratings or less.

b. The cross-sectional area of a stem is permitted to include a portion of the area in the flange, provided the width of the flange used in the calculation does not exceed three times the average width of the stem.

c. U-shaped or hooped stirrups spaced not to exceed the depth of the member and having a minimum cover of 1 inch shall be provided.

tendons located along the vertical centerline of the stem, the average cover shall be the distance from the bottom of the member to the centroid of the tendons. The actual cover for any individual tendon shall not be less than one-half the smaller value shown in Tables 721.2.3(4) and 721.2.3(5), or 1 inch (25 mm), whichever is greater.

721.2.4 Concrete columns. Concrete columns shall comply with this section.

TABLE 721.2.4
MINIMUM DIMENSION OF CONCRETE COLUMNS (inches)

TYPES OF CONCRETE	FIRE-RESISTANCE RATING (hours)				
	1	$1^1/_2$	2^a	3^a	4^b
Siliceous	8	9	10	12	14
Carbonate	8	9	10	11	12
Sand-lightweight	8	$8^1/_2$	9	$10^1/_2$	12

For SI: 1 inch = 25 mm.

a. The minimum dimension is permitted to be reduced to 8 inches for rectangular columns with two parallel sides at least 36 inches in length.

b. The minimum dimension is permitted to be reduced to 10 inches for rectangular columns with two parallel sides at least 36 inches in length.

721.2.4.1 Minimum size. The minimum overall dimensions of reinforced concrete columns for fire-resistance ratings of 1 hour to 4 hours shall comply with Table 721.2.4.

721.2.4.2 Minimum cover for R/C columns. The minimum thickness of concrete cover to the main longitudinal reinforcement in columns, regardless of the type of aggregate used in the concrete, shall not be less than 1 inch (25 mm) times the number of hours of required fire resistance or 2 inches (51 mm), whichever is less.

721.2.4.3 Columns built into walls. The minimum dimensions of Table 721.2.4 do not apply to a reinforced

concrete column that is built into a concrete or masonry wall provided all of the following are met:

1. The fire-resistance rating for the wall is equal to or greater than the required rating of the column;

2. The main longitudinal reinforcing in the column has cover not less than that required by Section 721.2.4.2; and

3. Openings in the wall are protected in accordance with Table 715.4.

Where openings in the wall are not protected as required by Section 715.4, the minimum dimension of columns required to have a fire-resistance rating of 3 hours or less shall be 8 inches (203 mm), and 10 inches (254 mm) for columns required to have a fire-resistance rating of 4 hours, regardless of the type of aggregate used in the concrete.

721.2.4.4 Precast cover units for steel columns. See Section 721.5.1.4.

721.3 Concrete masonry. The provisions of this section contain procedures by which the fire-resistance ratings of concrete masonry are established by calculations.

721.3.1 Equivalent thickness. The equivalent thickness of concrete masonry construction shall be determined in accordance with the provisions of this section.

721.3.1.1 Concrete masonry unit plus finishes. The equivalent thickness of concrete masonry assemblies, T_{ea}, shall be computed as the sum of the equivalent thickness of the concrete masonry unit, T_e, as determined by Section 721.3.1.2, 721.3.1.3, or 721.3.1.4, plus the equivalent thickness of finishes, T_{ef}, determined in accordance with Section 721.3.2:

$$T_{ea} = T_e + T_{ef} \qquad \textbf{(Equation 7-6)}$$

$T_e = V_n/LH =$ Equivalent thickness of concrete masonry unit (inch) (mm).

where:

V_n = Net volume of masonry unit (inch^3) (mm^3).

L = Specified length of masonry unit (inch) (mm).

H = Specified height of masonry unit (inch) (mm).

721.3.1.2 Ungrouted or partially grouted construction. T_e shall be the value obtained for the concrete masonry unit determined in accordance with ASTM C 140.

721.3.1.3 Solid grouted construction. The equivalent thickness, T_e, of solid grouted concrete masonry units is the actual thickness of the unit.

721.3.1.4 Airspaces and cells filled with loose-fill material. The equivalent thickness of completely filled hollow concrete masonry is the actual thickness of the unit when loose-fill materials are: sand, pea gravel, crushed stone, or slag that meet ASTM C 33 requirements; pumice, scoria, expanded shale, expanded clay, expanded slate, expanded slag, expanded fly ash, or cinders that comply with ASTM C 331; or perlite or vermiculite meeting the requirements of ASTM C 549 and ASTM C 516, respectively.

721.3.2 Concrete masonry walls. The fire-resistance rating of walls and partitions constructed of concrete masonry units shall be determined from Table 721.3.2. The rating shall be based on the equivalent thickness of the masonry and type of aggregate used.

721.3.2.1 Finish on nonfire-exposed side. Where plaster or gypsum wallboard is applied to the side of the wall not exposed to fire, the contribution of the finish to the total fire-resistance rating shall be determined as follows: The thickness of gypsum wallboard or plaster shall be corrected by multiplying the actual thickness of the finish by applicable factor determined from Table 721.2.1.4(1). This corrected thickness of finish shall be added to the equivalent thickness of masonry and the fire-resistance rating of the masonry and finish determined from Table 721.3.2.

721.3.2.2 Finish on fire-exposed side. Where plaster or gypsum wallboard is applied to the fire-exposed side of the wall, the contribution of the finish to the total fire-resistance rating shall be determined as follows: The time assigned to the finish as established by Table 721.2.1.4(2) shall be added to the fire-resistance rating determined in Section 721.3.2 for the masonry alone, or in Section 721.3.2.1 for the masonry and finish on the nonfire-exposed side.

721.3.2.3 Nonsymmetrical assemblies. For a wall having no finish on one side or having different types or thicknesses of finish on each side, the calculation procedures of this section shall be performed twice, assuming either side of the wall to be the fire-exposed side. The fire-resistance rating of the wall shall not exceed the lower of the two values calculated.

Exception: For exterior walls with more than 5 feet (1524 mm) of horizontal separation, the fire shall be assumed to occur on the interior side only.

721.3.2.4 Minimum concrete masonry fire-resistance rating. Where the finish applied to a concrete masonry wall contributes to its fire-resistance rating, the masonry alone shall provide not less than one-half the total required fire-resistance rating.

721.3.2.5 Attachment of finishes. Installation of finishes shall be as follows:

1. Gypsum wallboard and gypsum lath applied to concrete masonry or concrete walls shall be secured to wood or steel furring members spaced not more than 16 inches (406 mm) on center (o.c.).

2. Gypsum wallboard shall be installed with the long dimension parallel to the furring members and shall have all joints finished.

3. Other aspects of the installation of finishes shall comply with the applicable provisions of Chapters 7 and 25.

TABLE 721.3.2
MINIMUM EQUIVALENT THICKNESS (inches) OF BEARING OR NONBEARING CONCRETE MASONRY WALLS[a,b,c,d]

TYPE OF AGGREGATE	FIRE-RESISTANCE RATING (hours)														
	1/2	3/4	1	1 1/4	1 1/2	1 3/4	2	2 1/4	2 1/2	2 3/4	3	3 1/4	3 1/2	3 3/4	4
Pumice or expanded slag	1.5	1.9	2.1	2.5	2.7	3.0	3.2	3.4	3.6	3.8	4.0	4.2	4.4	4.5	4.7
Expanded shale, clay or slate	1.8	2.2	2.6	2.9	3.3	3.4	3.6	3.8	4.0	4.2	4.4	4.6	4.8	4.9	5.1
Limestone, cinders or unexpanded slag	1.9	2.3	2.7	3.1	3.4	3.7	4.0	4.3	4.5	4.8	5.0	5.2	5.5	5.7	5.9
Calcareous or siliceous gravel	2.0	2.4	2.8	3.2	3.6	3.9	4.2	4.5	4.8	5.0	5.3	5.5	5.8	6.0	6.2

For SI: 1 inch = 25.4 mm.

a. Values between those shown in the table can be determined by direct interpolation.

b. Where combustible members are framed into the wall, the thickness of solid material between the end of each member and the opposite face of the wall, or between members set in from opposite sides, shall not be less than 93 percent of the thickness shown in the table.

c. Requirements of ASTM C 55, ASTM C 73 or ASTM C 90 shall apply.

d. Minimum required equivalent thickness corresponding to the hourly fire-resistance rating for units with a combination of aggregate shall be determined by linear interpolation based on the percent by volume of each aggregate used in manufacture.

721.3.3 Multiwythe masonry walls. The fire-resistance rating of wall assemblies constructed of multiple wythes of masonry materials shall be permitted to be based on the fire-resistance rating period of each wythe and the continuous airspace between each wythe in accordance with the following formula:

$$R_A = (R_1^{0.59} + R_2^{0.59} + ... + R_n^{0.59} + A_1 + A_2 + ... + A_n)^{1.7}$$
(Equation 7-7)

where:

R_A = Fire endurance rating of the assembly (hours).

$R_1, R_2, ..., R_n$ = Fire endurance rating of wythes for 1, 2, n (hours), respectively.

$A_1, A_2, ..., A_n$ = 0.30, factor for each continuous airspace for 1, 2, ...n, respectively, having a depth of $\frac{1}{2}$ inch (12.7 mm) or more between wythes.

721.3.4 Concrete masonry lintels. Fire-resistance ratings for concrete masonry lintels shall be determined based upon the nominal thickness of the lintel and the minimum thickness of concrete masonry or concrete, or any combination thereof, covering the main reinforcing bars, as determined according to Table 721.3.4, or by approved alternate methods.

TABLE 721.3.4
MINIMUM COVER OF LONGITUDINAL
REINFORCEMENT IN FIRE-RESISTANCE-RATED
REINFORCED CONCRETE MASONRY LINTELS (inches)

NOMINAL WIDTH OF LINTEL (inches)	FIRE-RESISTANCE RATING (hours)			
	1	2	3	4
6	$1\frac{1}{2}$	2	—	—
8	$1\frac{1}{2}$	$1\frac{1}{2}$	$1\frac{3}{4}$	3
10 or greater	$1\frac{1}{2}$	$1\frac{1}{2}$	$1\frac{1}{2}$	$1\frac{3}{4}$

For SI: 1 inch = 25.4 mm.

721.3.5 Concrete masonry columns. The fire-resistance rating of concrete masonry columns shall be determined based upon the least plan dimension of the column in accordance with Table 721.3.5 or by approved alternate methods.

TABLE 721.3.5
MINIMUM DIMENSION OF
CONCRETE MASONRY COLUMNS (inches)

FIRE-RESISTANCE RATING (hours)			
1	2	3	4
8	10	12	14

For SI: 1 inch = 25.4 mm.

721.4 Clay brick and tile masonry. The provisions of this section contain procedures by which the fire-resistance ratings of clay brick and tile masonry are established by calculations.

721.4.1 Masonry walls. The fire-resistance rating of masonry walls shall be based upon the equivalent thickness as calculated in accordance with this section. The calculation shall take into account finishes applied to the wall and airspaces between wythes in multiwythe construction.

721.4.1.1 Equivalent thickness. The fire-resistance ratings of walls or partitions constructed of solid or hollow clay masonry units shall be determined from Table 721.4.1(1) or 721.4.1(2). The equivalent thickness of the clay masonry unit shall be determined by Equation 7-8 when using Table 721.4.1(1). The fire-resistance rating determined from Table 721.4.1(1) shall be permitted to be used in the calculated fire-resistance rating procedure in Section 721.4.2.

$$T_e = V_n/LH$$
(Equation 7-8)

where:

T_e = The equivalent thickness of the clay masonry unit (inches).

V_n = The net volume of the clay masonry unit (inch³).

L = The specified length of the clay masonry unit (inches).

H = The specified height of the clay masonry unit (inches).

721.4.1.1.1 Hollow clay units. The equivalent thickness, T_e, shall be the value obtained for hollow clay units as determined in accordance with ASTM C 67.

TABLE 721.4.1(1)
FIRE-RESISTANCE PERIODS OF CLAY MASONRY WALLS

MATERIAL TYPE	MINIMUM REQUIRED EQUIVALENT THICKNESS FOR FIRE RESISTANCE[a,b,c] (inches)			
	1 hour	2 hour	3 hour	4 hour
Solid brick of clay or shale[d]	2.7	3.8	4.9	6.0
Hollow brick or tile of clay or shale, unfilled	2.3	3.4	4.3	5.0
Hollow brick or tile of clay or shale, grouted or filled with materials specified in Section 721.4.1.1.3	3.0	4.4	5.5	6.6

For SI: 1 inch = 25.4 mm.

a. Equivalent thickness as determined from Section 721.4.1.1.

b. Calculated fire resistance between the hourly increments listed shall be determined by linear interpolation.

c. Where combustible members are framed in the wall, the thickness of solid material between the end of each member and the opposite face of the wall, or between members set in from opposite sides, shall be not less than 93 percent of the thickness shown.

d. For units in which the net cross-sectional area of cored brick in any plane parallel to the surface containing the cores is at least 75 percent of the gross cross-sectional area measured in the same plane.

TABLE 721.4.1(2)
FIRE-RESISTANCE RATINGS FOR BEARING STEEL FRAME
BRICK VENEER WALLS OR PARTITIONS

WALL OR PARTITION ASSEMBLY	PLASTER SIDE EXPOSED (hours)	BRICK FACED SIDE EXPOSED (hours)
Outside facing of steel studs: $^1/_2''$ wood fiberboard sheathing next to studs, $^3/_4''$ airspace formed with $^3/_4'' \times 1\,^5/_8''$ wood strips placed over the fiberboard and secured to the studs; metal or wire lath nailed to such strips, $3^3/_4''$ brick veneer held in place by filling $^3/_4''$ airspace between the brick and lath with mortar. Inside facing of studs: $^3/_4''$ unsanded gypsum plaster on metal or wire lath attached to $^5/_{16}''$ wood strips secured to edges of the studs.	1.5	4
Outside facing of steel studs: $1''$ insulation board sheathing attached to studs, $1''$ airspace, and $3^3/_4''$ brick veneer attached to steel frame with metal ties every 5th course. Inside facing of studs: $^7/_8''$ sanded gypsum plaster (1:2 mix) applied on metal or wire lath attached directly to the studs.	1.5	4
Same as above except use $^7/_8''$ vermiculite—gypsum plaster or $1''$ sanded gypsum plaster (1:2 mix) applied to metal or wire.	2	4
Outside facing of steel studs: $^1/_2''$ gypsum sheathing board, attached to studs, and $3^3/_4''$ brick veneer attached to steel frame with metal ties every 5th course. Inside facing of studs: $^1/_2''$ sanded gypsum plaster (1:2 mix) applied to $^1/_2''$ perforated gypsum lath securely attached to studs and having strips of metal lath 3 inches wide applied to all horizontal joints of gypsum lath.	2	4

For SI: 1 inch = 25.4 mm.

721.4.1.1.2 Solid grouted clay units. The equivalent thickness of solid grouted clay masonry units shall be taken as the actual thickness of the units.

721.4.1.1.3 Units with filled cores. The equivalent thickness of the hollow clay masonry units is the actual thickness of the unit when completely filled with loose-fill materials of: sand, pea gravel, crushed stone, or slag that meet ASTM C 33 requirements; pumice, scoria, expanded shale, expanded clay, expanded slate, expanded slag, expanded fly ash, or cinders in compliance with ASTM C 331; or perlite or vermiculite meeting the requirements of ASTM C 549 and ASTM C 516, respectively.

721.4.1.2 Plaster finishes. Where plaster is applied to the wall, the total fire-resistance rating shall be determined by the formula:

$$R = (R_n^{0.59} + pl)\,1.7 \qquad \textbf{(Equation 7-9)}$$

where:

R = The fire endurance of the assembly (hours).

R_n = The fire endurance of the individual wall (hours).

pl = Coefficient for thickness of plaster.

Values for $R_n^{0.59}$ for use in Equation 7-9 are given in Table 721.4.1(3). Coefficients for thickness of plaster shall be selected from Table 721.4.1(4) based on the actual thickness of plaster applied to the wall or partition and whether one or two sides of the wall are plastered.

721.4.1.3 Multiwythe walls with airspace. Where a continuous airspace separates multiple wythes of the wall or partition, the total fire-resistance rating shall be determined by the formula:

$$R = (R_1^{0.59} + R_2^{0.59} + ... + R_n^{0.59} + as)^{1.7} \qquad \textbf{(Equation 7-10)}$$

where:

R = The fire endurance of the assembly (hours).

R_1, R_2 and R_n = The fire endurance of the individual wythes (hours).

as = Coefficient for continuous airspace.

Values for $R_n^{0.59}$ for use in Equation 7-10 are given in Table 721.4.1(3). The coefficient for each continuous airspace of $^1/_2$ inch to $3^1/_2$ inches (12.7 to 89 mm) separating two individual wythes shall be 0.3.

721.4.1.4 Nonsymmetrical assemblies. For a wall having no finish on one side or having different types or thicknesses of finish on each side, the calculation procedures of this section shall be performed twice, assuming either side to be the fire-exposed side of the wall. The fire resistance of the wall shall not exceed the lower of the two values determined.

Exception: For exterior walls with more than 5 feet (1524 mm) of horizontal separation, the fire shall be assumed to occur on the interior side only.

TABLE 721.4.1(3)
VALUES OF $R_n^{0.59}$

$R_n^{0.59}$	R (hours)
1	1.0
2	1.50
3	1.91
4	2.27

TABLE 721.4.1(4)
COEFFICIENTS FOR PLASTER, pl [a]

THICKNESS OF PLASTER (inch)	ONE SIDE	TWO SIDE
$1/2$	0.3	0.6
$5/8$	0.37	0.75
$3/4$	0.45	0.90

For SI: 1 inch = 25.4 mm.

a. Values listed in table are for 1:3 sanded gypsum plaster.

TABLE 721.4.1(5)
REINFORCED MASONRY LINTELS

NOMINAL LINTEL WIDTH (inches)	MINIMUM LONGITUDINAL REINFORCEMENT COVER FOR FIRE RESISTANCE (inch)			
	1 hour	2 hour	3 hour	4 hour
6	$1^1/_2$	2	NP	NP
8	$1^1/_2$	$1^1/_2$	$1^3/_4$	3
10 or more	$1^1/_2$	$1^1/_2$	$1^1/_2$	$1^3/_4$

For SI: 1 inch = 25.4 mm.

NP = Not permitted.

TABLE 721.4.1(6)
REINFORCED CLAY MASONRY COLUMNS

COLUMN SIZE	FIRE-RESISTANCE RATING (hour)			
	1	2	3	4
Minimum column dimension (inches)	8	10	12	14

For SI: 1 inch = 25.4 mm.

721.4.2 Multiwythe walls. The fire-resistance rating for walls or partitions consisting of two or more dissimilar wythes shall be permitted to be determined by the formula:

$$R = (R_1^{0.59} + R_2^{0.59} + ... + R_n^{0.59})^{1.7} \qquad \textbf{(Equation 7-11)}$$

where:

R = The fire endurance of the assembly (hours).

R_1, R_2 and R_n = The fire endurance of the individual wythes (hours).

Values for $R_n^{0.59}$ for use in Equation 7-11 are given in Table 721.4.1(3).

721.4.2.1 Multiwythe walls of different material. For walls that consist of two or more wythes of different materials (concrete or concrete masonry units) in combination with clay masonry units, the fire-resistance rating of the different materials shall be permitted to be determined from Table 721.2.1.1 for concrete; Table 721.3.2 for concrete masonry units or Table 721.4.1(1) or 721.4.1(2) for clay and tile masonry units.

721.4.3 Reinforced clay masonry lintels. Fire-resistance ratings for clay masonry lintels shall be determined based on the nominal width of the lintel and the minimum covering for the longitudinal reinforcement in accordance with Table 721.4.1(5).

721.4.4 Reinforced clay masonry columns. The fire-resistance ratings shall be determined based on the last plan dimension of the column in accordance with Table 721.4.1(6). The minimum cover for longitudinal reinforcement shall be 2 inches (51 mm).

721.5 Steel assemblies. The provisions of this section contain procedures by which the fire-resistance ratings of steel assemblies are established by calculations.

721.5.1 Structural steel columns. The fire-resistance ratings of steel columns shall be based on the size of the element and the type of protection provided in accordance with this section.

721.5.1.1 General. These procedures establish a basis for determining the fire resistance of column assemblies as a function of the thickness of fire-resistant material and, the weight, W, and heated perimeter, D, of steel columns. As used in these sections, W is the average weight of a structural steel column in pounds per linear foot. The heated perimeter, D, is the inside perimeter of the fire-resistant material in inches as illustrated in Figure 721.5.1(1).

721.5.1.1.1 Nonload-bearing protection. The application of these procedures shall be limited to column assemblies in which the fire-resistant material is not designed to carry any of the load acting on the column.

721.5.1.1.2 Embedments. In the absence of substantiating fire-endurance test results, ducts, conduit, piping, and similar mechanical, electrical, and plumbing installations shall not be embedded in any required fire-resistant materials.

721.5.1.1.3 Weight-to-perimeter ratio. Table 721.5.1(1) contains weight-to-heated-perimeter ratios (W/D) for both contour and box fire-resistant profiles, for the wide flange shapes most often used as columns.

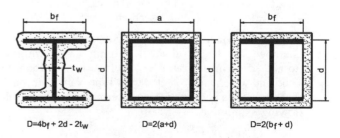

FIGURE 721.5.1(1)
DETERMINATION OF THE HEATED PERIMETER
OF STRUCTURAL STEEL COLUMNS

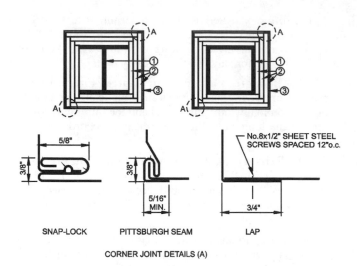

FIGURE 721.5.1(2)
GYPSUM WALLBOARD PROTECTED STRUCTURAL STEEL
COLUMNS WITH SHEET STEEL COLUMN COVERS

For SI: 1 inch = 25.4 mm, 1 foot = 305 mm.

1. Structural steel column, either wide flange or tubular shapes.

2. Type X gypsum wallboard in accordance with ASTM C 36. For single-layer applications, the wallboard shall be applied vertically with no horizontal joints. For multiple-layer applications, horizontal joints are permitted at a minimum spacing of 8 feet, provided that the joints in successive layers are staggered at least 12 inches. The total required thickness of wallboard shall be determined on the basis of the specified fire-resistance rating and the weight-to-heated-perimeter ratio (*W/D*) of the column. For fire-resistance ratings of 2 hours or less, one of the required layers of gypsum wallboard may be applied to the exterior of the sheet steel column covers with 1-inch long Type S screws spaced 1 inch from the wallboard edge and 8 inches on center. For such installations, 0.0149-inch minimum thickness galvanized steel corner beads with $1^1/_2$-inch legs shall be attached to the wallboard with Type S screws spaced 12 inches on center.

3. For fire-resistance ratings of 3 hours or less, the column covers shall be fabricated from 0.0239-inch minimum thickness galvanized or stainless steel. For 4-hour fire-resistance ratings, the column covers shall be fabricated from 0.0239-inch minimum thickness stainless steel. The column covers shall be erected with the Snap Lock or Pittsburgh joint details.

For fire-resistance ratings of 2 hours or less, column covers fabricated from 0.0269-inch minimum thickness galvanized or stainless steel shall be permitted to be erected with lap joints. The lap joints shall be permitted to be located anywhere around the perimeter of the column cover. The lap joints shall be secured with $^1/_2$-inch-long No. 8 sheet metal screws spaced 12 inches on center.

The column covers shall be provided with a minimum expansion clearance of $^1/_8$ inch per linear foot between the ends of the cover and any restraining construction.

For different fire-resistant protection profiles or column cross sections, the weight-to-heated-perimeter ratios (*W/D*) shall be determined in accordance with the definitions given in this section.

721.5.1.2 Gypsum wallboard protection. The fire resistance of structural steel columns with weight-to-heated-perimeter ratios (*W/D*) less than or equal to 3.65 and which are protected with Type X gypsum wallboard shall be permitted to be determined from the following expression:

$$R = 130\left[\frac{h(W'/D)}{2}\right]^{0.75} \qquad \textbf{(Equation 7-12)}$$

where:

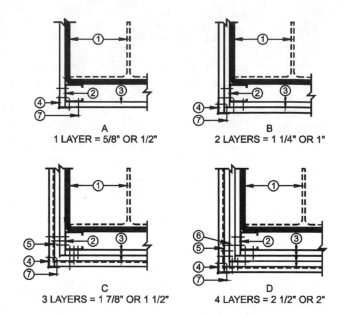

FIGURE 721.5.1(3)
GYPSUM WALLBOARD PROTECTED STRUCTURAL STEEL
COLUMNS WITH STEEL STUD/SCREW ATTACHMENT SYSTEM

For SI: 1 inch = 25.4 mm, 1 foot = -305 mm.

1. Structural steel column, either wide flange or tubular shapes.

2. $1^5/_8$-inch deep studs fabricated from 0.0179-inch minimum thickness galvanized steel with $1^5/_{16}$ or $1^7/_{16}$-inch legs. The length of the steel studs shall be $^1/_2$ inch less than the height of the assembly.

3. Type X gypsum wallboard in accordance with ASTM C 36. For single-layer applications, the wallboard shall be applied vertically with no horizontal joints. For multiple-layer applications, horizontal joints are permitted at a minimum spacing of 8 feet, provided that the joints in successive layers are staggered at least 12 inches. The total required thickness of wallboard shall be determined on the basis of the specified fire-resistance rating and the weight-to-heated-perimeter ratio (*W/D*) of the column.

4. Galvanized 0.0149-inch minimum thickness steel corner beads with $1^1/_2$-inch legs attached to the wallboard with 1-inch-long Type S screws spaced 12 inches on center.

5. No. 18 SWG steel tie wires spaced 24 inches on center.

6. Sheet metal angles with 2-inch legs fabricated from 0.0221-inch minimum thickness galvanized steel.

7. Type S screws, 1 inch long, shall be used for attaching the first layer of wallboard to the steel studs and the third layer to the sheet metal angles at 24 inches on center. Type S screws $1^3/_4$-inch long shall be used for attaching the second layer of wallboard to the steel studs and the fourth layer to the sheet metal angles at 12 inches on center. Type S screws $2^1/_4$ inches long shall be used for attaching the third layer of wallboard to the steel studs at 12 inches on center.

R = Fire resistance (minutes).

h = Total thickness of gypsum wallboard (inches).

D = Heated perimeter of the structural steel column (inches).

W' = Total weight of the structural steel column and gypsum wallboard protection (pounds per linear foot).

W' = $W + 50hD/144$.

721.5.1.2.1 Attachment. The gypsum wallboard shall be supported as illustrated in either Figure 721.5.1(2) for fire-resistance ratings of 4 hours or less, or Figure 721.5.1(3) for fire-resistance ratings of 3 hours or less.

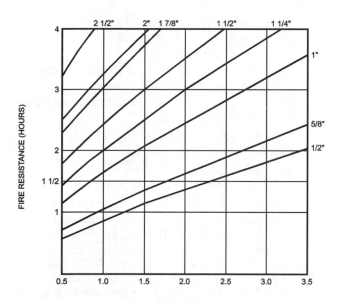

For SI: 1 inch = 25.4 mm, 1 pound per linear foot/inch = 0.059 kg/m/mm.

FIGURE 721.5.1(4)
FIRE RESISTANCE OF STRUCTURAL STEEL COLUMNS
PROTECTED WITH VARIOUS THICKNESSES OF
TYPE X GYPSUM WALLBOARD

a. The W/D ratios for typical wide flange columns are listed in Table 721.5.1(1). For other column shapes, the W/D ratios shall be determined in accordance with Section 720.5.1.1.

721.5.1.2.2 Gypsum wallboard equivalent to concrete. The determination of the fire resistance of structural steel columns from Figure 721.5.1(4) is permitted for various thicknesses of gypsum wallboard as a function of the weight-to-heated-perimeter ratio (*W/D*) of the column. For structural steel columns with weight-to-heated-perimeter ratios (*W/D*) greater than 3.65, the thickness of gypsum wallboard required for specified fire-resistance ratings shall be the same as the thickness determined for a *W*14 x 233 wide flange shape.

721.5.1.3 Spray-applied fire-resistant materials. The fire resistance of wide-flange structural steel columns protected with spray-applied fire-resistant materials, as illustrated in Figure 721.5.1(5), shall be permitted to be determined from the following expression:

$$R = \left[C_1 \left(W / D \right) + C_2 \right] h \qquad \textbf{(Equation 7-13)}$$

where:

R = Fire resistance (minutes).

h = Thickness of spray-applied fire-resistant material (inches).

D = Heated perimeter of the structural steel column (inches).

C_1 and C_2 = Material-dependent constants.

W = Weight of structural steel column (pounds per linear foot).

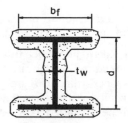

FIGURE 721.5.1(5)
WIDE FLANGE STRUCTURAL STEEL COLUMNS WITH
SPRAY-APPLIED FIRE-RESISTANT MATERIALS

721.5.1.3.1 Material-dependent constants. The material-dependent constants, C_1 and C_2, shall be determined for specific fire-resistant materials on the basis of standard fire endurance tests in accordance with Section 703.2. Unless evidence is submitted to the building official substantiating a broader application, this expression shall be limited to determining the fire resistance of structural steel columns with weight-to-heated-perimeter ratios (*W/D*) between the largest and smallest columns for which standard fire-endurance test results are available.

721.5.1.3.2 Spray-applied identification. Spray-applied fire-resistant materials shall be identified by density and thickness required for a given fire-resistance rating.

721.5.1.4 Concrete-protected columns. The fire resistance of structural steel columns protected with concrete, as illustrated in Figure 721.5.1(6) (a) and (b), shall be permitted to be determined from the following expression:

$$R = R_o \left(1 + 0.03m \right) \qquad \textbf{(Equation 7-14)}$$

where:

$$R_o = 10 \, (W/D)^{0.7} + 17 \, (h^{1.6}/k_c^{0.2}) \times (1 + 26 \, (H/p_c c_c h \, (L + h))^{0.8})$$

As used in these expressions:

R = Fire endurance at equilibrium moisture conditions (minutes).

R_o = Fire endurance at zero moisture content (minutes).

m = Equilibrium moisture content of the concrete by volume (percent).

W = Average weight of the steel column (pounds per linear foot).

D = Heated perimeter of the steel column (inches).

h = Thickness of the concrete cover (inches).

k_c = Ambient temperature thermal conductivity of the concrete (Btu/hr ft °F).

H = Ambient temperature thermal capacity of the steel column = 0.11W (Btu/ ft °F).

p_c = Concrete density (pounds per cubic foot).

c_c = Ambient temperature specific heat of concrete (Btu/lb °F).

L = Interior dimension of one side of a square concrete box protection (inches).

721.5.1.4.1 Reentrant space filled. For wide-flange steel columns completely encased in concrete with all reentrant spaces filled [Figure 721.5.1(6)(c)], the thermal capacity of the concrete within the reentrant spaces shall be permitted to be added to the thermal capacity of the steel column, as follows:

$$H = 0.11W + (p_c c_c/144)(b_f d - A_s) \qquad \textbf{(Equation 7-15)}$$

where:

b_f = Flange width of the steel column (inches).

d = Depth of the steel column (inches).

A_s = Cross-sectional area of the steel column (square inches).

721.5.1.4.2 Concrete properties unknown. If specific data on the properties of concrete are not available, the values given in Table 721.5.1(2) are permitted.

721.5.1.4.3 Minimum concrete cover. For structural steel column encased in concrete with all reentrant spaces filled, Figure 721.5.1(6)(c) and Tables 721.5.1(7) and 721.5.1(8) indicate the thickness of concrete cover required for various fire-resistance ratings for typical wide-flange sections. The thicknesses of concrete indicated in these tables also apply to structural steel columns larger than those listed.

721.5.1.4.4 Minimum precast concrete cover. For structural steel columns protected with precast concrete column covers as shown in Figure 721.5.1(6)(a), Tables 721.5.1(9) and 721.5.1(10) indicate the thickness of the column covers required for various fire-resistance ratings for typical wide-flange shapes. The thicknesses of concrete given in these tables also apply to structural steel columns larger than those listed.

721.5.1.4.5 Masonry protection. The fire resistance of structural steel columns protected with concrete masonry units or clay masonry units as illustrated in Figure 721.5.1(7), shall be permitted to be determined from the following expression:

$$R = 0.17 (W/D)^{0.7} + [0.285 (T_e^{1.6}/K^{0.2})]$$
$$[1.0 + 42.7 \{ (A_s/d_m T_e) / (0.25p + T_e) \}^{0.8}]$$
$$\textbf{(Equation 7-16)}$$

where:

R = Fire-resistance rating of column assembly (hours).

W = Average weight of steel column (pounds per foot).

D = Heated perimeter of steel column (inches) [see Figure 721.5.1(7)].

T_e = Equivalent thickness of concrete or clay masonry unit (inches) (see Table 721.3.2 Note a or Section 721.4.1).

K = Thermal conductivity of concrete or clay masonry unit (Btu/hr ft °F) [see Table 721.5.1(3)].

A_s = Cross-sectional area of steel column (square inches).

d_m = Density of the concrete or clay masonry unit (pounds per cubic foot).

p = Inner perimeter of concrete or clay masonry protection (inches) [see Figure 721.5.1(7)].

721.5.1.4.6 Equivalent concrete masonry thickness. For structural steel columns protected with concrete masonry, Table 721.5.1(5) gives the equivalent thickness of concrete masonry required for various fire-resistance ratings for typical column shapes. For structural steel columns protected with clay masonry, Table 721.5.1(6) gives the equivalent thickness of concrete masonry required for various fire-resistance ratings for typical column shapes.

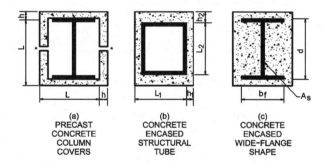

| (a) PRECAST CONCRETE COLUMN COVERS | (b) CONCRETE ENCASED STRUCTURAL TUBE | (c) CONCRETE ENCASED WIDE-FLANGE SHAPE |

FIGURE 721.5.1(6)
CONCRETE PROTECTED STRUCTURAL STEEL COLUMNS[a]

a. When the inside perimeter of the concrete protection is not square, L shall be taken as the average of L_1 and L_2. When the thickness of concrete cover is not constant, h shall be taken as the average of h_1 and h_2.

b. Joints shall be protected with a minimum 1 inch thickness of ceramic fiber blanket but in no case less than one-half the thickness of the column cover (see Section 720.2.1.3).

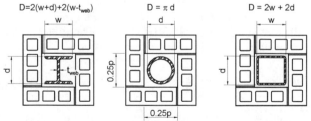

W SHAPE COLUMN   STEEL PIPE COLUMN   STRUCTURAL TUBE COLUMN

For SI: 1 inch = 25.4 mm.

FIGURE 721.5.1(7)
CONCRETE OR CLAY MASONRY PROTECTED STRUCTURAL STEEL COLUMNS

d = Depth of a wide flange column, outside diameter of pipe column, or outside dimension of structural tubing column (inches).

t_{web} = Thickness of web of wide flange column (inches).

w = Width of flange of wide flange column (inches).

TABLE 721.5.1(1)
***W/D* RATIOS FOR STEEL COLUMNS**

STRUCTURAL SHAPE	CONTOUR PROFILE	BOX PROFILE	STRUCTURAL SHAPE	CONTOUR PROFILE	BOX PROFILE
W14 × 233	2.49	3.65	W10 × 112	1.78	2.57
× 211	2.28	3.35	× 100	1.61	2.33
× 193	2.10	3.09	× 88	1.43	2.08
× 176	1.93	2.85	× 77	1.26	1.85
× 159	1.75	2.60	× 68	1.13	1.66
× 145	1.61	2.39	× 60	1.00	1.48
× 132	1.52	2.25	× 54	0.91	1.34
× 120	1.39	2.06	× 49	0.83	1.23
× 109	1.27	1.88	× 45	0.87	1.24
× 99	1.16	1.72	× 39	0.76	1.09
× 90	1.06	1.58	× 33	0.65	0.93
× 82	1.20	1.68			
× 74	1.09	1.53	W8 × 67	1.34	1.94
× 68	1.01	1.41	× 58	1.18	1.71
× 61	0.91	1.28	× 48	0.99S	1.44
× 53	0.89	1.21	× 40	0.83	1.23
× 48	0.81	1.10	× 35	0.73	1.08
× 43	0.73	0.99	× 31	0.65	0.97
			× 28	0.67	0.96
W12 × 190	2.46	3.51	× 24	0.58	0.83
× 170	2.22	3.20	× 21	0.57	0.77
× 152	2.01	2.90	× 18	0.49	0.67
× 136	1.82	2.63			
× 120	1.62	2.36	W6 × 25	0.69	1.00
× 106	1.44	2.11	× 20	0.56	0.82
× 96	1.32	1.93	× 16	0.57	0.78
× 87	1.20	1.76	× 15	0.42	0.63
× 79	1.10	1.61	× 12	0.43	0.60
× 72	1.00	1.48	× 9	0.33	0.46
× 65	0.91	1.35			
× 58	0.91	1.31	W5 × 19	0.64	0.93
× 53	0.84	1.20	× 16	0.54	0.80
× 50	0.89	1.23			
× 45	0.81	1.12	W4 × 13	0.54	0.79
× 40	0.72	1.00			

For SI: 1 pound per linear foot per inch = 0.059 kg/m/mm.

TABLE 721.5.1(2)
PROPERTIES OF CONCRETE

PROPERTY	NORMAL WEIGHT CONCRETE	STRUCTURAL LIGHTWEIGHT CONCRETE
Thermal conductivity (k_c)	0.95 Btu/hr ft °F	0.35 Btu/hr ft °F
Specific heat (c_c)	0.20 Btu/lb °F	0.20 Btu/lb °F
Density (P_c)	145 lb/ft³	110 lb/ft³
Equilibrium (free) moisture content (m) by volume	4%	5%

For SI: 1 inch = 25.4 mm, 1 foot = 304.8 mm, 1 lb/ft³ = 16.0185 kg/m³, Btu/hr ft °F = 1.731 W/(m · K)

TABLE 721.5.1(3)
THERMAL CONDUCTIVITY OF CONCRETE OR CLAY
MASONRY UNITS

DENSITY (d_m) OF UNITS (lb/ft³)	THERMAL CONDUCTIVITY (K) OF UNITS (Btu/hr ft °F)
Concrete Masonry Units	
80	0.207
85	0.228
90	0.252
95	0.278
100	0.308
105	0.340
110	0.376
115	0.416
120	0.459
125	0.508
130	0.561
135	0.620
140	0.685
145	0.758
150	0.837
Clay Masonry Units	
120	1.25
130	2.25

For SI: 1 pound per cubic foot = 16.0185 kg/m³, Btu per hour foot °F = 1.731 W/(m · K).

TABLE 721.5.1(4)
WEIGHT-TO-HEATED-PERIMETER RATIOS (*W/D*)
FOR TYPICAL WIDE FLANGE BEAM AND GIRDER SHAPES

STRUCTURAL SHAPE	CONTOUR PROFILE	BOX PROFILE	STRUCTURAL SHAPE	CONTOUR PROFILE	BOX PROFILE
W36 × 300	2.47	3.33	× 68	0.92	1.21
× 280	2.31	3.12	× 62	0.92	1.14
× 260	2.16	2.92	× 55	0.82	1.02
× 245	2.04	2.76			
× 230	1.92	2.61	W21 × 147	1.83	2.60
× 210	1.94	2.45	× 132	1.66	2.35
× 194	1.80	2.28	× 122	1.54	2.19
× 182	1.69	2.15	× 111	1.41	2.01
× 170	1.59	2.01	× 101	1.29	1.84
× 160	1.50	1.90	× 93	1.38	1.80
× 150	1.41	1.79	× 83	1.24	1.62
× 135	1.28	1.63	× 73	1.10	1.44
			× 68	1.03	1.35
W33 × 241	2.11	2.86	× 62	0.94	1.23
× 221	1.94	2.64	× 57	0.93	1.17
× 201	1.78	2.42	× 50	0.83	1.04
× 152	1.51	1.94	× 44	0.73	0.92
× 141	1.41	1.80			
× 130	1.31	1.67	W18 × 119	1.69	2.42
× 118	1.19	1.53	× 106	1.52	2.18
			× 97	1.39	2.01
W30 × 211	2.00	2.74	× 86	1.24	1.80
× 191	1.82	2.50	× 76	1.11	1.60
× 173	1.66	2.28	× 71	1.21	1.59
× 132	1.45	1.85	× 65	1.11	1.47
× 124	1.37	1.75	× 60	1.03	1.36
× 116	1.28	1.65	× 55	0.95	1.26
× 108	1.20	1.54	× 50	0.87	1.15
× 99	1.10	1.42	× 46	0.86	1.09
			× 40	0.75	0.96
W27 × 178	1.85	2.55	× 35	0.66	0.85
× 161	1.68	2.33			
× 146	1.53	2.12	W16 × 100	1.56	2.25
× 114	1.36	1.76	× 89	1.40	2.03
× 102	1.23	1.59	× 77	1.22	1.78
× 94	1.13	1.47	× 67	1.07	1.56
× 84	1.02	1.33	× 57	1.07	1.43
			× 50	0.94	1.26
			× 45	0.85	1.15
W24 × 162	1.85	2.57	× 40	0.76	1.03
× 146	1.68	2.34	× 36	0.69	0.93
× 131	1.52	2.12	× 31	0.65	0.83
× 117	1.36	1.91	× 26	0.55	0.70
× 104	1.22	1.71			
× 94	1.26	1.63	W14 × 132	1.83	3.00
× 84	1.13	1.47	× 120	1.67	2.75
× 76	1.03	1.34	× 109	1.53	2.52

(continued)

TABLE 721.5.1(4)—continued
WEIGHT-TO-HEATED-PERIMETER RATIOS (W/D)
FOR TYPICAL WIDE FLANGE BEAM AND GIRDER SHAPES

STRUCTURAL SHAPE	CONTOUR PROFILE	BOX PROFILE	STRUCTURAL SHAPE	CONTOUR PROFILE	BOX PROFILE
× 99	1.39	2.31	× 30	0.79	1.12
× 90	1.27	2.11	× 26	0.69	0.98
× 82	1.41	2.12	× 22	0.59	0.84
× 74	1.28	1.93	× 19	0.59	0.78
× 68	1.19	1.78	× 17	0.54	0.70
× 61	1.07	1.61	× 15	0.48	0.63
× 53	1.03	1.48	× 12	0.38	0.51
× 48	0.94	1.35			
× 43	0.85	1.22	W8 × 67	1.61	2.55
× 38	0.79	1.09	× 58	1.41	2.26
× 34	0.71	0.98	× 48	1.18	1.91
× 30	0.63	0.87	× 40	1.00	1.63
× 26	0.61	0.79	× 35	0.88	1.44
× 22	0.52	0.68	× 31	0.79	1.29
			× 28	0.80	1.24
W12 × 87	1.44	2.34	× 24	0.69	1.07
× 79	1.32	2.14	× 21	0.66	0.96
× 72	1.20	1.97	× 18	0.57	0.84
× 65	1.09	1.79	× 15	0.54	0.74
× 58	1.08	1.69	× 13	0.47	0.65
× 53	0.99	1.55	× 10	0.37	0.51
× 50	1.04	1.54			
× 45	0.95	1.40	W6 × 25	0.82	1.33
× 40	0.85	1.25	× 20	0.67	1.09
× 35	0.79	1.11	× 16	0.66	0.96
× 30	0.69	0.96	× 15	0.51	0.83
× 26	0.60	0.84	× 12	0.51	0.75
× 22	0.61	0.77	× 9	0.39	0.57
× 19	0.53	0.67			
× 16	0.45	0.57	W5 × 19	0.76	1.24
× 14	0.40	0.50	× 16	0.65	1.07
W10 × 112	2.14	3.38	W4 × 13	0.65	1.05
× 100	1.93	3.07			
× 88	1.7	2.75			
× 77	1.52	2.45			
× 68	1.35	2.20			
× 60	1.20	1.97			
× 54	1.09	1.79			
× 49	0.99	1.64			
× 45	1.03	1.59			
× 39	0.94	1.40			
× 33	0.77	1.20			

For SI: Pounds per linear foot per inch = 0.059 kg/m/mm.

TABLE 721.5.1(5)
FIRE RESISTANCE OF CONCRETE MASONRY PROTECTED STEEL COLUMNS

COLUMN SIZE	CONCRETE MASONRY DENSITY POUNDS PER CUBIC FOOT	MINIMUM REQUIRED EQUIVALENT THICKNESS FOR FIRE-RESISTANCE RATING OF CONCRETE. MASONRY PROTECTION ASSEMBLY T_e, (inches)				COLUMN SIZE	CONCRETE MASONRY DENSITY POUNDS PER CUBIC FOOT	MINIMUM REQUIRED EQUIVALENT THICKNESS FOR FIRE-RESISTANCE RATING OF CONCRETE. MASONRY PROTECTION ASSEMBLY T_e, (inches)			
		1-hour	2-hour	3-hour	4-hour			1-hour	2-hour	3-hour	4-hour
W14 × 82	80	0.74	1.61	2.36	3.04	W10 × 68	80	0.72	1.58	2.33	3.01
	100	0.89	1.85	2.67	3.40		100	0.87	1.83	2.65	3.38
	110	0.96	1.97	2.81	3.57		110	0.94	1.95	2.79	3.55
	120	1.03	2.08	2.95	3.73		120	1.01	2.06	2.94	3.72
W14 × 68	80	0.83	1.70	2.45	3.13	W10 × 54	80	0.88	1.76	2.53	3.21
	100	0.99	1.95	2.76	3.49		100	1.04	2.01	2.83	3.57
	110	1.06	2.06	2.91	3.66		110	1.11	2.12	2.98	3.73
	120	1.14	2.18	3.05	3.82		120	1.19	2.24	3.12	3.90
W14 × 53	80	0.91	1.81	2.58	3.27	W10 × 45	80	0.92	1.83	2.60	3.30
	100	1.07	2.05	2.88	3.62		100	1.08	2.07	2.90	3.64
	110	1.15	2.17	3.02	3.78		110	1.16	2.18	3.04	3.80
	120	1.22	2.28	3.16	3.94		120	1.23	2.29	3.18	3.96
W14 × 43	80	1.01	1.93	2.71	3.41	W10 × 33	80	1.06	2.00	2.79	3.49
	100	1.17	2.17	3.00	3.74		100	1.22	2.23	3.07	3.81
	110	1.25	2.28	3.14	3.90		110	1.30	2.34	3.20	3.96
	120	1.32	2.38	3.27	4.05		120	1.37	2.44	3.33	4.12
W12 × 72	80	0.81	1.66	2.41	3.09	W8 × 40	80	0.94	1.85	2.63	3.33
	100	0.91	1.88	2.70	3.43		100	1.10	2.10	2.93	3.67
	110	0.99	1.99	2.84	3.60		110	1.18	2.21	3.07	3.83
	120	1.06	2.10	2.98	3.76		120	1.25	2.32	3.20	3.99
W12 × 58	80	0.88	1.76	2.52	3.21	W8 × 31	80	1.06	2.00	2.78	3.49
	100	1.04	2.01	2.83	3.56		100	1.22	2.23	3.07	3.81
	110	1.11	2.12	2.97	3.73		110	1.29	2.33	3.20	3.97
	120	1.19	2.23	3.11	3.89		120	1.36	2.44	3.33	4.12
W12 × 50	80	0.91	1.81	2.58	3.27	W8 × 24	80	1.14	2.09	2.89	3.59
	100	1.07	2.05	2.88	3.62		100	1.29	2.31	3.16	3.90
	110	1.15	2.17	3.02	3.78		110	1.36	2.42	3.28	4.05
	120	1.22	2.28	3.16	3.94		120	1.43	2.52	3.41	4.20
W12 × 40	80	1.01	1.94	2.72	3.41	W8 × 18	110	1.22	2.20	3.01	3.72
	100	1.17	2.17	3.01	3.75		100	1.36	2.40	3.25	4.01
	110	1.25	2.28	3.14	3.90		110	1.42	2.50	3.37	4.14
	120	1.32	2.39	3.27	4.06		120	1.48	2.59	3.49	4.28

(continued)

TABLE 721.5.1(5)—continued
FIRE RESISTANCE OF CONCRETE MASONRY PROTECTED STEEL COLUMNS

NOMINAL TUBE SIZE (inches)	CONCRETE MASONRY DENSITY, POUNDS PER CUBIC FOOT	MINIMUM REQUIRED EQUIVALENT THICKNESS FOR FIRE-RESISTANCE RATING OF CONCRETE. MASONRY PROTECTION ASSEMBLY T_e, (inches)				NOMINAL PIPE SIZE (inches)	CONCRETE MASONRY DENSITY, POUNDS PER CUBIC FOOT	MINIMUM REQUIRED EQUIVALENT THICKNESS FOR FIRE-RESISTANCE RATING OF CONCRETE. MASONRY PROTECTION ASSEMBLY T_e, (inches)			
		1-hour	2-hour	3-hour	4-hour			1-hour	2-hour	3-hour	4-hour
4 × 4 × 1/2 wall thickness	80	0.93	1.90	2.71	3.43	4 double extra strong 0.674 wall thickness	80	0.80	1.75	2.56	3.28
	100	1.08	2.13	2.99	3.76		100	0.95	1.99	2.85	3.62
	110	1.16	2.24	3.13	3.91		110	1.02	2.10	2.99	3.78
	120	1.22	2.34	3.26	4.06		120	1.09	2.20	3.12	3.93
4 × 4 × 3/8 wall thickness	80	1.05	2.03	2.84	3.57	4 extra strong 0.337 wall thickness	80	1.12	2.11	2.93	3.65
	100	1.20	2.25	3.11	3.88		100	1.26	2.32	3.19	3.95
	110	1.27	2.35	3.24	4.02		110	1.33	2.42	3.31	4.09
	120	1.34	2.45	3.37	4.17		120	1.40	2.52	3.43	4.23
4 × 4 × 1/4 wall thickness	80	1.21	2.20	3.01	3.73	4 standard 0.237 wall thickness	80	1.26	2.25	3.07	3.79
	100	1.35	2.40	3.26	4.02		100	1.40	2.45	3.31	4.07
	110	1.41	2.50	3.38	4.16		110	1.46	2.55	3.43	4.21
	120	1.48	2.59	3.50	4.30		120	1.53	2.64	3.54	4.34
6 × 6 × 1/2 wall thickness	80	0.82	1.75	2.54	3.25	5 double extra strong 0.750 wall thickness	80	0.70	1.61	2.40	3.12
	100	0.98	1.99	2.84	3.59		100	0.85	1.86	2.71	3.47
	110	1.05	2.10	2.98	3.75		110	0.91	1.97	2.85	3.63
	120	1.12	2.21	3.11	3.91		120	0.98	2.02	2.99	3.79
6 × 6 × 3/8 wall thickness	80	0.96	1.91	2.71	3.42	5 extra strong 0.375 wall thickness	80	1.04	2.01	2.83	3.54
	100	1.12	2.14	3.00	3.75		100	1.19	2.23	3.09	3.85
	110	1.19	2.25	3.13	3.90		110	1.26	2.34	3.22	4.00
	120	1.26	2.35	3.26	4.05		120	1.32	2.44	3.34	4.14
6 × 6 × 1/4 wall thickness	80	1.14	2.11	2.92	3.63	5 standard 0.258 wall thickness	80	1.20	2.19	3.00	3.72
	100	1.29	2.32	3.18	3.93		100	1.34	2.39	3.25	4.00
	110	1.36	2.43	3.30	4.08		110	1.41	2.49	3.37	4.14
	120	1.42	2.52	3.43	4.22		120	1.47	2.58	3.49	4.28
8 × 8 × 1/2 wall thickness	80	0.77	1.66	2.44	3.13	6 double extra strong 0.864 wall thickness	80	0.59	1.46	2.23	2.92
	100	0.92	1.91	2.75	3.49		100	0.73	1.71	2.54	3.29
	110	1.00	2.02	2.89	3.66		110	0.80	1.82	2.69	3.47
	120	1.07	2.14	3.03	3.82		120	0.86	1.93	2.83	3.63
8 × 8 × 3/8 wall thickness	80	0.91	1.84	2.63	3.33	6 extra strong 0.432 wall thickness	80	0.94	1.90	2.70	3.42
	100	1.07	2.08	2.92	3.67		100	1.10	2.13	2.98	3.74
	110	1.14	2.19	3.06	3.83		110	1.17	2.23	3.11	3.89
	120	1.21	2.29	3.19	3.98		120	1.24	2.34	3.24	4.04
8 × 8 × 1/4 wall thickness	80	1.10	2.06	2.86	3.57	6 standard 0.280 wall thickness	80	1.14	2.12	2.93	3.64
	100	1.25	2.28	3.13	3.87		100	1.29	2.33	3.19	3.94
	110	1.32	2.38	3.25	4.02		110	1.36	2.43	3.31	4.08
	120	1.39	2.48	3.38	4.17		120	1.42	2.53	3.43	4.22

For SI: 1 inch = 25.4 mm, 1 pound per cubic feet = 16.02 kg/m^3.
Note: Tabulated values assume 1-inch air gap between masonry and steel section.

TABLE 721.5.1(6)
FIRE RESISTANCE OF CLAY MASONRY PROTECTED STEEL COLUMNS

COLUMN SIZE	CLAY MASONRY DENSITY, POUNDS PER CUBIC FOOT	MINIMUM REQUIRED EQUIVALENT THICKNESS FOR FIRE-RESISTANCE RATING OF CLAY. MASONRY PROTECTION ASSEMBLY T_e, (inches)				COLUMN SIZE	CLAY MASONRY DENSITY, POUNDS PER CUBIC FOOT	MINIMUM REQUIRED EQUIVALENT THICKNESS FOR FIRE-RESISTANCE RATING OF CLAY. MASONRY PROTECTION ASSEMBLY T_e, (inches)			
		1-hour	2-hour	3-hour	4-hour			1-hour	2-hour	3-hour	4-hour
W14 × 82	120	1.23	2.42	3.41	4.29	W10 × 68	120	1.27	2.46	3.26	4.35
	130	1.40	2.70	3.78	4.74		130	1.44	2.75	3.83	4.80
W14 × 68	120	1.34	2.54	3.54	4.43	W10 × 54	120	1.40	2.61	3.62	4.51
	130	1.51	2.82	3.91	4.87		130	1.58	2.89	3.98	4.95
W14 × 53	120	1.43	2.65	3.65	4.54	W10 × 45	120	1.44	2.66	3.67	4.57
	130	1.61	2.93	4.02	4.98		130	1.62	2.95	4.04	5.01
W14 × 43	120	1.54	2.76	3.77	4.66	W10 × 33	120	1.59	2.82	3.84	4.73
	130	1.72	3.04	4.13	5.09		130	1.77	3.10	4.20	5.13
W12 × 72	120	1.32	2.52	3.51	4.40	W8 × 40	120	1.47	2.70	3.71	4.61
	130	1.50	2.80	3.88	4.84		130	1.65	2.98	4.08	5.04
W12 × 58	120	1.40	2.61	3.61	4.50	W8 × 31	120	1.59	2.82	3.84	4.73
	130	1.57	2.89	3.98	4.94		130	1.77	3.10	4.20	5.17
W12 × 50	120	1.43	2.65	3.66	4.55	W8 × 24	120	1.66	2.90	3.92	4.82
	130	1.61	2.93	4.02	4.99		130	1.84	3.18	4.28	5.25
W12 × 40	120	1.54	2.77	3.78	4.67	W8 × 18	120	1.75	3.00	4.01	4.91
	130	1.72	3.05	4.14	5.10		130	1.93	3.27	4.37	5.34

Nominal tube size (inches)	Clay masonry density, pounds per cubic foot	Steel tubing — Minimum required equivalent thickness for fire-resistance rating of clay. Masonry protection assembly T_e, (inches)				Nominal pipe size (inches)	Clay masonry density, pounds per cubic foot	Steel pipe — Minimum required equivalent thickness for fire-resistance rating of clay. Masonry protection assembly T_e, (inches)			
		1-hour	2-hour	3-hour	4-hour			1-hour	2-hour	3-hour	4-hour
4 × 4 × 1/2 wall thickness	120	1.44	2.72	3.76	4.68	4 double extra strong 0.674 wall thickness	120	1.26	2.55	3.60	4.52
	130	1.62	3.00	4.12	5.11		130	1.42	2.82	3.96	4.95
4 × 4 × 3/8 wall thickness	120	1.56	2.84	3.88	4.78	4 extra strong 0.337 wall thickness	120	1.60	2.89	3.92	4.83
	130	1.74	3.12	4.23	5.21		130	1.77	3.16	4.28	5.25
4 × 4 × 1/4 wall thickness	120	1.72	2.99	4.02	4.92	4 standard 0.237 wall thickness	120	1.74	3.02	4.05	4.95
	130	1.89	3.26	4.37	5.34		130	1.92	3.29	4.40	5.37
6 × 6 × 1/2 wall thickness	120	1.33	2.58	3.62	4.52	5 double extra strong 0.750 wall thickness	120	1.17	2.44	3.48	4.40
	130	1.50	2.86	3.98	4.96		130	1.33	2.72	3.84	4.83
6 × 6 × 3/8 wall thickness	120	1.48	2.74	3.76	4.67	5 extra strong 0.375 wall thickness	120	1.55	2.82	3.85	4.76
	130	1.65	3.01	4.13	5.10		130	1.72	3.09	4.21	5.18
6 × 6 × 1/4 wall thickness	120	1.66	2.91	3.94	4.84	5 standard 0.258 wall thickness	120	1.71	2.97	4.00	4.90
	130	1.83	3.19	4.30	5.27		130	1.88	3.24	4.35	5.32
8 × 8 × 1/2 wall thickness	120	1.27	2.50	3.52	4.42	6 double extra strong 0.864 wall thickness	120	1.04	2.28	3.32	4.23
	130	1.44	2.78	3.89	4.86		130	1.19	2.60	3.68	4.67
8 × 8 × 3/8 wall thickness	120	1.43	2.67	3.69	4.59	6 extra strong 0.432 wall thickness	120	1.45	2.71	3.75	4.65
	130	1.60	2.95	4.05	5.02		130	1.62	2.99	4.10	5.08
8 × 8 × 1/4 wall thickness	120	1.62	2.87	3.89	4.78	6 standard 0.280 wall thickness	120	1.65	2.91	3.94	4.84
	130	1.79	3.14	4.24	5.21		130	1.82	3.19	4.30	5.27

TABLE 721.5.1(7)
MINIMUM COVER (inch) FOR STEEL COLUMNS ENCASED IN NORMAL-WEIGHT CONCRETE[a]
[FIGURE 721.5.1(6)(c)]

STRUCTURAL SHAPE	FIRE-RESISTANCE RATING (hours)				
	1	1½	2	3	4
W14 × 233				1½	2
× 176			1		
× 132		1			2½
× 90	1			2	
× 61		1½			
× 48					3
× 43		1½		2½	
W12 × 152			1		2½
× 96		1	2		
× 65	1				
× 50		1½			3
× 40		1½		2½	
W10 × 88	1			2	
× 49					3
× 45	1	1½	1½		
× 39				2½	3½
× 33			2		
W8 × 67		1			3
× 58		1½			
× 48	1			2½	
× 31		1½			3½
× 21			2		
× 18				3	4
W6 × 25		1½	2		3½
× 20				3	
× 16	1	2			4
× 15					
× 9	1½		2½	3½	

For SI: 1 inch = 25.4 mm.

a. The tabulated thicknesses are based upon the assumed properties of normal-weight concrete given in Table 721.5.1(2).

TABLE 721.5.1(8)
MINIMUM COVER (inch) FOR STEEL COLUMNS ENCASED IN STRUCTURAL LIGHTWEIGHT CONCRETE[a]
[FIGURE 721.5.1(6)(c)]

STRUCTURAL SHAPE	FIRE-RESISTANCE RATING (HOURS)				
	1	1½	2	3	4
W14 × 233				1	1½
× 193					
× 74	1	1	1	1½	2
× 61					
× 43			1½	2	2½
W12 × 65				1½	2
× 53	1	1	1		
× 40			1½	2	2½
W10 × 112					2
× 88	1		1	1½	
× 60		1			
× 33			1½	2	2½
W8 × 35					2½
× 28	1	1		2	
× 24			1½		3
× 18		1½		2½	

For SI: 1 inch = 25.4 mm.

a. The tabulated thicknesses are based upon the assumed properties of structural lightweight concrete given in Table 721.5.1(2).

**TABLE 721.5.1(9)
MINIMUM COVER (inch) FOR STEEL COLUMNS
IN NORMAL-WEIGHT PRECAST COVERS[a]
[FIGURE 721.5.1(6)(a)]**

STRUCTURAL SHAPE	FIRE-RESISTANCE RATING (hours)				
	1	1 1/2	2	3	4
W14 × 233			1 1/2		3
× 211				2 1/2	
× 176					3 1/2
× 145		1 1/2	2		
× 109	1 1/2			3	
× 99					
× 61					4
× 43		2	2 1/2	3 1/2	4 1/2
W12 × 190			1 1/2	2 1/2	3 1/2
× 152					
× 120		1 1/2	2		
× 96				3	
× 87	1 1/2				4
× 58					
× 40		2	2 1/2	3 1/2	4 1/2
W10 × 112					3 1/2
× 88		1 1/2	2	3	
× 77	1 1/2				4
× 54		2	2 1/2	3 1/2	
× 33					4 1/2
W8 × 67		1 1/2	2	3	
× 58					4
× 48	1 1/2	2	2 1/2	3 1/2	
× 28					
× 21					4 1/2
× 18		2 1/2	3	4	
W6 × 25		2	2 1/2	3 1/2	
× 20	1 1/2				4 1/2
× 16			3		
× 12	2	2 1/2		4	
× 9					5

For SI: 1 inch = 25.4 mm.

a. The tabulated thicknesses are based upon the assumed properties of normal-weight concrete given in Table 721.5.1(2).

**TABLE 721.5.1(10)
MINIMUM COVER (inch) FOR STEEL COLUMNS
IN STRUCTURAL LIGHTWEIGHT PRECAST COVERS[a]
[FIGURE 721.5.1(6)(a)]**

STRUCTURAL SHAPE	FIRE-RESISTANCE RATING (hours)				
	1	1 1/2	2	3	4
W14 × 233					2 1/2
× 176				2	
× 145			1 1/2		
× 132	1 1/2	1 1/2			3
× 109					
× 99				2 1/2	
× 68			2		
× 43				3	3 1/2
W12 × 190					2 1/2
× 152				2	
× 136			1 1/2		3
× 106					
× 96	1 1/2	1 1/2		2 1/2	
× 87					
× 65			2		
× 40				3	3 1/2
W10 × 112				2	
× 100			1 1/2		3
× 88					
× 77	1 1/2	1 1/2		2 1/2	
× 60			2		
× 39				3	3 1/2
× 33		2			
W8 × 67			1 1/2	2 1/2	3
× 48		1 1/2			
× 35	1 1/2		2		3 1/2
× 28				3	
× 18		2	2 1/2		4
W6 × 25			2	3	3 1/2
× 15	1 1/2	2			4
× 9			2 1/2	3 1/2	

For SI: 1 inch = 25.4 mm.

a. The tabulated thicknesses are based upon the assumed properties of structural lightweight concrete given in Table 721.5.1(2).

721.5.2 Structural steel beams and girders. The fire-resistance ratings of steel beams and girders shall be based upon the size of the element and the type of protection provided in accordance with this section.

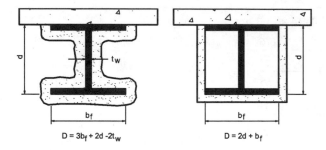

$D = 3b_f + 2d - 2t_w$ $D = 2d + b_f$

FIGURE 721.5.2
DETERMINATION OF THE HEATED PERIMETER
OF STRUCTURAL STEEL BEAMS AND GIRDERS

721.5.2.1 Determination of fire resistance. These procedures establish a basis for determining resistance of structural steel beams and girders which differ in size from that specified in approved fire-resistance-rated assemblies as a function of the thickness of fire-resistant material and the weight (W) and heated perimeter (D) of the beam or girder. As used in these sections, W is the average weight of a structural steel member in pounds per linear foot (plf). The heated perimeter, D, is the inside perimeter of the fire-resistant material in inches as illustrated in Figure 721.5.2.

721.5.2.1.1 Weight-to-heated perimeter. The weight-to-heated-perimeter ratios (W/D), for both contour and box fire-resistant protection profiles, for the wide flange shapes most often used as beams or girders are given in Table 721.5.1(4). For different shapes, the weight-to-heated-perimeter ratios (W/D) shall be determined in accordance with the definitions given in this section.

721.5.2.1.2 Beam and girder substitutions. Except as provided for in Section 721.5.2.2, structural steel beams in approved fire-resistance-rated assemblies shall be considered the minimum permissible size. Other beam or girder shapes shall be permitted to be substituted provided that the weight-to-heated-perimeter ratio (W/D) of the substitute beam is equal to or greater than that of the beam specified in the approved assembly.

721.5.2.2 Spray-applied fire-resistant materials. The provisions in this section apply to unrestrained structural steel beams and girders protected with spray-applied fire-resistance-rated materials. Larger or smaller unrestrained beam and girder shapes shall be permitted to be substituted for beams specified in approved unrestrained or restrained fire-resistance-rated assemblies provided that the thickness of the fire-resistant material is adjusted in accordance with the following expression:

$$h_2 = \left[\frac{(W_1 / D_1) + 0.60}{(W_2 / D_2) + 0.60} \right] h_1$$ **(Equation 7-17)**

where:

h = Thickness of spray-applied fire-resistant material in inches.

W = Weight of the structural steel beam or girder in pounds per linear foot.

D = Heated perimeter of the structural steel beam or girder in inches.

Subscript 1 refers to the beam and fire-resistant material thickness in the fire-resistance-rated assembly.

Subscript 2 refers to the substitute beam or girder and the required thickness of fire-resistant material.

721.5.2.2.1 Minimum thickness. Equation 7-17 is limited to beams with a weight-to-heated-perimeter ratio (W/D) of 0.37 or greater. The minimum thickness of fire-resistant material shall not be less than $^3/_8$ inch (9.5 mm).

721.5.2.3 Structural steel trusses. The fire resistance of structural steel trusses protected with fire-resistant materials spray applied to each of the individual truss elements shall be permitted to be determined in accordance with this section. The thickness of the fire-resistant material shall be determined in accordance with Section 721.5.1.3. The weight-to-heated-perimeter ratio (W/D) of truss elements that can be simultaneously exposed to fire on all sides shall be determined on the same basis as columns, as specified in Section 721.5.1.1. The weight-to-heated-perimeter ratio (W/D) of truss elements that directly support floor or roof construction shall be determined on the same basis as beams and girders, as specified in Section 721.5.2.1.

721.6 Wood assemblies. The provisions of this section contain procedures by which the fire-resistance ratings of wood assemblies are established by calculations.

721.6.1 General. This section contains procedures for calculating the fire-resistance ratings of walls, floor/ceiling and roof/ceiling assemblies based in part on the standard method of testing referenced in Section 703.2.

721.6.1.1 Maximum fire-resistance rating. Fire-resistance ratings calculated using the procedures in this section shall be used only for 1-hour rated assemblies.

721.6.1.2 Dissimilar membranes. Where dissimilar membranes are used on a wall assembly, the calculation shall be made from the least fire-resistant (weaker) side.

721.6.2 Walls, floors and roofs. These procedures apply to both load-bearing and nonload-bearing assemblies.

721.6.2.1 Fire-resistance rating of wood frame assemblies. The fire-resistance rating of a wood frame assembly is equal to the sum of the time assigned to the membrane on the fire-exposed side, the time assigned to the framing members and the time assigned for additional contribution by other protective measures such as insulation. The membrane on the unexposed side shall not be included in determining the fire resistance of the assembly.

721.6.2.2 Time assigned to membranes. Table 721.6.2(1) indicates the time assigned to membranes on the fire-exposed side.

721.6.2.3 Exterior walls. For an exterior wall having more than 5 feet (1524 mm) of horizontal separation, the wall is assigned a rating dependent on the interior membrane and the framing as described in Tables 721.6.2(1) and 721.6.2(2). The membrane on the outside of the nonfire-exposed side of exterior walls having more than 5 feet (1524 mm) of horizontal separation may consist of sheathing, sheathing paper, and siding as described in Table 721.6.2(3).

721.6.2.4 Floors and roofs. In the case of a floor or roof, the standard test provides only for testing for fire exposure from below. Except as noted in Section 703.3, Item 5, floor or roof assemblies of wood framing shall have an upper membrane consisting of a subfloor and finished floor conforming to Table 721.6.2(4) or any other membrane that has a contribution to fire resistance of at least 15 minutes in Table 721.6.2(1).

721.6.2.5 Additional protection. Table 721.6.2(5) indicates the time increments to be added to the fire resistance where glass fiber, rockwool, slag mineral wool, or cellulose insulation is incorporated in the assembly.

721.6.2.6 Fastening. Fastening of wood frame assemblies and the fastening of membranes to the wood framing members shall be done in accordance with Chapter 23.

TABLE 721.6.2(1)
TIME ASSIGNED TO WALLBOARD MEMBRANES[a,b,c,d]

DESCRIPTION OF FINISH	TIME[e] (minutes)
$^3/_8$-inch wood structural panel bonded with exterior glue	5
$^{15}/_{32}$-inch wood structural panel bonded with exterior glue	10
$^{19}/_{32}$-inch wood structural panel bonded with exterior glue	15
$^3/_8$-inch gypsum wallboard	10
$^1/_2$-inch gypsum wallboard	15
$^5/_8$-inch gypsum wallboard	30
$^1/_2$-inch Type X gypsum wallboard	25
$^5/_8$-inch Type X gypsum wallboard	40
Double $^3/_8$-inch gypsum wallboard	25
$^1/_2$- + $^3/_8$-inch gypsum wallboard	35
Double $^1/_2$-inch gypsum wallboard	40

For SI: 1 inch = 25.4 mm.

a. These values apply only when membranes are installed on framing members which are spaced 16 inches o.c.

b. Gypsum wallboard installed over framing or furring shall be installed so that all edges are supported, except $^5/_8$-inch Type X gypsum wallboard shall be permitted to be installed horizontally with the horizontal joints staggered 24 inches each side and unsupported but finished.

c. On wood frame floor/ceiling or roof/ceiling assemblies, gypsum board shall be installed with the long dimension perpendicular to framing members and shall have all joints finished.

d. The membrane on the unexposed side shall not be included in determining the fire resistance of the assembly. When dissimilar membranes are used on a wall assembly, the calculation shall be made from the least fire-resistant (weaker) side.

e. The time assigned is not a finished rating.

TABLE 721.6.2(2)
TIME ASSIGNED FOR CONTRIBUTION OF WOOD FRAME[a,b,c]

DESCRIPTION	TIME ASSIGNED TO FRAME (minutes)
Wood studs 16 inches o.c.	20
Wood floor and roof joists 16 inches o.c.	10

For SI: 1 inch = 25.4 mm.

a. This table does not apply to studs or joists spaced more than 16 inches o.c.

b. All studs shall be nominal 2 × 4 and all joists shall have a nominal thickness of at least 2 inches.

c. Allowable spans for joists shall be determined in accordance with Sections 2308.8, 2308.10.2 and 2308.10.3.

TABLE 721.6.2(3)
MEMBRANE[a] ON EXTERIOR FACE OF WOOD STUD WALLS

SHEATHING	PAPER	EXTERIOR FINISH
$^5/_8$-inch T & G lumber		Lumber siding
$^5/_{16}$-inch exterior glue plywood	Sheathing paper	Wood shingles and shakes
$^1/_2$-inch gypsum wallboard		$^1/_4$-inch wood structural panels—exterior type
$^5/_8$-inch gypsum wallboard		$^1/_4$-inch hardboard
$^1/_2$-inch fiberboard		Metal siding
		Stucco on metal lath
		Masonry veneer
None	—	$^3/_8$-inch exterior-grade wood structural panels

For SI: 1 pound/cubic feet = 16.0185 kg/m².

a. Any combination of sheathing, paper, and exterior finish is permitted.

TABLE 721.6.2(4)
FLOORING OR ROOFING OVER WOOD FRAMING[a]

ASSEMBLY	STRUCTURAL MEMBERS	SUBFLOOR OR ROOF DECK	FINISHED FLOORING OR ROOFING
Floor	Wood	$^{15}/_{32}$-inch wood structural panels or $^{11}/_{16}$ inch T & G softwood	Hardwood or softwood flooring on building paper resilient flooring, parquet floor felted-synthetic fiber floor coverings, carpeting, or ceramic tile on $^3/_8$-inch-thick panel-type underlay Ceramic tile on $1^1/_4$-inch mortar bed
Roof	Wood	$^{15}/_{32}$-inch wood structural panels or $^{11}/_{16}$ inch T & G softwood	Finished roofing material with or without insulation

For SI: 1 inch = 25.4 mm.
a. This table applies only to wood joist construction. It is not applicable to wood truss construction.

TABLE 721.6.2(5)
TIME ASSIGNED FOR ADDITIONAL PROTECTION

DESCRIPTION OF ADDITIONAL PROTECTION	FIRE RESISTANCE (minutes)
Add to the fire-resistance rating of wood stud walls if the spaces between the studs are completely filled with glass fiber mineral wool batts weighing not less than 2 pounds per cubic foot (0.6 pound per square foot of wall surface) or rockwool or slag material wool batts weighing not less than 3.3 pounds per cubic foot (1 pound per square foot of wall surface), or cellulose insulation having a nominal density not less than 2.6 pounds per cubic foot.	15

For SI: 1 pound/cubic foot = 16.0185 kg/m^3.

721.6.3 Design of fire-resistant exposed wood members.
The fire-resistance rating, in minutes, of timber beams and columns with a minimum nominal dimension of 6 inches (152 mm) is equal to:

Beams: $2.54Zb(4-2(b/d))$ for beams which may be exposed to fire on four sides.

(Equation 7-18)

$2.54Zb(4-(b/d))$ for beams which may be exposed to fire on three sides.

(Equation 7-19)

Columns: $2.54Zd(3-(d/b))$ for columns which may be exposed to fire on four sides

(Equation 7-20)

$2.54Zd(3-(d/2b))$ for columns which may be exposed to fire on three sides.

(Equation 7-21)

where:

b =   The breadth (width) of a beam or larger side of a column before exposure to fire (inches).

d =   The depth of a beam or smaller side of a column before exposure to fire (inches).

Z =   Load factor, based on Figure 720.6.3(1).

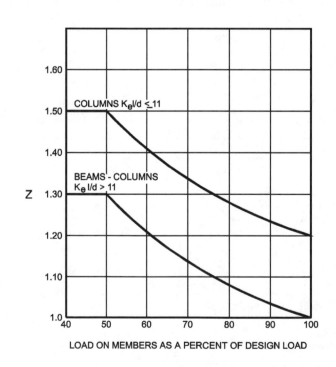

FIGURE 721.6.3(1)
LOAD FIGURE

K_e =   The effective length factor as noted in Figure 721.6.3(2).
l =   The unsupported length of columns (inches).

BUCKLING MODES						
THEORETICAL K_e VALUE	0.5	0.7	1.0	1.0	2.0	2.0
RECOMMENDED DESIGN K_e WHEN IDEAL CONDITIONS APPROXIMATED	0.65	0.80	1.2	1.0	2.10	2.4
END CONDITION CODE	ROTATION FIXED, TRANSLATION FIXED					
	ROTATION FREE, TRANSLATION FIXED					
	ROTATION FIXED, TRANSLATION FREE					
	ROTATION FREE, TRANSLATION FREE					

**FIGURE 721.6.3(2)
EFFECTIVE LENGTH FACTORS**

721.6.3.1 Equation 7-21. Equation 7-21 applies only where the unexposed face represents the smaller side of the column. If a column is recessed into a wall, its full dimension shall be used for the purpose of these calculations.

721.6.3.2 Allowable loads. Allowable loads on beams and columns are determined using design values given in ANSI/AF&PA NDS.

721.6.3.3 Fastener protection. Where minimum 1-hour fire resistance is required, connectors and fasteners shall be protected from fire exposure by $1^1/_2$ inches (38 mm) of wood, or other approved covering or coating for a 1-hour rating. Typical details for commonly used fasteners and connectors are shown in AITC Technical Note 7.

721.6.3.4 Minimum size. Wood members are limited to dimensions of 6 inches (152 mm) nominal or greater. Glued-laminated timber beams utilize standard laminating combinations except that a core lamination is removed. The tension zone is moved inward and the equivalent of an extra nominal 2-inch-thick (51 mm) outer tension lamination is added.

721.7 Other reference documents. Refer to Section 703.3, Item 1, and NBS BMS 71 and NBSTRBM-44 for fire-resistance ratings of materials and assemblies.

CHAPTER 8

INTERIOR FINISHES

SECTION 801
GENERAL

801.1 Scope. Provisions of this chapter shall govern the use of materials used as interior finishes, trim and decorative materials.

801.1.1 Interior finishes. These provisions shall limit the allowable flame spread and smoke development based on location and occupancy classification.

Exceptions:

1. Materials having a thickness less than 0.036 inch (0.9 mm) applied directly to the surface of walls or ceilings.

2. Exposed portions of structural members complying with the requirements for buildings of Type IV construction in Section 602.4 shall not be subject to interior finish requirements.

[F] 801.1.2 Decorative materials and trim. Decorative materials and trim shall be restricted by combustibility and flame resistance in accordance with Section 805.

801.1.3 Applicability. For buildings in flood hazard areas as established in Section 1612.3, interior finishes, trim and decorative materials below the design flood elevation shall be flood-damage-resistant materials.

801.2 Application. Combustible materials shall be permitted to be used as finish for walls, ceilings, floors and other interior surfaces of buildings.

801.2.1 Windows. Show windows in the first story of buildings shall be permitted to be of wood or of unprotected metal framing.

801.2.2 Foam plastics. Foam plastics shall not be used as interior finish or trim except as provided in Section 2603.7 or 2604.

SECTION 802
DEFINITIONS

802.1 General. The following words and terms shall, for the purposes of this chapter and as used elsewhere in this code, have the meanings shown herein.

EXPANDED VINYL WALL COVERING. Wall covering consisting of a woven textile backing, an expanded vinyl base coat layer and a nonexpanded vinyl skin coat. The expanded base coat layer is a homogeneous vinyl layer that contains a blowing agent. During processing, the blowing agent decomposes, causing this layer to expand by forming closed cells. The total thickness of the wall covering is approximately 0.055 inch to 0.070 inch (1.4 mm to 1.78 mm).

FLAME RESISTANCE. That property of materials or combinations of component materials that restricts the spread of flame in accordance with NFPA 701.

FLAME SPREAD. The propagation of flame over a surface.

FLAME SPREAD INDEX. The numerical value assigned to a material tested in accordance with ASTM E 84.

INTERIOR FINISH. Interior finish includes interior wall and ceiling finish and interior floor finish.

INTERIOR FLOOR FINISH. The exposed floor surfaces of buildings including coverings applied over a finished floor or stair, including risers.

INTERIOR WALL AND CEILING FINISH. The exposed interior surfaces of buildings including, but not limited to: fixed or movable walls and partitions; columns; ceilings; and interior wainscotting, paneling or other finish applied structurally or for decoration, acoustical correction, surface insulation, structural fire resistance or similar purposes, but not including trim.

SMOKE-DEVELOPED INDEX. The numerical value assigned to a material tested in accordance with ASTM E 84.

TRIM. Picture molds, chair rails, baseboards, handrails, door and window frames and similar decorative or protective materials used in fixed applications.

SECTION 803
WALL AND CEILING FINISHES

803.1 General. Interior wall and ceiling finishes shall be classified in accordance with ASTM E 84. Such interior finish materials shall be grouped in the following classes in accordance with their flame spread and smoke-developed indexes.

Class A: Flame spread 0-25; smoke-developed 0-450.

Class B: Flame spread 26-75; smoke-developed 0-450.

Class C: Flame spread 76-200; smoke-developed 0-450.

Exception: Materials, other than textiles, tested in accordance with Section 803.2.

803.2. Interior wall or ceiling finishes other than textiles. Interior wall or ceiling finishes, other than textiles, shall be permitted to be tested in accordance with NFPA 286. Finishes tested in accordance with NFPA 286 shall comply with Section 803.2.1.

803.2.1 Acceptance criteria. During the 40 kW exposure, the interior finish shall comply with Item 1. During the 160 kW exposure, the interior finish shall comply with Item 2. During the entire test, the interior finish shall comply with Item 3.

1. During the 40kW exposure, flames shall not spread to the ceiling.

2. During the 160 kW exposure, the interior finish shall comply with the following:

2.1. Flame shall not spread to the outer extremity of the sample on any wall or ceiling.

2.2. Flashover, as defined in NFPA 286, shall not occur.

3. The total smoke released throughout the NFPA 286 test shall not exceed 1,000 m².

803.3 Stability. Interior finish materials regulated by this chapter shall be applied or otherwise fastened in such a manner that such materials will not readily become detached where subjected to room temperatures of 200°F (93°C) for not less than 30 minutes.

803.4 Application. Where these materials are applied on walls, ceilings or structural elements required to have a fire-resistance rating or to be of noncombustible construction, they shall comply with the provisions of this section.

803.4.1 Direct attachment and furred construction. Where walls and ceilings are required by any provision in this code to be of fire-resistance-rated or noncombustible construction, the interior finish material shall be applied directly against such construction or to furring strips not exceeding 1.75 inches (44 mm) applied directly against such surfaces. The intervening spaces between such furring strips shall be filled with inorganic or Class A material or shall be fireblocked at a maximum of 8 feet (2438 mm) in any direction in accordance with Section 717.

803.4.2 Set-out construction. Where walls and ceilings are required to be of fire-resistance-rated or noncombustible construction and walls are set out or ceilings are dropped distances greater than specified in Section 803.4.1, Class A finish materials shall be used except where interior finish materials are protected on both sides by an automatic sprinkler system or attached to noncombustible backing or furring strips installed as specified in Section 803.4.1. The hangers and assembly members of such dropped ceilings that are below the main ceiling line shall be of noncombustible materials, except that in Type III and V construction, fire-retardant-treated wood shall be permitted. The construction of each set-out wall shall be of fire-resistance-rated construction as required elsewhere in this code.

803.4.3 Heavy timber construction. Wall and ceiling finishes of all classes as permitted in this chapter that are installed directly against the wood decking or planking of Type IV construction or to wood furring strips applied directly to the wood decking or planking shall be fireblocked as specified in Section 803.4.1.

803.4.4 Materials. An interior wall or ceiling finish that is not more than 0.25 inch (6.4 mm) thick shall be applied directly against a noncombustible backing.

Exceptions:

1. Class A materials.

2. Materials where the qualifying tests were made with the material suspended or furred out from the noncombustible backing.

803.5 Interior finish requirements based on group. Interior wall and ceiling finish shall have a flame spread index not greater than that specified in Table 803.5 for the group and location designated. Interior wall and ceiling finish materials, other than textiles, tested in accordance with NFPA 286 and meeting the acceptance criteria of Section 803.2.1, shall be permitted to be used where a Class A classification in accordance with ASTM E 84 is required.

803.6 Textiles. Where used as interior wall or ceiling finish materials, textiles, including materials having woven or nonwoven, napped, tufted, looped or similar surface, shall comply with the requirements of this section.

803.6.1 Textile wall coverings. Textile wall coverings shall have a Class A flame spread index in accordance with ASTM E 84 and be protected by automatic sprinklers installed in accordance with Section 903.3.1.1 or 903.3.1.2 or the covering shall meet the criteria of Section 803.6.1.1 or 803.6.1.2 when tested in the manner intended for use in accordance with NFPA 265 using the product mounting system, including adhesive.

803.6.1.1 Method A test protocol. During the Method A protocol, flame shall not spread to the ceiling during the 40 kW exposure. During the 150 kW exposure, the textile wall covering shall comply with all of the following:

1. Flame shall not spread to the outer extremity of the sample on the 8-foot by 12-foot (203 mm by 305 mm) wall.

2. The specimen shall not burn to the outer extremity of the 2-foot-wide (610 mm) samples mounted in the corner of the room.

3. Burning droplets deemed capable of igniting textile wall coverings or that burn for 30 seconds or more shall not form.

4. Flashover, as defined in NFPA 265, shall not occur.

5. The maximum net instantaneous peak heat release rate, determined by subtracting the burner output from the maximum heat release rate, does not exceed 300 kW.

803.6.1.2 Method B test protocol. During the Method B protocol, flames shall not spread to the ceiling at any time during the 40 kW exposure. During the 150 kW exposure, the textile wall covering shall comply with the following:

1. Flame shall not spread to the outer extremities of the samples on the 8-foot by 12-foot (203 mm by 305 mm) walls.

2. Flashover, as described in NFPA 265, shall not occur.

803.6.2 Textile ceiling finish. Where used as a ceiling finish, carpet and similar textile materials shall have a Class A flame spread index in accordance with ASTM E 84 and be protected by automatic sprinklers.

TABLE 803.5
INTERIOR WALL AND CEILING FINISH REQUIREMENTS BY OCCUPANCY[k]

GROUP	SPRINKLERED[l]			NONSPRINKLERED		
	Vertical exits and exit passageways[a, b]	Exit access corridors and other exitways	Rooms and enclosed spaces[c]	Vertical exits and exit passageways[a, b]	Exit access corridors and other exitways	Rooms and enclosed spaces[c]
A-1 & A-2	B	B	C	A	A[d]	B[e]
A-3[f], A-4, A-5	B	B	C	A	A[d]	C
B, E, M, R-1, R-4	B	C	C	A	B	C
F	C	C	C	B	C	C
H	B	B	C[g]	A	A	B
I-1	B	C	C	A	B	B
I-2	B	B	B[h, i]	A	A	B
I-3	A	A[j]	C	A	A	B
I-4	B	B	B[h, i]	A	A	B
R-2	C	C	C	B	B	C
R-3	C	C	C	C	C	C
S	C	C	C	B	B	C
U	No restrictions			No restrictions		

For SI: 1 inch = 25.4 mm, 1 square foot = 0.0929 m^2.

a. Class C interior finish materials shall be permitted for wainscotting or paneling of not more than 1,000 square feet of applied surface area in the grade lobby where applied directly to a noncombustible base or over furring strips applied to a noncombustible base and fireblocked as required by Section 803.4.1.

b. In vertical exits of buildings less than three stories in height of other than Group I-3, Class B interior finish for nonsprinklered buildings and Class C interior finish for sprinklered buildings shall be permitted.

c. Requirements for rooms and enclosed spaces shall be based upon spaces enclosed by partitions. Where a fire-resistance rating is required for structural elements, the enclosing partitions shall extend from the floor to the ceiling. Partitions that do not comply with this shall be considered enclosing spaces and the rooms or spaces on both sides shall be considered one. In determining the applicable requirements for rooms and enclosed spaces, the specific occupancy thereof shall be the governing factor regardless of the group classification of the building or structure.

d. Lobby areas in Group A-1, A-2 and A-3 occupancies shall not be less than Class B materials.

e. Class C interior finish materials shall be permitted in places of assembly with an occupant load of 300 persons or less.

f. For churches and places of worship, wood used for ornamental purposes, trusses, paneling or chancel furnishing shall be permitted.

g. Class B material is required where the building exceeds two stories.

h. Class C interior finish materials shall be permitted in administrative spaces.

i. Class C interior finish materials shall be permitted in rooms with a capacity of four persons or less.

j. Class B materials shall be permitted as wainscotting extending not more than 48 inches above the finished floor in exit access corridors.

k. Finish materials as provided for in other sections of this code.

l. Applies when the vertical exits, exit passageways, exit access corridors or exitways, or rooms and spaces are protected by a sprinkler system installed in accordance with Section 903.3.1.1 or 903.3.1.2.

803.7 Expanded vinyl wall coverings. Expanded vinyl wall coverings shall comply with the requirements for textile wall and ceiling materials and their use shall comply with Section 803.6.

> **Exception:** Expanded vinyl wall or ceiling coverings complying with Section 803.2 shall not be required to comply with Section 803.1 or 803.6.

803.8 Insulation. Thermal and acoustical insulation shall comply with Section 719.

803.9 Acoustical ceiling systems. The quality, design, fabrication and erection of metal suspension systems for acoustical tile and lay-in panel ceilings in buildings or structures shall conform with generally accepted engineering practice, the provisions of this chapter and other applicable requirements of this code.

> **803.9.1 Materials and installation.** Acoustical materials complying with the interior finish requirements of Section 803 shall be installed in accordance with the manufacturer's recommendations and applicable provisions for applying interior finish.
>
> > **803.9.1.1 Suspended acoustical ceilings.** Suspended acoustical ceiling systems shall be installed in accordance with the provisions of ASTM C 635 and ASTM C 636.
> >
> > **803.9.1.2 Fire-resistance-rated construction.** Acoustical ceiling systems that are part of fire-resistance-rated construction shall be installed in the same manner used in the assembly tested and shall comply with the provisions of Chapter 7.

SECTION 804
INTERIOR FLOOR FINISH

804.1 General. Interior floor finish and floor covering materials shall comply with this section.

> **Exception:** Floors and floor coverings of a traditional type, such as wood, vinyl, linoleum or terrazzo, and resilient floor covering materials which are not comprised of fibers.

804.2 Classification. Interior floor finish and floor covering materials required by Section 804.5.1 to be of Class I or II materials shall be classified in accordance with NFPA 253. The classification referred to herein corresponds to the classifications determined by NFPA 253 as follows: Class I, 0.45 watts/cm^2 or greater; Class II, 0.22 watts/cm^2 or greater.

804.3 Testing and identification. Floor covering materials shall be tested by an approved agency in accordance with NFPA 253 and identified by a hang tag or other suitable method so as to identify the manufacturer or supplier and style, and shall indicate the interior floor finish or floor covering classification according to Section 804.2. Carpet-type floor coverings shall be tested as proposed for use, including underlayment. Test reports confirming the information provided in the manufacturer's product identification shall be furnished to the building official upon request.

804.4 Application. Combustible materials installed in or on floors of buildings of Type I or II construction shall conform with the requirements of this section.

> **Exception:** Stages and platforms constructed in accordance with Sections 410.3 and 410.4, respectively.

804.4.1 Subfloor construction. Floor sleepers, bucks and nailing blocks shall not be constructed of combustible materials, unless the space between the fire-resistance-rated floor construction and the flooring is either solidly filled with approved noncombustible materials or fireblocked in accordance with Section 717, and provided that such open spaces shall not extend under or through permanent partitions or walls.

804.4.2 Wood finish flooring. Wood finish flooring is permitted to be attached directly to the embedded or fireblocked wood sleepers and shall be permitted where cemented directly to the top surface of approved fire-resistance-rated construction or directly to a wood subfloor attached to sleepers as provided for in Section 804.4.1.

804.4.3 Insulating boards. Combustible insulating boards not more than 0.5-inch (12.7 mm) thick and covered with approved finished flooring are permitted, where attached directly to a noncombustible floor assembly or to wood subflooring attached to sleepers as provided for in Section 804.4.1.

804.5 Interior floor finish requirements. In all occupancies, interior floor finish in vertical exits, exit passageways, exit access corridors and rooms or spaces not separated from exit access corridors by full-height partitions extending from the floor to the underside of the ceiling shall withstand a minimum critical radiant flux as specified in Section 804.5.1.

> **804.5.1 Minimum critical radiant flux.** Interior floor finish in vertical exits, exit passageways and exit access corridors shall not be less than Class I in Groups I-2 and I-3 and not less than Class II in Groups A, B, E, H, I-4, M, R-1, R-2 and S. In all other areas, the interior floor finish shall comply with the DOC FF-1 "pill test" (CPSC 16 CFR, Part 1630).
>
> > **Exception:** Where a building is equipped throughout with an automatic sprinkler system in accordance with Section 903.3.1.1, Class II materials are permitted in any area where Class I materials are required and materials complying with DOC FF-1 "pill test" (CPSC 16 CFR, Part 1630) are permitted in any area where Class II materials are required.

[F] SECTION 805
DECORATIONS AND TRIM

805.1 General. In occupancies of Groups A, E, I, R-1 and dormitories in Group R-2, curtains, draperies, hangings and other decorative materials suspended from walls or ceilings shall be flame resistant in accordance with Section 805.2 and NFPA 701 or noncombustible.

In Groups I-1 and I-2, combustible decorations shall be flame retardant unless the decorations, such as photographs and paintings, are of such limited quantities that a hazard of fire development or spread is not present. In Group I-3, combustible decorations are prohibited.

805.1.1 Noncombustible materials. The permissible amount of noncombustible decorative material shall not be limited.

805.1.2 Flame-resistant materials. The permissible amount of flame-resistant decorative materials shall not exceed 10 percent of the aggregate area of walls and ceilings.

> **Exception:** In auditoriums of Group A, the permissible amount of flame-resistant decorative material shall not exceed 50 percent of the aggregate area of walls and ceilings where the building is equipped throughout with an automatic sprinkler system in accordance with Section 903.3.1.1 and the material is installed in accordance with Section 803.4.

805.2 Acceptance criteria and reports. Where required to be flame resistant, decorative materials shall be tested by an approved agency and pass Test 1 or 2, as appropriate, described in NFPA 701 or such materials shall be noncombustible. Reports of test results shall be prepared in accordance with NFPA 701 and furnished to the building official upon request.

805.3 Foam plastic. Foam plastic used as trim in any occupancy shall comply with Section 2604.2.

805.4 Pyroxylin plastic. Imitation leather or other material consisting of or coated with a pyroxylin or similarly hazardous base shall not be used in Group A occupancies.

805.5 Trim. Material used as interior trim shall have minimum Class C flame spread and smoke-developed indexes. Combustible trim, excluding handrails and guardrails, shall not exceed 10 percent of the aggregate wall or ceiling area in which it is located.

CHAPTER 9

FIRE PROTECTION SYSTEMS

SECTION 901
GENERAL

901.1 Scope. The provisions of this chapter shall specify where fire protection systems are required and shall apply to the design, installation and operation of fire protection systems.

901.2 Fire protection systems. Fire protection systems shall be installed, repaired, operated and maintained in accordance with this code and the *International Fire Code*.

Any fire protection system for which an exception or reduction to the provisions of this code has been granted shall be considered to be a required system.

> **Exception:** Any fire protection system or portion thereof not required by this code shall be permitted to be installed for partial or complete protection provided that such system meets the requirements of this code.

901.3 Modifications. No person shall remove or modify any fire protection system installed or maintained under the provisions of this code or the *International Fire Code* without approval by the building official.

901.4 Threads. Threads provided for fire department connections to sprinkler systems, standpipes, yard hydrants or any other fire hose connection shall be compatible with the connections used by the local fire department.

901.5 Acceptance tests. Fire protection systems shall be tested in accordance with the requirements of this code and the *International Fire Code*. When required, the tests shall be conducted in the presence of the building official. Tests required by this code, the *International Fire Code* and the standards listed in this code shall be conducted at the expense of the owner or the owner's representative. It shall be unlawful to occupy portions of a structure until the required fire protection systems within that portion of the structure have been tested and approved.

901.6 Supervisory service. Where required, fire protection systems shall be monitored by an approved supervising station in accordance with NFPA 72.

> **901.6.1 Automatic sprinkler systems.** Automatic sprinkler systems shall be monitored by an approved supervising station.
>
> **Exceptions:**
> 1. A supervising station is not required for automatic sprinkler systems protecting one- and two-family dwellings.
> 2. Limited area systems serving fewer than 20 sprinklers.

901.6.2 Fire alarm systems. Fire alarm systems required by the provisions of Section 907.2 of this code and Section 907.2 of the *International Fire Code* shall be monitored by an approved supervising station in accordance with Section 907.14.

> **Exceptions:**
> 1. Single- and multiple-station smoke alarms required by Section 907.2.10.
> 2. Smoke detectors in Group I-3 occupancies.
> 3. Supervisory service is not required for automatic sprinkler systems in one- and two-family dwellings.

901.6.3 Group H. Manual fire alarm, automatic fire-extinguishing and emergency alarm systems in Group H occupancies shall be monitored by an approved supervising station.

> **Exception:** When approved by the building official, on-site monitoring at a constantly attended location shall be permitted provided that notifications to the fire department will be equal to those provided by an approved supervising station.

901.7 Fire areas. Where buildings, or portions thereof, are divided into fire areas so as not to exceed the limits established for requiring a fire protection system in accordance with this chapter, such fire areas shall be separated by fire barriers having a fire-resistance rating of not less than that determined in accordance with Section 706.3.7.

SECTION 902
DEFINITIONS

902.1 Definitions. The following words and terms shall, for the purposes of this chapter, and as used elsewhere in this code, have the meanings shown herein.

[F] ALARM NOTIFICATION APPLIANCE. A fire alarm system component such as a bell, horn, speaker, light or text display that provides audible, tactile or visible outputs, or any combination thereof.

[F] ALARM SIGNAL. A signal indicating an emergency requiring immediate action, such as a signal indicative of fire.

[F] ALARM VERIFICATION FEATURE. A feature of automatic fire detection and alarm systems to reduce unwanted alarms wherein smoke detectors report alarm conditions for a minimum period of time, or confirm alarm conditions within a given time period, after being automatically reset, in order to be accepted as a valid alarm-initiation signal.

[F] ANNUNCIATOR. A unit containing one or more indicator lamps, alphanumeric displays or other equivalent means in which each indication provides status information about a circuit, condition or location.

[F] AUDIBLE ALARM NOTIFICATION APPLIANCE. A notification appliance that alerts by the sense of hearing.

[F] AUTOMATIC. As applied to fire protection devices, is a device or system providing an emergency function without the necessity for human intervention and activated as a result of a predetermined temperature rise, rate of temperature rise or combustion products.

[F] AUTOMATIC FIRE-EXTINGUISHING SYSTEM. An approved system of devices and equipment which automatically detects a fire and discharges an approved fire-extinguishing agent onto or in the area of a fire.

[F] AUTOMATIC SPRINKLER SYSTEM. A sprinkler system, for fire protection purposes, is an integrated system of underground and overhead piping designed in accordance with fire protection engineering standards. The system includes a suitable water supply. The portion of the system above the ground is a network of specially sized or hydraulically designed piping installed in a structure or area, generally overhead, and to which automatic sprinklers are connected in a systematic pattern. The system is usually activated by heat from a fire and discharges water over the fire area.

[F] AVERAGE AMBIENT SOUND LEVEL. The root mean square, A-weighted sound pressure level measured over a 24-hour period.

[F] CARBON DIOXIDE EXTINGUISHING SYSTEMS. A system supplying carbon dioxide (CO_2) from a pressurized vessel through fixed pipes and nozzles. The system includes a manual- or automatic-actuating mechanism.

[F] CEILING LIMIT. The maximum concentration of an air-borne contaminant to which one may be exposed, as published in DOL 29 CFR Part 1910.1000.

[F] CLEAN AGENT. Electrically nonconducting, volatile or gaseous fire extinguishant that does not leave a residue upon evaporation.

[F] CONSTANTLY ATTENDED LOCATION. A designated location at a facility staffed by trained personnel on a continuous basis where alarm or supervisory signals are monitored and facilities are provided for notification of the fire department or other emergency services.

[F] DELUGE SYSTEM. A sprinkler system employing open sprinklers attached to a piping system connected to a water supply through a valve that is opened by the operation of a detection system installed in the same areas as the sprinklers. When this valve opens, water flows into the piping system and discharges from all sprinklers attached thereto.

[F] DETECTOR, HEAT. A fire detector that senses heat produced by burning substances. Heat is the energy produced by combustion that causes substances to rise in temperature.

[F] DRY-CHEMICAL EXTINGUISHING AGENT. A powder composed of small particles, usually of sodium bicarbonate, potassium bicarbonate, urea-potassium-based bicarbonate, potassium chloride or monoammonium phosphate, with added particulate material supplemented by special treatment to provide resistance to packing, resistance to moisture absorption (caking) and the proper flow capabilities.

[F] EMERGENCY ALARM SYSTEM. A system to provide indication and warning of emergency situations involving hazardous materials.

[F] EMERGENCY VOICE/ALARM COMMUNICATIONS. Dedicated manual or automatic facilities for originating and distributing voice instructions, as well as alert and evacuation signals pertaining to a fire emergency, to the occupants of a building.

[F] EXPLOSION. An effect produced by the sudden violent expansion of gases, that is accompanied by a shock wave or disruption of enclosing materials or structures, or both.

[F] FIRE ALARM BOX, MANUAL. See "Manual Fire Alarm Box."

[F] FIRE ALARM CONTROL UNIT. A system component that receives inputs from automatic and manual fire alarm devices and is capable of supplying power to detection devices and transponder(s) or off-premises transmitter(s). The control unit is capable of providing a transfer of power to the notification appliances and transfer of condition to relays or devices.

[F] FIRE ALARM SIGNAL. A signal initiated by a fire alarm-initiating device such as a manual fire alarm box, automatic fire detector, water flow switch, or other device whose activation is indicative of the presence of a fire or fire signature.

[F] FIRE ALARM SYSTEM. A system or portion of a combination system consisting of components and circuits arranged to monitor and annunciate the status of fire alarm or supervisory signal-initiating devices and to initiate the appropriate response to those signals.

[F] FIRE COMMAND CENTER. The principal attended or unattended location where the status of detection, alarm communications and control systems is displayed, and from which the system(s) can be manually controlled.

[F] FIRE DETECTOR, AUTOMATIC. A device designed to detect the presence of a fire signature and to initiate action.

[F] FIRE PROTECTION SYSTEM. Approved devices, equipment and systems or combinations of systems used to detect a fire, activate an alarm, extinguish or control a fire, control or manage smoke and products of a fire or any combination thereof.

[F] FIRE SAFETY FUNCTIONS. Building and fire control functions that are intended to increase the level of life safety for occupants or to control the spread of harmful effects of fire.

[F] FOAM-EXTINGUISHING SYSTEM. A special system discharging a foam made from concentrates, either mechanically or chemically, over the area to be protected.

[F] HALOGENATED EXTINGUISHING SYSTEM. A fire-extinguishing system using one or more atoms of an element from the halogen chemical series: fluorine, chlorine, bromine and iodine.

[F] INITIATING DEVICE. A system component that originates transmission of a change-of-state condition, such as in a smoke detector, manual fire alarm box or supervisory switch.

LISTED. Equipment, materials or services included in a list published by an organization acceptable to the building official and concerned with evaluation of products or services that maintains periodic inspection of production of listed equipment or materials or periodic evaluation of services and whose listing states either that the equipment, material or service meets identified standards or has been tested and found suitable for a specified purpose.

[F] MANUAL FIRE ALARM BOX. A manually operated device used to initiate an alarm signal.

[F] MULTIPLE-STATION ALARM DEVICE. Two or more single-station alarm devices that are capable of interconnection such that actuation of one causes all integral or separate audible alarms to operate. It also can consist of one single-station alarm device having connections to other detectors or to a manual fire alarm box.

[F] MULTIPLE-STATION SMOKE ALARM. Two or more single-station alarm devices that are capable of interconnection such that actuation of one causes all integral or separate audible alarms to operate.

[F] NUISANCE ALARM. An alarm caused by mechanical failure, malfunction, improper installation or lack of proper maintenance, or an alarm activated by a cause that cannot be determined.

[F] RECORD DRAWINGS. Drawings ("as builts") that document the location of all devices, appliances, wiring sequences, wiring methods and connections of the components of a fire alarm system as installed.

[F] SINGLE-STATION SMOKE ALARM. An assembly incorporating the detector, the control equipment and the alarm-sounding device in one unit, operated from a power supply either in the unit or obtained at the point of installation.

[F] SMOKE ALARM. A single- or multiple-station alarm responsive to smoke and not connected to a system.

[F] SMOKE DETECTOR. A listed device that senses visible or invisible particles of combustion.

SMOKEPROOF ENCLOSURE. An exit stairway designed and constructed so that the movement of the products of combustion produced by a fire occurring in any part of the building into the enclosure is limited.

[F] STANDPIPE SYSTEM, CLASSES OF. Standpipe classes are as follows:

Class I system. A system providing 2.5-inch (64 mm) hose connections to supply water for use by fire departments and those trained in handling heavy fire streams.

Class II system. A system providing 1.5-inch (38 mm) hose stations to supply water for use primarily by the building occupants or by the fire department during initial response.

Class III system. A system providing 1.5-inch (38 mm) hose stations to supply water for use by building occupants and 2.5-inch (64 mm) hose connections to supply a larger volume of water for use by fire departments and those trained in handling heavy fire streams.

[F] STANDPIPE, TYPES OF. Standpipe types are as follows:

Automatic dry. A dry standpipe system, normally filled with pressurized air, that is arranged through the use of a device, such as dry pipe valve, to admit water into the system piping automatically upon the opening of a hose valve. The water supply for an automatic dry standpipe system shall be capable of supplying the system demand.

Automatic wet. A wet standpipe system that has a water supply that is capable of supplying the system demand automatically.

Manual dry. A dry standpipe system that does not have a permanent water supply attached to the system. Manual dry standpipe systems require water from a fire department pumper to be pumped into the system through the fire department connection in order to meet the system demand.

Manual wet. A wet standpipe system connected to a water supply for the purpose of maintaining water within the system but does not have a water supply capable of delivering the system demand attached to the system. Manual-wet standpipe systems require water from a fire department pumper (or the like) to be pumped into the system in order to meet the system demand.

Semiautomatic dry. A dry standpipe system that is arranged through the use of a device, such as a deluge valve, to admit water into the system piping upon activation of a remote control device located at a hose connection. A remote control activation device shall be provided at each hose connection. The water supply for a semiautomatic dry standpipe system shall be capable of supplying the system demand.

[F] SUPERVISING STATION. A facility that receives signals and at which personnel are in attendance at all times to respond to these signals.

[F] SUPERVISORY SERVICE. The service required to monitor performance of guard tours and the operative condition of fixed suppression systems or other systems for the protection of life and property.

[F] SUPERVISORY SIGNAL. A signal indicating the need of action in connection with the supervision of guard tours, the fire suppression systems or equipment or the maintenance features of related systems.

[F] SUPERVISORY SIGNAL-INITIATING DEVICE. An initiation device, such as a valve supervisory switch, water-level indicator or low-air pressure switch on a dry-pipe sprinkler system, whose change of state signals an off-normal condition and its restoration to normal of a fire protection or life safety system, or a need for action in connection with guard tours, fire suppression systems or equipment or maintenance features of related systems.

[F] TIRES, BULK STORAGE OF. Storage of tires where the area available for storage exceeds 20,000 cubic feet (566 m^3).

[F] TROUBLE SIGNAL. A signal initiated by the fire alarm system or device indicative of a fault in a monitored circuit or component.

[F] VISIBLE ALARM NOTIFICATION APPLIANCE. A notification appliance that alerts by the sense of sight.

[F] WET-CHEMICAL EXTINGUISHING SYSTEM. A solution of water and potassium-carbonate-based chemical, potassium-acetate-based chemical or a combination thereof, forming an extinguishing agent.

[F] WIRELESS PROTECTION SYSTEM. A system or a part of a system that can transmit and receive signals without the aid of wire.

[F] ZONE. A defined area within the protected premises. A zone can define an area from which a signal can be received, an area to which a signal can be sent or an area in which a form of control can be executed.

SECTION 903
AUTOMATIC SPRINKLER SYSTEMS

[F] 903.1 General. Automatic sprinkler systems shall comply with this section.

[F] 903.1.1 Alternative protection. Alternative automatic fire-extinguishing systems complying with Section 904 shall be permitted in lieu of automatic sprinkler protection where recognized by the applicable standard and approved by the building official.

[F] 903.2 Where required. Approved automatic sprinkler systems in new buildings and structures shall be provided in the locations described in this section.

Exception: Spaces or areas in telecommunications buildings used exclusively for telecommunications equipment, associated electrical power distribution equipment, batteries and standby engines, provided those spaces or areas are equipped throughout with an automatic fire alarm system and are separated from the remainder of the building by a wall with a fire-resistance rating of not less than 1 hour and a floor/ceiling assembly with a fire-resistance rating of not less than 2 hours.

[F] 903.2.1 Group A. An automatic sprinkler system shall be provided throughout buildings and portions thereof used as Group A occupancies as provided in this section. For Group A-1, A-2, A-3 and A-4 occupancies, the automatic sprinkler system shall be provided throughout the floor area where the Group A-1, A-2, A-3 or A-4 occupancy is located, and in all floors between the Group A occupancy and the level of exit discharge. For Group A-5 occupancies, the automatic sprinkler system shall be provided in the spaces indicated in Section 903.2.1.5.

[F] 903.2.1.1 Group A-1. An automatic sprinkler system shall be provided for Group A-1 occupancies where one of the following conditions exists:

1. The fire area exceeds 12,000 square feet (1115 m^2).

2. The fire area has an occupant load of 300 or more.

3. The fire area is located on a floor other than the level of exit discharge.

4. The fire area contains a multitheater complex.

[F] 903.2.1.2 Group A-2. An automatic sprinkler system shall be provided for Group A-2 occupancies where one of the following conditions exists:

1. The fire area exceeds 5,000 square feet (464.5 m^2).

2. The fire area has an occupant load of 300 or more.

3. The fire area is located on a floor other than the level of exit discharge.

[F] 903.2.1.3 Group A-3. An automatic sprinkler system shall be provided for Group A-3 occupancies where one of the following conditions exists:

1. The fire area exceeds 12,000 square feet (1115 m^2).

2. The fire area has an occupant load of 300 or more.

3. The fire area is located on a floor other than the level of exit discharge.

Exception: Areas used exclusively as participant sports areas where the main floor area is located at the same level as the level of exit discharge of the main entrance and exit.

[F] 903.2.1.4 Group A-4. An automatic sprinkler system shall be provided for Group A-4 occupancies where one of the following conditions exists:

1. The fire area exceeds 12,000 square feet (1115 m^2).

2. The fire area has an occupant load of 300 or more.

3. The fire area is located on a floor other than the level of exit discharge.

Exception: Areas used exclusively as participant sports areas where the main floor area is located at the same level as the level of exit discharge of the main entrance and exit.

[F] 903.2.1.5 Group A-5. An automatic sprinkler system shall be provided in concession stands, retail areas, press boxes and other accessory use areas in excess of 1,000 square feet (93 m^2).

[F] 903.2.2 Group E. An automatic sprinkler system shall be provided for Group E occupancies as follows:

1. Throughout all Group E fire areas greater than 20,000 square feet (1858 m^2) in area.

2. Throughout every portion of educational buildings below the level of exit discharge.

   Exception: An automatic sprinkler system is not required in any fire area or area below the level of exit discharge where every classroom throughout the building has at least one exterior exit door at ground level.

[F] 903.2.3 Group F-1. An automatic sprinkler system shall be provided throughout all buildings containing a Group F-1 occupancy where one of the following conditions exists:

1. Where a Group F-1 fire area exceeds 12,000 square feet (1115 m^2);

2. Where a Group F-1 fire area is located more than three stories above grade; or

3. Where the combined area of all Group F-1 fire areas on all floors, including any mezzanines, exceeds 24,000 square feet (2230 m^2).

[F] 903.2.3.1 Woodworking operations. An automatic sprinkler system shall be provided throughout all Group F-1 occupancy fire areas that contain woodworking operations in excess of 2,500 square feet (232 m^2) in area which generate finely divided combustible waste or use finely divided combustible materials.

[F] 903.2.4 Group H. Automatic sprinkler systems shall be provided in high-hazard occupancies as required in Sections 903.2.4.1 through 903.2.4.3.

[F] 903.2.4.1 General. An automatic sprinkler system shall be installed in Group H occupancies.

[F] 903.2.4.2 Group H-5. An automatic sprinkler system shall be installed throughout buildings containing Group H-5 occupancies. The design of the sprinkler system shall not be less than that required by this code for the occupancy hazard classifications in accordance with Table 903.2.4.2. Where the design area of the sprinkler system consists of a corridor protected by one row of sprinklers, the maximum number of sprinklers required to be calculated is 13.

[F] TABLE 903.2.4.2
GROUP H-5 SPRINKLER DESIGN CRITERIA

LOCATION	OCCUPANCY HAZARD CLASSIFICATION
Fabrication areas	Ordinary Hazard Group 2
Service corridors	Ordinary Hazard Group 2
Storage rooms without dispensing	Ordinary Hazard Group 2
Storage rooms with dispensing	Extra Hazard Group 2
Corridors	Ordinary Hazard Group 2

[F] 903.2.4.3 Pyroxylin plastics. An automatic sprinkler system shall be provided in buildings, or portions thereof, where cellulose nitrate film or pyroxylin plastics are manufactured, stored or handled in quantities exceeding 100 pounds (45 kg).

[F] 903.2.5 Group I. An automatic sprinkler system shall be provided throughout buildings with a Group I fire area.

Exception: An automatic sprinkler system installed in accordance with Section 903.3.1.2 or 903.3.1.3 shall be allowed in Group I-1 facilities.

[F] 903.2.6 Group M. An automatic sprinkler system shall be provided throughout buildings containing a Group M occupancy where one of the following conditions exists:

1. Where a Group M fire area exceeds 12,000 square feet (1115 m^2);
2. Where a Group M fire area is located more than three stories above grade; or
3. Where the combined area of all Group M fire areas on all floors, including any mezzanines, exceeds 24,000 square feet (2230 m^2).

[F] 903.2.6.1 High-piled storage. An automatic sprinkler system shall be provided in accordance with the *International Fire Code* in all buildings of Group M where storage of merchandise is in high-piled or rack storage arrays.

[F] 903.2.7 Group R. An automatic sprinkler system installed in accordance with Section 903.3 shall be provided throughout all buildings with a Group R fire area.

[F] 903.2.8 Group S-1. An automatic sprinkler system shall be provided throughout all buildings containing a Group S-1 occupancy where one of the following conditions exists:

1. A Group S-1 fire area exceeds 12,000 square feet (1115 m^2);
2. A Group S-1 fire area is located more than three stories above grade; or
3. The combined area of all Group S-1 fire areas on all floors, including any mezzanines, exceeds 24,000 square feet (2230 m^2).

[F] 903.2.8.1 Repair garages. An automatic sprinkler system shall be provided throughout all buildings used as repair garages in accordance with Section 406, as shown:

1. Buildings two or more stories in height, including basements, with a fire area containing a repair garage exceeding 10,000 square feet (929 m^2).
2. One-story buildings with a fire area containing a repair garage exceeding 12,000 square feet (1115 m^2).
3. Buildings with a repair garage servicing vehicles parked in the basement.

[F] 903.2.8.2 Bulk storage of tires. Buildings and structures where the area for the storage of tires exceeds 20,000 cubic feet (566 m^3) shall be equipped throughout with an automatic sprinkler system in accordance with Section 903.3.1.1.

[F] 903.2.9 Group S-2. An automatic sprinkler system shall be provided throughout buildings classified as enclosed parking garages in accordance with Section 406.4 or where located beneath other groups.

Exception: Enclosed parking garages located beneath Group R-3 occupancies as applicable in Section 101.2.

[F] 903.2.9.1 Commercial parking garages. An automatic sprinkler system shall be provided throughout buildings used for storage of commercial trucks or buses where the fire area exceeds 5,000 square feet (464 m^2).

[F] 903.2.10 All occupancies except Groups R-3 and U. An automatic sprinkler system shall be installed in the locations set forth in Sections 903.2.10.1 through 903.2.10.1.3.

Exception: Group R-3 as applicable in Section 101.2 and Group U.

[F] 903.2.10.1 Stories and basements without openings. An automatic sprinkler system shall be installed throughout every story or basement of all buildings where the floor area exceeds 1,500 square feet (139.4 m^2) and where there is not provided at least one of the following types of exterior wall openings:

1. Openings below grade that lead directly to ground level by an exterior stairway complying with Section 1009 or an outside ramp complying with Section 1010. Openings shall be located in each 50 linear feet (15 240 mm), or fraction thereof, of exterior wall in the story on at least one side.
2. Openings entirely above the adjoining ground level totaling at least 20 square feet (1.86 m^2) in each 50 linear feet (15 240 mm), or fraction thereof, of exterior wall in the story on at least one side.

[F] 903.2.10.1.1 Opening dimensions and access. Openings shall have a minimum dimension of not less than 30 inches (762 mm). Such openings shall be accessible to the fire department from the exterior and shall not be obstructed in a manner that fire fighting or rescue cannot be accomplished from the exterior.

[F] 903.2.10.1.2 Openings on one side only. Where openings in a story are provided on only one side and the opposite wall of such story is more than 75 feet (22 860 mm) from such openings, the story shall be equipped throughout with an approved automatic sprinkler system, or openings as specified above shall be provided on at least two sides of the story.

[F] 903.2.10.1.3 Basements. Where any portion of a basement is located more than 75 feet (22 860 mm) from openings required by Section 903.2.10.1, the basement shall be equipped throughout with an approved automatic sprinkler system.

[F] 903.2.10.2 Rubbish and linen chutes. An automatic sprinkler system shall be installed at the top of rubbish and linen chutes and in their terminal rooms. Chutes extending through three or more floors shall have additional sprinkler heads installed within such chutes at alternate floors. Chute sprinklers shall be accessible for servicing.

[F] 903.2.10.3 Buildings over 55 feet in height. An automatic sprinkler system shall be installed throughout buildings with a floor level having an occupant load of 30 or more that is located 55 feet (16 764 mm) or more above the lowest level of fire department vehicle access.

Exceptions:

1. Airport control towers.
2. Open parking structures.
3. Occupancies in Group F-2.

[F] 903.2.11 During construction. Automatic sprinkler systems required during construction, alteration and demolition operations shall be provided in accordance with the *International Fire Code*.

[F] 903.2.12 Other hazards. Automatic sprinkler protection shall be provided for the hazards indicated in Sections 903.2.12.1 and 903.2.12.2.

[F] 903.2.12.1 Ducts conveying hazardous exhausts. Where required by the *International Mechanical Code*, automatic sprinklers shall be provided in ducts conveying hazardous exhaust, or flammable or combustible materials.

Exception: Ducts in which the largest cross-sectional diameter of the duct is less than 10 inches (254 mm).

[F] 903.2.12.2 Commercial cooking operations. An automatic sprinkler system shall be installed in commercial kitchen exhaust hood and duct system where an automatic sprinkler system is used to comply with Section 904.

[F] 903.2.13 Other required suppression systems. In addition to the requirements of Section 903.2, the provisions indicated in Table 903.2.13 also require the installation of a suppression system for certain buildings and areas.

[F] 903.3 Installation requirements. Automatic sprinkler systems shall be designed and installed in accordance with Sections 903.3.1 through 903.3.7.

[F] TABLE 903.2.13
ADDITIONAL REQUIRED SUPPRESSION SYSTEMS

SECTION	SUBJECT
402.8	Covered malls
403.2, 403.3	High-rise buildings
404.3	Atriums
405.3	Underground structures
407.5	Group I-2
410.6	Stages
411.4	Special amusement buildings
412.2.5, 412.2.6	Aircraft hangars
415.7.2.4	Group H-2
416.4	Flammable finishes
417.4	Drying rooms
507	Unlimited area buildings
IFC	Sprinkler requirements as set forth in Section 903.2.13 of the *International Fire Code*

[F] 903.3.1 Standards. Sprinkler systems shall be designed and installed in accordance with Section 903.3.1.1, 903.3.1.2 or 903.3.1.3.

[F] 903.3.1.1 NFPA 13 sprinkler systems. Where the provisions of this code require that a building or portion thereof be equipped throughout with an automatic sprinkler system in accordance with Section 903.3.1.1, sprinklers shall be installed throughout in accordance with NFPA 13 except as provided in Section 903.3.1.1.1.

[F] 903.3.1.1.1 Exempt locations. Automatic sprinklers shall not be required in the following rooms or areas where such rooms or areas are protected with an approved automatic fire detection system in accordance with Section 907.2 that will respond to visible or invisible particles of combustion. Sprinklers shall not be omitted from any room merely because it is damp, of fire-resistance-rated construction or contains electrical equipment.

1. Any room where the application of water, or flame and water, constitutes a serious life or fire hazard.
2. Any room or space where sprinklers are considered undesirable because of the nature of the contents, when approved by the building official.
3. Generator and transformer rooms separated from the remainder of the building by walls and floor/ceiling or roof/ceiling assemblies having a fire-resistance rating of not less than 2 hours.
4. In rooms or areas that are of noncombustible construction with wholly noncombustible contents.

[F] 903.3.1.2 NFPA 13R sprinkler systems. Where allowed in buildings of Group R, up to and including four stories in height, automatic sprinkler systems shall be installed throughout in accordance with NFPA 13R.

[F] 903.3.1.2.1 Balconies. Sprinkler protection shall be provided for exterior balconies and ground-floor patios of dwelling units where the building is of Type V construction. Sidewall sprinklers that are used to protect such areas shall be permitted to be located such that their deflectors are within 1 inch (25 mm) to 6 inches (152 mm) below the structural members, and a maximum distance of 14 inches (356 mm) below the deck of the exterior balconies that are constructed of open wood joist construction.

[F] 903.3.1.3 NFPA 13D sprinkler systems. Where allowed, automatic sprinkler systems in one- and two-family dwellings shall be installed throughout in accordance with NFPA 13D.

[F] 903.3.2 Quick-response and residential sprinklers. Where automatic sprinkler systems are required by this code, quick-response or residential automatic sprinklers shall be installed in the following areas in accordance with Section 903.3.1 and their listings:

1. Throughout all spaces within a smoke compartment containing patient sleeping units in Group I-2 in accordance with this code.
2. Dwelling units, and sleeping units in Group R and I-1 occupancies.
3. Light-hazard occupancies as defined in NFPA 13.

[F] 903.3.3 Obstructed locations. Automatic sprinklers shall be installed with due regard to obstructions that will delay activation or obstruct the water distribution pattern. Automatic sprinklers shall be installed in or under covered kiosks, displays, booths, concession stands, or equipment that exceeds 4 feet (1219 mm) in width. Not less than a 3-foot (914 mm) clearance shall be maintained between automatic sprinklers and the top of piles of combustible fibers.

Exception: Kitchen equipment under exhaust hoods protected with a fire-extinguishing system in accordance with Section 904.

[F] 903.3.4 Actuation. Automatic sprinkler systems shall be automatically actuated unless specifically provided for in this code.

[F] 903.3.5 Water supplies. Water supplies for automatic sprinkler systems shall comply with this section and the standards referenced in Section 903.3.1. The potable water supply shall be protected against backflow in accordance with the requirements of this section and the *International Plumbing Code.*

[F] 903.3.5.1 Domestic services. Where the domestic service provides the water supply for the automatic sprinkler system, the supply shall be in accordance with this section.

[F] 903.3.5.1.1 Limited area sprinkler systems. Limited area sprinkler systems serving fewer than 20 sprinklers on any single connection are permitted to be connected to the domestic service where a wet automatic standpipe is not available. Limited area sprinkler systems connected to domestic water supplies shall comply with each of the following requirements:

1. Valves shall not be installed between the domestic water riser control valve and the sprinklers.

   Exception: An approved indicating control valve supervised in the open position in accordance with Section 903.4.

2. The domestic service shall be capable of supplying the simultaneous domestic demand and the sprinkler demand required to be hydraulically calculated by NFPA 13, NFPA 13R or NFPA 13D.

[F] 903.3.5.1.2 Residential combination services. A single combination water supply shall be permitted provided that the domestic demand is added to the sprinkler demand as required by NFPA 13R.

[F] 903.3.5.2 Secondary water supply. A secondary on-site water supply equal to the hydraulically calculated sprinkler demand, including the hose stream requirement, shall be provided for high-rise buildings in Seismic Design Category C, D, E or F as determined by this code. The secondary water supply shall have a duration not less than 30 minutes as determined by the occupancy hazard classification in accordance with NFPA 13.

Exception: Existing buildings.

[F] 903.3.6 Hose threads. Fire hose threads used in connection with automatic sprinkler systems shall be approved and compatible with fire department hose threads.

[F] 903.3.7 Fire department connections. The location of fire department connections shall be approved by the building official.

[F] 903.3.7.1 Locking fire department connection (FDC) caps. The fire code official is authorized to require locking FDC caps on fire department connections for water-based fire protection systems where the responding fire department carries appropriate key wrenches for removal.

[F] 903.4 Sprinkler system monitoring and alarms. All valves controlling the water supply for automatic sprinkler systems, pumps, tanks, water levels and temperatures, critical air pressures and water-flow switches on all sprinkler systems shall be electrically supervised.

Exceptions:

1. Automatic sprinkler systems protecting one- and two-family dwellings.
2. Limited area systems serving fewer than 20 sprinklers.
3. Automatic sprinkler systems installed in accordance with NFPA 13R where a common supply main is used to supply both domestic water and the automatic sprinkler systems and a separate shutoff valve for the automatic sprinkler system is not provided.
4. Jockey pump control valves that are sealed or locked in the open position.

5. Control valves to commercial kitchen hoods, paint spray booths or dip tanks that are sealed or locked in the open position.

6. Valves controlling the fuel supply to fire pump engines that are sealed or locked in the open position.

7. Trim valves to pressure switches in dry, preaction and deluge sprinkler systems that are sealed or locked in the open position.

[F] 903.4.1 Signals. Alarm, supervisory and trouble signals shall be distinctly different and automatically transmitted to an approved central station, remote supervising station or proprietary supervising station as defined in NFPA 72 or, when approved by the building official, shall sound an audible signal at a constantly attended location.

Exceptions:

1. Underground key or hub valves in roadway boxes provided by the municipality or public utility are not required to be monitored.

2. Backflow prevention device test valves, located in limited area sprinkler system supply piping, shall be locked in the open position. In occupancies required to be equipped with a fire alarm system, the backflow preventer valves shall be electrically supervised by a tamper switch installed in accordance with NFPA 72 and separately annunciated.

[F] 903.4.2 Alarms. Approved audible devices shall be connected to every automatic sprinkler system. Such sprinkler water-flow alarm devices shall be activated by water flow equivalent to the flow of a single sprinkler of the smallest orifice size installed in the system. Alarm devices shall be provided on the exterior of the building in an approved location. Where a fire alarm system is installed, actuation of the automatic sprinkler system shall actuate the building fire alarm system.

[F] 903.4.3 Floor control valves. Approved supervised indicating control valves shall be provided at the point of connection to the riser on each floor in high-rise buildings.

[F] 903.5 Testing and maintenance. Sprinkler systems shall be tested and maintained in accordance with the *International Fire Code*.

SECTION 904
ALTERNATIVE AUTOMATIC
FIRE-EXTINGUISHING SYSTEMS

[F] 904.1 General. Automatic fire-extinguishing systems, other than automatic sprinkler systems, shall be designed, installed, inspected, tested and maintained in accordance with the provisions of this section and the applicable referenced standards.

[F] 904.2 Where required. Automatic fire-extinguishing systems installed as an alternative to the required automatic sprinkler systems of Section 903 shall be approved by the building official. Automatic fire-extinguishing systems shall not be considered alternatives for the purposes of exceptions or reductions permitted by other requirements of this code.

[F] 904.2.1 Hood system suppression. Each required commercial kitchen exhaust hood and duct system required by the *International Fire Code* or the *International Mechanical Code* to have a Type I hood shall be protected with an approved automatic fire-extinguishing system installed in accordance with this code.

[F] 904.3 Installation. Automatic fire-extinguishing systems shall be installed in accordance with this section.

[F] 904.3.1 Electrical wiring. Electrical wiring shall be in accordance with the ICC *Electrical Code*.

[F] 904.3.2 Actuation. Automatic fire-extinguishing systems shall be automatically actuated and provided with a manual means of actuation in accordance with Section 904.11.1.

[F] 904.3.3 System interlocking. Automatic equipment interlocks with fuel shutoffs, ventilation controls, door closers, window shutters, conveyor openings, smoke and heat vents and other features necessary for proper operation of the fire-extinguishing system shall be provided as required by the design and installation standard utilized for the hazard.

[F] 904.3.4 Alarms and warning signs. Where alarms are required to indicate the operation of automatic fire-extinguishing systems, distinctive audible and visible alarms and warning signs shall be provided to warn of pending agent discharge. Where exposure to automatic-extinguishing agents poses a hazard to persons and a delay is required to ensure the evacuation of occupants before agent discharge, a separate warning signal shall be provided to alert occupants once agent discharge has begun. Audible signals shall be in accordance with Section 907.9.2.

[F] 904.3.5 Monitoring. Where a building fire alarm system is installed, automatic fire-extinguishing systems shall be monitored by the building fire alarm system in accordance with NFPA 72.

[F] 904.4 Inspection and testing. Automatic fire-extinguishing systems shall be inspected and tested in accordance with the provisions of this section prior to acceptance.

[F] 904.4.1 Inspection. Prior to conducting final acceptance tests, the following items shall be inspected:

1. Hazard specification for consistency with design hazard.

2. Type, location and spacing of automatic- and manual-initiating devices.

3. Size, placement and position of nozzles or discharge orifices.

4. Location and identification of audible and visible alarm devices.

5. Identification of devices with proper designations.

6. Operating instructions.

[F] 904.4.2 Alarm testing. Notification appliances, connections to fire alarm systems and connections to approved supervising stations shall be tested in accordance with this section and Section 907 to verify proper operation.

[F] 904.4.2.1 Audible and visible signals. The audibility and visibility of notification appliances signaling agent discharge or system operation, where required, shall be verified.

[F] 904.4.3 Monitor testing. Connections to protected premises and supervising station fire alarm systems shall be tested to verify proper identification and retransmission of alarms from automatic fire-extinguishing systems.

[F] 904.5 Wet-chemical systems. Wet-chemical extinguishing systems shall be installed, maintained, periodically inspected and tested in accordance with NFPA 17A and their listing.

[F] 904.6 Dry-chemical systems. Dry-chemical extinguishing systems shall be installed, maintained, periodically inspected and tested in accordance with NFPA 17 and their listing.

[F] 904.7 Foam systems. Foam-extinguishing systems shall be installed, maintained, periodically inspected and tested in accordance with NFPA 11 and NFPA 16 and their listing.

[F] 904.8 Carbon dioxide systems. Carbon dioxide extinguishing systems shall be installed, maintained, periodically inspected and tested in accordance with NFPA 12 and their listing.

[F] 904.9 Halon systems. Halogenated extinguishing systems shall be installed, maintained, periodically inspected and tested in accordance with NFPA 12A and their listing.

[F] 904.10 Clean-agent systems. Clean-agent fire-extinguishing systems shall be installed, maintained, periodically inspected and tested in accordance with NFPA 2001 and their listing.

[F] 904.11 Commercial cooking systems. The automatic fire-extinguishing system for commercial cooking systems shall be of a type recognized for protection of commercial cooking equipment and exhaust systems of the type and arrangement protected. Preengineered automatic dry- and wet-chemical extinguishing systems shall be tested in accordance with UL 300 and listed and labeled for the intended application. Other types of automatic fire-extinguishing systems shall be listed and labeled for specific use as protection for commercial cooking operations. The system shall be installed in accordance with this code, its listing and the manufacturer's installation instructions. Automatic fire-extinguishing systems of the following types shall be installed in accordance with the referenced standard indicated, as shown:

1. Carbon dioxide extinguishing systems, NFPA 12.
2. Automatic sprinkler systems, NFPA 13.
3. Foam-water sprinkler system or foam-water spray systems, NFPA 16.
4. Dry-chemical extinguishing systems, NFPA 17.
5. Wet-chemical extinguishing systems, NFPA 17A.

Exception: Factory-built commercial cooking recirculating systems that are tested in accordance with UL 197, and listed, labeled and installed in accordance with Section 304.1 of the *International Mechanical Code.*

[F] 904.11.1 Manual system operation. A manual actuation device shall be located at or near a means of egress from the cooking area, a minimum of 10 feet (3048 mm) and a maximum of 20 feet (6096 mm) from the kitchen exhaust system. The manual actuation device shall be located a minimum of 4 feet (1219 mm) and a maximum of 5 feet (1524 mm) above the floor. The manual actuation shall require a maximum force of 40 pounds (178 N) and a maximum movement of 14 inches (356 mm) to actuate the fire suppression system.

Exception: Automatic sprinkler systems shall not be required to be equipped with manual actuation means.

[F] 904.11.2 System interconnection. The actuation of the fire suppression system shall automatically shut down the fuel or electrical power supply to the cooking equipment. The fuel and electrical supply reset shall be manual.

[F] 904.11.3 Carbon dioxide systems. When carbon dioxide systems are used, there shall be a nozzle at the top of the ventilating duct. Additional nozzles that are symmetrically arranged to give uniform distribution shall be installed within vertical ducts exceeding 20 feet (6096 mm) and horizontal ducts exceeding 50 feet (15 240 mm). Dampers shall be installed at either the top or the bottom of the duct and shall be arranged to operate automatically upon activation of the fire-extinguishing system. Where the damper is installed at the top of the duct, the top nozzle shall be immediately below the damper. Automatic carbon dioxide fire-extinguishing systems shall be sufficiently sized to protect against all hazards venting through a common duct simultaneously.

[F] 904.11.3.1 Ventilation system. Commercial-type cooking equipment protected by an automatic carbon dioxide-extinguishing system shall be arranged to shut off the ventilation system upon activation.

[F] 904.11.4 Special provisions for automatic sprinkler systems. Automatic sprinkler systems protecting commercial-type cooking equipment shall be supplied from a separate, readily accessible, indicating-type control valve that is identified.

[F] 904.11.4.1 Listed sprinklers. Sprinklers used for the protection of fryers shall be listed for that application and installed in accordance with their listing.

SECTION 905
STANDPIPE SYSTEMS

[F] 905.1 General. Standpipe systems shall be provided in new buildings and structures in accordance with this section. Fire hose threads used in connection with standpipe systems shall be approved and shall be compatible with fire department hose threads. The location of fire department hose connections shall be approved. In buildings used for high-piled combustible storage, fire protection shall be in accordance with the *International Fire Code.*

[F] 905.2 Installation standards. Standpipe systems shall be installed in accordance with this section and NFPA 14.

[F] 905.3 Required installations. Standpipe systems shall be installed where required by Sections 905.3.1 through 905.3.6 and in the locations indicated in Sections 905.4, 905.5 and

905.6. Standpipe systems are permitted to be combined with automatic sprinkler systems.

Exception: Standpipe systems are not required in Group R-3 occupancies as applicable in Section 101.2.

[F] 905.3.1 Building height. Class III standpipe systems shall be installed throughout buildings where the floor level of the highest story is located more than 30 feet (9144 mm) above the lowest level of fire department vehicle access, or where the floor level of the lowest story is located more than 30 feet (9144 mm) below the highest level of fire department vehicle access.

Exceptions:

1. Class I standpipes are allowed in buildings equipped throughout with an automatic sprinkler system in accordance with Section 903.3.1.1 or 903.3.1.2.

2. Class I manual standpipes are allowed in open parking garages where the highest floor is located not more than 150 feet (45 720 mm) above the lowest level of fire department vehicle access.

3. Class I manual dry standpipes are allowed in open parking garages that are subject to freezing temperatures, provided that the hose connections are located as required for Class II standpipes in accordance with Section 905.5.

4. Class I standpipes are allowed in basements equipped throughout with an automatic sprinkler system.

[F] 905.3.2 Group A. Class I automatic wet standpipes shall be provided in nonsprinklered Group A buildings having an occupant load exceeding 1,000 persons.

Exceptions:

1. Open-air-seating spaces without enclosed spaces.

2. Class I automatic dry and semiautomatic dry standpipes or manual wet standpipes are allowed in buildings where the highest floor surface used for human occupancy is 75 feet (22 860 mm) or less above the lowest level of fire department vehicle access.

[F] 905.3.3 Covered mall buildings. A covered mall building shall be equipped throughout with a standpipe system where required by Section 905.3. Covered mall buildings not required to be equipped with a standpipe system by Section 905.3 shall be equipped with Class I hose connections connected to a system sized to deliver 250 gallons per minute (946.4 L/min.) at the most hydraulically remote outlet. Hose connections shall be provided at each of the following locations:

1. Within the mall at the entrance to each exit passageway or corridor.

2. At each floor-level landing within enclosed stairways opening directly on the mall.

3. At exterior public entrances to the mall.

[F] 905.3.4 Stages. Stages greater than 1,000 square feet in area (93 m^2) shall be equipped with a Class III wet standpipe system with 1.5-inch and 2.5-inch (38 mm and 64 mm) hose connections on each side of the stage.

Exception: Where the building or area is equipped throughout with an automatic sprinkler system, the hose connections are allowed to be supplied from the automatic sprinkler system and shall have a flow rate of not less than that required by NFPA 14 for Class III standpipes.

[F] 905.3.4.1 Hose and cabinet. The 1.5-inch (38 mm) hose connections shall be equipped with sufficient lengths of 1.5-inch (38 mm) hose to provide fire protection for the stage area. Hose connections shall be equipped with an approved adjustable fog nozzle and be mounted in a cabinet or on a rack.

[F] 905.3.5 Underground buildings. Underground buildings shall be equipped throughout with a Class I automatic wet or manual wet standpipe system.

[F] 905.3.6 Helistops and heliports. Buildings with a helistop or heliport that are equipped with a standpipe shall extend the standpipe to the roof level on which the helistop or heliport is located in accordance with Section 1107.5 of the *International Fire Code*.

[F] 905.4 Location of Class I standpipe hose connections. Class I standpipe hose connections shall be provided in all of the following locations:

1. In every required stairway, a hose connection shall be provided for each floor level above or below grade. Hose connections shall be located at an intermediate floor level landing between floors, unless otherwise approved by the building official.

2. On each side of the wall adjacent to the exit opening of a horizontal exit.

3. In every exit passageway at the entrance from the exit passageway to other areas of a building.

4. In covered mall buildings, adjacent to each exterior public entrance to the mall and adjacent to each entrance from an exit passageway or exit corridor to the mall.

5. Where the roof has a slope less than four units vertical in 12 units horizontal (33.3-percent slope), each standpipe shall be provided with a hose connection located either on the roof or at the highest landing of stairways with stair access to the roof. An additional hose connection shall be provided at the top of the most hydraulically remote standpipe for testing purposes.

6. Where the most remote portion of a nonsprinklered floor or story is more than 150 feet (45 720 mm) from a hose connection or the most remote portion of a sprinklered floor or story is more than 200 feet (60 960 mm) from a hose connection, the building official is authorized to require that additional hose connections be provided in approved locations.

[F] 905.4.1 Protection. Risers and laterals of Class I standpipe systems not located within an enclosed stairway or pressurized enclosure shall be protected by a degree of fire resistance equal to that required for vertical enclosures in the building in which they are located.

> **Exception:** In buildings equipped throughout with an approved automatic sprinkler system, laterals that are not located within an enclosed stairway or pressurized enclosure are not required to be enclosed within fire-resistance-rated construction.

[F] 905.4.2 Interconnection. In buildings where more than one standpipe is provided, the standpipes shall be interconnected in accordance with NFPA 14.

[F] 905.5 Location of Class II standpipe hose connections. Class II standpipe hose connections shall be accessible and located so that all portions of the building are within 30 feet (9144 mm) of a nozzle attached to 100 feet (30 480 mm) of hose.

[F] 905.5.1 Groups A-1 and A-2. In Group A-1 and A-2 occupancies with occupant loads of more than 1,000, hose connections shall be located on each side of any stage, on each side of the rear of the auditorium, on each side of the balcony and on each tier of dressing rooms.

[F] 905.5.2 Protection. Fire-resistance-rated protection of risers and laterals of Class II standpipe systems is not required.

[F] 905.5.3 Class II system 1-inch hose. A minimum 1-inch (25 mm) hose shall be permitted to be used for hose stations in light-hazard occupancies where investigated and listed for this service and where approved by the building official.

[F] 905.6 Location of Class III standpipe hose connections. Class III standpipe systems shall have hose connections located as required for Class I standpipes in Section 905.4 and shall have Class II hose connections as required in Section 905.5.

[F] 905.6.1 Protection. Risers and laterals of Class III standpipe systems shall be protected as required for Class I systems in accordance with Section 905.4.1.

[F] 905.6.2 Interconnection. In buildings where more than one Class III standpipe is provided, the standpipes shall be interconnected at the bottom.

[F] 905.7 Cabinets. Cabinets containing fire-fighting equipment such as standpipes, fire hoses, fire extinguishers or fire department valves shall not be blocked from use or obscured from view.

[F] 905.7.1 Cabinet equipment identification. Cabinets shall be identified in an approved manner by a permanently attached sign with letters not less than 2 inches (51 mm) high in a color that contrasts with the background color, indicating the equipment contained therein.

> **Exceptions:**
>
> 1. Doors not large enough to accommodate a written sign shall be marked with a permanently attached pictogram of the equipment contained therein.
> 2. Doors that have either an approved visual identification clear glass panel or a complete glass door panel are not required to be marked.

[F] 905.7.2 Locking cabinet doors. Cabinets shall be unlocked.

> **Exceptions:**
>
> 1. Visual identification panels of glass or other approved transparent frangible material that is easily broken and allows access.
> 2. Approved locking arrangements.
> 3. Group I-3.

[F] 905.8 Dry standpipes. Dry standpipes shall not be installed.

> **Exception:** Where subject to freezing and in accordance with NFPA 14.

[F] 905.9 Valve supervision. Valves controlling water supplies shall be supervised in the open position so that a change in the normal position of the valve will generate a supervisory signal at the supervising station required by Section 903.4. Where a fire alarm system is provided, a signal shall also be transmitted to the control unit.

> **Exceptions:**
>
> 1. Valves to underground key or hub valves in roadway boxes provided by the municipality or public utility do not require supervision.
> 2. Valves locked in the normal position and inspected as provided in this code in buildings not equipped with a fire alarm system.

[F] 905.10 During construction. Standpipe systems required during construction and demolition operations shall be provided in accordance with Section 3311.

SECTION 906
PORTABLE FIRE EXTINGUISHERS

[F] 906.1 General. Portable fire extinguishers shall be provided in occupancies and locations as required by the *International Fire Code.*

SECTION 907
FIRE ALARM AND DETECTION SYSTEMS

[F] 907.1 General. This section covers the application, installation, performance and maintenance of fire alarm systems and their components.

[F] 907.1.1 Construction documents. Construction documents for fire alarm systems shall be submitted for review and approval prior to system installation. Construction documents shall include, but not be limited to, all of the following:

1. A floor plan which indicates the use of all rooms.
2. Locations of alarm-initiating and notification appliances.
3. Alarm control and trouble signaling equipment.
4. Annunciation.

5. Power connection.

6. Battery calculations.

7. Conductor type and sizes.

8. Voltage drop calculations.

9. Manufacturers, model numbers and listing information for equipment, devices and materials.

10. Details of ceiling height and construction.

11. The interface of fire safety control functions

[F] 907.1.2 Equipment. Systems and their components shall be listed and approved for the purpose for which they are installed.

[F] 907.2 Where required. An approved manual, automatic or manual and automatic fire alarm system shall be provided in accordance with Sections 907.2.1 through 907.2.23. Where automatic sprinkler protection, installed in accordance with Section 903.3.1.1 or 903.3.1.2, is provided and connected to the building fire alarm system, automatic heat detection required by this section shall not be required. An approved automatic fire detection system shall be installed in accordance with the provisions of this code and NFPA 72. Devices, combinations of devices, appliances and equipment shall comply with Section 907.1.2. The automatic fire detectors shall be smoke detectors, except that an approved alternative type of detector shall be installed in spaces such as boiler rooms where, during normal operation, products of combustion are present in sufficient quantity to actuate a smoke detector.

[F] 907.2.1 Group A. A manual fire alarm system shall be installed in accordance with NFPA 72 in Group A occupancies having an occupant load of 300 or more. Portions of Group E occupancies occupied for assembly purposes shall be provided with a fire alarm system as required for the Group E occupancy.

Exception: Manual fire alarm boxes are not required where the building is equipped throughout with an automatic sprinkler system and the notification appliances will activate upon sprinkler water flow.

[F] 907.2.1.1 System initiation in Group A occupancies with an occupant load of 1,000 or more. Activation of the fire alarm in Group A occupancies with an occupant load of 1,000 or more shall initiate a signal using an emergency voice/alarm communications system in accordance with NFPA 72.

Exception: Where approved, the prerecorded announcement is allowed to be manually deactivated for a period of time, not to exceed 3 minutes, for the sole purpose of allowing a live voice announcement from an approved, constantly attended location.

[F] 907.2.1.2 Emergency power. Emergency voice/alarm communications systems shall be provided with an approved emergency power source.

[F] 907.2.2 Group B. A manual fire alarm system shall be installed in Group B occupancies having an occupant load of 500 or more persons or more than 100 persons above or below the lowest level of exit discharge.

Exception: Manual fire alarm boxes are not required where the building is equipped throughout with an automatic sprinkler system and the alarm notification appliances will activate upon sprinkler water flow.

[F] 907.2.3 Group E. A manual fire alarm system shall be installed in Group E occupancies. When automatic sprinkler systems or smoke detectors are installed, such systems or detectors shall be connected to the building fire alarm system.

Exceptions:

1. Group E occupancies with an occupant load of less than 50.

2. Manual fire alarm boxes are not required in Group E occupancies where all the following apply:

   2.1. Interior corridors are protected by smoke detectors with alarm verification.

   2.2. Auditoriums, cafeterias, gymnasiums and the like are protected by heat detectors or other approved detection devices.

   2.3. Shops and laboratories involving dusts or vapors are protected by heat detectors or other approved detection devices.

   2.4. Off-premises monitoring is provided.

   2.5. The capability to activate the evacuation signal from a central point is provided.

   2.6. In buildings where normally occupied spaces are provided with a two-way communication system between such spaces and a constantly attended receiving station from where a general evacuation alarm can be sounded, except in locations specifically designated by the building official.

[F] 907.2.4 Group F. A manual fire alarm system shall be installed in Group F occupancies that are two or more stories in height and have an occupant load of 500 or more above or below the lowest level of exit discharge.

Exception: Manual fire alarm boxes are not required if the building is equipped throughout with an automatic sprinkler system and the notification appliances will activate upon sprinkler water flow.

[F] 907.2.5 Group H. A manual fire alarm system shall be installed in Group H-5 occupancies and in occupancies used for the manufacture of organic coatings. An automatic smoke detection system shall be installed for highly toxic gases, organic peroxides and oxidizers in accordance with Chapters 37, 39 and 40, respectively, of the *International Fire Code.*

[F] 907.2.6 Group I. A manual fire alarm system and an automatic fire detection system shall be installed in Group I occupancies. An electrically supervised, automatic smoke detection system shall be provided in waiting areas that are open to corridors.

Exception: Manual fire alarm boxes in patient sleeping areas of Group I-1 and I-2 occupancies shall not be required at exits if located at all nurses' control stations or other constantly attended staff locations, provided such stations are visible and continuously accessible and that travel distances required in Section 907.3.1 are not exceeded.

[F] 907.2.6.1 Group I-2. Corridors in nursing homes (both intermediate-care and skilled nursing facilities), detoxification facilities and spaces open to the corridors shall be equipped with an automatic fire detection system.

Exceptions:

1. Corridor smoke detection is not required in smoke compartments that contain patient sleeping rooms where patient sleeping units are provided with smoke detectors that comply with UL 268. Such detectors shall provide a visual display on the corridor side of each patient sleeping unit and an audible and visual alarm at the nursing station attending each unit.

2. Corridor smoke detection is not required in smoke compartments that contain patient sleeping rooms where patient sleeping unit doors are equipped with automatic door-closing devices with integral smoke detectors on the unit sides installed in accordance with their listing, provided that the integral detectors perform the required alerting function.

[F] 907.2.6.2 Group I-3. Group I-3 occupancies shall be equipped with a manual and automatic fire alarm system installed for alerting staff.

[F] 907.2.6.2.1 System initiation. Actuation of an automatic fire-extinguishing system, a manual fire alarm box or a fire detector shall initiate an approved fire alarm signal which automatically notifies staff. Presignal systems shall not be used.

[F] 907.2.6.2.2 Manual fire alarm boxes. Manual fire alarm boxes are not required to be located in accordance with Section 907.3 where the fire alarm boxes are provided at staff-attended locations having direct supervision over areas where manual fire alarm boxes have been omitted. Manual fire alarm boxes shall be permitted to be locked in areas occupied by detainees, provided that staff members are present within the subject area and have keys readily available to operate the manual fire alarm boxes.

[F] 907.2.6.2.3 Smoke detectors. An approved automatic smoke detection system shall be installed throughout resident housing areas, including sleeping areas and contiguous day rooms, group activity spaces and other common spaces normally accessible to residents.

Exceptions:

1. Other approved smoke detection arrangements providing equivalent protection including, but not limited to, placing detectors in exhaust ducts from cells or behind protective guards listed for the purpose are allowed when necessary to prevent damage or tampering.

2. Sleeping units in Use Conditions 2 and 3.

3. Smoke detectors are not required in sleeping units with four or fewer occupants in smoke compartments that are equipped throughout with an approved automatic sprinkler system.

[F] 907.2.7 Group M. A manual fire alarm system shall be installed in Group M occupancies, other than covered mall buildings complying with Section 402, having an occupant load of 500 or more persons or more than 100 persons above or below the lowest level of exit discharge.

Exception: Manual fire alarm boxes are not required if the building is equipped throughout with an automatic sprinkler system and the alarm notification appliances will activate upon sprinkler water flow.

[F] 907.2.7.1 Occupant notification. During times that the building is occupied, in lieu of the automatic activation of alarm notification appliances, the manual fire alarm system shall be allowed to activate an alarm signal at a constantly attended location from which evacuation instructions shall be initiated over an emergency voice/alarm communication system installed in accordance with Section 907.2.12.2. The emergency voice/alarm communication system shall be allowed to be used for other announcements, provided the manual fire alarm use takes precedence over any other use.

[F] 907.2.8 Group R-1. Fire alarm systems shall be installed in Group R-1 occupancies as required in Sections 907.2.8.1 through 907.2.8.3.

[F] 907.2.8.1 Manual fire alarm system. A manual fire alarm system shall be installed in Group R-1 occupancies.

Exceptions:

1. A manual fire alarm system is not required in buildings not over two stories in height where all individual guestrooms and contiguous attic and crawl spaces are separated from each other and public or common areas by at least 1-hour fire partitions and each individual guestroom has an exit directly to a public way, exit court or yard.

2. Manual fire alarm boxes are not required throughout the building when the following conditions are met:

2.1. The building is equipped throughout with an automatic sprinkler system installed in accordance with Section 903.3.1.1 or 903.3.1.2.

2.2. The notification appliances will activate upon sprinkler water flow, and

2.3. At least one manual fire alarm box is installed at an approved location.

[F] 907.2.8.2 Automatic fire alarm system. An automatic fire alarm system shall be installed throughout all interior corridors serving guestrooms.

Exception: An automatic fire detection system is not required in buildings that do not have interior corridors serving guestrooms and each guestroom has a means of egress door opening directly to an exterior exit access that leads directly to an exit.

[F] 907.2.8.3 Smoke alarms. Smoke alarms shall be installed as required by Section 907.2.10. In buildings that are not equipped throughout with an automatic sprinkler system installed in accordance with Section 903.3.1.1 or 903.3.1.2, the smoke alarms in guestrooms shall be connected to an emergency electrical system and shall be annunciated by guestroom at a constantly attended location from which the fire alarm system is capable of being manually activated

[F] 907.2.9 Group R-2. A manual fire alarm system shall be installed in Group R-2 occupancies where:

1. Any dwelling unit or sleeping unit is located three or more stories above the lowest level of exit discharge;

2. Any dwelling unit or sleeping unit is located more than one story below the highest level of exit discharge of exits serving the dwelling unit or sleeping unit; or

3. The building contains more than 16 dwelling units or sleeping units.

Exceptions:

1. A fire alarm system is not required in buildings not over two stories in height where all dwelling units or sleeping units and contiguous attic and crawl spaces are separated from each other and public or common areas by at least 1-hour fire partitions and each dwelling unit or sleeping unit has an exit directly to a public way, exit court or yard.

2. Manual fire alarm boxes are not required throughout the building when the following conditions are met:

   2.1. The building is equipped throughout with an automatic sprinkler system in accordance with Section 903.3.1.1 or 903.3.1.2.

   2.2. The notification appliances will activate upon sprinkler flow, and

   2.3. At least one manual fire alarm box is installed at an approved location.

3. A fire alarm system is not required in buildings that do not have interior corridors serving dwelling units and are protected by an approved automatic sprinkler system installed in accordance with Section 903.3.1.1 or 903.3.1.2, provided that dwelling units either have a means of egress door opening directly to an exterior exit access that leads directly to the exits or are served by open-ended corridors

designed in accordance with Section 1022.6, Exception 4.

[F] 907.2.10 Single- and multiple-station smoke alarms. Listed single- and multiple-station smoke alarms shall be installed in accordance with the provisions of this code and the household fire-warning equipment provisions of NFPA 72.

[F] 907.2.10.1 Where required. Single- or multiple-station smoke alarms shall be installed in the locations described in Sections 907.2.10.1.1 through 907.2.10.1.3.

[F] 907.2.10.1.1 Group R-1. Single- or multiple-station smoke alarms shall be installed in all of the following locations in Group R-1:

1. In sleeping areas.

2. In every room in the path of the means of egress from the sleeping area to the door leading from the sleeping unit.

3. In each story within the sleeping unit, including basements. For sleeping units with split levels and without an intervening door between the adjacent levels, a smoke alarm installed on the upper level shall suffice for the adjacent lower level provided that the lower level is less than one full story below the upper level.

[F] 907.2.10.1.2 Groups R-2, R-3, R-4 and I-1. Single- or multiple-station smoke alarms shall be installed and maintained in Groups R-2, R-3, R-4 and I-1, regardless of occupant load at all of the following locations:

1. On the ceiling or wall outside of each separate sleeping area in the immediate vicinity of bedrooms.

2. In each room used for sleeping purposes.

3. In each story within a dwelling unit, including basements but not including crawl spaces and uninhabitable attics. In dwellings or dwelling units with split levels and without an intervening door between the adjacent levels, a smoke alarm installed on the upper level shall suffice for the adjacent lower level provided that the lower level is less than one full story below the upper level.

[F] 907.2.10.1.3 Group I-1. Single- or multiple-station smoke alarms shall be installed and maintained in sleeping areas in occupancies in Group I-1. Single- or multiple-station smoke alarms shall not be required where the building is equipped throughout with an automatic fire detection system in accordance with Section 907.2.6.

[F] 907.2.10.2 Power source. In new construction, required smoke alarms shall receive their primary power from the building wiring where such wiring is served from a commercial source and shall be equipped with a battery backup. Smoke alarms shall emit a signal when the batteries are low. Wiring shall be permanent and without a disconnecting switch other than as required for overcurrent protection.

> **Exception:** Smoke alarms are not required to be equipped with battery backup in Group R-1 where they are connected to an emergency electrical system.

[F] 907.2.10.3 Interconnection. Where more than one smoke alarm is required to be installed within an individual dwelling unit in Group R-2, R-3 or R-4, or within an individual dwelling unit or sleeping unit in Group R-1, the smoke alarms shall be interconnected in such a manner that the activation of one alarm will activate all of the alarms in the individual unit. The alarm shall be clearly audible in all bedrooms over background noise levels with all intervening doors closed.

[F] 907.2.10.4 Acceptance testing. When the installation of the alarm devices is complete, each detector and interconnecting wiring for multiple-station alarm devices shall be tested in accordance with the household fire warning equipment provisions of NFPA 72.

[F] 907.2.11 Special amusement buildings. An approved automatic smoke detection system shall be provided in special amusement buildings in accordance with this section.

> **Exception:** In areas where ambient conditions will cause a smoke detection system to alarm, an approved alternative type of automatic detector shall be installed.

[F] 907.2.11.1 Alarm. Activation of any single smoke detector, the automatic sprinkler system or any other automatic fire detection device shall immediately sound an alarm at the building at a constantly attended location from which emergency action can be initiated, including the capability of manual initiation of requirements in Section 907.2.11.2.

[F] 907.2.11.2 System response. The activation of two or more smoke detectors, a single smoke detector with alarm verification, the automatic sprinkler system or other approved fire detection device shall automatically:

1. Cause illumination of the means of egress with light of not less than 1 foot-candle (11 lux) at the walking surface level;

2. Stop any conflicting or confusing sounds and visual distractions; and

3. Activate an approved directional exit marking that will become apparent in an emergency. Such system response shall also include activation of a prerecorded message, clearly audible throughout the special amusement building, instructing patrons to proceed to the nearest exit. Alarm signals used in conjunction with the prerecorded message shall produce a sound which is distinctive from other sounds used during normal operation. The wiring to the auxiliary devices and equipment used to accomplish the above fire safety functions shall be monitored for integrity in accordance with NFPA 72.

[F] 907.2.11.3 Emergency voice/alarm communication system. An emergency voice/alarm communication system, which is also allowed to serve as a public address system, shall be installed in accordance with NFPA 72, and shall be audible throughout the entire special amusement building.

[F] 907.2.12 High-rise buildings. Buildings having floors used for human occupancy located more than 75 feet (22 860 mm) above the lowest level of fire department vehicle access shall be provided with an automatic fire alarm system and an emergency voice/alarm communication system in accordance with Section 907.2.12.2.

> **Exceptions:**
> 1. Airport traffic control towers in accordance with Sections 412 and 907.2.22.
> 2. Open parking garages in accordance with Section 406.3.
> 3. Buildings with an occupancy in Group A-5.
> 4. Low-hazard special occupancies in accordance with Section 503.1.2.
> 5. Buildings with an occupancy in Group H-1, H-2 or H-3 in accordance with Section 415.

[F] 907.2.12.1 Automatic fire detection. Smoke detectors shall be provided in accordance with this section. Smoke detectors shall be connected to an automatic fire alarm system. The activation of any detector required by this section shall operate the emergency voice/alarm communication system. Smoke detectors shall be located as follows:

1. In each mechanical equipment, electrical, transformer, telephone equipment or similar room which is not provided with sprinkler protection, elevator machine rooms and in elevator lobbies.

2. In the main return air and exhaust air plenum of each air-conditioning system having a capacity greater than 2,000 cubic feet per minute (cfm) (0.94 m³/s). Such detectors shall be located in a serviceable area downstream of the last duct inlet.

3. At each connection to a vertical duct or riser serving two or more stories from a return air duct or plenum of an air-conditioning system. In Group R-1 and R-2 occupancies a listed smoke detector is allowed to be used in each return air riser carrying not more than 5,000 cfm (2.4 m³/s) and serving not more than 10 air inlet openings.

[F] 907.2.12.2 Emergency voice/alarm communication system. The operation of any automatic fire detector, sprinkler water-flow device or manual fire alarm box shall automatically sound an alert tone followed by voice instructions giving approved information and directions on a general or selective basis to the following terminal areas on a minimum of the alarming floor, the floor above and the floor below in accordance with the *International Fire Code*.

1. Elevator lobbies.
2. Corridors.
3. Rooms and tenant spaces exceeding 1,000 square feet (93 m²) in area.

4. Dwelling units or sleeping units in Group R-2 occupancies.

5. Sleeping units in Group R-1 occupancies.

6. Areas of refuge as defined in Section 1002.

Exception: In Group I-1 and I-2 occupancies, the alarm shall sound in a constantly attended area and a general occupant notification shall be broadcast over the overhead page.

[F] 907.2.12.2.1 Manual override. A manual override for emergency voice communication shall be provided for all paging zones.

[F] 907.2.12.2.2 Live voice messages. The emergency voice/alarm communication system shall also have the capability to broadcast live voice messages through speakers located in elevators, exit stairways and throughout a selected floor or floors.

[F] 907.2.12.2.3 Standard. The emergency voice/alarm communication system shall be designed and installed in accordance with NFPA 72.

[F] 907.2.12.3 Fire department communication system. An approved two-way, fire department communication system designed and installed in accordance with NFPA 72 shall be provided for fire department use. It shall operate between a fire command center complying with Section 911 and elevators, elevator lobbies, emergency and standby power rooms, fire pump rooms, areas of refuge and inside enclosed exit stairways. The fire department communication device shall be provided at each floor level within the enclosed stairway.

Exception: Fire department radio systems where approved by the fire department.

[F] 907.2.13 Atriums connecting more than two stories. A fire alarm system shall be installed in occupancies with an atrium that connects more than two stories. The system shall be activated in accordance with Section 907.6. Such occupancies in Group A, E or M shall be provided with an emergency voice/alarm communication system complying with the requirements of Section 907.2.12.2.

[F] 907.2.14 High-piled combustible storage areas. An automatic fire detection system shall be installed throughout high-piled combustible storage areas where required by the *International Fire Code.*

[F] 907.2.15 Delayed egress locks. Where delayed egress locks are installed on means of egress doors in accordance with Section 1008.1.8.6, an automatic smoke or heat detection system shall be installed as required by that section.

[F] 907.2.16 Aerosol storage uses. Aerosol storage rooms and general-purpose warehouses containing aerosols shall be provided with an approved manual fire alarm system where required by the *International Fire Code.*

[F] 907.2.17 Lumber, plywood and veneer mills. Lumber, plywood and veneer mills shall be provided with a manual fire alarm system.

[F] 907.2.18 Underground buildings with smoke exhaust system. Where a smoke exhaust system is installed in an underground building in accordance with this code, automatic fire detectors shall be provided in accordance with this section.

[F] 907.2.18.1 Smoke detectors. A minimum of one smoke detector listed for the intended purpose shall be installed in the following areas:

1. Mechanical equipment, electrical, transformer, telephone equipment, elevator machine or similar rooms.

2. Elevator lobbies.

3. The main return and exhaust air plenum of each air-conditioning system serving more than one story and located in a serviceable area downstream of the last duct inlet.

4. Each connection to a vertical duct or riser serving two or more floors from return air ducts or plenums of heating, ventilating and air-conditioning systems, except that in Group R occupancies, a listed smoke detector is allowed to be used in each return air riser carrying not more than 5,000 cfm (2.4 m³/s) and serving not more than 10 air inlet openings.

[F] 907.2.18.2 Alarm required. Activation of the smoke exhaust system shall activate an audible alarm at a constantly attended location.

[F] 907.2.19 Underground buildings. Where the lowest level of a structure is more than 60 feet (18 288 mm) below the lowest level of exit discharge, the structure shall be equipped throughout with a manual fire alarm system, including an emergency voice/alarm communication system installed in accordance with Section 907.2.12.2.

[F] 907.2.19.1 Public address system. Where a fire alarm system is not required by Section 907.2, a public address system shall be provided that shall be capable of transmitting voice communications to the highest level of exit discharge serving the underground portions of the structure and all levels below.

[F] 907.2.20 Covered mall buildings. Covered mall buildings exceeding 50,000 square feet (4645 m²) in total floor area shall be provided with an emergency voice/alarm communication system. An emergency voice/alarm communication system serving a mall, required or otherwise, shall be accessible to the fire department. The system shall be provided in accordance with Section 907.2.12.2.

[F] 907.2.21 Residential aircraft hangars. A minimum of one listed smoke alarm shall be installed within a residential aircraft hangar as defined in Section 412.3.1 and shall be interconnected into the residential smoke alarm or other sounding device to provide an alarm that will be audible in all sleeping areas of the dwelling.

[F] 907.2.22 Airport traffic control towers. An automatic fire detection system shall be provided in airport traffic control towers.

[F] 907.2.23 Battery rooms. An approved automatic smoke detection system shall be installed in areas containing stationary lead-acid battery systems having a liquid capacity of more than 50 gallons (189.3 L). The detection system shall be supervised by an approved central, proprietary or remote station service or a local alarm that will sound an audible signal at a constantly attended location.

[F] 907.3 Manual fire alarm boxes. Manual fire alarm boxes shall be installed in accordance with Sections 907.3.1 through 907.3.5.

[F] 907.3.1 Location. Manual fire alarm boxes shall be located not more than 5 feet (1524 mm) from the entrance to each exit. Additional manual fire alarm boxes shall be located so that travel distance to the nearest box does not exceed 200 feet (60 960 mm).

> **Exception:** Manual fire alarm boxes shall not be required in Group E occupancies where the building is equipped throughout with an approved automatic sprinkler system, the notification appliances will activate on sprinkler water flow and manual activation is provided from a normally occupied location.

[F] 907.3.2 Height. The height of the manual fire alarm boxes shall be a minimum of 42 inches (1067 mm) and a maximum of 48 inches (1219 mm), measured vertically, from the floor level to the activating handle or lever of the box.

[F] 907.3.3 Color. Manual fire alarm boxes shall be red in color.

[F] 907.3.4 Signs. Where fire alarm systems are not monitored by a supervising station, an approved permanent sign that reads: WHEN ALARM SOUNDS—CALL FIRE DEPARTMENT shall be installed adjacent to each manual fire alarm box.

> **Exception:** Where the manufacturer has permanently provided this information on the manual fire alarm box.

[F] 907.3.5 Protective covers. The building official is authorized to require the installation of listed manual fire alarm box protective covers to prevent malicious false alarms or provide the manual fire alarm box with protection from physical damage. The protective cover shall be transparent or red in color with a transparent face to permit visibility of the manual fire alarm box. Each cover shall include proper operating instructions. A protective cover that emits a local alarm signal shall not be installed unless approved.

[F] 907.4 Power supply. The primary and secondary power supplies for the fire alarm system shall be provided in accordance with NFPA 72.

[F] 907.5 Wiring. Wiring shall comply with the requirements of the ICC *Electrical Code* and NFPA 72. Wireless protection systems utilizing radio-frequency transmitting devices shall comply with the special requirements for supervision of low-power wireless systems in NFPA 72.

[F] 907.6 Activation. Where an alarm notification system is required by another section of this code, it shall be activated by:

1. A required automatic fire alarm system.
2. Sprinkler water-flow devices.
3. Required manual fire alarm boxes.

[F] 907.7 Presignal system. Presignal systems shall not be installed unless approved by the building official and the fire department. Where a presignal system is installed, 24-hour personnel supervision shall be provided at a location approved by the fire department, in order that the alarm signal can be actuated in the event of fire or other emergency.

[F] 907.8 Zones. Each floor shall be zoned separately and a zone shall not exceed 22,500 square feet (2090 m²). The length of any zone shall not exceed 300 feet (91 440 mm) in any direction.

> **Exception:** Automatic sprinkler system zones shall not exceed the area permitted by NFPA 13.

[F] 907.8.1 Zoning indicator panel. A zoning indicator panel and the associated controls shall be provided in an approved location. The visual zone indication shall lock in until the system is reset and shall not be canceled by the operation of an audible alarm-silencing switch.

[F] 907.8.2 High-rise buildings. In buildings used for human occupancy that have floors located more than 75 feet (22 860 mm) above the lowest level of fire department vehicle access, a separate zone by floor shall be provided for all of the following types of alarm-initiating devices where provided:

1. Smoke detectors.
2. Sprinkler water-flow devices.
3. Manual fire alarm boxes.
4. Other approved types of automatic fire detection devices or suppression systems.

[F] 907.9 Alarm notification appliances. Alarm notification appliances shall be provided and shall be listed for their purpose.

[F] 907.9.1 Visible alarms. Visible alarm notification appliances shall be provided in accordance with Sections 907.9.1.1 through 907.9.1.3.

> **Exceptions:**
> 1. Visible alarm notification appliances are not required in alterations, except where an existing fire alarm system is upgraded or replaced, or a new fire alarm system is installed.
> 2. Visible alarm notification appliances shall not be required in exits as defined in Section 1002.1.

[F] 907.9.1.1 Public and common areas. Visible alarm notification appliances shall be provided in public areas and common areas.

[F] 907.9.1.2 Employee work areas. Where employee work areas have audible alarm coverage, the wiring systems shall be designed so that visible alarm notification appliances can be integrated into the alarm system.

[F] 907.9.1.3 Groups I-1 and R-1. Group I-1 and R-1 sleeping units in accordance with Table 907.9.1.3 shall be provided with a visible alarm notification appliance, activated by both the in-room smoke alarm and the building fire alarm system.

[F] 907.9.1.4 Group R-2. In Group R-2 occupancies required by Section 907 to have a fire alarm system, all dwelling units and sleeping units shall be provided with the capability to support visible alarm notification appliances in accordance with ICC A117.1.

[F] 907.9.2 Audible alarms. Audible alarm notification appliances shall be provided and shall sound a distinctive sound that is not to be used for any purpose other than that of a fire alarm. The audible alarm notification appliances shall provide a sound pressure level of 15 decibels (dBA) above the average ambient sound level or 5 dBA above the maximum sound level having a duration of at least 60 seconds, whichever is greater, in every occupied space within the building. The minimum sound pressure levels shall be: 70 dBA in occupancies in Groups R and I-1; 90 dBA in mechanical equipment rooms and 60 dBA in other occupancies. The maximum sound pressure level for audible alarm notification appliances shall be 120 dBA at the minimum hearing distance from the audible appliance. Where the average ambient noise is greater than 105 dBA, visible alarm notification appliances shall be provided in accordance with NFPA 72 and audible alarm notification appliances shall not be required.

Exception: Visible alarm notification appliances shall be allowed in lieu of audible alarm notification appliances in critical-care areas of Group I-2 occupancies.

[F] TABLE 907.9.1.3
VISIBLE AND AUDIBLE ALARMS

NUMBER OF SLEEPING UNITS	SLEEPING UNITS WITH VISIBLE AND AUDIBLE ALARMS
6 to 25	2
26 to 50	4
51 to 75	7
76 to 100	9
101 to 150	12
151 to 200	14
201 to 300	17
301 to 400	20
401 to 500	22
501 to 1,000	5% of total
1,001 and over	50 plus 3 for each 100 over 1,000

[F] 907.10 Fire safety functions. Automatic fire detectors utilized for the purpose of performing fire safety functions shall be connected to the building's fire alarm control panel where a fire alarm system is required by Section 907.2. Detectors shall, upon actuation, perform the intended function and activate the alarm notification appliances or a visible and audible supervisory signal at a constantly attended location. In buildings not required to be equipped with a fire alarm system, the automatic fire detector shall be powered by normal electrical service and, upon actuation, perform the intended function. The detectors shall be located in accordance with NFPA 72.

[F] 907.11 Duct smoke detectors. Duct smoke detectors shall be connected to the building's fire alarm control panel when a fire alarm system is provided. Activation of a duct smoke detector shall initiate a visible and audible supervisory signal at a constantly attended location. Duct smoke detectors shall not be used as a substitute for required open-area detection.

Exceptions:
1. The supervisory signal at a constantly attended location is not required where duct smoke detectors activate the building's alarm notification appliances.
2. In occupancies not required to be equipped with a fire alarm system, actuation of a smoke detector shall activate a visible and audible signal in an approved location. Smoke detector trouble conditions shall activate a visible or audible signal in an approved location and shall be identified as air duct detector trouble.

[F] 907.12 Access. Access shall be provided to each detector for periodic inspection, maintenance and testing.

[F] 907.13 Fire-extinguishing systems. Automatic fire-extinguishing systems shall be connected to the building fire alarm system where a fire alarm system is required by another section of this code or is otherwise installed.

[F] 907.14 Monitoring. Where required by this chapter or the *International Fire Code*, an approved supervising station in accordance with NFPA 72 shall monitor fire alarm systems.

Exception: Supervisory service is not required for:
1. Single- and multiple-station smoke alarms required by Section 907.2.10.
2. Smoke detectors in Group I-3 occupancies.
3. Automatic sprinkler systems in one- and two-family dwellings.

[F] 907.15 Automatic telephone-dialing devices. Automatic telephone-dialing devices used to transmit an emergency alarm shall not be connected to any fire department telephone number unless approved by the fire chief.

[F] 907.16 Acceptance tests. Upon completion of the installation of the fire alarm system, alarm notification appliances and circuits, alarm-initiating devices and circuits, supervisory-signal initiating devices and circuits, signaling line circuits, and primary and secondary power supplies shall be tested in accordance with NFPA 72.

[F] 907.17 Record of completion. A record of completion in accordance with NFPA 72 verifying that the system has been installed in accordance with the approved plans and specifications shall be provided.

[F] 907.18 Instructions. Operating, testing and maintenance instructions, and record drawings ("as builts") and equipment specifications shall be provided at an approved location.

[F] 907.19 Inspection, testing and maintenance. The maintenance and testing schedules and procedures for fire alarm and fire detection systems shall be in accordance with the *International Fire Code*.

SECTION 908
EMERGENCY ALARM SYSTEMS

[F] 908.1 Group H occupancies. Emergency alarms for the detection and notification of an emergency condition in Group

H occupancies shall be provided in accordance with Section 414.7.

[F] 908.2 Group H-5 occupancy. Emergency alarms for notification of an emergency condition in an HPM facility shall be provided as required in Section 415.9.4.6. A continuous gas-detection system shall be provided for HPM gases in accordance with Section 415.9.7.

[F] 908.3 Highly toxic and toxic materials. A gas detection system shall be provided for indoor storage and use of highly toxic and toxic gases to detect the presence of gas at or below the permissible exposure limit (PEL) or ceiling limit of the gas for which detection is provided. The system shall be capable of monitoring the discharge from the treatment system at or below one-half the IDLH limit.

Exception: A gas detection system is not required for toxic gases when the physiological warning properties are at a level below the accepted PEL for the gas.

[F] 908.3.1 Alarms. The gas detection system shall initiate a local alarm and transmit a signal to a constantly attended control station when a short-term hazard condition is detected. The alarm shall be both visible and audible and shall provide warning both inside and outside the area where gas is detected. The audible alarm shall be distinct from all other alarms.

Exception: Signal transmission to a constantly attended control station is not required when not more than one cylinder of highly toxic or toxic gas is stored.

[F] 908.3.2 Shutoff of gas supply. The gas detection system shall automatically close the shutoff valve at the source on gas supply piping and tubing related to the system being monitored for whichever gas is detected.

Exception: Automatic shutdown is not required for reactors utilized for the production of highly toxic or toxic compressed gases where such reactors are:

1. Operated at pressures less than 15 pounds per square inch gauge (psig) (103.4 kPa).
2. Constantly attended.
3. Provided with readily accessible emergency shut-off valves.

[F] 908.3.3 Valve closure. The automatic closure of shutoff valves shall be in accordance with the following:

1. When the gas-detection sampling point initiating the gas detection system alarm is within a gas cabinet or exhausted enclosure, the shutoff valve in the gas cabinet or exhausted enclosure for the specific gas detected shall automatically close.
2. Where the gas-detection sampling point initiating the gas detection system alarm is within a gas room and compressed gas containers are not in gas cabinets or exhausted enclosures, the shutoff valves on all gas lines for the specific gas detected shall automatically close.
3. Where the gas-detection sampling point initiating the gas detection system alarm is within a piping distribu-

tion manifold enclosure, the shutoff valve for the compressed container of specific gas detected supplying the manifold shall automatically close.

Exception: When the gas-detection sampling point initiating the gas-detection system alarm is at a use location or within a gas valve enclosure of a branch line downstream of a piping distribution manifold, the shutoff valve in the gas valve enclosure for the branch line located in the piping distribution manifold enclosure shall automatically close.

[F] 908.4 Ozone gas-generator rooms. Ozone gas-generator rooms shall be equipped with a continuous gas-detection system that will shut off the generator and sound a local alarm when concentrations above the PEL occur.

[F] 908.5 Repair garages. A flammable-gas detection system shall be provided in repair garages for vehicles fueled by nonodorized gases in accordance with Section 406.6.6.

[F] 908.6 Refrigerant detector. Machinery rooms shall contain a refrigerant detector with an audible and visual alarm. The detector, or a sampling tube that draws air to the detector, shall be located in an area where refrigerant from a leak will concentrate. The alarm shall be actuated at a value not greater than the corresponding TLV-TWA values for the refrigerant classification indicated in the *International Mechanical Code*. Detectors and alarms shall be placed in approved locations.

Exception: Detectors are not required in ammonia system machinery rooms equipped with a vapor detector in accordance with the *International Mechanical Code*.

SECTION 909
SMOKE CONTROL SYSTEMS

909.1 Scope and purpose. This section applies to mechanical or passive smoke control systems when they are required by other provisions of this code. The purpose of this section is to establish minimum requirements for the design, installation and acceptance testing of smoke control systems that are intended to provide a tenable environment for the evacuation or relocation of occupants. These provisions are not intended for the preservation of contents, the timely restoration of operations or for assistance in fire suppression or overhaul activities. Smoke control systems regulated by this section serve a different purpose than the smoke- and heat-venting provisions found in Section 910. Mechanical smoke control systems shall not be considered exhaust systems under Chapter 5 of the *International Mechanical Code*.

909.2 General design requirements. Buildings, structures or parts thereof required by this code to have a smoke control system or systems shall have such systems designed in accordance with the applicable requirements of Section 909 and the generally accepted and well-established principles of engineering relevant to the design. The construction documents shall include sufficient information and detail to adequately describe the elements of the design necessary for the proper implementation of the smoke control systems. These documents shall be accompanied by sufficient information and analysis to demonstrate compliance with these provisions.

909.3 Special inspection and test requirements. In addition to the ordinary inspection and test requirements which buildings, structures and parts thereof are required to undergo, smoke control systems subject to the provisions of Section 909 shall undergo special inspections and tests sufficient to verify the proper commissioning of the smoke control design in its final installed condition. The design submission accompanying the construction documents shall clearly detail procedures and methods to be used and the items subject to such inspections and tests. Such commissioning shall be in accordance with generally accepted engineering practice and, where possible, based on published standards for the particular testing involved. The special inspections and tests required by this section shall be conducted under the same terms in Section 1704.

909.4 Analysis. A rational analysis supporting the types of smoke control systems to be employed, their methods of operation, the systems supporting them and the methods of construction to be utilized shall accompany the submitted construction documents and shall include, but not be limited to, the items indicated in Sections 909.4.1 through 909.4.6.

909.4.1 Stack effect. The system shall be designed such that the maximum probable normal or reverse stack effect will not adversely interfere with the system's capabilities. In determining the maximum probable stack effect, altitude, elevation, weather history and interior temperatures shall be used.

909.4.2 Temperature effect of fire. Buoyancy and expansion caused by the design fire in accordance with Section 909.9 shall be analyzed. The system shall be designed such that these effects do not adversely interfere with the system's capabilities.

909.4.3 Wind effect. The design shall consider the adverse effects of wind. Such consideration shall be consistent with the wind-loading provisions of Chapter 16.

909.4.4 HVAC systems. The design shall consider the effects of the heating, ventilating and air-conditioning (HVAC) systems on both smoke and fire transport. The analysis shall include all permutations of systems status. The design shall consider the effects of the fire on the HVAC systems.

909.4.5 Climate. The design shall consider the effects of low temperatures on systems, property and occupants. Air inlets and exhausts shall be located so as to prevent snow or ice blockage.

909.4.6 Duration of operation. All portions of active or passive smoke control systems shall be capable of continued operation after detection of the fire event for not less than 20 minutes.

909.5 Smoke barrier construction. Smoke barriers shall comply with Section 709, and shall be constructed and sealed to limit leakage areas exclusive of protected openings. The maximum allowable leakage area shall be the aggregate area calculated using the following leakage area ratios:

1. Walls: $A/A_w = 0.00100$
2. Exit enclosures: $A/A_w = 0.00035$
3. All other shafts: $A/A_w = 0.00150$

4. Floors and roofs: $A/A_F = 0.00050$

where:

A = Total leakage area, square feet (m^2).

A_F = Unit floor or roof area of barrier, square feet (m^2).

A_w = Unit wall area of barrier, square feet (m^2).

The leakage area ratios shown do not include openings due to doors, operable windows or similar gaps. These shall be included in calculating the total leakage area.

909.5.1 Leakage area. The total leakage area of the barrier is the product of the smoke barrier gross area monitored by the allowable leakage area ratio, plus the area of other openings such as gaps and operable windows. Compliance shall be determined by achieving the minimum air pressure difference across the barrier with the system in the smoke control mode for mechanical smoke control systems. Passive smoke control systems tested using other approved means such as door fan testing shall be as approved by the building official.

909.5.2 Opening protection. Openings in smoke barriers shall be protected by automatic-closing devices actuated by the required controls for the mechanical smoke control system. Door openings shall be protected by fire door assemblies complying with Section 715.3.3.

Exceptions:

1. Passive smoke control systems with automatic-closing devices actuated by spot-type smoke detectors listed for releasing service installed in accordance with Section 907.10.

2. Fixed openings between smoke zones which are protected utilizing the airflow method.

3. In Group I-2, where such doors are installed across corridors, a pair of opposite-swinging doors without a center mullion shall be installed having vision panels with approved fire-rated glazing materials in approved fire-rated frames, the area of which shall not exceed that tested. The doors shall be close fitting within operational tolerances and shall not have undercuts, louvers or grilles. The doors shall have head and jamb stops, astragals or rabbets at meeting edges, and automatic-closing devices. Positive-latching devices are not required.

4. Group I-3.

5. Openings between smoke zones with clear ceiling heights of 14 feet (4267 mm) or greater and bank-down capacity of greater than 20 minutes as determined by the design fire size.

909.5.2.1 Ducts and air transfer openings. Ducts and air transfer openings are required to be protected with a minimum Class II, 250°F (121°C) smoke damper complying with Section 716.

909.6 Pressurization method. The primary mechanical means of controlling smoke shall be by pressure differences across smoke barriers. Maintenance of a tenable environment is not required in the smoke control zone of fire origin.

909.6.1 Minimum pressure difference. The minimum pressure difference across a smoke barrier shall be 0.05-inch water gage (0.0124 kPa) in fully sprinklered buildings. In buildings permitted to be other than fully sprinklered, the smoke control system shall be designed to achieve pressure differences at least two times the maximum calculated pressure difference produced by the design fire.

909.6.2 Maximum pressure difference. The maximum air pressure difference across a smoke barrier shall be determined by required door-opening or closing forces. The actual force required to open exit doors when the system is in the smoke control mode shall be in accordance with Section 1008.1.2. Opening and closing forces for other doors shall be determined by standard engineering methods for the resolution of forces and reactions. The calculated force to set a side-hinged, swinging door in motion shall be determined by:

$$F = F_{dc} + K(WA\Delta P)/2(W - d) \qquad \textbf{(Equation 9-1)}$$

where:

A = Door area, square feet (m²).

d = Distance from door handle to latch edge of door, feet (m).

F = Total door opening force, pounds (N).

F_{dc} = Force required to overcome closing device, pounds (N).

K = Coefficient 5.2 (1.0).

W = Door width, feet (m).

ΔP = Design pressure difference, inches of water (Pa).

909.7 Airflow design method. When approved by the building official, smoke migration through openings fixed in a permanently open position, which are located between smoke control zones by the use of the airflow method, shall be permitted. The design airflow shall be in accordance with this section. Airflow shall be directed to limit smoke migration from the fire zone. The geometry of openings shall be considered to prevent flow reversal from turbulent effects.

909.7.1 Velocity. The minimum average velocity through a fixed opening shall not be less than:

$$v = 217.2 \, [h \, (T_f - T_o)/(T_f + 460)]^{1/2} \qquad \textbf{(Equation 9-2)}$$

For SI: $v = 119.9 \, [h \, (T_f - T_o)/T_f]^{1/2}$

where:

h = Height of opening, feet (m).

T_f = Temperature of smoke, °F (°K).

T_o = Temperature of ambient air, °F (°K).

v = Air velocity, feet per minute (m/minute).

909.7.2 Prohibited conditions. This method shall not be employed where either the quantity of air or the velocity of the airflow will adversely affect other portions of the smoke control system, unduly intensify the fire, disrupt plume dynamics or interfere with exiting. In no case shall airflow to-

ward the fire exceed 200 feet per minute (1.02 m/s). Where the formula in Section 909.7.1 requires airflow to exceed this limit, the airflow method shall not be used.

909.8 Exhaust method. When approved by the building official, mechanical smoke control for large enclosed volumes, such as in atriums or malls, shall be permitted to utilize the exhaust method. The design exhaust volumes shall be in accordance with this section.

909.8.1 Exhaust rate. The height of the lowest horizontal surface of the accumulating smoke layer shall be maintained at least 10 feet (3048 mm) above any walking surface which forms a portion of a required egress system within the smoke zone. The required exhaust rate for the zone shall be the largest of the calculated plume mass flow rates for the possible plume configurations. Provisions shall be made for natural or mechanical supply of air from outside or adjacent smoke zones to make up for the air exhausted. Makeup airflow rates, when measured at the potential fire location, shall not exceed 200 feet per minute (60 960 mm per minute) toward the fire. The temperature of the makeup air shall be such that it does not expose temperature-sensitive fire protection systems beyond their limits.

909.8.2 Axisymmetric plumes. The plume mass flow rate (m_p), in pounds per second (kg/s), shall be determined by placing the design fire center on the axis of the space being analyzed. The limiting flame height shall be determined by:

$$z_l = 0.533 Q_c^{2/5} \qquad \textbf{(Equation 9-3)}$$

For SI: $z_l = 0.166 Q_c^{2/5}$

where:

m_p = Plume mass flow rate, pounds per second (kg/s).

Q = Total heat output.

Q_c = Convective heat output, British thermal units per second (kW). (The value of Q_c shall not be taken as less than $0.70Q$).

z = Height from top of fuel surface to bottom of smoke layer, feet (m).

z_l = Limiting flame height, feet (m). The z_l value must be greater than the fuel equivalent diameter (see Section 909.9).

for $z > z_l$

$m_p = 0.022 Q_c^{1/3} z^{5/3} + 0.0042 Q_c$

For SI: $m_p = 0.071 \, Q_c^{1/3} z^{5/3} + 0.0018 Q_c$

for $z = z_l$

$m_p = 0.011 \, Q_c$

For SI: $m_p = 0.035 Q_c$

for $z < z_l$

$m_p = 0.0208 Q_c^{3/5} z$

For SI: $m_p = 0.032 Q_c^{3/5} z$

To convert m_p from pounds per second of mass flow to a volumetric rate, the following equation shall be used:

$$V = 60 \, m_p / \rho \qquad \textbf{(Equation 9-4)}$$

where:

V = Volumetric flow rate, cubic feet per minute (m³/s).

ρ = Density of air at the temperature of the smoke layer, pounds per cubic feet (T: in °F) [kg/m³ (T: in °C)].

909.8.3 Balcony spill plumes. The plume mass flow rate (m_p) for spill plumes shall be determined using the geometrically probable width based on architectural elements and projections in the following equation:

$$m_p = 0.124(QW^2)^{1/3}(z_b + 0.25H) \quad \textbf{(Equation 9-5)}$$

For SI: $m_p = 0.36(QW^2)^{1/3}(z_b + 0.25H)$

where:

H = Height above fire to underside of balcony, feet (m).

m_p = Plume mass flow rate, pounds per second (kg/s).

Q = Total heat output.

W = Plume width at point of spill, feet (m).

z_b = Height from balcony, feet (m).

909.8.4 Window plumes. The plume mass flow rate (m_p) shall be determined from:

$$m_p = 0.077(A_wH_w^{1/2})^{1/3}(z_w+a)^{5/3} + 0.18A_wH_w^{1/2} \quad \textbf{(Equation 9-6)}$$

For SI: $m_p = 0.68(A_wH_w^{1/2})^{1/3}(z_w + a)^{5/3} + 1.5A_wH_w^{1/2}$

where:

A_w = Area of the opening, square feet (m²).

H_w = Height of the opening, feet (m).

m_p = plume mass flow rate, pounds per second (kg/s).

z_w = Height from the top of the window or opening to the bottom of the smoke layer, feet (m).

a = $2.4A_w^{2/5}H_w^{1/5} - 2.1H_w$.

909.8.5 Plume contact with walls. When a plume contacts one or more of the surrounding walls, the mass flow rate shall be adjusted for the reduced entrainment resulting from the contact provided that the contact remains constant. Use of this provision requires calculation of the plume diameter, that shall be calculated by:

$$d = 0.48 [(T_c+460)/(T_a+460)]^{1/2}z \quad \textbf{(Equation 9-7)}$$

For SI: $d = 0.48 (T_c/T_a)^{1/2}z$

where:

d = Plume diameter, feet (m).

T_a = Ambient air temperature, °F (°K).

T_c = Plume centerline temperature, °F (°K).

= $0.60 (T_a + 460) Q_c^{2/3} z^{-5/3} + T_a$

z = Height at which T_c is determined, feet (m).

For SI: $T_c = 0.08 T_a Q_c^{2/3} z^{-5/3} + T_a$

909.9 Design fire. The design fire shall be based on a Q of not less than 5,000 Btu/s (5275 kW) unless a rational analysis is performed by the registered design professional and approved by the building official. The design fire shall be based on the analysis in accordance with Section 909.4 and this section.

909.9.1 Factors considered. The engineering analysis shall include the characteristics of the fuel, fuel load, effects included by the fire and whether the fire is likely to be steady or unsteady.

909.9.2 Separation distance. Determination of the design fire shall include consideration of the type of fuel, fuel spacing and configuration. The ratio of the separation distance to the fuel equivalent radius shall not be less than 4. The fuel equivalent radius shall be the radius of a circle of equal area to floor area of the fuel package. The design fire shall be increased if other combustibles are within the separation distance as determined by:

$$R = [Q/(12\pi q'')]^{1/2} \quad \textbf{(Equation 9-8)}$$

where:

q'' = Incident radiant heat flux required for nonpiloted ignition, Btu/ft² · s (W/m²).

Q = Heat release from fire, Btu/s (kW).

R = Separation distance from target to center of fuel package, feet (m).

909.9.3 Heat-release assumptions. The analysis shall make use of best available data from approved sources and shall not be based on excessively stringent limitations of combustible material.

909.9.4 Sprinkler effectiveness assumptions. A documented engineering analysis shall be provided for conditions that assume fire growth is halted at the time of sprinkler activation.

909.10 Equipment. Equipment such as, but not limited to, fans, ducts, automatic dampers and balance dampers, shall be suitable for its intended use, suitable for the probable exposure temperatures that the rational analysis indicates, and as approved by the building official.

909.10.1 Exhaust fans. Components of exhaust fans shall be rated and certified by the manufacturer for the probable temperature rise to which the components will be exposed. This temperature rise shall be computed by:

$$T_s = (Q_c/mc) + (T_a) \quad \textbf{(Equation 9-9)}$$

where:

c = Specific heat of smoke at smoke layer temperature, Btu/lb°F (kJ/kg · K).

m = Exhaust rate, pounds per second (kg/s).

Q_c = Convective heat output of fire, Btu/s (kW).

T_a = Ambient temperature, °F (°K).

T_s = Smoke temperature, °F (°K).

Exception: Reduced T_s as calculated based on the assurance of adequate dilution air.

909.10.2 Ducts. Duct materials and joints shall be capable of withstanding the probable temperatures and pressures to which they are exposed as determined in accordance with Section 909.10.1. Ducts shall be constructed and supported in accordance with the *International Mechanical Code.* Ducts shall be leak tested to 1.5 times the maximum design pressure in accordance with nationally accepted practices. Measured leakage shall not exceed 5 percent of design flow. Results of such testing shall be a part of the documentation procedure. Ducts shall be supported directly from fire-resistance-rated structural elements of the building by substantial, noncombustible supports.

> **Exception:** Flexible connections (for the purpose of vibration isolation) complying with the *International Mechanical Code*, that are constructed of approved fire-resistance-rated materials.

909.10.3 Equipment, inlets and outlets. Equipment shall be located so as to not expose uninvolved portions of the building to an additional fire hazard. Outside air inlets shall be located so as to minimize the potential for introducing smoke or flame into the building. Exhaust outlets shall be so located as to minimize reintroduction of smoke into the building and to limit exposure of the building or adjacent buildings to an additional fire hazard.

909.10.4 Automatic dampers. Automatic dampers, regardless of the purpose for which they are installed within the smoke control system, shall be listed and conform to the requirements of approved, recognized standards.

909.10.5 Fans. In addition to other requirements, belt-driven fans shall have 1.5 times the number of belts required for the design duty, with the minimum number of belts being two. Fans shall be selected for stable performance based on normal temperature and, where applicable, elevated temperature. Calculations and manufacturer's fan curves shall be part of the documentation procedures. Fans shall be supported and restrained by noncombustible devices in accordance with the requirements of Chapter 16. Motors driving fans shall not be operated beyond their nameplate horsepower (kilowatts), as determined from measurement of actual current draw, and shall have a minimum service factor of 1.15.

909.11 Power systems. The smoke control system shall be supplied with two sources of power. Primary power shall be the normal building power systems. Secondary power shall be from an approved standby source complying with the ICC *Electrical Code.* The standby power source and its transfer switches shall be in a separate room from the normal power transformers and switch gear and shall be enclosed in a room constructed of not less than 1-hour fire-resistance-rated fire barriers ventilated directly to and from the exterior. Power distribution from the two sources shall be by independent routes. Transfer to full standby power shall be automatic and within 60 seconds of failure of the primary power. The systems shall comply with the ICC *Electrical Code.*

909.11.1 Power sources and power surges. Elements of the smoke management system relying on volatile memories or the like shall be supplied with uninterruptable power sources of sufficient duration to span a 15-minute primary power interruption. Elements of the smoke management system susceptible to power surges shall be suitably protected by conditioners, suppressors or other approved means.

909.12 Detection and control systems. Fire detection systems providing control input or output signals to mechanical smoke control systems or elements thereof shall comply with the requirements of Section 907. Such systems shall be equipped with a control unit complying with UL 864 and listed as smoke control equipment.

Control systems for mechanical smoke control systems shall include provisions for verification. Verification shall include positive confirmation of actuation, testing, manual override, the presence of power downstream of all disconnects and, through a preprogrammed weekly test sequence report, abnormal conditions audibly, visually and by printed report.

909.12.1 Wiring. In addition to meeting requirements of the ICC *Electrical Code*, all wiring, regardless of voltage, shall be fully enclosed within continuous raceways.

[F] 909.12.2 Activation. Smoke control systems shall be activated in accordance with this section.

[F] 909.12.2.1 Pressurization, airflow or exhaust method. Mechanical smoke control systems using the pressurization, airflow or exhaust method shall have completely automatic control.

[F] 909.12.2.2 Passive method. Passive smoke control systems actuated by approved spot-type detectors listed for releasing service shall be permitted.

[F] 909.12.3 Automatic control. Where completely automatic control is required or used, the automatic-control sequences shall be initiated from an appropriately zoned automatic sprinkler system complying with Section 903.3.1.1, manual controls that are readily accessible to the fire department and any smoke detectors required by engineering analysis.

909.13 Control air tubing. Control air tubing shall be of sufficient size to meet the required response times. Tubing shall be flushed clean and dry prior to final connections and shall be adequately supported and protected from damage. Tubing passing through concrete or masonry shall be sleeved and protected from abrasion and electrolytic action.

909.13.1 Materials. Control air tubing shall be hard drawn copper, Type L, ACR in accordance with ASTM B 42, ASTM B 43, ASTM B 68, ASTM B 88, ASTM B 251 and ASTM B 280. Fittings shall be wrought copper or brass, solder type, in accordance with ASME B 16.18 or ASME B 16.22. Changes in direction shall be made with appropriate tool bends. Brass compression-type fittings shall be used at final connection to devices; other joints shall be brazed using a BCuP5 brazing alloy with solidus above 1,100°F (593°C) and liquids below 1,500°F (816°C). Brazing flux shall be used on copper-to-brass joints only.

> **Exception:** Nonmetallic tubing used within control panels and at the final connection to devices, provided that all of the following conditions are met:
>
> 1. Tubing shall be listed by an approved agency for flame and smoke characteristics.

2. Tubing and connected devices shall be completely enclosed within galvanized or paint-grade steel enclosure of not less than 0.030 inch (0.76 mm) (No. 22 galvanized sheet gage) thickness. Entry to the enclosure shall be by copper tubing with a protective grommet of neoprene or teflon or by suitable brass compression to male-barbed adapter.

3. Tubing shall be identified by appropriately documented coding.

4. Tubing shall be neatly tied and supported within enclosure. Tubing bridging cabinet and door or moveable device shall be of sufficient length to avoid tension and excessive stress. Tubing shall be protected against abrasion. Tubing serving devices on doors shall be fastened along hinges.

909.13.2 Isolation from other functions. Control tubing serving other than smoke control functions shall be isolated by automatic isolation valves or shall be an independent system.

909.13.3 Testing. Control air tubing shall be tested at three times the operating pressure for not less than 30 minutes without any noticeable loss in gauge pressure prior to final connection to devices.

909.14 Marking and identification. The detection and control systems shall be clearly marked at all junctions, accesses and terminations.

[F] 909.15 Control diagrams. Identical control diagrams showing all devices in the system and identifying their location and function shall be maintained current and kept on file with the building official, the fire department and in the fire command center in format and manner approved by the fire chief.

[F] 909.16 Fire-fighter's smoke control panel. A fire-fighter's smoke control panel for fire department emergency response purposes only shall be provided and shall include manual control or override of automatic control for mechanical smoke control systems. The panel shall be located in a fire command center complying with Section 911, and shall comply with Sections 909.16.1 through 909.16.3.

[F] 909.16.1 Smoke control systems. Fans within the building shall be shown on the fire-fighter's control panel. A clear indication of the direction of airflow and the relationship of components shall be displayed. Status indicators shall be provided for all smoke control equipment, annunciated by fan and zone, and by pilot-lamp-type indicators as follows:

1. Fans, dampers and other operating equipment in their normal status—WHITE.

2. Fans, dampers and other operating equipment in their off or closed status—RED.

3. Fans, dampers and other operating equipment in their on or open status—GREEN.

4. Fans, dampers and other operating equipment in a fault status—YELLOW/AMBER.

[F] 909.16.2 Smoke control panel. The fire-fighter's control panel shall provide control capability over the complete smoke-control system equipment within the building as follows:

1. ON-AUTO-OFF control over each individual piece of operating smoke control equipment that can also be controlled from other sources within the building. This includes stairway pressurization fans; smoke exhaust fans; supply, return and exhaust fans; elevator shaft fans and other operating equipment used or intended for smoke control purposes.

2. OPEN-AUTO-CLOSE control over individual dampers relating to smoke control and that are also controlled from other sources within the building.

3. ON-OFF or OPEN-CLOSE control over smoke control and other critical equipment associated with a fire or smoke emergency and that can only be controlled from the fire-fighter's control panel.

Exceptions:

1. Complex systems, where approved, where the controls and indicators are combined to control and indicate all elements of a single smoke zone as a unit.

2. Complex systems, where approved, where the control is accomplished by computer interface using approved, plain English commands.

[F] 909.16.3 Control action and priorities. The fire-fighter's control panel actions shall be as follows:

1. ON-OFF, OPEN-CLOSE control actions shall have the highest priority of any control point within the building. Once issued from the fire-fighter's control panel, no automatic or manual control from any other control point within the building shall contradict the control action. Where automatic means are provided to interrupt normal, nonemergency equipment operation or produce a specific result to safeguard the building or equipment (i.e., duct freezestats, duct smoke detectors, high-temperature cutouts, temperature-actuated linkage and similar devices), such means shall be capable of being overridden by the fire-fighter's control panel. The last control action as indicated by each fire-fighter's control panel switch position shall prevail. In no case shall control actions require the smoke control system to assume more than one configuration at any one time.

   Exception: Power disconnects required by the ICC *Electrical Code*.

2. Only the AUTO position of each three-position fire-fighter's control panel switch shall allow automatic or manual control action from other control points within the building. The AUTO position shall be the NORMAL, nonemergency, building control position. Where a fire-fighter's control panel is in the AUTO position, the actual status of the device (on, off, open, closed) shall continue to be indicated by the status indicator described above. When directed by an automatic signal to assume an emergency condition, the NORMAL position shall become the emergency condition for that device or group of devices within the zone. In no case shall control actions require the

smoke control system to assume more than one configuration at any one time.

[F] 909.17 System response time. Smoke-control system activation shall be initiated immediately after receipt of an appropriate automatic or manual activation command. Smoke control systems shall activate individual components (such as dampers and fans) in the sequence necessary to prevent physical damage to the fans, dampers, ducts and other equipment. For purposes of smoke control, the fire-fighter's control panel response time shall be the same for automatic or manual smoke control action initiated from any other building control point. The total response time, including that necessary for detection, shutdown of operating equipment and smoke control system startup, shall allow for full operational mode to be achieved before the conditions in the space exceed the design smoke condition. The system response time for each component and their sequential relationships shall be detailed in the required rational analysis and verification of their installed condition reported in the required final report.

[F] 909.18 Acceptance testing. Devices, equipment, components and sequences shall be individually tested. These tests, in addition to those required by other provisions of this code, shall consist of determination of function, sequence and, where applicable, capacity of their installed condition.

[F] 909.18.1 Detection devices. Smoke or fire detectors that are a part of a smoke control system shall be tested in accordance with Chapter 9 in their installed condition. When applicable, this testing shall include verification of airflow in both minimum and maximum conditions.

[F] 909.18.2 Ducts. Ducts that are part of a smoke control system shall be traversed using generally accepted practices to determine actual air quantities.

[F] 909.18.3 Dampers. Dampers shall be tested for function in their installed condition.

[F] 909.18.4 Inlets and outlets. Inlets and outlets shall be read using generally accepted practices to determine air quantities.

[F] 909.18.5 Fans. Fans shall be examined for correct rotation. Measurements of voltage, amperage, revolutions per minute (rpm) and belt tension shall be made.

[F] 909.18.6 Smoke barriers. Measurements using inclined manometers or other approved calibrated measuring devices shall be made of the pressure differences across smoke barriers. Such measurements shall be conducted for each possible smoke control condition.

[F] 909.18.7 Controls. Each smoke zone, equipped with an automatic-initiation device, shall be put into operation by the actuation of one such device. Each additional device within the zone shall be verified to cause the same sequence without requiring the operation of fan motors in order to prevent damage. Control sequences shall be verified throughout the system, including verification of override from the fire-fighter's control panel and simulation of standby power conditions.

[F] 909.18.8 Special inspections for smoke control. Smoke control systems shall be tested by a special inspector.

[F] 909.18.8.1 Scope of testing. Special inspections shall be conducted in accordance with the following:

1. During erection of ductwork and prior to concealment for the purposes of leakage testing and recording of device location.

2. Prior to occupancy and after sufficient completion for the purposes of pressure-difference testing, flow measurements, and detection and control verification.

[F] 909.18.8.2 Qualifications. Special inspection agencies for smoke control shall have expertise in fire protection engineering, mechanical engineering and certification as air balancers.

[F] 909.18.8.3 Reports. A complete report of testing shall be prepared by the special inspector or special inspection agency. The report shall include identification of all devices by manufacturer, nameplate data, design values, measured values and identification tag or mark. The report shall be reviewed by the responsible registered design professional and, when satisfied that the design intent has been achieved, the responsible registered design professional shall seal, sign and date the report.

[F] 909.18.8.3.1 Report filing. A copy of the final report shall be filed with the building official and an identical copy shall be maintained in an approved location at the building.

[F] 909.18.9 Identification and documentation. Charts, drawings and other documents identifying and locating each component of the smoke control system, and describing its proper function and maintenance requirements, shall be maintained on file at the building as an attachment to the report required by Section 909.18.8.3. Devices shall have an approved identifying tag or mark on them consistent with the other required documentation and shall be dated indicating the last time they were successfully tested and by whom.

[F] 909.19 System acceptance. Buildings, or portions thereof, required by this code to comply with this section shall not be issued a certificate of occupancy until such time that the building official determines that the provisions of this section have been fully complied with, and that the fire department has received satisfactory instruction on the operation, both automatic and manual, of the system.

Exception: In buildings of phased construction, a temporary certificate of occupancy, as approved by the building official, shall be permitted provided that those portions of the building to be occupied meet the requirements of this section and that the remainder does not pose a significant hazard to the safety of the proposed occupants or adjacent buildings.

909.20 Smokeproof enclosures. Where required by Section 1019.1.8, a smokeproof enclosure shall be constructed in accordance with this section. A smokeproof enclosure shall consist of an enclosed interior exit stairway that conforms to Section 1019.1 and an outside balcony or ventilated vestibule meeting the requirements of this section. Where access to the roof is required by the *International Fire Code*, such access shall be from the smokeproof enclosure where a smokeproof enclosure is required.

909.20.1 Access. Access to the stair shall be by way of a vestibule or an open exterior balcony. The minimum dimension of the vestibule shall not be less than the required width of the corridor leading to the vestibule but shall not have a width of less than 44 inches (1118 mm) and shall not have a length of less than 72 inches (1829 mm) in the direction of egress travel.

909.20.2 Construction. The smokeproof enclosure shall be separated from the remainder of the building by not less than a 2-hour fire-resistance-rated fire barrier without openings other than the required means of egress doors. The vestibule shall be separated from the stairway by not less than a 2-hour fire-resistance-rated fire barrier. The open exterior balcony shall be constructed in accordance with the fire-resistance-rating requirements for floor construction.

909.20.2.1 Door closers. Doors in a smokeproof enclosure shall be self-closing or automatic-closing by actuation of a smoke detector installed at the floor-side entrance to the smokeproof enclosure in accordance with Section 715.3.7. The actuation of the smoke detector on any door shall activate the closing devices on all doors in the smokeproof enclosure at all levels. Smoke detectors shall be installed in accordance with Section 907.10.

909.20.3 Natural ventilation alternative. The provisions of Sections 909.20.3.1 through 909.20.3.3 shall apply to ventilation of smokeproof enclosures by natural means.

909.20.3.1 Balcony doors. Where access to the stairway is by way of an open exterior balcony, the door assembly into the enclosure shall be a fire door in accordance with Section 715.3.

909.20.3.2 Vestibule doors. Where access to the stairway is by way of a vestibule, the door assembly into the vestibule shall be a fire door complying with Section 715.3. The door assembly from the vestibule to the stairway shall have not less than a 20-minute fire protection rating complying with Section 715.3.

909.20.3.3 Vestibule ventilation. Each vestibule shall have a minimum net area of 16 square feet (1.5 m²) of opening in a wall facing an outer court, yard or public way that is at least 20 feet (6096 mm) in width.

909.20.4 Mechanical ventilation alternative. The provisions of Sections 909.20.4.1 through 909.20.4.4 shall apply to ventilation of smokeproof enclosures by mechanical means.

909.20.4.1 Vestibule doors. The door assembly from the building into the vestibule shall be a fire door complying with Section 715.3. The door assembly from the vestibule to the stairway shall have not less than a 20-minute fire protection rating in accordance with Section 715.3. The door from the building into the vestibule shall be provided with gaskets or other provisions to minimize air leakage.

909.20.4.2 Vestibule ventilation. The vestibule shall be supplied with not less than one air change per minute and the exhaust shall not be less than 150 percent of supply.

Supply air shall enter and exhaust air shall discharge from the vestibule through separate, tightly constructed ducts used only for that purpose. Supply air shall enter the vestibule within 6 inches (152 mm) of the floor level. The top of the exhaust register shall be located at the top of the smoke trap but not more than 6 inches (152 mm) down from the top of the trap, and shall be entirely within the smoke trap area. Doors in the open position shall not obstruct duct openings. Duct openings with controlling dampers are permitted where necessary to meet the design requirements, but dampers are not otherwise required.

909.20.4.2.1 Engineered ventilation system. Where a specially engineered system is used, the system shall exhaust a quantity of air equal to not less than 90 air changes per hour from any vestibule in the emergency operation mode and shall be sized to handle three vestibules simultaneously. Smoke detectors shall be located at the floor-side entrance to each vestibule and shall activate the system for the affected vestibule. Smoke detectors shall be installed in accordance with Section 907.10.

909.20.4.3 Smoke trap. The vestibule ceiling shall be at least 20 inches (508 mm) higher than the door opening into the vestibule to serve as a smoke and heat trap and to provide an upward-moving air column. The height shall not be decreased unless approved and justified by design and test.

909.20.4.4 Stair shaft air movement system. The stair shaft shall be provided with a dampered relief opening and supplied with sufficient air to maintain a minimum positive pressure of 0.10 inch of water (25 Pa) in the shaft relative to the vestibule with all doors closed.

909.20.5 Stair pressurization alternative. Where the building is equipped throughout with an automatic sprinkler system in accordance with Section 903.3.1.1, the vestibule is not required, provided that interior exit stairways are pressurized to a minimum of 0.15 inch of water (37 Pa) and a maximum of 0.35 inch of water (87 Pa) in the shaft relative to the building measured with all stairway doors closed under maximum anticipated stack pressures.

909.20.6 Ventilating equipment. The activation of ventilating equipment required by the alternatives in Sections 909.20.4 and 909.20.5 shall be by smoke detectors installed at each floor level at an approved location at the entrance to the smokeproof enclosure. When the closing device for the stair shaft and vestibule doors is activated by smoke detection or power failure, the mechanical equipment shall activate and operate at the required performance levels. Smoke detectors shall be installed in accordance with Section 907.10.

909.20.6.1 Ventilation systems. Smokeproof enclosure ventilation systems shall be independent of other building ventilation systems. The equipment and ductwork shall comply with one of the following:

1. Equipment and ductwork shall be located exterior to the building and directly connected to the smokeproof enclosure or connected to the

smokeproof enclosure by ductwork enclosed by 2-hour fire-resistance-rated fire barriers.

2. Equipment and ductwork shall be located within the smokeproof enclosure with intake or exhaust directly from and to the outside or through ductwork enclosed by 2-hour fire-resistance-rated fire barriers.

3. Equipment and ductwork shall be located within the building if separated from the remainder of the building, including other mechanical equipment, by 2-hour fire-resistance-rated fire barriers.

909.20.6.2 Standby power. Mechanical vestibule and stair shaft ventilation systems and automatic fire detection systems shall be powered by an approved standby power system conforming to Section 403.10.1 and Chapter 27.

909.20.6.3 Acceptance and testing. Before the mechanical equipment is approved, the system shall be tested in the presence of the building official to confirm that the system is operating in compliance with these requirements.

909.21 Underground building smoke exhaust system. Where required in accordance with Section 405.5 for underground buildings, a smoke exhaust system shall be provided in accordance with this section.

909.21.1 Exhaust capability. Where compartmentation is required, each compartment shall have an independent, automatically activated smoke exhaust system capable of manual operation. The system shall have an air supply and smoke exhaust capability that will provide a minimum of six air changes per hour.

[F] 909.21.2 Operation. The smoke exhaust system shall be operated in the compartment of origin by the following, independently of each other:

1. Two cross-zoned smoke detectors within a single protected area of a single smoke detector monitored by an alarm verification zone or an approved equivalent method.

2. The automatic sprinkler system.

3. Manual controls that are readily accessible to the fire department.

[F] 909.21.3 Alarm required. Activation of the smoke exhaust system shall activate an audible alarm at a constantly attended location.

SECTION 910
SMOKE AND HEAT VENTS

[F] 910.1 General. Where required by this code or otherwise installed, smoke and heat vents or mechanical smoke exhaust systems and draft curtains shall conform to the requirements of this section.

Exception: Frozen-food warehouses used solely for storage of Class I and II commodities where protected by an approved automatic sprinkler system.

[F] 910.2 Where required. Approved smoke and heat vents shall be installed in the roofs of one-story buildings or portions thereof occupied for the uses set forth in Sections 910.2.1 through 910.2.4.

[F] 910.2.1 Groups F-1 and S-1. Buildings and portions thereof used as a Group F-1 or S-1 occupancy having more than 50,000 square feet (4645 m²) in undivided area.

Exception: Group S-1 aircraft repair hangars.

[F] 910.2.2 Group H. Buildings and portions thereof used as a Group H occupancy as shown:

1. In occupancies classified as Group H-2 or H-3, any of which are over 15,000 square feet (1394 m²) in single floor area.

   Exception: Buildings of noncombustible construction containing only noncombustible materials.

2. In areas of buildings in Group H used for storing Class 2, 3, and 4 liquid and solid oxidizers, Class 1 and unclassified detonable organic peroxides, Class 3 and 4 unstable (reactive) materials, or Class 2 or 3 water-reactive materials as required for a high-hazard commodity classification.

   Exception: Buildings of noncombustible construction containing only noncombustible materials.

[F] 910.2.3 High-piled combustible storage. Buildings and portions thereof containing high-piled combustible stock or rack storage in any occupancy group in accordance with Section 413 and the *International Fire Code*.

[F] 910.2.4 Exit access travel distance increase. Buildings and portions thereof used as a Group F-1 or S-1 occupancy where the maximum exit access travel distance is increased in accordance with Section 1015.2.

[F] 910.3 Design and installation. The design and installation of smoke and heat vents and draft curtains shall be as specified in this section and Table 910.3.

[F] 910.3.1 Vent operation. Smoke and heat vents shall be approved and labeled and shall be capable of being operated by approved automatic and manual means. Automatic operation of smoke and heat vents shall conform to the provisions of this section.

[F] 910.3.1.1 Gravity-operated drop-out vents. Automatic smoke and heat vents containing heat-sensitive glazing designed to shrink and drop out of the vent opening when exposed to fire shall fully open within 5 minutes after the vent cavity is exposed to a simulated fire, represented by a time-temperature gradient that reaches an air temperature of 500°F (260°C) within 5 minutes.

[F] 910.3.1.2 Sprinklered buildings. Where installed in buildings provided with an approved automatic sprinkler system, smoke and heat vents shall be designed to operate automatically.

[F] 910.3.1.3 Nonsprinklered buildings. Where installed in buildings not provided with an approved automatic sprinkler system, smoke and heat vents shall operate automatically by actuation of a heat-responsive

TABLE 910.3
REQUIREMENTS FOR DRAFT CURTAINS AND SMOKE AND HEAT VENTS[a]

OCCUPANCY GROUP AND COMMODITY CLASSIFICATION	DESIGNATED STORAGE HEIGHT (feet)	MINIMUM DRAFT CURTAIN DEPTH (feet)	MAXIMUM AREA FORMED BY DRAFT CURTAINS (square feet)	VENT AREA TO FLOOR AREA RATIO	MAXIMUM SPACING OF VENT CENTERS (feet)	MAXIMUM DISTANCE TO VENTS FROM WALL OR DRAFT CURTAINS[b] (feet)
Group F-1	—	$0.2 \times H$ but ≥ 4	50,000	1:100	120	60
Group S-1 I-IV (Option 1)	≤ 20	6	10,000	1:100	100	60
	$> 20 \leq 40$	6	8,000	1:75	100	55
Group S-1 I-IV (Option 2)	≤ 20	4	3,000	1:75	100	55
	$> 20 \leq 40$	4	3,000	1:50	100	50
Group S-1 High hazard (Option 1)	≤ 20	6	6,000	1:50	100	50
	$> 20 \leq 30$	6	6,000	1:40	90	45
Group S-1 High hazard (Option 2)	≤ 20	4	4,000	1:50	100	50
	$> 20 \leq 30$	4	2,000	1:30	75	40

For SI: 1 foot = 304.8 mm, 1 square foot = 0.0929 m^2.

a. Requirements for rack storage heights in excess of those indicated shall be in accordance with Chapter 23 of the *International Fire Code*. For solid-piled storage heights in excess of those indicated, an approved engineered design shall be used.

b. The distance specified is the maximum distance from any vent in a particular draft curtained area to walls or draft curtains which form the perimeter of the draft curtained area.

device rated at between 100°F (38°C) and 220°F (104°C) above ambient.

Exception: Gravity-operated drop-out vents complying with Section 910.3.1.1

[F] 910.3.2 Vent dimensions. The effective venting area shall not be less than 16 square feet (1.5 m^2) with no dimension less than 4 feet (1219 mm), excluding ribs or gutters having a total width not exceeding 6 inches (152 mm).

[F] 910.3.3 Vent locations. Smoke and heat vents shall be located 20 feet (6096 mm) or more from adjacent lot lines and fire walls and 10 feet (3048 mm) or more from fire barrier walls. Vents shall be uniformly located within the roof area above high-piled storage areas, with consideration given to roof pitch, draft curtain location, sprinkler location and structural members.

[F] 910.3.4 Draft curtains. Where required, draft curtains shall be provided in accordance with this section.

Exception: Where areas of buildings are equipped with early suppression fast-response (ESFR) sprinklers, draft curtains shall not be provided within these areas. Draft curtains shall only be provided at the separation between the ESFR sprinklers and the conventional sprinklers.

[F] 910.3.4.1 Construction. Draft curtains shall be constructed of sheet metal, lath and plaster, gypsum board or other approved materials which provide equivalent performance to resist the passage of smoke. Joints and connections shall be smoke tight.

[F] 910.3.4.2 Location and depth. The location and minimum depth of draft curtains shall be in accordance with Table 910.3.

[F] 910.4 Mechanical smoke exhaust. Where approved by the building official, engineered mechanical smoke exhaust shall be an acceptable alternate to smoke and heat vents.

[F] 910.4.1 Location. Exhaust fans shall be uniformly spaced within each draft-curtained area and the maximum distance between fans shall not be greater than 100 feet (30 480 mm).

[F] 910.4.2 Size. Fans shall have a maximum individual capacity of 30,000 cfm (14.2 m^3/s). The aggregate capacity of smoke exhaust fans shall be determined by the equation:

$$C = A \times 300 \qquad \text{(Equation 9-10)}$$

where:

C = Capacity of mechanical ventilation required, in cubic feet per minute (m^3/s).

A = Area of roof vents provided in square feet (m^2) in accordance with Table 910.3.

[F] 910.4.3 Operation. Mechanical smoke exhaust fans shall be automatically activated by the automatic sprinkler system or by heat detectors having operating characteristics equivalent to those described in Section 910.3.1. Individual manual controls of each fan unit shall also be provided.

[F] 910.4.4 Wiring and control. Wiring for operation and control of smoke exhaust fans shall be connected ahead of the main disconnect and protected against exposure to temperatures in excess of 1,000°F (538°C) for a period of not less than 15 minutes. Controls shall be located so as to be immediately accessible to the fire service from the exterior of the building and protected against interior fire exposure by fire barriers having a fire-resistance rating not less than 1 hour.

[F] 910.4.5 Supply air. Supply air for exhaust fans shall be provided at or near the floor level and shall be sized to provide a minimum of 50 percent of required exhaust. Openings for supply air shall be uniformly distributed around the periphery of the area served.

[F] 910.4.6 Interlocks. In combination comfort air-handling/smoke removal systems or independent comfort air-handling systems, fans shall be controlled to shut down in accordance with the approved smoke control sequence.

SECTION 911
FIRE COMMAND CENTER

[F] 911.1 Features. Where required by other sections of this code, a fire command center for fire department operations shall be provided. The location and accessibility of the fire command center shall be separated from the remainder of the building by not less than a 1-hour fire-resistance-rated fire barrier. The room shall be a minimum of 96 square feet (9 m^2) with a minimum dimension of 8 feet (2438 mm). A layout of the fire command center and all features required by the section to be contained therein shall be submitted for approval prior to installation. The fire command center shall comply with NFPA 72 and shall contain the following features.

1. The emergency voice/alarm communication system unit.

2. The fire department communications unit.

3. Fire detection and alarm system annunciator unit.

4. Annunciator unit visually indicating the location of the elevators and whether they are operational.

5. Status indicators and controls for air-handling systems.

6. The fire-fighter's control panel required by Section 909.16 for smoke control systems installed in the building.

7. Controls for unlocking stairway doors simultaneously.

8. Sprinkler valve and water-flow detector display panels.

9. Emergency and standby power status indicators.

10. A telephone for fire department use with controlled access to the public telephone system.

11. Fire pump status indicators.

12. Schematic building plans indicating the typical floor plan and detailing the building core, means of egress, fire protection systems, fire-fighting equipment and fire department access.

13. Worktable.

14. Generator supervision devices, manual start and transfer features.

15. Public address system, where specifically required by other sections of this code.

CHAPTER 10

MEANS OF EGRESS

User Note: See Preface page iv ("marginal markings") for Chapter 10 reorganization information.

SECTION 1001
ADMINISTRATION

1001.1 General. Buildings or portions thereof shall be provided with a means of egress system as required by this chapter. The provisions of this chapter shall control the design, construction and arrangement of means of egress components required to provide an approved means of egress from structures and portions thereof.

1001.2 Minimum requirements. It shall be unlawful to alter a building or structure in a manner that will reduce the number of exits or the capacity of the means of egress to less than required by this code.

[F] 1001.3 Maintenance. Means of egress shall be maintained in accordance with the *International Fire Code*.

SECTION 1002
DEFINITIONS

1002.1 Definitions. The following words and terms shall, for the purposes of this chapter and as used elsewhere in this code, have the meanings shown herein.

ACCESSIBLE MEANS OF EGRESS. A continuous and unobstructed way of egress travel from any point in a building or facility that provides an accessible route to an area of refuge, a horizontal exit or a public way.

AISLE ACCESSWAY. That portion of an exit access that leads to an aisle.

ALTERNATING TREAD DEVICE. A device that has a series of steps between 50 and 70 degrees (0.87 and 1.22 rad) from horizontal, usually attached to a center support rail in an alternating manner so that the user does not have both feet on the same level at the same time.

AREA OF REFUGE. An area where persons unable to use stairways can remain temporarily to await instructions or assistance during emergency evacuation.

BLEACHERS. Tiered seating facilities.

COMMON PATH OF EGRESS TRAVEL. That portion of exit access which the occupants are required to traverse before two separate and distinct paths of egress travel to two exits are available. Paths that merge are common paths of travel. Common paths of egress travel shall be included within the permitted travel distance.

CORRIDOR. An enclosed exit access component that defines and provides a path of egress travel to an exit.

DOOR, BALANCED. A door equipped with double-pivoted hardware so designed as to cause a semicounterbalanced swing action when opening.

EGRESS COURT. A court or yard which provides access to a public way for one or more exits.

EMERGENCY ESCAPE AND RESCUE OPENING. An operable window, door or other similar device that provides for a means of escape and access for rescue in the event of an emergency.

EXIT. That portion of a means of egress system which is separated from other interior spaces of a building or structure by fire-resistance-rated construction and opening protectives as required to provide a protected path of egress travel between the exit access and the exit discharge. Exits include exterior exit doors at ground level, exit enclosures, exit passageways, exterior exit stairs, exterior exit ramps and horizontal exits.

EXIT, HORIZONTAL. A path of egress travel from one building to an area in another building on approximately the same level, or a path of egress travel through or around a wall or partition to an area on approximately the same level in the same building, which affords safety from fire and smoke from the area of incidence and areas communicating therewith.

EXIT ACCESS. That portion of a means of egress system that leads from any occupied portion of a building or structure to an exit.

EXIT DISCHARGE. That portion of a means of egress system between the termination of an exit and a public way.

EXIT DISCHARGE, LEVEL OF. The horizontal plane located at the point at which an exit terminates and an exit discharge begins.

EXIT ENCLOSURE. An exit component that is separated from other interior spaces of a building or structure by fire-resistance-rated construction and opening protectives, and provides for a protected path of egress travel in a vertical or horizontal direction to the exit discharge or the public way.

EXIT PASSAGEWAY. An exit component that is separated from all other interior spaces of a building or structure by fire-resistance-rated construction and opening protectives, and provides for a protected path of egress travel in a horizontal direction to the exit discharge or the public way.

FIRE EXIT HARDWARE. Panic hardware that is listed for use on fire door assemblies.

FLOOR AREA, GROSS. The floor area within the inside perimeter of the exterior walls of the building under consideration, exclusive of vent shafts and courts, without deduction for corridors, stairways, closets, the thickness of interior walls, columns or other features. The floor area of a building, or portion thereof, not provided with surrounding exterior walls shall be the usable area under the horizontal projection of the roof or floor above. The gross floor area shall not include shafts with no openings or interior courts.

FLOOR AREA, NET. The actual occupied area not including unoccupied accessory areas such as corridors, stairways, toilet rooms, mechanical rooms and closets.

FOLDING AND TELESCOPIC SEATING. Tiered seating facilities having an overall shape and size that are capable of being reduced for purposes of moving or storing.

GRANDSTAND. Tiered seating facilities.

GUARD. A building component or a system of building components located at or near the open sides of elevated walking surfaces that minimizes the possibility of a fall from the walking surface to a lower level.

HANDRAIL. A horizontal or sloping rail intended for grasping by the hand for guidance or support.

MEANS OF EGRESS. A continuous and unobstructed path of vertical and horizontal egress travel from any occupied portion of a building or structure to a public way. A means of egress consists of three separate and distinct parts: the exit access, the exit and the exit discharge.

NOSING. The leading edge of treads of stairs and of landings at the top of stairway flights.

OCCUPANT LOAD. The number of persons for which the means of egress of a building or portion thereof is designed.

PANIC HARDWARE. A door-latching assembly incorporating a device that releases the latch upon the application of a force in the direction of egress travel.

PUBLIC WAY. A street, alley or other parcel of land open to the outside air leading to a street, that has been deeded, dedicated or otherwise permanently appropriated to the public for public use and which has a clear width and height of not less than 10 feet (3048 mm).

RAMP. A walking surface that has a running slope steeper than one unit vertical in 20 units horizontal (5-percent slope).

SCISSOR STAIR. Two interlocking stairways providing two separate paths of egress located within one stairwell enclosure.

SMOKE-PROTECTED ASSEMBLY SEATING. Seating served by means of egress that is not subject to smoke accumulation within or under a structure.

STAIR. A change in elevation, consisting of one or more risers.

STAIRWAY. One or more flights of stairs, either exterior or interior, with the necessary landings and platforms connecting them, to form a continuous and uninterrupted passage from one level to another.

STAIRWAY, EXTERIOR. A stairway that is open on at least one side, except for required structural columns, beams, handrails and guards. The adjoining open areas shall be either yards, courts or public ways. The other sides of the exterior stairway need not be open.

STAIRWAY, INTERIOR. A stairway not meeting the definition of an exterior stairway.

STAIRWAY, SPIRAL. A stairway having a closed circular form in its plan view with uniform section-shaped treads attached to and radiating from a minimum-diameter supporting column.

WINDER. A tread with nonparallel edges.

SECTION 1003
GENERAL MEANS OF EGRESS

1003.1 Applicability. The general requirements specified in Sections 1003 through 1012 shall apply to all three elements of the means of egress system, in addition to those specific requirements for the exit access, the exit and the exit discharge detailed elsewhere in this chapter.

1003.2 Ceiling height. The means of egress shall have a ceiling height of not less than 7 feet (2134 mm).

Exceptions:

1. Sloped ceilings in accordance with Section 1208.2.
2. Ceilings of dwelling units and sleeping units within residential occupancies in accordance with Section 1208.2.
3. Allowable projections in accordance with Section 1003.3.
4. Stair headroom in accordance with Section 1009.2.
5. Door height in accordance with Section 1008.1.1.

1003.3 Protruding objects. Protruding objects shall comply with the requirements of Sections 1003.3.1 through 1003.3.4.

1003.3.1 Headroom. Protruding objects are permitted to extend below the minimum ceiling height required by Section 1003.2 provided a minimum headroom of 80 inches (2032 mm) shall be provided for any walking surface, including walks, corridors, aisles and passageways. Not more than 50 percent of the ceiling area of a means of egress shall be reduced in height by protruding objects.

Exception: Door closers and stops shall not reduce headroom to less than 78 inches (1981 mm).

A barrier shall be provided where the vertical clearance is less than 80 inches (2032 mm) high. The leading edge of such a barrier shall be located 27 inches (686 mm) maximum above the floor.

1003.3.2 Free-standing objects. A free-standing object mounted on a post or pylon shall not overhang that post or pylon more than 12 inches (305 mm) where the lowest point of the leading edge is more than 27 inches (686 mm) and less than 80 inches (2032 mm) above the walking surface. Where a sign or other obstruction is mounted between posts or pylons and the clear distance between the posts or pylons is greater than 12 inches (305 mm), the lowest edge of such sign or obstruction shall be 27 inches (685 mm) maximum or 80 inches (2030 mm) minimum above the finish floor or ground.

Exception: This requirement shall not apply to sloping portions of handrails serving stairs and ramps.

1003.3.3 Horizontal projections. Structural elements, fixtures or furnishings shall not project horizontally from either side more than 4 inches (102 mm) over any walking surface between the heights of 27 inches (686 mm) and 80 inches (2032 mm) above the walking surface.

Exception: Handrails serving stairs and ramps are permitted to protrude 4.5 inches (114 mm) from the wall.

1003.3.4 Clear width. Protruding objects shall not reduce the minimum clear width of accessible routes as required in Section 1104.

1003.4 Floor surface. Walking surfaces of the means of egress shall have a slip-resistant surface and be securely attached.

1003.5 Elevation change. Where changes in elevation of less than 12 inches (305 mm) exist in the means of egress, sloped surfaces shall be used. Where the slope is greater than one unit vertical in 20 units horizontal (5-percent slope), ramps complying with Section 1010 shall be used. Where the difference in elevation is 6 inches (152 mm) or less, the ramp shall be equipped with either handrails or floor finish materials that contrast with adjacent floor finish materials.

Exceptions:

1. A single step with a maximum riser height of 7 inches (178 mm) is permitted for buildings with occupancies in Groups F, H, R-2 and R-3 as applicable in Section 101.2, and Groups S and U at exterior doors not required to be accessible by Chapter 11.

2. A stair with a single riser or with two risers and a tread is permitted at locations not required to be accessible by Chapter 11, provided that the risers and treads comply with Section 1009.3, the minimum depth of the tread is 13 inches (330 mm) and at least one handrail complying with Section 1009.11 is provided within 30 inches (762 mm) of the centerline of the normal path of egress travel on the stair.

3. An aisle serving seating that has a difference in elevation less than 12 inches (305 mm) is permitted at locations not required to be accessible by Chapter 11, provided that the risers and treads comply with Section 1024.11 and the aisle is provided with a handrail complying with Section 1024.13.

Any change in elevation in a corridor serving nonambulatory persons in a Group I-2 occupancy shall be by means of a ramp or sloped walkway.

1003.6 Means of egress continuity. The path of egress travel along a means of egress shall not be interrupted by any building element other than a means of egress component as specified in this chapter. Obstructions shall not be placed in the required width of a means of egress except projections permitted by this chapter. The required capacity of a means of egress system shall not be diminished along the path of egress travel.

1003.7 Elevators, escalators and moving walks. Elevators, escalators and moving walks shall not be used as a component of a required means of egress from any other part of the building.

Exception: Elevators used as an accessible means of egress in accordance with Section 1007.4.

SECTION 1004
OCCUPANT LOAD

1004.1 Design occupant load. In determining means of egress requirements, the number of occupants for whom means of egress facilities shall be provided shall be established by the largest number computed in accordance with Sections 1004.1.1 through 1004.1.3.

1004.1.1 Actual number. The actual number of occupants for whom each occupied space, floor or building is designed.

1004.1.2 Number by Table 1004.1.2. The number of occupants computed at the rate of one occupant per unit of area as prescribed in Table 1004.1.2.

TABLE 1004.1.2
MAXIMUM FLOOR AREA ALLOWANCES PER OCCUPANT

OCCUPANCY	FLOOR AREA IN SQ. FT. PER OCCUPANT
Agricultural building	300 gross
Aircraft hangars	500 gross

Airport terminal
    Baggage claim
    Baggage handling
    Concourse
    Waiting areas
	
	20 gross
	300 gross
	100 gross
	15 gross

Assembly Gaming floors (keno, slots, etc.)	 11 gross
Assembly with fixed seats	See Section 1004.7

Assembly without fixed seats
    Concentrated (chairs only—not fixed)
    Standing space
    Unconcentrated (tables and chairs)
	
	7 net
	5 net
	15 net

Bowling centers, allow 5 persons for each lane including 15 feet of runway, and for additional areas	7 net
Business areas	100 gross
Courtrooms—other than fixed seating areas	40 net
Dormitories	50 gross

Educational
    Classroom area
    Shops and other vocational room areas
	
	20 net
	50 net

Exercise rooms	50 gross
H-5 Fabrication and manufacturing areas	200 gross
Industrial areas	100 gross

Institutional areas
    Inpatient treatment areas
    Outpatient areas
    Sleeping areas
	
	240 gross
	100 gross
	120 gross

Kitchens, commercial	200 gross

Library
    Reading rooms
    Stack area
	
	50 net
	100 gross

Locker rooms	50 gross

Mercantile
    Areas on other floors
    Basement and grade floor areas
    Storage, stock, shipping areas
	
	60 gross
	30 gross
	300 gross

Parking garages	200 gross
Residential	200 gross

Skating rinks, swimming pools
    Rink and pool
    Decks
	
	50 gross
	15 gross

Stages and platforms	15 net
Accessory storage areas, mechanical equipment room	300 gross
Warehouses	500 gross

For SI: 1 square foot = 0.0929 m^2.

1004.1.3 Number by combination. Where occupants from accessory spaces egress through a primary area, the calculated occupant load for the primary space shall include the total occupant load of the primary space plus the number of occupants egressing through it from the accessory space.

1004.2 Increased occupant load. The occupant load permitted in any building or portion thereof is permitted to be increased from that number established for the occupancies in Table 1004.1.2 provided that all other requirements of the code are also met based on such modified number and the occupant load shall not exceed one occupant per 5 square feet (0.47 m²) of occupiable floor space. Where required by the building official, an approved aisle, seating or fixed equipment diagram substantiating any increase in occupant load shall be submitted. Where required by the building official, such diagram shall be posted.

1004.3 Posting of occupant load. Every room or space that is an assembly occupancy shall have the occupant load of the room or space posted in a conspicuous place, near the main exit or exit access doorway from the room or space. Posted signs shall be of an approved legible permanent design and shall be maintained by the owner or authorized agent.

1004.4 Exiting from multiple levels. Where exits serve more than one floor, only the occupant load of each floor considered individually shall be used in computing the required capacity of the exits at that floor, provided that the exit capacity shall not decrease in the direction of egress travel.

1004.5 Egress convergence. Where means of egress from floors above and below converge at an intermediate level, the capacity of the means of egress from the point of convergence shall not be less than the sum of the two floors.

1004.6 Mezzanine levels. The occupant load of a mezzanine level with egress onto a room or area below shall be added to that room or area's occupant load, and the capacity of the exits shall be designed for the total occupant load thus established.

1004.7 Fixed seating. For areas having fixed seats and aisles, the occupant load shall be determined by the number of fixed seats installed therein.

For areas having fixed seating without dividing arms, the occupant load shall not be less than the number of seats based on one person for each 18 inches (457 mm) of seating length.

The occupant load of seating booths shall be based on one person for each 24 inches (610 mm) of booth seat length measured at the backrest of the seating booth.

1004.8 Outdoor areas. Yards, patios, courts and similar outdoor areas accessible to and usable by the building occupants shall be provided with means of egress as required by this chapter. The occupant load of such outdoor areas shall be assigned by the building official in accordance with the anticipated use. Where outdoor areas are to be used by persons in addition to the occupants of the building, and the path of egress travel from the outdoor areas passes through the building, means of egress requirements for the building shall be based on the sum of the occupant loads of the building plus the outdoor areas.

Exceptions:

1. Outdoor areas used exclusively for service of the building need only have one means of egress.
2. Both outdoor areas associated with Group R-3 and individual dwelling units of Group R-2, as applicable in Section 101.2.

1004.9 Multiple occupancies. Where a building contains two or more occupancies, the means of egress requirements shall apply to each portion of the building based on the occupancy of that space. Where two or more occupancies utilize portions of the same means of egress system, those egress components shall meet the more stringent requirements of all occupancies that are served.

SECTION 1005
EGRESS WIDTH

1005.1 Minimum required egress width. The means of egress width shall not be less than required by this section. The total width of means of egress in inches (mm) shall not be less than the total occupant load served by the means of egress multiplied by the factors in Table 1005.1 and not less than specified elsewhere in this code. Multiple means of egress shall be sized such that the loss of any one means of egress shall not reduce the available capacity to less than 50 percent of the required capacity. The maximum capacity required from any story of a building shall be maintained to the termination of the means of egress.

Exception: Means of egress complying with Section 1024.

TABLE 1005.1
EGRESS WIDTH PER OCCUPANT SERVED

OCCUPANCY	WITHOUT SPRINKLER SYSTEM		WITH SPRINKLER SYSTEM[a]	
	Stairways (inches per occupant)	Other egress components (inches per occupant)	Stairways (inches per occupant)	Other egress components (inches per occupant)
Occupancies other than those listed below	0.3	0.2	0.2	0.15
Hazardous: H-1, H-2, H-3 and H-4	0.7	0.4	0.3	0.2
Institutional: I-2	NA	NA	0.3	0.2

For SI: 1 inch = 25.4 mm. NA = Not applicable.

a. Buildings equipped throughout with an automatic sprinkler system in accordance with Section 903.3.1.1 or 903.3.1.2.

1005.2 Door encroachment. Doors opening into the path of egress travel shall not reduce the required width to less than one-half during the course of the swing. When fully open, the door shall not project more than 7 inches (178 mm) into the required width.

Exception: The restrictions on a door swing shall not apply to doors within individual dwelling units and sleeping units of Group R-2 and dwelling units of Group R-3.

SECTION 1006
MEANS OF EGRESS ILLUMINATION

1006.1 Illumination required. The means of egress, including the exit discharge, shall be illuminated at all times the building space served by the means of egress is occupied.

Exceptions:

1. Occupancies in Group U.
2. Aisle accessways in Group A.
3. Dwelling units and sleeping units in Groups R-1, R-2 and R-3.
4. Sleeping units of Group I occupancies.

1006.2 Illumination level. The means of egress illumination level shall not be less than 1 foot-candle (11 lux) at the floor level.

Exception: For auditoriums, theaters, concert or opera halls and similar assembly occupancies, the illumination at the floor level is permitted to be reduced during performances to not less than 0.2 foot-candle (2.15 lux) provided that the required illumination is automatically restored upon activation of a premise's fire alarm system where such system is provided.

1006.3 Illumination emergency power. The power supply for means of egress illumination shall normally be provided by the premise's electrical supply.

In the event of power supply failure, an emergency electrical system shall automatically illuminate the following areas:

1. Exit access corridors, passageways and aisles in rooms and spaces which require two or more means of egress.
2. Exit access corridors and exit stairways located in buildings required to have two or more exits.
3. Exterior egress components at other than the level of exit discharge until exit discharge is accomplished for buildings required to have two or more exits.
4. Interior exit discharge elements, as permitted in Section 1023.1, in buildings required to have two or more exits.
5. The portion of the exterior exit discharge immediately adjacent to exit discharge doorways in buildings required to have two or more exits.

The emergency power system shall provide power for a duration of not less than 90 minutes and shall consist of storage batteries, unit equipment or an on-site generator. The installation of the emergency power system shall be in accordance with Section 2702.

1006.4 Performance of system. Emergency lighting facilities shall be arranged to provide initial illumination that is at least an average of 1 foot-candle (11 lux) and a minimum at any point of 0.1 foot-candle (1 lux) measured along the path of egress at floor level. Illumination levels shall be permitted to decline to 0.6 foot-candle (6 lux) average and a minimum at any point of 0.06 foot-candle (0.6 lux) at the end of the emergency lighting time duration. A maximum-to-minimum illumination uniformity ratio of 40 to 1 shall not be exceeded.

SECTION 1007
ACCESSIBLE MEANS OF EGRESS

1007.1 Accessible means of egress required. Accessible means of egress shall comply with this section. Accessible spaces shall be provided with not less than one accessible means of egress. Where more than one means of egress is required by Section 1014.1 or 1018.1 from any accessible space, each accessible portion of the space shall be served by not less than two accessible means of egress.

Exceptions:

1. Accessible means of egress are not required in alterations to existing buildings.
2. One accessible means of egress is required from an accessible mezzanine level in accordance with Section 1007.3 or 1007.4.
3. In assembly spaces with sloped floors, one accessible means of egress is required from a space where the common path of travel of the accessible route for access to the wheelchair spaces meets the requirements in Section 1024.8.

1007.2 Continuity and components. Each required accessible means of egress shall be continuous to a public way and shall consist of one or more of the following components:

1. Accessible routes complying with Section 1104.
2. Stairways within exit enclosures complying with Sections 1007.3 and 1019.1.
3. Elevators complying with Section 1007.4.
4. Platform lifts complying with Section 1007.5.
5. Horizontal exits.
6. Smoke barriers.

Exceptions:

1. Where the exit discharge is not accessible, an exterior area for assisted rescue must be provided in accordance with Section 1007.8.
2. Where the exit stairway is open to the exterior, the accessible means of egress shall include either an area of refuge in accordance with Section 1007.6 or an exterior area for assisted rescue in accordance with Section 1007.8.

1007.2.1 Buildings with four or more stories. In buildings where a required accessible floor is four or more stories above or below a level of exit discharge, at least one required accessible means of egress shall be an elevator complying with Section 1007.4.

Exceptions:

1. In buildings equipped throughout with an automatic sprinkler system installed in accordance with Section 903.3.1.1 or 903.3.1.2, the elevator shall not be required on floors provided with a horizontal exit and located at or above the level of exit discharge.
2. In buildings equipped throughout with an automatic sprinkler system installed in accordance with Section 903.3.1.1 or 903.3.1.2, the elevator

shall not be required on floors provided with a ramp conforming to the provisions of Section 1010.

1007.3 Enclosed exit stairways. An enclosed exit stairway, to be considered part of an accessible means of egress, shall have a clear width of 48 inches (1219 mm) minimum between handrails and shall either incorporate an area of refuge within an enlarged floor-level landing or shall be accessed from either an area of refuge complying with Section 1007.6 or a horizontal exit.

Exceptions:

1. Open exit stairways as permitted by Section 1019.1 are permitted to be considered part of an accessible means of egress.

2. The area of refuge is not required at open stairways that are permitted by Section 1019.1 in buildings or facilities that are equipped throughout with an automatic sprinkler system installed in accordance with Section 903.3.1.1.

3. The clear width of 48 inches (1219 mm) between handrails and the area of refuge is not required at exit stairways in buildings or facilities equipped throughout with an automatic sprinkler system installed in accordance with Section 903.3.1.1 or 903.3.1.2.

4. The clear width of 48 inches (1219 mm) between handrails is not required for enclosed exit stairways accessed from a horizontal exit.

5. Areas of refuge are not required at exit stairways serving open parking garages.

1007.4 Elevators. An elevator to be considered part of an accessible means of egress shall comply with the emergency operation and signaling device requirements of Section 2.27 of ASME A17.1. Standby power shall be provided in accordance with Sections 2702 and 3003. The elevator shall be accessed from either an area of refuge complying with Section 1007.6 or a horizontal exit.

Exceptions:

1. Elevators are not required to be accessed from an area of refuge or horizontal exit in open parking garages.

2. Elevators are not required to be accessed from an area of refuge or horizontal exit in buildings and facilities equipped throughout with an automatic sprinkler system installed in accordance with Section 903.3.1.1 or 903.3.1.2.

1007.5 Platform lifts. Platform (wheelchair) lifts shall not serve as part of an accessible means of egress, except where allowed as part of a required accessible route in Section 1109.7. Platform lifts in accordance with Section 2702 shall be installed in accordance with ASME A18.1. Standby power shall be provided for platform lifts permitted to serve as part of a means of egress.

1007.6 Areas of refuge. Every required area of refuge shall be accessible from the space it serves by an accessible means of egress. The maximum travel distance from any accessible space to an area of refuge shall not exceed the travel distance permitted for the occupancy in accordance with Section

1015.1. Every required area of refuge shall have direct access to an enclosed stairway complying with Sections 1007.3 and 1019.1 or an elevator complying with Section 1007.4. Where an elevator lobby is used as an area of refuge, the shaft and lobby shall comply with Section 1019.1.8 for smokeproof enclosures except where the elevators are in an area of refuge formed by a horizontal exit or smoke barrier.

1007.6.1 Size. Each area of refuge shall be sized to accommodate one wheelchair space of 30 inches by 48 inches (762 mm by 1219 mm) for each 200 occupants or portion thereof, based on the occupant load of the area of refuge and areas served by the area of refuge. Such wheelchair spaces shall not reduce the required means of egress width. Access to any of the required wheelchair spaces in an area of refuge shall not be obstructed by more than one adjoining wheelchair space.

1007.6.2 Separation. Each area of refuge shall be separated from the remainder of the story by a smoke barrier complying with Section 709. Each area of refuge shall be designed to minimize the intrusion of smoke.

Exceptions:

1. Areas of refuge located within a stairway enclosure.

2. Areas of refuge where the area of refuge and areas served by the area of refuge are equipped throughout with an automatic sprinkler system installed in accordance with Section 903.3.1.1 or 903.3.1.2.

1007.6.3 Two-way communication. Areas of refuge shall be provided with a two-way communication system between the area of refuge and a central control point. If the central control point is not constantly attended, the area of refuge shall also have controlled access to a public telephone system. Location of the central control point shall be approved by the fire department. The two-way communication system shall include both audible and visible signals.

1007.6.4 Instructions. In areas of refuge that have a two-way emergency communications system, instructions on the use of the area under emergency conditions shall be posted adjoining the communications system. The instructions shall include all of the following:

1. Directions to find other means of egress.

2. Persons able to use the exit stairway do so as soon as possible, unless they are assisting others.

3. Information on planned availability of assistance in the use of stairs or supervised operation of elevators and how to summon such assistance.

4. Directions for use of the emergency communications system.

1007.6.5 Identification. Each door providing access to an area of refuge from an adjacent floor area shall be identified by a sign complying with ICC A117.1, stating: AREA OF REFUGE, and including the International Symbol of Accessibility. Where exit sign illumination is required by Section 1011.2, the area of refuge sign shall be illuminated. Additionally, tactile signage complying with ICC A117.1 shall be located at each door to an area of refuge.

1007.7 Signage. At exits and elevators serving a required accessible space but not providing an approved accessible means of egress, signage shall be installed indicating the location of accessible means of egress.

1007.8 Exterior area for assisted rescue. The exterior area for assisted rescue must be open to the outside air and meet the requirements of Section 1007.6.1. Separation walls shall comply with the requirements of Section 704 for exterior walls. Where walls or openings are between the area for assisted rescue and the interior of the building, the building exterior walls within 10 feet (3048 mm) horizontally of a nonrated wall or unprotected opening shall be constructed as required for a minimum 1-hour fire-resistance rating with $^3/_4$-hour opening protectives. This construction shall extend vertically from the ground to a point 10 feet (3048 mm) above the floor level of the area for assisted rescue or to the roof line, whichever is lower.

1007.8.1 Openness. The exterior area for assisted rescue shall be at least 50 percent open, and the open area above the guards shall be so distributed as to minimize the accumulation of smoke or toxic gases.

1007.8.2 Exterior exit stairway. Exterior exit stairways that are part of the means of egress for the exterior area for assisted rescue shall provide a clear width of 48 inches (1219 mm) between handrails.

1007.8.3 Identification. Exterior areas for assisted rescue shall have identification as required for area of refuge that complies with Section 1007.6.5.

SECTION 1008
DOORS, GATES AND TURNSTILES

1008.1 Doors. Means of egress doors shall meet the requirements of this section. Doors serving a means of egress system shall meet the requirements of this section and Section 1017.2. Doors provided for egress purposes in numbers greater than required by this code shall meet the requirements of this section.

Means of egress doors shall be readily distinguishable from the adjacent construction and finishes such that the doors are easily recognizable as doors. Mirrors or similar reflecting materials shall not be used on means of egress doors. Means of egress doors shall not be concealed by curtains, drapes, decorations or similar materials.

1008.1.1 Size of doors. The minimum width of each door opening shall be sufficient for the occupant load thereof and shall provide a clear width of not less than 32 inches (813 mm). Clear openings of doorways with swinging doors shall be measured between the face of the door and the stop, with the door open 90 degrees (1.57 rad). Where this section requires a minimum clear width of 32 inches (813 mm) and a door opening includes two door leaves without a mullion, one leaf shall provide a clear opening width of 32 inches (813 mm). The maximum width of a swinging door leaf shall be 48 inches (1219 mm) nominal. Means of egress doors in an occupancy in Group I-2 used for the movement of beds shall provide a clear width not less than $41^1/_2$ inches

(1054 mm). The height of doors shall not be less than 80 inches (2032 mm).

Exceptions:

1. The minimum and maximum width shall not apply to door openings that are not part of the required means of egress in occupancies in Groups R-2 and R-3 as applicable in Section 101.2.
2. Door openings to resident sleeping units in occupancies in Group I-3 shall have a clear width of not less than 28 inches (711 mm).
3. Door openings to storage closets less than 10 square feet (0.93 m²) in area shall not be limited by the minimum width.
4. Width of door leafs in revolving doors that comply with Section 1008.1.3.1 shall not be limited.
5. Door openings within a dwelling unit or sleeping unit shall not be less than 78 inches (1981 mm) in height.
6. Exterior door openings in dwelling units and sleeping units, other than the required exit door, shall not be less than 76 inches (1930 mm) in height.
7. Interior egress doors within a dwelling unit or sleeping unit which is not required to be adaptable or accessible.
8. Door openings required to be accessible within Type B dwelling units shall have a minimum clear width of $31^3/_4$ inches (806 mm).

1008.1.1.1 Projections into clear width. There shall not be projections into the required clear width lower than 34 inches (864 mm) above the floor or ground. Projections into the clear opening width between 34 inches (864 mm) and 80 inches (2032 mm) above the floor or ground shall not exceed 4 inches (102 mm).

1008.1.2 Door swing. Egress doors shall be side-hinged swinging.

Exceptions:

1. Private garages, office areas, factory and storage areas with an occupant load of 10 or less.
2. Group I-3 occupancies used as a place of detention.
3. Doors within or serving a single dwelling unit in Groups R-2 and R-3 as applicable in Section 101.2.
4. In other than Group H occupancies, revolving doors complying with Section 1008.1.3.1.
5. In other than Group H occupancies, horizontal sliding doors complying with Section 1008.1.3.3 are permitted in a means of egress.
6. Power-operated doors in accordance with Section 1008.1.3.2.

Doors shall swing in the direction of egress travel where serving an occupant load of 50 or more persons or a Group H occupancy.

The opening force for interior side-swinging doors without closers shall not exceed a 5-pound (22 N) force. For other side-swinging, sliding and folding doors, the door latch shall release when subjected to a 15-pound (67 N) force. The door shall be set in motion when subjected to a 30-pound (133 N) force. The door shall swing to a full-open position when subjected to a 15-pound (67 N) force. Forces shall be applied to the latch side.

1008.1.3 Special doors. Special doors and security grilles shall comply with the requirements of Sections 1008.1.3.1 through 1008.1.3.5.

1008.1.3.1 Revolving doors. Revolving doors shall comply with the following:

1. Each revolving door shall be capable of collapsing into a bookfold position with parallel egress paths providing an aggregate width of 36 inches (914 mm).

2. A revolving door shall not be located within 10 feet (3048 mm) of the foot of or top of stairs or escalators. A dispersal area shall be provided between the stairs or escalators and the revolving doors.

3. The revolutions per minute (rpm) for a revolving door shall not exceed those shown in Table 1008.1.3.1.

4. Each revolving door shall have a side-hinged swinging door which complies with Section 1008.1 in the same wall and within 10 feet (3048 mm) of the revolving door.

TABLE 1008.1.3.1
REVOLVING DOOR SPEEDS

INSIDE DIAMETER (feet-inches)	POWER-DRIVEN-TYPE SPEED CONTROL (rpm)	MANUAL-TYPE SPEED CONTROL (rpm)
6-6	11	12
7-0	10	11
7-6	9	11
8-0	9	10
8-6	8	9
9-0	8	9
9-6	7	8
10-0	7	8

For SI: 1 inch = 25.4 mm, 1 foot = 304.8 mm.

1008.1.3.1.1 Egress component. A revolving door used as a component of a means of egress shall comply with Section 1008.1.3.1 and the following three conditions:

1. Revolving doors shall not be given credit for more than 50 percent of the required egress capacity.

2. Each revolving door shall be credited with no more than a 50-person capacity.

3. Each revolving door shall be capable of being collapsed when a force of not more than 130 pounds (578 N) is applied within 3 inches (76 mm) of the outer edge of a wing.

1008.1.3.1.2 Other than egress component. A revolving door used as other than a component of a means of egress shall comply with Section 1008.1.3.1. The collapsing force of a revolving door not used as a component of a means of egress shall not be more than 180 pounds (801 N).

Exception: A collapsing force in excess of 180 pounds (801 N) is permitted if the collapsing force is reduced to not more than 130 pounds (578 N) when at least one of the following conditions is satisfied:

1. There is a power failure or power is removed to the device holding the door wings in position.

2. There is an actuation of the automatic sprinkler system where such system is provided.

3. There is an actuation of a smoke detection system which is installed in accordance with Section 907 to provide coverage in areas within the building which are within 75 feet (22 860 mm) of the revolving doors.

4. There is an actuation of a manual control switch, in an approved location and clearly defined, which reduces the holding force to below the 130-pound (578 N) force level.

1008.1.3.2 Power-operated doors. Where means of egress doors are operated by power, such as doors with a photoelectric-actuated mechanism to open the door upon the approach of a person, or doors with power-assisted manual operation, the design shall be such that in the event of power failure, the door is capable of being opened manually to permit means of egress travel or closed where necessary to safeguard means of egress. The forces required to open these doors manually shall not exceed those specified in Section 1008.1.2, except that the force to set the door in motion shall not exceed 50 pounds (220 N). The door shall be capable of swinging from any position to the full width of the opening in which such door is installed when a force is applied to the door on the side from which egress is made. Full-power-operated doors shall comply with BHMA A156.10. Power-assisted and low-energy doors shall comply with BHMA A156.19.

Exceptions:

1. Occupancies in Group I-3.

2. Horizontal sliding doors complying with Section 1008.1.3.3.

3. For a biparting door in the emergency breakout mode, a door leaf located within a multiple-leaf opening shall be exempt from the minimum 32-inch (813 mm) single-leaf requirement of Section 1008.1.1, provided a minimum 32-inch (813 mm) clear opening is provided when the

two biparting leaves meeting in the center are broken out.

1008.1.3.3 Horizontal sliding doors. In other than Group H occupancies, horizontal sliding doors permitted to be a component of a means of egress in accordance with Exception 5 to Section 1008.1.2 shall comply with all of the following criteria:

1. The doors shall be power operated and shall be capable of being operated manually in the event of power failure.

2. The doors shall be openable by a simple method from both sides without special knowledge or effort.

3. The force required to operate the door shall not exceed 30 pounds (133 N) to set the door in motion and 15 pounds (67 N) to close the door or open it to the minimum required width.

4. The door shall be openable with a force not to exceed 15 pounds (67 N) when a force of 250 pounds (1100 N) is applied perpendicular to the door adjacent to the operating device.

5. The door assembly shall comply with the applicable fire protection rating and, where rated, shall be self-closing or automatic-closing by smoke detection, shall be installed in accordance with NFPA 80 and shall comply with Section 715.

6. The door assembly shall have an integrated standby power supply.

7. The door assembly power supply shall be electrically supervised.

8. The door shall open to the minimum required width within 10 seconds after activation of the operating device.

1008.1.3.4 Access-controlled egress doors. The entrance doors in a means of egress in buildings with an occupancy in Group A, B, E, M, R-1 or R-2 and entrance doors to tenant spaces in occupancies in Groups A, B, E, M, R-1 and R-2 are permitted to be equipped with an approved entrance and egress access control system which shall be installed in accordance with all of the following criteria:

1. A sensor shall be provided on the egress side arranged to detect an occupant approaching the doors. The doors shall be arranged to unlock by a signal from or loss of power to the sensor.

2. Loss of power to that part of the access control system which locks the doors shall automatically unlock the doors.

3. The doors shall be arranged to unlock from a manual unlocking device located 40 inches to 48 inches (1016 mm to 1219 mm) vertically above the floor and within 5 feet (1524 mm) of the secured doors. Ready access shall be provided to the manual unlocking device and the device shall be clearly identified by a sign that reads "PUSH TO EXIT." When operated, the manual unlocking device shall result

in direct interruption of power to the lock—independent of the access control system electronics—and the doors shall remain unlocked for a minimum of 30 seconds.

4. Activation of the building fire alarm system, if provided, shall automatically unlock the doors, and the doors shall remain unlocked until the fire alarm system has been reset.

5. Activation of the building automatic sprinkler or fire detection system, if provided, shall automatically unlock the doors. The doors shall remain unlocked until the fire alarm system has been reset.

6. Entrance doors in buildings with an occupancy in Group A, B, E or M shall not be secured from the egress side during periods that the building is open to the general public.

1008.1.3.5 Security grilles. In Groups B, F, M and S, horizontal sliding or vertical security grilles are permitted at the main exit and shall be openable from the inside without the use of a key or special knowledge or effort during periods that the space is occupied. The grilles shall remain secured in the full-open position during the period of occupancy by the general public. Where two or more means of egress are required, not more than one-half of the exits or exit access doorways shall be equipped with horizontal sliding or vertical security grilles.

1008.1.4 Floor elevation. There shall be a floor or landing on each side of a door. Such floor or landing shall be at the same elevation on each side of the door. Landings shall be level except for exterior landings, which are permitted to have a slope not to exceed 0.25 unit vertical in 12 units horizontal (2-percent slope).

Exceptions:

1. Doors serving individual dwelling units in Groups R-2 and R-3 as applicable in Section 101.2 where the following apply:

   1.1. A door is permitted to open at the top step of an interior flight of stairs, provided the door does not swing over the top step.

   1.2. Screen doors and storm doors are permitted to swing over stairs or landings.

2. Exterior doors as provided for in Section 1003.5, Exception 1, and Section 1017.2, which are not on an accessible route.

3. In Group R-3 occupancies, the landing at an exterior doorway shall not be more than $7^3/_4$ inches (197 mm) below the top of the threshold, provided the door, other than an exterior storm or screen door, does not swing over the landing.

4. Variations in elevation due to differences in finish materials, but not more than 0.5 inch (12.7 mm).

5. Exterior decks, patios or balconies that are part of Type B dwelling units and have impervious surfaces, and that are not more than 4 inches (102

mm) below the finished floor level of the adjacent interior space of the dwelling unit.

1008.1.5 Landings at doors. Landings shall have a width not less than the width of the stairway or the door, whichever is the greater. Doors in the fully open position shall not reduce a required dimension by more than 7 inches (178 mm). When a landing serves an occupant load of 50 or more, doors in any position shall not reduce the landing to less than one-half its required width. Landings shall have a length measured in the direction of travel of not less than 44 inches (1118 mm).

> **Exception:** Landing length in the direction of travel in Group R-3 as applicable in Section 101.2 and Group U and within individual units of Group R-2 as applicable in Section 101.2, need not exceed 36 inches (914 mm).

1008.1.6 Thresholds. Thresholds at doorways shall not exceed 0.75 inch (19.1 mm) in height for sliding doors serving dwelling units or 0.5 inch (12.7 mm) for other doors. Raised thresholds and floor level changes greater than 0.25 inch (6.4 mm) at doorways shall be beveled with a slope not greater than one unit vertical in two units horizontal (50-percent slope).

> **Exception:** The threshold height shall be limited to 7 $^3/_4$ inches (197 mm) where the occupancy is Group R-2 or R-3 as applicable in Section 101.2, the door is an exterior door that is not a component of the required means of egress and the doorway is not on an accessible route.

1008.1.7 Door arrangement. Space between two doors in series shall be 48 inches (1219 mm) minimum plus the width of a door swinging into the space. Doors in series shall swing either in the same direction or away from the space between doors.

> **Exceptions:**
> 1. The minimum distance between horizontal sliding power-operated doors in a series shall be 48 inches (1219 mm).
> 2. Storm and screen doors serving individual dwelling units in Groups R-2 and R-3 as applicable in Section 101.2 need not be spaced 48 inches (1219 mm) from the other door.
> 3. Doors within individual dwelling units in Groups R-2 and R-3 as applicable in Section 101.2 other than within Type A dwelling units.

1008.1.8 Door operations. Except as specifically permitted by this section egress doors shall be readily openable from the egress side without the use of a key or special knowledge or effort.

1008.1.8.1 Hardware. Door handles, pulls, latches, locks and other operating devices on doors required to be accessible by Chapter 11 shall not require tight grasping, tight pinching or twisting of the wrist to operate.

1008.1.8.2 Hardware height. Door handles, pulls, latches, locks and other operating devices shall be installed 34 inches (864 mm) minimum and 48 inches (1219 mm) maximum above the finished floor. Locks used only for security purposes and not used for normal operation are permitted at any height.

1008.1.8.3 Locks and latches. Locks and latches shall be permitted to prevent operation of doors where any of the following exists:

1. Places of detention or restraint.
2. In buildings in occupancy Group A having an occupant load of 300 or less, Groups B, F, M and S, and in churches, the main exterior door or doors are permitted to be equipped with key-operated locking devices from the egress side provided:
   2.1. The locking device is readily distinguishable as locked,
   2.2. A readily visible durable sign is posted on the egress side on or adjacent to the door stating: THIS DOOR TO REMAIN UNLOCKED WHEN BUILDING IS OCCUPIED. The sign shall be in letters 1 inch (25 mm) high on a contrasting background,
   2.3. The use of the key-operated locking device is revokable by the building official for due cause.
3. Where egress doors are used in pairs, approved automatic flush bolts shall be permitted to be used, provided that the door leaf having the automatic flush bolts has no doorknob or surface-mounted hardware.
4. Doors from individual dwelling or sleeping units of Group R occupancies having an occupant load of 10 or less are permitted to be equipped with a night latch, dead bolt or security chain, provided such devices are openable from the inside without the use of a key or tool.

1008.1.8.4 Bolt locks. Manually operated flush bolts or surface bolts are not permitted.

> **Exceptions:**
> 1. On doors not required for egress in individual dwelling units or sleeping units.
> 2. Where a pair of doors serves a storage or equipment room, manually operated edge- or surface-mounted bolts are permitted on the inactive leaf.

1008.1.8.5 Unlatching. The unlatching of any leaf shall not require more than one operation.

> **Exception:** More than one operation is permitted for unlatching doors in the following locations:
> 1. Places of detention or restraint.
> 2. Where manually operated bolt locks are permitted by Section 1008.1.8.4.
> 3. Doors with automatic flush bolts as permitted by Section 1008.1.8.3, Exception 3.
> 4. Doors from individual dwelling units and guestrooms of Group R occupancies as permitted by Section 1008.1.8.3, Exception 4.

1008.1.8.6 Delayed egress locks. Approved, listed, delayed egress locks shall be permitted to be installed on doors serving any occupancy except Group A, E and H occupancies in buildings that are equipped throughout with an automatic sprinkler system in accordance with Section 903.3.1.1 or an approved automatic smoke or heat detection system installed in accordance with Section 907, provided that the doors unlock in accordance with Items 1 through 6 below. A building occupant shall not be required to pass through more than one door equipped with a delayed egress lock before entering an exit.

1. The doors unlock upon actuation of the automatic sprinkler system or automatic fire detection system.

2. The doors unlock upon loss of power controlling the lock or lock mechanism.

3. The door locks shall have the capability of being unlocked by a signal from the fire command center.

4. The initiation of an irreversible process which will release the latch in not more than 15 seconds when a force of not more than 15 pounds (67 N) is applied for 1 second to the release device. Initiation of the irreversible process shall activate an audible signal in the vicinity of the door. Once the door lock has been released by the application of force to the releasing device, relocking shall be by manual means only.

   Exception: Where approved, a delay of not more than 30 seconds is permitted.

5. A sign shall be provided on the door located above and within 12 inches (305 mm) of the release device reading: PUSH UNTIL ALARM SOUNDS. DOOR CAN BE OPENED IN 15 [30] SECONDS.

6. Emergency lighting shall be provided at the door.

1008.1.8.7 Stairway doors. Interior stairway means of egress doors shall be openable from both sides without the use of a key or special knowledge or effort.

Exceptions:

1. Stairway discharge doors shall be openable from the egress side and shall only be locked from the opposite side.

2. This section shall not apply to doors arranged in accordance with Section 403.12.

3. In stairways serving not more than four stories, doors are permitted to be locked from the side opposite the egress side, provided they are openable from the egress side.

1008.1.9 Panic and fire exit hardware. Where panic and fire exit hardware is installed, it shall comply with the following:

1. The actuating portion of the releasing device shall extend at least one-half of the door leaf width.

2. A maximum unlatching force of 15 pounds (67 N).

Each door in a means of egress from an occupancy of Group A or E having an occupant load of 100 or more and any occupancy of Group H-1, H-2, H-3 or H-5 shall not be provided with a latch or lock unless it is panic hardware or fire exit hardware.

If balanced doors are used and panic hardware is required, the panic hardware shall be the push-pad type and the pad shall not extend more than one-half the width of the door measured from the latch side.

1008.2 Gates. Gates serving the means of egress system shall comply with the requirements of this section. Gates used as a component in a means of egress shall conform to the applicable requirements for doors.

Exception: Horizontal sliding or swinging gates exceeding the 4-foot (1219 mm) maximum leaf width limitation are permitted in fences and walls surrounding a stadium.

1008.2.1 Stadiums. Panic hardware is not required on gates surrounding stadiums where such gates are under constant immediate supervision while the public is present, and further provided that safe dispersal areas based on 3 square feet (0.28 m²) per occupant are located between the fence and enclosed space. Such required safe dispersal areas shall not be located less than 50 feet (15 240 mm) from the enclosed space. See Section 1017 for means of egress from safe dispersal areas.

1008.3 Turnstiles. Turnstiles or similar devices that restrict travel to one direction shall not be placed so as to obstruct any required means of egress.

Exception: Each turnstile or similar device shall be credited with no more than a 50-person capacity where all of the following provisions are met:

1. Each device shall turn free in the direction of egress travel when primary power is lost, and upon the manual release by an employee in the area.

2. Such devices are not given credit for more than 50 percent of the required egress capacity.

3. Each device is not more than 39 inches (991 mm) high.

4. Each device has at least 16.5 inches (419 mm) clear width at and below a height of 39 inches (991 mm) and at least 22 inches (559 mm) clear width at heights above 39 inches (991 mm).

   Where located as part of an accessible route, turnstiles shall have at least 36 inches (914 mm) clear at and below a height of 34 inches (864 mm), at least 32 inches (813 mm) clear width between 34 inches (864 mm) and 80 inches (2032 mm) and shall consist of a mechanism other than a revolving device.

1008.3.1 High turnstile. Turnstiles more than 39 inches (991 mm) high shall meet the requirements for revolving doors.

1008.3.2 Additional door. Where serving an occupant load greater than 300, each turnstile that is not portable shall have a side-hinged swinging door which conforms to Section 1008.1 within 50 feet (15 240 mm).

SECTION 1009
STAIRWAYS AND HANDRAILS

1009.1 Stairway width. The width of stairways shall be determined as specified in Section 1005.1, but such width shall not be less than 44 inches (1118 mm). See Section 1007.3 for accessible means of egress stairways.

Exceptions:

1. Stairways serving an occupant load of 50 or less shall have a width of not less than 36 inches (914 mm).

2. Spiral stairways as provided for in Section 1009.9.

3. Aisle stairs complying with Section 1024.

4. Where a stairway lift is installed on stairways serving occupancies in Group R-3, or within dwelling units in occupancies in Group R-2, both as applicable in Section 101.2, a clear passage width not less than 20 inches (508 mm) shall be provided. If the seat and platform can be folded when not in use, the distance shall be measured from the folded position.

1009.2 Headroom. Stairways shall have a minimum headroom clearance of 80 inches (2032 mm) measured vertically from a line connecting the edge of the nosings. Such headroom shall be continuous above the stairway to the point where the line intersects the landing below, one tread depth beyond the bottom riser. The minimum clearance shall be maintained the full width of the stairway and landing.

Exception: Spiral stairways complying with Section 1009.9 are permitted a 78-inch (1981 mm) headroom clearance.

1009.3 Stair treads and risers. Stair riser heights shall be 7 inches (178 mm) maximum and 4 inches (102 mm) minimum. Stair tread depths shall be 11 inches (279 mm) minimum. The riser height shall be measured vertically between the leading edges of adjacent treads. The greatest riser height within any flight of stairs shall not exceed the smallest by more than 0.375 inch (9.5 mm). The tread depth shall be measured horizontally between the vertical planes of the foremost projection of adjacent treads and at right angle to the tread's leading edge. The greatest tread depth within any flight of stairs shall not exceed the smallest by more than 0.375 inch (9.5 mm). Winder treads shall have a minimum tread depth of 11 inches (279 mm) measured at a right angle to the tread's leading edge at a point 12 inches (305 mm) from the side where the treads are narrower and a minimum tread depth of 10 inches (254 mm). The greatest winder tread depth at the 12-inch (305 mm) walk line within any flight of stairs shall not exceed the smallest by more than 0.375 inch (9.5 mm).

Exceptions:

1. Circular stairways in accordance with Section 1009.7.

2. Winders in accordance with Section 1009.8.

3. Spiral stairways in accordance with Section 1009.9.

4. Aisle stairs in assembly seating areas where the stair pitch or slope is set, for sightline reasons, by the slope of the adjacent seating area in accordance with Section 1024.11.2.

5. In occupancies in Group R-3, as applicable in Section 101.2, within dwelling units in occupancies in Group R-2, as applicable in Section 101.2, and in occupancies in Group U, which are accessory to an occupancy in Group R-3, as applicable in Section 101.2, the maximum riser height shall be 7.75 inches (197 mm) and the minimum tread depth shall be 10 inches (254 mm), the minimum winder tread depth at the walk line shall be 10 inches (254 mm), and the minimum winder tread depth shall be 6 inches (152 mm). A nosing not less than 0.75 inch (19.1 mm) but not more than 1.25 inches (32 mm) shall be provided on stairways with solid risers where the tread depth is less than 11 inches (279 mm).

6. See the *International Existing Building Code* for the replacement of existing stairways.

1009.3.1 Dimensional uniformity. Stair treads and risers shall be of uniform size and shape. The tolerance between the largest and smallest riser or between the largest and smallest tread shall not exceed 0.375 inch (9.5 mm) in any flight of stairs.

Exceptions:

1. Nonuniform riser dimensions of aisle stairs complying with Section 1024.11.2.

2. Consistently shaped winders, complying with Section 1009.8, differing from rectangular treads in the same stairway flight.

Where the bottom or top riser adjoins a sloping public way, walkway or driveway having an established grade and serving as a landing, the bottom or top riser is permitted to be reduced along the slope to less than 4 inches (102 mm) in height with the variation in height of the bottom or top riser not to exceed one unit vertical in 12 units horizontal (8-percent slope) of stairway width. The nosings or leading edges of treads at such nonuniform height risers shall have a distinctive marking stripe, different from any other nosing marking provided on the stair flight. The distinctive marking stripe shall be visible in descent of the stair and shall have a slip-resistant surface. Marking stripes shall have a width of at least 1 inch (25 mm) but not more than 2 inches (51 mm).

1009.3.2 Profile. The radius of curvature at the leading edge of the tread shall be not greater than 0.5 inch (12.7 mm). Beveling of nosings shall not exceed 0.5 inch (12.7 mm). Risers shall be solid and vertical or sloped from the underside of the leading edge of the tread above at an angle not more than 30 degrees (0.52 rad) from the vertical. The leading edge (nosings) of treads shall project not more than 1.25 inches (32 mm) beyond the tread below and all projections of the leading edges shall be of uniform size, including the leading edge of the floor at the top of a flight.

Exceptions:

1. Solid risers are not required for stairways that are not required to comply with Section 1007.3, provided that the opening between treads does not permit the passage of a sphere with a diameter of 4 inches (102 mm).

2. Solid risers are not required for occupancies in Group I-3.

1009.4 Stairway landings. There shall be a floor or landing at the top and bottom of each stairway. The width of landings shall not be less than the width of stairways they serve. Every landing shall have a minimum dimension measured in the direction of travel equal to the width of the stairway. Such dimension need not exceed 48 inches (1219 mm) where the stairway has a straight run.

Exceptions:

1. Aisle stairs complying with Section 1024.

2. Doors opening onto a landing shall not reduce the landing to less than one-half the required width. When fully open, the door shall not project more than 7 inches (178 mm) into a landing.

1009.5 Stairway construction. All stairways shall be built of materials consistent with the types permitted for the type of construction of the building, except that wood handrails shall be permitted for all types of construction.

1009.5.1 Stairway walking surface. The walking surface of treads and landings of a stairway shall not be sloped steeper than one unit vertical in 48 units horizontal (2-percent slope) in any direction. Stairway treads and landings shall have a solid surface. Finish floor surfaces shall be securely attached.

Exception: In Group F, H and S occupancies, other than areas of parking structures accessible to the public, openings in treads and landings shall not be prohibited provided a sphere with a diameter of $1^1/_8$ inches (29 mm) cannot pass through the opening.

1009.5.2 Outdoor conditions. Outdoor stairways and outdoor approaches to stairways shall be designed so that water will not accumulate on walking surfaces. In other than occupancies in Group R-3, and occupancies in Group U that are accessory to an occupancy in Group R-3, treads, platforms and landings that are part of exterior stairways in climates subject to snow or ice shall be protected to prevent the accumulation of same.

1009.6 Vertical rise. A flight of stairs shall not have a vertical rise greater than 12 feet (3658 mm) between floor levels or landings.

Exception: Aisle stairs complying with Section 1024.

1009.7 Circular stairways. Circular stairways shall have a minimum tread depth and a maximum riser height in accordance with Section 1009.3 and the smaller radius shall not be less than twice the width of the stairway. The minimum tread depth measured 12 inches (305 mm) from the narrower end of the tread shall not be less than 11 inches (279 mm). The minimum tread depth at the narrow end shall not be less than 10 inches (254 mm).

Exception: For occupancies in Group R-3, and within individual dwelling units in occupancies in Group R-2, both as applicable in Section 101.2.

1009.8 Winders. Winders are not permitted in means of egress stairways except within a dwelling unit.

1009.9 Spiral stairways. Spiral stairways are permitted to be used as a component in the means of egress only within dwelling units or from a space not more than 250 square feet (23 m²) in area and serving not more than five occupants, or from galleries, catwalks and gridirons in accordance with Section 1014.6.

A spiral stairway shall have a 7.5-inch (191 mm) minimum clear tread depth at a point 12 inches (305 mm) from the narrow edge. The risers shall be sufficient to provide a headroom of 78 inches (1981 mm) minimum, but riser height shall not be more than 9.5 inches (241 mm). The minimum stairway width shall be 26 inches (660 mm).

1009.10 Alternating tread devices. Alternating tread devices are limited to an element of a means of egress in buildings of Groups F, H and S from a mezzanine not more than 250 square feet (23 m²) in area and which serves not more than five occupants; in buildings of Group I-3 from a guard tower, observation station or control room not more than 250 square feet (23 m²) in area and for access to unoccupied roofs.

1009.10.1 Handrails of alternating tread devices. Handrails shall be provided on both sides of alternating tread devices and shall conform to Section 1009.11.

1009.10.2 Treads of alternating tread devices. Alternating tread devices shall have a minimum projected tread of 5 inches (127 mm), a minimum tread depth of 8.5 inches (216 mm), a minimum tread width of 7 inches (178 mm) and a maximum riser height of 9.5 inches (241 mm). The initial tread of the device shall begin at the same elevation as the platform, landing or floor surface.

Exception: Alternating tread devices used as an element of a means of egress in buildings from a mezzanine area not more than 250 square feet (23 m²) in area which serves not more than five occupants shall have a minimum projected tread of 8.5 inches (216 mm) with a minimum tread depth of 10.5 inches (267 mm). The rise to the next alternating tread surface should not be more than 8 inches (203 mm).

1009.11 Handrails. Stairways shall have handrails on each side. Handrails shall be adequate in strength and attachment in accordance with Section 1607.7. Handrails for ramps, where required by Section 1010.8, shall comply with this section.

Exceptions:

1. Aisle stairs complying with Section 1024 provided with a center handrail need not have additional handrails.

2. Stairways within dwelling units, spiral stairways and aisle stairs serving seating only on one side are permitted to have a handrail on one side only.

3. Decks, patios and walkways that have a single change in elevation where the landing depth on each side of the change of elevation is greater than what is required for a landing do not require handrails.

4. In Group R-3 occupancies, a change in elevation consisting of a single riser at an entrance or egress door does not require handrails.

5. Changes in room elevations of only one riser within dwelling units and sleeping units in Group R-2 and R-3 occupancies do not require handrails.

1009.11.1 Height. Handrail height, measured above stair tread nosings, or finish surface of ramp slope, shall be uniform, not less than 34 inches (864 mm) and not more than 38 inches (965 mm).

1009.11.2 Intermediate handrails. Intermediate handrails are required so that all portions of the stairway width required for egress capacity are within 30 inches (762 mm) of a handrail. On monumental stairs, handrails shall be located along the most direct path of egress travel.

1009.11.3 Handrail graspability. Handrails with a circular cross section shall have an outside diameter of at least 1.25 inches (32 mm) and not greater than 2 inches (51 mm) or shall provide equivalent graspability. If the handrail is not circular, it shall have a perimeter dimension of at least 4 inches (102 mm) and not greater than 6.25 inches (160 mm) with a maximum cross-section dimension of 2.25 inches (57 mm). Edges shall have a minimum radius of 0.01 inch (0.25 mm).

1009.11.4 Continuity. Handrail-gripping surfaces shall be continuous, without interruption by newel posts or other obstructions.

Exceptions:

1. Handrails within dwelling units are permitted to be interrupted by a newel post at a stair landing.

2. Within a dwelling unit, the use of a volute, turnout or starting easing is allowed on the lowest tread.

3. Handrail brackets or balusters attached to the bottom surface of the handrail that do not project horizontally beyond the sides of the handrail within 1.5 inches (38 mm) of the bottom of the handrail shall not be considered to be obstructions and provided further that for each 0.5 inch (13 mm) of additional handrail perimeter dimension above 4 inches (102 mm), the vertical clearance dimension of 1.5 inches (38 mm) shall be permitted to be reduced by 0.125 inch (3 mm).

1009.11.5 Handrail extensions. Handrails shall return to a wall, guard or the walking surface or shall be continuous to the handrail of an adjacent stair flight. Where handrails are not continuous between flights, the handrails shall extend horizontally at least 12 inches (305 mm) beyond the top riser and continue to slope for the depth of one tread beyond the bottom riser.

Exceptions:

1. Handrails within a dwelling unit that is not required to be accessible need extend only from the top riser to the bottom riser.

2. Aisle handrails in Group A occupancies in accordance with Section 1024.13.

1009.11.6 Clearance. Clear space between a handrail and a wall or other surface shall be a minimum of 1.5 inches (38 mm). A handrail and a wall or other surface adjacent to the handrail shall be free of any sharp or abrasive elements.

1009.11.7 Stairway projections. Projections into the required width at each handrail shall not exceed 4.5 inches (114 mm) at or below the handrail height. Projections into the required width shall not be limited above the minimum headroom height required in Section 1009.2.

1009.12 Stairway to roof. In buildings four or more stories in height above grade, one stairway shall extend to the roof surface, unless the roof has a slope steeper than four units vertical in 12 units horizontal (33-percent slope). In buildings without an occupied roof, access to the roof from the top story shall be permitted to be by an alternating tread device.

1009.12.1 Roof access. Where a stairway is provided to a roof, access to the roof shall be provided through a penthouse complying with Section 1509.2.

Exception: In buildings without an occupied roof, access to the roof shall be permitted to be a roof hatch or trap door not less than 16 square feet (1.5 m²) in area and having a minimum dimension of 2 feet (610 mm).

SECTION 1010
RAMPS

1010.1 Scope. The provisions of this section shall apply to ramps used as a component of a means of egress.

Exceptions:

1. Other than ramps that are part of the accessible routes providing access in accordance with Sections 1108.2.2 through 1108.2.4.1, ramped aisles within assembly rooms or spaces shall conform with the provisions in Section 1024.11.

2. Curb ramps shall comply with ICC A117.1.

3. Vehicle ramps in parking garages for pedestrian exit access shall not be required to comply with Sections 1010.3 through 1010.9 when they are not an accessible route serving accessible parking spaces, other required accessible elements or part of an accessible means of egress.

1010.2 Slope. Ramps used as part of a means of egress shall have a running slope not steeper than one unit vertical in 12 units horizontal (8-percent slope). The slope of other ramps shall not be steeper than one unit vertical in eight units horizontal (12.5-percent slope).

Exception: Aisle ramp slope in occupancies of Group A shall comply with Section 1024.11.

1010.3 Cross slope. The slope measured perpendicular to the direction of travel of a ramp shall not be steeper than one unit vertical in 48 units horizontal (2-percent slope).

1010.4 Vertical rise. The rise for any ramp run shall be 30 inches (762 mm) maximum.

1010.5 Minimum dimensions. The minimum dimensions of means of egress ramps shall comply with Sections 1010.5.1 through 1010.5.3.

1010.5.1 Width. The minimum width of a means of egress ramp shall not be less than that required for corridors by Section 1016.2. The clear width of a ramp and the clear width between handrails, if provided, shall be 36 inches (914 mm) minimum.

1010.5.2 Headroom. The minimum headroom in all parts of the means of egress ramp shall not be less than 80 inches (2032 mm).

1010.5.3 Restrictions. Means of egress ramps shall not reduce in width in the direction of egress travel. Projections into the required ramp and landing width are prohibited. Doors opening onto a landing shall not reduce the clear width to less than 42 inches (1067 mm).

1010.6 Landings. Ramps shall have landings at the bottom and top of each ramp, points of turning, entrance, exits and at doors. Landings shall comply with Sections 1010.6.1 through 1010.6.5.

1010.6.1 Slope. Landings shall have a slope not steeper than one unit vertical in 48 units horizontal (2-percent slope) in any direction. Changes in level are not permitted.

1010.6.2 Width. The landing shall be at least as wide as the widest ramp run adjoining the landing.

1010.6.3 Length. The landing length shall be 60 inches (1525 mm) minimum.

> **Exception:** Landings in nonaccessible Group R-2 and R-3 individual dwelling units, as applicable in Section 101.2, are permitted to be 36 inches (914 mm) minimum.

1010.6.4 Change in direction. Where changes in direction of travel occur at landings provided between ramp runs, the landing shall be 60 inches by 60 inches (1524 mm by 1524 mm) minimum.

> **Exception:** Landings in nonaccessible Group R-2 and R-3 individual dwelling units, as applicable in Section 101.2, are permitted to be 36 inches by 36 inches (914 mm by 914 mm) minimum.

1010.6.5 Doorways. Where doorways are located adjacent to a ramp landing, maneuvering clearances required by ICC A117.1 are permitted to overlap the required landing area.

1010.7 Ramp construction. All ramps shall be built of materials consistent with the types permitted for the type of construction of the building; except that wood handrails shall be permitted for all types of construction. Ramps used as an exit shall conform to the applicable requirements of Sections 1019.1 and 1019.1.1 through 1019.1.3 for vertical exit enclosures.

1010.7.1 Ramp surface. The surface of ramps shall be of slip-resistant materials that are securely attached.

1010.7.2 Outdoor conditions. Outdoor ramps and outdoor approaches to ramps shall be designed so that water will not accumulate on walking surfaces. In other than occupancies in Group R-3, and occupancies in Group U that are accessory to an occupancy in Group R-3, surfaces and landings which are part of exterior ramps in climates subject to snow or ice shall be designed to minimize the accumulation of same.

1010.8 Handrails. Ramps with a rise greater than 6 inches (152 mm) shall have handrails on both sides complying with Section 1009.11.

1010.9 Edge protection. Edge protection complying with Section 1010.9.1 or 1010.9.2 shall be provided on each side of ramp runs and at each side of ramp landings.

> **Exceptions:**
> 1. Edge protection is not required on ramps not required to have handrails, provided they have flared sides that comply with the ICC A117.1 curb ramp provisions.
> 2. Edge protection is not required on the sides of ramp landings serving an adjoining ramp run or stairway.
> 3. Edge protection is not required on the sides of ramp landings having a vertical dropoff of not more than 0.5 inch (13 mm) within 10 inches (254 mm) horizontally of the required landing area.

1010.9.1 Railings. A rail shall be mounted below the handrail 17 inches to 19 inches (432 mm to 483 mm) above the ramp or landing surface.

1010.9.2 Curb or barrier. A curb or barrier shall be provided that prevents the passage of a 4-inch-diameter (102 mm) sphere, where any portion of the sphere is within 4 inches (102 mm) of the floor or ground surface.

1010.10 Guards. Guards shall be provided where required by Section 1012 and shall be constructed in accordance with Section 1012.

SECTION 1011
EXIT SIGNS

1011.1 Where required. Exits and exit access doors shall be marked by an approved exit sign readily visible from any direction of egress travel. Access to exits shall be marked by readily visible exit signs in cases where the exit or the path of egress travel is not immediately visible to the occupants. Exit sign placement shall be such that no point in an exit access corridor is more than 100 feet (30 480 mm) or the listed viewing distance for the sign, whichever is less, from the nearest visible exit sign.

> **Exceptions:**
> 1. Exit signs are not required in rooms or areas which require only one exit or exit access.
> 2. Main exterior exit doors or gates which obviously and clearly are identifiable as exits need not have exit signs where approved by the building official.
> 3. Exit signs are not required in occupancies in Group U and individual sleeping units or dwelling units in Group R-1, R-2 or R-3.
> 4. Exit signs are not required in sleeping areas in occupancies in Group I-3.
> 5. In occupancies in Groups A-4 and A-5, exit signs are not required on the seating side of vomitories or openings into seating areas where exit signs are provided in the concourse that are readily apparent from the vomitories. Egress lighting is provided to identify each

vomitory or opening within the seating area in an emergency.

1011.2 Illumination. Exit signs shall be internally or externally illuminated.

> **Exception:** Tactile signs required by Section 1011.3 need not be provided with illumination.

1011.3 Tactile exit signs. A tactile sign stating EXIT and complying with ICC A117.1 shall be provided adjacent to each door to an egress stairway, an exit passageway and the exit discharge.

1011.4 Internally illuminated exit signs. Internally illuminated exit signs shall be listed and labeled and shall be installed in accordance with the manufacturer's instructions and Section 2702. Exit signs shall be illuminated at all times.

1011.5 Externally illuminated exit signs. Externally illuminated exit signs shall comply with Sections 1011.5.1 through 1011.5.3.

1011.5.1 Graphics. Every exit sign and directional exit sign shall have plainly legible letters not less than 6 inches (152 mm) high with the principal strokes of the letters not less than 0.75 inch (19.1 mm) wide. The word "EXIT" shall have letters having a width not less than 2 inches (51 mm) wide except the letter "I," and the minimum spacing between letters shall not be less than 0.375 inch (9.5 mm). Signs larger than the minimum established in this section shall have letter widths, strokes and spacing in proportion to their height.

The word "EXIT" shall be in high contrast with the background and shall be clearly discernible when the exit sign illumination means is or is not energized. If an arrow is provided as part of the exit sign, the construction shall be such that the arrow direction cannot be readily changed.

1011.5.2 Exit sign illumination. The face of an exit sign illuminated from an external source shall have an intensity of not less than 5 foot-candles (54 lux).

1011.5.3 Power source. Exit signs shall be illuminated at all times. To ensure continued illumination for a duration of not less than 90 minutes in case of primary power loss, the sign illumination means shall be connected to an emergency power system provided from storage batteries, unit equipment or an on-site generator. The installation of the emergency power system shall be in accordance with Section 2702.

> **Exception:** Approved exit sign illumination means that provide continuous illumination independent of external power sources for a duration of not less than 90 minutes, in case of primary power loss, are not required to be connected to an emergency electrical system.

SECTION 1012
GUARDS

1012.1 Where required. Guards shall be located along open-sided walking surfaces, mezzanines, industrial equipment platforms, stairways, ramps and landings which are located more than 30 inches (762 mm) above the floor or grade below. Guards shall be adequate in strength and attachment in accordance with Section 1607.7. Guards shall also be located along glazed sides of stairways, ramps and landings that are located more than 30 inches (762 mm) above the floor or grade below where the glazing provided does not meet the strength and attachment requirements in Section 1607.7.

> **Exception:** Guards are not required for the following locations:
>
> 1. On the loading side of loading docks or piers.
> 2. On the audience side of stages and raised platforms, including steps leading up to the stage and raised platforms.
> 3. On raised stage and platform floor areas such as runways, ramps and side stages used for entertainment or presentations.
> 4. At vertical openings in the performance area of stages and platforms.
> 5. At elevated walking surfaces appurtenant to stages and platforms for access to and utilization of special lighting or equipment.
> 6. Along vehicle service pits not accessible to the public.
> 7. In assembly seating where guards in accordance with Section 1024.14 are permitted and provided.

1012.2 Height. Guards shall form a protective barrier not less than 42 inches (1067 mm) high, measured vertically above the leading edge of the tread, adjacent walking surface or adjacent seatboard.

> **Exceptions:**
>
> 1. For occupancies in Group R-3, and within individual dwelling units in occupancies in Group R-2, both as applicable in Section 101.2, guards whose top rail also serves as a handrail shall have a height not less than 34 inches (864 mm) and not more than 38 inches (965 mm) measured vertically from the leading edge of the stair tread nosing.
> 2. The height in assembly seating areas shall be in accordance with Section 1024.14.

1012.3 Opening limitations. Open guards shall have balusters or ornamental patterns such that a 4-inch-diameter (102 mm) sphere cannot pass through any opening up to a height of 34 inches (864 mm). From a height of 34 inches (864 mm) to 42 inches (1067 mm) above the adjacent walking surfaces, a sphere 8 inches (203 mm) in diameter shall not pass.

> **Exceptions:**
>
> 1. The triangular openings formed by the riser, tread and bottom rail at the open side of a stairway shall be of a maximum size such that a sphere of 6 inches (152 mm) in diameter cannot pass through the opening.
> 2. At elevated walking surfaces for access to and use of electrical, mechanical or plumbing systems or equipment, guards shall have balusters or be of solid materials such that a sphere with a diameter of 21 inches (533 mm) cannot pass through any opening.
> 3. In areas which are not open to the public within occupancies in Group I-3, F, H or S, balusters, horizontal intermediate rails or other construction shall not per-

mit a sphere with a diameter of 21 inches (533 mm) to pass through any opening.

4. In assembly seating areas, guards at the end of aisles where they terminate at a fascia of boxes, balconies and galleries shall have balusters or ornamental patterns such that a 4-inch-diameter (102 mm) sphere cannot pass through any opening up to a height of 26 inches (660 mm). From a height of 26 inches (660 mm) to 42 inches (1067 mm) above the adjacent walking surfaces, a sphere 8 inches (203 mm) in diameter shall not pass.

1012.4 Screen porches. Porches and decks which are enclosed with insect screening shall be provided with guards where the walking surface is located more than 30 inches (762 mm) above the floor or grade below.

1012.5 Mechanical equipment. Guards shall be provided where appliances, equipment, fans or other components that require service are located within 10 feet (3048 mm) of a roof edge or open side of a walking surface and such edge or open side is located more than 30 inches (762 mm) above the floor, roof or grade below. The guard shall be constructed so as to prevent the passage of a 21-inch-diameter (533 mm) sphere.

SECTION 1013
EXIT ACCESS

1013.1 General. The exit access arrangement shall comply with Sections 1013 through 1016 and the applicable provisions of Sections 1003 through 1012.

1013.2 Egress through intervening spaces. Egress from a room or space shall not pass through adjoining or intervening rooms or areas, except where such adjoining rooms or areas are accessory to the area served; are not a high-hazard occupancy and provide a discernible path of egress travel to an exit. Egress shall not pass through kitchens, storage rooms, closets or spaces used for similar purposes. An exit access shall not pass through a room that can be locked to prevent egress. Means of egress from dwelling units or sleeping areas shall not lead through other sleeping areas, toilet rooms or bathrooms.

Exceptions:

1. Means of egress are not prohibited through a kitchen area serving adjoining rooms constituting part of the same dwelling unit or sleeping unit.

2. Means of egress are not prohibited through adjoining or intervening rooms or spaces in a Group H occupancy when the adjoining or intervening rooms or spaces are the same or a lesser hazard occupancy group.

1013.2.1 Multiple tenants. Where more than one tenant occupies any one floor of a building or structure, each tenant space, dwelling unit and sleeping unit shall be provided with access to the required exits without passing through adjacent tenant spaces, dwelling units and sleeping units.

1013.2.2 Group I-2. Habitable rooms or suites in Group I-2 occupancies shall have an exit access door leading directly to an exit access corridor.

Exceptions:

1. Rooms with exit doors opening directly to the outside at ground level.

2. Patient sleeping rooms are permitted to have one intervening room if the intervening room is not used as an exit access for more than eight patient beds.

3. Special nursing suites are permitted to have one intervening room where the arrangement allows for direct and constant visual supervision by nursing personnel.

4. For rooms other than patient sleeping rooms, suites of rooms are permitted to have one intervening room if the travel distance within the suite to the exit access door is not greater than 100 feet (30 480 mm) and are permitted to have two intervening rooms where the travel distance within the suite to the exit access door is not greater than 50 feet (15 240 mm).

Suites of sleeping rooms shall not exceed 5,000 square feet (465 m²). Suites of rooms, other than patient sleeping rooms, shall not exceed 10,000 square feet (929 m²). Any patient sleeping room, or any suite that includes patient sleeping rooms, of more than 1,000 square feet (93 m²) shall have at least two exit access doors remotely located from each other. Any room or suite of rooms, other than patient sleeping rooms, of more than 2,500 square feet (232 m²) shall have at least two access doors remotely located from each other. The travel distance between any point in a Group I-2 occupancy and an exit access door in the room shall not exceed 50 feet (15 240 mm). The travel distance between any point in a suite of sleeping rooms and an exit access door of that suite shall not exceed 100 feet (30 480 mm).

1013.3 Common path of egress travel. In occupancies other than Groups H-1, H-2 and H-3, the common path of egress travel shall not exceed 75 feet (22 860 mm). In occupancies in Groups H-1, H-2, and H-3, the common path of egress travel shall not exceed 25 feet (7620 mm).

Exceptions:

1. The length of a common path of egress travel in an occupancy in Groups B, F and S shall not be more than 100 feet (30 480 mm), provided that the building is equipped throughout with an automatic sprinkler system installed in accordance with Section 903.3.1.1.

2. Where a tenant space in an occupancy in Groups B, S and U has an occupant load of not more than 30, the length of a common path of egress travel shall not be more than 100 feet (30 480 mm).

3. The length of a common path of egress travel in occupancies in Group I-3 shall not be more than 100 feet (30 480 mm).

1013.4 Aisles. Aisles serving as a portion of the exit access in the means of egress system shall comply with the requirements of this section. Aisles shall be provided from all occupied portions of the exit access which contain seats, tables, furnishings, displays and similar fixtures or equipment. Aisles serving assembly areas, other than seating at tables, shall comply with

Section 1024. Aisles serving reviewing stands, grandstands and bleachers shall also comply with Section 1024.

The required width of aisles shall be unobstructed.

Exception: Doors, when fully opened, and handrails shall not reduce the required width by more than 7 inches (178 mm). Doors in any position shall not reduce the required width by more than one-half. Other nonstructural projections such as trim and similar decorative features are permitted to project into the required width 1.5 inches (38 mm) from each side.

1013.4.1 Groups B and M. In Group B and M occupancies, the minimum clear aisle width shall be determined by Section 1005.1 for the occupant load served, but shall not be less than 36 inches (914 mm).

Exception: Nonpublic aisles serving less than 50 people, and not required to be accessible by Chapter 11 need not exceed 28 inches (711 mm) in width.

1013.4.2 Seating at tables. Where seating is located at a table or counter and is adjacent to an aisle or aisle accessway, the measurement of required clear width of the aisle or aisle accessway shall be made to a line 19 inches (483 mm) away from and parallel to the edge of the table or counter. The 19-inch (483 mm) distance shall be measured perpendicular to the side of the table or counter. In the case of other side boundaries for aisle or aisle accessways, the clear width shall be measured to walls, edges of seating and tread edges, except that handrail projections are permitted.

Exception: Where tables or counters are served by fixed seats, the width of the aisle accessway shall be measured from the back of the seat.

1013.4.2.1 Aisle accessway for tables and seating. Aisle accessways serving arrangements of seating at tables or counters shall have sufficient clear width to conform to the capacity requirements of Section 1005.1 but shall not have less than the appropriate minimum clear width specified in Section 1013.4.1.

1013.4.2.2 Table and seating accessway width. Aisle accessways shall provide a minimum of 12 inches (305 mm) of width plus 0.5 inch (12.7 mm) of width for each additional 1 foot (305 mm), or fraction thereof, beyond 12 feet (3658 mm) of aisle accessway length measured from the center of the seat farthest from an aisle.

Exception: Portions of an aisle accessway having a length not exceeding 6 feet (1829 mm) and used by a total of not more than four persons.

1013.4.2.3 Table and seating aisle accessway length. The length of travel along the aisle accessway shall not exceed 30 feet (9144 mm) from any seat to the point where a person has a choice of two or more paths of egress travel to separate exits.

1013.5 Egress balconies. Balconies used for egress purposes shall conform to the same requirements as corridors for width, headroom, dead ends and projections. Exterior balconies shall be designed to minimize accumulation of snow or ice that impedes the means of egress.

Exception: Exterior balconies and concourses in outdoor stadiums shall be exempt from the design requirement to protect against the accumulation of snow or ice.

1013.5.1 Wall separation. Exterior egress balconies shall be separated from the interior of the building by walls and opening protectives as required for corridors.

Exception: Separation is not required where the exterior egress balcony is served by at least two stairs and a dead-end travel condition does not require travel past an unprotected opening to reach a stair.

1013.5.2 Openness. The long side of an egress balcony shall be at least 50 percent open, and the open area above the guards shall be so distributed as to minimize the accumulation of smoke or toxic gases.

SECTION 1014
EXIT AND EXIT ACCESS DOORWAYS

1014.1 Exit or exit access doorways required. Two exits or exit access doorways from any space shall be provided where one of the following conditions exists:

1. The occupant load of the space exceeds the values in Table 1014.1.
2. The common path of egress travel exceeds the limitations of Section 1013.3.
3. Where required by Sections 1014.3, 1014.4 and 1014.5.

Exception: Group I-2 occupancies shall comply with Section 1013.2.2.

TABLE 1014.1
SPACES WITH ONE MEANS OF EGRESS

OCCUPANCY	MAXIMUM OCCUPANT LOAD
A, B, E, F, M, U	50
H-1, H-2, H-3	3
H-4, H-5, I-1, I-3, I-4, R	10
S	30

1014.1.1 Three or more exits. Access to three or more exits shall be provided from a floor area where required by Section 1018.1.

1014.2 Exit or exit access doorway arrangement. Required exits shall be located in a manner that makes their availability obvious. Exits shall be unobstructed at all times. Exit and exit access doorways shall be arranged in accordance with Sections 1014.2.1 and 1014.2.2.

1014.2.1 Two exits or exit access doorways. Where two exits or exit access doorways are required from any portion of the exit access, the exit doors or exit access doorways shall be placed a distance apart equal to not less than one-half of the length of the maximum overall diagonal dimension of the building or area to be served measured in a straight line between exit doors or exit access doorways. Interlocking or scissor stairs shall be counted as one exit stairway.

Exceptions:

1. Where exit enclosures are provided as a portion of the required exit and are interconnected by a 1-hour fire-resistance-rated corridor conforming to the requirements of Section 1016, the required exit separation shall be measured along the shortest direct line of travel within the corridor.

2. Where a building is equipped throughout with an automatic sprinkler system in accordance with Section 903.3.1.1 or 903.3.1.2, the separation distance of the exit doors or exit access doorways shall not be less than one-third of the length of the maximum overall diagonal dimension of the area served.

1014.2.2 Three or more exits or exit access doorways. Where access to three or more exits is required, at least two exit doors or exit access doorways shall be placed a distance apart equal to not less than one-half of the length of the maximum overall diagonal dimension of the area served measured in a straight line between such exit doors or exit access doorways. Additional exits or exit access doorways shall be arranged a reasonable distance apart so that if one becomes blocked, the others will be available.

Exception: Where a building is equipped throughout with an automatic sprinkler system in accordance with Section 903.3.1.1 or 903.3.1.2, the separation distance of at least two of the exit doors or exit access doorways shall not be less than one-third of the length of the maximum overall diagonal dimension of the area served.

1014.3 Boiler, incinerator and furnace rooms. Two exit access doorways are required in boiler, incinerator and furnace rooms where the area is over 500 square feet (46 m²) and any fuel-fired equipment exceeds 400,000 British thermal units (Btu) (422 000 KJ) input capacity. Where two exit access doorways are required, one is permitted to be a fixed ladder or an alternating tread device. Exit access doorways shall be separated by a horizontal distance equal to one-half the maximum horizontal dimension of the room.

1014.4 Refrigeration machinery rooms. Machinery rooms larger than 1,000 square feet (93 m²) shall have not less than two exits or exit access doors. Where two exit access doorways are required, one such doorway is permitted to be served by a fixed ladder or an alternating tread device. Exit access doorways shall be separated by a horizontal distance equal to one-half the maximum horizontal dimension of room.

All portions of machinery rooms shall be within 150 feet (45 720 mm) of an exit or exit access doorway. An increase in travel distance is permitted in accordance with Section 1015.1.

Doors shall swing in the direction of egress travel, regardless of the occupant load served. Doors shall be tight fitting and self-closing.

1014. 5 Refrigerated rooms or spaces. Rooms or spaces having a floor area of 1,000 square feet (93 m²) or more, containing a refrigerant evaporator and maintained at a temperature below 68°F (20°C), shall have access to not less than two exits or exit access doors.

Travel distance shall be determined as specified in Section 1015.1, but all portions of a refrigerated room or space shall be within 150 feet (45 720 mm) of an exit or exit access door where such rooms are not protected by an approved automatic sprinkler system. Egress is allowed through adjoining refrigerated rooms or spaces.

Exception: Where using refrigerants in quantities limited to the amounts based on the volume set forth in the *International Mechanical Code*.

1014.6 Stage means of egress. Where two means of egress are required, based on the stage size or occupant load, one means of egress shall be provided on each side of the stage.

1014.6.1 Gallery, gridiron and catwalk means of egress. The means of egress from lighting and access catwalks, galleries and gridirons shall meet the requirements for occupancies in Group F-2.

Exceptions:

1. A minimum width of 22 inches (559 mm) is permitted for lighting and access catwalks.

2. Spiral stairs are permitted in the means of egress.

3. Stairways required by this subsection need not be enclosed.

4. Stairways with a minimum width of 22 inches (559 mm), ladders, or spiral stairs are permitted in the means of egress.

5. A second means of egress is not required from these areas where a means of escape to a floor or to a roof is provided. Ladders, alternating tread devices or spiral stairs are permitted in the means of escape.

6. Ladders are permitted in the means of egress.

SECTION 1015
EXIT ACCESS TRAVEL DISTANCE

1015.1 Travel distance limitations. Exits shall be so located on each story such that the maximum length of exit access travel, measured from the most remote point within a story to the entrance to an exit along the natural and unobstructed path of egress travel, shall not exceed the distances given in Table 1015.1.

Where the path of exit access includes unenclosed stairways or ramps within the exit access or includes unenclosed exit ramps or stairways as permitted in Section 1019.1, the distance of travel on such means of egress components shall also be included in the travel distance measurement. The measurement along stairways shall be made on a plane parallel and tangent to the stair tread nosings in the center of the stairway.

Exceptions:

1. Travel distance in open parking garages is permitted to be measured to the closest riser of open stairs.

2. In outdoor facilities with open exit access components and open exterior stairs or ramps, travel distance is permitted to be measured to the closest riser of a stair or the closest slope of the ramp.

3. Where an exit stair is permitted to be unenclosed in accordance with Exception 8 or 9 of Section 1019.1, the travel distance shall be measured from the most remote point within a building to an exit discharge.

TABLE 1015.1
EXIT ACCESS TRAVEL DISTANCE[a]

OCCUPANCY	WITHOUT SPRINKLER SYSTEM (feet)	WITH SPRINKLER SYSTEM (feet)
A, E, F-1, I-1, M, R, S-1	200	250[b]
B	200	300[c]
F-2, S-2, U	300	400[b]
H-1	Not Permitted	75[c]
H-2	Not Permitted	100[c]
H-3	Not Permitted	150[c]
H-4	Not Permitted	175[c]
H-5	Not Permitted	200[c]
I-2, I-3, I-4	150	200[c]

For SI: 1 foot = 304.8 mm.

a. See the following sections for modifications to exit access travel distance requirements:
Section 402: For the distance limitation in malls.
Section 404: For the distance limitation through an atrium space.
Section 1015.2: For increased limitation in Groups F-1 and S-1.
Section 1024.7: For increased limitation in assembly seating.
Section 1024.7: For increased limitation for assembly open-air seating.
Section 1018.2: For buildings with one exit.
Chapter 31: For the limitation in temporary structures.

b. Buildings equipped throughout with an automatic sprinkler system in accordance with Section 903.3.1.1 or 903.3.1.2. See Section 903 for occupancies where sprinkler systems according to Section 903.3.1.2 are permitted.

c. Buildings equipped throughout with an automatic sprinkler system in accordance with Section 903.3.1.1.

1015.2 Roof vent increase. In buildings which are one story in height, equipped with automatic heat and smoke roof vents complying with Section 910 and equipped throughout with an automatic sprinkler system in accordance with Section 903.3.1.1, the maximum exit access travel distance shall be 400 feet (122 m) for occupancies in Group F-1 or S.

1015.3 Exterior egress balcony increase. Travel distances specified in Section 1015.1 shall be increased up to an additional 100 feet (30 480 mm) provided the last portion of the exit access leading to the exit occurs on an exterior egress balcony constructed in accordance with Section 1013.5. The length of such balcony shall not be less than the amount of the increase taken.

SECTION 1016
CORRIDORS

1016.1 Construction. Corridors shall be fire-resistance rated in accordance with Table 1016.1. The corridor walls required to be fire-resistance rated shall comply with Section 708 for fire partitions.

Exceptions:

1. A fire-resistance rating is not required for corridors in an occupancy in Group E where each room that is used for instruction has at least one door directly to the exterior and rooms for assembly purposes have at least one-half of the required means of egress doors opening directly to the exterior. Exterior doors specified in this exception are required to be at ground level.

2. A fire-resistance rating is not required for corridors contained within a dwelling or sleeping unit in an occupancy in Group R.

3. A fire-resistance rating is not required for corridors in open parking garages.

4. A fire-resistance rating is not required for corridors in an occupancy in Group B which is a space requiring only a single means of egress complying with Section 1014.1.

TABLE 1016.1
CORRIDOR FIRE-RESISTANCE RATING

OCCUPANCY	OCCUPANT LOAD SERVED BY CORRIDOR	REQUIRED FIRE-RESISTANCE RATING (hours)	
		Without sprinkler system	With sprinkler system[c]
H-1, H-2, H-3	All	Not Permitted	1
H-4, H-5	Greater than 30	Not Permitted	1
A, B, E, F, M, S, U	Greater than 30	1	0
R	Greater than 10	1	0.5
I-2[a], I-4	All	Not Permitted	0
I-1, I-3	All	Not Permitted	1[b]

a. For requirements for occupancies in Group I-2, see Section 407.3.

b. For a reduction in the fire-resistance rating for occupancies in Group I-3, see Section 408.7.

c. Buildings equipped throughout with an automatic sprinkler system in accordance with Section 903.3.1.1 or 903.3.1.2 where allowed.

1016.2 Corridor width. The minimum corridor width shall be as determined in Section 1005.1, but not less than 44 inches (1118 mm).

Exceptions:

1. Twenty-four inches (610 mm)—For access to and utilization of electrical, mechanical or plumbing systems or equipment.
2. Thirty-six inches (914 mm)—With a required occupant capacity of 50 or less.
3. Thirty-six inches (914 mm)—Within a dwelling unit.
4. Seventy-two inches (1829 mm)—In Group E with a corridor having a required capacity of 100 or more.
5. Seventy-two inches (1829 mm)—In corridors serving surgical Group I, health care centers for ambulatory patients receiving outpatient medical care, which causes the patient to be not capable of self-preservation.
6. Ninety-six inches (2438 mm)—In Group I-2 in areas where required for bed movement.

1016.3 Dead ends. Where more than one exit or exit access doorway is required, the exit access shall be arranged such that there are no dead ends in corridors more than 20 feet (6096 mm) in length.

Exceptions:

1. In occupancies in Group I-3 of Occupancy Condition 2, 3 or 4 (see Section 308.4), the dead end in a corridor shall not exceed 50 feet (15 240 mm).
2. In occupancies in Groups B and F where the building is equipped throughout with an automatic sprinkler system in accordance with Section 903.3.1.1, the length of dead-end corridors shall not exceed 50 feet (15 240 mm).
3. A dead-end corridor shall not be limited in length where the length of the dead-end corridor is less than 2.5 times the least width of the dead-end corridor.

1016.4 Air movement in corridors. Exit access corridors shall not serve as supply, return, exhaust, relief or ventilation air ducts.

Exceptions:

1. Use of a corridor as a source of makeup air for exhaust systems in rooms that open directly onto such corridors, including toilet rooms, bathrooms, dressing rooms, smoking lounges and janitor closets, shall be permitted provided that each such corridor is directly supplied with outdoor air at a rate greater than the rate of makeup air taken from the corridor.
2. Where located within a dwelling unit, the use of corridors for conveying return air shall not be prohibited.
3. Where located within tenant spaces of 1,000 square feet (93 m²) or less in area, utilization of corridors for conveying return air is permitted.

1016.4.1 Corridor ceiling. Use of the space between the corridor ceiling and the floor or roof structure above as a re-

turn air plenum is permitted for one or more of the following conditions:

1. The corridor is not required to be of fire-resistance-rated construction;
2. The corridor is separated from the plenum by fire-resistance-rated construction;
3. The air-handling system serving the corridor is shut down upon activation of the air-handling unit smoke detectors required by the *International Mechanical Code*.
4. The air-handling system serving the corridor is shut down upon detection of sprinkler waterflow where the building is equipped throughout with an automatic sprinkler system; or
5. The space between the corridor ceiling and the floor or roof structure above the corridor is used as a component of an approved engineered smoke control system.

1016.5 Corridor continuity. Fire-resistance-rated corridors shall be continuous from the point of entry to an exit, and shall not be interrupted by intervening rooms.

Exception: Foyers, lobbies or reception rooms constructed as required for corridors shall not be construed as intervening rooms.

SECTION 1017
EXITS

1017.1 General. Exits shall comply with Sections 1017 through 1022 and the applicable requirements of Sections 1003 through 1012. An exit shall not be used for any purpose that interferes with its function as a means of egress. Once a given level of exit protection is achieved, such level of protection shall not be reduced until arrival at the exit discharge.

1017.2 Exterior exit doors. Buildings or structures used for human occupancy shall have at least one exterior door that meets the requirements of Section 1008.1.1.

1017.2.1 Detailed requirements. Exterior exit doors shall comply with the applicable requirements of Section 1008.1.

1017.2.2 Arrangement. Exterior exit doors shall lead directly to the exit discharge or the public way.

SECTION 1018
NUMBER OF EXITS AND CONTINUITY

1018.1 Minimum number of exits. All rooms and spaces within each story shall be provided with and have access to the minimum number of approved independent exits as required by Table 1018.1 based on the occupant load, except as modified in Section 1014.1 or 1018.2. For the purposes of this chapter, occupied roofs shall be provided with exits as required for stories. The required number of exits from any story, basement or individual space shall be maintained until arrival at grade or the public way.

TABLE 1018.1
MINIMUM NUMBER OF EXITS FOR OCCUPANT LOAD

OCCUPANT LOAD	MINIMUM NUMBER OF EXITS
1-500	2
501-1,000	3
More than 1,000	4

1018.1.1 Open parking structures. Parking structures shall not have less than two exits from each parking tier, except that only one exit is required where vehicles are mechanically parked. Unenclosed vehicle ramps shall not be considered as required exits unless pedestrian facilities are provided.

1018.1.2 Helistops. The means of egress from helistops shall comply with the provisions of this chapter, provided that landing areas located on buildings or structures shall have two or more exits. For landing platforms or roof areas less than 60 feet (18 288 mm) long, or less than 2,000 square feet (186 m²) in area, the second means of egress is permitted to be a fire escape or ladder leading to the floor below.

1018.2 Buildings with one exit. Only one exit shall be required in buildings as described below:

1. Buildings described in Table 1018.2, provided that the building has not more than one level below the first story above grade plane.

2. Buildings of Group R-3 occupancy.

3. Single-level buildings with the occupied space at the level of exit discharge provided that the story or space complies with Section 1014.1 as a space with one means of egress.

TABLE 1018.2
BUILDINGS WITH ONE EXIT

OCCUPANCY	MAXIMUM HEIGHT OF BUILDING ABOVE GRADE PLANE	MAXIMUM OCCUPANTS (OR DWELLING UNITS) PER FLOOR AND TRAVEL DISTANCE
A, B[d], E, F, M, U	1 Story	50 occupants and 75 feet travel distance
H-2, H-3	1 Story	3 occupants and 25 feet travel distance
H-4, H-5, I, R	1 Story	10 occupants and 75 feet travel distance
S[a]	1 Story	30 occupants and 100 feet travel distance
B[b], F, M, S[a]	2 Stories	30 occupants and 75 feet travel distance
R-2	2 Stories[c]	4 dwelling units and 50 feet travel distance

For SI: 1 foot = 304.8 mm.

a. For the required number of exits for open parking structures, see Section 1018.1.1.

b. For the required number of exits for air traffic control towers, see Section 412.1.

c. Buildings classified as Group R-2 equipped throughout with an automatic sprinkler system in accordance with Section 903.3.1.1 or 903.3.1.2 and provided with emergency escape and rescue openings in accordance with Section 1025 shall have a maximum height of three stories above grade.

d. Buildings equipped throughout with an automatic sprinkler system in accordance with Section 903.3.1.1 with an occupancy in Group B shall have a maximum travel distance of 100 feet.

1018.3 Exit continuity. Exits shall be continuous from the point of entry into the exit to the exit discharge.

1018.4 Exit door arrangement. Exit door arrangement shall meet the requirements of Sections 1014.2 through 1014.2.2.

SECTION 1019
VERTICAL EXIT ENCLOSURES

1019.1 Enclosures required. Interior exit stairways and interior exit ramps shall be enclosed with fire barriers. Exit enclosures shall have a fire-resistance rating of not less than 2 hours where connecting four stories or more and not less than 1 hour where connecting less than four stories. The number of stories connected by the shaft enclosure shall include any basements but not any mezzanines. An exit enclosure shall not be used for any purpose other than means of egress. Enclosures shall be constructed as fire barriers in accordance with Section 706.

Exceptions:

1. In other than Group H and I occupancies, a stairway serving an occupant load of less than 10 not more than one story above the level of exit discharge is not required to be enclosed.

2. Exits in buildings of Group A-5 where all portions of the means of egress are essentially open to the outside need not be enclosed.

3. Stairways serving and contained within a single residential dwelling unit or sleeping unit in occupancies in Group R-2 or R-3 and sleeping units in occupancies in Group R-1 are not required to be enclosed.

4. Stairways that are not a required means of egress element are not required to be enclosed where such stairways comply with Section 707.2.

5. Stairways in open parking structures which serve only the parking structure are not required to be enclosed.

6. Stairways in occupancies in Group I-3 as provided for in Section 408.3.6 are not required to be enclosed.

7. Means of egress stairways as required by Section 410.5.4 are not required to be enclosed.

8. In other than occupancy Groups H and I, a maximum of 50 percent of egress stairways serving one adjacent floor are not required to be enclosed, provided at least two means of egress are provided from both floors served by the unenclosed stairways. Any two such interconnected floors shall not be open to other floors.

9. In other than occupancy Groups H and I, interior egress stairways serving only the first and second stories of a building equipped throughout with an automatic sprinkler system in accordance with Section 903.3.1.1 are not required to be enclosed, provided at least two means of egress are provided from both floors served by the unenclosed stairways. Such interconnected stories shall not be open to other stories.

1019.1.1 Openings and penetrations. Exit enclosure opening protectives shall be in accordance with the requirements of Section 715.

Except as permitted in Section 402.4.6, openings in exit enclosures other than unexposed exterior openings shall be limited to those necessary for exit access to the enclosure from normally occupied spaces and for egress from the enclosure.

Where interior exit enclosures are extended to the exterior of a building by an exit passageway, the door assembly from the exit enclosure to the exit passageway shall be protected by a fire door conforming to the requirements in Section 715.3. Fire door assemblies in exit enclosures shall comply with Section 715.3.4.

1019.1.2 Penetrations. Penetrations into and openings through an exit enclosure are prohibited except for required exit doors, equipment and ductwork necessary for independent pressurization, sprinkler piping, standpipes, electrical raceway for fire department communication and electrical raceway serving the exit enclosure and terminating at a steel box not exceeding 16 square inches (0.010 m²). Such penetrations shall be protected in accordance with Section 712. There shall be no penetrations or communication openings, whether protected or not, between adjacent exit enclosures.

1019.1.3 Ventilation. Equipment and ductwork for exit enclosure ventilation shall comply with one of the following items:

1. Such equipment and ductwork shall be located exterior to the building and shall be directly connected to the exit enclosure by ductwork enclosed in construction as required for shafts.

2. Where such equipment and ductwork is located within the exit enclosure, the intake air shall be taken directly from the outdoors and the exhaust air shall be discharged directly to the outdoors, or such air shall be conveyed through ducts enclosed in construction as required for shafts.

3. Where located within the building, such equipment and ductwork shall be separated from the remainder of the building, including other mechanical equipment, with construction as required for shafts.

In each case, openings into the fire-resistance-rated construction shall be limited to those needed for maintenance and operation and shall be protected by self-closing fire-resistance-rated devices in accordance with Chapter 7 for enclosure wall opening protectives.

Exit enclosure ventilation systems shall be independent of other building ventilation systems.

1019.1.4 Vertical enclosure exterior walls. Exterior walls of a vertical exit enclosure shall comply with the requirements of Section 704 for exterior walls. Where nonrated walls or unprotected openings enclose the exterior of the stairway and the walls or openings are exposed by other parts of the building at an angle of less than 180 degrees (3.14 rad), the building exterior walls within 10 feet (3048 mm) horizontally of a nonrated wall or unprotected opening shall be constructed as required for a minimum 1-hour fire-resistance rating with ³/₄-hour opening protectives. This construction shall extend vertically from the ground to a point 10 feet (3048 mm) above the topmost landing of the stairway or to the roof line, whichever is lower.

1019.1.5 Enclosures under stairways. The walls and soffits within enclosed usable spaces under enclosed and unenclosed stairways shall be protected by 1-hour fire-resistance-rated construction, or the fire-resistance rating of the stairway enclosure, whichever is greater. Access to the enclosed usable space shall not be directly from within the stair enclosure.

Exception: Spaces under stairways serving and contained within a single residential dwelling unit in Group R-2 or R-3 as applicable in Section 101.2.

There shall be no enclosed usable space under exterior exit stairways unless the space is completely enclosed in 1-hour fire-resistance-rated construction. The open space under exterior stairways shall not be used for any purpose.

1019.1.6 Discharge identification. A stairway in an exit enclosure shall not continue below the level of exit discharge unless an approved barrier is provided at the level of exit discharge to prevent persons from unintentionally continuing into levels below. Directional exit signs shall be provided as specified in Section 1011.

1019.1.7 Stairway floor number signs. A sign shall be provided at each floor landing in interior vertical exit enclosures connecting more than three stories designating the floor level, the terminus of the top and bottom of the stair enclosure and the identification of the stair. The signage shall also state the story of, and the direction to the exit discharge and the availability of roof access from the stairway for the fire department. The sign shall be located 5 feet (1524 mm) above the floor landing in a position which is readily visible when the doors are in the open and closed positions.

1019.1.8 Smokeproof enclosures. In buildings required to comply with Section 403 or 405, each of the exits of a building that serves stories where the floor surface is located more than 75 feet (22 860 mm) above the lowest level of fire department vehicle access or more than 30 feet (9144 mm) below the level of exit discharge serving such floor levels shall be a smokeproof enclosure or pressurized stairway in accordance with Section 909.20.

1019.1.8.1 Enclosure exit. A smokeproof enclosure or pressurized stairway shall exit into a public way or into an exit passageway, yard or open space having direct access to a public way. The exit passageway shall be without other openings and shall be separated from the remainder of the building by 2-hour fire-resistance-rated construction.

Exceptions:

1. Openings in the exit passageway serving a smokeproof enclosure are permitted where the exit passageway is protected and pressurized in the same manner as the smokeproof enclosure, and openings are protected as required for access from other floors.

2. Openings in the exit passageway serving a pressurized stairway are permitted where the exit

passageway is protected and pressurized in the same manner as the pressurized stairway.

1019.1.8.2 Enclosure access. Access to the stairway within a smokeproof enclosure shall be by way of a vestibule or an open exterior balcony.

Exception: Access is not required by way of a vestibule or exterior balcony for stairways using the pressurization alternative complying with Section 909.20.5.

SECTION 1020
EXIT PASSAGEWAYS

1020.1 Exit passageway. Exit passageways serving as an exit component in a means of egress system shall comply with the requirements of this section. An exit passageway shall not be used for any purpose other than as a means of egress.

1020.2 Width. The width of exit passageways shall be determined as specified in Section 1005.1 but such width shall not be less than 44 inches (1118 mm), except that exit passageways serving an occupant load of less than 50 shall not be less than 36 inches (914 mm) in width.

The required width of exit passageways shall be unobstructed.

Exception: Doors, when fully opened, and handrails, shall not reduce the required width by more than 7 inches (178 mm). Doors in any position shall not reduce the required width by more than one-half. Other nonstructural projections such as trim and similar decorative features are permitted to project into the required width 1.5 inches (38 mm) on each side.

1020.3 Construction. Exit passageway enclosures shall have walls, floors and ceilings of not less than 1-hour fire-resistance rating, and not less than that required for any connecting exit enclosure. Exit passageways shall be constructed as fire barriers in accordance with Section 706.

1020.4 Openings and penetrations. Exit passageway opening protectives shall be in accordance with the requirements of Section 715.

Except as permitted in Section 402.4.6, openings in exit passageways other than unexposed exterior openings shall be limited to those necessary for exit access to the exit passageway from normally occupied spaces and for egress from the exit passageway.

Where interior exit enclosures are extended to the exterior of a building by an exit passageway, the door assembly from the exit enclosure to the exit passageway shall be protected by a fire door conforming to the requirements in Section 715.3. Fire door assemblies in exit passageways shall comply with Section 715.3.4.

Elevators shall not open into an exit passageway.

1020.5 Penetrations. Penetrations into and openings through an exit passageway are prohibited except for required exit doors, equipment and ductwork necessary for independent pressurization, sprinkler piping, standpipes, electrical raceway for fire department communication and electrical raceway serving the exit passageway and terminating at a steel box not exceeding 16 square inches (0.010 m²). Such penetrations shall be protected in accordance with Section 712. There shall be no penetrations or communicating openings, whether protected or not, between adjacent exit passageways.

SECTION 1021
HORIZONTAL EXITS

1021.1 Horizontal exits. Horizontal exits serving as an exit in a means of egress system shall comply with the requirements of this section. A horizontal exit shall not serve as the only exit from a portion of a building, and where two or more exits are required, not more than one-half of the total number of exits or total exit width shall be horizontal exits.

Exceptions:

1. Horizontal exits are permitted to comprise two-thirds of the required exits from any building or floor area for occupancies in Group I-2.

2. Horizontal exits are permitted to comprise 100 percent of the exits required for occupancies in Group I-3. At least 6 square feet (0.6 m²) of accessible space per occupant shall be provided on each side of the horizontal exit for the total number of people in adjoining compartments.

Every fire compartment for which credit is allowed in connection with a horizontal exit shall not be required to have a stairway or door leading directly outside, provided the adjoining fire compartments have stairways or doors leading directly outside and are so arranged that egress shall not require the occupants to return through the compartment from which egress originates.

The area into which a horizontal exit leads shall be provided with exits adequate to meet the occupant requirements of this chapter, but not including the added occupant capacity imposed by persons entering it through horizontal exits from another area. At least one of its exits shall lead directly to the exterior or to an exit enclosure.

1021.2 Separation. The separation between buildings or areas of refuge connected by a horizontal exit shall be provided by a fire wall complying with Section 705 or a fire barrier complying with Section 706 and having a fire-resistance rating of not less than 2 hours. Opening protectives in horizontal exit walls shall also comply with Section 715. The horizontal exit separation shall extend vertically through all levels of the building unless floor assemblies are of 2-hour fire resistance with no unprotected openings.

Exception: A fire-resistance rating is not required at horizontal exits between a building area and an above-grade pedestrian walkway constructed in accordance with Section 3104, provided that the distance between connected buildings is more than 20 feet (6096 mm).

Horizontal exit walls constructed as fire barriers shall be continuous from exterior wall to exterior wall so as to divide completely the floor served by the horizontal exit.

1021.3 Opening protectives. Fire doors in horizontal exits shall be self-closing or automatic-closing when activated by a smoke detector installed in accordance with Section 907.10. Opening protectives in horizontal exits shall be consistent with the fire-resistance rating of the wall. Such doors where located in a cross-corridor condition shall be automatic-closing by activation of a smoke detector installed in accordance with Section 907.10.

1021.4 Capacity of refuge area. The refuge area of a horizontal exit shall be spaces occupied by the same tenant or public areas and each such area of refuge shall be adequate to house the original occupant load of the refuge space plus the occupant load anticipated from the adjoining compartment. The anticipated occupant load from the adjoining compartment shall be based on the capacity of the horizontal exit doors entering the area of refuge. The capacity of areas of refuge shall be computed on a net floor area allowance of 3 square feet (0.2787 m²) for each occupant to be accommodated therein, not including areas of stairways, elevators and other shafts or courts.

Exception: The net floor area allowable per occupant shall be as follows for the indicated occupancies:

1. Six square feet (0.6 m²) per occupant for occupancies in Group I-3.

2. Fifteen square feet (1.4 m²) per occupant for ambulatory occupancies in Group I-2.

3. Thirty square feet (2.8 m²) per occupant for nonambulatory occupancies in Group I-2.

SECTION 1022
EXTERIOR EXIT RAMPS AND STAIRWAYS

1022.1 Exterior exit ramps and stairways. Exterior exit ramps and stairways serving as an element of a required means of egress shall comply with this section.

Exception: Exterior exit ramps and stairways for outdoor stadiums complying with Section 1019.1, Exception 2.

1022.2 Use in a means of egress. Exterior exit ramps and stairways shall not be used as an element of a required means of egress for occupancies in Group I-2. For occupancies in other than Group I-2, exterior exit ramps and stairways shall be permitted as an element of a required means of egress for buildings not exceeding six stories or 75 feet (22 860 mm) in height.

1022.3 Open side. Exterior exit ramps and stairways serving as an element of a required means of egress shall be open on at least one side. An open side shall have a minimum of 35 square feet (3.3 m²) of aggregate open area adjacent to each floor level and the level of each intermediate landing. The required open area shall be located not less than 42 inches (1067 mm) above the adjacent floor or landing level.

1022.4 Side yards. The open areas adjoining exterior exit ramps or stairways shall be either yards, courts or public ways; the remaining sides are permitted to be enclosed by the exterior walls of the building.

1022.5 Location. Exterior exit ramps and stairways shall be located in accordance with Section 1023.3.

1022.6 Exterior ramps and stairway protection. Exterior exit ramps and stairways shall be separated from the interior of the building as required in Section 1019.1. Openings shall be limited to those necessary for egress from normally occupied spaces.

Exceptions:

1. Separation from the interior of the building is not required for occupancies, other than those in Group R-1 or R-2, in buildings that are no more than two stories above grade where the level of exit discharge is the first story above grade.

2. Separation from the interior of the building is not required where the exterior ramp or stairway is served by an exterior ramp and/or balcony that connects two remote exterior stairways or other approved exits, with a perimeter that is not less than 50 percent open. To be considered open, the opening shall be a minimum of 50 percent of the height of the enclosing wall, with the top of the openings no less than 7 feet (2134 mm) above the top of the balcony.

3. Separation from the interior of the building is not required for an exterior ramp or stairway located in a building or structure that is permitted to have unenclosed interior stairways in accordance with Section 1019.1.

4. Separation from the interior of the building is not required for exterior ramps or stairways connected to open-ended corridors, provided that Items 4.1 through 4.4 are met:

   4.1. The building, including corridors and ramps and/or stairs, shall be equipped throughout with an automatic sprinkler system in accordance with Section 903.3.1.1 or 903.3.1.2.

   4.2. The open-ended corridors comply with Section 1016.

   4.3. The open-ended corridors are connected on each end to an exterior exit ramp or stairway complying with Section 1022.

   4.4. At any location in an open-ended corridor where a change of direction exceeding 45 degrees (0.79 rad) occurs, a clear opening of not less than 35 square feet (3.3 m²) or an exterior ramp or stairway shall be provided. Where clear openings are provided, they shall be located so as to minimize the accumulation of smoke or toxic gases.

SECTION 1023
EXIT DISCHARGE

1023.1 General. Exits shall discharge directly to the exterior of the building. The exit discharge shall be at grade or shall provide direct access to grade. The exit discharge shall not reenter a building.

Exceptions:

1. A maximum of 50 percent of the number and capacity of the exit enclosures is permitted to egress through

areas on the level of discharge provided all of the following are met:

1.1. Such exit enclosures egress to a free and unobstructed way to the exterior of the building, which way is readily visible and identifiable from the point of termination of the exit enclosure.

1.2. The entire area of the level of discharge is separated from areas below by construction conforming to the fire-resistance rating for the exit enclosure.

1.3. The egress path from the exit enclosure on the level of discharge is protected throughout by an approved automatic sprinkler system. All portions of the level of discharge with access to the egress path shall either be protected throughout with an automatic sprinkler system installed in accordance with Section 903.3.1.1 or 903.3.1.2, or separated from the egress path in accordance with the requirements for the enclosure of exits.

2. A maximum of 50 percent of the number and capacity of the exit enclosures is permitted to egress through a vestibule provided all of the following are met:

2.1. The entire area of the vestibule is separated from areas below by construction conforming to the fire-resistance rating for the exit enclosure.

2.2. The depth from the exterior of the building is not greater than 10 feet (3048 mm) and the length is not greater than 30 feet (9144 mm).

2.3. The area is separated from the remainder of the level of exit discharge by construction providing protection at least the equivalent of approved wired glass in steel frames.

2.4. The area is used only for means of egress and exits directly to the outside.

3. Stairways in open parking garages complying with Section 1019.1, Exception 5, are permitted to egress through the open parking garage at the level of exit discharge.

1023.2 Exit discharge capacity. The capacity of the exit discharge shall be not less than the required discharge capacity of the exits being served.

1023.3 Exit discharge location. Exterior balconies, stairways and ramps shall be located at least 10 feet (3048 mm) from adjacent lot lines and from other buildings on the same lot unless the adjacent building exterior walls and openings are protected in accordance with Section 704 based on fire separation distance.

1023.4 Exit discharge components. Exit discharge components shall be sufficiently open to the exterior so as to minimize the accumulation of smoke and toxic gases.

1023.5 Egress courts. Egress courts serving as a portion of the exit discharge in the means of egress system shall comply with the requirements of Section 1023.

1023.5.1 Width. The width of egress courts shall be determined as specified in Section 1005.1, but such width shall not be less than 44 inches (1118 mm), except as specified herein. Egress courts serving occupancies in Group R-3 applicable in Section 101.2 and Group U shall not be less than 36 inches (914 mm) in width.

The required width of egress courts shall be unobstructed to a height of 7 feet (2134 mm).

Exception: Doors, when fully opened, and handrails shall not reduce the required width by more than 7 inches (178 mm). Doors in any position shall not reduce the required width by more than one-half. Other nonstructural projections such as trim and similar decorative features are permitted to project into the required width 1.5 inches (38 mm) from each side.

Where an egress court exceeds the minimum required width and the width of such egress court is then reduced along the path of exit travel, the reduction in width shall be gradual. The transition in width shall be affected by a guard not less than 36 inches (914 mm) in height and shall not create an angle of more than 30 degrees (0.52 rad) with respect to the axis of the egress court along the path of egress travel. In no case shall the width of the egress court be less than the required minimum.

1023.5.2 Construction and openings. Where an egress court serving a building or portion thereof is less than 10 feet (3048 mm) in width, the egress court walls shall be not less than 1-hour fire-resistance-rated exterior walls complying with Section 704 for a distance of 10 feet (3048 mm) above the floor of the court, and openings therein shall be equipped with fixed or self-closing, $^3/_4$-hour opening protective assemblies.

Exceptions:

1. Egress courts serving an occupant load of less than 10.

2. Egress courts serving Group R-3 as applicable in Section 101.2.

1023.6 Access to a public way. The exit discharge shall provide a direct and unobstructed access to a public way.

Exception: Where access to a public way cannot be provided, a safe dispersal area shall be provided where all of the following are met:

1. The area shall be of a size to accommodate at least 5 square feet (0.28 m²) for each person.

2. The area shall be located on the same property at least 50 feet (15 240 mm) away from the building requiring egress.

3. The area shall be permanently maintained and identified as a safe dispersal area.

4. The area shall be provided with a safe and unobstructed path of travel from the building.

SECTION 1024
ASSEMBLY

1024.1 General. Occupancies in Group A which contain seats, tables, displays, equipment or other material shall comply with this section.

1024.1.1 Bleachers. Bleachers, grandstands, and folding and telescopic seating shall comply with ICC 300.

1024.2 Assembly main exit. Group A occupancies that have an occupant load of greater than 300 shall be provided with a main exit. The main exit shall be of sufficient width to accommodate not less than one-half of the occupant load, but such width shall not be less than the total required width of all means of egress leading to the exit. Where the building is classified as a Group A occupancy, the main exit shall front on at least one street or an unoccupied space of not less than 10 feet (3048 mm) in width that adjoins a street or public way.

Exception: In assembly occupancies where there is no well-defined main exit or where multiple main exits are provided, exits shall be permitted to be distributed around the perimeter of the building provided that the total width of egress is not less than 100 percent of the required width.

1024.3 Assembly other exits. In addition to having access to a main exit, each level of an occupancy in Group A having an occupant load of greater than 300 shall be provided with additional exits that shall provide an egress capacity for at least one-half of the total occupant load served by that level and comply with Section 1014.2.

Exception: In assembly occupancies where there is no well-defined main exit or where multiple main exits are provided, exits shall be permitted to be distributed around the perimeter of the building provided that the total width of egress is not less than 100 percent of the required width.

1024.4 Foyers and lobbies. In Group A-1 occupancies, where persons are admitted to the building at times when seats are not available and are allowed to wait in a lobby or similar space, such use of lobby or similar space shall not encroach upon the required clear width of the means of egress. Such waiting areas shall be separated from the required means of egress by substantial permanent partitions or by fixed rigid railings not less than 42 inches (1067 mm) high. Such foyer, if not directly connected to a public street by all the main entrances or exits, shall have a straight and unobstructed corridor or path of travel to every such main entrance or exit.

1024.5 Interior balcony and gallery means of egress. For balconies or galleries having a seating capacity of over 50 located in Group A occupancies, at least two means of egress shall be provided, one from each side of every balcony or gallery, with at least one leading directly to an exit.

1024.5.1 Enclosure of balcony openings. Interior stairways and other vertical openings shall be enclosed in a vertical exit enclosure as provided in Section 1019.1, except that stairways are permitted to be open between the balcony and the main assembly floor in occupancies such as theaters, churches and auditoriums. At least one accessible means of egress is required from a balcony or gallery level containing accessible seating locations in accordance with Section 1007.3 or 1007.4.

1024.6 Width of means of egress for assembly. The clear width of aisles and other means of egress shall comply with Section 1024.6.1 where smoke-protected seating is not provided and with Section 1024.6.2 or 1024.6.3 where smoke-protected seating is provided. The clear width shall be measured to walls, edges of seating and tread edges except for permitted projections.

1024.6.1 Without smoke protection. The clear width of the means of egress shall provide sufficient capacity in accordance with all of the following, as applicable:

1. At least 0.3 inch (7.6 mm) of width for each occupant served shall be provided on stairs having riser heights 7 inches (178 mm) or less and tread depths 11 inches (279 mm) or greater, measured horizontally between tread nosing.

2. At least 0.005 inch (0.127 mm) of additional stair width for each occupant shall be provided for each 0.10 inch (2.5 mm) of riser height above 7 inches (178 mm).

3. Where egress requires stair descent, at least 0.075 inch (1.9 mm) of additional width for each occupant shall be provided on those portions of stair width having no handrail within a horizontal distance of 30 inches (762 mm).

4. Ramped means of egress, where slopes are steeper than one unit vertical in 12 units horizontal (8-percent slope), shall have at least 0.22 inch (5.6 mm) of clear width for each occupant served. Level or ramped means of egress, where slopes are not steeper than one unit vertical in 12 units horizontal (8-percent slope), shall have at least 0.20 inch (5.1 mm) of clear width for each occupant served.

1024.6.2 Smoke-protected seating. The clear width of the means of egress for smoke-protected assembly seating shall be not less than the occupant load served by the egress element multiplied by the appropriate factor in Table 1024.6.2. The total number of seats specified shall be those within a single assembly space and exposed to the same smoke-protected environment. Interpolation is permitted between the specific values shown. A life safety evaluation, complying with NFPA 101, shall be done for a facility utilizing the reduced width requirements of Table 1024.6.2 for smoke-protected assembly seating.

Exception: For an outdoor smoke-protected assembly with an occupant load not greater than 18,000, the clear width shall be determined using the factors in Section 1024.6.3.

1024.6.2.1 Smoke control. Means of egress serving a smoke-protected assembly seating area shall be provided with a smoke control system complying with Section 909 or natural ventilation designed to maintain the smoke level at least 6 feet (1829 mm) above the floor of the means of egress.

TABLE 1024.6.2
WIDTH OF AISLES FOR SMOKE-PROTECTED ASSEMBLY

TOTAL NUMBER OF SEATS IN THE SMOKE-PROTECTED ASSEMBLY OCCUPANCY	INCHES OF CLEAR WIDTH PER SEAT SERVED			
	Stairs and aisle steps with handrails within 30 inches	Stairs and aisle steps without handrails within 30 inches	Passageways, doorways and ramps not steeper than 1 in 10 in slope	Ramps steeper than 1 in 10 in slope
Equal to or less than 5,000	0.200	0.250	0.150	0.165
10,000	0.130	0.163	0.100	0.110
15,000	0.096	0.120	0.070	0.077
20,000	0.076	0.095	0.056	0.062
Equal to or greater than 25,000	0.060	0.075	0.044	0.048

For SI: 1 inch = 25.4 mm.

1024.6.2.2 Roof height. A smoke-protected assembly seating area with a roof shall have the lowest portion of the roof deck not less than 15 feet (4572 mm) above the highest aisle or aisle accessway.

Exception: A roof canopy in an outdoor stadium shall be permitted to be less than 15 feet (4572 mm) above the highest aisle or aisle accessway provided that there are no objects less than 80 inches (2032 mm) above the highest aisle or aisle accessway.

1024.6.2.3 Automatic sprinklers. Enclosed areas with walls and ceilings in buildings or structures containing smoke-protected assembly seating shall be protected with an approved automatic sprinkler system in accordance with Section 903.3.1.1.

Exceptions:

1. The floor area used for contests, performances or entertainment provided the roof construction is more than 50 feet (15 240 mm) above the floor level and the use is restricted to low fire hazard uses.

2. Press boxes and storage facilities less than 1,000 square feet (93 m²) in area.

3. Outdoor seating facilities where seating and the means of egress in the seating area are essentially open to the outside.

1024.6.3 Width of means of egress for outdoor smoke-protected assembly. The clear width in inches (mm) of aisles and other means of egress shall be not less than the total occupant load served by the egress element multiplied by 0.08 (2.0 mm) where egress is by aisles and stairs and multiplied by 0.06 (1.52 mm) where egress is by ramps, corridors, tunnels or vomitories.

Exception: The clear width in inches (mm) of aisles and other means of egress shall be permitted to comply with Section 1024.6.2 for the number of seats in the outdoor smoke-protected assembly where Section 1024.6.2 permits less width.

1024.7 Travel distance. Exits and aisles shall be so located that the travel distance to an exit door shall not be greater than 200 feet (60 960 mm) measured along the line of travel in nonsprinklered buildings. Travel distance shall not be more than 250 feet (76 200 mm) in sprinklered buildings. Where aisles are provided for seating, the distance shall be measured along the aisles and aisle accessway without travel over or on the seats.

Exceptions:

1. Smoke-protected assembly seating: The travel distance from each seat to the nearest entrance to a vomitory or concourse shall not exceed 200 feet (60 960 mm). The travel distance from the entrance to the vomitory or concourse to a stair, ramp or walk on the exterior of the building shall not exceed 200 feet (60 960 mm).

2. Open-air seating: The travel distance from each seat to the building exterior shall not exceed 400 feet (122 m). The travel distance shall not be limited in facilities of Type I or II construction.

1024.8 Common path of travel. The common path of travel shall not exceed 30 feet (9144 mm) from any seat to a point where a person has a choice of two paths of egress travel to two exits.

Exceptions:

1. For areas serving not more than 50 occupants, the common path of travel shall not exceed 75 feet (22 860 mm).

2. For smoke-protected assembly seating, the common path of travel shall not exceed 50 feet (15 240 mm).

1024.8.1 Path through adjacent row. Where one of the two paths of travel is across the aisle through a row of seats to another aisle, there shall be not more than 24 seats between the two aisles, and the minimum clear width between rows for the row between the two aisles shall be 12 inches (305 mm) plus 0.6 inch (15.2 mm) for each additional seat above seven in the row between aisles.

Exception: For smoke-protected assembly seating there shall not be more than 40 seats between the two aisles and the minimum clear width shall be 12 inches (305 mm) plus 0.3 inch (7.6 mm) for each additional seat.

1024.9 Assembly aisles are required. Every occupied portion of any occupancy in Group A that contains seats, tables, displays, similar fixtures or equipment shall be provided with aisles leading to exits or exit access doorways in accordance with this section. Aisle accessways for tables and seating shall comply with Section 1013.4.2.

1024.9.1 Minimum aisle width. The minimum clear width of aisles shall be as shown:

1. Forty-eight inches (1219 mm) for aisle stairs having seating on each side.

   Exception: Thirty-six inches (914 mm) where the aisle does not serve more than 50 seats.

2. Thirty-six inches (914 mm) for aisle stairs having seating on only one side.

3. Twenty-three inches (584 mm) between an aisle stair handrail or guard and seating where the aisle is subdivided by a handrail.

4. Forty-two inches (1067 mm) for level or ramped aisles having seating on both sides.

   Exceptions:

   1. Thirty-six inches (914 mm) where the aisle does not serve more than 50 seats.

   2. Thirty inches (762 mm) where the aisle does not serve more than 14 seats.

5. Thirty-six inches (914 mm) for level or ramped aisles having seating on only one side.

   Exception: Thirty inches (762 mm) where the aisle does not serve more than 14 seats.

6. Twenty-three inches (584 mm) between an aisle stair handrail and seating where an aisle does not serve more than five rows on one side.

1024.9.2 Aisle width. The aisle width shall provide sufficient egress capacity for the number of persons accommodated by the catchment area served by the aisle. The catchment area served by an aisle is that portion of the total space that is served by that section of the aisle. In establishing catchment areas, the assumption shall be made that there is a balanced use of all means of egress, with the number of persons in proportion to egress capacity.

1024.9.3 Converging aisles. Where aisles converge to form a single path of egress travel, the required egress capacity of that path shall not be less than the combined required capacity of the converging aisles.

1024.9.4 Uniform width. Those portions of aisles, where egress is possible in either of two directions, shall be uniform in required width.

1024.9.5 Assembly aisle termination. Each end of an aisle shall terminate at cross aisle, foyer, doorway, vomitory or concourse having access to an exit.

Exceptions:

1. Dead-end aisles shall not be greater than 20 feet (6096 mm) in length.

2. Dead-end aisles longer than 20 feet (6096 mm) are permitted where seats beyond the 20-foot (6096 mm) dead-end aisle are no more than 24 seats from another aisle, measured along a row of seats having a minimum clear width of 12 inches (305 mm) plus 0.6 inch (15.2 mm) for each additional seat above seven in the row.

3. For smoke-protected assembly seating, the dead-end aisle length of vertical aisles shall not exceed a distance of 21 rows.

4. For smoke-protected assembly seating, a longer dead-end aisle is permitted where seats beyond the 21-row dead-end aisle are not more than 40 seats from another aisle, measured along a row of seats having an aisle accessway with a minimum clear width of 12 inches (305 mm) plus 0.3 inch (7.6 mm) for each additional seat above seven in the row.

1024.9.6 Assembly aisle obstructions. There shall be no obstructions in the required width of aisles except for handrails as provided in Section 1024.13.

1024.10 Clear width of aisle accessways serving seating. Where seating rows have 14 or fewer seats, the minimum clear aisle accessway width shall not be less than 12 inches (305 mm) measured as the clear horizontal distance from the back of the row ahead and the nearest projection of the row behind. Where chairs have automatic or self-rising seats, the measurement shall be made with seats in the raised position. Where any chair in the row does not have an automatic or self-rising seat, the measurements shall be made with the seat in the down position. For seats with folding tablet arms, row spacing shall be determined with the tablet arm down.

1024.10.1 Dual access. For rows of seating served by aisles or doorways at both ends, there shall not be more than 100 seats per row. The minimum clear width of 12 inches (305 mm) between rows shall be increased by 0.3 inch (7.6 mm) for every additional seat beyond 14 seats, but the minimum clear width is not required to exceed 22 inches (559 mm).

Exception: For smoke-protected assembly seating, the row length limits for a 12-inch-wide (305 mm) aisle accessway, beyond which the aisle accessway minimum clear width shall be increased, are in Table 1024.10.1.

TABLE 1024.10.1
SMOKE-PROTECTED
ASSEMBLY AISLE ACCESSWAYS

TOTAL NUMBER OF SEATS IN THE SMOKE-PROTECTED ASSEMBLY OCCUPANCY	MAXIMUM NUMBER OF SEATS PER ROW PERMITTED TO HAVE A MINIMUM 12-INCH CLEAR WIDTH AISLE ACCESSWAY	
	Aisle or doorway at both ends of row	Aisle or doorway at one end of row only
Less than 4,000	14	7
4,000	15	7
7,000	16	8
10,000	17	8
13,000	18	9
16,000	19	9
19,000	20	10
22,000 and greater	21	11

For SI: 1 inch = 25.4 mm.

1024.10.2 Single access. For rows of seating served by an aisle or doorway at only one end of the row, the minimum clear width of 12 inches (305 mm) between rows shall be increased by 0.6 inch (15.2 mm) for every additional seat beyond seven seats, but the minimum clear width is not required to exceed 22 inches (559 mm).

Exception: For smoke-protected assembly seating, the row length limits for a 12-inch-wide (305 mm) aisle accessway, beyond which the aisle accessway minimum clear width shall be increased, are in Table 1024.10.1.

1024.11 Assembly aisle walking surfaces. Aisles with a slope not exceeding one unit vertical in eight units horizontal (12.5-percent slope) shall consist of a ramp having a slip-resistant walking surface. Aisles with a slope exceeding one unit vertical in eight units horizontal (12.5-percent slope) shall consist of a series of risers and treads that extends across the full width of aisles and complies with Sections 1024.11.1 through 1024.11.3.

1024.11.1 Treads. Tread depths shall be a minimum of 11 inches (279 mm) and shall have dimensional uniformity.

Exception: The tolerance between adjacent treads shall not exceed 0.188 inch (4.8 mm).

1024.11.2 Risers. Where the gradient of aisle stairs is to be the same as the gradient of adjoining seating areas, the riser height shall not be less than 4 inches (102 mm) nor more than 8 inches (203 mm) and shall be uniform within each flight.

Exceptions:

1. Riser height nonuniformity shall be limited to the extent necessitated by changes in the gradient of the adjoining seating area to maintain adequate sightlines. Where nonuniformities exceed 0.188 inch (4.8 mm) between adjacent risers, the exact location of such nonuniformities shall be indicated with a distinctive marking stripe on each tread at the nosing or leading edge adjacent to the nonuniform risers. Such stripe shall be a minimum of 1 inch (25 mm), and a maximum of 2 inches (51 mm), wide. The edge marking stripe shall be distinctively different from the contrasting marking stripe.

2. Riser heights not exceeding 9 inches (229 mm) shall be permitted where they are necessitated by the slope of the adjacent seating areas to maintain sightlines.

1024.11.3 Tread contrasting marking stripe. A contrasting marking stripe shall be provided on each tread at the nosing or leading edge such that the location of each tread is readily apparent when viewed in descent. Such stripe shall be a minimum of 1 inch (25 mm), and a maximum of 2 inches (51 mm), wide.

Exception: The contrasting marking stripe is permitted to be omitted where tread surfaces are such that the location of each tread is readily apparent when viewed in descent.

1024.12 Seat stability. In places of assembly, the seats shall be securely fastened to the floor.

Exceptions:

1. In places of assembly or portions thereof without ramped or tiered floors for seating and with 200 or fewer seats, the seats shall not be required to be fastened to the floor.

2. In places of assembly or portions thereof with seating at tables and without ramped or tiered floors for seating, the seats shall not be required to be fastened to the floor.

3. In places of assembly or portions thereof without ramped or tiered floors for seating and with greater than 200 seats, the seats shall be fastened together in groups of not less than three or the seats shall be securely fastened to the floor.

4. In places of assembly where flexibility of the seating arrangement is an integral part of the design and function of the space and seating is on tiered levels, a maximum of 200 seats shall not be required to be fastened to the floor. Plans showing seating, tiers and aisles shall be submitted for approval.

5. Groups of seats within a place of assembly separated from other seating by railings, guards, partial height walls or similar barriers with level floors and having no more than 14 seats per group shall not be required to be fastened to the floor.

6. Seats intended for musicians or other performers and separated by railings, guards, partial height walls or similar barriers shall not be required to be fastened to the floor.

1024.13 Handrails. Ramped aisles having a slope exceeding one unit vertical in 15 units horizontal (6.7-percent slope) and aisle stairs shall be provided with handrails located either at the side or within the aisle width.

Exceptions:

1. Handrails are not required for ramped aisles having a gradient no greater than one unit vertical in eight units horizontal (12.5-percent slope) and seating on both sides.

2. Handrails are not required if, at the side of the aisle, there is a guard that complies with the graspability requirements of handrails.

1024.13.1 Discontinuous handrails. Where there is seating on both sides of the aisle, the handrails shall be discontinuous with gaps or breaks at intervals not exceeding five rows to facilitate access to seating and to permit crossing from one side of the aisle to the other. These gaps or breaks shall have a clear width of at least 22 inches (559 mm) and not greater than 36 inches (914 mm), measured horizontally, and the handrail shall have rounded terminations or bends.

1024.13.2 Intermediate handrails. Where handrails are provided in the middle of aisle stairs, there shall be an additional intermediate handrail located approximately 12 inches (305 mm) below the main handrail.

1024.14 Assembly guards. Assembly guards shall comply with Sections 1024.14.1 through 1024.14.3.

1024.14.1 Cross aisles. Cross aisles located more than 30 inches (762 mm) above the floor or grade below shall have guards in accordance with Section 1012.

Where an elevation change of 30 inches (762 mm) or less occurs between a cross aisle and the adjacent floor or grade below, guards not less than 26 inches (660 mm) above the aisle floor shall be provided.

Exception: Where the backs of seats on the front of the cross aisle project 24 inches (610 mm) or more above the adjacent floor of the aisle, a guard need not be provided.

1024.14.2 Sightline-constrained guard heights. Unless subject to the requirements of Section 1024.14.3, a fascia or railing system in accordance with the guard requirements of Section 1012 and having a minimum height of 26 inches (660 mm) shall be provided where the floor or footboard elevation is more than 30 inches (762 mm) above the floor or grade below and the fascia or railing would otherwise interfere with the sightlines of immediately adjacent seating. At bleachers, a guard must be provided where the floor or footboard elevation is more than 24 inches (610 mm) above the floor or grade below and the fascia or railing would otherwise interfere with the sightlines of the immediately adjacent seating.

1024.14.3 Guards at the end of aisles. A fascia or railing system complying with the guard requirements of Section 1012 shall be provided for the full width of the aisle where the foot of the aisle is more than 30 inches (762 mm) above the floor or grade below. The fascia or railing shall be a minimum of 36 inches (914 mm) high and shall provide a minimum 42 inches (1067 mm) measured diagonally between the top of the rail and the nosing of the nearest tread.

1024.15 Bench seating. Where bench seating is used, the number of persons shall be based on one person for each 18 inches (457 mm) of length of the bench.

SECTION 1025
EMERGENCY ESCAPE AND RESCUE

1025.1 General. In addition to the means of egress required by this chapter, provisions shall be made for emergency escape and rescue in Group R as applicable in Section 101.2 and Group I-1 occupancies. Basements and sleeping rooms below the fourth story above grade plane shall have at least one exterior emergency escape and rescue opening in accordance with this section. Where basements contain one or more sleeping rooms, emergency egress and rescue openings shall be required in each sleeping room, but shall not be required in adjoining areas of the basement. Such opening shall open directly into a public street, public alley, yard or court.

Exceptions:

1. In other than Group R-3 occupancies as applicable in Section 101.2, buildings equipped throughout with an approved automatic sprinkler system in accordance with Section 903.3.1.1 or 903.3.1.2.

2. In other than Group R-3 occupancies as applicable in Section 101.2, sleeping rooms provided with a door to a fire-resistance-rated corridor having access to two remote exits in opposite directions.

3. The emergency escape and rescue opening is permitted to open onto a balcony within an atrium in accordance with the requirements of Section 404, provided the balcony provides access to an exit and the dwelling unit or sleeping unit has a means of egress that is not open to the atrium.

4. Basements with a ceiling height of less than 80 inches (2032 mm) shall not be required to have emergency escape and rescue windows.

5. High-rise buildings in accordance with Section 403.

6. Emergency escape and rescue openings are not required from basements or sleeping rooms which have an exit door or exit access door that opens directly into a public street, public alley, yard, egress court or to an exterior exit balcony that opens to a public street, public alley, yard or egress court.

7. Basements without habitable spaces and having no more than 200 square feet (18.6 square meters) in floor area shall not be required to have emergency escape windows.

1025.2 Minimum size. Emergency escape and rescue openings shall have a minimum net clear opening of 5.7 square feet (0.53 m²).

Exception: The minimum net clear opening for emergency escape and rescue grade-floor openings shall be 5 square feet (0.46 m²).

1025.2.1 Minimum dimensions. The minimum net clear opening height dimension shall be 24 inches (610 mm). The minimum net clear opening width dimension shall be 20 inches (508 mm). The net clear opening dimensions shall be the result of normal operation of the opening.

1025.3 Maximum height from floor. Emergency escape and rescue openings shall have the bottom of the clear opening not greater than 44 inches (1118 mm) measured from the floor.

1025.4 Operational constraints. Emergency escape and rescue openings shall be operational from the inside of the room without the use of keys or tools. Bars, grilles, grates or similar devices are permitted to be placed over emergency escape and rescue openings provided the minimum net clear opening size complies with Section 1025.2 and such devices shall be releasable or removable from the inside without the use of a key, tool or force greater than that which is required for normal operation of the escape and rescue opening. Where such bars, grilles, grates or similar devices are installed in existing buildings, smoke alarms shall be installed in accordance with Section 907.2.10 regardless of the valuation of the alteration.

1025.5 Window wells. An emergency escape and rescue opening with a finished sill height below the adjacent ground level shall be provided with a window well in accordance with Sections 1025.5.1 and 1025.5.2.

1025.5.1 Minimum size. The minimum horizontal area of the window well shall be 9 square feet (0.84 m²), with a minimum dimension of 36 inches (914 mm). The area of the window well shall allow the emergency escape and rescue opening to be fully opened.

1025.5.2 Ladders or steps. Window wells with a vertical depth of more than 44 inches (1118 mm) shall be equipped with an approved permanently affixed ladder or steps. Ladders or rungs shall have an inside width of at least 12 inches (305 mm), shall project at least 3 inches (76 mm) from the wall and shall be spaced not more than 18 inches (457 mm) on center (o.c.) vertically for the full height of the window well. The ladder or steps shall not encroach into the required dimensions of the window well by more than 6 inches (152 mm). The ladder or steps shall not be obstructed by the emergency escape and rescue opening. Ladders or steps required by this section are exempt from the stairway requirements of Section 1009.

CHAPTER 11
ACCESSIBILITY

SECTION 1101
GENERAL

1101.1 Scope. The provisions of this chapter shall control the design and construction of facilities for accessibility to physically disabled persons.

1101.2 Design. Buildings and facilities shall be designed and constructed to be accessible in accordance with this code and ICC A117.1.

SECTION 1102
DEFINITIONS

1102.1 Definitions. The following words and terms shall, for the purposes of this chapter and as used elsewhere in the code, have the following meanings:

ACCESSIBLE. A site, building, facility or portion thereof, that complies with this chapter.

ACCESSIBLE ROUTE. A continuous, unobstructed path that complies with this chapter.

ACCESSIBLE UNIT. A dwelling unit or sleeping unit that complies with this code and Chapters 1 through 9 of ICC A117.1.

CIRCULATION PATH. An exterior or interior way of passage from one place to another for pedestrians.

COMMON USE. Interior or exterior circulation paths, rooms, spaces or elements that are not for public use and are made available for the shared use of two or more people.

DETECTABLE WARNING. A standardized surface feature built in or applied to walking surfaces or other elements to warn visually impaired persons of hazards on a circulation path.

DWELLING UNIT OR SLEEPING UNIT, MULTI-STORY. A dwelling unit or sleeping unit with habitable space located on more than one story.

DWELLING UNIT OR SLEEPING UNIT, TYPE A. A dwelling unit or sleeping unit designed and constructed for accessibility in accordance with ICC A117.1.

DWELLING UNIT OR SLEEPING UNIT, TYPE B. A dwelling unit or sleeping unit designed and constructed for accessibility in accordance with ICC A117.1, consistent with the design and construction requirements of the federal Fair Housing Act.

EMPLOYEE WORK AREA. All or any portion of a space used only by employees and only for work. Corridors, toilet rooms, kitchenettes and break rooms are not employee areas.

FACILITY. All or any portion of buildings, structures, site improvements, elements and pedestrian or vehicular routes located on a site.

INTENDED TO BE OCCUPIED AS A RESIDENCE. This refers to a dwelling unit or sleeping unit that can or will be used all or part of the time as the occupant's place of abode.

MULTILEVEL ASSEMBLY SEATING. Seating that is arranged in distinct levels where each level is comprised of either multiple rows, or a single row of box seats accessed from a separate level.

PUBLIC ENTRANCE. An entrance that is not a service entrance or a restricted entrance.

PUBLIC-USE AREAS. Interior or exterior rooms or spaces that are made available to the general public.

RESTRICTED ENTRANCE. An entrance that is made available for common use on a controlled basis, but not public use, and that is not a service entrance.

SELF-SERVICE STORAGE FACILITY. Real property designed and used for the purpose of renting or leasing individual storage spaces to customers for the purpose of storing and removing personal property on a self-service basis.

SERVICE ENTRANCE. An entrance intended primarily for delivery of goods or services.

SITE. A parcel of land bounded by a property line or a designated portion of a public right-of-way.

WHEELCHAIR SPACE. A space for a single wheelchair and its occupant.

SECTION 1103
SCOPING REQUIREMENTS

1103.1 Where required. Buildings and structures, temporary or permanent, including their associated sites and facilities, shall be accessible to persons with physical disabilities.

1103.2 General exceptions. Sites, buildings, facilities and elements shall be exempt from this chapter to the extent specified in this section.

1103.2.1 Specific requirements. Accessibility is not required in buildings and facilities, or portions thereof, to the extent permitted by Sections 1104 through 1110.

1103.2.2 Existing buildings. Existing buildings shall comply with Section 3409.

1103.2.3 Employee work areas. Spaces and elements within employee work areas shall only be required to comply with Sections 907.9.1.2, 1007 and 1104.3.1 and shall be designed and constructed so that individuals with disabilities can approach, enter and exit the work area. Work areas, or portions of work areas, that are less than 150 square feet (14 m^2) in area and elevated 7 inches (178 mm) or more above the ground or finish floor where the elevation is essential to the function of the space shall be exempt from all requirements.

1103.2.4 Detached dwellings. Detached one- and two-family dwellings and accessory structures, and their associated sites and facilities as applicable in Section 101.2, are not required to be accessible.

1103.2.5 Utility buildings. Occupancies in Group U are exempt from the requirements of this chapter other than the following:

1. In agricultural buildings, access is required to paved work areas and areas open to the general public.

2. Private garages or carports that contain required accessible parking.

1103.2.6 Construction sites. Structures, sites and equipment directly associated with the actual processes of construction including, but not limited to, scaffolding, bridging, materials hoists, materials storage or construction trailers are not required to be accessible.

1103.2.7 Raised areas. Raised areas used primarily for purposes of security, life safety or fire safety including, but not limited to, observation galleries, prison guard towers, fire towers or lifeguard stands are not required to be accessible or to be served by an accessible route.

1103.2.8 Limited access spaces. Nonoccupiable spaces accessed only by ladders, catwalks, crawl spaces, freight elevators or very narrow passageways are not required to be accessible.

1103.2.9 Equipment spaces. Spaces frequented only by personnel for maintenance, repair or monitoring of equipment are not required to be accessible. Such spaces include, but are not limited to, elevator pits, elevator penthouses, mechanical, electrical or communications equipment rooms, piping or equipment catwalks, water or sewage treatment pump rooms and stations, electric substations and transformer vaults, and highway and tunnel utility facilities.

1103.2.10 Single-occupant structures. Single-occupant structures accessed only by passageways below grade or elevated above grade including, but not limited to, toll booths that are accessed only by underground tunnels, are not required to be accessible.

1103.2.11 Residential Group R-1. Buildings of Group R-1 containing not more than five sleeping units for rent or hire that are also occupied as the residence of the proprietor are not required to be accessible.

1103.2.12 Day care facilities. Where a day care facility (Groups A-3, E, I-4 and R-3) is part of a dwelling unit, only the portion of the structure utilized for the day care facility is required to be accessible.

1103.2.13 Detention and correctional facilities. In detention and correctional facilities, common use areas that are used only by inmates or detainees and security personnel, and that do not serve holding cells or housing cells required to be accessible, are not required to be accessible or to be served by an accessible route.

1103.2.14 Fuel-dispensing systems. The operable parts on fuel-dispensing devices shall comply with ICC A117.1, Section 308.2.1 or 308.3.1.

SECTION 1104
ACCESSIBLE ROUTE

1104.1 Site arrival points. Accessible routes within the site shall be provided from public transportation stops, accessible parking and accessible passenger loading zones and public streets or sidewalks to the accessible building entrance served.

Exception: An accessible route shall not be required between site arrival points and the building or facility entrance if the only means of access between them is a vehicular way not providing for pedestrian access.

1104.2 Within a site. At least one accessible route shall connect accessible buildings, accessible facilities, accessible elements and accessible spaces that are on the same site.

Exception: An accessible route is not required between accessible buildings, accessible facilities, accessible elements and accessible spaces that have, as the only means of access between them, a vehicular way not providing for pedestrian access.

1104.3 Connected spaces. When a building, or portion of a building, is required to be accessible, an accessible route shall be provided to each portion of the building, to accessible building entrances connecting accessible pedestrian walkways and the public way. Where only one accessible route is provided, the accessible route shall not pass through kitchens, storage rooms, restrooms, closets or similar spaces.

Exceptions:

1. In assembly areas with fixed seating required to be accessible, an accessible route shall not be required to serve fixed seating where wheelchair spaces or designated aisle seats required to be on an accessible route are not provided.

2. Accessible routes shall not be required to mezzanines provided that the building or facility has no more than one story, or where multiple stories are not connected by an accessible route as permitted by Section 1104.4.

3. A single accessible route is permitted to pass through a kitchen or storage room in an accessible dwelling unit.

1104.3.1 Employee work areas. Common use circulation paths within employee work areas shall be accessible routes.

Exceptions:

1. Common use circulation paths, located within employee work areas that are less than 300 square feet (27.9 m^2) in size and defined by permanently in-

stalled partitions, counters, casework or furnishings, shall not be required to be accessible routes.

2. Common use circulation paths, located within employee work areas, that are an integral component of equipment, shall not be required to be accessible routes.

3. Common use circulation paths, located within exterior employee work areas that are fully exposed to the weather, shall not be required to be accessible routes.

1104.3.2 Press boxes. Press boxes in assembly areas shall be on an accessible route.

Exceptions:

1. An accessible route shall not be required to press boxes in bleachers that have points of entry at only one level, provided that the aggregate area of all press boxes is 500 square feet (46 m²) maximum.

2. An accessible route shall not be required to free-standing press boxes that are elevated above grade 12 feet (3660 mm) minimum provided that the aggregate area of all press boxes is 500 square feet (46 m²) maximum.

1104.4 Multilevel buildings and facilities. At least one accessible route shall connect each accessible level, including mezzanines, in multilevel buildings and facilities.

Exceptions:

1. An accessible route is not required to stories and mezzanines above and below accessible levels that have an aggregate area of not more than 3,000 square feet (278.7 m²). This exception shall not apply to:

1.1. Multiple tenant facilities of Group M occupancies containing five or more tenant spaces;

1.2. Levels containing offices of health care providers (Group B or I); or

1.3. Passenger transportation facilities and airports (Group A-3 or B).

2. In Group A, I, R and S occupancies, levels that do not contain accessible elements or other spaces required by Section 1107 or 1108 are not required to be served by an accessible route from an accessible level.

3. In air traffic control towers, an accessible route is not required to serve the cab and the floor immediately below the cab.

4. Where a two-story building or facility has one story with an occupant load of five or fewer persons that does not contain public use space, that story shall not be required to be connected by an accessible route to the story above or below.

1104.5 Location. Accessible routes shall coincide with or be located in the same area as a general circulation path. Where the circulation path is interior, the accessible route shall also be interior.

Exception: Accessible routes from parking garages contained within and serving Type B dwelling units are not required to be interior.

1104.6 Security barriers. Security barriers including, but not limited to, security bollards and security check points shall not obstruct a required accessible route or accessible means of egress.

Exception: Where security barriers incorporate elements that cannot comply with these requirements, such as certain metal detectors, fluoroscopes or other similar devices, the accessible route shall be permitted to be provided adjacent to security screening devices. The accessible route shall permit persons with disabilities passing around security barriers to maintain visual contact with their personal items to the same extent provided others passing through the security barrier.

SECTION 1105
ACCESSIBLE ENTRANCES

1105.1 Public entrances. In addition to accessible entrances required by Sections 1105.1.1 through 1105.1.6, at least 50 percent of all public entrances shall be accessible.

Exceptions:

1. An accessible entrance is not required to areas not required to be accessible.

2. Loading and service entrances that are not the only entrance to a tenant space.

1105.1.1 Parking garage entrances. Where provided, direct access for pedestrians from parking structures to buildings or facility entrances shall be accessible.

1105.1.2 Entrances from tunnels or elevated walkways. Where direct access is provided for pedestrians from a pedestrian tunnel or elevated walkway to a building or facility, at least one entrance to the building or facility from each tunnel or walkway shall be accessible.

1105.1.3 Restricted entrances. Where restricted entrances are provided to a building or facility, at least one restricted entrance to the building or facility shall be accessible.

1105.1.4 Entrances for inmates or detainees. Where entrances used only by inmates or detainees and security personnel are provided at judicial facilities, detention facilities or correctional facilities, at least one such entrance shall be accessible.

1105.1.5 Service entrances. If a service entrance is the only entrance to a building or a tenant space in a facility, that entrance shall be accessible.

1105.1.6 Tenant spaces, dwelling units and sleeping units. At least one accessible entrance shall be provided to each tenant, dwelling unit and sleeping unit in a facility.

Exceptions:

1. An accessible entrance is not required to tenants that are not required to be accessible.

2. An accessible entrance is not required to dwelling units and sleeping units that are not required to be Accessible units, Type A units or Type B units.

SECTION 1106
PARKING AND PASSENGER LOADING FACILITIES

1106.1 Required. Where parking is provided, accessible parking spaces shall be provided in compliance with Table 1106.1, except as required by Sections 1106.2 through 1106.4. The number of accessible parking spaces shall be determined based on the total number of parking spaces provided for the facility.

Exception: This section does not apply to parking spaces used exclusively for buses, trucks, other delivery vehicles, law enforcement vehicles or vehicular impound and motor pools where lots accessed by the public are provided with an accessible passenger loading zone.

TABLE 1106.1
ACCESSIBLE PARKING SPACES

TOTAL PARKING SPACES PROVIDED	REQUIRED MINIMUM NUMBER OF ACCESSIBLE SPACES
1 to 25	1
26 to 50	2
51 to 75	3
76 to 100	4
101 to 150	5
151 to 200	6
201 to 300	7
301 to 400	8
401 to 500	9
501 to 1,000	2% of total
More than 1,000	20, plus one for each 100 over 1,000

1106.2 Groups R-2 and R-3. Two percent, but not less than one, of each type of parking space provided for occupancies in Groups R-2 and R-3, which are required to have Accessible, Type A or Type B dwelling or sleeping units, shall be accessible. Where parking is provided within or beneath a building, accessible parking spaces shall also be provided within or beneath the building.

1106.3 Hospital outpatient facilities. Ten percent of patient and visitor parking spaces provided to serve hospital outpatient facilities shall be accessible.

1106.4 Rehabilitation facilities and outpatient physical therapy facilities. Twenty percent, but not less than one, of the portion of patient and visitor parking spaces serving rehabilitation facilities and outpatient physical therapy facilities shall be accessible.

1106.5 Van spaces. For every six or fraction of six accessible arking spaces, at least one shall be a van-accessible parking space.

1106.6 Location. Accessible parking spaces shall be located on the shortest accessible route of travel from adjacent parking to an accessible building entrance. Accessible parking spaces shall be dispersed among the various types of parking facilities provided. In parking facilities that do not serve a particular building, accessible parking spaces shall be located on the shortest route to an accessible pedestrian entrance to the parking facility. Where buildings have multiple accessible entrances with adjacent parking, accessible parking spaces shall be dispersed and located near the accessible entrances.

Exception: In multilevel parking structures, van-accessible parking spaces are permitted on one level.

1106.7 Passenger loading zones. Passenger loading zones shall be designed and constructed in accordance with ICC A117.1.

1106.7.1 Continuous loading zones. Where passenger loading zones are provided, one passenger loading zone in every continuous 100 linear feet (30.4 m) maximum of loading zone space shall be accessible.

1106.7.2 Medical facilities. A passenger loading zone shall be provided at an accessible entrance to licensed medical and long-term care facilities where people receive physical or medical treatment or care and where the period of stay exceeds 24 hours.

1106.7.3 Valet parking. A passenger loading zone shall be provided at valet parking services.

SECTION 1107
DWELLING UNITS AND SLEEPING UNITS

1107.1 General. In addition to the other requirements of this chapter, occupancies having dwelling units or sleeping units shall be provided with accessible features in accordance with this section.

1107.2 Design. Dwelling units and sleeping units which are required to be Accessible units shall comply with this code and the applicable portions of Chapters 1 through 9 of ICC A117.1. Type A and Type B units shall comply with the applicable portions of Chapter 10 of ICC A117.1. Units required to be Type A units are permitted to be designed and constructed as Accessible units. Units required to be Type B units are permitted to be designed and constructed as Accessible units or as Type A units.

1107.3 Accessible spaces. Rooms and spaces available to the general public or available for use by residents and serving Accessible units, Type A units or Type B units shall be accessible. Accessible spaces shall include toilet and bathing rooms, kitchen, living and dining areas and any exterior spaces, including patios, terraces and balconies.

Exception: Recreational facilities in accordance with Section 1109.14.

1107.4 Accessible route. At least one accessible route shall connect accessible building or facility entrances with the primary entrance of each Accessible unit, Type A unit and Type B unit within the building or facility and with those exterior and interior spaces and facilities that serve the units.

Exceptions:

1. If the slope of the finished ground level between accessible facilities and buildings exceeds one unit vertical in 12 units horizontal (1:12), or where physical barriers prevent the installation of an accessible route, a vehicular route with parking that complies with Section 1106 at each public or common use facility or building is permitted in place of the accessible route.

2. Exterior decks, patios or balconies that are part of Type B units and have impervious surfaces, and that are not more than 4 inches (102 mm) below the finished floor level of the adjacent interior space of the unit.

1107.5 Group I. Occupancies in Group I shall be provided with accessible features in accordance with Sections 1107.5.1 through 1107.5.5.

1107.5.1 Group I-1. Group I-1 occupancies shall be provided with accessible features in accordance with Sections 1107.5.1.1 and 1107.5.1.2.

1107.5.1.1 Accessible units. At least 4 percent, but not less than one, of the dwelling units and sleeping units shall be Accessible units.

1107.5.1.2 Type B units. In structures with four or more dwelling or sleeping units intended to be occupied as a residence, every dwelling and sleeping unit intended to be occupied as a residence shall be a Type B unit.

Exception: The number of Type B units is permitted to be reduced in accordance with Section 1107.7.

1107.5.2 Group I-2 Nursing homes. Nursing homes of Group I-2 shall be provided with accessible features in accordance with Sections 1107.5.2.1 and 1107.5.2.2.

1107.5.2.1 Accessible units. At least 50 percent, but not less than one, of the dwelling units and sleeping units shall be Accessible units.

1107.5.2.2 Type B units. In structures with four or more dwelling or sleeping units intended to be occupied as a residence, every dwelling and sleeping unit intended to be occupied as a residence shall be a Type B unit.

Exception: The number of Type B units is permitted to be reduced in accordance with Section 1107.7.

1107.5.3 Group I-2 Hospitals. General-purpose hospitals, psychiatric facilities, detoxification facilities and residential care/assisted living facilities of Group I-2 shall be provided with accessible features in accordance with Sections 1107.5.3.1 and 1107.5.3.2.

1107.5.3.1 Accessible units. At least 10 percent, but not less than one, of the dwelling units and sleeping units shall be Accessible units.

1107.5.3.2 Type B units. In structures with four or more dwelling or sleeping units intended to be occupied as a residence, every dwelling and sleeping unit intended to be occupied as a residence shall be a Type B unit.

Exception: The number of Type B units is permitted to be reduced in accordance with Section 1107.7.

1107.5.4 Group I-2 Rehabilitation facilities. In hospitals and rehabilitation facilities of Group I-2 which specialize in treating conditions that affect mobility, or units within either which specialize in treating conditions that affect mobility, 100 percent of the dwelling units and sleeping units shall be Accessible units.

1107.5.5 Group I-3. Buildings, facilities or portions thereof with Group I-3 occupancies shall comply with Sections 1107.5.5.1 through 1107.5.5.3.

1107.5.5.1 Group I-3 sleeping units. In occupancies in Group I-3, at least 2 percent, but not less than one, of the dwelling units and sleeping units shall be Accessible units.

1107.5.5.2 Special holding cells and special housing cells or rooms. In addition to the units required to be accessible by Section 1107.5.5.1, where special holding cells or special housing cells or rooms are provided, at least one serving each purpose shall be accessible. Cells or rooms subject to this requirement include, but are not limited to, those used for purposes of orientation, protective custody, administrative or disciplinary detention or segregation, detoxification and medical isolation.

Exception: Cells or rooms specially designed without protrusions and that are used solely for purposes of suicide prevention shall not be required to include grab bars.

1107.5.5.3 Medical care facilities. Patient sleeping units or cells required to be accessible in medical care facilities shall be provided in addition to any medical isolation cells required to comply with Section 1107.5.5.2.

1107.6 Group R. Occupancies in Group R shall be provided with accessible features in accordance with Sections 1107.6.1 through 1107.6.4.

1107.6.1 Group R-1. Group R-1 occupancies shall be provided with accessible features in accordance with Sections 1107.6.1.1 and 1107.6.1.2.

1107.6.1.1 Accessible units. In occupancies in Group R-1, Accessible dwelling units and sleeping units shall be provided in accordance with Table 1107.6.1.1. All facilities on a site shall be considered to determine the total number of Accessible units. Accessible units shall be dispersed among the various classes of units. Roll-in showers provided in Accessible units shall include a permanently mounted folding shower seat.

TABLE 1107.6.1.1
ACCESSIBLE DWELLING AND SLEEPING UNITS

TOTAL NUMBER OF UNITS PROVIDED	MINIMUM REQUIRED NUMBER OF ACCESSIBLE UNITS ASSOCIATED WITH ROLL-IN SHOWERS	TOTAL NUMBER OF REQUIRED ACCESSIBLE UNITS
1 to 25	0	1
26 to 50	0	2
51 to 75	1	4
76 to 100	1	5
101 to 150	2	7
151 to 200	2	8
201 to 300	3	10
301 to 400	4	12
401 to 500	4	13
501 to 1,000	1% of total	3% of total
Over 1,000	10, plus 1 for each 100 over 1,000	30, plus 2 for each 100 over 1,000

1107.6.1.2 Type B units. In structures with four or more dwelling or sleeping units intended to be occupied as a residence, every dwelling and sleeping unit intended to be occupied as a residence shall be a Type B unit.

Exception: The number of Type B units is permitted to be reduced in accordance with Section 1107.7.

1107.6.2 Group R-2. Accessible units, Type A units and Type B units shall be provided in occupancies in Group R-2 in accordance with Sections 1107.6.2.1 and 1107.6.2.2.

1107.6.2.1 Apartment houses, monasteries and convents. Type A and Type B units shall be provided in apartment houses, monasteries and convents in accordance with Sections 1107.6.2.1.1 and 1107.6.2.1.2.

1107.6.2.1.1 Type A units. In occupancies in Group R-2 containing more than 20 dwelling units or sleeping units, at least 2 percent, but not less than one, of the units shall be a Type A unit. All units on a site shall be considered to determine the total number of units and the required number of Type A units. Type A units shall be dispersed among the various classes of units.

Exceptions:

1. The number of Type A units is permitted to be reduced in accordance with Section 1107.7.
2. Existing structures on a site shall not contribute to the total number of units on a site.

1107.6.2.1.2 Type B units. Where there are four or more dwelling units or sleeping units intended to be occupied as a residence in a single structure, every dwelling unit and sleeping unit intended to be occupied as a residence shall be a Type B unit.

Exception: The number of Type B units is permitted to be reduced in accordance with Section 1107.7.

1107.6.2.2 Boarding houses, dormitories, fraternity houses and sorority houses. Accessible units and Type B dwelling units shall be provided in boarding houses, dormitories, fraternity houses and sorority houses in accordance with Sections 1107.6.2.2.1 and 1107.6.2.2.2.

1107.6.2.2.1 Accessible units. Accessible dwelling units and sleeping units shall be provided in accordance with Table 1107.6.1.1.

1107.6.2.2.2 Type B units. Where there are four or more dwelling units or sleeping units intended to be occupied as a residence in a single structure, every dwelling unit and every sleeping unit intended to be occupied as a residence shall be a Type B unit.

Exception: The number of Type B units is permitted to be reduced in accordance with Section 1107.7.

1107.6.3 Group R-3. In occupancies in Group R-3 where there are four or more dwelling units or sleeping units intended to be occupied as a residence in a single structure, every dwelling and sleeping unit intended to be occupied as a residence shall be a Type B unit.

Exception: The number of Type B units is permitted to be reduced in accordance with Section 1107.7.

1107.6.4 Group R-4. Group R-4 occupancies shall be provided with accessible features in accordance with Sections 1107.6.4.1 and 1107.6.4.2.

1107.6.4.1 Accessible units. At least one of the dwelling or sleeping units shall be an Accessible unit.

1107.6.4.2 Type B units. In structures with four or more dwelling or sleeping units intended to be occupied as a residence, every dwelling and sleeping unit intended to be occupied as a residence shall be a Type B unit.

Exception: The number of Type B units is permitted to be reduced in accordance with Section 1107.7.

1107.7 General exceptions. Where specifically permitted by Section 1107.5 or 1107.6, the required number of Type A and Type B units is permitted to be reduced in accordance with Sections 1107.7.1 through 1107.7.5.

1107.7.1 Buildings without elevator service. Where no elevator service is provided in a building, only the dwelling and sleeping units that are located on stories indicated in Sections 1107.7.1.1 and 1107.7.1.2 are required to be Type A and Type B units. The number of Type A units shall be determined in accordance with Section 1107.6.2.1.1.

1107.7.1.1 One story with Type B units required. At least one story containing dwelling units or sleeping units intended to be occupied as a residence shall be provided with an accessible entrance from the exterior of the building and all units intended to be occupied as a residence on that story shall be Type B units.

1107.7.1.2 Additional stories with Type B units. On all other stories that have a building entrance in proximity to arrival points intended to serve units on that story, as indicated in Items 1 and 2, all dwelling units and sleeping units intended to be occupied as a residence served by that entrance on that story shall be Type B units.

1. Where the slopes of the undisturbed site measured between the planned entrance and all vehicular or pedestrian arrival points within 50 feet of the planned entrance are 10 percent or less, and

2. Where the slopes of the planned finished grade measured between the entrance and all vehicular or pedestrian arrival points within 50 feet of the planned entrance are 10 percent or less.

Where no such arrival points are within 50 feet (15 240 mm) of the entrance, the closest arrival point shall be used unless that arrival point serves the story required by Section 1107.7.1.1.

1107.7.2 Multistory units. A multistory dwelling or sleeping unit which is not provided with elevator service is not required to be a Type B unit. Where a multistory unit is provided with external elevator service to only one floor, the floor provided with elevator service shall be the primary entry to the unit, shall comply with the requirements for a Type B unit and a toilet facility shall be provided on that floor.

1107.7.3 Elevator service to the lowest story with units. Where elevator service in the building provides an accessible route only to the lowest story containing dwelling or sleeping units intended to be occupied as a residence, only the units on that story which are intended to be occupied as a residence are required to be Type B units.

1107.7.4 Site impracticality. On a site with multiple nonelevator buildings, the number of units required by Section 1107.7.1 to be Type B units is permitted to be reduced to a percentage which is equal to the percentage of the entire site having grades, prior to development, which are less than 10 percent, provided that all of the following conditions are met:

1. Not less than 20 percent of the units required by Section 1107.7.1 on the site are Type B units;

2. Units required by Section 1107.7.1, where the slope between the building entrance serving the units on that story and a pedestrian or vehicular arrival point is no greater than 8.33 percent, are Type B units;

3. Units required by Section 1107.7.1, where an elevated walkway is planned between a building entrance serving the units on that story and a pedestrian or vehicular arrival point and the slope between them is 10 percent or less are Type B units; and

4. Units served by an elevator in accordance with Section 1107.7.3 are Type B units.

1107.7.5 Design flood elevation. The required number of Type A and Type B units shall not apply to a site where the lowest floor or the lowest structural building members of nonelevator buildings are required to be at or above the design flood elevation resulting in:

1. A difference in elevation between the minimum required floor elevation at the primary entrances and vehicular and pedestrian arrival points within 50 feet (15 240 mm) exceeding 30 inches (762 mm), and

2. A slope exceeding 10 percent between the minimum required floor elevation at the primary entrances and vehicular and pedestrian arrival points within 50 feet (15 240 mm).

Where no such arrival points are within 50 feet (15 240 mm) of the primary entrances, the closest arrival point shall be used.

SECTION 1108
SPECIAL OCCUPANCIES

1108.1 General. In addition to the other requirements of this chapter, the requirements of Sections 1108.2 through 1108.4 shall apply to specific occupancies.

1108.2 Assembly area seating. Assembly areas with fixed seating shall comply with Sections 1108.2.1 through 1108.2.8. Dining areas shall comply with Section 1108.2.9.

1108.2.1 Services. Services and facilities provided in areas not required to be accessible shall be provided on an accessible level and shall be accessible.

1108.2.2 Wheelchair spaces. In theaters, bleachers, grandstands, stadiums, arenas and other fixed seating assembly areas, accessible wheelchair spaces complying with ICC A117.1 shall be provided in accordance with Sections 1108.2.2.1 through 1108.2.2.5.

1108.2.2.1 General seating. Wheelchair spaces shall be provided in accordance with Table 1108.2.2.1

TABLE 1108.2.2.1
ACCESSIBLE WHEELCHAIR SPACES

CAPACITY OF SEATING IN ASSEMBLY AREAS	MINIMUM REQUIRED NUMBER OF WHEELCHAIR SPACES
4 to 25	1
26 to 50	2
51 to 100	4
101 to 300	5
301 to 500	6
501 to 5,000	6, plus 1 for each 150, or fraction thereof, between 501 through 5,000
5,001 and over	36 plus 1 for each 200, or fraction thereof, over 5,000

1108.2.2.2 Luxury boxes, club boxes and suites. In each luxury box, club box, and suite within arenas, stadiums and grandstands, wheelchair spaces shall be provided in accordance with Table 1108.2.2.1.

1108.2.2.3 Other boxes. In boxes other than those required to comply with Section 1108.2.2.2, the total number of wheelchair spaces provided shall be determined in accordance with Table 1108.2.2.1. Wheelchair spaces shall be located in not less than 20 percent of all boxes provided.

1108.2.3 Integration. Wheelchair spaces shall be an integral part of the seating plan.

1108.2.4 Dispersion of wheelchair spaces. Dispersion of wheelchair spaces shall be based on the availability of accessible routes to various seating areas including seating at various levels in multilevel facilities.

1108.2.4.1 Multilevel assembly seating areas. In multilevel assembly seating areas, wheelchair spaces shall be provided on the main floor level and on one of each two additional floor or mezzanine levels. Wheelchair spaces shall be provided in each luxury box, club box and suite within assembly facilities.

Exceptions:

1. In multilevel assembly spaces utilized for worship services, where the second floor or mezza-
nine level contains 25 percent or less of the total seating capacity, wheelchair spaces shall be permitted to all be located on the main level.

2. In multilevel assembly seating where the second floor or mezzanine level provides 25 percent or less of the total seating capacity and 300 or fewer seats, wheelchair spaces shall be permitted to all be located on the main level.

1108.2.5 Companion seats. At least one companion seat complying with ICC A117.1 shall be provided for each wheelchair space required by Section 1108.2.2.

1108.2.6 Designated aisle seats. At least five percent, but not less than one, of the total number of aisle seats provided shall be designated aisle seats.

1108.2.7 Assistive listening systems. Each assembly area where audible communications are integral to the use of the space shall have an assistive listening system.

Exception: Other than in courtrooms, an assistive listening system is not required where there is no audio amplification system.

1108.2.7.1 Receivers. Receivers shall be provided for assistive listening systems in accordance with Table 1108.2.7.1.

Exception: Where a building contains more than one assembly area, the total number of required receivers shall be permitted to be calculated according to the total number of seats in the assembly areas in the building provided that all receivers are usable with all systems, and if assembly areas required to provide assistive listening are under one management.

1108.2.7.2 Public address systems. Where stadiums, arenas and grandstands provide audible public announcements, they shall also provide equivalent text information regarding events and facilities in compliance with Sections 1108.2.7.2.1 and 1108.2.7.2.2.

1108.2.7.2.1 Prerecorded text messages. Where electronic signs are provided and have the capability to display prerecorded text messages containing information that is the same, or substantially equivalent, to informa-

TABLE 1108.2.7.1
RECEIVERS FOR ASSISTIVE LISTENING SYSTEMS

CAPACITY OF SEATING IN ASSEMBLY AREAS	MINIMUM REQUIRED NUMBER OF RECEIVERS	MINIMUM NUMBER OF RECEIVERS TO BE HEARING-AID COMPATIBLE
50 or less	2	2
51 to 200	2, plus 1 per 25 seats over 50 seats*	2
201 to 500	2, plus 1 per 25 seats over 50 seats.*	1 per 4 receivers*
501 to 1,000	20, plus 1 per 33 seats over 500 seats*	1 per 4 receivers*
1,001 to 2,000	35, plus 1 per 50 seats over 1,000 seats*	1 per 4 receivers*
Over 2,000	55, plus 1 per 100 seats over 2,000 seats*	1 per 4 receivers*

NOTE: * = or fraction thereof

tion that is provided audibly, signs shall display text that is equivalent to audible announcements.

> **Exception:** Announcements that cannot be prerecorded in advance of the event shall not be required to be displayed.

1108.2.7.2.2 Real-time messages. Where electronic signs are provided and have the capability to display real-time messages containing information that is the same, or substantially equivalent, to information that is provided audibly, signs shall display text that is equivalent to audible announcements.

1108.2.8 Performance areas. An accessible route shall directly connect the performance area to the assembly seating area where a circulation path directly connects a performance area to an assembly seating area. An accessible route shall be provided from performance areas to ancillary areas or facilities used by performers.

1108.2.9 Dining areas. In dining areas, the total floor area allotted for seating and tables shall be accessible.

Exceptions:

1. In buildings or facilities not required to provide an accessible route between levels, an accessible route to a mezzanine seating area is not required, provided that the mezzanine contains less than 25 percent of the total area and the same services are provided in the accessible area.

2. In sports facilities, tiered dining areas providing seating required to be accessible shall be required to have accessible routes serving at least 25 percent of the dining area, provided that accessible routes serve accessible seating and where each tier is provided with the same services.

1108.2.9.1 Dining surfaces. Where dining surfaces for the consumption of food or drink are provided, at least 5 percent, but not less than one, of the seating and standing spaces at the dining surfaces shall be accessible and be distributed throughout the facility.

1108.3 Self-service storage facilities. Self-service storage facilities shall provide accessible individual self-storage spaces in accordance with Table 1108.3.

TABLE 1108.3
ACCESSIBLE SELF-SERVICE STORAGE FACILITIES

TOTAL SPACES IN FACILITY	MINIMUM NUMBER OF REQUIRED ACCESSIBLE SPACES
1 to 200	5%, but not less than 1
Over 200	10, plus 2% of total number of units over 200

1108.3.1 Dispersion. Accessible individual self-service storage spaces shall be dispersed throughout the various classes of spaces provided. Where more classes of spaces are provided than the number of required accessible spaces, the number of accessible spaces shall not be required to exceed that required by Table 1108.3. Accessible spaces are permitted to be dispersed in a single building of a multibuilding facility.

1108.4 Judicial facilities. Judicial facilities shall comply with Sections 1108.4.1 through 1108.4.3.

1108.4.1 Courtrooms. Each courtroom shall be accessible.

1108.4.2 Holding cells. Where provided, central holding cells and court-floor holding cells shall comply with Sections 1108.4.2.1 and 1108.4.2.2.

1108.4.2.1 Central holding cells. Where separate central holding cells are provided for adult males, juvenile males, adult females or juvenile females, one of each type shall be accessible. Where central holding cells are provided and are not separated by age or sex, at least one accessible cell shall be provided.

1108.4.2.2 Court-floor holding cells. Where separate court-floor holding cells are provided for adult males, juvenile males, adult females or juvenile females, each courtroom shall be served by one accessible cell of each type. Where court-floor holding cells are provided and are not separated by age or sex, courtrooms shall be served by at least one accessible cell. Accessible cells shall be permitted to serve more than one courtroom.

1108.4.3 Visiting areas. Visiting areas shall comply with Sections 1108.4.3.1 and 1108.4.3.2.

1108.4.3.1 Cubicles and counters. At least 5 percent, but no fewer than one, of cubicles shall be accessible on both the visitor and detainee sides. Where counters are provided, at least one shall be accessible on both the visitor and detainee sides.

> **Exception:** This requirement shall not apply to the detainee side of cubicles or counters at noncontact visiting areas not serving holding cells.

1108.4.3.2 Partitions. Where solid partitions or security glazing separate visitors from detainees, at least one of each type of cubicle or counter partition shall be accessible.

SECTION 1109
OTHER FEATURES AND FACILITIES

1109.1 General. Accessible building features and facilities shall be provided in accordance with Sections 1109.2 through 1109.15.

> **Exception:** Type A and Type B dwelling and sleeping units shall comply with ICC A117.1.

1109.2 Toilet and bathing facilities. Toilet rooms and bathing facilities shall be accessible. Where a floor level is not required to be connected by an accessible route, the only toilet rooms or bathing facilities provided within the facility shall not be located on the inaccessible floor. At least one of each type of fixture, element, control or dispenser in each accessible toilet room and bathing facility shall be accessible.

Exceptions:

1. In toilet rooms or bathing facilities accessed only through a private office, not for common or public use, and intended for use by a single occupant, any of the following alternatives are allowed:

1.1. Doors are permitted to swing into the clear floor space provided the door swing can be reversed to meet the requirements in ICC A117.1,

1.2. The height requirements for the water closet in ICC A117.1 are not applicable,

1.3. Grab bars are not required to be installed in a toilet room, provided that reinforcement has been installed in the walls and located so as to permit the installation of such grab bars, and

1.4. The requirement for height, knee and toe clearance shall not apply to a lavatory.

2. This section is not applicable to toilet and bathing facilities that serve dwelling units or sleeping units that are not required to be accessible by Section 1107.

3. Where multiple single-user toilet rooms or bathing facilities are clustered at a single location and contain fixtures in excess of the minimum required number of plumbing fixtures, at least 5 percent, but not less than one room for each use at each cluster, shall be accessible.

4. Toilet room fixtures that are in excess of those required by the *International Plumbing Code* and that are designated for use by children in day care and primary school occupancies.

5. Where no more than one urinal is provided in a toilet room or bathing facility, the urinal is not required to be accessible.

6. Toilet rooms that are part of critical-care or intensive-care patient sleeping rooms are not required to be accessible.

1109.2.1 Unisex toilet and bathing rooms. In assembly and mercantile occupancies, an accessible unisex toilet room shall be provided where an aggregate of six or more male and female water closets is required. In buildings of mixed occupancy, only those water closets required for the assembly or mercantile occupancy shall be used to determine the unisex toilet room requirement. In recreational facilities where separate-sex bathing rooms are provided, an accessible unisex bathing room shall be provided. Fixtures located within unisex toilet and bathing rooms shall be included in determining the number of fixtures provided in an occupancy.

Exception: Where each separate-sex bathing room has only one shower or bathtub fixture, a unisex bathing room is not required.

1109.2.1.1 Standard. Unisex toilet and bathing rooms shall comply with Sections 1109.2.1.2 through 1109.2.1.7 and ICC A117.1.

1109.2.1.2 Unisex toilet rooms. Unisex toilet rooms shall include only one water closet and only one lavatory. A unisex bathing room in accordance with Section 1109.2.1.3 shall be considered a unisex toilet room.

Exception: A urinal is permitted to be provided in addition to the water closet in a unisex toilet room.

1109.2.1.3 Unisex bathing rooms. Unisex bathing rooms shall include only one shower or bathtub fixture. Unisex bathing rooms shall also include one water closet and one lavatory. Where storage facilities are provided for separate-sex bathing rooms, accessible storage facilities shall be provided for unisex bathing rooms.

1109.2.1.4 Location. Unisex toilet and bathing rooms shall be located on an accessible route. Unisex toilet rooms shall be located not more than one story above or below separate-sex toilet rooms. The accessible route from any separate-sex toilet room to a unisex toilet room shall not exceed 500 feet (152 m).

1109.2.1.5 Prohibited location. In passenger transportation facilities and airports, the accessible route from separate-sex toilet rooms to a unisex toilet room shall not pass through security checkpoints.

1109.2.1.6 Clear floor space. Where doors swing into a unisex toilet or bathing room, a clear floor space not less than 30 inches by 48 inches (762 mm by 1219 mm) shall be provided, within the room, beyond the area of the door swing.

1109.2.1.7 Privacy. Doors to unisex toilet and bathing rooms shall be securable from within the room.

1109.2.2 Water closet compartment. Where water closet compartments are provided in a toilet room or bathing facility, at least one wheelchair-accessible compartment shall be provided. Where the combined total water closet compartments and urinals provided in a toilet room or bathing facility is six or more, at least one ambulatory-accessible water closet compartment shall be provided in addition to the wheelchair-accessible compartment. Wheelchair-accessible and ambulatory-accessible compartments shall comply with ICC A117.1.

1109.3 Sinks. Where sinks are provided, at least 5 percent, but not less than one, provided in accessible spaces shall comply with ICC A117.1.

Exceptions:

1. Mop or service sinks are not required to be accessible.

2. Sinks designated for use by children in day care and primary school occupancies.

1109.4 Kitchens and kitchenettes. Where kitchens and kitchenettes are provided in accessible spaces or rooms, they shall be accessible in accordance with ICC A117.1.

1109.5 Drinking fountains. On floors where drinking fountains are provided, at least 50 percent, but not less than one fountain, shall be accessible.

1109.6 Elevators. Passenger elevators on an accessible route shall be accessible and comply with Section 3001.3.

1109.7 Lifts. Platform (wheelchair) lifts are permitted to be a part of a required accessible route in new construction where indicated in Items 1 through 7. Platform (wheelchair) lifts shall be installed in accordance with ASME A18.1.

1. An accessible route to a performing area and speakers' platforms in occupancies in Group A.

2. An accessible route to wheelchair spaces required to comply with the wheelchair space dispersion requirements of Section 1108.2.2 through 1108.2.4.

3. An accessible route to spaces that are not open to the general public with an occupant load of not more than five.

4. An accessible route within a dwelling or sleeping unit.

5. An accessible route to wheelchair seating spaces located in outdoor dining terraces in A-5 occupancies where the means of egress from the dining terraces to a public way are open to the outdoors.

6. An accessible route to raised judges' benches, clerks' stations, jury boxes, witness stands and other raised or depressed areas in a court.

7. An accessible route where existing exterior site constraints make use of a ramp or elevator infeasible.

1109.8 Storage. Where fixed or built-in storage elements such as cabinets, shelves, medicine cabinets, closets and drawers are provided in required accessible spaces, at least one of each type shall contain storage space complying with ICC A117.1.

1109.8.1 Lockers. Where lockers are provided in accessible spaces, at least five percent, but not less than one, of each type shall be accessible.

1109.8.2 Shelving and display units. Self-service shelves and display units shall be located on an accessible route. Such shelving and display units shall not be required to comply with reach-range provisions.

1109.8.3 Coat hooks and folding shelves. Where coat hooks and folding shelves are provided in toilet rooms, toilet compartments, or in dressing, fitting or locker rooms, at least one of each type shall be accessible and shall be provided in accessible toilet rooms without toilet compartments, accessible toilet compartments and accessible dressing, fitting and locker rooms.

1109.9 Detectable warnings. Passenger transit platform edges bordering a drop-off and not protected by platform screens or guards shall have a detectable warning.

Exception: Detectable warnings are not required at bus stops.

1109.10 Assembly area seating. Assembly areas with fixed seating in every occupancy shall comply with Section 1108.2 for accessible seating and assistive listening devices.

1109.11 Seating at tables, counters and work surfaces. Where seating or standing space at fixed or built-in tables, counters or work surfaces is provided in accessible spaces, at least 5 percent of the seating and standing spaces, but not less than one, shall be accessible. In Group I-3 occupancy visiting areas at least 5 percent, but not less than one, cubicle or counter shall be accessible on both the visitor and detainee sides.

Exceptions:

1. Check-writing surfaces at check-out aisles not required to comply with Section 1109.12.2 are not required to be accessible.

2. In Group I-3 occupancies, the counter or cubicle on the detainee side is not required to be accessible at noncontact visiting areas or in areas not serving accessible holding cells or sleeping units.

1109.11.1 Dispersion. Accessible fixed or built-in seating at tables, counters or work surfaces shall be distributed throughout the space or facility containing such elements.

1109.12 Service facilities. Service facilities shall provide for accessible features in accordance with Sections 1109.12.1 through 1109.12.5.

1109.12.1 Dressing, fitting and locker rooms. Where dressing rooms, fitting rooms or locker rooms are provided, at least 5 percent, but not less than one, of each type of use in each cluster provided shall be accessible.

1109.12.2 Check-out aisles. Where check-out aisles are provided, accessible check-out aisles shall be provided in accordance with Table 1109.12.2. Where check-out aisles serve different functions, at least one accessible check-out aisle shall be provided for each function. Where checkout aisles serve different functions, accessible check-out aisles shall be provided in accordance with Table 1109.12.2 for each function. Where check-out aisles are dispersed throughout the building or facility, accessible check-out aisles shall also be dispersed. Traffic control devices, security devices and turnstiles located in accessible check-out aisles or lanes shall be accessible.

Exception: Where the area of the selling space is less than 5,000 square feet (465 m^2), only one check-out aisle is required to be accessible.

TABLE 1109.12.2
ACCESSIBLE CHECK-OUT AISLES

TOTAL CHECK-OUT AISLES OF EACH FUNCTION	MINIMUM NUMBER OF ACCESSIBLE CHECK-OUT AISLES OF EACH FUNCTION
1 to 4	1
5 to 8	2
9 to 15	3
Over 15	3, plus 20% of additional aisles

1109.12.3 Point of sale and service counters. Where counters are provided for sales or distribution of goods or services, at least one of each type provided shall be accessible. Where such counters are dispersed throughout the building or facility, accessible counters shall also be dispersed.

1109.12.4 Food service lines. Food service lines shall be accessible. Where self-service shelves are provided, at least 50 percent, but not less than one, of each type provided shall be accessible.

1109.12.5 Queue and waiting lines. Queue and waiting lines servicing accessible counters or check-out aisles shall be accessible.

1109.13 Controls, operating mechanisms and hardware. Controls, operating mechanisms and hardware intended for operation by the occupant, including switches that control lighting and ventilation, and electrical convenience outlets, in accessible spaces, along accessible routes or as parts of accessible elements shall be accessible.

Exceptions:

1. Operable parts that are intended for use only by service or maintenance personnel shall not be required to be accessible.

2. Electrical or communication receptacles serving a dedicated use shall not be required to be accessible.

3. Where two or more outlets are provided in a kitchen above a length of counter top that is uninterrupted by a sink or appliance, one outlet shall not be required to be accessible.

4. Floor electrical receptacles shall not be required to be accessible.

5. HVAC diffusers shall not be required to be accessible.

6. Except for light switches, where redundant controls are provided for a single element, one control in each space shall not be required to be accessible.

1109.13.1 Operable windows. Where operable windows are provided in rooms that are required to be accessible in accordance with Sections 1107.5.1.1, 1107.5.2.1, 1107.5.3.1, 1107.5.4, 1107.6.1.1, 1107.6.2.2.1 and 1107.6.4.1, at least one window in each room shall be accessible and each required operable window shall be accessible.

Exception: Accessible windows are not required in bathrooms or kitchens.

1109.14 Recreational facilities. Recreational facilities shall be provided with accessible features in accordance with Sections 1109.14.1 through 1109.14.3.

1109.14.1 Facilities serving a single building. In Group R-2 and R-3 occupancies where recreational facilities are provided serving a single building containing Type A or Type B units, 25 percent, but not less than one, of each type of recreational facility shall be accessible. Every recreational facility of each type on a site shall be considered to determine the total number of each type that is required to be accessible.

1109.14.2 Facilities serving multiple buildings. In Group R-2 and R-3 occupancies on a single site where multiple buildings containing Type A or Type B units are served by recreational facilities, 25 percent, but not less than one, of each type of recreational facility serving each building shall be accessible. The total number of each type of recreational facility that is required to be accessible shall be determined by considering every recreational facility of each type serving each building on the site.

1109.14.3 Other occupancies. All recreational facilities not falling within the purview of Section 1109.14.1 or 1109.14.2 shall be accessible.

1109.15 Stairways. Stairways located along accessible routes connecting floor levels that are not connected by an elevator shall be designed and constructed to comply with ICC A117.1 and Chapter 10.

SECTION 1110
SIGNAGE

1110.1 Signs. Required accessible elements shall be identified by the International Symbol of Accessibility at the following locations:

1. Accessible parking spaces required by Section 1106.1 except where the total number of parking spaces provided is four or less.

2. Accessible passenger loading zones.

3. Accessible areas of refuge required by Section 1007.6.

4. Accessible rooms where multiple single-user toilet or bathing rooms are clustered at a single location.

5. Accessible entrances where not all entrances are accessible.

6. Accessible check-out aisles where not all aisles are accessible. The sign, where provided, shall be above the check-out aisle in the same location as the check-out aisle number or type of check-out identification.

7. Unisex toilet and bathing rooms.

8. Accessible dressing, fitting and locker rooms where not all such rooms are accessible.

1110.2 Directional signage. Directional signage indicating the route to the nearest like accessible element shall be provided at the following locations. These directional signs shall include the International Symbol of Accessibility:

1. Inaccessible building entrances.

2. Inaccessible public toilets and bathing facilities.

3. Elevators not serving an accessible route.

4. At each separate-sex toilet and bathing room indicating the location of the nearest unisex toilet or bathing room where provided in accordance with Section 1109.2.1.

5. At exits and elevators serving a required accessible space, but not providing an approved accessible means of egress, signage shall be provided in accordance with Section 1007.7.

1110.3 Other signs. Signage indicating special accessibility provisions shall be provided as shown:

1. Each assembly area required to comply with Section 1108.2.7 shall provide a sign notifying patrons of the availability of assistive listening systems.

   Exception: Where ticket offices or windows are provided, signs are not required at each assembly area provided that signs are displayed at each ticket office or window informing patrons of the availability of assistive listening systems.

2. At each door to an egress stairway, exit passageway and exit discharge, signage shall be provided in accordance with Section 1011.3.

3. At areas of refuge, signage shall be provided in accordance with Sections 1007.6.3 through 1007.6.5.

4. At areas for assisted rescue, signage shall be provided in accordance with Section 1007.8.3.

CHAPTER 12

INTERIOR ENVIRONMENT

SECTION 1201
GENERAL

1201.1 Scope. The provisions of this chapter shall govern ventilation, temperature control, lighting, yards and courts, sound transmission, room dimensions, surrounding materials and rodent proofing associated with the interior spaces of buildings.

SECTION 1202
DEFINITIONS

1202.1 General. The following words and terms shall, for the purposes of this chapter and as used elsewhere in this code, have the meanings shown herein.

SUNROOM ADDITION. A one-story addition added to an existing building with a glazing area in excess of 40 percent of the gross area of the structure's exterior walls and roof.

THERMAL ISOLATION. A separation of conditioned spaces, between a sunroom addition and a dwelling unit, consisting of existing or new wall(s), doors and/or windows.

SECTION 1203
VENTILATION

1203.1 General. Buildings shall be provided with natural ventilation in accordance with Section 1203.4, or mechanical ventilation in accordance with the *International Mechanical Code*.

1203.2 Attic spaces. Enclosed attics and enclosed rafter spaces formed where ceilings are applied directly to the underside of roof framing members shall have cross ventilation for each separate space by ventilating openings protected against the entrance of rain and snow. Blocking and bridging shall be arranged so as not to interfere with the movement of air. A minimum of 1 inch (25 mm) of airspace shall be provided between the insulation and the roof sheathing. The net free ventilating area shall not be less than 1/150 of the area of the space ventilated, with 50 percent of the required ventilating area provided by ventilators located in the upper portion of the space to be ventilated at least 3 feet (914 mm) above eave or cornice vents with the balance of the required ventilation provided by eave or cornice vents.

Exception: The minimum required net free ventilating area shall be $^1/_{300}$ of the area of the space ventilated, provided a vapor retarder having a transmission rate not exceeding 1 perm in accordance with ASTM E 96 is installed on the warm side of the attic insulation and provided 50 percent of the required ventilating area provided by ventilators located in the upper portion of the space to be ventilated at least 3 feet (914 mm) above eave or cornice vents, with the balance of the required ventilation provided by eave or cornice vents.

1203.2.1 Openings into attic. Exterior openings into the attic space of any building intended for human occupancy shall be covered with corrosion-resistant wire cloth screening, hardware cloth, perforated vinyl or similar material that will prevent the entry of birds, squirrels, rodents, snakes and other similar creatures. The openings therein shall be a minimum of $^1/_8$ inch (3.2 mm) and shall not exceed $^1/_4$ inch (6.4 mm). Where combustion air is obtained from an attic area, it shall be in accordance with Chapter 7 of the *International Mechanical Code*.

1203.3 Under-floor ventilation. The space between the bottom of the floor joists and the earth under any building except spaces occupied by a basement or cellar shall be provided with ventilation openings through foundation walls or exterior walls. Such openings shall be placed so as to provide cross ventilation of the under-floor space.

1203.3.1 Openings for under-floor ventilation. The minimum net area of ventilation openings shall not be less than 1 square foot for each 150 square feet (0.67 m² for each 100 m²) of crawl-space area. Ventilation openings shall be covered for their height and width with any of the following materials, provided that the least dimension of the covering shall not exceed $^1/_4$ inch (6 mm):

1. Perforated sheet metal plates not less than 0.070 inch (1.8 mm) thick.

2. Expanded sheet metal plates not less than 0.047 inch (1.2 mm) thick.

3. Cast-iron grills or gratings.

4. Extruded load-bearing vents.

5. Hardware cloth of 0.035 inch (0.89 mm) wire or heavier.

6. Corrosion-resistant wire mesh, with the least dimension not exceeding $^1/_8$ inch (3.2 mm).

1203.3.2 Exceptions. The following are exceptions to Sections 1203.3 and 1203.3.1:

1. Where warranted by climatic conditions, ventilation openings to the outdoors are not required if ventilation openings to the interior are provided.

2. The total area of ventilation openings is permitted to be reduced to $^1/_{1,500}$ of the under-floor area where the ground surface is treated with an approved vapor retarder material and the required openings are placed so as to provide cross ventilation of the space. The installation of operable louvers shall not be prohibited.

3. Ventilation openings are not required where continuously operated mechanical ventilation is provided at a rate of 1.0 cubic foot per minute (cfm) for each 50 square feet (1.02 L/s for each 10 m²) of crawl-space floor area and the ground surface is covered with an approved vapor retarder.

4. Ventilation openings are not required when the ground surface is covered with an approved vapor retarder, the perimeter walls are insulated and the space

is conditioned in accordance with the *International Energy Conservation Code.*

5. For buildings in flood hazard areas as established in Section 1612.3, the openings for under-floor ventilation shall be deemed as meeting the flood opening requirements of ASCE 24 provided that the ventilation openings are designed and installed in accordance with ASCE 24.

1203.4 Natural ventilation. Natural ventilation of an occupied space shall be through windows, doors, louvers or other openings to the outdoors. The operating mechanism for such openings shall be provided with ready access so that the openings are readily controllable by the building occupants.

1203.4.1 Ventilation area required. The minimum openable area to the outdoors shall be 4 percent of the floor area being ventilated.

1203.4.1.1 Adjoining spaces. Where rooms and spaces without openings to the outdoors are ventilated through an adjoining room, the opening to the adjoining room shall be unobstructed and shall have an area of not less than 8 percent of the floor area of the interior room or space, but not less than 25 square feet (2.3 m²). The minimum openable area to the outdoors shall be based on the total floor area being ventilated.

Exception: Exterior openings required for ventilation shall be permitted to open into a thermally isolated sunroom addition or patio cover provided that the openable area between the sunroom addition or patio cover and the interior room shall have an area of not less than 8 percent of the floor area of the interior room or space, but not less than 20 square feet (1.86 m²). The minimum openable area to the outdoors shall be based on the total floor area being ventilated.

1203.4.1.2 Openings below grade. Where openings below grade provide required natural ventilation, the outside horizontal clear space measured perpendicular to the opening shall be one and one-half times the depth of the opening. The depth of the opening shall be measured from the average adjoining ground level to the bottom of the opening.

1203.4.2 Contaminants exhausted. Contaminant sources in naturally ventilated spaces shall be removed in accordance with the *International Mechanical Code* and the *International Fire Code.*

1203.4.2.1 Bathrooms. Rooms containing bathtubs, showers, spas and similar bathing fixtures shall be mechanically ventilated in accordance with the *International Mechanical Code.*

1203.4.3 Openings on yards or courts. Where natural ventilation is to be provided by openings onto yards or courts, such yards or courts shall comply with Section 1206.

1203.5 Other ventilation and exhaust systems. Ventilation and exhaust systems for occupancies and operations involving flammable or combustible hazards or other contaminant sources as covered in the *International Mechanical Code* or the *International Fire Code* shall be provided as required by both codes.

SECTION 1204
TEMPERATURE CONTROL

1204.1 Equipment and systems. Interior spaces intended for human occupancy shall be provided with active or passive space-heating systems capable of maintaining a minimum indoor temperature of 68°F (20°C) at a point 3 feet (914 mm) above the floor on the design heating day.

Exception: Interior spaces where the primary purpose is not associated with human comfort.

SECTION 1205
LIGHTING

1205.1 General. Every space intended for human occupancy shall be provided with natural light by means of exterior glazed openings in accordance with Section 1205.2 or shall be provided with artificial light in accordance with Section 1205.3. Exterior glazed openings shall open directly onto a public way or onto a yard or court in accordance with Section 1206.

1205.2 Natural light. The minimum net glazed area shall not be less than 8 percent of the floor area of the room served.

1205.2.1 Adjoining spaces. For the purpose of natural lighting, any room is permitted to be considered as a portion of an adjoining room where one-half of the area of the common wall is open and unobstructed and provides an opening of not less than one-tenth of the floor area of the interior room or 25 square feet (2.32 m²), whichever is greater.

Exception: Openings required for natural light shall be permitted to open into a thermally isolated sunroom addition or patio cover where the common wall provides a glazed area of not less than one-tenth of the floor area of the interior room or 20 square feet (1.86 m²), whichever is greater.

1205.2.2 Exterior openings. Exterior openings required by Section 1205.2 for natural light shall open directly onto a public way, yard or court, as set forth in Section 1206.

Exceptions:

1. Required exterior openings are permitted to open into a roofed porch where the porch:

    1.1. Abuts a public way, yard or court.

    1.2. Has a ceiling height of not less than 7 feet (2134 mm).

    1.3. Has a longer side at least 65 percent open and unobstructed.

2. Skylights are not required to open directly onto a public way, yard or court.

1205.3 Artificial light. Artificial light shall be provided that is adequate to provide an average illumination of 10 foot-candles (107 lux) over the area of the room at a height of 30 inches (762 mm) above the floor level.

1205.4 Stairway illumination. Stairways within dwelling units and exterior stairways serving a dwelling unit shall have an illumination level on tread runs of not less than 1 foot-candle (11 lux). Stairs in other occupancies shall be governed by Chapter 10.

1205.4.1 Controls. The control for activation of the required stairway lighting shall be in accordance with the ICC *Electrical Code.*

1205.5 Emergency egress lighting. The means of egress shall be illuminated in accordance with Section 1006.1.

SECTION 1206
YARDS OR COURTS

1206.1 General. This section shall apply to yards and courts adjacent to exterior openings that provide natural light or ventilation. Such yards and courts shall be on the same property as the building.

1206.2 Yards. Yards shall not be less than 3 feet (914 mm) in width for one- and two-story buildings. For buildings more than two stories in height, the minimum width of the yard shall be increased at the rate of 1 foot (305 mm) for each additional story. For buildings exceeding 14 stories in height, the required width of the yard shall be computed on the basis of 14 stories.

1206.3 Courts. Courts shall not be less than 3 feet (914 mm) in width. Courts having windows opening on opposite sides shall not be less than 6 feet (1829 mm) in width. Courts shall not be less than 10 feet (3048 mm) in length unless bounded on one end by a public way or yard. For buildings more than two stories in height, the court shall be increased 1 foot (305 mm) in width and 2 feet (310 mm) in length for each additional story. For buildings exceeding 14 stories in height, the required dimensions shall be computed on the basis of 14 stories.

1206.3.1 Court access. Access shall be provided to the bottom of courts for cleaning purposes.

1206.3.2 Air intake. Courts more than two stories in height shall be provided with a horizontal air intake at the bottom not less than 10 square feet (0.93 m²) in area and leading to the exterior of the building unless abutting a yard or public way.

1206.3.3 Court drainage. The bottom of every court shall be properly graded and drained to a public sewer or other approved disposal system complying with the *International Plumbing Code.*

SECTION 1207
SOUND TRANSMISSION

1207.1 Scope. This section shall apply to common interior walls, partitions and floor/ceiling assemblies between adjacent dwelling units or between dwelling units and adjacent public areas such as halls, corridors, stairs or service areas.

1207.2 Air-borne sound. Walls, partitions and floor/ceiling assemblies separating dwelling units from each other or from public or service areas shall have a sound transmission class (STC) of not less than 50 (45 if field tested) for air-borne noise when tested in accordance with ASTM E 90. Penetrations or openings in construction assemblies for piping; electrical devices; recessed cabinets; bathtubs; soffits; or heating, ventilating or exhaust ducts shall be sealed, lined, insulated or otherwise treated to maintain the required ratings. This require-

ment shall not apply to dwelling unit entrance doors; however, such doors shall be tight fitting to the frame and sill.

1207.3 Structure-borne sound. Floor/ceiling assemblies between dwelling units or between a dwelling unit and a public or service area within the structure shall have an impact insulation class (IIC) rating of not less than 50 (45 if field tested) when tested in accordance with ASTM E 492.

SECTION 1208
INTERIOR SPACE DIMENSIONS

1208.1 Minimum room widths. Habitable spaces, other than a kitchen, shall not be less than 7 feet (2134 mm) in any plan dimension. Kitchens shall have a clear passageway of not less than 3 feet (914 mm) between counter fronts and appliances or counter fronts and walls.

1208.2 Minimum ceiling heights. Occupiable spaces, habitable spaces and corridors shall have a ceiling height of not less than 7 feet 6 inches (2286 mm). Bathrooms, toilet rooms, kitchens, storage rooms and laundry rooms shall be permitted to have a ceiling height of not less than 7 feet (2134 mm).

Exceptions:

1. In one- and two-family dwellings, beams or girders spaced not less than 4 feet (1219 mm) on center and projecting not more than 6 inches (152 mm) below the required ceiling height.

2. If any room in a building has a sloped ceiling, the prescribed ceiling height for the room is required in one-half the area thereof. Any portion of the room measuring less than 5 feet (1524 mm) from the finished floor to the ceiling shall not be included in any computation of the minimum area thereof.

3. Mezzanines constructed in accordance with Section 505.1.

1208.2.1 Furred ceiling. Any room with a furred ceiling shall be required to have the minimum ceiling height in two-thirds of the area thereof, but in no case shall the height of the furred ceiling be less than 7 feet (2134 mm).

1208.3 Room area. Every dwelling unit shall have at least one room that shall have not less than 120 square feet (13.9 m²) of net floor area. Other habitable rooms shall have a net floor area of not less than 70 square feet (6.5 m²).

Exception: Every kitchen in a one- and two-family dwelling shall have not less than 50 square feet (4.64 m²) of gross floor area.

1208.4 Efficiency dwelling units. An efficiency living unit shall conform to the requirements of the code except as modified herein:

1. The unit shall have a living room of not less than 220 square feet (20.4 m) of floor area. An additional 100 square feet (9.3 m) of floor area shall be provided for each occupant of such unit in excess of two.

2. The unit shall be provided with a separate closet.

3. The unit shall be provided with a kitchen sink, cooking appliance and refrigeration facilities, each having a clear working space of not less than 30 inches (762

mm) in front. Light and ventilation conforming to this code shall be provided.

4. The unit shall be provided with a separate bathroom containing a water closet, lavatory and bathtub or shower.

SECTION 1209
ACCESS TO UNOCCUPIED SPACES

1209.1 Crawl spaces. Crawl spaces shall be provided with a minimum of one access opening not less than 18 inches by 24 inches (457 mm by 610 mm).

1209.2 Attic spaces. An opening not less than 20 inches by 30 inches (559 mm by 762 mm) shall be provided to any attic area having a clear height of over 30 inches (762 mm). A 30-inch (762 mm) minimum clear headroom in the attic space shall be provided at or above the access opening.

1209.3 Mechanical appliances. Access to mechanical appliances installed in under-floor areas, in attic spaces and on roofs or elevated structures shall be in accordance with the *International Mechanical Code*.

SECTION 1210
SURROUNDING MATERIALS

1210.1 Floors. In other than dwelling units, toilet and bathing room floors shall have a smooth, hard, nonabsorbent surface that extends upward onto the walls at least 6 inches (152 mm).

1210.2 Walls. Walls within 2 feet (610 mm) of urinals and water closets shall have a smooth, hard, nonabsorbent surface, to a height of 4 feet (1219 mm) above the floor, and except for structural elements, the materials used in such walls shall be of a type that is not adversely affected by moisture.

Exceptions:

1. Dwelling units and sleeping units.
2. Toilet rooms that are not accessible to the public and which have not more than one water closet.

Accessories such as grab bars, towel bars, paper dispensers and soap dishes, provided on or within walls, shall be installed and sealed to protect structural elements from moisture.

1210.3 Showers. Shower compartments and walls above bathtubs with installed shower heads shall be finished with a smooth, nonabsorbent surface to a height not less than 70 inches (1778 mm) above the drain inlet.

1210.4 Waterproof joints. Built-in tubs with showers shall have waterproof joints between the tub and adjacent wall.

1210.5 Toilet rooms. Toilet rooms shall not open directly into a room used for the preparation of food for service to the public.

CHAPTER 13

ENERGY EFFICIENCY

SECTION 1301
GENERAL

1301.1 Scope. This chapter governs the design and construction of buildings for energy efficiency.

1301.1.1 Criteria. Buildings shall be designed and constructed in accordance with the *International Energy Conservation Code*.

CHAPTER 14
EXTERIOR WALLS

SECTION 1401
GENERAL

1401.1 Scope. The provisions of this chapter shall establish the minimum requirements for exterior walls, exterior wall coverings, exterior wall openings, exterior windows and doors, architectural trim, balconies and bay windows.

SECTION 1402
DEFINITIONS

1402.1 General. The following words and terms shall, for the purposes of this chapter and as used elsewhere in this code, have the meanings shown herein.

ADHERED MASONRY VENEER. Veneer secured and supported through the adhesion of an approved bonding material applied to an approved backing.

ANCHORED MASONRY VENEER. Veneer secured with approved mechanical fasteners to an approved backing.

BACKING. The wall or surface to which the veneer is secured.

EXTERIOR WALL. A wall, bearing or nonbearing, that is used as an enclosing wall for a building, other than a fire wall, and that has a slope of 60 degrees (1.05 rad) or greater with the horizontal plane.

EXTERIOR WALL COVERING. A material or assembly of materials applied on the exterior side of exterior walls for the purpose of providing a weather-resisting barrier, insulation or for aesthetics, including but not limited to, veneers, siding, exterior insulation and finish systems, architectural trim and embellishments such as cornices, soffits, facias, gutters and leaders.

EXTERIOR WALL ENVELOPE. A system or assembly of exterior wall components, including exterior wall finish materials, that provides protection of the building structural members, including framing and sheathing materials, and conditioned interior space, from the detrimental effects of the exterior environment.

FIBER CEMENT SIDING. A manufactured, fiber-reinforcing product made with an inorganic hydraulic or calcium silicate binder formed by chemical reaction and reinforced with organic or inorganic nonasbestos fibers, or both. Additives that enhance manufacturing or product performance are permitted. Fiber cement siding products have either smooth or textured faces and are intended for exterior wall and related applications.

METAL COMPOSITE MATERIAL (MCM). A factory-manufactured panel consisting of metal skins bonded to both faces of a plastic core.

METAL COMPOSITE MATERIAL (MCM) SYSTEM. An exterior wall finish system fabricated using MCM in a specific assembly including joints, seams, attachments, substrate, framing and other details as appropriate to a particular design.

VENEER. A facing attached to a wall for the purpose of providing ornamentation, protection or insulation, but not counted as adding strength to the wall.

SECTION 1403
PERFORMANCE REQUIREMENTS

1403.1 General. The provisions of this section shall apply to exterior walls, wall coverings and components thereof.

1403.2 Weather protection. Exterior walls shall provide the building with a weather-resistant exterior wall envelope. The exterior wall envelope shall include flashing, as described in Section 1405.3. The exterior wall envelope shall be designed and constructed in such a manner as to prevent the accumulation of water within the wall assembly by providing a water-resistive barrier behind the exterior veneer, as described in Section 1404.2 and a means for draining water that enters the assembly to the exterior of the veneer, unless it is determined that penetration of water behind the veneer shall not be detrimental to the building performance. Protection against condensation in the exterior wall assembly shall be provided in accordance with the *International Energy Conservation Code*.

Exceptions:

1. A weather-resistant exterior wall envelope shall not be required over concrete or masonry walls designed in accordance with Chapters 19 and 21, respectively.

2. Compliance with the requirements for a means of drainage, and the requirements of Sections 1405.2 and 1405.3, shall not be required for an exterior wall envelope that has been demonstrated through testing to resist wind-driven rain, including joints, penetrations and intersections with dissimilar materials, in accordance with ASTM E 331 under the following conditions:

   2.1. Exterior wall envelope test assemblies shall include at least one opening, one control joint, one wall/eave interface and one wall sill. All tested openings and penetrations shall be representative of the intended end-use configuration.

   2.2. Exterior wall envelope test assemblies shall be at least 4 feet by 8 feet (1219 mm by 2438 mm) in size.

2.3. Exterior wall envelope assemblies shall be tested at a minimum differential pressure of 6.24 pounds per square foot (psf) (0.297 kN/m^2).

2.4. Exterior wall envelope assemblies shall be subjected to a minimum test exposure duration of 2 hours.

The exterior wall envelope design shall be considered to resist wind-driven rain where the results of testing indicate that water did not penetrate control joints in the exterior wall envelope, joints at the perimeter of openings or intersections of terminations with dissimilar materials.

1403.3 Vapor retarder. An approved vapor retarder shall be provided.

Exceptions:

1. Where other approved means to avoid condensation and leakage of moisture are provided.

2. Plain and reinforced concrete or masonry exterior walls designed and constructed in accordance with Chapter 19 or 21, respectively.

1403.4 Structural. Exterior walls, and the associated openings, shall be designed and constructed to resist safely the superimposed loads required by Chapter 16.

1403.5 Fire resistance. Exterior walls shall be fire-resistance rated as required by other sections of this code with opening protection as required by Chapter 7.

1403.6 Flood resistance. For buildings in flood hazard areas as established in Section 1612.3, exterior walls extending below the design flood elevation shall be resistant to water damage. Wood shall be pressure-preservative treated in accordance with AWPA C1, C2, C3, C4, C9, C15, C18, C22, C23, C24, C28, P1, P2 and P3, or decay-resistant heartwood of redwood, black locust or cedar.

1403.7 Flood resistance for high-velocity wave action areas. For buildings in flood hazard areas subject to high-velocity wave action as established in Section 1612.3, electrical, mechanical and plumbing system components shall not be mounted on or penetrate through exterior walls that are designed to break away under flood loads.

SECTION 1404
MATERIALS

1404.1 General. Materials used for the construction of exterior walls shall comply with the provisions of this section. Materials not prescribed herein shall be permitted, provided that any such alternative has been approved.

1404.2 Water-resistive barrier. A minimum of one layer of No. 15 asphalt felt, complying with ASTM D 226 for Type 1 felt, shall be attached to the sheathing, with flashing as described in Section 1405.3, in such a manner as to provide a continuous water-resistive barrier behind the exterior wall veneer.

1404.3 Wood. Exterior walls of wood construction shall be designed and constructed in accordance with Chapter 23.

1404.3.1 Basic hardboard. Basic hardboard shall conform to the requirements of AHA A135.4.

1404.3.2 Hardboard siding. Hardboard siding shall conform to the requirements of AHA A135.6 and, where used structurally, shall be so identified by the label of an approved agency.

1404.4 Masonry. Exterior walls of masonry construction shall be designed and constructed in accordance with this section and Chapter 21. Masonry units, mortar and metal accessories used in anchored and adhered veneer shall meet the physical requirements of Chapter 21. The backing of anchored and adhered veneer shall be of concrete, masonry, steel framing or wood framing.

1404.5 Metal. Exterior walls of formed steel construction, structural steel or lightweight metal alloys shall be designed in accordance with Chapters 22 and 20, respectively.

1404.5.1 Aluminum siding. Aluminum siding shall conform to the requirements of AAMA 1402.

1404.6 Concrete. Exterior walls of concrete construction shall be designed and constructed in accordance with Chapter 19.

1404.7 Glass-unit masonry. Exterior walls of glass-unit masonry shall be designed and constructed in accordance with Chapter 21.

1404.8 Plastics. Plastic panel, apron or spandrel walls as defined in this code shall not be limited in thickness, provided that such plastics and their assemblies conform to the requirements of Chapter 26 and are constructed of approved weather-resistant materials of adequate strength to resist the wind loads for cladding specified in Chapter 16.

1404.9 Vinyl siding. Vinyl siding shall conform to the requirements of ASTM D 3679.

1404.10 Fiber cement siding. Fiber cement siding shall conform to the requirements of ASTM C 1186 and shall be so identified on labeling listing an approved quality control agency.

SECTION 1405
INSTALLATION OF WALL COVERINGS

1405.1 General. Exterior wall coverings shall be designed and constructed in accordance with the applicable provisions of this section.

1405.2 Weather protection. Exterior walls shall provide weather protection for the building. The materials of the minimum nominal thickness specified in Table 1405.2 shall be acceptable as approved weather coverings.

TABLE 1405.2
MINIMUM THICKNESS OF WEATHER COVERINGS

COVERING TYPE	MINIMUM THICKNESS (inches)
Adhered masonry veneer	0.25
Anchored masonry veneer	2.625
Aluminum siding	0.019
Asbestos-cement boards	0.125
Asbestos shingles	0.156
Cold-rolled copper[d]	0.0216 nominal
Copper shingles[d]	0.0162 nominal
Exterior plywood (with sheathing)	0.313
Exterior plywood (without sheathing)	See Section 2304.6
Fiberboard siding	0.5
Fiber cement lap siding	0.25[c]
Fiber cement panel siding	0.25[c]
Glass-fiber reinforced concrete panels	0.375
Hardboard siding[c]	0.25
High-yield copper[d]	0.0162 nominal
Lead-coated copper[d]	0.0216 nominal
Lead-coated high-yield copper	0.0162 nominal
Marble slabs	1
Particleboard (with sheathing)	See Section 2304.6
Particleboard (without sheathing)	See Section 2304.6
Precast stone facing	0.625
Steel (approved corrosion resistant)	0.0149
Stone (cast artificial)	1.5
Stone (natural)	2
Structural glass	0.344
Stucco or exterior portland cement plaster	
Three-coat work over:	
Metal plaster base	0.875[b]
Unit masonry	0.625[b]
Cast-in-place or precast concrete	0.625[b]
Two-coat work over:	
Unit masonry	0.5[b]
Cast-in-place or precast concrete	0.375[b]
Terra cotta (anchored)	1
Terra cotta (adhered)	0.25
Vinyl siding	0.035
Wood shingles	0.375
Wood siding (without sheathing)[a]	0.5

For SI: 1 inch = 25.4 mm.

a. Wood siding of thicknesses less than 0.5 inch shall be placed over sheathing that conforms to Section 2304.6.
b. Exclusive of texture.
c. As measured at the bottom of decorative grooves.
d. 16 ounces per square foot for cold-rolled copper and lead-coated copper, 12 ounces per square foot for copper shingles, high-yield copper and lead-coated high-yield copper.

1405.3 Flashing. Flashing shall be installed in such a manner so as to prevent moisture from entering the wall or to redirect it to the exterior. Flashing shall be installed at the perimeters of exterior door and window assemblies, penetrations and terminations of exterior wall assemblies, exterior wall intersections with roofs, chimneys, porches, decks, balconies and similar projections and at built-in gutters and similar locations where moisture could enter the wall. Flashing with projecting flanges shall be installed on both sides and the ends of copings, under sills and continuously above projecting trim.

1405.3.1 Exterior wall pockets. In exterior walls of buildings or structures, wall pockets or crevices in which moisture can accumulate shall be avoided or protected with caps or drips, or other approved means shall be provided to prevent water damage.

1405.3.2 Masonry. Flashing and weepholes shall be located in the first course of masonry above finished ground level above the foundation wall or slab, and other points of support, including structural floors, shelf angles and lintels where anchored veneers are designed in accordance with Section 1405.5.

1405.4 Wood veneers. Wood veneers on exterior walls of buildings of Type I, II, III and IV construction shall be not less than 1-inch (25 mm) nominal thickness, 0.438-inch (11.1 mm) exterior hardboard siding or 0.375-inch (9.5 mm) exterior-type wood structural panels or particleboard and shall conform to the following:

1. The veneer does not exceed three stories in height, measured from grade, except where fire-retardant-treated wood is used, the height shall not exceed four stories.

2. The veneer is attached to or furred from a noncombustible backing that is fire-resistance rated as required by other provisions of this code.

3. Where open or spaced wood veneers (without concealed spaces) are used, they shall not project more than 24 inches (610 mm) from the building wall.

1405.5 Anchored masonry veneer. Anchored masonry veneer shall comply with the provisions of Sections 1405.5, 1405.6, 1405.7 and 1405.8 and Sections 6.1 and 6.2 of ACI 530/ASCE 5/TMS 402.

1405.5.1 Tolerances. Anchored masonry veneers in accordance with Chapter 14 are not required to meet the tolerances in Article 3.3 G1 of ACI 530.1/ASCE 6/TMS 602.

1405.5.2 Seismic requirements. Anchored masonry veneer located in Seismic Design Category C, D, E or F shall conform to the requirements of Section 6.2.2.10 of ACI 530/ASCE 5/TMS 402.

1405.6 Stone veneer. Stone veneer units not exceeding 10 inches (254 mm) in thickness shall be anchored directly to masonry, concrete or to stud construction by one of the following methods:

1. With concrete or masonry backing, anchor ties shall be not less than 0.1055-inch (2.68 mm) corrosion-resistant wire, or approved equal, formed beyond the base of the backing. The legs of the loops shall be not less than 6 inches (152 mm) in length bent at right angles and laid in the mortar joint, and spaced so that the eyes or loops are

12 inches (305 mm) maximum on center (o.c.) in both directions. There shall be provided not less than a 0.1055-inch (2.68 mm) corrosion-resistant wire tie, or approved equal, threaded through the exposed loops for every 2 square feet (0.2 m²) of stone veneer. This tie shall be a loop having legs not less than 15 inches (381 mm) in length bent so that it will lie in the stone veneer mortar joint. The last 2 inches (51 mm) of each wire leg shall have a right-angle bend. One-inch (25 mm) minimum thickness of cement grout shall be placed between the backing and the stone veneer.

2. With stud backing, a 2-inch by 2-inch (51 by 51 mm) 0.0625-inch (1.59 mm) corrosion-resistant wire mesh with two layers of waterproofed paper backing in accordance with Section 1403.3 shall be applied directly to wood studs spaced a maximum of 16 inches (406 mm) o.c. On studs, the mesh shall be attached with 2-inch-long (51 mm) corrosion-resistant steel wire furring nails at 4 inches (102 mm) o.c. providing a minimum 1.125-inch (29 mm) penetration into each stud and with 8d common nails at 8 inches (203 mm) o.c. into top and bottom plates or with equivalent wire ties. There shall be not less than a 0.1055-inch (2.68 mm) corrosion-resistant wire, or approved equal, looped through the mesh for every 2 square feet (0.2 m²) of stone veneer. This tie shall be a loop having legs not less than 15 inches (381 mm) in length, so bent that it will lie in the stone veneer mortar joint. The last 2 inches (51 mm) of each wire leg shall have a right-angle bend. One-inch (25 mm) minimum thickness of cement grout shall be placed between the backing and the stone veneer.

1405.7 Slab-type veneer. Slab-type veneer units not exceeding 2 inches (51 mm) in thickness shall be anchored directly to masonry, concrete or stud construction. For veneer units of marble, travertine, granite or other stone units of slab form ties of corrosion-resistant dowels in drilled holes located in the middle third of the edge of the units spaced a maximum of 24 inches (610 m) apart around the periphery of each unit with not less than four ties per veneer unit. Units shall not exceed 20 square feet (1.9 m²) in area. If the dowels are not tight fitting, the holes shall be drilled not more than 0.063 inch (1.6 mm) larger in diameter than the dowel, with the hole countersunk to a diameter and depth equal to twice the diameter of the dowel in order to provide a tight-fitting key of cement mortar at the dowel locations when the mortar in the joint has set. Veneer ties shall be corrosion-resistant metal capable of resisting, in tension or compression, a force equal to two times the weight of the attached veneer. If made of sheet metal, veneer ties shall be not smaller in area than 0.0336 by 1 inch (0.853 by 25 mm) or, if made of wire, not smaller in diameter than 0.1483-inch (3.76 mm) wire.

1405.8 Terra cotta. Anchored terra cotta or ceramic units not less than 1.625 inches (41 mm) thick shall be anchored directly to masonry, concrete or stud construction. Tied terra cotta or ceramic veneer units shall be not less than 1.625 inches (41 mm) thick with projecting dovetail webs on the back surface spaced approximately 8 inches (203 mm) o.c. The facing shall be tied to the backing wall with corrosion-resistant metal anchors of not less than No. 8 gage wire installed at the top of each piece in horizontal bed joints not less than 12 inches (305 mm) nor more than 18 inches (457 mm) o.c.; these anchors shall be secured to 0.25-inch (6.4 mm) corrosion-resistant pencil rods that pass through the vertical aligned loop anchors in the backing wall. The veneer ties shall have sufficient strength to support the full weight of the veneer in tension. The facing shall be set with not less than a 2-inch (51 mm) space from the backing wall and the space shall be filled solidly with portland cement grout and pea gravel. Immediately prior to setting, the backing wall and the facing shall be drenched with clean water and shall be distinctly damp when the grout is poured.

1405.9 Adhered masonry veneer. Adhered masonry veneer shall comply with the applicable requirements in Section 1405.9.1 and Sections 6.1 and 6.3 of ACI 530/ASCE 5/TMS 402.

1405.9.1 Interior adhered masonry veneers. Interior adhered masonry veneers shall have a maximum weight of 20 psf (0.958 kg/m²) and shall be installed in accordance with Section 1405.9. Where the interior adhered masonry veneer is supported by wood construction, the supporting members shall be designed to limit deflection to 1/600 of the span of the supporting members.

1405.10 Metal veneers. Veneers of metal shall be fabricated from approved corrosion-resistant materials or shall be protected front and back with porcelain enamel, or otherwise be treated to render the metal resistant to corrosion. Such veneers shall not be less than 0.0149-inch (0.378 mm) nominal thickness sheet steel mounted on wood or metal furring strips or approved sheathing on the wood construction.

1405.10.1 Attachment. Exterior metal veneer shall be securely attached to the supporting masonry or framing members with corrosion-resistant fastenings, metal ties or by other approved devices or methods. The spacing of the fastenings or ties shall not exceed 24 inches (610 mm) either vertically or horizontally, but where units exceed 4 square feet (0.4 m²) in area there shall be not less than four attachments per unit. The metal attachments shall have a cross-sectional area not less than provided by W 1.7 wire. Such attachments and their supports shall be capable of resisting a horizontal force in accordance with the wind loads specified in Section 1609, but in no case less than 20 psf (0.958 kg/m²).

1405.10.2 Weather protection. Metal supports for exterior metal veneer shall be protected by painting, galvanizing or by other equivalent coating or treatment. Wood studs, furring strips or other wood supports for exterior metal veneer shall be approved pressure-treated wood or protected as required in Section 1403.2. Joints and edges exposed to the weather shall be caulked with approved durable waterproofing material or by other approved means to prevent penetration of moisture.

1405.10.3 Backup. Masonry backup shall not be required for metal veneer except as is necessary to meet the fire-resistance requirements of this code.

1405.10.4 Grounding. Grounding of metal veneers on buildings shall comply with the requirements of Chapter 27 and the ICC *Electrical Code.*

1405.11 Glass veneer. The area of a single section of thin exterior structural glass veneer shall not exceed 10 square feet (0.93 m²) where it is not more than 15 feet (4572 mm) above the level of the sidewalk or grade level directly below, and shall not exceed 6 square feet (0.56 m²) where it is more than 15 feet (4572 mm) above that level.

1405.11.1 Length and height. The length or height of any section of thin exterior structural glass veneer shall not exceed 48 inches (1219 mm).

1405.11.2 Thickness. The thickness of thin exterior structural glass veneer shall be not less than 0.344 inch (8.7 mm).

1405.11.3 Application. Thin exterior structural glass veneer shall be set only after backing is thoroughly dry and after application of an approved bond coat uniformly over the entire surface of the backing so as to effectively seal the surface. Glass shall be set in place with an approved mastic cement in sufficient quantity so that at least 50 percent of the area of each glass unit is directly bonded to the backing by mastic not less than 0.25 inch (6.4 mm) thick and not more than 0.625 inch (15.9 mm) thick. The bond coat and mastic shall be evaluated for compatibility and shall bond firmly together.

1405.11.4 Installation at sidewalk level. Where glass extends to a sidewalk surface, each section shall rest in an approved metal molding, and be set at least 0.25 inch (6.4 mm) above the highest point of the sidewalk. The space between the molding and the sidewalk shall be thoroughly caulked and made water tight.

1405.11.4.1 Installation above sidewalk level. Where thin exterior structural glass veneer is installed above the level of the top of a bulkhead facing, or at a level more than 36 inches (914 mm) above the sidewalk level, the mastic cement binding shall be supplemented with approved nonferrous metal shelf angles located in the horizontal joints in every course. Such shelf angles shall be not less than 0.0478-inch (12 mm) thick and not less than 2 inches (51 mm) long and shall be spaced at approved intervals, with not less than two angles for each glass unit. Shelf angles shall be secured to the wall or backing with expansion bolts, toggle bolts or by other approved methods.

1405.11.5 Joints. Unless otherwise specifically approved by the building official, abutting edges of thin exterior structural glass veneer shall be ground square. Mitered joints shall not be used except where specifically approved for wide angles. Joints shall be uniformly buttered with an approved jointing compound and horizontal joints shall be held to not less than 0.063 inch (1.6 mm) by an approved

nonrigid substance or device. Where thin exterior structural glass veneer abuts nonresilient material at sides or top, expansion joints not less than 0.25 inch (6.4 mm) wide shall be provided.

1405.11.6 Mechanical fastenings. Thin exterior structural glass veneer installed above the level of the heads of show windows and veneer installed more than 12 feet (3658 mm) above sidewalk level shall, in addition to the mastic cement and shelf angles, be held in place by the use of fastenings at each vertical or horizontal edge, or at the four corners of each glass unit. Fastenings shall be secured to the wall or backing with expansion bolts, toggle bolts or by other methods. Fastenings shall be so designed as to hold the glass veneer in a vertical plane independent of the mastic cement. Shelf angles providing both support and fastenings shall be permitted.

1405.11.7 Flashing. Exposed edges of thin exterior structural glass veneer shall be flashed with overlapping corrosion-resistant metal flashing and caulked with a waterproof compound in a manner to effectively prevent the entrance of moisture between the glass veneer and the backing.

1405.12 Exterior windows and doors. Windows and doors installed in exterior walls shall conform to the testing and performance requirements of Section 1714.5.

1405.12.1 Installation. Windows and doors shall be installed in accordance with approved manufacturer's instructions. Fastener size and spacing shall be provided in such instructions and shall be calculated based on maximum loads and spacing used in the tests.

1405.13 Vinyl siding. Vinyl siding conforming to the requirements of this section and complying with ASTM D 3679 shall be permitted on exterior walls of buildings of Type V construction located in areas where the basic wind speed specified in Chapter 16 does not exceed 100 miles per hour (161 km/h) and the building height is less than or equal to 40 feet (12 192 mm) in Exposure C. Where construction is located in areas where the basic wind speed exceeds 100 miles per hour (161 km/h), or building heights are in excess of 40 feet (12 192 mm), tests or calculations indicating compliance with Chapter 16 shall be submitted. Vinyl siding shall be secured to the building so as to provide weather protection for the exterior walls of the building.

1405.13.1 Application. The siding shall be applied over sheathing or materials listed in Section 2304.6. Siding shall be applied to conform with the weather-resistant barrier requirements in Section 1403. Siding and accessories shall be installed in accordance with approved manufacturer's instructions. Unless otherwise specified in the approved manufacturer's instructions, nails used to fasten the siding and accessories shall have a minimum 0.313-inch (7.9 mm) head diameter and 0.125-inch (3.18 mm) shank diameter. The nails shall be corrosion resistant and shall be long enough to penetrate the studs or nailing strip at least 0.75 inch (19 mm). Where the siding is installed horizontally, the fastener spacing shall not exceed 16 inches (406 mm) horizontally and 12 inches (305 mm) vertically. Where the siding is installed vertically, the fastener spacing shall not exceed 12 inches (305 mm) horizontally and 12 inches (305 mm) vertically.

1405.14 Cement plaster. Cement plaster applied to exterior walls shall conform to the requirements specified in Chapter 25.

1405.15 Fiber cement siding. Fiber cement siding complying with Section 1404.10 shall be permitted on exterior walls of Type I, II, III, IV and V construction for wind pressure resistance or wind speed exposures as indicated in the manufacturer's compliance report and approved installation instructions. Where specified, the siding shall be installed over sheathing or materials listed in Section 2304.6 and shall be installed to conform to the weather-resistant barrier requirements in Section 1403. Siding and accessories shall be installed in accordance with approved manufacturer's instructions. Unless otherwise specified in the approved manufacturer's instructions, nails used to fasten the siding to wood studs shall be corrosion-resistant round head smooth shank and shall be long enough to penetrate the studs at least 1 inch (25 mm). For metal framing, all-weather screws shall be used and shall penetrate the metal framing at least three full threads.

1405.16 Fastening. Weather boarding and wall coverings shall be securely fastened with aluminum, copper, zinc, zinc-coated or other approved corrosion-resistant fasteners in accordance with the nailing schedule in Table 2304.9.1 or the approved manufacturer's installation instructions. Shingles and other weather coverings shall be attached with appropriate standard-shingle nails to furring strips securely nailed to studs, or with approved mechanically bonding nails, except where sheathing is of wood not less than 1-inch (25 mm) nominal thickness or of wood structural panels as specified in Table 2308.9.3(3).

1405.17 Fiber cement siding.

1405.17.1 Panel siding. Panels shall be installed with the long dimension parallel to framing. Vertical joints shall occur over framing members and shall be sealed with caulking or covered with battens. Horizontal joints shall be flashed with Z-flashing and blocked with solid wood framing.

1405.17.2 Horizontal lap siding. Lap siding shall be lapped a minimum of $1^1/_4$ inches (32 mm) and shall have the ends sealed with caulking, covered with an H-section joint cover or located over a strip of flashing. Lap siding courses shall be permitted to be installed with the fastener heads exposed or concealed, according to approved manufacturers' instructions.

SECTION 1406
COMBUSTIBLE MATERIALS ON THE EXTERIOR SIDE OF EXTERIOR WALLS

1406.1 General. This section shall apply to exterior wall coverings, balconies and similar appendages, and bay and oriel windows constructed of combustible materials.

1406.2 Combustible exterior wall coverings. Combustible exterior wall coverings shall comply with this section.

Exception: Plastics complying with Chapter 26.

1406.2.1 Ignition resistance. Combustible exterior wall coverings shall be tested in accordance with NFPA 268.

Exceptions:

1. Wood or wood-based products.

2. Other combustible materials covered with an exterior covering other than vinyl sidings listed in Table 1405.2.

3. Aluminum having a minimum thickness of 0.019 inch (0.48 mm).

4. Exterior wall coverings on exterior walls of Type V construction.

1406.2.1.1 Fire separation 5 feet or less. Where installed on exterior walls having a fire separation distance of 5 feet (1524 mm) or less, combustible exterior wall coverings shall not exhibit sustained flaming as defined in NFPA 268.

1406.2.1.2 Fire separation greater than 5 feet. For fire separation distances greater than 5 feet (1524 mm), an assembly shall be permitted that has been exposed to a reduced level of incident radiant heat flux in accordance with the NFPA 268 test method without exhibiting sustained flaming. The minimum fire separation distance required for the assembly shall be determined from Table 1406.2.1.2 based on the maximum tolerable level of incident radiant heat flux that does not cause sustained flaming of the assembly.

1406.2.2 Architectural trim. In buildings of Type I, II, III and IV construction that do not exceed three stories or 40 feet (12 192 mm) in height above grade plane, exterior wall coverings shall be permitted to be constructed of wood where permitted by Section 1405.4 or other equivalent combustible material. Combustible exterior wall coverings, other than fire-retardant-treated wood complying with Section 2303.2 for exterior installation, shall not exceed 10 percent of an exterior wall surface area where the fire separation distance is 5 feet (1524 mm) or less. Architectural trim that exceeds 40 feet (12 192 mm) in height above grade plane shall be constructed of approved noncombustible materials and shall be secured to the wall with metal or other approved noncombustible brackets.

TABLE 1406.2.1.2
MINIMUM FIRE SEPARATION FOR COMBUSTIBLE VENEERS

FIRE SEPARATION DISTANCE (feet)	TOLERABLE LEVEL INCIDENT RADIANT HEAT ENERGY(kW/m²)	FIRE SEPARATION DISTANCE (feet)	TOLERABLE LEVEL INCIDENT RADIANT HEAT ENERGY(kW/m²)
5	12.5	16	5.9
6	11.8	17	5.5
7	11.0	18	5.2
8	10.3	19	4.9
9	9.6	20	4.6
10	8.9	21	4.4
11	8.3	22	4.1
12	7.7	23	3.9
13	7.2	24	3.7
14	6.7	25	3.5
15	6.3		

For SI: 1 foot = 304.8 mm, 1 Btu/H² ×°F = .0057 kW/m² × K.

1406.2.3 Location. Where combustible exterior wall covering is located along the top of exterior walls, such trim shall be completely backed up by the exterior wall and shall not extend over or above the top of exterior walls.

1406.2.4 Fireblocking. Where the combustible exterior wall covering is furred from the wall and forms a solid surface, the distance between the back of the covering and the wall shall not exceed 1.625 inches (41 mm) and the space thereby created shall be fireblocked in accordance with Section 717 so that there will be no open space exceeding 100 square feet (9.3 m²). Where wood furring strips are used, they shall be of approved wood of natural decay resistance or preservative-treated wood.

Exceptions:

1. Fireblocking of cornices is not required in single-family dwellings.

2. Fireblocking shall not be required where installed on noncombustible framing and the face of the exterior wall finish exposed to the concealed space is covered by one of the following materials:

   2.1. Aluminum having a minimum thickness of 0.019 inch (0.5 mm);

   2.2. Corrosion-resistant steel having a base metal thickness not less than 0.016 inch (0.4 mm) at any point; or

   2.3. Other approved noncombustible materials.

1406.3 Balconies and similar projections. Balconies and similar projections of combustible construction, other than fire-retardant-treated wood, shall afford the fire-resistance rating required by Table 601 for floor construction or shall be of Type IV construction as described in Section 602.4, and the aggregate length shall not exceed 50 percent of the building perimeter on each floor.

Exceptions:

1. On buildings of Type I and II construction, three stories or less in height, fire-retardant-treated wood shall be permitted for balconies, porches, decks and exterior stairways not used as required exits.

2. Untreated wood is permitted for pickets and rails, or similar guardrail devices that are limited to 42 inches (1067 mm) in height.

3. Balconies and similar appendages on buildings of Type III, IV and V construction shall be permitted to be of Type V construction, and shall not be required to have a fire-resistance rating where sprinkler protection is extended to these areas.

4. Where sprinkler protection is extended to the balcony areas, the aggregate length of the balcony on each floor shall not be limited.

1406.4 Bay windows and oriel windows. Bay and oriel windows shall conform to the type of construction required for the building to which they are attached.

Exception: Fire-retardant-treated wood shall be permitted on buildings three stories or less of Type I, II, III and IV construction.

SECTION 1407
METAL COMPOSITE MATERIALS (MCM)

1407.1 General. The provisions of this section shall govern the materials, construction and quality of metal composite materials (MCM) for use as exterior wall coverings in addition to other applicable requirements of Chapters 14 and 16.

1407.2 Exterior wall finish. MCM used as exterior wall finish or as elements of balconies and similar appendages and bay and oriel windows to provide cladding or weather resistance shall comply with Sections 1407.4 through 1407.13.

1407.3 Architectural trim and embellishments. MCM used as architectural trim or embellishments shall comply with Sections 1407.7 through 1407.13.

1407.4 Structural design. MCM systems shall be designed and constructed to resist wind loads as required by Chapter 16 for components and cladding.

1407.5 Approval. Results of approved tests or an engineering analysis shall be submitted to the building official to verify compliance with the requirements of Chapter 16 for wind loads.

1407.6 Weather resistance. MCM systems shall comply with Section 1403 and shall be designed and constructed to resist wind and rain in accordance with this section and the manufacturer's installation instructions.

1407.7 Durability. MCM systems shall be constructed of approved materials that maintain the performance characteristics required in Section 1407 for the duration of use.

1407.8 Fire-resistance rating. Where MCM systems are used on exterior walls required to have a fire-resistance rating in accordance with Section 704, evidence shall be submitted to the building official that the required fire-resistance rating is maintained.

1407.9 Surface-burning characteristics. Unless otherwise specified, MCM shall have a flame spread index of 75 or less and a smoke-developed index of 450 or less when tested as an assembly in the maximum thickness intended for use in accordance with ASTM E84.

1407.10 Type I, II, III and IV construction. Where installed on buildings of Type I, II, III and IV construction, MCM systems shall comply with Sections 1407.10.1 through 1407.10.4, or 1407.11.

1407.10.1 Surface-burning characteristics. MCM shall have a flame spread index of not more than 25 and a smoke-developed index of not more than 450 when tested as an assembly in the maximum thickness intended for use in accordance with ASTM E 84.

1407.10.2 Thermal barriers. MCM shall be separated from the interior of a building by an approved thermal barrier consisting of 0.5-inch (12.7 mm) gypsum wallboard or equivalent thermal barrier material that will limit the average temperature rise of the unexposed surface to not more than 250°F (121°C) after 15 minutes of fire exposure in accordance with the standard time-temperature curve of ASTM E 119. The thermal barrier shall be installed in such a manner that it will remain in place for not less than 15 minutes based on a test conducted in accordance with UL 1715.

1407.10.3 Thermal barrier not required. The thermal barrier specified for MCM in Section 1407.10.2 is not required where:

1. The MCM system is specifically approved based on tests conducted in accordance with UL 1040 or UL 1715. Such testing shall be performed with the MCM in the maximum thickness intended for use. The MCM system shall include seams, joints and other typical details used in the installation and shall be tested in the manner intended for use.

2. The MCM is used as elements of balconies and similar appendages, architectural trim or embellishments.

1407.10.4 Full-scale tests. The MCM exterior wall assembly shall be tested in accordance with, and comply with, the acceptance criteria of NFPA 285. Such testing shall be performed on the MCM system with the MCM in the maximum thickness intended for use.

1407.11 Alternate conditions. MCM and MCM systems shall not be required to comply with Sections 1407.10.1 through 1407.10.4 provided such systems comply with Section 1407.11.1 or 1407.11.2.

1407.11.1 Installations up to 40 feet in height. MCM shall not be installed more than 40 feet (12 190 mm) in height above the grade plane where installed in accordance with Sections 1407.11.1.1 and 1407.11.1.2.

1407.11.1.1 Fire separation distance of 5 feet or less. Where the fire separation distance is 5 feet (1524 mm) or less, the area of MCM shall not exceed 10 percent of the exterior wall surface.

1407.11.1.2 Fire separation distance greater than 5 feet. Where the fire separation distance is greater than 5 feet (1524 mm), there shall be no limit on the area of exterior wall surface coverage using MCM.

1407.11.2 Installations up to 50 feet in height. MCM shall not be installed more than 50 feet (15 240 mm) in height above the grade plane where installed in accordance with Sections 1407.11.2.1 and 1407.11.2.2.

1407.11.2.1 Self ignition temperature. MCM shall have a self-ignition temperature of 650°F (343°C) or greater when tested in accordance with ASTM D 1929.

1407.11.2.2 Limitations. Sections of MCM shall not exceed 300 square feet (27.9 m²) in area and shall be separated by a minimum of 4 feet (1219 mm) vertically.

1407.12 Type V construction. MCM shall be permitted to be installed on buildings of Type V construction.

1407.13 Labeling. MCM shall be labeled in accordance with Section 1703.5.

CHAPTER 15

ROOF ASSEMBLIES AND ROOFTOP STRUCTURES

SECTION 1501
GENERAL

1501.1 Scope. The provisions of this chapter shall govern the design, materials, construction and quality of roof assemblies, and rooftop structures.

SECTION 1502
DEFINITIONS

1502.1 General. The following words and terms shall, for the purposes of this chapter and as used elsewhere in this code, have the meanings shown herein.

BUILT-UP ROOF COVERING. Two or more layers of felt cemented together and surfaced with a cap sheet, mineral aggregate, smooth coating or similar surfacing material.

➡ **INTERLAYMENT.** A layer of felt or nonbituminous saturated felt not less than 18 inches (457 mm) wide, shingled between each course of a wood-shake roof covering.

MECHANICAL EQUIPMENT SCREEN. A partially enclosed rooftop structure used to aesthetically conceal heating, ventilating and air conditioning (HVAC) electrical or mechanical equipment from view.

METAL ROOF PANEL. An interlocking metal sheet having a minimum installed weather exposure of 3 square feet (.279 m²) per sheet.

METAL ROOF SHINGLE. An interlocking metal sheet having an installed weather exposure less than 3 square feet (.279 m²) per sheet.

MODIFIED BITUMEN ROOF COVERING. One or more layers of polymer-modified asphalt sheets. The sheet materials shall be fully adhered or mechanically attached to the substrate or held in place with an approved ballast layer.

PENTHOUSE. An enclosed, unoccupied structure above the roof of a building, other than a tank, tower, spire, dome cupola or bulkhead, occupying not more than one-third of the roof area.

POSITIVE ROOF DRAINAGE. The drainage condition in which consideration has been made for all loading deflections of the roof deck, and additional slope has been provided to ensure drainage of the roof within 48 hours of precipitation.

REROOFING. The process of recovering or replacing an existing roof covering. See "Roof recover" and "Roof replacement."

ROOF ASSEMBLY. A system designed to provide weather protection and resistance to design loads. The system consists of a roof covering and roof deck or a single component serving as both the roof covering and the roof deck. A roof assembly includes the roof deck, vapor retarder, substrate or thermal barrier, insulation, vapor retarder and roof covering.

ROOF COVERING. The covering applied to the roof deck for weather resistance, fire classification or appearance.

ROOF COVERING SYSTEM. See "Roof assembly."

ROOF DECK. The flat or sloped surface not including its supporting members or vertical supports.

ROOF RECOVER. The process of installing an additional roof covering over a prepared existing roof covering without removing the existing roof covering.

ROOF REPAIR. Reconstruction or renewal of any part of an existing roof for the purposes of its maintenance.

ROOF REPLACEMENT. The process of removing the existing roof covering, repairing any damaged substrate and installing a new roof covering.

ROOF VENTILATION. The natural or mechanical process of supplying conditioned or unconditioned air to, or removing such air from, attics, cathedral ceilings or other enclosed spaces over which a roof assembly is installed.

ROOFTOP STRUCTURE. An enclosed structure on or above the roof of any part of a building.

SCUPPER. An opening in a wall or parapet that allows water to drain from a roof.

SINGLE-PLY MEMBRANE. A roofing membrane that is field applied using one layer of membrane material (either homogeneous or composite) rather than multiple layers.

UNDERLAYMENT. One or more layers of felt, sheathing paper, nonbituminous saturated felt or other approved material over which a steep-slope roof covering is applied.

SECTION 1503
WEATHER PROTECTION

1503.1 General. Roof decks shall be covered with approved roof coverings secured to the building or structure in accordance with the provisions of this chapter. Roof coverings shall be designed, installed and maintained in accordance with this code and the approved manufacturer's instructions such that the roof covering shall serve to protect the building or structure.

1503.2 Flashing. Flashing shall be installed in such a manner so as to prevent moisture entering the wall and roof through joints in copings, through moisture-permeable materials and at intersections with parapet walls and other penetrations through the roof plane.

1503.2.1 Locations. Flashing shall be installed at wall and roof intersections, at gutters, wherever there is a change in roof slope or direction and around roof openings. Where flashing is of metal, the metal shall be corrosion resistant with a thickness of not less than 0.019 inch (.483 mm) (No. 26 galvanized sheet).

1503.3 Coping. Parapet walls shall be properly coped with noncombustible, weatherproof materials of a width no less than the thickness of the parapet wall.

[P] 1503.4 Roof drainage. Design and installation of roof drainage systems shall comply with the *International Plumbing Code.*

1503.4.1 Gutters. Gutters and leaders placed on the outside of buildings, other than Group R-3 as applicable in Section 101.2, private garages and buildings of Type V construction, shall be of noncombustible material or a minimum of Schedule 40 plastic pipe.

1503.5 Roof ventilation. Intake and exhaust vents shall be provided in accordance with Section 1203.2 and the manufacturer's installation instructions.

SECTION 1504
PERFORMANCE REQUIREMENTS

1504.1 Wind resistance of roofs. Roof decks and roof coverings shall be designed for wind loads in accordance with Chapter 16 and Sections 1504.2, 1504.3 and 1504.4.

1504.1.1 Wind resistance of asphalt shingles. Asphalt shingles shall be designed for wind speeds in accordance with Section 1507.2.7.

1504.2 Wind resistance of clay and concrete tile. Clay and concrete tile roof coverings shall be connected to the roof deck in accordance with Chapter 16.

1504.3 Wind resistance of nonballasted roofs. Roof coverings installed on roofs in accordance with Section 1507 that are mechanically attached or adhered to the roof deck shall be designed to resist the design wind load pressures for cladding in Chapter 16.

1504.3.1 Other roof systems. Roof systems with built-up, modified bitumen, fully adhered or mechanically attached single-ply through fastened metal panel roof systems, and other types of membrane roof coverings shall also be tested in accordance with FM 4450, FM 4470, UL 580 or UL 1897.

1504.3.2 Metal panel roof systems. Metal panel roof systems through fastened or standing seam shall be tested in accordance with UL 580 or ASTM E 1592.

1504.4 Ballasted low-slope roof systems. Ballasted low-slope (roof slope < 2:12) single-ply roof system coverings installed in accordance with Section 1507 shall be designed in accordance with ANSI/SPRI RP-4.

1504.5 Edge securement for low-slope roofs. Low-slope membrane roof systems metal edge securement, except gutters, installed in accordance with Section 1507, shall be designed in accordance with ANSI/SPRI ES-1, except the basic wind speed shall be determined from Figure 1609.

1504.6 Physical properties. Roof coverings installed on low-slope roofs (roof slope < 2:12) in accordance with Section 1507 shall demonstrate physical integrity over the working life of the roof based upon 2,000 hours of exposure to accelerated weathering tests conducted in accordance with ASTM G 152, ASTM G 155 or ASTM G 154. Those roof coverings that are subject to cyclical flexural response due to wind loads shall not demonstrate any significant loss of tensile strength for unreinforced membranes or breaking strength for reinforced membranes when tested as herein required.

1504.7 Impact resistance. Roof coverings installed on low-slope roofs (roof slope < 2:12) in accordance with Section 1507 shall resist impact damage based on the results of tests conducted in accordance with ASTM D 3746, ASTM D 4272, CGSB 37-GP-52M or FM 4470.

SECTION 1505
FIRE CLASSIFICATION

1505.1 General. Roof assemblies shall be divided into the classes defined below. Class A, B and C roof assemblies and roof coverings required to be listed by this section shall be tested in accordance with ASTM E 108 or UL 790. In addition, fire-retardant-treated wood roof coverings shall be tested in accordance with ASTM D 2898. The minimum roof coverings installed on buildings shall comply with Table 1505.1 based on the type of construction of the building.

TABLE 1505.1[a,b]
MINIMUM ROOF COVERING CLASSIFICATION
FOR TYPES OF CONSTRUCTION

IA	IB	IIA	IIB	IIIA	IIIB	IV	VA	VB
B	B	B	C[c]	B	C[c]	B	B	C[c]

For SI: 1 foot = 304.8 mm, 1 square foot = 0.0929 m².

a. Unless otherwise required in accordance with the *International Urban Wildland Interface Code* or due to the location of the building within a fire district in accordance with Appendix D.

b. Nonclassified roof coverings shall be permitted on buildings of Group R-3, as applicable in Section 101.2, and Group U occupancies, where there is a minimum fire-separation distance of 6 feet measured from the leading edge of the roof.

c. Buildings that are not more than two stories in height and having not more than 6,000 square feet of projected roof area and where there is a minimum 10-foot fire-separation distance from the leading edge of the roof to a lot line on all sides of the building, except for street fronts or public ways, shall be permitted to have roofs of No. 1 cedar or redwood shakes and No. 1 shingles.

1505.2 Class A roof assemblies. Class A roof assemblies are those that are effective against severe fire test exposure. Class A roof assemblies and roof coverings shall be listed and identified as Class A by an approved testing agency. Class A roof assemblies shall be permitted for use in buildings or structures of all types of construction.

Exception: Class A roof assemblies include those with coverings of brick, masonry, slate, clay or concrete roof tile, exposed concrete roof deck, ferrous or copper shingles or sheets.

1505.3 Class B roof assemblies. Class B roof assemblies are those that are effective against moderate fire-test exposure. Class B roof assemblies and roof coverings shall be listed and identified as Class B by an approved testing agency.

Exception: Class B roof assemblies include those with coverings of metal sheets and shingles.

1505.4 Class C roof assemblies. Class C roof assemblies are those that are effective against light fire-test exposure. Class C roof assemblies and roof coverings shall be listed and identified as Class C by an approved testing agency.

1505.5 Nonclassified roofing. Nonclassified roofing is approved material that is not listed as a Class A, B or C roof covering.

1505.6 Fire-retardant-treated wood shingles and shakes. Fire-retardant-treated wood shakes and shingles shall be treated by impregnation with chemicals by the full-cell vacuum-pressure process, in accordance with AWPA C1. Each bundle shall be marked to identify the manufactured unit and the manufacturer, and shall also be labeled to identify the classification of the material in accordance with the testing required in Section 1505.1, the treating company and the quality control agency.

1505.7 Special purpose roofs. Special purpose wood shingle or wood shake roofing shall conform with the grading and application requirements of Section 1507.8 or 1507.9. In addition, an underlayment of 0.625-inch (15.9 mm) Type X water-resistant gypsum backing board or gypsum sheathing shall be placed under minimum nominal 0.5-inch-thick (12.7 mm) wood structural panel solid sheathing or 1-inch (25 mm) nominal spaced sheathing.

SECTION 1506
MATERIALS

1506.1 Scope. The requirements set forth in this section shall apply to the application of roof-covering materials specified herein. Roof coverings shall be applied in accordance with this chapter and the manufacturer's installation instructions. Installation of roof coverings shall comply with the applicable provisions of Section 1507.

1506.2 Compatibility of materials. Roofs and roof coverings shall be of materials that are compatible with each other and with the building or structure to which the materials are applied.

1506.3 Material specifications and physical characteristics. Roof-covering materials shall conform to the applicable standards listed in this chapter. In the absence of applicable standards or where materials are of questionable suitability, testing by an approved agency shall be required by the building official to determine the character, quality and limitations of application of the materials.

1506.4 Product identification. Roof-covering materials shall be delivered in packages bearing the manufacturer's identifying marks and approved testing agency labels required in accordance with Section 1505. Bulk shipments of materials shall be accompanied with the same information issued in the form of a certificate or on a bill of lading by the manufacturer.

SECTION 1507
REQUIREMENTS FOR ROOF COVERINGS

1507.1 Scope. Roof coverings shall be applied in accordance with the applicable provisions of this section and the manufacturer's installation instructions.

1507.2 Asphalt shingles. The installation of asphalt shingles shall comply with the provisions of this section and Table 1507.2.

1507.2.1 Deck requirements. Asphalt shingles shall be fastened to solidly sheathed decks.

1507.2.2 Slope. Asphalt shingles shall only be used on roof slopes of two units vertical in 12 units horizontal (17-percent slope) or greater. For roof slopes from two units vertical in 12 units horizontal (17-percent slope) up to four units vertical in 12 units horizontal (33-percent slope), double underlayment application is required in accordance with Section 1507.2.8.

1507.2.3 Underlayment. Unless otherwise noted, required underlayment shall conform to ASTM D 226, Type I, or ASTM D 4869, Type I.

1507.2.4 Self-adhering polymer modified bitumen sheet. Self-adhering polymer modified bitumen sheet shall comply with ASTM D 1970.

1507.2.5 Asphalt shingles. Asphalt shingles shall have self-seal strips or be interlocking, and comply with ASTM D 225 or ASTM D 3462.

1507.2.6 Fasteners. Fasteners for asphalt shingles shall be galvanized, stainless steel, aluminum or copper roofing nails, minimum 12 gage [0.105 inch (2.67 mm)] shank with a minimum 0.375 inch-diameter (9.5 mm) head, of a length to penetrate through the roofing materials and a minimum of 0.75 inch (19.1 mm) into the roof sheathing. Where the roof sheathing is less than 0.75 inch (19.1 mm) thick, the nails shall penetrate through the sheathing. Fasteners shall comply with ASTM F 1667.

1507.2.7 Attachment. Asphalt shingles shall have the minimum number of fasteners required by the manufacturer and Section 1504.1. Asphalt shingles shall be secured to the roof with not less than four fasteners per strip shingle or two fasteners per individual shingle. Where the roof slope exceeds 20 units vertical in 12 units horizontal (166-percent slope), special methods of fastening are required. For roofs located where the basic wind speed in accordance with Figure 1609 is 110 mph or greater, special methods of fastening are required. Special fastening methods shall be tested in accordance with ASTM D 3161, modified to use a wind speed of 110 mph.

1507.2.8 Underlayment application. For roof slopes from two units vertical in 12 units horizontal (17-percent slope), up to four units vertical in 12 units horizontal (33-percent slope), underlayment shall be two layers applied in the following manner. Apply a minimum 19-inch-wide (483 mm) strip of underlayment felt parallel with and starting at the eaves, fastened sufficiently to hold in place. Starting at the eave, apply 36-inch-wide (914 mm) sheets of underlayment overlapping successive sheets 19 inches (483 mm) and fastened sufficiently to hold in place. For roof slopes of four units vertical in 12 units horizontal (33-percent slope) or greater, underlayment shall be one layer applied in the following manner. Underlayment shall be applied shingle fashion, parallel to and starting from the eave and lapped 2 inches (51 mm), fastened only as necessary to hold in place.

1507.2.8.1 High wind attachment. Underlayment applied in areas subject to high winds (greater than 110 mph in accordance with Figure 1609) shall be applied

TABLE 1507.2
ASPHALT SHINGLE APPLICATION

COMPONENT	INSTALLATION REQUIREMENT
1. Roof slope	Asphalt shingles shall only be used on roof slopes of two units vertical in 12 units horizontal (2:12) or greater. For roof slopes from two units vertical in 12 units horizontal (2:12) up to four units vertical in 12 units horizontal (4:12), double underlayment application is required in accordance with Section 1507.2.8.
2. Deck requirement	Asphalt shingles shall be fastened to solidly sheathed roofs.
3. Underlayment	Underlayment shall conform with ASTM D 226, Type 1, or ASTM D 4869, Type 1.
For roof slopes from two units vertical in 12 units horizontal (2:12), up to four units vertical in 12 units horizontal (4:12)	Underlayment shall be two layers applied in the following manner. Apply a minimum 19-inch strip or underlayment felt parallel to and starting at the eaves, fastened sufficiently to hold in place. Starting at the eave, apply 35-inch-wide sheets of underlayment overlapping successive sheets 19 inches and fastened sufficiently to hold in place.
For roof slopes from four units vertical in 12 units horizontal (4:12) or greater	Underlayment shall be one layer applied in the following manner. Underlayment shall be applied shingle fashion, parallel to and starting from the eave and lapped 2 inches, fastened only as necessary to hold in place.
In areas where the average daily temperature in January is 25°F or less or where there is a possibility of ice forming along the eaves causing a backup of water	A membrane that consists of at least two layers of underlayment cemented together or of a self-adhering polymer-modified bitumen sheet shall be used in lieu of normal underlayment and extend from the eave's edge to a point at least 24 inches inside the exterior wall line of the building.
4. Application	—
Attachment	Asphalt shingles shall have the minimum number of fasteners required by the manufacturer and Section 1504.1. Asphalt shingles shall be secured to the roof with not less than four fasteners per strip shingle or two fasteners per individual shingle. Where the roof slope exceeds 20 units vertical in 12 units horizontal (20:12), special methods of fastening are required.
Fasteners	Galvanized, stainless steel, aluminum or copper roofing nails, minimum 12-gage (0.105 inch) shank with a minimum $^3/_8$-inch diameter head. Fasteners shall be long enough to penetrate into the sheathing $^3/_4$ inch or through the thickness of the sheathing.
Flashings	In accordance with Section 1507.2.9.

For SI: 1 inch = 25.4 mm, 1 foot = 304.8 mm, °C = [(°F) - 32]/1.8, 1 mile per hour = 1.609 km/h.

with corrosion-resistant fasteners in accordance with the manufacturer's instructions. Fasteners are to be applied along the overlap at a maximum spacing of 36 inches (914 mm) on center.

1507.2.8.2 Ice dam membrane. In areas where the average daily temperature in January is 25°F (-4°C) or less or where there is a possibility of ice forming along the eaves causing a backup of water, a membrane that consists of at least two layers of underlayment cemented together or of a self-adhering polymer modified bitumen sheet shall be used in lieu of normal underlayment and extend from the eave's edge to a point at least 24 inches (610 mm) inside the exterior wall line of the building.

Exception: Detached accessory structures that contain no conditioned floor area.

1507.2.9 Flashings. Flashing for asphalt shingles shall comply with this section. Flashing shall be applied in accordance with this section and the asphalt shingle manufacturer's printed instructions.

1507.2.9.1 Base and cap flashing. Base and cap flashing shall be installed in accordance with the manufacturer's instructions. Base flashing shall be of either corrosion-resistant metal of minimum nominal 0.019-inch (0.483 mm) thickness or mineral-surfaced roll roofing weighing a minimum of 77 pounds per 100 square feet (3.76 kg/m²). Cap flashing shall be corrosion-resistant metal of minimum nominal 0.019-inch (0.483 mm) thickness.

1507.2.9.2 Valleys. Valley linings shall be installed in accordance with the manufacturer's instructions before applying shingles. Valley linings of the following types shall be permitted:

1. For open valleys (valley lining exposed) lined with metal, the valley lining shall be at least 16 inches (406 mm) wide and of any of the corrosion-resistant metals in Table 1507.2.9.2.

2. For open valleys, valley lining of two plies of mineral-surfaced roll roofing shall be permitted. The

bottom layer shall be 18 inches (457 mm) and the top layer a minimum of 36 inches (914 mm) wide.

3. For closed valleys (valleys covered with shingles), valley lining of one ply of smooth roll roofing complying with ASTM D 224 and at least 36 inches (914 mm) wide or types as described in Items 1 and 2 above shall be permitted. Specialty underlayment shall comply with ASTM D 1970.

TABLE 1507.2.9.2
VALLEY LINING MATERIAL

MATERIAL	MINIMUM THICKNESS	GAGE	WEIGHT
Copper	—	—	16 oz
Aluminum	0.024 in.	—	—
Stainless steel	—	28	—
Galvanized steel	0.0179 in.	26 (zinc-coated G90)	—
Zinc alloy	0.027 in.	—	—
Lead	—	—	2.5 pounds
Painted terne	—	—	20 pounds

For SI: 1 inch = 25.4 mm, 1 pound = 0.454 kg, 1 ounce = 28.35 g.

1507.2.9.3 Drip edge. Provide drip edge at eaves and gables of shingle roofs. Overlap to be a minimum of 2 inches (51 mm). Eave drip edges shall extend 0.25 inch (6.4 mm) below sheathing and extend back on the roof a minimum of 2 inches (51 mm). Drip edge shall be mechanically fastened a maximum of 12 inches (305 mm) o.c. A cricket or saddle shall be installed on the ridge side of any chimney greater than 30 inches (762 mm) wide. Cricket or saddle coverings shall be sheet metal or of the same material as the roof covering.

1507.3 Clay and concrete tile. The installation of clay and concrete tile shall comply with the provisions of this section.

1507.3.1 Deck requirements. Concrete and clay tile shall be installed only over solid sheathing or spaced structural sheathing boards.

1507.3.2 Deck slope. Clay and concrete roof tile shall be installed on roof slopes of $2^1/_2$ units vertical in 12 units horizontal (21-percent slope) or greater. For roof slopes from $2^1/_2$ units vertical in 12 units horizontal (21-percent slope) to four units vertical in 12 units horizontal (33-percent slope), double underlayment application is required in accordance with Section 1507.3.3.

1507.3.3 Underlayment. Unless otherwise noted, required underlayment shall conform to: ASTM D 226, Type II; ASTM D 2626 or ASTM D 249 Type I mineral-surfaced roll roofing.

1507.3.3.1 Low-slope roofs. For roof slopes from $2^1/_2$ units vertical in 12 units horizontal (21-percent slope), up to four units vertical in 12 units horizontal (33-percent

slope), underlayment shall be a minimum of two layers applied as follows:

1. Starting at the eave, a 19-inch (483 mm) strip of underlayment shall be applied parallel with the eave and fastened sufficiently in place.

2. Starting at the eave, 36-inch-wide (914 mm) strips of underlayment felt shall be applied overlapping successive sheets 19 inches (483 mm) and fastened sufficiently in place.

1507.3.3.2 High-slope roofs. For roof slopes of four units vertical in 12 units horizontal (33-percent slope) or greater, underlayment shall be a minimum of one layer of underlayment felt applied shingle fashion, parallel to, and starting from the eaves and lapped 2 inches (51 mm), fastened only as necessary to hold in place.

1507.3.4 Clay tile. Clay roof tile shall comply with ASTM C 1167.

1507.3.5 Concrete tile. Concrete roof tiles shall be in accordance with the physical test requirements as follows:

1. The transverse strength of tiles shall be determined according to Section 6.3 of ASTM C 1167 and in accordance with Table 1507.3.5.

2. The absorption of concrete roof tiles shall be according to Section 8 of ASTM C 140. Roof tiles shall absorb not more than 15 percent of the dry weight of the tile during a 24-hour immersion test.

3. Roof tiles shall be tested for freeze/thaw resistance according to Section 8 of ASTM C67. Roof tiles shall show no breakage and not have more than 1 percent loss in dry weight of any individual concrete roof tile.

TABLE 1507.3.5
TRANSVERSE BREAKING STRENGTH
OF CONCRETE ROOF TILE (lbs.)

TILE PROFILE	DRY	
	Average of five tiles	Individual tile
High profile	400	350
Medium profile	300	250
Flat profile	300	250

For SI: 1 pound = 4.45 N.

1507.3.6 Fasteners. Tile fasteners shall be corrosion resistant and not less than 11 gage, $5/_{16}$-inch (8.0 mm) head, and of sufficient length to penetrate the deck a minimum of 0.75 inch (19.1 mm) or through the thickness of the deck, whichever is less. Attaching wire for clay or concrete tile shall not be smaller than 0.083 inch (2.1 mm). Perimeter fastening areas include three tile courses but not less than 36 inches (914 mm) from either side of hips or ridges and edges of eaves and gable rakes.

1507.3.7 Attachment. Clay and concrete roof tiles shall be fastened in accordance with Table 1507.3.7.

TABLE 1507.3.7
CLAY AND CONCRETE TILE ATTACHMENT[a, b, c]

GENERAL — CLAY OR CONCRETE ROOF TILE			
Maximum basic wind speed (mph)	Mean roof height (feet)	Roof slope up to < 3:12	Roof slope 3:12 and over
85	0-60	One fastener per tile. Flat tile without vertical laps, two fasteners per tile.	Two fasteners per tile. Only one fastener on slopes of 7:12 and less for tiles with installed weight exceeding 7.5 lbs./sq. ft. having a width no greater than 16 inches.
100	0-40		
100	> 40-60	The head of all tiles shall be nailed. The nose of all eave tiles shall be fastened with approved clips. All rake tiles shall be nailed with two nails. The nose of all ridge, hip and rake tiles shall be set in a bead of roofer's mastic.	
110	0-60	The fastening system shall resist the wind forces in Section 1609.7.2.	
120	0-60	The fastening system shall resist the wind forces in Section 1609.7.2.	
130	0-60	The fastening system shall resist the wind forces in Section 1609.7.2.	
All	> 60	The fastening system shall resist the wind forces in Section 1609.7.2.	

INTERLOCKING CLAY OR CONCRETE ROOF TILE WITH PROJECTING ANCHOR LUGS[d, e] (Installations on spaced/solid sheathing with battens or spaced sheathing)				
Maximum basic wind speed (mph)	Mean roof height (feet)	Roof slope up to < 5:12	Roof slope 5:12 < 12:12	Roof slope 12:12 and over
85	0-60	Fasteners are not required. Tiles with installed weight less than 9 lbs./sq. ft. require a minimum of one fastener per tile.	One fastener per tile every other row. All perimeter tiles require one fastener. Tiles with installed weight less than 9 lbs./sq. ft. require a minimum of one fastener per tile.	One fastener required for every tile. Tiles with installed weight less than 9 lbs./sq. ft. require a minimum of one fastener per tile.
100	0-40			
100	> 40-60	The head of all tiles shall be nailed. The nose of all eave tiles shall be fastened with approved clips. All rake tiles shall be nailed with two nails The nose of all ridge, hip and rake tiles shall be set in a bead of roofers's mastic.		
110	0-60	The fastening system shall resist the wind forces in Section 1609.7.2.		
120	0-60	The fastening system shall resist the wind forces in Section 1609.7.2.		
130	0-60	The fastening system shall resist the wind forces in Section 1609.7.2.		
All	> 60	The fastening system shall resist the wind forces in Section 1609.7.2.		

INTERLOCKING CLAY OR CONCRETE ROOF TILE WITH PROJECTING ANCHOR LUGS (Installations on solid sheathing without battens)		
Maximum basic wind speed (mph)	Mean roof height (feet)	All roof slopes
85	0-60	One fastener per tile.
100	0-40	One fastener per tile.
100	> 40-60	The head of all tiles shall be nailed. The nose of all eave tiles shall be fastened with approved clips. All rake tiles shall be nailed with two nails The nose of all ridge, hip and rake tiles shall be set in a bead of roofers's mastic.
110	0-60	The fastening system shall resist the wind forces in Section 1609.7.2.
120	0-60	The fastening system shall resist the wind forces in Section 1609.7.2.
130	0-60	The fastening system shall resist the wind forces in Section 1609.7.2.
All	> 60	The fastening system shall resist the wind forces in Section 1609.7.2.

For SI: 1 inch = 25.4 mm, 1 foot = 304.8 mm, 1 mile per hour = 1.609 km/h, 1 pound per square foot = 0.0478 kn/m^2.

a. Minimum fastener size. Corrosion-resistant nails not less than No. 11 gage with $5/16$-inch head. Fasteners shall be long enough to penetrate into the sheathing 0.75 inch or through the thickness of the sheathing, whichever is less. Attaching wire for clay and concrete tile shall not be smaller than 0.083 inch.

b. Snow areas. A minimum of two fasteners per tile are required or battens and one fastener.

c. Roof slopes greater than 24:12. The nose of all tiles shall be securely fastened.

d. Horizontal battens. Battens shall be not less than 1inch by 2 inch nominal. Provisions shall be made for drainage by a minimum of $1/8$-inch riser at each nail or by 4-foot-long battens with at least a 0.5-inch separation between battens. Horizontal battens are required for slopes over 7:12.

e. Perimeter fastening areas include three tile courses but not less than 36 inches from either side of hips or ridges and edges of eaves and gable rakes.

1507.3.8 Application. Tile shall be applied according to the manufacturer's installation instructions, based on the following:

1. Climatic conditions.
2. Roof slope.
3. Underlayment system.
4. Type of tile being installed.

1507.3.9 Flashing. At the juncture of the roof vertical surfaces, flashing and counterflashing shall be provided in accordance with the manufacturer's installation instructions, and where of metal, shall not be less than 0.019-inch (0.48 mm) (No. 26 galvanized sheet gage) corrosion-resistant. The valley flashing shall extend at least 11 inches (279 mm) from the centerline each way and have a splash diverter rib not less than 1 inch (25 mm) high at the flow line formed as part of the flashing. Sections of flashing shall have an end lap of not less than 4 inches (102 mm). For roof slopes of three units vertical in 12 units horizontal (25-percent slope) and over, the valley flashing shall have a 36-inch-wide (914 mm) underlayment of one layer of Type I underlayment running the full length of the valley, in addition to other required underlayment. In areas where the average daily temperature in January is 25°F (-4°C) or less or where there is a possibility of ice forming along the eaves causing a backup of water, the metal valley flashing underlayment shall be solid cemented to the roofing underlayment for slopes under seven units vertical in 12 units horizontal (58-percent slope) or of self-adhering polymer modified bitumen sheet.

1507.4 Metal roof panels. The installation of metal roof panels shall comply with the provisions of this section.

1507.4.1 Deck requirements. Metal roof panel roof coverings shall be applied to a solid or closely fitted deck, except where the roof covering is specifically designed to be applied to spaced supports.

1507.4.2 Deck slope. The minimum slope for lapped, nonsoldered seam metal roofs without applied lap sealant shall be three units vertical in 12 units horizontal (25-percent slope). The minimum slope for lapped, nonsoldered seam metal roofs with applied lap sealant shall be one-half vertical unit in 12 units horizontal (4-percent slope). The minimum slope for standing seam of roof systems shall be one-quarter unit vertical in 12 units horizontal (2-percent slope).

1507.4.3 Material standards. Metal-sheet roof covering systems that incorporate supporting structural members shall be designed in accordance with Chapter 22. Metal-sheet roof coverings installed over structural decking shall comply with Table 1507.4.3.

1507.4.4 Attachment. Metal roofing fastened directly to steel framing shall be attached by approved manufacturers' fasteners. In the absence of manufacturer recommendations, all of the following fasteners shall be used:

1. Galvanized fasteners shall be used for galvanized roofs.
2. 300 series stainless-steel fasteners shall be used for copper roofs.

3. Stainless-steel fasteners are acceptable for all types of metal roofs.

TABLE 1507.4.3
METAL ROOF COVERINGS

ROOF COVERING TYPE	STANDARD APPLICATION RATE/THICKNESS
Aluminum	ASTM B 209, 0.024 inch minimum thickness for roll-formed panels and 0.019 inch minimum thickness for press-formed shingles.
Aluminum-zinc alloy coated steel	ASTM A 792 AZ 50
Copper	16 oz./sq. ft. for metal-sheet roof-covering systems; 12 oz./sq. ft. for preformed metal shingle systems.
Galvanized steel	ASTM A 653 G-90 zinc-coated, 0.013-inch-thick minimum
Lead-coated copper	ASTM B 101
Hard lead	2 lbs./sq. ft.
Soft lead	3 lbs./sq. ft.
Prepainted steel	ASTM A 755
Terne (tin) and terne-coated stainless	Terne coating of 40 lbs. per double base box, field painted where applicable in accordance with manufacturer's installation instructions.

For SI: 1 ounce per square foot = 0.0026 kg/m², 1 pound per square foot = 4.882 kg/m², 1 inch = 25.4 mm, 1 pound = 0.454 kg.

1507.5 Metal roof shingles. The installation of metal roof shingles shall comply with the provisions of this section.

1507.5.1 Deck requirements. Metal roof shingles shall be applied to a solid or closely fitted deck, except where the roof covering is specifically designed to be applied to spaced sheathing.

1507.5.2 Deck slope. Metal roof shingles shall not be installed on roof slopes below three units vertical in 12 units horizontal (25-percent slope).

1507.5.3 Underlayment. Underlayment shall conform to ASTM D 226, Type I. In areas where the average daily temperature in January is 25°F (-4°C) or less or where there is a possibility of ice forming along the eaves causing a backup of water, an ice barrier that consists of at least two layers of underlayment cemented together or of a self-adhering polymer-modified bitumen sheet, shall be used in lieu of normal underlayment and extend from the eave's edge to a point at least 24 inches (610 mm) inside the exterior wall line of the building.

Exception: Detached accessory structures that contain no conditioned floor area.

1507.5.4 Material standards. Metal roof shingle roof coverings shall comply with Table 1507.4.3.

1507.5.5 Attachment. Metal roof shingles shall be secured to the roof in accordance with the approved manufacturer's installation instructions.

1507.5.6 Flashing. Roof valley flashing shall be of corrosion-resistant metal of the same material as the roof covering or shall comply with the standards in Table 1507.4.3. The valley flashing shall extend at least 8 inches (203 mm) from the centerline each way and shall have a splash diverter rib not less than 0.75 inch (19.1 mm) high at the flow line formed as part of the flashing. Sections of flashing shall have an end lap of not less than 4 inches (102 mm). In areas where the average daily temperature in January is 25°F (-4°C) or less or where there is a possibility of ice forming along the eaves causing a backup of water, the metal valley flashing shall have a 36-inch-wide (914 mm) underlayment directly under it consisting of one layer of underlayment running the full length of the valley, in addition to underlayment required for metal roof shingles. The metal valley flashing underlayment shall be solid cemented to the roofing underlayment for roof slopes under seven units vertical in 12 units horizontal (58-percent slope) or of self-adhering polymer-modified bitumen sheet.

1507.6 Mineral-surfaced roll roofing. The installation of mineral-surfaced roll roofing shall comply with this section.

1507.6.1 Deck requirements. Mineral-surfaced roll roofing shall be fastened to solidly sheathed roofs.

1507.6.2 Deck slope. Mineral-surfaced roll roofing shall not be applied on roof slopes below one unit vertical in 12 units horizontal (8-percent slope).

1507.6.3 Underlayment. Underlayment shall conform to ASTM D 226, Type I. In areas where the average daily temperature in January is 25°F (-4°C) or less or where there is a possibility of ice forming along the eaves causing a backup of water, an ice barrier that consists of at least two layers of underlayment cemented together or of a self-adhering polymer-modified bitumen sheet, shall extend from the eave's edge to a point at least 24 inches (610 mm) inside the exterior wall line of the building.

Exception: Detached accessory structures that contain no conditioned floor area.

1507.6.4 Material standards. Mineral-surfaced roll roofing shall conform to ASTM D 224, ASTM D 249, ASTM D 371 or ASTM D 3909.

1507.7 Slate shingles. The installation of slate shingles shall comply with the provisions of this section.

1507.7.1 Deck requirements. Slate shingles shall be fastened to solidly sheathed roofs.

1507.7.2 Deck slope. Slate shingles shall only be used on slopes of four units vertical in 12 units horizontal (4:12) or greater.

1507.7.3 Underlayment. Underlayment shall comply with ASTM D 226, Type II. In areas where the average daily temperature in January is 25°F (-4°C) or less or where there is a possibility of ice forming along the eaves causing a backup of water, an ice barrier that consists of at least two layers of underlayment cemented together or of a self-adhering polymer-modified bitumen sheet, shall extend from the eave's edge to a point at least 24 inches (610 mm) inside the exterior wall line of the building.

Exception: Detached accessory structures that contain no conditioned floor area.

1507.7.4 Material standards. Slate shingles shall comply with ASTM C 406.

1507.7.5 Application. Minimum headlap for slate shingles shall be in accordance with Table 1507.7.5. Slate shingles shall be secured to the roof with two fasteners per slate.

TABLE 1507.7.5
SLATE SHINGLE HEADLAP

SLOPE	HEADLAP (inches)
4:12 < slope < 8:12	4
8:12 < slope < 20:12	3
slope ≥ 20:12	2

For SI: 1 inch = 25.4 mm.

1507.7.6 Flashing. Flashing and counterflashing shall be made with sheet metal. Valley flashing shall be a minimum of 15 inches (381 mm) wide. Valley and flashing metal shall be a minimum uncoated thickness of 0.0179-inch (0.455 mm) zinc-coated G90. Chimneys, stucco or brick walls shall have a minimum of two plies of felt for a cap flashing consisting of a 4-inch-wide (102 mm) strip of felt set in plastic cement and extending 1 inch (25 mm) above the first felt and a top coating of plastic cement. The felt shall extend over the base flashing 2 inches (51 mm).

1507.8 Wood shingles. The installation of wood shingles shall comply with the provisions of this section and Table 1507.8.

1507.8.1 Deck requirements. Wood shingles shall be installed on solid or spaced sheathing. Where spaced sheathing is used, sheathing boards shall not be less than 1-inch by 4-inch (25 mm by 102 mm) nominal dimensions and shall be spaced on centers equal to the weather exposure to coincide with the placement of fasteners.

1507.8.1.1 Solid sheathing required. Solid sheathing is required in areas where the average daily temperature in January is 25°F (-4°C) or less or where there is a possibility of ice forming along the eaves causing a backup of water.

1507.8.2 Deck slope. Wood shingles shall be installed on slopes of three units vertical in 12 units horizontal (25-percent slope) or greater.

1507.8.3 Underlayment. Underlayment shall comply with ASTM D 226, Type I. In areas where the average daily temperature in January is 25°F (-4°C) or less or where there is a possibility of ice forming along the eaves causing a backup of water, an ice barrier that consists of at least two layers of underlayment cemented together or of a self-adhering polymer-modified bitumen sheet shall extend from the eave's edge to a point at least 24 inches (610 mm) inside the exterior wall line of the building.

Exception: Detached accessory structures that contain no conditioned floor area.

TABLE 1507.8
WOOD SHINGLE AND SHAKE INSTALLATION

ROOF ITEM	WOOD SHINGLES	WOOD SHAKES
1. Roof slope	Wood shingles shall be installed on slopes of three units vertical in 12 units horizontal (3:12) or greater.	Wood shakes shall be installed on slopes of four units vertical in 12 units horizontal (4:12) or greater.
2. Deck requirement	—	—
Temperate climate	Shingles shall be applied to roofs with solid or spaced sheathing. Where spaced sheathing is used, sheathing boards shall not be 4 less than 1″ × 4″ nominal dimensions and shall be spaced on center equal to the weather exposure to coincide with the placement of fasteners.	Shakes shall be applied to roofs with solid or spaced sheathing. Where spaced sheathing is used, sheathing boards shall not be less than 1″ × 4″ nominal dimensions and shall be spaced on center equal to the weather exposure to coincide with the placement of fasteners. When 1″ × 4″ spaced sheathing is installed at 10 inches, boards must be installed between the sheathing boards.
In areas where the average daily temperature in January is 25°F or less or where there is a possibility of ice forming along the eaves causing a backup of water.	Solid sheathing required.	Solid sheathing is required.
3. Interlayment	No requirements.	Interlayment shall comply with ASTM D 226, Type 1.
4. Underlayment	—	—
Temperate climate	Underlayment shall comply with ASTM D 226, Type 1.	Underlayment shall comply with ASTM D 226, Type 1.
In areas where the average daily temperature in January is 25°F or less or where there is a possibility of ice forming along the eaves causing a backup of water.	An ice shield that consists of at least two layers of underlayment cemented together or of a self-adhering polymer-modified bitumen sheet shall extend from the eave's edge to a point at least 24 inches inside the exterior wall line of the building.	An ice shield that consists of at least two layers of underlayment cemented together or of a self-adhering polymer-modified bitumen sheet shall extend from the eave's edge to a point at least 24 inches inside the exterior wall line of the building.
5. Application	—	—
Attachment	Fasteners for wood shingles shall be corrosion resistant with a minimum penetration of 0.75 inch into the sheathing. For sheathing less than 0.5 inch thick, the fasteners shall extend through the sheathing.	Fasteners for wood shakes shall be corrosion resistant with a minimum penetration of 0.75 inch into the sheathing. For sheathing less than 0.5 inch thick, the fasteners shall extend through the sheathing.
No. of fasteners	Two per shingle.	Two per shake.
Exposure	Weather exposures shall not exceed those set forth in Table 1507.8.6	Weather exposures shall not exceed those set forth in Table 1507.9.7
Method	Shingles shall be laid with a side lap of not less than 1.5 inches between joints in courses, and no two joints in any three adjacent courses shall be in direct alignment. Spacing between shingles shall be 0.25 to 0.375 inch.	Shakes shall be laid with a side lap of not less than 1.5 inches between joints in adjacent courses. Spacing between shakes shall not be less than 0.375 inch or more than 0.625 inch for shakes and tapersawn shakes of naturally durable wood and shall be 0.25 to 0.375 inch for preservative taper sawn shakes.
Flashing	In accordance with Section 1507.8.7.	In accordance with Section 1507.9.8.

For SI: 1 inch = 25.4 mm, °C = [(°F) - 32]/1.8.

1507.8.4 Material standards. Wood shingles shall be of naturally durable wood and comply with the requirements of Table 1507.8.4.

TABLE 1507.8.4
WOOD SHINGLE MATERIAL REQUIREMENTS

MATERIAL	APPLICABLE MINIMUM GRADES	GRADING RULES
Wood shingles of naturally durable wood	1, 2 or 3	CSSB

CSSB = Cedar Shake and Shingle Bureau

1507.8.5 Attachment. Fasteners for wood shingles shall be corrosion resistant with a minimum penetration of 0.75 inch (19.1 mm) into the sheathing. For sheathing less than 0.5 inch (12.7 mm) in thickness, the fasteners shall extend through the sheathing. Each shingle shall be attached with a minimum of two fasteners.

1507.8.6 Application. Wood shingles shall be laid with a side lap not less than 1.5 inches (38 mm) between joints in adjacent courses, and not be in direct alignment in alternate courses. Spacing between shingles shall be 0.25 to 0.375 inches (6.4 to 9.5 mm). Weather exposure for wood shingles shall not exceed that set in Table 1507.8.6.

TABLE 1507.8.6
WOOD SHINGLE WEATHER EXPOSURE AND ROOF SLOPE

ROOFING MATERIAL	LENGTH (inches)	GRADE	EXPOSURE (inches) 3:12 pitch to < 4:12	4:12 pitch or steeper
Shingles of naturally durable wood	16	No. 1	3.75	5
		No. 2	3.5	4
		No. 3	3	3.5
	18	No. 1	4.25	5.5
		No. 2	4	4.5
		No. 3	3.5	4
	24	No. 1	5.75	7.5
		No. 2	5.5	6.5
		No. 3	5	5.5

For SI: 1 inch = 25.4 mm.

1507.8.7 Flashing. At the juncture of the roof and vertical surfaces, flashing and counterflashing shall be provided in accordance with the manufacturer's installation instructions, and where of metal, shall not be less than 0.019-inch (0.48 mm) (No. 26 galvanized sheet gage) corrosion-resistant metal. The valley flashing shall extend at least 11 inches (279 mm) from the centerline each way and have a splash diverter rib not less than 1 inch (25 mm) high at the flow line formed as part of the flashing. Sections of flashing shall have an end lap of not less than 4 inches (102 mm). For roof slopes of three units vertical in 12 units horizontal (25-percent slope) and over, the valley flashing shall have a 36-inch-wide (914 mm) underlayment of one layer of Type I underlayment running the full length of the valley, in addition to other required underlayment. In areas where the average daily temperature in January is 25°F (-4°C) or less or where there is a possibility of ice forming along the eaves causing a backup of water, the metal valley flashing underlayment shall be solid cemented to the roofing underlayment for slopes under seven units vertical in 12 units horizontal (58-percent slope).

1507.9 Wood shakes. The installation of wood shakes shall comply with the provisions of this section and Table 1507.8.

1507.9.1 Deck requirements. Wood shakes shall only be used on solid or spaced sheathing. Where spaced sheathing is used, sheathing boards shall not be less than 1-inch by 4-inch (25 mm by 102 mm) nominal dimensions and shall be spaced on centers equal to the weather exposure to coincide with the placement of fasteners. Where 1-inch by 4-inch (25 mm by 102 mm) spaced sheathing is installed at 10 inches (254 mm) o.c., additional 1-inch by 4-inch (25 mm by 102 mm) boards shall be installed between the sheathing boards.

1507.9.1.1 Solid sheathing required. Solid sheathing is required in areas where the average daily temperature in January is 25°F (-4°C) or less or where there is a possibility of ice forming along the eaves causing a backup of water.

1507.9.2 Deck slope. Wood shakes shall only be used on slopes of four units vertical in 12 units horizontal (33-percent slope) or greater.

1507.9.3 Underlayment. Underlayment shall comply with ASTM D 226, Type I. In areas where the average daily temperature in January is 25°F (-4°C) or less or where there is a possibility of ice forming along the eaves causing a backup of water, an ice barrier that consists of at least two layers of underlayment cemented together or a self-adhering polymer-modified bitumen sheet shall extend from the edge of the eave to a point at least 24 inches (610 mm) inside the exterior wall line of the building.

Exception: Detached accessory structures that contain no conditioned floor area.

1507.9.4 Interlayment. Interlayment shall comply with ASTM D 226, Type I.

1507.9.5 Material standards. Wood shakes shall comply with the requirements of Table 1507.9.5.

TABLE 1507.9.5
WOOD SHAKE MATERIAL REQUIREMENTS

MATERIAL	MINIMUM GRADES	APPLICABLE GRADING RULES
Wood shakes of naturally durable wood	1	CSSB
Taper sawn shakes of naturally durable wood	1 or 2	CSSB
Preservative-treated shakes and shingles of naturally durable wood	1	CSSB
Fire-retardant-treated shakes and shingles of naturally durable wood	1	CSSB
Preservative-treated taper sawn shakes of Southern yellow pine treated in accordance with AWPA Standard C2	1 or 2	TFS

CSSB = Cedar Shake and Shingle Bureau.
TFS = Forest Products Laboratory of the Texas Forest Services.

1507.9.6 Attachment. Fasteners for wood shakes shall be corrosion resistant with a minimum penetration of 0.75 inch (19.1 mm) into the sheathing. For sheathing less than 0.5 inch (12.7 mm) in thickness, the fasteners shall extend through the sheathing. Each shake shall be attached with a minimum of two fasteners.

1507.9.7 Application. Wood shakes shall be laid with a side lap not less than 1.5 inches (38 mm) between joints in adjacent courses. Spacing between shakes in the same course shall be 0.375 to 0.625 inches (9.5 to 15.9 mm) for shakes and taper sawn shakes of naturally durable wood and shall be 0.25 to 0.375 inch (6.4 to 9.5 mm) for preservative taper sawn shakes. Weather exposure for wood shakes shall not exceed those set in Table 1507.9.7.

TABLE 1507.9.7
WOOD SHAKE WEATHER EXPOSURE
AND ROOF SLOPE

ROOFING MATERIAL	LENGTH (inches)	GRADE	EXPOSURE (inches) 4:12 PITCH OR STEEPER
Shakes of naturally durable wood	18	No. 1	7.5
	24	No. 1	10a
Preservative-treated taper sawn shakes of Southern yellow pine	18	No. 1	7.5
	24	No. 1	10
	18	No. 2	5.5
	24	No. 2	7.5
Taper sawn shakes of naturally durable wood	18	No. 1	7.5
	24	No. 1	10
	18	No. 2	5.5
	24	No. 2	7.5

For SI: 1 inch = 25.4 mm.
a. For 24-inch by 0.375-inch handsplit shakes, the maximum exposure is 7.5 inches.

1507.9.8 Flashing. At the juncture of the roof and vertical surfaces, flashing and counterflashing shall be provided in accordance with the manufacturer's installation instructions, and where of metal, shall not be less than 0.019-inch (0.48 mm) (No. 26 galvanized sheet gage) corrosion-resistant metal. The valley flashing shall extend at least 11 inches (279 mm) from the centerline each way and have a splash diverter rib not less than 1 inch (25 mm) high at the flow line formed as part of the flashing. Sections of flashing shall have an end lap of not less than 4 inches (102 mm). For roof slopes of 3 units vertical in 12 units horizontal (25-percent slope) and over, the valley flashing shall have a 36-inch-wide (914 mm) underlayment of one layer of Type I underlayment running the full length of the valley, in addition to other required underlayment. In areas where the average daily temperature in January is 25°F (-4°C) or less or where there is a possibility of ice forming along the eaves causing a backup of water, the metal valley flashing underlayment shall be solid cemented to the roofing underlayment for slopes under seven units vertical in 12 units horizontal (58-percent slope).

1507.10 Built-up roofs. The installation of built-up roofs shall comply with the provisions of this section.

1507.10.1 Slope. Built-up roofs shall have a design slope of a minimum of one-fourth unit vertical in 12 units horizontal (2-percent slope) for drainage, except for coal-tar built-up roofs that shall have a design slope of a minimum one-eighth unit vertical in 12 units horizontal (1-percent slope).

1507.10.2 Material standards. Built-up roof covering materials shall comply with the standards in Table 1507.10.2.

1507.11 Modified bitumen roofing. The installation of modified bitumen roofing shall comply with the provisions of this section.

TABLE 1507.10.2
BUILT-UP ROOFING MATERIAL STANDARDS

MATERIAL STANDARD	STANDARD
Acrylic coatings used in roofing	ASTM D 6083
Aggregate surfacing	ASTM D 1863
Asphalt adhesive used in roofing	ASTM D 3747
Asphalt cements used in roofing	ASTM D 3019; D 2822; D 4586
Asphalt-coated glass fiber base sheet	ASTM D 4601
Asphalt coatings used in roofing	ASTM D1227; D 2823; D 4479
Asphalt glass felt	ASTM D 2178
Asphalt primer used in roofing	ASTM D 41
Asphalt-saturated and asphalt-coated organic felt base sheet	ASTM D 2626
Asphalt-saturated organic felt (perforated)	ASTM D 226
Asphalt used in roofing	ASTM D 312
Coal-tar cements used in roofing	ASTM D 4022; D 5643
Coal-tar saturated organic felt	ASTM D 227
Coal-tar pitch used in roofing	ASTM D 450; Type I or II
Coal-tar primer used in roofing, damproofing and waterproofing	ASTM D 43
Glass mat, coal tar	ASTM D 4990
Glass mat, venting type	ASTM D 4897
Mineral-surfaced inorganic cap sheet	ASTM D 3909
Thermoplastic fabrics used in roofing	ASTM D 5665, D 5726

1507.11.1 Slope. Modified bitumen membrane roofs shall have a design slope of a minimum of one-fourth unit vertical in 12 units horizontal (2-percent slope) for drainage.

1507.11.2 Material standards. Modified bitumen roof coverings shall comply with CGSB 37-GP-56M, ASTM D 6162, ASTM D 6163, ASTM D 6164, ASTM D 6222, ASTM D 6223 and ASTM D 6298.

1507.12 Thermoset single-ply roofing. The installation of thermoset single-ply roofing shall comply with the provisions of this section.

1507.12.1 Slope. Thermoset single-ply membrane roofs shall have a design slope of a minimum of one-fourth unit vertical in 12 units horizontal (2-percent slope) for drainage.

1507.12.2 Material standards. Thermoset single-ply roof coverings shall comply with RMA RP-1, RP-2 or RP-3, or ASTM D 4637, ASTM D 5019 or CGSB 37-GP-52M.

1507.13 Thermoplastic single-ply roofing. The installation of thermoplastic single-ply roofing shall comply with the provisions of this section.

1507.13.1 Slope. Thermoplastic single-ply membrane roofs shall have a design slope of a minimum of one-fourth unit vertical in 12 units horizontal (2-percent slope).

1507.13.2 Material standards. Thermoplastic single-ply roof coverings shall comply with ASTM D 4434 or CGSB 37-GP-54M.

1507.14 Sprayed polyurethane foam roofing. The installation of sprayed polyurethane foam roofing shall comply with the provisions of this section.

1507.14.1 Slope. Sprayed polyurethane foam roofs shall have a design slope of a minimum of one-fourth unit vertical in 12 units horizontal (2-percent slope) for drainage.

1507.14.2 Material standards. Spray-applied polyurethane foam insulation shall comply with ASTM C 1029.

1507.14.3 Application. Foamed-in-place roof insulation shall be installed in accordance with the manufacturer's instructions. A liquid-applied protective coating that complies with Section 1507.15 shall be applied no less than 2 hours nor more than 72 hours following the application of the foam.

1507.14.4 Foam plastics. Foam plastic materials and installation shall comply with Chapter 26.

1507.15 Liquid-applied coatings. The installation of liquid-applied coatings shall comply with the provisions of this section.

1507.15.1 Slope. Liquid-applied roofs shall have a design slope of a minimum of one-fourth unit vertical in 12 units horizontal (2-percent slope).

1507.15.2 Material standards. Liquid-applied roof coatings shall comply with ASTM C 836, ASTM C 957, ASTM D 6083, ASTM D 1227 or ASTM D 3468.

SECTION 1508
ROOF INSULATION

1508.1 General. The use of above-deck thermal insulation shall be permitted provided such insulation is covered with an approved roof covering and passes the tests of FM 4450 or UL 1256 when tested as an assembly.

Exception: Foam plastic roof insulation shall conform to the material and installation requirements of Chapter 26.

1508.1.1 Cellulosic fiberboard. Cellulosic fiberboard roof insulation shall conform to the material and installation requirements of Chapter 23.

SECTION 1509
ROOFTOP STRUCTURES

1509.1 General. The provisions of this section shall govern the construction of rooftop structures.

1509.2 Penthouses. A penthouse or other projection above the roof in structures of other than Type I construction shall not exceed 28 feet (8534 mm) above the roof where used as an enclosure for tanks or for elevators that run to the roof and in all other cases shall not extend more than 18 feet (8534 mm) above the roof. The aggregate area of penthouses and other rooftop structures shall not exceed one-third the area of the supporting roof. A penthouse, bulkhead or any other similar projection above the roof shall not be used for purposes other than shelter of mechanical equipment or shelter of vertical shaft openings in the roof. Provisions such as louvers, louver blades or flashing shall be made to protect the mechanical equipment and the building interior from the elements. Penthouses or bulkheads used for purposes other than permitted by this section shall conform to the requirements of this code for an additional story. The restrictions of this section shall not prohibit the placing of wood flagpoles or similar structures on the roof of any building.

1509.2.1 Type of construction. Penthouses shall be constructed with walls, floors and roof as required for the building.

Exceptions:

1. On buildings of Type I and II construction, the exterior walls and roofs of penthouses with a fire separation distance of more than 5 feet (1524 mm) and less than 20 feet (6096 mm) shall be of at least 1-hour fire-resistance-rated noncombustible construction. Walls and roofs with a fire separation distance of 20 feet (6096 mm) or greater shall be of noncombustible construction. Interior framing and walls shall be of noncombustible construction.

2. On buildings of Type III, IV and V construction, the exterior walls of penthouses with a fire separation distance of more than 5 feet (1524 mm) and less than 20 feet (6096 mm) shall be at least 1-hour fire-resistance-rated construction. Walls with a fire separation distance of 20 feet (6096 mm) or greater from a common property line shall be of Type IV or noncombustible construction. Roofs shall be constructed of materials and fire-resistance rated as required in Table 601. Interior framing and walls shall be Type IV or noncombustible construction.

3. Unprotected noncombustible enclosures housing only mechanical equipment and located with a minimum fire separation distance of 20 feet (6096 mm) shall be permitted.

4. On one-story buildings, combustible unroofed mechanical equipment screens, fences or similar enclosures are permitted where located with a fire separation distance of at least 20 feet (6096 mm) from adjacent property lines and where not exceeding 4 feet (1219 mm) in height above the roof surface.

5. Dormers shall be of the same type of construction as the roof on which they are placed, or of the exterior walls of the building.

1509.3 Tanks. Tanks having a capacity of more than 500 gallons (2 m³) placed in or on a building shall be supported on masonry, reinforced concrete, steel or Type IV construction provided that, where such supports are located in the building above the lowest story, the support shall be fire-resistance rated as required for Type IA construction.

1509.3.1 Valve. Such tanks shall have in the bottom or on the side near the bottom, a pipe or outlet, fitted with a suitable quick opening valve for discharging the contents in an emergency through an adequate drain.

1509.3.2 Location. Such tanks shall not be placed over or near a line of stairs or an elevator shaft, unless there is a solid roof or floor underneath the tank.

1509.3.3 Tank cover. Unenclosed roof tanks shall have covers sloping toward the outer edges.

1509.4 Cooling towers. Cooling towers in excess of 250 square feet (23.2 m²) in base area or in excess of 15 feet (4572 mm) high where located on buildings more than 50 feet (15 240 mm) high shall be of noncombustible construction. Cooling towers shall not exceed one-third of the supporting roof area.

Exception: Drip boards and the enclosing construction of wood not less than 1 inch (25 mm) nominal thickness, provided the wood is covered on the exterior of the tower with noncombustible material.

1509.5 Towers, spires, domes and cupolas. Any tower, spire, dome or cupola shall be of a type of construction not less in fire-resistance rating than required for the building to which it is attached except that any such tower, spire, dome or cupola that exceeds 85 feet (25 908 mm) in height above grade, or exceeds 200 square feet (18.6 m²) in horizontal area or is used for any purpose other than a belfry or an architectural embellishment shall be constructed of and supported on Type I or II construction.

1509.5.1 Noncombustible construction required. Any tower, spire, dome or cupola that exceeds 60 feet (18 288) in height above the highest point at which it comes in contact with the roof, or that exceeds 200 square feet (18.6 m²) in area at any horizontal section, or which is intended to be used for any purpose other than a belfry or architectural embellishment, shall be entirely constructed of and supported by noncombustible materials. Such structures shall be separated from the building below by construction having a fire-resistance rating of not less than 1.5 hours with openings protected with a minimum 1.5-hour fire-protection rating. Structures, except aerial supports 12 feet (3658 mm) high or less, flagpoles, water tanks and cooling towers, placed above the roof of any building more than 50 feet (15 240 mm) in height, shall be of noncombustible material and shall be supported by construction of noncombustible material.

1509.5.2 Towers and spires. Towers and spires where enclosed shall have exterior walls as required for the building to which they are attached. The roof covering of spires shall be of a class of roof covering as required for the main roof of the rest of the structure.

SECTION 1510
REROOFING

1510.1 General. Materials and methods of application used for recovering or replacing an existing roof covering shall comply with the requirements of Chapter 15.

Exception: Reroofing shall not be required to meet the minimum design slope requirement of one-quarter unit vertical in 12 units horizontal (2-percent slope) in Section 1507 for roofs that provide positive roof drainage.

1510.2 Structural and construction loads. Structural roof components shall be capable of supporting the roof-covering system and the material and equipment loads that will be encountered during installation of the system.

1510.3 Recovering versus replacement. New roof coverings shall not be installed without first removing all existing layers of roof coverings where any of the following conditions occur:

1. Where the existing roof or roof covering is water soaked or has deteriorated to the point that the existing roof or roof covering is not adequate as a base for additional roofing.
2. Where the existing roof covering is wood shake, slate, clay, cement or asbestos-cement tile.
3. Where the existing roof has two or more applications of any type of roof covering.

Exceptions:

1. Complete and separate roofing systems, such as standing-seam metal roof systems, that are designed to transmit the roof loads directly to the building's structural system and that do not rely on existing roofs and roof coverings for support, shall not require the removal of existing roof coverings.
2. Metal panel, metal shingle, and concrete and clay tile roof coverings shall be permitted to be installed over existing wood shake roofs when applied in accordance with Section 1510.4.

1510.4 Roof recovering. Where the application of a new roof covering over wood shingle or shake roofs creates a combustible concealed space, the entire existing surface shall be covered with gypsum board, mineral fiber, glass fiber or other approved materials securely fastened in place.

1510.5 Reinstallation of materials. Existing slate, clay or cement tile shall be permitted for reinstallation, except that damaged, cracked or broken slate or tile shall not be reinstalled. Existing vent flashing, metal edgings, drain outlets, collars and metal counterflashings shall not be reinstalled where rusted, damaged or deteriorated. Aggregate surfacing materials shall not be reinstalled.

1510.6 Flashings. Flashings shall be reconstructed in accordance with approved manufacturer's installation instructions. Metal flashing to which bituminous materials are to be adhered shall be primed prior to installation.

CHAPTER 16

STRUCTURAL DESIGN

SECTION 1601
GENERAL

1601.1 Scope. The provisions of this chapter shall govern the structural design of buildings, structures and portions thereof regulated by this code.

SECTION 1602
DEFINITIONS

1602.1 Definitions. The following words and terms shall, for the purposes of this chapter, have the meanings shown herein.

ALLOWABLE STRESS DESIGN. A method of proportioning structural members, such that elastically computed stresses produced in the members by nominal loads do not exceed specified allowable stresses (also called "working stress design").

BALCONY, EXTERIOR. An exterior floor projecting from and supported by a structure without additional independent supports.

BASE SHEAR. Total design lateral force or shear at the base.

BASIC SEISMIC-FORCE-RESISTING SYSTEMS.

Bearing wall system. A structural system without a complete vertical load-carrying space frame. Bearing walls or bracing elements provide support for substantial vertical loads. Seismic lateral force resistance is provided by shear walls or braced frames.

Building frame system. A structural system with an essentially complete space frame providing support for vertical loads. Seismic lateral force resistance is provided by shear walls or braced frames.

Dual system. A structural system with an essentially complete space frame providing support for vertical loads. Seismic lateral force resistance is provided by a moment frame and shear walls or braced frames.

Inverted pendulum system. A structure with a large portion of its mass concentrated at the top; therefore, having essentially one degree of freedom in horizontal translation. Seismic lateral force resistance is provided by the columns acting as cantilevers.

Moment-resisting frame system. A structural system with an essentially complete space frame providing support for vertical loads. Seismic lateral force resistance is provided by moment frames.

Shear wall-frame interactive system. A structural system which uses combinations of shear walls and frames designed to resist seismic lateral forces in proportion to their rigidities, considering interaction between shear walls and frames on all levels. Support of vertical loads is provided by the same shear walls and frames.

BOUNDARY MEMBERS. Strengthened portions along shear wall and diaphragm edges (also called "boundary elements").

Boundary element. In light-frame construction, diaphragms and shear wall boundary members to which sheathing transfers forces. Boundary elements include chords and drag struts at diaphragm and shear wall perimeters, interior openings, discontinuities and reentrant corners.

CANTILEVERED COLUMN SYSTEM. A structural system relying on column elements that cantilever from a fixed base and have minimal rotational resistance capacity at the top with lateral forces applied essentially at the top and are used for lateral resistance.

COLLECTOR ELEMENTS. Members that serve to transfer forces between floor diaphragms and members of the lateral-force-resisting system.

CONFINED REGION. The portion of a reinforced concrete component in which the concrete is confined by closely spaced special transverse reinforcement restraining the concrete in directions perpendicular to the applied stress.

DEAD LOADS. The weight of materials of construction incorporated into the building, including but not limited to walls, floors, roofs, ceilings, stairways, built-in partitions, finishes, cladding and other similarly incorporated architectural and structural items, and fixed service equipment, including the weight of cranes. All dead loads are considered permanent loads.

DECK. An exterior floor supported on at least two opposing sides by an adjacent structure, and/or posts, piers or other independent supports.

DEFORMABILITY. The ratio of the ultimate deformation to the limit deformation.

High deformability element. An element whose deformability is not less than 3.5 when subjected to four fully reversed cycles at the limit deformation.

Limited deformability element. An element that is neither a low deformability or a high deformability element.

Low deformability element. An element whose deformability is 1.5 or less.

DEFORMATION.

Limit deformation. Two times the initial deformation that occurs at a load equal to 40 percent of the maximum strength.

Ultimate deformation. The deformation at which failure occurs and which shall be deemed to occur if the sustainable load reduces to 80 percent or less of the maximum strength.

DESIGN STRENGTH. The product of the nominal strength and a resistance factor (or strength reduction factor).

DIAPHRAGM. A horizontal or sloped system acting to transmit lateral forces to the vertical-resisting elements. When the term "diaphragm" is used, it shall include horizontal bracing systems.

> **Diaphragm, blocked.** In light-frame construction, a diaphragm in which all sheathing edges not occurring on a framing member are supported on and fastened to blocking.

> **Diaphragm boundary.** In light-frame construction, a location where shear is transferred into or out of the diaphragm sheathing. Transfer is either to a boundary element or to another force-resisting element.

> **Diaphragm chord.** A diaphragm boundary element perpendicular to the applied load that is assumed to take axial stresses due to the diaphragm moment.

> **Diaphragm, flexible.** A diaphragm is flexible for the purpose of distribution of story shear and torsional moment when the computed maximum in-plane deflection of the diaphragm itself under lateral load is more than two times the average drift of adjoining vertical elements of the lateral-force-resisting system of the associated story under equivalent tributary lateral load (see Section 1617.5.3).

> **Diaphragm, rigid.** A diaphragm is rigid for the purpose of distribution of story shear and torsional moment when the lateral deformation of the diaphragm is less than or equal to two times the average story drift.

DURATION OF LOAD. The period of continuous application of a given load, or the aggregate of periods of intermittent applications of the same load.

ELEMENT.

> **Ductile element.** An element capable of sustaining large cyclic deformations beyond the attainment of its nominal strength without any significant loss of strength.

> **Limited ductile element.** An element that is capable of sustaining moderate cyclic deformations beyond the attainment of nominal strength without significant loss of strength.

> **Nonductile element.** An element having a mode of failure that results in an abrupt loss of resistance when the element is deformed beyond the deformation corresponding to the development of its nominal strength. Nonductile elements cannot reliably sustain significant deformation beyond that attained at their nominal strength.

EQUIPMENT SUPPORT. Those structural members or assemblies of members or manufactured elements, including braces, frames, lugs, snuggers, hangers or saddles, that transmit gravity load and operating load between the equipment and the structure.

ESSENTIAL FACILITIES. Buildings and other structures that are intended to remain operational in the event of extreme environmental loading from flood, wind, snow or earthquakes.

FACTORED LOAD. The product of a nominal load and a load factor.

FLEXIBLE EQUIPMENT CONNECTIONS. Those connections between equipment components that permit rotational

and/or translational movement without degradation of performance.

FRAME.

> **Braced frame.** An essentially vertical truss, or its equivalent, of the concentric or eccentric type that is provided in a building frame system or dual system to resist lateral forces.

> **Concentrically braced frame (CBF).** A braced frame in which the members are subjected primarily to axial forces.

> **Eccentrically braced frame (EBF).** A diagonally braced frame in which at least one end of each brace frames into a beam a short distance from a beam-column or from another diagonal brace.

> **Ordinary concentrically braced frame (OCBF).** A steel concentrically braced frame in which members and connections are designed in accordance with the provisions of AISC Seismic without modification.

> **Special concentrically braced frame (SCBF).** A steel or composite steel and concrete concentrically braced frame in which members and connections are designed for ductile behavior.

> **Moment frame.** A frame in which members and joints resist lateral forces by flexure as well as along the axis of the members. Moment frames are categorized as "intermediate moment frames" (IMF), "ordinary moment frames" (OMF), and "special moment frames" (SMF).

GUARD. See Section 1002.1.

IMPACT LOAD. The load resulting from moving machinery, elevators, craneways, vehicles and other similar forces and kinetic loads, pressure and possible surcharge from fixed or moving loads.

JOINT. A portion of a column bounded by the highest and lowest surfaces of the other members framing into it.

LIMIT STATE. A condition beyond which a structure or member becomes unfit for service and is judged to be no longer useful for its intended function (serviceability limit state) or to be unsafe (strength limit state).

LIVE LOADS. Those loads produced by the use and occupancy of the building or other structure and do not include construction or environmental loads such as wind load, snow load, rain load, earthquake load, flood load or dead load.

LIVE LOADS (ROOF). Those loads produced (1) during maintenance by workers, equipment and materials; and (2) during the life of the structure by movable objects such as planters and by people.

LOAD AND RESISTANCE FACTOR DESIGN (LRFD). A method of proportioning structural members and their connections using load and resistance factors such that no applicable limit state is reached when the structure is subjected to appropriate load combinations. The term "LRFD" is used in the design of steel and wood structures.

LOAD FACTOR. A factor that accounts for deviations of the actual load from the nominal load, for uncertainties in the analysis that transforms the load into a load effect, and for the probability that more than one extreme load will occur simultaneously.

LOADS. Forces or other actions that result from the weight of building materials, occupants and their possessions, environmental effects, differential movement and restrained dimensional changes. Permanent loads are those loads in which variations over time are rare or of small magnitude, such as dead loads. All other loads are variable loads (see also "Nominal loads").

LOADS EFFECTS. Forces and deformations produced in structural members by the applied loads.

NOMINAL LOADS. The magnitudes of the loads specified in this chapter (dead, live, soil, wind, snow, rain, flood and earthquake).

NOTATIONS.

D = Dead load.

E = Combined effect of horizontal and vertical earthquake-induced forces as defined in Sections 1616.4.1 and 1617.1.

E_m = Maximum seismic load effect of horizontal and vertical seismic forces as set forth in Sections 1616.4.1 and 1617.1.

F = Load due to fluids.

F_a = Flood load.

H = Load due to lateral pressure of soil and water in soil.

L = Live load, except roof live load, including any permitted live load reduction.

L_r = Roof live load including any permitted live load reduction.

P = Ponding load.

R = Rain load.

S = Snow load.

T = Self-straining force arising from contraction or expansion resulting from temperature change, shrinkage, moisture change, creep in component materials, movement due to differential settlement or combinations thereof.

W = Load due to wind pressure.

OTHER STRUCTURES. Structures, other than buildings, for which loads are specified in this chapter.

P-DELTA EFFECT. The second order effect on shears, axial forces and moments of frame members induced by axial loads on a laterally displaced building frame.

PANEL (PART OF A STRUCTURE). The section of a floor, wall or roof comprised between the supporting frame of two adjacent rows of columns and girders or column bands of floor or roof construction.

RESISTANCE FACTOR. A factor that accounts for deviations of the actual strength from the nominal strength and the manner and consequences of failure (also called "strength reduction factor").

SHALLOW ANCHORS. Shallow anchors are those with embedment length-to-diameter ratios of less than eight.

SHEAR PANEL. A floor, roof or wall component sheathed to act as a shear wall or diaphragm.

SHEAR WALL. A wall designed to resist lateral forces parallel to the plane of the wall.

SPACE FRAME. A structure composed of interconnected members, other than bearing walls, that is capable of supporting vertical loads and that also may provide resistance to seismic lateral forces.

SPECIAL TRANSVERSE REINFORCEMENT. Reinforcement composed of spirals, closed stirrups or hoops and supplementary cross ties provided to restrain the concrete and qualify the portion of the component, where used, as a confined region.

STRENGTH, NOMINAL. The capacity of a structure or member to resist the effects of loads, as determined by computations using specified material strengths and dimensions and equations derived from accepted principles of structural mechanics or by field tests or laboratory tests of scaled models, allowing for modeling effects and differences between laboratory and field conditions.

STRENGTH, REQUIRED. Strength of a member, cross section or connection required to resist factored loads or related internal moments and forces in such combinations as stipulated by these provisions.

STRENGTH DESIGN. A method of proportioning structural members such that the computed forces produced in the members by factored loads do not exceed the member design strength [also called "load and resistance factor design" (LRFD)]. The term "strength design" is used in the design of concrete and masonry structural elements.

WALL, LOAD BEARING. Any wall meeting either of the following classifications:

1. Any metal or wood stud wall that supports more than 100 pounds per linear foot (plf) (1459 N/m) of vertical load in addition to its own weight.

2. Any masonry or concrete wall that supports more than 200 plf (2919 N/m) of vertical load in addition to its own weight.

WALL, NONLOAD BEARING. Any wall that is not a load-bearing wall.

SECTION 1603
CONSTRUCTION DOCUMENTS

1603.1 General. Construction documents shall show the size, section and relative locations of structural members with floor levels, column centers and offsets fully dimensioned. The design loads and other information pertinent to the structural design required by Sections 1603.1.1 through 1603.1.8 shall be clearly indicated on the construction documents for parts of the building or structure.

Exception: Construction documents for buildings constructed in accordance with the conventional light-frame construction provisions of Section 2308 shall indicate the following structural design information:

1. Floor and roof live loads.

STRUCTURAL DESIGN

2. Ground snow load, P_g.

3. Basic wind speed (3-second gust), miles per hour (mph) (km/hr) and wind exposure.

4. Seismic design category and site class.

1603.1.1 Floor live load. The uniformly distributed, concentrated and impact floor live load used in the design shall be indicated for floor areas. Live load reduction of the uniformly distributed floor live loads, if used in the design, shall be indicated.

1603.1.2 Roof live load. The roof live load used in the design shall be indicated for roof areas (Section 1607.11).

1603.1.3 Roof snow load. The ground snow load, P_g, shall be indicated. In areas where the ground snow load, P_g, exceeds 10 pounds per square foot (psf) (0.479 kN/m^2), the following additional information shall also be provided, regardless of whether snow loads govern the design of the roof:

1. Flat-roof snow load, P_f.

2. Snow exposure factor, C_e.

3. Snow load importance factor, I_s.

4. Thermal factor, C_t.

1603.1.4 Wind design data. The following information related to wind loads shall be shown, regardless of whether wind loads govern the design of the lateral-force-resisting system of the building:

1. Basic wind speed (3-second gust), miles per hour (km/hr).

2. Wind importance factor, I_w, and building category.

3. Wind exposure, if more than one wind exposure is utilized, the wind exposure and applicable wind direction shall be indicated.

4. The applicable internal pressure coefficient.

5. Components and cladding. The design wind pressures in terms of psf (kN/m^2) to be used for the design of exterior component and cladding materials not specifically designed by the registered design professional.

1603.1.5 Earthquake design data. The following information related to seismic loads shall be shown, regardless of whether seismic loads govern the design of the lateral-force-resisting system of the building:

1. Seismic importance factor, I_E, and seismic use group.

2. Mapped spectral response accelerations S_S and S_1.

3. Site class.

4. Spectral response coefficients S_{DS} and S_{D1}.

5. Seismic design category.

6. Basic seismic-force-resisting system(s).

7. Design base shear.

8. Seismic response coefficient(s), C_S.

9. Response modification factor(s), R.

10. Analysis procedure used.

1603.1.6 Flood load. For buildings located in flood hazard areas as established in Section 1612.3, the following information, referenced to the datum on the community's Flood Insurance Rate Map (FIRM), shall be shown, regardless of whether flood loads govern the design of the building:

1. In flood hazard areas not subject to high-velocity wave action, the elevation of proposed lowest floor, including basement.

2. In flood hazard areas not subject to high-velocity wave action, the elevation to which any nonresidential building will be dry floodproofed.

3. In flood hazard areas subject to high-velocity wave action, the proposed elevation of the bottom of the lowest horizontal structural member of the lowest floor, including basement.

1603.1.7 Special loads. Special loads that are applicable to the design of the building, structure or portions thereof shall be indicated along with the specified section of this code that addresses the special loading condition.

1603.1.8 System and components requiring special inspections for seismic resistance. Construction documents or specifications shall be prepared for those systems and components requiring special inspection for seismic resistance as specified in Section 1707.1 by the registered design professional responsible for their design and shall be submitted for approval in accordance with Section 106.1. Reference to seismic standards in lieu of detailed drawings is acceptable.

1603.2 Restrictions on loading. It shall be unlawful to place, or cause or permit to be placed, on any floor or roof of a building, structure or portion thereof, a load greater than is permitted by these requirements.

1603.3 Live loads posted. Where the live loads for which each floor or portion thereof of a commercial or industrial building is or has been designed to exceed 50 psf (2.40 kN/m^2), such design live loads shall be conspicuously posted by the owner in that part of each story in which they apply, using durable signs. It shall be unlawful to remove or deface such notices.

1603.4 Occupancy permits for changed loads. Construction documents for other than residential buildings filed with the building official with applications for permits shall show on each drawing the live loads per square foot (m^2) of area covered for which the building is designed. Occupancy permits for buildings hereafter erected shall not be issued until the floor load signs, required by Section 1603.3, have been installed.

SECTION 1604
GENERAL DESIGN REQUIREMENTS

1604.1 General. Building, structures and parts thereof shall be designed and constructed in accordance with strength design, load and resistance factor design, allowable stress design, empirical design or conventional construction methods, as permitted by the applicable material chapters.

1604.2 Strength. Buildings and other structures, and parts thereof, shall be designed and constructed to support safely the factored loads in load combinations defined in this code without exceeding the appropriate strength limit states for the materials of construction. Alternatively, buildings and other structures, and parts thereof, shall be designed and constructed to support safely the nominal loads in load combinations defined in this code without exceeding the appropriate specified allowable stresses for the materials of construction.

Loads and forces for occupancies or uses not covered in this chapter shall be subject to the approval of the building official.

1604.3 Serviceability. Structural systems and members thereof shall be designed to have adequate stiffness to limit deflections and lateral drift. See Section 1617.3 for drift limits applicable to earthquake loading.

1604.3.1 Deflections. The deflections of structural members shall not exceed the more restrictive of the limitations of Sections 1604.3.2 through 1604.3.5 or that permitted by Table 1604.3.

1604.3.2 Reinforced concrete. The deflection of reinforced concrete structural members shall not exceed that permitted by ACI 318.

1604.3.3 Steel. The deflection of steel structural members shall not exceed that permitted by AISC LRFD, AISC HSS, AISC 335, AISI -NASPEC, AISI-General, AISI-Truss, ASCE 3, ASCE 8-SSD-LRFD/ASD, and the standard specifications of SJI Standard Specifications, Load Tables and Weight Tables for Steel Joists and Joist Girders as applicable.

1604.3.4 Masonry. The deflection of masonry structural members shall not exceed that permitted by ACI 530/ASCE 5/TMS 402.

1604.3.5 Aluminum. The deflection of aluminum structural members shall not exceed that permitted by AA-94.

1604.3.6 Limits. Deflection of structural members over span, l, shall not exceed that permitted by Table 1604.3.

1604.4 Analysis. Load effects on structural members and their connections shall be determined by methods of structural analysis that take into account equilibrium, general stability, geometric compatibility and both short- and long-term material properties.

Members that tend to accumulate residual deformations under repeated service loads shall have included in their analysis the added eccentricities expected to occur during their service life.

Any system or method of construction to be used shall be based on a rational analysis in accordance with well-established principles of mechanics. Such analysis shall result in a system that provides a complete load path capable of transferring loads from their point of origin to the load-resisting elements.

The total lateral force shall be distributed to the various vertical elements of the lateral-force-resisting system in proportion to their rigidities considering the rigidity of the horizontal bracing system or diaphragm. Rigid elements that are assumed not to be a part of the lateral-force-resisting system shall be permitted to be incorporated into buildings provided that their effect

on the action of the system is considered and provided for in design. Provisions shall be made for the increased forces induced on resisting elements of the structural system resulting from torsion due to eccentricity between the center of application of the lateral forces and the center of rigidity of the lateral-force-resisting system.

Every structure shall be designed to resist the overturning effects caused by the lateral forces specified in this chapter. See Section 1609 for wind, Section 1610 for lateral soil loads and Sections 1613 through 1623 for earthquake.

TABLE 1604.3
DEFLECTION LIMITS[a, b, c, h, i]

CONSTRUCTION	L	S or W[f]	$D + L$[d,g]
Roof members:[e]			
Supporting plaster ceiling	$l/360$	$l/360$	$l/240$
Supporting nonplaster ceiling	$l/240$	$l/240$	$l/180$
Not supporting ceiling	$l/180$	$l/180$	$l/120$
Floor members	$l/360$	—	$l/240$
Exterior walls and interior partitions:			
With brittle finishes	—	$l/240$	—
With flexible finishes	—	$l/120$	—
Farm buildings	—	—	$l/180$
Greenhouses	—	—	$l/120$

For SI: 1 foot = 304.8 mm.

a. For structural roofing and siding made of formed metal sheets, the total load deflection shall not exceed $l/60$. For secondary roof structural members supporting formed metal roofing, the live load deflection shall not exceed $l/150$. For secondary wall members supporting formed metal siding, the design wind load deflection shall not exceed $l/90$. For roofs, this exception only applies when the metal sheets have no roof covering.

b. Interior partitions not exceeding 6 feet in height and flexible, folding and portable partitions are not governed by the provisions of this section. The deflection criterion for interior partitions is based on the horizontal load defined in Section 1607.13.

c. See Section 2403 for glass supports.

d. For wood structural members having a moisture content of less than 16 percent at time of installation and used under dry conditions, the deflection resulting from $L + 0.5D$ is permitted to be substituted for the deflection resulting from $L + D$.

e. The above deflections do not ensure against ponding. Roofs that do not have sufficient slope or camber to assure adequate drainage shall be investigated for ponding. See Section 1611 for rain and ponding requirements and Section 1503.4 for roof drainage requirements.

f. The wind load is permitted to be taken as 0.7 times the "component and cladding" loads for the purpose of determining deflection limits herein.

g. For steel structural members, the dead load shall be taken as zero.

h. For aluminum structural members or aluminum panels used in roofs or walls of sunroom additions or patio covers, not supporting edge of glass or aluminum sandwich panels, the total load deflection shall not exceed $l/60$. For aluminum sandwich panels used in roofs or walls of sunroom additions or patio covers, the total load deflection shall not exceed $l/120$.

i. For cantilever members, l shall be taken as twice the length of the cantilever.

1604.5 Importance factors. The value for snow load, wind load and seismic load importance factors shall be determined in accordance with Table 1604.5.

1604.6 In-situ load tests. The building official is authorized to require an engineering analysis or a load test, or both, of any construction whenever there is reason to question the safety of the construction for the intended occupancy. Engineering analysis and load tests shall be conducted in accordance with Section 1713.

TABLE 1604.5
CLASSIFICATION OF BUILDINGS AND OTHER STRUCTURES FOR IMPORTANCE FACTORS

CATEGORY[a]	NATURE OF OCCUPANCY	SEISMIC FACTOR I_E	SNOW FACTOR I_S	WIND FACTOR I_W
I	Buildings and other structures that represent a low hazard to human life in the event of failure including, but not limited to: • Agricultural facilities • Certain temporary facilities • Minor storage facilities	1.00	0.8	0.87[b]
II	Buildings and other structures except those listed in Categories I, III and IV	1.00	1.0	1.00
III	Buildings and other structures that represent a substantial hazard to human life in the event of failure including, but not limited to: • Buildings and other structures where more than 300 people congregate in one area • Buildings and other structures with elementary school, secondary school or day care facilities with an occupant load greater than 250 • Buildings and other structures with an occupant load greater than 500 for colleges or adult education facilities • Health care facilities with an occupant load of 50 or more resident patients but not having surgery or emergency treatment facilities • Jails and detention facilities • Any other occupancy with an occupant load greater than 5,000 • Power-generating stations, water treatment for potable water, waste water treatment facilities and other public utility facilities not included in Category IV • Buildings and other structures not included in Category IV containing sufficient quantities of toxic or explosive substances to be dangerous to the public if released	1.25	1.1	1.15
IV	Buildings and other structures designed as essential facilities including, but not limited to: • Hospitals and other health care facilities having surgery or emergency treatment facilities • Fire, rescue and police stations and emergency vehicle garages • Designated earthquake, hurricane or other emergency shelters • Designated emergency preparedness, communication, and operation centers and other facilities required for emergency response • Power-generating stations and other public utility facilities required as emergency backup facilities for Category IV structures • Structures containing highly toxic materials as defined by Section 307 where the quantity of the material exceeds the maximum allowable quantities of Table 307.7(2) • Aviation control towers, air traffic control centers and emergency aircraft hangars • Buildings and other structures having critical national defense functions • Water treatment facilities required to maintain water pressure for fire suppression	1.50	1.2	1.15

a. For the purpose of Section 1616.2, Categories I and II are considered Seismic Use Group I, Category III is considered Seismic Use Group II and Category IV is equivalent to Seismic Use Group III.

b. In hurricane-prone regions with $V > 100$ miles per hour, I_w shall be 0.77.

1604.7 Preconstruction load tests. Materials and methods of construction that are not capable of being designed by approved engineering analysis or that do not comply with the applicable material design standards listed in Chapter 35, or alternative test procedures in accordance with Section 1711, shall be load tested in accordance with Section 1714.

1604.8 Anchorage.

1604.8.1 General. Anchorage of the roof to walls and columns, and of walls and columns to foundations, shall be provided to resist the uplift and sliding forces that result from the application of the prescribed loads.

1604.8.2 Concrete and masonry walls. Concrete and masonry walls shall be anchored to floors, roofs and other structural elements that provide lateral support for the wall. Such anchorage shall provide a positive direct connection capable of resisting the horizontal forces specified in this chapter but not less than a minimum strength design horizontal force of 280 plf (4.10 kN/m) of wall, substituted for "E" in the load combinations of Section 1605.2 or 1605.3. Walls shall be designed to resist bending between anchors where the anchor spacing exceeds 4 feet (1219 mm). Required anchors in masonry walls of hollow units or cavity walls shall be embedded in a reinforced grouted structural element of the wall. See Sections 1609.6.2.2 and 1620 for wind and earthquake design requirements.

1604.8.3 Decks. Where supported by attachment to an exterior wall, decks shall be positively anchored to the primary structure and designed for both vertical and lateral loads as applicable. Such attachment shall not be accomplished by the use of toenails or nails subject to withdrawal. Where positive connection to the primary building structure cannot be verified during inspection, decks shall be self-supporting. For decks with cantilevered framing members, connections to exterior walls or other framing members shall be designed and constructed to resist uplift resulting from the full live load specified in Table 1607.1 acting on the cantilevered portion of the deck.

SECTION 1605
LOAD COMBINATIONS

1605.1 General. Buildings and other structures and portions thereof shall be designed to resist the load combinations specified in Section 1605.2 or 1605.3 and Chapters 18 through 23, and the special seismic load combinations of Section 1605.4 where required by Section 1620.2.6, 1620.2.9 or 1620.4.4 or Section 9.5.2.6.2.11 or 9.5.2.6.3.1 of ASCE 7. Applicable loads shall be considered, including both earthquake and wind, in accordance with the specified load combinations. Each load combination shall also be investigated with one or more of the variable loads set to zero.

1605.2 Load combinations using strength design or load and resistance factor design.

1605.2.1 Basic load combinations. Where strength design or load and resistance factor design is used, structures and portions thereof shall resist the most critical effects from the following combinations of factored loads:

$$1.4D \qquad \text{(Equation 16-1)}$$

$$1.2D + 1.6L + 0.5(L_r \text{ or } S \text{ or } R) \qquad \text{(Equation 16-2)}$$

$$1.2D + 1.6(L_r \text{ or } S \text{ or } R) + (f_1 L \text{ or } 0.8W) \qquad \text{(Equation 16-3)}$$

$$1.2D + 1.6W + f_1 L + 0.5(L_r \text{ or } S \text{ or } R) \qquad \text{(Equation 16-4)}$$

$$1.2D + 1.0E + f_1 L + f_2 S \qquad \text{(Equation 16-5)}$$

$$0.9D + (1.0E \text{ or } 1.6W) \qquad \text{(Equation 16-6)}$$

where:

f_1 = 1.0 for floors in places of public assembly, for live loads in excess of 100 pounds per square foot (4.79 kN/m²), and for parking garage live load.

f_1 = 0.5 for other live loads.

f_2 = 0.7 for roof configurations (such as saw tooth) that do not shed snow off the structure.

f_2 = 0.2 for other roof configurations.

Exception: Where other factored load combinations are specifically required by the provisions of this code, such combinations shall take precedence.

1605.2.2 Other loads. Where F, H, P or T is to be considered in design, each applicable load shall be added to the above combinations in accordance with Section 2.3.2 of ASCE 7. Where F_a is to be considered in design, the load combinations of Section 2.3.3 of ASCE 7 shall be used.

1605.3 Load combinations using allowable stress design.

1605.3.1 Basic load combinations. Where allowable stress design (working stress design), as permitted by this code, is used, structures and portions thereof shall resist the most critical effects resulting from the following combinations of loads:

$$D \qquad \text{(Equation 16-7)}$$

$$D + L \qquad \text{(Equation 16-8)}$$

$$D + L + (L_r \text{ or } S \text{ or } R) \qquad \text{(Equation 16-9)}$$

$$D + (W \text{ or } 0.7E) + L + (L_r \text{ or } S \text{ or } R) \qquad \text{(Equation 16-10)}$$

$$0.6D + W \qquad \text{(Equation 16-11)}$$

$$0.6D + 0.7E \qquad \text{(Equation 16-12)}$$

Exceptions:

1. Crane hook loads need not be combined with roof live load or with more than three-fourths of the snow load or one-half of the wind load.

2. Flat roof snow loads of 30 psf (1.44 kN/m²) or less need not be combined with seismic loads. Where flat roof snow loads exceed 30 psf (1.44 kN/m²), 20 percent shall be combined with seismic loads.

1605.3.1.1 Load reduction. It is permitted to multiply the combined effect of two or more variable loads by 0.75 and add to the effect of dead load. The combined load used in design shall not be less than the sum of the effects of dead load and any one of the variable loads. The 0.7 factor on E does not apply for this provision.

Increases in allowable stresses specified in the appropriate materials section of this code or referenced standard shall not be used with the load combinations of Section 1605.3.1 except that a duration of load increase shall be permitted in accordance with Chapter 23.

1605.3.1.2 Other loads. Where F, H, P or T are to be considered in design, the load combinations of Section 2.4.1 of ASCE 7 shall be used. Where F_a is to be considered in design, the load combinations of Section 2.4.2 of ASCE 7 shall be used.

1605.3.2 Alternative basic load combinations. In lieu of the basic load combinations specified in Section 1605.3.1, structures and portions thereof shall be permitted to be designed for the most critical effects resulting from the following combinations. When using these alternate basic load combinations that include wind or seismic loads, allowable stresses are permitted to be increased or load combinations reduced, where permitted by the material section of this code or referenced standard. Where wind loads are calculated in accordance with Section 1609.6 or Section 6 of ASCE 7, the coefficient ω in the following equations shall be taken as 1.3. For other wind loads ω shall be taken as 1.0.

$$D + L + (L_r \text{ or } S \text{ or } R) \qquad \textbf{(Equation 16-13)}$$

$$D + L + (\omega W) \qquad \textbf{(Equation 16-14)}$$

$$D + L + \omega W + S/2 \qquad \textbf{(Equation 16-15)}$$

$$D + L + S + \omega W/2 \qquad \textbf{(Equation 16-16)}$$

$$D + L + S + E/1.4 \qquad \textbf{(Equation 16-17)}$$

$$0.9D + E/1.4 \qquad \textbf{(Equation 16-18)}$$

Exceptions:

1. Crane hook loads need not be combined with roof live load or with more than three-fourths of the snow load or one-half of the wind load.

2. Flat roof snow loads of 30 pounds per square foot (1.44 kN/m²) or less need not be combined with seismic loads. Where flat roof snow loads exceed 30 psf (1.44 kN/m²), 20 percent shall be combined with seismic loads.

1605.3.2.1 Other loads. Where F, H, P or T are to be considered in design, 1.0 times each applicable load shall be added to the combinations specified in Section 1605.3.2.

1605.4 Special seismic load combinations. For both allowable stress design and strength design methods, where specifically required by Sections 1613 through 1622 or by Chapters 18 through 23, elements and components shall be designed to resist the forces calculated using Equation 16-19 when the effects of the seismic ground motion are additive to gravity forces and those calculated using Equation 16-20 when the effects of the seismic ground motion counteract gravity forces.

$$1.2D + f_1 L + E_m \qquad \textbf{(Equation 16-19)}$$

$$0.9D + E_m \qquad \textbf{(Equation 16-20)}$$

where:

E_m = The maximum effect of horizontal and vertical forces as set forth in Section 1617.1.

f_1 = 1.0 for floors in places of public assembly, for live loads in excess of 100 psf (4.79 kN/m²) and for parking garage live load.

f_1 = 0.5 for other live loads.

1605.5 Heliports and helistops. Heliport and helistop landing or touchdown areas shall be designed for the following loads, combined in accordance with Section 1605:

1. Dead load, D, plus the gross weight of the helicopter, D_h, plus snow load, S.

2. Dead load, D, plus two single concentrated impact loads, L, approximately 8 feet (2438 mm) apart applied anywhere on the touchdown pad (representing each of the helicopter's two main landing gear, whether skid type or wheeled type), having a magnitude of 0.75 times the gross weight of the helicopter. Both loads acting together total 1.5 times the gross weight of the helicopter.

3. Dead load, D, plus a uniform live load, L, of 100 psf (4.79 kN/m²).

SECTION 1606
DEAD LOADS

1606.1 Weights of materials and construction. In determining dead loads for purposes of design, the actual weights of materials and construction shall be used. In the absence of definite information, values used shall be subject to the approval of the building official.

1606.2 Weights of fixed service equipment. In determining dead loads for purposes of design, the weight of fixed service equipment, such as plumbing stacks and risers, electrical feeders, heating, ventilating and air-conditioning systems (HVAC) and fire sprinkler systems, shall be included.

SECTION 1607
LIVE LOADS

1607.1 General. Live loads are those loads defined in Section 1602.1.

1607.2 Loads not specified. For occupancies or uses not designated in Table 1607.1, the live load shall be determined in accordance with a method approved by the building official.

1607.3 Uniform live loads. The live loads used in the design of buildings and other structures shall be the maximum loads expected by the intended use or occupancy but shall in no case be less than the minimum uniformly distributed unit loads required by Table 1607.1.

1607.4 Concentrated loads. Floors and other similar surfaces shall be designed to support the uniformly distributed live loads prescribed in Section 1607.3 or the concentrated load, in pounds (kilonewtons), given in Table 1607.1, whichever produces the greater load effects. Unless otherwise specified, the indicated concentration shall be assumed to be uniformly distributed over an area 2.5 feet by 2.5 feet [6.25 ft² (0.58 m²)] and shall be located so as to produce the maximum load effects in the structural members.

TABLE 1607.1
MINIMUM UNIFORMLY DISTRIBUTED LIVE LOADS AND MINIMUM CONCENTRATED LIVE LOADS[g]

OCCUPANCY OR USE	UNIFORM (psf)	CONCENTRATED (lbs.)
1. Apartments (see residential)	—	—
2. Access floor systems		
Office use	50	2,000
Computer use	100	2,000
3. Armories and drill rooms	150	—
4. Assembly areas and theaters		
Fixed seats (fastened to floor)	60	
Lobbies	100	
Movable seats	100	
Stages and platforms	125	—
Follow spot, projections and control rooms	50	
Catwalks	40	
5. Balconies (exterior)	100	
On one- and two-family residences only, and not exceeding 100 ft.[2]	60	
6. Decks	Same as occupancy served[h]	—
7. Bowling alleys	75	—
8. Cornices	60	—
9. Corridors, except as otherwise indicated	100	—
10. Dance halls and ballrooms	100	—
11. Dining rooms and restaurants	100	—
12. Dwellings (see residential)	—	—
13. Elevator machine room grating (on area of 4 in.[2])	—	300
14. Finish light floor plate construction (on area of 1 in.[2])	—	200
15. Fire escapes	100	
On single-family dwellings only	40	—
16. Garages (passenger vehicles only)	40	Note a
Trucks and buses	See Section 1607.6	
17. Grandstands (see stadium and arena bleachers)	—	—
18. Gymnasiums, main floors and balconies	100	—
19. Handrails, guards and grab bars	See Section 1607.7	
20. Hospitals		
Operating rooms, laboratories	60	1,000
Private rooms	40	1,000
Wards	40	1,000
Corridors above first floor	80	1,000
21. Hotels (see residential)	—	—
22. Libraries		
Reading rooms	60	1,000
Stack rooms	150[b]	1,000
Corridors above first floor	80	1,000
23. Manufacturing		
Light	125	2,000
Heavy	250	3,000
24. Marquees	75	—
25. Office buildings		
File and computer rooms shall be designed for heavier loads based on anticipated occupancy		
Lobbies and first-floor corridors	100	2,000
Offices	50	2,000
Corridors above first floor	80	2,000
26. Penal institutions		
Cell blocks	40	—
Corridors	100	
27. Residential		
One- and two-family dwellings		
Uninhabitable attics without storage	10	
Uninhabitable attics with storage	20	
Habitable attics and sleeping areas	30	
All other areas except balconies and decks	40	—
Hotels and multifamily dwellings		
Private rooms and corridors serving them	40	
Public rooms and corridors serving them	100	
28. Reviewing stands, grandstands and bleachers	Note c	—
29. Roofs	See Section 1607.11	
30. Schools		
Classrooms	40	1,000
Corridors above first floor	80	1,000
First-floor corridors	100	1,000
31. Scuttles, skylight ribs and accessible ceilings	—	200
32. Sidewalks, vehicular driveways and yards, subject to trucking	250[d]	8,000[e]
33. Skating rinks	100	—
34. Stadiums and arenas		
Bleachers	100[c]	—
Fixed seats (fastened to floor)	60[c]	
35. Stairs and exits	100	Note f
One- and two-family dwellings	40	
All other	100	
36. Storage warehouses (shall be designed for heavier loads if required for anticipated storage)		—
Light	125	
Heavy	250	
37. Stores		
Retail		
First floor	100	1,000
Upper floors	75	1,000
Wholesale, all floors	125	1,000
38. Vehicle barriers	See Section 1607.7	
39. Walkways and elevated platforms (other than exitways)	60	—
40. Yards and terraces, pedestrians	100	—

(continued)

Notes to Table 1607.1

For SI: 1 inch = 25.4 mm, 1 square inch = 645.16 mm^2, 1 pound per square foot = 0.0479 kN/m^2, 1 pound = 0.004448 kN. 1 pound per cubic foot = 16 kg/m^3

a. Floors in garages or portions of buildings used for the storage of motor vehicles shall be designed for the uniformly distributed live loads of Table 1607.1 or the following concentrated loads: (1) for garages restricted to vehicles accommodating not more than nine passengers, 3,000 pounds acting on an area of 4.5 inches by 4.5 inches; (2) for mechanical parking structures without slab or deck which are used for storing passenger vehicles only, 2,250 pounds per wheel.

b. The loading applies to stack room floors that support nonmobile, double-faced library bookstacks, subject to the following limitations:
  1. The nominal bookstack unit height shall not exceed 90 inches;
  2. The nominal shelf depth shall not exceed 12 inches for each face; and
  3. Parallel rows of double-faced bookstacks shall be separated by aisles not less than 36 inches wide.

c. Design in accordance with the ICC *Standard on Bleachers, Folding and Telescopic Seating and Grandstands.*

d. Other uniform loads in accordance with an approved method which contains provisions for truck loadings shall also be considered where appropriate.

e. The concentrated wheel load shall be applied on an area of 20 square inches.

f. Minimum concentrated load on stair treads (on area of 4 square inches) is 300 pounds.

g. Where snow loads occur that are in excess of the design conditions, the structure shall be designed to support the loads due to the increased loads caused by drift buildup or a greater snow design determined by the building official (see Section 1608). For special-purpose roofs, see Section 1607.11.2.2.

h. See Section 1604.8.3 for decks attached to exterior walls.

1607.5 Partition loads. In office buildings and in other buildings where partition locations are subject to change, provision for partition weight shall be made, whether or not partitions are shown on the construction documents, unless the specified live load exceeds 80 psf (3.83 kN/m^2). Such partition load shall not be less than a uniformly distributed live load of 20 psf (0.96kN/m^2).

1607.6 Truck and bus garages. Minimum live loads for garages having trucks or buses shall be as specified in Table 1607.6, but shall not be less than 50 psf (2.40 kN/m^2), unless other loads are specifically justified and approved by the building official. Actual loads shall be used where they are greater than the loads specified in the table.

1607.6.1 Truck and bus garage live load application. The concentrated load and uniform load shall be uniformly distributed over a 10-foot (3048 mm) width on a line normal to the centerline of the lane placed within a 12-foot-wide (3658 mm) lane. The loads shall be placed within their individual lanes so as to produce the maximum stress in each structural member. Single spans shall be designed for the uniform load in Table 1607.6 and one simultaneous concentrated load positioned to produce the maximum effect. Multiple spans shall be designed for the uniform load in Table 1607.6 on the spans and two simultaneous concentrated loads in two spans positioned to produce the maximum negative moment effect. Multiple span design loads, for other effects, shall be the same as for single spans.

1607.7 Loads on handrails, guards, grab bars and vehicle barriers. Handrails, guards, grab bars as designed in ICC A117.1 and vehicle barriers shall be designed and constructed to the structural loading conditions set forth in this section.

TABLE 1607.6
UNIFORM AND CONCENTRATED LOADS

LOADING CLASS[a]	UNIFORM LOAD (pounds/linear foot of lane)	CONCENTRATED LOAD (pounds)[b]	
		For moment design	For shear design
H20-44 and HS20-44	640	18,000	26,000
H15-44 and HS15-44	480	13,500	19,500

For SI: 1 pound per linear foot = 0.01459 kN/m, 1 pound = 0.004448 kN, 1 ton = 8.90 kN.

a. An H loading class designates a two-axle truck with a semitrailer. An HS loading class designates a tractor truck with a semitrailer. The numbers following the letter classification indicate the gross weight in tons of the standard truck and the year the loadings were instituted.

b. See Section 1607.6.1 for the loading of multiple spans.

1607.7.1 Handrails and guards. Handrail assemblies and guards shall be designed to resist a load of 50 plf (0.73 kN/m) applied in any direction at the top and to transfer this load through the supports to the structure.

Exceptions:
  1. For one- and two-family dwellings, only the single, concentrated load required by Section 1607.7.1.1 shall be applied.
  2. In Group I-3, F, H and S occupancies, for areas that are not accessible to the general public and that have an occupant load no greater than 50, the minimum load shall be 20 pounds per foot (0.29 kN/m).

1607.7.1.1 Concentrated load. Handrail assemblies and guards shall be able to resist a single concentrated load of 200 pounds (0.89 kN), applied in any direction at any point along the top, and have attachment devices and supporting structure to transfer this loading to appropriate structural elements of the building. This load need not be assumed to act concurrently with the loads specified in the preceding paragraph.

1607.7.1.2 Components. Intermediate rails (all those except the handrail), balusters and panel fillers shall be designed to withstand a horizontally applied normal load of 50 pounds (0.22 kN) on an area equal to 1 square foot (0.093m^2), including openings and space between rails. Reactions due to this loading are not required to be superimposed with those of Section 1607.7.1 or 1607.7.1.1.

1607.7.1.3 Stress increase. Where handrails and guards are designed in accordance with the provisions for allowable stress design (working stress design) exclusively for the loads specified in Section 1607.7.1, the allowable stress for the members and their attachments are permitted to be increased by one-third.

1607.7.2 Grab bars, shower seats and dressing room bench seats. Grab bars, shower seats and dressing room bench seat systems shall be designed to resist a single concentrated load of 250 pounds (1.11 kN) applied in any direction at any point.

1607.7.3 Vehicle barriers. Vehicle barrier systems for passenger cars shall be designed to resist a single load of 6,000 pounds (26.70 kN) applied horizontally in any direction to the barrier system and shall have anchorage or attachment capable of transmitting this load to the structure. For design of the system, the load shall be assumed to act at a minimum height of 1 foot, 6 inches (457 mm) above the floor or ramp surface on an area not to exceed 1 square foot (305 mm²), and is not required to be assumed to act concurrently with any handrail or guard loadings specified in the preceding paragraphs of Section 1607.7.1. Garages accommodating trucks and buses shall be designed in accordance with an approved method that contains provision for traffic railings.

1607.8 Impact loads. The live loads specified in Section 1607.2 include allowance for impact conditions. Provisions shall be made in the structural design for uses and loads that involve unusual vibration and impact forces.

1607.8.1 Elevators. Elevator loads shall be increased by 100 percent for impact and the structural supports shall be designed within the limits of deflection prescribed by ASME A17.1.

1607.8.2 Machinery. For the purpose of design, the weight of machinery and moving loads shall be increased as follows to allow for impact: (1) elevator machinery, 100 percent; (2) light machinery, shaft- or motor-driven, 20 percent; (3) reciprocating machinery or power-driven units, 50 percent; (4) hangers for floors or balconies, 33 percent. Percentages shall be increased where specified by the manufacturer.

1607.9 Reduction in live loads. The minimum uniformly distributed live loads, L_o, in Table 1607.1 are permitted to be reduced according to the following provisions.

1607.9.1 General. Subject to the limitations of Sections 1607.9.1.1 through 1607.9.1.4, members for which a value of $K_{LL}A_T$ is 400 square feet (37.16 m²) or more are permitted to be designed for a reduced live load in accordance with the following equation:

$$L = L_o\left(0.25 + \frac{15}{\sqrt{K_{LL}A_T}}\right) \qquad \textbf{(Equation 16-21)}$$

For SI: $L = L_o\left(0.25 + \frac{4.57}{\sqrt{K_{LL}A_T}}\right)$

where:

L = Reduced design live load per square foot (meter) of area supported by the member.

L_o = Unreduced design live load per square foot (meter) of area supported by the member (see Table 1607.1).

K_{LL} = Live load element factor (see Table 1607.9.1).

A_T = Tributary area, in square feet (square meters). L shall not be less than $0.50L_o$ for members supporting one floor and L shall not be less than $0.40L_o$ for members supporting two or more floors.

TABLE 1607.9.1
LIVE LOAD ELEMENT FACTOR, K_{LL}

ELEMENT	K_{LL}
Interior columns	4
Exterior columns without cantilever slabs	4
Edge columns with cantilever slabs	3
Corner columns with cantilever slabs	2
Edge beams without cantilever slabs	2
Interior beams	2

All other members not identified above including:
    Edge beams with cantilever slabs
    Cantilever beams
    Two-way slabs
    Members without provisions for continuous shear
    transfer normal to their span
	1

1607.9.1.1 Heavy live loads. Live loads that exceed 100 psf (4.79 kN/m²) shall not be reduced except the live loads for members supporting two or more floors are permitted to be reduced by a maximum of 20 percent, but the live load shall not be less than L as calculated in Section 1607.9.1.

1607.9.1.2 Passenger vehicle garages. The live loads shall not be reduced in passenger vehicle garages except the live loads for members supporting two or more floors are permitted to be reduced by a maximum of 20 percent, but the live load shall not be less than L as calculated in Section 1607.9.1.

1607.9.1.3 Special occupancies. Live loads of 100 psf (4.79 kN/m²) or less shall not be reduced in public assembly occupancies.

1607.9.1.4 Special structural elements. Live loads shall not be reduced for one-way slabs except as permitted in Section 1607.9.1.1. Live loads of 100 psf (4.79 kN/m²) or less shall not be reduced for roof members except as specified in Section 1607.11.2.

1607.9.2 Alternate floor live load reduction. As an alternative to Section 1607.9.1, floor live loads are permitted to be reduced in accordance with the following provisions. Such reductions shall apply to slab systems, beams, girders, columns, piers, walls and foundations.

1. A reduction shall not be permitted in Group A occupancies.

2. A reduction shall not be permitted when the live load exceeds 100 psf (4.79 kN/m²) except that the design live load for columns may be reduced by 20 percent.

3. For live loads not exceeding 100 psf (4.79 kN/m²), the design live load for any structural member supporting 150 square feet (13.94 m²) or more is permitted to be reduced in accordance with the following equation:

$$R = r(A - 150) \qquad \textbf{(Equation 16-22)}$$

For SI: $R = r(A - 13.94)$

Such reduction shall not exceed 40 percent for horizontal members, 60 percent for vertical members, nor R as determined by the following equation:

$$R = 23.1\,(1 + D/L_o)\qquad\text{(Equation 16-23)}$$

where:

A = Area of floor or roof supported by the member, square feet (m²).

D = Dead load per square foot (m²) of area supported.

L_o = Unreduced live load per square foot (m²) of area supported.

R = Reduction in percent.

r = Rate of reduction equal to 0.08 percent for floors.

1607.10 Distribution of floor loads. Where uniform floor live loads are involved in the design of structural members arranged so as to create continuity, the minimum applied loads shall be the full dead loads on all spans in combination with the floor live loads on spans selected to produce the greatest effect at each location under consideration. It shall be permitted to reduce floor live loads in accordance with Section 1607.9.

1607.11 Roof loads. The structural supports of roofs and marquees shall be designed to resist wind and, where applicable, snow and earthquake loads, in addition to the dead load of construction and the appropriate live loads as prescribed in this section, or as set forth in Table 1607.1. The live loads acting on a sloping surface shall be assumed to act vertically on the horizontal projection of that surface.

1607.11.1 Distribution of roof loads. Where uniform roof live loads are involved in the design of structural members arranged so as to create continuity, the minimum applied loads shall be the full dead loads on all spans in combination with the roof live loads on adjacent spans or on alternate spans, whichever produces the greatest effect. See Section 1607.11.2 for minimum roof live loads and Section 1608.5 for partial snow loading.

1607.11.2 Minimum roof live loads. Minimum roof loads shall be determined for the specific conditions in accordance with Sections 1607.11.2.1 through 1607.11.2.4.

1607.11.2.1 Flat, pitched and curved roofs. Ordinary flat, pitched and curved roofs shall be designed for the live loads specified in the following equation or other controlling combinations of loads in Section 1605, whichever produces the greater load. In structures where special scaffolding is used as a work surface for workers and materials during maintenance and repair operations, a lower roof load than specified in the following equation shall not be used unless approved by the building official. Greenhouses shall be designed for a minimum roof live load of 10 psf (0.479 kN/m²).

$$L_r = 20R_1R_2\qquad\text{(Equation 16-24)}$$

where: $12 \le L_r \le 20$

For SI: $L_r = 0.96\,R_1R_2$

where: $0.58 \le L_r \le 0.96$

L_r = Roof live load per square foot (m²) of horizontal projection in pounds per square foot (kN/m²).

The reduction factors R_1 and R_2 shall be determined as follows:

$R_1 = 1$ for $A_t \le 200$ square feet (18.58 m²)

$$\text{(Equation 16-25)}$$

$R_1 = 1.2 - 0.001A_t$ for 200 square feet $< A_t <$ 600 square feet

$$\text{(Equation 16-26)}$$

For SI: $1.2 - 0.011A_t$ for 18.58 square meters $< A_t <$ 55.74 square meters

$R_1 = 0.6$ for $A_t \ge 600$ square feet (55.74 m²)

$$\text{(Equation 16-27)}$$

where:

A_t = Tributary area (span length multiplied by effective width) in square feet (m²) supported by any structural member, and

F = for a sloped roof, the number of inches of rise per foot (for SI: $F = 0.12 \times$ slope, with slope expressed in percentage points), and

F = for an arch or dome, rise-to-span ratio multiplied by 32, and

$R_2 = 1$	for $F \le 4$	**(Equation 16-28)**
$R_2 = 1.2 - 0.05\,F$	for $4 < F < 12$	**(Equation 16-29)**
$R_2 = 0.6$	for $F \ge 12$	**(Equation 16-30)**

1607.11.2.2 Special-purpose roofs. Roofs used for promenade purposes shall be designed for a minimum live load of 60 psf (2.87 kN/m²). Roofs used for roof gardens or assembly purposes shall be designed for a minimum live load of 100 psf (4.79 kN/m²). Roofs used for other special purposes shall be designed for appropriate loads, as directed or approved by the building official.

1607.11.2.3 Landscaped roofs. Where roofs are to be landscaped, the uniform design live load in the landscaped area shall be 20 psf (0.958 kN/m²). The weight of the landscaping materials shall be considered as dead load and shall be computed on the basis of saturation of the soil.

1607.11.2.4 Awnings and canopies. Awnings and canopies shall be designed for a uniform live load of 5 psf (0.240 kN/m²) as well as for snow loads and wind loads as specified in Sections 1608 and 1609.

1607.12 Crane loads. The crane live load shall be the rated capacity of the crane. Design loads for the runway beams, including connections and support brackets, of moving bridge cranes and monorail cranes shall include the maximum wheel loads of the crane and the vertical impact, lateral and longitudinal forces induced by the moving crane.

1607.12.1 Maximum wheel load. The maximum wheel loads shall be the wheel loads produced by the weight of the bridge, as applicable, plus the sum of the rated capacity and the weight of the trolley with the trolley positioned on its runway at the location where the resulting load effect is maximum.

1607.12.2 Vertical impact force. The maximum wheel loads of the crane shall be increased by the percentages shown below to determine the induced vertical impact or vibration force:

Monorail cranes (powered) · · · · · · · · 25 percent

Cab-operated or remotely operated
bridge cranes (powered) · · · · · · · · · · 25 percent

Pendant-operated bridge cranes
(powered) · · · · · · · · · · · · · · · · · 10 percent

Bridge cranes or monorail cranes with
hand-geared bridge, trolley and hoist · · · · · 0 percent

1607.12.3 Lateral force. The lateral force on crane runway beams with electrically powered trolleys shall be calculated as 20 percent of the sum of the rated capacity of the crane and the weight of the hoist and trolley. The lateral force shall be assumed to act horizontally at the traction surface of a runway beam, in either direction perpendicular to the beam, and shall be distributed according to the lateral stiffness of the runway beam and supporting structure.

1607.12.4 Longitudinal force. The longitudinal force on crane runway beams, except for bridge cranes with hand-geared bridges, shall be calculated as 10 percent of the maximum wheel loads of the crane. The longitudinal force shall be assumed to act horizontally at the traction surface of a runway beam, in either direction parallel to the beam.

1607.13 Interior walls and partitions. Interior walls and partitions that exceed 6 feet (1829 mm) in height, including their finish materials, shall have adequate strength to resist the loads to which they are subjected but not less than a horizontal load of 5 psf (0.240 kN/m²).

SECTION 1608
SNOW LOADS

1608.1 General. Design snow loads shall be determined in accordance with Section 7 of ASCE 7, but the design roof load shall not be less than that determined by Section 1607.

1608.2 Ground snow loads. The ground snow loads to be used in determining the design snow loads for roofs are given in Figure 1608.2 for the contiguous United States and Table 1608.2 for Alaska. Site-specific case studies shall be made in areas designated CS in Figure 1608.2. Ground snow loads for sites at elevations above the limits indicated in Figure 1608.2 and for all sites within the CS areas shall be approved. Ground snow load determination for such sites shall be based on an extreme value statistical analysis of data available in the vicinity of the site using a value with a 2-percent annual probability of being exceeded (50-year mean recurrence interval). Snow loads are zero for Hawaii, except in mountainous regions as approved by the building official.

1608.3 Flat roof snow loads. The flat roof snow load, p_f, on a roof with a slope equal to or less than 5 degrees (0.09 rad) (1 inch per foot = 4.76 degrees) shall be calculated in accordance with Section 7.3 of ASCE 7.

1608.3.1 Exposure factor. The value for the snow exposure factor, C_e, used in the calculation of p_f shall be determined from Table 1608.3.1.

1608.3.2 Thermal factor. The value for the thermal factor, C_t, used in the calculation of p_f shall be determined from Table 1608.3.2.

TABLE 1608.2
GROUND SNOW LOADS, p_g, FOR ALASKAN LOCATIONS

LOCATION	POUNDS PER SQUARE FOOT	LOCATION	POUNDS PER SQUARE FOOT	LOCATION	POUNDS PER SQUARE FOOT
Adak	30	Galena	60	Petersburg	150
Anchorage	50	Gulkana	70	St. Paul Islands	40
Angoon	70	Homer	40	Seward	50
Barrow	25	Juneau	60	Shemya	25
Barter Island	35	Kenai	70	Sitka	50
Bethel	40	Kodiak	30	Talkeetna	120
Big Delta	50	Kotzebue	60	Unalakleet	50
Cold Bay	25	McGrath	70	Valdez	160
Cordova	100	Nenana	80	Whittier	300
Fairbanks	60	Nome	70	Wrangell	60
Fort Yukon	60	Palmer	50	Yakutat	150

For SI: 1 pound per square foot = 0.0479 kN/m².

In CS areas, site-specific Case Studies are required to
establish ground snow loads. Extreme local variations
in ground snow loads in these areas preclude mapping
at this scale.

Numbers in parentheses represent the upper elevation
limits in feet for the ground snow load values presented
below. Site-specific case studies are required to establish
ground snow loads at elevations not covered.

To convert lb/sq ft to kNm^2, multiply by 0.0479.

To convert feet to meters, multiply by 0.3048.

```
0        100        200        300 miles
```

FIGURE 1608.2
GROUND SNOW LOADS, p_g, FOR THE UNITED STATES (psf)

FIGURE 1608.2–continued
GROUND SNOW LOADS, p_g, FOR THE UNITED STATES (psf)

TABLE 1608.3.1
SNOW EXPOSURE FACTOR, C_e

TERRAIN CATEGORY[a]	EXPOSURE OF ROOF[a,b]		
	Fully exposed[c]	Partially exposed	Sheltered
A (see Section 1609.4)	N/A	1.1	1.3
B (see Section 1609.4)	0.9	1.0	1.2
C (see Section 1609.4)	0.9	1.0	1.1
D (see Section 1609.4)	0.8	0.9	1.0
Above the treeline in windswept mountainous areas	0.7	0.8	N/A
In Alaska, in areas where trees do not exist within a 2-mile radius of the site	0.7	0.8	N/A

For SI: 1 mile = 1609 m.

a. The terrain category and roof exposure condition chosen shall be representative of the anticipated conditions during the life of the structure. An exposure factor shall be determined for each roof of a structure.

b. Definitions of roof exposure are as follows:
   1. Fully exposed shall mean roofs exposed on all sides with no shelter afforded by terrain, higher structures or trees. Roofs that contain several large pieces of mechanical equipment, parapets which extend above the height of the balanced snow load, h_b, or other obstructions are not in this category.
   2. Partially exposed shall include all roofs except those designated as "fully exposed" or "sheltered."
   3. Sheltered roofs shall mean those roofs located tight in among conifers that qualify as "obstructions."

c. Obstructions within a distance of $10\,h_o$ provide "shelter," where h_o is the height of the obstruction above the roof level. If the only obstructions are a few deciduous trees that are leafless in winter, the "fully exposed" category shall be used except for terrain category "A." Note that these are heights above the roof. Heights used to establish the terrain category in Section 1609.4 are heights above the ground.

TABLE 1608.3.2
THERMAL FACTOR, C_t

THERMAL CONDITION[a]	C_t
All structures except as indicated below	1.0
Structures kept just above freezing and others with cold, ventilated roofs in which the thermal resistance (R-value) between the ventilated space and the heated space exceeds $25\text{h} \cdot \text{ft}^2 \cdot {}^\circ\text{F/Btu}$	1.1
Unheated structures	1.2
Continuously heated greenhouses[b] with a roof having a thermal resistance (R-value) less than $2.0\text{h} \cdot \text{ft}^2 \cdot {}^\circ\text{F/Btu}$	0.85

For SI: $1\,\text{h} \cdot \text{ft}^2 \cdot {}^\circ\text{F/Btu} = 0.176\,\text{m}^2 \cdot \text{K/W}$.

a. The thermal condition shall be representative of the anticipated conditions during winters for the life of the structure.

b. A continuously heated greenhouse shall mean a greenhouse with a constantly maintained interior temperature of 50°F or more during winter months. Such greenhouse shall also have a maintenance attendant on duty at all times or a temperature alarm system to provide warning in the event of a heating system failure.

1608.3.3 Snow load importance factor. The value for the snow load importance factor, I_s, used in the calculation of p_f shall be determined in accordance with Table 1604.5. Greenhouses that are occupied for growing plants on production or research basis, without public access, shall be included in Importance Category I.

1608.3.4 Rain-on-snow surcharge load. Roofs with a slope less than $1/2$ inch per foot (2.38 degrees) shall be designed for a rain-on-snow surcharge load determined in accordance with Section 7.10 of ASCE 7.

1608.3.5 Ponding instability. For roofs with a slope less than $1/4$ inch per foot (1.19 degrees), the design calculations shall include verification of the prevention of ponding instability in accordance with Section 7.11 of ASCE 7.

1608.4 Sloped roof snow loads. The snow load, p_s, on a roof with a slope greater than 5 degrees (0.09 rad) (1 inch per foot = 4.76 degrees) shall be calculated in accordance with Section 7.4 of ASCE 7.

1608.5 Partial loading. The effect of not having the balanced snow load over the entire loaded roof area shall be analyzed in accordance with Section 7.5 of ASCE 7.

1608.6 Unbalanced snow loads. Unbalanced roof snow loads shall be determined in accordance with Section 7.6 of ASCE 7. Winds from all directions shall be accounted for when establishing unbalanced snow loads.

1608.7 Drifts on lower roofs. In areas where the ground snow load, p_g, as determined by Section 1608.2, is equal to or greater than 5 psf (0.240 kN/m²), roofs shall be designed to sustain localized loads from snow drifts in accordance with Section 7.7 of ASCE 7.

1608.8 Roof projections. Drift loads due to mechanical equipment, penthouses, parapets and other projections above the roof shall be determined in accordance with Section 7.8 of ASCE 7.

1608.9 Sliding snow. The extra load caused by snow sliding off a sloped roof onto a lower roof shall be determined in accordance with Section 7.9 of ASCE 7.

SECTION 1609
WIND LOADS

1609.1 Applications. Buildings, structures and parts thereof shall be designed to withstand the minimum wind loads prescribed herein. Decreases in wind loads shall not be made for the effect of shielding by other structures.

1609.1.1 Determination of wind loads. Wind loads on every building or structure shall be determined in accordance with Section 6 of ASCE 7. Wind shall be assumed to come from any horizontal direction and wind pressures shall be assumed to act normal to the surface considered.

Exceptions:

1. Wind loads determined by the provisions of Section 1609.6.

2. Subject to the limitations of Section 1609.1.1.1, the provisions of *SBCCI SSTD 10 Standard for Hurricane Resistant Residential Construction* shall be permitted for applicable Group R2 and R3 buildings.

3. Subject to the limitations of Section 1609.1.1.1, residential structures using the provisions of the *AF&PA Wood Frame Construction Manual for One- and Two-Family Dwellings.*

4. Designs using *NAAMM FP 1001 Guide Specification for Design of Metal Flagpoles.*

5. Designs using TIA/EIA-222 for antenna-supporting structures and antennas.

1609.1.1.1 Applicability. The provisions of SSTD 10 are applicable only to buildings located within Exposure, B or C as defined in Section 1609.4. The provisions of SSTD 10 and the *AF&PA Wood Frame construction Manual for One- and Two-Family Dwellings* shall not apply to buildings sited on the upper half of an isolated hill, ridge or escarpment meeting the following conditions:

1. The hill, ridge or escarpment is 60 feet (18 288 mm) or higher if located in Exposure B or 30 feet (9144 mm) or higher if located in Exposure C;

2. The maximum average slope of the hill exceeds 10 percent; and

3. The hill, ridge or escarpment is unobstructed upwind by other such topographic features for a distance from the high point of 50 times the height of the hill or 1 mile (1.61 km), whichever is greater.

1609.1.2 Minimum wind loads. The wind loads used in the design of the main wind-force-resisting system shall not be less than 10 psf (0.479 kN/m^2) multiplied by the area of the building or structure projected on a vertical plane normal to the wind direction. In the calculation of design wind loads for components and cladding for buildings, the algebraic sum of the pressures acting on opposite faces shall be taken into account. The design pressure for components and cladding of buildings shall not be less than 10 psf (0.479 kN/m^2) acting in either direction normal to the surface. The design force for open buildings and other structures shall not be less than 10 psf (0.479 kN/m^2) multiplied by the area A_f.

1609.1.3 Anchorage against overturning, uplift and sliding. Structural members and systems and components and cladding in a building or structure shall be anchored to resist wind-induced overturning, uplift and sliding and to provide continuous load paths for these forces to the foundation. Where a portion of the resistance to these forces is provided by dead load, the dead load, including the weight of soils and foundations, shall be taken as the minimum dead load likely to be in place during a design wind event. Where the alternate basic load combinations of Section 1605.3.2 are used, only two-thirds of the minimum dead load likely to be in place during a design wind event shall be used.

1609.1.4 Protection of openings. In wind-borne debris regions, glazing that receives positive external pressure in the lower 60 feet (18 288 mm) in buildings shall be assumed to be openings unless such glazing is impact resistant or protected with an impact-resistant covering meeting the requirements of an approved impact-resisting standard or ASTM E 1996 and of ASTM E 1886 referenced therein as follows:

1. Glazed openings located within 30 feet (9144 mm) of grade shall meet the requirements of the Large Missile Test of ASTM E 1996.

2. Glazed openings located more than 30 feet (9144 mm) above grade shall meet the provisions of the Small Missile Test of ASTM E 1996.

Exceptions:

1. Wood structural panels with a minimum thickness of $^7/_{16}$ inch (11.1 mm) and maximum panel span of 8 feet (2438 mm) shall be permitted for opening protection in one- and two-story buildings. Panels shall be precut to cover the glazed openings with attachment hardware provided. Attachments shall be designed to resist the components and cladding loads determined in accordance with the provisions of Section 1609.6.1.2. Attachment in accordance with Table 1609.1.4 is permitted for buildings with a mean roof height of 33 feet (10 058 mm) or less where wind speeds do not exceed 130 mph (57.2 m/s).

2. Buildings in Category I as defined in Table 1604.5, including production greenhouses as defined in Section 1608.3.3.

TABLE 1609.1.4
WIND-BORNE DEBRIS PROTECTION FASTENING SCHEDULE FOR WOOD STRUCTURAL PANELS[a,b,c]

FASTENER TYPE	FASTENER SPACING (inches)			
	Panel span ≤ 2 feet	2 feet < Panel span ≤ 4 feet	4 feet < Panel span ≤ 6 feet	6 feet < Panel span ≤ 8 feet
2$^1/_2$ No. 6 Wood screws	16	16	12	9
2$^1/_2$ No. 8 Wood screws	16	16	16	12

For SI: 1 inch = 25.4 mm, 1 foot = 304.8 mm, 1 pound = 4.4 N, 1 mile per hour = 0.44 m/s.

a. This table is based on a maximum wind speed (3-second gust) of 130 mph and mean roof height of 33 feet or less.

b. Fasteners shall be installed at opposing ends of the wood structural panel.

c. Where screws are attached to masonry or masonry/stucco, they shall be attached utilizing vibration-resistant anchors having a minimum withdrawal capacity of 490 pounds.

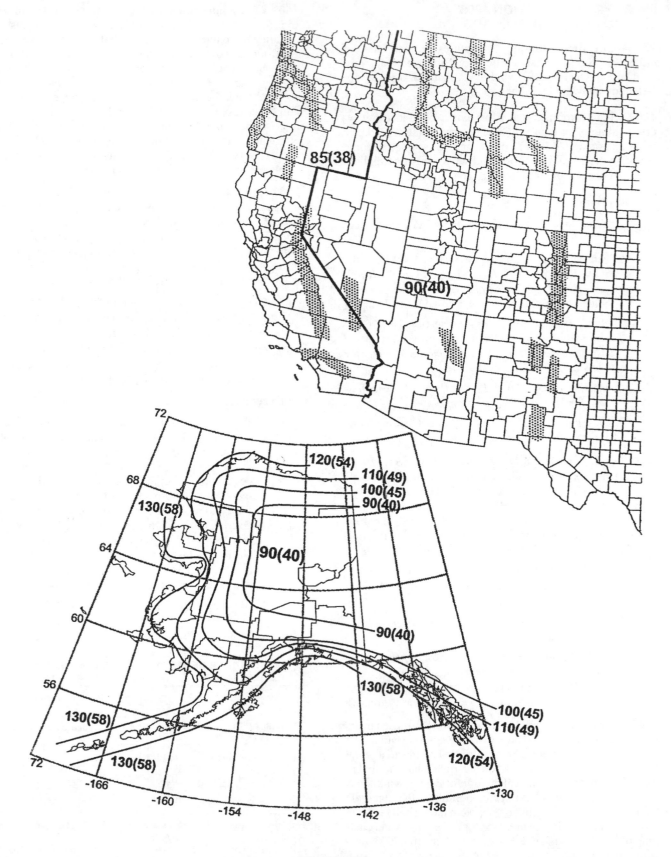

FIGURE 1609
BASIC WIND SPEED (3-SECOND GUST)

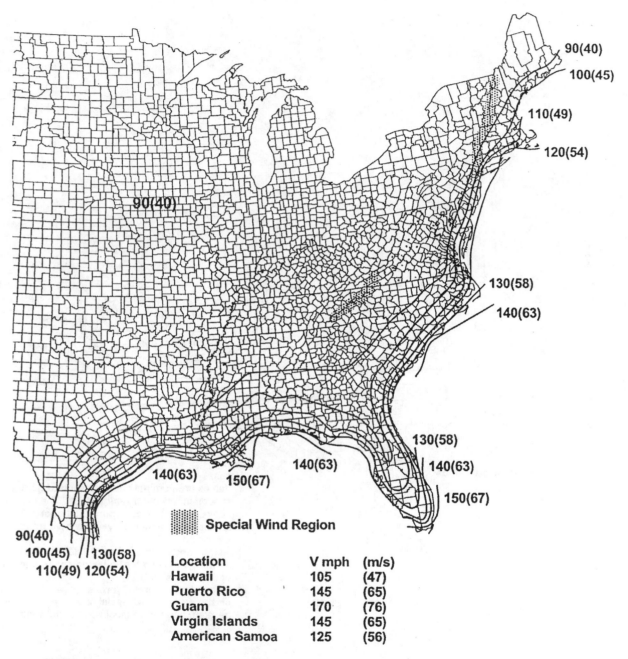

90(40)
100(45)
110(49)
120(54)
130(58)
140(63)
130(58)
140(63)
150(67)
140(63)
150(67)
140(63)
150(67)
90(40)
90(40)
100(45) 130(58)
110(49) 120(54)

Special Wind Region

Location	V mph	(m/s)
Hawaii	105	(47)
Puerto Rico	145	(65)
Guam	170	(76)
Virgin Islands	145	(65)
American Samoa	125	(56)

Notes:
1. Values are nominal design 3-second gust wind speeds in miles per hour (m/s) at 33 ft (10 m) above ground for Exposure C category.
2. Linear interpolation between wind contours is permitted.
3. Islands and coastal areas outside the last contour shall use the last wind speed contour of the coastal area.
4. Mountainous terrain, gorges, ocean promontories, and special wind regions shall be examined for unusual wind conditions.

FIGURE 1609—continued
BASIC WIND SPEED (3-SECOND GUST)

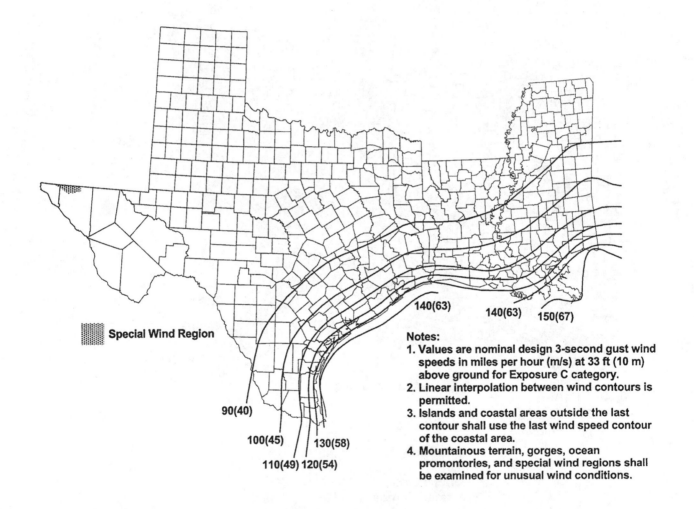

Special Wind Region

140(63)　　140(63)　　150(67)

90(40)

100(45)　　130(58)

110(49) 120(54)

Notes:
1. Values are nominal design 3-second gust wind speeds in miles per hour (m/s) at 33 ft (10 m) above ground for Exposure C category.
2. Linear interpolation between wind contours is permitted.
3. Islands and coastal areas outside the last contour shall use the last wind speed contour of the coastal area.
4. Mountainous terrain, gorges, ocean promontories, and special wind regions shall be examined for unusual wind conditions.

FIGURE 1609–continued
BASIC WIND SPEED (3-SECOND GUST)
WESTERN GULF OF MEXICO HURRICANE COASTLINE

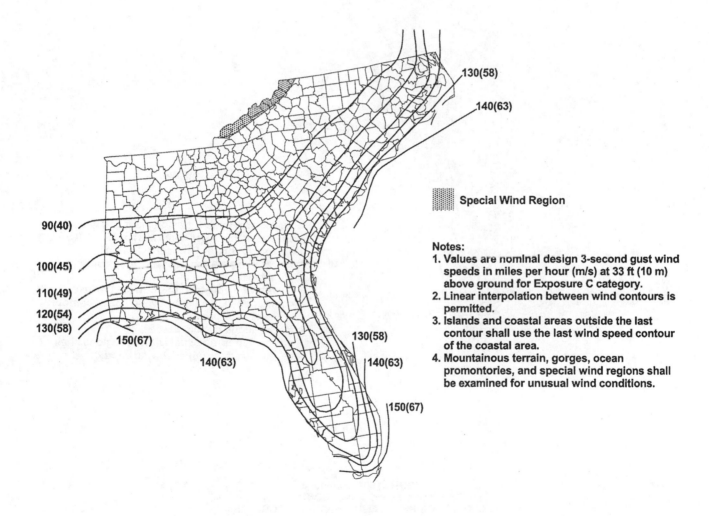

130(58)

140(63)

Special Wind Region

90(40)

100(45)

110(49)

120(54)

130(58)

150(67)

140(63)

130(58)

140(63)

150(67)

Notes:
1. Values are nominal design 3-second gust wind speeds in miles per hour (m/s) at 33 ft (10 m) above ground for Exposure C category.
2. Linear interpolation between wind contours is permitted.
3. Islands and coastal areas outside the last contour shall use the last wind speed contour of the coastal area.
4. Mountainous terrain, gorges, ocean promontories, and special wind regions shall be examined for unusual wind conditions.

FIGURE 1609–continued
BASIC WIND SPEED (3-SECOND GUST)
EASTERN GULF OF MEXICO AND SOUTHEASTERN U.S. HURRICANE COASTLINE

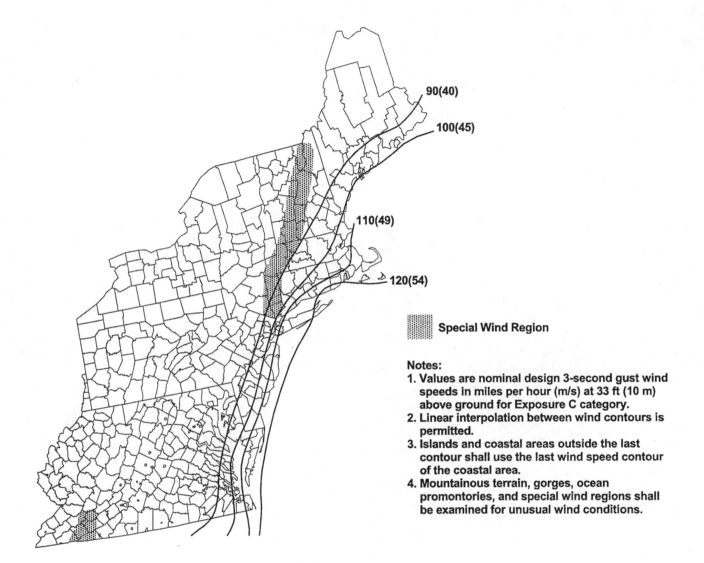

90(40)

100(45)

110(49)

120(54)

Special Wind Region

Notes:
1. Values are nominal design 3-second gust wind speeds in miles per hour (m/s) at 33 ft (10 m) above ground for Exposure C category.
2. Linear interpolation between wind contours is permitted.
3. Islands and coastal areas outside the last contour shall use the last wind speed contour of the coastal area.
4. Mountainous terrain, gorges, ocean promontories, and special wind regions shall be examined for unusual wind conditions.

FIGURE 1609–continued
BASIC WIND SPEED (3-SECOND GUST)
MID AND NORTHERN ATLANTIC HURRICANE COASTLINE

STRUCTURAL DESIGN

1609.1.4.1 Building with openings. Where glazing is assumed to be an opening in accordance with Section 1609.1.4, the building shall be evaluated to determine if the openings are of sufficient area to constitute an open or partially enclosed building as defined in Section 1609.2. Open and partially enclosed buildings shall be designed in accordance with the applicable provisions of ASCE 7.

1609.1.5 Wind and seismic detailing. Lateral-force-resisting systems shall meet seismic detailing requirements and limitations prescribed in this code, even when wind code prescribed load effects are greater than seismic load effects.

1609.2 Definitions. The following words and terms shall, for the purposes of Section 1609.6, have the meanings shown herein.

BUILDINGS AND OTHER STRUCTURES, FLEXIBLE. Slender buildings and other structures that have a fundamental natural frequency less than 1 Hz.

BUILDING, ENCLOSED. A building that does not comply with the requirements for open or partially enclosed buildings.

BUILDING, LOW-RISE. Enclosed or partially enclosed buildings that comply with the following conditions:

1. Mean roof height, h, less than or equal to 60 feet (18 288 mm).

2. Mean roof height, h, does not exceed least horizontal dimension.

BUILDING, OPEN. A building having each wall at least 80 percent open. This condition is expressed for each wall by the equation:

$$A_o \geq 0.8 A_g \qquad \text{(Equation 16-31)}$$

where:

A_o = Total area of openings in a wall that receives positive external pressure, in square feet (m²).

A_g = The gross area of that wall in which A_o is identified, in square feet (m²).

BUILDING, PARTIALLY ENCLOSED. A building that complies with both of the following conditions:

1. The total area of openings in a wall that receives positive external pressure exceeds the sum of the areas of openings in the balance of the building envelope (walls and roof) by more than 10 percent; and

2. The total area of openings in a wall that receives positive external pressure exceeds 4 square feet (0.37 m²) or 1 percent of the area of that wall, whichever is smaller, and the percentage of openings in the balance of the building envelope does not exceed 20 percent.

These conditions are expressed by the following equations:

$$A_o > 1.10 A_{oi} \qquad \text{(Equation 16-32)}$$

$A_o > 4$ square feet (0.37 m²) or $> 0.01 A_g$, whichever is smaller, and $A_{oi}/A_{gi} \leq 0.20$ **(Equation 16-33)**

where:

A_o, A_g are as defined for an open building.

A_{oi} = The sum of the areas of openings in the building envelope (walls and roof) not including A_o, in square feet (m²).

A_{gi} = The sum of the gross surface areas of the building envelope (walls and roof) not including A_g, in square feet (m²).

BUILDING, SIMPLE DIAPHRAGM. A building in which wind loads are transmitted through floor and roof diaphragms to the vertical lateral-force-resisting systems.

COMPONENTS AND CLADDING. Elements of the building envelope that do not qualify as part of the main windforce-resisting system.

EFFECTIVE WIND AREA. The area used to determine GC_p. For component and cladding elements, the effective wind area in Tables 1609.6.2.1(2) and 1609.6.2.1(3) is the span length multiplied by an effective width that need not be less than one-third the span length. For cladding fasteners, the effective wind area shall not be greater than the area that is tributary to an individual fastener.

HURRICANE-PRONE REGIONS. Areas vulnerable to hurricanes defined as:

1. The U.S. Atlantic Ocean and Gulf of Mexico coasts where the basic wind speed is greater than 90 mph (39.6 m/s) and

2. Hawaii, Puerto Rico, Guam, Virgin Islands and American Samoa.

IMPORTANCE FACTOR, I_w. A factor that accounts for the degree of hazard to human life and damage to property.

MAIN WINDFORCE-RESISTING SYSTEM. An assemblage of structural elements assigned to provide support and stability for the overall structure. The system generally receives wind loading from more than one surface.

MEAN ROOF HEIGHT. The average of the roof eave height and the height to the highest point on the roof surface, except that eave height shall be used for roof angle of less than or equal to 10 degrees (0.1745 rad).

WIND-BORNE DEBRIS REGION. Areas within hurricane-prone regions within 1 mile (1.61 km) of the coastal mean high water line where the basic wind speed is 110 mph (48.4 m/s) or greater; or where the basic wind speed is 120 mph (52.8 m/s) or greater; or Hawaii.

1609.3 Basic wind speed. The basic wind speed, in mph, for the determination of the wind loads shall be determined by Figure 1609 or by ASCE 7 Figure 6-1 when using the provisions of ASCE 7. Basic wind speed for the special wind regions indicated, near mountainous terrain, and near gorges, shall be in accordance with local jurisdiction requirements. Basic wind speeds determined by the local jurisdiction shall be in accordance with Section 6.5.4 of ASCE 7.

In nonhurricane-prone regions, when the basic wind speed is estimated from regional climatic data, the basic wind speed shall be not less than the wind speed associated with an annual probability of 0.02 (50-year mean recurrence interval), and the estimate shall be adjusted for equivalence to a 3-second gust wind speed at 33 feet (10 m) above ground in exposure Category C. The data analysis shall be performed in accordance with Section 6.5.4 of ASCE 7.

1609.3.1 Wind speed conversion. When required, the 3-second gust wind velocities of Figure 1609 shall be converted to fastest-mile wind velocities using Table 1609.3.1.

1609.4 Exposure category. For each wind direction considered, an exposure category that adequately reflects the characteristics of ground surface irregularities shall be determined for the site at which the building or structure is to be constructed. For a site located in the transition zone between categories, the category resulting in the largest wind forces shall apply. Account shall be taken of variations in ground surface roughness that arise from natural topography and vegetation as well as from constructed features. For any given wind direction, the exposure in which a specific building or other structure is sited shall be assessed as being one of the following categories. When applying the simplified wind load method of Section 1609.6, a single exposure category shall be used based upon the most restrictive for any given wind direction.

1. **Exposure A.** This exposure category is no longer used in ASCE 7.

2. **Exposure B.** Urban and suburban areas, wooded areas or other terrain with numerous closely spaced obstructions having the size of single-family dwellings or larger. Exposure B shall be assumed unless the site meets the definition of another type of exposure.

3. **Exposure C.** Open terrain with scattered obstructions, including surface undulations or other irregularities, having heights generally less than 30 feet (9144 mm) extending more than 1,500 feet (457.2 m) from the building site in any quadrant. This exposure shall also apply to any building located within Exposure B-type terrain where the building is directly adjacent to open areas of Exposure C-type terrain in any quadrant for a distance of more than 600 feet (182.9 m). This category includes flat open country, grasslands and shorelines in hurricane-prone regions.

4. **Exposure D.** Flat, unobstructed areas exposed to wind flowing over open water (excluding shorelines in hurricane-prone regions) for a distance of at least 1 mile (1.61 km). Shorelines in Exposure D include inland waterways, the Great Lakes and coastal areas of California, Oregon, Washington and Alaska. This exposure shall apply only to those buildings and other structures exposed to the wind coming from over the water. Exposure D extends inland from the shoreline a distance of 1,500 feet (460 m) or 10 times the height of the building or structure, whichever is greater.

1609.5 Importance factor. Buildings and other structures shall be assigned a wind load importance factor, I_w, in accordance with Table 1604.5.

1609.6 Simplified wind load method.

1609.6.1 Scope. The procedures in Section 1609.6 shall be permitted to be used for determining and applying wind pressures in the design of enclosed buildings with flat, gabled and hipped roofs and having a mean roof height not exceeding the least horizontal dimension or 60 feet (18 288 mm), whichever is less, subject to the limitations of Sections 1609.6.1.1 and 1609.6.1.2. If a building qualifies only under Section 1609.6.1.2 for design of its components and cladding, then its main windforce-resisting system shall be designed in accordance with Section 1609.1.1.

Exception: The provisions of Section 1609.6 shall not apply to buildings sited on the upper half of an isolated hill or escarpment meeting all of the following conditions:

1. The hill or escarpment is 60 feet (18 288 mm) or higher if located in Exposure B or 30 feet (9144 mm) or higher if located in Exposure C.

2. The maximum average slope of the hill exceeds 10 percent.

3. The hill or escarpment is unobstructed upwind by other such topographic features for a distance from the high point of 50 times the height of the hill or 1 mile (1.61 km), whichever is less.

TABLE 1609.3.1
EQUIVALENT BASIC WIND SPEEDS[a,b,c]

V_{3S}	85	90	100	105	110	120	125	130	140	145	150	160	170
V_{fm}	70	75	80	85	90	100	105	110	120	125	130	140	150

For SI: 1 mile per hour = 0.44 m/s.
a. Linear interpolation is permitted.
b. V_{3S} is the 3-second gust wind speed (mph).
c. V_{fm} is the fastest mile wind speed (mph).

1609.6.1.1 Main windforce-resisting systems. For the design of main windforce- resisting systems, the building must meet all of the following conditions:

1. The building is a simple diaphragm building as defined in Section 1609.2.

2. The building is not classified as a flexible building as defined in Section 1609.2.

3. The building does not have response characteristics making it subject to across wind loading, vortex shedding, instability due to galloping or flutter; and does not have a site location for which channeling effects or buffeting in the wake of upwind obstructions warrant special consideration.

4. The building structure has no expansion joints or separations.

5. The building is regular shaped and has an approximately symmetrical cross section in each direction with roof slopes not exceeding 45 degrees (0.78 rad.).

1609.6.1.2 Components and cladding. For the design of components and cladding, the building must meet all of the following conditions:

1. The building does not have response characteristics making it subject to across wind loading, vortex shedding, instability due to galloping or flutter; and does not have a site location for which channeling effects or buffeting in the wake of upwind obstructions warrant special consideration.

2. The building is regular shaped with roof slopes not exceeding 45 degrees (0.78 rad.) for gable roofs, or 27 degrees (0.47 rad.) for hip roofs.

1609.6.2 Design procedure.

1. The basic wind speed, V, shall be determined in accordance with Section 1609.3. The wind shall be assumed to come from any horizontal direction.

2. An importance factor I_w shall be determined in accordance with Section 1609.5.

3. An exposure category shall be determined in accordance with Section 1609.4.

4. A height and exposure adjustment coefficient, λ, shall be determined from Table 1609.6.2.1(4).

1609.6.2.1 Main windforce-resisting system. Simplified design wind pressures, p_s, for the main windforce-resisting systems represent the net pressures (sum of internal and external) to be applied to the horizontal and vertical projections of building surfaces as shown in Figure 1609.6.2.1. For the horizontal pressures (Zones A, B, C, D), p_s is the combination of the windward and leeward net pressures. p_s shall be determined from Equation 16-34).

$$p_s = \lambda\, I_w\, p_{s30} \qquad \textbf{(Equation 16-34)}$$

where:

λ = Adjustment factor for building height and exposure from Table 1609.6.2.1(4).

I_w = Importance factor as defined in Section 1609.5

p_{s30} = Simplified design wind pressure for Exposure B, at h = 30 feet (9144 mm), and for I_w = 1.0, from Table 1609.6.2.1(1).

1609.6.2.1.1 Minimum pressures. The load effects of the design wind pressures from Section 1609.6.2.1 shall not be less than assuming the pressures , p_s, for Zones A, B, C and D all equal to +10 psf (0.48 kN/m²), while assuming Zones E, F, G, and H all equal to 0 psf.

1609.6.2.2 Components and cladding. Net design wind pressures, p_{net}, for the components and cladding of buildings represent the net pressures (sum of internal and external) to be applied normal to each building surface as shown in Figure 1609.6.2.2. The net design wind pressure, p_{net}, shall be determined from Equation 16-35:

$$p_{net} = \lambda I_w p_{net30} \qquad \textbf{(Equation 16-35)}$$

where:

λ = Adjustment factor for building height and exposure from Table 1609.6.2.1(4).

I_w = Importance factor as defined in Section 1609.5.

p_{net30} = Net design wind pressure for Exposure B, at h = 30 feet (9144 mm), and for I_w = 1.0, from Tables 1609.6.2.1(2) and 1609.6.2.1(3).

1609.6.2.2.1 Minimum pressures. The positive design wind pressures, p_{net}, from Section 1609.6.2.2 shall not be less than +10 psf (0.48 kN/m²), and the negative design wind pressures, p_{net}, from Section 1609.6.2.2 shall not be less than -10 psf (-0.48 kN/m²).

1609.6.2.3 Load case. Members that act as both part of the main windforce-resisting system and as components and cladding shall be designed for each separate load case.

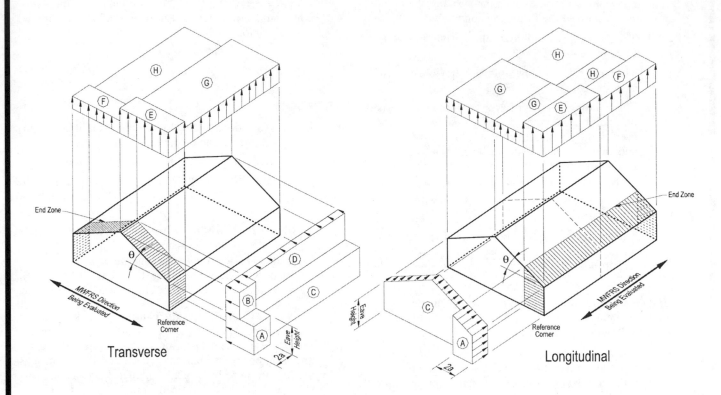

FIGURE 1609.6.2.1
MAIN WINDFORCE LOADING DIAGRAM

For SI: 1 foot = 304.8 mm, 1 degree = 0.0174 rad.

Notes:

1. Pressures are applied to the horizontal and vertical projections for Exposure B, at h = 30 feet, for I_w = 1.0. Adjust to other exposures and heights with adjustment factor λ.

2. The load patterns shown shall be applied to each corner of the building in turn as the reference corner.

3. For the design of the longitudinal MWFRS, use θ = 0°, and locate the Zone E/F, G/H boundary at the mid-length of the building.

4. Load Cases 1 and 2 must be checked for 25° < θ ≤ 45°. Load Case 2 at 25° is provided only for interpolation between 25° to 30°.

5. Plus and minus signs signify pressures acting toward and away from the projected surfaces, respectively.

6. For roof slopes other than those shown, linear interpolation is permitted.

7. The total horizontal load shall not be less than that determined by assuming p_S = 0 in Zones B and D.

8. The zone pressures represent the following:

   Horizontal pressure zones — Sum of the windward and leeward net (sum of internal and external) pressures on vertical projection of:

A – End zone of wall	C – Interior zone of wall
B – End zone of roof	D – Interior zone of roof

   Vertical pressure zones — Net (sum of internal and external) pressures on horizontal projection of:

E – End zone of windward roof	G – Interior zone of windward roof
F – End zone of leeward roof	H – Interior zone of leeward roof

9. Where Zone E or G falls on a roof overhang on the windward side of the building, use E_{OH} and G_{OH} for the pressure on the horizontal projection of the overhang. Overhangs on the leeward and side edges shall have the basic zone pressure applied.

10. Notation:

   a: 10 percent of least horizontal dimension or 0.4h, whichever is smaller, but not less than either 4 percent of least horizontal dimension or 3 feet.

   h: Mean roof height, in feet (meters), except that eave height shall be used for roof angles <10°.

   θ: Angle of plane of roof from horizontal, in degrees.

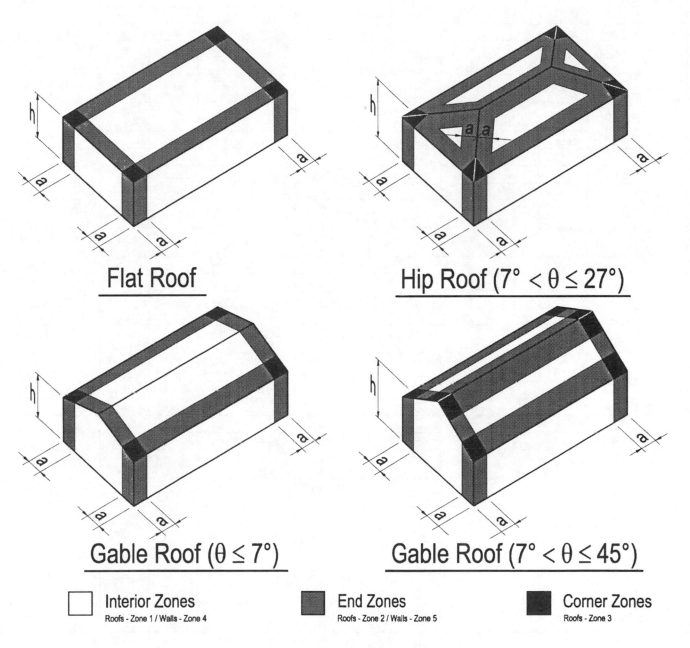

Flat Roof

Hip Roof (7° < θ ≤ 27°)

Gable Roof (θ ≤ 7°)

Gable Roof (7° < θ ≤ 45°)

☐ Interior Zones
Roofs - Zone 1 / Walls - Zone 4

▨ End Zones
Roofs - Zone 2 / Walls - Zone 5

▮ Corner Zones
Roofs - Zone 3

FIGURE 1609.6.2.2
COMPONENT AND CLADDING PRESSURE

For SI: 1 foot = 304.8 mm, 1 degree = 0.0174 rad.

Notes:

1. Pressures are applied normal to the surface for Exposure B, at $h = 30$ feet, for $I_w = 1.0$. Adjust to other exposures and heights with adjustment factor λ.
2. Plus and minus signs signify pressures acting toward and away from the surfaces, respectively.
3. For hip roofs with $\theta \leq 25°$, Zone 3 shall be treated as Zone 2.
4. For effective areas between those given, the value is permitted to be interpolated, otherwise use the value associated with the lower effective area.
5. Notation:

   a: 10 percent of least horizontal dimension or 0.4h, whichever is smaller, but not less than either 4 percent of least horizontal dimension or 3 feet.

   h: Mean roof height, in feet (meters), except that eave height shall be used for roof angles <10°.

   θ: Angle of plane of roof from horizontal, in degrees.

TABLE 1609.6.2.1(1)
SIMPLIFIED DESIGN WIND PRESSURE (MAIN WINDFORCE-RESISTING SYSTEM), p_{s30} (Exposure B at h = 30 feet with I_w = 1.0) (psf)

BASIC WIND SPEED (mph)	ROOF ANGLE (degrees)	ROOF RISE IN 12"	LOAD CASE	ZONES									
				Horizontal Pressures				Vertical Pressures				Overhangs	
				A	B	C	D	E	F	G	H	E_{OH}	G_{OH}
85	0 to 5°	Flat	1	11.5	-5.9	7.6	-3.5	-13.8	-7.8	-9.6	-6.1	-19.3	-15.1
	10°	2	1	12.9	-5.4	8.6	-3.1	-13.8	-8.4	-9.6	-6.5	-19.3	-15.1
	15°	3	1	14.4	-4.8	9.6	-2.7	-13.8	-9.0	-9.6	-6.9	-19.3	-15.1
	20°	4	1	15.9	-4.2	10.6	-2.3	-13.8	-9.6	-9.6	-7.3	-19.3	-15.1
	25°	6	1	14.4	2.3	10.4	2.4	-6.4	-8.7	-4.6	-7.0	-11.9	-10.1
			2	—	—	—	—	-2.4	-4.7	-0.7	-3.0	—	—
	30° to 45°	7 to 12	1	12.9	8.8	10.2	7.0	1.0	-7.8	0.3	-6.7	-4.5	-5.2
			2	12.9	8.8	10.2	7.0	5.0	-3.9	4.3	-2.8	-4.5	-5.2
90	0 to 5°	Flat	1	12.8	-6.7	8.5	-4.0	-15.4	-8.8	-10.7	-6.8	-21.6	-16.9
	10°	2	1	14.5	-6.0	9.6	-3.5	-15.4	-9.4	-10.7	-7.2	-21.6	-16.9
	15°	3	1	16.1	-5.4	10.7	-3.0	-15.4	-10.1	-10.7	-7.7	-21.6	-16.9
	20°	4	1	17.8	-4.7	11.9	-2.6	-15.4	-10.7	-10.7	-8.1	-21.6	-16.9
	25°	6	1	16.1	2.6	11.7	2.7	-7.2	-9.8	-5.2	-7.8	-13.3	-11.4
			2	—	—	—	—	-2.7	-5.3	-0.7	-3.4	—	—
	30° to 45°	7 to 12	1	14.4	9.9	11.5	7.9	1.1	-8.8	0.4	-7.5	-5.1	-5.8
			2	14.4	9.9	11.5	7.9	5.6	-4.3	4.8	-3.1	-5.1	-5.8
100	0 to 5°	Flat	1	15.9	-8.2	10.5	-4.9	-19.1	-10.8	-13.3	-8.4	-26.7	-20.9
	10°	2	1	17.9	-7.4	11.9	-4.3	-19.1	-11.6	-13.3	-8.9	-26.7	-20.9
	15°	3	1	19.9	-6.6	13.3	-3.8	-19.1	-12.4	-13.3	-9.5	-26.7	-20.9
	20°	4	1	22.0	-5.8	14.6	-3.2	-19.1	-13.3	-13.3	-10.1	-26.7	-20.9
	25°	6	1	19.9	3.2	14.4	3.3	-8.8	-12.0	-6.4	-9.7	-16.5	-14.0
			2	—	—	—	—	-3.4	-6.6	-0.9	-4.2	—	—
	30° to 45°	7 to 12	1	17.8	12.2	14.2	9.8	1.4	-10.8	0.5	-9.3	-6.3	-7.2
			2	17.8	12.2	14.2	9.8	6.9	-5.3	5.9	-3.8	-6.3	-7.2
110	0 to 5°	Flat	1	19.2	-10.0	12.7	-5.9	-23.1	-13.1	-16.0	-10.1	-32.3	-25.3
	10°	2	1	21.6	-9.0	14.4	-5.2	-23.1	-14.1	-16.0	-10.8	-32.3	-25.3
	15°	3	1	24.1	-8.0	16.0	-4.6	-23.1	-15.1	-16.0	-11.5	-32.3	-25.3
	20°	4	1	26.6	-7.0	17.7	-3.9	-23.1	-16.0	-16.0	-12.2	-32.3	-25.3
	25°	6	1	24.1	3.9	17.4	4.0	-10.7	-14.6	-7.7	-11.7	-19.9	-17.0
			2	—	—	—	—	-4.1	-7.9	-1.1	-5.1	—	—
	30° to 45°	7 to 12	1	21.6	14.8	17.2	11.8	1.7	-13.1	0.6	-11.3	-7.6	-8.7
			2	21.6	14.8	17.2	11.8	8.3	-6.5	7.2	-4.6	-7.6	-8.7
120	0 to 5°	Flat	1	22.8	-11.9	15.1	-7.0	-27.4	-15.6	-19.1	-12.1	-38.4	-30.1
	10°	2	1	25.8	-10.7	17.1	-6.2	-27.4	-16.8	-19.1	-12.9	-38.4	-30.1
	15°	3	1	28.7	-9.5	19.1	-5.4	-27.4	-17.9	-19.1	-13.7	-38.4	-30.1
	20°	4	1	31.6	-8.3	21.1	-4.6	-27.4	-19.1	-19.1	-14.5	-38.4	-30.1
	25°	6	1	28.6	4.6	20.7	4.7	-12.7	-17.3	-9.2	-13.9	-23.7	-20.2
			2	—	—	—	—	-4.8	-9.4	-1.3	-6.0	—	—
	30° to 45°	7 to 12	1	25.7	17.6	20.4	14.0	2.0	-15.6	0.7	-13.4	-9.0	-10.3
			2	25.7	17.6	20.4	14.0	9.9	-7.7	8.6	-5.5	-9.0	-10.3
130	0 to 5°	Flat	1	26.8	-13.9	17.8	-8.2	-32.2	-18.3	-22.4	-14.2	-45.1	-35.3
	10°	2	1	30.2	-12.5	20.1	-7.3	-32.2	-19.7	-22.4	-15.1	-45.1	-35.3
	15°	3	1	33.7	-11.2	22.4	-6.4	-32.2	-21.0	-22.4	-16.1	-45.1	-35.3
	20°	4	1	37.1	-9.8	24.7	-5.4	-32.2	-22.4	-22.4	-17.0	-45.1	-35.3
	25°	6	1	33.6	5.4	24.3	5.5	-14.9	-20.4	-10.8	-16.4	-27.8	-23.7
			2	—	—	—	—	-5.7	-11.1	-1.5	-7.1	—	—
	30° to 45°	7 to 12	1	30.1	20.6	24.0	16.5	2.3	-18.3	0.8	-15.7	-10.6	-12.1
			2	30.1	20.6	24.0	16.5	11.6	-9.0	10.0	-6.4	-10.6	-12.1

continued

TABLE 1609.6.2.1(1)-continued
SIMPLIFIED DESIGN WIND PRESSURE (MAIN WINDFORCE-RESISTING SYSTEM), p_{s30} (Exposure B at h = 30 feet with I_w = 1.0) (psf)

BASIC WIND SPEED (mph)	ROOF ANGLE (degrees)	ROOF RISE IN 12"	LOAD CASE	ZONES									
				Horizontal Pressures				Vertical Pressures				Overhangs	
				A	B	C	D	E	F	G	H	E_{OH}	G_{OH}
140	0 to 5°	Flat	1	31.1	-16.1	20.6	-9.6	-37.3	-21.2	-26.0	-16.4	-52.3	-40.9
	10°	2	1	35.1	-14.5	23.3	-8.5	-37.3	-22.8	-26.0	-17.5	-52.3	-40.9
	15°	3	1	39.0	-12.9	26.0	-7.4	-37.3	-24.4	-26.0	-18.6	-52.3	-40.9
	20°	4	1	43.0	-11.4	28.7	-6.3	-37.3	-26.0	-26.0	-19.7	-52.3	-40.9
	25°	6	1	39.0	6.3	28.2	6.4	-17.3	-23.6	-12.5	-19.0	-32.3	-27.5
			2	—	—	—	—	-6.6	-12.8	-1.8	-8.2	—	—
	30° to 45°	7 to 12	1	35.0	23.9	27.8	19.1	2.7	-21.2	0.9	-18.2	-12.3	-14.0
			2	35.0	23.9	27.8	19.1	13.4	-10.5	11.7	-7.5	-12.3	-14.0
150	0 to 5°	Flat	1	35.7	-18.5	23.7	-11.0	-42.9	-24.4	-29.8	-18.9	-60.0	-47.0
	10°	2	1	40.2	-16.7	26.8	-9.7	-42.9	-26.2	-29.8	-20.1	-60.0	-47.0
	15°	3	1	44.8	-14.9	29.8	-8.5	-42.9	-28.0	-29.8	-21.4	-60.0	-47.0
	20°	4	1	49.4	-13.0	32.9	-7.2	-42.9	-29.8	-29.8	-22.6	-60.0	-47.0
	25°	6	1	44.8	7.2	32.4	7.4	-19.9	-27.1	-14.4	-21.8	-37.0	-31.6
			2	—	—	—	—	-7.5	-14.7	-2.1	-9.4	—	—
	30° to 45°	7 to 12	1	40.1	27.4	31.9	22.0	3.1	-24.4	1.0	-20.9	-14.1	-16.1
			2	40.1	27.4	31.9	22.0	15.4	-12.0	13.4	-8.6	-14.1	-16.1
170	0 to 5°	Flat	1	45.8	-23.8	30.4	-14.1	-55.1	-31.3	-38.3	-24.2	-77.1	-60.4
	10°	2	1	51.7	-21.4	34.4	-12.5	-55.1	-33.6	-38.3	-25.8	-77.1	-60.4
	15°	3	1	57.6	-19.1	38.3	-10.9	-55.1	-36.0	-38.3	-27.5	-77.1	-60.4
	20°	4	1	63.4	-16.7	42.3	-9.3	-55.1	-38.3	-38.3	-29.1	-77.1	-60.4
	25°	6	1	57.5	9.3	41.6	9.5	-25.6	-34.8	-18.5	-28.0	-47.6	-40.5
			2	—	—	—	—	-9.7	-18.9	-2.6	-12.1	—	—
	30° to 45°	7 to 12	1	51.5	35.2	41.0	28.2	4.0	-31.3	1.3	-26.9	-18.1	-20.7
			2	51.5	35.2	41.0	28.2	19.8	-15.4	17.2	-11.0	-18.1	-20.7

For SI: 1 inch = 25.4 mm, 1 foot = 304.8 mm, 1 degree = 0.0174 rad, 1 mile per hour = 0.44 m/s, 1 pound per square foot = 47.9 N/m².

TABLE 1609.6.2.1(2)
NET DESIGN WIND PRESSURE (COMPONENT AND CLADDING), p_{net30} (Exposure B at h = 30 feet with I_w = 1.0) (psf)

	ZONE	EFFECTIVE WIND AREA	BASIC WIND SPEED V (mph—3-second gust)															
			85		90		100		110		120		130		140		150		170	
Roof 0 to 7 degrees	1	10	5.3	-13.0	5.9	-14.6	7.3	-18.0	8.9	-21.8	10.5	-25.9	12.4	-30.4	14.3	-35.3	16.5	-40.5	21.1	-52.0
	1	20	5.0	-12.7	5.6	-14.2	6.9	-17.5	8.3	-21.2	9.9	-25.2	11.6	-29.6	13.4	-34.4	15.4	-39.4	19.8	-50.7
	1	50	4.5	-12.2	5.1	-13.7	6.3	-16.9	7.6	-20.5	9.0	-24.4	10.6	-28.6	12.3	-33.2	14.1	-38.1	18.1	-48.9
	1	100	4.2	-11.9	4.7	-13.3	5.8	-16.5	7.0	-19.9	8.3	-23.7	9.8	-27.8	11.4	-32.3	13.0	-37.0	16.7	-47.6
	2	10	5.3	-21.8	5.9	-24.4	7.3	-30.2	8.9	-36.5	10.5	-43.5	12.4	-51.0	14.3	-59.2	16.5	-67.9	21.1	-87.2
	2	20	5.0	-19.5	5.6	-21.8	6.9	-27.0	8.3	-32.6	9.9	-38.8	11.6	-45.6	13.4	-52.9	15.4	-60.7	19.8	-78.0
	2	50	4.5	-16.4	5.1	-18.4	6.3	-22.7	7.6	-27.5	9.0	-32.7	10.6	-38.4	12.3	-44.5	14.1	-51.1	18.1	-65.7
	2	100	4.2	-14.1	4.7	-15.8	5.8	-19.5	7.0	-23.6	8.3	-28.1	9.8	-33.0	11.4	-38.2	13.0	-43.9	16.7	-56.4
	3	10	5.3	-32.8	5.9	-36.8	7.3	-45.4	8.9	-55.0	10.5	-65.4	12.4	-76.8	14.3	-89.0	16.5	-102.2	21.1	-131.3
	3	20	5.0	-27.2	5.6	-30.5	6.9	-37.6	8.3	-45.5	9.9	-54.2	11.6	-63.6	13.4	-73.8	15.4	-84.7	19.8	-108.7
	3	50	4.5	-19.7	5.1	-22.1	6.3	-27.3	7.6	-33.1	9.0	-39.3	10.6	-46.2	12.3	-53.5	14.1	-61.5	18.1	-78.9
	3	100	4.2	-14.1	4.7	-15.8	5.8	-19.5	7.0	-23.6	8.3	-28.1	9.8	-33.0	11.4	-38.2	13.0	-43.9	16.7	-56.4
Roof > 7 to 27 degrees	1	10	7.5	-11.9	8.4	-13.3	10.4	-16.5	12.5	-19.9	14.9	-23.7	17.5	-27.8	20.3	-32.3	23.3	-37.0	30.0	-47.6
	1	20	6.8	-11.6	7.7	-13.0	9.4	-16.0	11.4	-19.4	13.6	-23.0	16.0	-27.0	18.5	-31.4	21.3	-36.0	27.3	-46.3
	1	50	6.0	-11.1	6.7	-12.5	8.2	-15.4	10.0	-18.6	11.9	-22.2	13.9	-26.0	16.1	-30.2	18.5	-34.6	23.8	-44.5
	1	100	5.3	-10.8	5.9	-12.1	7.3	-14.9	8.9	-18.1	10.5	-21.5	12.4	-25.2	14.3	-29.3	16.5	-33.6	21.1	-43.2
	2	10	7.5	-20.7	8.4	-23.2	10.4	-28.7	12.5	-34.7	14.9	-41.3	17.5	-48.4	20.3	-56.2	23.3	-64.5	30.0	-82.8
	2	20	6.8	-19.0	7.7	-21.4	9.4	-26.4	11.4	-31.9	13.6	-38.0	16.0	-44.6	18.5	-51.7	21.3	-59.3	27.3	-76.2
	2	50	6.0	-16.9	6.7	-18.9	8.2	-23.3	10.0	-28.2	11.9	-33.6	13.9	-39.4	16.1	-45.7	18.5	-52.5	23.8	-67.4
	2	100	5.3	-15.2	5.9	-17.0	7.3	-21.0	8.9	-25.5	10.5	-30.3	12.4	-35.6	14.3	-41.2	16.5	-47.3	21.1	-60.8
	3	10	7.5	-30.6	8.4	-34.3	10.4	-42.4	12.5	-51.3	14.9	-61.0	17.5	-71.6	20.3	-83.1	23.3	-95.4	30.0	-122.5
	3	20	6.8	-28.6	7.7	-32.1	9.4	-39.6	11.4	-47.9	13.6	-57.1	16.0	-67.0	18.5	-77.7	21.3	-89.2	27.3	-114.5
	3	50	6.0	-26.0	6.7	-29.1	8.2	-36.0	10.0	-43.5	11.9	-51.8	13.9	-60.8	16.1	-70.5	18.5	-81.0	23.8	-104.0
	3	100	5.3	-24.0	5.9	-26.9	7.3	-33.2	8.9	-40.2	10.5	-47.9	12.4	-56.2	14.3	-65.1	16.5	-74.8	21.1	-96.0
Roof > 27 to 45 degrees	1	10	11.9	-13.0	13.3	-14.6	16.5	-18.0	19.9	-21.8	23.7	-25.9	27.8	-30.4	32.3	-35.3	37.0	-40.5	47.6	-52.0
	1	20	11.6	-12.3	13.0	-13.8	16.0	-17.1	19.4	-20.7	23.0	-24.6	27.0	-28.9	31.4	-33.5	36.0	-38.4	46.3	-49.3
	1	50	11.1	-11.5	12.5	-12.8	15.4	-15.9	18.6	-19.2	22.2	-22.8	26.0	-26.8	30.2	-31.1	34.6	-35.7	44.5	-45.8
	1	100	10.8	-10.8	12.1	-12.1	14.9	-14.9	18.1	-18.1	21.5	-21.5	25.2	-25.2	29.3	-29.3	33.6	-33.6	43.2	-43.2
	2	10	11.9	-15.2	13.3	-17.0	16.5	-21.0	19.9	-25.5	23.7	-30.3	27.8	-35.6	32.3	-41.2	37.0	-47.3	47.6	-60.8
	2	20	11.6	-14.5	13.0	-16.3	16.0	-20.1	19.4	-24.3	23.0	-29.0	27.0	-34.0	31.4	-39.4	36.0	-45.3	46.3	-58.1
	2	50	11.1	-13.7	12.5	-15.3	15.4	-18.9	18.6	-22.9	22.2	-27.2	26.0	-32.0	30.2	-37.1	34.6	-42.5	44.5	-54.6
	2	100	10.8	-13.0	12.1	-14.6	14.9	-18.0	18.1	-21.8	21.5	-25.9	25.2	-30.4	29.3	-35.3	33.6	-40.5	43.2	-52.0
	3	10	11.9	-15.2	13.3	-17.0	16.5	-21.0	19.9	-25.5	23.7	-30.3	27.8	-35.6	32.3	-41.2	37.0	-47.3	47.6	-60.8
	3	20	11.6	-14.5	13.0	-16.3	16.0	-20.1	19.4	-24.3	23.0	-29.0	27.0	-34.0	31.4	-39.4	36.0	-45.3	46.3	-58.1
	3	50	11.1	-13.7	12.5	-15.3	15.4	-18.9	18.6	-22.9	22.2	-27.2	26.0	-32.0	30.2	-37.1	34.6	-42.5	44.5	-54.6
	3	100	10.8	-13.0	12.1	-14.6	14.9	-18.0	18.1	-21.8	21.5	-25.9	25.2	-30.4	29.3	-35.3	33.6	-40.5	43.2	-52.0
Wall	4	10	13.0	-14.1	14.6	-15.8	18.0	-19.5	21.8	-23.6	25.9	-28.1	30.4	-33.0	35.3	-38.2	40.5	-43.9	52.0	-56.4
	4	20	12.4	-13.5	13.9	-15.1	17.2	-18.7	20.8	-22.6	24.7	-26.9	29.0	-31.6	33.7	-36.7	38.7	-42.1	49.6	-54.1
	4	50	11.6	-12.7	13.0	-14.3	16.1	-17.6	19.5	-21.3	23.2	-25.4	27.2	-29.8	31.6	-34.6	36.2	-39.7	46.6	-51.0
	4	100	11.1	-12.2	12.4	-13.6	15.3	-16.8	18.5	-20.4	22.0	-24.2	25.9	-28.4	30.0	-33.0	34.4	-37.8	44.2	-48.6
	4	500	9.7	-10.8	10.9	-12.1	13.4	-14.9	16.2	-18.1	19.3	-21.5	22.7	-25.2	26.3	-29.3	30.2	-33.6	38.8	-43.2
	5	10	13.0	-17.4	14.6	-19.5	18.0	-24.1	21.8	-29.1	25.9	-34.7	30.4	-40.7	35.3	-47.2	40.5	-54.2	52.0	-69.6
	5	20	12.4	-16.2	13.9	-18.2	17.2	-22.5	20.8	-27.2	24.7	-32.4	29.0	-38.0	33.7	-44.0	38.7	-50.5	49.6	-64.9
	5	50	11.6	-14.7	13.0	-16.5	16.1	-20.3	19.5	-24.6	23.2	-29.3	27.2	-34.3	31.6	-39.8	36.2	-45.7	46.6	-58.7
	5	100	11.1	-13.5	12.4	-15.1	15.3	-18.7	18.5	-22.6	22.0	-26.9	25.9	-31.6	30.0	-36.7	34.4	-42.1	44.2	-54.1
	5	500	9.7	-10.8	10.9	-12.1	13.4	-14.9	16.2	-18.1	19.3	-21.5	22.7	-25.2	26.3	-29.3	30.2	-33.6	38.8	-43.2

For SI: 1 foot = 304.8 mm, 1 degree = 0.0174 rad, 1 mile per hour = 0.44 m/s, 1 pound per square foot = 47.9 N/m².

Note: For effective areas between those given above, the load is permitted to be interpolated, otherwise use the load associated with the lower effective area.

TABLE 1609.6.2.1(3)
ROOF OVERHANG NET DESIGN WIND PRESSURE (COMPONENT AND CLADDING), p_{net30} **(Exposure B at** h **= 30 feet with** I_w **= 1.0) (psf)**

	ZONE	EFFECTIVE WIND AREA (sq. ft.)	BASIC WIND SPEED V (mph—3-second gust)							
			90	100	110	120	130	140	150	170
Roof 0 to 7 degrees	2	10	-21.0	-25.9	-31.4	-37.3	-43.8	-50.8	-58.3	-74.9
	2	20	-20.6	-25.5	-30.8	-36.7	-43.0	-49.9	-57.3	-73.6
	2	50	-20.1	-24.9	-30.1	-35.8	-42.0	-48.7	-55.9	-71.8
	2	100	-19.8	-24.4	-29.5	-35.1	-41.2	-47.8	-54.9	-70.5
	3	10	-34.6	-42.7	-51.6	-61.5	-72.1	-83.7	-96.0	-123.4
	3	20	-27.1	-33.5	-40.5	-48.3	-56.6	-65.7	-75.4	-96.8
	3	50	-17.3	-21.4	-25.9	-30.8	-36.1	-41.9	-48.1	-61.8
	3	100	-10.0	-12.2	-14.8	-17.6	-20.6	-23.9	-27.4	-35.2
Roof > 7 to 27 degrees	2	10	-27.2	-33.5	-40.6	-48.3	-56.7	-65.7	-75.5	-96.9
	2	20	-27.2	-33.5	-40.6	-48.3	-56.7	-65.7	-75.5	-96.9
	2	50	-27.2	-33.5	-40.6	-48.3	-56.7	-65.7	-75.5	-96.9
	2	100	-27.2	-33.5	-40.6	-48.3	-56.7	-65.7	-75.5	-96.9
	3	10	-45.7	-56.4	-68.3	-81.2	-95.3	-110.6	-126.9	-163.0
	3	20	-41.2	-50.9	-61.6	-73.3	-86.0	-99.8	-114.5	-147.1
	3	50	-35.3	-43.6	-52.8	-62.8	-73.7	-85.5	-98.1	-126.1
	3	100	-30.9	-38.1	-46.1	-54.9	-64.4	-74.7	-85.8	-110.1
Roof > 27 to 45 degrees	2	10	-24.7	-30.5	-36.9	-43.9	-51.5	-59.8	-68.6	-88.1
	2	20	-24.0	-29.6	-35.8	-42.6	-50.0	-58.0	-66.5	-85.5
	2	50	-23.0	-28.4	-34.3	-40.8	-47.9	-55.6	-63.8	-82.0
	2	100	-22.2	-27.4	-33.2	-39.5	-46.4	-53.8	-61.7	-79.3
	3	10	-24.7	-30.5	-36.9	-43.9	-51.5	-59.8	-68.6	-88.1
	3	20	-24.0	-29.6	-35.8	-42.6	-50.0	-58.0	-66.5	-85.5
	3	50	-23.0	-28.4	-34.3	-40.8	-47.9	-55.5	-63.8	-82.2
	3	100	-22.2	-27.4	-33.2	-39.5	-46.4	-53.8	-61.7	-79.3

For SI: 1 foot = 304.8 mm, 1 degree = 0.0174 rad, 1 mile per hour = 0.45 m/s, 1 pound per square foot = 47.9 N/m^2.
Note: For effective areas between those given above, the load is permitted to be interpolated, otherwise use the load associated with the lower effective area.

TABLE 1609.6.2.1(4)
ADJUSTMENT FACTOR FOR BUILDING HEIGHT AND EXPOSURE, (λ)

MEAN ROOF HEIGHT (feet)	EXPOSURE		
	B	C	D
15	1.00	1.21	1.47
20	1.00	1.29	1.55
25	1.00	1.35	1.61
30	1.00	1.40	1.66
35	1.05	1.45	1.70
40	1.09	1.49	1.74
45	1.12	1.53	1.78
50	1.16	1.56	1.81
55	1.19	1.59	1.84
60	1.22	1.62	1.87

For SI: 1 foot = 304.8 mm.
a. All table values shall be adjusted for other exposures and heights by multiplying by the above coefficients.

1609.7 Roof systems.

1609.7.1 Roof deck. The roof deck shall be designed to withstand the wind pressures determined under either the provisions of Section 1609.6 for buildings with a mean roof height not exceeding 60 feet (18 288 mm) or Section 1609.1.1 for buildings of any height.

1609.7.2 Roof coverings. Roof coverings shall comply with Section 1609.7.1.

> **Exception:** Rigid tile roof coverings that are air permeable and installed over a roof deck complying with Section 1609.7.1 are permitted to be designed in accordance with Section 1609.7.3.

1609.7.3 Rigid tile. Wind loads on rigid tile roof coverings shall be determined in accordance with the following equation:

$$M_a = q_h C_L b L L_a [1.0 - Gc_p] \qquad \text{(Equation 16-36)}$$

For SI: $M_a = \dfrac{q_h C_L b L L_a [1.0 - Gc_p]}{1,000}$

where:

b = Exposed width, feet (mm) of the roof tile.

C_L = Lift coefficient. The lift coefficient for concrete and clay tile shall be 0.2 or shall be determined by test in accordance with Section 1715.2.

GC_p = Roof pressure coefficient for each applicable roof zone determined from Section 6 of ASCE 7. Roof coefficients shall not be adjusted for internal pressure.

L = Length, feet (mm) of the roof tile.

L_a = Moment arm, feet (mm) from the axis of rotation to the point of uplift on the roof tile. The point of uplift shall be taken at 0.76L from the head of the tile and the middle of the exposed width. For roof tiles with nails or screws (with or without a tail clip), the axis of rotation shall be taken as the head of the tile for direct deck application or as the top edge of the batten for battened applications. For roof tiles fastened only by a nail or screw along the side of the tile, the axis of rotation shall be determined by testing. For roof tiles installed with battens and fastened only by a clip near the tail of the tile, the moment arm shall be determined about the top edge of the batten with consideration given for the point of rotation of the tiles based on straight bond or broken bond and the tile profile.

M_a = Aerodynamic uplift moment, feet-pounds (N-mm) acting to raise the tail of the tile.

q_h = Wind velocity pressure, psf (kN/m²) determined from Section 6.5.10 of ASCE 7.

Concrete and clay roof tiles complying with the following limitations shall be designed to withstand the aerodynamic uplift moment as determined by this section.

1. The roof tiles shall be either loose laid on battens, mechanically fastened, mortar set or adhesive set.

2. The roof tiles shall be installed on solid sheathing which has been designed as components and cladding.

3. An underlayment shall be installed in accordance with Chapter 15.

4. The tile shall be single lapped interlocking with a minimum head lap of not less than 2 inches (51 mm).

5. The length of the tile shall be between 1.0 and 1.75 feet (305 mm and 533 mm).

6. The exposed width of the tile shall be between 0.67 and 1.25 feet (204 mm and 381 mm).

7. The maximum thickness of the tail of the tile shall not exceed 1.3 inches (33 mm).

8. Roof tiles using mortar set or adhesive set systems shall have at least two-thirds of the tile's area free of mortar or adhesive contact.

SECTION 1610
SOIL LATERAL LOAD

1610.1 General. Basement, foundation and retaining walls shall be designed to resist lateral soil loads. Soil loads specified in Table 1610.1 shall be used as the minimum design lateral soil loads unless specified otherwise in a soil investigation report approved by the building official. Basement walls and other walls in which horizontal movement is restricted at the top shall be designed for at-rest pressure. Retaining walls free to move and rotate at the top are permitted to be designed for active pressure. Design lateral pressure from surcharge loads shall be added to the lateral earth pressure load. Design lateral pressure shall be increased if soils with expansion potential are present at the site.

> **Exception:** Basement walls extending not more than 8 feet (2438 mm) below grade and supporting flexible floor systems shall be permitted to be designed for active pressure.

SECTION 1611
RAIN LOADS

1611.1 Design rain loads. Each portion of a roof shall be designed to sustain the load of rainwater that will accumulate on it if the primary drainage system for that portion is blocked plus the uniform load caused by water that rises above the inlet of the secondary drainage system at its design flow.

$$R = 5.2 (d_s + d_h) \qquad \text{(Equation 16-37)}$$

For SI: $R = 0.0098 (d_s + d_h)$

where:

d_h = Additional depth of water on the undeflected roof above the inlet of secondary drainage system at its design flow (i.e., the hydraulic head), in inches (mm).

d_s = Depth of water on the undeflected roof up to the inlet of secondary drainage system when the primary drainage system is blocked (i.e., the static head), in inches (mm).

R = Rain load on the undeflected roof, in psf (kN/m²). When the phrase "undeflected roof" is used, deflections from loads (including dead loads) shall not be considered when determining the amount of rain on the roof.

TABLE 1610.1
SOIL LATERAL LOAD

DESCRIPTION OF BACKFILL MATERIAL[c]	UNIFIED SOIL CLASSIFICATION	DESIGN LATERAL SOIL LOAD[a] (pound per square foot per foot of depth)	
		Active pressure	At-rest pressure
Well-graded, clean gravels; gravel-sand mixes	GW	30	60
Poorly graded clean gravels; gravel-sand mixes	GP	30	60
Silty gravels, poorly graded gravel-sand mixes	GM	40	60
Clayey gravels, poorly graded gravel-and-clay mixes	GC	45	60
Well-graded, clean sands; gravelly sand mixes	SW	30	60
Poorly graded clean sands; sand-gravel mixes	SP	30	60
Silty sands, poorly graded sand-silt mixes	SM	45	60
Sand-silt clay mix with plastic fines	SM-SC	45	100
Clayey sands, poorly graded sand-clay mixes	SC	60	100
Inorganic silts and clayey silts	ML	45	100
Mixture of inorganic silt and clay	ML-CL	60	100
Inorganic clays of low to medium plasticity	CL	60	100
Organic silts and silt clays, low plasticity	OL	Note b	Note b
Inorganic clayey silts, elastic silts	MH	Note b	Note b
Inorganic clays of high plasticity	CH	Note b	Note b
Organic clays and silty clays	OH	Note b	Note b

For SI: 1 pound per square foot per foot of depth = 0.157 kPa/m, 1 foot = 304.8 mm.

a. Design lateral soil loads are given for moist conditions for the specified soils at their optimum densities. Actual field conditions shall govern. Submerged or saturated soil pressures shall include the weight of the buoyant soil plus the hydrostatic loads.

b. Unsuitable as backfill material.

c. The definition and classification of soil materials shall be in accordance with ASTM D 2487.

1611.2 Ponding instability. Ponding refers to the retention of water due solely to the deflection of relatively flat roofs. Roofs with a slope less than one-fourth unit vertical in 12 units horizontal (2-percent slope) shall be investigated by structural analysis to ensure that they possess adequate stiffness to preclude progressive deflection (i.e., instability) as rain falls on them or meltwater is created from snow on them. The larger of snow load or rain load shall be used in this analysis. The primary drainage system within an area subjected to ponding shall be considered to be blocked in this analysis.

1611.3 Controlled drainage. Roofs equipped with hardware to control the rate of drainage shall be equipped with a secondary drainage system at a higher elevation that limits accumulation of water on the roof above that elevation. Such roofs shall be designed to sustain the load of rainwater that will accumulate on them to the elevation of the secondary drainage system plus the uniform load caused by water that rises above the inlet of the secondary drainage system at its design flow determined from Section 1611.1. Such roofs shall also be checked for ponding instability in accordance with Section 1611.2.

SECTION 1612
FLOOD LOADS

1612.1 General. Within flood hazard areas as established in Section 1612.3, all new construction of buildings, structures and portions of buildings and structures, including substantial improvements and restoration of substantial damage to buildings and structures, shall be designed and constructed to resist the effects of flood hazards and flood loads.

1612.2 Definitions. The following words and terms shall, for the purposes of this section, have the meanings shown herein.

BASE FLOOD. The flood having a 1-percent chance of being equaled or exceeded in any given year.

BASE FLOOD ELEVATION. The elevation of the base flood, including wave height, relative to the National Geodetic Vertical Datum (NGVD), North American Vertical Datum (NAVD) or other datum specified on the Flood Insurance Rate Map (FIRM).

BASEMENT. The portion of a building having its floor subgrade (below ground level) on all sides.

DESIGN FLOOD. The flood associated with the greater of the following two areas:

1. Area with a flood plain subject to a 1-percent or greater chance of flooding in any year; or

2. Area designated as a flood hazard area on a community's flood hazard map, or otherwise legally designated.

DESIGN FLOOD ELEVATION. The elevation of the "design flood," including wave height, relative to the datum specified on the community's legally designated flood hazard map. In areas designated as Zone AO, the design flood elevation shall be the elevation of the highest existing grade of the building's perimeter plus the depth number (in feet) specified on the flood hazard map. In areas designated as Zone AO where a depth number is not specified on the map, the depth number shall be taken as being equal to 2 feet (610 mm).

DRY FLOODPROOFING. A combination of design modifications that results in a building or structure, including the attendant utility and sanitary facilities, being water tight with walls substantially impermeable to the passage of water and with structural components having the capacity to resist loads as identified in ASCE 7.

EXISTING CONSTRUCTION. Any buildings and structures for which the "start of construction" commenced before the effective date of the community's first flood plain management code, ordinance or standard. "Existing construction" is also referred to as "existing structures."

EXISTING STRUCTURE. See "Existing construction."

FLOOD or FLOODING. A general and temporary condition of partial or complete inundation of normally dry land from:

1. The overflow of inland or tidal waters.
2. The unusual and rapid accumulation or runoff of surface waters from any source.

FLOOD DAMAGE-RESISTANT MATERIALS. Any construction material capable of withstanding direct and prolonged contact with floodwaters without sustaining any damage that requires more than cosmetic repair.

FLOOD HAZARD AREA. The greater of the following two areas:

1. The area within a flood plain subject to a 1-percent or greater chance of flooding in any year.
2. The area designated as a flood hazard area on a community's flood hazard map, or otherwise legally designated.

FLOOD HAZARD AREA SUBJECT TO HIGH VELOCITY WAVE ACTION. Area within the flood hazard area that is subject to high velocity wave action, and shown on a Flood Insurance Rate Map (FIRM) or other flood hazard map as Zone V, VO, VE or V1-30.

FLOOD INSURANCE RATE MAP (FIRM). An official map of a community on which the Federal Emergency Management Agency (FEMA) has delineated both the special flood hazard areas and the risk premium zones applicable to the community.

FLOOD INSURANCE STUDY. The official report provided by the Federal Emergency Management Agency containing the Flood Insurance Rate Map (FIRM), the Flood Boundary and Floodway Map (FBFM), the water surface elevation of the base flood and supporting technical data.

FLOODWAY. The channel of the river, creek or other watercourse and the adjacent land areas that must be reserved in order to discharge the base flood without cumulatively increasing the water surface elevation more than a designated height.

LOWEST FLOOR. The floor of the lowest enclosed area, including basement, but excluding any unfinished or flood-resistant enclosure, usable solely for vehicle parking, building access or limited storage provided that such enclosure

is not built so as to render the structure in violation of this section.

SPECIAL FLOOD HAZARD AREA. The land area subject to flood hazards and shown on a Flood Insurance Rate Map or other flood hazard map as Zone A, AE, A1-30, A99, AR, AO, AH, V, VO, VE or V1-30.

START OF CONSTRUCTION. The date of permit issuance for new construction and substantial improvements to existing structures, provided the actual start of construction, repair, reconstruction, rehabilitation, addition, placement or other improvement is within 180 days after the date of issuance. The actual start of construction means the first placement of permanent construction of a building (including a manufactured home) on a site, such as the pouring of a slab or footings, installation of pilings or construction of columns.

Permanent construction does not include land preparation (such as clearing, excavation, grading or filling), the installation of streets or walkways, excavation for a basement, footings, piers or foundations, the erection of temporary forms or the installation of accessory buildings such as garages or sheds not occupied as dwelling units or not part of the main building. For a substantial improvement, the actual "start of construction" means the first alteration of any wall, ceiling, floor or other structural part of a building, whether or not that alteration affects the external dimensions of the building.

SUBSTANTIAL DAMAGE. Damage of any origin sustained by a structure whereby the cost of restoring the structure to its before-damaged condition would equal or exceed 50 percent of the market value of the structure before the damage occurred.

SUBSTANTIAL IMPROVEMENT. Any repair, reconstruction, rehabilitation, addition or improvement of a building or structure, the cost of which equals or exceeds 50 percent of the market value of the structure before the improvement or repair is started. If the structure has sustained substantial damage, any repairs are considered substantial improvement regardless of the actual repair work performed. The term does not, however, include either:

1. Any project for improvement of a building required to correct existing health, sanitary or safety code violations identified by the building official and that are the minimum necessary to assure safe living conditions.
2. Any alteration of a historic structure provided that the alteration will not preclude the structure's continued designation as a historic structure.

1612.3 Establishment of flood hazard areas. To establish flood hazard areas, the governing body shall adopt a flood hazard map and supporting data. The flood hazard map shall include, at a minimum, areas of special flood hazard as identified by the Federal Emergency Management Agency in an engineering report entitled "The Flood Insurance Study for [INSERT NAME OF JURISDICTION]," dated [INSERT DATE OF ISSUANCE], as amended or revised with the accompanying Flood Insurance Rate Map (FIRM) and Flood Boundary and Floodway Map (FBFM) and related supporting data along with any revisions thereto. The adopted flood hazard map and supporting data are hereby adopted by reference and declared to be part of this section.

1612.4 Design and construction. The design and construction of buildings and structures located in flood hazard areas, including flood hazard areas subject to high velocity wave action, shall be in accordance with ASCE 24.

1612.5 Flood hazard documentation. The following documentation shall be prepared and sealed by a registered design professional and submitted to the building official:

1. For construction in flood hazard areas not subject to high-velocity wave action:

    1.1. The elevation of the lowest floor, including basement, as required by the lowest floor elevation inspection in Section 109.3.3.

    1.2. For fully enclosed areas below the design flood elevation where provisions to allow for the automatic entry and exit of floodwaters do not meet the minimum requirements in Section 2.6.1.1, ASCE 24, construction documents shall include a statement that the design will provide for equalization of hydrostatic flood forces in accordance with Section 2.6.1.2, ASCE 24.

    1.3. For dry floodproofed nonresidential buildings, construction documents shall include a statement that the dry floodproofing is designed in accordance with ASCE 24.

2. For construction in flood hazard areas subject to high-velocity wave action:

    2.1. The elevation of the bottom of the lowest horizontal structural member as required by the lowest floor elevation inspection in Section 109.3.3.

    2.2. Construction documents shall include a statement that the building is designed in accordance with ASCE 24, including that the pile or column foundation and building or structure to be attached thereto is designed to be anchored to resist flotation, collapse and lateral movement due to the effects of wind and flood loads acting simultaneously on all building components, and other load requirements of Chapter 16.

    2.3. For breakaway walls designed to resist a nominal load of less than 10 psf (0.48 kN/m^2) or more than 20 psf (0.96 kN/m^2), construction documents shall include a statement that the breakaway wall is designed in accordance with ASCE 24.

SECTION 1613
EARTHQUAKE LOADS DEFINITIONS

1613.1 Definitions. The following words and terms shall, for the purposes of this section, have the meanings shown herein.

ACTIVE FAULT/ACTIVE FAULT TRACE. A fault for which there is an average historic slip rate of 1 mm per year or more and geologic evidence of seismic activity within Holocene (past 11,000 years) times. Active fault traces are designated by the appropriate regulatory agency and/or registered design professional subject to identification by a geologic report.

ATTACHMENTS, SEISMIC. Means by which components and their supports are secured or connected to the seismic-force-resisting system of the structure. Such attachments include anchor bolts, welded connections and mechanical fasteners.

BASE. The level at which the horizontal seismic ground motions are considered to be imparted to the structure.

BOUNDARY ELEMENTS. Chords and collectors at diaphragm and shear wall edges, interior openings, discontinuities and reentrant corners.

BRITTLE. Systems, members, materials and connections that do not exhibit significant energy dissipation capacity in the inelastic range.

COLLECTOR. A diaphragm or shear wall element parallel to the applied load that collects and transfers shear forces to the vertical-force-resisting elements or distributes forces within a diaphragm or shear wall.

COMPONENT. A part or element of an architectural, electrical, mechanical or structural system.

Component, equipment. A mechanical or electrical component or element that is part of a mechanical and/or electrical system within or without a building system.

Component, flexible. Component, including its attachments, having a fundamental period greater than 0.06 second.

Component, rigid. Component, including its attachments, having a fundamental period less than or equal to 0.06 second.

DESIGN EARTHQUAKE. The earthquake effects that buildings and structures are specifically proportioned to resist in Sections 1613 through 1622.

DESIGNATED SEISMIC SYSTEM. Those architectural, electrical and mechanical systems and their components that require design in accordance with Section 1621 that have a component importance factor, I_p, greater than one.

DISPLACEMENT.

Design displacement. The design earthquake lateral displacement, excluding additional displacement due to actual and accidental torsion, required for design of the isolation system.

Total design displacement. The design earthquake lateral displacement, including additional displacement due to actual and accidental torsion, required for design of the isolation system.

Total maximum displacement. The maximum considered earthquake lateral displacement, including additional displacement due to actual and accidental torsion, required for verification of the stability of the isolation system or elements thereof, design of building separations and vertical load testing of isolator unit prototype.

DISPLACEMENT RESTRAINT SYSTEM. A collection of structural elements that limits lateral displacement of seismically isolated structures due to the maximum considered earthquake.

EFFECTIVE DAMPING. The value of equivalent viscous damping corresponding to energy dissipated during cyclic response of the isolation system.

EFFECTIVE STIFFNESS. The value of the lateral force in the isolation system, or an element thereof, divided by the corresponding lateral displacement.

HAZARDOUS CONTENTS. A material that is highly toxic or potentially explosive and in sufficient quantity to pose a significant life-safety threat to the general public if an uncontrolled release were to occur.

INVERTED PENDULUM-TYPE STRUCTURES. Structures that have a large portion of their mass concentrated near the top, and thus have essentially one degree of freedom in horizontal translation. The structures are usually T-shaped with a single column supporting the beams or framing at the top.

ISOLATION INTERFACE. The boundary between the upper portion of the structure, which is isolated, and the lower portion of the structure, which moves rigidly with the ground.

ISOLATION SYSTEM. The collection of structural elements that includes individual isolator units, structural elements that transfer force between elements of the isolation system and connections to other structural elements.

ISOLATOR UNIT. A horizontally flexible and vertically stiff structural element of the isolation system that permits large lateral deformations under design seismic load. An isolator unit is permitted to be used either as part of or in addition to the weight-supporting system of the building.

LOAD.

> **Gravity load (*W*).** The total dead load and applicable portions of other loads as defined in Sections 1613 through 1622.

MAXIMUM CONSIDERED EARTHQUAKE. The most severe earthquake effects considered by this code.

NONBUILDING STRUCTURE. A structure, other than a building, constructed of a type included in Section 1622.

OCCUPANCY IMPORTANCE FACTOR. A factor assigned to each structure according to its seismic use group as prescribed in Table 1604.5.

SEISMIC DESIGN CATEGORY. A classification assigned to a structure based on its seismic use group and the severity of the design earthquake ground motion at the site.

SEISMIC-FORCE-RESISTING SYSTEM. The part of the structural system that has been considered in the design to provide the required resistance to the seismic forces prescribed herein.

SEISMIC FORCES. The assumed forces prescribed herein, related to the response of the structure to earthquake motions, to be used in the design of the structure and its components.

SEISMIC USE GROUP. A classification assigned to a building based on its use as defined in Section 1616.2.

SHEAR WALL. A wall designed to resist lateral forces parallel to the plane of the wall.

SHEAR WALL-FRAME INTERACTIVE SYSTEM. A structural system that uses combinations of shear walls and frames designed to resist lateral forces in proportion to their rigidities, considering interaction between shear walls and frames on all levels.

SITE CLASS. A classification assigned to a site based on the types of soils present and their engineering properties as defined in Section 1615.1.5.

SITE COEFFICIENTS. The values of, F_a, and, F_v, indicated in Tables 1615.1.2(1) and 1615.1.2(2), respectively.

STORY DRIFT RATIO. The story drift divided by the story height.

TORSIONAL FORCE DISTRIBUTION. The distribution of horizontal seismic forces through a rigid diaphragm when the center of mass of the structure at the level under consideration does not coincide with the center of rigidity (sometimes referred to as a "diaphragm rotation").

TOUGHNESS. The ability of a material to absorb energy without losing significant strength.

WIND-RESTRAINT SEISMIC SYSTEM. The collection of structural elements that provides restraint of the seismic-isolated structure for wind loads. The wind-restraint system may be either an integral part of isolator units or a separate device.

SECTION 1614
EARTHQUAKE LOADS—GENERAL

1614.1 Scope. Every structure, and portion thereof, shall as a minimum, be designed and constructed to resist the effects of earthquake motions and assigned a seismic design category as set forth in Section 1616.3. Structures determined to be in Seismic Design Category A need only comply with Section 1616.4.

Exceptions:

1. Structures designed in accordance with the provisions of Sections 9.1 through 9.6, 9.13 and 9.14 of ASCE 7 shall be permitted.

2. Detached one- and two-family dwellings as applicable in Section 101.2 in Seismic Design Categories A, B and C, or located where the mapped short-period spectral response acceleration, S_S, is less than 0.4 g, are exempt from the requirements of Sections 1613 through 1622.

3. The seismic-force-resisting system of wood frame buildings that conform to the provisions of Section 2308 are not required to be analyzed as specified in Section 1616.1.

4. Agricultural storage structures intended only for incidental human occupancy are exempt from the requirements of Sections 1613 through 1623.

5. Structures located where mapped short-period spectral response acceleration, S_S, determined in accordance with Section 1615.1, is less than or equal to 0.15g and where the mapped spectral response acceleration at 1-second period, S_1, determined in accordance with Section 1615.1, is less than or equal to 0.04g shall be categorized as Seismic Design Category A. Seismic Design Category A structures need only comply with Section 1616.4.

6. Structures located where the short-period design spectral response acceleration, S_{DS}, determined in accordance with Section 1615.1, is less than or equal to 0.167g and the design spectral response acceleration at 1-second period, S_{D1}, determined in accordance with Section 1615.1, is less than or equal to 0.067g, shall be categorized as Seismic Design Category A and need only comply with Section 1616.4.

[EB] 1614.1.1 Additions to existing buildings. An addition that is structurally independent from an existing structure shall be designed and constructed as required for a new structure in accordance with the seismic requirements for new structures. An addition that is not structurally independent from an existing structure shall be designed and constructed such that the entire structure conforms to the seismic-force resistance requirements for new structures unless the following conditions are satisfied:

1. The addition conforms with the requirements for new structures,

2. The addition does not increase the seismic forces in any structural element of the existing structure by more than 5 percent, unless the element has the capacity to resist the increased forces determined in accordance with Sections 1613 through 1622, and

3. Additions do not decrease the seismic resistance of any structural element of the existing structure by more than 5 percent cumulative since the original construction, unless the element has the capacity to resist the forces determined in accordance with Sections 1613 through 1622.

[EB] 1614.2 Change of occupancy. When a change of occupancy results in a structure being reclassified to a higher seismic use group, the structure shall conform to the seismic requirements for a new structure.

Exceptions:

1. Specific detailing provisions required for a new structure are not required to be met where it can be shown an equivalent level of performance and seismic safety contemplated for a new structure is obtained. Such analysis shall consider the regularity, overstrength, redundancy and ductility of the structure within the context of the specific detailing provided.

2. When a change of use results in a structure being reclassified from Seismic Use Group I to Seismic Use Group II and the structure is located in a seismic map area where $S_{DS} < 0.33$, compliance with this section is not required.

[EB] 1614.3 Alterations. Alterations are permitted to be made to any structure without requiring the structure to comply with Sections 1613 through 1623 provided the alterations conform to the requirements for a new structure. Alterations that increase the seismic force in any existing structural element by more than 5 percent or decrease the design strength of any existing structural element to resist seismic forces by more than 5 percent shall not be permitted unless the entire seismic-force-resisting system is determined to conform to Sections 1613 through 1623 for a new structure.

Exception: Alterations to existing structural elements or additions of new structural elements that are not required by Sections 1613 through 1623 and are initiated for the purpose of increasing the strength or stiffness of the seismic-force-resisting system of an existing structure need not be designed for forces conforming to Sections 1613 through 1623 provided that an engineering analysis is submitted indicating the following:

1. The design strength of existing structural elements required to resist seismic forces is not reduced.

2. The seismic force to required existing structural elements is not increased beyond their design strength.

3. New structural elements are detailed and connected to the existing structural elements as required by this chapter.

4. New or relocated nonstructural elements are detailed and connected to existing or new structural elements as required by this chapter.

5. The alterations do not create a structural irregularity as defined in Section 1616.5 or make an existing structural irregularity more severe.

6. The alterations do not result in the creation of an unsafe condition.

1614.4 Quality assurance. A quality assurance plan shall be provided where required by Chapter 17.

1614.5 Seismic and wind. When the code-prescribed wind design produces greater effects, the wind design shall govern, but detailing requirements and limitations prescribed in this and referenced sections shall be followed.

SECTION 1615
EARTHQUAKE LOADS—SITE GROUND MOTION

1615.1 General procedure for determining maximum considered earthquake and design spectral response accelerations. Ground motion accelerations, represented by response spectra and coefficients derived from these spectra, shall be determined in accordance with the general procedure of Section 1615.1, or the site-specific procedure of Section 1615.2. The site-specific procedure of Section 1615.2 shall be used for structures on sites classified as Site Class F, in accordance with Section 1615.1.1.

The mapped maximum considered earthquake spectral response acceleration at short periods (S_S) and at 1-second period (S_1) shall be determined from Figures 1615(1) through (10). Where a site is between contours, straight-line interpolation or the value of the higher contour shall be used.

The site class shall be determined in accordance with Section 1615.1.1. The maximum considered earthquake spectral response accelerations at short period and 1-second period adjusted for site class effects, S_{MS} and S_{M1}, shall be determined in accordance with Section 1615.1.2. The design spectral response accelerations at short period, S_{DS}, and at 1-second period, S_{D1}, shall be determined in accordance with Section 1615.1.3. The general response spectrum shall be determined in accordance with Section 1615.1.4.

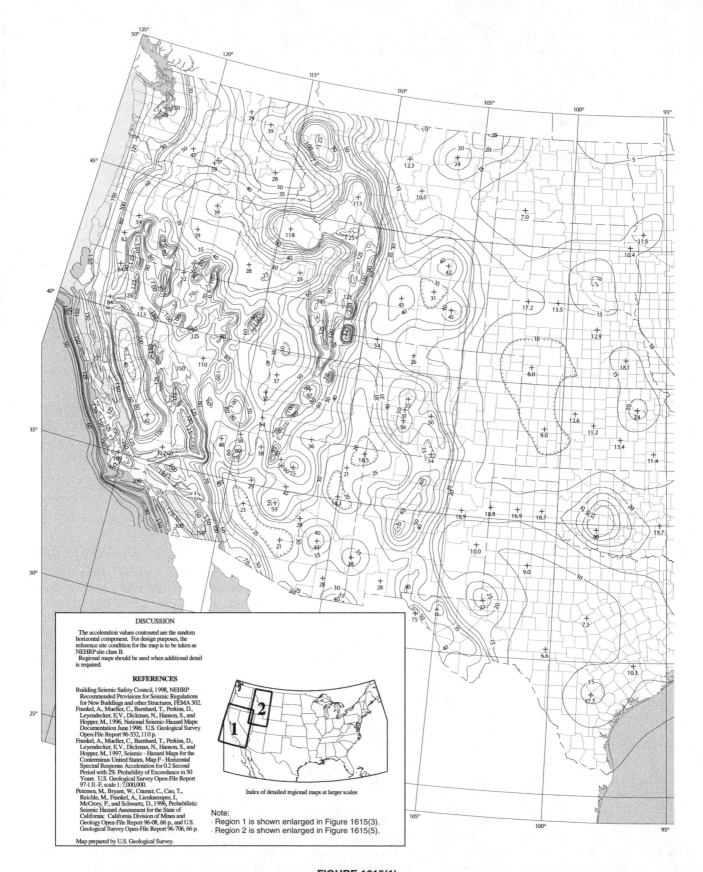

Index of detailed regional maps at larger scales

Note:
· Region 1 is shown enlarged in Figure 1615(3).
· Region 2 is shown enlarged in Figure 1615(5).

FIGURE 1615(1)
MAXIMUM CONSIDERED EARTHQUAKE GROUND MOTION FOR THE CONTERMINOUS UNITED STATES
OF 0.2 SEC SPECTRAL RESPONSE ACCELERATION (5 PERCENT OF CRITICAL DAMPING), SITE CLASS B

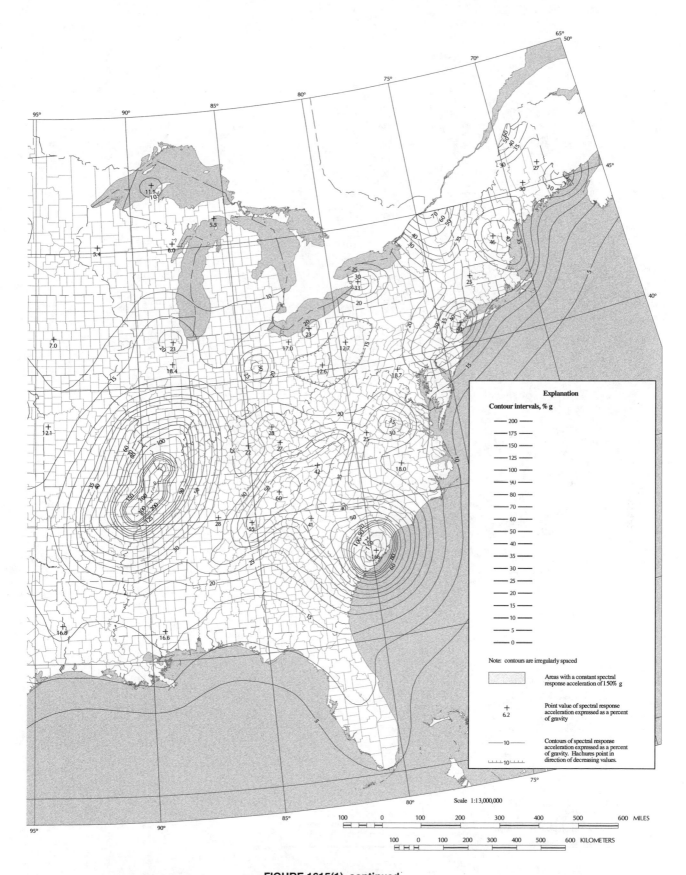

FIGURE 1615(1)–continued
**MAXIMUM CONSIDERED EARTHQUAKE GROUND MOTION FOR THE CONTERMINOUS UNITED STATES
OF 0.2 SEC SPECTRAL RESPONSE ACCELERATION (5 PERCENT OF CRITICAL DAMPING), SITE CLASS B**

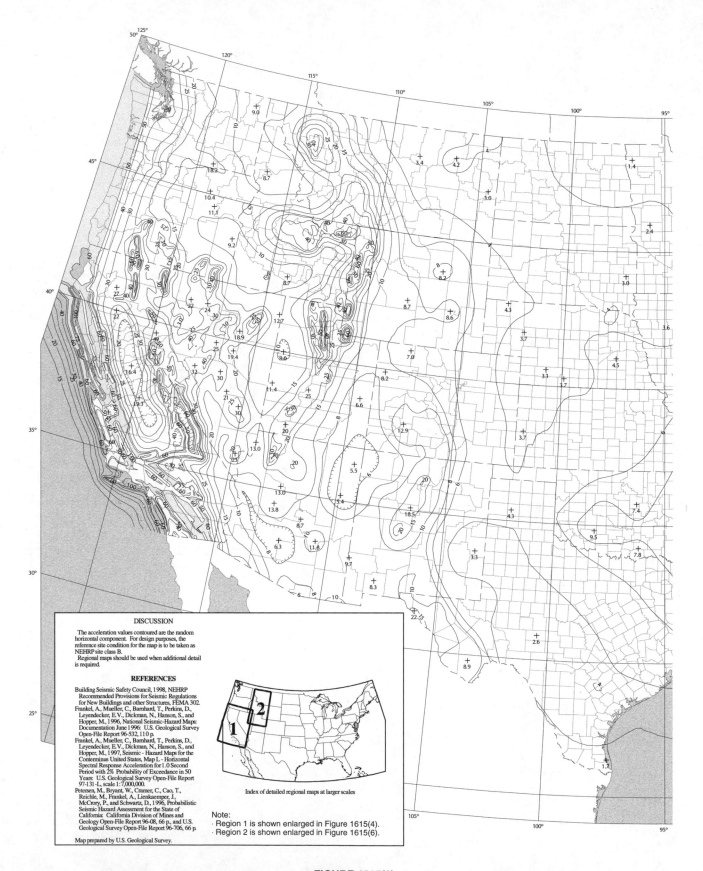

Index of detailed regional maps at larger scales

Note:
· Region 1 is shown enlarged in Figure 1615(4).
· Region 2 is shown enlarged in Figure 1615(6).

FIGURE 1615(2)
MAXIMUM CONSIDERED EARTHQUAKE GROUND MOTION FOR THE CONTERMINOUS UNITED STATES OF 1.0 SEC SPECTRAL RESPONSE ACCELERATION (5 PERCENT OF CRITICAL DAMPING), SITE CLASS B

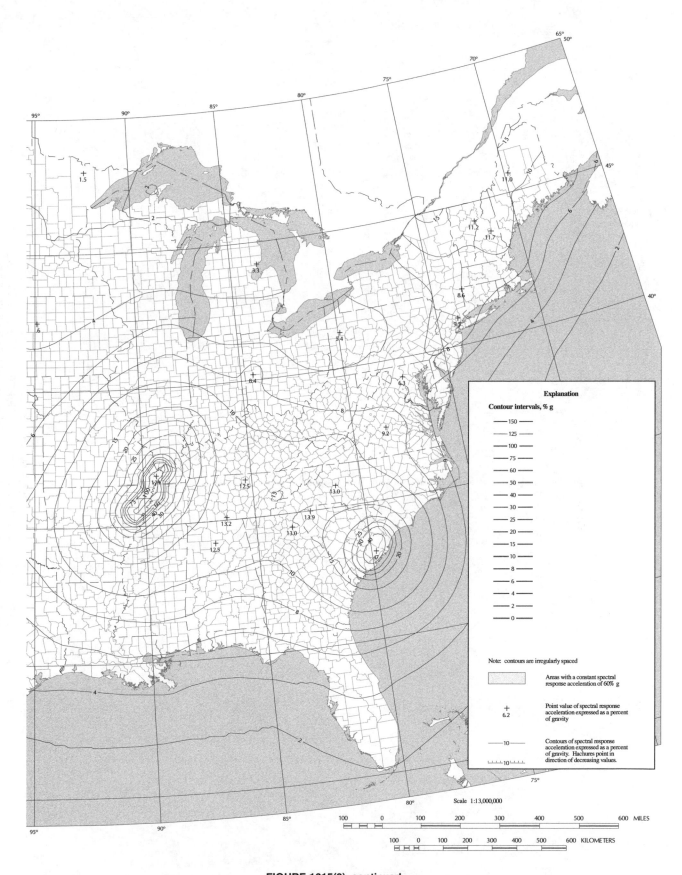

FIGURE 1615(2)–continued
MAXIMUM CONSIDERED EARTHQUAKE GROUND MOTION FOR THE CONTERMINOUS UNITED STATES
OF 1.0 SEC SPECTRAL RESPONSE ACCELERATION (5 PERCENT OF CRITICAL DAMPING), SITE CLASS B

FIGURE 1615(3)
MAXIMUM CONSIDERED EARTHQUAKE GROUND MOTION FOR REGION 1 OF 0.2 SEC SPECTRAL
RESPONSE ACCELERATION (5 PERCENT OF CRITICAL DAMPING), SITE CLASS B

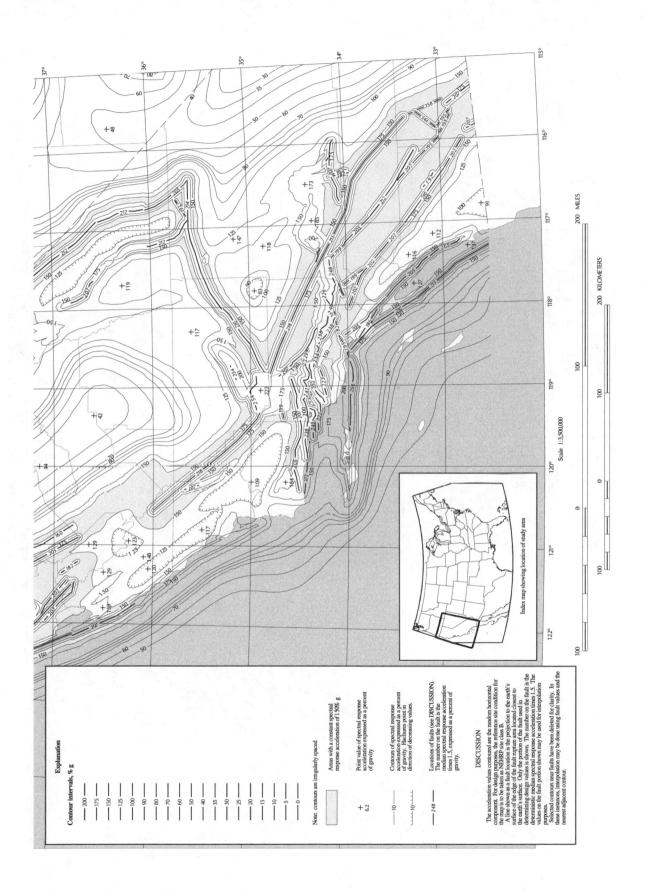

FIGURE 1615(3)–continued
MAXIMUM CONSIDERED EARTHQUAKE GROUND MOTION FOR REGION 1 OF 0.2 SEC SPECTRAL
RESPONSE ACCELERATION (5 PERCENT OF CRITICAL DAMPING), SITE CLASS B

FIGURE 1615(4)
MAXIMUM CONSIDERED EARTHQUAKE GROUND MOTION FOR REGION 1 OF 1.0 SEC SPECTRAL
RESPONSE ACCELERATION (5 PERCENT OF CRITICAL DAMPING), SITE CLASS B

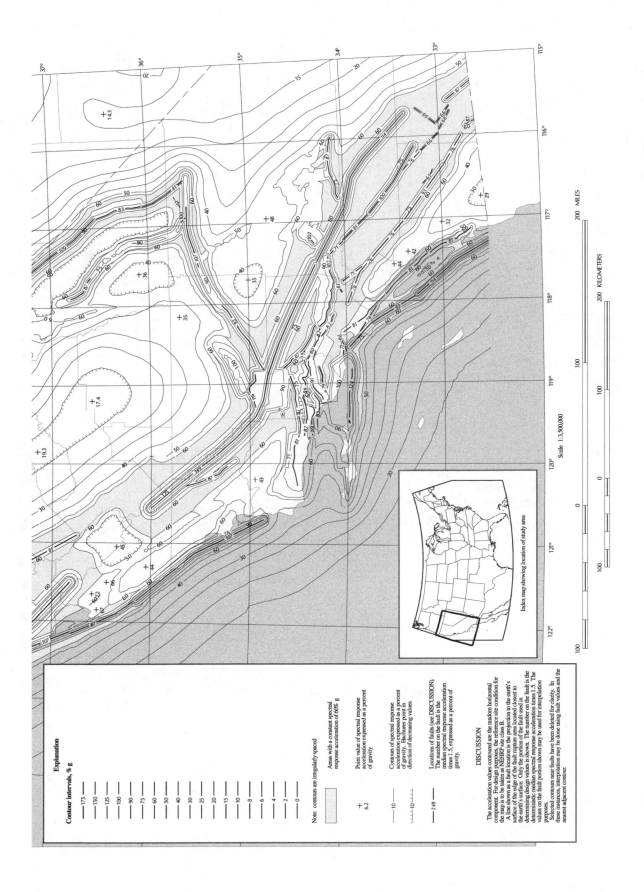

FIGURE 1615(4)—continued
MAXIMUM CONSIDERED EARTHQUAKE GROUND MOTION FOR REGION 1 OF 1.0 SEC SPECTRAL
RESPONSE ACCELERATION (5 PERCENT OF CRITICAL DAMPING), SITE CLASS B

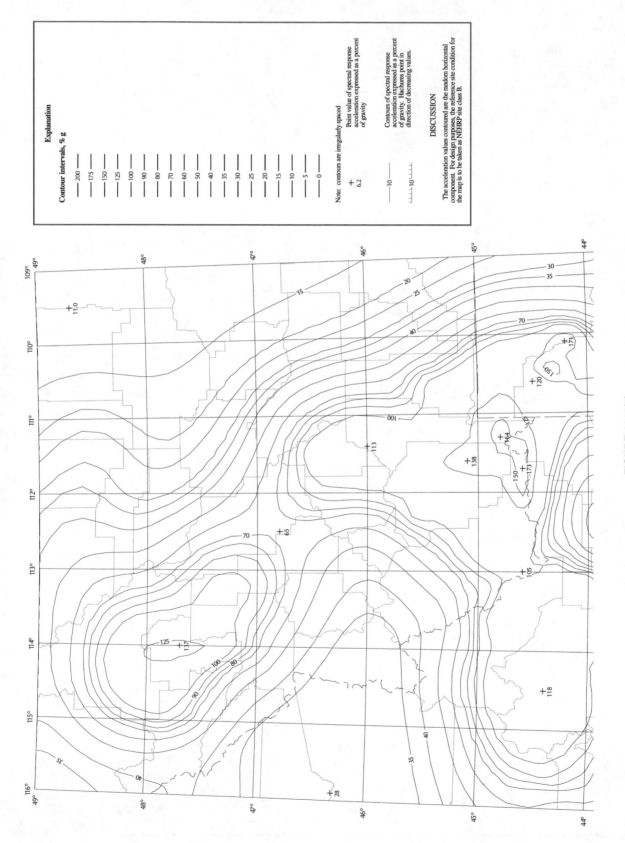

FIGURE 1615(5)
MAXIMUM CONSIDERED EARTHQUAKE GROUND MOTION FOR REGION 2 OF 0.2 SEC SPECTRAL RESPONSE ACCELERATION (5 PERCENT OF CRITICAL DAMPING), SITE CLASS B

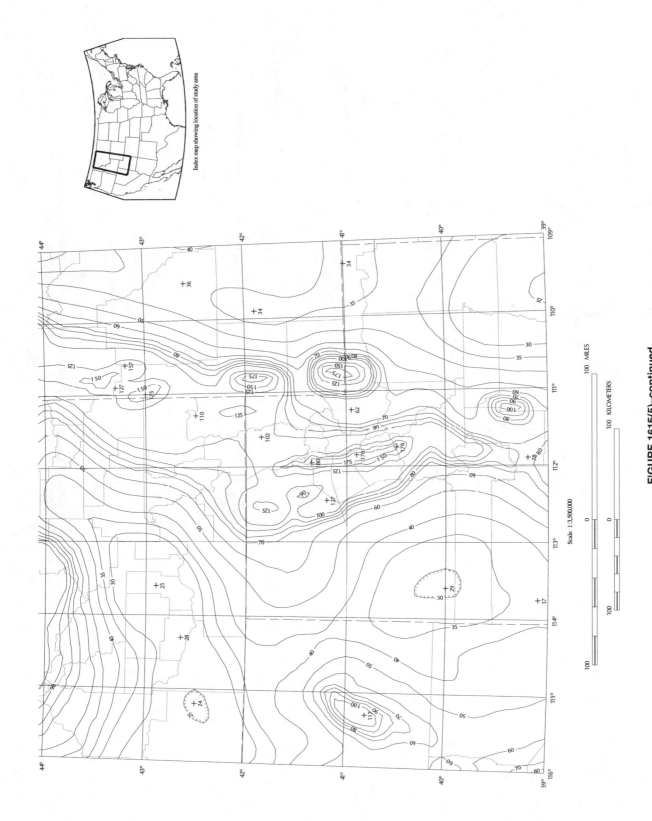

FIGURE 1615(5)—continued
**MAXIMUM CONSIDERED EARTHQUAKE GROUND MOTION FOR REGION 2 OF 0.2 SEC SPECTRAL
RESPONSE ACCELERATION (5 PERCENT OF CRITICAL DAMPING), SITE CLASS B**

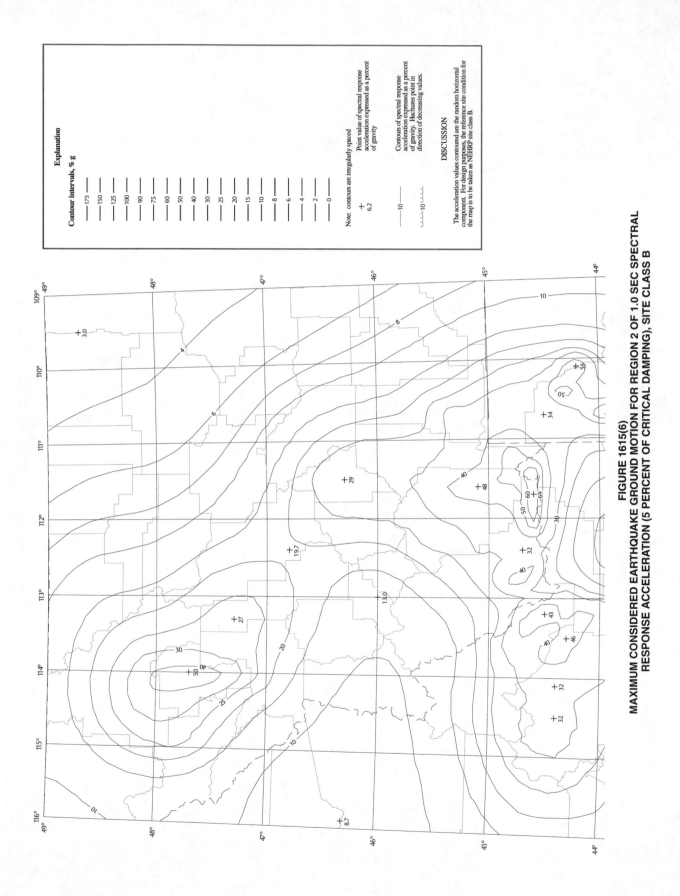

FIGURE 1615(6)
MAXIMUM CONSIDERED EARTHQUAKE GROUND MOTION FOR REGION 2 OF 1.0 SEC SPECTRAL RESPONSE ACCELERATION (5 PERCENT OF CRITICAL DAMPING), SITE CLASS B

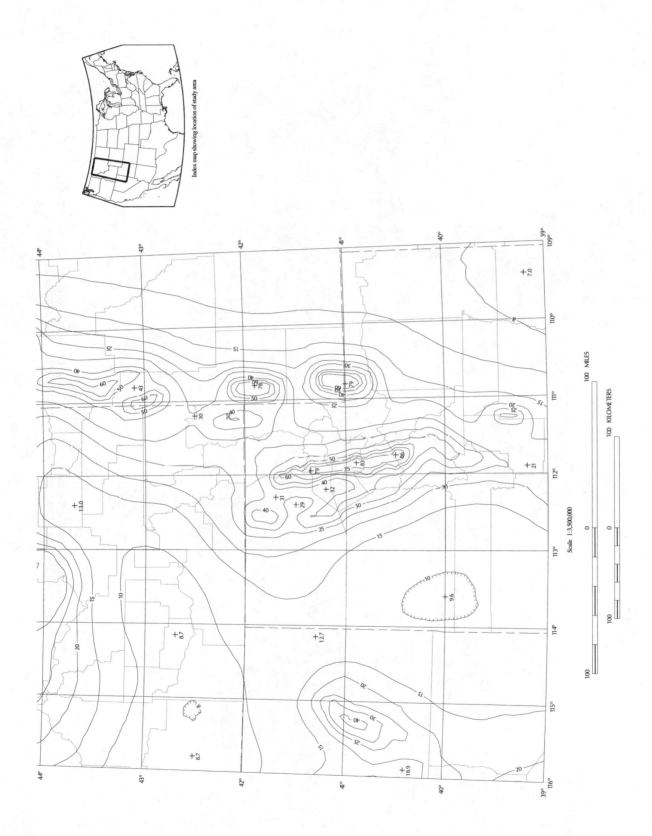

FIGURE 1615(6)—continued
MAXIMUM CONSIDERED EARTHQUAKE GROUND MOTION FOR REGION 2 OF 1.0 SEC SPECTRAL
RESPONSE ACCELERATION (5 PERCENT OF CRITICAL DAMPING), SITE CLASS B

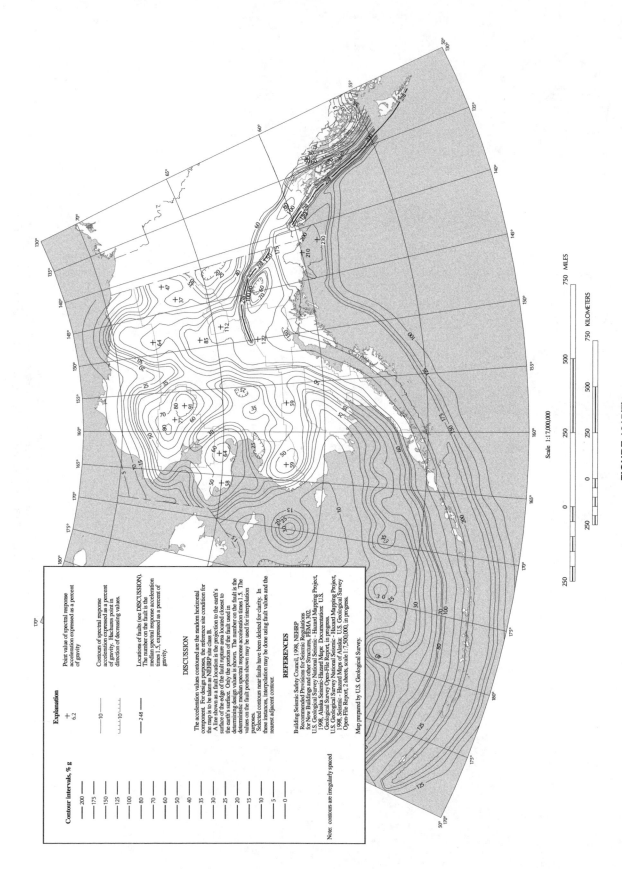

FIGURE 1615(7)
MAXIMUM CONSIDERED EARTHQUAKE GROUND MOTION FOR ALASKA OF 0.2 SEC SPECTRAL
RESPONSE ACCELERATION (5 PERCENT OF CRITICAL DAMPING), SITE CLASS B

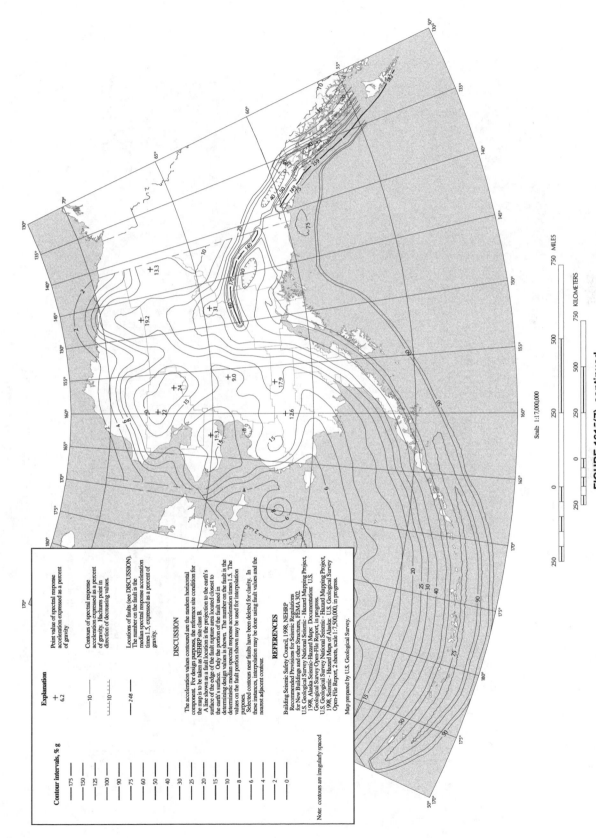

Explanation

+
6.2
Point value of spectral response acceleration expressed as a percent of gravity

—10—
Contours of spectral response acceleration expressed as a percent of gravity. Hachures point in direction of decreasing values.

⊥10⊥
Locations of faults (see DISCUSSION). The number on the fault is the median spectral response acceleration times 1.5, expressed as a percent of gravity.

248
Contour intervals, % g

—175—
—150—
—125—
—100—
—90—
—75—
—60—
—50—
—40—
—30—
—25—
—20—
—15—
—10—
—8—
—6—
—4—
—2—
—0—

Note: contours are irregularly spaced

DISCUSSION

The acceleration values contoured are the random horizontal component. For design purposes, the reference site condition for the map is to be taken as NEHRP site class B.
A line shown as a fault location is the projection to the earth's surface of the edge of the fault rupture area located closest to the earth's surface. Only the portion of the fault used in determining design values is shown. The number on the fault is the deterministic median spectral response acceleration times 1.5. The values on the fault portion shown may be used for interpolation purposes.
Selected contours near faults have been deleted for clarity. In these instances, interpolation may be done using fault values and the nearest adjacent contour.

REFERENCES

Building Seismic Safety Council, 1998. NEHRP Recommended Provisions for Seismic Regulations for New Buildings and other Structures, FEMA 302.
U.S. Geological Survey National Seismic - Hazard Mapping Project, 1998, Alaska Seismic-Hazard Maps: Documentation: U.S. Geological Survey Open-File Report, in progress.
U.S. Geological Survey National Seismic - Hazard Mapping Project, 1998, Seismic - Hazard Maps of Alaska: U.S. Geological Survey Open-File Report, 2 sheets, scale 1:7,500,000, in progress.

Map prepared by U.S. Geological Survey.

Scale 1:17,000,000

250   0   250   500   750 MILES

250 0 250 500 750 KILOMETERS

FIGURE 1615(7)—continued
MAXIMUM CONSIDERED EARTHQUAKE GROUND MOTION FOR ALASKA OF 1.0 SEC SPECTRAL
RESPONSE ACCELERATION (5 PERCENT OF CRITICAL DAMPING), SITE CLASS B

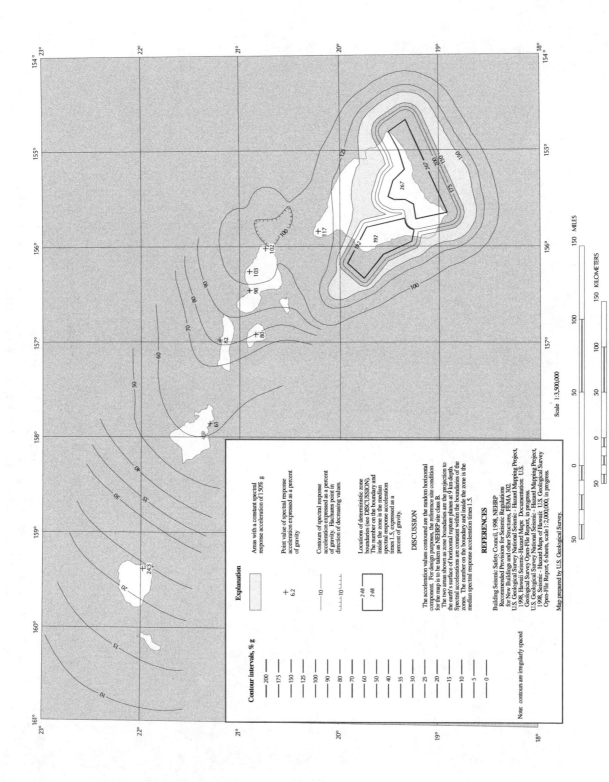

FIGURE 1615(8)
MAXIMUM CONSIDERED EARTHQUAKE GROUND MOTION FOR HAWAII OF 0.2 SEC SPECTRAL
RESPONSE ACCELERATION (5 PERCENT OF CRITICAL DAMPING), SITE CLASS B

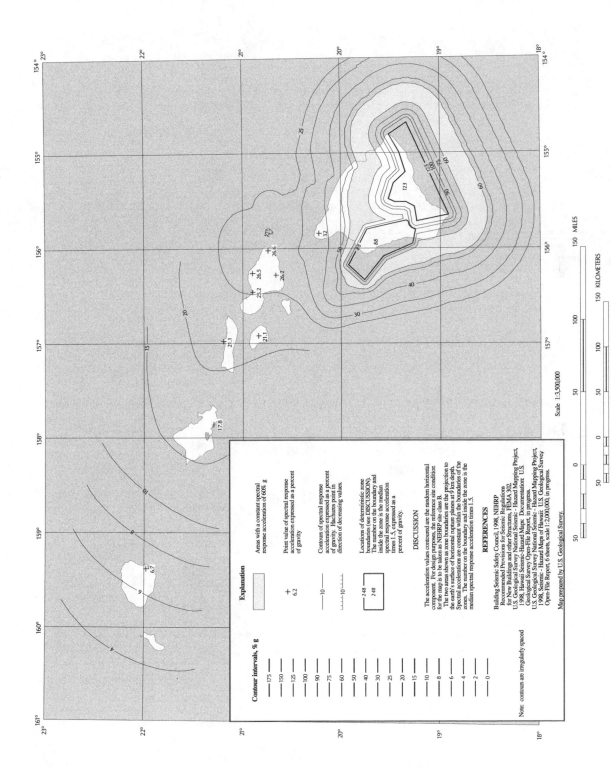

FIGURE 1615(8)—continued
MAXIMUM CONSIDERED EARTHQUAKE GROUND MOTION FOR HAWAII OF 1.0 SEC SPECTRAL
RESPONSE ACCELERATION (5 PERCENT OF CRITICAL DAMPING), SITE CLASS B

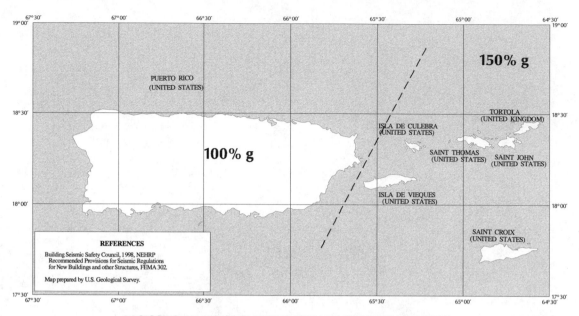

0.2 SEC SPECTRAL RESPONSE ACCELERATION (5% OF CRITICAL DAMPING)

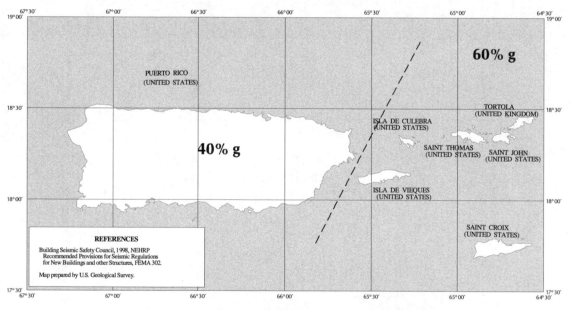

1.0 SEC SPECTRAL RESPONSE ACCELERATION (5% OF CRITICAL DAMPING)

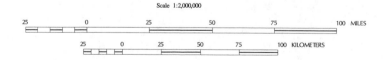

FIGURE 1615(9)
MAXIMUM CONSIDERED EARTHQUAKE GROUND MOTION FOR PUERTO RICO,
CULEBRA, VIEQUES, ST. THOMAS, ST. JOHN, AND ST. CROIX OF 0.2 AND 1.0 SEC SPECTRAL
RESPONSE ACCELERATION (5 PERCENT OF CRITICAL DAMPING), SITE CLASS B

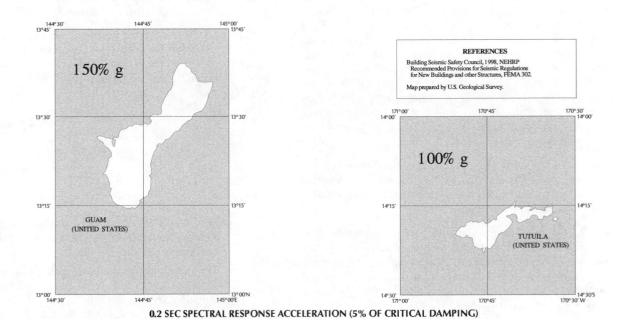

0.2 SEC SPECTRAL RESPONSE ACCELERATION (5% OF CRITICAL DAMPING)

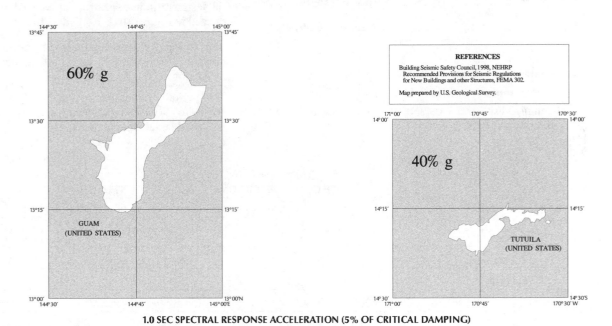

1.0 SEC SPECTRAL RESPONSE ACCELERATION (5% OF CRITICAL DAMPING)

Scale 1:1,000,000

25   0   25   MILES

25   0   25   KILOMETERS

**FIGURE 1615(10)
MAXIMUM CONSIDERED EARTHQUAKE GROUND MOTION FOR GUAM AND TUTUILLA
OF 0.2 AND 1.0 SEC SPECTRAL RESPONSE ACCELERATION
(5 PERCENT OF CRITICAL DAMPING), SITE CLASS B**

1615.1.1 Site class definitions. The site shall be classified as one of the site classes defined in Table 1615.1.1. Where the soil shear wave velocity, $\bar{v}_s$, is not known, site class shall be determined, as permitted in Table 1615.1.1, from standard penetration resistance, $\bar{N}$, or from soil undrained shear strength, $\bar{s}_u$, calculated in accordance with Section 1615.1.5. Where site-specific data are not available to a depth of 100 feet (30 480 mm), appropriate soil properties are permitted to be estimated by the registered design professional preparing the soils report based on known geologic conditions.

When the soil properties are not known in sufficient detail to determine the site class, Site Class D shall be used unless the building official determines that Site Class E or F soil is likely to be present at the site.

1615.1.2 Site coefficients and adjusted maximum considered earthquake spectral response acceleration parameters. The maximum considered earthquake spectral response acceleration for short periods, S_{MS}, and at 1-second period, S_{M1}, adjusted for site class effects, shall be determined by Equations 16-38 and 16-39, respectively:

$$S_{MS} = F_a S_s \qquad \textbf{(Equation 16-38)}$$

$$S_{M1} = F_v S_1 \qquad \textbf{(Equation 16-39)}$$

where:

F_a = Site coefficient defined in Table 1615.1.2(1).

F_v = Site coefficient defined in Table 1615.1.2(2).

S_S = The mapped spectral accelerations for short periods as determined in Section 1615.1.

S_1 = The mapped spectral accelerations for a 1-second period as determined in Section 1615.1.

1615.1.3 Design spectral response acceleration parameters. Five-percent damped design spectral response acceleration at short periods, S_{DS}, and at 1-second period, S_{D1}, shall be determined from Equations 16-40 and 16-41, respectively:

$$S_{DS} = \frac{2}{3} S_{MS} \qquad \textbf{(Equation 16-40)}$$

$$S_{D1} = \frac{2}{3} S_{M1} \qquad \textbf{(Equation 16-41)}$$

where:

S_{MS} = The maximum considered earthquake spectral response accelerations for short period as determined in Section 1615.1.2.

S_{M1} = The maximum considered earthquake spectral response accelerations for 1-second period as determined in Section 1615.1.2.

1615.1.4 General procedure response spectrum. The general design response spectrum curve shall be developed as indicated in Figure 1615.1.4 and as follows:

1. For periods less than or equal to T_O, the design spectral response acceleration, S_a, shall be determined by Equation 16-42.

2. For periods greater than or equal to T_O and less than or equal to T_S, the design spectral response acceleration, S_a, shall be taken equal to S_{DS}.

TABLE 1615.1.1
SITE CLASS DEFINITIONS

SITE CLASS	SOIL PROFILE NAME	AVERAGE PROPERTIES IN TOP 100 feet, AS PER SECTION 1615.1.5		
		Soil shear wave velocity, $\bar{v}_s$, (ft/s)	Standard penetration resistance, N	Soil undrained shear strength, $\bar{s}_u$, (psf)
A	Hard rock	$\bar{v}_s > 5{,}000$	N/A	N/A
B	Rock	$2{,}500 < \bar{v}_s \le 5{,}000$	N/A	N/A
C	Very dense soil and soft rock	$1{,}200 < \bar{v}_s \le 2{,}500$	$\bar{N} > 50$	$\bar{s}_u \ge 2{,}000$
D	Stiff soil profile	$600 \le \bar{v}_s \le 1{,}200$	$15 \le \bar{N} \le 50$	$1{,}000 \le \bar{s}_u \le 2{,}000$
E	Soft soil profile	$\bar{v}_s < 600$	$\bar{N} < 15$	$\bar{s}_u < 1{,}000$
E	—	Any profile with more than 10 feet of soil having the following characteristics: 1. Plasticity index $PI > 20$, 2. Moisture content $w \ge 40\%$, and 3. Undrained shear strength $\bar{s}_u < 500$ psf		
F	—	Any profile containing soils having one or more of the following characteristics: 1. Soils vulnerable to potential failure or collapse under seismic loading such as liquefiable soils, quick and highly sensitive clays, collapsible weakly cemented soils. 2. Peats and/or highly organic clays ($H > 10$ feet of peat and/or highly organic clay where H = thickness of soil) 3. Very high plasticity clays ($H > 25$ feet with plasticity index $PI > 75$) 4. Very thick soft/medium stiff clays ($H > 120$ feet)		

For SI: 1 foot = 304.8 mm, 1 square foot = 0.0929 m², 1 pound per square foot = 0.0479 kPa. N/A = Not applicable

TABLE 1615.1.2(1)
VALUES OF SITE COEFFICIENT F_a AS A FUNCTION OF SITE CLASS
AND MAPPED SPECTRAL RESPONSE ACCELERATION AT SHORT PERIODS (S_S)[a]

SITE CLASS	MAPPED SPECTRAL RESPONSE ACCELERATION AT SHORT PERIODS				
	$S_s \leq 0.25$	$S_s = 0.50$	$S_s = 0.75$	$S_s = 1.00$	$S_s \geq 1.25$
A	0.8	0.8	0.8	0.8	0.8
B	1.0	1.0	1.0	1.0	1.0
C	1.2	1.2	1.1	1.0	1.0
D	1.6	1.4	1.2	1.1	1.0
E	2.5	1.7	1.2	0.9	0.9
F	Note b	Note b	Note b	Note b	Note b

a. Use straight-line interpolation for intermediate values of mapped spectral response acceleration at short period, S_s.

b. Site-specific geotechnical investigation and dynamic site response analyses shall be performed to determine appropriate values, except that for structures with periods of vibration equal to or less than 0.5 second, values of F_a for liquefiable soils are permitted to be taken equal to the values for the site class determined without regard to liquefaction in Section 1615.1.5.1.

TABLE 1615.1.2(2)
VALUES OF SITE COEFFICIENT F_V AS A FUNCTION OF SITE CLASS
AND MAPPED SPECTRAL RESPONSE ACCELERATION AT 1-SECOND PERIOD (S_1)[a]

SITE CLASS	MAPPED SPECTRAL RESPONSE ACCELERATION AT SHORT PERIODS				
	$S_1 \leq 0.1$	$S_1 = 0.2$	$S_1 = 0.3$	$S_1 = 0.4$	$S_1 \geq 0.5$
A	0.8	0.8	0.8	0.8	0.8
B	1.0	1.0	1.0	1.0	1.0
C	1.7	1.6	1.5	1.4	1.3
D	2.4	2.0	1.8	1.6	1.5
E	3.5	3.2	2.8	2.4	2.4
F	Note b	Note b	Note b	Note b	Note b

a. Use straight-line interpolation for intermediate values of mapped spectral response acceleration at 1-second period, S_1.

b. Site-specific geotechnical investigation and dynamic site response analyses shall be performed to determine appropriate values, except that for structures with periods of vibration equal to or less than 0.5 second, values of F_v for liquefiable soils are permitted to be taken equal to the values for the site class determined without regard to liquefaction in Section 1615.1.5.1.

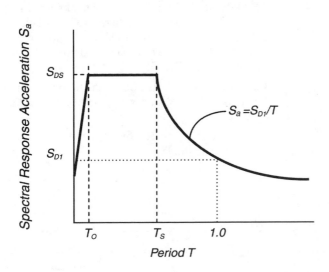

FIGURE 1615.1.4
DESIGN RESPONSE SPECTRUM

3. For periods greater than T_S, the design spectral response acceleration, S_a, shall be determined by Equation 16-43.

$$S_a = 0.6 \frac{S_{DS}}{T_O} T + 0.4 S_{DS} \qquad \textbf{(Equation 16-42)}$$

$$S_a = \frac{S_{DI}}{T} \qquad \textbf{(Equation 16-43)}$$

where:

S_{DS} = The design spectral response acceleration at short periods as determined in Section 1615.1.3.

S_{DI} = The design spectral response acceleration at 1-second period as determined in Section 1615.1.3.

T = Fundamental period (in seconds) of the structure (see Section 9.5.5.3 of ASCE 7).

T_O = 0.2 S_{DI}/S_{DS}

T_S = S_{DI}/S_{DS}

STRUCTURAL DESIGN

1615.1.5 Site classification for seismic design. Site classification for Site Class C, D or E shall be determined from Table 1615.1.5.

The notations presented below apply to the upper 100 feet (30 480 mm) of the site profile. Profiles containing distinctly different soil layers shall be subdivided into those layers designated by a number that ranges from 1 to n at the bottom where there is a total of n distinct layers in the upper 100 feet (30 480 mm). The symbol, i, then refers to any one of the layers between 1 and n.

where:

v_{si} = The shear wave velocity in feet per second (m/s).

d_i = The thickness of any layer between 0 and 100 feet (30 480 mm).

$$\bar{v}_s = \frac{\sum_{i=1}^{n} d_i}{\sum_{i=1}^{n} \frac{d_i}{v_{si}}} \qquad \text{(Equation 16-44)}$$

$$\sum_{i=1}^{n} d_i = 100 \text{ feet} (30\,480\text{ mm})$$

N_i is the Standard Penetration Resistance (ASTM D 1586) not to exceed 100 blows/foot (mm) as directly measured in the field without corrections.

$$\bar{N} = \frac{\sum_{i=1}^{n} d_i}{\sum_{i=1}^{n} \frac{d_i}{N_i}} \qquad \text{(Equation 16-45)}$$

$$\bar{N}_{ch} = \frac{d_s}{\sum_{i=1}^{m} \frac{d_i}{N_i}} \qquad \text{(Equation 16-46)}$$

where:

$$\sum_{i=1}^{m} d_i = d_s$$

Use only d_i and N_i for cohesionless soils.

d_s = The total thickness of cohesionless soil layers in the top 100 feet (30 480 mm).

s_{ui} = The undrained shear strength in psf (kPa), not to exceed 5,000 psf (240 kPa), ASTM D 2166 or D 2850.

$$\bar{s}_u = \frac{d_c}{\sum_{i=1}^{k} \frac{d_i}{s_{ui}}} \qquad \text{(Equation 16-47)}$$

where:

$$\sum_{i=1}^{k} d_i = d_c$$

d_c = The total thickness $(100 - d_s)$ (For SI: $30\,480 - d_s$) of cohesive soil layers in the top 100 feet (30 480 mm).

PI = The plasticity index, ASTM D 4318.

w = The moisture content in percent, ASTM D 2216.

The shear wave velocity for rock, Site Class B, shall be either measured on site or estimated by a geotechnical engineer or engineering geologist/seismologist for competent rock with moderate fracturing and weathering. Softer and more highly fractured and weathered rock shall either be measured on site for shear wave velocity or classified as Site Class C.

The hard rock, Site Class A, category shall be supported by shear wave velocity measurements either on site or on profiles of the same rock type in the same formation with an equal or greater degree of weathering and fracturing. Where hard rock conditions are known to be continuous to a depth of 100 feet (30 480 mm), surficial shear wave velocity measurements are permitted to be extrapolated to assess $\bar{v}_s$.

The rock categories, Site Classes A and B, shall not be used if there is more than 10 feet (3048 mm) of soil between the rock surface and the bottom of the spread footing or mat foundation.

1615.1.5.1 Steps for classifying a site.

1. Check for the four categories of Site Class F requiring site-specific evaluation. If the site corresponds to any of these categories, classify the site as Site Class F and conduct a site-specific evaluation.

2. Check for the existence of a total thickness of soft clay > 10 feet (3048 mm) where a soft clay layer is defined by: $\bar{s}_u < 500$ psf (25 kPa), $w \geq 40$ percent, and $PI > 20$. If these criteria are satisfied, classify the site as Site Class E.

3. Categorize the site using one of the following three methods with $\bar{v}_s$, $\bar{N}$, and $\bar{s}_u$ computed in all cases as specified.

   3.1. $\bar{v}_s$ for the top 100 feet (30 480 mm) ($\bar{v}_s$ method).

TABLE 1615.1.5
SITE CLASSIFICATION[a]

SITE CLASS	$\bar{v}_s$	$\bar{N}$ or $\bar{N}_{ch}$	$\bar{s}_u$
E	< 600 ft/s	< 15	< 1,000 psf
D	600 to 1,200 ft/s	15 to 50	1,000 to 2,000 psf
C	1,200 to 2,500 ft/s	> 50	> 2,000

For SI: 1 foot per second = 304.8 mm per second, 1 pound per square foot = 0.0479 kN/m².
a. If the $\bar{s}_u$ method is used and the $\bar{N}_{ch}$ and $\bar{s}_u$ criteria differ, select the category with the softer soils (for example, use Site Class E instead of D).

3.2. $\overline{N}$ for the top 100 feet (30 480 mm) ($\overline{N}$ method).

3.3. $\overline{N}_{ch}$ for cohesionless soil layers ($PI < 20$) in the top 100 feet (30 480 mm) and average, $\overline{s}_u$, for cohesive soil layers ($PI > 20$) in the top 100 feet (30 480 mm) ($\overline{s}_u$ method).

1615.2 Site-specific procedure for determining ground motion accelerations. A site-specific study shall account for the regional seismicity and geology; the expected recurrence rates and maximum magnitudes of events on known faults and source zones; the location of the site with respect to these; near source effects if any and the characteristics of subsurface site conditions.

1615.2.1 Probabilistic maximum considered earthquake. Where site-specific procedures are used as required or permitted by Section 1615, the maximum considered earthquake ground motion shall be taken as that motion represented by an acceleration response spectrum having a 2-percent probability of exceedance within a 50-year period. The maximum considered earthquake spectral response acceleration at any period, S_{aM}, shall be taken from the 2-percent probability of exceedance within a 50-year period spectrum.

> **Exception:** Where the spectral response ordinates at 0.2 second or 1 second for a 5-percent damped spectrum having a 2-percent probability of exceedance within a 50-year period exceed the corresponding ordinates of the deterministic limit of Section 1615.2.2, the maximum considered earthquake ground motion spectrum shall be taken as the lesser of the probabilistic maximum considered earthquake ground motion or the deterministic maximum considered earthquake ground motion spectrum of Section 1615.2.3, but shall not be taken as less than the deterministic limit ground motion of Section 1615.2.2.

1615.2.2 Deterministic limit on maximum considered earthquake ground motion. The deterministic limit for the maximum considered earthquake ground motion shall be the response spectrum determined in accordance with Figure 1615.2.2, where site coefficients, F_a and F_v, are determined in accordance with Section 1615.1.2, with the value of the mapped short-period spectral response acceleration, S_S, taken as 1.5g and the value of the mapped spectral response acceleration at 1 second, S_1, taken as 0.6g.

1615.2.3 Deterministic maximum considered earthquake ground motion. The deterministic maximum considered earthquake ground motion response spectrum shall be calculated as 150 percent of the median spectral response accelerations, S_{aM}, at all periods resulting from a characteristic earthquake on any known active fault within the region.

1615.2.4 Site-specific design ground motion. Where site-specific procedures are used to determine the maximum considered earthquake ground motion response spectrum, the design spectral response acceleration, S_a, at any period shall be determined from Equation 16-48:

$$S_a = \frac{2}{3} S_{aM}$$

(Equation 16-48)

and shall be greater than or equal to 80 percent of the design spectral response acceleration, S_a, determined by the general response spectrum in Section 1615.1.4.

1615.2.5 Design spectral response coefficients. Where the site-specific procedure is used to determine the design ground motion in accordance with Section 1615.2.4, the parameter S_{DS} shall be taken as the spectral acceleration, S_a, obtained from the site-specific spectra at a period of 0.2 second, except that it shall not be taken as less than 90 percent of the peak spectral acceleration, S_a, at any period. The parameter S_{D1} shall be taken as the greater of the spectral acceleration, S_a, at a period of 1 second or two times the spectral acceleration, S_a, at a period of 2 seconds. The parameters S_{MS} and S_{M1} shall be taken as 1.5 times S_{DS} and S_{D1}, respectively. The values so obtained shall not be taken as less than 80 percent of the values obtained from the general procedures of Section 1615.1.

SECTION 1616
EARTHQUAKE LOADS—CRITERIA SELECTION

1616.1 Structural design criteria. Each structure shall be assigned to a seismic design category in accordance with Section 1616.3. Seismic design categories are used in this code to determine permissible structural systems, limitations on height and irregularity, those components of the structure that must be designed for seismic resistance and the types of lateral force analysis that must be performed. Each structure shall be provided with complete lateral- and vertical-force-resisting systems capable of providing adequate strength, stiffness and energy dissipation capacity to withstand the design earthquake ground motions determined in accordance with Section 1615 within the prescribed deformation limits of Section 1617.3. The design ground motions shall be assumed to occur along any horizontal direction of a structure. A continuous load path, or paths, with adequate strength and stiffness to transfer forces induced by the design earthquake ground motions from the points of application to the final point of resistance shall be provided.

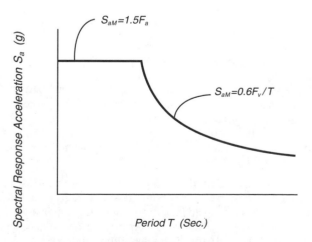

FIGURE 1615.2.2
DETERMINISTIC LIMIT ON MAXIMUM CONSIDERED
EARTHQUAKE RESPONSE SPECTRUM

Allowable stress design is permitted to be used to evaluate sliding, overturning and soil bearing at the soil-structure interface regardless of the approach used in the design of the structure, provided load combinations of Section 1605.3 are utilized. When using allowable stress design for proportioning foundations, the value of $0.2 S_{DS}D$ in Equations 16-50, 16-51, 16-52 and 16-53 or Equations 9.5.2.7-1, 9.5.2.7-2, 9.5.2.7.1-1 and 9.5.2.7.1-2 of ASCE 7 is permitted to be taken equal to zero. When the load combinations of Section 1605.3.2 are utilized, a one-third increase in soil allowable stresses is permitted for all load combinations that include W or E.

1616.2 Seismic use groups and occupancy importance factors. Each structure shall be assigned a seismic use group and a corresponding occupancy importance factor (I_E) as indicated in Table 1604.5.

1616.2.1 Seismic Use Group I. Seismic Use Group I structures are those not assigned to either Seismic Use Group II or III.

1616.2.2 Seismic Use Group II. Seismic Use Group II structures are those, the failure of which would result in a substantial public hazard due to occupancy or use as indicated by Table 1604.5, or as designated by the building official.

1616.2.3 Seismic Use Group III. Seismic Use Group III structures are those having essential facilities that are required for postearthquake recovery and those containing substantial quantities of hazardous substances, as indicated in Table 1604.5, or as designated by the building official.

Where operational access to a Seismic Use Group III structure is required through an adjacent structure, the adjacent structure shall conform to the requirements for Seismic Use Group III structures. Where operational access is less than 10 feet (3048 mm) from an interior lot line or less than 10 feet (3048 mm) from another structure, access protection from potential falling debris shall be provided by the owner of the Seismic Use Group III structure.

1616.2.4 Multiple occupancies. Where a structure is occupied for two or more occupancies not included in the same seismic use group, the structure shall be assigned the classification of the highest seismic use group corresponding to the various occupancies.

Where structures have two or more portions that are structurally separated in accordance with Section 1620, each portion shall be separately classified. Where a structurally separated portion of a structure provides required access to, required egress from or shares life safety components with another portion having a higher seismic use group, both portions shall be assigned the higher seismic use group.

1616.3 Determination of seismic design category. All structures shall be assigned to a seismic design category based on their seismic use group and the design spectral response acceleration coefficients, S_{DS} and S_{D1}, determined in accordance with Section 1615.1.3 or 1615.2.5. Each building and structure shall be assigned to the most severe seismic design category in accordance with Table 1616.3(1) or 1616.3(2), irrespective of the fundamental period of vibration of the structure, T.

Exception: The seismic design category is permitted to be determined from Table 1616.3(1) alone when all of the following apply:

1. The approximate fundamental period of the structure, T_a, in each of the two orthogonal directions determined in accordance with Section 9.5.5.3.2 of ASCE 7, is less than 0.8 T_s determined in accordance with Section 1615.1.4,

2. Equation 9.5.5.2.1-1 of ASCE 7 is used to determine the seismic response coefficient, C_s, and

3. The diaphragms are rigid as defined in Section 1602.

TABLE 1616.3(1)
SEISMIC DESIGN CATEGORY BASED ON
SHORT-PERIOD RESPONSE ACCELERATIONS

VALUE OF S_{DS}	SEISMIC USE GROUP		
	I	II	III
$S_{DS} < 0.167g$	A	A	A
$0.167g \leq S_{DS} < 0.33g$	B	B	C
$0.33g \leq S_{DS} < 0.50g$	C	C	D
$0.50g \leq S_{DS}$	D[a]	D[a]	D[a]

a. Seismic Use Group I and II structures located on sites with mapped maximum considered earthquake spectral response acceleration at 1-second period, S_1, equal to or greater than 0.75g, shall be assigned to Seismic Design Category E, and Seismic Use Group III structures located on such sites shall be assigned to Seismic Design Category F.

TABLE 1616.3(2)
SEISMIC DESIGN CATEGORY BASED ON
1-SECOND PERIOD RESPONSE ACCELERATION

VALUE OF S_{D1}	SEISMIC USE GROUP		
	I	II	III
$S_{D1} < 0.067g$	A	A	A
$0.067g \leq S_{D1} < 0.133g$	B	B	C
$0.133g \leq S_{D1} < 0.20g$	C	C	D
$0.20g \leq S_{D1}$	D[a]	D[a]	D[a]

a. Seismic Use Group I and II structures located on sites with mapped maximum considered earthquake spectral response acceleration at 1-second period, S_1, equal to or greater than 0.75g, shall be assigned to Seismic Design Category E, and Seismic Use Group III structures located on such sites shall be assigned to Seismic Design Category F.

1616.3.1 Site limitation for Seismic Design Category E or F. A structure assigned to Seismic Design Category E or F shall not be sited over an identified active fault trace.

Exception: Detached Group R-3 as applicable in Section 101.2 of light-frame construction.

1616.4 Design requirements for Seismic Design Category A. Structures assigned to Seismic Design Category A need only comply with the requirements of Sections 1616.4.1 through 1616.4.5.

1616.4.1 Minimum lateral force. Structures shall be provided with a complete lateral-force-resisting system designed to resist the minimum lateral force, F_x, applied simultaneously at each floor level given by Equation 16-49:

$$F_x = 0.01 w_x \qquad \textbf{(Equation 16-49)}$$

where:

F_x = The design lateral force applied at Level x.

w_x = The portion of the total gravity load of the structure, W, located or assigned to Level x.

W = The total dead load and other loads listed below:

1. In areas used for storage, a minimum of 25 percent of the reduced floor live load (floor live load in public garages and open parking structures need not be included).

2. Where an allowance for partition load is included in the floor load design, the actual partition weight or a minimum weight of 10 psf (0.479 kN/m²) of floor area, whichever is greater.

3. Total operating weight of permanent equipment.

4. Twenty percent of flat roof snow load where flat roof snow load exceeds 30 psf (1.44 kN/m²).

The direction of application of seismic forces used in design shall be that which will produce the most critical load effect in each component. The design seismic forces are permitted to be applied separately in each of two orthogonal directions and orthogonal effects are permitted to be neglected.

The effect of this lateral force shall be taken as E in the load combinations prescribed in Section 1605.2 for strength or load and resistance factor design methods, or Section 1605.3 for allowable stress design methods. Special seismic load combinations that include E_m need not be considered.

1616.4.2 Connections. All parts of the structure between separation joints shall be interconnected, and the connections shall be capable of transmitting the seismic force, F_p, induced in the connection by the parts being connected. Any smaller portion of the structure shall be tied to the remainder of the structure for F_p equal to 0.05 times the weight of the smaller portion. A positive connection for resisting horizontal forces acting on the member shall be provided for each beam, girder or truss to its support. The connection shall have strength sufficient to resist 5 percent of the dead and live load vertical reaction applied horizontally.

1616.4.3 Anchorage of concrete or masonry walls. See Section 1604.8.2.

1616.4.4 Conventional light-frame construction. Buildings constructed in compliance with Section 2308 are deemed to comply with Sections 1616.4.1, 1616.4.2 and 1616.4.3.

1616.4.5 Tank freeboard. Tanks in Seismic Use Group III according to Table 9.14.5.1.2 of ASCE 7 shall also comply with the freeboard requirements of Section 9.14.7.3.6.1.2 of ASCE 7.

1616.5 Building configuration. Buildings shall be classified as regular or irregular based on the criteria in Section 9.5.2.3 of ASCE 7.

Exception: Buildings designed using the simplified analysis procedure in Section 1617.5 shall be classified in accordance with Section 1616.5.1.

1616.5.1 Building configuration (for use in the simplified analysis procedure of Section 1617.5). Buildings designed using the simplified analysis procedure in Section 1617.5 shall be classified as regular or irregular based on the criteria in this section. Such classification shall be based on the plan and vertical configuration. Buildings shall not exceed the limitations of Section 1616.6.1.

1616.5.1.1 Plan irregularity. Buildings having one or more of the features listed in Table 1616.5.1.1 shall be designated as having plan structural irregularity and shall comply with the requirements in the sections referenced in that table.

1616.5.1.2 Vertical irregularity. Buildings having one or more of the features listed in Table 1616.5.1.2 shall be designated as having vertical irregularity and shall comply with the requirements in the sections referenced in that table.

Exceptions:

1. Structural irregularities of Type 1a, 1b or 2 in Table 1616.5.1.2 do not apply where no story drift ratio under design lateral load is greater than 130 percent of the story drift ratio of the next story above. Torsional effects need not be considered in the calculation of story drifts for the purpose of this determination. The story drift ratio relationship for the top two stories of the building is not required to be evaluated.

2. Irregularities of Types 1a, 1b and 2 of Table 1616.5.1.2 are not required to be considered for one-story buildings in any seismic design category or for two-story buildings in Seismic Design Category A, B, C or D.

1616.6 Analysis procedures. A structural analysis conforming to one of the types permitted in Section 9.5.2.5.1 of ASCE 7 or to the simplified procedure in Section 1617.5 shall be made for all structures. The analysis shall form the basis for determining the seismic forces, E and E_m, to be applied in the load combinations of Section 1605 and shall form the basis for determining the design drift as required by Section 9.5.2.8 of ASCE 7 or Section 1617.3.

Exceptions:

1. Structures assigned to Seismic Design Category A.

2. Design drift need not be evaluated in accordance with Section 1617.3 when the simplified analysis method of Section 1617.5 is used.

1616.6.1 Simplified analysis. A simplified analysis, in accordance with Section 1617.5, shall be permitted to be used for any structure in Seismic Use Group I, subject to the following limitations, or a more rigorous analysis shall be made:

1. Buildings of light-framed construction not exceeding three stories in height, excluding basements.

2. Buildings of any construction other than light-framed construction, not exceeding two stories in height, excluding basements, with flexible diaphragms at every level as defined in Section 1602.

TABLE 1616.5.1.1
PLAN STRUCTURAL IRREGULARITIES

	IRREGULARITY TYPE AND DESCRIPTION	REFERENCE SECTION	SEISMIC DESIGN CATEGORY[a] APPLICATION
1a	Torsional Irregularity—to be considered when diaphragms are not flexible as determined in Section 1602.1.1 Torsional irregularity shall be considered to exist when the maximum story drift, computed including accidental torsion, at one end of the structure transverse to an axis is more than 1.2 times the average of the story drifts at the two ends of the structure.	9.5.5.5.2 of ASCE 7 1620.4.1 9.5.2.5.1 of ASCE 7 9.5.5.7.1 of ASCE 7	C, D, E and F D, E and F D, E and F C, D, E and F
1b	Extreme Torsional Irregularity—to be considered when diaphragms are not flexible as determined in Section 1602.1. Extreme torsional irregularity shall be considered to exist when the maximum story drift, computed and including accidental torsion, at one end of the structure transverse to an axis is more than 1.4 times the average of the story drifts at the two ends of the structure.	9.5.5.5.2 of ASCE 7 1620.4.1 1620.5.1 9.5.2.5.1 of ASCE 7 9.5.5.7.1 of ASCE 7	C, D, E and F D E and F D, E and F C, D, E and F
2	Reentrant Corners Plan configurations of a structure and its lateral-force-resisting system contain reentrant corners where both projections of the structure beyond a reentrant corner are greater than 15 percent of the plan dimension of the structure in the given direction.	1620.4.1	D, E and F
3	Diaphragm Discontinuity Diaphragms with abrupt discontinuities or variations in stiffness, including those having cutout or open areas greater than 50 percent of the gross enclosed diaphragm area, or changes in effective diaphragm stiffness of more than 50 percent from one story to the next.	1620.4.1	D, E and F
4	Out-of-Plane Offsets Discontinuities in a lateral-force-resistance path, such as out-of-plane offsets of the vertical elements.	1620.4.1 9.5.2.5.1 of ASCE 7 1620.2.9	D, E and F D, E and F B, C, D, E and F
5	Nonparallel Systems The vertical lateral-force-resisting elements are not parallel to or symmetric about the major orthogonal axes of the lateral-force-resisting system.	1620.3.2	C, D, E and F

a. Seismic design category is determined in accordance with Section 1616.

TABLE 1616.5.1.2
VERTICAL STRUCTURAL IRREGULARITIES

	IRREGULARITY TYPE AND DESCRIPTION	REFERENCE SECTION	SEISMIC DESIGN CATEGORY[a] APPLICATION
1a	Stiffness Irregularity—Soft Story A soft story is one in which the lateral stiffness is less than 70 percent of that in the story above or less than 80 percent of the average stiffness of the three stories above.	9.5.2.5.1 of ASCE 7	D, E, and F
1b	Stiffness Irregularity—Extreme Soft Story An extreme soft story is one in which the lateral stiffness is less than 60 percent of that in the story above or less than 70 percent of the average stiffness of the three stories above.	1620.5.1 9.5.2.5.1 of ASCE 7	E and F D, E and F
2	Weight (Mass) Irregularity Mass irregularity shall be considered to exist where the effective mass of any story is more than 150 percent of the effective mass of an adjacent story. A roof that is lighter than the floor below need not be considered.	9.5.2.5.1 of ASCE 7	D, E and F
3	Vertical Geometric Irregularity Vertical geometric irregularity shall be considered to exist where the horizontal dimension of the lateral-force-resisting system in any story is more than 130 percent of that in an adjacent story.	9.5.2.5.1 of ASCE 7	D, E and F
4	In-plane Discontinuity in Vertical Lateral-Force-Resisting Elements An in-plane offset of the lateral-force-resisting elements greater than the length of those elements or a reduction in stiffness of the resisting element in the story below.	1620.4.1 9.5.2.5.1 of ASCE 7 1620.2.9	D, E and F D, E and F B, C, D, E and F
5	Discontinuity in Capacity—Weak Story A weak story is one in which the story lateral strength is less than 80 percent of that in the story above. The story strength is the total strength of seismic-resisting elements sharing the story shear for the direction under consideration.	1620.2.3 9.5.2.5.1 of ASCE 7 1620.5.1	B, C, D, E and F D, E and F E and F

a. Seismic design category is determined in accordance with Section 1616.

SECTION 1617
EARTHQUAKE LOADS—MINIMUM DESIGN
LATERAL FORCE AND RELATED EFFECTS

1617.1 Seismic load effect E and E_m. The seismic load effect, E, for use in the basic load combinations of Sections 1605.2 and 1605.3 shall be determined from Section 9.5.2.7 of ASCE 7. The maximum seismic load effect, E_m, for use in the special seismic load combination of Section 1605.4 shall be the special seismic load determined from Section 9.5.2.7.1 of ASCE 7.

Exception: For structures designed using the simplified analysis procedure in Section 1617.5, the seismic load effects, E and E_m, shall be determined from Section 1617.1.1.

1617.1.1 Seismic load effects, E and E_m (for use in the simplified analysis procedure of Section 1617.5). Seismic load effects, E and E_m, for use in the load combinations of Section 1605 for structures designed using the simplified analysis procedure in Section 1617.5 shall be determined as follows.

1617.1.1.1 Seismic load effect, E. Where the effects of gravity and the seismic ground motion are additive, seismic load, E, for use in Equations 16-5, 16-10 and 16-17, shall be defined by Equation 16-50:

$$E = \rho Q_E + 0.2 S_{DS} D \qquad \textbf{(Equation 16-50)}$$

where:

D = The effect of dead load.

E = The combined effect of horizontal and vertical earthquake-induced forces.

ρ = A redundancy coefficient obtained in accordance with Section 1617.2.

Q_E = The effect of horizontal seismic forces.

S_{DS} = The design spectral response acceleration at short periods obtained from Section 1615.1.3 or 1615.2.5.

Where the effects of gravity and seismic ground motion counteract, the seismic load, E, for use in Equations 16-6, 16-12 and 16-18 shall be defined by Equation 16-51.

$$E = \rho Q_E - 0.2 S_{DS} D \qquad \textbf{(Equation 16-51)}$$

Design shall use the load combinations prescribed in Section 1605.2 for strength or load and resistance factor design methodologies, or Section 1605.3 for allowable stress design methods.

1617.1.1.2 Maximum seismic load effect, E_m. The maximum seismic load effect, E_m, shall be used in the special seismic load combinations in Section 1605.4.

Where the effects of the seismic ground motion and gravity loads are additive, seismic load, E_m, for use in Equation 16-19, shall be defined by Equation 16-52.

$$E_m = \Omega_0 Q_E + 0.2 S_{DS} D \qquad \textbf{(Equation 16-52)}$$

Where the effects of the seismic ground and gravity loads counteract, seismic load, E_m, for use in Equation 16-20, shall be defined by Equation 16-53.

$$E_m = \Omega_0 Q_E - 0.2 S_{DS} D \qquad \textbf{(Equation 16-53)}$$

where E, Q_E, S_{DS} are as defined above and Ω_0 is the system overstrength factor as given in Table 1617.6.2.

The term $\Omega_0 Q_E$ need not exceed the maximum force that can be transferred to the element by the other elements of the lateral-force-resisting system.

Where allowable stress design methodologies are used with the special load combinations of Section 1605.4, design strengths are permitted to be determined using an allowable stress increase of 1.7 and a resistance factor, ϕ, of 1.0. This increase shall not be combined with increases in allowable stresses or load combination reductions otherwise permitted by this code or the material reference standard except that combination with the duration of load increases in Chapter 23 is permitted.

1617.2 Redundancy. The provisions given in Section 9.5.2.4 of ASCE 7 shall be used.

Exception: Structures designed using the simplified analysis procedure in Section 1617.5 shall use the redundancy provisions in Sections 1617.2.2.

1617.2.1 ASCE 7, Sections 9.5.2.4.2 and 9.5.2.4.3. Modify Sections 9.5.2.4.2 and 9.5.2.4.3 as follows:

9.5.2.4.2 Seismic Design Category D: For structures in Seismic Design Category D, ρ shall be taken as the largest of the values of ρ_x calculated at each story "x" of the structure in accordance with Equation 9.5.2.4.2-1 as follows:

$$\rho_x = 2 - \frac{20}{r_{max_x} \sqrt{A_x}}$$

where:

r_{max_x} = The ratio of the design story shear resisted by the single element carrying the most shear force in the story to the total story shear, for a given direction of loading. For braced frames, the value of r_{max_x} is equal to the lateral force component in the most heavily loaded brace element divided by the story shear. For moment frames, r_{max_x} shall be taken as the maximum of the sum of the shears in any two adjacent columns in the plane of a moment frame divided by the story shear. For columns common to two bays with moment-resisting connections on opposite sides at the level under consideration, 70 percent of the shear in that column is permitted to be used in the column shear summation. For shear walls, r_{max_x} shall be taken equal to shear in the most heavily loaded wall or wall pier multiplied by $10/l_w$ (the metric coefficient is $3.3/l_w$), divided by the story shear, where l_w is the wall or wall pier length in feet (m). The value of the ratio of $10/l_w$ need not be greater than 1.0 for buildings of light-framed construction. For dual systems, r_{max_x} shall be taken as the maximum value defined above, considering all lateral-load-resisting elements in the story. The lateral loads shall be distributed to elements based on relative rigidities considering the interaction of the dual system. For dual systems, the value of ρ need not exceed 80 percent of the value calculated above.

A_x = The floor area in square feet of the diaphragm level immediately above the story.

Calculation of r_{max_x} need not consider the effects of accidental torsion and any dynamic amplification of torsion required by Section 9.5.5.5.2.

For a story with a flexible diaphragm immediately above, r_{max_x} shall be permitted to be calculated from an analysis that assumes rigid diaphragm behavior and ρ_x, need not exceed 1.25.

The value of ρ need not exceed 1.5, which is permitted to be used for any structure. The value of ρ shall not be taken as less than 1.0.

Exception: For structures with seismic-force-resisting systems in any direction comprised solely of special moment frames, the seismic-force-resisting system shall be configured such that the value of ρ calculated in accordance with this section does not exceed 1.25. The calculated value of ρ is permitted to exceed this limit when the design story drift, Δ, as determined in Section 9.5.5.7, does not exceed Δ_a/ρ for any story where Δ_a is the allowable story drift from Table 9.5.2.8.

The metric equivalent of Equation 9.5.2.4.2-1 is:

$$\rho_x = 2 - \frac{6.1}{r_{max_x} \sqrt{A_x}}$$

where: A_x is in square meters.

The value ρ shall be permitted to be taken equal to 1.0 in the following circumstances:

1. When calculating displacements for dynamic amplification of torsion in Section 9.5.5.5.2.

2. When calculating deflections, drifts and seismic shear forces related to Sections 9.5.5.7.1 and 9.5.5.7.2.

3. For design calculations required by Section 9.5.2.6, 9.6 or 9.14.

For structures with vertical combinations of seismic-force-resisting systems, the value of ρ shall be determined independently for each seismic-force-resisting system. The redundancy coefficient of the lower portion shall not be less than the following:

$$\rho_L = \frac{R_L \rho_u}{R_u}$$

where:

ρ_L = ρ of lower portion.

R_L = R of lower portion.

ρ_u = ρ of upper portion.

R_u = R of upper portion.

9.5.2.4.3 Seismic Design Categories E and F. For structures in Seismic Design Categories E and F, the value of ρ shall be calculated as indicated in Section 9.5.2.4.2, above.

Exception: For structures with lateral-force-resisting systems in any direction consisting solely of special moment frames, the lateral-force-resisting system shall be configured such that the value of ρ calculated in accordance with Section 9.5.2.4.2 does not exceed 1.1. The calculated value of ρ is permitted to exceed this limit when the design story drift, Δ, as determined in Section 9.5.5.7, does not exceed Δ_a/ρ for any story where Δ_a is the allowable story drift from Table 9.5.2.8.

1617.2.2 Redundancy (for use in the simplified analysis procedure of Section 1617.5). A redundancy coefficient, ρ, shall be assigned to each structure designed using the simplified analysis procedure in Section 1617.5 in accordance with this section. Buildings shall not exceed the limitations of Section 1616.6.1.

1617.2.2.1 Seismic Design Category A, B or C. For structures assigned to Seismic Design Category A, B or C (see Section 1616), the value of the redundancy coefficient ρ is 1.0.

1617.2.2.2 Seismic Design Category D, E or F. For structures in Seismic Design Category D, E or F (see Section 1616), the redundancy coefficient, ρ, shall be taken as the largest of the values of, ρ_i, calculated at each story "i" of the structure in accordance with Equation 16-54, as follows:

$$\rho_i = 2 - \frac{20}{r_{max_i} \sqrt{A_i}} \qquad \textbf{(Equation 16-54)}$$

For SI:

$$\rho_i = 2 - \frac{6.1}{r_{max_i} \sqrt{A_i}}$$

where:

r_{max_i} = The ratio of the design story shear resisted by the most heavily loaded single element in the story to the total story shear, for a given direction of loading.

r_{max_i} = For braced frames, the value r_{max_i}, is equal to the horizontal force component in the most heavily loaded brace element divided by the story shear.

r_{max_i} = For moment frames, r_{max_i}, shall be taken as the maximum of the sum of the shears in any two adjacent columns in a moment frame divided by the story shear. For columns common to two bays with moment-resisting connections on opposite sides at the level under consideration, it is permitted to use 70 percent of the shear in that column in the column shear summation.

r_{max_i} = For shear walls, r_{max_i}, shall be taken as the maximum value of the product of the shear in the wall or wall pier and $10/l_w$ ($3.3/l_w$ for SI), divided by the story shear, where l_w is the length of the wall or wall pier in feet (m). In light-framed construction, the value of the ratio of $10/l_w$ need not be greater than 1.0.

r_{max_i} = For dual systems, r_{max_i}, shall be taken as the maximum value defined above, considering all lateral-load-resisting elements in the story. The lateral loads shall be distributed to elements based on relative rigidities considering the interaction of the dual system. For dual systems, the value of ρ need not exceed 80 percent of the value calculated above.

A_i = The floor area in square feet of the diaphragm level immediately above the story.

For a story with a flexible diaphragm immediately above, r_{max_i} shall be permitted to be calculated from an analysis that assumes rigid diaphragm behavior and ρ need not exceed 1.25.

The value, ρ, shall be not less than 1.0, and need not exceed 1.5.

Calculation of r_{max_i} need not consider the effects of accidental torsion and any dynamic amplification of torsion required by Section 9.5.5.5.2 of ASCE 7.

For structures with seismic-force-resisting systems in any direction comprised solely of special moment frames, the seismic-force-resisting system shall be configured such that the value of ρ calculated in accordance with this section does not exceed 1.25 for structures assigned to Seismic Design Category D, and does not exceed 1.1 for structures assigned to Seismic Design Category E or F.

Exception: The calculated value of ρ is permitted to exceed these limits when the design story drift, Δ, as determined in Section 1617.5.4, does not exceed Δ_a/ρ for any story where Δ_a is the allowable story drift from Table 1617.3.1.

The value ρ shall be permitted to be taken equal to 1.0 in the following circumstances:

1. When calculating displacements for dynamic amplification of torsion in Section 9.5.5.5.2 of ASCE 7.

2. When calculating deflections, drifts and seismic shear forces related to Sections 9.5.5.7.1 and 9.5.5.7.2 of ASCE 7.

3. For design calculations required by Section 1620, 1621 or 1622.

For structures with vertical combinations of seismic-force-resisting systems, the value, ρ, shall be determined independently for each seismic-force-resisting system. The redundancy coefficient of the lower portion shall be not less than the following:

$$\rho_L = \frac{R_L \rho_u}{R_u}$$ (Equation 16-55)

where:

ρ_L = ρ of lower portion.
R_L = R of lower portion.
ρ_u = ρ of upper portion.
R_u = R of upper portion.

1617.3 Deflection and drift limits. The provisions given in Section 9.5.2.8 of ASCE 7 shall be used.

Exception: Structures designed using the simplified analysis procedure in Section 1617.5 shall meet the provisions in Section 1617.3.1.

1617.3.1 Deflection and drift limits (for use in the simplified analysis procedure of Section 1617.5). The design story drift, Δ, as determined in Section 1617.5.4, shall not exceed the allowable story drift, Δ_a, as obtained from Table 1617.3.1 for any story. All portions of the building shall be designed to act as an integral unit in resisting seismic forces unless separated structurally by a distance sufficient to avoid damaging contact under total deflection as determined in Section 1617.5.4. Buildings shall not exceed the limitations of Section 1616.6.1.

1617.4 Equivalent lateral force procedure for seismic design of buildings. The provisions given in Section 9.5.5 of ASCE 7 shall be used.

1617.5 Simplified analysis procedure for seismic design of buildings. See Section 1616.6.1 for limitations on the use of this procedure. For purposes of this analytical procedure, a building is considered to be fixed at the base.

1617.5.1 Seismic base shear. The seismic base shear, V, in a given direction shall be determined in accordance with the following equation:

$$V = \frac{1.2 S_{DS}}{R} W$$ (Equation 16-56)

where:

S_{DS} = The design elastic response acceleration at short period as determined in accordance with Section 1615.1.3.

R = The response modification factor from Table 1617.6.2.

W = The effective seismic weight of the structure, including the total dead load and other loads listed below:

1. In areas used for storage, a minimum of 25 percent of the reduced floor live load (floor live load in public garages and open parking structures need not be included).

2. Where an allowance for partition load is included in the floor load design, the actual partition weight or a minimum weight of 10 psf of floor area, whichever is greater (0.48 kN/m²).

3. Total weight of permanent operating equipment.

4. 20 percent of flat roof snow load where flat snow load exceeds 30 psf (1.44 kN/m²).

TABLE 1617.3.1
ALLOWABLE STORY DRIFT, Δ_a (inches)[a]

BUILDING	SEISMIC USE GROUP		
	I	II	III
Buildings, other than masonry shear wall or masonry wall frame buildings, four stories or less in height with interior walls, partitions, ceilings and exterior wall systems that have been designed to accommodate the story drifts	$0.025\ h_{sx}$[b]	$0.020\ h_{sx}$	$0.015\ h_{sx}$
Masonry cantilever shear wall buildings[c]	$0.010\ h_{sx}$	$0.010\ h_{sx}$	$0.010\ h_{sx}$
Other masonry shear wall buildings	$0.007\ h_{sx}$	$0.007\ h_{sx}$	$0.007\ h_{sx}$
Masonry wall frame buildings	$0.013\ h_{sx}$	$0.013\ h_{sx}$	$0.010\ h_{sx}$
All other buildings	$0.020\ h_{sx}$	$0.015\ h_{sx}$	$0.010\ h_{sx}$

For SI: 1 inch = 25.4 mm.

a. There shall be no drift limit for single-story buildings with interior walls, partitions, ceilings and exterior wall systems that have been designed to accommodate the story drifts.

b. h_{sx} is the story height below Level x.

c. Buildings in which the basic structural system consists of masonry shear walls designed as vertical elements cantilevered from their base or foundation support which are so constructed that moment transfer between shear walls (coupling) is negligible.

1617.5.2 Vertical distribution. The forces at each level shall be calculated using the following equation:

$$F_x = \frac{1.2 S_{DS}}{R} w_x \qquad \textbf{(Equation 16-57)}$$

where:

w_x = The portion of the effective seismic weight of the structure, W, at Level x.

1617.5.3 Horizontal distribution. Diaphragms constructed of untopped steel decking or wood structural panels or similar light-framed construction are permitted to be considered as flexible.

1617.5.4 Design drift. For the purposes of Sections 1617.3.1 and 1620.4.6, the design story drift, Δ, shall be taken as 1 percent of the story height unless a more exact analysis is provided.

1617.6 Seismic-force-resisting systems. The provisions given in Section 9.5.2.2 of ASCE 7 shall be used except as modified in Section 1617.6.1.

Exception: For structures designed using the simplified analysis procedure in Section 1617.5, the provisions of Section 1617.6.2 shall be used.

1617.6.1 Modifications to ASCE 7, Section 9.5.2.2.

1617.6.1.1 ASCE 7, Table 9.5.2.2. Modify Table 9.5.2.2 as follows:

1. Bearing wall systems: Ordinary reinforced masonry shear walls shall use a response modification coefficient of $2^1/_2$. Light-framed walls sheathed with wood structural panels rated for shear resistance or steel sheets shall use a response modification coefficient of $6^1/_2$. Table 1617.6.2 entries for ordinary plain prestressed masonry shear walls, intermediate prestressed masonry shear walls and special prestressed masonry shear walls shall apply.

2. Building frame systems: Ordinary reinforced masonry shear walls shall use a response modification coefficient of 3. Light-framed walls sheathed with wood structural panels rated for shear resistance or steel sheets shall use a response modification coefficient of 7. Table 1617.6.2 entries for ordinary plain prestressed masonry shear walls, intermediate prestressed masonry shear walls and special prestressed masonry shear walls shall apply.

3. Dual systems with intermediate moment frames capable of resisting at least 25 percent of prescribed seismic forces. Special steel concentrically braced frames shall use a deflection amplification factor of 4.

4. The table column titled Detailing Reference Section in Table 1617.6.2 shall apply.

1617.6.1.2 ASCE 7, Section 9.5.2.2.2.1. Modify Section 9.5.2.2.2.1 by adding Exception 3 as follows:

3. The following two-stage static analysis procedure is permitted to be used for structures having a flexible upper portion supported on a rigid lower portion where both portions of the structure considered separately can be classified as being regular, the average story stiffness of the lower portion is at least 10 times the average story stiffness of the upper portion and the period of the entire structure is not greater than 1.1 times the period of the upper portion considered as a separate structure fixed at the base:

3.1. The flexible upper portion shall be designed as a separate structure using the appropriate values of R and ρ.

3.2. The rigid lower portion shall be designed as a separate structure using the appropriate values of R and ρ. The reactions from the upper portion shall be those determined from the analysis of the upper portion amplified by the ratio of the R/ρ of the upper portion over R/ρ of the lower portion. This ratio shall not be less than 1.0.

1617.6.1.3 ASCE 7, Section 9.5.2.2.4.3. Modify Section 9.5.2.2.4.3 by changing exception to read as follows:

Exception: Reinforced concrete frame members not designed as part of the seismic-force-resisting system and slabs shall comply with Section 21.11 of Ref. 9.9-1.

1617.6.2 Seismic-force-resisting systems (for use in the Simplified analysis procedure of Section 1617.5). The basic lateral and vertical seismic-force-resisting systems shall conform to one of the types indicated in Table 1617.6.2 subject to the limitations on height indicated in the table based on seismic design category as determined in Section 1616. The appropriate response modification coefficient, R, system overstrength factor, Ω_0, and deflection amplification factor, C_d, indicated in Table 1617.6.2 shall be used in determining the base shear, element design forces and design story drift. For seismic-force-resisting systems not listed in Table 1617.6.2, analytical and test data shall be submitted that establish the dynamic characteristics and demonstrate the lateral-force resistance and energy dissipation capacity to be equivalent to the structural systems listed in Table 1617.6.2 for equivalent response modification coefficient, R, system overstrength coefficient, Ω_0, and deflection amplification factor, C_d, values. Buildings shall not exceed the limitations of Section 1616.6.1.

Exception: Structures assigned to Seismic Design Category A.

1617.6.2.1 Dual systems. For a dual system, the moment frame shall be capable of resisting at least 25 percent of the design forces. The total seismic force resistance is to be provided by the combination of the moment frame and the shear walls or braced frames in proportion to their stiffness.

1617.6.2.2 Combination along the same axis. For other than dual systems and shear wall-frame interactive systems, where a combination of different structural systems is utilized to resist lateral forces in the same direction, the value, R, used for design in that direction shall not be greater than the least value for any of the systems utilized in that same direction.

Exception: For light-framed, flexible diaphragm buildings, of Seismic Use Group I and two stories or less in height: Resisting elements are permitted to be designed using the least value of R for the different structural systems found on each independent line of resistance. The value of R used for design of diaphragms in such structures shall not be greater than the least value for any of the systems utilized in that same direction.

1617.6.2.3 Combinations of framing systems. Where different seismic-force-resisting systems are used along the two orthogonal axes of the structure, the appropriate response modification coefficient, R, system overstrength factor, Ω_0, and deflection amplification factor, C_d, indicated in Table 1617.6.2 for each system shall be used.

1617.6.2.3.1 Combination framing factor. The response modification coefficient, R, in the direction under consideration at any story shall not exceed the lowest response modification coefficient, R, for the seismic-force-resisting system in the same direction considered above that story, excluding penthouses. The system overstrength factor, Ω_0, in the direction under consideration at any story, shall not be less than the largest value of this factor for the seismic-force-resisting system in the same direction considered above that story. In structures assigned to Seismic Design Category D, E or F, if a system with a response modification coefficient, R, with a value less than five is used as part of the seismic-force-resisting system in any direction of the structure, the lowest such value shall be used for the entire structure.

Exceptions:

1. Detached one- and two-family dwellings constructed of light framing.

2. The response modification coefficient, R, and system overstrength factor, Ω_0, for supported structural systems with a weight equal to or less than 10 percent of the weight of the structure are permitted to be determined independent of the values of these parameters for the structure as a whole.

3. The following two-stage static analysis procedure is permitted to be used for structures having a flexible upper portion supported on a rigid lower portion where both portions of the structure considered separately can be classified as being regular, the average story stiffness of the lower portion is at least 10 times the average story stiffness of the upper portion and the period of the entire structure is not greater than 1.1 times the period of the upper portion considered as a separate structure fixed at the base:

   3.1. The flexible upper portion shall be designed as a separate structure using the appropriate values of R and ρ.

   3.2. The rigid lower portion shall be designed as a separate structure using the appropriate values of R and ρ. The reactions from the upper portion shall be those determined from the analysis of the upper portion amplified by the ratio of, R/ρ, of the upper portion over, R/ρ, of the lower portion. This ratio shall not be less than 1.0.

1617.6.2.3.2 Combination framing detailing requirements. The detailing requirements of Section 1620 required by the higher response modification coefficient, R, shall be used for structural components common to systems having different response modification coefficients.

1617.6.2.4 System limitations for Seismic Design Category D, E or F. In addition to the system limitation indicated in Table 1617.6.2, structures assigned to Seismic Design Category D, E or F shall be subject to the following.

TABLE 1617.6.2
DESIGN COEFFICIENTS AND FACTORS FOR BASIC SEISMIC-FORCE-RESISTING SYSTEMS

BASIC SEISMIC-FORCE-RESISTING SYSTEM	DETAILING REFERENCE SECTION	RESPONSE MODIFICATION COEFFICIENT, R^a	SYSTEM OVERSTRENGTH FACTOR, Ω_o^g	DEFLECTION AMPLIFICATION FACTOR, C_d^b	SYSTEM LIMITATIONS AND BUILDING HEIGHT LIMITATIONS (FEET) BY SEISMIC DESIGN CATEGORY AS DETERMINED IN SECTION 1616.3[c]				
					A or B	C	D^d	E^e	F^e
1. Bearing Wall Systems									
A. Ordinary steel braced frames in light-frame construction	2211	4	2	$3^1/_2$	NL	NL	65	65	65
B. Special reinforced concrete shear walls	1910.2.4	$5^1/_2$	$2^1/_2$	5	NL	NL	160	160	100
C. Ordinary reinforced concrete shear walls	1910.2.3	$4^1/_2$	$2^1/_2$	4	NL	NL	NP	NP	NP
D. Detailed plain concrete shear walls	1910.2.2	$2^1/_2$	$2^1/_2$	2	NL	NP	NP	NP	NP
E. Ordinary plain concrete shear walls	1910.2.1	$1^1/_2$	$2^1/_2$	$1^1/_2$	NL	NP	NP	NP	NP
F. Special reinforced masonry shear walls	1.13.2.2.5[o]	5	$2^1/_2$	$3^1/_2$	NL	NL	160	160	100
G. Intermediate reinforced masonry shear walls	1.13.2.2.4[o]	$3^1/_2$	$2^1/_2$	$2^1/_4$	NL	NL	NP	NP	NP
H. Ordinary reinforced masonry shear walls	1.13.2.2.3[o]	$2^1/_2$	$2^1/_2$	$1^3/_4$	NL	160	NP	NP	NP
I. Detailed plain masonry shear walls	1.13.2.2.2[o]	2	$2^1/_2$	$1^3/_4$	NL	NP	NP	NP	NP
J. Ordinary plain masonry shear walls	1.13.2.2.1[o]	$1^1/_2$	$2^1/_2$	$1^1/_4$	NL	NP	NP	NP	NP
K. Light frame walls with shear panels—wood structural panels/sheet steel panels	2306.4.1/2211	$6^1/_2$	3	4	NL	NL	65	65	65
L. Light framed walls with shear panels—all other materials	2306.4.5/2211	2	$2^1/_2$	2	NL	NL	35	NP	NP
M. Ordinary plain prestressed masonry shear walls	2106.1.1.1	$1^1/_2$	$2^1/_2$	$1^1/_4$	NL	NP	NP	NP	NP
N. Intermediate prestressed masonry shear walls	2106.1.1.2, 1.13.2.2.4[o]	$2^1/_2$	$2^1/_2$	$2^1/_2$	NL	35	NP	NP	NP
O. Special prestressed masonry shear walls	2106.1.1.3, 1.13.2.2.5[o]	$4^1/_2$	$2^1/_2$	$3^1/_2$	NL	35	35	35	35
2. Building Frame Systems									
A. Steel eccentrically braced frames, moment-resisting, connections at columns away from links	(15)[j]	8	2	4	NL	NL	160	160	100
B. Steel eccentrically braced frames, nonmoment resisting, connections at columns away from links	(15)[j]	7	2	4	NL	NL	160	160	100
C. Special steel concentrically braced frames	(13)[j]	6	2	5	NL	NL	160	160	100
D. Ordinary steel concentrically braced frames	(14)[j]	5	2	$4^1/_2$	NL	NL	35^n	35^n	NP[n]
E. Special reinforced concrete shear walls	1910.2.4	6	$2^1/_2$	5	NL	NL	160	160	100
F. Ordinary reinforced concrete shear walls	1910.2.3	5	$2^1/_2$	$4^1/_2$	NL	NL	NP	NP	NP
G. Detailed plain concrete shear walls	1910.2.2	3	$2^1/_2$	$2^1/_2$	NL	NP	NP	NP	NP

(continued)

TABLE 1617.6.2—continued
DESIGN COEFFICIENTS AND FACTORS FOR BASIC SEISMIC-FORCE-RESISTING SYSTEMS

BASIC SEISMIC-FORCE-RESISTING SYSTEM	DETAILING REFERENCE SECTION	RESPONSE MODIFICATION COEFFICIENT, R^a	SYSTEM OVERSTRENGTH FACTOR, Ω_o^g	DEFLECTION AMPLIFICATION FACTOR, C_d^b	SYSTEM LIMITATIONS AND BUILDING HEIGHT LIMITATIONS (FEET) BY SEISMIC DESIGN CATEGORY[c] AS DETERMINED IN SECTION 1616.3[c]				
					A or B	C	D^d	E^e	F^e
H. Ordinary plain concrete shear walls	1910.2.1	2	$2\frac{1}{2}$	2	NL	NP	NP	NP	NP
I. Composite eccentrically braced frames	(14)[k]	8	2	4	NL	NL	160	160	100
J. Composite concentrically braced frames	(13)[k]	5	2	$4\frac{1}{2}$	NL	NL	160	160	100
K. Ordinary composite braced frames	(12)[k]	3	2	3	NL	NL	NP	NP	NP
L. Composite steel plate shear walls	(17)[k]	$6\frac{1}{2}$	$2\frac{1}{2}$	$5\frac{1}{2}$	NL	NL	160	160	100
M. Special composite reinforced concrete shear walls with steel elements	(16)[k]	6	$2\frac{1}{2}$	5	NL	NL	160	160	100
N. Ordinary composite reinforced concrete shear walls with steel elements	(15)[k]	5	$2\frac{1}{2}$	$4\frac{1}{2}$	NL	NL	NP	NP	NP
O. Special reinforced masonry shear walls	1.13.2.2.5[o]	$5\frac{1}{2}$	$2\frac{1}{2}$	4	NL	NL	160	160	100
P. Intermediate reinforced masonry shear walls	1.13.2.2.4[o]	4	$2\frac{1}{2}$	$2\frac{1}{2}$	NL	NL	NP	NP	NP
Q. Ordinary reinforced masonry shear walls	1.13.2.2.3[o]	3	$2\frac{1}{2}$	$2\frac{1}{4}$	NL	160	NP	NP	NP
R. Detailed plain masonry shear walls	1.13.2.2.2[o]	$2\frac{1}{2}$	$2\frac{1}{2}$	$2\frac{1}{4}$	NL	NP	NP	NP	NP
S. Ordinary plain masonry shear walls	1.13.2.2.1[o]	$1\frac{1}{2}$	$2\frac{1}{2}$	$1\frac{1}{4}$	NL	NP	NP	NP	NP
T. Light frame walls with shear panels—wood structural panels/sheet steel panels	2306.4.1/ 2211	7	$2\frac{1}{2}$	$4\frac{1}{2}$	NL	NL	65	65	65
U. Light framed walls with shear panels—all other materials	2306.4.5/ 2211	$2\frac{1}{2}$	$2\frac{1}{2}$	$2\frac{1}{2}$	NL	NL	35	NP	NP
V. Ordinary plain prestressed masonry shear walls	2106.1.1.1	$1\frac{1}{2}$	$2\frac{1}{2}$	$1\frac{1}{4}$	NL	NP	NP	NP	NP
W. Intermediate prestressed masonry shear walls	2106.1.1.2; 1.13.2.2.4[o]	3	$2\frac{1}{2}$	$2\frac{1}{2}$	NL	35	NP	NP	NP
X. Special prestressed masonry shear walls	2106.1.1.3; 1.13.2.2.5[o]	$4\frac{1}{2}$	$2\frac{1}{2}$	4	NL	35	35	35	35
3. Moment-resisting Frame Systems									
A. Special steel moment frames	(9)[j]	8	3	$5\frac{1}{2}$	NL	NL	NL	NL	NL
B. Special steel truss moment frames	(12)[j]	7	3	$5\frac{1}{2}$	NL	NL	160	100	NP
C. Intermediate steel moment frames	(10)[j]	$4\frac{1}{2}$	3	4	NL	NL	35^h	$NP^{h,i}$	$NP^{h,i}$
D. Ordinary steel moment frames	(11)[j]	$3\frac{1}{2}$	3	3	NL	NL	$NP^{h,i}$	$NP^{h,i}$	$NP^{h,i}$
E. Special reinforced concrete moment frames	(21.1)[l]	8	3	$5\frac{1}{2}$	NL	NL	NL	NL	NL

(continued)

TABLE 1617.6.2—continued
DESIGN COEFFICIENTS AND FACTORS FOR BASIC SEISMIC-FORCE-RESISTING SYSTEMS

BASIC SEISMIC-FORCE-RESISTING SYSTEM	DETAILING REFERENCE SECTION	RESPONSE MODIFICATION COEFFICIENT, R^a	SYSTEM OVERSTRENGTH FACTOR, Ω_o^g	DEFLECTION AMPLIFICATION FACTOR, C_d^b	SYSTEM LIMITATIONS AND BUILDING HEIGHT LIMITATIONS (FEET) BY SEISMIC DESIGN CATEGORY AS DETERMINED IN SECTION 1616.3[c]				
					A or B	C	D^d	E^e	F^e
F. Intermediate reinforced concrete moment frames	$(21.1)^l$	5	3	$4\frac{1}{2}$	NL	NL	NP	NP	NP
G. Ordinary reinforced concrete moment frames	$(21.1)^l$	3	3	$2\frac{1}{2}$	NL	NP	NP	NP	NP
H. Special composite moment frames	$(9)^k$	8	3	$5\frac{1}{2}$	NL	NL	NL	NL	NL
I. Intermediate composite moment frames	$(10)^k$	5	3	$4\frac{1}{2}$	NL	NL	NP	NP	NP
J. Composite partially restrained moment frames	$(8)^k$	6	3	$5\frac{1}{2}$	160	160	100	NP	NP
K. Ordinary composite moment frames	$(11)^k$	3	3	$2\frac{1}{2}$	NL	NP	NP	NP	NP
L. Masonry wall frames	2106	$5\frac{1}{2}$	3	5	NL	NL	160	160	100
4. Dual Systems with Special Moment Frames									
A. Steel eccentrically braced frames, moment-resisting connections, at columns away from links	$(15)^j$	8	$2\frac{1}{2}$	4	NL	NL	NL	NL	NL
B. Steel eccentrically braced frames, nonmoment-resisting connections, at columns away from links	$(15)^j$	7	$2\frac{1}{2}$	4	NL	NL	NL	NL	NL
C. Special steel concentrically braced frames	$(13)^j$	8	$2\frac{1}{2}$	$6\frac{1}{2}$	NL	NL	NL	NL	NL
D. Special reinforced concrete shear walls	1910.2.4	8	$2\frac{1}{2}$	$6\frac{1}{2}$	NL	NL	NL	NL	NL
E. Ordinary reinforced concrete shear walls	1910.2.3	7	$2\frac{1}{2}$	6	NL	NL	NP	NP	NP
F. Composite eccentrically braced frames	$(14)^k$	8	$2\frac{1}{2}$	4	NL	NL	NL	NL	NL
G. Composite concentrically braced frames	$(13)^k$	6	$2\frac{1}{2}$	5	NL	NL	NL	NL	NL
H. Composite steel plate shear walls	$(17)^k$	8	$2\frac{1}{2}$	$6\frac{1}{2}$	NL	NL	NL	NL	NL
I. Special composite reinforced concrete shear walls with steel elements	$(16)^k$	8	$2\frac{1}{2}$	$6\frac{1}{2}$	NL	NL	NL	NL	NL
J. Ordinary composite reinforced concrete shear walls with steel elements	$(15)^k$	7	$2\frac{1}{2}$	6	NL	NL	NP	NP	NP
K. Special reinforced masonry shear walls	$1.13.2.2.5^o$	7	3	$6\frac{1}{2}$	NL	NL	NL	NL	NL
L. Intermediate reinforced masonry shear walls	$1.13.2.2.4^o$	$6\frac{1}{2}$	3	$5\frac{1}{2}$	NL	NL	NP	NP	NP
5. Dual Systems with Intermediate Moment Frames[m]									
A. Special steel concentrically braced frames[f]	$(13)^j$	$4\frac{1}{2}$	$2\frac{1}{2}$	4	NL	NL	35^h	$NP^{h,i}$	NP
B. Special reinforced concrete shear walls	1910.2.4	6	$2\frac{1}{2}$	5	NL	NL	160	100	100
C. Ordinary reinforced concrete shear walls	1910.2.3	$5\frac{1}{2}$	$2\frac{1}{2}$	$4\frac{1}{2}$	NL	NL	NP	NP	NP

(continued)

TABLE 1617.6.2-continued
DESIGN COEFFICIENTS AND FACTORS FOR BASIC SEISMIC-FORCE-RESISTING SYSTEMS

BASIC SEISMIC-FORCE-RESISTING SYSTEM	DETAILING REFERENCE SECTION	RESPONSE MODIFICATION COEFFICIENT, R [a]	SYSTEM OVERSTRENGTH FACTOR, Ω_o [g]	DEFLECTION AMPLIFICATION FACTOR, C_d [b]	SYSTEM LIMITATIONS AND BUILDING HEIGHT LIMITATIONS (FEET) BY SEISMIC DESIGN CATEGORY AS DETERMINED IN SECTION 1616.3 [c]				
					A or B	C	D [d]	E [e]	F [e]
D. Ordinary reinforced masonry shear walls	1.13.2.2.3 [o]	3	3	$2\frac{1}{2}$	NL	160	NP	NP	NP
E. Intermediate reinforced masonry shear walls	1.13.2.2.4 [o]	5	3	$4\frac{1}{2}$	NL	NL	NP	NP	NP
F. Composite concentrically braced frames	(13) [k]	5	$2\frac{1}{2}$	$4\frac{1}{2}$	NL	NL	160	100	NP
G. Ordinary composite braced frames	(12) [k]	4	$2\frac{1}{2}$	3	NL	NL	NP	NP	NP
H. Ordinary composite reinforced concrete shear walls with steel elements	(15) [k]	$5\frac{1}{2}$	$2\frac{1}{2}$	$4\frac{1}{2}$	NL	NL	NP	NP	NP
6. Shear Wall-frame Interactive System with Ordinary Reinforced Concrete Moment Frames and Ordinary Reinforced Concrete Shear Walls	21.1 [l] 1910.2.3	$5\frac{1}{2}$	$2\frac{1}{2}$	5	NL	NP	NP	NP	NP
7. Inverted Pendulum Systems									
A. Cantilevered column systems		$2\frac{1}{2}$	2	$2\frac{1}{2}$	NL	NL	35	35	35
B. Special steel moment frames	(9) [j]	$2\frac{1}{2}$	2	$2\frac{1}{2}$	NL	NL	NL	NL	NL
C. Ordinary steel moment frames	(11) [j]	$1\frac{1}{4}$	2	$2\frac{1}{2}$	NL	NL	NP	NP	NL
D. Special reinforced concrete moment frames	21.1 [l]	$2\frac{1}{2}$	2	$1\frac{1}{4}$	NL	NL	NL	NL	NL
8. Structural Steel Systems Not Specifically Detailed for Seismic Resistance	AISC—335 AISC—LRFD AISI AISC—HSS	3	3	3	NL	NL	NP	NP	NP

For SI: 1 foot = 304.8 mm, 1 pound per square foot = 0.0479 KN/m².

a. Response modification coefficient, R, for use throughout.
b. Deflection amplification factor, C_d.
c. NL = Not limited and NP = Not permitted.
d. See Section 1617.6.2.4.1 for a description of building systems limited to buildings with a height of 240 feet or less.
e. See Section 1617.6.2.4.1 for building systems limited to buildings with a height of 160 feet or less.
f. Ordinary moment frame is permitted to be used in lieu of intermediate moment frame in Seismic Design Categories B and C.
g. The tabulated value of the overstrength factor, Ω_o, is permitted to be reduced by subtracting $\frac{1}{2}$ for structures with flexible diaphragms but shall not be taken as less than 2.0 for any structure.
h. Steel ordinary moment frames and intermediate moment frames are permitted in single-story buildings up to a height of 60 feet, when the moment joints of field connections are constructed of bolted end plates and the dead load of the roof does not exceed 15 pounds per square foot. The dead weight of the portion of walls more than 35 feet above the base shall not exceed 15 pounds per square foot.
i. Steel ordinary moment frames are permitted in buildings up to a height of 35 feet, where the dead load of the walls, floors and roof does not exceed 15 pounds per square foot.
j. AISC 341 Part I or Part III section number.
k. AISC 341 Part II section number.
l. ACI 318, Section number.
m. Steel intermediate moment resisting frames as part of a dual system are not permitted in Seismic Design Categories D, E, and F.
n. Steel ordinary concentrically braced frames are permitted in penthouse structures and in single-story buildings up to a height of 60 feet when the dead load of the roof does not exceed 15 pounds per square foot.
o. ACI 530/ASCE 5/TMS 402 section number.

1617.6.2.4.1 Limited building height. For buildings that have steel-braced frames or concrete cast-in-place shear walls, the height limits in Table 1617.6.2 for Seismic Design Category D or E are increased to 240 feet (73 152 mm) and for Seismic Design Category F to 160 feet (48 768 mm) provided that the buildings are configured such that the braced frames or shear walls arranged in any one plane conform to the following:

1. The braced frames or shear walls in any one plane shall resist no more than 50 percent of the total seismic forces in each direction, neglecting torsional effects.

2. The seismic force in the braced frames or shear walls in any one plane resulting from torsional effects shall not exceed 20 percent of the total seismic force in the braced frames or shear walls.

1617.6.2.4.2 Interaction effects. Moment-resisting frames that are enclosed or adjoined by stiffer elements not considered to be part of the seismic-force-resisting system shall be designed so that the action or failure of those elements will not impair the vertical load and seismic-force-resisting capability of the frame. The design shall consider and provide for the effect of these rigid elements on the structural system at deformations corresponding to the design story drift, Δ, as determined in Section 1617.5.4. In addition, the effects of these elements shall be considered when determining whether a structure has one or more of the irregularities defined in Section 1616.5.1.

1617.6.2.4.3 Deformational compatibility. Every structural component not included in the seismic-force-resisting system in the direction under consideration shall be designed to be adequate for vertical load-carrying capacity and the induced moments and shears resulting from the design story drift, Δ, as determined in accordance with Section 1617.5.4. Where allowable stress design is used, Δ shall be computed without dividing the earthquake force by 1.4. The moments and shears induced in components that are not included in the seismic-force-resisting system in the direction under consideration shall be calculated including the stiffening effects of adjoining rigid structural and nonstructural elements.

> **Exception:** Reinforced concrete frame members not designed as part of the seismic-force-resisting system and slabs shall comply with Section 21.11 of ACI 318.

1617.6.2.4.4 Special moment frames. A special moment frame that is used but not required by Table 1617.6.2 is permitted to be discontinued and supported by a stiffer system with a lower response modification coefficient, R, provided the requirements of Sections 1620.2.3 and 1620.4.1 are met. Where a special moment frame is required by Table 1617.6.2, the frame shall be continuous to the foundation.

SECTION 1618
DYNAMIC ANALYSIS PROCEDURE FOR THE SEISMIC DESIGN OF BUILDINGS

1618.1 Dynamic analysis procedures. The following dynamic analysis procedures are permitted to be used in lieu of the equivalent lateral force procedure of Section 1617.4:

1. Modal Response Spectral Analysis.

2. Linear Time-history Analysis.

3. Nonlinear Time-history Analysis.

The dynamic analysis procedures listed above shall be performed in accordance with the requirements of Sections 9.5.6, 9.5.7 and 9.5.8, respectively, of ASCE 7.

SECTION 1619
EARTHQUAKE LOADS
SOIL-STRUCTURE INTERACTION EFFECTS

1619.1 Analysis procedure. If soil-structure interaction is considered in the determination of seismic design forces and corresponding displacements in the structure, the procedure given in Section 9.5.9 of ASCE 7 shall be used.

SECTION 1620
EARTHQUAKE LOADS—DESIGN, DETAILING REQUIREMENTS AND STRUCTURAL COMPONENT LOAD EFFECTS

1620.1 Structural component design and detailing. The design and detailing of the components of the seismic-force-resisting system shall comply with the requirements of Section 9.5.2.6 of ASCE 7 in addition to the nonseismic requirements of this code except as modified in Sections 1620.1.1, 1620.1.2 and 1620.1.3.

> **Exception:** For structures designed using the simplified analysis procedure in Section 1617.5, the provisions of Sections 1620.2 through 1620.5 shall be used.

> **1620.1.1 ASCE 7, Section 9.5.2.6.2.5.** Section 9.5.2.6.2.5 of ASCE 7 shall not apply.

> **1620.1.2 ASCE 7, Section 9.5.2.6.2.11.** Modify ASCE 7, Section 9.5.2.6.2.11, to read as follows:

> 9.5.2.6.2.11 Elements supporting discontinuous walls or frames. Columns, beams, trusses or slabs supporting discontinuous walls or frames of structures and the connections of the discontinuous element to the supporting member having plan irregularity Type 4 of Table 9.5.2.3.2 or vertical irregularity Type 4 of Table 9.5.2.3.3 shall have the design strength to resist the maximum axial force that can develop in accordance with the special seismic loads of Section 9.5.2.7.1.

> **Exceptions:**

> 1. The quantity E in Section 9.5.2.7.1 need not exceed the maximum force that can be transmitted to the element by the lateral-force-resisting system at yield.

> 2. Concrete slabs supporting light-framed walls.

1620.1.3 ASCE 7, Section 9.5.2.6.3. Modify ASCE 7, Section 9.5.2.6.3, to read as follows:

9.5.2.6.3 Seismic Design Category C. Structures assigned to Category C shall conform to the requirements of Section 9.5.2.6.2 for Category B and to the requirements of this section. Structures that have plan structural irregularity Type 1a or 1b of Table 9.5.2.3.2 along both principal plan axes, or plan structural irregularity Type 5 of Table 9.5.2.3.2, shall be analyzed for seismic forces in compliance with Section 9.5.2.5.2.2. When the square root of the sum of the squares method of combining directional effects is used, each term computed shall be assigned the sign that will yield the most conservative result.

The orthogonal combination procedure of Section 9.5.2.5.2.2, Item a, shall be required for any column or wall that forms part of two or more intersecting seismic-force-resisting systems and is subjected to axial load due to seismic forces acting along either principal plan axis equaling or exceeding 20 percent of the axial load design strength of the column or wall.

1620.2 Structural component design and detailing (for use in the simplified analysis procedure of Section 1617.5). The design and detailing of the components of the seismic-force-resisting system for structures designed using the simplified analysis procedure in Section 1617.5 shall comply with the requirements of Sections 1620.2 through 1620.5 in addition to the nonseismic requirements of this code. Buildings shall not exceed the limitations of Section 1616.6.1.

Exception: Structures assigned to Seismic Design Category A.

Structures assigned to Seismic Design Category B (see Section 1616) shall conform to Sections 1620.2.1 through 1620.2.10.

1620.2.1 Second-order load effects. Where θ exceeds 0.10 as determined in Section 9.5.5.7.2 in ASCE 7, second-order load effects shall be included in the evaluation of component and connection strengths.

1620.2.2 Openings. Where openings occur in shear walls, diaphragms or other plate-type elements, reinforcement at the edges of the openings shall be designed to transfer the stresses into the structure. The edge reinforcement shall extend into the body of the wall or diaphragm a distance sufficient to develop the force in the reinforcement.

1620.2.3 Discontinuities in vertical system. Structures with a discontinuity in lateral capacity, vertical irregularity Type 5, as defined in Table 1616.5.1.2, shall not be over two stories or 30 feet (9144 mm) in height where the "weak" story has a calculated strength of less than 65 percent of the story above.

Exception: Where the "weak" story is capable of resisting a total seismic force equal to the overstrength factor, Ω_0, as given in Table 1617.6.2, multiplied by the design force prescribed in Section 1617.5, the height limitation does not apply.

1620.2.4 Connections. All parts of the structure, except at separation joints, shall be interconnected and the connec-

tions shall be designed to resist the seismic force, F_p, induced by the parts being connected. Any smaller portion of the structure shall be tied to the remainder of the structure for the greater of:

$$F_p = 0.133 S_{DS} w_p \qquad \text{(Equation 16-58)}$$

or

$$F_p = 0.05 w_p \qquad \text{(Equation 16-59)}$$

where:

S_{DS} = The design, 5-percent damped, spectral response acceleration at short periods as defined in Section 1615.

w_p = The weight of the smaller portion.

A positive connection for resisting a horizontal force acting parallel to the member shall be provided for each beam, girder or truss to its support for a force not less than 5 percent of the dead plus live load reaction.

1620.2.5 Diaphragms. Permissible deflection shall be that deflection up to which the diaphragm and any attached distributing or resisting element will maintain its structural integrity under design load conditions, such that the resisting element will continue to support design loads without danger to occupants of the structure.

Floor and roof diaphragms shall be designed to resist F_p as follows:

$$F_p = 0.2 I_E S_{DS} w_p + V_{px} \qquad \text{(Equation 16-60)}$$

where:

F_p = The seismic force induced by the parts.

I_E = Occupancy importance factor (Table 1604.5).

S_{DS} = The short-period site design spectral response acceleration coefficient (Section 1615).

w_p = The weight of the diaphragm and other elements of the structure attached to the diaphragm.

V_{px} = The portion of the seismic shear force at the level of the diaphragm, required to be transferred to the components of the vertical seismic-force-resisting system because of the offsets or changes in stiffness of the vertical components above or below the diaphragm.

Diaphragms shall provide for both shear and bending stresses resulting from these forces. Diaphragms shall have ties or struts to distribute the wall anchorage forces into the diaphragm. Diaphragm connections shall be positive, mechanical or welded-type connections.

1620.2.6 Collector elements. Collector elements shall be provided that are capable of transferring the seismic forces originating in other portions of the structure to the element providing the resistance to those forces. Collector elements, splices and their connections to resisting elements shall have the design strength to resist the special load combinations of Section 1605.4.

Exception: In structures or portions thereof braced entirely by light-framed shear walls, collector elements, splices and connections to resisting elements need only have the strength to resist the load combinations of Section 1605.2 or 1605.3.

1620.2.7 Bearing walls and shear walls. Bearing walls and shear walls and their anchorage shall be designed for an out-of-plane force, F_p, that is the greater of 10 percent of the weight of the wall, or the quantity given by Equation 16-61:

$$F_p = 0.40 I_E S_{DS} w_w \qquad \text{(Equation 16-61)}$$

where:

I_E = Occupancy importance factor (Table 1604.5).

S_{DS} = The short-period site design spectral response acceleration coefficient (Section 1615.1.3 or 1615.2.5).

w_w = The weight of the wall.

In addition, concrete and masonry walls shall be anchored to the roof and floors and members that provide lateral support for the wall or that are supported by the wall. The anchorage shall provide a direct connection between the wall and the supporting construction capable of resisting the greater of the force, F_p, as given by Equation 16-61 or $(400 S_{DS} I_E)$ pounds per linear foot of wall. For SI: 5838 $S_{DS} I_E$ N/m. Walls shall be designed to resist bending between anchors where the anchor spacing exceeds 4 feet (1219 mm). Parapets shall conform to the requirements of Section 9.6.2.2 of ASCE 7.

1620.2.8 Inverted pendulum-type structures. Supporting columns or piers of inverted pendulum-type structures shall be designed for the bending moment calculated at the base determined using the procedures given in Section 1617.4 and varying uniformly to a moment at the top equal to one-half the calculated bending moment at the base.

1620.2.9 Elements supporting discontinuous walls or frames. Columns or other elements subject to vertical reactions from discontinuous walls or frames of structures having plan irregularity Type 4 of Table 1616.5.1.1 or vertical irregularity Type 4 of Table 1616.5.1.2 shall have the design strength to resist special seismic load combinations of Section 1605.4. The connections from the discontinuous walls or frames to the supporting elements need not have the design strength to resist the special seismic load combinations of Section 1605.4.

Exceptions:

1. The quantity, E_m, in Section 1617.1.1.2 need not exceed the maximum force that can be transmitted to the element by the lateral-force-resisting system at yield.

2. Concrete slabs supporting light-framed walls.

1620.2.10 Direction of seismic load. The direction of application of seismic forces used in design shall be that which will produce the most critical load effect in each component. The requirement will be deemed satisfied if the design seismic forces are applied separately and independently in each of the two orthogonal directions.

1620.3 Seismic Design Category C. Structures assigned to Seismic Design Category C (see Section 1616) shall conform to the requirements of Section 1620.2 for Seismic Design Category B and to Sections 1620.3.1 through 1620.3.2.

1620.3.1 Anchorage of concrete or masonry walls. Concrete or masonry walls shall be anchored to floors and roofs and members that provide out-of-plane lateral support for the wall or that are supported by the wall. The anchorage shall provide a positive direct connection between the wall and floor or roof capable of resisting the horizontal forces specified in Equation 16-62 for structures with flexible diaphragms or in Section 9.6.1.3 of ASCE 7 (using a_p of 1.0 and R_p of 2.5) for structures with diaphragms that are not flexible.

$$F_p = 0.8 S_{DS} I_E w_w \qquad \text{(Equation 16-62)}$$

where:

F_p = The design force in the individual anchors.

I_E = Occupancy importance factor in accordance with Section 1616.2.

S_{DS} = The design earthquake spectral response acceleration at short period in accordance with Section 1615.1.3.

w_w = The weight of the wall tributary to the anchor.

Diaphragms shall be provided with continuous ties or struts between diaphragm chords to distribute these anchorage forces into the diaphragms. Where added chords are used to form subdiaphragms, such chords shall transmit the anchorage forces to the main cross ties. The maximum length-to-width ratio of the structural subdiaphragm shall be $2^1/_2$ to 1. Connections and anchorages capable of resisting the prescribed forces shall be provided between the diaphragm and the attached components. Connections shall extend into the diaphragms a sufficient distance to develop the force transferred into the diaphragm.

The strength design forces for steel elements of the wall anchorage system shall be 1.4 times the force otherwise required by this section.

In wood diaphragms, the continuous ties shall be in addition to the diaphragm sheathing. Anchorage shall not be accomplished by use of toenails or nails subject to withdrawal, nor shall wood ledgers or framing be used in cross-grain bending or cross-grain tension. The diaphragm sheathing shall not be considered effective as providing the ties or struts required by this section.

In metal deck diaphragms, the metal deck shall not be used as the continuous ties required by this section in the direction perpendicular to the deck span.

Diaphragm-to-wall anchorage using embedded straps shall be attached to or hooked around the reinforcing steel or otherwise terminated so as to directly transfer force to the reinforcing steel.

1620.3.2 Direction of seismic load. For structures that have plan structural irregularity Type 1a or 1b of Table 1616.5.1.1 along both principal plan axes, or plan structural irregularity Type 5 in Table 1616.5.1.1, the critical direction requirement of Section 1620.2.10 shall be deemed satisfied if components and their foundations are designed for the following orthogonal combination of prescribed loads.

One hundred percent of the forces for one direction plus 30 percent of the forces for the perpendicular direction. The combination requiring the maximum component strength shall be used. Alternatively, the effects of the two orthogonal directions are permitted to be combined on a square root of the sum of the squares (SRSS) basis. When the SRSS method of combining directional effects is used, each term computed shall be assigned the sign that will result in the most conservative result.

The orthogonal combination procedure above shall be required for any column or wall that forms part of two or more intersecting seismic-force-resisting systems and is subjected to axial load due to seismic forces acting along either principal plan axis equaling or exceeding 20 percent of the axial load design strength of the column or wall.

1620.4 Seismic Design Category D. Structures assigned to Seismic Design Category D shall conform to the requirements of Section 1620.3 for Seismic Design Category C and to Sections 1620.4.1 through 1620.4.6.

1620.4.1 Plan or vertical irregularities. For buildings having a plan structural irregularity of Type 1a, 1b, 2, 3 or 4 in Table 1616.5.1.1 or a vertical structural irregularity of Type 4 in Table 1616.5.1.2, the design forces determined from Section 1617.5 shall be increased 25 percent for connections of diaphragms to vertical elements and to collectors, and for connections of collectors to the vertical elements.

Exception: When connection design forces are determined using the special seismic load combinations of Section 1605.4

1620.4.2 Vertical seismic forces. In addition to the applicable load combinations of Section 1605, horizontal cantilever and horizontal prestressed components shall be designed to resist a minimum net upward force of 0.2 times the dead load.

1620.4.3 Diaphragms. Floor and roof diaphragms shall be designed to resist design seismic forces determined in accordance with Equation 16-63 as follows:

$$F_{px} = \frac{\sum_{i=x}^{n} F_i}{\sum_{i=x}^{n} w_i} w_{px} \qquad \textbf{(Equation 16-63)}$$

where:

F_i = The design force applied to Level i.

F_{px} = The diaphragm design force.

w_i = The weight tributary to Level i.

w_{px} = The weight tributary to the diaphragm at Level x.

The force determined from Equation 16-63 need not exceed $0.4S_{DS}I_E w_{px}$ but shall not be less than $0.2S_{DS}I_E w_{px}$ where S_{DS} is the design spectral response acceleration at short period determined in Section 1615.1.3 and I_E is the occupancy importance factor determined in Section 1616.2. When the diaphragm is required to transfer design seismic force from the vertical-resisting elements above the diaphragm to other vertical-resisting elements below the diaphragm due to offsets in the placement of the elements or to changes in rela-

tive lateral stiffness in the vertical elements, these forces shall be added to those determined from Equation 16-63 and to the upper and lower limits on that equation.

1620.4.4 Collector elements. Collector elements shall be provided that are capable of transferring the seismic forces originating in other portions of the structure to the element providing resistance to those forces.

Collector elements, splices and their connections to resisting elements shall resist the forces determined in accordance with Equation 16-63. In addition, collector elements, splices and their connections to resisting elements shall have the design strength to resist the earthquake loads as defined in the special load combinations of Section 1605.4.

Exception: In structures, or portions thereof, braced entirely by light-framed shear walls, collector elements, splices and their connections to resisting elements need only be designed to resist forces in accordance with Equation 16-63.

1620.4.5 Building separations. All structures shall be separated from adjoining structures. Separations shall allow for the displacement δ_M. Adjacent buildings on the same property shall be separated by at least δ_{MT} where

$$\delta_{MT} = \sqrt{(\delta_{M1})^2 + (\delta_{M2})^2} \qquad \textbf{(Equation 16-64)}$$

and δ_{M1} and δ_{M2} are the displacements of the adjacent buildings.

When a structure adjoins a property line not common to a public way, that structure shall also be set back from the property line by at least the displacement, δ_M, of that structure.

Exception: Smaller separations or property line setbacks shall be permitted when justified by rational analyses based on maximum expected ground motions.

1620.4.6 Anchorage of concrete or masonry walls to flexible diaphragms. In addition to the requirements of Section 1620.3.1, concrete and masonry walls shall be anchored to flexible diaphragms based on the following:

1. When elements of the wall anchorage system are not loaded concentrically or are not perpendicular to the wall, the system shall be designed to resist all components of the forces induced by the eccentricity.

2. When pilasters are present in the wall, the anchorage force at the pilasters shall be calculated considering the additional load transferred from the wall panels to the pilasters. The minimum anchorage at a floor or roof shall not be less than that specified in Item 1.

1620.5 Seismic Design Category E or F. Structures assigned to Seismic Design Category E or F (Section 1616) shall conform to the requirements of Section 1620.4 for Seismic Design Category D and to Section 1620.5.1.

1620.5.1 Plan or vertical irregularities. Structures having plan irregularity Type 1b of Table 1616.5.1.1 or vertical irregularities Type 1b or 5 of Table 1616.5.1.2 shall not be permitted.

SECTION 1621
ARCHITECTURAL, MECHANICAL AND ELECTRICAL COMPONENT SEISMIC DESIGN REQUIREMENTS

1621.1 Component design. Architectural, mechanical, electrical and nonstructural systems, components and elements permanently attached to structures, including supporting structures and attachments (hereinafter referred to as "components"), and nonbuilding structures that are supported by other structures, shall meet the requirements of Section 9.6 of ASCE 7 except as modified in Sections 1621.1.1, 1621.1.2 and 1621.1.3, excluding Section 9.6.3.11.2, of ASCE 7, as amended in this section.

1621.1.1 ASCE 7, Section 9.6.3.11.2: Section 9.6.3.11.2 of ASCE 7 shall not apply.

1621.1.2 ASCE 7, Section 9.6.2.8.1. Modify ASCE 7, Section 9.6.2.8.1, to read as follows:

9.6.2.8.1 General. Partitions that are tied to the ceiling and all partitions greater than 6 feet (1829 mm) in height shall be laterally braced to the building structure. Such bracing shall be independent of any ceiling splay bracing. Bracing shall be spaced to limit horizontal deflection at the partition head to be compatible with ceiling deflection requirements as determined in Section 9.6.2.6 for suspended ceilings and Section 9.6.2.6 for other systems.

Exception: Partitions not taller than 9 feet (2743 mm) when the horizontal seismic load does not exceed 5 psf (0.240 KN/m^2) required in Section 1607.13.

1621.1.3 ASCE 7, Section 9.6.3.13. Modify ASCE 7, Section 9.6.3.13, to read as follows:

9.6.3.13 Mechanical equipment, attachments and supports. Attachments and supports for mechanical equipment not covered in Sections 9.6.3.8 through 9.6.3.12 or Section 9.6.3.16 shall be designed to meet the force and displacement provisions of Section 9.6.1.3 and 9.6.1.4 and the additional provisions of this section. In addition to their attachments and supports, such mechanical equipment designated as having an I_p = 1.5, which contains hazardous or flammable materials in quantities that exceed the maximum allowable quantities for an open system listed in Chapter 4, shall, itself, be designed to meet the force and displacement provisions of Sections 9.6.1.3 and 9.6.1.4 and the additional provisions of this section. The seismic design of mechanical equipment, attachments and their supports shall include analysis of the following: the dynamic effects of the equipment, its contents and, when appropriate, its supports. The interaction between the equipment and the supporting structures, including other mechanical and electrical equipment, shall also be considered.

SECTION 1622
NONBUILDING STRUCTURES SEISMIC DESIGN REQUIREMENTS

1622.1 Nonbuilding structures. The requirements of Section 9.14 of ASCE 7 shall apply to nonbuilding structures except as modified by Sections 1622.1.1, 1622.1.2 and 1622.1.3.

1622.1.1 ASCE 7, Section 9.14.5.1. Modify Section 9.14.5.1, Item 9, to read as follows:

9. Where an approved national standard provides a basis for the earthquake-resistant design of a particular type of nonbuilding structure covered by Section 9.14, such a standard shall not be used unless the following limitations are met:

1. The seismic force shall not be taken as less than 80 percent of that given by the remainder of Section 9.14.5.1.

2. The seismic ground acceleration, and seismic coefficient, shall be in conformance with the requirements of Sections 9.4.1 and 9.4.1.2.5, respectively.

3. The values for total lateral force and total base overturning moment used in design shall not be less than 80 percent of the base shear value and overturning moment, each adjusted for the effects of soil structure interaction that is obtained by using this standard.

1622.1.2 ASCE 7, Section 9.14.7.2.1. Modify Section 9.14.7.2.1 to read as follows:

9.14.7.2.1 General. This section applies to all earth-retaining walls. The applied seismic forces shall be determined in accordance with Section 9.7.5.1 with a geotechnical analysis prepared by a registered design professional.

The seismic use group shall be determined by the proximity of the retaining wall to other nonbuilding structures or buildings. If failure of the retaining wall would affect an adjacent structure, the seismic use group shall not be less than that of the adjacent structure, as determined in Section 9.1.3. Earth-retaining walls are permitted to be designed for seismic loads as either yielding or nonyielding walls. Cantilevered reinforced concrete retaining walls shall be assumed to be yielding walls and shall be designed as simple flexural wall elements.

1622.1.3 ASCE 7, Section 9.14.7.9. Add a new Section 9.14.7.9 to read as follows:

9.14.7.9 Buried structures. As used in this section, the term "buried structures" means subgrade structures such as tanks, tunnels and pipes. Buried structures that are designated as Seismic Use Group II or III, as determined in Section 9.1.3, or are of such a size or length as to warrant special seismic design as determined by the registered design professional, shall be identified in the geotechnical report. Buried structures shall be designed to resist seismic lateral forces determined from a substantiated analysis using standards approved by the building official. Flexible couplings shall be provided for buried structures where changes in the support system, configurations or soil condition occur.

SECTION 1623
SEISMICALLY ISOLATED STRUCTURES

1623.1 Design requirements. Every seismically isolated structure and every portion thereof shall be designed and constructed in accordance with the requirements of Section 9.13 of ASCE 7, except as modified in Section 1623.1.1.

1623.1.1 ASCE 7, Section 9.13.6.2.3. Modify ASCE 7, Section 9.13.6.2.3, to read as follows:

9.13.6.2.3 Fire resistance. Fire-resistance ratings for the isolation system shall comply with Section 713.7 of the *International Building Code.*

CHAPTER 17

STRUCTURAL TESTS AND SPECIAL INSPECTIONS

SECTION 1701
GENERAL

1701.1 Scope. The provisions of this chapter shall govern the quality, workmanship and requirements for materials covered. Materials of construction and tests shall conform to the applicable standards listed in this code.

1701.2 New materials. New building materials, equipment, appliances, systems or methods of construction not provided for in this code, and any material of questioned suitability proposed for use in the construction of a building or structure, shall be subjected to the tests prescribed in this chapter and in the approved rules to determine character, quality and limitations of use.

1701.3 Used materials. The use of second-hand materials that meet the minimum requirements of this code for new materials shall be permitted.

SECTION 1702
DEFINITIONS

1702.1 General. The following words and terms shall, for the purposes of this chapter and as used elsewhere in this code, have the meanings shown herein.

APPROVED AGENCY. An established and recognized agency regularly engaged in conducting tests or furnishing inspection services, when such agency has been approved.

APPROVED FABRICATOR. An established and qualified person, firm or corporation approved by the building official pursuant to Chapter 17 of this code.

CERTIFICATE OF COMPLIANCE. A certificate stating that materials and products meet specified standards or that work was done in compliance with approved construction documents.

FABRICATED ITEM. Structural, load-bearing or lateral load-resisting assemblies consisting of materials assembled prior to installation in a building or structure, or subjected to operations such as heat treatment, thermal cutting, cold working or reforming after manufacture and prior to installation in a building or structure. Materials produced in accordance with standard specifications referenced by this code, such as rolled structural steel shapes, steel-reinforcing bars, masonry units and plywood sheets, shall not be considered "fabricated items."

INSPECTION CERTIFICATE. An identification applied on a product by an approved agency containing the name of the manufacturer, the function and performance characteristics, and the name and identification of an approved agency that indicates that the product or material has been inspected and evaluated by an approved agency (see Section 1703.5 and "Label," "Manufacturer's designation" and "Mark").

LABEL. An identification applied on a product by the manufacturer that contains the name of the manufacturer, the func-

tion and performance characteristics of the product or material, and the name and identification of an approved agency and that indicates that the representative sample of the product or material has been tested and evaluated by an approved agency (see Section 1703.5 and "Inspection certificate," "Manufacturer's designation" and "Mark").

MANUFACTURER'S DESIGNATION. An identification applied on a product by the manufacturer indicating that a product or material complies with a specified standard or set of rules (see also "Inspection certificate," "Label" and "Mark").

MARK. An identification applied on a product by the manufacturer indicating the name of the manufacturer and the function of a product or material (see also "Inspection certificate," "Label" and "Manufacturer's designation").

SPECIAL INSPECTION. Inspection as herein required of the materials, installation, fabrication, erection or placement of components and connections requiring special expertise to ensure compliance with approved construction documents and referenced standards (see Section 1704).

SPECIAL INSPECTION, CONTINUOUS. The full-time observation of work requiring special inspection by an approved special inspector who is present in the area where the work is being performed.

SPECIAL INSPECTION, PERIODIC. The part-time or intermittent observation of work requiring special inspection by an approved special inspector who is present in the area where the work has been or is being performed and at the completion of the work.

SPRAYED FIRE-RESISTANT MATERIALS. Cementitious or fibrous materials that are spray applied to provide fire-resistant protection of the substrates.

STRUCTURAL OBSERVATION. The visual observation of the structural system by a registered design professional for general conformance to the approved construction documents at significant construction stages and at completion of the structural system. Structural observation does not include or waive the responsibility for the inspection required by Section 109, 1704 or other sections of this code.

SECTION 1703
APPROVALS

1703.1 Approved agency. An approved agency shall provide all information as necessary for the building official to determine that the agency meets the applicable requirements.

1703.1.1 Independent. An approved agency shall be objective and competent. The agency shall also disclose possible conflicts of interest so that objectivity can be confirmed.

1703.1.2 Equipment. An approved agency shall have adequate equipment to perform required tests. The equipment shall be periodically calibrated.

1703.1.3 Personnel. An approved agency shall employ experienced personnel educated in conducting, supervising and evaluating tests and/or inspections.

1703.2 Written approval. Any material, appliance, equipment, system or method of construction meeting the requirements of this code shall be approved in writing after satisfactory completion of the required tests and submission of required test reports.

1703.3 Approved record. For any material, appliance, equipment, system or method of construction that has been approved, a record of such approval, including the conditions and limitations of the approval, shall be kept on file in the building official's office and shall be open to public inspection at appropriate times.

1703.4 Performance. Specific information consisting of test reports conducted by an approved testing agency in accordance with standards referenced in Chapter 35, or other such information as necessary, shall be provided for the building official to determine that the material meets the applicable code requirements.

1703.4.1 Research and investigation. Sufficient technical data shall be submitted to the building official to substantiate the proposed use of any material or assembly. If it is determined that the evidence submitted is satisfactory proof of performance for the use intended, the building official shall approve the use of the material or assembly subject to the requirements of this code. The cost offsets, reports and investigations required under these provisions shall be paid by the permit applicant.

1703.4.2 Research reports. Supporting data, where necessary to assist in the approval of materials or assemblies not specifically provided for in this code, shall consist of valid research reports from approved sources.

1703.5 Labeling. Where materials or assemblies are required by this code to be labeled, such materials and assemblies shall be labeled by an approved agency in accordance with Section 1703. Products and materials required to be labeled shall be labeled in accordance with the procedures set forth in Sections 1703.5.1 through 1703.5.3.

1703.5.1 Testing. An approved agency shall test a representative sample of the product or material being labeled to the relevant standard or standards. The approved agency shall maintain a record of the tests performed. The record shall provide sufficient detail to verify compliance with the test standard.

1703.5.2 Inspection and identification. The approved agency shall periodically perform an inspection, which shall be in-plant if necessary, of the product or material that is to be labeled. The inspection shall verify that the labeled product or material is representative of the product or material tested.

1703.5.3 Label information. The label shall contain the manufacturer's or distributor's identification, model number, serial number or definitive information describing the product or material's performance characteristics and approved agency's identification.

1703.6 Heretofore approved materials. The use of any material already fabricated or of any construction already erected, which conformed to requirements or approvals heretofore in effect, shall be permitted to continue, if not detrimental to life, health or safety to the public.

1703.7 Evaluation and follow-up inspection services. Where structural components or other items regulated by this code are not visible for inspection after completion of a prefabricated assembly, the permit applicant shall submit a report of each prefabricated assembly. The report shall indicate the complete details of the assembly, including a description of the assembly and its components, the basis upon which the assembly is being evaluated, test results and similar information and other data as necessary for the building official to determine conformance to this code. Such a report shall be approved by the building official.

1703.7.1 Follow-up inspection. The permit applicant shall provide for special inspections of fabricated items in accordance with Section 1704.2.

1703.7.2 Test and inspection records. Copies of necessary test and inspection records shall be filed with the building official.

SECTION 1704
SPECIAL INSPECTIONS

1704.1 General. Where application is made for construction as described in this section, the owner or the registered design professional in responsible charge acting as the owner's agent shall employ one or more special inspectors to provide inspections during construction on the types of work listed under Section 1704. The special inspector shall be a qualified person who shall demonstrate competence, to the satisfaction of the building official, for inspection of the particular type of construction or operation requiring special inspection. These inspections are in addition to the inspections specified in Section 109.

Exceptions:

1. Special inspections are not required for work of a minor nature or as warranted by conditions in the jurisdiction as approved by the building official.

2. Special inspections are not required for building components unless the design involves the practice of professional engineering or architecture as defined by applicable state statutes and regulations governing the professional registration and certification of engineers or architects.

3. Unless otherwise required by the building official, special inspections are not required for occupancies in Group R-3 as applicable in Section 101.2 and occupancies in Group U that are accessory to a residential occupancy including, but not limited to, those listed in Section 312.1.

1704.1.1 Building permit requirement. The permit applicant shall submit a statement of special inspections prepared by the registered design professional in responsible charge in accordance with Section 106.1 as a condition for permit issuance. This statement shall include a complete list of materials and work requiring special inspections by this section, the inspections to be performed and a list of the individuals, approved agencies or firms intended to be retained for conducting such inspections.

1704.1.2 Report requirement. Special inspectors shall keep records of inspections. The special inspector shall furnish inspection reports to the building official, and to the registered design professional in responsible charge. Reports shall indicate that work inspected was done in conformance to approved construction documents. Discrepancies shall be brought to the immediate attention of the contractor for correction. If the discrepancies are not corrected, the discrepancies shall be brought to the attention of the building official and to the registered design professional in responsible charge prior to the completion of that phase of the work. A final report documenting required special inspections and correction of any discrepancies noted in the inspections shall be submitted at a point in time agreed upon by the permit applicant and the building official prior to the start of work.

1704.2 Inspection of fabricators. Where fabrication of structural load-bearing members and assemblies is being performed on the premises of a fabricator's shop, special inspection of the fabricated items shall be required by this section and as required elsewhere in this code.

1704.2.1 Fabrication and implementation procedures. The special inspector shall verify that the fabricator maintains detailed fabrication and quality control procedures that provide a basis for inspection control of the workmanship and the fabricator's ability to conform to approved construction documents and referenced standards. The special inspector shall review the procedures for completeness and adequacy relative to the code requirements for the fabricator's scope of work.

> **Exception:** Special inspections as required by Section 1704.2 shall not be required where the fabricator is approved in accordance with Section 1704.2.2.

1704.2.2 Fabricator approval. Special inspections required by this code are not required where the work is done on the premises of a fabricator registered and approved to perform such work without special inspection. Approval shall be based upon review of the fabricator's written procedural and quality control manuals and periodic auditing of fabrication practices by an approved special inspection agency. At completion of fabrication, the approved fabricator shall submit a certificate of compliance to the building official stating that the work was performed in accordance with the approved construction documents.

1704.3 Steel construction. The special inspections for steel elements of buildings and structures shall be as required by Section 1704.3 and Table 1704.3. Where required, special inspection of steel shall also comply with Section 1715.

Exceptions:

1. Special inspection of the steel fabrication process shall not be required where the fabricator does not perform any welding, thermal cutting or heating operation of any kind as part of the fabrication process. In such cases, the fabri-

cator shall be required to submit a detailed procedure for material control that demonstrates the fabricator's ability to maintain suitable records and procedures such that, at any time during the fabrication process, the material specification, grade and mill test reports for the main stress-carrying elements are capable of being determined.

2. The special inspector need not be continuously present during welding of the following items, provided the materials, welding procedures and qualifications of welders are verified prior to the start of the work; periodic inspections are made of the work in progress and a visual inspection of all welds is made prior to completion or prior to shipment of shop welding.

   2.1. Single-pass fillet welds not exceeding $^5/_{16}$ inch (7.9 mm) in size.

   2.2. Floor and roof deck welding.

   2.3. Welded studs when used for structural diaphragm.

   2.4. Welded sheet steel for cold-formed steel framing members such as studs and joists.

   2.5. Welding of stairs and railing systems.

1704.3.1 Welding. Welding inspection shall be in compliance with AWS D1.1. The basis for welding inspector qualification shall be AWS D1.1.

1704.3.2 Details. The special inspector shall perform an inspection of the steel frame to verify compliance with the details shown on the approved construction documents, such as bracing, stiffening, member locations and proper application of joint details at each connection.

1704.3.3 High-strength bolts. Installation of high-strength bolts shall be periodically inspected in accordance with AISC specifications.

1704.3.3.1 General. While the work is in progress, the special inspector shall determine that the requirements for bolts, nuts, washers and paint; bolted parts and installation and tightening in such standards are met. For bolts requiring pretensioning, the special inspector shall observe the preinstallation testing and calibration procedures when such procedures are required by the installation method or by project plans or specifications; determine that all plies of connected materials have been drawn together and properly snugged and monitor the installation of bolts to verify that the selected procedure for installation is properly used to tighten bolts. For joints required to be tightened only to the snug-tight condition, the special inspector need only verify that the connected materials have been drawn together and properly snugged.

1704.3.3.2 Periodic monitoring. Monitoring of bolt installation for pretensioning is permitted to be performed on a periodic basis when using the turn-of-nut method with matchmarking techniques, the direct tension indicator method or the alternate design fastener (twist-off bolt) method. Joints designated as snug tight need be inspected only on a periodic basis.

TABLE 1704.3
REQUIRED VERIFICATION AND INSPECTION OF STEEL CONSTRUCTION

VERIFICATION AND INSPECTION	CONTINUOUS	PERIODIC	REFERENCED STANDARD[a]	IBC REFERENCE
1. Material verification of high-strength bolts, nuts and washers:				
a. Identification markings to conform to ASTM standards specified in the approved construction documents.	—	X	Applicable ASTM material specifications; AISC 335, Section A3.4; AISC LRFD, Section A3.3	—
b. Manufacturer's certificate of compliance required.	—	X	—	—
2. Inspection of high-strength bolting:				
a. Bearing-type connections.	—	X	AISC LRFD Section M2.5	1704.3.3
b. Slip-critical connections.	X	X		
3. Material verification of structural steel:				
a. Identification markings to conform to ASTM standards specified in the approved construction documents.	—	—	ASTM A 6 orASTM A 568	1708.4
b. Manufacturers' certified mill test reports.	—	—	ASTM A 6 or ASTM A 568	
4. Material verification of weld filler materials:				
a. Identification markings to conform to AWS specification in the approved construction documents.	—	—	AISC, ASD, Section A3.6; AISC LRFD, Section A3.5	—
b. Manufacturer's certificate of compliance required.	—	—	—	—
5. Inspection of welding: a. Structural steel:	—	—		
1) Complete and partial penetration groove welds.	X	—	AWS D1.1	1704.3.1
2) Multipass fillet welds.	X	—		
3) Single-pass fillet welds $> {}^5/_{16}''$	X	—		
4) Single-pass fillet welds $\leq {}^5/_{16}''$	—	X		
5) Floor and deck welds.	—	X	AWS D1.3	—
b. Reinforcing steel:	—	—		
1) Verification of weldability of reinforcing steel other than ASTM A 706.	—	X	AWS D1.4 ACI 318: 3.5.2	1903.5.2
2) Reinforcing steel-resisting flexural and axial forces in intermediate and special moment frames, and boundary elements of special reinforced concrete shear walls and shear reinforcement.	X	—		
3) Shear reinforcement.	X	—		
4) Other reinforcing steel.	—	X		

6. Inspection of steel frame joint details for compliance with approved construction documents:
   a. Details such as bracing and stiffening.
   b. Member locations.
   c. Application of joint details at each connection.
	
	—
	—
	—
		X
		
			—	1704.3.2

For SI: 1 inch = 25.4 mm.

a. Where applicable, see also Section 1707.1, Special inspection for seismic resistance.

1704.3.3.3 Continuous monitoring. Monitoring of bolt installation for pretensioning using the calibrated wrench method or the turn-of-nut method without matchmarking shall be performed on a continuous basis.

1704.4 Concrete construction. The special inspections and verifications for concrete construction shall be as required by this section and Table 1704.4.

Exception: Special inspections shall not be required for:

1. Isolated spread concrete footings of buildings three stories or less in height that are fully supported on earth or rock.

2. Continuous concrete footings supporting walls of buildings three stories or less in height that are fully supported on earth or rock where:

   2.1. The footings support walls of light frame construction;

   2.2. The footings are designed in accordance with Table 1805.4.2; or

   2.3. The structural design is based on a f'_c no greater than 2,500 pounds per square inch (psi) (17.2 Mpa).

3. Nonstructural concrete slabs supported directly on the ground, including prestressed slabs on grade, where the effective prestress in the concrete is less than 150 psi (1.03 Mpa).

4. Concrete foundation walls constructed in accordance with Table 1805.5(1), 1805.5(2), 1805.5(3) or 1805.5(4).

5. Concrete patios, driveways and sidewalks, on grade.

1704.4.1 Materials. In the absence of sufficient data or documentation providing evidence of conformance to quality standards for materials in Chapter 3 of ACI 318, the building official shall require testing of materials in accordance with the appropriate standards and criteria for the material in Chapter 3 of ACI 318. Weldability of reinforcement, except that which conforms to ASTM A 706, shall be determined in accordance with the requirements of Section 1903.5.2.

TABLE 1704.4
REQUIRED VERIFICATION AND INSPECTION OF CONCRETE CONSTRUCTION

VERIFICATION AND INSPECTION	CONTINUOUS	PERIODIC	REFERENCED STANDARD[a]	IBC REFERENCE
1. Inspection of reinforcing steel, including prestressing tendons, and placement.	—	X	ACI 318: 3.5, 7.1-7.7	1903.5, 1907.1, 1907.7, 1914.4
2. Inspection of reinforcing steel welding in accordance with Table 1704.3, Item 5B.	—	—	AWS D1.4 ACI 318: 3.5.2	1903.5.2
3. Inspect bolts to be installed in concrete prior to and during placement of concrete where allowable loads have been increased.	X	—	—	1912.5
4. Verifying use of required design mix.	—	X	ACI 318: Ch. 4, 5.2-5.4	1904, 1905.2-1905.4, 1914.2, 1914.3
5. At the time fresh concrete is sampled to fabricate specimens for strength tests, perform slump and air content tests, and determine the temperature of the concrete.	X	—	ASTM C 172 ASTM C 31 ACI 318: 5.6, 5.8	1905.6, 1914.10
6. Inspection of concrete and shotcrete placement for proper application techniques.	X	—	ACI 318: 5.9, 5.10	1905.9, 1905.10, 1914.6, 1914.7, 1914.8
7. Inspection for maintenance of specified curing temperature and techniques.	—	X	ACI 318: 5.11-5.13	1905.11, 1905.13, 1914.9
8. Inspection of prestressed concrete: a. Application of prestressing forces. b. Grouting of bonded prestressing tendons in the seismic-force-resisting system.	X X	—	ACI 318: 18.20 ACI 318: 18.18.4	—
9. Erection of precast concrete members.	—	X	ACI 318: Ch. 16	—
10. Verification of in-situ concrete strength, prior to stressing of tendons in posttensioned concrete and prior to removal of shores and forms from beams and structural slabs.	—	X	ACI 318: 6.2	1906.2

For SI: 1 inch = 25.4 mm.

a. Where applicable, see also Section 1707.1, Special inspection for seismic resistance.

1704.5 Masonry construction. Masonry construction shall be inspected and evaluated in accordance with the requirements of this section, depending on the classification of the building or structure or nature of occupancy, as defined by this code (see Table 1604.5 and Section 1617.2).

> **Exception:** Special inspections shall not be required for:
>
> 1. Empirically designed masonry, glass unit masonry or masonry veneer designed by Section 2109, 2110 or ACI 530/ASCE 5/TMS 402, Chapters 5, 6 or 7, when they are part of nonessential buildings (see Table 1604.5 and Section 1617.2).
>
> 2. Masonry foundation walls constructed in accordance with Table 1805.5(1), 1805.5(2), 1805.5(3) or 1805.5(4).

1704.5.1 Empirically designed masonry, glass unit masonry and masonry veneer in essential facilities. The minimum inspection program for masonry designed by Chapter 14, Section 2109 or 2110, or by Chapter 5, 6 or 7 of ACI 530/ASCE 5/TMS 402, in essential facilities (see Table 1604.5 and Section 1616.2) shall comply with Table 1704.5.1.

1704.5.2 Engineered masonry in nonessential facilities. The minimum special inspection program for masonry designed by Section 2106, 2107 or 2108, or by chapters other than Chapters 5, 6 or 7 of ACI 530/ASCE 5/TMS 402, in nonessential facilities (see Table 1604.5 and Section 1617.2), shall comply with Table 1704.5.1.

1704.5.3 Engineered masonry in essential facilities. The minimum special inspection program for masonry designed by Section 2106, 2107 or 2108, or by chapters other than Chapters 5, 6 or 7 of ACI 530/ASCE 5/TMS 402, in essential facilities (see Table 1604.5 and Section 1616.2), shall comply with Table 1704.5.3.

1704.6 Wood construction. Special inspections of the fabrication process of prefabricated wood structural elements and assemblies shall be in accordance with Section 1704.2. Special inspections of site-built assemblies shall be in accordance with Section 1704.1.

1704.6.1 Fabrication of high-load diaphragms. High-load diaphragms using values from Table 2306.3.2 shall be installed with special inspections as indicated in Section 1704.1. The special inspector shall inspect the wood structural panel sheathing to ascertain whether it is of the grade and thickness shown on the approved building plans. Additionally, the special inspector must verify the nominal size of framing members at adjoining panel edges, the nail or staple diameter and length, the number of fastener lines and that spacing between fasteners in each line and at edge margins agrees with the approved building plans.

1704.7 Soils. The special inspections for existing site soil conditions, fill placement and load-bearing requirements shall follow Sections 1704.7.1 through 1704.7.3. The approved soils report, required by Section 1802.2, shall be used to determine compliance.

> **Exception:** Special inspections not required during placement of fill less than 12 inches (305 mm) deep.

1704.7.1 Site preparation. Prior to placement of the prepared fill, the special inspector shall determine that the site has been prepared in accordance with the approved soils report.

1704.7.2 During fill placement. During placement and compaction of the fill material, the special inspector shall determine that the material being used and the maximum lift thickness comply with the approved report, as specified in Section 1803.5.

1704.7.3 Evaluation of in-place density. The special inspector shall determine, at the approved frequency, that the in-place dry density of the compacted fill complies with the approved report.

1704.8 Pile foundations. A special inspector shall be present when pile foundations are being installed and during tests. The special inspector shall make and submit to the building official records of the installation of each pile and results of load tests. Records shall include the cutoff and tip elevation of each pile relative to a permanent reference.

1704.9 Pier foundations. Special inspection is required for pier foundations for buildings assigned to Seismic Design Category C, D, E or F in accordance with Section 1616.3.

1704.10 Wall panels and veneers. Special inspection is required for exterior and interior architectural wall panels and the anchoring of veneers for buildings assigned to Seismic Design Category E or F in accordance with Section 1616.3. Special inspection of such masonry veneer shall be in accordance with Section 1704.5.

1704.11 Sprayed fire-resistant materials. Special inspections for sprayed fire-resistant materials applied to structural elements and decks shall be in accordance with Sections 1704.11.1 through 1704.11.5. Special inspections shall be based on the fire-resistance design as designated in the approved construction documents.

1704.11.1 Structural member surface conditions. The surfaces shall be prepared in accordance with the approved fire-resistance design and the approved manufacturer's written instructions. The prepared surface of structural members to be sprayed shall be inspected before the application of the sprayed fire-resistant material.

1704.11.2 Application. The substrate shall have a minimum ambient temperature before and after application as specified in the approved manufacturer's written instructions. The area for application shall be ventilated during and after application as required by the approved manufacturer's written instructions.

TABLE 1704.5.1
LEVEL 1 SPECIAL INSPECTION

INSPECTION TASK	FREQUENCY OF INSPECTION		REFERENCE FOR CRITERIA		
	Continuous during task listed	Periodically during task listed	IBC section	ACI 530/ASCE 5/TMS 402[a]	ACI 530.1/ASCE 6/TMS 602[a]
1. As masonry construction begins, the following shall be verified to ensure compliance:					
a. Proportions of site-prepared mortar.		X			Art. 2.6A
b. Construction of mortar joints.	—	X	—	—	Art. 3.3B
c. Location of reinforcement and connectors.		X			Art. 3.4, 3.6A
d. Prestressing technique.	—	X			Art. 3.6B
e. Grade and size of prestressing tendons and anchorages.	—	X	—	—	Art. 2.4B, 2.4H
2. The inspection program shall verify:					
a. Size and location of structural elements.	—	X	—	—	Art. 3.3G
b. Type, size and location of anchors, including other details of anchorage of masonry to structural members, frames or other construction.	—	X	—	Sec. 1.2.2(e), 2.1.4, 3.1.6	—
c. Specified size, grade and type of reinforcement.	—	X	—	Sec. 1.12	Art. 2.4, 3.4
d. Welding of reinforcing bars.	X	—	—	Sec. 2.1.10.6.2, 3.2.3.4(b)	—
e. Protection of masonry during cold weather (temperature below 40°F) or hot weather (temperature above 90°F).	—	X	Sec. 2104.3, 2104.4	—	Art. 1.8C, 1.8D
f. Application and measurement of prestressing force.	—	X	—	—	Art. 3.6B
3. Prior to grouting, the following shall be verified to ensure compliance:					
a. Grout space is clean.		X		—	Art. 3.2D
b. Placement of reinforcement and connectors and prestressing tendons and anchorages.		X		Sec. 1.12	Art. 3.4
c. Proportions of site-prepared grout and prestressing grout for bonded tendons.	—	X	—	—	Art. 2.6B
d. Construction of mortar joints.		X		—	Art. 3.3B
4. Grout placement shall be verified to ensure compliance with code and construction document provisions.	X	—	—	—	Art 3.5
a. Grouting of prestressing bonded tendons.	X	—	—	—	Art. 3.6C
5. Preparation of any required grout specimens, mortar specimens and/or prisms shall be observed.	X	—	Sec. 2105.2.2, 2105.3	—	Art. 1.4
6. Compliance with required inspection provisions of the construction documents and the approved submittals shall be verified.	—	X	—	—	Art. 1.5

For SI: °C = (°F - 32)/1.8.

a. The specific standards referenced are those listed in Chapter 35.

TABLE 1704.5.3
LEVEL 2 SPECIAL INSPECTION

INSPECTION TASK	FREQUENCY OF INSPECTION		REFERENCE FOR CRITERIA		
	Continuous during task listed	Periodically during task listed	IBC section	ACI 530/ ASCE 5/ TMS 402[a]	ACI 530.1/ ASCE 6/ TMS 602[a]
1. From the beginning of masonry construction, the following shall be verified to ensure compliance:					
a. Proportions of site-prepared mortar, grout and prestressing grout for bonded tendons.	—	X	—	—	Art. 2.6A
b. Placement of masonry units and construction of mortar joints.	—	X	—	—	Art. 3.3B
c. Placement of reinforcement, connectors and prestressing tendons and anchorages.	—	X	—	Sec. 1.12	Art. 3.4, 3.6A
d. Grout space prior to grouting.	X	—	—	—	Art. 3.2D
e. Placement of grout.	X	—	—	—	Art. 3.5
f. Placement of prestressing grout.	X	—	—	—	Art. 3.6C
2. The inspection program shall verify:					
a. Size and location of structural elements.	—	X	—	—	Art. 3.3G
b. Type, size and location of anchors, including other details of anchorage of masonry to structural members, frames or other construction.	X	—	—	Sec. 1.2.2(e), 2.1.4, 3.1.6	—
c. Specified size, grade and type of reinforcement.	—	X	—	Sec. 1.12	Art. 2.4, 3.4
d. Welding of reinforcment.	X	—	—	Sec. 2.1.10.6.2, 3.2.3.4(b)	—
e. Protection of masonry during cold weather (temperature below 40°F) or hot weather (temperature above 90°F).	—	X	Sec. 2104.3, 2104.4	—	Art. 1.8C, 1.8D
f. Application and measurement of prestressing force.	X	—	—	—	Art. 3.6B
3. Preparation of any required grout specimens, mortar specimens and/or prisms shall be observed.	X	—	Sec. 2105.2.2, 2105.3	—	Art. 1.4
4. Compliance with required inspection provisions of the construction documents and the approved submittals shall be verified.	—	X	—	—	Art. 1.5

For SI: °C = (°F - 32)/1.8.

a. The specific standards referenced are those listed in Chapter 35.

1704.11.3 Thickness. The average thickness of the sprayed fire-resistant materials applied to structural elements shall not be less than the thickness required by the approved fire-resistant design. Individual measured thickness, which exceeds the thickness specified in a design by $^1/_4$ inch (6.4 mm) or more, shall be recorded as the thickness specified in the design plus $^1/_4$ inch (6.4 mm). For design thicknesses 1 inch (25 mm) or greater, the minimum allowable individual thickness shall be the design thickness minus $^1/_4$ inch (6.4 mm). For design thicknesses less than 1 inch (25 mm), the minimum allowable individual thickness shall be the design thickness minus 25 percent. Thickness shall be determined in accordance with ASTM E 605. Samples of the sprayed fire-resistant materials shall be selected in accordance with Sections 1704.11.3.1 and 1704.11.3.2.

1704.11.3.1 Floor, roof and wall assemblies. The thickness of the sprayed fire-resistant material applied to floor, roof and wall assemblies shall be determined in accordance with ASTM E 605, taking the average of not less than four measurements for each 1,000 square feet (93 m²) of the sprayed area on each floor or part thereof.

1704.11.3.2 Structural framing members. The thickness of the sprayed fire-resistant material applied to structural members shall be determined in accordance with ASTM E 605. Thickness testing shall be performed on not less than 25 percent of the structural members on each floor.

1704.11.4 Density. The density of the sprayed fire-resistant material shall not be less than the density specified in the approved fire-resistant design. Density of the sprayed fire-resistant material shall be determined in accordance with ASTM E 605.

1704.11.5 Bond strength. The cohesive/adhesive bond strength of the cured sprayed fire-resistant material applied to structural elements shall not be less than 150 pounds per square foot (psf) (7.18 kN/m²). The cohesive/adhesive bond strength shall be determined in accordance with the field test specified in ASTM E 736 by testing in-place samples of the sprayed fire-resistant material selected in accordance with Sections 1704.11.5.1 and 1704.11.5.2.

1704.11.5.1 Floor, roof and wall assemblies. The test samples for determining the cohesive/adhesive bond strength of the sprayed fire-resistant materials shall be selected from each floor, roof and wall assembly at the rate of not less than one sample for every 10,000 square feet (929 m²) or part thereof of the sprayed area in each story.

1704.11.5.2 Structural framing members. The test samples for determining the cohesive/adhesive bond strength of the sprayed fire-resistant materials shall be selected from beams, girders, joists, trusses and columns at the rate of not less than one sample for each type of structural framing member for each 10,000 square feet (929 m²) of floor area or part thereof in each story.

1704.12 Exterior insulation and finish systems (EIFS). Special inspections shall be required for all EIFS applications.

Exceptions:

1. Special inspections shall not be required for EIFS applications installed over a water-resistive barrier with a means of draining moisture to the exterior.

2. Special inspections shall not be required for EIFS applications installed over masonry or concrete walls.

1704.13 Special cases. Special inspections shall be required for proposed work that is, in the opinion of the building official, unusual in its nature, such as, but not limited to, the following examples:

1. Construction materials and systems that are alternatives to materials and systems prescribed by this code.

2. Unusual design applications of materials described in this code.

3. Materials and systems required to be installed in accordance with additional manufacturer's instructions that prescribe requirements not contained in this code or in standards referenced by this code.

1704.14 Special inspection for smoke control. Smoke control systems shall be tested by a special inspector.

1704.14.1 Testing scope. The test scope shall be as follows:

1. During erection of ductwork and prior to concealment for the purposes of leakage testing and recording of device location.

2. Prior to occupancy and after sufficient completion for the purposes of pressure difference testing, flow measurements and detection and control verification.

1704.14.2 Qualifications. Special inspection agencies for smoke control shall have expertise in fire protection engineering, mechanical engineering and certification as air balancers.

SECTION 1705
QUALITY ASSURANCE FOR SEISMIC RESISTANCE

1705.1 Scope. A quality assurance plan for seismic requirements shall be provided in accordance with Section 1705.2 for the following:

1. The seismic-force-resisting systems in structures assigned to Seismic Design Category C, D, E or F, in accordance with Section 1616.

2. Designated seismic systems in structures assigned to Seismic Design Category D, E or F.

3. The following additional systems in structures assigned to Seismic Design Category C:

    3.1. Heating, ventilating and air-conditioning (HVAC) ductwork containing hazardous materials and anchorage of such ductwork.

    3.2. Piping systems and mechanical units containing flammable, combustible or highly toxic materials.

    3.3. Anchorage of electrical equipment used for emergency or standby power systems.

4. The following additional systems in structures assigned to Seismic Design Category D:

    4.1. Systems required for Seismic Design Category C.

    4.2. Exterior wall panels and their anchorage.

    4.3. Suspended ceiling systems and their anchorage.

    4.4. Access floors and their anchorage.

    4.5. Steel storage racks and their anchorage, where the factor, Ip, determined in Section 9.6.1.5 of ASCE 7, is equal to 1.5.

5. The following additional systems in structures assigned to Seismic Design Category E or F:

    5.1. Systems required for Seismic Design Categories C and D.

    5.2. Electrical equipment.

Exceptions:

1. A quality assurance plan is not required for structures designed and constructed in accordance with the conventional construction provisions of Section 2308.

2. A quality assurance plan is not required for structures designed and constructed in accordance with the following:

    2.1. The structure is constructed of light wood framing or light framed cold-formed steel; the design spectral response acceleration at short periods, S_{DS}, as determined in Section 1615.1, does not exceed 0.5g, and the height of the structure does not exceed 35 feet (10 668 mm) above grade plane; or

2.2. The structure is constructed using a reinforced masonry structural system or reinforced concrete structural system; the design spectral response acceleration at short periods, S_{DS}, as determined in Section 1615.1, does not exceed 0.5g, and the height of the structure does not exceed 25 feet (7620 mm) above grade plane; or

2.3. The structure is a detached one- or two-family dwelling not exceeding two stories in height; and

    2.3.1. The structure is classified as Seismic Use Group I, as determined in Section 1616.2; and

    2.3.2. The structure does not have any of the following plan or vertical irregularities as defined in Section 1616.5:

        a. Torsional irregularity.

        b. Nonparallel systems.

        c. Stiffness irregularity–extreme soft story and soft story.

        d. Discontinuity in capacity–weak story.

1705.2 Quality assurance plan preparation. The design of each designated seismic system shall include a quality assurance plan prepared by a registered design professional. The quality assurance plan shall identify the following:

1. The designated seismic systems and seismic-force-resisting systems that are subject to quality assurance in accordance with Section 1705.1.

2. The special inspections and testing to be provided as required by Sections 1704 and 1708 and other applicable sections of this code, including the applicable standards referenced by this code.

3. The type and frequency of testing required.

4. The type and frequency of special inspections required.

5. The required frequency and distribution of testing and special inspection reports.

6. The structural observations to be performed.

7. The required frequency and distribution of structural observation reports.

1705.3 Contractor responsibility. Each contractor responsible for the construction of a seismic-force-resisting system, designated seismic system, or component listed in the quality assurance plan shall submit a written contractor's statement of responsibility to the building official and to the owner prior to the commencement of work on the system or component. The contractor's statement of responsibility shall contain the following:

1. Acknowledgment of awareness of the special requirements contained in the quality assurance plan.

2. Acknowledgment that control will be exercised to obtain conformance with the construction documents approved by the building official.

3. Procedures for exercising control within the contractor's organization, the method and frequency of reporting and the distribution of the reports.

4. Identification and qualifications of the person(s) exercising such control and their position(s) in the organization.

SECTION 1706
QUALITY ASSURANCE FOR WIND REQUIREMENTS

1706.1 Scope. A quality assurance plan shall be provided in accordance with Section 1706.1.1.

1706.1.1 Where required. A quality assurance plan for wind requirements shall be provided for all structures constructed in the following areas:

1. In wind exposure Categories A and B, where the 3-second-gust basic wind speed is 120 miles per hour (mph) (52.8 m/sec) or greater.

2. In wind exposure Categories C and D, where the 3-second-gust basic wind speed is 110 mph (49 m/sec) or greater.

Exception: A quality assurance plan is not required for structures designed and constructed in accordance with the *International Residential Code* or the conventional construction provisions of Section 2308 of this code, provided that all of the applicable items listed in Section 1706.1.2 are inspected during construction by a qualified person approved by the building official.

1706.1.2 Detailed requirements. Where required by Section 1706.1.1, a quality assurance plan shall be provided for the following:

1. Roof cladding and roof framing connections.

2. Wall connections to roof and floor diaphragms and framing.

3. Roof and floor diaphragm systems, including collectors, drag struts and boundary elements.

4. Vertical windforce-resisting systems, including braced frames, moment frames and shear walls.

5. Windforce-resisting system connections to the foundation.

6. Fabrication and installation of components and assemblies required to meet the impact-resistance requirements of Section 1609.1.4.

Exception: Fabrication of manufactured components and assemblies that have a label indicating compliance with the wind-load and impact-resistance requirements of this code.

1706.2 Quality assurance plan preparation. The design of each main windforce-resisting system and each wind-resisting component shall include a quality assurance plan prepared by a registered design professional.

Exception: For construction that is not required to be designed by a registered design professional, the quality assurance plan may be prepared by a qualified person approved by the building official.

The quality assurance plan shall identify the following:

1. The main windforce-resisting systems and wind-resisting components that are subject to quality assurance in accordance with Section 1706.1.

2. The special inspections and testing to be provided as required by Section 1704 and other applicable sections of this code, including the applicable standards referenced by this code.

3. The type and frequency of testing required.

4. The type and frequency of special inspections required.

5. The required frequency and distribution of testing and special inspection reports.

6. The structural observations to be performed.

7. The required frequency and distribution of structural observation reports.

1706.3 Contractor responsibility. Each contractor responsible for the construction of a main windforce-resisting system or a wind-resisting component listed in the quality assurance plan shall submit a written statement of responsibility to the building official and the owner prior to the commencement of work on the system or component. The contractor's statement of responsibility shall contain the following:

1. Acknowledgment of awareness of the special requirements contained in the quality assurance plan;

2. Acknowledgment that control will be exercised to obtain conformance with the construction documents approved by the building official;

3. Procedures for exercising control within the contractor's organization, the method and frequency of reporting and the distribution of the reports; and

4. Identification and qualifications of the person(s) exercising such control and their position(s) in the organization.

SECTION 1707
SPECIAL INSPECTIONS FOR
SEISMIC RESISTANCE

1707.1 Special inspections for seismic resistance. Special inspection as specified in this section is required for the following, where required in Section 1704.1. Special inspections itemized in Sections 1707.2 through 1707.8 are required for the following:

1. The seismic-force-resisting systems in structures assigned to Seismic Design Category C, D, E or F, as determined in Section 1616.

2. Designated seismic systems in structures assigned to Seismic Design Category D, E or F.

3. Architectural, mechanical and electrical components in structures assigned to Seismic Design Category C, D, E or F that are required in Sections 1707.6 and 1707.7.

1707.2 Structural steel. Continuous special inspection for structural welding in accordance with AISC 341.

Exceptions:

1. Single-pass fillet welds not exceeding $^5/_{16}$ inch (7.9 mm) in size.

2. Floor and roof deck welding.

1707.3 Structural wood. Continuous special inspection during field gluing operations of elements of the seismic-force-resisting system. Periodic special inspections for nailing, bolting, anchoring and other fastening of components within the seismic-force-resisting system, including drag struts, braces and hold-downs.

Exception: Fastening of wood sheathing used for wood shear walls, shear panels and diaphragms where the fastener spacing is more than 4 inches (102 mm) on center (o.c.).

1707.4 Cold-formed steel framing. Periodic special inspections during welding operations of elements of the seismic-force-resisting system. Periodic special inspections for screw attachment, bolting, anchoring and other fastening of components within the seismic-force-resisting system, including struts, braces, and hold-downs.

1707.5 Storage racks and access floors. Periodic special inspection during the anchorage of access floors and storage racks 8 feet (2438 mm) or greater in height in structures assigned to Seismic Design Category D, E or F.

1707.6 Architectural components. Periodic special inspection during the erection and fastening of exterior cladding, interior and exterior nonbearing walls and interior and exterior veneer in structures assigned to Seismic Design Category D, E or F.

Exceptions:

1. Special inspection is not required for architectural components in structures 30 feet (9144 mm) or less in height.

2. Special inspection is not required for cladding and veneer weighing 5 psf ($24.5N/m^2$) or less.

3. Special inspection is not required for interior nonbearing walls weighing 15 psf ($73.5 N/m^2$) or less.

1707.7 Mechanical and electrical components. Periodic special inspection is required during the anchorage of electrical equipment for emergency or standby power systems in structures assigned to Seismic Design Category C, D, E or F. Periodic special inspection is required during the installation of anchorage of other electrical equipment in structures assigned to Seismic Design Category E or F. Periodic special inspection is required during installation of piping systems intended to carry flammable, combustible or highly toxic contents and their associated mechanical units in structures assigned to Seismic Design Category C, D, E or F. Periodic special inspection is required during the installation of HVAC ductwork that will contain hazardous materials in structures assigned to Seismic Design Category C, D, E or F.

1707.7.1 Component inspection. Special inspection is required for the installation of the following components, where the component has a Component Importance Factor of 1.0 or 1.5 in accordance with Section 9.6.1.5 of ASCE 7. Evidence of the quality control program shall be permanently identified on each piece of equipment by a label.

1. Equipment using combustible energy sources.
2. Electrical motors, transformers, switchgear unit substations and motor control centers.
3. Reciprocating and rotating-type machinery.
4. Piping distribution systems 3 inches (76 mm) and larger.
5. Tanks, heat exchangers and pressure vessels.

1707.7.2 Component and attachment testing. The component manufacturer shall test or analyze the component and the component mounting system or anchorage for the design forces in Chapter 16 for those components having a Component Importance Factor of 1.0 or 1.5 in accordance with Chapter 16. The manufacturer shall submit a certificate of compliance for review and acceptance by the registered design professional responsible for the design, and for approval by the building official. The basis of certification shall be by test on a shaking table, by three-dimensional shock tests, by an analytical method using dynamic characteristics and forces from Chapter 16 or by more rigorous analysis. The special inspector shall inspect the component and verify that the label, anchorage or mounting conforms to the certificate of compliance.

1707.7.3 Component manufacturer certification. Each manufacturer of equipment to be placed in a building assigned to Seismic Design Categories E or F, in accordance with Chapter 16, where the equipment has a Component Importance Factor of 1.0 or 1.5 in accordance with Chapter 16, shall maintain an approved quality control program. Evidence of the quality control program shall be permanently identified on each piece of equipment by a label.

1707.8 Seismic isolation system. Provide periodic special inspection during the fabrication and installation of isolator units and energy dissipation devices if used as part of the seismic isolation system.

SECTION 1708
STRUCTURAL TESTING FOR
SEISMIC RESISTANCE

1708.1 Masonry. Testing and verification of masonry materials and assemblies prior to construction shall comply with the requirements of this section, depending on the classification of building or structure or nature of occupancy, as defined in this code (see Table 1604.5 or Section 1616.2).

1708.1.1 Empirically designed masonry and glass unit masonry in nonessential facilities. For masonry designed by Section 2109 or 2110, or by Chapter 5 or 7 of ACI 530/ASCE 5/TMS 402, in nonessential facilities (see Table 1604.5 or Section 1616.2), certificates of compliance used in masonry construction shall be verified prior to construction.

1708.1.2 Empirically designed masonry and glass unit masonry in essential facilities. The minimum testing and verification prior to construction for masonry designed by Section 2109 or 2110, or by Chapter 5 or 7 of ACI 530/ASCE 5/TMS 402, in essential facilities (see Table 1604.5 or Section 1616.2), shall comply with the requirements of Table 1708.1.2, Level 1 Quality Assurance.

TABLE 1708.1.2
LEVEL 1 QUALITY ASSURANCE

MINIMUM TESTS AND SUBMITTALS
Certificates of compliance used in masonry construction.
Verification of f'_m prior to construction, except where specifically exempted by this code.

1708.1.3 Engineered masonry in nonessential facilities. The minimum testing and verification prior to construction for masonry designed by Section 2107 or 2108, or by chapters other than Chapter 5, 6 or 7 of ACI 530/ASCE 5/TMS 402, in nonessential facilities (see Table 1604.5 or Section 1616.2), shall comply with Table 1708.1.2, Level 1 Quality Assurance.

1708.1.4 Engineered masonry in essential facilities. The minimum testing and verification prior to construction for masonry designed by Section 2107 or 2108, or by chapters other than Chapter 5, 6 or 7 of ACI 530/ASCE 5/TMS 402, in essential facilities (see Table 1604.5 or Section 1616.2), shall comply with Table 1708.1.4, Level 2 Quality Assurance.

TABLE 1708.1.4
LEVEL 2 QUALITY ASSURANCE

MINIMUM TESTS AND SUBMITTALS
Certificates of compliance used in masonry construction.
Verification of f'_m prior to construction and every 5,000 square feet during construction.
Verification of proportions of materials in mortar and grout as delivered to the site.

For SI: 1 square foot = 0.0929 m^2.

1708.2 Testing for seismic resistance. The tests specified in Sections 1708.3 through 1708.6 are required for the following:

1. The seismic-force-resisting systems in structures assigned to Seismic Design Category C, D, E or F, as determined in Section 1616.
2. Designated seismic systems in structures assigned to Seismic Design Category D, E or F.
3. Architectural, mechanical and electrical components in structures assigned to Seismic Design Category C, D, E or F that are required in Section 1708.5.

1708.3 Reinforcing and prestressing steel. Certified mill test reports shall be provided for each shipment of reinforcing steel used to resist flexural, shear and axial forces in reinforced concrete intermediate frames, special moment frames and boundary elements of special reinforced concrete or reinforced masonry shear walls. Where ASTM A 615 reinforcing steel is used to resist earthquake-induced flexural and axial forces in special moment frames and in wall boundary elements of shear walls in structures assigned to Seismic Design Category D, E or F, as determined in Section 1616, the testing requirements of ACI 318 shall be met. Where ASTM A 615 reinforcing steel is to be welded, chemical tests shall be performed to determine weldability in accordance with Section 1903.5.2.

1708.4 Structural steel. The testing contained in the quality assurance plan shall be as required by AISC 341 and the additional requirements herein. The acceptance criteria for nondestructive testing shall be as required in AWS D1.1 as specified by the registered design professional.

Base metal thicker than 1.5 inches (38 mm), where subject to through-thickness weld shrinkage strains, shall be ultrasonically tested for discontinuities behind and adjacent to such welds after joint completion. Any material discontinuities shall be accepted or rejected on the basis of ASTM A 435 or ASTM A 898 (Level 1 criteria) and criteria as established by the registered design professional(s) in responsible charge and the construction documents.

1708.5 Mechanical and electrical equipment. Each manufacturer of designated seismic system components shall test or analyze the component and its mounting system or anchorage and submit a certificate of compliance for review and acceptance by the registered design professional in responsible charge of the design of the designated seismic system and for approval by the building official. The evidence of compliance shall be by actual test on a shake table, by three-dimensional shock tests, by an analytical method using dynamic characteristics and forces, by the use of experience data (i.e., historical data demonstrating acceptable seismic performance) or by more rigorous analysis providing for equivalent safety. The special inspector shall examine the designated seismic system and determine whether the anchorages and label conform with the evidence of compliance.

1708.6 Seismically isolated structures. For required system tests, see Section 9.13.9 of ASCE 7.

SECTION 1709
STRUCTURAL OBSERVATIONS

1709.1 Structural observations. Structural observations shall be provided for those structures included in Seismic Design Category D, E or F, as determined in Section 1616, where one or more of the following conditions exist:

1. The structure is included in Seismic Use Group II or III,
2. The height of the structure is greater than 75 feet (22 860 mm) above the base,
3. The structure is in Seismic Design Category E and Seismic Use Group I and greater than two stories in height,
4. When so designated by the registered design professional in responsible charge of the design,
5. When such observation is specifically required by the building official.

Structural observations shall also be provided for those structures sited where the basic wind speed exceeds 110 mph (49 m/sec) determined from Figure 1609, where one or more of the following conditions exist:

1. The structure is included in Category III or IV according to Table 1604.5.

2. The height of the structure is greater than 75 feet (22 860 mm).
3. When so designated by the registered design professional in responsible charge of the design,
4. When such observation is specifically required by the building official.

The owner shall employ a registered design professional to perform structural observations as defined in Section 1702.

Deficiencies shall be reported in writing to the owner and the building official. At the conclusion of the work included in the permit, the structural observer shall submit to the building official a written statement that the site visits have been made and identify any reported deficiencies which, to the best of the structural observer's knowledge, have not been resolved.

SECTION 1710
DESIGN STRENGTHS OF MATERIALS

1710.1 Conformance to standards. The design strengths and permissible stresses of any structural material that are identified by a manufacturer's designation as to manufacture and grade by mill tests, or the strength and stress grade is otherwise confirmed to the satisfaction of the building official, shall conform to the specifications and methods of design of accepted engineering practice or the approved rules in the absence of applicable standards.

1710.2 New materials. For materials that are not specifically provided for in this code, the design strengths and permissible stresses shall be established by tests as provided for in Section 1711.

SECTION 1711
ALTERNATIVE TEST PROCEDURE

1711.1 General. In the absence of approved rules or other approved standards, the building official shall make, or cause to be made, the necessary tests and investigations; or the building official shall accept duly authenticated reports from approved agencies in respect to the quality and manner of use of new materials or assemblies as provided for in Section 104.11. The cost of all tests and other investigations required under the provisions of this code shall be borne by the permit applicant.

SECTION 1712
TEST SAFE LOAD

1712.1 Where required. Where proposed construction is not capable of being designed by approved engineering analysis, or where proposed construction design method does not comply with the applicable material design standard, the system of construction or the structural unit and the connections shall be subjected to the tests prescribed in Section 1714. The building official shall accept certified reports of such tests conducted by an approved testing agency, provided that such tests meet the requirements of this code and approved procedures.

SECTION 1713
IN-SITU LOAD TESTS

1713.1 General. Whenever there is a reasonable doubt as to the stability or load-bearing capacity of a completed building, structure or portion thereof for the expected loads, an engineering assessment shall be required. The engineering assessment shall involve either a structural analysis or an in-situ load test, or both. The structural analysis shall be based on actual material properties and other as-built conditions that affect stability or load-bearing capacity, and shall be conducted in accordance with the applicable design standard. If the structural assessment determines that the load-bearing capacity is less than that required by the code, load tests shall be conducted in accordance with Section 1713.2. If the building, structure or portion thereof is found to have inadequate stability or load-bearing capacity for the expected loads, modifications to ensure structural adequacy or the removal of the inadequate construction shall be required.

1713.2 Test standards. Structural components and assemblies shall be tested in accordance with the appropriate material standards listed in Chapter 35. In the absence of a standard that contains an applicable load test procedure, the test procedure shall be developed by a registered design professional and approved. The test procedure shall simulate loads and conditions of application that the completed structure or portion thereof will be subjected to in normal use.

1713.3 In-situ load tests. In-situ load tests shall be conducted in accordance with Section 1713.3.1 or 1713.3.2 and shall be supervised by a registered design professional. The test shall simulate the applicable loading conditions specified in Chapter 16 as necessary to address the concerns regarding structural stability of the building, structure or portion thereof.

1713.3.1 Load test procedure specified. Where a standard listed in Chapter 35 contains an applicable load test procedure and acceptance criteria, the test procedure and acceptance criteria in the standard shall apply. In the absence of specific load factors or acceptance criteria, the load factors and acceptance criteria in Section 1713.3.2 shall apply.

1713.3.2 Load test procedure not specified. In the absence of applicable load test procedures contained within a standard referenced by this code or acceptance criteria for a specific material or method of construction, such existing structure shall be subjected to a test procedure developed by a registered design professional that simulates applicable loading and deformation conditions. For components that are not a part of the seismic-load-resisting system, the test load shall be equal to two times the unfactored design loads. The test load shall be left in place for a period of 24 hours. The structure shall be considered to have successfully met the test requirements where the following criteria are satisfied:

1. Under the design load, the deflection shall not exceed the limitations specified in Section 1604.3.

2. Within 24 hours after removal of the test load, the structure shall have recovered not less than 75 percent of the maximum deflection.

3. During and immediately after the test, the structure shall not show evidence of failure.

SECTION 1714
PRECONSTRUCTION LOAD TESTS

1714.1 General. In evaluating the physical properties of materials and methods of construction that are not capable of being designed by approved engineering analysis or do not comply with applicable material design standards listed in Chapter 35, the structural adequacy shall be predetermined based on the load test criteria established in this section.

1714.2 Load test procedures specified. Where specific load test procedures, load factors and acceptance criteria are included in the applicable design standards listed in Chapter 35, such test procedures, load factors and acceptance criteria shall apply. In the absence of specific test procedures, load factors or acceptance criteria, the corresponding provisions in Section 1714.3 shall apply.

1714.3 Load test procedures not specified. Where load test procedures are not specified in the applicable design standards listed in Chapter 35, the load-bearing and deformation capacity of structural components and assemblies shall be determined on the basis of a test procedure developed by a registered design professional that simulates applicable loading and deformation conditions. For components and assemblies that are not a part of the seismic-load-resisting system, the test shall be as specified in Section 1714.3.1. Load tests shall simulate the applicable loading conditions specified in Chapter 16.

1714.3.1 Test procedure. The test assembly shall be subjected to an increasing superimposed load equal to not less than two times the superimposed design load. The test load shall be left in place for a period of 24 hours. The tested assembly shall be considered to have successfully met the test requirements if the assembly recovers not less than 75 percent of the maximum deflection within 24 hours after the removal of the test load. The test assembly shall then be reloaded and subjected to an increasing superimposed load until either structural failure occurs or the superimposed load is equal to two and one-half times the load at which the deflection limitations specified in Section 1714.3.2 were reached, or the load is equal to two and one-half times the superimposed design load. In the case of structural components and assemblies for which deflection limitations are not specified in Section 1714.3.2, the test specimen shall be subjected to an increasing superimposed load until structural failure occurs or the load is equal to two and one-half times the desired superimposed design load. The allowable superimposed design load shall be taken as the lesser of:

1. The load at the deflection limitation given in Section 1714.3.2.

2. The failure load divided by 2.5.

3. The maximum load applied divided by 2.5.

1714.3.2 Deflection. The deflection of structural members under the design load shall not exceed the limitations in Section 1604.3.

1714.4 Wall and partition assemblies. Load-bearing wall and partition assemblies shall sustain the test load both with and without window framing. The test load shall include all design load components. Wall and partition assemblies shall be tested both with and without door and window framing.

1714.5 Exterior window and door assemblies. The design pressure rating of exterior windows and doors in buildings shall be determined in accordance with Section 1714.5.1 or 1714.5.2.

Exception: Structural wind load design pressures for window units smaller than the size tested in accordance with Section 1714.5.1 or 1714.5.2 shall be permitted to be higher than the design value of the tested unit provided such higher pressures are determined by accepted engineering analysis. All components of the small unit shall be the same as the tested unit. Where such calculated design pressures are used, they shall be validated by an additional test of the window unit having the highest allowable design pressure.

1714.5.1 Aluminum, vinyl and wood exterior windows and glass doors. Aluminum, vinyl and wood exterior windows and glass doors shall be labeled as conforming to AAMA/NWWDA 101/I.S.2 or 101/I.S.2/NAFS. The label shall state the name of the manufacturer, the approved labeling agency and the product designation as specified in AAMA/NWWDA 101/I.S.2 or 101/I.S.2/NAFS. Products tested and labeled as conforming to AAMA/NWWDA 101/I.S.2 or 101/I.S.2/NAFS shall not be subject to the requirements of Sections 2403.2 and 2403.3.

1714.5.2 Exterior windows and door assemblies not provided for in Section 1714.5.1. Exterior window and door assemblies shall be tested in accordance with ASTM E 330. Exterior window and door assemblies containing glass shall comply with Section 2403. The design pressure for testing shall be calculated in accordance with Chapter 16. Each assembly shall be tested for 10 seconds at a load equal to 1.5 times the design pressure.

1714.6 Test specimens. Test specimens and construction shall be representative of the materials, workmanship and details normally used in practice. The properties of the materials used to construct the test assembly shall be determined on the basis of tests on samples taken from the load assembly or on representative samples of the materials used to construct the load test assembly. Required tests shall be conducted or witnessed by an approved agency.

SECTION 1715
MATERIAL AND TEST STANDARDS

1715.1 Test standards for joist hangers and connectors.

1715.1.1 Test standards for joist hangers. The vertical load-bearing capacity, torsional moment capacity and deflection characteristics of joist hangers shall be determined in accordance with ASTM D 1761, using lumber having a specific gravity of 0.49 or greater, but not greater than 0.55, as determined in accordance with AFPA NDS for the joist and hangers.

1715.1.2 Vertical load capacity for joist hangers. The vertical load capacity for the joist hanger shall be determined by testing three joist hanger assemblies as specified in ASTM D 1761. If the ultimate vertical load for any one of the tests varies more than 20 percent from the average ultimate vertical load, at least three additional tests shall be conducted. The allowable vertical load for a normal duration of

loading of the joist hanger shall be the lowest value determined from the following:

1. The lowest ultimate vertical load from any test divided by three (where three tests are conducted and each ultimate vertical load does not vary more than 20 percent from the average ultimate vertical load).
2. The average ultimate vertical load for all tests divided by six (where six or more tests are conducted).
3. The vertical load at which the vertical movement of the joist with respect to the header is 0.125 inch (3.2 mm) in any test.
4. The allowable design load for nails or other fasteners utilized to secure the joist hanger to the wood members.
5. The allowable design load for the wood members forming the connection.

1715.1.3 Torsional moment capacity for joist hangers. The torsional moment capacity for the joist hanger shall be determined by testing at least three joist hanger assemblies as specified in ASTM D 1761. The allowable torsional moment for normal duration of loading of the joist hanger shall be the average torsional moment at which the lateral movement of the top or bottom of the joist with respect to the original position of the joist is 0.125 inch (3.2 mm).

1715.1.4 Design value modifications for joist hangers. Allowable design values for joist hangers that are determined by Item 4 or 5 in Section 1715.1.2 shall be permitted to be modified by the appropriate duration of loading factors as specified in AFPA NDS but shall not exceed the direct loads as determined by Item 1, 2 or 3 in Section 1715.1.2. Allowable design values determined by Item 1, 2 or 3 in Sections 1715.1.2 and 2305.1 shall not be modified by duration of loading factors.

1715.2 Concrete and clay roof tiles.

1715.2.1 Overturning resistance. Concrete and clay roof tiles shall be tested to determine their resistance to overturning due to wind in accordance with SBCCI SSTD 11 and Chapter 15.

1715.2.2 Wind tunnel testing. When roof tiles do not satisfy the limitations in Chapter 16 for rigid tile, a wind tunnel test shall be used to determine the wind characteristics of the concrete or clay tile roof covering in accordance with SBCCI SSTD 11 and Chapter 15.

CHAPTER 18

SOILS AND FOUNDATIONS

SECTION 1801
GENERAL

1801.1 Scope. The provisions of this chapter shall apply to building and foundation systems in those areas not subject to scour or water pressure by wind and wave action. Buildings and foundations subject to such scour or water pressure loads shall be designed in accordance with Chapter 16.

1801.2 Design. Allowable bearing pressures, allowable stresses and design formulas provided in this chapter shall be used with the allowable stress design load combinations specified in Section 1605.3. The quality and design of materials used structurally in excavations, footings and foundations shall conform to the requirements specified in Chapters 16, 19, 21, 22 and 23 of this code. Excavations and fills shall also comply with Chapter 33.

1801.2.1 Foundation design for seismic overturning. Where the foundation is proportioned using the strength design load combinations of Section 1605.2, the seismic overturning moment need not exceed 75 percent of the value computed from Section 9.5.5.6 of ASCE 7 for the equivalent lateral force method, or Section 1618 for the modal analysis method.

SECTION 1802
FOUNDATION AND SOILS INVESTIGATIONS

1802.1 General. Foundation and soils investigations shall be conducted in conformance with Sections 1802.2 through 1802.6. Where required by the building official, the classification and investigation of the soil shall be made by a registered design professional.

1802.2 Where required. The owner or applicant shall submit a foundation and soils investigation to the building official where required in Sections 1802.2.1 through 1802.2.7.

Exception: The building official need not require a foundation or soils investigation where satisfactory data from adjacent areas is available that demonstrates an investigation is not necessary for any of the conditions in Sections 1802.2.1 through 1802.2.6.

1802.2.1 Questionable soil. Where the safe-sustaining power of the soil is in doubt, or where a load-bearing value superior to that specified in this code is claimed, the building official shall require that the necessary investigation be made. Such investigation shall comply with the provisions of Sections 1802.4 through 1802.6.

1802.2.2 Expansive soils. In areas likely to have expansive soil, the building official shall require soil tests to determine where such soils do exist.

1802.2.3 Ground-water table. A subsurface soil investigation shall be performed to determine whether the existing ground-water table is above or within 5 feet (1524 mm) below the elevation of the lowest floor level where such floor is located below the finished ground level adjacent to the foundation.

Exception: A subsurface soil investigation shall not be required where waterproofing is provided in accordance with Section 1807.

1802.2.4 Pile and pier foundations. Pile and pier foundations shall be designed and installed on the basis of a foundation investigation and report as specified in Sections 1802.4 through 1802.6 and Section 1808.2.1.

1802.2.5 Rock strata. Where subsurface explorations at the project site indicate variations or doubtful characteristics in the structure of the rock upon which foundations are to be constructed, a sufficient number of borings shall be made to a depth of not less than 10 feet (3048 mm) below the level of the foundations to provide assurance of the soundness of the foundation bed and its load-bearing capacity.

1802.2.6 Seismic Design Category C. Where a structure is determined to be in Seismic Design Category C in accordance with Section 1616, an investigation shall be conducted, and shall include an evaluation of the following potential hazards resulting from earthquake motions: slope instability, liquefaction and surface rupture due to faulting or lateral spreading.

1802.2.7 Seismic Design Category D, E or F. Where the structure is determined to be in Seismic Design Category D, E or F, in accordance with Section 1616, the soils investigation requirements for Seismic Design Category C, given in Section 1802.2.6, shall be met, in addition to the following. The investigation shall include:

1. A determination of lateral pressures on basement and retaining walls due to earthquake motions.

2. An assessment of potential consequences of any liquefaction and soil strength loss, including estimation of differential settlement, lateral movement or reduction in foundation soil-bearing capacity, and shall address mitigation measures. Such measures shall be given consideration in the design of the structure and can include, but are not limited to, ground stabilization, selection of appropriate foundation type and depths, selection of appropriate structural systems to accommodate anticipated displacements or any combination of these measures. The potential for liquefaction and soil strength loss shall be evaluated for site peak ground acceleration magnitudes and source characteristics consistent with the design earthquake ground motions. Peak ground acceleration shall be

determined from a site-specific study taking into account soil amplification effects, as specified in Section 1615.2.

Exception: A site-specific study need not be performed provided that peak ground acceleration equal to $S_{DS}/2.5$ is used, where S_{DS} is determined in accordance with Section 1615.2.1.

1802.3 Soil classification. Where required, soils shall be classified in accordance with Section 1802.3.1 or 1802.3.2.

1802.3.1 General. For the purposes of this chapter, the definition and classification of soil materials for use in Table 1804.2 shall be in accordance with ASTM D 2487.

1802.3.2 Expansive soils. Soils meeting all four of the following provisions shall be considered expansive, except that tests to show compliance with Items 1, 2 and 3 shall not be required if the test prescribed in Item 4 is conducted:

1. Plasticity index (PI) of 15 or greater, determined in accordance with ASTM D 4318.

2. More than 10 percent of the soil particles pass a No. 200 sieve (75 µm), determined in accordance with ASTM D 422.

3. More than 10 percent of the soil particles are less than 5 micrometers in size, determined in accordance with ASTM D 422.

4. Expansion index greater than 20, determined in accordance with ASTM D 4829.

1802.4 Investigation. Soil classification shall be based on observation and any necessary tests of the materials disclosed by borings, test pits or other subsurface exploration made in appropriate locations. Additional studies shall be made as necessary to evaluate slope stability, soil strength, position and adequacy of load-bearing soils, the effect of moisture variation on soil-bearing capacity, compressibility, liquefaction and expansiveness.

1802.4.1 Exploratory boring. The scope of the soil investigation including the number and types of borings or soundings, the equipment used to drill and sample, the in-situ testing equipment and the laboratory testing program shall be determined by a registered design professional.

1802.5 Soil boring and sampling. The soil boring and sampling procedure and apparatus shall be in accordance with generally accepted engineering practice. The registered design professional shall have a fully qualified representative on the site during all boring and sampling operations.

1802.6 Reports. The soil classification and design load-bearing capacity shall be shown on the construction document. Where required by the building official, a written report of the investigation shall be submitted that includes, but need not be limited to, the following information:

1. A plot showing the location of test borings and/or excavations.

2. A complete record of the soil samples.

3. A record of the soil profile.

4. Elevation of the water table, if encountered.

5. Recommendations for foundation type and design criteria, including but not limited to: bearing capacity of natural or compacted soil; provisions to mitigate the effects of expansive soils; mitigation of the effects of liquefaction, differential settlement and varying soil strength; and the effects of adjacent loads.

6. Expected total and differential settlement.

7. Pile and pier foundation information in accordance with Section 1808.2.2.

8. Special design and construction provisions for footings or foundations founded on expansive soils, as necessary.

9. Compacted fill material properties and testing in accordance with Section 1803.5.

SECTION 1803
EXCAVATION, GRADING AND FILL

1803.1 Excavations near footings or foundations. Excavations for any purpose shall not remove lateral support from any footing or foundation without first underpinning or protecting the footing or foundation against settlement or lateral translation.

1803.2 Placement of backfill. The excavation outside the foundation shall be backfilled with soil that is free of organic material, construction debris, cobbles and boulders or a controlled low-strength material (CLSM). The backfill shall be placed in lifts and compacted, in a manner that does not damage the foundation or the waterproofing or dampproofing material.

Exception: Controlled low-strength material need not be compacted.

1803.3 Site grading. The ground immediately adjacent to the foundation shall be sloped away from the building at a slope of not less than one unit vertical in 20 units horizontal (5-percent slope) for a minimum distance of 10 feet (3048 mm) measured perpendicular to the face of the wall or an approved alternate method of diverting water away from the foundation shall be used.

Exception: Where climatic or soil conditions warrant, the slope of the ground away from the building foundation is permitted to be reduced to not less than one unit vertical in 48 units horizontal (2-percent slope).

The procedure used to establish the final ground level adjacent to the foundation shall account for additional settlement of the backfill.

1803.4 Grading and fill in floodways. In floodways shown on the flood hazard map established in Section 1612.3, grading and/or fill shall not be approved unless it has been demonstrated through hydrologic and hydraulic analyses performed by a registered design professional in accordance with standard engineering practice that the proposed grading or fill, or both, will not result in any increase in flood levels during the occurrence of the design flood.

1803.5 Compacted fill material. Where footings will bear on compacted fill material, the compacted fill shall comply with the provisions of an approved report, which shall contain the following:

1. Specifications for the preparation of the site prior to placement of compacted fill material.

2. Specifications for material to be used as compacted fill.

3. Test method to be used to determine the maximum dry density and optimum moisture content of the material to be used as compacted fill.

4. Maximum allowable thickness of each lift of compacted fill material.

5. Field test method for determining the in-place dry density of the compacted fill.

6. Minimum acceptable in-place dry density expressed as a percentage of the maximum dry density determined in accordance with Item 3.

7. Number and frequency of field tests required to determine compliance with Item 6.

Exception: Compacted fill material less than 12 inches (305 mm) in depth need not comply with an approved report, provided it has been compacted to a minimum of 90 percent Modified Proctor in accordance with ASTM D 1557. The compaction shall be verified by a qualified inspector approved by the building official.

1803.6 Controlled low-strength material (CLSM). Where footings will bear on controlled low-strength material (CLSM), the CLSM shall comply with the provisions of an approved report, which shall contain the following:

1. Specifications for the preparation of the site prior to placement of the CLSM.

2. Specifications for the CLSM.

3. Laboratory or field test method(s) to be used to determine the compressive strength or bearing capacity of the CLSM.

4. Test methods for determining the acceptance of the CLSM in the field.

5. Number and frequency of field tests required to determine compliance with Item 4.

SECTION 1804
ALLOWABLE LOAD-BEARING VALUES OF SOILS

1804.1 Design. The presumptive load-bearing values provided in Table 1804.2 shall be used with the allowable stress design load combinations specified in Section 1605.3.

1804.2 Presumptive load-bearing values. The maximum allowable foundation pressure, lateral pressure or lateral sliding resistance values for supporting soils at or near the surface shall not exceed the values specified in Table 1804.2 unless data to substantiate the use of a higher value are submitted and approved.

Presumptive load-bearing values shall apply to materials with similar physical characteristics and dispositions.

Mud, organic silt, organic clays, peat or unprepared fill shall not be assumed to have a presumptive load-bearing capacity unless data to substantiate the use of such a value are submitted.

Exception: A presumptive load-bearing capacity is permitted to be used where the building official deems the load-bearing capacity of mud, organic silt or unprepared fill is adequate for the support of lightweight and temporary structures.

1804.3 Lateral sliding resistance. The resistance of structural walls to lateral sliding shall be calculated by combining the values derived from the lateral bearing and the lateral sliding resis-

TABLE 1804.2
ALLOWABLE FOUNDATION AND LATERAL PRESSURE

CLASS OF MATERIALS	ALLOWABLE FOUNDATION PRESSURE (psf)[d]	LATERAL BEARING (psf/f below natural grade)[d]	LATERAL SLIDING	
			Coefficient of friction[a]	Resistance (psf)[b]
1. Crystalline bedrock	12,000	1,200	0.70	—
2. Sedimentary and foliated rock	4,000	400	0.35	—
3. Sandy gravel and/or gravel (GW and GP)	3,000	200	0.35	—
4. Sand, silty sand, clayey sand, silty gravel and clayey gravel (SW, SP, SM, SC, GM and GC)	2,000	150	0.25	—
5. Clay, sandy clay, silty clay, clayey silt, silt and sandy silt (CL, ML, MH and CH)	1,500[c]	100	—	130

For SI: 1 pound per square foot = 0.0479 kPa, 1 pound per square foot per foot = 0.157 kPa/m.

a. Coefficient to be multiplied by the dead load.

b. Lateral sliding resistance value to be multiplied by the contact area, as limited by Section 1804.3.

c. Where the building official determines that in-place soils with an allowable bearing capacity of less than 1,500 psf are likely to be present at the site, the allowable bearing capacity shall be determined by a soils investigation.

d. An increase of one-third is permitted when using the alternate load combinations in Section 1605.3.2 that include wind or earthquake loads.

tance shown in Table 1804.2 unless data to substantiate the use of higher values are submitted for approval.

For clay, sandy clay, silty clay and clayey silt, in no case shall the lateral sliding resistance exceed one-half the dead load.

1804.3.1 Increases in allowable lateral sliding resistance. The resistance values derived from the table are permitted to be increased by the tabular value for each additional foot (305 mm) of depth to a maximum of 15 times the tabular value.

Isolated poles for uses such as flagpoles or signs and poles used to support buildings that are not adversely affected by a $1/_2$-inch (12.7 mm) motion at the ground surface due to short-term lateral loads are permitted to be designed using lateral-bearing values equal to two times the tabular values.

SECTION 1805
FOOTINGS AND FOUNDATIONS

1805.1 General. Footings and foundations shall be designed and constructed in accordance with Sections 1805.1 through 1805.9. Footings and foundations shall be built on undisturbed soil, compacted fill material or CLSM. Compacted fill material shall be placed in accordance with Section 1803.5. CLSM shall be placed in accordance with Section 1803.6.

The top surface of footings shall be level. The bottom surface of footings is permitted to have a slope not exceeding one unit vertical in 10 units horizontal (10-percent slope). Footings shall be stepped where it is necessary to change the elevation of the top surface of the footing or where the surface of the ground slopes more than one unit vertical in 10 units horizontal (10-percent slope).

1805.2 Depth of footings. The minimum depth of footings below the undisturbed ground surface shall be 12 inches (305 mm). Where applicable, the depth of footings shall also conform to Sections 1805.2.1 through 1805.2.3.

1805.2.1 Frost protection. Except where otherwise protected from frost, foundation walls, piers and other permanent supports of buildings and structures shall be protected from frost by one or more of the following methods:

1. Extending below the frost line of the locality;

2. Constructing in accordance with ASCE-32; or

3. Erecting on solid rock.

Exception: Free-standing buildings meeting all of the following conditions shall not be required to be protected:

1. Classified in Importance Category I (see Table 1604.5);

2. Area of 400 square feet (37 m²) or less; and

3. Eave height of 10 feet (3048 mm) or less.

Footings shall not bear on frozen soil unless such frozen condition is of a permanent character.

1805.2.2 Isolated footings. Footings on granular soil shall be so located that the line drawn between the lower edges of adjoining footings shall not have a slope steeper than 30 degrees (0.52 rad) with the horizontal, unless the material supporting the higher footing is braced or retained or otherwise laterally supported in an approved manner or a greater slope has been properly established by engineering analysis.

1805.2.3 Shifting or moving soils. Where it is known that the shallow subsoils are of a shifting or moving character, footings shall be carried to a sufficient depth to ensure stability.

1805.3 Footings on or adjacent to slopes. The placement of buildings and structures on or adjacent to slopes steeper than one unit vertical in three units horizontal (33.3-percent slope) shall conform to Sections 1805.3.1 through 1805.3.5.

1805.3.1 Building clearance from ascending slopes. In general, buildings below slopes shall be set a sufficient distance from the slope to provide protection from slope drainage, erosion and shallow failures. Except as provided for in Section 1805.3.5 and Figure 1805.3.1, the following criteria will be assumed to provide this protection. Where the existing slope is steeper than one unit vertical in one unit horizontal (100-percent slope), the toe of the slope shall be assumed to be at the intersection of a horizontal plane drawn from the top of the foundation and a plane drawn tangent to the slope at an angle of 45 degrees (0.79 rad) to the horizontal. Where a retaining wall is constructed at the toe of the slope, the height of the slope shall be measured from the top of the wall to the top of the slope.

1805.3.2 Footing setback from descending slope surface. Footings on or adjacent to slope surfaces shall be founded in firm material with an embedment and set back from the slope surface sufficient to provide vertical and lateral support for the footing without detrimental settlement. Except as provided for in Section 1805.3.5 and Figure 1805.3.1, the following setback is deemed adequate to meet the criteria. Where the slope is steeper than 1 unit vertical in 1 unit horizontal (100-percent slope), the required setback shall be measured from an imaginary plane 45 degrees (0.79 rad) to the horizontal, projected upward from the toe of the slope.

1805.3.3 Pools. The setback between pools regulated by this code and slopes shall be equal to one-half the building footing setback distance required by this section. That portion of the pool wall within a horizontal distance of 7 feet (2134 mm) from the top of the slope shall be capable of supporting the water in the pool without soil support.

1805.3.4 Foundation elevation. On graded sites, the top of any exterior foundation shall extend above the elevation of the street gutter at point of discharge or the inlet of an approved drainage device a minimum of 12 inches (305 mm) plus 2 percent. Alternate elevations are permitted subject to the approval of the building official, provided it can be demonstrated that required drainage to the point of discharge and away from the structure is provided at all locations on the site.

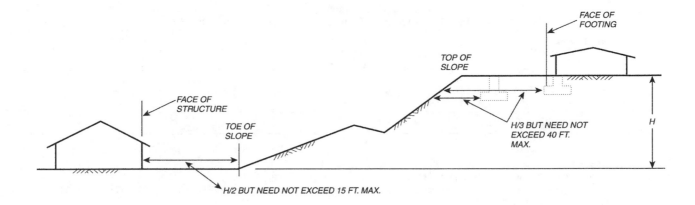

For SI: 1 foot = 304.8 mm.

FIGURE 1805.3.1
FOUNDATION CLEARANCES FROM SLOPES

1805.3.5 Alternate setback and clearance. Alternate setbacks and clearances are permitted, subject to the approval of the building official. The building official is permitted to require an investigation and recommendation of a registered design professional to demonstrate that the intent of this section has been satisfied. Such an investigation shall include consideration of material, height of slope, slope gradient, load intensity and erosion characteristics of slope material.

1805.4 Footings. Footings shall be designed and constructed in accordance with Sections 1805.4.1 through 1805.4.6.

1805.4.1 Design. Footings shall be so designed that the allowable bearing capacity of the soil is not exceeded, and that differential settlement is minimized. The minimum width of footings shall be 12 inches (305 mm).

Footings in areas with expansive soils shall be designed in accordance with the provisions of Section 1805.8.

1805.4.1.1 Design loads. Footings shall be designed for the most unfavorable effects due to the combinations of loads specified in Section 1605.3. The dead load shall include the weight of foundations, footings and overlying fill. Reduced live loads, as specified in Section 1607.9, are permitted to be used in designing footings.

1805.4.1.2 Vibratory loads. Where machinery operations or other vibrations are transmitted through the foundation, consideration shall be given in the footing design to prevent detrimental disturbances of the soil.

1805.4.2 Concrete footings. The design, materials and construction of concrete footings shall comply with Sections 1805.4.2.1 through 1805.4.2.6 and the provisions of Chapter 19.

Exception: Where a specific design is not provided, concrete footings supporting walls of light-frame construction are permitted to be designed in accordance with Table 1805.4.2.

1805.4.2.1 Concrete strength. Concrete in footings shall have a specified compressive strength (f'_c) of not less than 2,500 pounds per square inch (psi) (17 237 kPa) at 28 days.

1805.4.2.2 Footing seismic ties. Where a structure is assigned to Seismic Design Category D, E or F in accordance with Section 1616, individual spread footings founded on soil defined in Section 1615.1.1 as Site Class E or F shall be interconnected by ties. Ties shall be capable of carrying, in tension or compression, a force equal to the product of the larger footing load times the seismic coefficient S_{DS} divided by 10 unless it is demonstrated that equivalent restraint is provided by reinforced concrete beams within slabs on grade or reinforced concrete slabs on grade.

1805.4.2.3 Plain concrete footings. The edge thickness of plain concrete footings supporting walls of other than light-frame construction shall not be less than 8 inches (203 mm) where placed on soil.

Exception: For plain concrete footings supporting Group R-3 occupancies, the edge thickness is permitted to be 6 inches (152 mm), provided that the footing does not extend beyond a distance greater than the thickness of the footing on either side of the supported wall.

1805.4.2.4 Placement of concrete. Concrete footings shall not be placed through water unless a tremie or other method approved by the building official is used. Where placed under or in the presence of water, the concrete shall be deposited by approved means to ensure minimum segregation of the mix and negligible turbulence of the water.

1805.4.2.5 Protection of concrete. Concrete footings shall be protected from freezing during depositing and for a period of not less than five days thereafter. Water shall not be allowed to flow through the deposited concrete.

TABLE 1805.4.2
FOOTINGS SUPPORTING WALLS OF LIGHT-FRAME CONSTRUCTION[a, b, c, d, e]

NUMBER OF FLOORS SUPPORTED BY THE FOOTING[f]	WIDTH OF FOOTING (inches)	THICKNESS OF FOOTING (inches)
1	12	6
2	15	6
3	18	8[g]

For SI: 1 inch = 25.4 mm, 1 foot = 304.8 mm.

a. Depth of footings shall be in accordance with Section 1805.2.

b. The ground under the floor is permitted to be excavated to the elevation of the top of the footing.

c. Interior-stud-bearing walls are permitted to be supported by isolated footings. The footing width and length shall be twice the width shown in this table, and footings shall be spaced not more than 6 feet on center.

d. See Section 1910 for additional requirements for footings of structures assigned to Seismic Design Category C, D, E or F.

e. For thickness of foundation walls, see Section 1805.5.

f. Footings are permitted to support a roof in addition to the stipulated number of floors. Footings supporting roof only shall be as required for supporting one floor.

g. Plain concrete footings for Group R-3 occupancies are permitted to be 6 inches thick.

1805.4.2.6 Forming of concrete. Concrete footings are permitted to be cast against the earth where, in the opinion of the building official, soil conditions do not require forming. Where forming is required, it shall be in accordance with Chapter 6 of ACI 318.

1805.4.3 Masonry-unit footings. The design, materials and construction of masonry-unit footings shall comply with Sections 1805.4.3.1 and 1805.4.3.2, and the provisions of Chapter 21.

Exception: Where a specific design is not provided, masonry-unit footings supporting walls of light-frame construction are permitted to be designed in accordance with Table 1805.4.2.

1805.4.3.1 Dimensions. Masonry-unit footings shall be laid in Type M or S mortar complying with Section 2103.7 and the depth shall not be less than twice the projection beyond the wall, pier or column. The width shall not be less than 8 inches (203 mm) wider than the wall supported thereon.

1805.4.3.2 Offsets. The maximum offset of each course in brick foundation walls stepped up from the footings shall be $1^1/_2$ inches (38 mm) where laid in single courses, and 3 inches (76 mm) where laid in double courses.

1805.4.4 Steel grillage footings. Grillage footings of structural steel shapes shall be separated with approved steel spacers and be entirely encased in concrete with at least 6 inches (152 mm) on the bottom and at least 4 inches (102 mm) at all other points. The spaces between the shapes shall be completely filled with concrete or cement grout.

1805.4.5 Timber footings. Timber footings are permitted for buildings of Type V construction and as otherwise approved by the building official. Such footings shall be treated in accordance with AWPA C2 or C3. Treated timbers are not required where placed entirely below permanent water level, or where used as capping for wood piles that project above the water level over submerged or marsh lands. The compressive stresses perpendicular to grain in untreated timber footings supported upon piles shall not exceed 70 percent of the allowable stresses for the species and grade of timber as specified in the AFPA NDS.

1805.4.6 Wood foundations. Wood foundation systems shall be designed and installed in accordance with AFPA Technical Report No. 7. Lumber and plywood shall be treated in accordance with AWPA C22 and shall be identified in accordance with Section 2303.1.8.1.

1805.5 Foundation walls. Concrete and masonry foundation walls shall be designed in accordance with Chapter 19 or 21. Foundation walls that are laterally supported at the top and bottom and within the parameters of Tables 1805.5(1) through 1805.5(4) are permitted to be designed and constructed in accordance with Sections 1805.5.1 through 1805.5.5.

1805.5.1 Foundation wall thickness. The minimum thickness of concrete and masonry foundation walls shall comply with Sections 1805.5.1.1 through 1805.5.1.3.

1805.5.1.1 Thickness based on walls supported. The thickness of foundation walls shall not be less than the thickness of the wall supported, except that foundation walls of at least 8 inch (203 mm) nominal width are permitted to support brick-veneered frame walls and 10-inch-wide (254 mm) cavity walls provided the requirements of Section 1805.5.1.2 are met. Corbeling of masonry shall be in accordance with Section 2104.2. Where an 8-inch (203 mm) wall is corbeled, the top corbel shall be a full course of headers at least 6 inches (152 mm) in length, extending not higher than the bottom of the floor framing.

TABLE 1805.5(1)
PLAIN MASONRY AND PLAIN CONCRETE FOUNDATION WALLS[a, b, c]

		PLAIN MASONRY		
		MINIMUM NOMINAL WALL THICKNESS (inches)		
		Soil classes and lateral soil load[a] (psf per foot below natural grade)		
WALL HEIGHT (feet)	HEIGHT OF UNBALANCED BACKFILL (feet)	GW, GP, SW and SP soils 30	GM, GC, SM, SM-SC and ML soils 45	SC, MH, ML-CL and Inorganic CL soils 60
7	4 (or less)	8	8	8
	5	8	10	10
	6	10	12	10 (solid[c])
	7	12	10 (solid[c])	10 (solid[c])
8	4 (or less)	8	8	8
	5	8	10	12
	6	10	12	12 (solid[c])
	7	12	12 (solid[c])	Note d
	8	10 (solid[c])	12 (solid[c])	Note d
9	4 (or less)	8	8	8
	5	8	10	12
	6	12	12	12 (solid[c])
	7	12 (solid[c])	12 (solid[c])	Note d
	8	12 (solid[c])	Note d	Note d
	9	Note d	Note d	Note d

		PLAIN CONCRETE		
		MINIMUM NOMINAL WALL THICKNESS (inches)		
		Soil classes and lateral soil load[a] (psf per foot below natural grade)		
WALL HEIGHT (feet)	HEIGHT OF UNBALANCED BACKFILL (feet)	GW, GP, SW and SP soils 30	GM, GC, SM, SM-SC and ML soils 45	SC, MH, ML-CL and Inorganic CL soils 60
7	4 (or less)	$7^{1}/_{2}$	$7^{1}/_{2}$	$7^{1}/_{2}$
	5	$7^{1}/_{2}$	$7^{1}/_{2}$	$7^{1}/_{2}$
	6	$7^{1}/_{2}$	$7^{1}/_{2}$	8
	7	$7^{1}/_{2}$	8	10
8	4 (or less)	$7^{1}/_{2}$	$7^{1}/_{2}$	$7^{1}/_{2}$
	5	$7^{1}/_{2}$	$7^{1}/_{2}$	$7^{1}/_{2}$
	6	$7^{1}/_{2}$	$7^{1}/_{2}$	10
	7	$7^{1}/_{2}$	10	10
	8	10	10	12
9	4 (or less)	$7^{1}/_{2}$	$7^{1}/_{2}$	$7^{1}/_{2}$
	5	$7^{1}/_{2}$	$7^{1}/_{2}$	$7^{1}/_{2}$
	6	$7^{1}/_{2}$	$7^{1}/_{2}$	10
	7	$7^{1}/_{2}$	10	10
	8	10	10	12
	9	10	12	Note e

For SI: 1 inch = 25.4 mm, 1 foot = 304.8 mm, 1 pound per square foot per foot = 0.157 kPa/m.

a. For design lateral soil loads, see Section 1610. Soil classes are in accordance with the Unified Soil Classification System and design lateral soil loads are for moist soil conditions without hydrostatic pressure.

b. Provisions for this table are based on construction requirements specified in Section 1805.5.2.

c. Solid grouted hollow units or solid masonry units.

d. A design in compliance with Chapter 21 or reinforcement in accordance with Table 1805.5(2) is required.

e. A design in compliance with Chapter 19 is required.

TABLE 1805.5(2)
8-INCH CONCRETE AND MASONRY FOUNDATION WALLS WITH REINFORCING WHERE _d_ ≥ 5 INCHES[a, b, c]

WALL HEIGHT (feet)	HEIGHT OF UNBALANCED BACKFILL (feet)	VERTICAL REINFORCEMENT		
		Soil classes and lateral soil load[a] (psf per foot below natural grade)		
		GW, GP, SW and SP soils 30	GM, GC, SM, SM-SC and ML soils 45	SC, MH, ML-CL and Inorganic CL soils 60
7	4 (or less)	#4 at 48″ o.c.	#4 at 48″ o.c.	#4 at 48″ o.c.
	5	#4 at 48″ o.c.	#4 at 48″ o.c.	#4 at 40″ o.c.
	6	#4 at 48″ o.c.	#5 at 48″ o.c.	#5 at 40″ o.c.
	7	#4 at 40″ o.c.	#5 at 40″ o.c.	#6 at 48″ o.c.
8	4 (or less)	#4 at 48″ o.c.	#4 at 48″ o.c.	#4 at 48″ o.c.
	5	#4 at 48″ o.c.	#4 at 48″ o.c.	#4 at 40″ o.c.
	6	#4 at 48″ o.c.	#5 at 48″ o.c.	#5 at 40″ o.c.
	7	#5 at 48″ o.c.	#6 at 48″ o.c.	#6 at 40″ o.c.
	8	#5 at 40″ o.c.	#6 at 40″ o.c.	#7 at 40″ o.c.
9	4 (or less)	#4 at 48″ o.c.	#4 at 48″ o.c.	#4 at 48″ o.c.
	5	#4 at 48″ o.c.	#4 at 48″ o.c.	#5 at 48″ o.c.
	6	#4 at 48″ o.c.	#5 at 48″ o.c.	#6 at 48″ o.c.
	7	#5 at 48″ o.c.	#6 at 48″ o.c.	#7 at 48″ o.c.
	8	#5 at 40″ o.c.	#7 at 48″ o.c.	#8 at 48″ o.c.
	9	#6 at 40″ o.c.	#8 at 48″ o.c.	#8 at 32″ o.c.

For SI: 1 inch = 25.4 mm, 1 foot = 304.8 mm, 1 pound per square foot per foot = 0.157 kPa/m.

a. For design lateral soil loads, see Section 1610. Soil classes are in accordance with the Unified Soil Classification System and design lateral soil loads are for moist soil conditions without hydrostatic pressure.

b. Provisions for this table are based on construction requirements specified in Section 1805.5.2.

c. For alternative reinforcement, see Section 1805.5.3.

TABLE 1805.5(3)
10-INCH CONCRETE AND MASONRY FOUNDATION WALLS WITH REINFORCING WHERE _d_ ≥ 6.75 INCHES[a, b, c]

WALL HEIGHT (feet)	HEIGHT OF UNBALANCED BACKFILL (feet)	VERTICAL REINFORCEMENT		
		Soil classes and lateral soil load[a] (psf per foot below natural grade)		
		GW, GP, SW and SP soils 30	GM, GC, SM, SM-SC and ML soils 45	SC, MH, ML-CL and Inorganic CL soils 60
7	4 (or less)	#4 at 56″ o.c.	#4 at 56″ o.c.	#4 at 56″ o.c.
	5	#4 at 56″ o.c.	#4 at 56″ o.c.	#4 at 56″ o.c.
	6	#4 at 56″ o.c.	#4 at 48″ o.c.	#4 at 40″ o.c.
	7	#4 at 56″ o.c.	#5 at 56″ o.c.	#5 at 40″ o.c.
8	4 (or less)	#4 at 56″ o.c.	#4 at 56″ o.c.	#4 at 56″ o.c.
	5	#4 at 56″ o.c.	#4 at 56″ o.c.	#4 at 48″ o.c.
	6	#4 at 56″ o.c.	#4 at 48″ o.c.	#5 at 56″ o.c.
	7	#4 at 48″ o.c.	#4 at 32″ o.c.	#6 at 56″ o.c.
	8	#5 at 56″ o.c.	#5 at 40″ o.c.	#7 at 56″ o.c.
9	4 (or less)	#4 at 56″ o.c.	#4 at 56″ o.c.	#4 at 56″ o.c.
	5	#4 at 56″ o.c.	#4 at 56″ o.c.	#4 at 48″ o.c.
	6	#4 at 56″ o.c.	#4 at 40″ o.c.	#4 at 32″ o.c.
	7	#4 at 40″ o.c.	#5 at 48″ o.c.	#6 at 48″ o.c.
	8	#4 at 32″ o.c.	#6 at 48″ o.c.	#4 at 16″ o.c.
	9	#5 at 40″ o.c.	#6 at 40″ o.c.	#7 at 40″ o.c.

For SI: 1 inch = 25.4 mm, 1 foot = 304.8 mm, 1 pound per square foot per foot = 0.157 kPa/m.

a. For design lateral soil loads, see Section 1610. Soil classes are in accordance with the Unified Soil Classification System and design lateral soil loads are for moist soil conditions without hydrostatic pressure.

b. Provisions for this table are based on construction requirements specified in Section 1805.5.2.

c. For alternative reinforcement, see Section 1805.5.3.

TABLE 1805.5(4)
12-INCH CONCRETE AND MASONRY FOUNDATION WALLS WITH REINFORCING WHERE $d \geq 8.75$ INCHES[a, b, c]

WALL HEIGHT (feet)	HEIGHT OF UNBALANCED BACKFILL (feet)	VERTICAL REINFORCEMENT		
		Soil classes and lateral soil load[a] (psf per foot below natural grade)		
		GW, GP, SW and SP soils 30	GM, GC, SM, SM-SC and ML soils 45	SC, MH, ML-CL and Inorganic CL soils 60
7	4 (or less)	#4 at 72″ o.c.	#4 at 72″ o.c.	#4 at 72″ o.c.
	5	#4 at 72″ o.c.	#4 at 72″ o.c.	#4 at 72″ o.c.
	6	#4 at 72″ o.c.	#4 at 64″ o.c.	#4 at 48″ o.c.
	7	#4 at 72″ o.c.	#4 at 48″ o.c.	#5 at 56″ o.c.
8	4 (or less)	#4 at 72″ o.c.	#4 at 72″ o.c.	#4 at 72″ o.c.
	5	#4 at 72″ o.c.	#4 at 72″ o.c.	#4 at 72″ o.c.
	6	#4 at 72″ o.c.	#4 at 56″ o.c.	#5 at 72″ o.c.
	7	#4 at 64″ o.c.	#5 at 64″ o.c.	#4 at 32″ o.c.
	8	#4 at 48″ o.c.	#4 at 32″ o.c.	#5 at 40″ o.c.
9	4 (or less)	#4 at 72″ o.c.	#4 at 72″ o.c.	#4 at 72″ o.c.
	5	#4 at 72″ o.c.	#4 at 72″ o.c.	#4 at 64″ o.c.
	6	#4 at 72″ o.c.	#4 at 56″ o.c.	#5 at 64″ o.c.
	7	#4 at 56″ o.c.	#4 at 40″ o.c.	#6 at 64″ o.c.
	8	#4 at 64″ o.c.	#6 at 64″ o.c.	#6 at 48″ o.c.
	9	#5 at 56″ o.c.	#7 at 72″ o.c.	#6 at 40″ o.c.

For SI: 1 inch = 25.4 mm, 1 foot = 304.8 mm, 1 pound per square foot per foot = 0.157 kPa/m.

a. For design lateral soil loads, see Section 1610. Soil classes are in accordance with the Unified Soil Classification System and design lateral soil loads are for moist soil conditions without hydrostatic pressure.

b. Provisions for this table are based on construction requirements specified in Section 1805.5.2.

c. For alternative reinforcement, see Section 1805.5.3.

1805.5.1.2 Thickness based on soil loads, unbalanced backfill height and wall height. The thickness of foundation walls shall comply with the requirements of Table 1805.5(1) for plain masonry and plain concrete walls or Table 1805.5(2), 1805.5(3) or 1805.5(4) for reinforced concrete and masonry walls. When using the tables, masonry shall be laid in running bond and the mortar shall be Type M or S.

Unbalanced backfill height is the difference in height of the exterior and interior finish ground levels. Where an interior concrete slab is provided, the unbalanced backfill height shall be measured from the exterior finish ground level to the top of the interior concrete slab.

1805.5.1.3 Rubble stone. Foundation walls of rough or random rubble stone shall not be less than 16 inches (406 mm) thick. Rubble stone shall not be used for foundations for structures in Seismic Design Category C, D, E or F.

1805.5.2 Foundation wall materials. Foundation walls constructed in accordance with Table 1805.5(1), 1805.5(2), 1805.5(3) or 1805.5(4) shall comply with the following:

1. Vertical reinforcement shall have a minimum yield strength of 60,000 psi (414 Mpa).

2. The specified location of the reinforcement shall equal or exceed the effective depth distance, d, noted in Tables 1805.5(2), 1805.5(3) and 1805.5(4) and shall be measured from the face of the soil side of the wall to the center of vertical reinforcement. The reinforcement shall be placed within the toler-

ances specified in ACI 530.1/ASCE 6/TMS 402, Article 3.4 B7 of the specified location.

3. Concrete shall have a specified compressive strength of not less than 2,500 psi (17.2 MPa) at 28 days.

4. Grout shall have a specified compressive strength of not less than 2,000 psi (13.8 MPa) at 28 days.

5. Hollow masonry units shall comply with ASTM C 90 and be installed with Type M or S mortar.

1805.5.3 Alternative foundation wall reinforcement. In lieu of the reinforcement provisions in Table 1805.5(2), 1805.5(3) or 1805.5(4), alternative reinforcing bar sizes and spacings having an equivalent cross-sectional area of reinforcement per linear foot (mm) of wall are permitted to be used, provided the spacing of reinforcement does not exceed 72 inches (1829 mm) and reinforcing bar sizes do not exceed No. 11.

1805.5.4 Hollow masonry walls. At least 4 inches (102 mm) of solid masonry shall be provided at girder supports at the top of hollow masonry unit foundation walls.

1805.5.5 Seismic requirements. Tables 1805.5(1) through 1805.5(4) shall be subject to the following limitations in Sections 1805.5.5.1 and 1805.5.5.2 based on the seismic design category assigned to the structure as defined in Section 1616.

1805.5.5.1 Seismic requirements for concrete foundation walls. Concrete foundation walls designed using

Tables 1805.5(1) through 1805.5(4) shall be subject to the following limitations:

1. Seismic Design Categories A and B. No limitations, except provide not less than two No. 5 bars around window and door openings. Such bars shall extend at least 24 inches (610 mm) beyond the corners of the openings

2. Seismic Design Category C. Tables shall not be used except as allowed for plain concrete members in Section 1910.4.

3. Seismic Design Categories D, E and F. Tables shall not be used except as allowed for plain concrete members in ACI 318, Section 22.10.

1805.5.5.2 Seismic requirements for masonry foundation walls. Masonry foundation walls designed using Tables 1805.5(1) through 1805.5(4) shall be subject to the following limitations:

1. Seismic Design Categories A and B. No additional seismic requirements.

2. Seismic Design Category C. A design using Tables 1805.5(1) through 1805.5(4) subject to the seismic requirements of Section 2106.4.

3. Seismic Design Category D. A design using Tables 1805.2(2) through 1805.5(4) subject to the seismic requirements of Section 2106.5.

4. Seismic Design Categories E and F. A design using Tables 1805.2(2) through 1805.5(4) subject to the seismic requirements of Section 2106.6.

1805.5.6 Foundation wall drainage. Foundation walls shall be designed to support the weight of the full hydrostatic pressure of undrained backfill unless a drainage system is installed in accordance with Sections 1807.4.2 and 1807.4.3.

1805.5.7 Pier and curtain wall foundations. Except in Seismic Design Categories D, E and F, pier and curtain wall foundations are permitted to be used to support light-frame construction not more than two stories in height, provided the following requirements are met:

1. All load-bearing walls shall be placed on continuous concrete footings bonded integrally with the exterior wall footings.

2. The minimum actual thickness of a load-bearing masonry wall shall not be less than 4 inches (102 mm) nominal or $3^5/_8$ inches (92 mm) actual thickness, and shall be bonded integrally with piers spaced 6 feet (1829 mm) on center (o.c.).

3. Piers shall be constructed in accordance with Chapter 21 and the following:

    3.1. The unsupported height of the masonry piers shall not exceed 10 times their least dimension.

    3.2. Where structural clay tile or hollow concrete masonry units are used for piers supporting beams and girders, the cellular spaces shall be filled solidly with concrete or Type M or S mortar.

    Exception: Unfilled hollow piers are permitted where the unsupported height of the pier is not more than four times its least dimension.

    3.3. Hollow piers shall be capped with 4 inches (102 mm) of solid masonry or concrete or the cavities of the top course shall be filled with concrete or grout.

4. The maximum height of a 4-inch (102 mm) load-bearing masonry foundation wall supporting wood frame walls and floors shall not be more than 4 feet (1219 mm) in height.

5. The unbalanced fill for 4-inch (102 mm) foundation walls shall not exceed 24 inches (610 mm) for solid masonry, nor 12 inches (305 mm) for hollow masonry.

1805.6 Foundation plate or sill bolting. Wood foundation plates or sills shall be bolted or strapped to the foundation or foundation wall as provided in Chapter 23.

1805.7 Designs employing lateral bearing. Designs to resist both axial and lateral loads employing posts or poles as columns embedded in earth or embedded in concrete footings in the earth shall conform to the requirements of Sections 1805.7.1 through 1805.7.3.

1805.7.1 Limitations. The design procedures outlined in this section are subject to the following limitations:

1. The frictional resistance for structural walls and slabs on silts and clays shall be limited to one-half of the normal force imposed on the soil by the weight of the footing or slab.

2. Posts embedded in earth shall not be used to provide lateral support for structural or nonstructural materials such as plaster, masonry or concrete unless bracing is provided that develops the limited deflection required.

Wood poles shall be treated in accordance with AWPA C2 or C4.

1805.7.2 Design criteria. The depth to resist lateral loads shall be determined by the design criteria established in Sections 1805.7.2.1 through 1805.7.2.3, or by other methods approved by the building official.

1805.7.2.1 Nonconstrained. The following formula shall be used in determining the depth of embedment required to resist lateral loads where no constraint is provided at the ground surface, such as rigid floor or rigid ground surface pavement, and where no lateral constraint is provided above the ground surface, such as a structural diaphragm.

$$d = 0.5A\{1 + [1 + (4.36h/A)]^{1/2}\} \qquad \textbf{(Equation 18-1)}$$

where:

$A = 2.34P/S_1 b.$

b = Diameter of round post or footing or diagonal dimension of square post or footing, feet (m).

d = Depth of embedment in earth in feet (m) but not over 12 feet (3658 mm) for purpose of computing lateral pressure.

h = Distance in feet (m) from ground surface to point of application of "P."

P = Applied lateral force in pounds (kN).

S_1 = Allowable lateral soil-bearing pressure as set forth in Section 1804.3 based on a depth of one-third the depth of embedment in pounds per square foot (psf) (kPa).

1805.7.2.2 Constrained. The following formula shall be used to determine the depth of embedment required to resist lateral loads where constraint is provided at the ground surface, such as a rigid floor or pavement.

$$d^2 = 4.25(Ph/S_3 b) \qquad \text{(Equation 18-2)}$$

or alternatively

$$d^2 = 4.25 (M_g/S_3 b) \qquad \text{(Equation 18-3)}$$

where:

M_g = moment in the post at grade, in foot-pounds (kN-m).

S_3 = Allowable lateral soil-bearing pressure as set forth in Section 1804.3 based on a depth equal to the depth of embedment in pounds per square foot (kPa).

1805.7.2.3 Vertical load. The resistance to vertical loads shall be determined by the allowable soil-bearing pressure set forth in Table 1804.2.

1805.7.3 Backfill. The backfill in the annular space around columns not embedded in poured footings shall be by one of the following methods:

1. Backfill shall be of concrete with an ultimate strength of 2,000 psi (13.8 MPa) at 28 days. The hole shall not be less than 4 inches (102 mm) larger than the diameter of the column at its bottom or 4 inches (102 mm) larger than the diagonal dimension of a square or rectangular column.

2. Backfill shall be of clean sand. The sand shall be thoroughly compacted by tamping in layers not more than 8 inches (203 mm) in depth.

3. Backfill shall be of controlled low-strength material (CLSM).

1805.8 Design for expansive soils. Footings or foundations for buildings and structures founded on expansive soils shall be designed in accordance with Section 1805.8.1 or 1805.8.2.

Footing or foundation design need not comply with Section 1805.8.1 or 1805.8.2 where the soil is removed in accordance with Section 1805.8.3, nor where the building official approves stabilization of the soil in accordance with Section 1805.8.4.

1805.8.1 Foundations. Footings or foundations placed on or within the active zone of expansive soils shall be designed to resist differential volume changes and to prevent structural damage to the supported structure. Deflection and racking of the supported structure shall be limited to that which will not interfere with the usability and serviceability of the structure.

Foundations placed below where volume change occurs or below expansive soil shall comply with the following provisions:

1. Foundations extending into or penetrating expansive soils shall be designed to prevent uplift of the supported structure.

2. Foundations penetrating expansive soils shall be designed to resist forces exerted on the foundation due to soil volume changes or shall be isolated from the expansive soil.

1805.8.2 Slab-on-ground foundations. Slab-on-ground, mat or raft foundations on expansive soils shall be designed and constructed in accordance with WRI/CRSI *Design of Slab-on-Ground Foundations* or PTI *Design and Construction of Post-Tensioned Slabs-On-Ground.*

Exception: Slab-on-ground systems that have performed adequately in soil conditions similar to those encountered at the building site are permitted subject to the approval of the building official.

1805.8.3 Removal of expansive soil. Where expansive soil is removed in lieu of designing footings or foundations in accordance with Section 1805.8.1 or 1805.8.2, the soil shall be removed to a depth sufficient to ensure a constant moisture content in the remaining soil. Fill material shall not contain expansive soils and shall comply with Section 1803.5 or 1803.6.

Exception: Expansive soil need not be removed to the depth of constant moisture, provided the confining pressure in the expansive soil created by the fill and supported structure exceeds the swell pressure.

1805.8.4 Stabilization. Where the active zone of expansive soils is stabilized in lieu of designing footings or foundations in accordance with Section 1805.8.1 or 1805.8.2, the soil shall be stabilized by chemical, dewatering, presaturation or equivalent techniques.

1805.9 Seismic requirements. See Section 1910 for additional requirements for footings and foundations of structures assigned to Seismic Design Category C, D, E or F.

For structures assigned to Seismic Design Category D, E or F, provisions of ACI 318, Sections 21.10.1 to 21.10.3, shall apply when not in conflict with the provisions of Section 1805. Concrete shall have a specified compressive strength of not less than 3,000 psi (20.68 MPa) at 28 days.

Exceptions:

1. Group R or U occupancies of light-framed construction and two stories or less in height are permitted to

use concrete with a specified compressive strength of not less than 2,500 psi (17.2 MPa) at 28 days.

2. Detached one- and two-family dwellings of light-frame construction and two stories or less in height are not required to comply with the provisions of ACI 318, Sections 21.10.1 to 21.10.3.

SECTION 1806
RETAINING WALLS

1806.1 General. Retaining walls shall be designed to ensure stability against overturning, sliding, excessive foundation pressure and water uplift. Retaining walls shall be designed for a safety factor of 1.5 against lateral sliding and overturning.

SECTION 1807
DAMPPROOFING AND WATERPROOFING

1807.1 Where required. Walls or portions thereof that retain earth and enclose interior spaces and floors below grade shall be waterproofed and dampproofed in accordance with this section, with the exception of those spaces containing groups other than residential and institutional where such omission is not detrimental to the building or occupancy.

Ventilation for crawl spaces shall comply with Section 1203.4.

1807.1.1 Story above grade. Where a basement is considered a story above grade and the finished ground level adjacent to the basement wall is below the basement floor elevation for 25 percent or more of the perimeter, the floor and walls shall be dampproofed in accordance with Section 1807.2 and a foundation drain shall be installed in accordance with Section 1807.4.2. The foundation drain shall be installed around the portion of the perimeter where the basement floor is below ground level. The provisions of Sections 1802.2.3, 1807.3 and 1807.4.1 shall not apply in this case.

1807.1.2 Under-floor space. The finished ground level of an under-floor space such as a crawl space shall not be located below the bottom of the footings. Where there is evidence that the ground-water table rises to within 6 inches (152 mm) of the ground level at the outside building perimeter, or that the surface water does not readily drain from the building site, the ground level of the under-floor space shall be as high as the outside finished ground level, unless an approved drainage system is provided. The provisions of Sections 1802.2.3, 1807.2, 1807.3 and 1807.4 shall not apply in this case.

1807.1.2.1 Flood hazard areas. For buildings and structures in flood hazard areas as established in Section 1612.3, the finished ground level of an under-floor space such as a crawl space shall be equal to or higher than the outside finished ground level.

Exception: Under-floor spaces of Group R-3 buildings that meet the requirements of FEMA/FIA-TB-11.

1807.1.3 Ground-water control. Where the ground-water table is lowered and maintained at an elevation not less than 6 inches (152 mm) below the bottom of the lowest floor, the floor and walls shall be dampproofed in accordance with Section 1807.2. The design of the system to lower the ground-water table shall be based on accepted principles of engineering that shall consider, but not necessarily be limited to, permeability of the soil, rate at which water enters the drainage system, rated capacity of pumps, head against which pumps are to operate and the rated capacity of the disposal area of the system.

1807.2 Dampproofing required. Where hydrostatic pressure will not occur as determined by Section 1802.2.3, floors and walls for other than wood foundation systems shall be dampproofed in accordance with this section. Wood foundation systems shall be constructed in accordance with AFPA TR7.

1807.2.1 Floors. Dampproofing materials for floors shall be installed between the floor and the base course required by Section 1807.4.1, except where a separate floor is provided above a concrete slab.

Where installed beneath the slab, dampproofing shall consist of not less than 6-mil (0.006 inch; 0.152 mm) polyethylene with joints lapped not less than 6 inches (152 mm), or other approved methods or materials. Where permitted to be installed on top of the slab, dampproofing shall consist of mopped-on bitumen, not less than 4-mil (0.004 inch; 0.102 mm) polyethylene, or other approved methods or materials. Joints in the membrane shall be lapped and sealed in accordance with the manufacturer's installation instructions.

1807.2.2 Walls. Dampproofing materials for walls shall be installed on the exterior surface of the wall, and shall extend from the top of the footing to above ground level.

Dampproofing shall consist of a bituminous material, 3 pounds per square yard (16 N/m²) of acrylic modified cement, $^1/_8$-inch (3.2 mm) coat of surface-bonding mortar complying with ASTM C 887, any of the materials permitted for waterproofing by Section 1807.3.2 or other approved methods or materials.

1807.2.2.1 Surface preparation of walls. Prior to application of dampproofing materials on concrete walls, holes and recesses resulting from the removal of form ties shall be sealed with a bituminous material or other approved methods or materials. Unit masonry walls shall be parged on the exterior surface below ground level with not less than $^3/_8$ inch (9.5 mm) of portland cement mortar. The parging shall be coved at the footing.

Exception: Parging of unit masonry walls is not required where a material is approved for direct application to the masonry.

1807.3 Waterproofing required. Where the ground-water investigation required by Section 1802.2.3 indicates that a hydrostatic pressure condition exists, and the design does not include a ground-water control system as described in Section 1807.1.3, walls and floors shall be waterproofed in accordance with this section.

1807.3.1 Floors. Floors required to be waterproofed shall be of concrete, designed and constructed to withstand the hydrostatic pressures to which the floors will be subjected.

Waterproofing shall be accomplished by placing a membrane of rubberized asphalt, butyl rubber, or not less than 6-mil (0.006 inch; 0.152 mm) polyvinyl chloride with joints lapped not less than 6 inches (152 mm) or other approved materials under the slab. Joints in the membrane shall be lapped and sealed in accordance with the manufacturer's installation instructions.

1807.3.2 Walls. Walls required to be waterproofed shall be of concrete or masonry and shall be designed and constructed to withstand the hydrostatic pressures and other lateral loads to which the walls will be subjected.

Waterproofing shall be applied from the bottom of the wall to not less than 12 inches (305 mm) above the maximum elevation of the ground-water table. The remainder of the wall shall be dampproofed in accordance with Section 1807.2.2. Waterproofing shall consist of two-ply hot-mopped felts, not less than 6-mil (0.006 inch; 0.152 mm) polyvinyl chloride, 40-mil (0.040 inch; 1.02 mm) polymer-modified asphalt, 6-mil (0.006 inch; 0.152 mm) polyethylene or other approved methods or materials capable of bridging nonstructural cracks. Joints in the membrane shall be lapped and sealed in accordance with the manufacturer's installation instructions.

1807.3.2.1 Surface preparation of walls. Prior to the application of waterproofing materials on concrete or masonry walls, the walls shall be prepared in accordance with Section 1807.2.2.1.

1807.3.3 Joints and penetrations. Joints in walls and floors, joints between the wall and floor and penetrations of the wall and floor shall be made water-tight utilizing approved methods and materials.

1807.4 Subsoil drainage system. Where a hydrostatic pressure condition does not exist, dampproofing shall be provided and a base shall be installed under the floor and a drain installed around the foundation perimeter. A subsoil drainage system designed and constructed in accordance with Section 1807.1.3 shall be deemed adequate for lowering the ground-water table.

1807.4.1 Floor base course. Floors of basements, except as provided for in Section 1807.1.1, shall be placed over a floor base course not less than 4 inches (102 mm) in thickness that consists of gravel or crushed stone containing not more than 10 percent of material that passes through a No. 4 (4.75 mm) sieve.

Exception: Where a site is located in well-drained gravel or sand/gravel mixture soils, a floor base course is not required.

1807.4.2 Foundation drain. A drain shall be placed around the perimeter of a foundation that consists of gravel or crushed stone containing not more than 10-percent material that passes through a No. 4 (4.75 mm) sieve. The drain shall extend a minimum of 12 inches (305 mm) beyond the outside edge of the footing. The thickness shall be such that the bottom of the drain is not higher than the bottom of the base under the floor, and that the top of the drain is not less than 6 inches (152 mm) above the top of the footing. The top of the drain shall be covered with an approved filter membrane material. Where a drain tile or perforated pipe is used, the invert of the pipe or tile shall not be higher than the floor elevation. The top of joints or the top of perforations shall be protected with an approved filter membrane material. The pipe or tile shall be placed on not less than 2 inches (51 mm) of gravel or crushed stone complying with Section 1807.4.1, and shall be covered with not less than 6 inches (152 mm) of the same material.

1807.4.3 Drainage discharge. The floor base and foundation perimeter drain shall discharge by gravity or mechanical means into an approved drainage system that complies with the *International Plumbing Code*.

Exception: Where a site is located in well-drained gravel or sand/gravel mixture soils, a dedicated drainage system is not required.

SECTION 1808
PIER AND PILE FOUNDATIONS

1808.1 Definitions. The following words and terms shall, for the purposes of this section, have the meanings shown herein.

FLEXURAL LENGTH. Flexural length is the length of the pile from the first point of zero lateral deflection to the underside of the pile cap or grade beam.

PIER FOUNDATIONS. Pier foundations consist of isolated masonry or cast-in-place concrete structural elements extending into firm materials. Piers are relatively short in comparison to their width, with lengths less than or equal to 12 times the least horizontal dimension of the pier. Piers derive their load-carrying capacity through skin friction, through end bearing, or a combination of both.

Belled piers. Belled piers are cast-in-place concrete piers constructed with a base that is larger than the diameter of the remainder of the pier. The belled base is designed to increase the load-bearing area of the pier in end bearing.

PILE FOUNDATIONS. Pile foundations consist of concrete, wood or steel structural elements either driven into the ground or cast in place. Piles are relatively slender in comparison to their length, with lengths exceeding 12 times the least horizontal dimension. Piles derive their load-carrying capacity through skin friction, through end bearing, or a combination of both.

Augered uncased piles. Augered uncased piles are constructed by depositing concrete into an uncased augered hole, either during or after the withdrawal of the auger.

Caisson piles. Caisson piles are cast-in-place concrete piles extending into bedrock. The upper portion of a caisson pile consists of a cased pile that extends to the bedrock. The lower portion of the caisson pile consists of an uncased socket drilled into the bedrock.

Concrete-filled steel pipe and tube piles. Concrete-filled steel pipe and tube piles are constructed by driving a steel pipe or tube section into the soil and filling the pipe or tube section with concrete. The steel pipe or tube section is left in place during and after the deposition of the concrete.

Driven uncased piles. Driven uncased piles are constructed by driving a steel shell into the soil to shore an unexcavated hole that is later filled with concrete. The steel casing is lifted out of the hole during the deposition of the concrete.

Enlarged base piles. Enlarged base piles are cast-in-place concrete piles constructed with a base that is larger than the diameter of the remainder of the pile. The enlarged base is designed to increase the load-bearing area of the pile in end bearing.

Steel-cased piles. Steel-cased piles are constructed by driving a steel shell into the soil to shore an unexcavated hole. The steel casing is left permanently in place and filled with concrete.

1808.2 Piers and piles—general requirements.

1808.2.1 Design. Piles are permitted to be designed in accordance with provisions for piers in Section 1808 and Sections 1812.3 through 1812.10 where either of the following conditions exists, subject to the approval of the building official:

1. Group R-3 and U occupancies not exceeding two stories of light-frame construction, or

2. Where the surrounding foundation materials furnish adequate lateral support for the pile.

1808.2.2 General. Pier and pile foundations shall be designed and installed on the basis of a foundation investigation as defined in Section 1802, unless sufficient data upon which to base the design and installation is available.

The investigation and report provisions of Section 1802 shall be expanded to include, but not be limited to, the following:

1. Recommended pier or pile types and installed capacities.

2. Recommended center-to-center spacing of piers or piles.

3. Driving criteria.

4. Installation procedures.

5. Field inspection and reporting procedures (to include procedures for verification of the installed bearing capacity where required).

6. Pier or pile load test requirements.

7. Durability of pier or pile materials.

8. Designation of bearing stratum or strata.

9. Reductions for group action, where necessary.

1808.2.3 Special types of piles. The use of types of piles not specifically mentioned herein is permitted, subject to the approval of the building official, upon the submission of acceptable test data, calculations and other information relating to the structural properties and load capacity of such piles. The allowable stresses shall not in any case exceed the limitations specified herein.

1808.2.4 Pile caps. Pile caps shall be of reinforced concrete, and shall include all elements to which piles are connected, including grade beams and mats. The soil immediately below the pile cap shall not be considered as carrying any vertical load. The tops of piles shall be embedded not less than 3 inches (76 mm) into pile caps and the caps shall extend at least 4 inches (102 mm) beyond the edges of piles. The tops of piles shall be cut back to sound material before capping.

1808.2.5 Stability. Piers or piles shall be braced to provide lateral stability in all directions. Three or more piles connected by a rigid cap shall be considered braced, provided that the piles are located in radial directions from the centroid of the group not less than 60 degrees (1 rad) apart. A two-pile group in a rigid cap shall be considered to be braced along the axis connecting the two piles. Methods used to brace piers or piles shall be subject to the approval of the building official.

Piles supporting walls shall be driven alternately in lines spaced at least 1 foot (305 mm) apart and located symmetrically under the center of gravity of the wall load carried, unless effective measures are taken to provide for eccentricity and lateral forces, or the wall piles are adequately braced to provide for lateral stability. A single row of piles without lateral bracing is permitted for one- and two-family dwellings and lightweight construction not exceeding two stories or 35 feet (10 668 mm) in height, provided the centers of the piles are located within the width of the foundation wall.

1808.2.6 Structural integrity. Piers or piles shall be installed in such a manner and sequence as to prevent distortion or damage to piles being installed or already in place to the extent that such distortion or damage affects the structural integrity of the piles.

1808.2.7 Splices. Splices shall be constructed so as to provide and maintain true alignment and position of the component parts of the pier or pile during installation and subsequent thereto and shall be of adequate strength to transmit the vertical and lateral loads and moments occurring at the location of the splice during driving and under service loading. Splices shall develop not less than 50 percent of the least capacity of the pier or pile in bending. In addition, splices occurring in the upper 10 feet (3048 mm) of the embedded portion of the pier or pile shall be capable of resisting at allowable working stresses the moment and shear that would result from an assumed eccentricity of the pier or pile load of 3 inches (76 mm), or the pier or pile shall be braced in accordance with Section 1808.2.5 to other piers or piles that do not have splices in the upper 10 feet (3048 mm) of embedment.

1808.2.8 Allowable pier or pile loads.

1808.2.8.1 Determination of allowable loads. The allowable axial and lateral loads on piers or piles shall be determined by an approved formula, load tests or method of analysis.

1808.2.8.2 Driving criteria. The allowable compressive load on any pile where determined by the application of an approved driving formula shall not exceed 40 tons (356 kN). For allowable loads above 40 tons (356 kN), the wave equation method of analysis shall be used to estimate pile driveability of both driving stresses and net displacement per blow at the ultimate load. Allowable loads shall be verified by load tests in accordance with Section 1808.2.8.3. The formula or wave equation load shall be determined for gravity-drop or power-actuated hammers and the hammer energy used shall be the maximum consistent with the size, strength and weight of the driven piles. The use of a follower is permitted only with the approval of the building official. The introduction of fresh hammer cushion or pile cushion material just prior to final penetration is not permitted.

1808.2.8.3 Load tests. Where design compressive loads per pier or pile are greater than those permitted by Section 1808.2.10, or where the design load for any pier or pile foundation is in doubt, control test piers or piles shall be tested in accordance with ASTM D 1143 or ASTM D 4945. At least one pier or pile shall be test loaded in each area of uniform subsoil conditions. Where required by the building official, additional piers or piles shall be load tested where necessary to establish the safe design capacity. The resulting allowable loads shall not be more than one-half of the ultimate load capacity of the test pier or pile as assessed by one of the published methods listed in Section 1808.2.8.3.1 with consideration for the test type, duration and subsoil. The ultimate load capacity shall be determined by a registered design professional, but shall be no greater than two times the test load that produces a settlement of 0.3 inches (7.6 mm). In subsequent installation of the balance of foundation piles, all piles shall be deemed to have a supporting capacity equal to the control pile where such piles are of the same type, size and relative length as the test pile; are installed using the same or comparable methods and equipment as the test pile; are installed in similar subsoil conditions as the test pile; and, for driven piles, where the rate of penetration (e.g., net displacement per blow) of such piles is equal to or less than that of the test pile through a comparable driving distance.

1808.2.8.3.1 Load test evaluation. It shall be permitted to evaluate pile load tests with any of the following methods:

1. Davisson Offset Limit.
2. Brinch-Hansen 90% Criterion.
3. Chin-Konder Extrapolation.

4. Other methods approved by the building official.

1808.2.8.4 Allowable frictional resistance. The assumed frictional resistance developed by any pier or uncased cast-in-place pile shall not exceed one-sixth of the bearing value of the soil material at minimum depth as set forth in Table 1804.2, up to a maximum of 500 psf (24 kPa), unless a greater value is allowed by the building official after a soil investigation as specified in Section 1802 is submitted. Frictional resistance and bearing resistance shall not be assumed to act simultaneously unless recommended by a soil investigation as specified in Section 1802.

1808.2.8.5 Uplift capacity. Where required by the design, the uplift capacity of a single pier or pile shall be determined by an approved method of analysis based on a minimum factor of safety of three or by load tests conducted in accordance with ASTM D 3689. The maximum allowable uplift load shall not exceed the ultimate load capacity as determined in Section 1808.2.8.3 divided by a factor of safety of two. For pile groups subjected to uplift, the allowable working uplift load for the group shall be the lesser of:

1. The proposed individual pile uplift working load times the number of piles in the group.
2. Two-thirds of the effective weight of the pile group and the soil contained within a block defined by the perimeter of the group and the length of the pile.

1808.2.8.6 Load-bearing capacity. Piers, individual piles and groups of piles shall develop ultimate load capacities of at least twice the design working loads in the designated load-bearing layers. Analysis shall show that no soil layer underlying the designated load-bearing layers causes the load-bearing capacity safety factor to be less than two.

1808.2.8.7 Bent piers or piles. The load-bearing capacity of piers or piles discovered to have a sharp or sweeping bend shall be determined by an approved method of analysis or by load testing a representative pier or pile.

1808.2.8.8 Overloads on piers or piles. The maximum compressive load on any pier or pile due to mislocation shall not exceed 110 percent of the allowable design load.

1808.2.9 Lateral support.

1808.2.9.1 General. Any soil other than fluid soil shall be deemed to afford sufficient lateral support to the pier or pile to prevent buckling and to permit the design of the pier or pile in accordance with accepted engineering practice and the applicable provisions of this code.

1808.2.9.2 Unbraced piles. Piles standing unbraced in air, water or in fluid soils shall be designed as columns in accordance with the provisions of this code. Such piles driven into firm ground can be considered fixed and laterally supported at 5 feet (1524 mm) below the ground surface and in soft material at 10 feet (3048 mm) below the ground surface unless otherwise prescribed by the building official after a foundation investigation by an approved agency.

1808.2.9.3 Allowable lateral load. Where required by the design, the lateral load capacity of a pier, a single pile or a pile group shall be determined by an approved method of analysis or by lateral load tests to at least twice the proposed design working load. The resulting allowable load shall not be more than one-half of that test load that produces a gross lateral movement of 1 inch (25 mm) at the ground surface.

1808.2.10 Use of higher allowable pier or pile stresses. Allowable stresses greater than those specified for piers or for each pile type in Sections 1808 and 1809 are permitted where supporting data justifying such higher stresses is filed with the building official. Such substantiating data shall include:

1. A soils investigation in accordance with Section 1802.

2. Pier or pile load tests in accordance with Section 1808.2.8.3, regardless of the load supported by the pier or pile.

The design and installation of the pier or pile foundation shall be under the direct supervision of a registered design professional knowledgeable in the field of soil mechanics and pier or pile foundations who shall certify to the building official that the piers or piles as installed satisfy the design criteria.

1808.2.11 Piles in subsiding areas. Where piles are driven through subsiding fills or other subsiding strata and derive support from underlying firmer materials, consideration shall be given to the downward frictional forces that may be imposed on the piles by the subsiding upper strata.

Where the influence of subsiding fills is considered as imposing loads on the pile, the allowable stresses specified in this chapter are permitted to be increased where satisfactory substantiating data are submitted.

1808.2.12 Settlement analysis. The settlement of piers, individual piles or groups of piles shall be estimated based on approved methods of analysis. The predicted settlement shall cause neither harmful distortion of, nor instability in, the structure, nor cause any stresses to exceed allowable values.

1808.2.13 Preexcavation. The use of jetting, augering or other methods of preexcavation shall be subject to the approval of the building official. Where permitted, preexcavation shall be carried out in the same manner as used for piers or piles subject to load tests and in such a manner that will not impair the carrying capacity of the piers or piles already in place or damage adjacent structures. Pile tips shall be driven below the preexcavated depth until the required resistance or penetration is obtained.

1808.2.14 Installation sequence. Piles shall be installed in such sequence as to avoid compacting the surrounding soil to the extent that other piles cannot be installed properly, and to prevent ground movements that are capable of damaging adjacent structures.

1808.2.15 Use of vibratory drivers. Vibratory drivers shall only be used to install piles where the pile load capacity is verified by load tests in accordance with Section 1808.2.8.3. The installation of production piles shall be controlled according to power consumption, rate of penetration or other approved means that ensure pile capacities equal or exceed those of the test piles.

1808.2.16 Pile driveability. Pile cross sections shall be of sufficient size and strength to withstand driving stresses without damage to the pile, and to provide sufficient stiffness to transmit the required driving forces.

1808.2.17 Protection of pile materials. Where boring records or site conditions indicate possible deleterious action on pier or pile materials because of soil constituents, changing water levels or other factors, the pier or pile materials shall be adequately protected by materials, methods or processes approved by the building official. Protective materials shall be applied to the piles so as not to be rendered ineffective by driving. The effectiveness of such protective measures for the particular purpose shall have been thoroughly established by satisfactory service records or other evidence.

1808.2.18 Use of existing piers or piles. Piers or piles left in place where a structure has been demolished shall not be used for the support of new construction unless satisfactory evidence is submitted to the building official, which indicates that the piers or piles are sound and meet the requirements of this code. Such piers or piles shall be load tested or redriven to verify their capacities. The design load applied to such piers or piles shall be the lowest allowable load as determined by tests or redriving data.

1808.2.19 Heaved piles. Piles that have heaved during the driving of adjacent piles shall be redriven as necessary to develop the required capacity and penetration, or the capacity of the pile shall be verified by load tests in accordance with Section 1808.2.8.3.

1808.2.20 Identification. Pier or pile materials shall be identified for conformity to the specified grade with this identity maintained continuously from the point of manufacture to the point of installation or shall be tested by an approved agency to determine conformity to the specified grade. The approved agency shall furnish an affidavit of compliance to the building official.

1808.2.21 Pier or pile location plan. A plan showing the location and designation of piers or piles by an identification system shall be filed with the building official prior to installation of such piers or piles. Detailed records for piers or individual piles shall bear an identification corresponding to that shown on the plan.

1808.2.22 Special inspection. Special inspections in accordance with Sections 1704.8 and 1704.9 shall be provided for piles and piers, respectively.

1808.2.23 Seismic design of piers or piles.

1808.2.23.1 Seismic Design Category C. Where a structure is assigned to Seismic Design Category C in accordance with Section 1616, the following shall apply. Individual pile caps, piers or piles shall be interconnected by ties. Ties shall be capable of carrying, in tension and compression, a force equal to the product of the larger pile cap or column load times the seismic coefficient, S_{DS}, divided by 10 unless it can be demonstrated that equivalent restraint is provided by reinforced concrete beams within slabs on grade or reinforced concrete slabs on grade or confinement by competent rock, hard cohesive soils or very dense granular soils.

Exception: Piers supporting foundation walls, isolated interior posts detailed so the pier is not subject to lateral loads, lightly loaded exterior decks and patios, of Group R-3 and U occupancies not exceeding two stories of light-frame construction, are not subject to interconnection if it can be shown the soils are of adequate stiffness, subject to the approval of the building official.

1808.2.23.1.1 Connection to pile cap. Concrete piles and concrete-filled steel pipe piles shall be connected to the pile cap by embedding the pile reinforcement or field-placed dowels anchored in the concrete pile in the pile cap for a distance equal to the development length. For deformed bars, the development length is the full development length for compression or tension, in the case of uplift, without reduction in length for excess area. Alternative measures for laterally confining concrete and maintaining toughness and ductile-like behavior at the top of the pile will be permitted provided the design is such that any hinging occurs in the confined region.

Ends of hoops, spirals and ties shall be terminated with seismic hooks, as defined in Section 21.1 of ACI 318, turned into the confined concrete core. The minimum transverse steel ratio for confinement shall not be less than one-half of that required for columns.

For resistance to uplift forces, anchorage of steel pipe (round HSS sections), concrete-filled steel pipe or H-piles to the pile cap shall be made by means other than concrete bond to the bare steel section.

Exception: Anchorage of concrete-filled steel pipe piles is permitted to be accomplished using deformed bars developed into the concrete portion of the pile.

Splices of pile segments shall develop the full strength of the pile, but the splice need not develop the nominal strength of the pile in tension, shear and bending when it has been designed to resist axial and shear forces and moments from the load combinations of Section 1605.4.

1808.2.23.1.2 Design details. Pier or pile moments, shears and lateral deflections used for design shall be established considering the nonlinear interaction of the shaft and soil, as recommended by a registered design professional. Where the ratio of the depth of embedment of the pile-to-pile diameter or width is less than or equal to six, the pile may be assumed to be rigid.

Pile group effects from soil on lateral pile nominal strength shall be included where pile center-to-center spacing in the direction of lateral force is less than eight pile diameters. Pile group effects on vertical nominal strength shall be included where pile center-to-center spacing is less than three pile diameters. The pile uplift soil nominal strength shall be taken as the pile uplift strength as limited by the frictional force developed between the soil and the pile.

Where a minimum length for reinforcement or the extent of closely spaced confinement reinforcement is specified at the top of the pier or pile, provisions shall be made so that those specified lengths or extents are maintained after pier or pile cutoff.

1808.2.23.2 Seismic Design Category D, E or F. Where a structure is assigned to Seismic Design Category D, E or F in accordance with Section 1616, the requirements for Seismic Design Category C given in Section 1808.2.23.1 shall be met, in addition to the following. Provisions of ACI 318, Section 21.10.4, shall apply when not in conflict with the provisions of Sections 1808 through 1812. Concrete shall have a specified compressive strength of not less than 3,000 psi (20.68 MPa) at 28 days.

Exceptions:

1. Group R or U occupancies of light-framed construction and two stories or less in height are permitted to use concrete with a specified compressive strength of not less than 2,500 psi (17.2 MPa) at 28 days.

2. Detached one- and two-family dwellings of light-frame construction and two stories or less in height are not required to comply with the provisions of ACI 318, Section 21.10.4.

3. Section 21.10.4.4(a) of ACI 318 need not apply to concrete piles.

1808.2.23.2.1 Design details for piers, piles and grade beams. Piers or piles shall be designed and constructed to withstand maximum imposed curvatures from earthquake ground motions and structure response. Curvatures shall include free-field soil strains modified for soil-pile-structure interaction coupled with pier or pile deformations induced by lateral pier or pile resistance to structure seismic forces. Concrete piers or piles on Site Class E or F sites, as determined in Section 1615.1.1, shall be designed and detailed in accordance with Sections 21.4.4.1, 21.4.4.2 and 21.4.4.3 of ACI 318 within seven pile diameters of the pile cap and the interfaces of soft to medium stiff clay or liquefiable strata. For precast prestressed concrete piles, detailing provisions as given in Sections 1809.2.3.2.1 and 1809.2.3.2.2 shall apply.

Grade beams shall be designed as beams in accordance with ACI 318, Chapter 21. When grade beams have the capacity to resist the forces from the load combinations in Section 1605.4, they need not conform to ACI 318, Chapter 21.

1808.2.23.2.2 Connection to pile cap. For piles required to resist uplift forces or provide rotational restraint, design of anchorage of piles into the pile cap shall be provided considering the combined effect of axial forces due to uplift and bending moments due to fixity to the pile cap. Anchorage shall develop a minimum of 25 percent of the strength of the pile in tension. Anchorage into the pile cap shall be capable of developing the following:

1. In the case of uplift, the lesser of the nominal tensile strength of the longitudinal reinforcement in a concrete pile, or the nominal tensile strength of a steel pile, or the pile uplift soil nominal strength factored by 1.3 or the axial tension force resulting from the load combinations of Section 1605.4.

2. In the case of rotational restraint, the lesser of the axial and shear forces, and moments resulting from the load combinations of Section 1605.4 or development of the full axial, bending and shear nominal strength of the pile.

1808.2.23.2.3 Flexural strength. Where the vertical lateral-force-resisting elements are columns, the grade beam or pile cap flexural strengths shall exceed the column flexural strength.

The connection between batter piles and grade beams or pile caps shall be designed to resist the nominal strength of the pile acting as a short column. Batter piles and their connection shall be capable of resisting forces and moments from the load combinations of Section 1605.4.

SECTION 1809
DRIVEN PILE FOUNDATIONS

1809.1 Timber piles. Timber piles shall be designed in accordance with the AFPA NDS.

1809.1.1 Materials. Round timber piles shall conform to ASTM D 25. Sawn timber piles shall conform to DOC PS-20.

1809.1.2 Preservative treatment. Timber piles used to support permanent structures shall be treated in accordance with this section unless it is established that the tops of the untreated timber piles will be below the lowest ground-water level assumed to exist during the life of the structure. Preservative and minimum final retention shall be in accordance with AWPA C3 for round timber piles and AWPA C24 for sawn timber piles. Preservative-treated timber piles shall be subject to a quality control program administered by an approved agency. Pile cutoffs shall be treated in accordance with AWPA M4.

1809.1.3 End-supported piles. Any sudden decrease in driving resistance of an end-supported timber pile shall be investigated with regard to the possibility of damage. If the sudden decrease in driving resistance cannot be correlated to load-bearing data, the pile shall be removed for inspection or rejected.

1809.2 Precast concrete piles.

1809.2.1 General. The materials, reinforcement and installation of precast concrete piles shall conform to Sections 1809.2.1.1 through 1809.2.1.4.

1809.2.1.1 Design and manufacture. Piles shall be designed and manufactured in accordance with accepted engineering practice to resist all stresses induced by handling, driving and service loads.

1809.2.1.2 Minimum dimension. The minimum lateral dimension shall be 8 inches (203 mm). Corners of square piles shall be chamfered.

1809.2.1.3 Reinforcement. Longitudinal steel shall be arranged in a symmetrical pattern and be laterally tied with steel ties or wire spiral spaced not more than 4 inches (102 mm) apart, center to center, for a distance of 2 feet (610 mm) from the ends of the pile; and not more than 6 inches (152 mm) elsewhere except that at the ends of each pile, the first five ties or spirals shall be spaced 1 inch (25 mm) center to center. The gage of ties and spirals shall be as follows:

For piles having a diameter of 16 inches (406 mm) or less, wire shall not be smaller than 0.22 inch (5.6 mm) (No. 5 gage).

For piles having a diameter of more than 16 inches (406 mm) and less than 20 inches (508 mm), wire shall not be smaller than 0.238 inch (6 mm) (No. 4 gage).

For piles having a diameter of 20 inches (508 mm) and larger, wire shall not be smaller than $^1/_4$ inch (6.4 mm) round or 0.259 inch (6.6 mm) (No. 3 gage).

1809.2.1.4 Installation. Piles shall be handled and driven so as not to cause injury or overstressing, which affects durability or strength.

1809.2.2 Precast nonprestressed piles. Precast nonprestressed concrete piles shall conform to Sections 1809.2.2.1 through 1809.2.2.5.

1809.2.2.1 Materials. Concrete shall have a 28-day specified compressive strength (f'_c) of not less than 3,000 psi (20.68 MPa).

1809.2.2.2 Minimum reinforcement. The minimum amount of longitudinal reinforcement shall be 0.8 percent of the concrete section and shall consist of at least four bars.

1809.2.2.2.1 Seismic reinforcement in Seismic Design Category C. Where a structure is assigned to Seismic Design Category C in accordance with Section 1616, the following shall apply. Longitudinal reinforcement with a minimum steel ratio of 0.01 shall be provided throughout the length of precast concrete piles. Within three pile diameters of the bottom of the pile cap, the longitudinal reinforcement shall be confined with closed ties or spirals of a minimum $^3/_8$ inch (9.5 mm) diameter. Ties or spirals shall be provided at a maximum spacing of eight times the diameter of the smallest longitudinal bar, not to exceed 6 inches (152 mm). Throughout the remainder of the pile, the closed ties or spirals shall have a maximum spacing of 16 times the smallest longitudinal-bar diameter, not to exceed 8 inches (203 mm).

1809.2.2.2.2 Seismic reinforcement in Seismic Design Category D, E or F. Where a structure is assigned to Seismic Design Category D, E or F in accordance with Section 1616, the requirements for Seismic Design Category C in Section 1809.2.2.2.1 shall apply except as modified by this section. Transverse confinement reinforcement consisting of closed ties or equivalent spirals shall be provided in accordance with Sections 21.4.4.1, 21.4.4.2 and 21.4.4.3 of ACI 318 within three pile diameters of the bottom of the pile cap. For other than Site Class E or F, or liquefiable sites and where spirals are used as the transverse reinforcement, it shall be permitted to use a volumetric ratio of spiral reinforcement of not less than one-half that required by Section 21.4.4.1(a) of ACI 318.

1809.2.2.3 Allowable stresses. The allowable compressive stress in the concrete shall not exceed 33 percent of the 28-day specified compressive strength (f'_c) applied to the gross cross-sectional area of the pile. The allowable compressive stress in the reinforcing steel shall not exceed 40 percent of the yield strength of the steel (f_y) or a maximum of 30,000 psi (207 MPa). The allowable tensile stress in the reinforcing steel shall not exceed 50 percent of the yield strength of the steel (f_y) or a maximum of 24,000 psi (165 MPa).

1809.2.2.4 Installation. A precast concrete pile shall not be driven before the concrete has attained a compressive strength of at least 75 percent of the 28-day specified compressive strength (f'_c), but not less than the strength sufficient to withstand handling and driving forces.

1809.2.2.5 Concrete cover. Reinforcement for piles that are not manufactured under plant conditions shall have a concrete cover of not less than 2 inches (51 mm).

Reinforcement for piles manufactured under plant control conditions shall have a concrete cover of not less than $1^1/_4$ inches (32 mm) for No. 5 bars and smaller, and not less than $1^1/_2$ inches (38 mm) for No. 6 through No. 11 bars except that longitudinal bars spaced less than $1^1/_2$ inches (38 mm) clear distance apart shall be considered bundled bars for which the minimum concrete cover shall be equal to that for the equivalent diameter of the bundled bars.

Reinforcement for piles exposed to seawater shall have a concrete cover of not less than 3 inches (76 mm).

1809.2.3 Precast prestressed piles. Precast prestressed concrete piles shall conform to the requirements of Sections 1809.2.3.1 through 1809.2.3.5.

1809.2.3.1 Materials. Prestressing steel shall conform to ASTM A 416. Concrete shall have a 28-day specified compressive strength (f'_c) of not less than 5,000 psi (34.48 MPa).

1809.2.3.2 Design. Precast prestressed piles shall be designed to resist stresses induced by handling and driving as well as by loads. The effective prestress in the pile shall not be less than 400 psi (2.76 MPa) for piles up to 30 feet (9144 mm) in length, 550 psi (3.79 MPa) for piles up to 50 feet (15 240 mm) in length and 700 psi (4.83 MPa) for piles greater than 50 feet (15 240 mm) in length.

Effective prestress shall be based on an assumed loss of 30,000 psi (207 MPa) in the prestressing steel. The tensile stress in the prestressing steel shall not exceed the values specified in ACI 318.

1809.2.3.2.1 Design in Seismic Design Category C. Where a structure is assigned to Seismic Design Category C in accordance with Section 1616, the following shall apply. The minimum volumetric ratio of spiral reinforcement shall not be less than 0.007 or the amount required by the following formula for the upper 20 feet (6096 mm) of the pile.

$$\rho_s = 0.12 f'_c / f_{yh} \qquad \textbf{(Equation 18-4)}$$

where:

f'_c = Specified compressive strength of concrete, psi (MPa)

f_{yh} = Yield strength of spiral reinforcement $\leq$ 85,000 psi (586 MPa).

ρ_s = Spiral reinforcement index (vol. spiral/vol. core).

At least one-half the volumetric ratio required by Equation 18-4 shall be provided below the upper 20 feet (6096 mm) of the pile.

The pile cap connection by means of dowels as indicated in Section 1808.2.23.1 is permitted. Pile cap connection by means of developing pile reinforcing strand is permitted provided that the pile reinforcing strand results in a ductile connection.

1809.2.3.2.2 Design in Seismic Design Category D, E or F. Where a structure is assigned to Seismic Design Category D, E or F in accordance with Section 1616, the requirements for Seismic Design Category C in Section 1809.2.3.2.1 shall be met, in addition to the following:

1. Requirements in ACI 318, Chapter 21, need not apply, unless specifically referenced.

2. Where the total pile length in the soil is 35 feet (10 668 mm) or less, the lateral transverse reinforcement in the ductile region shall occur through the length of the pile. Where the pile length exceeds 35 feet (10 668 mm), the ductile pile region shall be taken as the greater of 35 feet (10 668 mm) or the distance from the underside of the pile cap to the point of zero curvature plus three times the least pile dimension.

3. In the ductile region, the center-to-center spacing of the spirals or hoop reinforcement shall not exceed one-fifth of the least pile dimension, six times the diameter of the longtitudinal strand, or 8 inches (203 mm), whichever is smaller.

4. Circular spiral reinforcement shall be spliced by lapping one full turn and bending the end of the spiral to a 90-degree hook or by use of a mechanical or welded splice complying with Sec. 12.14.3 of ACI 318.

5. Where the transverse reinforcement consists of circular spirals, the volumetric ratio of spiral transverse reinforcement in the ductile region shall comply with the following:

$$\rho_s = 0.25(f'_c/f_{yh})(A_g/A_{ch} - 1.0)[0.5 + 1.4P/(f'_c A_g)]$$

(Equation 18-5)

but not less than:

$$\rho_s = 0.12(f'_c/f_{yh})[0.5 + 1.4P/(f'_c A_g)]$$

(Equation 18-6)

and need not exceed:

$$\rho_s = 0.021 \qquad \textbf{(Equation 18-7)}$$

where:

A_g = Pile cross-sectional area, square inches (mm²).

A_{ch} = Core area defined by spiral outside diameter, square inches (mm²).

f'_c = Specified compressive strength of concrete, psi (MPa)

f_{yh} = Yield strength of spiral reinforcement ≤ 85,000 psi (586 MPa).

P = Axial load on pile, pounds (kN), as determined from Equations 16-5 and 16-6.

ρ_s = Volumetric ratio (vol. spiral/ vol. core).

This required amount of spiral reinforcement is permitted to be obtained by providing an inner and outer spiral.

6. When transverse reinforcement consists of rectangular hoops and cross ties, the total cross-sectional area of lateral transverse reinforcement in the ductile region with spacings, and perpendicular to dimension, h_c, shall conform to:

$$A_{sh} = 0.3sh_c\,(f'_c/f_{yh})(A_g/A_{ch} - 1.0)[0.5 + 1.4P/(f'_c A_g)]$$

(Equation 18-8)

but not less than:

$$A_{sh} = 0.12sh_c\,(f'_c/f_{yh})[0.5 + 1.4P/(f'_c A_g)]$$

(Equation 18-9)

where:

f_{yh} = ≤ 70,000 psi (483 MPa).

h_c = Cross-sectional dimension of pile core measured center to center of hoop reinforcement, inch (mm).

s = Spacing of transverse reinforcement measured along length of pile, inch (mm).

A_{sh} = Cross-sectional area of tranverse reinforcement, square inches (mm²)

f'_c = Specified compressive strength of concrete, psi (MPa)

The hoops and cross ties shall be equivalent to deformed bars not less than No. 3 in size. Rectangular hoop ends shall terminate at a corner with seismic hooks.

Outside of the length of the pile requiring transverse confinement reinforcing, the spiral or hoop reinforcing with a volumetric ratio not less than one-half of that required for transverse confinement reinforcing shall be provided.

1809.2.3.3 Allowable stresses. The maximum allowable design compressive stress, f'_c, in concrete shall be determined as follows:

$$f_c = 0.33\,f'_c - 0.27f_{pc} \qquad \textbf{(Equation 18-10)}$$

where:

f'_c = The 28-day specified compressive strength of the concrete.

f_{pc} = The effective prestress stress on the gross section.

1809.2.3.4 Installation. A prestressed pile shall not be driven before the concrete has attained a compressive strength of at least 75 percent of the 28-day specified compressive strength (f'_c), but not less than the strength sufficient to withstand handling and driving forces.

1809.2.3.5 Concrete cover. Prestressing steel and pile reinforcement shall have a concrete cover of not less than $1^1/_4$ inches (32 mm) for square piles of 12 inches (305 mm) or smaller size and $1^1/_2$ inches (38 mm) for larger piles, except that for piles exposed to seawater, the minimum protective concrete cover shall not be less than $2^1/_2$ inches (64 mm).

1809.3 Structural steel piles. Structural steel piles shall conform to the requirements of Sections 1809.3.1 through 1809.3.5.

1809.3.1 Materials. Structural steel piles, steel pipe and fully welded steel piles fabricated from plates shall conform to ASTM A 36, ASTM A 252, ASTM A 283, ASTM A 572, ASTM A 588 or ASTM A 913.

1809.3.2 Allowable stresses. The allowable axial stresses shall not exceed 35 percent of the minimum specified yield strength (F_y).

Exception: Where justified in accordance with Section 1808.2.10, the allowable axial stress is permitted to be increased above $0.35F_y$, but shall not exceed $0.5F_y$.

1809.3.3 Dimensions of H-piles. Sections of H-piles shall comply with the following:

1. The flange projections shall not exceed 14 times the minimum thickness of metal in either the flange or the web and the flange widths shall not be less than 80 percent of the depth of the section.

2. The nominal depth in the direction of the web shall not be less than 8 inches (203 mm).

3. Flanges and web shall have a minimum nominal thickness of $3/_8$ inch (9.5 mm).

1809.3.4 Dimensions of steel pipe piles. Steel pipe piles driven open ended shall have a nominal outside diameter of not less than 8 inches (203 mm). The pipe shall have a minimum of 0.34 square inch (219 mm^2) of steel in cross section to resist each 1,000 foot-pounds (1356 N×m) of pile hammer energy or the equivalent strength for steels having a yield strength greater than 35,000 psi (241 MPa). Where pipe wall thickness less than 0.188 inch (4.8 mm) is driven open ended, a suitable cutting shoe shall be provided.

SECTION 1810
CAST-IN-PLACE CONCRETE PILE FOUNDATIONS

1810.1 General. The materials, reinforcement and installation of cast-in-place concrete piles shall conform to Sections 1810.1.1 through 1810.1.3.

1810.1.1 Materials. Concrete shall have a 28-day specified compressive strength (f'_c) of not less than 2,500 psi (17.24 MPa). Where concrete is placed through a funnel hopper at the top of the pile, the concrete mix shall be designed and proportioned so as to produce a cohesive workable mix having a slump of not less than 4 inches (102 mm) and not more than 6 inches (152 mm). Where concrete is to be pumped, the mix design including slump shall be adjusted to produce a pumpable concrete.

1810.1.2 Reinforcement. Except for steel dowels embedded 5 feet (1524 mm) or less in the pile and as provided in Section 1810.3.4, reinforcement where required shall be assembled and tied together and shall be placed in the pile as a unit before the reinforced portion of the pile is filled with concrete except in augered uncased cast-in-place piles. Tied reinforcement in augered uncased cast-in-place piles shall be placed after piles are concreted, while the concrete is still in a semifluid state.

1810.1.2.1 Reinforcement in Seismic Design Category C. Where a structure is assigned to Seismic Design Category C in accordance with Section 1616, the following shall apply. A minimum longitudinal reinforcement ratio of 0.0025 shall be provided for uncased cast-in-place concrete drilled or augered piles, piers or caissons in the top one-third of the pile length, a minimum length of 10 feet (3048 mm) below the ground or that required by analysis, whichever length is greatest. The minimum reinforcement ratio, but no less than that ratio required by rational analysis, shall be continued throughout the flexural length of the pile. There shall be a minimum of four longitudinal bars with closed ties (or equivalent spirals) of a minimum $3/_8$ inch (9 mm) diameter provided at 16-longitudinal-bar diameter maximum spacing. Transverse confinement reinforcing with a maximum spacing of 6 inches (152 mm) or 8-longitudinal-bar diameters, whichever is less, shall be provided within a distance equal to three times the least pile dimension of the bottom of the pile cap.

1810.1.2.2 Reinforcement in Seismic Design Category D, E or F. Where a structure is assigned to Seismic Design Category D, E or F in accordance with Section 1616, the requirements for Seismic Design Category C given above shall be met, in addition to the following. A minimum longitudinal reinforcement ratio of 0.005 shall be provided for uncased cast-in-place drilled or augered concrete piles, piers or caissons in the top one-half of the pile length, a minimum length of 10 feet (3048 mm) below ground or throughout the flexural length of the pile, whichever length is greatest. The flexural length shall be taken as the length of the pile to a point where the concrete section cracking moment strength multiplied by 0.4 exceeds the required moment strength at that point. There shall be a minimum of four longitudinal bars with transverse confinement reinforcing provided in the pile in accordance with Sections 21.4.4.1, 21.4.4.2 and 21.4.4.3 of ACI 318 within three times the least pile dimension of the bottom of the pile cap. It shall be permitted to use a transverse spiral reinforcing ratio of not less than one-half of that required in Section 21.4.4.1(a) of ACI 318 for other than Class E, F or liquefiable sites. Tie spacing throughout the remainder of the concrete section shall not exceed 12-longitudinal-bar diameters, one-half the least dimension of the section, nor 12 inches (305 mm). Ties shall be a minimum of No. 3 bars for piles with a least dimension up to 20 inches (508 mm), and No. 4 bars for larger piles.

1810.1.3 Concrete placement. Concrete shall be placed in such a manner as to ensure the exclusion of any foreign matter and to secure a full-sized shaft. Concrete shall not be placed through water except where a tremie or other approved method is used. When depositing concrete from the top of the pile, the concrete shall not be chuted directly into the pile but shall be poured in a rapid and continuous operation through a funnel hopper centered at the top of the pile.

1810.2 Enlarged base piles. Enlarged base piles shall conform to the requirements of Sections 1810.2.1 through 1810.2.5.

1810.2.1 Materials. The maximum size for coarse aggregate for concrete shall be $^3/_4$ inch (19.1 mm). Concrete to be compacted shall have a zero slump.

1810.2.2 Allowable stresses. The maximum allowable design compressive stress for concrete not placed in a permanent steel casing shall be 25 percent of the 28-day specified compressive strength (f'_c). Where the concrete is placed in a permanent steel casing, the maximum allowable concrete stress shall be 33 percent of the 28-day specified compressive strength (f'_c).

1810.2.3 Installation. Enlarged bases formed either by compacting concrete or driving a precast base shall be formed in or driven into granular soils. Piles shall be constructed in the same manner as successful prototype test piles driven for the project. Pile shafts extending through peat or other organic soil shall be encased in a permanent steel casing. Where a cased shaft is used, the shaft shall be adequately reinforced to resist column action or the annular space around the pile shaft shall be filled sufficiently to reestablish lateral support by the soil. Where pile heave occurs, the pile shall be replaced unless it is demonstrated that the pile is undamaged and capable of carrying twice its design load.

1810.2.4 Load-bearing capacity. Pile load-bearing capacity shall be verified by load tests in accordance with Section 1808.2.8.3.

1810.2.5 Concrete cover. The minimum concrete cover shall be $2^1/_2$ inches (64 mm) for uncased shafts and 1 inch (25 mm) for cased shafts.

1810.3 Drilled or augered uncased piles. Drilled or augered uncased piles shall conform to Sections 1810.3.1 through 1810.3.5.

1810.3.1 Allowable stresses. The allowable design stress in the concrete of drilled uncased piles shall not exceed 33 percent of the 28-day specified compressive strength (f'_c). The allowable design stress in the concrete of augered cast-in-place piles shall not exceed 25 percent of the 28-day specified compressive strength (f'_c). The allowable compressive stress of reinforcement shall not exceed 34 percent of the yield strength of the steel or 25,500 psi (175.8 Mpa).

1810.3.2 Dimensions. The pile length shall not exceed 30 times the average diameter. The minimum diameter shall be 12 inches (305 mm).

> **Exception:** The length of the pile is permitted to exceed 30 times the diameter, provided that the design and installation of the pile foundation are under the direct supervision of a registered design professional knowledgeable in the field of soil mechanics and pile foundations. The registered design professional shall certify to the building official that the piles were installed in compliance with the approved construction documents.

1810.3.3 Installation. Where pile shafts are formed through unstable soils and concrete is placed in an open-drilled hole, a steel liner shall be inserted in the hole prior to placing the concrete. Where the steel liner is withdrawn during concreting, the level of concrete shall be maintained above the bottom of the liner at a sufficient height to offset any hydrostatic or lateral soil pressure.

Where concrete is placed by pumping through a hollow-stem auger, the auger shall be permitted to rotate in a clockwise direction during withdrawal. The auger shall be withdrawn in a continuous manner in increments of about 12 inches (305 mm) each. Concreting pumping pressures shall be measured and maintained high enough at all times to offset hydrostatic and lateral earth pressures. Concrete volumes shall be measured to ensure that the volume of concrete placed in each pile is equal to or greater than the theoretical volume of the hole created by the auger. Where the installation process of any pile is interrupted or a loss of concreting pressure occurs, the pile shall be redrilled to 5 feet (1524 mm) below the elevation of the tip of the auger when the installation was interrupted or concrete pressure was lost and reformed. Augered cast-in-place piles shall not be installed within six pile diameters center to center of a pile filled with concrete less than 12 hours old, unless approved by the building official. If the concrete level in any completed pile drops during installation of an adjacent pile, the pile shall be replaced.

1810.3.4 Reinforcement. For piles installed with a hollow-stem auger, where full-length longitudinal steel reinforcement is placed without lateral ties, the reinforcement shall be placed through ducts in the auger prior to filling the pile with concrete. All pile reinforcement shall have a concrete cover of not less than $2^1/_2$ inches (64 mm).

Exception: Where physical constraints do not allow the placement of the longitudinal reinforcement prior to filling the pile with concrete or where partial-length longitudinal reinforcement is placed without lateral ties, the reinforcement is allowed to be placed after the piles are completely concreted but while concrete is still in a semifluid state.

1810.3.5 Reinforcement in Seismic Design Category C, D, E or F. Where a structure is assigned to Seismic Design Category C, D, E or F in accordance with Section 1616, the corresponding requirements of Sections 1810.1.2.1 and 1810.1.2.2 shall be met.

1810.4 Driven uncased piles. Driven uncased piles shall conform to Sections 1810.4.1 through 1810.4.4.

1810.4.1 Allowable stresses. The allowable design stress in the concrete shall not exceed 25 percent of the 28-day specified compressive strength (f'_c) applied to a cross-sectional area not greater than the inside area of the drive casing or mandrel.

1810.4.2 Dimensions. The pile length shall not exceed 30 times the average diameter. The minimum diameter shall be 12 inches (305 mm).

Exception: The length of the pile is permitted to exceed 30 times the diameter, provided that the design and installation of the pile foundation is under the direct supervision of a registered design professional knowledgeable in the field of soil mechanics and pile foundations. The registered design professional shall certify to the building official that the piles were installed in compliance with the approved design.

1810.4.3 Installation. Piles shall not be driven within six pile diameters center to center in granular soils or within one-half the pile length in cohesive soils of a pile filled with concrete less than 48 hours old unless approved by the building official. If the concrete surface in any completed pile rises or drops, the pile shall be replaced. Piles shall not be installed in soils that could cause pile heave.

1810.4.4 Concrete cover. Pile reinforcement shall have a concrete cover of not less than $2^1/_2$ inches (64 mm), measured from the inside face of the drive casing or mandrel.

1810.5 Steel-cased piles. Steel-cased piles shall comply with the requirements of Sections 1810.5.1 through 1810.5.4.

1810.5.1 Materials. Pile shells or casings shall be of steel and shall be sufficiently strong to resist collapse and sufficiently water tight to exclude any foreign materials during the placing of concrete. Steel shells shall have a sealed tip with a diameter of not less than 8 inches (203 mm).

1810.5.2 Allowable stresses. The allowable design compressive stress in the concrete shall not exceed 33 percent of the 28-day specified compressive strength (f'_c). The allowable concrete compressive stress shall be 0.40 (f'_c) for that portion of the pile meeting the conditions specified in Sections 1810.5.2.1 through 1810.5.2.4.

1810.5.2.1 Shell thickness. The thickness of the steel shell shall not be less than manufacturer's standard gage No. 14 gage (0.068 inch) (1.75 mm) minimum.

1810.5.2.2 Shell type. The shell shall be seamless or provided with seams of strength equal to the basic material and be of a configuration that will provide confinement to the cast-in-place concrete.

1810.5.2.3 Strength. The ratio of steel yield strength (f_y) to 28-day specified compressive strength (f'_c) shall not be less than six.

1810.5.2.4 Diameter. The nominal pile diameter shall not be greater than 16 inches (406 mm).

1810.5.3 Installation. Steel shells shall be mandrel driven their full length in contact with the surrounding soil.

The steel shells shall be driven in such order and with such spacing as to ensure against distortion of or injury to piles already in place. A pile shall not be driven within four and one-half average pile diameters of a pile filled with concrete less than 24 hours old unless approved by the building official. Concrete shall not be placed in steel shells within heave range of driving.

1810.5.4 Reinforcement. Reinforcement shall not be placed within 1 inch (25 mm) of the steel shell. Reinforcing shall be required for unsupported pile lengths or where the pile is designed to resist uplift or unbalanced lateral loads.

1810.5.4.1 Seismic reinforcement. Where a structure is assigned to Seismic Design Category C, D, E or F in accordance with Section 1616, the reinforcement requirements for drilled or augered uncased piles in Section 1810.3.5 shall be met.

Exception: A spiral-welded metal casing of a thickness not less than manufacturer's standard gage No. 14 gage (0.068 inch) is permitted to provide concrete confinement in lieu of the closed ties or equivalent spirals required in an uncased concrete pile. Where used as such, the metal casing shall be protected against possible deleterious action due to soil constituents, changing water levels or other factors indicated by boring records of site conditions.

1810.6 Concrete-filled steel pipe and tube piles. Concrete-filled steel pipe and tube piles shall conform to the requirements of Sections 1810.6.1 through 1810.6.5.

1810.6.1 Materials. Steel pipe and tube sections used for piles shall conform to ASTM A 252 or ASTM A 283. Concrete shall conform to Section 1810.1.1. The maximum coarse aggregate size shall be $3/_4$ inch (19.1 mm).

1810.6.2 Allowable stresses. The allowable design compressive stress in the concrete shall not exceed 33 percent of the 28-day specified compressive strength (f'_c). The allowable design compressive stress in the steel shall not exceed 35 percent of the minimum specified yield strength of the steel (F_y), provided F_y shall not be assumed greater than 36,000 psi (248 MPa) for computational purposes.

Exception: Where justified in accordance with Section 1810.2.10, the allowable stresses are permitted to be increased to 0.50 F_y.

1810.6.3 Minimum dimensions. Piles shall have a nominal outside diameter of not less than 8 inches (203 mm) and a minimum wall thickness in accordance with Section 1809.3.4. For mandrel-driven pipe piles, the minimum wall thickness shall be $^1/_{10}$ inch (2.5 mm).

1810.6.4 Reinforcement. Reinforcement steel shall conform to Section 1810.1.2. Reinforcement shall not be placed within 1 inch (25 mm) of the steel casing.

1810.6.4.1 Seismic reinforcement. Where a structure is assigned to Seismic Design Category C, D, E or F in accordance with Section 1616, the following shall apply. Minimum reinforcement no less than 0.01 times the cross-sectional area of the pile concrete shall be provided in the top of the pile with a length equal to two times the required cap embedment anchorage into the pile cap, but not less than the tension development length of the reinforcement. The wall thickness of the steel pipe shall not be less than $^3/_{16}$ inch (5 mm).

1810.6.5 Placing concrete. The placement of concrete shall conform to Section 1810.1.3.

1810.7 Caisson piles. Caisson piles shall conform to the requirements of Sections 1810.7.1 through 1810.7.6.

1810.7.1 Construction. Caisson piles shall consist of a shaft section of concrete-filled pipe extending to bedrock with an uncased socket drilled into the bedrock and filled with concrete. The caisson pile shall have a full-length structural steel core or a stub core installed in the rock socket and extending into the pipe portion a distance equal to the socket depth.

1810.7.2 Materials. Pipe and steel cores shall conform to the material requirements in Section 1809.3. Pipes shall have a minimum wall thickness of $^3/_8$ inch (9.5 mm) and shall be fitted with a suitable steel-driving shoe welded to the bottom of the pipe. Concrete shall have a 28-day specified compressive strength (f'_c) of not less than 4,000 psi (27.58 MPa). The concrete mix shall be designed and proportioned so as to produce a cohesive workable mix with a slump of 4 inches to 6 inches (102 mm to 152 mm).

1810.7.3 Design. The depth of the rock socket shall be sufficient to develop the full load-bearing capacity of the caisson pile with a minimum safety factor of two, but the depth shall not be less than the outside diameter of the pipe. The design of the rock socket is permitted to be predicated on the sum of the allowable load-bearing pressure on the bottom of the socket plus bond along the sides of the socket. The mini-

mum outside diameter of the caisson pile shall be 18 inches (457 mm), and the diameter of the rock socket shall be approximately equal to the inside diameter of the pile.

1810.7.4 Structural core. The gross cross-sectional area of the structural steel core shall not exceed 25 percent of the gross area of the caisson. The minimum clearance between the structural core and the pipe shall be 2 inches (51 mm). Where cores are to be spliced, the ends shall be milled or ground to provide full contact and shall be full-depth welded.

1810.7.5 Allowable stresses. The allowable design compressive stresses shall not exceed the following: concrete, $0.33 f'_c$; steel pipe, $0.35 F_y$ and structural steel core, $0.50 F_y$.

1810.7.6 Installation. The rock socket and pile shall be thoroughly cleaned of foreign materials before filling with concrete. Steel cores shall be bedded in cement grout at the base of the rock socket. Concrete shall not be placed through water except where a tremie or other approved method is used.

SECTION 1811
COMPOSITE PILES

1811.1 General. Composite piles shall conform to the requirements of Sections 1811.2 through 1811.5.

1811.2 Design. Composite piles consisting of two or more approved pile types shall be designed to meet the conditions of installation.

1811.3 Limitation of load. The maximum allowable load shall be limited by the capacity of the weakest section incorporated in the pile.

1811.4 Splices. Splices between concrete and steel or wood sections shall be designed to prevent separation both before and after the concrete portion has set, and to ensure the alignment and transmission of the total pile load. Splices shall be designed to resist uplift caused by upheaval during driving of adjacent piles, and shall develop the full compressive strength and not less than 50 percent of the tension and bending strength of the weaker section.

1811.5 Seismic reinforcement. Where a structure is assigned to Seismic Design Category C, D, E or F in accordance with Section 1616, the following shall apply. Where concrete and steel are used as part of the pile assembly, the concrete reinforcement shall comply with that given in Sections 1810.1.2.1 and 1810.1.2.2 or the steel section shall comply with Section 1809.3.5 or 1810.6.4.1.

SECTION 1812
PIER FOUNDATIONS

1812.1 General. Isolated and multiple piers used as foundations shall conform to the requirements of Sections 1812.2 through 1812.10, as well as the applicable provisions of Section 1808.2.

1812.2 Lateral dimensions and height. The minimum dimension of isolated piers used as foundations shall be 2 feet (610 mm), and the height shall not exceed 12 times the least horizontal dimension.

1812.3 Materials. Concrete shall have a 28-day specified compressive strength (f'_c) of not less than 2,500 psi (17.24 MPa). Where concrete is placed through a funnel hopper at the top of the pier, the concrete mix shall be designed and proportioned so as to produce a cohesive workable mix having a slump of not less than 4 inches (102 mm) and not more than 6 inches (152 mm). Where concrete is to be pumped, the mix design including slump shall be adjusted to produce a pumpable concrete.

1812.4 Reinforcement. Except for steel dowels embedded 5 feet (1524 mm) or less in the pier, reinforcement where required shall be assembled and tied together and shall be placed in the pier hole as a unit before the reinforced portion of the pier is filled with concrete.

> **Exception:** Reinforcement is permitted to be wet set and the $2^1/_2$- inch (64 mm) concrete cover requirement be reduced to 2 inches (51 mm) for Group R-3 and U occupancies not exceeding two stories of light-frame construction, provided the construction method can be demonstrated to the satisfaction of the building official.

Reinforcement shall conform to the requirements of Sections 1810.1.2.1 and 1810.1.2.2.

Exceptions:

1. Isolated piers supporting posts of Group R-3 and U occupancies not exceeding two stories of light-frame construction are permitted to be reinforced as required by rational analysis but not less than a minimum of one No. 4 bar, without ties or spirals, when detailed so the pier is not subject to lateral loads and the soil is determined to be of adequate stiffness.

2. Isolated piers supporting posts and bracing from decks and patios appurtenant to Group R-3 and U occupancies not exceeding two stories of light-frame construction are permitted to be reinforced as required by rational analysis but not less than one No. 4 bar, without ties or spirals, when the lateral load, E, to the top of the pier does not exceed 200 pounds (890 N) and the soil is determined to be of adequate stiffness.

3. Piers supporting the concrete foundation wall of Group R-3 and U occupancies not exceeding two stories of light-frame construction are permitted to be reinforced as required by rational analysis but not less than two No. 4 bars, without ties or spirals, when it can be shown the concrete pier will not rupture when designed for the maximum seismic load, E_m, and the soil is determined to be of adequate stiffness.

4. Closed ties or spirals where required by Section 1810.1.2.2 are permitted to be limited to the top 3 feet (914 mm) of the piers 10 feet (3048 mm) or less in depth supporting Group R-3 and U occupancies of Seismic Design Category D, not exceeding two stories of light-frame construction.

1812.5 Concrete placement. Concrete shall be placed in such a manner as to ensure the exclusion of any foreign matter and to secure a full-sized shaft. Concrete shall not be placed through water except where a tremie or other approved method is used. When depositing concrete from the top of the pier, the concrete shall not be chuted directly into the pier but shall be poured in a rapid and continuous operation through a funnel hopper centered at the top of the pier.

1812.6 Belled bottoms. Where pier foundations are belled at the bottom, the edge thickness of the bell shall not be less than that required for the edge of footings. Where the sides of the bell slope at an angle less than 60 degrees (1 rad) from the horizontal, the effects of vertical shear shall be considered.

1812.7 Masonry. Where the unsupported height of foundation piers exceeds six times the least dimension, the allowable working stress on piers of unit masonry shall be reduced in accordance with ACI 530/ASCE 5/TMS 402.

1812.8 Concrete. Where adequate lateral support is not provided, and the unsupported height to least lateral dimension does not exceed three, piers of plain concrete shall be designed and constructed as pilasters in accordance with ACI 318. Where the unsupported height to least lateral dimension exceeds three, piers shall be constructed of reinforced concrete, and shall conform to the requirements for columns in ACI 318.

> **Exception:** Where adequate lateral support is furnished by the surrounding materials as defined in Section 1808.2.9, piers are permitted to be constructed of plain or reinforced concrete. The requirements of ACI 318 for bearing on concrete shall apply.

1812.9 Steel shell. Where concrete piers are entirely encased with a circular steel shell, and the area of the shell steel is considered reinforcing steel, the steel shall be protected under the conditions specified in Section 1808.2.17. Horizontal joints in the shell shall be spliced to comply with Section 1808.2.7.

1812.10 Dewatering. Where piers are carried to depths below water level, the piers shall be constructed by a method that will provide accurate preparation and inspection of the bottom, and the depositing or construction of sound concrete or other masonry in the dry.

CHAPTER 19

CONCRETE

SECTION 1901
GENERAL

1901.1 Scope. The provisions of this chapter shall govern the materials, quality control, design and construction of concrete used in structures.

1901.2 Plain and reinforced concrete. Structural concrete shall be designed and constructed in accordance with the requirements of this chapter and ACI 318 as amended in Section 1908 of this code. Except for the provisions of Sections 1904 and 1911, the design and construction of slabs on grade shall not be governed by this chapter unless they transmit vertical loads or lateral forces from other parts of the structure to the soil.

1901.3 Source and applicability. The contents of Sections 1902 through 1907 of this chapter are patterned after, and in general conformity with, the provisions for structural concrete in ACI 318. Where sections within Chapters 2 through 7 of ACI 318 are referenced in other chapters and appendices of ACI 318, the provisions of Sections 1902 through 1907 of this code shall apply.

1901.4 Construction documents. The construction documents for structural concrete construction shall include:

1. The specified compressive strength of concrete at the stated ages or stages of construction for which each concrete element is designed.
2. The specified strength or grade of reinforcement.
3. The size and location of structural elements, reinforcement, and anchors.
4. Provision for dimensional changes resulting from creep, shrinkage and temperature.
5. The magnitude and location of prestressing forces.
6. Anchorage length of reinforcement and location and length of lap splices.
7. Type and location of mechanical and welded splices of reinforcement.
8. Details and location of contraction or isolation joints specified for plain concrete.
9. Minimum concrete compressive strength at time of posttensioning.
10. Stressing sequence for posttensioning tendons.
11. For structures assigned to Seismic Design Category D, E or F, a statement if slab on grade is designed as a structural diaphragm (see Section 21.10.3.4 of ACI 318).

1901.5 Special inspection. The special inspection of concrete elements of buildings and structures and concreting operations shall be as required by Chapter 17.

SECTION 1902
DEFINITIONS

1902.1 General. The following words and terms shall, for the purposes of this chapter and as used elsewhere in this code, have the meanings shown herein.

ADMIXTURE. Material other than water, aggregate or hydraulic cement, used as an ingredient of concrete and added to concrete before or during its mixing to modify its properties.

AGGREGATE. Granular material, such as sand, gravel, crushed stone and iron blast-furnace slag, used with a cementing medium to form a hydraulic cement concrete or mortar.

AGGREGATE, LIGHTWEIGHT. Aggregate with a dry, loose weight of 70 pounds per cubic foot (pcf) (1120 kg/m³) or less.

CEMENTITIOUS MATERIALS. Materials as specified in Section 1903 that have cementing value when used in concrete either by themselves, such as portland cement, blended hydraulic cements and expansive cement, or such materials in combination with fly ash, other raw or calcined natural pozzolans, silica fume, and/or ground granulated blast-furnace slag.

COLUMN. A member with a ratio of height-to-least-lateral dimension exceeding three, used primarily to support axial compressive load.

CONCRETE. A mixture of portland cement or any other hydraulic cement, fine aggregate, coarse aggregate and water, with or without admixtures.

CONCRETE, SPECIFIED COMPRESSIVE STRENGTH OF, (f'_c). The compressive strength of concrete used in design and evaluated in accordance with the provisions of Section 1905, expressed in pounds per square inch (psi) (MPa). Whenever the quantity f'_c is under a radical sign, the square root of the numerical value only is intended, and the result has units of psi (MPa).

CONTRACTION JOINT. Formed, sawed or tooled groove in a concrete structure to create a weakened plane and regulate the location of cracking resulting from the dimensional change of different parts of the structure.

DEFORMED REINFORCEMENT. Deformed reinforcing bars, bar mats, deformed wire, welded plain wire fabric and welded deformed wire fabric conforming to ACI 318, Section 3.5.3.

DUCT. A conduit (plain or corrugated) to accommodate prestressing steel for post-tensioned installation.

EFFECTIVE DEPTH OF SECTION (d). The distance measured from extreme compression fiber to the centroid of tension reinforcement.

CONCRETE

ISOLATION JOINT. A separation between adjoining parts of a concrete structure, usually a vertical plane, at a designed location such as to interfere least with performance of the structure, yet to allow relative movement in three directions and avoid formation of cracks elsewhere in the concrete and through which all or part of the bonded reinforcement is interrupted.

PEDESTAL. An upright compression member with a ratio of unsupported height-to-average-least-lateral dimension of three or less.

PLAIN CONCRETE. Structural concrete with no reinforcement or with less reinforcement than the minimum amount specified for reinforced concrete.

PLAIN REINFORCEMENT. Reinforcement that does not conform to the definition of "Deformed reinforcement" (see ACI 318, Section 3.5.4).

POSTTENSIONING. Method of prestressing in which prestressing steel is tensioned after concrete has hardened.

PRECAST CONCRETE. A structural concrete element cast elsewhere than its final position in the structure.

PRESTRESSED CONCRETE. Structural concrete in which internal stresses have been introduced to reduce potential tensile stresses in concrete resulting from loads.

PRESTRESSING STEEL. High-strength steel element such as wire, bar or strand, or a bundle of such elements, used to impart prestress forces to concrete.

PRETENSIONING. Method of prestressing in which prestressing steel is tensioned before concrete is placed.

REINFORCED CONCRETE. Structural concrete reinforced with no less than the minimum amounts of prestressing steel or nonprestressed reinforcement specified in ACI 318, Chapters 1 through 21 and ACI 318 Appendices A through C.

REINFORCEMENT. Material that conforms to Section 1903.5, excluding prestressing steel unless specifically included.

RESHORES. Shores placed snugly under a concrete slab or other structural member after the original forms and shores have been removed from a larger area, thus requiring the new slab or structural member to deflect and support its own weight and existing construction loads applied prior to the installation of the reshores.

SHORES. Vertical or inclined support members designed to carry the weight of the formwork, concrete and construction loads above.

SPIRAL REINFORCEMENT. Continuously wound reinforcement in the form of a cylindrical helix.

STIRRUP. Reinforcement used to resist shear and torsion stresses in a structural member; typically bars, wires or welded wire fabric (plain or deformed) either single leg or bent into L, U or rectangular shapes and located perpendicular, or at an angle to, longitudinal reinforcement. (The term "stirrups" is usually applied to lateral reinforcement in flexural members and the term "ties" to those in compression members.)

STRUCTURAL CONCRETE. Concrete used for structural purposes, including plain and reinforced concrete.

TENDON. In pretensioning applications, the tendon is the prestressing steel. In posttensioned applications, the tendon is a complete assembly consisting of anchorages, prestressing steel and sheathing with coating for unbonded applications or ducts with grout for bonded applications.

SECTION 1903
SPECIFICATIONS FOR TESTS AND MATERIALS

1903.1 General. Materials used to produce concrete and testing thereof shall comply with the applicable standards listed in ACI 318 and this section. *Tests of concrete and the materials used in concrete shall be in accordance with ACI 318, Section 3.8. Where required, special inspections and tests shall be in accordance with Chapter 17.*

1903.2 Cement. Cement used to produce concrete shall comply with ACI 318, Section 3.2.

1903.3 Aggregates. Aggregates used in concrete shall comply with ACI 318, Section 3.3.

1903.4 Water. Water used in mixing concrete shall be clean and free from injurious amounts of oils, acids, alkalis, salts, organic materials or other substances that are deleterious to concrete or steel reinforcement and shall comply with ACI 318, Section 3.4.

1903.5 Steel reinforcement. Reinforcement and welding of reinforcement to be placed in concrete construction shall conform to the requirements of this section.

1903.5.1 Reinforcement type. Reinforcement shall be deformed reinforcement, except plain reinforcement is permitted for spirals or prestressing steel, and reinforcement consisting of structural steel, steel pipe or steel tubing is permitted where specified in ACI 318. Reinforcement shall comply with ACI 318, Section 3.5.

1903.5.2 Welding. Welding of reinforcing bars shall conform to AWS D1.4. Type and location of welded splices and other required welding of reinforcing bars shall be indicated on the design drawings or in the project specifications. The ASTM reinforcing bar specifications, except for ASTM A 706, shall be supplemented to require a report of material properties necessary to conform to the requirements in AWS D1.4.

1903.6 Admixtures. Admixtures to be used in concrete shall be subject to prior approval by the registered design professional and shall comply with ACI 318, Section 3.6.

1903.7 Storage of materials. The storage of materials for use in concrete shall comply with the provisions of Sections 1903.7.1 and 1903.7.2.

1903.7.1 Manner of storage. Cementitious materials and aggregates shall be stored in such a manner as to prevent deterioration or intrusion of foreign matter.

1903.7.2 Unacceptable material. Any material that has deteriorated or has been contaminated shall not be used for concrete.

1903.8 Glass fiber reinforced concrete. *Glass fiber reinforced concrete (GFRC) and the materials used in such concrete shall be in accordance with the PCI MNL 128 standard.*

SECTION 1904
DURABILITY REQUIREMENTS

1904.1 Water-cementitious materials ratio. The water-cementitious materials ratios specified in Tables 1904.2.2 and 1904.3 shall be calculated using the weight of cement meeting ASTM C 150, ASTM C 595, ASTM C 845 or ASTM C 1157, plus the weight of fly ash and other pozzolans meeting ASTM C 618, slag meeting ASTM C 989 and silica fume meeting ASTM C 1240, if any, except that where concrete is exposed to deicing chemicals, Section 1904.2.3 further limits the amount of fly ash, pozzolans, silica fume, slag or the combination of these materials.

1904.2 Freezing and thawing exposures. Concrete that will be exposed to freezing and thawing or deicing chemicals shall comply with Sections 1904.2.1 through 1904.2.3.

1904.2.1 Air entrainment. Normal-weight and lightweight concrete exposed to freezing and thawing or deicing chemicals shall be air entrained with air content indicated in Table 1904.2.1. Tolerance on air content as delivered shall be ±1.5 percent. For specified compressive strength (f'_c) greater than 5,000 psi (34.47 MPa), reduction of air content indicated in Table 1904.2.1 by 1.0 percent is permitted.

TABLE 1904.2.1
TOTAL AIR CONTENT FOR FROST-RESISTANT CONCRETE

NOMINAL MAXIMUM AGGREGATE SIZE[a] (inches)	AIR CONTENT (percent)	
	Severe exposure[b]	Moderate exposure[b]
$3/_8$	$7^1/_2$	6
$1/_2$	7	$5^1/_2$
$3/_4$	6	5
1	6	$4^1/_2$
$1^1/_2$	$5^1/_2$	$4^1/_2$
2[c]	5	4
3[c]	$4^1/_2$	$3^1/_2$

For SI: 1 inch = 25.4 mm.

a. See ASTM C 33 for tolerance on oversize for various nominal maximum size designations.

b. The severe and moderate exposures referenced in this table are not based on the weathering regions shown in Figure 1904.2.2. For the purposes of this section, severe and moderate exposures shall be defined as follows:
 1. Severe exposure occurs where concrete will be in almost continuous contact with moisture prior to freezing, or where deicing salts are used. Examples are pavements, bridge decks, sidewalks, parking garages and water tanks.
 2. Moderate exposure occurs where concrete will be only occasionally exposed to moisture prior to freezing, and where deicing salts are not used. Examples are certain exterior walls, beams, girders and slabs not in direct contact with soil.

c. These air contents apply to total mix, as for the preceding aggregate sizes. When testing these concretes, however, aggregate larger than $1^1/_2$ inches is removed by hand picking or sieving and air content is determined on the minus $1^1/_2$-inch fraction of the mix (tolerance on air content as delivered applies to this value). Air content of total mix is computed from value determined on the minus $1^1/_2$-inch fraction.

1904.2.2 Concrete properties. Concrete that will be subject to the exposures given in Table 1904.2.2(1) shall conform to the corresponding maximum water-cementitious materials ratios and minimum specified concrete compressive strength requirements of that table. In addition, concrete that will be exposed to deicing chemicals shall conform to the limitations of Section 1904.2.3.

> *Exception: For occupancies and appurtenances thereto in Group R occupancies that are in buildings less than four stories in height, normal-weight aggregate concrete that is subject to weathering (freezing and thawing), as determined from Figure 1904.2.2, or deicer chemicals shall comply with the requirements of Table 1904.2.2(2).*

1904.2.3 Deicing chemicals. For concrete exposed to deicing chemicals, the maximum weight of fly ash, other pozzolans, silica fume or slag that is included in the concrete shall not exceed the percentages of the total weight of cementitious materials given in Table 1904.2.3.

1904.3 Sulfate exposures. Where concrete will be exposed to sulfate-containing solutions, it shall comply with the provisions of Sections 1904.3.1 and 1904.3.2.

1904.3.1 Concrete quality. Concrete to be exposed to sulfate-containing solutions or soils shall conform to the requirements of Table 1904.3 or shall be concrete made with a cement that provides sulfate resistance and that has a maximum water-cementitious materials ratio and minimum compressive strength from Table 1904.3.

1904.3.2 Calcium chloride. Calcium chloride as an admixture shall not be used in concrete to be exposed to severe or very severe sulfate-containing solutions as defined in Table 1904.3.

1904.4 Corrosion protection of reinforcement. Reinforcement in concrete shall be protected from corrosion and exposure to chlorides as provided by Sections 1904.4.1 and 1904.4.2.

1904.4.1 General. For corrosion protection of reinforcement in concrete, the maximum water-soluble chloride ion concentrations in hardened concrete at ages from 28 to 42 days contributed from the ingredients including water, aggregates, cementitious materials and admixtures shall not exceed the limits of Table 1904.4.1. When testing is performed to determine water-soluble chloride ion content, test procedures shall conform to ASTM C 1218.

1904.4.2 Exposure to chlorides. Where concrete with reinforcement will be exposed to chlorides from deicing chemicals, salt, saltwater, brackish water, seawater or spray from these sources, the requirements of Table 1904.2.2(1) for water-cementitious materials ratio and concrete strength, and the minimum concrete cover requirements of Section 1907.7, shall be satisfied. See ACI 318, Section 18.16, for corrosion protection of unbonded tendons.

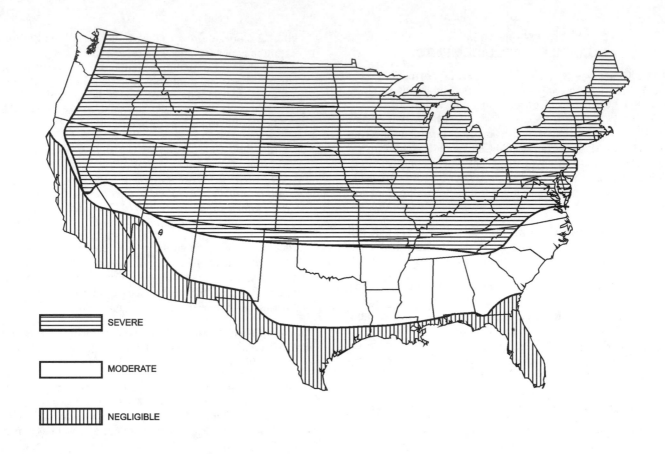

FIGURE 1904.2.2
WEATHERING PROBABILITY MAP FOR CONCRETE[a, b, c]

a. Lines defining areas are approximate only. Local areas can be more or less severe than indicated by the region classification.

b. A "severe" classification is where weather conditions encourage or require the use of deicing chemicals or where there is potential for a continuous presence of moisture during frequent cycles of freezing and thawing. A "moderate" classification is where weather conditions occasionally expose concrete in the presence of moisture to freezing and thawing, but where deicing chemicals are not generally used. A "negligible" classification is where weather conditions rarely expose concrete in the presence of moisture to freezing and thawing.

c. Alaska and Hawaii are classified as severe and negligible, respectively.

TABLE 1904.2.2(1)
REQUIREMENTS FOR SPECIAL EXPOSURE CONDITIONS

EXPOSURE CONDITION	MAXIMUM WATER-CEMENTITIOUS MATERIALS RATIO, BY WEIGHT, NORMAL-WEIGHT AGGREGATE CONCRETE	MINIMUM f'_c, NORMAL-WEIGHT AND LIGHTWEIGHT AGGREGATE CONCRETE (psi)
Concrete intended to have low permeability when exposed to water	0.50	4,000
Concrete exposed to freezing and thawing in a moist condition or to deicing chemicals	0.45	4,500
For corrosion protection of reinforcement in concrete exposed to chlorides from deicing chemicals, salt, saltwater, brackish water, seawater or spray from these sources	0.40	5,000

For SI: 1 pound per square inch = 0.00689 MPa.

TABLE 1904.2.2(2)
MINIMUM SPECIFIED COMPRESSIVE STRENGTH (f'_c)

TYPE OR LOCATION OF CONCRETE CONSTRUCTION	MINIMUM SPECIFIED COMPRESSIVE STRENGTH (f'_c at 28 days, psi)		
	Negligible exposure	Moderate exposure	Severe exposure
Basement walls[c] and foundations not exposed to the weather	2,500	2,500	2,500[a]
Basement slabs and interior slabs on grade, except garage floor slabs	2,500	2,500	2,500[a]
Basement walls[c], foundation walls, exterior walls and other vertical concrete surfaces exposed to the weather	2,500	3,000[b]	3,000[b]
Driveways, curbs, walks, patios, porches, carport slabs, steps and other flatwork exposed to the weather, and garage floor slabs	2,500	3,000[b]	3,500[b]

For SI: 1 pound per square inch = 0.00689 MPa.

a. Concrete in these locations that can be subjected to freezing and thawing during construction shall be of air-entrained concrete in accordance with Table 1904.2.1.

b. Concrete shall be air entrained in accordance with Table 1904.2.1.

c. Structural plain concrete basement walls are exempt from the requirements for special exposure conditions of Section 1904.2.2 (see Section 1909.1.1).

TABLE 1904.2.3
REQUIREMENTS FOR CONCRETE EXPOSED TO DEICING CHEMICALS

CEMENTITIOUS MATERIALS	MAXIMUM PERCENT OF TOTAL CEMENTITIOUS MATERIALS BY WEIGHT[a, b]
Fly ash or other pozzolans conforming to ASTM C 618	25
Slag conforming to ASTM C 989	50
Silica fume conforming to ASTM C 1240	10
Total of fly ash or other pozzolans, slag and silica fume	50[c]
Total of fly ash or other pozzolans and silica fume	35[c]

a. The total cementitious material also includes ASTM C 150, ASTM C 595, ASTM C 845 and ASTM C 1157 cement.

b. The maximum percentages shall include:
  1. Fly ash or other pozzolans present in Type IP or I (PM) blended cement, ASTM C 595, or ASTM C 1157.
  2. Slag used in the manufacture of an IS or I (SM) blended cement, ASTM C 595, or ASTM C 1157.
  3. Silica fume, ASTM C 1240, present in a blended cement.

c. Fly ash or other pozzolans and silica fume shall constitute no more than 25 and 10 percent, respectively, of the total weight of the cementitious materials.

TABLE 1904.3
REQUIREMENTS FOR CONCRETE EXPOSED TO SULFATE-CONTAINING SOLUTIONS

SULFATE EXPOSURE	WATER SOLUBLE SULFATE (SO$_4$) IN SOIL, PERCENT BY WEIGHT	SULFATE (SO$_4$) IN WATER (ppm)	CEMENT TYPE			MAXIMUM WATER-CEMENTITIOUS MATERIALS RATIO, BY WEIGHT, NORMAL-WEIGHT AGGREGATE CONCRETE[a]	MINIMUM f'_c, NORMAL-WEIGHT AND LIGHTWEIGHT AGGREGATE CONCRETE (psi)[a]
			ASTM C 150	ASTM C 595	*ASTM C 1157*		
Negligible	0.00 - 0.10	0 - 150	—	—	—	—	—
Moderate[b]	0.10 - 0.20	150 - 1,500	II	II, IP (MS), IS (MS), P (MS), I (PM)(MS), I (SM)(MS)	*MS*	0.50	4,000
Severe	0.20 - 2.00	1,500 - 10,000	V	—	*HS*	0.45	4,500
Very severe	Over 2.00	Over 10,000	V plus pozzolan[c]	—	*HS plus pozzolan[d]*	0.45	4,500

For SI: 1 pound per square inch = 0.00689 MPa.

a. A lower water-cementitious materials ratio or higher strength may be required for low permeability or for protection against corrosion of embedded items or freezing and thawing (see Table 1904.2.2).

b. Seawater.

c. Pozzolan that has been determined by test or service record to improve sulfate resistance when used in concrete containing Type V cement.

d. *Pozzolan that has been determined by test or service record to improve sulfate resistance when used in concrete containing Type HS blended cement.*

TABLE 1904.4.1
MAXIMUM CHLORIDE ION CONTENT FOR CORROSION PROTECTION OF REINFORCEMENT

TYPE OF MEMBER	MAXIMUM WATER SOLUBLE CHLORIDE ION (CI) IN CONCRETE, PERCENT BY WEIGHT OF CEMENT
Prestressed concrete	0.06
Reinforced concrete exposed to chloride in service	0.15
Reinforced concrete that will be dry or protected from moisture in service	1.00
Other reinforced concrete construction	0.30

SECTION 1905
CONCRETE QUALITY, MIXING AND PLACING

1905.1 General. The required strength and durability of concrete shall be determined by compliance with the proportioning, testing, mixing and placing provisions of Sections 1905.1.1 through 1905.13.

1905.1.1 Strength. Concrete shall be proportioned to provide an average compressive strength as prescribed in Section 1905.3, and shall satisfy the durability criteria of Section 1904. Concrete shall be produced to minimize the frequency of strengths below f'_c as prescribed in Section 1905.6.3.3. *For concrete designed and constructed in accordance with this chapter, f'_c shall not be less than 2,500 psi (17.22 MPa).* No maximum specified compressive strength shall apply unless restricted by a specific provision of this code or ACI 318.

1905.1.2 Cylinder tests. Requirements for f'_c shall be based on tests of cylinders made and tested as prescribed in Section 1905.6.3.

1905.1.3 Basis of f'_c. Unless otherwise specified, f'_c shall be based on 28-day tests. If other than 28 days, test age for f'_c shall be as indicated in construction documents.

1905.1.4 Lightweight aggregate concrete. Where design criteria in ACI 318, Sections 9.5.2.3, 11.2 and 12.2.4, provide for use of a splitting tensile strength value of concrete (f_{ct}), laboratory tests shall be made in accordance with ASTM C 330 to establish the value of f_{ct} corresponding to the specified value of f'_c.

1905.1.5 Field acceptance. Splitting tensile strength tests shall not be used as a basis for field acceptance of concrete.

1905.2 Selection of concrete proportions. Concrete proportions shall be determined in accordance with the provisions of Sections 1905.2.1 through 1905.2.3.

1905.2.1 General. Proportions of materials for concrete shall be established to provide:

1. Workability and consistency to permit concrete to be worked readily into forms and around reinforcement under the conditions of placement to be employed, without segregation or excessive bleeding.
2. Resistance to special exposures as required by Section 1904.
3. Conformance with the strength test requirements of Section 1905.6.

1905.2.2 Different materials. Where different materials are to be used for different portions of proposed work, each combination shall be evaluated.

1905.2.3 Basis of proportions. Concrete proportions shall be established in accordance with Section 1905.3 or Section 1905.4, and shall comply with the applicable requirements of Section 1904.

1905.3 Proportioning on the basis of field experience and/or trial mixtures. Concrete proportioning determined on the basis of field experience and/or trial mixtures shall be done in accordance with ACI 318, Section 5.3.

1905.4 Proportioning without field experience or trial mixtures. Concrete proportioning determined without field experience or trial mixtures shall be done in accordance with ACI 318, Section 5.4.

1905.5 Average strength reduction. As data become available during construction, it is permissible to reduce the amount by which the average compressive strength (f'_c) is required to exceed the specified value of f'_c in accordance with ACI 318, Section 5.5.

1905.6 Evaluation and acceptance of concrete. The criteria for evaluation and acceptance of concrete shall be as specified in Sections 1905.6.2 through 1905.6.5.5.

1905.6.1 Qualified technicians. Concrete shall be tested in accordance with the requirements in Sections 1905.6.2 through 1905.6.5. Qualified field testing technicians shall perform tests on fresh concrete at the job site, prepare specimens required for curing under field conditions, prepare specimens required for testing in the laboratory and record the temperature of the fresh concrete when preparing specimens for strength tests. Qualified laboratory technicians shall perform all required laboratory tests.

1905.6.2 Frequency of testing. The frequency of conducting strength tests of concrete shall be as specified in Sections 1905.6.2.1 through 1905.6.2.4.

1905.6.2.1 Minimum frequency. Samples for strength tests of each class of concrete placed each day shall be taken not less than once a day, nor less than once for each 150 cubic yards (115 m³) of concrete, nor less than once for each 5,000 square feet (465 m²) of surface area for slabs or walls.

1905.6.2.2 Minimum number. On a given project, if the total volume of concrete is such that the frequency of testing required by Section 1905.6.2.1 would provide less than five strength tests for a given class of concrete, tests shall be made from at least five randomly selected batches or from each batch if fewer than five batches are used.

1905.6.2.3 Small volume. When the total volume of a given class of concrete is less than 50 cubic yards (38 m³), strength tests are not required when evidence of satisfactory strength is submitted to and approved by the building official.

1905.6.2.4 Strength test. A strength test shall be the average of the strengths of two cylinders made from the same sample of concrete and tested at 28 days or at the test age designated for the determination of f'_c.

1905.6.3 Laboratory-cured specimens. Laboratory-cured specimens shall comply with the provisions of Sections 1905.6.3.1 through 1905.6.3.4.

1905.6.3.1 Sampling. Samples for strength tests shall be taken in accordance with ASTM C 172.

1905.6.3.2 Cylinders. Cylinders for strength tests shall be molded and laboratory cured in accordance with ASTM C 31 and tested in accordance with ASTM C 39.

1905.6.3.3 Acceptance of results. The strength level of an individual class of concrete shall be considered satisfactory if both of the following requirements are met:

1. Every arithmetic average of any three consecutive strength tests equals or exceeds f'_c.

2. No individual strength test (average of two cylinders) falls below f'_c by more than 500 psi (3.45 MPa) when f'_c is 5,000 psi (34.45 MPa) or less, or by more than $0.10 f'_c$ when f'_c is more than 5,000 psi.

1905.6.3.4 Correction. If either of the requirements of Section 1905.6.3.3 is not met, steps shall be taken to increase the average of subsequent strength test results. The requirements of Section 1905.6.5 shall be observed if the requirement of Section 1905.6.3.3, Item 2, is not met.

1905.6.4 Field-cured specimens. Field-cured specimens shall comply with the provisions of Sections 1905.6.4.1 through 1905.6.4.4.

1905.6.4.1 When required. Where required by the building official, the results of strength tests of cylinders cured under field conditions shall be provided.

1905.6.4.2 Curing. Field-cured cylinders shall be cured under field conditions in accordance with ASTM C 31.

1905.6.4.3 Sampling. Field-cured test cylinders shall be molded at the same time and from the same samples as laboratory-cured test cylinders.

1905.6.4.4 Correction. Procedures for protecting and curing concrete shall be improved when the strength of field-cured cylinders at the test age designated for determination of f'_c is less than 85 percent of that of companion laboratory-cured cylinders. The 85-percent limitation shall not apply if the field-cured strength exceeds f'_c by more than 500 psi (3.45 MPa).

1905.6.5 Low-strength test results. The investigation of low-strength test results shall be in accordance with the provisions of Sections 1905.6.5.1 through 1905.6.5.5.

1905.6.5.1 Precaution. If any strength test (see Section 1905.6.2.4) of laboratory-cured cylinders falls below the specified value of f'_c by more than the values given in Section 1905.6.3.3, Item 2, or if tests of field-cured cylinders indicate deficiencies in protection and curing (see Section 1905.6.4.4), steps shall be taken to assure that the load-carrying capacity of the structure is not jeopardized.

1905.6.5.2 Core tests. If the likelihood of low-strength concrete is confirmed and calculations indicate that load-carrying capacity is significantly reduced, tests of cores drilled from the area in question in accordance with ASTM C 42 are permitted. In such cases, three cores shall be taken for each strength test that falls below the values given in Section 1905.6.3.3, Item 2.

1905.6.5.3 Condition of cores. Cores shall be prepared for transport and storage by wiping drilling water from their surfaces and placing the cores in water-tight bags or containers immediately after drilling. Cores shall be tested no earlier than 48 hours and not later than seven days after coring unless approved by the registered design professional.

1905.6.5.4 Test results. Concrete in an area represented by core tests shall be considered structurally adequate if the average of three cores is equal to at least 85 percent of f'_c and if no single core is less than 75 percent of f'_c. Additional testing of cores extracted from locations represented by erratic core strength results is permitted.

1905.6.5.5 Strength evaluation. If the criteria of Section 1905.6.5.4 are not met and the structural adequacy remains in doubt, the building official is permitted to order a strength evaluation in accordance with ACI 318, Chapter 20, for the questionable portion of the structure, or take other appropriate action.

1905.7 Preparation of equipment and place of deposit. Preparation before concrete placement shall include the following:

1. Equipment for mixing and transporting concrete shall be clean.

2. Debris and ice shall be removed from spaces to be occupied by concrete.

3. Forms shall be properly coated.

4. Masonry filler units that will be in contact with concrete shall be well drenched.

5. Reinforcement shall be thoroughly clean of ice or other deleterious coatings.

6. Water shall be removed from the place of deposit before concrete is placed unless a tremie is to be used or unless otherwise permitted by the building official.

7. Laitance and other unsound material shall be removed before additional concrete is placed against hardened concrete.

1905.8 Mixing. Mixing of concrete shall be performed in accordance with Sections 1905.8.1 through 1905.8.3.

1905.8.1 General. Concrete shall be mixed until there is a uniform distribution of materials and shall be discharged completely before the mixer is recharged.

1905.8.2 Ready-mixed concrete. Ready-mixed concrete shall be mixed and delivered in accordance with the requirements of ASTM C 94 or ASTM C 685.

1905.8.3 Job-mixed concrete. Job-mixed concrete shall comply with ACI 318, Section 5.8.3.

1905.9 Conveying. The method and equipment for conveying concrete to the place of deposit shall comply with Sections 1905.9.1 and 1905.9.2.

1905.9.1 Method of conveyance. Concrete shall be conveyed from the mixer to the place of final deposit by methods that will prevent separation or loss of materials.

1905.9.2 Conveying equipment. The conveying equipment shall be capable of providing a supply of concrete at the site of placement without separation of ingredients and without interruptions sufficient to permit the loss of plasticity between successive increments.

1905.10 Depositing. The depositing of concrete shall comply with the provisions of Sections 1905.10.1 through 1905.10.8.

1905.10.1 Segregation. Concrete shall be deposited as nearly as practicable to its final position to avoid segregation due to rehandling or flowing.

1905.10.2 Placement timing. Concreting operations shall be carried on at such a rate that the concrete is at all times plastic and flows readily into spaces between reinforcement.

1905.10.3 Unacceptable concrete. Concrete that has partially hardened or been contaminated by foreign materials shall not be deposited in the structure.

1905.10.4 Retempering. Retempered concrete or concrete that has been remixed after initial set shall not be used unless approved by the registered design professional.

1905.10.5 Continuous operation. After concreting has started, it shall be carried on as a continuous operation until placing of a panel or section, as defined by its boundaries or predetermined joints, is completed, except as permitted or prohibited by Section 1906.4.

1905.10.6 Placement in vertical lifts. The top surfaces of vertically formed lifts shall be generally level.

1905.10.7 Construction joints. When construction joints are required, they shall be made in accordance with Section 1906.4.

1905.10.8 Consolidation. Concrete shall be thoroughly consolidated by suitable means during placement and shall be thoroughly worked around reinforcement and embedded fixtures and into corners of the forms.

1905.11 Curing. The curing of concrete shall be in accordance with Sections 1905.11.1 through 1905.11.3.

1905.11.1 Regular. Concrete (other than high early strength) shall be maintained above 50°F (10°C) and in a moist condition for at least the first seven days after placement, except when cured in accordance with Section 1905.11.3.

1905.11.2 High early strength. High-early-strength concrete shall be maintained above 50°F (10°C) and in a moist condition for at least the first three days, except when cured in accordance with Section 1905.11.3.

1905.11.3 Accelerated curing. Accelerated curing of concrete shall comply with ACI 318, Section 5.11.3.

1905.12 Cold weather requirements. Concrete that is to be placed during freezing or near-freezing weather shall comply with the following:

1. Adequate equipment shall be provided for heating concrete materials and protecting concrete during freezing or near-freezing weather.

2. Concrete materials and reinforcement, forms, fillers and ground with which concrete is to come in contact shall be free from frost.

3. Frozen materials or materials containing ice shall not be used.

1905.13 Hot weather requirements. During hot weather, proper attention shall be given to ingredients, production methods, handling, placing, protection and curing to prevent excessive concrete temperatures or water evaporation that could impair the required strength or serviceability of the member or structure.

SECTION 1906
FORMWORK, EMBEDDED PIPES AND CONSTRUCTION JOINTS

1906.1 Formwork. The design, fabrication and erection of forms shall comply with Sections 1906.1.1 through 1906.1.6.

1906.1.1 General. Forms shall result in a final structure that conforms to shapes, lines and dimensions of the members as required by the construction documents.

1906.1.2 Strength. Forms shall be substantial and sufficiently tight to prevent leakage of mortar.

1906.1.3 Bracing. Forms shall be properly braced or tied together to maintain position and shape.

1906.1.4 Placement. Forms and their supports shall be designed so as not to damage previously placed structures.

1906.1.5 Design. Design of formwork shall comply with ACI 318, Section 6.1.5.

1906.1.6 Forms for prestressed concrete. Forms for prestressed concrete members shall be designed and constructed to permit movement of the member without damage during application of the prestressing force.

1906.2 Removal of forms, shores and reshores. The removal of forms and shores and the installation of reshores shall comply with Sections 1906.2.1 through 1906.2.2.3.

1906.2.1 Removal of forms. Forms shall be removed in such a manner so as not to impair safety and serviceability of the structure. Concrete to be exposed by form removal shall have sufficient strength not to be damaged by the removal operation.

1906.2.2 Removal of shores and reshores. The provisions of Sections 1906.2.2.1 through 1906.2.2.3 shall apply to slabs and beams, except where cast on the ground.

1906.2.2.1 Removal schedule. Before starting construction, the contractor shall develop a procedure and schedule for removal of shores and installation of reshores and for calculating the loads transferred to the structure during the process.

1. The structural analysis and concrete strength data used in planning and implementing form removal and shoring shall be furnished by the contractor to the building official when so requested.

2. No construction loads shall be supported on, nor any shoring removed from, any part of the structure under construction except when that portion of the structure in combination with the remaining forming and shoring system has sufficient strength to support safely its weight and the loads placed thereon.

3. Sufficient strength shall be demonstrated by structural analysis considering the proposed loads, the strength of the forming and shoring system and concrete strength data. Concrete strength data shall be based on tests of field-cured cylinders or, when approved by the building official, on other procedures to evaluate concrete strength.

1906.2.2.2 Construction loads. No construction loads exceeding the combination of superimposed dead load plus specified live load shall be supported on any unshored portion of the structure under construction, unless analysis indicates adequate strength to support such additional loads.

1906.2.2.3 Prestressed members. Form supports for prestressed concrete members shall not be removed until sufficient prestressing has been applied to enable prestressed members to carry their dead load and anticipated construction loads.

1906.3 Conduits and pipes embedded in concrete. Conduits, pipes and sleeves of any material not harmful to concrete and within the limitations of ACI 318, Section 6.3, are permitted to be embedded in concrete with approval of the registered design professional.

1906.4 Construction joints. Construction joints shall comply with the provisions of Sections 1906.4.1 through 1906.4.6.

1906.4.1 Surface cleaning. The surface of concrete construction joints shall be cleaned and laitance removed.

1906.4.2 Joint treatment. Immediately before new concrete is placed, construction joints shall be wetted and standing water removed.

1906.4.3 Location for force transfer. Construction joints shall be so made and located as not to impair the strength of the structure. Provision shall be made for the transfer of

shear and other forces through construction joints (see ACI 318, Section 11.7.9).

1906.4.4 Location in slabs, beams and girders. Construction joints in floors shall be located within the middle third of spans of slabs, beams and girders. Joints in girders shall be offset a minimum distance of two times the width of intersecting beams.

1906.4.5 Vertical support. Beams, girders or slabs supported by columns or walls shall not be cast or erected until concrete in the vertical support members is no longer plastic.

1906.4.6 Monolithic placement. Beams, girders, haunches, drop panels and capitals shall be placed monolithically as part of a slab system, unless otherwise shown in the design drawings or specifications.

SECTION 1907
DETAILS OF REINFORCEMENT

1907.1 Hooks. Standard hooks on reinforcing bars used in concrete construction shall comply with ACI 318, Section 7.1.

1907.2 Minimum bend diameters. Minimum reinforcement bend diameters utilized in concrete construction shall comply with ACI 318, Section 7.2.

1907.3 Bending. The bending of reinforcement shall comply with Sections 1907.3.1 and 1907.3.2.

1907.3.1 Cold bending. Reinforcement shall be bent cold, unless otherwise permitted by the registered design professional.

1907.3.2 Embedded reinforcement. Reinforcement partially embedded in concrete shall not be field bent, except as shown on the construction documents or permitted by the registered design professional.

1907.4 Surface conditions of reinforcement. The surface conditions of reinforcement shall comply with the provisions of Sections 1907.4.1 through 1907.4.3.

1907.4.1 Coatings. At the time concrete is placed, reinforcement shall be free from mud, oil or other nonmetallic coatings that decrease bond. Epoxy coatings of steel reinforcement in accordance with ACI 318, Sections 3.5.3.7 and 3.5.3.8, are permitted.

1907.4.2 Rust or mill scale. Except for prestressing steel, steel reinforcement with rust, mill scale or a combination of both, shall be considered satisfactory, provided the minimum dimensions, including height of deformations and weight of a hand-wire-brushed test specimen, comply with applicable ASTM specifications (see Section 1903.5).

1907.4.3 Prestressing steel. Prestressing steel shall be clean and free of oil, dirt, scale, pitting and excessive rust. A light coating of rust is permitted.

1907.5 Placing reinforcement. The placement of concrete reinforcement shall comply with the provisions of Sections 1907.5.1 through 1907.5.4.

1907.5.1 Support. Reinforcement, including tendons, and posttensioning ducts shall be accurately placed and adequately supported before concrete is placed, and shall be secured against displacement within tolerances permitted in Section 1907.5.2. Where approved by the registered design professional, embedded items (such as dowels or inserts) that either protrude from precast concrete members or remain exposed for inspection are permitted to be embedded while the concrete is in a plastic state, provided the following conditions are met:

1. Embedded items are not required to be hooked or tied to reinforcement within the concrete.

2. Embedded items are maintained in the correct position while the concrete remains plastic.

3. The concrete is properly consolidated around the embedded item.

1907.5.2 Tolerances. Unless otherwise specified by the registered design professional, reinforcement, including tendons, and posttensioning ducts shall be placed within the tolerances specified in Sections 1907.5.2.1 and 1907.5.2.2.

1907.5.2.1 Depth and cover. Tolerance for depth, d, and minimum concrete cover in flexural members, walls and compression members shall be as shown in Table 1907.5.2.1, except that tolerance for the clear distance to formed soffits shall be minus $1/4$ inch (6.4 mm) and tolerance for cover shall not exceed minus one-third the minimum concrete cover required in the design drawings or specifications.

1907.5.2.2 Bends and ends. Tolerance for longitudinal location of bends and ends of reinforcement shall be ± 2 inches (± 51 mm) except the tolerance shall be ± $1/2$ inch (± 12.7 mm) at the discontinuous ends of brackets and corbels, and ± 1 inch (25 mm) at the discontinuous ends of other members. The tolerance for minimum concrete cover of Section 1907.5.2.1 shall also apply at discontinuous ends of members.

TABLE 1907.5.2.1
TOLERANCES

DEPTH (d) (inches)	TOLERANCE ON d (inch)	TOLERANCE ON MINIMUM CONCRETE COVER (inch)
$d > 8$	± $3/8$	− $3/8$
$d > 8$	± $1/2$	− $1/2$

For SI: 1 inch = 25.4 mm.

1907.5.3 Welded wire fabric. Welded wire fabric with wire size not greater than W5 or D5 used in slabs not exceeding 10 feet (3048 mm) in span is permitted to be curved from a point near the top of the slab over the support to a point near the bottom of the slab at midspan, provided such reinforcement is either continuous over, or securely anchored at support.

1907.5.4 Welding. Welding of crossing bars shall not be permitted for assembly of reinforcement unless authorized by the registered design professional.

1907.6 Spacing limits for reinforcement. The clear distance between reinforcing bars, bundled bars, tendons and ducts shall comply with ACI 318, Section 7.6.

1907.7 Concrete protection for reinforcement. The minimum concrete cover for reinforcement shall comply with Sections 1907.7.1 through 1907.7.7.

1907.7.1 Cast-in-place concrete (nonprestressed). Minimum concrete cover shall be provided for reinforcement in nonprestressed, cast-in-place concrete construction in accordance with Table 1907.7.1, but shall not be less than required by Sections 1907.7.5 and 1907.7.7.

TABLE 1907.7.1
MINIMUM CONCRETE COVER

CONCRETE EXPOSURE	MINIMUM COVER (inches)
1. Concrete cast against and permanently exposed to earth	3

2. Concrete exposed to earth or weather
    No. 6 through No. 18 bar
    No. 5 bar, W31 or D31 wire, and smaller
	
	2
	$1^1/_2$

3. Concrete not exposed to weather or in contact with ground
    Slabs, walls, joists:
        No. 14 and No. 18 bars
        No. 11 bar and smaller
    Beams, columns:
        Primary reinforcement, ties, stirrups, spirals
    Shells, folded plate members:
        No. 6 bar and larger
        No. 5 bar, W31 or D31 wire, and smaller
	
	$1^1/_2$
	$3/_4$
	
	$1^1/_2$
	
	$3/_4$
	$1/_2$

For SI: 1 inch = 25.4 mm.

1907.7.2 Cast-in-place concrete (prestressed). The minimum concrete cover for prestressed and nonprestressed reinforcement, ducts and end fittings in cast-in-place prestressed concrete shall comply with ACI 318, Section 7.7.2.

1907.7.3 Precast concrete (manufactured under plant control conditions). The minimum concrete cover for prestressed and nonprestressed reinforcement, ducts and end fittings in precast concrete manufactured under plant control conditions shall comply with ACI 318, Section 7.7.3.

1907.7.4 Bundled bars. The minimum concrete cover for bundled bars shall comply with ACI 318, Section 7.7.4.

1907.7.5 Corrosive environments. In corrosive environments or other severe exposure conditions, prestressed and nonprestressed reinforcement shall be provided with additional protection in accordance with ACI 318, Section 7.7.5.

1907.7.6 Future extensions. Exposed reinforcement, inserts and plates intended for bonding with future extensions shall be protected from corrosion.

1907.7.7 Fire protection. When this code requires a thickness of cover for fire protection greater than the minimum concrete cover specified in Section 1907.7, such greater thickness shall be used.

1907.8 Special reinforcement details for columns. Offset bent longitudinal bars in columns and load transfer in structural steel cores of composite compression members shall comply with the provisions of ACI 318, Section 7.8.

1907.9 Connections. Connections between concrete framing members shall comply with the provisions of ACI 318, Section 7.9.

1907.10 Lateral reinforcement for compression members. Lateral reinforcement for concrete compression members shall comply with the provisions of ACI 318, Section 7.10.

1907.11 Lateral reinforcement for flexural members. Lateral reinforcement for compression reinforcement in concrete flexural members shall comply with the provisions of ACI 318, Section 7.11.

1907.12 Shrinkage and temperature reinforcement. Reinforcement for shrinkage and temperature stresses in concrete members shall comply with the provisions of ACI 318, Section 7.12.

1907.13 Requirements for structural integrity. The detailing of reinforcement and connections between concrete members shall comply with the provisions of ACI 318, Section 7.13, to improve structural integrity.

SECTION 1908
MODIFICATIONS TO ACI 318

1908.1 General. The text of ACI 318 shall be modified as indicated in Sections 1908.1.1 through 1908.1.9.

1908.1.1 ACI 318, Section 21.1. Modify existing definitions and add the following definitions to ACI 318, Section 21.1.

DESIGN DISPLACEMENT. Total lateral displacement expected for the design-basis earthquake, *as specified by Section 9.5.5.7 of ASCE 7 or 1617.5.4 of the International Building Code.*

STORY DRIFT RATIO. *The design displacement over a story divided by the story height.*

WALL PIER. *A wall segment with a horizontal length-to-thickness ratio of at least 2.5, but not exceeding six, whose clear height is at least two times its horizontal length.*

1908.1.2 ACI 318, Section 21.2.1. Modify Sections 21.2.1.2, 21.2.1.3 and 21.2.1.4 to read as follows:

21.2.1.2 For structures assigned to Seismic Design Category A or B, provisions of Chapters 1 through 18 and 22 shall apply except as modified by the provisions of this chapter. Where the seismic design loads are computed using provisions for intermediate or special concrete systems, the requirements of Chapter 21 for intermediate or special systems, as applicable, shall be satisfied.

21.2.1.3 For structures assigned to Seismic Design Category C, intermediate or special moment frames, *or ordinary or special reinforced concrete structural walls* shall be used to resist *seismic* forces induced by earthquake motions. Where the design seismic loads are computed using provisions for special concrete systems, the requirements of Chapter 21 for special systems, as applicable, shall be satisfied.

21.2.1.4 For structures assigned to Seismic Design Category D, E or F, special moment frames, *special reinforced concrete structural walls,* diaphragms and trusses *and foundations* complying with Sections 21.2 through 21.10 shall be used to resist forces induced by earthquake motions. Frame members not proportioned to resist earthquake forces shall comply with Section 21.11.

1908.1.3 ACI 318, Section 21.2.5. Modify ACI 318, Section 21.2.5, by renumbering as Section 21.2.5.1 and adding new Sections 21.2.5.2, 21.2.5.3 and 21.2.5.4 to read as follows:

21.2.5 Reinforcement in members resisting earthquake-induced forces.

21.2.5.1 Except as permitted in Sections 21.2.5.2 through 21.2.5.4, reinforcement resisting earthquake-induced flexural and axial forces in frame members and in structural wall boundary elements shall comply with ASTM A 706. ASTM 615, Grades 40 and 60 reinforcement, shall be permitted in these members if (a) the actual yield strength based on mill tests does not exceed the specified yield strength by more than 18,000 psi (retests shall not exceed this value by more than an additional 3,000 psi), and (b) the ratio of the actual ultimate tensile strength to the actual tensile yield strength is not less than 1.25.

21.2.5.2 Prestressing steel shall be permitted in flexural members of frames, provided the average prestress, f_{pc}, calculated for an area equal to the member's shortest cross-sectional dimension multiplied by the perpendicular dimension shall be the lesser of 700 psi (4.83 MPa) or $f'_c/6$ at locations of nonlinear action where prestressing steel is used in members of frames.

21.2.5.3 Unless the seismic-force-resisting frame is qualified for use through structural testing as required by the ACI T1.1, for members in which prestressing steel is used together with mild reinforcement to resist earthquake-induced forces, prestressing steel shall not provide more than one-quarter of the strength for either positive or negative moments at the nonlinear action location and shall be anchored at the exterior face of the joint or beyond.

21.2.5.4 Anchorages for tendons must be demonstrated to perform satisfactorily for seismic loadings. Anchorage assemblies shall withstand, without failure, a minimum of 50 cycles of loading ranging between 40 and 85 percent of the minimum specified tensile strength of the prestressing steel.

1908.1.4 ACI 318, Section 21.7. Modify ACI 318, Section 21.7, by adding a new Section 21.7.10 to read as follows:

21.7.10 Wall piers and wall segments.

21.7.10.1 Wall piers not designed as a part of a special moment frame shall have transverse reinforcement designed to satisfy the requirements in Section 21.7.10.2.

Exceptions:

1. *Wall piers that satisfy Section 21.11.*

2. *Wall piers along a wall line within a story where other shear wall segments provide lateral support to the wall piers, and such segments have a*

total stiffness of at least six times the sum of the stiffness of all the wall piers.

21.7.10.2 Transverse reinforcement shall be designed to resist the shear forces determined from Sections 21.3.4.2 and 21.4.5.1. Where the axial compressive force, including earthquake effects, is less than $A_g f'_c/20$, transverse reinforcement in wall piers is permitted to have standard hooks at each end in lieu of hoops. Spacing of transverse reinforcement shall not exceed 6 inches (152 mm). Transverse reinforcement shall be extended beyond the pier clear height for at least the development length of the largest longitudinal reinforcement in the wall pier.

21.7.10.3 Wall segments with a horizontal length-to-thickness ratio less than 2.5 shall be designed as columns.

1908.1.5 ACI 318, Section 21.10.1.1. Modify ACI 318, Section 21.10.1.1, to read as follows:

21.10.1.1 Foundations resisting earthquake-induced forces or transferring earthquake-induced forces between a structure and the ground shall comply with the requirements of Section 21.10 and other applicable provisions of ACI 318 *unless modified by Chapter 18 of the International Building Code.*

1908.1.6 ACI 318, Section 21.11. Modify ACI Sections 21.11.1 and 21.11.2.2 and add Sections 21.11.5 through 21.11.7 as follows:

21.11.1 Frame members assumed not to contribute to lateral resistance shall be detailed according to Section 21.11.2 or 21.11.3 depending on the magnitude of moments induced in those members when subjected to the design displacement. If effects of design displacements are not explicitly checked, it shall be permitted to apply the requirements of Section 21.11.3. *Slab-column connections shall comply with Sections 21.11.5 through 21.11.7. Conformance to Section 21.11 satisfies the deformation compatibility requirements of Section 9.5.2.2.4.3 of ASCE 7.*

21.11.2.2 Members with factored gravity axial forces exceeding ($A_g f'_c/10$) shall satisfy Sections 21.4.3, 21.4.4.1(c), 21.4.4.3 and 21.4.5. The maximum longitudinal spacing of ties shall be, s_o, for the full column height. The spacing, s_o, shall be not more than six diameters of the smallest longitudinal bar enclosed or 6 inches (152 mm), whichever is smaller. *Lap splices of longitudinal reinforcement in such members need not satisfy Section 21.4.3.2 in structures where the seismic-force-resisting system does not include special moment frames.*

21.11.5 Reinforcement to resist punching shear shall be provided in accordance with Sections 21.11.5.1 and 21.11.5.2 at slab column connections where story drift ratio exceeds [$0.035 - 0.05 (V_u/\phi V_c)$] except that Sections 21.11.4.1 and 21.11.4.2 need not be satisfied where $V_u/\phi V_c$ is less than 0.2 or where the story drift ratio is less than 0.005. V_u equals the factored punching shear from gravity load excluding shear stress from unbalanced moment. V_u is calculated for the load combination 1.2D +

1.0L + 0.2S. The load factor on L is permitted to be reduced to 0.5 in accordance with Section 9.2.1(a). In no case shall shear reinforcement be less than that required in Section 11.12 for loads without consideration of seismic effects.

21.11.5.1 — The slab shear reinforcement shall provide V_s not less than $3.5\sqrt{f'_c}$.

21.11.5.2 — Slab shear reinforcement shall extend not less than five times the slab thickness from the face of column.

21.11.6 — Bottom bars or wires within the column strip shall conform to Section 13.3.8.5 except that splices shall be Class B.

21.11.7 — Within the effective slab width defined in Section 13.5.3.2, the ratio of nonprestressed bottom reinforcement to gross concrete area shall not be less than 0.004. Where bottom reinforcement is not required to be continuous, such reinforcement shall extend a minimum of five times the slab thickness plus one development length beyond the face of the column or terminated at the slab edge with a standard hook.

1908.1.7 ACI 318, Section 21.13.2. Modify ACI 318, Section 21.13.2, to read as follows:

21.13.2 In connections between wall panels, or between wall panels and the foundation, *yielding shall be restricted to reinforcement.*

SECTION 1909
STRUCTURAL PLAIN CONCRETE

1909.1 Scope. The design and construction of structural plain concrete, both cast-in-place and precast, shall comply with the minimum requirements of Section 1909 and ACI 318, Chapter 22.

1909.1.1 Special structures. For special structures, such as arches, underground utility structures, gravity walls and shielding walls, the provisions of this section shall govern where applicable.

1909.2 Limitations. The use of structural plain concrete shall be limited to:

1. Members that are continuously supported by soil, such as walls and footings, or by other structural members capable of providing continuous vertical support.

2. Members for which arch action provides compression under all conditions of loading.

3. Walls and pedestals.

The use of structural plain concrete columns and structural plain concrete footings on piles is not permitted. See Section 1910 for additional limitations on the use of structural plain concrete.

1909.3 Joints. Contraction or isolation joints shall be provided to divide structural plain concrete members into flexurally discontinuous elements in accordance with ACI 318, Section 22.3.

1909.4 Design. Structural plain concrete walls, footings and pedestals shall be designed for adequate strength in accordance with ACI 318, Sections 22.4 through 22.8.

Exception: For Group R-3 as applicable in Section 101.2 occupancies and buildings of other occupancies less than two stories in height of light-frame construction, the required edge thickness of ACI 318 is permitted to be reduced to 6 inches (152 mm), provided that the footing does not extend more than 4 inches (102 mm) on either side of the supported wall.

1909.5 Precast members. The design, fabrication, transportation and erection of precast, structural plain concrete elements shall be in accordance with ACI 318, Section 22.9.

1909.6 Walls. In addition to the requirements of this section, structural plain concrete walls shall comply with the applicable requirements of ACI 318, Chapter 22.

1909.6.1 Basement walls. The thickness of exterior basement walls and foundation walls shall be not less than $7^1/_2$ inches (191 mm). Structural plain concrete exterior basement walls shall be exempt from the requirements for special exposure conditions of Section 1904.2.2.

1909.6.2 Other walls. Except as provided for in Section 1909.6.1, the thickness of bearing walls shall be not less than $^1/_{24}$ the unsupported height or length, whichever is shorter, but not less than $5^1/_2$ inches (140 mm).

1909.6.3 Openings in walls. Not less than two No. 5 bars shall be provided around window and door openings. Such bars shall extend at least 24 inches (610 mm) beyond the corners of openings.

SECTION 1910
SEISMIC DESIGN PROVISIONS

1910.1 General. The design and construction of concrete components that resist seismic forces shall conform to the requirements of this section and to ACI 318 except as modified by Section 1908.

1910.2 Classification of shear walls. Structural concrete shear walls that resist seismic forces shall be classified in accordance with Sections 1910.2.1 through 1910.2.4.

1910.2.1 Ordinary plain concrete shear walls. Ordinary plain concrete shear walls are walls conforming to the requirements of Chapter 22 of ACI 318.

1910.2.2 Detailed plain concrete shear walls. Detailed plain concrete shear walls are walls conforming to the requirements for ordinary plain concrete shear walls and shall have reinforcement as follows: Vertical reinforcement of at least 0.20 square inch (129 mm²) in cross-sectional area shall be provided continuously from support to support at each corner, at each side of each opening and at the ends of walls. The continuous vertical bar required beside an opening is permitted to substitute for one of the two No. 5 bars required by Section 22.6.6.5 of ACI 318. Horizontal reinforcement at least 0.20 square inch (129 mm²) in cross-sectional area shall be provided:

1. Continuously at structurally connected roof and floor levels and at the top of walls;
2. At the bottom of load-bearing walls or in the top of foundations where doweled to the wall; and
3. At a maximum spacing of 120 inches (3048 mm).

Reinforcement at the top and bottom of openings, where used in determining the maximum spacing specified in Item 3 above, shall be continuous in the wall.

1910.2.3 Ordinary reinforced concrete shear walls. Ordinary reinforced concrete shear walls are walls conforming to the requirements of ACI 318 for ordinary reinforced concrete structural walls.

1910.2.4 Special reinforced concrete shear walls. Special reinforced concrete shear walls are walls conforming to the requirements of ACI 318 for special reinforced concrete structural walls or special precast structural walls.

1910.3 Seismic Design Category B. Structures assigned to Seismic Design Category B, as determined in Section 1616, shall conform to the requirements for Seismic Design Category A and to the additional requirements for Seismic Design Category B of this section.

1910.3.1 Ordinary moment frames. In flexural members of ordinary moment frames forming part of the seismic-force-resisting system, at least two main flexural reinforcing bars shall be provided continuously top and bottom throughout the beams, through or developed within exterior columns or boundary elements.

Columns of ordinary moment frames having a clear height-to-maximum-plan-dimension ratio of five or less shall be designed for shear in accordance with Section 21.12.3 of ACI 318.

1910.4 Seismic Design Category C. Structures assigned to Seismic Design Category C, as determined in Section 1616, shall conform to the requirements for Seismic Design Category B and to the additional requirements for Seismic Design Category C of this section.

1910.4.1 Seismic-force-resisting systems. Moment frames used to resist seismic forces shall be intermediate moment frames or special moment frames. Shear walls used to resist seismic forces shall be ordinary reinforced concrete shear walls or special reinforced concrete shear walls. Ordinary reinforced concrete shear walls constructed of precast concrete elements shall comply with the additional requirements of Section 21.13 of ACI 318 for intermediate precast concrete structural walls, as modified by Section 1908.1.7.

1910.4.2 Discontinuous members. Columns supporting reactions from discontinuous stiff members, such as walls, shall be designed for the special load combinations in Section 1605.4 and shall be provided with transverse reinforcement at the spacing, s_o, as defined in Section 21.12.5.2 of ACI 318 over their full height beneath the level at which the discontinuity occurs. This transverse reinforcement shall be extended above and below the column as required in Section 21.4.4.5 of ACI 318.

1910.4.3 Plain concrete. Structural plain concrete members in structures assigned to Seismic Design Category C

CONCRETE

shall conform to ACI 318 and with Sections 1910.4.3.1 through 1910.4.3.3.

1910.4.3.1 Walls. Structural plain concrete walls are not permitted in structures assigned to Seismic Design Category C.

Exception: Structural plain concrete basement, foundation or other walls below the base are permitted in detached one- and two-family dwellings constructed with stud-bearing walls. Such walls shall have reinforcement in accordance with Section 22.6.6.5 of ACI 318.

1910.4.3.2 Footings. Isolated footings of plain concrete supporting pedestals or columns are permitted provided the projection of the footing beyond the face of the supported member does not exceed the footing thickness.

Exception: In detached one- and two-family dwellings three stories or less in height, the projection of the footing beyond the face of the supported member is permitted to exceed the footing thickness.

Plain concrete footings supporting walls shall be provided with not less than two continuous longitudinal reinforcing bars. Bars shall not be smaller than No. 4 and shall have a total area of not less than 0.002 times the gross cross-sectional area of the footing. For footings which exceed 8 inches (203 mm) in thickness, a minimum of one bar shall be provided at the top and bottom of the footing. For foundation systems consisting of a plain concrete footing and a plain concrete stemwall, a minimum of one bar shall be provided at the top of the stemwall and at the bottom of the footing. Continuity of reinforcement shall be provided at corners and intersections.

Exceptions:

1. In detached one- and two-family dwellings three stories or less in height and constructed with stud-bearing walls, plain concrete footings supporting walls are permitted without longitudinal reinforcement.

2. Where a slab-on-ground is cast monolithically with the footing, one No. 5 bar is permitted to be located at either the top or bottom of the footing.

1910.4.3.3 Pedestals. Plain concrete pedestals shall not be used to resist lateral seismic forces.

1910.5 Seismic Design Category D, E or F. Structures assigned to Seismic Design Category D, E or F, as determined in Section 1616, shall conform to the requirements for Seismic Design Category C and to the additional requirements of this section.

1910.5.1 Seismic-force-resisting systems. Moment frames used to resist seismic forces shall be special moment frames. Shear walls used to resist seismic forces shall be special reinforced concrete shear walls.

1910.5.2 Frame members not proportioned to resist forces induced by earthquake motions. Frame components assumed not to contribute to lateral force resistance shall conform to ACI 318, Section 21.11, as modified by Section 1908.1.6 of this chapter.

SECTION 1911
MINIMUM SLAB PROVISIONS

1911.1 General. The thickness of concrete floor slabs supported directly on the ground shall not be less than $3^1/_2$ inches (89 mm). A 6-mil (0.006 inch; 152 mm) polyethylene vapor retarder with joints lapped not less than 6 inches (152 mm) shall be placed between the base course or subgrade and the concrete floor slab, or other approved equivalent methods or materials shall be used to retard vapor transmission through the floor slab.

Exception: A vapor retarder is not required:

1. For detached structures accessory to occupancies in Group R-3 as applicable in Section 101.2, such as garages, utility buildings or other unheated facilities.

2. For unheated storage rooms having an area of less than 70 square feet (6.5 m²) and carports attached to occupancies in Group R-3 as applicable in Section 101.2.

3. For buildings of other occupancies where migration of moisture through the slab from below will not be detrimental to the intended occupancy of the building.

4. For driveways, walks, patios and other flatwork which will not be enclosed at a later date.

5. Where approved based on local site conditions.

SECTION 1912
ANCHORAGE TO CONCRETE—
ALLOWABLE STRESS DESIGN

1912.1 Scope. The provisions of this section shall govern the allowable stress design of headed bolts and headed stud anchors cast in normal-weight concrete for purposes of transmitting structural loads from one connected element to the other. These provisions do not apply to anchors installed in hardened concrete or where load combinations include earthquake loads or effects. The bearing area of headed anchors shall be not less than one and one-half times the shank area. Where strength design is used, or where load combinations include earthquake loads or effects, the design strength of anchors shall be determined in accordance with Section 1913. Bolts shall conform to ASTM A 307 or an approved equivalent.

1912.2 Allowable service load. The allowable service load for headed anchors in shear or tension shall be as indicated in Table 1912.2. Where anchors are subject to combined shear and tension, the following relationship shall be satisfied:

$$(P_s/P_t)^{5/3} + (V_s/V_t)^{5/3} \le 1 \qquad \textbf{(Equation 19-1)}$$

where:

P_s = Applied tension service load, pounds (newtons).

P_t = Allowable tension service load from Table 1912.2, pounds (newtons).

V_s = Applied shear service load, pounds (newtons).

V_t = Allowable shear service load from Table 1912.2, pounds (newtons).

TABLE 1912.2
ALLOWABLE SERVICE LOAD ON EMBEDDED BOLTS (pounds)

BOLT DIAMETER (inches)	MINIMUM EMBEDMENT (inches)	EDGE DISTANCE (inches)	SPACING (inches)	MINIMUM CONCRETE STRENGTH (psi)					
				$f'_c = 2,500$		$f'_c = 3,000$		$f'_c = 4,000$	
				Tension	Shear	Tension	Shear	Tension	Shear
$^1/_4$	$2^1/_2$	$1^1/_2$	3	200	500	200	500	200	500
$^3/_8$	3	$2^1/_4$	$4^1/_2$	500	1,100	500	1,100	500	1,100
$^1/_2$	4	3	6	950	1,250	950	1,250	950	1,250
	4	5	5	1,450	1,600	1,500	1,650	1,550	1,750
$^5/_8$	$4^1/_2$	$3^3/_4$	$7^1/_2$	1,500	2,750	1,500	2,750	1,500	2,750
	$4^1/_2$	$6^1/_4$	$7^1/_2$	2,125	2,950	2,200	3,000	2,400	3,050
$^3/_4$	5	$4^1/_2$	9	2,250	3,250	2,250	3,560	2,250	3,560
	5	$7^1/_2$	9	2,825	4,275	2,950	4,300	3,200	4,400
$^7/_8$	6	$5^1/_4$	$10^1/_2$	2,550	3,700	2,550	4,050	2,550	4,050
1	7	6	12	3,050	4,125	3,250	4,500	3,650	5,300
$1^1/_8$	8	$6^3/_4$	$13^1/_2$	3,400	4,750	3,400	4,750	3,400	4,750
$1^1/_4$	9	$7^1/_2$	15	4,000	5,800	4,000	5,800	4,000	5,800

For SI: 1 inch = 25.4 mm, 1 pound per square inch = 0.00689 MPa, 1 pound = 4.45 N.

1912.3 Required edge distance and spacing. The allowable service loads in tension and shear specified in Table 1912.2 are for the edge distance and spacing specified. The edge distance and spacing are permitted to be reduced to 50 percent of the values specified with an equal reduction in allowable service load. Where edge distance and spacing are reduced less than 50 percent, the allowable service load shall be determined by linear interpolation.

1912.4 Increase in allowable load. Increase of the values in Table 1912.2 by one-third is permitted where the provisions of Section 1605.3.2 permit an increase in allowable stress for wind loading.

1912.5 Increase for special inspection. Where special inspection is provided for the installation of anchors, a 100-percent increase in the allowable tension values of Table 1912.2 is permitted. No increase in shear value is permitted.

SECTION 1913
ANCHORAGE TO CONCRETE—
STRENGTH DESIGN

1913.1 Scope. The provisions of this section shall govern the strength design of anchors installed in concrete for purposes of transmitting structural loads from one connected element to the other. Headed bolts, headed studs and hooked (J- or L-) bolts cast in concrete and expansion anchors and undercut anchors installed in hardened concrete shall be designed in accordance with Appendix D of ACI 318, provided they are within the scope of Appendix D.

Exception: Where the basic concrete breakout strength in tension of a single anchor, N_b, is determined in accordance with Equation (D-7), the concrete breakout strength requirements of Section D.4.2.2 shall be considered satisfied by the design procedures of Sections D.5.2 and D.6.2 for anchors exceeding 2 inches (51 mm) in diameter or 25 inches (635 mm) tensile embedment depth.

The strength design of anchors that are not within the scope of Appendix D of ACI 318, and as amended above, shall be in accordance with an approved procedure.

SECTION 1914
SHOTCRETE

1914.1 General. Shotcrete is mortar or concrete that is pneumatically projected at high velocity onto a surface. Except as specified in this section, shotcrete shall conform to the requirements of this chapter for plain or reinforced concrete.

1914.2 Proportions and materials. Shotcrete proportions shall be selected that allow suitable placement procedures using the delivery equipment selected and shall result in finished in-place hardened shotcrete meeting the strength requirements of this code.

1914.3 Aggregate. Coarse aggregate, if used, shall not exceed $^3/_4$ inch (19.1 mm).

1914.4 Reinforcement. Reinforcement used in shotcrete construction shall comply with the provisions of Sections 1914.4.1 through 1914.4.4.

1914.4.1 Size. The maximum size of reinforcement shall be No. 5 bars unless it is demonstrated by preconstruction tests that adequate encasement of larger bars will be achieved.

1914.4.2 Clearance. When No. 5 or smaller bars are used, there shall be a minimum clearance between parallel reinforcement bars of $2^1/_2$ inches (64 mm). When bars larger than No. 5 are permitted, there shall be a minimum clearance between parallel bars equal to six diameters of the bars used. When two curtains of steel are provided, the curtain nearer the nozzle shall have a minimum spacing equal to 12 bar diameters and the remaining curtain shall have a minimum spacing of six bar diameters.

> **Exception:** Subject to the approval of the building official, required clearances shall be reduced where it is demonstrated by preconstruction tests that adequate encasement of the bars used in the design will be achieved.

1914.4.3 Splices. Lap splices of reinforcing bars shall utilize the noncontact lap splice method with a minimum clearance of 2 inches (51 mm) between bars. The use of contact lap splices necessary for support of the reinforcing is permitted when approved by the building official, based on satisfactory preconstruction tests that show that adequate encasement of the bars will be achieved, and provided that the splice is oriented so that a plane through the center of the spliced bars is perpendicular to the surface of the shotcrete.

1914.4.4 Spirally tied columns. Shotcrete shall not be applied to spirally tied columns.

1914.5 Preconstruction tests. When required by the building official, a test panel shall be shot, cured, cored or sawn, examined and tested prior to commencement of the project. The sample panel shall be representative of the project and simulate job conditions as closely as possible. The panel thickness and reinforcing shall reproduce the thickest and most congested area specified in the structural design. It shall be shot at the same angle, using the same nozzleman and with the same concrete mix design that will be used on the project. The equipment used in preconstruction testing shall be the same equipment used in the work requiring such testing, unless substitute equipment is approved by the building official.

1914.6 Rebound. Any rebound or accumulated loose aggregate shall be removed from the surfaces to be covered prior to placing the initial or any succeeding layers of shotcrete. Rebound shall not be used as aggregate.

1914.7 Joints. Except where permitted herein, unfinished work shall not be allowed to stand for more than 30 minutes unless edges are sloped to a thin edge. For structural elements that will be under compression and for construction joints shown on the approved construction documents, square joints are permitted. Before placing additional material adjacent to previously applied work, sloping and square edges shall be cleaned and wetted.

1914.8 Damage. In-place shotcrete that exhibits sags, sloughs, segregation, honeycombing, sand pockets or other obvious defects shall be removed and replaced. Shotcrete above sags and sloughs shall be removed and replaced while still plastic.

1914.9 Curing. During the curing periods specified herein, shotcrete shall be maintained above 40°F (4°C) and in moist condition.

1914.9.1 Initial curing. Shotcrete shall be kept continuously moist for 24 hours after shotcreting is complete or shall be sealed with an approved curing compound.

1914.9.2 Final curing. Final curing shall continue for seven days after shotcreting, or for three days if high-early-strength cement is used, or until the specified strength is obtained. Final curing shall consist of the initial curing process or the shotcrete shall be covered with an approved moisture-retaining cover.

1914.9.3 Natural curing. Natural curing shall not be used in lieu of that specified in this section unless the relative humidity remains at or above 85 percent, and is authorized by the registered design professional and approved by the building official.

1914.10 Strength tests. Strength tests for shotcrete shall be made by an approved agency on specimens that are representative of the work and which have been water soaked for at least 24 hours prior to testing. When the maximum-size aggregate is larger than $^3/_8$ inch (9.5 mm), specimens shall consist of not less than three 3-inch-diameter (76 mm) cores or 3-inch (76 mm) cubes. When the maximum-size aggregate is $^3/_8$ inch (9.5 mm) or smaller, specimens shall consist of not less than 2-inch-diameter (51 mm) cores or 2-inch (51 mm) cubes.

1914.10.1 Sampling. Specimens shall be taken from the in-place work or from test panels, and shall be taken at least once each shift, but not less than one for each 50 cubic yards (38.2 m^3) of shotcrete.

1914.10.2 Panel criteria. When the maximum-size aggregate is larger than $^3/_8$ inch (9.5 mm), the test panels shall have minimum dimensions of 18 inches by 18 inches (457 mm by 457 mm). When the maximum size aggregate is $^3/_8$ inch (9.5 mm) or smaller, the test panels shall have minimum dimensions of 12 inches by 12 inches (305 mm by 305 mm). Panels shall be shot in the same position as the work, during the course of the work and by the nozzlemen doing the work. The conditions under which the panels are cured shall be the same as the work.

1914.10.3 Acceptance criteria. The average compressive strength of three cores from the in-place work or a single test panel shall equal or exceed $0.85 f'_c$ with no single core less than $0.75 f'_c$. The average compressive strength of three cubes taken from the in-place work or a single test panel shall equal or exceed f'_c with no individual cube less than $0.88 f'_c$. To check accuracy, locations represented by erratic core or cube strengths shall be retested.

SECTION 1915
REINFORCED GYPSUM CONCRETE

1915.1 General. Reinforced gypsum concrete shall comply with the requirements of ASTM C 317 and ASTM C 956.

1915.2 Minimum thickness. The minimum thickness of reinforced gypsum concrete shall be 2 inches (51 mm) except the minimum required thickness shall be reduced to $1^1/_2$ inches (38 mm), provided the following conditions are satisfied:

1. The overall thickness, including the formboard, is not less than 2 inches (51 mm).

2. The clear span of the gypsum concrete between supports does not exceed 33 inches (838 mm).

3. Diaphragm action is not required.

4. The design live load does not exceed 40 pounds per square foot (psf) (1915 Pa).

SECTION 1916
CONCRETE-FILLED PIPE COLUMNS

1916.1 General. Concrete-filled pipe columns shall be manufactured from standard, extra-strong or double-extra-strong steel pipe or tubing that is filled with concrete so placed and manipulated as to secure maximum density and to ensure complete filling of the pipe without voids.

1916.2 Design. The safe supporting capacity of concrete-filled pipe columns shall be computed in accordance with the approved rules or as determined by a test.

1916.3 Connections. Caps, base plates and connections shall be of approved types and shall be positively attached to the shell and anchored to the concrete core. Welding of brackets without mechanical anchorage shall be prohibited. Where the pipe is slotted to accommodate webs of brackets or other connections, the integrity of the shell shall be restored by welding to ensure hooping action of the composite section.

1916.4 Reinforcement. To increase the safe load-supporting capacity of concrete-filled pipe columns, the steel reinforcement shall be in the form of rods, structural shapes or pipe embedded in the concrete core with sufficient clearance to ensure the composite action of the section, but not nearer than 1 inch (25 mm) to the exterior steel shell. Structural shapes used as reinforcement shall be milled to ensure bearing on cap and base plates.

1916.5 Fire-resistance-rating protection. Pipe columns shall be of such size or so protected as to develop the required fire-resistance ratings specified in Table 601. Where an outer steel shell is used to enclose the fire-resistant covering, the shell shall not be included in the calculations for strength of the column section. The minimum diameter of pipe columns shall be 4 inches (102 mm) except that in structures of Type V construction not exceeding three stories or 40 feet (12 192 mm) in height, pipe columns used in the basement and as secondary steel members shall have a minimum diameter of 3 inches (76 mm).

1916.6 Approvals. Details of column connections and splices shall be shop fabricated by approved methods and shall be approved only after tests in accordance with the approved rules. Shop-fabricated concrete-filled pipe columns shall be inspected by the building official or by an approved representative of the manufacturer at the plant.

CHAPTER 20

ALUMINUM

SECTION 2001
GENERAL

2001.1 Scope. This chapter shall govern the quality, design, fabrication and erection of aluminum.

SECTION 2002
MATERIALS

2002.1 General. Aluminum used for structural purposes in buildings and structures shall comply with AA ASM 35 and Parts 1-A and 1-B of the Aluminum Design Manual. The nominal loads shall be the minimum design loads required by Chapter 16.

CHAPTER 21

MASONRY

SECTION 2101
GENERAL

2101.1 Scope. This chapter shall govern the materials, design, construction and quality of masonry.

2101.2 Design methods. Masonry shall comply with the provisions of one of the following design methods in this chapter as well as the requirements of Sections 2101 through 2104. Masonry designed by the working stress design provisions of Section 2101.2.1, the strength design provisions of Section 2101.2.2 or the prestressed masonry provisions of Section 2101.2.3 shall comply with Section 2105.

2101.2.1 Working stress design. Masonry designed by the working stress design method shall comply with the provisions of Sections 2106 and 2107.

2101.2.2 Strength design. Masonry designed by the strength design method shall comply with the provisions of Sections 2106 and 2108.

2101.2.3 Prestressed masonry. Prestressed masonry shall be designed in accordance with Chapters 1 and 4 of ACI 530/ASCE 5/TMS 402 and Section 2106. Special inspection during construction shall be provided as set forth in Section 1704.5.

2101.2.4 Empirical design. Masonry designed by the empirical design method shall comply with the provisions of Sections 2106 and 2109 or Chapter 5 of ACI 530/ASCE 5/TMS 402.

2101.2.5 Glass masonry. Glass masonry shall comply with the provisions of Section 2110 or with the requirements of Chapter 7 of ACI 530/ASCE 5/TMS 402.

2101.2.6 Masonry veneer. Masonry veneer shall comply with the provisions of Chapter 14.

2101.3 Construction documents. The construction documents shall show all of the items required by this code including the following:

1. Specified size, grade, type and location of reinforcement, anchors and wall ties.
2. Reinforcing bars to be welded and welding procedure.
3. Size and location of structural elements.
4. Provisions for dimensional changes resulting from elastic deformation, creep, shrinkage, temperature and moisture.

2101.3.1 Fireplace drawings. The construction documents shall describe in sufficient detail the location, size and construction of masonry fireplaces. The thickness and characteristics of materials and the clearances from walls, partitions and ceilings shall be clearly indicated.

SECTION 2102
DEFINITIONS AND NOTATIONS

2102.1 General. The following words and terms shall, for the purposes of this chapter and as used elsewhere in this code, have the meanings shown herein.

ADOBE CONSTRUCTION. Construction in which the exterior load-bearing and nonload-bearing walls and partitions are of unfired clay masonry units, and floors, roofs and interior framing are wholly or partly of wood or other approved materials.

> **Adobe, stabilized.** Unfired clay masonry units to which admixtures, such as emulsified asphalt, are added during the manufacturing process to limit the units' water absorption so as to increase their durability.

> **Adobe, unstabilized.** Unfired clay masonry units that do not meet the definition of "Adobe, stabilized."

ANCHOR. Metal rod, wire or strap that secures masonry to its structural support.

ARCHITECTURAL TERRA COTTA. Plain or ornamental hard-burned modified clay units, larger in size than brick, with glazed or unglazed ceramic finish.

AREA.

> **Bedded.** The area of the surface of a masonry unit that is in contact with mortar in the plane of the joint.

> **Gross cross-sectional.** The area delineated by the out-to-out specified dimensions of masonry in the plane under consideration.

> **Net cross-sectional.** The area of masonry units, grout and mortar crossed by the plane under consideration based on out-to-out specified dimensions.

BED JOINT. The horizontal layer of mortar on which a masonry unit is laid.

BOND BEAM. A horizontal grouted element within masonry in which reinforcement is embedded.

BOND REINFORCING. The adhesion between steel reinforcement and mortar or grout.

BRICK.

> **Calcium silicate (sand lime brick).** A masonry unit made of sand and lime.

> **Clay or shale.** A masonry unit made of clay or shale, usually formed into a rectangular prism while in the plastic state and burned or fired in a kiln.

> **Concrete.** A masonry unit having the approximate shape of a rectangular prism and composed of inert aggregate particles embedded in a hardened cementitious matrix.

BUTTRESS. A projecting part of a masonry wall built integrally therewith to provide lateral stability.

CAST STONE. A building stone manufactured from portland cement concrete precast and used as a trim, veneer or facing on or in buildings or structures.

CELL. A void space having a gross cross-sectional area greater than $1^1/_2$ square inches (967 mm^2).

CHIMNEY. A primarily vertical enclosure containing one or more passageways for conveying flue gases to the outside atmosphere.

CHIMNEY TYPES.

High-heat appliance type. An approved chimney for removing the products of combustion from fuel-burning, high-heat appliances producing combustion gases in excess of 2,000°F (1093°C) measured at the appliance flue outlet (see Section 2113.11.3).

Low-heat appliance type. An approved chimney for removing the products of combustion from fuel-burning, low-heat appliances producing combustion gases not in excess of 1,000°F (538°C) under normal operating conditions, but capable of producing combustion gases of 1,400°F (760°C) during intermittent forces firing for periods up to 1 hour. Temperatures shall be measured at the appliance flue outlet.

Masonry type. A field-constructed chimney of solid masonry units or stones.

Medium-heat appliance type. An approved chimney for removing the products of combustion from fuel-burning, medium-heat appliances producing combustion gases not exceeding 2,000°F (1093°C) measured at the appliance flue outlet (see Section 2113.11.2).

CLEANOUT. An opening to the bottom of a grout space of sufficient size and spacing to allow the removal of debris.

COLLAR JOINT. Vertical longitudinal joint between wythes of masonry or between masonry and backup construction that is permitted to be filled with mortar or grout.

COLUMN, MASONRY. An isolated vertical member whose horizontal dimension measured at right angles to its thickness does not exceed three times its thickness and whose height is at least four times its thickness.

COMPOSITE ACTION. Transfer of stress between components of a member designed so that in resisting loads, the combined components act together as a single member.

COMPOSITE MASONRY. Multiwythe masonry members acting with composite action.

COMPRESSIVE STRENGTH OF MASONRY. Maximum compressive force resisted per unit of net cross-sectional area of masonry, determined by the testing of masonry prisms or a function of individual masonry units, mortar and grout.

CONNECTOR. A mechanical device for securing two or more pieces, parts or members together, including anchors, wall ties and fasteners.

COVER. Distance between surface of reinforcing bar and edge of member.

DIAPHRAGM. A roof or floor system designed to transmit lateral forces to shear walls or other lateral-load-resisting elements.

DIMENSIONS.

Actual. The measured dimension of a masonry unit or element.

Nominal. A dimension equal to a specified dimension plus an allowance for the joints with which the units are to be laid. Thickness is given first, followed by height and then length.

Specified. The dimensions specified for the manufacture or construction of masonry, masonry units, joints or any other component of a structure.

EFFECTIVE HEIGHT. For braced members, the effective height is the clear height between lateral supports and is used for calculating the slenderness ratio. The effective height for unbraced members is calculated in accordance with engineering mechanics.

FIREPLACE. A hearth and fire chamber or similar prepared place in which a fire may be made and which is built in conjunction with a chimney.

FIREPLACE THROAT. The opening between the top of the firebox and the smoke chamber.

GROUTED MASONRY.

Grouted hollow-unit masonry. That form of grouted masonry construction in which certain designated cells of hollow units are continuously filled with grout.

Grouted multiwythe masonry. That form of grouted masonry construction in which the space between the wythes is solidly or periodically filled with grout.

HEAD JOINT. Vertical mortar joint placed between masonry units within the wythe at the time the masonry units are laid.

HEADER (Bonder). A masonry unit that connects two or more adjacent wythes of masonry.

HEIGHT, WALLS. The vertical distance from the foundation wall or other immediate support of such wall to the top of the wall.

MASONRY. A built-up construction or combination of building units or materials of clay, shale, concrete, glass, gypsum, stone or other approved units bonded together with or without mortar or grout or other accepted method of joining.

Ashlar masonry. Masonry composed of various sized rectangular units having sawed, dressed or squared bed surfaces, properly bonded and laid in mortar.

Coursed ashlar. Ashlar masonry laid in courses of stone of equal height for each course, although different courses shall be permitted to be of varying height.

Glass unit masonry. Nonload-bearing masonry composed of glass units bonded by mortar.

Plain masonry. Masonry in which the tensile resistance of the masonry is taken into consideration and the effects of stresses in reinforcement are neglected.

Random ashlar. Ashlar masonry laid in courses of stone set without continuous joints and laid up without drawn patterns. When composed of material cut into modular heights, discontinuous but aligned horizontal joints are discernible.

Reinforced masonry. Masonry construction in which reinforcement acting in conjunction with the masonry is used to resist forces.

Solid masonry. Masonry consisting of solid masonry units laid contiguously with the joints between the units filled with mortar.

MASONRY UNIT. Brick, tile, stone, glass block or concrete block conforming to the requirements specified in Section 2103.

Clay. A building unit larger in size than a brick, composed of burned clay, shale, fired clay or mixtures thereof.

Concrete. A building unit or block larger in size than 12 inches by 4 inches by 4 inches (305 mm by 102 mm by 102 mm) made of cement and suitable aggregates.

Hollow. A masonry unit whose net cross-sectional area in any plane parallel to the load-bearing surface is less than 75 percent of its gross cross-sectional area measured in the same plane.

Solid. A masonry unit whose net cross-sectional area in every plane parallel to the load-bearing surface is 75 percent or more of its gross cross-sectional area measured in the same plane.

MEAN DAILY TEMPERATURE. The average daily temperature of temperature extremes predicted by a local weather bureau for the next 24 hours.

MORTAR. A plastic mixture of approved cementitious materials, fine aggregates and water used to bond masonry or other structural units.

MORTAR, SURFACE-BONDING. A mixture to bond concrete masonry units that contains hydraulic cement, glass fiber reinforcement with or without inorganic fillers or organic modifiers and water.

PLASTIC HINGE. The zone in a structural member in which the yield moment is anticipated to be exceeded under loading combinations that include earthquakes.

PRESTRESSED MASONRY. Masonry in which internal stresses have been introduced to counteract potential tensile stresses in masonry resulting from applied loads.

PRISM. An assemblage of masonry units and mortar with or without grout used as a test specimen for determining properties of the masonry.

RUBBLE MASONRY. Masonry composed of roughly shaped stones.

Coursed rubble. Masonry composed of roughly shaped stones fitting approximately on level beds and well bonded.

Random rubble. Masonry composed of roughly shaped stones laid without regularity of coursing but well bonded and fitted together to form well-divided joints.

Rough or ordinary rubble. Masonry composed of unsquared field stones laid without regularity of coursing but well bonded.

RUNNING BOND. The placement of masonry units such that head joints in successive courses are horizontally offset at least one-quarter the unit length.

SHEAR WALL.

Detailed plain masonry shear wall. A masonry shear wall designed to resist lateral forces neglecting stresses in reinforcement, and designed in accordance with Section 2106.1.1.

Intermediate prestressed masonry shear wall. A prestressed masonry shear wall designed to resist lateral forces considering stresses in reinforcement, and designed in accordance with Section 2106.1.1.2.

Intermediate reinforced masonry shear wall. A masonry shear wall designed to resist lateral forces considering stresses in reinforcement, and designed in accordance with Section 2106.1.1.

Ordinary plain masonry shear wall. A masonry shear wall designed to resist lateral forces neglecting stresses in reinforcement, and designed in accordance with Section 2106.1.1.

Ordinary plain prestressed masonry shear wall. A prestressed masonry shear wall designed to resist lateral forces considering stresses in reinforcement, and designed in accordance with Section 2106.1.1.1.

Ordinary reinforced masonry shear wall. A masonry shear wall designed to resist lateral forces considering stresses in reinforcement, and designed in accordance with Section 2106.1.1.

Special prestressed masonry shear wall. A prestressed masonry shear wall designed to resist lateral forces considering stresses in reinforcement and designed in accordance with Section 2106.1.1.3 except that only grouted, laterally restrained tendons are used.

Special reinforced masonry shear wall. A masonry shear wall designed to resist lateral forces considering stresses in reinforcement, and designed in accordance with Section 2106.1.1.

SHELL. The outer portion of a hollow masonry unit as placed in masonry.

SPECIFIED. Required by construction documents.

SPECIFIED COMPRESSIVE STRENGTH OF MASONRY, f'_m. Minimum compressive strength, expressed as force per unit of net cross-sectional area, required of the masonry used in construction by the construction documents, and upon which the project design is based. Whenever the quantity f'_m is under the radical sign, the square root of numerical value only is intended and the result has units of pounds per square inch (psi) (Mpa).

STACK BOND. The placement of masonry units in a bond pattern is such that head joints in successive courses are vertically aligned. For the purpose of this code, requirements for stack bond shall apply to masonry laid in other than running bond.

STONE MASONRY. Masonry composed of field, quarried or cast stone units bonded by mortar.

Ashlar stone masonry. Stone masonry composed of rectangular units having sawed, dressed or squared bed surfaces and bonded by mortar.

Rubble stone masonry. Stone masonry composed of irregular-shaped units bonded by mortar.

STRENGTH.

Design strength. Nominal strength multiplied by a strength reduction factor.

Nominal strength. Strength of a member or cross section calculated in accordance with these provisions before application of any strength-reduction factors.

Required strength. Strength of a member or cross section required to resist factored loads.

TIE, LATERAL. Loop of reinforcing bar or wire enclosing longitudinal reinforcement.

TIE, WALL. A connector that connects wythes of masonry walls together.

TILE. A ceramic surface unit, usually relatively thin in relation to facial area, made from clay or a mixture of clay or other ceramic materials, called the body of the tile, having either a "glazed" or "unglazed" face and fired above red heat in the course of manufacture to a temperature sufficiently high enough to produce specific physical properties and characteristics.

TILE, STRUCTURAL CLAY. A hollow masonry unit composed of burned clay, shale, fire clay or mixture thereof, and having parallel cells.

WALL. A vertical element with a horizontal length-to-thickness ratio greater than three, used to enclose space.

Cavity wall. A wall built of masonry units or of concrete, or a combination of these materials, arranged to provide an airspace within the wall, and in which the inner and outer parts of the wall are tied together with metal ties.

Composite wall. A wall built of a combination of two or more masonry units bonded together, one forming the backup and the other forming the facing elements.

Dry-stacked, surface-bonded walls. A wall built of concrete masonry units where the units are stacked dry, without mortar on the bed or head joints, and where both sides of the wall are coated with a surface-bonding mortar.

Masonry-bonded hollow wall. A wall built of masonry units so arranged as to provide an airspace within the wall, and in which the facing and backing of the wall are bonded together with masonry units.

Parapet wall. The part of any wall entirely above the roof line.

WEB. An interior solid portion of a hollow masonry unit as placed in masonry.

WYTHE. Each continuous, vertical section of a wall, one masonry unit in thickness.

NOTATIONS.

A_n = Net cross-sectional area of masonry, square inches (mm^2).

b = Effective width of rectangular member or width of flange for T and I sections, inches (mm).

d_b = Diameter of reinforcement, inches (mm).

f_r = Modulus of rupture, psi (MPa).

f_y = Specified yield stress of the reinforcement or the anchor bolt, psi (MPa).

f'_m = Specified compressive strength of masonry at age of 28 days, psi (MPa).

K = The lesser of the masonry cover, clear spacing between adjacent reinforcement, or five times d_b, inches (mm).

L_s = Distance between supports, inches (mm).

L_w = Length of wall, inches (mm).

l_d = Required development length of reinforcement, inches (mm).

l_{de} = Embedment length of reinforcement, inches (mm).

P_w = Weight of wall tributary to section under consideration, pounds (N).

t = Specified wall thickness dimension or the least lateral dimension of a column, inches (mm).

V_n = Nominal shear strength, pounds (N).

V_u = Required shear strength due to factored loads, pounds (N).

W = Wind load, or related internal moments in forces.

γ = Reinforcement size factor.

ρ_n = Ratio of distributed shear reinforcement on plane perpendicular to plane of A_{mv}.

ρ_{max} = Maximum reinforcement ratio.

ϕ = Strength reduction factor.

SECTION 2103
MASONRY CONSTRUCTION MATERIALS

2103.1 Concrete masonry units. Concrete masonry units shall conform to the following standards: ASTM C 55 for concrete brick; ASTM C 73 for calcium silicate face brick; ASTM C 90 for load-bearing concrete masonry units or ASTM C 744 for prefaced concrete and calcium silicate masonry units.

2103.2 Clay or shale masonry units. Clay or shale masonry units shall conform to the following standards: ASTM C 34 for structural clay load-bearing wall tile; ASTM C 56 for structural clay nonload-bearing wall tile; ASTM C 62 for building brick (solid masonry units made from clay or shale); ASTM C 1088 for solid units of thin veneer brick; ASTM C 126 for ceramic-glazed structural clay facing tile, facing brick and solid masonry units; ASTM C 212 for structural clay facing tile; ASTM C 216 for facing brick (solid masonry units made from clay or shale) and ASTM C 652 for hollow brick (hollow masonry units made from clay or shale).

Exception: Structural clay tile for nonstructural use in fireproofing of structural members and in wall furring shall not be required to meet the compressive strength specifications. The fire-resistance rating shall be determined in accordance with ASTM E 119 and shall comply with the requirements of Table 602.

2103.3 Stone masonry units. Stone masonry units shall conform to the following standards: ASTM C 503 for marble building stone (exterior): ASTM C 568 for limestone building stone; ASTM C 615 for granite building stone; ASTM C 616 for sandstone building stone or ASTM C 629 for slate building stone.

2103.4 Ceramic tile. Ceramic tile shall be as defined in, and shall conform to the requirements of, ANSI A137.1.

2103.5 Glass unit masonry. Hollow glass units shall be partially evacuated and have a minimum average glass face thickness of $^3/_{16}$ inch (4.8 mm). Solid glass-block units shall be provided when required. The surfaces of units intended to be in contact with mortar shall be treated with a polyvinyl butyral coating or latex-based paint. Reclaimed units shall not be used.

2103.6 Second-hand units. Second-hand masonry units shall not be reused unless they conform to the requirements of new units. The units shall be of whole, sound materials and free from cracks and other defects that will interfere with proper laying or use. Old mortar shall be cleaned from the unit before reuse.

2103.7 Mortar. Mortar for use in masonry construction shall conform to ASTM C 270 and shall conform to the proportion specifications of Table 2103.7(1) or the property specifications of Table 2103.7(2). Type S or N mortar shall be used for glass unit masonry. The amount of water used in mortar for glass unit masonry shall be adjusted to account for the lack of absorption. Retempering of mortar for glass unit masonry shall not be permitted after initial set. Unused mortar shall be discarded within $2^1/_2$ hours after initial mixing except that unused mortar for glass unit masonry shall be discarded within $1^1/_2$ hours after initial mixing.

2103.8 Surface-bonding mortar. Surface-bonding mortar shall comply with ASTM C 887. Surface bonding of concrete masonry units shall comply with ASTM C 946.

2103.9 Mortars for ceramic wall and floor tile. Portland cement mortars for installing ceramic wall and floor tile shall comply with ANSI A108.1A and ANSI A108.1B and be of the compositions indicated in Table 2103.9.

TABLE 2103.7(1)
MORTAR PROPORTIONS

MORTAR	TYPE	PROPORTIONS BY VOLUME (cementitious materials)							HYDRATED LIME[e] OR LIME PUTTY	AGGREGATE MEASURED IN A DAMP, LOOSE CONDITION
		Portland cement[a] or blended cement[b]	Masonry cement[c]			Mortar cement[d]				
			M	S	N	M	S	N		
Cement-lime	M	1	—	—	—	—	—	—	$^1/_4$	
	S	1	—	—	—	—	—	—	over $^1/_4$ to $^1/_2$	
	N	1	—	—	—	—	—	—	over $^1/_2$ to $1^1/_4$	
	O	1	—	—	—	—	—	—	over $1^1/_4$ to $2^1/_2$	
Mortar cement	M	1	—	—	—	—	—	1	—	Not less than $2^1/_4$ and not more than 3 times the sum of the separate volumes of cementitious materials
	M	—	—	—	—	1	—	—	—	
	S	$^1/_2$	—	—	—	—	—	1	—	
	S	—	—	—	—	—	1	—	—	
	N	—	—	—	—	—	—	1	—	
	O	—	—	—	—	—	—	1	—	
Masonry cement	M	1	—	—	1	—	—	—	—	
	M	—	1	—	—	—	—	—	—	
	S	$^1/_2$	—	—	1	—	—	—	—	
	S	—	—	1	—	—	—	—	—	
	N	—	—	—	1	—	—	—	—	
	O	—	—	—	1	—	—	—	—	

a. Portland cement conforming to the requirements of ASTM C 150.
b. Blended cement conforming to the requirements of ASTM C 595.
c. Masonry cement conforming to the requirements of ASTM C 91.
d. Mortar cement conforming to the requirements of ASTM C 1329.
e. Hydrated lime conforming to the requirements of ASTM C 207.

TABLE 2103.7(2)
MORTAR PROPERTIES[a]

MORTAR	TYPE	AVERAGE COMPRESSIVE[b] STRENGTH AT 28 DAYS minimum (psi)	WATER RETENTION minimum (%)	AIR CONTENT maximum (%)
Cement-lime	M	2,500	75	12
	S	1,800	75	12
	N	750	75	14[c]
	O	350	75	14[c]
Mortar cement	M	2,500	75	12
	S	1,800	75	12
	N	750	75	14[c]
	O	350	75	14[c]
Masonry cement	M	2,500	75	18
	S	1,800	75	18
	N	750	75	20[d]
	O	350	75	20[d]

For SI: 1 inch = 25.4 mm, 1 pound per square inch = 6.895 kPa.

a. This aggregate ratio (measured in damp, loose condition) shall not be less than $2^1/_4$ and not more than 3 times the sum of the separate volumes of cementitious materials.

b. Average of three 2-inch cubes of laboratory-prepared mortar, in accordance with ASTM C 270.

c. When structural reinforcement is incorporated in cement-lime or mortar cement mortars, the maximum air content shall not exceed 12 percent.

d. When structural reinforcement is incorporated in masonry cement mortar, the maximum air content shall not exceed 18 percent.

TABLE 2103.9
CERAMIC TILE MORTAR COMPOSITIONS

LOCATION	MORTAR	COMPOSITION
Walls	Scratchcoat	1 cement; $^1/_5$ hydrated lime; 4 dry or 5 damp sand
	Setting bed and leveling coat	1 cement; $^1/_2$ hydrated lime; 5 damp sand to 1 cement 1 hydrated lime, 7 damp sand
Floors	Setting bed	1 cement; $^1/_{10}$ hydrated lime; 5 dry or 6 damp sand; or 1 cement; 5 dry or 6 damp sand
Ceilings	Scratchcoat and sand bed	1 cement; $^1/_2$ hydrated lime; $2^1/_2$ dry sand or 3 damp sand

2103.9.1 Dry-set portland cement mortars. Premixed prepared portland cement mortars, which require only the addition of water and are used in the installation of ceramic tile, shall comply with ANSI A118.1. The shear bond strength for tile set in such mortar shall be as required in accordance with ANSI A118.1. Tile set in dry-set portland cement mortar shall be installed in accordance with ANSI A108.5.

2103.9.2 Electrically conductive dry-set mortars. Premixed prepared portland cement mortars, which require only the addition of water and comply with ANSI A118.2, shall be used in the installation of electrically conductive ceramic tile. Tile set in electrically conductive dry-set mortar shall be installed in accordance with ANSI A108.7.

2103.9.3 Latex-modified portland cement mortar. Latex-modified portland cement thin-set mortars in which latex is added to dry-set mortar as a replacement for all or part of the gauging water that are used for the installation of ceramic tile shall comply with ANSI A118.4. Tile set in latex-modified portland cement shall be installed in accordance with ANSI A108.5.

2103.9.4 Epoxy mortar. Ceramic tile set and grouted with chemical-resistant epoxy shall comply with ANSI A118.3. Tile set and grouted with epoxy shall be installed in accordance with ANSI A108.6.

2103.9.5 Furan mortar and grout. Chemical-resistant furan mortar and grout that are used to install ceramic tile shall comply with ANSI A118.5. Tile set and grouted with furan shall be installed in accordance with ANSI A108.8.

2103.9.6 Modified epoxy-emulsion mortar and grout. Modified epoxy-emulsion mortar and grout that are used to install ceramic tile shall comply with ANSI A118.8. Tile set and grouted with modified epoxy-emulsion mortar and grout shall be installed in accordance with ANSI A108.9.

2103.9.7 Organic adhesives. Water-resistant organic adhesives used for the installation of ceramic tile shall comply with ANSI A136.1. The shear bond strength after water immersion shall not be less than 40 psi (275 kPa) for Type I adhesive, and not less than 20 psi (138 kPa) for Type II adhesive, when tested in accordance with ANSI A136.1. Tile set in organic adhesives shall be installed in accordance with ANSI A108.4.

2103.9.8 Portland cement grouts. Portland cement grouts used for the installation of ceramic tile shall comply with ANSI A118.6. Portland cement grouts for tile work shall be installed in accordance with ANSI A108.10.

2103.10 Grout. Grout shall conform to Table 2103.10 or to ASTM C 476. When grout conforms to ASTM C 476, the grout shall be specified by proportion requirements or property requirements.

TABLE 2103.10
GROUT PROPORTIONS BY VOLUME FOR
MASONRY CONSTRUCTION

TYPE	PARTS BY VOLUME OF PORTLAND CEMENT OR BLENDED CEMENT	PARTS BY VOLUME OF HYDRATED LIME OR LIME PUTTY	AGGREGATE, MEASURED IN A DAMP, LOOSE CONDITION	
			Fine	Coarse
Fine grout	1	$0-{}^1/_{10}$	$2^1/_4$-3 times the sum of the volumes of the cementitious materials	—
Coarse grout	1	$0-{}^1/_{10}$	$2^1/_4$-3 times the sum of the volumes of the cementitious materials	1-2 times the sum of the volumes of the cementitious materials

2103.11 Metal reinforcement and accessories. Metal reinforcement and accessories shall conform to Sections 2103.11.1 through 2103.11.7.

2103.11.1 Deformed reinforcing bars. Deformed reinforcing bars shall conform to one of the following standards: ASTM A 615 for deformed and plain billet-steel bars for concrete reinforcement; ASTM A 706 for low-alloy steel deformed bars for concrete reinforcement; ASTM A 767 for zinc-coated reinforcing steel bars; ASTM A 775 for epoxy-coated reinforcing steel bars and ASTM A 996 for rail steel and axle steel deformed bars for concrete reinforcement.

2103.11.2 Joint reinforcement. Joint reinforcement shall comply with ASTM A 951. The maximum spacing of crosswires in ladder-type joint reinforcement and of point of connection of cross wires to longitudinal wires of truss-type reinforcement shall be 16 inches (400 mm).

2103.11.3 Deformed reinforcing wire. Deformed reinforcing wire shall conform to ASTM A 496.

2103.11.4 Wire fabric. Wire fabric shall conform to ASTM A 185 for plain steel-welded wire fabric for concrete reinforcement or ASTM A 496 for welded deformed steel wire fabric for concrete reinforcement.

2103.11.5 Anchors, ties and accessories. Anchors, ties and accessories shall conform to the following standards: ASTM A 36 for structural steel; ASTM A 82 for plain steel wire for concrete reinforcement; ASTM A 185 for plain steel-welded wire fabric for concrete reinforcement; ASTM A 167, Type 304, for stainless and heat-resisting chromium-nickel steel plate, sheet and strip and ASTM A 366 for cold-rolled carbon steel sheet, commercial quality.

2103.11.6 Prestressing tendons. Prestressing tendons shall conform to one of the following standards:

a. Wire . . . . ASTM A 421

b. Low-relaxation wire . . ASTM A 421

c. Strand . . . . ASTM A 416

d. Low-relaxation strand. . ASTM A 416

e. Bar . . . . ASTM A 722

Exceptions:

1. Wire, strands and bars not specifically listed in ASTM A 421, ASTM A 416 or ASTM A 722 are permitted, provided they conform to the minimum requirements in ASTM A 421, ASTM A 416, or ASTM A 722 and are approved by the architect/engineer.

2. Bars and wires of less than 150 kips per square inch (ksi) (1034 MPa) tensile strength and conforming to ASTM A 82, ASTM A 510, ASTM A 615, ASTM A 616, ASTM A 996 or ASTM A 706/A 706 M are permitted to be used as prestressed tendons provided that:

   2.1. The stress relaxation properties have been assessed by tests according to ASTM E 328 for the maximum permissible stress in the tendon.

   2.2. Other nonstress-related requirements of ACI 530/ASCE 5/TMS 402, Chapter 4, addressing prestressing tendons are met.

2103.11.7 Corrosion protection. Corrosion protection for prestressing tendons, prestressing anchorages, couplers and end block shall comply with the requirements of ACI 530.1/ASCE 6/TMS 602, Article 2.4G. Corrosion protection for carbon steel accessories used in exterior wall construction or interior walls exposed to a mean relative humidity exceeding 75 percent shall comply with either Section 2103.11.7.1 or 2103.11.7.2. Corrosion protection for carbon steel accessories used in interior walls exposed to a mean relative humidity equal to or less than 75 percent shall comply with either Section 2103.11.7.1, 2103.11.7.2 or 2103.11.7.3.

2103.11.7.1 Hot-dipped galvanized. Apply a hot-dipped galvanized coating after fabrication as follows:

1. For joint reinforcement, wall ties, anchors and inserts, apply a minimum coating of 1.5 ounces per square foot (psf) (458 g/m²) complying with the requirements of ASTM A 153, Class B.

2. For sheet metal ties and sheet metal anchors, comply with the requirements of ASTM A 153, Class B.

3. For steel plates and bars, comply with the requirements of either ASTM A 123 or ASTM A 153, Class B.

2103.11.7.2 Epoxy coatings. Carbon steel accessories shall be epoxy coated as follows:

1. For joint reinforcement, comply with the requirements of ASTM A 884 Class B, Type 2 – 18 mils (457 μm).

2. For wire ties and anchors, comply with the requirements of ASTM A 899 Class C —20 mils (508 μm).

3. For sheet metal ties and anchors, provide a minimum thickness of 20 mils (508 μm) or in accordance with the manufacturer's specification.

2103.11.7.3 Mill galvanized. Apply a mill galvanized coating as follows:

1. For joint reinforcement, wall ties, anchors and inserts, apply a minimum coating of 0.1 ounce psf

(31g/m²) complying with the requirements of ASTM A 641.

2. For sheet metal ties and sheet metal anchors, apply a minimum coating complying with Coating Designation G-60 according to the requirements of ASTM A 653.

3. For anchor bolts, steel plates or bars not exposed to the earth, weather or a mean relative humidity exceeding 75 percent, a coating is not required.

2103.11.8 Tests. Where unidentified reinforcement is approved for use, not less than three tension and three bending tests shall be made on representative specimens of the reinforcement from each shipment and grade of reinforcing steel proposed for use in the work.

SECTION 2104
CONSTRUCTION

2104.1 Masonry construction. Masonry construction shall comply with the requirements of Sections 2104.1.1 through 2104.5 and with ACI 530.1/ASCE 6/TMS 602.

2104.1.1 Tolerances. Masonry, except masonry veneer, shall be constructed within the tolerances specified in ACI 530.1/ASCE 6/TMS 602.

2104.1.2 Placing mortar and units. Placement of mortar and units shall comply with Sections 2104.1.2.1 through 2104.1.2.5.

2104.1.2.1 Bed and head joints. Unless otherwise required or indicated on the construction documents, head and bed joints shall be $^3/_8$ inch (9.5 mm) thick, except that the thickness of the bed joint of the starting course placed over foundations shall not be less than $^1/_4$ inch (6.4 mm) and not more than $^3/_4$ inch (19.1 mm).

2104.1.2.1.1 Open-end units. Open-end units with beveled ends shall be fully grouted. Head joints of open-end units with beveled ends need not be mortared. The beveled ends shall form a grout key that permits grouts within $^5/_8$ inch (15.9 mm) of the face of the unit. The units shall be tightly butted to prevent leakage of the grout.

2104.1.2.2 Hollow units. Hollow units shall be placed such that face shells of bed joints are fully mortared. Webs shall be fully mortared in all courses of piers, columns, pilasters, in the starting course on foundations where adjacent cells or cavities are to be grouted, and where otherwise required. Head joints shall be mortared a minimum distance from each face equal to the face shell thickness of the unit.

2104.1.2.3 Solid units. Unless otherwise required or indicated on the construction documents, solid units shall be placed in fully mortared bed and head joints. The ends of the units shall be completely buttered. Head joints shall not be filled by slushing with mortar. Head joints shall be constructed by shoving mortar tight against the adjoining unit. Bed joints shall not be furrowed deep enough to produce voids.

2104.1.2.4 Glass unit masonry. Glass units shall be placed so head and bed joints are filled solidly. Mortar shall not be furrowed.

Unless otherwise required, head and bed joints of glass unit masonry shall be $^1/_4$ inch (6.4 mm) thick, except that vertical joint thickness of radial panels shall not be less than $^1/_8$ inch (3.2 mm). The bed joint thickness tolerance shall be minus $^1/_{16}$ inch (1.6 mm) and plus $^1/_8$ inch (3.2 mm). The head joint thickness tolerance shall be plus or minus $^1/_8$ inch (3.2 mm).

2104.1.2.5 All units. Units shall be placed while the mortar is soft and plastic. Any unit disturbed to the extent that the initial bond is broken after initial positioning shall be removed and relaid in fresh mortar.

2104.1.3 Installation of wall ties. The ends of wall ties shall be embedded in mortar joints. Wall tie ends shall engage outer face shells of hollow units by at least $^1/_2$ inch (12.7 mm). Wire wall ties shall be embedded at least $1^1/_2$ inches (38 mm) into the mortar bed of solid masonry units or solid-grouted hollow units. Wall ties shall not be bent after being embedded in grout or mortar.

2104.1.4 Chases and recesses. Chases and recesses shall be constructed as masonry units are laid. Masonry directly above chases or recesses wider than 12 inches (305 mm) shall be supported on lintels.

2104.1.5 Lintels. The design for lintels shall be in accordance with the masonry design provisions of either Section 2107 or 2108. Minimum length of end support shall be 4 inches (102 mm).

2104.1.6 Support on wood. Masonry shall not be supported on wood girders or other forms of wood construction except as permitted in Section 2304.12.

2104.1.7 Masonry protection. The top of unfinished masonry work shall be covered to protect the masonry from the weather.

2104.1.8 Weep holes. Weep holes provided in the outside wythe of masonry walls shall be at a maximum spacing of 33 inches (838 mm) on center (o.c.). Weep holes shall not be less than $^3/_{16}$ inch (4.8 mm) in diameter.

2104.2 Corbeled masonry. The maximum corbeled projection beyond the face of the wall shall not be more than one-half of the wall thickness nor one-half the wythe thickness for hollow walls. The maximum projection of one unit shall neither exceed one-half the height of the unit nor one-third the thickness at right angles to the wall.

2104.2.1 Molded cornices. Unless structural support and anchorage are provided to resist the overturning moment, the center of gravity of projecting masonry or molded cornices shall lie within the middle one-third of the supporting wall. Terra cotta and metal cornices shall be provided with a structural frame of approved noncombustible material anchored in an approved manner.

2104.3 Cold weather construction. The cold weather construction provisions of ACI 530.1/ASCE 6/TMS 602, Article 1.8 C, or the following procedures shall be implemented when either the ambient temperature falls below 40°F (4°C) or the temperature of masonry units is below 40°F (4°C).

2104.3.1 Preparation.

1. Temperatures of masonry units shall not be less than 20°F (-7°C) when laid in the masonry. Masonry units containing frozen moisture, visible ice or snow on their surface shall not be laid.

2. Visible ice and snow shall be removed from the top surface of existing foundations and masonry to receive new construction. These surfaces shall be heated to above freezing, using methods that do not result in damage.

2104.3.2 Construction. The following requirements shall apply to work in progress and shall be based on ambient temperature.

2104.3.2.1 Construction requirements for temperatures between 40°F (4°C) and 32°F (0°F). The following construction requirements shall be met when the ambient temperature is between 40°F (4°C) and 32°F (0°C):

1. Glass unit masonry shall not be laid.

2. Water and aggregates used in mortar and grout shall not be heated above 140°F (60°C).

3. Mortar sand or mixing water shall be heated to produce mortar temperatures between 40°F (4°C) and 120°F (49°C) at the time of mixing. When water and aggregates for grout are below 32°F(0°C), they shall be heated.

2104.3.2.2 Construction requirements for temperatures between 32°F (0°C) and 25°F (-4°C). The requirements of Section 2104.3.2.1 and the following construction requirements shall be met when the ambient temperature is between 32°F (0°C) and 25°F (-4°C):

1. The mortar temperature shall be maintained above freezing until used in masonry.

2. Aggregates and mixing water for grout shall be heated to produce grout temperature between 70°F (21°C) and 120°F (49°C) at the time of mixing. Grout temperature shall maintained above 70°F (21°C) at the time of grout placement.

2104.3.2.3 Construction requirements for temperatures between 25°F (-4°C) and 20°F (-7°C). The requirements of Sections 2104.3.2.1 and 2104.3.2.2 and the following construction requirements shall be met when the ambient temperature is between 25°F (-4°C) and 20°F (-7°C):

1. Masonry surfaces under construction shall be heated to 40°F (4°C).

2. Wind breaks or enclosures shall be provided when the wind velocity exceeds 15 miles per hour (mph) (24 km/h).

3. Prior to grouting, masonry shall be heated to a minimum of 40°F (4°C).

2104.3.2.4. Construction requirements for temperatures below 20°F (-7°C). The requirements of Sections

2104.3.2.1, 2104.3.2.2 and 2104.3.2.3 and the following construction requirement shall be met when the ambient temperature is below 20°F (-7°C): Enclosures and auxiliary heat shall be provided to maintain air temperature within the enclosure to above 32°F (0°C).

2104.3.3 Protection. The requirements of this section and Sections 2104.3.3.1 through 2104.3.3.4 apply after the masonry is placed and shall be based on anticipated minimum daily temperature for grouted masonry and anticipated mean daily temperature for ungrouted masonry.

2104.3.3.1 Glass unit masonry. The temperature of glass unit masonry shall be maintained above 40°F (4°C) for 48 hours after construction.

2104.3.3.2 Protection requirements for temperatures between 40°F (4°C) and 25°F (-4°C). When the temperature is between 40°F (4°C) and 25°F (-4°C), newly constructed masonry shall be covered with a weather-resistive membrane for 24 hours after being completed.

2104.3.3.3 Protection requirements for temperatures between 25°F (-4°C) and 20°F (-7°C). When the temperature is between 25°F (-4°C) and 20°F (-7°C), newly constructed masonry shall be completely covered with weather-resistive insulating blankets, or equal protection, for 24 hours after being completed. The time period shall be extended to 48 hours for grouted masonry, unless the only cement in the grout is Type III portland cement.

2104.3.3.4 Protection requirements for temperatures below 20°F (-7°C). When the temperature is below 20°F (-7°C), newly constructed masonry shall be maintained at a temperature above 32°F (0°C) for at least 24 hours after being completed by using heated enclosures, electric heating blankets, infrared lamps or other acceptable methods. The time period shall be extended to 48 hours for grouted masonry, unless the only cement in the grout is Type III portland cement.

2104.4 Hot weather construction. The hot weather construction provisions of ACI 530.1/ASCE 6/TMS 602, Article 1.8 D, or the following procedures shall be implemented when the temperature or the temperature and wind-velocity limits of this section are exceeded.

2104.4.1 Preparation. The following requirements shall be met prior to conducting masonry work.

2104.4.1.1 Temperature. When the ambient temperature exceeds 100°F (38°C), or exceeds 90°F (32°C) with a wind velocity greater than 8 mph (13 km/h):

1. Necessary conditions and equipment shall be provided to produce mortar having a temperature below 120°F (49°C).

2. Sand piles shall be maintained in a damp, loose condition.

2104.4.1.2 Special conditions. When the ambient temperature exceeds 115°F (46°C), or 105°F (40°C) with a wind velocity greater than 8 mph (13 km/h), the requirements of Section 2104.4.1.1 shall be implemented, and materials and mixing equipment shall be shaded from direct sunlight.

2104.4.2 Construction. The following requirements shall be met while masonry work is in progress.

2104.4.2.1 Temperature. When the ambient temperature exceeds 100°F (38°C), or exceeds 90°F (32°C) with a wind velocity greater than 8 mph (13 km/h):

1. The temperature of mortar and grout shall be maintained below 120°F (49°C).

2. Mixers, mortar transport containers and mortar boards shall be flushed with cool water before they come into contact with mortar ingredients or mortar.

3. Mortar consistency shall be maintained by retempering with cool water.

4. Mortar shall be used within 2 hours of initial mixing.

2104.4.2.2 Special conditions. When the ambient temperature exceeds 115°F (46°C), or exceeds 105°F (40°C) with a wind velocity greater than 8 mph (13 km/h), the requirements of Section 2104.4.2.1 shall be implemented and cool mixing water shall be used for mortar and grout. The use of ice shall be permitted in the mixing water prior to use. Ice shall not be permitted in the mixing water when added to the other mortar or grout materials.

2104.4.3 Protection. When the mean daily temperature exceeds 100°F (38°C), or exceeds 90°F (32°C) with a wind velocity greater than 8 mph (13 km/h), newly constructed masonry shall be fog sprayed until damp at least three times a day until the masonry is three days old.

2104.5 Wetting of brick. Brick (clay or shale) at the time of laying shall require wetting if the unit's initial rate of water absorption exceeds 30 grams per 30 square inches (19 355 mm²) per minute or 0.035 ounce psi (1 g/645 mm²), as determined by ASTM C 67.

SECTION 2105
QUALITY ASSURANCE

2105.1 General. A quality assurance program shall be used to ensure that the constructed masonry is in compliance with the construction documents.

The quality assurance program shall comply with the inspection and testing requirements of Chapter 17.

2105.2 Acceptance relative to strength requirements.

2105.2.1 Compliance with f'_m. Compressive strength of masonry shall be considered satisfactory if the compressive strength of each masonry wythe and grouted collar joint equals or exceeds the value of f'_m.

2105.2.2 Determination of compressive strength. The compressive strength for each wythe shall be determined by the unit strength method or by the prism test method as specified herein.

2105.2.2.1 Unit strength method.

2105.2.2.1.1 Clay masonry. The compressive strength of masonry shall be determined based on the strength of the units and the type of mortar specified using Table 2105.2.2.1.1, provided:

1. Units conform to ASTM C 62, ASTM C 216 or ASTM C 652 and are sampled and tested in accordance with ASTM C 67.

2. Thickness of bed joints does not exceed $^5/_8$ inch (15.9 mm).

3. For grouted masonry, the grout meets one of the following requirements:

   3.1. Grout conforms to ASTM C 476.

   3.2. Minimum grout compressive strength equals f'_m but not less than 2,000 psi (13.79 MPa). The compressive strength of grout shall be determined in accordance with ASTM C 1019.

TABLE 2105.2.2.1.1
COMPRESSIVE STRENGTH OF CLAY MASONRY

NET AREA COMPRESSIVE STRENGTH OF CLAY MASONRY UNITS (psi)		NET AREA COMPRESSIVE STRENGTH OF MASONRY (psi)
Type M or S mortar	Type N mortar	
1,700	2,100	1,000
3,350	4,150	1,500
4,950	6,200	2,000
6,600	8,250	2,500
8,250	10,300	3,000
9,900	—	3,500
13,200	—	4,000

For SI: 1 pound per square inch = 0.00689 Mpa.

2105.2.2.1.2 Concrete masonry. The compressive strength of masonry shall be determined based on the strength of the unit and type of mortar specified using Table 2105.2.2.1.2, provided:

1. Units conform to ASTM C 55 or ASTM C 90 and are sampled and tested in accordance with ASTM C 140.

2. Thickness of bed joints does not exceed $^5/_8$ inch (15.9 mm).

3. For grouted masonry, the grout meets one of the following requirements:

   3.1. Grout conforms to ASTM C 476.

   3.2. Minimum grout compressive strength equals f'_m but not less than 2,000 psi (13.79 MPa). The compressive strength of grout shall be determined in accordance with ASTM C 1019.

TABLE 2105.2.2.1.2
COMPRESSIVE STRENGTH OF CONCRETE MASONRY

NET AREA COMPRESSIVE STRENGTH OF CONCRETE MASONRY UNITS (psi)		NET AREA COMPRESSIVE STRENGTH OF MASONRY (psi)[a]
Type M or S mortar	Type N mortar	
1,250	1,300	1,000
1,900	2,150	1,500
2,800	3,050	2,000
3,750	4,050	2,500
4,800	5,250	3,000

For SI: 1 inch = 25.4 mm, 1 pound per square inch = 0.00689 MPa.
a. For units less than 4 inches in height, 85 percent of the values listed.

2105.2.2.2 Prism test method.

2105.2.2.2.1 General. The compressive strength of masonry shall be determined by the prism test method:

1. Where specified in the construction documents.

2. Where masonry does not meet the requirements for application of the unit strength method in Section 2105.2.2.1.

2105.2.2.2.2 Number of prisms per test. A prism test shall consist of three prisms constructed and tested in accordance with ASTM C 1314.

2105.3 Testing prisms from constructed masonry. When approved by the building official, acceptance of masonry that does not meet the requirements of Section 2105.2.2.1 or 2105.2.2.2 shall be permitted to be based on tests of prisms cut from the masonry construction in accordance with Sections 2105.3.1, 2105.3.2 and 2105.3.3.

2105.3.1 Prism sampling and removal. A set of three masonry prisms that are at least 28 days old shall be saw cut from the masonry for each 5,000 square feet (465 m²) of the wall area that is in question but not less than one set of three masonry prisms for the project. The length, width and height dimensions of the prisms shall comply with the requirements of ASTM C 1314. Transporting, preparation and testing of prisms shall be in accordance with ASTM C 1314.

2105.3.2 Compressive strength calculations. The compressive strength of prisms shall be the value calculated in accordance ASTM C 1314, except that the net cross-sectional area of the prism shall be based on the net mortar bedded area.

2105.3.3 Compliance. Compliance with the requirement for the specified compressive strength of masonry, f'_m, shall be considered satisfied provided the modified compressive strength equals or exceeds the specified f'_m. Additional testing of specimens cut from locations in question shall be permitted.

SECTION 2106
SEISMIC DESIGN

2106.1 Seismic design requirements for masonry. Masonry structures and components shall comply with the requirements in Section 1.13.2.2 of ACI 530/ASCE 5/TMS 402 and Section 1.13.3, 1.13.4, 1.13.5, 1.13.6 or 1.13.7 of ACI 530/ASCE 5/TMS 402 depending on the structure's seismic design category as determined in Section 1616.3. All masonry walls, unless isolated on three edges from in-plane motion of the basic structural systems, shall be considered to be part of the seismic-force-resisting system. In addition, the following requirements shall be met.

2106.1.1 Basic seismic-force-resisting system. Buildings relying on masonry shear walls as part of the basic seismic-force-resisting system shall comply with Section 1.13.2.2 of ACI 530/ASCE 5/TMS 402 or with Section 2106.1.1.1, 2106.1.1.2 or 2106.1.1.3.

2106.1.1.1 Ordinary plain prestressed masonry shear walls. Ordinary plain prestressed masonry shear walls shall comply with the requirements of Chapter 4 of ACI 530/ASCE 5/TMS 402.

2106.1.1.2 Intermediate prestressed masonry shear walls. Intermediate prestressed masonry shear walls shall comply with the requirements of Section 1.13.2.2.4 of ACI 530/ASCE 5/TMS 402 and shall be designed by Chapter 4, Section 4.5.3.3, of ACI 530/ASCE 5/TMS 402 for flexural strength and by Section 3.2.4.1.2 of ACI 530/ASCE 5/TMS 402 for shear strength. Sections 1.13.2.2.5(a), 3.2.3.5 and 3.2.4.3.2(c) of ACI 530/ASCE 5/TMS 402 shall be applicable for reinforcement. Flexural elements subjected to load reversals shall be symmetrically reinforced. The nominal moment strength at any section along a member shall not be less than one-fourth the maximum moment strength. The cross-sectional area of bonded tendons shall be considered to contribute to the minimum reinforcement in Section 1.13.2.2.4 of ACI 530/ASCE 5/TMS 402. Tendons shall be located in cells that are grouted the full height of the wall.

2106.1.1.3 Special prestressed masonry shear walls. Special prestressed masonry shear walls shall comply with the requirements of Section 1.13.2.2.5 of ACI 530/ASCE 5/TMS 402 and shall be designed by Chapter 4, Section 4.5.3.3, of ACI 530/ASCE 5/TMS 402 for flexural strength and by Section 3.2.4.1.2 of ACI 530/ASCE 5/TMS 402 for shear strength. Sections 1.13.2.2.5(a), 3.2.3.5 and 3.2.4.3.2(c) of ACI 530/ASCE 5/TMS 402 shall be applicable for reinforcement. Flexural elements subjected to load reversals shall be symmetrically reinforced. The nominal moment strength at any section along a member shall not be less than one-fourth the maximum moment strength. The cross-sectional area of bonded tendons shall be considered to contribute to the minimum reinforcement in Section 1.13.2.2.5 of ACI 530/ASCE 5/TMS 402. Special prestressed masonry shear walls shall also comply with the requirements of Section 3.2.3.5 of ACI 530/ASCE 5/TMS 402.

2106.1.1.3.1 Prestressing tendons. Prestressing tendons shall consist of bars conforming to ASTM A 722.

2106.1.1.3.2 Grouting. All cells of the masonry wall shall be grouted.

2106.2 Anchorage of masonry walls. Masonry walls shall be anchored to the roof and floors that provide lateral support for the wall in accordance with Section 1604.8.2.

2106.3 Seismic Design Category B. Structures assigned to Seismic Design Category B shall conform to the requirements of Section 1.13.4 of ACI 530/ASCE 5/TMS 402 and to the additional requirements of this section.

2106.3.1 Masonry walls not part of the lateral-force-resisting system. Masonry partition walls, masonry screen walls and other masonry elements that are not designed to resist vertical or lateral loads, other than those induced by their own mass, shall be isolated from the structure so that the vertical and lateral forces are not imparted to these elements. Isolation joints and connectors between these elements and the structure shall be designed to accommodate the design story drift.

2106.4 Additional requirements for structures in Seismic Design Category C. Structures assigned to Seismic Design Category C shall conform to the requirements of Section 1.13.5 of ACI 530/ASCE 5/TMS 402 and the additional requirements of this section.

2106.4.1 Design of discontinuous members that are part of the lateral-force-resisting system. Columns and pilasters that are part of the lateral-force-resisting system and that support reactions from discontinuous stiff members such as walls shall be provided with transverse reinforcement spaced at no more than one-fourth of the least nominal dimension of the column or pilaster. The minimum transverse reinforcement ratio shall be 0.0015. Beams supporting reactions from discontinuous walls or frames shall be provided with transverse reinforcement spaced at no more than one-half of the nominal depth of the beam. The minimum transverse reinforcement ratio shall be 0.0015.

2106.5 Additional requirements for structures in Seismic Design Category D. Structures assigned to Seismic Design Category D shall conform to the requirements of Section 2106.4, Section 1.13.6 of ACI 530/ASCE 5/TMS 402 and the additional requirements of this section.

2106.5.1 Loads for shear walls designed by the working stress design method. When calculating in-plane shear or diagonal tension stresses by the working stress design method, shear walls that resist seismic forces shall be designed to resist 1.5 times the seismic forces required by Chapter 16. The 1.5 multiplier need not be applied to the overturning moment.

2106.5.2 Shear wall shear strength. For a shear wall whose nominal shear strength exceeds the shear corresponding to development of its nominal flexural strength, two shear regions exist.

For all cross sections within a region defined by the base of the shear wall and a plane at a distance L_w above the base of the shear wall, the nominal shear strength shall be determined by Equation 21-1.

$$V_n = A_n \rho_n f_y \qquad \text{(Equation 21-1)}$$

The required shear strength for this region shall be calculated at a distance $L_w/2$ above the base of the shear wall, but not to exceed one-half story height.

For the other region, the nominal shear strength of the shear wall shall be determined from Section 2108.

2106.6 Additional requirements for structures in Seismic Design Category E or F. Structures assigned to Seismic Design Category E or F shall conform to the requirements of Section 2106.5 and Section 1.13.7 of ACI 530/ASCE 5/TMS 402.

SECTION 2107
WORKING STRESS DESIGN

2107.1 General. The design of masonry structures using working stress design shall comply with Section 2106 and the requirements of Chapters 1 and 2, except Section 2.1.2.1 and 2.1.3.3 of ACI 530/ASCE 5/TMS 402. The text of ACI 530/ASCE 5/TMS 402 shall be modified as follows.

2107.2 Modifications to ACI 530/ASCE 5/TMS 402.

2107.2.1 ACI 530/ASCE 5/TMS 402, Chapter 2. Special inspection during construction shall be provided as set forth in Section 1704.5.

2107.2.2 ACI 530/ASCE 5/TMS 402, Section 2.1.6. Masonry columns used only to support light-frame roofs of carports, porches, sheds or similar structures with a maximum area of 450 square feet (41.8 m²) assigned to Seismic Design Category A, B or C are permitted to be designed and constructed as follows:

1. Concrete masonry materials shall be in accordance with Section 2103.1. Clay or shale masonry units shall be in accordance with Section 2103.2.
2. The nominal cross-sectional dimension of columns shall not be less than 8 inches (203 mm).
3. Columns shall be reinforced with not less than one No. 4 bar centered in each cell of the column.
4. Columns shall be grouted solid.
5. Columns shall not exceed 12 feet (3658 mm) in height.
6. Roofs shall be anchored to the columns. Such anchorage shall be capable of resisting the design loads specified in Chapter 16.
7. Where such columns are required to resist uplift loads, the columns shall be anchored to their footings with two No. 4 bars extending a minimum of 24 inches (610 mm) into the columns and bent horizontally a minimum of 15 inches (381 mm) in opposite directions into the footings. One of these bars is permitted to be the reinforcing bar specified in Item 3 above. The total weight of a column and its footing shall not be less than 1.5 times the design uplift load.

2107.2.3 ACI 530/ASCE 5/TMS 402, Section 2.1.10.6.1.1, lap splices. The minimum length of lap splices for reinforcing bars in tension or compression, l_{ld}, shall be calculated by Equation 21-2, but shall not be less than 15 inches (380 mm).

$$l_{ld} = \frac{0.16d_b^2 f_y \gamma}{K\sqrt{f'_m}}$$ **(Equation 21-2)**

For SI: $l_{ld} = \frac{1.95d_b^2 f_y \gamma}{K\sqrt{f'_m}}$

where:

d_b = Diameter of reinforcement, inches (mm).

f_y = Specified yield stress of the reinforcement or the anchor bolt, psi (MPa).

f'_m = Specified compressive strength of masonry at age of 28 days, psi (MPa).

l_{ld} = Minimum lap splice length, inches (mm).

K = The lesser of the masonry cover, clear spacing between adjacent reinforcement or five times d_b, inches (mm).

γ = 1.0 for No. 3 through No. 5 reinforcing bars. 1.4 for No. 6 and No. 7 reinforcing bars. 1.5 for No. 8 through No. 9 reinforcing bars.

2107.2.4 ACI 530/ASCE 5/TMS 402, maximum bar size. The bar diameter shall not exceed one-eighth of the nominal wall thickness and shall not exceed one-quarter of the least dimension of the cell, course or collar joint in which it is placed.

2107.2.5 ACI 530/ASCE 5/TMS 402, splices for large bars. Reinforcing bars larger than No. 9 in size shall be spliced using mechanical connectors in accordance with ACI 530/ASCE 5/TMS 402, Section 2.1.10.6.3.

2107.2.6 ACI 530/ASCE 5/TMS 402, Maximum reinforcement percentage. Special reinforced masonry shear walls having a shear span ratio, M/Vd, equal to or greater than 1.0 and having an axial load, P greater than $0.05 f'_m A_n$ which are subjected to in-plane forces, shall have a maximum reinforcement ratio, ρ_{max}, not greater than that computed as follows:

$$\rho_{max} = \frac{nf'_m}{2f_y\left(n + \dfrac{f_y}{f'_m}\right)}$$ **(Equation 21-3)**

SECTION 2108
STRENGTH DESIGN OF MASONRY

2108.1 General. The design of masonry structures using strength design shall comply with Section 2106 and the requirements of Chapters 1 and 3 of ACI 530/ASCE 5/TMS 402.

The minimum nominal thickness for hollow clay masonry in accordance with Section 3.2.5.5 of ACI 530/ASCE 5/TMS 402 shall be 4 inches (102 mm).

2108.2 ACI 530/ASCE 5/TMS 402, Section 3.2.2(g). Modify Section 3.2.2(g) as follows:

3.2.2(g). The relationship between masonry compressive stress and masonry strain shall be assumed to be defined by the following:

Masonry stress of 0.80 f'_m shall be assumed uniformly distributed over an equivalent compression zone bounded by edges of the cross section and a straight line located parallel to the neutral axis at a distance, $a = 0.80\ c$, from the fiber of maximum compressive strain. The distance, c, from the fiber of maximum strain to the neutral axis shall be measured perpendicular to that axis. For out-of-plane bending, the width of the equivalent stress block shall not be taken greater than six times the nominal thickness of the masonry wall or the spacing between reinforcement, whichever is less. For in-plane bending of flanged walls, the effective flange width shall not exceed six times the thickness of the flange.

2108.3 ACI 530/ASCE 5/TMS 402, Section 3.2.3.4. Modify Section 3.2.3.4 (b) and (c) as follows:

3.2.3.4 (b). A welded splice shall have the bars butted and welded to develop at least 125 percent of the yield strength, f_y, of the bar in tension or compression, as required. Welded splices shall be of ASTM A 706 steel reinforcement. Welded splices shall not be permitted in plastic hinge zones of intermediate or special reinforced walls or special moment frames of masonry.

3.2.3.4 (c). Mechanical splices shall be classified as Type 1 or 2 according to Section 21.2.6.1 of ACI 318. Type 1 mechanical splices shall not be used within a plastic hinge zone or within a beam-column joint of intermediate or special reinforced masonry shear walls or special moment frames. Type 2 mechanical splices are permitted in any location within a member.

2108.4 ACI 530/ASCE 5/TMS 402, Section 3.2.3.5.1. Add the following text to Section 3.2.3.5.1:

For special prestressed masonry shear walls, strain in all prestressing steel shall be computed to be compatible with a strain in the extreme tension reinforcement equal to five times the strain associated with the reinforcement yield stress, f_y. The calculation of the maximum reinforcement shall consider forces in the prestressing steel that correspond to these calculated strains.

SECTION 2109
EMPIRICAL DESIGN OF MASONRY

2109.1 General. Empirically designed masonry shall conform to this chapter or Chapter 5 of ACI 530/ASCE 5/TMS 402.

2109.1.1 Limitations. Empirical masonry design shall not be utilized for any of the following conditions:

1. The design or construction of masonry in buildings assigned to Seismic Design Category D, E or F as specified in Section 1616, and the design of the seismic-force-resisting system for buildings assigned to Seismic Design Category B or C.

2. The design or construction of masonry structures located in areas where the basic wind speed exceeds 110 mph (177 km/hr).

3. Buildings more than 35 feet (10 668 mm) in height which have masonry wall lateral-force-resisting systems.

In buildings that exceed one or more of the above limitations, masonry shall be designed in accordance with the engineered design provisions of Section 2107 or 2108, or the foundation wall provisions of Section 1805.5.

2109.2 Lateral stability.

2109.2.1 Shear walls. Where the structure depends upon masonry walls for lateral stability, shear walls shall be provided parallel to the direction of the lateral forces resisted.

2109.2.1.1 Shear wall thickness. Minimum nominal thickness of masonry shear walls shall be 8 inches (203 mm).

Exception: Shear walls of one-story buildings are permitted to be a minimum nominal thickness of 6 inches (152 mm).

2109.2.1.2 Cumulative length of shear walls. In each direction in which shear walls are required for lateral stability, shear walls shall be positioned in two separate planes. The minimum cumulative length of shear walls provided shall be 0.4 times the long dimension of the building. Cumulative length of shear walls shall not include openings or any element whose length is less than one-half its height.

2109.2.1.3 Maximum diaphragm ratio. Masonry shear walls shall be spaced so that the length-to-width ratio of each diaphragm transferring lateral forces to the shear walls does not exceed the values given in Table 2109.2.1.3.

TABLE 2109.2.1.3
DIAPHRAGM LENGTH-TO-WIDTH RATIOS

FLOOR OR ROOF DIAPHRAGM CONSTRUCTION	MAXIMUM LENGTH-TO-WIDTH RATIO OF DIAPHRAGM PANEL
Cast-in-place concrete	5:1
Precast concrete	4:1
Metal deck with concrete fill	3:1
Metal deck with no fill	2:1
Wood	2:1

2109.2.2 Roofs. The roof construction shall be designed so as not to impart out-of-plane lateral thrust to the walls under roof gravity load.

2109.2.3 Surface-bonded walls. Dry-stacked, surface-bonded concrete masonry walls shall comply with the requirements of this code for masonry wall construction, except where otherwise noted in this section.

2109.2.3.1 Strength. Dry-stacked, surface-bonded concrete masonry walls shall be of adequate strength and proportions to support all superimposed loads without exceeding the allowable stresses listed in Table 2109.2.3.1. Allowable stresses not specified in Table 2109.2.3.1 shall comply with the requirements of ACI 530/ASCE 5/TMS 402.

TABLE 2109.2.3.1
ALLOWABLE STRESS GROSS CROSS-SECTIONAL AREA FOR DRY-STACKED, SURFACE-BONDED CONCRETE MASONRY WALLS

DESCRIPTION	MAXIMUM ALLOWABLE STRESS (psi)
Compression standard block	45
Shear	10

Flexural tension
  Vertical span
  Horizontal span
	
	18
	30

For SI:1 pound per square inch = 0.006895 Mpa.

2109.2.3.2 Construction. Construction of dry-stacked, surface-bonded masonry walls, including stacking and leveling of units, mixing and application of mortar and curing and protection shall comply with ASTM C 946.

2109.3 Compressive stress requirements.

2109.3.1 Calculations. Compressive stresses in masonry due to vertical dead plus live loads, excluding wind or seismic loads, shall be determined in accordance with Section 2109.3.2.1. Dead and live loads shall be in accordance with Chapter 16, with live load reductions as permitted in Section 1607.9.

2109.3.2 Allowable compressive stresses. The compressive stresses in masonry shall not exceed the values given in Table 2109.3.2. Stress shall be calculated based on specified rather than nominal dimensions.

2109.3.2.1 Calculated compressive stresses. Calculated compressive stresses for single wythe walls and for multiwythe composite masonry walls shall be determined by dividing the design load by the gross cross-sectional area of the member. The area of openings, chases or recesses in walls shall not be included in the gross cross-sectional area of the wall.

2109.3.2.2 Multiwythe walls. The allowable stress shall be as given in Table 2109.3.2 for the weakest combination of the units used in each wythe.

TABLE 2109.3.2
ALLOWABLE COMPRESSIVE STRESSES FOR EMPIRICAL DESIGN OF MASONRY

CONSTRUCTION; COMPRESSIVE STRENGTH OF UNIT GROSS AREA (psi)	ALLOWABLE COMPRESSIVE STRESSES[a] GROSS CROSS-SECTIONAL AREA (psi)	
	Type M or S mortar	Type N mortar
Solid masonry of brick and other solid units of clay or shale; sand-lime or concrete brick:		
8,000 or greater	350	300
4,500	225	200
2,500	160	140
1,500	115	100
Grouted masonry, of clay or shale; sand-lime or concrete:		
4,500 or greater	225	200
2,500	160	140
1,500	115	100
Solid masonry of solid concrete masonry units:		
3,000 or greater	225	200
2,000	160	140
1,200	115	100
Masonry of hollow load-bearing units:		
2,000 or greater	140	120
1,500	115	100
1,000	75	70
700	60	55
Hollow walls (noncomposite masonry bonded)[b] Solid units:		
2,500 or greater	160	140
1,500	115	100
Hollow units	75	70
Stone ashlar masonry:		
Granite	720	640
Limestone or marble	450	400
Sandstone or cast stone	360	320
Rubble stone masonry Coursed, rough or random	120	100

For SI: 1 pound per square inch = 0.006895 MPa.

a. Linear interpolation for determining allowable stresses for masonry units having compressive strengths which are intermediate between those given in the table is permitted.

b. Where floor and roof loads are carried upon one wythe, the gross cross-sectional area is that of the wythe under load; if both wythes are loaded, the gross cross-sectional area is that of the wall minus the area of the cavity between the wythes. Walls bonded with metal ties shall be considered as noncomposite walls unless collar joints are filled with mortar or grout.

2109.4 Lateral support.

2109.4.1 Intervals. Masonry walls shall be laterally supported in either the horizontal or vertical direction at intervals not exceeding those given in Table 2109.4.1.

TABLE 2109.4.1
WALL LATERAL SUPPORT REQUIREMENTS

CONSTRUCTION	MAXIMUM WALL LENGTH TO THICKNESS OR WALL HEIGHT TO THICKNESS
Bearing walls Solid units or fully grouted All others	20 18
Nonbearing walls Exterior Interior	18 36

2109.4.2 Thickness. Except for cavity walls and cantilever walls, the thickness of a wall shall be its nominal thickness measured perpendicular to the face of the wall. For cavity walls, the thickness shall be determined as the sum of the nominal thicknesses of the individual wythes. For cantilever walls, except for parapets, the ratio of height-to-nominal thickness shall not exceed six for solid masonry or four for hollow masonry. For parapets, see Section 2109.5.5.

2109.4.3 Support elements. Lateral support shall be provided by cross walls, pilasters, buttresses or structural frame members when the limiting distance is taken horizontally, or by floors, roofs acting as diaphragms or structural frame members when the limiting distance is taken vertically.

2109.5 Thickness of masonry. Minimum thickness requirements shall be based on nominal dimensions of masonry.

2109.5.1 Thickness of walls. The thickness of masonry walls shall conform to the requirements of Section 2109.5.

2109.5.2 Minimum thickness. The minimum thickness of masonry bearing walls more than one story high shall be 8 inches (203 mm). Bearing walls of one-story buildings shall not be less than 6 inches (152 mm) thick.

2109.5.3 Rubble stone walls. The minimum thickness of rough or random or coursed rubble stone walls shall be 16 inches (406 mm).

2109.5.4 Change in thickness. Where walls of masonry of hollow units or masonry bonded hollow walls are decreased in thickness, a course or courses of solid masonry shall be interposed between the wall below and the thinner wall above, or special units or construction shall be used to transmit the loads from face shells or wythes above to those below.

2109.5.5 Parapet walls.

2109.5.5.1 Minimum thickness. Unreinforced parapet walls shall be at least 8 inches (203 mm) thick, and their height shall not exceed three times their thickness.

2109.5.5.2 Additional provisions. Additional provisions for parapet walls are contained in Sections 1503.2 and 1503.3.

2109.5.6 Foundation walls. Foundation walls shall comply with the requirements of Sections 2109.5.6.1 and 2109.5.6.2.

2109.5.6.1 Minimum thickness. Minimum thickness for foundation walls shall comply with the requirements of Table 2109.5.6.1. The provisions of Table 2109.5.6.1 are only applicable where the following conditions are met:

1. The foundation wall does not exceed 8 feet (2438 mm) in height between lateral supports,

2. The terrain surrounding foundation walls is graded to drain surface water away from foundation walls,

3. Backfill is drained to remove ground water away from foundation walls,

4. Lateral support is provided at the top of foundation walls prior to backfilling,

5. The length of foundation walls between perpendicular masonry walls or pilasters is a maximum of three times the basement wall height,

6. The backfill is granular and soil conditions in the area are nonexpansive, and

7. Masonry is laid in running bond using Type M or S mortar.

2109.5.6.2 Design requirements. Where the requirements of Section 2109.5.6.1 are not met, foundation walls shall be designed in accordance with Section 1805.5.

TABLE 2109.5.6.1
FOUNDATION WALL CONSTRUCTION

WALL CONSTRUCTION	NOMINAL WALL THICKNESS (inches)	MAXIMUM DEPTH OF UNBALANCED BACKFILL (feet)
Hollow unit masonry	8 10 12	5 6 7
Solid unit masonry	8 10 12	5 7 7
Fully grouted masonry	8 10 12	7 8 8

For SI: 1 inch = 25.4 mm, 1 foot = 304.8 mm.

2109.6 Bond.

2109.6.1 General. The facing and backing of multiwythe masonry walls shall be bonded in accordance with Section 2109.6.2, 2109.6.3 or 2109.6.4.

2109.6.2 Bonding with masonry headers.

2109.6.2.1 Solid units. Where the facing and backing (adjacent wythes) of solid masonry construction are bonded by means of masonry headers, no less than 4 percent of the wall surface of each face shall be composed of headers extending not less than 3 inches (76 mm) into the backing. The distance between adjacent full-length headers shall not exceed 24 inches (610 mm) either vertically or horizontally. In walls in which a single header does not extend through the wall, headers from the opposite sides shall overlap at least 3 inches (76 mm), or headers from opposite sides shall be covered with another header course overlapping the header below at least 3 inches (76 mm).

2109.6.2.2 Hollow units. Where two or more hollow units are used to make up the thickness of a wall, the stretcher courses shall be bonded at vertical intervals not exceeding 34 inches (864 mm) by lapping at least 3 inches (76 mm) over the unit below, or by lapping at vertical intervals not exceeding 17 inches (432 mm) with units that are at least 50 percent greater in thickness than the units below.

2109.6.2.3 Masonry bonded hollow walls. In masonry bonded hollow walls, the facing and backing shall be bonded so that not less than 4 percent of the wall surface of each face is composed of masonry bonded units extending not less than 3 inches (76 mm) into the backing. The distance between adjacent bonders shall not exceed 24 inches (610 mm) either vertically or horizontally.

2109.6.3 Bonding with wall ties or joint reinforcement.

2109.6.3.1 Bonding with wall ties. Except as required by Section 2109.6.3.1.1, where the facing and backing (adjacent wythes) of masonry walls are bonded with wire size W2.8 (MW18) wall ties or metal wire of equivalent stiffness embedded in the horizontal mortar joints, there shall be at least one metal tie for each $4^1/_2$ square feet (0.42 m²) of wall area. The maximum vertical distance between ties shall not exceed 24 inches (610 mm), and the maximum horizontal distance shall not exceed 36 inches (914 mm). Rods or ties bent to rectangular shape shall be used with hollow masonry units laid with the cells vertical. In other walls, the ends of ties shall be bent to 90-degree (1.57 rad) angles to provide hooks no less than 2 inches (51 mm) long. Wall ties shall be without drips. Additional bonding ties shall be provided at all openings, spaced not more than 36 inches (914 mm) apart around the perimeter and within 12 inches (305 mm) of the opening.

2109.6.3.1.1 Bonding with adjustable wall ties. Where the facing and backing (adjacent wythes) of masonry are bonded with adjustable wall ties, there shall be at least one tie for each 1.77 square feet (0.164 m²) of wall area. Neither the vertical nor horizontal spacing of the adjustable wall ties shall exceed 16 inches (406 mm). The maximum vertical offset of bed joints from one wythe to the other shall be $1^1/_4$ inches (32 mm). The maximum clearance between connecting parts of the ties shall be $^1/_{16}$ inch (1.6 mm). When pintle legs are used, ties shall have at least two wire size W2.8 (MW18) legs.

2109.6.3.2 Bonding with prefabricated joint reinforcement. Where the facing and backing (adjacent wythes) of masonry are bonded with prefabricated joint reinforcement, there shall be at least one cross wire serving as a tie for each $2^2/_3$ square feet (0.25 m²) of wall area. The vertical spacing of the joint reinforcing shall not exceed 24 inches (610 mm). Cross wires on prefabricated

joint reinforcement shall not be less than W1.7 (MW11) and shall be without drips. The longitudinal wires shall be embedded in the mortar.

2109.6.4 Bonding with natural or cast stone.

2109.6.4.1 Ashlar masonry. In ashlar masonry, bonder units, uniformly distributed, shall be provided to the extent of not less than 10 percent of the wall area. Such bonder units shall extend not less than 4 inches (102 mm) into the backing wall.

2109.6.4.2 Rubble stone masonry. Rubble stone masonry 24 inches (610 mm) or less in thickness shall have bonder units with a maximum spacing of 36 inches (914 mm) vertically and 36 inches (914 mm) horizontally, and if the masonry is of greater thickness than 24 inches (610 mm), shall have one bonder unit for each 6 square feet (0.56 m²) of wall surface on both sides.

2109.6.5 Masonry bonding pattern.

2109.6.5.1 Masonry laid in running bond. Each wythe of masonry shall be laid in running bond, head joints in successive courses shall be offset by not less than one-fourth the unit length or the masonry walls shall be reinforced longitudinally as required in Section 2109.6.5.2.p

2109.6.5.2 Masonry laid in stack bond. Where unit masonry is laid with less head joint offset than in Section 2109.6.5.1, the minimum area of horizontal reinforcement placed in mortar bed joints or in bond beams spaced not more than 48 inches (1219 mm) apart, shall be 0.0003 times the vertical cross-sectional area of the wall.

2109.7 Anchorage.

2109.7.1 General. Masonry elements shall be anchored in accordance with Sections 2109.7.2 through 2109.7.4.

2109.7.2 Intersecting walls. Masonry walls depending upon one another for lateral support shall be anchored or bonded at locations where they meet or intersect by one of the methods indicated in Sections 2109.7.2.1 through 2109.7.2.5.

2109.7.2.1 Bonding pattern. Fifty percent of the units at the intersection shall be laid in an overlapping masonry bonding pattern, with alternate units having a bearing of not less than 3 inches (76 mm) on the unit below.

2109.7.2.2 Steel connectors. Walls shall be anchored by steel connectors having a minimum section of $^1/_4$ inch (6.4 mm) by $1^1/_2$ inches (38 mm), with ends bent up at least 2 inches (51 mm) or with cross pins to form anchorage. Such anchors shall be at least 24 inches (610 mm) long and the maximum spacing shall be 48 inches (1219 mm).

2109.7.2.3 Joint reinforcement. Walls shall be anchored by joint reinforcement spaced at a maximum distance of 8 inches (203 mm). Longitudinal wires of such reinforcement shall be at least wire size W1.7 (MW 11) and shall extend at least 30 inches (762 mm) in each direction at the intersection.

2109.7.2.4 Interior nonload-bearing walls. Interior nonload-bearing walls shall be anchored at their intersection, at vertical intervals of not more than 16 inches (406 mm) with joint reinforcement or $^1/_4$-inch (6.4 mm) mesh galvanized hardware cloth.

2109.7.2.5 Ties, joint reinforcement or anchors. Other metal ties, joint reinforcement or anchors, if used, shall be spaced to provide equivalent area of anchorage to that required by this section.

2109.7.3 Floor and roof anchorage. Floor and roof diaphragms providing lateral support to masonry shall comply with the live loads in Section 1607.3 and shall be connected to the masonry in accordance with Sections 2109.7.3.1 through 2109.7.3.3.

2109.7.3.1 Wood floor joists. Wood floor joists bearing on masonry walls shall be anchored to the wall at intervals not to exceed 72 inches (1829 mm) by metal strap anchors. Joists parallel to the wall shall be anchored with metal straps spaced not more than 72 inches (1829 mm) o.c. extending over or under and secured to at least three joists. Blocking shall be provided between joists at each strap anchor.

2109.7.3.2 Steel floor joists. Steel floor joists bearing on masonry walls shall be anchored to the wall with $^3/_8$-inch (9.5 mm) round bars, or their equivalent, spaced not more than 72 inches (1829 mm) o.c. Where joists are parallel to the wall, anchors shall be located at joist bridging.

2109.7.3.3 Roof diaphragms. Roof diaphragms shall be anchored to masonry walls with $^1/_2$-inch-diameter (12.7 mm) bolts, 72 inches (1829 mm) o.c. or their equivalent. Bolts shall extend and be embedded at least 15 inches (381 mm) into the masonry, or be hooked or welded to not less than 0.20 square inch (129 mm²) of bond beam reinforcement placed not less than 6 inches (152 mm) from the top of the wall.

2109.7.4 Walls adjoining structural framing. Where walls are dependent upon the structural frame for lateral support, they shall be anchored to the structural members with metal anchors or otherwise keyed to the structural members. Metal anchors shall consist of $^1/_2$-inch (12.7 mm) bolts spaced at 48 inches (1219 mm) o.c. embedded 4 inches (102 mm) into the masonry, or their equivalent area.

2109.8 Adobe construction. Adobe construction shall comply with this section and shall be subject to the requirements of this code for Type V construction.

2109.8.1 Unstabilized adobe.

2109.8.1.1 Compressive strength. Adobe units shall have an average compressive strength of 300 psi (2068 kPa) when tested in accordance with ASTM C 67. Five samples shall be tested and no individual unit is permitted to have a compressive strength of less than 250 psi (1724 kPa).

2109.8.1.2 Modulus of rupture. Adobe units shall have an average modulus of rupture of 50 psi (345 kPa) when tested in accordance with the following procedure. Five samples shall be tested and no individual unit shall have a modulus of rupture of less than 35 psi (241 kPa).

2109.8.1.2.1 Support conditions. A cured unit shall be simply supported by 2-inch-diameter (51 mm) cylindrical supports located 2 inches (51 mm) in from each end and extending the full width of the unit.

2109.8.1.2.2 Loading conditions. A 2-inch-diameter (51 mm) cylinder shall be placed at midspan parallel to the supports.

2109.8.1.2.3 Testing procedure. A vertical load shall be applied to the cylinder at the rate of 500 pounds per minute (37 N/s) until failure occurs.

2109.8.1.2.4 Modulus of rupture determination. The modulus of rupture shall be determined by the equation:

$$f_r = 3WL_s/2bt^2 \qquad \text{(Equation 21-4)}$$

where, for the purposes of this section only:

b = Width of the test specimen measured parallel to the loading cylinder, inches (mm).

f_r = Modulus of rupture, psi (MPa).

L_s = Distance between supports, inches (mm).

t = Thickness of the test specimen measured parallel to the direction of load, inches (mm).

W = The applied load at failure, pounds (N).

2109.8.1.3 Moisture content requirements. Adobe units shall have a moisture content not exceeding 4 percent by weight.

2109.8.1.4 Shrinkage cracks. Adobe units shall not contain more than three shrinkage cracks and any single shrinkage crack shall not exceed 3 inches (76 mm) in length or $^1/_8$ inch (3.2 mm) in width.

2109.8.2 Stabilized adobe.

2109.8.2.1 Material requirements. Stabilized adobe shall comply with the material requirements of unstabilized adobe in addition to Sections 2109.8.2.1.1 and 2109.8.2.1.2.

2109.8.2.1.1 Soil requirements. Soil used for stabilized adobe units shall be chemically compatible with the stabilizing material.

2109.8.2.1.2 Absorption requirements. A 4-inch (102 mm) cube, cut from a stabilized adobe unit dried to a constant weight in a ventilated oven at 212°F to 239°F (100°C to 115°C), shall not absorb more than $2^1/_2$- percent moisture by weight when placed upon a constantly water-saturated, porous surface for seven days. A minimum of five specimens shall be tested and each specimen shall be cut from a separate unit.

2109.8.3 Working stress. The allowable compressive stress based on gross cross-sectional area of adobe shall not exceed 30 psi (207 kPa).

2109.8.3.1 Bolts. Bolt values shall not exceed those set forth in Table 2109.8.3.1.

TABLE 2109.8.3.1
ALLOWABLE SHEAR ON BOLTS IN ADOBE MASONRY

DIAMETER OF BOLTS (inches)	MINIMUM EMBEDMENT (inches)	SHEAR (pounds)
$^1/_2$	—	—
$^5/_8$	12	200
$^3/_4$	15	300
$^7/_8$	18	400
1	21	500
$1^1/_8$	24	600

For SI: 1 inch = 25.4 mm, 1 pound = 4.448 N.

2109.8.4 Construction.

2109.8.4.1 General.

2109.8.4.1.1 Height restrictions. Adobe construction shall be limited to buildings not exceeding one story, except that two-story construction is allowed when designed by a registered design professional.

2109.8.4.1.2 Mortar restrictions. Mortar for stabilized adobe units shall comply with Chapter 21 or adobe soil. Adobe soil used as mortar shall comply with material requirements for stabilized adobe. Mortar for unstabilized adobe shall be portland cement mortar.

2109.8.4.1.3 Mortar joints. Adobe units shall be laid with full head and bed joints and in full running bond.

2109.8.4.1.4 Parapet walls. Parapet walls constructed of adobe units shall be waterproofed.

2109.8.4.2 Wall thickness. The minimum thickness of exterior walls in one-story buildings shall be 10 inches (254 mm). The walls shall be laterally supported at intervals not exceeding 24 feet (7315 mm). The minimum thickness of interior load-bearing walls shall be 8 inches (203 mm). In no case shall the unsupported height of any wall constructed of adobe units exceed 10 times the thickness of such wall.

2109.8.4.3 Foundations.

2109.8.4.3.1 Foundation support. Walls and partitions constructed of adobe units shall be supported by foundations or footings that extend not less than 6 inches (152 mm) above adjacent ground surfaces and are constructed of solid masonry (excluding adobe) or concrete. Footings and foundations shall comply with Chapter 18.

2109.8.4.3.2 Lower course requirements. Stabilized adobe units shall be used in adobe walls for the first 4 inches (102 mm) above the finished first-floor elevation.

2109.8.4.4 Isolated piers or columns. Adobe units shall not be used for isolated piers or columns in a load-bearing capacity. Walls less than 24 inches (610 mm) in length shall be considered isolated piers or columns.

2109.8.4.5 Tie beams. Exterior walls and interior load-bearing walls constructed of adobe units shall have a continuous tie beam at the level of the floor or roof bearing and meeting the following requirements.

2109.8.4.5.1 Concrete tie beams. Concrete tie beams shall be a minimum depth of 6 inches (152 mm) and a minimum width of 10 inches (254 mm). Concrete tie beams shall be continuously reinforced with a minimum of two No. 4 reinforcing bars. The ultimate compressive strength of concrete shall be at least 2,500 psi (17.2 MPa) at 28 days.

2109.8.4.5.2 Wood tie beams. Wood tie beams shall be solid or built up of lumber having a minimum nominal thickness of 1 inch (25 mm), and shall have a minimum depth of 6 inches (152 mm) and a minimum width of 10 inches (254 mm). Joints in wood tie beams shall be spliced a minimum of 6 inches (152 mm). No splices shall be allowed within 12 inches (305 mm) of an opening. Wood used in tie beams shall be approved naturally decay-resistant or pressure-treated wood.

2109.8.4.6 Exterior finish. Exterior walls constructed of unstabilized adobe units shall have their exterior surface covered with a minimum of two coats of portland cement plaster having a minimum thickness of $^3/_4$ inch (19.1 mm) and conforming to ANSI A42.2. Lathing shall comply with ANSI A42.3. Fasteners shall be spaced at 16 inches (406 mm) o.c. maximum. Exposed wood surfaces shall be treated with an approved wood preservative or other protective coating prior to lath application.

2109.8.4.7 Lintels. Lintels shall be considered structural members and shall be designed in accordance with the applicable provisions of Chapter 16.

SECTION 2110
GLASS UNIT MASONRY

2110.1 Scope. This section covers the empirical requirements for nonload-bearing glass unit masonry elements in exterior or interior walls.

2110.1.1 Limitations. Solid or hollow approved glass block shall not be used in fire walls, party walls, fire barriers or fire partitions, or for load-bearing construction. Such blocks shall be erected with mortar and reinforcement in metal channel-type frames, structural frames, masonry or concrete recesses, embedded panel anchors as provided for both exterior and interior walls or other approved joint materials. Wood strip framing shall not be used in walls required to have a fire-resistance rating by other provisions of this code.

Exceptions:

1. Glass-block assemblies having a fire protection rating of not less than $^3/_4$ hour shall be permitted as opening protectives in accordance with Section

715 in fire barriers and fire partitions that have a required fire-resistance rating of 1 hour or less and do not enclose exit stairways or exit passageways.

2. Glass-block assemblies as permitted in Section 404.5, Exception 2.

2110.2 Units. Hollow or solid glass-block units shall be standard or thin units.

2110.2.1 Standard units. The specified thickness of standard units shall be $3^7/_8$ inches (98 mm).

2110.2.2 Thin units. The specified thickness of thin units shall be $3^1/_8$ inches (79 mm) for hollow units or 3 inches (76 mm) for solid units.

2110.3 Panel size.

2110.3.1 Exterior standard-unit panels. The maximum area of each individual exterior standard-unit panel shall be 144 square feet (13.4 m²) when the design wind pressure is 20 psf (958 N/m²). The maximum panel dimension between structural supports shall be 25 feet (7620 mm) in width or 20 feet (6096 mm) in height. The panel areas are permitted to be adjusted in accordance with Figure 2110.3.1 for other wind pressures.

2110.3.2 Exterior thin-unit panels. The maximum area of each individual exterior thin-unit panel shall be 85 square feet (7.9 m²). The maximum dimension between structural supports shall be 15 feet (4572 mm) in width or 10 feet (3048 mm) in height. Thin units shall not be used in applications where the design wind pressure exceeds 20 psf (958 N/m²).

2110.3.3 Interior panels. The maximum area of each individual standard-unit panel shall be 250 square feet (23.2 m²). The maximum area of each thin-unit panel shall be 150 square feet (13.9 m²). The maximum dimension between structural supports shall be 25 feet (7620 mm) in width or 20 feet (6096 mm) in height.

2110.3.4 Solid units. The maximum area of solid glass-block wall panels in both exterior and interior walls shall not be more than 100 square feet (9.3 m²).

2110.3.5 Curved panels. The width of curved panels shall conform to the requirements of Sections 2110.3.1, 2110.3.2 and 2110.3.3, except additional structural supports shall be provided at locations where a curved section joins a straight section, and at inflection points in multicurved walls.

2110.4 Support.

2110.4.1 Isolation. Glass unit masonry panels shall be isolated so that in-plane loads are not imparted to the panel.

2110.4.2 Vertical. Maximum total deflection of structural members supporting glass unit masonry shall not exceed $^1/_{600}$.

2110.4.3 Lateral. Glass unit masonry panels more than one unit wide or one unit high shall be laterally supported along their tops and sides. Lateral support shall be provided by panel anchors along the top and sides spaced not more than 16 inches (406 mm) o.c. or by channel-type restraints. Glass unit masonry panels shall be recessed at least 1 inch (25 mm) within channels and chases. Channel-type restraints shall be oversized to accommodate expansion material in the opening and packing and sealant between the framing restraints and the glass unit masonry perimeter units. Lateral supports for glass unit masonry panels shall be designed to resist applied loads, or a minimum of 200 pounds per lineal feet (plf) (2919 N/m) of panel, whichever is greater.

Exceptions:

1. Lateral support at the top of glass unit masonry panels that are no more than one unit wide shall not be required.

2. Lateral support at the sides of glass unit masonry panels that are no more than one unit high shall not be required.

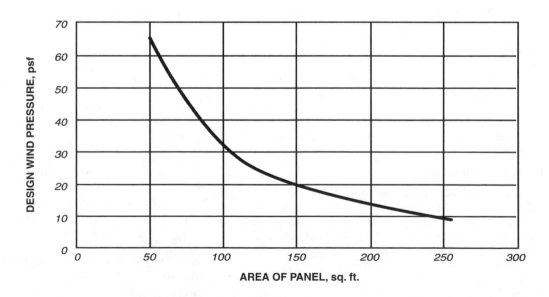

For SI: 1 square foot = 0.0929 m², 1 pound per square foot = 47.g N/m².

FIGURE 2110.3.1
GLASS MASONRY DESIGN WIND LOAD RESISTANCE

2110.4.3.1 Single unit panels. Single unit glass unit masonry panels shall conform to the requirements of Section 2110.4.3, except lateral support shall not be provided by panel anchors.

2110.5 Expansion joints. Glass unit masonry panels shall be provided with expansion joints along the top and sides at all structural supports. Expansion joints shall have sufficient thickness to accommodate displacements of the supporting structure, but shall not be less than $^3/_8$ inch (9.5 mm) in thickness. Expansion joints shall be entirely free of mortar or other debris and shall be filled with resilient material. The sills of glass-block panels shall be coated with approved water-based asphaltic emulsion, or other elastic waterproofing material, prior to laying the first mortar course.

2110.6 Mortar. Mortar for glass unit masonry shall comply with Section 2103.7.

2110.7 Reinforcement. Glass unit masonry panels shall have horizontal joint reinforcement spaced not more than 16 inches (406 mm) on center, located in the mortar bed joint, and extending the entire length of the panel but not across expansion joints. Longitudinal wires shall be lapped a minimum of 6 inches (152 mm) at splices. Joint reinforcement shall be placed in the bed joint immediately below and above openings in the panel. The reinforcement shall have not less than two parallel longitudinal wires of size W1.7 (MW11), and have welded cross wires of size W1.7 (MW11).

SECTION 2111
MASONRY FIREPLACES

2111.1 Definition. A masonry fireplace is a fireplace constructed of concrete or masonry. Masonry fireplaces shall be constructed in accordance with this section, Table 2111.1 and Figure 2111.1.

2111.2 Footings and foundations. Footings for masonry fireplaces and their chimneys shall be constructed of concrete or solid masonry at least 12 inches (305 mm) thick and shall extend at least 6 inches (153 mm) beyond the face of the fireplace or foundation wall on all sides. Footings shall be founded on natural undisturbed earth or engineered fill below frost depth. In areas not subjected to freezing, footings shall be at least 12 inches (305 mm) below finished grade.

2111.2.1 Ash dump cleanout. Cleanout openings, located within foundation walls below fireboxes, when provided, shall be equipped with ferrous metal or masonry doors and frames constructed to remain tightly closed, except when in use. Cleanouts shall be accessible and located so that ash removal will not create a hazard to combustible materials.

2111.3 Seismic reinforcing. Masonry or concrete fireplaces shall be constructed, anchored, supported and reinforced as required in this chapter. In Seismic Design Category D, masonry and concrete fireplaces shall be reinforced and anchored as detailed in Sections 2111.3.1, 2111.3.2, 2111.4 and 2111.4.1 for chimneys serving fireplaces. In Seismic Design Category A, B or C, reinforcement and seismic anchorage is not required. In Seismic Design Category E or F, masonry and concrete chimneys shall be reinforced in accordance with the requirements of Sections 2101 through 2109.

2111.3.1 Vertical reinforcing. For fireplaces with chimneys up to 40 inches (1016 mm) wide, four No. 4 continuous vertical bars, anchored in the foundation, shall be placed in the concrete, between wythes of solid masonry or within the cells of hollow unit masonry and grouted in accordance with Section 2103.10. For fireplaces with chimneys greater than 40 inches (1016 mm) wide, two additional No. 4 vertical bars shall be provided for each additional 40 inches (1016 mm) in width or fraction thereof.

2111.3.2 Horizontal reinforcing. Vertical reinforcement shall be placed enclosed within $^1/_4$-inch (6.4 mm) ties or other reinforcing of equivalent net cross-sectional area, spaced not to exceed 18 inches (457 mm) on center in concrete; or placed in the bed joints of unit masonry at a minimum of every 18 inches (457 mm) of vertical height. Two such ties shall be provided at each bend in the vertical bars.

2111.4 Seismic anchorage. Masonry and concrete chimneys in Seismic Design Category D shall be anchored at each floor, ceiling or roof line more than 6 feet (1829 mm) above grade, except where constructed completely within the exterior walls. Anchorage shall conform to the following requirements.

2111.4.1 Anchorage. Two $^3/_{16}$-inch by 1-inch (4.8 mm by 25.4 mm) straps shall be embedded a minimum of 12 inches (305 mm) into the chimney. Straps shall be hooked around the outer bars and extend 6 inches (152 mm) beyond the bend. Each strap shall be fastened to a minimum of four floor joists with two $^1/_2$-inch (12.7 mm) bolts.

2111.5 Firebox walls. Masonry fireboxes shall be constructed of solid masonry units, hollow masonry units grouted solid, stone or concrete. When a lining of firebrick at least 2 inches (51 mm) in thickness or other approved lining is provided, the minimum thickness of back and sidewalls shall each be 8 inches (203 mm) of solid masonry, including the lining. The width of joints between firebricks shall not be greater than $^1/_4$ inch (6.4 mm). When no lining is provided, the total minimum thickness of back and sidewalls shall be 10 inches (254 mm) of solid masonry. Firebrick shall conform to ASTM C 27 or ASTM C 1261 and shall be laid with medium-duty refractory mortar conforming to ASTM C 199.

2111.5.1 Steel fireplace units. Steel fireplace units are permitted to be installed with solid masonry to form a masonry fireplace provided they are installed according to either the requirements of their listing or the requirements of this section. Steel fireplace units incorporating a steel firebox lining shall be constructed with steel not less than $^1/_4$ inch (6.4 mm) in thickness, and an air-circulating chamber which is ducted to the interior of the building. The firebox lining shall be encased with solid masonry to provide a total thickness at the back and sides of not less than 8 inches (203 mm), of which not less than 4 inches (102 mm) shall be of solid masonry or concrete. Circulating air ducts employed with steel fireplace units shall be constructed of metal or masonry.

TABLE 2111.1
SUMMARY OF REQUIREMENTS FOR MASONRY FIREPLACES AND CHIMNEYS[a]

ITEM	LETTER	REQUIREMENTS	SECTION
Hearth and hearth extension thickness	A	4-inch minimum thickness for hearth, 2-inch minimum thickness for hearth extension.	2111.9
Hearth extension (each side of opening)	B	8 inches for fireplace opening less than 6 square feet. 12 inches for fireplace opening greater than or equal to 6 square feet.	2111.10
Hearth extension (front of opening)	C	16 inches for fireplace opening less than 6 square feet. 20 inches for fireplace opening greater than or equal to 6 square feet.	2111.10
Firebox dimensions	—	20-inch minimum firebox depth. 12-inch minimum firebox depth for Rumford fireplaces.	2111.6
Hearth and hearth extension reinforcing	D	Reinforced to carry its own weight and all imposed loads.	2111.9
Thickness of wall of firebox	E	10 inches solid masonry or 8 inches where firebrick lining is used.	2111.5
Distance from top of opening to throat	F	8 inches minimum.	2111.7 2111.7.1
Smoke chamber wall thickness dimensions	G	6 inches lined; 8 inches unlined. Not taller than opening width; walls not inclined more than 45 degrees from vertical for prefabricated smoke chamber linings or 30 degrees from vertical for corbeled masonry.	2111.8
Chimney vertical reinforcing	H	Four No. 4 full-length bars for chimney up to 40 inches wide. Add two No. 4 bars for each additional 40 inches or fraction of width, or for each additional flue.	2111.3.1, 2113.3.1
Chimney horizontal reinforcing	J	$^1/_4$-inch ties at each 18 inches, and two ties at each bend in vertical steel.	2111.3.2, 2113.3.2
Fireplace lintel	L	Noncombustible material with 4-inch bearing length of each side of opening.	2111.7
Chimney walls with flue lining	M	4-inch-thick solid masonry with $^5/_8$-inch fireclay liner or equivalent. $^1/_2$-inch grout or airspace between fireclay liner and wall.	2113.11.1
Effective flue area (based on area of fireplace opening and chimney)	P	See Section 2113.16.	2113.16
Clearances From chimney From fireplace From combustible trim or materials Above roof	R	 2 inches interior, 1 inch exterior or 12 inches from lining. 2 inches back or sides or 12 inches from lining. 6 inches from opening 3 feet above roof penetration, 2 feet above part of structure within 10 feet.	 2113.19 2111.11 2111.12 2113.9
Anchorage strap Number required Embedment into chimney Fasten to Number of bolts	S	$^3/_{16}$ inch by 1 inch Two 12 inches hooked around outer bar with 6-inch extension. 4 joists Two $^1/_2$-inch diameter.	 2111.4 2113.4.1
Footing Thickness Width	T	 12-inch minimum. 6 inches each side of fireplace wall.	 2111.2

For SI: 1 inch = 25.4 mm, 1 foot = 304.8 mm, 1 square foot = 0.0929 m², 1 degree = 0.017 rad.

a. This table provides a summary of major requirements for the construction of masonry chimneys and fireplaces. Letter references are to Figure 2111.1, which shows examples of typical construction. This table does not cover all requirements, nor does it cover all aspects of the indicated requirements. For the actual mandatory requirements of the code, see the indicated section of text.

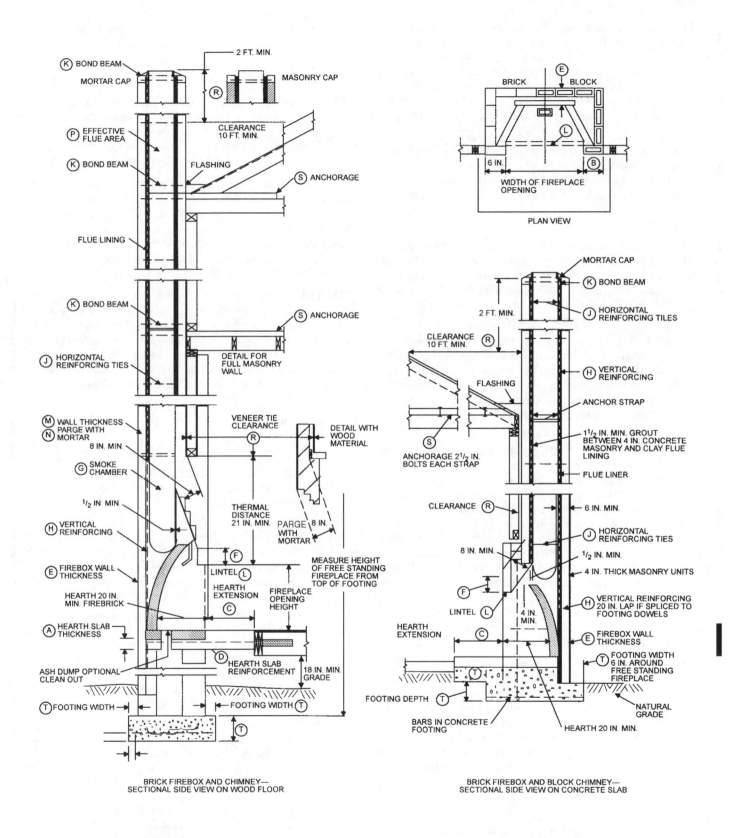

For SI: 1 inch = 25.4 mm, 1 foot = 304.8 mm.

FIGURE 2111.1
FIREPLACE AND CHIMNEY DETAILS

2111.6 Firebox dimensions. The firebox of a concrete or masonry fireplace shall have a minimum depth of 20 inches (508 mm). The throat shall not be less than 8 inches (203 mm) above the fireplace opening. The throat opening shall not be less than 4 inches (102 mm) in depth. The cross-sectional area of the passageway above the firebox, including the throat, damper and smoke chamber, shall not be less than the cross-sectional area of the flue.

Exception: Rumsford fireplaces shall be permitted provided that the depth of the fireplace is at least 12 inches (305 mm) and at least one-third of the width of the fireplace opening, and the throat is at least 12 inches (305 mm) above the lintel, and at least $1/20$ the cross-sectional area of the fireplace opening.

2111.7 Lintel and throat. Masonry over a fireplace opening shall be supported by a lintel of noncombustible material. The minimum required bearing length on each end of the fireplace opening shall be 4 inches (102 mm). The fireplace throat or damper shall be located a minimum of 8 inches (203 mm) above the top of the fireplace opening.

2111.7.1 Damper. Masonry fireplaces shall be equipped with a ferrous metal damper located at least 8 inches (203 mm) above the top of the fireplace opening. Dampers shall be installed in the fireplace or at the top of the flue venting the fireplace, and shall be operable from the room containing the fireplace. Damper controls shall be permitted to be located in the fireplace.

2111.8 Smoke chamber walls. Smoke chamber walls shall be constructed of solid masonry units, hollow masonry units grouted solid, stone or concrete. Corbeling of masonry units shall not leave unit cores exposed to the inside of the smoke chamber. The inside surface of corbeled masonry shall be parged smooth. Where no lining is provided, the total minimum thickness of front, back and sidewalls shall be 8 inches (203 mm) of solid masonry. When a lining of firebrick at least 2 inches (51 mm) thick, or a lining of vitrified clay at least $5/8$ inch (15.9 mm) thick, is provided, the total minimum thickness of front, back and sidewalls shall be 6 inches (152 mm) of solid masonry, including the lining. Firebrick shall conform to ASTM C 27 or ASTM C 1261 and shall be laid with refractory mortar conforming to ASTM C 199.

2111.8.1 Smoke chamber dimensions. The inside height of the smoke chamber from the fireplace throat to the beginning of the flue shall not be greater than the inside width of the fireplace opening. The inside surface of the smoke chamber shall not be inclined more than 45 degrees (0.76 rad) from vertical when prefabricated smoke chamber linings are used or when the smoke chamber walls are rolled or sloped rather than corbeled. When the inside surface of the smoke chamber is formed by corbeled masonry, the walls shall not be corbeled more than 30 degrees (0.52 rad) from vertical.

2111.9 Hearth and hearth extension. Masonry fireplace hearths and hearth extensions shall be constructed of concrete or masonry, supported by noncombustible materials, and reinforced to carry their own weight and all imposed loads. No combustible material shall remain against the underside of hearths or hearth extensions after construction.

2111.9.1 Hearth thickness. The minimum thickness of fireplace hearths shall be 4 inches (102 mm).

2111.9.2 Hearth extension thickness. The minimum thickness of hearth extensions shall be 2 inches (51 mm).

Exception: When the bottom of the firebox opening is raised at least 8 inches (203 mm) above the top of the hearth extension, a hearth extension of not less than $3/8$-inch-thick (9.5 mm) brick, concrete, stone, tile or other approved noncombustible material is permitted.

2111.10 Hearth extension dimensions. Hearth extensions shall extend at least 16 inches (406 mm) in front of, and at least 8 inches (203 mm) beyond, each side of the fireplace opening. Where the fireplace opening is 6 square feet (0.557 m²) or larger, the hearth extension shall extend at least 20 inches (508 mm) in front of, and at least 12 inches (305 mm) beyond, each side of the fireplace opening.

2111.11 Fireplace clearance. Any portion of a masonry fireplace located in the interior of a building or within the exterior wall of a building shall have a clearance to combustibles of not less than 2 inches (51 mm) from the front faces and sides of masonry fireplaces and not less than 4 inches (102 mm) from the back faces of masonry fireplaces. The airspace shall not be filled, except to provide fireblocking in accordance with Section 2111.13.

Exceptions:

1. Masonry fireplaces listed and labeled for use in contact with combustibles in accordance with UL 127, and installed in accordance with the manufacturer's installation instructions, are permitted to have combustible material in contact with their exterior surfaces.

2. When masonry fireplaces are constructed as part of masonry or concrete walls, combustible materials shall not be in contact with the masonry or concrete walls less than 12 inches (306 mm) from the inside surface of the nearest firebox lining.

3. Exposed combustible trim and the edges of sheathing materials, such as wood siding, flooring and drywall, are permitted to abut the masonry fireplace sidewalls and hearth extension, in accordance with Figure 2111.11, provided such combustible trim or sheathing is a minimum of 12 inches (306 mm) from the inside surface of the nearest firebox lining.

4. Exposed combustible mantels or trim is permitted to be placed directly on the masonry fireplace front surrounding the fireplace opening provided such combustible materials shall not be placed within 6 inches (153 mm) of a fireplace opening. Combustible material within 12 inches (306 mm) of the fireplace opening shall not project more than $1/8$ inch (3.2 mm) for each 1-inch (25 mm) distance from such opening.

2111.12 Mantel and trim. Woodwork or other combustible materials shall not be placed within 6 inches (152 mm) of a fireplace opening. Combustible material within 12 inches (305 mm) of the fireplace opening shall not project more than $1/8$ inch (3.2 mm) for each 1-inch (25 mm) distance from such opening.

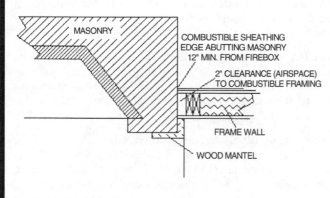

MASONRY

COMBUSTIBLE SHEATHING
EDGE ABUTTING MASONRY
12" MIN. FROM FIREBOX

2" CLEARANCE (AIRSPACE)
TO COMBUSTIBLE FRAMING

FRAME WALL

WOOD MANTEL

For SI: 1 inch = 25.4 mm

**FIGURE 2111.11
ILLUSTRATION OF EXCEPTION TO
FIREPLACE CLEARANCE PROVISION**

2111.13 Fireplace fireblocking. All spaces between fireplaces and floors and ceilings through which fireplaces pass shall be fireblocked with noncombustible material securely fastened in place. The fireblocking of spaces between wood joists, beams or headers shall be to a depth of 1 inch (25 mm) and shall only be placed on strips of metal or metal lath laid across the spaces between combustible material and the chimney.

2111.14 Exterior air. Factory-built or masonry fireplaces covered in this section shall be equipped with an exterior air supply to ensure proper fuel combustion unless the room is mechanically ventilated and controlled so that the indoor pressure is neutral or positive.

> **2111.14.1 Factory-built fireplaces.** Exterior combustion air ducts for factory-built fireplaces shall be listed components of the fireplace, and installed according to the fireplace manufacturer's instructions.
>
> **2111.14.2 Masonry fireplaces.** Listed combustion air ducts for masonry fireplaces shall be installed according to the terms of their listing and manufacturer's instructions.
>
> **2111.14.3 Exterior air intake.** The exterior air intake shall be capable of providing all combustion air from the exterior of the dwelling. The exterior air intake shall not be located within the garage, attic, basement or crawl space of the dwelling nor shall the air intake be located at an elevation higher than the firebox. The exterior air intake shall be covered with a corrosion-resistant screen of $^1/_4$-inch (6.4 mm) mesh.
>
> **2111.14.4 Clearance.** Unlisted combustion air ducts shall be installed with a minimum 1-inch (25 mm) clearance to combustibles for all parts of the duct within 5 feet (1524 mm) of the duct outlet.
>
> **2111.14.5 Passageway.** The combustion air passageway shall be a minimum of 6 square inches (3870 mm^2) and not more than 55 square inches (0.035 m^2), except that combustion air systems for listed fireplaces or for fireplaces tested for emissions shall be constructed according to the fireplace manufacturer's instructions.

2111.14.6 Outlet. The exterior air outlet is permitted to be located in the back or sides of the firebox chamber or within 24 inches (610 mm) of the firebox opening on or near the floor. The outlet shall be closable and designed to prevent burning material from dropping into concealed combustible spaces.

SECTION 2112
MASONRY HEATERS

2112.1 Definition. A masonry heater is a heating appliance constructed of concrete or solid masonry, hereinafter referred to as "masonry," having a mass of at least 1,760 pounds (800 kg), excluding the chimney and foundation, which is designed to absorb and store heat from a solid fuel fire built in the firebox by routing the exhaust gases through internal heat exchange channels in which the flow path downstream of the firebox includes at least one 180-degree (3.14 rad) change in flow direction before entering the chimney, and that delivers heat by radiation from the masonry surface of the heater that shall not exceed 230°F (110°C) except within 8 inches (203 mm) surrounding the fuel loading door(s).

2112.2 Installation. Masonry heaters shall be listed or installed in accordance with ASTM E 1602.

2112.3 Seismic reinforcing. Seismic reinforcing shall not be required within the body of a masonry heater whose height is equal to or less than 2.5 times its body width and where the masonry chimney serving the heater is not supported by the body of the heater. Where the masonry chimney shares a common wall with the facing of the masonry heater, the chimney portion of the structure shall be reinforced in accordance with Sections 2113 and 2113.4.

2112.4 Masonry heater clearance. Wood or other combustible framing shall not be placed within 4 inches (102 mm) of the outside surface of a masonry heater, provided the wall thickness of the firebox is not less than 8 inches (203 mm) and the wall thickness of the heat exchange channels is not less than 5 inches (127 mm). A clearance of at least 8 inches (203 mm) shall be provided between the gas-tight capping slab of the heater and a combustible ceiling. The required space between the heater and combustible material shall be fully vented to permit the free flow of air around all heater surfaces.

SECTION 2113
MASONRY CHIMNEYS

2113.1 General. A masonry chimney is a chimney constructed of concrete or masonry, hereinafter referred to as "masonry." Masonry chimneys shall be constructed, anchored, supported and reinforced as required in this chapter.

2113.2 Footings and foundations. Foundations for masonry chimneys shall be constructed of concrete or solid masonry at least 12 inches (305 mm) thick and shall extend at least 6 inches (152 mm) beyond the face of the foundation or support wall on all sides. Footings shall be founded on natural undisturbed earth or engineered fill below frost depth. In areas not subjected to freezing, footings shall be at least 12 inches (305 mm) below finished grade.

2113.3 Seismic reinforcing. Masonry or concrete chimneys shall be constructed, anchored, supported and reinforced as required in this chapter. In Seismic Design Category D, masonry and concrete chimneys shall be reinforced and anchored as detailed in Sections 2113.3.1, 2113.3.2 and 2113.4. In Seismic Design Category A, B or C, reinforcement and seismic anchorage is not required. In Seismic Design Category E or F, masonry and concrete chimneys shall be reinforced in accordance with the requirements of Sections 2101 through 2108.

2113.3.1 Vertical reinforcing. For chimneys up to 40 inches (1016 mm) wide, four No. 4 continuous vertical bars anchored in the foundation shall be placed in the concrete, between wythes of solid masonry or within the cells of hollow unit masonry and grouted in accordance with Section 2103.10. Grout shall be prevented from bonding with the flue liner so that the flue liner is free to move with thermal expansion. For chimneys greater than 40 inches (1016 mm) wide, two additional No. 4 vertical bars shall be provided for each additional 40 inches (1016 mm) in width or fraction thereof.

2113.3.2 Horizontal reinforcing. Vertical reinforcement shall be placed enclosed within $^1/_4$-inch (6.4 mm) ties, or other reinforcing of equivalent net cross-sectional area, spaced not to exceed 18 inches (457 mm) o.c. in concrete, or placed in the bed joints of unit masonry, at a minimum of every 18 inches (457 mm) of vertical height. Two such ties shall be provided at each bend in the vertical bars.

2113.4 Seismic anchorage. Masonry and concrete chimneys and foundations in Seismic Design Category D shall be anchored at each floor, ceiling or roof line more than 6 feet (1829 mm) above grade, except where constructed completely within the exterior walls. Anchorage shall conform to the following requirements.

2113.4.1 Anchorage. Two $^3/_{16}$-inch by 1-inch (4.8 mm by 25 mm) straps shall be embedded a minimum of 12 inches (305 mm) into the chimney. Straps shall be hooked around the outer bars and extend 6 inches (152 mm) beyond the bend. Each strap shall be fastened to a minimum of four floor joists with two $^1/_2$-inch (12.7 mm) bolts.

2113.5 Corbeling. Masonry chimneys shall not be corbeled more than half of the chimney's wall thickness from a wall or foundation, nor shall a chimney be corbeled from a wall or foundation that is less than 12 inches (305 mm) in thickness unless it projects equally on each side of the wall, except that on the second story of a two-story dwelling, corbeling of chimneys on the exterior of the enclosing walls is permitted to equal the wall thickness. The projection of a single course shall not exceed one-half the unit height or one-third of the unit bed depth, whichever is less.

2113.6 Changes in dimension. The chimney wall or chimney flue lining shall not change in size or shape within 6 inches (152 mm) above or below where the chimney passes through floor components, ceiling components or roof components.

2113.7 Offsets. Where a masonry chimney is constructed with a fireclay flue liner surrounded by one wythe of masonry, the maximum offset shall be such that the centerline of the flue above the offset does not extend beyond the center of the chimney wall below the offset. Where the chimney offset is supported by masonry below the offset in an approved manner, the maximum offset limitations shall not apply. Each individual corbeled masonry course of the offset shall not exceed the projection limitations specified in Section 2113.5.

2113.8 Additional load. Chimneys shall not support loads other than their own weight unless they are designed and constructed to support the additional load. Masonry chimneys are permitted to be constructed as part of the masonry walls or concrete walls of the building.

2113.9 Termination. Chimneys shall extend at least 2 feet (610 mm) higher than any portion of the building within 10 feet (3048 mm), but shall not be less than 3 feet (914 mm) above the highest point where the chimney passes through the roof.

2113.9.1 Spark arrestors. Where a spark arrestor is installed on a masonry chimney, the spark arrestor shall meet all of the following requirements:

1. The net free area of the arrestor shall not be less than four times the net free area of the outlet of the chimney flue it serves.
2. The arrestor screen shall have heat and corrosion resistance equivalent to 19-gage galvanized steel or 24-gage stainless steel.
3. Openings shall not permit the passage of spheres having a diameter greater than $^1/_2$ inch (13 mm) nor block the passage of spheres having a diameter less than $^3/_8$ inch (11 mm).
4. The spark arrestor shall be accessible for cleaning and the screen or chimney cap shall be removable to allow for cleaning of the chimney flue.

2113.10 Wall thickness. Masonry chimney walls shall be constructed of concrete, solid masonry units or hollow masonry units grouted solid with not less than 4 inches (102 mm) nominal thickness.

2113.11 Flue lining (material). Masonry chimneys shall be lined. The lining material shall be appropriate for the type of appliance connected, according to the terms of the appliance listing and the manufacturer's instructions.

2113.11.1 Residential-type appliances (general). Flue lining systems shall comply with one of the following:

1. Clay flue lining complying with the requirements of ASTM C 315, or equivalent.
2. Listed chimney lining systems complying with UL 1777.
3. Factory-built chimneys or chimney units listed for installation within masonry chimneys.
4. Other approved materials that will resist corrosion, erosion, softening or cracking from flue gases and condensate at temperatures up to 1,800°F (982°C).

2113.11.1.1 Flue linings for specific appliances. Flue linings other than those covered in Section 2113.11.1 intended for use with specific appliances shall comply with Sections 2113.11.1.2 through 2113.11.1.4 and Sections 2113.11.2 and 2113.11.3.

2113.11.1.2 Gas appliances. Flue lining systems for gas appliances shall be in accordance with the *International Fuel Gas Code.*

2113.11.1.3 Pellet fuel-burning appliances. Flue lining and vent systems for use in masonry chimneys with pellet fuel-burning appliances shall be limited to flue lining systems complying with Section 2113.11.1 and pellet vents listed for installation within masonry chimneys (see Section 2113.11.1.5 for marking).

2113.11.1.4 Oil-fired appliances approved for use with L-vent. Flue lining and vent systems for use in masonry chimneys with oil-fired appliances approved for use with Type L vent shall be limited to flue lining systems complying with Section 2113.11.1 and listed chimney liners complying with UL 641 (see Section 2113.11.1.5 for marking).

2113.11.1.5 Notice of usage. When a flue is relined with a material not complying with Section 2113.11.1, the chimney shall be plainly and permanently identified by a label attached to a wall, ceiling or other conspicuous location adjacent to where the connector enters the chimney. The label shall include the following message or equivalent language: "This chimney is for use only with (type or category of appliance) that burns (type of fuel). Do not connect other types of appliances."

2113.11.2 Concrete and masonry chimneys for medium-heat appliances.

2113.11.2.1 General. Concrete and masonry chimneys for medium-heat appliances shall comply with Sections 2113.1 through 2113.5.

2113.11.2.2 Construction. Chimneys for medium-heat appliances shall be constructed of solid masonry units or of concrete with walls a minimum of 8 inches (203 mm) thick, or with stone masonry a minimum of 12 inches (305 mm) thick.

2113.11.2.3 Lining. Concrete and masonry chimneys shall be lined with an approved medium-duty refractory brick a minimum of $4^1/_2$ inches (114 mm) thick laid on the $4^1/_2$-inch bed (114 mm) in an approved medium-duty refractory mortar. The lining shall start 2 feet (610 mm) or more below the lowest chimney connector entrance. Chimneys terminating 25 feet (7620 mm) or less above a chimney connector entrance shall be lined to the top.

2113.11.2.4 Multiple passageway. Concrete and masonry chimneys containing more than one passageway shall have the liners separated by a minimum 4-inch-thick (102 mm) concrete or solid masonry wall.

2113.11.2.5 Termination height. Concrete and masonry chimneys for medium-heat appliances shall extend a minimum of 10 feet (3048 mm) higher than any portion of any building within 25 feet (7620 mm).

2113.11.2.6 Clearance. A minimum clearance of 4 inches (102 mm) shall be provided between the exterior surfaces of a concrete or masonry chimney for medium-heat appliances and combustible material.

2113.11.3 Concrete and masonry chimneys for high-heat appliances.

2113.11.3.1 General. Concrete and masonry chimneys for high-heat appliances shall comply with Sections 2113.1 through 2113.5.

2113.11.3.2 Construction. Chimneys for high-heat appliances shall be constructed with double walls of solid masonry units or of concrete, each wall to be a minimum of 8 inches (203 mm) thick with a minimum airspace of 2 inches (51 mm) between the walls.

2113.11.3.3 Lining. The inside of the interior wall shall be lined with an approved high-duty refractory brick, a minimum of $4^1/_2$ inches (114 mm) thick laid on the $4^1/_2$-inch bed (114 mm) in an approved high-duty refractory mortar. The lining shall start at the base of the chimney and extend continuously to the top.

2113.11.3.4 Termination height. Concrete and masonry chimneys for high-heat appliances shall extend a minimum of 20 feet (6096 mm) higher than any portion of any building within 50 feet (15 240 mm).

2113.11.3.5 Clearance. Concrete and masonry chimneys for high-heat appliances shall have approved clearance from buildings and structures to prevent overheating combustible materials, permit inspection and maintenance operations on the chimney and prevent danger of burns to persons.

2113.12 Flue lining (installation). Flue liners shall be installed in accordance with ASTM C 1283 and extend from a point not less than 8 inches (203 mm) below the lowest inlet or, in the case of fireplaces, from the top of the smoke chamber, to a point above the enclosing walls. The lining shall be carried up vertically, with a maximum slope no greater than 30 degrees (0.52 rad) from the vertical.

Fireclay flue liners shall be laid in medium-duty refractory mortar conforming to ASTM C 199, with tight mortar joints left smooth on the inside and installed to maintain an airspace or insulation not to exceed the thickness of the flue liner separating the flue liners from the interior face of the chimney masonry walls. Flue lining shall be supported on all sides. Only enough mortar shall be placed to make the joint and hold the liners in position.

2113.13 Additional requirements.

2113.13.1 Listed materials. Listed materials used as flue linings shall be installed in accordance with the terms of their listings and the manufacturer's instructions.

2113.13.2 Space around lining. The space surrounding a chimney lining system or vent installed within a masonry chimney shall not be used to vent any other appliance.

Exception: This shall not prevent the installation of a separate flue lining in accordance with the manufacturer's instructions.

2113.14 Multiple flues. When two or more flues are located in the same chimney, masonry wythes shall be built between adjacent flue linings. The masonry wythes shall be at least 4 inches (102 mm) thick and bonded into the walls of the chimney.

Exception: When venting only one appliance, two flues are permitted to adjoin each other in the same chimney with only the flue lining separation between them. The joints of the adjacent flue linings shall be staggered at least 4 inches (102 mm).

2113.15 Flue area (appliance). Chimney flues shall not be smaller in area than the area of the connector from the appliance. Chimney flues connected to more than one appliance shall not be less than the area of the largest connector plus 50 percent of the areas of additional chimney connectors.

Exceptions:

1. Chimney flues serving oil-fired appliances sized in accordance with NFPA 31.

2. Chimney flues serving gas-fired appliances sized in accordance with the *International Fuel Gas Code*.

2113.16 Flue area (masonry fireplace). Flue sizing for chimneys serving fireplaces shall be in accordance with Section 2113.16.1 or 2113.16.2.

2113.16.1 Minimum area. Round chimney flues shall have a minimum net cross-sectional area of at least $^1/_{12}$ of the fireplace opening. Square chimney flues shall have a minimum net cross-sectional area of at least $^1/_{10}$ of the fireplace opening. Rectangular chimney flues with an aspect ratio less than 2 to 1 shall have a minimum net cross-sectional area of at least $^1/_{10}$ of the fireplace opening. Rectangular chimney flues with an aspect ratio of 2 to 1 or more shall have a minimum net cross-sectional area of at least $^1/_8$ of the fireplace opening.

2113.16.2 Determination of minimum area. The minimum net cross-sectional area of the flue shall be determined in accordance with Figure 2113.16. A flue size providing at least the equivalent net cross-sectional area shall be used. Cross-sectional areas of clay flue linings are as provided in Tables 2113.16(1) and 2113.16(2) or as provided by the manufacturer or as measured in the field. The height of the chimney shall be measured from the firebox floor to the top of the chimney flue.

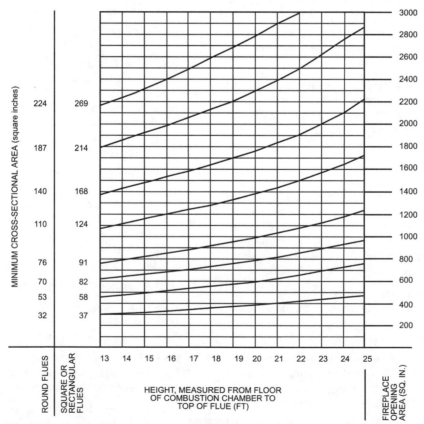

For SI: 1 inch = 25.4 mm, 1 square inch = 645 mm^2.

FIGURE 2113.16
FLUE SIZES FOR MASONRY CHIMNEYS

TABLE 2113.16(1)
NET CROSS-SECTIONAL AREA OF ROUND FLUE SIZES[a]

FLUE SIZE, INSIDE DIAMETER (inches)	CROSS-SECTIONAL AREA (square inches)
6	28
7	38
8	50
10	78
$10^3/_4$	90
12	113
15	176
18	254

For SI: 1 inch = 25.4 mm, 1 square inch = 645.16 mm².

a. Flue sizes are based on ASTM C 315.

TABLE 2113.16(2)
NET CROSS-SECTIONAL AREA OF SQUARE AND RECTANGULAR FLUE SIZES[a]

FLUE SIZE, INSIDE DIMENSION (inches)	CROSS-SECTIONAL AREA (square inches)
$4^1/_2 \times 13$	34
$7^1/_2 \times 7^1/_2$	37
$8^1/_2 \times 8^1/_2$	47
$7^1/_2 \times 11^1/_2$	58
$8^1/_2 \times 13$	74
$7^1/_2 \times 15^1/_2$	82
$11^1/_2 \times 11^1/_2$	91
$8^1/_2 \times 17^1/_2$	101
13×13	122
$11^1/_2 \times 15^1/_2$	124
$13 \times 17^1/_2$	165
$15^1/_2 \times 15^1/_2$	168
$15^1/_2 \times 19^1/_2$	214
$17^1/_2 \times 17^1/_2$	226
$19^1/_2 \times 19^1/_2$	269
20×20	286

For SI: 1 inch = 25.4 mm, 1 square inch = 645.16 mm².

a. Flue sizes are based on ASTM C 315.

2113.17 Inlet. Inlets to masonry chimneys shall enter from the side. Inlets shall have a thimble of fireclay, rigid refractory material or metal that will prevent the connector from pulling out of the inlet or from extending beyond the wall of the liner.

2113.18 Masonry chimney cleanout openings. Cleanout openings shall be provided within 6 inches (152 mm) of the base of each flue within every masonry chimney. The upper edge of the cleanout shall be located at least 6 inches (152 mm) below the lowest chimney inlet opening. The height of the opening shall be at least 6 inches (152 mm). The cleanout shall be provided with a noncombustible cover.

Exception: Chimney flues serving masonry fireplaces, where cleaning is possible through the fireplace opening.

2113.19 Chimney clearances. Any portion of a masonry chimney located in the interior of the building or within the exterior wall of the building shall have a minimum airspace clearance to combustibles of 2 inches (51 mm). Chimneys located entirely outside the exterior walls of the building, including chimneys that pass through the soffit or cornice, shall have a minimum airspace clearance of 1 inch (25 mm). The airspace shall not be filled, except to provide fireblocking in accordance with Section 2113.20.

Exceptions:

1. Masonry chimneys equipped with a chimney lining system listed and labeled for use in chimneys in contact with combustibles in accordance with UL 1777, and installed in accordance with the manufacturer's instructions, are permitted to have combustible material in contact with their exterior surfaces.

2. Where masonry chimneys are constructed as part of masonry or concrete walls, combustible materials shall not be in contact with the masonry or concrete wall less than 12 inches (306 mm) from the inside surface of the nearest flue lining.

3. Exposed combustible trim and the edges of sheathing materials, such as wood siding, are permitted to abut the masonry chimney sidewalls, in accordance with Figure 2113.19, provided such combustible trim or sheathing is a minimum of 12 inches (306 mm) from the inside surface of the nearest flue lining. Combustible material and trim shall not overlap the corners of the chimney by more than 1 inch (25 mm).

2113.20 Chimney fireblocking. All spaces between chimneys and floors and ceilings through which chimneys pass shall be fireblocked with noncombustible material securely fastened in place. The fireblocking of spaces between wood joists, beams or headers shall be to a depth of 1 inch (25 mm) and shall only be placed on strips of metal or metal lath laid across the spaces between combustible material and the chimney.

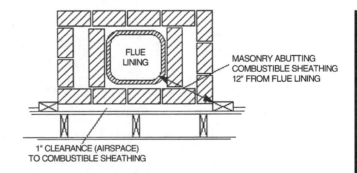

For SI: 1 inch = 25.4 mm

FIGURE 2113.19
ILLUSTRATION OF EXCEPTION TO CHIMNEY CLEARANCE PROVISION

CHAPTER 22

STEEL

SECTION 2201
GENERAL

2201.1 Scope. The provisions of this chapter govern the quality, design, fabrication and erection of steel used structurally in buildings or structures.

SECTION 2202
DEFINITIONS AND NOMENCLATURE

2202.1 Definitions. The following words and terms shall, for the purposes of this chapter and as used elsewhere in this code, have the meaning shown herein.

ADJUSTED SHEAR RESISTANCE. In Type II shear walls, the unadjusted shear resistance multiplied by the shear resistance adjustment factors of Table 2211.3.

STEEL CONSTRUCTION, COLD-FORMED. That type of construction made up entirely or in part of steel structural members cold formed to shape from sheet or strip steel such as roof deck, floor and wall panels, studs, floor joists, roof joists and other structural elements.

STEEL JOIST. Any steel structural member of a building or structure made of hot-rolled or cold-formed solid or open-web sections, or riveted or welded bars, strip or sheet steel members, or slotted and expanded, or otherwise deformed rolled sections.

STEEL MEMBER, STRUCTURAL. Any steel structural member of a building or structure consisting of a rolled steel structural shape other than cold-formed steel, or steel joist members.

TYPE I SHEAR WALL. A wall designed to resist in-plane lateral forces that is fully sheathed and provided with hold-down anchors at each end of the wall segment. Type I walls are permitted to have openings where detailing for force transfer around the openings is provided (see Figure 2202.1).

TYPE II SHEAR WALL. A wall designed to resist in-plane lateral forces that is sheathed with wood structural panel or sheet steel that contains openings, that have not been specifically designed and detailed for force transfer around wall openings. Hold-down anchors for Type II shear walls are only required at the ends of the wall (see Figure 2202.1).

TYPE II SHEAR WALL SEGMENT. A section of shear wall with full-height sheathing and which meets the aspect ratio limits of Section 2211.3.2(3).

UNADJUSTED SHEAR RESISTANCE. In Type II walls, the unadjusted shear resistance is based on the design shear and the limitations of Section 2211.3.1.

2202.2 Nomenclature. The following symbols shall, for the purposes of this chapter and as used elsewhere in this code, have the meanings shown herein.

ϕ = Resistance factor (see Section 2211.2.1).

Ω = Factor of safety (see Section 2211.2.1).

Ω_o = System overstrength factor (see Section 1617.6).

C_o = Shear resistance adjustment factor from Table 2211.3.

ΣL_i = Sum of widths of Type II shear wall segments, feet (mm/1,000).

C = Compression chord uplift force, lbs (kN).

V = Shear force in Type II shear wall, lbs (kN).

h = The height of a shear wall measured as:

    1. The maximum clear height from top of foundation to bottom of diaphragm framing above or,

    2. The maximum clear height from top of a diaphragm to bottom of diaphragm framing above.

v = Unit shear force, plf (kN/m).

w = The width of a shear wall or wall pier in the direction of application of force measured as the sheathed dimension of the shear wall.

SECTION 2203
IDENTIFICATION AND PROTECTION
OF STEEL FOR STRUCTURAL PURPOSES

2203.1 Identification. Steel furnished for structural load-carrying purposes shall be properly identified for conformity to the ordered grade in accordance with the specified ASTM standard or other specification and the provisions of this chapter. Steel that is not readily identifiable as to grade from marking and test records shall be tested to determine conformity to such standards.

2203.2 Protection. Painting of structural steel shall comply with the requirements contained in either the *AISC Load and Resistance Factor Design Specification for Structural Steel Buildings* (AISC-LRFD), *AISC Specification for Structural Steel Buildings—Allowable Stress Design* (AISC 335) or *AISC Specification for the Design of Steel Hollow Structural Sections* (AISC-HSS). Individual structural members and assembled panels of cold-formed steel construction, except where fabricated of approved corrosion-resistant steel or of steel having a corrosion resistant or other approved coating, shall be protected against corrosion with an approved coat of paint, enamel or other approved protection.

SECTION 2204
CONNECTIONS

2204.1 Welding. The details of design, workmanship and technique for welding, inspection of welding and qualification of welding operators shall conform to the requirements of the specifications listed in Sections 2205, 2206, 2207, 2209 and 2210. Special inspection of welding shall be provided where required by Section 1704.

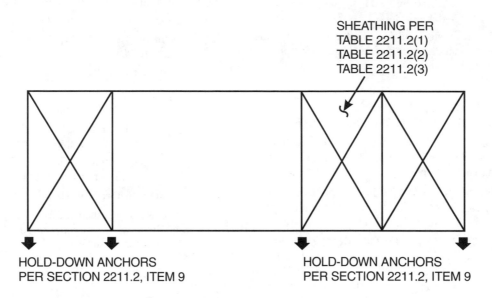

SHEATHING PER
TABLE 2211.2(1)
TABLE 2211.2(2)
TABLE 2211.2(3)

HOLD-DOWN ANCHORS
PER SECTION 2211.2, ITEM 9

HOLD-DOWN ANCHORS
PER SECTION 2211.2, ITEM 9

TYPE I SHEAR WALL

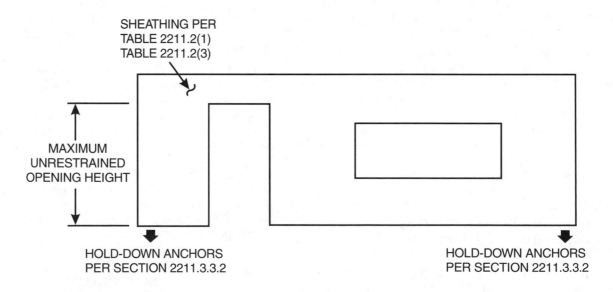

SHEATHING PER
TABLE 2211.2(1)
TABLE 2211.2(3)

MAXIMUM
UNRESTRAINED
OPENING HEIGHT

HOLD-DOWN ANCHORS
PER SECTION 2211.3.3.2

HOLD-DOWN ANCHORS
PER SECTION 2211.3.3.2

TYPE II SHEAR WALL

FIGURE 2202.1
TYPE I AND TYPE II SHEAR WALLS

2204.2 Bolting. The design, installation and inspection of bolts shall be in accordance with the requirements of the specifications listed in Sections 2205, 2206, 2209 and 2210. Special inspection of the installation of high-strength bolts shall be provided where required by Section 1704.

2204.2.1 Anchor rods. Anchor rods shall be set accurately to the pattern and dimensions called for on the plans. The protrusion of the threaded ends through the connected material shall be sufficient to fully engage the threads of the nuts, but shall not be greater than the length of the threads on the bolts.

SECTION 2205
STRUCTURAL STEEL

2205.1 General. The design, fabrication and erection of structural steel for buildings and structures shall be in accordance with either the AISC-LRFD, AISC 335 or AISC-HSS. Where required, the seismic design of steel structures shall be in accordance with the additional provisions of Section 2205.2.

2205.2 Seismic requirements for steel structures. The design of structural steel structures to resist seismic forces shall be in accordance with the provisions of Section 2205.2.1 or 2205.2.2 for the appropriate seismic design category.

2205.2.1 Seismic Design Category A, B or C. Structural steel structures assigned to Seismic Design Category A, B or C, in accordance with Section 1616, shall be of any construction permitted in Section 2205. An R factor as set forth in Section 1617.6 for the appropriate steel system is permitted where the structure is designed and detailed in accordance with the provisions of AISC 341, Parts I and III. Systems not detailed in accordance with the above shall use the R factor in Section 1617.6 designated for "steel systems not detailed for seismic."

2205.2.2 Seismic Design Category D, E or F. Structural steel structures assigned to Seismic Design Category D, E or F shall be designed and detailed in accordance with AISC 341, Part I.

2205.3 Seismic requirements for composite construction. The design, construction and quality of composite steel and concrete components that resist seismic forces shall conform to the requirements of the AISC LRFD and ACI 318. An R factor as set forth in Section 1617.6 for the appropriate composite steel and concrete system is permitted where the structure is designed and detailed in accordance with the provisions of AISC 341, Part II. In Seismic Design Category B or above, the design of such systems shall conform to the requirements of AISC 341, Part II.

2205.3.1 Seismic Design Categories D, E and F. Composite structures are permitted in Seismic Design Categories D, E and F, subject to the limitations in Section 1617.6, where substantiating evidence is provided to demonstrate that the proposed system will perform as intended by AISC 341, Part II. The substantiating evidence shall be subject to building official approval. Where composite elements or connections are required to sustain inelastic deformations, the substantiating evidence shall be based on cyclic testing.

SECTION 2206
STEEL JOISTS

2206.1 General. The design, manufacturing and use of open web steel joists and joist girders shall be in accordance with one of the following Steel Joist Institute specifications:

1. *Standard Specifications for Open Web Steel Joists, K Series.*
2. *Standard Specifications for Longspan Steel Joists, LH Series and Deep Longspan Steel Joists, DLH Series.*
3. *Standard Specifications for Joist Girders.*

Where required, the seismic design of buildings shall be in accordance with the additional provisions of Section 2205.2 or 2211.

SECTION 2207
STEEL CABLE STRUCTURES

2207.1 General. The design, fabrication and erection including related connections, and protective coatings of steel cables for buildings shall be in accordance with ASCE 19.

2207.2 Seismic requirements for steel cable. The design strength of steel cables shall be determined by the provisions of ASCE 19 except as modified by these provisions.

1. A load factor of 1.1 shall be applied to the prestress force included in T_3 and T_4 as defined in Section 3.12.
2. In Section 3.2.1, Item (c) shall be replaced with "1.5 T_3" and Item (d) shall be replaced with "1.5 T_4"

SECTION 2208
STEEL STORAGE RACKS

2208.1 Storage racks. The design, testing and utilization of industrial steel storage racks shall be in accordance with the *RMI Specification for the Design, Testing and Utilization of Industrial Steel Storage Racks*. Racks in the scope of this specification include industrial pallet racks, movable shelf racks and stacker racks, and does not apply to other types of racks, such as drive-in and drive-through racks, cantilever racks, portable racks or rack buildings. Where required, the seismic design of storage racks shall be in accordance with the provisions of Section 9.6.2.9 of ASCE 7.

SECTION 2209
COLD-FORMED STEEL

2209.1 General. The design of cold-formed carbon and low-alloy steel structural members shall be in accordance with the *North American Specification for the Design of Cold-Formed Steel Structural Members* (AISI-NASPEC). The design of cold-formed stainless-steel structural members shall be in accordance with ASCE 8. Cold-formed steel light-framed construction shall comply with Section 2210.

2209.2 Composite slabs on steel decks. Composite slabs of concrete and steel deck shall be designed and constructed in accordance with ASCE 3.

SECTION 2210
COLD-FORMED STEEL
LIGHT-FRAMED CONSTRUCTION

2210.1 General. The design, installation and construction of cold-formed carbon or low-alloy steel, structural and nonstructural steel framing, shall be in accordance with the *Standard for Cold-Formed Steel Framing—General Provisions, American Iron and Steel Institute* (AISI-General) *and* AISI-NASPEC.

2210.2 Headers. The design and installation of cold-formed steel box and back-to-back headers, and double L-headers used in single-span conditions for load-carrying purposes shall be in accordance with the *Standard for Cold-Formed Steel Framing—Header Design*, American Iron and Steel Institute (AISI-Header), subject to the limitations therein.

2210.3 Trusses. The design, quality assurance, installation and testing of cold-formed steel trusses shall be in accordance with the *Standard for Cold-Formed Steel Framing–Trusses*, American Iron and Steel Institute (AISI-Truss), subject to the limitations therein.

SECTION 2211
COLD-FORMED STEEL
LIGHT-FRAMED SHEAR WALLS

2211.1 General. In addition to the requirements of Section 2210, the design of cold-formed steel light-framed shear walls, to resist wind and seismic loads shall be in accordance with the requirements of Section 2211.2 for Type I (segmented) shear walls or Section 2211.3 for Type II (perforated) shear walls.

Light-framed structures assigned to Seismic Design Categories A, B and C, in accordance with Section 1616, shall be of any construction permitted in Section 2210. An *R* factor as set forth in Section 1617.6 for the appropriate steel system is permitted where the lateral design of the structure is in accordance with the provisions of Section 2211.4. Systems not detailed in accordance with Section 2211.4 shall use the *R* factor in Section 1617.6 designated for "steel systems not detailed for seismic."

In Seismic Design Categories D, E and F, the lateral design of light-framed structures shall also comply with the requirements in Section 2211.4

2211.2 Type I shear walls. The design of Type I shear walls, of cold-formed steel light-framed construction, to resist wind and seismic loads, shall be in accordance the requirements of this section.

1. The nominal shear value for Type I shear walls, as shown in Table 2211.2(1) for wind loads, Table 2211.2(2) for wind or seismic loads or Table 2211.2(3) for seismic loads, is permitted to establish allowable shear values or design shear values.

2. Boundary members, chords, collectors and connections thereto shall be proportioned to transmit the induced forces.

3. As an alternative to the values in Tables 2211.2(1), 2211.2(2) and 2211.2(3), shear values are permitted to be calculated by the principles of mechanics by using ap-

proved fastener values and shear values appropriate for the sheathing material attached.

4. Type I shear walls sheathed with wood structural or sheet steel panels are permitted to have window openings, between hold-down anchors at each end of a wall segment, where details are provided to account for force transfer around openings.

5. The aspect ratio limitations of Section 2211.2.2, Item 5, shall apply to the entire Type I segment and to each wall pier at the side of each opening.

6. The height of the wall pier (*h*) shall be defined as the clear height of the pier at the side of an opening.

7. The width of a pier (*w*) shall be defined as the sheathed width of the pier.

8. The width of wall piers shall not be less than 24 inches (102 mm).

9. Hold-down anchors shall be provided at each end of a Type I shear wall capable of resisting the design forces.

2211.2.1 Design shear determination. Where allowable stress design (ASD) is used, the allowable shear value shall be determined by dividing the nominal shear value, shown in Tables 2211.2(1), 2211.2(2) and 2211.2(3), by a factor of safety (Ω) of 2.5.

Where load and resistance factor design (LRFD) is used, the design shear value shall be determined by multiplying the nominal shear value, shown in Tables 2211.2(1), 2211.2(2) and 2211.2(3), by a resistance factor (ϕ) of 0.55.

2211.2.2 Limitations for systems. The lateral-resistant systems listed in Tables 2211.2(1), 2211.2(2) and 2211.2(3) shall conform to the following requirements:

1. Studs shall be a minimum $1^5/_8$ inches (41.3 mm) by $3^1/_2$ inches (89 mm) with a $3/_8$-inch (9.5 mm) return lip. As a minimum, studs shall be doubled (back to back) at shear wall ends.

2. Track shall be a minimum $1^1/_4$ inches (31.8 mm) by $3^1/_2$ inches (89 mm).

3. Both studs and track shall have a minimum uncoated base metal thickness of 33 mils (0.84 mm) and shall be of the following grades of structural quality steel: ASTM A 653 SS Grade 33, ASTM A 792 SS Grade 33 or ASTM A 875 SS Grade 33.

4. Fasteners along the edges in shear panels shall be placed not less than $3/_8$ inch (9.5 mm) in from panel edges.

5. The height-to-width shear wall aspect ratio (*h/w*) of wall systems shall not exceed the values in Tables 2211.2(1), 2211.2(2) and 2211.2(3). Where the limiting ratio of *h/w* is greater than 2:1, the shear values shall be multiplied by 2*w/h*.

6. Panel thicknesses shown are minimums. Panels less than 12 inches (305 mm) wide shall not be used. All panel edges shall be fully blocked.

7. Where horizontal strap blocking is used to provide edge blocking, it shall be a minimum $1^1/_2$ inches (38 mm) wide and of the same material and equal or greater thickness as the track and studs.

8. The design shear values for shear panels with different nominal shear values applied to the same side of a wall are not cumulative except as permitted in Tables 2211.2(1), 2211.2(2) and 2211.2(3). For walls with material applied to both faces of the same wall, the design shear value of material of the same capacity is cumulative. Where the material nominal shear values are not equal, the design shear value shall be either two times the design shear value of the material with the smaller values or shall be taken as the value of the stronger side, whichever is greater. Summing shear values of dissimilar material applied to opposite faces or to the same wall line is not allowed unless permitted by Table 2211.2(1).

2211.2.2.1 Sheet steel sheathing. Steel sheets, attached to cold-formed steel framing, are permitted to resist horizontal forces produced by wind or seismic loads.

1. Steel sheets shall have a minimum base metal thickness as shown in Table 2211.2(1) or 2211.2(3), and shall be of the following grades of structural quality steel: ASTM A653 SS Grade 33, ASTM A792 SS Grade 33 or ASTM A 875 SS Grade 33.

2. Nominal shear values, used to establish the allowable shear value or design shear value, are given in Tables 2211.2(1) for wind loads and 2211.2(3) for seismic loads.

3. Steel sheets are permitted to be applied either parallel or perpendicular to framing. All edges of steel sheets shall be attached to framing members, strap blocking or shall be overlapped and attached to each other with screw spacing as required for edges.

4. Screws used to attach steel sheets shall be a minimum No. 8 modified truss head.

TABLE 2211.2(1)
NOMINAL SHEAR VALUES FOR WIND FORCES IN POUNDS PER FOOT FOR SHEAR WALLS FRAMED WITH COLD-FORMED STEEL STUDS[a]

ASSEMBLY DESCRIPTION	MAXIMUM HEIGHT/LENGTH RATIO h/w	FASTENER SPACING AT PANEL EDGES[b] (inches)				MAXIMUM FRAMING SPACING (inches o.c.)
		6	4	3	2	
$^{15}/_{32}$-inch structural 1 sheathing (4-ply) plywood one side	2:1	1,065[c]	—	—	—	24
$^{7}/_{16}$-inch rated sheathing (OSB), one side	2:1	910[c]	1,410	1,735	1,910	24
$^{7}/_{16}$-inch rated sheathing (OSB), one side, oriented perpendicular to framing	2:1	1,020[c]	—	—	—	24
$^{7}/_{16}$-inch rated sheathing (OSB), one side	4:1[d]	—	1,025	1,425	1,825	24
0.018-inch steel sheet, one side	2:1	485	—	—	—	24
0.027-inch steel sheet, one side	4:1[d]	—	1,000	—	—	24

For SI: 1 inch = 25.4 mm, 1 pound per foot = 14.5939 N/m.

a. Nominal shear values shall be multiplied by the resistance factor ϕ to determine design strength or divided by the safety factor Ω to determine allowable shear values as set forth in Section 2211.2.1.

b. Screws shall be attached to intermediate supports at 12 inches on center unless otherwise shown.

c. Where fully blocked gypsum board is applied to the opposite side of this assembly, in accordance with Table 2211.2(2) with screw spacing at 7 inches o.c. edge and 7 inches o.c. field, these nominal values are permitted to be increased by 30 percent.

d. Where aspect ratio (h/w) is greater that 2:1, the design shear shall be reduced as required by Section 2211.2.2, item 5.

TABLE 2211.2(2)
NOMINAL SHEAR VALUES FOR WIND AND SEISMIC FORCES IN POUNDS PER FOOT FOR SHEAR WALLS FRAMED WITH COLD-FORMED STEEL STUDS AND FACED WITH GYPSUM BOARD[a,b]

WALL CONSTRUCTION	MAXIMUM HEIGHT/LENGTH RATIO h/w	ORIENTATION	SCREW SPACING (inches)		NOMINAL SHEAR VALUE (plf)
			Edge	Field	
$^{1}/_{2}$-inch gypsum board on both sides of wall; Studs maximum 24 inches o.c.	2:1	Gypsum board applied perpendicular to framing with strap blocking behind the horizontal joint and with solid blocking between the first two end studs	7	7	585
			4	4	850

For SI: 1 inch = 25.4 mm, 1 pound per foot = 14.5939 N/m.

a. Nominal shear values shall be multiplied by the resistance factor ϕ to determine design strength or divided by the safety factor Ω to determine allowable shear values as set forth in Section 2211.2.1.

b. Walls resisting seismic loads shall be subject to the limitations in Section 1617.6.

TABLE 2211.2(3)
NOMINAL SHEAR VALUES FOR SEISMIC FORCES IN POUNDS PER FOOT FOR SHEAR WALLS
FRAMED WITH COLD-FORMED STEEL STUDS[a]

ASSEMBLY DESCRIPTION	MAXIMUM HEIGHT/LENGTH RATIO h/w	FASTENER SPACING AT PANEL EDGES[b] (inches)				MAXIMUM FRAMING SPACING (inches o.c.)
		6	4	3	2	
$^{15}/_{32}$-inch Structural 1 Sheathing (4-ply) plywood one side	2:1[c]	780	990	1,465	1,625	24
$^{15}/_{32}$-inch Structural 1 Sheathing (4-ply) plywood one side; end studs 0.043 inch minimum thickness	2:1	—	—	1,775	2,190	24
$^{15}/_{32}$-inch Structural 1 Sheathing (4-ply) plywood one side; all studs and track 0.043 inch minimum thickness	2:1	890	1,330	1,775	2,190	24
$^{7}/_{16}$-inch OSB one side	2:1[c]	700	915	1,275	1,625	24
$^{7}/_{16}$-inch OSB one side end studs, 0.043 inch minimum thickness	2:1	—	—	1,520	2,060	24
0.018-inch minimum thickness steel sheet one side	2:1	390	—	—	—	24
0.027-inch minimum thickness steel sheet one side	2:1[c]	—	1,000	1,085	1,170	24

For SI: 1 inch = 25.4 mm, 1 pound per foot = 14.5939 N/m.

a. Nominal shear values shall be multiplied by the resistance factor φ to determine design strength or divided by the safety factor Ω to determine allowable shear values as set forth in Section 2211.2.1.

b. Screws shall be attached to intermediate supports at 12 inches o.c. unless otherwise shown.

c. In Seismic Design Category A, B and C the aspect ratio (h/w) is permitted to be 4:1 where the design shear is reduced as required by Section 2211.2.2, item 5.

2211.2.2.2 Wood structural panel sheathing. Cold-formed steel framed wall systems, sheathed with wood structural panels, are permitted to resist horizontal forces produced by wind or seismic loads subject to the following:

1. Nominal shear values, used to establish the allowable shear value or design shear value, are given in Tables 2211.2(1), for wind loads, and 2211.2(3), for seismic loads.

2. Wood structural panels shall comply with DOC PS 1 or PS 2 and shall be manufactured using exterior glue.

3. Wood structural panels shall be attached to steel framing with flat-head self-drilling tapping screws with a minimum head diameter of 0.292 inch (8 mm).

4. Where $^{7}/_{16}$-inch oriented strand board (OSB) is specified, $^{15}/_{32}$-inch structural 1 sheathing (plywood) is permitted.

5. Structural panels are permitted to be applied either parallel or perpendicular to framing.

6. Increases of the nominal loads shown in Tables 2211.2(1) and 2211.2(3) shall not be permitted for duration of load as permitted in Chapter 23.

2211.2.2.3 Gypsum board panel sheathing. Cold-formed steel framed wall systems, sheathed with gypsum board, are permitted to resist horizontal forces produced by wind or seismic loads subject to the following:

1. Nominal shear values, used to establish the allowable shear value or design shear value, are given in Table 2211.2(2).

2. The shear values listed in Table 2211.2(2) shall not be cumulative with the shear values of other materials applied to the same wall unless otherwise permitted herein.

3. The nominal shear values shown are for gypsum board that is applied to both sides of the wall.

4. Where gypsum board is only applied to one side of the wall, the nominal shear values shall be taken as one-half of the value shown.

5. Where gypsum board is applied perpendicular to studs, end joints of adjacent courses of gypsum board sheets shall not occur over the same stud.

6. Screws used to attach gypsum board shall be a minimum No. 6 in accordance with ASTM C 954.

7. Walls resisting seismic loads shall be subject to the limitations in Section 1617.6.

2211.3 Type II shear walls. Type II shear walls sheathed with wood structural panels or sheet steel are permitted to resist wind and seismic loads when designed in accordance with this section. Type II walls shall meet the requirements for Type I walls except as revised by this section.

2211.3.1 Limitations. The following limitations shall apply to the use of Type II shear walls:

1. A Type II shear wall segment, meeting the minimum aspect ratio (h/w) of Section 2211.3.2, Item 3, shall be located at each end of a Type II shear wall. Openings shall be permitted to occur beyond the ends of the Type II shear wall; however, the width of such openings shall not be included in the width of the perforated shear wall.

2. In Seismic Design Categories B, C, D, E and F, the nominal shear values shall be based upon edge screw spacing not less than 4 inches o.c.

3. A Type II shear wall shall not have out-of-plane (horizontal) offsets. Where out-of-plane offsets occur, portions of the wall on each side of the offset shall be considered as separate perforated shear walls.

4. Collectors for shear transfer shall be provided through the full length of the Type II shear wall.

5. A Type II shear wall shall have uniform top of wall and bottom of wall elevations. Type II shear walls not having uniform elevations shall be designed by other methods.

6. Type II shear wall height, h, shall not exceed 20 feet (6096 mm).

2211.3.2 Type II shear wall resistance. The Type II shear wall resistance shall be equal to the adjusted shear resistance multiplied by the sum of the widths (ΣL_i) of the Type II shear wall segments and shall be calculated in accordance with the following:

1. The percent of full-height sheathing shall be calculated as the sum of widths (ΣL_i) of Type II shear wall segments divided by the total width of the Type II shear wall including openings.

2. The maximum opening height ratio shall be calculated by dividing the maximum opening clear height by the shear wall height, h.

3. The unadjusted shear resistance shall be the design shear values calculated in accordance with Section 2211.2.1 based upon the values in Tables 2211.2(1) and 2211.2(3). The aspect ratio of all Type II shear wall segments used in calculations shall not exceed 2:1.

   Exception: Where permitted by Tables 2211.2.1(1) and 2211.2(3), the aspect ratio (h/w) of Type II wall segments greater than 2:1, but in no case greater than 4:1, is permitted to be included in the calculation of the unadjusted shear resistance for the wall, provided the values are multiplied by $2w/h$.

4. The adjusted shear resistance shall be calculated by multiplying the unadjusted shear resistance by the shear resistance adjustment factors of Table 2211.3. For intermediate percentages of full-height sheathing, the values are permitted to be determined by interpolation.

2211.3.3 Anchorage and load path. Design of Type II shear wall anchorage and load path shall conform to the requirements of this section, or shall be calculated using principles of mechanics.

2211.3.3.1 Anchorage for in-plane shear. The unit shear force, v, transmitted into the top and out of the base of the Type II shear wall full-height sheathing segments,

TABLE 2211.3
SHEAR RESISTANCE ADJUSTMENT FACTOR—C_o

WALL HEIGHT (h)	MAXIMUM OPENING HEIGHT RATIO[a] AND HEIGHT				
	h/3	h/2	2h/3	5h/6	h
8'0"	2'8"	4'0"	5'4"	6'8"	8'0"
10'0"	3'4"	5'0"	6'8"	8'4"	10'0"
Percent full-height sheathing[b]	Shear Resistance Adjustment Factor				
10%	1.00	0.69	0.53	0.43	0.36
20%	1.00	0.71	0.56	0.45	0.38
30%	1.00	0.74	0.59	0.49	0.42
40%	1.00	0.77	0.63	0.53	0.45
50%	1.00	0.80	0.67	0.57	0.50
60%	1.00	0.83	0.71	0.63	0.56
70%	1.00	0.87	0.77	0.69	0.63
80%	1.00	0.91	0.83	0.77	0.71
90%	1.00	0.95	0.91	0.87	0.83
100%	1.00	1.00	1.00	1.00	1.00

a. See Section 2211.3.2, item 2.
b. See Section 2211.3.2, item 1.

and into collectors (drag struts) connecting shear wall segments, shall be calculated in accordance with the following:

$$v = \frac{V}{C_o \Sigma L_i}$$ (Equation 22-1)

where:

v = Unit shear force, plf (kN/m).

V = Shear force in Type II shear wall, lbs (kN).

C_o = Shear resistance adjustment factor from Table 2211.3.

ΣL_i = Sum of widths of Type II shear wall segments, feet (mm/1,000).

2211.3.3.2 Uplift anchorage at Type II shear wall ends. Anchorage for uplift forces due to overturning shall be provided at each end of the Type II shear wall. Where seismic loads govern, the uplift anchorage shall be determined in accordance with the requirements of Section 2211.4.3.

2211.3.3.3. Uplift anchorage between perforated shear wall ends. In addition to the requirements of Section 2211.3.3.1, Type II shear wall bottom plates at full-height sheathing shall be anchored for a uniform uplift force, t, equal to the unit shear force, v, determined in Section 2211.3.3.1.

2211.3.3.4. Compression chords. Vertical elements at each end of each Type II shear wall segment shall be designed for a compression force, C, from each story calculated in accordance with the following:

$$C = \frac{Vh}{C_o \Sigma L_i}$$ (Equation 22-2)

where:

C = Compression chord uplift force, lbs (kN).

V = Shear force in Type II shear wall, lbs (kN).

h = Shear wall height feet, (mm/1,000).

C_o = Shear resistance adjustment factor from Table 2211.3.

ΣL_i = Sum of widths of Type II shear wall segments, feet (mm/1,000).

2211.3.3.5. Load path. A load path to the foundation shall be provided for the uplift shear and compression forces as determined from Sections 2211.3.3.1 through 2211.3.3.4, inclusive. Elements resisting shear wall forces contributed by multiple stories shall be designed for the sum of forces contributed by each story.

2211.4 Seismic Design Categories D, E and F.

2211.4.1 General. In addition to the requirements of Sections 2211.2 and 2211.3, light-framed cold-formed steel wall systems, that resist seismic loads, in buildings assigned to Seismic Design Category D, E or F, shall comply with the requirements of this section.

2211.4.2 Connections. Connections for diagonal bracing members, top chord splices, boundary members and collectors shall be designed to develop the lesser of the nominal

tensile strength of the member or the design seismic force multiplied by the seismic overstrength factor, Ω_o, from Section 1617.6. The pull-out resistance of screws shall not be used to resist design seismic forces.

2211.4.3 Anchorage of braced wall segments. Studs or other vertical boundary members at the ends of wall segments, that resist seismic loads, braced with either sheathing or diagonal braces, shall be anchored such that the bottom track is not required to resist uplift by bending of the track web. Both flanges of the studs shall be braced to prevent lateral torsional buckling. Studs or other vertical boundary members and anchorage thereto shall have the nominal strength to resist design seismic force multiplied by the seismic overstrength factor, Ω_o, from Section 1617.6.

2211.4.4 Sheet steel sheathing. Where steel sheathing provides lateral resistance, the design and construction of such walls shall be in accordance with the additional requirements of this section. Perimeter members at openings shall be provided and shall be detailed to distribute the shearing stresses. Wall studs and track shall have a minimum uncoated base metal thickness of 33 mils (0.84 mm) and shall not have an uncoated base metal thickness greater than 48 mils (1.10 mm). The nominal shear value for light-framed wall systems for buildings in Seismic Design Category D, E or F shall be based upon values from Table 2211.2(3).

2211.4.5 Wood structural panel sheathing. Where wood structural panels provide lateral resistance, the design and construction of such walls shall be in accordance with the additional requirements of this section. Perimeter members at openings shall be provided and shall be detailed to distribute the shearing stresses. Wood sheathing shall not be used to splice these members. Wall studs and track shall have a minimum uncoated base metal thickness of 33 mils (0.84 mm) and shall not have an uncoated base metal thickness greater than 48 mils (1.10 mm). The nominal shear value for light-framed wall systems for buildings in Seismic Design Category D, E or F shall be based upon values from Table 2211.2(3).

2211.4.6 Diagonal bracing. Where diagonal bracing is provided for lateral resistance, provisions shall be made for pretensioning or other methods of installing tension-only bracing shall be used to guard against loose diagonal straps. The l/r of the brace is permitted to exceed 200.

2211.4.7 Gypsum board panel sheathing. Gypsum board panel sheathing is permitted to resist seismic loads, subject to the limitations in Table 2211.2(2) and Section 1617.6.

CHAPTER 23
WOOD

SECTION 2301
GENERAL

2301.1 Scope. The provisions of this chapter shall govern the materials, design, construction and quality of wood members and their fasteners.

2301.2 General design requirements. The design of structural elements or systems, constructed partially or wholly of wood or wood-based products, shall be based on one of the following methods.

2301.2.1 Allowable stress design. Design using allowable stress design methods shall resist the applicable load combinations of Chapter 16 in accordance with the provisions of Sections 2304, 2305 and 2306.

2301.2.2 Load and resistance factor design (LRFD). Design using load and resistance factor design (LRFD) methods shall resist the applicable load combinations of Chapter 16 in accordance with the provisions of Sections 2304, 2305 and 2307.

2301.2.3 Conventional light-frame wood construction. The design and construction of conventional light-frame wood construction shall be accordance with the provisions of Sections 2304 and 2308.

> **Exception:** Buildings designed in accordance with the provisions of the *AF&PA Wood Frame Construction Manual for One- and Two-Family Dwellings* shall be deemed to meet the requirements of the provisions of Section 2308.

2301.3 Nominal sizes. For the purposes of this chapter, where dimensions of lumber are specified, they shall be deemed to be nominal dimensions unless specifically designated as actual dimensions (see Section 2304.2).

SECTION 2302
DEFINITIONS

2302.1 Definitions. The following words and terms shall, for the purposes of this chapter, have the meanings shown herein.

ACCREDITATION BODY. An approved, third-party organization that is independent of the grading and inspection agencies, and the lumber mills, and that initially accredits and subsequently monitors, on a continuing basis, the competency and performance of a grading or inspection agency related to carrying out specific tasks.

ADJUSTED SHEAR RESISTANCE. The unadjusted shear resistance multiplied by the shear resistance adjustment factors of Table 2305.3.7.2.

BRACED WALL LINE. A series of braced wall panels in a single story that meets the requirements of Section 2308.3 or 2308.12.4.

BRACED WALL PANEL. A section of wall braced in accordance with Section 2308.9.3 or 2308.12.4.

COLLECTOR. A horizontal diaphragm element parallel and in line with the applied force that collects and transfers diaphragm shear forces to the vertical elements of the lateral-force-resisting system and/or distributes forces within the diaphragm.

CONVENTIONAL LIGHT-FRAME WOOD CONSTRUCTION. A type of construction whose primary structural elements are formed by a system of repetitive wood-framing members. See Section 2308 for conventional light-frame wood construction provisions.

CRIPPLE WALL. A framed stud wall extending from the top of the foundation to the underside of floor framing for the lowest occupied floor level.

DIAPHRAGM, UNBLOCKED. A diaphragm that has edge nailing at supporting members only. Blocking between supporting structural members at panel edges is not included. Diaphragm panels are field nailed to supporting members.

DRAG STRUT. See "Collector."

FIBERBOARD. A fibrous, homogeneous panel made from lignocellulosic fibers (usually wood or cane) and having a density of less than 31 pounds per cubic foot (pcf) (497 kg/m³) but more than 10 pcf (160 kg/m³).

GLUED BUILT-UP MEMBER. A structural element, the section of which is composed of built-up lumber, wood structural panels or wood structural panels in combination with lumber, all parts bonded together with structural adhesives.

GRADE (LUMBER). The classification of lumber in regard to strength and utility in accordance with American Softwood Lumber Standard DOC PS 20 and the grading rules of an approved lumber rules-writing agency.

HARDBOARD. A fibrous-felted, homogeneous panel made from lignocellulosic fibers consolidated under heat and pressure in a hot press to a density not less than 31 pcf (497 kg/m³).

NAILING, BOUNDARY. A special nailing pattern required by design at the boundaries of diaphragms.

NAILING, EDGE. A special nailing pattern required by design at the edges of each panel within the assembly of a diaphragm or shear wall.

NAILING, FIELD. Nailing required between the sheathing panels and framing members at locations other than boundary nailing and edge nailing.

NATURALLY DURABLE WOOD. The heartwood of the following species with the exception that an occasional piece with corner sapwood is permitted if 90 percent or more of the width of each side on which it occurs is heartwood.

> **Decay resistant.** Redwood, cedar, black locust and black walnut.

> **Termite resistant.** Redwood and Eastern red cedar.

NOMINAL SIZE (LUMBER). The commercial size designation of width and depth, in standard sawn lumber and glued-laminated lumber grades; somewhat larger than the standard net size of dressed lumber, in accordance with DOC PS 20 for sawn lumber and with the *National Design Specification for Wood Construction* (NDS) for glued-laminated lumber.

PARTICLEBOARD. A generic term for a panel primarily composed of cellulosic materials (usually wood), generally in the form of discrete pieces or particles, as distinguished from fibers. The cellulosic material is combined with synthetic resin or other suitable bonding system by a process in which the interparticle bond is created by the bonding system under heat and pressure.

PERFORATED SHEAR WALL. A wood structural panel sheathed wall with openings, that has not been specifically designed and detailed for force transfer around openings.

PERFORATED SHEAR WALL SEGMENT. A section of shear wall with full-height sheathing that meets the aspect ratio limits of Section 2305.3.3.

PRESERVATIVE-TREATED WOOD. Wood (including plywood) pressure treated with preservatives in accordance with Section 2303.1.8.

REFERENCE RESISTANCE (D). The resistance (force or moment as appropriate) of a member or connection computed at the reference end use conditions.

SHEAR WALL. A wall designed to resist lateral forces parallel to the plane of a wall.

STRUCTURAL GLUED-LAMINATED TIMBER. Any member comprising an assembly of laminations of lumber in which the grain of all laminations is approximately parallel longitudinally, in which the laminations are bonded with adhesives.

SUBDIAPHRAGM. A portion of a larger wood diaphragm designed to anchor and transfer local forces to primary diaphragm struts and the main diaphragm.

TIE-DOWN (HOLD-DOWN). A device used to resist uplift of the chords of shear walls.

TREATED WOOD. Wood impregnated under pressure with compounds that reduce its susceptibility to flame spread or to deterioration caused by fungi, insects or marine borers.

UNADJUSTED SHEAR RESISTANCE. The allowable shear set forth in Table 2306.4.1 where the aspect ratio of any perforated shear wall segment used in calculation of perforated shear wall resistance does not exceed 2:1. Where the aspect ratio of any perforated shear wall segment used in calculation of perforated shear wall resistance is greater than 2:1, but not exceeding 3.5:1, the unadjusted shear resistance shall be the allowable shear set forth in Table 2306.4.1, multiplied by $2w/h$.

WOOD SHEAR PANEL. A wood floor, roof or wall component sheathed to act as a shear wall or diaphragm.

WOOD STRUCTURAL PANEL. A panel manufactured from veneers, or wood strands or wafers, or a combination of veneer and wood strands or wafers, bonded together with waterproof synthetic resins or other suitable bonding systems. Examples of wood structural panels are:

Composite panels. A structural panel that is made of layers of veneer and wood-based material;

Oriented strand board (OSB). A wood structural panel that is a mat-formed product composed of thin rectangular wood strands or wafers arranged in oriented layers; or

Plywood. A wood structural panel comprised of plies of wood veneer arranged in cross-aligned layers.

SECTION 2303
MINIMUM STANDARDS AND QUALITY

2303.1 General. Structural lumber, end-jointed lumber, prefabricated I-joists, structural glued-laminated timber, wood structural panels, fiberboard sheathing (when used structurally), hardboard siding (when used structurally), particleboard, preservative-treated wood, fire-retardant-treated wood, hardwood, plywood, trusses and joist hangers shall conform to the applicable provisions of this section.

2303.1.1 Lumber. Lumber used for load-supporting purposes, including end-jointed or edge-glued lumber, machine stress-rated or machine evaluated lumber, shall be identified by the grade mark of a lumber grading or inspection agency that has been approved by an accreditation body that complies with DOC PS 20 or equivalent. Grading practices and identification shall comply with rules published by an agency approved in accordance with the procedures of DOC PS 20 or equivalent procedures. In lieu of a grade mark on the material, a certificate of inspection as to species and grade issued by a lumber-grading or inspection agency meeting the requirements of this section is permitted to be accepted for precut, remanufactured or rough-sawn lumber, and for sizes larger than 3 inches (76 mm) nominal thickness.

Approved end-jointed lumber is permitted to be used interchangeably with solid-sawn members of the same species and grade.

2303.1.2 Prefabricated wood I-joists. Structural capacities and design provisions for prefabricated wood I-joists shall be established and monitored in accordance with ASTM D 5055.

2303.1.3 Structural glued-laminated timber. Glued-laminated timbers shall be manufactured and identified as required in AITC A190.1 and ASTM D 3737.

2303.1.4 Wood structural panels. Wood structural panels, when used structurally (including those used for siding, roof and wall sheathing, subflooring, diaphragms and built-up members), shall conform to the requirements for their type in DOC PS 1 or PS 2. Each panel or member shall be identified for grade and glue type by the trademarks of an approved testing and grading agency. Wood structural panel components shall be designed and fabricated in accordance with the applicable standards listed in Section 2306.1 and identified by the trademarks of an approved testing and inspection agency indicating conformance with the applicable standard. In addition, wood structural panels when permanently exposed in outdoor applications shall be of exterior type, except that wood structural panel roof sheathing exposed to the outdoors on the underside is permitted to be interior type bonded with exterior glue, Exposure 1.

2303.1.5 Fiberboard. Fiberboard for its various uses shall conform to ANSI/AHA A194.1 or ASTM C 208. Fiberboard sheathing, when used structurally, shall be so identified by an approved agency as conforming to ANSI/AHA A194.1 or ASTM C 208.

2303.1.5.1 Jointing. To ensure tight-fitting assemblies, edges shall be manufactured with square, shiplapped, beveled, tongue-and-groove or U-shaped joints.

2303.1.5.2 Roof insulation. Where used as roof insulation in all types of construction, fiberboard shall be protected with an approved roof covering.

2303.1.5.3 Wall insulation. Where installed and fireblocked to comply with Chapter 7, fiberboards are permitted as wall insulation in all types of construction. In fire walls and fire barriers, unless treated to comply with Section 803.1 for Class A materials, the boards shall be cemented directly to the concrete, masonry or other noncombustible base and shall be protected with an approved noncombustible veneer anchored to the base without intervening airspaces.

2303.1.5.3.1 Protection. Fiberboard wall insulation applied on the exterior of foundation walls shall be protected below ground level with a bituminous coating.

2303.1.5.4 Insulating roof deck. Where used as roof decking in open beam construction, fiberboard insulation roof deck shall have a nominal thickness of not less than 1 inch (25 mm).

2303.1.6 Hardboard. Hardboard siding used structurally shall be identified by an approved agency conforming to AHA A135.6. Hardboard underlayment shall meet the strength requirements of $^7/_{32}$-inch (5.6 mm) or $^1/_4$-inch (6.4 mm) service class hardboard planed or sanded on one side to a uniform thickness of not less than 0.200 inch (5.1 mm). Prefinished hardboard paneling shall meet the requirements of AHA A135.5. Other basic hardboard products shall meet the requirements of AHA A135.4. Hardboard products shall be installed in accordance with manufacturer's recommendations.

2303.1.7 Particleboard. Particleboard shall conform to ANSI A208.1. Particleboard shall be identified by the grade mark or certificate of inspection issued by an approved agency. Particleboard shall not be utilized for applications other than indicated in this section unless the particleboard complies with the provisions of Section 2306.4.3.

2303.1.7.1 Floor underlayment. Particleboard floor underlayment shall conform to Type PBU of ANSI A208.1. Type PBU underlayment shall not be less than $^1/_4$-inch (6.4 mm) thick and shall be installed in accordance with the instructions of the Composite Panel Association.

2303.1.8 Preservative-treated wood. Lumber, timber, plywood, piles and poles supporting permanent structures required by Section 2304.11 to be preservative treated shall conform to the requirements of the applicable AWPA Standard C1, C2, C3, C4, C9, C14, C15, C16, C22, C23, C24, C28, C31, C33 and M4, for the species, product, preservative and end use. Preservatives shall conform to AWPA P1/P13, P2, P5, P8 and P9. Lumber and plywood used in wood foundation systems shall conform to Chapter 18.

2303.1.8.1 Identification. Wood required by Section 2304.11 to be preservative treated shall bear the quality mark of an inspection agency that maintains continuing supervision, testing and inspection over the quality of the reservative-treated wood. Inspection agencies for preservative-treated wood shall be listed by an accreditation body that complies with the requirements of the American Lumber Standards Treated Wood Program, or equivalent. The quality mark shall be on a stamp or label affixed to the preservative-treated wood, and shall include the following information:

1. Identification of treating manufacturer.
2. Type of preservative used.
3. Minimum preservative retention (pcf).
4. End use for which the product is treated.
5. AWPA standard to which the product was treated.
6. Identity of the accredited inspection agency.

2303.1.8.2 Moisture content. Where preservative-treated wood is used in enclosed locations where drying in service cannot readily occur, such wood shall be at a moisture content of 19 percent or less before being covered with insulation, interior wall finish, floor covering or other materials.

2303.1.9 Structural composite lumber. Structural capacities for structural composite lumber shall be established and monitored in accordance with ASTM D 5456.

2303.2 Fire-retardant-treated wood. Fire-retardant-treated wood is any wood product which, when impregnated with chemicals by a pressure process or other means during manufacture, shall have, when tested in accordance with ASTM E 84, a listed flame spread index of 25 or less and show no evidence of significant progressive combustion when the test is continued for an additional 20-minute period. In addition, the flame front shall not progress more than 10.5 feet (3200 mm) beyond the centerline of the burners at any time during the test.

2303.2.1 Labeling. Fire-retardant-treated lumber and wood structural panels shall be labeled. The label shall contain the following items:

1. The identification mark of an approved agency in accordance with Section 1703.5.
2. Identification of the treating manufacturer.
3. The name of the fire-retardant treatment.
4. The species of wood treated.
5. Flame spread and smoke-developed index.
6. Method of drying after treatment.
7. Conformance with appropriate standards in accordance with Sections 2303.2.2 through 2303.2.5.
8. For fire-retardant-treated wood exposed to weather, damp or wet locations, include the words "No in-

crease in the listed classification when subjected to the Standard Rain Test" (ASTM D 2898).

2303.2.2 Strength adjustments. Design values for untreated lumber and wood structural panels, as specified in Section 2303.1, shall be adjusted for fire-retardant-treated wood. Adjustments to design values shall be based on an approved method of investigation that takes into consideration the effects of the anticipated temperature and humidity to which the fire-retardant-treated wood will be subjected, the type of treatment and redrying procedures.

2303.2.2.1 Wood structural panels. The effect of treatment and the method of redrying after treatment, and exposure to high temperatures and high humidities on the flexure properties of fire-retardant-treated softwood plywood shall be determined in accordance with ASTM D 5516. The test data developed by ASTM D 5516 shall be used to develop adjustment factors, maximum loads and spans, or both, for untreated plywood design values in accordance with ASTM D 6305. Each manufacturer shall publish the allowable maximum loads and spans for service as floor and roof sheathing for its treatment.

2303.2.2.2 Lumber. For each species of wood treated, the effect of the treatment and the method of redrying after treatment and exposure to high temperatures and high humidities on the allowable design properties of fire-retardant-treated lumber shall be determined in accordance with ASTM D 5664. The test data developed by ASTM D 5664 shall be used to develop modification factors for use at or near room temperature and at elevated temperatures and humidity in accordance with an approved method of investigation. Each manufacturer shall publish the modification factors for service at temperatures of not less than 80°F (26.7°C) and for roof framing. The roof framing modification factors shall take into consideration the climatological location.

2303.2.3 Exposure to weather, damp or wet locations. Where fire-retardant-treated wood is exposed to weather, or damp or wet locations, it shall be identified as "Exterior" to indicate there is no increase in the listed flame spread index as defined in Section 2303.2 when subjected to ASTM D 2898.

2303.2.4 Interior applications. Interior fire-retardant-treated wood shall have moisture content of not over 28 percent when tested in accordance with ASTM D 3201 procedures at 92-percent relative humidity. Interior fire-retardant-treated wood shall be tested in accordance with Section 2303.2.2.1 or 2303.2.2.2. Interior fire-retardant-treated wood designated as Type A shall be tested in accordance with the provisions of this section.

2303.2.5 Moisture content. Fire-retardant-treated wood shall be dried to a moisture content of 19 percent or less for lumber and 15 percent or less for wood structural panels before use. For wood kiln dried after treatment (KDAT), the kiln temperatures shall not exceed those used in kiln drying the lumber and plywood submitted for the tests described in Section 2303.2.2.1 for plywood and 2303.2.2.2 for lumber.

2303.2.6 Type I and II construction applications. See Section 603.1 for limitations on the use of fire-retardant-treated wood in buildings of Type I or II construction.

2303.3 Hardwood plywood. Hardwood and decorative plywood shall be manufactured and identified as required in HPVA HP-1.

2303.4 Trusses. Metal-plate-connected wood trusses shall be manufactured as required by TPI 1. Each manufacturer of trusses using metal plate connectors shall retain an approved agency to make unscheduled inspections of truss manufacturing and delivery operations. The inspection shall cover all phases of truss operations, including lumber storage, handling, cutting fixtures, presses or rollers, manufacturing, bundling and banding.

2303.4.1 Truss design drawings. Truss construction documents shall be prepared by a registered design professional and shall be provided to the building official and approved prior to installation. These construction documents shall include, at a minimum, the information specified below. Truss shop drawings shall be provided with the shipment of trusses delivered to the job site.

1. Slope or depth, span and spacing;
2. Location of joints;
3. Required bearing widths;
4. Design loads as applicable;
5. Top chord live load (including snow loads);
6. Top chord dead load;
7. Bottom chord live load;
8. Bottom chord dead load;
9. Concentrated loads and their points of application;
10. Controlling wind and earthquake loads;
11. Adjustments to lumber and metal connector plate design value for conditions of use;
12. Each reaction force and direction;
13. Metal connector plate type, size, thickness or gage, and the dimensioned location of each metal connector plate except where symmetrically located relative to the joint interface;
14. Lumber size, species and grade for each member;
15. Connection requirements for:
    15.1. Truss to truss girder;
    15.2. Truss ply to ply; and
    15.3. Field splices.
16. Calculated deflection ratio or maximum deflection for live and total load;
17. Maximum axial compression forces in the truss members to design the size, connections and anchorage of the permanent continuous lateral bracing. Forces shall be shown on the truss construction documents or on supplemental documents; and
18. Required permanent truss member bracing location.

2303.5 Test standard for joist hangers and connectors. For the required test standards for joist hangers and connectors, see Section 1715.1.

2303.6 Nails and staples. Nails and staples shall conform to requirements of ASTM F 1667. Nails used for framing and sheathing connections shall have minimum average bending yield strengths as follows: 80 kips per square inch (ksi) (551 MPa) for shank diameters larger than 0.177 inch (4.50 mm) but not larger than 0.254 inch (6.45 mm), 90 ksi (620 MPa) for shank diameters larger than 0.142 inch (3.61 mm) but not larger than 0.177 inch (4.50 mm) and 100 ksi (689 MPa) for shank diameters of 0.142 inch (3.61 mm) or less.

2303.7 Shrinkage. Consideration shall be given in design to the possible effect of cross-grain dimensional changes considered vertically which may occur in lumber fabricated in a green condition.

SECTION 2304
GENERAL CONSTRUCTION REQUIREMENTS

2304.1 General. The provisions of this section apply to design methods specified in Section 2301.2.

2304.2 Size of structural members. Computations to determine the required sizes of members shall be based on the net dimensions (actual sizes) and not nominal sizes.

2304.3 Wall framing. The framing of exterior and interior walls shall be in accordance with the provisions specified in Section 2308 unless a specific design is furnished.

2304.3.1 Bottom plates. Studs shall have full bearing on a 2 inch-thick (actual 1¹/₂-inch, 38 mm) or larger plate or sill having a width at least equal to the width of the studs.

2304.3.2 Framing over openings. Headers, double joists, trusses or other approved assemblies that are of adequate size to transfer loads to the vertical members shall be provided over window and door openings in load-bearing walls and partitions.

2304.3.3. Shrinkage. Wood walls and bearing partitions shall not support more than two floors and a roof unless an analysis satisfactory to the building official shows that shrinkage of the wood framing will not have adverse effects

on the structure or any plumbing, electrical or mechanical systems, or other equipment installed therein due to excessive shrinkage or differential movements caused by shrinkage. The analysis shall also show that the roof drainage system and the foregoing systems or equipment will not be adversely affected or, as an alternate, such systems shall be designed to accommodate the differential shrinkage or movements.

2304.4 Floor and roof framing. The framing of wood-joisted floors and wood framed roofs shall be in accordance with the provisions specified in Section 2308 unless a specific design is furnished.

2304.5 Framing around flues and chimneys. Combustible framing shall be a minimum of 2 inches (51 mm), but shall not be less than the distance specified in Sections 2111 and 2113 and the *International Mechanical Code*, from flues, chimneys and fireplaces, and 6 inches (152 mm) away from flue openings.

2304.6 Wall sheathing. Except as provided for in Section 1405 for weatherboarding or where stucco construction that complies with Section 2510 is installed, enclosed buildings shall be sheathed with one of the materials of the nominal thickness specified in Table 2304.6 or any other approved material of equivalent strength or durability.

2304.6.1 Wood structural panel sheathing. Where wood structural panel sheathing is used as the exposed finish on the exterior of outside walls, it shall have an exterior exposure durability classification. Where wood structural panel sheathing is used on the exterior of outside walls but not as the exposed finish, it shall be of a type manufactured with exterior glue (Exposure 1 or Exterior). Where wood structural panel sheathing is used elsewhere, it shall be of a type manufactured with intermediate or exterior glue.

2304.6.2 Interior paneling. Softwood wood structural panels used for interior paneling shall conform with the provisions of Chapter 8 and shall be installed in accordance with Table 2304.9.1. Panels shall comply with DOC PS 1 or PS 2. Prefinished hardboard paneling shall meet the requirements of AHA A135.5, *Prefinished Hardboard Paneling*. Hardwood plywood shall conform to HPVA

TABLE 2304.6
MINIMUM THICKNESS OF WALL SHEATHING

SHEATHING TYPE	MINIMUM THICKNESS	MAXIMUM WALL STUD SPACING
Wood boards	⁵/₈ inch	24 inches on center
Fiberboard	¹/₂ inch	16 inches on center
Wood structural panel	In accordance with Tables 2308.9.3(2) and 2308.9.3(3)	—
M-S "Exterior Glue" and M-2 "Exterior Glue" Particleboard	In accordance with Tables 2306.4.3 and 2308.9.3(5)	—
Gypsum sheathing	¹/₂ inch	16 inches on center
Gypsum wallboard	¹/₂ inch	24 inches on center
Reinforced cement mortar	1 inch	24 inches on center

For SI: 1 inch = 25.4 mm.

HP-1, *The American National Standard for Hardwood and Decorative Plywood.*

2304.7 Floor and roof sheathing.

2304.7.1 Structural floor sheathing. Structural floor sheathing shall be designed in accordance with the general provisions of this code and the special provisions in this section.

Floor sheathing conforming to the provisions of Table 2304.7(1), 2304.7(2), 2304.7(3) or 2304.7(4) shall be deemed to meet the requirements of this section.

2304.7.2 Structural roof sheathing. Structural roof sheathing shall be designed in accordance with the general provisions of this code and the special provisions in this section.

Roof sheathing conforming to the provisions of Table 2304.7(1), 2304.7(2), 2304.7(3) or 2304.7(5) shall be deemed to meet the requirements of this section. Wood structural panel roof sheathing shall be bonded by exterior glue.

2304.8 Mechanically laminated floors and decks.

2304.8.1 General. A laminated lumber floor or deck built up of wood members set on edge, when meeting the following requirements, is permitted to be designed as a solid floor or roof deck of the same thickness, and continuous spans are permitted to be designed on the basis of the full cross section using the simple span moment coefficient.

Nail lengths shall not be less than two and one-half times the net thickness of each lamination. Where deck supports are 4 feet (1219 mm) on center (o.c.) or less, side nails shall be spaced not more than 30 inches (762 mm) o.c. alternately near top and bottom edges, and staggered one-third of the spacing in adjacent laminations. Where supports are spaced more than 4 feet (1219 mm) o.c., side nails shall be spaced not more than 18 inches (457 mm) o.c. alternately near top and bottom edges, and staggered one-third of the spacing in adjacent laminations. Two side nails shall be used at each end of butt-jointed pieces.

Laminations shall be toenailed to supports with 20d or larger common nails. Where the supports are 4 feet (1219 mm) o.c. or less, alternate laminations shall be toenailed to alternate supports; where supports are spaced more than 4 feet (1219 mm) o.c., alternate laminations shall be toenailed to every support. A single-span deck shall have all laminations full length. A continuous deck of two spans shall not have more than every fourth lamination spliced within quarter points adjoining supports. Joints shall be closely butted over supports or staggered across the deck but within the adjoining quarter spans. No lamination shall be spliced more than twice in any span.

2304.9 Connections and fasteners.

2304.9.1 Fastener requirements. Connections for wood members shall be designed in accordance with the appropriate methodology in Section 2301.2. The number and size of nails connecting wood members shall not be less than that set forth in Table 2304.9.1.

2304.9.2 Sheathing fasteners. Sheathing nails or other approved sheathing connectors shall be driven so that their head or crown is flush with the surface of the sheathing.

2304.9.3 Joist hangers and framing anchors. Connections depending on joist hangers or framing anchors, ties

TABLE 2304.7(1)
ALLOWABLE SPANS FOR LUMBER FLOOR AND ROOF SHEATHING[a,b]

	MINIMUM NET THICKNESS (inches) OF LUMBER PLACED			
	Perpendicular to supports		Diagonally to supports	
SPAN (inches)	Surfaced dry[c]	Surfaced unseasoned	Surfaced dry[c]	Surfaced unseasoned
Floors				
24	$3/4$	$25/32$	$3/4$	$25/32$
16	$5/8$	$11/16$	$5/8$	$11/16$
Roofs				
24	$5/8$	$11/16$	$3/4$	$25/32$

For SI: 1 inch = 25.4 mm.
a. Installation details shall conform to Sections 2304.6.1 and 2304.6.2 for floor and roof sheathing, respectively.
b. Floor or roof sheathing conforming with this table shall be deemed to meet the design criteria of Section 2304.6.
c. Maximum 19-percent moisture content.

TABLE 2304.7(2)
SHEATHING LUMBER, MINIMUM GRADE REQUIREMENTS: BOARD GRADE

SOLID FLOOR OR ROOF SHEATHING	SPACED ROOF SHEATHING	GRADING RULES
Utility	Standard	NLGA, WCLIB, WWPA
4 common or utility	3 common or standard	NLGA, WCLIB, WWPA, NSLB or NELMA
No. 3	No. 2	SPIB
Merchantable	Construction common	RIS

TABLE 2304.7(3)
**ALLOWABLE SPANS AND LOADS FOR WOOD STRUCTURAL PANEL SHEATHING AND
SINGLE-FLOOR GRADES CONTINUOUS OVER TWO OR MORE SPANS WITH
STRENGTH AXIS PERPENDICULAR TO SUPPORTS[a,b]**

SHEATHING GRADES		ROOF[c]				FLOOR[d]
		Maximum span (inches)		Load[e] (psf)		Maximum span (inches)
Panel span rating roof/floor span	Panel thickness (inches)	With edge support[f]	Without edge support	Total load	Live load	
12/0	$5/16$	12	12	40	30	0
16/0	$5/16, 3/8$	16	16	40	30	0
20/0	$5/16, 3/8$	20	20	40	30	0
24/0	$3/8, 7/16, 1/2$	24	20[g]	40	30	0
24/16	$7/16, 1/2$	24	24	50	40	16
32/16	$15/32, 1/2, 5/8$	32	28	40	30	16[h]
40/20	$19/32, 5/8, 3/4, 7/8$	40	32	40	30	20[h,i]
48/24	$23/32, 3/4, 7/8$	48	36	45	35	24
54/32	$7/8, 1$	54	40	45	35	32
60/32	$7/8, 1 1/8$	60	48	45	35	32
SINGLE FLOOR GRADES		ROOF[c]				FLOOR[d]
		Maximum span (inches)		Load[e] (psf)		Maximum span (inches)
Panel span rating	Panel thickness (inches)	With edge support[f]	Without edge support	Total load	Live load	
16 o.c.	$1/2, 19/32, 5/8$	24	24	50	40	16[h]
20 o.c.	$19/32, 5/8, 3/4$	32	32	40	30	20[h,i]
24 o.c.	$23/32, 3/4$	48	36	35	25	24
32 o.c.	$7/8, 1$	48	40	50	40	32
48 o.c.	$1 3/32, 1 1/8$	60	48	50	40	48

For SI: 1 inch = 25.4 mm, 1 pound per square foot = 0.0479 kN/m².

a. Applies to panels 24 inches or wider.

b. Floor and roof sheathing conforming with this table shall be deemed to meet the design criteria of Section 2304.7.

c. Uniform load deflection limitations $1/180$ of span under live load plus dead load, $1/240$ under live load only.

d. Panel edges shall have approved tongue-and-groove joints or shall be supported with blocking unless $1/4$-inch minimum thickness underlayment or $1 1/2$ inches of approved cellular or lightweight concrete is placed over the subfloor, or finish floor is $3/4$-inch wood strip. Allowable uniform load based on deflection of $1/360$ of span is 100 pounds per square foot except the span rating of 48 inches on center is based on a total load of 65 pounds per square foot.

e. Allowable load at maximum span.

f. Tongue-and-groove edges, panel edge clips (one midway between each support, except two equally spaced between supports 48 inches on center), lumber blocking or other. Only lumber blocking shall satisfy blocked diaphragm requirements.

g. For $1/2$-inch panel, maximum span shall be 24 inches.

h. Span is permitted to be 24 inches on center where $3/4$-inch wood strip flooring is installed at right angles to joist.

i. Span is permitted to be 24 inches on center for floors where $1 1/2$ inches of cellular or lightweight concrete is applied over the panels.

TABLE 2304.7(4)
ALLOWABLE SPAN FOR WOOD STRUCTURAL PANEL COMBINATION SUBFLOOR-UNDERLAYMENT (SINGLE FLOOR)[a,b]
(Panels Continuous Over Two or More Spans and Strength Axis Perpendicular to Supports)

IDENTIFICATION	MAXIMUM SPACING OF JOISTS (inches)				
	16	20	24	32	48
Species group[c]	Thickness (inches)				
1	$^1/_2$	$^5/_8$	$^3/_4$	—	—
2, 3	$^5/_8$	$^3/_4$	$^7/_8$	—	—
4	$^3/_4$	$^7/_8$	1	—	—
Single floor span rating[d]	16 o.c.	20 o.c.	24 o.c.	32 o.c.	48 o.c.

For SI: 1 inch = 25.4 mm, 1 pound per square foot = 0.0479 kN/m^2.

a. Spans limited to value shown because of possible effects of concentrated loads. Allowable uniform loads based on deflection of $^1/_{360}$ of span is 100 pounds per square foot except allowable total uniform load for $1^1/_8$-inch wood structural panels over joists spaced 48 inches on center is 65 pounds per square foot. Panel edges shall have approved tongue-and-groove joints or shall be supported with blocking, unless $^1/_4$-inch minimum thickness underlayment or $1^1/_2$ inches of approved cellular or lightweight concrete is placed over the subfloor, or finish floor is $^3/_4$-inch wood strip.

b. Floor panels conforming with this table shall be deemed to meet the design criteria of Section 2304.7.

c. Applicable to all grades of sanded exterior-type plywood. See DOC PS 1 for plywood species groups.

d. Applicable to Underlayment grade, C-C (Plugged) plywood, and Single Floor grade wood structural panels.

TABLE 2304.7(5)
ALLOWABLE LOAD (PSF) FOR WOOD STRUCTURAL PANEL ROOF SHEATHING CONTINUOUS OVER TWO OR MORE SPANS AND STRENGTH AXIS PARALLEL TO SUPPORTS
(Plywood Structural Panels Are Five-Ply, Five-Layer Unless Otherwise Noted)[a,b]

PANEL GRADE	THICKNESS (inch)	MAXIMUM SPAN (inches)	LOAD AT MAXIMUM SPAN (psf)	
			Live	Total
Structural I sheathing	$^7/_{16}$	24	20	30
	$^{15}/_{32}$	24	35[c]	45[c]
	$^1/_2$	24	40[c]	50[c]
	$^{19}/_{32}, ^5/_8$	24	70	80
	$^{23}/_{32}, ^3/_4$	24	90	100
Sheathing, other grades covered in DOC PS 1 or DOC PS 2	$^7/_{16}$	16	40	50
	$^{15}/_{32}$	24	20	25
	$^1/_2$	24	25	30
	$^{19}/_{32}$	24	40[c]	50[c]
	$^5/_8$	24	45[c]	55[c]
	$^{23}/_{32}, ^3/_4$	24	60[c]	65[c]

For SI: 1 inch = 25.4 mm, 1 pound per square foot = 0.0479 kN/m^2.

a. Roof sheathing conforming with this table shall be deemed to meet the design criteria of Section 2304.7.

b. Uniform load deflection limitations $^1/_{180}$ of span under live load plus dead load, $^1/_{240}$ under live load only. Edges shall be blocked with lumber or other approved type of edge supports.

c. For composite and four-ply plywood structural panel, load shall be reduced by 15 pounds per square foot.

TABLE 2304.9.1
FASTENING SCHEDULE

CONNECTION	FASTENING[a,m]	LOCATION
1. Joist to sill or girder	3 - 8d common 3 - 3″ × 0.131″ nails 3 - 3″ 14 gage staples	toenail
2. Bridging to joist	2 - 8d common 2 - 3″ × 0.131″ nails 2 - 3″14 gage staples	toenail each end
3. 1″ × 6″ subfloor or less to each joist	2 - 8d common	face nail
4. Wider than 1″ × 6″ subfloor to each joist	3 - 8d common	face nail
5. 2″ subfloor to joist or girder	2 - 16d common	blind and face nail
6. Sole plate to joist or blocking	16d at 16″ o.c. 3″ × 0.131″ nails at 8″ o.c. 3″ 14 gage staples at 12″ o.c.	typical face nail
Sole plate to joist or blocking at braced wall panel	3 - 16d at 16″ 4 - 3″ × 0.131″ nails at 16″ 4 - 3″ 14 gage staples per 16″	braced wall panels
7. Top plate to stud	2 - 16d common 3 - 3″ × 0.131″ nails 3 - 3″ 14 gage staples	end nail
8. Stud to sole plate	4 - 8d common 4 - 3″ × 0.131″ nails 3 - 3″ 14 gage staples	toenail
	2 - 16d common 3 - 3″ × 0.131″ nails 3 - 3″ 14 gage staples	end nail
9. Double studs	16d at 24″ o.c. 3″ × 0.131″ nail at 8″ o.c. 3″ 14 gage staple at 8″ o.c.	face nail
10. Double top plates	16d at 16″ o.c. 3″ × 0.131″ nail at 12″ o.c. 3″ 14 gage staple at 12″ o.c.	typical face nail
Double top plates	8-16d common 12 - 3″ × 0.131″ nails 12 - 3″ 14 gage staples typical face nail	lap splice
11. Blocking between joists or rafters to top plate	3 - 8d common 3 - 3″ × 0.131″ nails 3 - 3″ 14 gage staples	toenail
12. Rim joist to top plate	8d at 6″ (152 mm) o.c. 3″ × 0.131″ nail at 6″ o.c. 3″ 14 gage staple at 6″ o.c.	toenail
13. Top plates, laps and intersections	2 - 16d common 3 - 3″ × 0.131″ nails 3 - 3″ 14 gage staples	face nail
14. Continuous header, two pieces	16d common	16″ o.c. along edge
15. Ceiling joists to plate	3 - 8d common 5 - 3″ × 0.131″ nails 5 - 3″ 14 gage staples	toenail
16. Continuous header to stud	4 - 8d common	toenail

(continued)

TABLE 2304.9.1—continued
FASTENING SCHEDULE

CONNECTION	FASTENING[a,m]	LOCATION
17. Ceiling joists, laps over partitions (see Section 2308.10.4.1, Table 2308.10.4.1)	3 - 16d common minimum, Table 2308.10.4.1 4 - 3″ × 0.131″ nails 4 - 3″ 14 gage staples	face nail
18. Ceiling joists to parallel rafters (see Section 2308.10.4.1, Table 2308.10.4.1)	3 - 16d common minimum, Table 2308.10.4.1 4 - 3″ × 0.131″ nails 4 - 3″ 14 gage staples	face nail
19. Rafter to plate (see Section 2308.10.1, Table 2308.10.1)	3 - 8d common 3 - 3″ × 0.131″ nails 3 - 3″ 14 gage staples	toenail
20. 1″ diagonal brace to each stud and plate	2 - 8d common 2 - 3″ × 0.131″ nails 2 - 3″ 14 gage staples face nail	face nail
21. 1″ × 8″ sheathing to each bearing wall	2 - 8d common	face nail
22. Wider than 1″× 8″ sheathing to each bearing	3 - 8d common	face nail
23. Built-up corner studs	16d common 3″ × 0.131″ nails 3″ 14 gage staples	24″ o.c. 16″ o.c. 16″ o.c.
24. Built-up girder and beams	20d common 32″ o.c. 3″ × 0.131″ nail at 24″ o.c. 3″ 14 gage staple at 24″ o.c. 2 - 20d common 3 - 3″ × 0.131″ nails 3 - 3″ 14 gage staples	face nail at top and bottom staggered on opposite sides face nail at ends and at each splice
25. 2″ planks	16d common	at each bearing
26. Collar tie to rafter	3 - 10d common 4 - 3″ × 0.131″ nails 4 - 3″ 14 gage staples face nail	face nail
27. Jack rafter to hip	3 - 10d common 4 - 3″ × 0.131″nails 4 - 3″ 14 gage staples 2 - 16d common 3 - 3″ × 0.131″ nails 3 - 3″ 14 gage staples	toenail face nail
28. Roof rafter to 2-by ridge beam	2 - 16d common 3 - 3″ × 0.131″ nails 3 - 3″ 14 gage staples 2 - 16d common 3 - 3″ × 0.131″ nails 3 - 3″ 14 gage staples	toenail face nail
29. Joist to band joist	3 - 16d common 5 - 3″ × 0.131″ nails 5 - 3″ 14 gage staples	face nail

(continued)

TABLE 2304.9.1—continued
FASTENING SCHEDULE

CONNECTION	FASTENING[a,m]		LOCATION
30. Ledger strip	3 - 16d common 4 - 3″ × 0.131″ nails 4 - 3″ 14 gage staples		face nail
31. Wood structural panels and particleboard:[b] Subfloor, roof and wall sheathing (to framing):	$^1/_2$″ and less	6d[c,l] 2 $^3/_8$″ × 0.113″ nail[n] 1 $^3/_4$″ 16 gage[o]	
	$^{19}/_{32}$″ to $^3/_4$″	8d[d] or 6d[e] 2 $^3/_8$″ × 0.113″ nail[p] 2″ 16 gage[p]	
	$^7/_8$″ to 1″	8d[c]	
Single Floor (combination subfloor-underlayment to framing):	1 $^1/_8$″ to 1 $^1/_4$″	10d[d] or 8d[e]	
	$^3/_4$″ and less	6d[e]	
	$^7/_8$″ to 1″	8d[e]	
	1 $^1/_8$″ to 1 $^1/_4$″	10d[d] or 8d[e]	
32. Panel siding (to framing)	$^1/_2$″ or less	6d[f]	
	$^5/_8$″	8d[f]	
33. Fiberboard sheathing:[g]	$^1/_2$″	No. 11 gage roofing nail[h] 6d common nail No. 16 gage staple[i]	
	$^{25}/_{32}$″	No. 11 gage roofing nail[h] 8d common nail No. 16 gage staple[i]	
34. Interior paneling	$^1/_4$″	4d[j]	
	$^3/_8$″	6d[k]	

For SI: 1 inch = 25.4 mm.

a. Common or box nails are permitted to be used except where otherwise stated.
b. Nails spaced at 6 inches on center at edges, 12 inches at intermediate supports except 6 inches at supports where spans are 48 inches or more. For nailing of wood structural panel and particleboard diaphragms and shear walls, refer to Section 2305. Nails for wall sheathing are permitted to be common, box or casing.
c. Common or deformed shank.
d. Common.
e. Deformed shank.
f. Corrosion-resistant siding or casing nail.
g. Fasteners spaced 3 inches on center at exterior edges and 6 inches on center at intermediate supports.
h. Corrosion-resistant roofing nails with $^7/_{16}$-inch-diameter head and 1$^1/_2$-inch length for $^1/_2$-inch sheathing and 1 $^3/_4$-inch length for $^{25}/_{32}$-inch sheathing.
i. Corrosion-resistant staples with nominal $^7/_{16}$-inch crown and 1 $^1/_8$-inch length for $^1/_2$-inch sheathing and 1 $^1/_2$-inch length for $^{25}/_{32}$-inch sheathing. Panel supports at 16 inches (20 inches if strength axis in the long direction of the panel, unless otherwise marked).
j. Casing or finish nails spaced 6 inches on panel edges, 12 inches at intermediate supports.
k. Panel supports at 24 inches. Casing or finish nails spaced 6 inches on panel edges, 12 inches at intermediate supports.
l. For roof sheathing applications, 8d nails are the minimum required for wood structural panels.
m. Staples shall have a minimum crown width of $^7/_{16}$ inch.
n. For roof sheathing applications, fasteners spaced 4 inches on center at edges, 8 inches at intermediate supports.
o. Fasteners spaced 4 inches on center at edges, 8 inches at intermediate supports for subfloor and wall sheathing and 3 inches on center at edges, 6 inches at intermediate supports for roof sheathing.
p. Fasteners spaced 4 inches on center at edges, 8 inches at intermediate supports.

and other mechanical fastenings not otherwise covered are permitted where approved. The vertical load-bearing capacity, torsional moment capacity and deflection characteristics of joist hangers shall be determined in accordance with Section 1715.1.

2304.9.4 Other fasteners. Clips, staples, glues and other approved methods of fastening are permitted where approved.

2304.9.5 Fasteners in preservative-treated and fire-retardant-treated wood. Fasteners for preservative-treated and fire-retardant-treated wood shall be of hot-dipped zinc-coated galvanized steel, stainless steel, silicon bronze or copper. Fastenings for wood foundations shall be as required in AF&PA Technical Report No. 7.

2304.9.6 Load path. Where wall framing members are not continuous from foundation sill to roof, the members shall be secured to ensure a continuous load path. Where required, sheet metal clamps, ties or clips shall be formed of galvanized steel or other approved corrosion-resistant material not less than 0.040 inch (1.01 mm) nominal thickness.

2304.9.7 Framing requirements. Wood columns and posts shall be framed to provide full end bearing. Alternatively, column-and-post end connections shall be designed to resist the full compressive loads, neglecting end-bearing capacity. Column-and-post end connections shall be fastened to resist lateral and net induced uplift forces.

2304.10 Heavy timber construction.

2304.10.1 Columns. Columns shall be continuous or superimposed throughout all stories by means of reinforced concrete or metal caps with brackets, or shall be connected by properly designed steel or iron caps, with pintles and base plates, or by timber splice plates affixed to the columns by metal connectors housed within the contact faces, or by other approved methods.

2304.10.1.1 Column connections. Girders and beams shall be closely fitted around columns and adjoining ends shall be cross tied to each other, or intertied by caps or ties, to transfer horizontal loads across joints. Wood bolsters shall not be placed on tops of columns unless the columns support roof loads only.

2304.10.2 Floor framing. Approved wall plate boxes or hangers shall be provided where wood beams, girders or trusses rest on masonry or concrete walls. Where intermediate beams are used to support a floor, they shall rest on top of girders, or shall be supported by ledgers or blocks securely fastened to the sides of the girders, or they shall be supported by an approved metal hanger into which the ends of the beams shall be closely fitted.

2304.10.3 Roof framing. Every roof girder and at least every alternate roof beam shall be anchored to its supporting member; and every monitor and every sawtooth construction shall be anchored to the main roof construction. Such anchors shall consist of steel or iron bolts of sufficient strength to resist vertical uplift of the roof.

2304.10.4 Floor decks. Floor decks and covering shall not extend closer than $^1/_2$ inch (12.7 mm) to walls. Such $^1/_2$-inch (12.7 mm) spaces shall be covered by a molding fastened to the wall either above or below the floor and arranged such that the molding will not obstruct the expansion or contraction movements of the floor. Corbeling of masonry walls under floors is permitted in place of such molding.

2304.10.5 Roof decks. Where supported by a wall, roof decks shall be anchored to walls to resist uplift forces determined in accordance with Chapter 16. Such anchors shall consist of steel or iron bolts of sufficient strength to resist vertical uplift of the roof.

2304.11 Protection against decay and termites.

2304.11.1 General. Where required by this section, protection from decay and termites shall be provided by the use of naturally durable or preservative-treated wood.

2304.11.2 Wood used above ground. Wood installed above ground in the locations specified in Sections 2304.11.2.1 through 2304.11.2.6 shall be naturally durable wood or preservative-treated wood that uses water-borne preservatives, and shall be treated in accordance with AWPA C2 or C9 or applicable AWPA standards for above-ground use.

2304.11.2.1 Joists, girders and subfloor. Where wood joists or the bottom of a wood structural floor without joists are closer than 18 inches (457 mm), or wood girders are closer than 12 inches (305 mm) to the exposed ground in crawl spaces or unexcavated areas located within the perimeter of the building foundation, the floor assembly (including posts, girders, joists and subfloor) shall be of naturally durable or preservative-treated wood.

2304.11.2.2 Framing. Wood framing members, including wood sheathing, which rest on exterior foundation walls and are less than 8 inches (203 mm) from exposed earth shall be of naturally durable or preservative-treated wood. Wood framing members and furring strips attached directly to the interior of exterior masonry or concrete walls below grade shall be of approved naturally durable or preservative-treated wood.

2304.11.2.3 Sleepers and sills. Sleepers and sills on a concrete or masonry slab that is in direct contact with earth shall be of naturally durable or preservative-treated wood.

2304.11.2.4 Girder ends. The ends of wood girders entering exterior masonry or concrete walls shall be provided with a $^1/_2$-inch (12.7 mm) air space on top, sides and end, unless naturally durable or preservative-treated wood is used.

2304.11.2.5 Wood siding. Clearance between wood siding and earth on the exterior of a building shall not be less than 6 inches (152 mm) except where siding, sheathing and wall framing are of naturally durable or preservative-treated wood.

2304.11.2.6 Posts or columns. Posts or columns supporting permanent structures and supported by a concrete or masonry slab or footing that is in direct contact with the earth shall be of naturally durable or preservative-treated wood.

Exceptions:

1. Posts or columns that are either exposed to the weather or located in basements or cellars, supported by concrete piers or metal pedestals projected at least 1 inch (25 mm) above the slab or deck and 6 inches (152 mm) above exposed earth, and are separated therefrom by an impervious moisture barrier.

2. Posts or columns in enclosed crawl spaces or unexcavated areas located within the periphery of the building, supported by a concrete pier or metal pedestal at a height greater than 8 inches (203 mm) from exposed ground, and are separated therefrom by an impervious moisture barrier.

2304.11.3 Laminated timbers. The portions of glued-laminated timbers that form the structural supports of a building or other structure and are exposed to weather and not properly protected by a roof, eave or similar covering shall be pressure treated with preservative, or be manufactured from naturally durable or preservative-treated wood.

2304.11.4 Wood in contact with the ground or fresh water. Wood in contact with the ground (exposed earth) that supports permanent structures shall be of naturally durable (species for both decay and termite resistance) or preservative-treated wood using water-borne preservatives and shall be treated in accordance with AWPA C2, C9 or other applicable AWPA standard for soil or fresh water contact, where used in the locations specified in Sections 2304.11.4.1 and 2304.11.4.2.

Exception: Untreated wood is permitted where such wood is continuously and entirely below the ground-water level or submerged in fresh water.

2304.11.4.1 Posts or columns. Posts and columns supporting permanent structures that are embedded in concrete in direct contact with the earth or embedded in concrete exposed to the weather, or in direct contact with the earth, shall be of preservative-treated wood.

2304.11.4.2 Wood structural members. Wood structural members that support moisture-permeable floors or roofs that are exposed to the weather, such as concrete or masonry slabs, shall be of naturally durable or preservative-treated wood unless separated from such floors or roofs by an impervious moisture barrier.

2304.11.5 Supporting member for permanent appurtenances. Naturally durable or preservative-treated wood shall be utilized for those portions of wood members that form the structural supports of buildings, balconies, porches or similar permanent building appurtenances where such members are exposed to the weather without adequate protection from a roof, eave, overhang or other covering to prevent moisture or water accumulation on the surface or at joints between members.

Exception: When a building is located in a geographical region where experience has demonstrated that climatic conditions preclude the need to use durable materials where the structure is exposed to the weather.

2304.11.6 Termite protection. In geographical areas where the hazard of termite damage is known to be very heavy, the floor framing shall be of naturally durable or preservative-treated wood, or provided with approved methods of termite protection.

2304.11.7 Wood used in retaining walls and cribs. Wood installed in retaining or crib walls shall be of preservative-treated wood treated in accordance with AWPA C2 or C9 for soil and fresh water contact.

2304.11.8 Attic ventilation. For attic ventilation, see Section 1203.2.

2304.11.9 Under-floor ventilation (crawl space). For under-floor ventilation (crawl space), see Section 1203.3.

2304.12 Wood supporting masonry or concrete. Wood members shall not be used to permanently support the dead load of any masonry or concrete.

Exceptions:

1. Masonry or concrete nonstructural floor or roof surfacing not more than 4 inches (102 mm) thick is permitted to be supported by wood members.

2. Any structure is permitted to rest upon wood piles constructed in accordance with the requirements of Chapter 18.

3. Veneer of brick, concrete or stone applied as specified in Section 1405.5 having an installed weight of 40 pounds per square foot (psf) (1.9 kN/m^2) or less is permitted to be supported by an approved treated wood foundation when the maximum height of veneer does not exceed 30 feet (9144 mm) above the foundation. Such veneer used as an interior wall finish is permitted to be supported on wood floor construction. The wood floor construction shall be designed to support the additional weight of the veneer plus any other loads and to limit the deflection and shrinkage to $^1/_{600}$ of the span of the supporting members.

4. Glass unit masonry having an installed weight of 20 psf (0.96 kN/m^2) or less is permitted to be installed in accordance with the provisions of Section 2110. The wood construction supporting the glass unit masonry shall be designed for dead and live loads to limit deflection and shrinkage to $^1/_{600}$ of the span of the supporting members.

SECTION 2305
GENERAL DESIGN REQUIREMENTS FOR LATERAL-FORCE-RESISTING SYSTEMS

2305.1 General. Structures using wood shear walls and diaphragms to resist wind, seismic and other lateral loads shall be designed and constructed in accordance with the provisions of this section.

2305.1.1 Shear resistance based on principles of mechanics. Shear resistance of diaphragms and shear walls are permitted to be calculated by principles of mechanics using values of fastener strength and sheathing shear resistance.

2305.1.2 Framing. Boundary elements shall be provided to transmit tension and compression forces. Perimeter members at openings shall be provided and shall be detailed to distribute the shearing stresses. Diaphragm and shear wall sheathing shall not be used to splice boundary elements. Diaphragm chords and collectors shall be placed in, or tangent to, the plane of the diaphragm framing unless it can be demonstrated that the moments, shears and deformations, considering eccentricities resulting from other configurations can be tolerated without exceeding the adjusted resistance and drift limits.

2305.1.2.1 Framing members. Framing members shall be at least 2 inch (51 mm) nominal width. In general, adjoining panel edges shall bear and be attached to the framing members and butt along their centerlines. Nails shall be placed not less than $^3/_8$ inch (9.5 mm) from the panel edge, not more than 12 inches (305 mm) apart along intermediate supports, and 6 inches (152 mm) along panel edge bearings, and shall be firmly driven into the framing members.

2305.1.3 Openings in shear panels. Openings in shear panels that materially affect their strength shall be fully detailed on the plans, and shall have their edges adequately reinforced to transfer all shearing stresses.

2305.1.4 Shear panel connections. Positive connections and anchorages, capable of resisting the design forces, shall be provided between the shear panel and the attached components. In Seismic Design Category D, E or F, toenails shall not be used to transfer lateral forces in excess of 150 pounds per foot (2189 N/m) from diaphragms to shear walls, drag struts (collectors) or other elements, or from shear walls to other elements.

2305.1.5 Wood members resisting horizontal seismic forces contributed by masonry and concrete. Wood shear walls, diaphragms, horizontal trusses and other members shall not be used to resist horizontal seismic forces contributed by masonry or concrete construction in structures over one story in height.

Exceptions:

1. Wood floor and roof members are permitted to be used in horizontal trusses and diaphragms to resist horizontal seismic forces contributed by masonry or concrete construction (including those due to masonry veneer, fireplaces and chimneys) provided such forces do not result in torsional force distribution through the truss or diaphragm.

2. Wood structural panel sheathed shear walls are permitted to be used to provide resistance to seismic forces contributed by masonry or concrete construction in two-story structures of masonry or concrete construction, provided the following requirements are met:

2.1. Story-to-story wall heights shall not exceed 12 feet (3658 mm).

2.2. Diaphragms shall not be designed to transmit lateral forces by rotation. Diaphragms shall not cantilever past the outermost supporting shear wall.

2.3. Combined deflections of diaphragms and shear walls shall not permit story drift of supported masonry or concrete walls to exceed the limit of Section 1617.3.

2.4. Wood structural panel sheathing in diaphragms shall have unsupported edges blocked. Wood structural panel sheathing for both stories of shear walls shall have unsupported edges blocked and, for the lower story, shall have a minimum thickness of $^{15}/_{32}$ inch (11.9 mm).

2.5. There shall be no out-of-plane horizontal offsets between the first and second stories of wood structural panel shear walls.

2305.2 Design of wood diaphragms.

2305.2.1 General. Wood diaphragms are permitted to be used to resist horizontal forces provided the deflection in the plane of the diaphragm, as determined by calculations, tests or analogies drawn therefrom, does not exceed the permissible deflection of attached distributing or resisting elements. Connections shall extend into the diaphragm a sufficient distance to develop the force transferred into the diaphragm.

2305.2.2 Deflection. Permissible deflection shall be that deflection up to which the diaphragm and any attached distributing or resisting element will maintain its structural integrity under design load conditions, such that the resisting element will continue to support design loads without danger to occupants of the structure. Calculations for diaphragm deflection shall account for the usual bending and shear components as well as any other factors, such as nail deformation, which will contribute to deflection.

The deflection (Δ) of a blocked wood structural panel diaphragm uniformly nailed throughout is permitted to be calculated by using the following formula. If not uniformly nailed, the constant 0.188 (For SI: 1/1627) in the third term must be modified accordingly.

$$\Delta = \frac{5vL^3}{8EAb} + \frac{vL}{4Gt} + 0.188Le_n + \frac{\Sigma(\Delta_c X)}{2b} \quad \textbf{(Equation 23-1)}$$

For SI: $\Delta = \dfrac{0.052L^3}{EAb} + \dfrac{vL}{4Gt} + \dfrac{Le_n}{1627} + \dfrac{\Sigma(\Delta_c X)}{2b}$

where:

A = Area of chord cross section, in square inches (mm²).

b = Diaphragm width, in feet (mm).

E = Elastic modulus of chords, in pounds per square inch (N/mm²).

e_n = Nail deformation, in inches (mm).

G = Modulus of rigidity of wood structural panel, in pounds per square inch (N/mm^2).

L = Diaphragm length, in feet (mm).

t = Effective thickness of wood structural panel for shear, in inches (mm).

v = Maximum shear due to design loads in the direction under consideration, in pounds per linear foot (plf) (N/mm).

Δ = The calculated deflection, in inches (mm).

$\sum(\Delta_c X)$ = Sum of individual chord-splice values on both sides of the diaphragm, each multiplied by its distance to the nearest support.

2305.2.3 Diaphragm aspect ratios. Size and shape of diaphragms shall be limited as set forth in Table 2305.2.3.

TABLE 2305.2.3
MAXIMUM DIAPHRAGM DIMENSION RATIOS
HORIZONTAL AND SLOPED DIAPHRAGM

TYPE	MAXIMUM LENGTH - WIDTH RATIO
Wood structural panel, nailed all edges	4:1
Wood structural panel, blocking omitted at intermediate joints	3:1
Diagonal sheathing, single	3:1
Diagonal sheathing, double	4:1

2305.2.4 Construction. Shear panels shall be constructed of wood structural panels, manufactured with exterior glue, not less than 4 feet by 8 feet (1219 mm by 2438 mm), except at boundaries and changes in framing. Boundary elements shall be connected at corners. Wood structural panel thickness for horizontal diaphragms shall not be less than set forth in Tables 2304.7(3) and 2304.7(5) for corresponding joist spacing and loads, except that $^1/_4$ inch (6.4 mm) is permitted to be used where perpendicular loads permit. Sheet-type sheathing shall be arranged so that the width of a sheet in a shear wall shall not be less than 2 feet (610 mm).

2305.2.4.1 Seismic Design Category F. Structures assigned to Seismic Design Category F shall conform to the requirements in Section 1620.5 or Section 9.5.2.6.5 of ASCE 7, and to the additional requirements of this section.

Wood structural panel sheathing used for diaphragms and shear walls that are part of the seismic-force-resisting system shall be applied directly to the framing members.

Exception: Wood structural panel sheathing in a diaphragm is permitted to be fastened over solid lumber planking or laminated decking provided the panel joints and lumber planking or laminated decking joints do not coincide.

2305.2.5 Rigid diaphragms. Design of structures with rigid diaphragms shall conform to the structure configuration requirements of Section 9.5.2.3 of ASCE 7 and the horizontal shear distribution requirements of Section 9.5.5.5 of ASCE 7.

Open front structures with rigid wood diaphragms resulting in torsional force distribution are permitted provided the length, l, of the diaphragm normal to the open side does not exceed 25 feet (7620 mm), the diaphragm sheathing conforms to Section 2305.2.4, and the l/w ratio [as shown in Figure 2305.2.5(1)] is less than 1.0 for one-story structures or 0.67 for structures over one story in height.

Exception: Where calculations show that diaphragm deflections can be tolerated, the length, l, normal to the open end is permitted to be increased to a l/w ratio not greater than 1.5 where sheathed in compliance with Section 2305.2.4 or to 1.0 where sheathed in compliance with Section 2306.3.4 or 2306.3.5.

Rigid wood diaphragms are permitted to cantilever past the outermost supporting shear wall (or other vertical resisting element) a length, l, of not more than 25 feet (7620 mm) or two-thirds of the diaphragm width, w, whichever is the smaller. Figure 2305.2.5(2) illustrates the dimensions of l and w for a cantilevered diaphragm.

Structures with rigid wood diaphragms having a torsional irregularity in accordance with Table 1616.5.1.1, Item 1, shall meet the following requirements: The l/w ratio shall not exceed 1.0 for one-story structures or 0.67 for structures over one story in height, where l is the dimension parallel to the load direction for which the irregularity exists.

Exception: Where calculations demonstrate that the diaphragm deflections can be tolerated, the width is permitted to be increased and the l/w ratio is permitted to be increased to 1.5 where sheathed in compliance with Section 2305.2.4 or 1.0 where sheathed in compliance with Section 2306.3.4 or 2306.3.5.

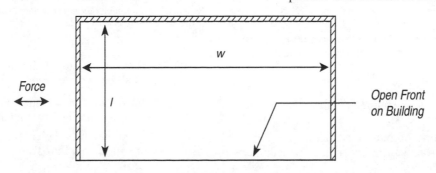

FIGURE 2305.2.5(1)
DIAPHRAGM LENGTH AND WIDTH FOR PLAN VIEW OF OPEN FRONT BUILDING

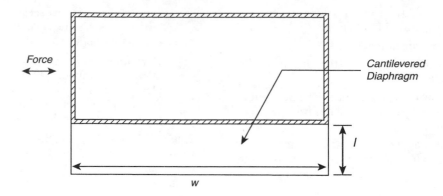

FIGURE 2305.2.5(2)
DIAPHRAGM LENGTH AND WIDTH FOR PLAN VIEW OF CANTILEVERED DIAPHRAGM

2305.3 Design of wood shear walls.

2305.3.1 General. Wood shear walls are permitted to resist horizontal forces in vertical distributing or resisting elements, provided the deflection in the plane of the shear wall, as determined by calculations, tests or analogies drawn therefrom, does not exceed the more restrictive of the permissible deflection of attached distributing or resisting elements or the drift limits of Section 1617.3. Shear wall sheathing other than wood structural panels shall not be permitted in Seismic Design Category E or F (see Section 1617.6).

2305.3.2 Deflection. Permissible deflection shall be that deflection up to which the shear wall and any attached distributing or resisting element will maintain its structural integrity under design load conditions, i.e., continue to support design loads without danger to occupants of the structure.

The deflection (Δ) of a blocked wood structural panel shear wall uniformly fastened throughout is permitted to be calculated by the use of the following formula:

$$\Delta = \frac{8vh^3}{EAb} + \frac{vh}{Gt} + 0.75he_n + d_a \qquad \textbf{(Equation 23-2)}$$

For SI: $\Delta = \dfrac{vh^3}{3EAb} + \dfrac{vh}{Gt} + \dfrac{he_n}{406.7} + d_a$

where:

A = Area of boundary element cross section in square inches (mm^2) (vertical member at shear wall boundary).

b = Wall width, in feet (mm).

d_a = Deflection due to anchorage details (rotation and slip at tie-down bolts).

E = Elastic modulus of boundary element (vertical member at shear wall boundary), in pounds per square inch (N/mm^2).

e_n = Deformation of mechanically fastened connections, in inches (mm^2).

G = Modulus of rigidity of wood structural panel, in pounds per square inch (N/mm^2).

h = Wall height, in feet (mm).

t = Effective thickness of wood structural panel for shear, in inches (mm).

v = Maximum shear due to design loads at the top of the wall, in pounds per linear foot (N/mm).

Δ = The calculated deflection, in inches (mm).

2305.3.3 Shear wall aspect ratios. Size and shape of shear walls and shear wall segments within shear walls containing openings shall be limited as set forth in Table 2305.3.3.

TABLE 2305.3.3
MAXIMUM SHEAR WALL ASPECT RATIOS

TYPE	MAXIMUM HEIGHT-WIDTH RATIO
Wood structural panels or particleboard, nailed edges	For other than seismic: $3^1/_2$:1 For seismic: 2:1[a]
Diagonal sheathing, single	2:1
Fiberboard	$1^1/_2$:1
Gypsum board, gypsum lath, cement plaster	$1^1/_2$:1[b]

a. For design to resist seismic forces, shear wall height-width ratios greater than 2:1, but not exceeding $3^1/_2$:1, are permitted provided the allowable shear resistance values in Table 2306.4.1 are multiplied by $2w/h$.

b. Ratio shown is for unblocked construction. Aspect ratio is permitted to be 2:1 where the wall is installed as blocked construction in accordance with Section 2306.4.5.1.2.

2305.3.4 Shear wall height definition. The height of a shear wall shall be defined as:

1. The maximum clear height from top of foundation to bottom of diaphragm framing above; or

2. The maximum clear height from top of diaphragm to bottom of diaphragm framing above [see Figure 2305.3.4(a)].

2305.3.5 Shear wall width definition. The width of a shear wall shall be defined as the sheathed dimension of the shear

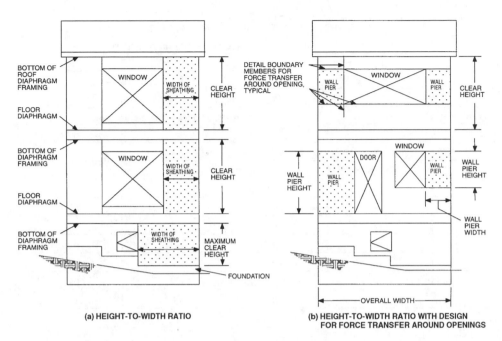

(a) HEIGHT-TO-WIDTH RATIO

(b) HEIGHT-TO-WIDTH RATIO WITH DESIGN FOR FORCE TRANSFER AROUND OPENINGS

FIGURE 2305.3.4
GENERAL DEFINITION OF SHEAR WALL HEIGHT, WIDTH AND HEIGHT-TO-WIDTH RATIO

wall in the direction of application of force [see Figure 2305.3.4(a)].

2305.3.5.1 Shear wall segment width definition. The width of full-height sheathing adjacent to unrestrained openings in a shear wall.

2305.3.6 Overturning restraint. Where the dead load stabilizing moment in accordance with Chapter 16 allowable stress design load combinations is not sufficient to prevent uplift due to overturning moments on the wall, an anchoring device shall be provided. Anchoring devices shall maintain a continuous load path to the foundation.

2305.3.7 Shear walls with openings. The provisions of this section shall apply to the design of shear walls with openings. Where framing and connections around the openings are designed for force transfer around the openings, the provisions of Section 2305.3.7.1 shall apply. Where framing and connections around the openings are not designed for force transfer around the openings, the provisions of Section 2305.3.7.2 shall apply.

2305.3.7.1 Force transfer around openings. Where shear walls with openings are designed for force transfer around the openings, the limitations of Table 2305.3.3 shall apply to the overall shear wall including openings and to each wall pier at the side of an opening. The height of a wall pier shall be defined as the clear height of the pier at the side of an opening. The width of a wall pier shall be defined as the sheathed width of the pier at the side of an opening. Design for force transfer shall be based on a rational analysis. Detailing of boundary elements around the opening shall be provided in accordance with the provisions of this section [see Figure 2305.3.4(b)].

2305.3.7.2 Perforated shear walls. The provisions of Section 2305.3.7.2 shall be permitted to be used for the design of perforated shear walls.

2305.3.7.2.1 Limitations. The following limitations shall apply to the use of Section 2305.3.7.2:

1. A perforated shear wall segment shall be located at each end of a perforated shear wall. Openings shall be permitted to occur beyond the ends of the perforated shear wall; however, the width of such openings shall not be included in the width of the perforated shear wall.

2. The allowable shear set forth in Table 2306.4.1 shall not exceed 490 plf (7150 N/m).

3. Where out-of-plane offsets occur, portions of the wall on each side of the offset shall be considered as separate perforated shear walls.

4. Collectors for shear transfer shall be provided through the full length of the perforated shear wall.

5. A perforated shear wall shall have uniform top of wall and bottom of wall elevations. Perforated shear walls not having uniform elevations shall be designed by other methods.

6. Perforated shear wall height, h, shall not exceed 20 feet (6096 mm).

2305.3.7.2.2 Perforated shear wall resistance. The resistance of a perforated shear wall shall be calculated in accordance with the following:

1. The percent of full-height sheathing shall be calculated as the sum of the widths of perforated shear wall segments divided by the total

TABLE 2305.3.7.2
SHEAR RESISTANCE ADJUSTMENT FACTOR, C_o

WALL HEIGHT, H	MAXIMUM OPENING HEIGHT[a]				
	H/3	H/2	2H/3	5H/6	H
8' wall	2'-8"	4'-0"	5'-4"	6'-8"	8'-0"
10' wall	3'-4"	5'-0"	6'-8"	8'-4"	10'-0"
Percent full-height sheathing[b]	Shear resistance adjustment factor				
10%	1.00	0.69	0.53	0.43	0.36
20%	1.00	0.71	0.56	0.45	0.38
30%	1.00	0.74	0.59	0.49	0.42
40%	1.00	0.77	0.63	0.53	0.45
50%	1.00	0.80	0.67	0.57	0.50
60%	1.00	0.83	0.71	0.63	0.56
70%	1.00	0.87	0.77	0.69	0.63
80%	1.00	0.91	0.83	0.77	0.71
90%	1.00	0.95	0.91	0.87	0.83
100%	1.00	1.00	1.00	1.00	1.00

For SI: 1 inch = 25.4 mm, 1 foot = 304.8 mm.
a. See Section 2305.3.7.2.2, Item 2.
b. See Section 2305.3.7.2.2, Item 1.

width of the perforated shear wall including openings.

2. The maximum opening height shall be taken as the maximum opening clear height. Where areas above and below an opening remain unsheathed, the height of opening shall be defined as the height of the wall.

3. The adjusted shear resistance shall be calculated by multiplying the unadjusted shear resistance by the shear resistance adjustment factors of Table 2305.3.7.2. For intermediate percentages of full-height sheathing, the values in Table 2305.3.7.2 are permitted to be interpolated.

4. The perforated shear wall resistance shall be equal to the adjusted shear resistance times the sum of the widths of the perforated shear wall segments.

2305.3.7.2.3 Anchorage and load path. Design of perforated shear wall anchorage and load path shall conform to the requirements of Sections 2305.3.7.2.4 through 2305.3.7.2.8, or shall be calculated using principles of mechanics. Except as modified by these sections, wall framing, sheathing, sheathing attachment and fastener schedules shall conform to the requirements of Section 2305.2.4 and Table 2306.4.1.

2305.3.7.2.4 Uplift anchorage at perforated shear wall ends. Anchorage for uplift forces due to overturning shall be provided at each end of the perforated shear wall. The uplift anchorage shall conform to the requirements of Section 2305.3.6 except that for each story the minimum tension chord uplift force, T, shall be calculated in accordance with the following:

$$T = \frac{Vh}{C_o \Sigma L_i}$$ **(Equation 23-3)**

where:

T = Tension chord uplift force, pounds (N).

V = Shear force in perforated shear wall, pounds (N).

h = Shear wall height, feet (mm).

C_o = Shear resistance adjustment factor from Table 2305.3.7.2.

ΣL_i= Sum of widths of perforated shear wall segments, feet (mm).

2305.3.7.2.5 Anchorage for in-plane shear. The unit shear force, v, transmitted into the top of a perforated shear wall, out of the base of the perforated shear wall at full-height sheathing and into collectors (drag struts) connecting shear wall segments, shall be calculated in accordance with the following:

$$v = \frac{V}{C_o \Sigma L_i}$$ **(Equation 23-4)**

where:

v = Unit shear force, pounds per lineal feet (N/m).

V = Shear force in perforated shear wall, pounds (N).

C_o = Shear resistance adjustment factor from Table 2305.3.7.2.

ΣL_i= Sum of widths of perforated shear wall segments, feet (mm).

2305.3.7.2.6 Uplift anchorage between perforated shear wall ends. In addition to the requirements of Section 2305.3.7.2.4, perforated shear wall bottom plates at full-height sheathing shall be anchored for a uniform uplift force, t, equal to the unit shear force, v, determined in Section 2305.3.7.2.5.

2305.3.7.2.7 Compression chords. Each end of each perforated shear wall segment shall be designed for a compression chord force, C, equal to the tension chord uplift force, T, calculated in Section 2305.3.7.2.4.

2305.3.7.2.8 Load path. A load path to the foundation shall be provided for each uplift force, T and t, for each shear force, V and v, and for each compression chord force, C. Elements resisting shear wall forces contributed by multiple stories shall be designed for the sum of forces contributed by each story.

2305.3.7.2.9 Deflection of shear walls with openings. The controlling deflection of a blocked shear wall with openings uniformly nailed throughout shall be taken as the maximum individual deflection of the shear wall segments calculated in accordance with Section 2305.3.2, divided by the appropriate shear resistance adjustment factors of Table 2305.3.7.2.

2305.3.8 Summing shear capacities. The shear values for shear panels of different capacities applied to the same side of the wall are not cumulative except as allowed in Table 2306.4.1.

The shear values for material of the same type and capacity applied to both faces of the same wall are cumulative. Where the material capacities are not equal, the allowable shear shall be either two times the smaller shear capacity or the capacity of the stronger side, whichever is greater.

Summing shear capacities of dissimilar materials applied to opposite faces or to the same wall line is not allowed.

Exception: For wind design, the allowable shear capacity of shear wall segments sheathed with a combination of wood structural panels and gypsum wallboard on opposite faces, fiberboard structural sheathing and gypsum wallboard on opposite faces or hardboard panel siding and gypsum wallboard on opposite faces shall equal the sum of the sheathing capacities of each face separately.

2305.3.9 Adhesives. Adhesive attachment of shear wall sheathing is not permitted as a substitute for mechanical fasteners, and shall not be used in shear wall strength calculations alone, or in combination with mechanical fasteners in Seismic Design Category D, E or F.

2305.3.10 Sill plate size and anchorage in Seismic Design Category D, E or F. Two-inch (51 mm) nominal wood sill plates for shear walls shall include steel plate washers, a minimum of $^3/_{16}$ inch by 2 inches by 2 inches (4.76 mm by 51 mm by 51 mm) in size, between the sill plate and nut. Sill plates resisting a design load greater than 490 plf (LRFD) (7154 N/m) or 350 plf (ASD) (5110 N/m) shall not be less than a 3-inch (76 mm) nominal member. Where a single 3-inch (76 mm) nominal sill plate is used,

2-20d box end nails shall be substituted for 2-16d common end nails found in Line 8 of Table 2304.9.1.

Exception: In shear walls where the design load is less than 840 plf (LRFD) (12 264 N/m) or 600 plf (ASD) (8760 N/m), the sill plate is permitted to be a 2-inch (51 mm) nominal member if the sill plate is anchored by two times the number of bolts required by design and $^3/_{16}$ inch by 2 inch by 2 inch (4.76 mm by 51 mm by 51 mm) plate washers are used.

SECTION 2306
ALLOWABLE STRESS DESIGN

2306.1 Allowable stress design. The structural analysis and construction of wood elements in structures using allowable design methods shall be in accordance with the following applicable standards:

American Forest & Paper Association.

NDS National Design Specification for Wood Construction

American Institute of Timber Construction.

AITC 104	Typical Construction Details
AITC 110	Standard Appearance Grades for Structural Glued Laminated Timber
AITC 112	Standard for Tongue-and-Groove Heavy Timber Roof Decking
AITC 113	Standard for Dimensions of Structural Glued Laminated Timber
AITC 117	Standard Specifications for Structural Glued Laminated Timber of Softwood Species
AITC 119	Structural Standard Specifications for Glued Laminated Timber of Hardwood Species
AITC A190.1	Structural Glued Laminated Timber
AITC 200	Inspection Manual
AITC 500	Determination of Design Values for Structural Glued Laminated Timber

Truss Plate Institute, Inc.

TPI 1 National Design Standard for Metal Plate Connected Wood Truss Construction

American Society of Agricultural Engineers.

ASAE EP 484.2	Diaphragm Design of Metal-Clad, Post-Frame Rectangular Buildings
ASAE EP 486.1	Shallow Post Foundation Design
ASAE 559	Design Requirements and Bending Properties for Mechanically Laminated Columns

APA—The Engineered Wood Association.

Plywood Design Specification

Plywood Design Specification Supplement 1 - Design & Fabrication of Plywood Curved Panels.

Plywood Design Specification Supplement 2 -
Design & Fabrication of Glued Plywood-Lumber beams.

Plywood Design Specification Supplement 3 -
Design & Fabrication of Plywood Stressed-Skin Panels.

Plywood Design Specification Supplement 4 -
Design & Fabrication of Plywood Sandwich Panels.

Plywood Design Specification Supplement 5 -
Design & Fabrication of All-Plywood Beams.

EWS T300	Glulam Connection Details
EWS S560	Field Notching and Drilling of Glued Laminated Timber Beams
EWS S475	Glued Laminated Beam Design Tables
EWS X450	Glulam in Residential Construction
EWS X440	Product and Application Guide: Glulam
EWS R540	Builders Tips: Proper Storage and Handling of Glulam Beams

2306.1.1 Joists and rafters. The design of rafter spans is permitted to be in accordance with the *AF&PA Span Tables for Joists and Rafters*.

2306.1.2 Plank and beam flooring. The design of plank and beam flooring is permitted to be in accordance with the *AF&PA Wood Construction Data No. 4*.

2306.1.3 Treated wood stress adjustments. The allowable unit stresses for preservative-treated wood need no adjustment for treatment, but are subject to other adjustments.

The allowable unit stresses for fire-retardant-treated wood, including fastener values, shall be developed from an approved method of investigation that considers the effects of anticipated temperature and humidity to which the fire-retardant-treated wood will be subjected, the type of treatment and the redrying process. Other adjustments are applicable except that the impact load duration shall not apply.

2306.2 Wind provisions for walls.

2306.2.1 Wall stud bending stress increase. The NDS fiber stress in bending (F_b) design values for wood studs resisting wind shall be increased by the factors in Table 2306.2.1, in lieu of the 1.15 repetitive member factor, to take into consideration the load sharing and composite actions provided by the wood structural panels as defined in Section 2302.1, where the studs are designed for bending in accordance with Section 1609.6 spaced no more than 16 inches (406 mm) o.c, covered on the inside with a minimum of $^1/_2$-inch (12.7 mm) gypsum board fastened in accordance with Table 2306.4.5, and sheathed on the exterior with a minimum of $^3/_8$-inch (9.5 mm) wood structural panel sheathing that is attached to the studs using a minimum of 8d common nails spaced a maximum of 6 inches o.c. (152 mm) at panel edges and 12 inches o.c. (305 mm) in the field of the panels.

2306.3 Wood diaphragms.

2306.3.1 Shear capacities modifications. The allowable shear capacities in Table 2306.3.1 for horizontal wood structural panel diaphragms shall be increased 40 percent for wind design.

2306.3.2 Wood structural panel diaphragms. Structural panel diaphragms with wood structural panels are permitted to be used to resist horizontal forces not exceeding those set forth in Table 2306.3.1 or 2306.3.2 or calculated by principles of mechanics without limitations by using values for fastener strength in the NDS structural design properties for wood structural panels based on DOC PS-1 and DOC PS-2 or plywood design properties given in the APA Plywood Design Specification.

TABLE 2306.2.1
WALL STUD BENDING STRESS INCREASE FACTORS

STUD SIZE	SYSTEM FACTOR
2 × 4	1.5
2 × 6	1.4
2 × 8	1.3
2 × 10	1.2
2 × 12	1.15

2306.3.3 Diagonally sheathed lumber diaphragms. Diagonally sheathed lumber diaphragms shall be nailed in accordance with Table 2306.3.3.

2306.3.4 Single diagonally sheathed lumber diaphragms. Single diagonally sheathed lumber diaphragms shall be constructed of minimum 1-inch (25 mm) thick nominal sheathing boards laid at an angle of approximately 45 degrees (0.78 rad) to the supports. The shear capacity for single diagonally sheathed lumber diaphragms of southern pine or Douglas fir-larch shall not exceed 300 plf (4378 N/m) of width. The shear capacities shall be adjusted by reduction factors of 0.82 for framing members of species with a specific gravity equal to or greater than 0.42 but less than 0.49 and 0.65 for species with a specific gravity of less than 0.42, as contained in the NDS.

2306.3.4.1 End joints. End joints in adjacent boards shall be separated by at least one stud or joist space and there shall be at least two boards between joints on the same support.

2306.3.4.2 Single diagonally sheathed lumber diaphragms. Single diagonally sheathed lumber diaphragms made up of 2-inch (51 mm) nominal diagonal lumber sheathing fastened with 16d nails shall be designed with the same shear capacities as shear panels using 1-inch (25 mm) boards fastened with 8d nails, provided there are not splices in adjacent boards on the same support and the supports are not less than 4 inch (102 mm) nominal depth or 3 inch (76 mm) nominal thickness.

2306.3.5 Double diagonally sheathed lumber diaphragms. Double diagonally sheathed lumber diaphragms shall be constructed of two layers of diagonal sheathing boards at 90 degrees (1.57 rad) to each other on

TABLE 2306.3.1
RECOMMENDED SHEAR (POUNDS PER FOOT) FOR WOOD STRUCTURAL PANEL DIAPHRAGMS WITH FRAMING OF DOUGLAS-FIR-LARCH, OR SOUTHERN PINE[a] FOR WIND OR SEISMIC LOADING

PANEL GRADE	COMMON NAIL SIZE OR STAPLE[f] LENGTH AND GAGE	MINIMUM FASTENER PENETRATION IN FRAMING (inches)	MINIMUM NOMINAL PANEL THICKNESS (inch)	MINIMUM NOMINAL WIDTH OF FRAMING MEMBER (inches)	BLOCKED DIAPHRAGMS — Fastener spacing (inches) at diaphragm boundaries (all cases), at continuous panel edges parallel to load (Cases 3, 4), and at all panel edges (Cases 5 and 6)[b] / Fastener spacing (inches) at other panel edges (Cases 1, 2, 3 and 4)[b]				UNBLOCKED DIAPHRAGMS — Fasteners spaced 6" max. At supported edges[b]	
					6 / 6	4 / 6	2½[c] / 4	2[c] / 3	Case 1 (No unblocked edges or continuous joints parallel to load)	All other configurations (Cases 2, 3, 4, 5 and 6)
Structural I Grades	6d[e]	1¼	5/16	2	185	250	375	420	165	125
				3	210	280	420	475	185	140
	1½ 16 Gage	1		2	155	205	310	350	135	105
				3	175	230	345	390	155	115
	8d	1⅜	3/8	2	270	360	530	600	240	180
				3	300	400	600	675	265	200
	1½ 16 Gage	1		2	175	235	350	400	155	115
				3	200	265	395	450	175	130
	10d[d]	1½	15/32	2	320	425	640	730	285	215
				3	360	480	720	820	320	240
	1½ 16 Gage	1		2	175	235	350	400	155	120
				3	200	265	395	450	175	130
Sheathing, single floor and other grades covered in DOC PS 1 and PS 2	6d[e]	1¼	5/16	2	170	225	335	380	150	110
				3	190	250	380	430	170	125
	1½ 16 Gage	1		2	140	185	275	315	125	90
				3	155	205	310	350	140	105
	6d[e]	1¼	3/8	2	185	250	375	420	165	125
				3	210	280	420	475	185	140
	8d	1⅜		2	240	320	480	545	215	160
				3	270	360	540	610	240	180

(continued)

TABLE 2306.3.1—continued
RECOMMENDED SHEAR (POUNDS PER FOOT) FOR WOOD STRUCTURAL PANEL DIAPHRAGMS WITH FRAMING OF DOUGLAS-FIR-LARCH, OR SOUTHERN PINE[a] FOR WIND OR SEISMIC LOADING

PANEL GRADE	COMMON NAIL SIZE OR STAPLE[f] LENGTH AND GAGE	MINIMUM FASTENER PENETRATION IN FRAMING (inches)	MINIMUM NOMINAL PANEL THICKNESS (inch)	MINIMUM NOMINAL WIDTH OF FRAMING MEMBER (inches)	BLOCKED DIAPHRAGMS								UNBLOCKED DIAPHRAGMS	
					Fastener spacing (inches) at diaphragm boundaries (all cases), at continuous panel edges parallel to load (Cases 3, 4), and at all panel edges (Cases 5 and 6)[b]				Fastener spacing (inches) at other panel edges (Cases 1, 2, 3 and 4)[b]				Fasteners spaced 6" max. at supported edges[b]	
					6	4	2½[c]	2[c]	6	4	2½[c]	2[c]	Case 1 (No unblocked edges or continuous joints parallel to load)	All other configurations (Cases 2, 3, 4, 5 and 6)
					6	4	2½	2	6	4	2½	2		
Sheathing, single floor and other grades covered in DOC PS 1 and PS 2 (continued)	1½ 16 Gage	1	3/8	2	160	210	315	360					140	105
				3	180	235	355	400					160	120
	8d	1 3/8	7/16	2	255	340	505	575					230	170
				3	285	380	570	645					255	190
	1½ 16 Gage	1		2	165	225	335	380					150	110
				3	190	250	375	425					165	125
	8d	1 3/8	15/32	2	270	360	530	600					240	180
				3	300	400	600	675					265	200
	10d[d]	1½		2	290	385	575	655					255	190
				3	325	430	650	735					290	215
	1½ 16 Gage	1	19/32	2	160	210	315	360					140	105
				3	180	235	355	405					160	120
	10d[d]	1½		2	320	425	640	730					285	215
				3	360	480	720	820					320	240
	1¾ 16 Gage	1		2	175	235	350	400					155	115
				3	200	265	395	450					175	130

(continued)

**TABLE 2306.3.1—continued
RECOMMENDED SHEAR (POUNDS PER FOOT) FOR WOOD STRUCTURAL
PANEL DIAPHRAGMS WITH FRAMING OF DOUGLAS-FIR-LARCH,
OR SOUTHERN PINE[a] FOR WIND OR SEISMIC LOADING**

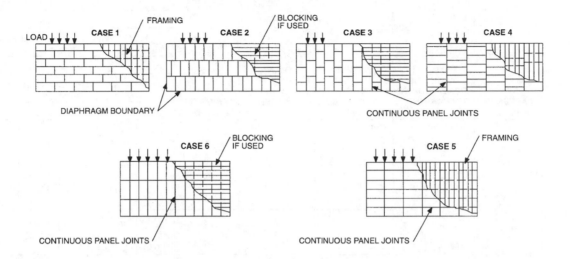

For SI: 1 inch = 25.4 mm, 1 pound per foot = 14.5939 N/m.

a. For framing of other species: (1) Find specific gravity for species of lumber in AFPA National Design Specification. (2) For staples find shear value from table above for Structural I panels (regardless of actual grade) and multiply value by 0.82 for species with specific gravity of 0.42 or greater, or 0.65 for all other species. (3) For nails find shear value from table above for nail size for actual grade and multiply value by the following adjustment factor: Specific Gravity Adjustment Factor = [1-(0.5 - SG)], where SG = Specific Gravity of the framing lumber. This adjustment factor shall not be greater than 1.

b. Space fasteners maximum 12 inches o.c. along intermediate framing members (6 inches o.c. where supports are spaced 48 inches o.c.).

c. Framing at adjoining panel edges shall be 3 inches nominal or wider, and nails shall be staggered where nails are spaced 2 inches o.c. or 2 1/2 inches o.c.

d. Framing at adjoining panel edges shall be 3 inches nominal or wider, and nails shall be staggered where both of the following conditions are met: (1) 10d nails having penetration into framing of more than 1 1/2 inches and (2) nails are spaced 3 inches o.c. or less.

e. 8d is recommended minimum for roofs due to negative pressures of high winds.

f. Staples shall have a minimum crown width of 7/16 inch.

the same face of the supporting members. Each chord shall be considered as a beam with uniform load per foot equal to 50 percent of the unit shear due to diaphragm action. The load shall be assumed as acting normal to the chord in the plan of the diaphragm in either direction. The span of the chord or portion thereof shall be the distance between framing members of the diaphragm, such as the joists, studs and blocking that serve to transfer the assumed load to the sheathing. The shear capacity of double diagonally sheathed diaphragms of Southern pine or Douglas fir-larch shall not exceed 600 plf (8756 kN/m) of width. The shear capacity shall be adjusted by reduction factors of 0.82 for framing members of species with a specific gravity equal to or greater than 0.42 but less than 0.49 and 0.65 for species with a specific gravity of less than 0.42, as contained in the NDS. Nailing of diagonally sheathed lumber diaphragms shall be in accordance with Table 2306.3.3.

2306.3.6 Gypsum board diaphragm ceilings. Gypsum board diaphragm ceilings shall be in accordance with Section 2508.5.

2306.4 Shear walls. Panel sheathing joints in shear walls shall occur over studs or blocking. Adjacent panel sheathing joints shall occur over and be nailed to common framing members (see Section 2305.3.1 for limitations on shear wall bracing materials).

2306.4.1 Wood structural panel shear walls. The allowable shear capacities for wood structural panel shear walls shall be in accordance with Table 2306.4.1. These capacities are permitted to be increased 40 percent for wind design. Shear walls are permitted to be calculated by principles of mechanics without limitations by using values for nail strength given in the NDS and wood structural panel design properties given in the APA/PDS.

2306.4.2 Lumber sheathed shear walls. Single and double diagonally sheathed lumber diaphragms are permitted using the construction and allowable load provisions of Sections 2306.3.4 and 2306.3.5.

2306.4.3 Particleboard shear walls. The design shear capacity of particleboard shear walls shall be in accordance with Table 2306.4.3. Shear panels shall be constructed with particleboard sheets not less than 4 feet by 8 feet (1219 mm by 2438 mm), except at boundaries and changes in framing.

WOOD

TABLE 2306.3.2
ALLOWABLE SHEAR IN POUNDS PER FOOT FOR HORIZONTAL BLOCKED DIAPHRAGMS UTILIZING MULTIPLE ROWS OF FASTENERS (HIGH LOAD DIAPHRAGMS) WITH FRAMING OF DOUGLAS FIR, LARCH OR SOUTHERN PINE[a] FOR WIND OR SEISMIC LOADING[b]

PANEL GRADE[c]	FASTENER AND SIZE	MINIMUM FASTENING PENETRATION IN FRAMING (inches)	MINIMUM NOMINAL PANEL THICKNESS (inch)	MINIMUM NOMINAL WIDTH OF FRAMING MEMBER[e] (inches)	LINES OF FASTENERS	BLOCKED DIAPHRAGMS Cases 1 and 2[d]					
						Fastener Spacing Per Line at Boundaries (inches)					
						4		2½		2	
						Fastener Spacing Per Line at Other Panel Edges (inches)					
						6	4	4	3	3	2
Structural I grades	10d common nails	1½	15/32	3	2	605	815	875	1,150	—	—
				4	2	700	915	1,005	1,290	—	—
				4	3	875	1,220	1,285	1,395	—	—
			19/32	3	2	670	880	965	1,255	—	—
				4	2	780	990	1,110	1,440	—	—
				4	3	965	1,320	1,405	1,790	—	—
			23/32	3	2	730	955	1,050	1,365	—	—
				4	2	855	1,070	1,210	1,565	—	—
				4	3	1,050	1,430	1,525	1,800	—	—
	14 gage staples	2	15/32	3	2	600	600	860	960	1,060	1,200
				4	3	860	900	1,160	1,295	1,295	1,400
			19/32	3	2	600	600	875	960	1,075	1,200
				4	3	875	900	1,175	1,440	1,475	1,795
Sheathing single floor and other grades covered in DOC PS 1 and PS 2	10d common nails	1½	15/32	3	2	525	725	765	1,010	—	—
				4	2	605	815	875	1,105	—	—
				4	3	765	1,085	1,130	1,195	—	—
			19/32	3	2	650	860	935	1,225	—	—
				4	2	755	965	1,080	1,370	—	—
				4	3	935	1,290	1,365	1,485	—	—
			23/32	3	2	710	935	1,020	1,335	—	—
				4	2	825	1,050	1,175	1,445	—	—
				4	3	1,020	1,400	1,480	1,565	—	—
	14 gage staples	2	15/32	3	2	540	540	735	865	915	1,080
				4	3	735	810	1,005	1,105	1,105	1,195
			19/32	3	2	600	600	865	960	1,065	1,200
				4	3	865	900	1,130	1,430	1,370	1,485
			23/32	4	3	865	900	1,130	1,490	1,430	1,545

For SI: 1 inch = 25.4 mm, 1 pound per foot = 14.5939 N/m.

a. For framing of the other species: (1) Find specific gravity for species of framing lumber in AFPA National Design Specification, (2) Find shear value from table above for nail size of actual grade, and (3) Multiply value by the following adjustment factor = [1 - (0.5 - SG)], where SG = Specific gravity of the framing lumber. This adjustment factor shall not be greater than 1.

b. Fastening along intermediate framing members: Space nails 12 inches on center, except 6 inches on center for spans greater than 32 inches.

c. Panels conforming to PS 1 or PS 2.

d. This table gives shear values for Cases 1 and 2 as shown in Table 2306.3.1. The values shown are applicable to Cases 3, 4, 5 and 6 as shown in Table 2306.3.1, providing fasteners at all continuous panel edges are spaced in accordance with the boundary fastener spacing.

e. The minimum depth of framing members shall be 3 inches.

TABLE 2306.3.3
DIAGONALLY SHEATHED LUMBER DIAPHRAGM NAILING SCHEDULE

SHEATHING NOMINAL DIMENSION	NAILING TO INTERMEDIATE AND END-BEARING STUDS		NAILING AT THE SHEAR PANEL BOUNDARIES	
	Type, size and number of nails per board			
	Common nails	Box nails	Common nails	Box nails
1 × 6	2 - 8d	3 - 8d	3 - 8d	5 - 8d
1 × 8	3 - 8d	4 - 8d	4 - 8d	6 - 8d
2 × 6	2 - 16d	3 - 16d	3 - 16d	5 - 16d
2 × 8	3 - 16d	4 - 16d	4 - 16d	6 - 16d

TABLE 2306.4.1
ALLOWABLE SHEAR (POUNDS PER FOOT) FOR WOOD STRUCTURAL PANEL SHEAR WALLS WITH FRAMING OF DOUGLAS-FIR-LARCH, OR SOUTHERN PINE[a] FOR WIND OR SEISMIC LOADING[b, h, i, j]

PANEL GRADE	MINIMUM NOMINAL PANEL THICKNESS (inch)	MINIMUM FASTENER PENETRATION IN FRAMING (inches)	PANELS APPLIED DIRECT TO FRAMING — NAIL (common or galvanized box) or staple size[k]	Fastener spacing at panel edges (inches) 6	4	3	2[e]	PANELS APPLIED OVER $\frac{1}{2}$" OR $\frac{5}{8}$" GYPSUM SHEATHING — NAIL (common or galvanized box) or staple size[k]	Fastener spacing at panel edges (inches) 6	4	3	2[e]
Structural I Sheathing	$\frac{5}{16}$	$1\frac{1}{4}$	6d	200	300	390	510	8d	200	300	390	510
		1	$1\frac{1}{2}$ 16 Gage	165	245	325	415	2 16 Gage	125	185	245	315
	$\frac{3}{8}$	$1\frac{3}{8}$	8d	230[d]	360[d]	460[d]	610[d]	10d	280	430	550[f]	730
		1	$1\frac{1}{2}$ 16 Gage	155	235	315	400	2 16 Gage	155	235	310	400
	$\frac{7}{16}$	$1\frac{3}{8}$	8d	255[d]	395[d]	505[d]	670[d]	10d	280	430	550[f]	730
		1	$1\frac{1}{2}$ 16 Gage	170	260	345	440	2 16 Gage	155	235	310	400
	$\frac{15}{32}$	$1\frac{3}{8}$	8d	280	430	550	730	10d	280	430	550[f]	730
		1	$1\frac{1}{2}$ 16 Gage	185	280	375	475	2 16 Gage	155	235	300	400
		$1\frac{1}{2}$	10d	340	510	665[f]	870	10d	—	—	—	—
Sheathing, plywood siding[g] except Group 5 Species	$\frac{5}{16}$ or $\frac{1}{4}$[c]	$1\frac{1}{4}$	6d	180	270	350	450	8d	180	270	350	450
		1	$1\frac{1}{2}$ 16 Gage	145	220	295	375	2 16 Gage	110	165	220	285
	$\frac{3}{8}$	$1\frac{1}{4}$	6d	200	300	390	510	8d	200	300	390	510
		$1\frac{3}{8}$	8d	220[d]	320[d]	410[d]	530[d]	10d	260	380	490[f]	640
		1	$1\frac{1}{2}$ 16 Gage	140	210	280	360	2 16 Gage	140	210	280	360
	$\frac{7}{16}$	$1\frac{3}{8}$	8d	240[d]	350[d]	450[d]	585[d]	10d	260	380	490[f]	640
		1	$1\frac{1}{2}$ 16 Gage	155	230	310	395	2 16 Gage	140	210	280	360
	$\frac{15}{32}$	$1\frac{3}{8}$	8d	260	380	490	640	10d	260	380	490[f]	640
		$1\frac{1}{2}$	10d	310	460	600[f]	770	—	—	—	—	—
		1	$1\frac{1}{2}$ 16 Gage	170	255	335	430	2 16 Gage	140	210	280	360
	$\frac{19}{32}$	$1\frac{1}{2}$	10d	340	510	665[f]	870	—	—	—	—	—
		1	$1\frac{3}{4}$ 16 Gage	185	280	375	475	—	—	—	—	—
			Nail Size (galvanized casing)					Nail Size (galvanized casing)				
	$\frac{5}{16}$[c]	$1\frac{1}{4}$	6d	140	210	275	360	8d	140	210	275	360
	$\frac{3}{8}$	$1\frac{3}{8}$	8d	160	240	310	410	10d	160	240	310[f]	410

For SI: 1 inch = 25.4 mm, 1 pound per foot = 14.5939 N/m.

a. For framing of other species: (1) Find specific gravity for species of lumber in AF&PA National Design Specification. (2) For staples find shear value from table above for Structural I panels (regardless of actual grade) and multiply value by 0.82 for species with specific gravity of 0.42 or greater, or 0.65 for all other species. (3) For nails find shear value from table above for nail size for actual grade and multiply value by the following adjustment factor: Specific Gravity Adjustment Factor = [1-(0.5 - SG)], where SG = Specific Gravity of the framing lumber. This adjustment factor shall not be greater than 1.

b. Panel edges backed with 2-inch nominal or wider framing. Install panels either horizontally or vertically. Space fasteners maximum 6 inches on center along intermediate framing members for $\frac{3}{8}$-inch and $\frac{7}{16}$-inch panels installed on studs spaced 24 inches on center. For other conditions and panel thickness, space fasteners maximum 12 inches on center on intermediate supports.

c. $\frac{3}{8}$-inch panel thickness or siding with a span rating of 16 inches on center is the minimum recommended where applied direct to framing as exterior siding.

d. Shears are permitted to be increased to values shown for $\frac{15}{32}$-inch sheathing with same nailing provided (a) studs are spaced a maximum of 16 inches on center, or (b) if panels are applied with long dimension across studs.

e. Framing at adjoining panel edges shall be 3 inches nominal or wider, and nails shall be staggered where nails are spaced 2 inches on center.

f. Framing at adjoining panel edges shall be 3 inches nominal or wider, and nails shall be staggered where both of the following conditions are met: (1) 10d nails having penetration into framing of more than 1 $\frac{1}{2}$ inches and (2) nails are spaced 3 inches on center.

g. Values apply to all-veneer plywood. Thickness at point of fastening on panel edges governs shear values.

h. Where panels are applied on both faces of a wall and nail spacing is less than 6 inches o.c. on either side, panel joints shall be offset to fall on different framing members. Or framing shall be 3 inch nominal or thicker and nails on each side shall be staggered.

i. In Seismic Design Category D, E or F, where shear design values exceed 490 pounds per lineal foot (LRFD) or 350 pounds per lineal foot (ASD) all framing members receiving edge nailing from abutting panels shall not be less than a single 3-inch nominal member. Plywood joint and sill plate nailing shall be staggered in all cases. See Section 2305.3.10 for sill plate size and anchorage requirements.

j. Galvanized nails shall be hot dipped or tumbled.

k. Staples shall have a minimum crown width of $\frac{7}{16}$-inch.

Particleboard panels shall be designed to resist shear only, and chords, collector members and boundary elements shall be connected at all corners. Panel edges shall be backed with 2-inch (51 mm) nominal or wider framing. Sheets are permitted to be installed either horizontally or vertically. For $^3/_8$-inch (9.5 mm) particleboard sheets installed with the long dimension parallel to the studs spaced 24 inches (610 mm) o.c, nails shall be spaced at 6 inches (152 mm) o.c. along intermediate framing members. For all other conditions, nails of the same size shall be spaced at 12 inches (305 mm) o.c. along intermediate framing members. Particleboard panels less than 12 inches (305 mm) wide shall be blocked. Particleboard shall not be used to resist seismic forces in structures in Seismic Design Category D, E or F.

2306.4.4 Fiberboard shear walls. The design shear capacity of fiberboard shear walls shall be in accordance with Table 2308.9.3(4). The fiberboard sheathing shall be applied vertically or horizontally to wood studs not less than 2 inch (51 mm) nominal thickness spaced 16 inches (406 mm) o.c. Blocking not less than 2 inch (51 mm) nominal in thickness shall be provided at horizontal joints. Fiberboard shall not be used to resist seismic forces in structures in Seismic Design Category D, E or F.

2306.4.5 Shear walls sheathed with other materials. Shear capacities for walls sheathed with lath and plaster, and gypsum board shall be in accordance with Table 2306.4.5. Shear walls sheathed with lath, plaster and gypsum board shall be constructed in accordance with Chapter 25 and Section 2306.4.5.1. Walls resisting seismic loads shall be subject to the limitations in Section 1617.6.

2306.4.5.1 Application of gypsum board or lath and plaster to wood framing.

2306.4.5.1.1 Joint staggering. End joints of adjacent courses of gypsum board shall not occur over the same stud.

2306.4.5.1.2 Blocking. Where required in Table 2306.4.5, wood blocking having the same cross-sectional dimensions as the studs shall be provided at joints that are perpendicular to the studs.

2306.4.5.1.3 Nailing. Studs, top and bottom plates and blocking shall be nailed in accordance with Table 2304.9.1.

2306.4.5.1.4 Fasteners. The size and spacing of nails shall be set forth in Table 2306.4.5. Nails shall be spaced not less than $^3/_8$ inch (9.5 mm) from edges and ends of gypsum boards or sides of studs, blocking and top and bottom plates.

2306.4.5.1.5 Gypsum lath. Gypsum lath shall be applied perpendicular to the studs. Maximum allowable shear values shall be as set forth in Table 2306.4.5.

2306.4.5.1.6 Gypsum sheathing. Four-foot-wide (1219 mm) pieces of gypsum sheathing shall be applied parallel or perpendicular to studs. Two-foot-wide (610 mm) pieces of gypsum sheathing shall be applied perpendicular to the studs. Maximum allowable shear values shall be as set forth in Table 2306.4.5.

2306.4.5.1.7 Other gypsum boards. Gypsum board shall be applied parallel or perpendicular to studs. Maximum allowable shear values shall be set forth in Table 2306.4.5.

SECTION 2307
LOAD AND RESISTANCE FACTOR DESIGN

2307.1 Load and resistance factor design (LRFD). The structural analysis and construction of wood elements and structures using load and resistance factor design (LRFD) methods shall be in accordance with ASCE 16.

TABLE 2306.4.3
ALLOWABLE SHEAR FOR PARTICLEBOARD SHEAR WALL SHEATHING

PANEL GRADE	MINIMUM NOMINAL PANEL THICKNESS (inch)	MINIMUM NAIL PENETRATION IN FRAMING (inches)	PANELS APPLIED DIRECT TO FRAMING				
			Nail size (common or galvanized box)	Allowable shear (pounds per foot) nail spacing at panel edges (inches)[a]			
				6	4	3	2
M-S "Exterior Glue" and M-2 "Exterior Glue"	$^3/_8$	$1\,^1/_2$	6d	120	180	230	300
	$^3/_8$	$1\,^1/_2$	8d	130	190	240	315
	$^1/_2$			140	210	270	350
	$^1/_2$	$1\,^5/_8$	10d	185	275	360	460
	$^5/_8$			200	305	395	520

For SI: 1 inch = 25.4 mm, 1 pound per foot = 14.5939 N/m.

a. Values are not permitted in Seismic Design Category D, E or F.

TABLE 2306.4.5
ALLOWABLE SHEAR FOR WIND OR SEISMIC FORCES FOR SHEAR WALLS OF LATH
AND PLASTER OR GYPSUM BOARD WOOD FRAMED WALL ASSEMBLIES

TYPE OF MATERIAL	THICKNESS OF MATERIAL	WALL CONSTRUCTION	FASTENER SPACING[b] MAXIMUM (inches)	SHEAR VALUE[a,e] (plf)	MINIMUM FASTENER SIZE[c,d,j,k]
1. Expanded metal or woven wire lath and portland cement plaster	$7/8''$	Unblocked	6	180	No. 11 gage $1^1/_2''$ long, $7/_{16}''$ head 16 Ga. Galv. Staple, $7/_8''$ legs
2. Gypsum lath, plain or perforated	$3/_8''$ lath and $1/_2''$ plaster	Unblocked	5	100	No. 13 gage, $1^1/_8''$ long, $19/_{64}''$ head, plasterboard nail 16 Ga. Galv. Staple, $1^1/_8''$ long 0.120'' Nail, min. $3/_8''$ head, $1^1/_4''$ long
3. Gypsum sheathing	$1/_2'' \times 2' \times 8'$	Unblocked	4	75	No. 11 gage, $1^3/_4''$ long, $7/_{16}''$ head, diamond-point, galvanized 16 Ga. Galv. Staple, $1^3/_4''$ long
	$1/_2'' \times 4'$	Blocked[f]	4	175	
		Unblocked	7	100	
	$5/_8'' \times 4'$	Blocked	4'' edge/ 7'' field	200	6d galvanized 0.120'' Nail, min. $3/_8''$ head, $1^3/_4''$ long
4. Gypsum board, gypsum veneer base, or water-resistant gypsum backing board	$1/_2''$	Unblocked[f]	7	75	5d cooler or wallboard 0.120'' Nail, min. $3/_8''$ head, $1^1/_2''$ long 16 Gage Staple, $1^1/_2''$ long
		Unblocked[f]	4	110	
		Unblocked	7	100	
		Unblocked	4	125	
		Blocked[g]	7	125	
		Blocked[g]	4	150	
		Unblocked	8/12[h]	60	No. 6-$1^1/_4''$ screws[i]
		Blocked[g]	4/16[h]	160	
		Blocked[g]	4/12[h]	155	
		Blocked[f, g]	8/12[h]	70	
		Blocked[g]	6/12[h]	90	
	$5/_8''$	Unblocked[f]	7	115	6d cooler or wallboard 0.120'' Nail, min. $3/_8''$ head, $1^3/_4''$ long 16 Gage Staple, $1^1/_2''$ legs, $1^5/_8''$ long
			4	145	
		Blocked[g]	7	145	
			4	175	
		Blocked[g] Two-ply	Base ply: 9 Face ply: 7	250	Base ply—6d cooler or wallboard $1^3/_4'' \times 0.120''$ Nail, min. $3/_8''$ head $1^5/_8''$ 16 Ga. Galv. Staple Face ply—8d cooler or wallboard 0.120'' Nail, min. $3/_8''$ head, $2^3/_8''$ long 15 Ga. Galv. Staple, $2^1/_4''$ long
		Unblocked	8/12[h]	70	No. 6-$1^1/_4''$ screws[i]
		Blocked[g]	8/12[h]	90	

For SI: 1 inch = 25.4 mm, 1 foot = 304.8 mm, 1 pound per foot = 14.5939 N/m.

a. These shear walls shall not be used to resist loads imposed by masonry or concrete construction (see Section 2305.1.5). Values shown are for short-term loading due to wind or seismic loading in Seismic Design Categories A, B and C. Walls resisting seismic loads shall be subject to the limitations in Section 1617.6. Values shown shall be reduced 25 percent for normal loading.

b. Applies to nailing at studs, top and bottom plates and blocking.

c. Alternate nails are permitted to be used if their dimensions are not less than the specified dimensions. Drywall screws are permitted to be substituted for the 5d, 6d (cooler) nails listed above. $1^1/_4$ inches Type S or W, No. 6 for 6d (cooler) nails.

d. For properties of cooler nails, see ASTM C 514.

e. Except as noted, shear values are based on a maximum framing spacing of 16 inches on center.

f. Maximum framing spacing of 24 inches on center.

g. All edges are blocked, and edge nailing is provided at all supports and all panel edges.

h. First number denotes fastener spacing at the edges; second number denotes fastener spacing in the field.

i. Screws are Type W or S.

j. Staples shall have a minimum crown width of $7/_{16}$ inch, measured outside the legs.

k. Staples for the attachment of gypsum lath and woven-wire lath shall have a minimum crown width of $3/_4$ inch, measured outside the legs.

SECTION 2308
CONVENTIONAL LIGHT-FRAME CONSTRUCTION

2308.1 General. The requirements of this section are intended for conventional light-frame construction. Other methods are permitted to be used provided a satisfactory design is submitted showing compliance with other provisions of this code. Interior nonload-bearing partitions, ceilings and curtain walls of conventional light-frame construction are not subject to the limitations of this section. Alternatively, compliance with the following standard shall be permitted subject to the limitations therein and the limitations of this code: *American Forest and Paper Association (AF&PA) Wood Frame Construction Manual for One- and Two-Family Dwellings* (WCFM).

2308.2 Limitations. Buildings are permitted to be constructed in accordance with the provisions of conventional light-frame construction, subject to the following limitations, and to further limitations of Sections 2308.11 and 2308.12.

1. Buildings shall be limited to a maximum of three stories above grade. For the purposes of this section, for buildings in Seismic Design Category D or E as determined in Section 1616, cripple stud walls shall be considered to be a story.

   Exception: Solid blocked cripple walls not exceeding 14 inches (356 mm) in height need not be considered a story.

2. Bearing wall floor-to-floor heights shall not exceed 10 feet (3048 mm).

3. Loads as determined in Chapter 16 shall not exceed the following:

   3.1. Average dead loads shall not exceed 15 psf (718 N/m²) for roofs and exterior walls, floors and partitions.

   3.2. Live loads shall not exceed 40 psf (1916 N/m²) for floors.

   3.3. Ground snow loads shall not exceed 50 psf (2395 N/m²).

4. Wind speeds shall not exceed 100 miles per hour (mph) (44 m/s) (3-second gust).

   Exception: Wind speeds shall not exceed 110 mph (48.4 m/s) 3-second gust for buildings in Exposure Category A or B.

5. Roof trusses and rafters shall not span more than 40 feet (12 192 mm) between points of vertical support.

6. The use of the provisions for conventional light-frame construction in this section shall not be permitted for buildings in Seismic Design Category B, C, D, E or F for Seismic Use Group III, as determined in Section 1616.

7. Conventional light-frame construction is limited in irregular structures in Seismic Design Category D or E, as specified in Section 2308.12.6.

2308.2.1 Basic wind speed greater than 100 mph (3-second gust). Where the basic wind speed exceeds 100 mph (3-second gust), the provisions of either the *AF&PA Wood Frame Construction Manual for One- and Two-Family Dwellings* (WFCM), or the SBCCI *Standard for Hurri-*

cane-Resistant Residential Construction (SSTD 10), are permitted to be used.

2308.2.2 Buildings in Seismic Design Category B, C, D or E. Buildings of conventional light-frame construction in Seismic Design Category B or C, as determined in Section 1616, shall comply with the additional requirements in Section 2308.11.

Exceptions:

1. Detached one- and two-family dwellings as applicable in Section 101.2 in Seismic Design Category B.

2. Detached one- and two-family dwellings as applicable in Section 101.2 in Seismic Design Category C where masonry veneer is limited to the first two stories above grade.

Buildings of conventional light-frame construction in Seismic Design Category D or E, as determined in Section 1616, shall comply with the additional requirements in Section 2308.12.

2308.3 Braced wall lines. Buildings shall be provided with exterior and interior braced wall lines as described in Section 2308.9.3 and installed in accordance with Sections 2308.3.1 through 2308.3.4.

2308.3.1 Spacing. Spacing of braced wall lines shall not exceed 35 feet (10 668 mm) o.c. in both the longitudinal and transverse directions in each story.

2308.3.2 Braced wall panel connections. Forces shall be transferred from the roofs and floors to braced wall panels and from the braced wall panels in upper stories to the braced wall panels in the story below by the following:

1. Braced wall panel top and bottom plates shall be fastened to joists, rafters or full-depth blocking. Braced wall panels shall be extended and fastened to roof framing at intervals not to exceed 50 feet (15 240 mm) between parallel braced wall lines.

   Exception: Where roof trusses are used, lateral forces shall be transferred from the roof diaphragm to the braced wall by blocking of the ends of the trusses or by other approved methods.

2. Bottom plate fastening to joist or blocking below shall be with not less than 3-16d nails at 16 inches (406 mm) o.c.

3. Blocking shall be nailed to the top plate below with not less than 3-8d toenails per block.

4. Joists parallel to the top plates shall be nailed to the top plate with not less than 8d toenails at 6 inches (152 mm) o.c.

In addition, top plate laps shall be nailed with not less than 8-16d face nails on each side of each break in the top plate.

2308.3.3 Sill anchorage. Where foundations are required by Section 2308.3.4, braced wall line sills shall be anchored to concrete or masonry foundations. Such anchorage shall conform to the requirements of Section 2308.6 except that such anchors shall be spaced at not more than 4 feet (1219 mm) o.c. for structures over two stories in height. The anchors shall be distributed along the length of the braced wall line. Other anchorage devices having equivalent capacity are permitted.

2308.3.3.1 Anchorage to all-wood foundations. Where all-wood foundations are used, the force transfer from the braced wall lines shall be determined based on calculation and shall have a capacity greater than or equal to the connections required by Section 2308.3.3.

2308.3.4 Braced wall line support. Braced wall lines shall be supported by continuous foundations.

Exception: For structures with a maximum plan dimension not over 50 feet (15 240 mm), continuous foundations are required at exterior walls only.

2308.4 Design of portions. Where a building of otherwise conventional construction contains nonconventional structural elements, those elements shall be designed to resist the forces specified in Chapter 16. The extent of such design need only demonstrate compliance of nonconventional elements with other applicable provisions of this code, and shall be compatible with the performance of the conventional framed system.

2308.5 Connections and fasteners. Connections and fasteners used in conventional construction shall comply with the requirements of Section 2304.9.

2308.6 Foundation plates or sills. Foundations and footings shall be as specified in Chapter 18. Foundation plates or sills resting on concrete or masonry foundations shall comply with Section 2304.3.1. Foundation plates or sills shall be bolted or anchored to the foundation with not less than $^1/_2$-inch-diameter (12.7 mm) steel bolts or approved anchors. Bolts shall be embedded at least 7 inches (178 mm) into concrete or masonry, and spaced not more than 6 feet (1829 mm) apart. There shall be a minimum of two bolts or anchor straps per piece with one bolt or anchor strap located not more than 12 inches (305 mm) or less than 4 inches (102 mm) from each end of each piece. A properly sized nut and washer shall be tightened on each bolt to the plate.

2308.7 Girders. Girders for single-story construction or girders supporting loads from a single floor shall not be less than 4 inches by 6 inches (102 mm by 152 mm) for spans 6 feet (1829 mm) or less, provided that girders are spaced not more than 8 feet (2438 mm) o.c. Spans for built-up 2-inch (51 mm) girders shall be in accordance with Table 2308.9.5 or 2308.9.6. Other girders shall be designed to support the loads specified in this code. Girder end joints shall occur over supports.

Where a girder is spliced over a support, an adequate tie shall be provided. The ends of beams or girders supported on masonry or concrete shall not have less than 3 inches (76 mm) of bearing.

2308.8 Floor joists. Spans for floor joists shall be in accordance with Table 2308.8(1) or 2308.8(2). For other grades and or species, refer to the *AF&PA Span Tables for Joists and Rafters*.

2308.8.1 Bearing. Except where supported on a 1-inch by 4-inch (25.4 mm by 102 mm) ribbon strip and nailed to the adjoining stud, the ends of each joist shall not have less than $1^1/_2$ inches (38 mm) of bearing on wood or metal, or less than 3 inches (76 mm) on masonry.

2308.8.2 Framing details. Joists shall be supported laterally at the ends and at each support by solid blocking except where the ends of the joists are nailed to a header, band or rim joist or to an adjoining stud or by other means. Solid blocking shall be not less than 2 inches (51mm) in thickness and the full depth of the joist. Notches on the ends of joists shall not exceed one-fourth the joist depth. Holes bored in joists shall not be within 2 inches (51 mm) of the top or bottom of the joist, and the diameter of any such hole shall not exceed one-third the depth of the joist. Notches in the top or bottom of joists shall not exceed one-sixth the depth and shall not be located in the middle third of the span.

Joist framing from opposite sides of a beam, girder or partition shall be lapped at least 3 inches (76 mm) or the opposing joists shall be tied together in an approved manner.

Joists framing into the side of a wood girder shall be supported by framing anchors or on ledger strips not less than 2 inches by 2 inches (51 mm by 51 mm).

2308.8.2.1 Engineered wood products. Cuts, notches and holes bored in trusses, laminated veneer lumber, glue-laminated members or I-joists are not permitted unless the effects of such penetrations are specifically considered in the design of the member.

2308.8.3 Framing around openings. Trimmer and header joists shall be doubled, or of lumber of equivalent cross section, where the span of the header exceeds 4 feet (1219 mm). The ends of header joists more than 6 feet (1829 mm) long shall be supported by framing anchors or joist hangers unless bearing on a beam, partition or wall. Tail joists over 12 feet (3658 mm) long shall be supported at the header by framing anchors or on ledger strips not less than 2 inches by 2 inches (51 mm by 51 mm).

2308.8.4 Supporting bearing partitions. Bearing partitions parallel to joists shall be supported on beams, girders, doubled joists, walls or other bearing partitions. Bearing partitions perpendicular to joists shall not be offset from supporting girders, walls or partitions more than the joist depth unless such joists are of sufficient size to carry the additional load.

2308.8.5 Lateral support. Floor, attic and roof framing with a nominal depth-to-thickness ratio greater than or equal to 5:1 shall have one edge held in line for the entire span. Where the nominal depth-to-thickness ratio of the framing member exceeds 6:1, there shall be one line of bridging for each 8 feet (2438 mm) of span, unless both edges of the member are held in line. The bridging shall consist of not less than 1-inch by 3-inch (25 mm by 76 mm) lumber, double nailed at each end, of equivalent metal bracing of equal rigidity, full-depth solid blocking or other approved means. A line of bridging shall also be required at supports where equivalent lateral support is not otherwise provided.

2308.8.6 Structural floor sheathing. Structural floor sheathing shall comply with the provisions of Section 2304.7.1.

2308.8.7 Under-floor ventilation. For under-floor ventilation, see Section 1203.3.

TABLE 2308.8(1)
FLOOR JOIST SPANS FOR COMMON LUMBER SPECIES
(Residential Sleeping Areas, Live Load = 30 psf, L/Δ = 360)

JOIST SPACING (inches)	SPECIES AND GRADE		DEAD LOAD = 10 psf				DEAD LOAD = 20 psf			
			2x6	2x8	2x10	2x12	2x6	2x8	2x10	2x12
			Maximum floor joist spans							
			(ft. - in.)	(ft. - in.)	(ft. - in.)	(ft. - in.)	(ft. - in.)	(ft. - in.)	(ft. - in.)	(ft. - in.)
12	Douglas Fir-Larch	SS	12-6	16-6	21-0	25-7	12-6	16-6	21-0	25-7
	Douglas Fir-Larch	#1	12-0	15-10	20-3	24-8	12-0	15-7	19-0	22-0
	Douglas Fir-Larch	#2	11-10	15-7	19-10	23-0	11-6	14-7	17-9	20-7
	Douglas Fir-Larch	#3	9-8	12-4	15-0	17-5	8-8	11-0	13-5	15-7
	Hem-Fir	SS	11-10	15-7	19-10	24-2	11-10	15-7	19-10	24-2
	Hem-Fir	#1	11-7	15-3	19-5	23-7	11-7	15-2	18-6	21-6
	Hem-Fir	#2	11-0	14-6	18-6	22-6	11-0	14-4	17-6	20-4
	Hem-Fir	#3	9-8	12-4	15-0	17-5	8-8	11-0	13-5	15-7
	Southern Pine	SS	12-3	16-2	20-8	25-1	12-3	16-2	20-8	25-1
	Southern Pine	#1	12-0	15-10	20-3	24-8	12-0	15-10	20-3	24-8
	Southern Pine	#2	11-10	15-7	19-10	24-2	11-10	15-7	18-7	21-9
	Southern Pine	#3	10-5	13-3	15-8	18-8	9-4	11-11	14-0	16-8
	Spruce-Pine-Fir	SS	11-7	15-3	19-5	23-7	11-7	15-3	19-5	23-7
	Spruce-Pine-Fir	#1	11-3	14-11	19-0	23-0	11-3	14-7	17-9	20-7
	Spruce-Pine-Fir	#2	11-3	14-11	19-0	23-0	11-3	14-7	17-9	20-7
	Spruce-Pine-Fir	#3	9-8	12-4	15-0	17-5	8-8	11-0	13-5	15-7
16	Douglas Fir-Larch	SS	11-4	15-0	19-1	23-3	11-4	15-0	19-1	23-0
	Douglas Fir-Larch	#1	10-11	14-5	18-5	21-4	10-8	13-6	16-5	19-1
	Douglas Fir-Larch	#2	10-9	14-1	17-2	19-11	9-11	12-7	15-5	17-10
	Douglas Fir-Larch	#3	8-5	10-8	13-0	15-1	7-6	9-6	11-8	13-6
	Hem-Fir	SS	10-9	14-2	18-0	21-11	10-9	14-2	18-0	21-11
	Hem-Fir	#1	10-6	13-10	17-8	20-9	10-4	13-1	16-0	18-7
	Hem-Fir	#2	10-0	13-2	16-10	19-8	9-10	12-5	15-2	17-7
	Hem-Fir	#3	8-5	10-8	13-0	15-1	7-6	9-6	11-8	13-6
	Southern Pine	SS	11-2	14-8	18-9	22-10	11-2	14-8	18-9	22-10
	Southern Pine	#1	10-11	14-5	18-5	22-5	10-11	14-5	17-11	21-4
	Southern Pine	#2	10-9	14-2	18-0	21-1	10-5	13-6	16-1	18-10
	Southern Pine	#3	9-0	11-6	13-7	16-2	8-1	10-3	12-2	14-6
	Spruce-Pine-Fir	SS	10-6	13-10	17-8	21-6	10-6	13-10	17-8	21-4
	Spruce-Pine-Fir	#1	10-3	13-6	17-2	19-11	9-11	12-7	15-5	17-10
	Spruce-Pine-Fir	#2	10-3	13-6	17-2	19-11	9-11	12-7	15-5	17-10
	Spruce-Pine-Fir	#3	8-5	10-8	13-0	15-1	7-6	9-6	11-8	13-6

(continued)

TABLE 2308.8(1)—continued
FLOOR JOIST SPANS FOR COMMON LUMBER SPECIES
(Residential Sleeping Areas, Live Load = 30 psf, L/Δ = 360)

JOIST SPACING (inches)	SPECIES AND GRADE		DEAD LOAD = 10 psf				DEAD LOAD = 20 psf			
			2x6	2x8	2x10	2x12	2x6	2x8	2x10	2x12
			\[Maximum floor joist spans\]							
			(ft. - in.)	(ft. - in.)	(ft. - in.)	(ft. - in.)	(ft. - in.)	(ft. - in.)	(ft. - in.)	(ft. - in.)
19.2	Douglas Fir-Larch	SS	10-8	14-1	18-0	21-10	10-8	14-1	18-0	21-0
	Douglas Fir-Larch	#1	10-4	13-7	16-9	19-6	9-8	12-4	15-0	17-5
	Douglas Fir-Larch	#2	10-1	12-10	15-8	18-3	9-1	11-6	14-1	16-3
	Douglas Fir-Larch	#3	7-8	9-9	11-10	13-9	6-10	8-8	10-7	12-4
	Hem-Fir	SS	10-1	13-4	17-0	20-8	10-1	13-4	17-0	20-7
	Hem-Fir	#1	9-10	13-0	16-4	19-0	9-6	12-0	14-8	17-0
	Hem-Fir	#2	9-5	12-5	15-6	17-1	8-11	11-4	13-10	16-1
	Hem-Fir	#3	7-8	9-9	11-10	13-9	6-10	8-8	10-7	12-4
	Southern Pine	SS	10-6	13-10	17-8	21-6	10-6	13-10	17-8	21-6
	Southern Pine	#1	10-4	13-7	17-4	21-1	10-4	13-7	16-4	19-6
	Southern Pine	#2	10-1	13-4	15-5	19-3	9-6	12-4	14-8	17-2
	Southern Pine	#3	8-3	10-6	12-5	14-9	7-4	9-5	11-1	13-2
	Spruce-Pine-Fir	SS	9-10	13-0	16-7	20-2	9-10	13-0	16-7	19-6
	Spruce-Pine-Fir	#1	9-8	12-9	15-8	18-3	9-1	11-6	14-1	16-3
	Spruce-Pine-Fir	#2	9-8	12-9	15-8	18-3	9-1	11-6	14-1	16-3
	Spruce-Pine-Fir	#3	7-8	9-9	11-10	13-9	6-10	8-8	10-7	12-4
24	Douglas Fir-Larch	SS	9-11	13-1	16-8	20-3	9-11	13-1	16-2	18-9
	Douglas Fir-Larch	#1	9-7	12-4	15-0	17-5	8-8	11-0	13-5	15-7
	Douglas Fir-Larch	#2	9-1	11-6	14-1	16-3	8-1	10-3	12-7	14-7
	Douglas Fir-Larch	#3	6-10	8-8	10-7	12-4	6-2	7-9	9-6	11-0
	Hem-Fir	SS	9-4	12-4	15-9	19-2	9-4	12-4	15-9	18-5
	Hem-Fir	#1	9-2	12-0	14-8	17-0	8-6	10-9	13-1	15-2
	Hem-Fir	#2	8-9	11-4	13-10	16-1	8-0	10-2	12-5	14-4
	Hem-Fir	#3	6-10	8-8	10-7	12-4	6-2	7-9	9-6	11-0
	Southern Pine	SS	9-9	12-10	16-5	19-11	9-9	12-10	16-5	19-11
	Southern Pine	#1	9-7	12-7	16-1	19-6	9-7	12-4	14-7	17-5
	Southern Pine	#2	9-4	12-4	14-8	17-2	8-6	11-0	13-1	15-5
	Southern Pine	#3	7-4	9-5	11-1	13-2	6-7	8-5	9-11	11-10
	Spruce-Pine-Fir	SS	9-2	12-1	15-5	18-9	9-2	12-1	15-0	17-5
	Spruce-Pine-Fir	#1	8-11	11-6	14-1	16-3	8-1	10-3	12-7	14-7
	Spruce-Pine-Fir	#2	8-11	11-6	14-1	16-3	8-1	10-3	12-7	14-7
	Spruce-Pine-Fir	#3	6-10	8-8	10-7	12-4	6-2	7-9	9-6	11-0

Check sources for availability of lumber in lengths greater than 20 feet.

For SI: 1 inch = 25.4 mm, 1 foot = 304.8 mm, 1 pound per square foot = 47.8 N/m².

TABLE 2308.8(2)
FLOOR JOIST SPANS FOR COMMON LUMBER SPECIES
(Residential Living Areas, Live Load = 40 psf, L/Δ = 360)

JOIST SPACING (inches)	SPECIES AND GRADE		DEAD LOAD = 10 psf				DEAD LOAD = 20 psf			
			2x6	2x8	2x10	2x12	2x6	2x8	2x10	2x12
			Maximum floor joist spans							
			(ft. - in.)	(ft. - in.)	(ft. - in.)	(ft. - in.)	(ft. - in.)	(ft. - in.)	(ft. - in.)	(ft. - in.)
12	Douglas Fir-Larch	SS	11-4	15-0	19-1	23-3	11-4	15-0	19-1	23-3
	Douglas Fir-Larch	#1	10-11	14-5	18-5	22-0	10-11	14-2	17-4	20-1
	Douglas Fir-Larch	#2	10-9	14-2	17-9	20-7	10-6	13-3	16-3	18-10
	Douglas Fir-Larch	#3	8-8	11-0	13-5	15-7	7-11	10-0	12-3	14-3
	Hem-Fir	SS	10-9	14-2	18-0	21-11	10-9	14-2	18-0	21-11
	Hem-Fir	#1	10-6	13-10	17-8	21-6	10-6	13-10	16-11	19-7
	Hem-Fir	#2	10-0	13-2	16-10	20-4	10-0	13-1	16-0	18-6
	Hem-Fir	#3	8-8	11-0	13-5	15-7	7-11	10-0	12-3	14-3
	Southern Pine	SS	11-2	14-8	18-9	22-10	11-2	14-8	18-9	22-10
	Southern Pine	#1	10-11	14-5	18-5	22-5	10-11	14-5	18-5	22-5
	Southern Pine	#2	10-9	14-2	18-0	21-9	10-9	14-2	16-11	19-10
	Southern Pine	#3	9-4	11-11	14-0	16-8	8-6	10-10	12-10	15-3
	Spruce-Pine-Fir	SS	10-6	13-10	17-8	21-6	10-6	13-10	17-8	21-6
	Spruce-Pine-Fir	#1	10-3	13-6	17-3	20-7	10-3	13-3	16-3	18-10
	Spruce-Pine-Fir	#2	10-3	13-6	17-3	20-7	10-3	13-3	16-3	18-10
	Spruce-Pine-Fir	#3	8-8	11-0	13-5	15-7	7-11	10-0	12-3	14-3
16	Douglas Fir-Larch	SS	10-4	13-7	17-4	21-1	10-4	13-7	17-4	21-0
	Douglas Fir-Larch	#1	9-11	13-1	16-5	19-1	9-8	12-4	15-0	17-5
	Douglas Fir-Larch	#2	9-9	12-7	15-5	17-10	9-1	11-6	14-1	16-3
	Douglas Fir-Larch	#3	7-6	9-6	11-8	13-6	6-10	8-8	10-7	12-4
	Hem-Fir	SS	9-9	12-10	16-5	19-11	9-9	12-10	16-5	19-11
	Hem-Fir	#1	9-6	12-7	16-0	18-7	9-6	12-0	14-8	17-0
	Hem-Fir	#2	9-1	12-0	15-2	17-7	8-11	11-4	13-10	16-1
	Hem-Fir	#3	7-6	9-6	11-8	13-6	6-10	8-8	10-7	12-4
	Southern Pine	SS	10-2	13-4	17-0	20-9	10-2	13-4	17-0	20-9
	Southern Pine	#1	9-11	13-1	16-9	20-4	9-11	13-1	16-4	19-6
	Southern Pine	#2	9-9	12-10	16-1	18-10	9-6	12-4	14-8	17-2
	Southern Pine	#3	8-1	10-3	12-2	14-6	7-4	9-5	11-1	13-2
	Spruce-Pine-Fir	SS	9-6	12-7	16-0	19-6	9-6	12-7	16-0	19-6
	Spruce-Pine-Fir	#1	9-4	12-3	15-5	17-10	9-1	11-6	14-1	16-3
	Spruce-Pine-Fir	#2	9-4	12-3	15-5	17-10	9-1	11-6	14-1	16-3
	Spruce-Pine-Fir	#3	7-6	9-6	11-8	13-6	6-10	8-8	10-7	12-4

(continued)

TABLE 2308.8(2)—continued
FLOOR JOIST SPANS FOR COMMON LUMBER SPECIES
(Residential Living Areas, Live Load = 40 psf, L/Δ = 360)

JOIST SPACING (inches)	SPECIES AND GRADE		DEAD LOAD = 10 psf				DEAD LOAD = 20 psf			
			2x6	2x8	2x10	2x12	2x6	2x8	2x10	2x12
			Maximum floor joist spans							
			(ft. - in.)	(ft. - in.)	(ft. - in.)	(ft. - in.)	(ft. - in.)	(ft. - in.)	(ft. - in.)	(ft. - in.)
19.2	Douglas Fir-Larch	SS	9-8	12-10	16-4	19-10	9-8	12-10	16-4	19-2
	Douglas Fir-Larch	#1	9-4	12-4	15-0	17-5	8-10	11-3	13-8	15-11
	Douglas Fir-Larch	#2	9-1	11-6	14-1	16-3	8-3	10-6	12-10	14-10
	Douglas Fir-Larch	#3	6-10	8-8	10-7	12-4	6-3	7-11	9-8	11-3
	Hem-Fir	SS	9-2	12-1	15-5	18-9	9-2	12-1	15-5	18-9
	Hem-Fir	#1	9-0	11-10	14-8	17-0	8-8	10-11	13-4	15-6
	Hem-Fir	#2	8-7	11-3	13-10	16-1	8-2	10-4	12-8	14-8
	Hem-Fir	#3	6-10	8-8	10-7	12-4	6-3	7-11	9-8	11-3
	Southern Pine	SS	9-6	12-7	16-0	19-6	9-6	12-7	16-0	19-6
	Southern Pine	#1	9-4	12-4	15-9	19-2	9-4	12-4	14-11	17-9
	Southern Pine	#2	9-2	12-1	14-8	17-2	8-8	11-3	13-5	15-8
	Southern Pine	#3	7-4	9-5	11-1	13-2	6-9	8-7	10-1	12-1
	Spruce-Pine-Fir	SS	9-0	11-10	15-1	18-4	9-0	11-10	15-1	17-9
	Spruce-Pine-Fir	#1	8-9	11-6	14-1	16-3	8-3	10-6	12-10	14-10
	Spruce-Pine-Fir	#2	8-9	11-6	14-1	16-3	8-3	10-6	12-10	14-10
	Spruce-Pine-Fir	#3	6-10	8-8	10-7	12-4	6-3	7-11	9-8	11-3
24	Douglas Fir-Larch	SS	9-0	11-11	15-2	18-5	9-0	11-11	14-9	17-1
	Douglas Fir-Larch	#1	8-8	11-0	13-5	15-7	7-11	10-0	12-3	14-3
	Douglas Fir-Larch	#2	8-1	10-3	12-7	14-7	7-5	9-5	11-6	13-4
	Douglas Fir-Larch	#3	6-2	7-9	9-6	11-0	5-7	7-1	8-8	10-1
	Hem-Fir	SS	8-6	11-3	14-4	17-5	8-6	11-3	14-4	16-10[a]
	Hem-Fir	#1	8-4	10-9	13-1	15-2	7-9	9-9	11-11	13-10
	Hem-Fir	#2	7-11	10-2	12-5	14-4	7-4	9-3	11-4	13-1
	Hem-Fir	#3	6-2	7-9	9-6	11-0	5-7	7-1	8-8	10-1
	Southern Pine	SS	8-10	11-8	14-11	18-1	8-10	11-8	14-11	18-1
	Southern Pine	#1	8-8	11-5	14-7	17-5	8-8	11-3	13-4	15-11
	Southern Pine	#2	8-6	11-0	13-1	15-5	7-9	10-0	12-0	14-0
	Southern Pine	#3	6-7	8-5	9-11	11-10	6-0	7-8	9-1	10-9
	Spruce-Pine-Fir	SS	8-4	11-0	14-0	17-0	8-4	11-0	13-8	15-11
	Spruce-Pine-Fir	#1	8-1	10-3	12-7	14-7	7-5	9-5	11-6	13-4
	Spruce-Pine-Fir	#2	8-1	10-3	12-7	14-7	7-5	9-5	11-6	13-4
	Spruce-Pine-Fir	#3	6-2	7-9	9-6	11-0	5-7	7-1	8-8	10-1

Check sources for availability of lumber in lengths greater than 20 feet.

For SI: 1 inch = 25.4 mm, 1 foot = 304.8 mm, 1 pound per square foot = 47.8 N/m².

a. End bearing length shall be increased to 2 inches.

2308.9 Wall framing.

2308.9.1 Size, height and spacing. The size, height and spacing of studs shall be in accordance with Table 2308.9.1 except that utility-grade studs shall not be spaced more than 16 inches (406 mm) o.c., or support more than a roof and ceiling, or exceed 8 feet (2438 mm) in height for exterior walls and load-bearing walls or 10 feet (3048 mm) for interior nonload-bearing walls.

2308.9.2 Framing details. Studs shall be placed with their wide dimension perpendicular to the wall. Not less than three studs shall be installed at each corner of an exterior wall.

Exception: At corners, two studs are permitted, provided wood spacers or backup cleats of $^3/_8$-inch-thick (9.5 mm) wood structural panel, $^3/_8$-inch (9.5 mm) Type M "Exterior Glue" particleboard, 1-inch-thick (25 mm) lumber or other approved devices that will serve as an adequate backing for the attachment of facing materials are used. Where fire-resistance ratings or shear values are involved, wood spacers, backup cleats or other devices shall not be used unless specifically approved for such use.

2308.9.2.1 Top plates. Bearing and exterior wall studs shall be capped with double top plates installed to provide overlapping at corners and at intersections with other partitions. End joints in double top plates shall be offset at least 48 inches (1219 mm), and shall be nailed with not less than eight 16d face nails on each side of the joint. Plates shall be a nominal 2 inches (51 mm) in depth and have a width at least equal to the width of the studs.

Exception: A single top plate is permitted, provided the plate is adequately tied at joints, corners and intersecting walls by at least the equivalent of 3-inch by 6-inch (76 mm by 152 mm) by 0.036-inch-thick (0.914 mm) galvanized steel that is nailed to each wall or segment of wall by six 8d nails or equivalent, provided the rafters, joists or trusses are centered over the studs with a tolerance of no more than 1 inch (25 mm).

2308.9.2.2 Top plates for studs spaced at 24 inches (610 mm). Where bearing studs are spaced at 24-inch (610 mm) intervals and top plates are less than two 2-inch by 6-inch (51 mm by 152 mm) or two 3-inch by 4-inch (76 mm by 102 mm) members and where the floor joists, floor trusses or roof trusses that they support are spaced at more than 16-inch (406 mm) intervals, such joists or trusses shall bear within 5 inches (127 mm) of the studs beneath or a third plate shall be installed.

2308.9.2.3 Nonbearing walls and partitions. In nonbearing walls and partitions, studs shall be spaced not more than 28 inches (711 mm) o.c. and are permitted to be set with the long dimension parallel to the wall. Interior nonbearing partitions shall be capped with no less than a single top plate installed to provide overlapping at corners and at intersections with other walls and partitions. The plate shall be continuously tied at joints by solid blocking at least 16 inches (406 mm) in length and equal in size to the plate or by $^1/_2$-inch by $1^1/_2$-inch (12.7 mm by 38 mm) metal ties with spliced sections fastened with two 16d nails on each side of the joint.

2308.9.2.4 Plates or sills. Studs shall have full bearing on a plate or sill not less than 2 inches (51 mm) in thickness having a width not less than that of the wall studs.

2308.9.3 Bracing. Braced wall lines shall consist of braced wall panels that meet the requirements for location, type and amount of bracing as shown in Figure 2308.9.3, specified in Table 2308.9.3(1), and are in line or offset from each other by not more than 4 feet (1219 mm). Braced wall panels shall start not more than 8 feet (2438 mm) from each end of a braced wall line. A designed collector shall be provided if the bracing begins more than 12.5 feet (3810 mm) from an end of a braced wall line. Braced wall panels shall be clearly indicated on the plans. Construction of braced wall panels shall be by one of the following methods:

1. Nominal 1-inch by 4-inch (25 mm by 102 mm) continuous diagonal braces let into top and bottom plates and intervening studs, placed at an angle not more than 60 degrees (1.0 rad) or less than 45 degrees (0.79 rad) from the horizontal and attached to the framing in conformance with Table 2304.9.1.

TABLE 2308.9.1
SIZE, HEIGHT AND SPACING OF WOOD STUDS

STUD SIZE (inches)	BEARING WALLS				NONBEARING WALLS	
	Laterally unsupported stud height[a] (feet)	Supporting roof and ceiling only	Supporting one floor, roof and ceiling	Supporting two floors, roof and ceiling	Laterally unsupported stud height[a] (feet)	Spacing (inches)
	Spacing (inches)					
2 × 3[b]	—	—	—	—	10	16
2 × 4	10	24	16	—	14	24
3 × 4	10	24	24	16	14	24
2 × 5	10	24	24	—	16	24
2 × 6	10	24	24	16	20	24

For SI: 1 inch = 25.4 mm, 1 foot = 304.8 mm.

a. Listed heights are distances between points of lateral support placed perpendicular to the plane of the wall. Increases in unsupported height are permitted where justified by an analysis.

b. Shall not be used in exterior walls.

2. Wood boards of $^5/_8$-inch (15.9 mm) net minimum thickness applied diagonally on studs spaced not over 24 inches (610 mm) o.c.

3. Wood structural panel sheathing with a thickness not less than $^5/_{16}$ inch (7.9 mm) for a 16-inch (406 mm) stud spacing and not less than $^3/_8$ inch (9.5 mm) for a 24-inch (610 mm) stud spacing in accordance with Tables 2308.9.3(2) and 2308.9.3(3).

4. Fiberboard sheathing panels not less than $^1/_2$ inch (12.7 mm) thick applied vertically or horizontally on studs spaced not over 16 inches (406 mm) o.c. where installed with fasteners in accordance with Section 2306.4.4 and Table 2308.9.3(4).

5. Gypsum board [sheathing $^1/_2$ inch (12.7 mm) thick by 4 feet (1219 mm) wide wallboard or veneer base] on studs spaced not over 24 inches (610 mm) o.c. and nailed at 7 inches (178 mm) o.c. with nails as required by Table 2306.4.5.

6. Particleboard wall sheathing panels where installed in accordance with Table 2308.9.3(5).

7. Portland cement plaster on studs spaced 16 inches (406 mm) o.c. installed in accordance with Section 2510.

8. Hardboard panel siding where installed in accordance with Section 2303.1.6 and Table 2308.9.3(6).

For cripple wall bracing, see Section 2308.9.4.1. For Methods 2, 3, 4, 6, 7 and 8, each panel must be at least 48 inches (1219 mm) in length, covering three stud spaces where studs are spaced 16 inches (406 mm) apart and covering two stud spaces where studs are spaced 24 inches (610 mm) apart.

For Method 5, each panel must be at least 96 inches (2438 mm) in length where applied to one face of a panel and 48 inches (1219 mm) where applied to both faces.

All vertical joints of panel sheathing shall occur over studs and adjacent panel joints shall be nailed to common framing members. Horizontal joints shall occur over blocking or other framing equal in size to the studding except where waived by the installation requirements for the specific sheathing materials.

Sole plates shall be nailed to the floor framing and top plates shall be connected to the framing above in accordance with Section 2308.3.2. Where joists are perpendicular to braced wall lines above, blocking shall be provided under and in line with the braced wall panels.

2308.9.3.1 Alternative bracing. Any bracing required by Section 2308.9.3 is permitted to be replaced by the following:

1. In one-story buildings, each panel shall have a length of not less than 2 feet 8 inches (813 mm) and a height of not more than 10 feet (3048 mm). Each panel shall be sheathed on one face with $^3/_8$-inch-minimum-thickness (9.5 mm) wood structural panel sheathing nailed with 8d common or galvanized box nails in accordance with Table 2304.9.1 and blocked at wood structural panel edges. Two anchor bolts installed in accordance with Section 2308.6 shall be provided in each panel. Anchor bolts shall be placed at each panel outside quarter points. Each panel end stud shall have a tie-down device fastened to the foundation, capable of providing an approved uplift capacity of not less than 1,800 pounds (8006 N). The tie-down device shall be installed in accordance with the manufacturer's recommendations. The panels shall be supported directly on a foundation or on floor framing supported directly on a foundation that is continuous across the entire length of the braced wall line. This foundation shall be reinforced with not less than one No. 4 bar top and bottom.

Where the continuous foundation is required to have a depth greater than 12 inches (305 mm), a minimum 12-inch by 12-inch (305 mm by 305 mm) continuous footing or turned down slab edge is permitted at door openings in the braced wall line. This continuous footing or turned down slab edge shall be reinforced with not less than one No. 4 bar top and bottom. This reinforcement shall be lapped 15 inches (381 mm) with the reinforcement required in the continuous foundation located directly under the braced wall line.

2. In the first story of two-story buildings, each wall panel shall be braced in accordance with Section 2308.9.3.1, Item 1, except that the wood structural panel sheathing shall be provided on both faces, three anchor bolts shall be placed at one-quarter points, and tie-down device uplift capacity shall not be less than 3,000 pounds (13 344 N).

2308.9.4 Cripple walls. Foundation cripple walls shall be framed of studs not less in size than the studding above with a minimum length of 14 inches (356 mm), or shall be framed of solid blocking. Where exceeding 4 feet (1219 mm) in height, such walls shall be framed of studs having the size required for an additional story.

2308.9.4.1 Bracing. For the purposes of this section, cripple walls having a stud height exceeding 14 inches (356 mm) shall be considered a story and shall be braced in accordance with Table 2308.9.3(1) for Seismic Design Category A, B or C. See Section 2308.12.4 for Seismic Design Category D or E.

2308.9.4.2 Nailing of bracing. Spacing of edge nailing for required wall bracing shall not exceed 6 inches (152 mm) o.c. along the foundation plate and the top plate of the cripple wall. Nail size, nail spacing for field nailing and more restrictive boundary nailing requirements shall be as required elsewhere in the code for the specific bracing material used.

2308.9.5 Openings in exterior walls.

SEISMIC DESIGN CATEGORY	MAXIMUM WALL SPACING (feet)	REQUIRED BRACING LENGTH, b
A, B, and C	35'-0"	Table 2308.9.3(1) and Section 2308.9.3
D and E	25'-0"	Table 2308.12.4

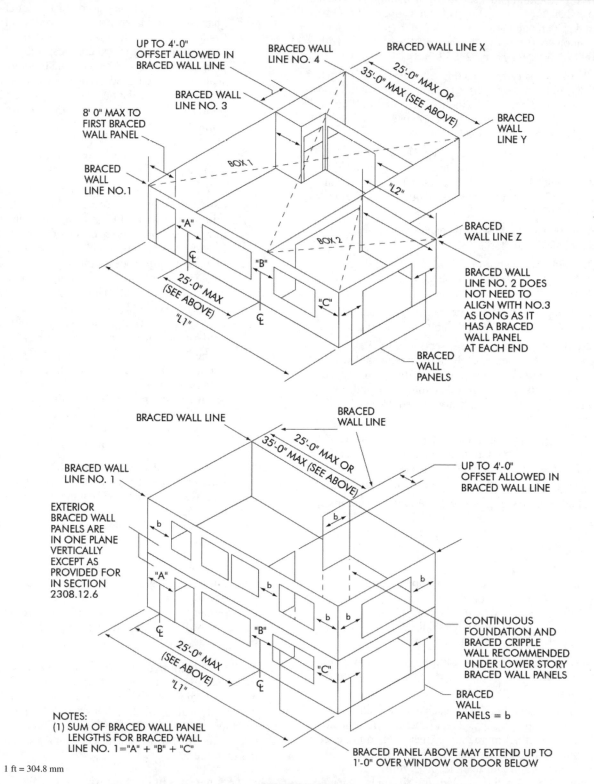

For SI: 1 ft = 304.8 mm

FIGURE 2308.9.3
BASIC COMPONENTS OF THE LATERAL BRACING SYSTEM

TABLE 2308.9.3(1)
BRACED WALL PANELS[a]

SEISMIC DESIGN CATEGORY	CONDITION	CONSTRUCTION METHODS[b,c]								BRACED PANEL LOCATION AND LENGTH[d]
		1	2	3	4	5	6	7	8	
A and B	One story, top of two or three story	X	X	X	X	X	X	X	X	Each end and not more than 25 feet on center
	First story of two story or second story of three story	X	X	X	X	X	X	X	X	
	First story of three story	—	X	X	X	X[e]	X	X	X	
C	One story, top of two or three story	—	X	X	X	X	X	X	X	Each end and not more than 25 feet on center
	First story of two story or second story of three story	—	X	X	X	X[e]	X	X	X	Each end and not more than 25 feet on center but total length shall not be less than 25% of building length[f]
	First story of three story	—	X	X	X	X[e]	X	X	X	Each end and not more than 25 feet on center but total length shall not be less than 40% of building length[f]

For SI: 1 inch = 25.4 mm, 1 foot = 304.8 mm.

a. This table specifies minimum requirements for braced panels that form interior or exterior braced wall lines.
b. See Section 2308.9.3 for full description.
c. See Section 2308.9.3.1 for alternative braced panel requirement.
d. Building length is the dimension parallel to the braced wall length.
e. Gypsum wallboard applied to framing supports that are spaced at 16 inches on center.
f. The required lengths shall be doubled for gypsum board applied to only one face of a braced wall panel.

TABLE 2308.9.3(2)
EXPOSED PLYWOOD PANEL SIDING

MINIMUM THICKNESS[a] (inch)	MINIMUM NUMBER OF PLIES	STUD SPACING (inches) Plywood siding applied directly to studs or over sheathing
$^3/_8$	3	16[b]
$^1/_2$	4	24

For SI: 1 inch = 25.4 mm.

a. Thickness of grooved panels is measured at bottom of grooves.
b. Spans are permitted to be 24 inches if plywood siding applied with face grain perpendicular to studs or over one of the following: (1) 1-inch board sheathing, (2) $^7/_{16}$-inch wood structural panel sheathing or (3) $^3/_8$-inch wood structural panel sheathing with strength axis (which is the long direction of the panel unless otherwise marked) of sheathing perpendicular to studs.

TABLE 2308.9.3(3)
WOOD STRUCTURAL PANEL WALL SHEATHING[b]
(Not Exposed to the Weather, Strength Axis Parallel or Perpendicular to Studs Except as Indicated Below)

MINIMUM THICKNESS (inch)	PANEL SPAN RATING	STUD SPACING (inches)		
			Nailable sheathing	
		Siding nailed to studs	Sheathing parallel to studs	Sheathing perpendicular to studs
$^5/_{16}$	12/0, 16/0, 20/0 Wall–16″ o.c.	16	—	16
$^3/_8$, $^{15}/_{32}$, $^1/_2$	16/0, 20/0, 24/0, 32/16 Wall–24″ o.c.	24	16	24
$^7/_{16}$, $^{15}/_{32}$, $^1/_2$	24/0, 24/16, 32/16 Wall–24″ o.c.	24	24[a]	24

For SI: 1 inch = 25.4 mm.

a. Plywood shall consist of four or more plies.

b. Blocking of horizontal joints shall not be required except as specified in Sections 2306.4 and 2308.12.4.

TABLE 2308.9.3(4)
ALLOWABLE SHEAR VALUES (plf) FOR WIND OR SEISMIC LOADING ON
VERTICAL DIAPHRAGMS OF FIBERBOARD SHEATHING BOARD CONSTRUCTION
FOR TYPE V CONSTRUCTION ONLY[a, b, c, d, e, f, g, h]

THICKNESS AND GRADE	FASTENER SIZE	SHEAR VALUE (pounds per linear foot) 3-INCH NAIL SPACING AROUND PERIMETER AND 6-INCH AT INTERMEDIATE POINTS
$^1/_2$″ Structural	No. 11 gage galvanized roofing nail $1^1/_2$″ long, $^7/_{16}$″ head	125[g]
$^{25}/_{32}$″ Structural	No. 11 gage galvanized roofing nail $1^3/_4$″ long, $^7/_{16}$″ head	175[g]

For SI: 1 inch = 25.4 mm, 1 pound per foot = 14.5939 N/m.

a. Fiberboard sheathing diaphragms shall not be used to brace concrete or masonry walls.

b. Panel edges shall be backed with 2 inch or wider framing of Douglas fir-larch or Southern pine.

c. Fiberboard sheathing on one side only.

d. Fiberboard panels are installed with their long dimension parallel or perpendicular to studs.

e. Fasteners shall be spaced 6 inches on center along intermediate framing members.

f. For framing of other species: (1) Find specific gravity for species of lumber in AF&PA National Design Specification, and (2) Multiply the shear value from the above table by 0.82 for species with specific gravity of 0.42 or greater, or 0.65 for all other species.

g. The same values can be applied when staples are used as described in Table 2304.9.1.

h. Values are not permitted in Seismic Design Category D, E or F.

TABLE 2308.9.3(5)
ALLOWABLE SPANS FOR PARTICLEBOARD WALL SHEATHING
(Not Exposed to the Weather, Long Dimension of the Panel Parallel or Perpendicular to Studs)

GRADE	THICKNESS (inch)	STUD SPACING (inches)	
		Siding nailed to studs	Sheathing under coverings specified in Section 2308.9.3 parallel or perpendicular to studs
M-S "Exterior Glue" and M-2 "Exterior Glue"	$^3/_8$	16	—
	$^1/_2$	16	16

For SI: 1 inch = 25.4 mm.

TABLE 2308.9.3(6)
HARDBOARD SIDING

SIDING	MINIMUM NOMINAL THICKNESS (inch)	2 × 4 FRAMING MAXIMUM SPACING	NAIL SIZE[a,b,d]	NAIL SPACING	
				General	Bracing panels[c]
1. Lap siding					
Direct to studs	$^3/_8$	16″ o.c.	8d	16″ o.c.	Not applicable
Over sheathing	$^3/_8$	16″ o.c.	10d	16″ o.c.	Not applicable
2. Square edge panel siding					
Direct to studs	$^3/_8$	24″ o.c.	6d	6″ o.c. edges; 12″ o.c. at intermediate supports	4″ o.c. edges; 8″ o.c. at intermediate supports
Over sheathing	$^3/_8$	24″ o.c.	8d	6″ o.c. edges; 12″ o.c. at intermediate supports	4″ o.c. edges; 8″ o.c. at intermediate supports
3. Shiplap edge panel siding					
Direct to studs	$^3/_8$	16″ o.c.	6d	6″ o.c. edges; 12″ o.c. at intermediate supports	4″ o.c. edges; 8″ o.c. at intermediate supports
Over sheathing	$^3/_8$	16″ o.c.	8d	6″ o.c. edges; 12″ o.c. At intermediate supports	4″ o.c. edges; 8″ o.c. at intermediate supports

For SI: 1 inch = 25.4 mm.

a. Nails shall be corrosion resistant.

b. Minimum acceptable nail dimensions:

	Panel Siding (inch)	Lap Siding (inch)
Shank diameter	0.092	0.099
Head diameter	0.225	0.240

c. Where used to comply with Section 2308.9.3.

d. Nail length must accommodate the sheathing and penetrate framing $1^1/_2$ inches.

2308.9.5.1 Headers. Headers shall be provided over each opening in exterior-bearing walls. The spans in Table 2308.9.5 are permitted to be used for one- and two-family dwellings. Headers for other buildings shall be designed in accordance with Section 2301.2.1 or 2301.2.2. Headers shall be of two pieces of nominal 2-inch (51 mm) framing lumber set on edge as permitted by Table 2308.9.5 and nailed together in accordance with Table 2304.9.1 or of solid lumber of equivalent size.

2308.9.5.2 Header support. Wall studs shall support the ends of the header in accordance with Table 2308.9.5. Each end of a lintel or header shall have a length of bearing of not less than $1^1/_2$ inches (38 mm) for the full width of the lintel.

2308.9.6 Openings in interior bearing partitions. Headers shall be provided over each opening in interior bearing partitions as required in Section 2308.9.5. The spans in Table 2308.9.6 are permitted to be used for one- and two-family dwellings. Wall studs shall support the ends of the header in accordance with Table 2308.9.5 or 2308.9.6 as appropriate.

2308.9.7 Openings in interior nonbearing partitions. Openings in nonbearing partitions are permitted to be framed with single studs and headers. Each end of a lintel or header shall have a length of bearing of not less than $1^1/_2$ inches (38 mm) for the full width of the lintel.

2308.9.8 Pipes in walls. Stud partitions containing plumbing, heating or other pipes shall be so framed and the joists underneath so spaced as to give proper clearance for the piping. Where a partition containing such piping runs parallel to the floor joists, the joists underneath such partitions shall be doubled and spaced to permit the passage of such pipes and shall be bridged. Where plumbing, heating or other pipes are placed in or partly in a partition, necessitating the cutting of the soles or plates, a metal tie not less than 0.058 inch (1.47 mm) (16 galvanized gage) and $1^1/_2$ inches (38 mm) wide shall be fastened to each plate across and to each side of the opening with not less than six 16d nails.

2308.9.9 Bridging. Unless covered by interior or exterior wall coverings or sheathing meeting the minimum requirements of this code, stud partitions or walls with studs having a height-to-least-thickness ratio exceeding 50 shall have

TABLE 2308.9.5
HEADER AND GIRDER SPANS[a] FOR EXTERIOR BEARING WALLS
(Maximum Spans for Douglas Fir-Larch, Hem-Fir, Southern Pine and Spruce-Pine-Fir[b] and Required Number of Jack Studs)

HEADERS SUPPORTING	SIZE	GROUND SNOW LOAD (psf)[e]											
		30						50					
		Building width[c] (feet)											
		20		28		36		20		28		36	
		Span	NJ[d]	Span	NJ[d]	Span	NJ[d]	Span	NJ[d]	Span	NJ[d]	Span	NJ[d]
Roof & Ceiling	2-2 × 4	3-6	1	3-2	1	2-10	1	3-2	1	2-9	1	2-6	1
	2-2 × 6	5-5	1	4-8	1	4-2	1	4-8	1	4-1	1	3-8	2
	2-2 × 8	6-10	1	5-11	2	5-4	2	5-11	2	5-2	2	4-7	2
	2-2 × 10	8-5	2	7-3	2	6-6	2	7-3	2	6-3	2	5-7	2
	2-2 × 12	9-9	2	8-5	2	7-6	2	8-5	2	7-3	2	6-6	2
	3-2 × 8	8-4	1	7-5	1	6-8	1	7-5	1	6-5	2	5-9	2
	3-2 × 10	10-6	1	9-1	2	8-2	2	9-1	2	7-10	2	7-0	2
	3-2 × 12	12-2	2	10-7	2	9-5	2	10-7	2	9-2	2	8-2	2
	4-2 × 8	9-2	1	8-4	1	7-8	1	8-4	1	7-5	1	6-8	1
	4-2 × 10	11-8	1	10-6	1	9-5	2	10-6	1	9-1	2	8-2	2
	4-2 × 12	14-1	1	12-2	2	10-11	2	12-2	2	10-7	2	9-5	2
Roof Ceiling & 1 Center-Bearing Floor	2-2 × 4	3-1	1	2-9	1	2-5	1	2-9	1	2-5	1	2-2	1
	2-2 × 6	4-6	1	4-0	1	3-7	1	4-1	1	3-7	2	3-3	2
	2-2 × 8	5-9	2	5-0	2	4-6	2	5-2	2	4-6	2	4-1	2
	2-2 × 10	7-0	2	6-2	2	5-6	2	6-4	2	5-6	2	5-0	2
	2-2 × 12	8-1	2	7-1	2	6-5	2	7-4	2	6-5	2	5-9	3
	3-2 × 8	7-2	1	6-3	1	5-8	2	6-5	2	5-8	2	5-1	2
	3-2 × 10	8-9	2	7-8	2	6-11	2	7-11	2	6-11	2	6-3	2
	3-2 × 12	10-2	2	8-11	2	8-0	2	9-2	2	8-0	2	7-3	2
	4-2 × 8	8-1	1	7-3	1	6-7	1	7-5	1	6-6	1	5-11	1
	4-2 × 10	10-1	1	8-10	2	8-0	2	9-1	2	8-0	2	7-2	2
	4-2 × 12	11-9	2	10-3	2	9-3	2	10-7	2	9-3	2	8-4	2
Roof Ceiling & 1 Clear Span Floor	2-2 × 4	2-8	1	2-4	1	2-1	1	2-7	1	2-3	1	2-0	1
	2-2 × 6	3-11	1	3-5	2	3-0	2	3-10	2	3-4	2	3-0	2
	2-2 × 8	5-0	2	4-4	2	3-10	2	4-10	2	4-2	2	3-9	2
	2-2 × 10	6-1	2	5-3	2	4-8	2	5-11	2	5-1	2	4-7	3
	2-2 × 12	7-1	2	6-1	3	5-5	3	6-10	2	5-11	3	5-4	3
	3-2 × 8	6-3	2	5-5	2	4-10	2	6-1	2	5-3	2	4-8	2
	3-2 × 10	7-7	2	6-7	2	5-11	2	7-5	2	6-5	2	5-9	2
	3-2 × 12	8-10	2	7-8	2	6-10	2	8-7	2	7-5	2	6-8	2
	4-2 × 8	7-2	1	6-3	2	5-7	2	7-0	1	6-1	2	5-5	2
	4-2 × 10	8-9	2	7-7	2	6-10	2	8-7	2	7-5	2	6-7	2
	4-2 × 12	10-2	2	8-10	2	7-11	2	9-11	2	8-7	2	7-8	2

(continued)

TABLE 2308.9.5—continued
HEADER AND GIRDER SPANS[a] FOR EXTERIOR BEARING WALLS
(Maximum Spans for Douglas Fir-Larch, Hem-Fir, Southern Pine and Spruce-Pine-Fir[b] and Required Number of Jack Studs)

HEADERS SUPPORTING	SIZE	GROUND SNOW LOAD (psf)[e]											
		30						50					
		Building width[c] (feet)											
		20		28		36		20		28		36	
		Span	NJ[d]	Span	NJ[d]	Span	NJ[d]	Span	NJ[d]	Span	NJ[d]	Span	NJ[d]
Roof Ceiling & 2 Center-Bearing Floors	2-2 × 4	2-7	1	2-3	1	2-0	1	2-6	1	2-2	1	1-11	1
	2-2 × 6	3-9	2	3-3	2	2-11	2	3-8	2	3-2	2	2-10	2
	2-2 × 8	4-9	2	4-2	2	3-9	2	4-7	2	4-0	2	3-8	2
	2-2 × 10	5-9	2	5-1	2	4-7	2	5-8	2	4-11	2	4-5	3
	2-2 × 12	6-8	2	5-10	3	5-3	3	6-6	2	5-9	3	5-2	3
	3-2 × 8	5-11	2	5-2	2	4-8	2	5-9	2	5-1	2	4-7	2
	3-2 × 10	7-3	2	6-4	2	5-8	2	7-1	2	6-2	2	5-7	2
	3-2 × 12	8-5	2	7-4	2	6-7	2	8-2	2	7-2	2	6-5	3
	4-2 × 8	6-10	1	6-0	2	5-5	2	6-8	1	5-10	2	5-3	2
	4-2 × 10	8-4	2	7-4	2	6-7	2	8-2	2	7-2	2	6-5	2
	4-2 × 12	9-8	2	8-6	2	7-8	2	9-5	2	8-3	2	7-5	2
Roof, Ceiling & 2 Clear Span Floors	2-2 × 4	2-1	1	1-8	1	1-6	1	2-0	1	1-8	1	1-5	1
	2-2 × 6	3-1	2	2-8	2	2-4	2	3-0	2	2-7	2	2-3	2
	2-2 × 8	3-10	2	3-4	2	3-0	2	3-10	2	3-4	2	2-11	3
	2-2 × 10	4-9	2	4-1	3	3-8	3	4-8	3	4-0	3	3-7	3
	2-2 × 12	5-6	3	4-9	3	4-3	3	5-5	3	4-8	3	4-2	3
	3-2 × 8	4-10	2	4-2	2	3-9	2	4-9	2	4-1	2	3-8	2
	3-2 × 10	5-11	2	5-1	2	4-7	2	5-10	2	5-0	2	4-6	3
	3-2 × 12	6-10	2	5-11	3	5-4	3	6-9	2	5-10	3	5-3	3
	4-2 × 8	5-7	2	4-10	2	4-4	2	5-6	2	4-9	2	4-3	2
	4-2 × 10	6-10	2	5-11	2	5-3	2	6-9	2	5-10	2	5-2	2
	4-2 × 12	7-11	2	6-10	2	6-2	2	7-9	2	6-9	2	6-0	3

For SI: 1 inch = 25.4 mm, 1 foot = 304.8 mm, 1 pound per square foot = 47.8 N/m².

a. Spans are given in feet and inches (ft-in).

b. Tabulated values are for No. 2 grade lumber.

c. Building width is measured perpendicular to the ridge. For widths between those shown, spans are permitted to be interpolated.

d. NJ - Number of jack studs required to support each end. Where the number of required jack studs equals one, the header is permitted to be supported by an approved framing anchor attached to the full-height wall stud and to the header.

e. Use 30 pounds per square foot ground snow load for cases in which ground snow load is less than 30 pounds per square foot and the roof live load is equal to or less than 20 pounds per square foot.

WOOD

TABLE 2308.9.6
HEADER AND GIRDER SPANS[a] FOR INTERIOR BEARING WALLS
(Maximum Spans for Douglas Fir-Larch, Hem-Fir, Southern Pine and Spruce-Pine-Fir[b] and Required Number of Jack Studs)

HEADERS AND GIRDERS SUPPORTING	SIZE	BUILDING WIDTH[c] (feet)					
		20		28		36	
		Span	NJ[d]	Span	NJ[d]	Span	NJ[d]
One Floor Only	2-2 × 4	3-1	1	2-8	1	2-5	1
	2-2 × 6	4-6	1	3-11	1	3-6	1
	2-2 × 8	5-9	1	5-0	2	4-5	2
	2-2 × 10	7-0	2	6-1	2	5-5	2
	2-2 × 12	8-1	2	7-0	2	6-3	2
	3-2 × 8	7-2	1	6-3	1	5-7	2
	3-2 × 10	8-9	1	7-7	2	6-9	2
	3-2 × 12	10-2	2	8-10	2	7-10	2
	4-2 × 8	9-0	1	7-8	1	6-9	1
	4-2 × 10	10-1	1	8-9	1	7-10	2
	4-2 × 12	11-9	1	10-2	2	9-1	2
Two Floors	2-2 × 4	2-2	1	1-10	1	1-7	1
	2-2 × 6	3-2	2	2-9	2	2-5	2
	2-2 × 8	4-1	2	3-6	2	3-2	2
	2-2 × 10	4-11	2	4-3	2	3-10	3
	2-2 × 12	5-9	2	5-0	3	4-5	3
	3-2 × 8	5-1	2	4-5	2	3-11	2
	3-2 × 10	6-2	2	5-4	2	4-10	2
	3-2 × 12	7-2	2	6-3	2	5-7	3
	4-2 × 8	6-1	1	5-3	2	4-8	2
	4-2 × 10	7-2	2	6-2	2	5-6	2
	4-2 × 12	8-4	2	7-2	2	6-5	2

For SI: 1 inch = 25.4 mm, 1 foot = 304.8 mm.

a. Spans are given in feet and inches (ft-in).

b. Tabulated values are for No. 2 grade lumber.

c. Building width is measured perpendicular to the ridge. For widths between those shown, spans are permitted to be interpolated.

d. NJ - Number of jack studs required to support each end. Where the number of required jack studs equals one, the headers are permitted to be supported by an approved framing anchor attached to the full-height wall stud and to the header.

bridging not less than 2 inches (51 mm) in thickness and of the same width as the studs fitted snugly and nailed thereto to provide adequate lateral support. Bridging shall be placed in every stud cavity and at a frequency such that no stud so braced shall have a height-to-least-thickness ratio exceeding 50 with the height of the stud measured between horizontal framing and bridging or between bridging, whichever is greater.

2308.9.10 Cutting and notching. In exterior walls and bearing partitions, any wood stud is permitted to be cut or notched to a depth not exceeding 25 percent of its width. Cutting or notching of studs to a depth not greater than 40 percent of the width of the stud is permitted in nonbearing partitions supporting no loads other than the weight of the partition.

2308.9.11 Bored holes. A hole not greater in diameter than 40 percent of the stud width is permitted to be bored in any wood stud. Bored holes not greater than 60 percent of the width of the stud are permitted in nonbearing partitions or in any wall where each bored stud is doubled, provided not more than two such successive doubled studs are so bored.

In no case shall the edge of the bored hole be nearer than $^5/_8$ inch (15.9 mm) to the edge of the stud.

Bored holes shall not be located at the same section of stud as a cut or notch.

2308.10 Roof and ceiling framing. The framing details required in this section apply to roofs having a minimum slope of three units vertical in 12 units horizontal (25-percent slope) or greater. Where the roof slope is less than three units vertical in 12 units horizontal (25-percent slope), members supporting rafters and ceiling joists such as ridge board, hips and valleys shall be designed as beams.

2308.10.1 Wind uplift. Roof assemblies shall have rafter and truss ties to the wall below. Resultant uplift loads shall be transferred to the foundation using a continuous load path. The rafter or truss to wall connection shall comply with Tables 2304.9.1 and 2308.10.1.

2308.10.2 Ceiling joist spans. Allowable spans for ceiling joists shall be in accordance with Table 2308.10.2(1) or 2308.10.2(2). For other grades and species, refer to the *AF&PA Span Tables for Joists and Rafters.*

2308.10.3 Rafter spans. Allowable spans for rafters shall be in accordance with Table 2308.10.3(1), 2308.10.3(2), 2308.10.3(3), 2308.10.3(4), 2308.10.3(5) or 2308.10.3(6). For other grades and species, refer to the *AF&PA Span Tables for Joists and Rafters.*

TABLE 2308.10.1
REQUIRED RATING OF APPROVED UPLIFT CONNECTORS (pounds)[a,b,c,e,f,g,h]

BASIC WIND SPEED (3-second gust)	ROOF SPAN (feet)							OVERHANGS (pounds/feet)[d]
	12	20	24	28	32	36	40	
85	-72	-120	-145	-169	-193	-217	-241	-38.55
90	-91	-151	-181	-212	-242	-272	-302	-43.22
100	-131	-281	-262	-305	-349	-393	-436	-53.36
110	-175	-292	-351	-409	-467	-526	-584	-64.56

For SI: 1 inch = 25.4 mm, 1 foot = 304.8 mm, 1 mile per hour = 1.61 km/hr, 1 pound = 0.454 Kg, 1 pound/foot = 14.5939 N/m.

a. The uplift connection requirements are based on a 30-foot mean roof height located in Exposure B. For Exposure C or D and for other mean roof heights, multiply the above loads by the adjustment coefficients in Table 1609.6.2.1(4).

b. The uplift connection requirements are based on the framing being spaced 24 inches on center. Multiply by 0.67 for framing spaced 16 inches on center and multiply by 0.5 for framing spaced 12 inches on center.

c. The uplift connection requirements include an allowance for 10 pounds of dead load.

d. The uplift connection requirements do not account for the effects of overhangs. The magnitude of the above loads shall be increased by adding the overhang loads found in the table. The overhang loads are also based on framing spaced 24 inches on center. The overhang loads given shall be multiplied by the overhang projection and added to the roof uplift value in the table.

e. The uplift connection requirements are based upon wind loading on end zones as defined in Section 1609.6.3. Connection loads for connections located a distance of 20 percent of the least horizontal dimension of the building from the corner of the building are permitted to be reduced by multiplying the table connection value by 0.7 and multiplying the overhang load by 0.8.

f. For wall-to-wall and wall-to-foundation connections, the capacity of the uplift connector is permitted to be reduced by 100 pounds for each full wall above. (For example, if a 500-pound rated connector is used on the roof framing, a 400-pound rated connector is permitted at the next floor level down.)

g. Interpolation is permitted for intermediate values of basic wind speeds and roof spans.

h. The rated capacity of approved tie-down devices is permitted to include up to a 60-percent increase for wind effects where allowed by material specifications.

TABLE 2308.10.2(1)
CEILING JOIST SPANS FOR COMMON LUMBER SPECIES
(Uninhabitable Attics Without Storage, Live Load = 10 pounds psf, $L/\Delta = 240$)

CEILING JOIST SPACING (inches)	SPECIES AND GRADE		DEAD LOAD = 5 pounds per square foot			
			Maximum ceiling joist spans			
			2 × 4	2 × 6	2 × 8	2 × 10
			(ft. - in.)	(ft. - in.)	(ft. - in.)	(ft. - in.)
12	Douglas Fir-Larch	SS	13-2	20-8	Note a	Note a
	Douglas Fir-Larch	#1	12-8	19-11	Note a	Note a
	Douglas Fir-Larch	#2	12-5	19-6	25-8	Note a
	Douglas Fir-Larch	#3	10-10	15-10	20-1	24-6
	Hem-Fir	SS	12-5	19-6	25-8	Note a
	Hem-Fir	#1	12-2	19-1	25-2	Note a
	Hem-Fir	#2	11-7	18-2	24-0	Note a
	Hem-Fir	#3	10-10	15-10	20-1	24-6
	Southern Pine	SS	12-11	20-3	Note a	Note a
	Southern Pine	#1	12-8	19-11	Note a	Note a
	Southern Pine	#2	12-5	19-6	25-8	Note a
	Southern Pine	#3	11-6	17-0	21-8	25-7
	Spruce-Pine-Fir	SS	12-2	19-1	25-2	Note a
	Spruce-Pine-Fir	#1	11-10	18-8	24-7	Note a
	Spruce-Pine-Fir	#2	11-10	18-8	24-7	Note a
	Spruce-Pine-Fir	#3	10-10	15-10	20-1	24-6
16	Douglas Fir-Larch	SS	11-11	18-9	24-8	Note a
	Douglas Fir-Larch	#1	11-6	18-1	23-10	Note a
	Douglas Fir-Larch	#2	11-3	17-8	23-0	Note a
	Douglas Fir-Larch	#3	9-5	13-9	17-5	21-3
	Hem-Fir	SS	11-3	17-8	23-4	Note a
	Hem-Fir	#1	11-0	17-4	22-10	Note a
	Hem-Fir	#2	10-6	16-6	21-9	Note a
	Hem-Fir	#3	9-5	13-9	17-5	21-3
	Southern Pine	SS	11-9	18-5	24-3	Note a
	Southern Pine	#1	11-6	18-1	23-1	Note a
	Southern Pine	#2	11-3	17-8	23-4	Note a
	Southern Pine	#3	10-0	14-9	18-9	22-2
	Spruce-Pine-Fir	SS	11-0	17-4	22-10	Note a
	Spruce-Pine-Fir	#1	10-9	16-11	22-4	Note a
	Spruce-Pine-Fir	#2	10-9	16-11	22-4	Note a
	Spruce-Pine-Fir	#3	9-5	13-9	17-5	21-3

(continued)

TABLE 2308.10.2(1)—continued
CEILING JOIST SPANS FOR COMMON LUMBER SPECIES
(Uninhabitable Attics Without Storage, Live Load = 10 pounds psf, L/Δ = 240)

CEILING JOIST SPACING (inches)	SPECIES AND GRADE		DEAD LOAD = 5 pounds per square foot			
			Maximum ceiling joist spans			
			2 × 4	2 × 6	2 × 8	2 × 10
			(ft. - in.)	(ft. - in.)	(ft. - in.)	(ft. - in.)
19.2	Douglas Fir-Larch	SS	11-3	17-8	23-3	Note a
	Douglas Fir-Larch	#1	10-10	17-0	22-5	Note a
	Douglas Fir-Larch	#2	10-7	16-7	21-0	25-8
	Douglas Fir-Larch	#3	8-7	12-6	15-10	19-5
	Hem-Fir	SS	10-7	16-8	21-11	Note a
	Hem-Fir	#1	10-4	16-4	21-6	Note a
	Hem-Fir	#2	9-11	15-7	20-6	25-3
	Hem-Fir	#3	8-7	12-6	15-10	19-5
	Southern Pine	SS	11-0	17-4	22-10	Note a
	Southern Pine	#1	10-10	17-0	22-5	Note a
	Southern Pine	#2	10-7	16-8	21-11	Note a
	Southern Pine	#3	9-1	13-6	17-2	20-3
	Spruce-Pine-Fir	SS	10-4	16-4	21-6	Note a
	Spruce-Pine-Fir	#1	10-2	15-11	21-0	25-8
	Spruce-Pine-Fir	#2	10-2	15-11	21-0	25-8
	Spruce-Pine-Fir	#3	8-7	12-6	15-10	19-5
24	Douglas Fir-Larch	SS	10-5	16-4	21-7	Note a
	Douglas Fir-Larch	#1	10-0	15-9	20-1	24-6
	Douglas Fir-Larch	#2	9-10	14-10	18-9	22-11
	Douglas Fir-Larch	#3	7-8	11-2	14-2	17-4
	Hem-Fir	SS	9-10	15-6	20-5	Note a
	Hem-Fir	#1	9-8	15-2	19-7	23-11
	Hem-Fir	#2	9-2	14-5	18-6	22-7
	Hem-Fir	#3	7-8	11-2	14-2	17-4
	Southern Pine	SS	10-3	16-1	21-2	Note a
	Southern Pine	#1	10-0	15-9	20-10	Note a
	Southern Pine	#2	9-10	15-6	20-1	23-11
	Southern Pine	#3	8-2	12-0	15-4	18-1
	Spruce-Pine-Fir	SS	9-8	15-2	19-11	25-5
	Spruce-Pine-Fir	#1	9-5	14-9	18-9	22-11
	Spruce-Pine-Fir	#2	9-5	14-9	18-9	22-11
	Spruce-Pine-Fir	#3	7-8	11-2	14-2	17-4

For SI: 1 inch = 25.4 mm, 1 foot = 304.8 mm, 1 pound per square foot = 47.8 N/m².

a. Span exceeds 26 feet in length. Check sources for availability of lumber in lengths greater than 20 feet.

TABLE 2308.10.2(2)
CEILING JOIST SPANS FOR COMMON LUMBER SPECIES
(Uninhabitable Attics With Limited Storage, Live Load = 20 pounds per square foot, L/Δ = 240)

CEILING JOIST SPACING (inches)	SPECIES AND GRADE		DEAD LOAD = 10 pounds per square foot			
			Maximum ceiling joist spans			
			2 × 4	2 × 6	2 × 8	2 × 10
			(ft. - in.)	(ft. - in.)	(ft. - in.)	(ft. - in.)
12	Douglas Fir-Larch	SS	10-5	16-4	21-7	Note a
	Douglas Fir-Larch	#1	10-0	15-9	20-1	24-6
	Douglas Fir-Larch	#2	9-10	14-10	18-9	22-11
	Douglas Fir-Larch	#3	7-8	11-2	14-2	17-4
	Hem-Fir	SS	9-10	15-6	20-5	Note a
	Hem-Fir	#1	9-8	15-2	19-7	23-11
	Hem-Fir	#2	9-2	14-5	18-6	22-7
	Hem-Fir	#3	7-8	11-2	14-2	17-4
	Southern Pine	SS	10-3	16-1	21-2	Note a
	Southern Pine	#1	10-0	15-9	20-10	Note a
	Southern Pine	#2	9-10	15-6	20-1	23-11
	Southern Pine	#3	8-2	12-0	15-4	18-1
	Spruce-Pine-Fir	SS	9-8	15-2	19-11	25-5
	Spruce-Pine-Fir	#1	9-5	14-9	18-9	22-11
	Spruce-Pine-Fir	#2	9-5	14-9	18-9	22-11
	Spruce-Pine-Fir	#3	7-8	11-2	14-2	17-4
16	Douglas Fir-Larch	SS	9-6	14-11	19-7	25-0
	Douglas Fir-Larch	#1	9-1	13-9	17-5	21-3
	Douglas Fir-Larch	#2	8-9	12-10	16-3	19-10
	Douglas Fir-Larch	#3	6-8	9-8	12-4	15-0
	Hem-Fir	SS	8-11	14-1	18-6	23-8
	Hem-Fir	#1	8-9	13-5	16-10	20-8
	Hem-Fir	#2	8-4	12-8	16-0	19-7
	Hem-Fir	#3	6-8	9-8	12-4	15-0
	Southern Pine	SS	9-4	14-7	19-3	24-7
	Southern Pine	#1	9-1	14-4	18-11	23-1
	Southern Pine	#2	8-11	13-6	17-5	20-9
	Southern Pine	#3	7-1	10-5	13-3	15-8
	Spruce-Pine-Fir	SS	8-9	13-9	18-1	23-1
	Spruce-Pine-Fir	#1	8-7	12-10	16-3	19-10
	Spruce-Pine-Fir	#2	8-7	12-10	16-3	19-10
	Spruce-Pine-Fir	#3	6-8	9-8	12-4	15-0

(continued)

TABLE 2308.10.2(2)—continued
CEILING JOIST SPANS FOR COMMON LUMBER SPECIES
(Uninhabitable Attics With Limited Storage, Live Load = 20 pounds per square foot, $L/\Delta = 240$)

CEILING JOIST SPACING (inches)	SPECIES AND GRADE		DEAD LOAD = 10 pounds per square foot			
			Maximum ceiling joist spans			
			2 × 4	2 × 6	2 × 8	2 × 10
			(ft. - in.)	(ft. - in.)	(ft. - in.)	(ft. - in.)
19.2	Douglas Fir-Larch	SS	8-11	14-0	18-5	23-4
	Douglas Fir-Larch	#1	8-7	12-6	15-10	19-5
	Douglas Fir-Larch	#2	8-0	11-9	14-10	18-2
	Douglas Fir-Larch	#3	6-1	8-10	11-3	13-8
	Hem-Fir	SS	8-5	13-3	17-5	22-3
	Hem-Fir	#1	8-3	12-3	15-6	18-11
	Hem-Fir	#2	7-10	11-7	14-8	17-10
	Hem-Fir	#3	6-1	8-10	11-3	13-8
	Southern Pine	SS	8-9	13-9	18-1	23-1
	Southern Pine	#1	8-7	13-6	17-9	21-1
	Southern Pine	#2	8-5	12-3	15-10	18-11
	Southern Pine	#3	6-5	9-6	12-1	14-4
	Spruce-Pine-Fir	SS	8-3	12-11	17-1	21-8
	Spruce-Pine-Fir	#1	8-0	11-9	14-10	18-2
	Spruce-Pine-Fir	#2	8-0	11-9	14-10	18-2
	Spruce-Pine-Fir	#3	6-1	8-10	11-3	13-8
24	Douglas Fir-Larch	SS	8-3	13-0	17-1	20-11
	Douglas Fir-Larch	#1	7-8	11-2	14-2	17-4
	Douglas Fir-Larch	#2	7-2	10-6	13-3	16-3
	Douglas Fir-Larch	#3	5-5	7-11	10-0	12-3
	Hem-Fir	SS	7-10	12-3	16-2	20-6
	Hem-Fir	#1	7-6	10-11	13-10	16-11
	Hem-Fir	#2	7-1	10-4	13-1	16-0
	Hem-Fir	#3	5-5	7-11	10-0	12-3
	Southern Pine	SS	8-1	12-9	16-10	21-6
	Southern Pine	#1	8-0	12-6	15-10	18-10
	Southern Pine	#2	7-8	11-0	14-2	16-11
	Southern Pine	#3	5-9	8-6	10-10	12-10
	Spruce-Pine-Fir	SS	7-8	12-0	15-10	19-5
	Spruce-Pine-Fir	#1	7-2	10-6	13-3	16-3
	Spruce-Pine-Fir	#2	7-2	10-6	13-3	16-3
	Spruce-Pine-Fir	#3	5-5	7-11	10-0	12-3

For SI: 1 inch = 25.4 mm, 1 foot = 304.8 mm, 1 pound per square foot = 47.8 N/m².

a. Span exceeds 26 feet in length. Check sources for availability of lumber in lengths greater than 20 feet.

TABLE 2308.10.3(1)
RAFTER SPANS FOR COMMON LUMBER SPECIES
(Roof Live Load = 20 pounds per square foot, Ceiling Not Attached to Rafters, $L/\Delta = 180$)

RAFTER SPACING (inches)	SPECIES AND GRADE		DEAD LOAD = 10 pounds per square foot					DEAD LOAD = 20 pounds per square foot				
			Maximum rafter spans									
			2 × 4	2 × 6	2 × 8	2 × 10	2 × 12	2 × 4	2 × 6	2 × 8	2 × 10	2 × 12
			(ft. - in.)	(ft. - in.)	(ft. - in.)	(ft. - in.)	(ft. - in.)	(ft. - in.)	(ft. - in.)	(ft. - in.)	(ft. - in.)	(ft. - in.)
12	Douglas Fir-Larch	SS	11-6	18-0	23-9	Note a	Note a	11-6	18-0	23-5	Note a	Note a
	Douglas Fir-Larch	#1	11-1	17-4	22-5	Note a	Note a	10-6	15-4	19-5	23-9	Note a
	Douglas Fir-Larch	#2	10-10	16-7	21-0	25-8	Note a	9-10	14-4	18-2	22-3	25-9
	Douglas Fir-Larch	#3	8-7	12-6	15-10	19-5	22-6	7-5	10-10	13-9	16-9	19-6
	Hem-Fir	SS	10-10	17-0	22-5	Note a	Note a	10-10	17-0	22-5	Note a	Note a
	Hem-Fir	#1	10-7	16-8	21-10	Note a	Note a	10-3	14-11	18-11	23-2	Note a
	Hem-Fir	#2	10-1	15-11	20-8	25-3	Note a	9-8	14-2	17-11	21-11	25-5
	Hem-Fir	#3	8-7	12-6	15-10	19-5	22-6	7-5	10-10	13-9	16-9	19-6
	Southern Pine	SS	11-3	17-8	23-4	Note a	Note a	11-3	17-8	23-4	Note a	Note a
	Southern Pine	#1	11-1	17-4	22-11	Note a	Note a	11-1	17-3	21-9	25-10	Note a
	Southern Pine	#2	10-10	17-0	22-5	Note a	Note a	10-6	15-1	19-5	23-2	25-9
	Southern Pine	#3	9-1	13-6	17-2	20-3	24-1	7-11	11-8	14-10	17-6	20-11
	Spruce-Pine-Fir	SS	10-7	16-8	21-11	Note a	Note a	10-7	16-8	21-9	Note a	Note a
	Spruce-Pine-Fir	#1	10-4	16-3	21-0	25-8	Note a	9-10	14-4	18-2	22-3	25-9
	Spruce-Pine-Fir	#2	10-4	16-3	21-0	25-8	Note a	9-10	14-4	18-2	22-3	25-9
	Spruce-Pine-Fir	#3	8-7	12-6	15-10	19-5	22-6	7-5	10-10	13-9	16-9	19-6
16	Douglas Fir-Larch	SS	10-5	16-4	21-7	Note a	Note a	10-5	16-0	20-3	24-9	Note a
	Douglas Fir-Larch	#1	10-0	15-4	19-5	23-9	Note a	9-1	13-3	16-10	20-7	23-10
	Douglas Fir-Larch	#2	9-10	14-4	18-2	22-3	25-9	8-6	12-5	15-9	19-3	22-4
	Douglas Fir-Larch	#3	7-5	10-10	13-9	16-9	19-6	6-5	9-5	11-11	14-6	16-10
	Hem-Fir	SS	9-10	15-6	20-5	Note a	Note a	9-10	15-6	19-11	24-4	Note a
	Hem-Fir	#1	9-8	14-11	18-11	23-2	Note a	8-10	12-11	16-5	20-0	23-3
	Hem-Fir	#2	9-2	14-2	17-11	21-11	25-5	8-5	12-3	15-6	18-11	22-0
	Hem-Fir	#3	7-5	10-10	13-9	16-9	19-6	6-5	9-5	11-11	14-6	16-10
	Southern Pine	SS	10-3	16-1	21-2	Note a	Note a	10-3	16-1	21-2	Note a	Note a
	Southern Pine	#1	10-0	15-9	20-10	25-10	Note a	10-0	15-0	18-10	22-4	Note a
	Southern Pine	#2	9-10	15-1	19-5	23-2	Note a	9-1	13-0	16-10	20-1	23-7
	Southern Pine	#3	7-11	11-8	14-10	17-6	20-11	6-10	10-1	12-10	15-2	18-1
	Spruce-Pine-Fir	SS	9-8	15-2	19-11	25-5	Note a	9-8	14-10	18-10	23-0	Note a
	Spruce-Pine-Fir	#1	9-5	14-4	18-2	22-3	25-9	8-6	12-5	15-9	19-3	22-4
	Spruce-Pine-Fir	#2	9-5	14-4	18-2	22-3	25-9	8-6	12-5	15-9	19-3	22-4
	Spruce-Pine-Fir	#3	7-5	10-10	13-9	16-9	19-6	6-5	9-5	11-11	14-6	16-10

(continued)

TABLE 2308.10.3(1)—continued
RAFTER SPANS FOR COMMON LUMBER SPECIES
(Roof Live Load = 20 pounds per square foot, Ceiling Not Attached to Rafters, $L/\Delta = 180$)

RAFTER SPACING (inches)	SPECIES AND GRADE		DEAD LOAD = 10 pounds per square foot					DEAD LOAD = 20 pounds per square foot				
			2 × 4	2 × 6	2 × 8	2 × 10	2 × 12	2 × 4	2 × 6	2 × 8	2 × 10	2 × 12
			Maximum rafter spans									
			(ft. - in.)	(ft. - in.)	(ft. - in.)	(ft. - in.)	(ft. - in.)	(ft. - in.)	(ft. - in.)	(ft. - in.)	(ft. - in.)	(ft. - in.)
19.2	Douglas Fir-Larch	SS	9-10	15-5	20-4	25-11	Note a	9-10	14-7	18-6	22-7	Note a
	Douglas Fir-Larch	#1	9-5	14-0	17-9	21-8	25-2	8-4	12-2	15-4	18-9	21-9
	Douglas Fir-Larch	#2	8-11	13-1	16-7	20-3	23-6	7-9	11-4	14-4	17-7	20-4
	Douglas Fir-Larch	#3	6-9	9-11	12-7	15-4	17-9	5-10	8-7	10-10	13-3	15-5
	Hem-Fir	SS	9-3	14-7	19-2	24-6	Note a	9-3	14-4	18-2	22-3	25-9
	Hem-Fir	#1	9-1	13-8	17-4	21-1	24-6	8-1	11-10	15-0	18-4	21-3
	Hem-Fir	#2	8-8	12-11	16-4	20-0	23-2	7-8	11-2	14-2	17-4	20-1
	Hem-Fir	#3	6-9	9-11	12-7	15-4	17-9	5-10	8-7	10-10	13-3	15-5
	Southern Pine	SS	9-8	15-2	19-11	25-5	Note a	9-8	15-2	19-11	25-5	Note a
	Southern Pine	#1	9-5	14-10	19-7	23-7	Note a	9-3	13-8	17-2	20-5	24-4
	Southern Pine	#2	9-3	13-9	17-9	21-2	24-10	8-4	11-11	15-4	18-4	21-6
	Southern Pine	#3	7-3	10-8	13-7	16-0	19-1	6-3	9-3	11-9	13-10	16-6
	Spruce-Pine-Fir	SS	9-1	14-3	18-9	23-11	Note a	9-1	13-7	17-2	21-0	24-4
	Spruce-Pine-Fir	#1	8-10	13-1	16-7	20-3	23-6	7-9	11-4	14-4	17-7	20-4
	Spruce-Pine-Fir	#2	8-10	13-1	16-7	20-3	23-6	7-9	11-4	14-4	17-7	20-4
	Spruce-Pine-Fir	#3	6-9	9-11	12-7	15-4	17-9	5-10	8-7	10-10	13-3	15-5
24	Douglas Fir-Larch	SS	9-1	14-4	18-10	23-4	Note b	8-11	13-1	16-7	20-3	23-5
	Douglas Fir-Larch	#1	8-7	12-6	15-10	19-5	22-6	7-5	10-10	13-9	16-9	19-6
	Douglas Fir-Larch	#2	8-0	11-9	14-10	18-2	21-0	6-11	10-2	12-10	15-8	18-3
	Douglas Fir-Larch	#3	6-1	8-10	11-3	13-8	15-11	5-3	7-8	9-9	11-10	13-9
	Hem-Fir	SS	8-7	13-6	17-10	22-9	Note b	8-7	12-10	16-3	19-10	23-0
	Hem-Fir	#1	8-4	12-3	15-6	18-11	21-11	7-3	10-7	13-5	16-4	19-0
	Hem-Fir	#2	7-11	11-7	14-8	17-10	20-9	6-10	10-0	12-8	15-6	17-11
	Hem-Fir	#3	6-1	8-10	11-3	13-8	15-11	5-3	7-8	9-9	11-10	13-9
	Southern Pine	SS	8-11	14-1	18-6	23-8	Note b	8-11	14-1	18-6	22-11	Note a
	Southern Pine	#1	8-9	13-9	17-9	21-1	25-2	8-3	12-3	15-4	18-3	21-9
	Southern Pine	#2	8-7	12-3	15-10	18-11	22-2	7-5	10-8	13-9	16-5	19-3
	Southern Pine	#3	6-5	9-6	12-1	14-4	17-1	5-7	8-3	10-6	12-5	14-9
	Spruce-Pine-Fir	SS	8-5	13-3	17-5	21-8	25-2	8-4	12-2	15-4	18-9	21-9
	Spruce-Pine-Fir	#1	8-0	11-9	14-10	18-2	21-0	6-11	10-2	12-10	15-8	18-3
	Spruce-Pine-Fir	#2	8-0	11-9	14-10	18-2	21-0	6-11	10-2	12-10	15-8	18-3
	Spruce-Pine-Fir	#3	6-1	8-10	11-3	13-8	15-11	5-3	7-8	9-9	11-10	13-9

For SI: 1 inch = 25.4 mm, 1 foot = 304.8 mm, 1 pound per square foot = 47.9 N/m².
a. Span exceeds 26 feet in length. Check sources for availability of lumber in lengths greater than 26 feet.

TABLE 2308.10.3(2)
RAFTER SPANS FOR COMMON LUMBER SPECIES
(Roof Live Load = 20 pounds per square foot, Ceiling Not Attached to Rafters, $L/\Delta = 240$)

RAFTER SPACING (inches)	SPECIES AND GRADE		DEAD LOAD = 10 pounds per square foot					DEAD LOAD = 20 pounds per square foot				
			2 × 4	2 × 6	2 × 8	2 × 10	2 × 12	2 × 4	2 × 6	2 × 8	2 × 10	2 × 12
			(ft. - in.)	(ft. - in.)	(ft. - in.)	(ft. - in.)	(ft. - in.)	(ft. - in.)	(ft. - in.)	(ft. - in.)	(ft. - in.)	(ft. - in.)
			Maximum rafter spans									
12	Douglas Fir-Larch	SS	10-5	16-4	21-7	Note a	Note a	10-5	16-4	21-7	Note a	Note a
	Douglas Fir-Larch	#1	10-0	15-9	20-10	Note a	Note a	10-0	15-4	19-5	23-9	Note a
	Douglas Fir-Larch	#2	9-10	15-6	20-5	25-8	Note a	9-10	14-4	18-2	22-3	25-9
	Douglas Fir-Larch	#3	8-7	12-6	15-10	19-5	22-6	7-5	10-10	13-9	16-9	19-6
	Hem-Fir	SS	9-10	15-6	20-5	Note a	Note a	9-10	15-6	20-5	Note a	Note a
	Hem-Fir	#1	9-8	15-2	19-11	25-5	Note a	9-8	14-11	18-11	23-2	Note a
	Hem-Fir	#2	9-2	14-5	19-0	24-3	Note a	9-2	14-2	17-11	21-11	25-5
	Hem-Fir	#3	8-7	12-6	15-10	19-5	22-6	7-5	10-10	13-9	16-9	19-6
	Southern Pine	SS	10-3	16-1	21-2	Note a	Note a	10-3	16-1	21-2	Note a	Note a
	Southern Pine	#1	10-0	15-9	20-10	Note a	Note a	10-0	15-9	20-10	25-10	Note a
	Southern Pine	#2	9-10	15-6	20-5	Note a	Note a	9-10	15-1	19-5	23-2	Note a
	Southern Pine	#3	9-1	13-6	17-2	20-3	24-1	7-11	11-8	14-10	17-6	20-11
	Spruce-Pine-Fir	SS	9-8	15-2	19-11	25-5	Note a	9-8	15-2	19-11	25-5	Note a
	Spruce-Pine-Fir	#1	9-5	14-9	19-6	24-10	Note a	9-5	14-4	18-2	22-3	25-9
	Spruce-Pine-Fir	#2	9-5	14-9	19-6	24-10	Note a	9-5	14-4	18-2	22-3	25-9
	Spruce-Pine-Fir	#3	8-7	12-6	15-10	19-5	22-6	7-5	10-10	13-9	16-9	19-6
16	Douglas Fir-Larch	SS	9-6	14-11	19-7	25-0	Note a	9-6	14-11	19-7	24-9	Note a
	Douglas Fir-Larch	#1	9-1	14-4	18-11	23-9	Note a	9-1	13-3	16-10	20-7	23-10
	Douglas Fir-Larch	#2	8-11	14-1	18-2	22-3	25-9	8-6	12-5	15-9	19-3	22-4
	Douglas Fir-Larch	#3	7-5	10-10	13-9	16-9	19-6	6-5	9-5	11-11	14-6	16-10
	Hem-Fir	SS	8-11	14-1	18-6	23-8	Note a	8-11	14-1	18-6	23-8	Note a
	Hem-Fir	#1	8-9	13-9	18-1	23-1	Note a	8-9	12-11	16-5	20-0	23-3
	Hem-Fir	#2	8-4	13-1	17-3	21-11	25-5	8-4	12-3	15-6	18-11	22-0
	Hem-Fir	#3	7-5	10-10	13-9	16-9	19-6	6-5	9-5	11-11	14-6	16-10
	Southern Pine	SS	9-4	14-7	19-3	24-7	Note a	9-4	14-7	19-3	24-7	Note a
	Southern Pine	#1	9-1	14-4	18-11	24-1	Note a	9-1	14-4	18-10	22-4	Note a
	Southern Pine	#2	8-11	14-1	18-6	23-2	Note a	8-11	13-0	16-10	20-1	23-7
	Southern Pine	#3	7-11	11-8	14-10	17-6	20-11	6-10	10-1	12-10	15-2	18-1
	Spruce-Pine-Fir	SS	8-9	13-9	18-1	23-1	Note a	8-9	13-9	18-1	23-0	Note a
	Spruce-Pine-Fir	#1	8-7	13-5	17-9	22-3	25-9	8-6	12-5	15-9	19-3	22-4
	Spruce-Pine-Fir	#2	8-7	13-5	17-9	22-3	25-9	8-6	12-5	15-9	19-3	22-4
	Spruce-Pine-Fir	#3	7-5	10-10	13-9	16-9	19-6	6-5	9-5	11-11	14-6	16-10

(continued)

TABLE 2308.10.3(2)—continued
RAFTER SPANS FOR COMMON LUMBER SPECIES
(Roof Live Load = 20 pounds per square foot, Ceiling Not Attached to Rafters, $L/\Delta = 240$)

RAFTER SPACING (inches)	SPECIES AND GRADE		DEAD LOAD = 10 pounds per square foot					DEAD LOAD = 20 pounds per square foot				
			2 × 4	2 × 6	2 × 8	2 × 10	2 × 12	2 × 4	2 × 6	2 × 8	2 × 10	2 × 12
			\(Maximum rafter spans\)									
			(ft.-in.)	(ft.-in.)	(ft.-in.)	(ft.-in.)	(ft.-in.)	(ft.-in.)	(ft.-in.)	(ft.-in.)	(ft.-in.)	(ft.-in.)
19.2	Douglas Fir-Larch	SS	8-11	14-0	18-5	23-7	Note a	8-11	14-0	18-5	22-7	Note a
	Douglas Fir-Larch	#1	8-7	13-6	17-9	21-8	25-2	8-4	12-2	15-4	18-9	21-9
	Douglas Fir-Larch	#2	8-5	13-1	16-7	20-3	23-6	7-9	11-4	14-4	17-7	20-4
	Douglas Fir-Larch	#3	6-9	9-11	12-7	15-4	17-9	5-10	8-7	10-10	13-3	15-5
	Hem-Fir	SS	8-5	13-3	17-5	22-3	Note a	8-5	13-3	17-5	22-3	25-9
	Hem-Fir	#1	8-3	12-11	17-1	21-1	24-6	8-1	11-10	15-0	18-4	21-3
	Hem-Fir	#2	7-10	12-4	16-3	20-0	23-2	7-8	11-2	14-2	17-4	20-1
	Hem-Fir	#3	6-9	9-11	12-7	15-4	17-9	5-10	8-7	10-10	13-3	15-5
	Southern Pine	SS	8-9	13-9	18-1	23-1	Note a	8-9	13-9	18-1	23-1	Note a
	Southern Pine	#1	8-7	13-6	17-9	22-8	Note a	8-7	13-6	17-2	20-5	24-4
	Southern Pine	#2	8-5	13-3	17-5	21-2	24-10	8-4	11-11	15-4	18-4	21-6
	Southern Pine	#3	7-3	10-8	13-7	16-0	19-1	6-3	9-3	11-9	13-10	16-6
	Spruce-Pine-Fir	SS	8-3	12-11	17-1	21-9	Note a	8-3	12-11	17-1	21-0	24-4
	Spruce-Pine-Fir	#1	8-1	12-8	16-7	20-3	23-6	7-9	11-4	14-4	17-7	20-4
	Spruce-Pine-Fir	#2	8-1	12-8	16-7	20-3	23-6	7-9	11-4	14-4	17-7	20-4
	Spruce-Pine-Fir	#3	6-9	9-11	12-7	15-4	17-9	5-10	8-7	10-10	13-3	15-5
24	Douglas Fir-Larch	SS	8-3	13-0	17-2	21-10	Note a	8-3	13-0	16-7	20-3	23-5
	Douglas Fir-Larch	#1	8-0	12-6	15-10	19-5	22-6	7-5	10-10	13-9	16-9	19-6
	Douglas Fir-Larch	#2	7-10	11-9	14-10	18-2	21-0	6-11	10-2	12-10	15-8	18-3
	Douglas Fir-Larch	#3	6-1	8-10	11-3	13-8	15-11	5-3	7-8	9-9	11-10	13-9
	Hem-Fir	SS	7-10	12-3	16-2	20-8	25-1	7-10	12-3	16-2	19-10	23-0
	Hem-Fir	#1	7-8	12-0	15-6	18-11	21-11	7-3	10-7	13-5	16-4	19-0
	Hem-Fir	#2	7-3	11-5	14-8	17-10	20-9	6-10	10-0	12-8	15-6	17-11
	Hem-Fir	#3	6-1	8-10	11-3	13-8	15-11	5-3	7-8	9-9	11-10	13-9
	Southern Pine	SS	8-1	12-9	16-10	21-6	Note a	8-1	12-9	16-10	21-6	Note a
	Southern Pine	#1	8-0	12-6	16-6	21-1	25-2	8-0	12-3	15-4	18-3	21-9
	Southern Pine	#2	7-10	12-3	15-10	18-11	22-2	7-5	10-8	13-9	16-5	19-3
	Southern Pine	#3	6-5	9-6	12-1	14-4	17-1	5-7	8-3	10-6	12-5	14-9
	Spruce-Pine-Fir	SS	7-8	12-0	15-10	20-2	24-7	7-8	12-0	15-4	18-9	21-9
	Spruce-Pine-Fir	#1	7-6	11-9	14-10	18-2	21-0	6-11	10-2	12-10	15-8	18-3
	Spruce-Pine-Fir	#2	7-6	11-9	14-10	18-2	21-0	6-11	10-2	12-10	15-8	18-3
	Spruce-Pine-Fir	#3	6-1	8-10	11-3	13-8	15-11	5-3	7-8	9-9	11-10	13-9

For SI: 1 inch = 25.4 mm, 1 foot = 304.8 mm, 1 pound per square foot = 47.9 N/m².
a. Span exceeds 26 feet in length. Check sources for availability of lumber in lengths greater than 20 feet.

TABLE 2308.10.3(3)
RAFTER SPANS FOR COMMON LUMBER SPECIES
(Ground Snow Load = 30 pounds per square foot, Ceiling Not Attached to Rafters, $L/\Delta = 180$)

RAFTER SPACING (inches)	SPECIES AND GRADE		DEAD LOAD = 10 pounds per square foot					DEAD LOAD = 20 pounds per square foot				
			2 × 4	2 × 6	2 × 8	2 × 10	2 × 12	2 × 4	2 × 6	2 × 8	2 × 10	2 × 12
			(ft. - in.)	(ft. - in.)	(ft. - in.)	(ft. - in.)	(ft. - in.)	(ft. - in.)	(ft. - in.)	(ft. - in.)	(ft. - in.)	(ft. - in.)
							Maximum rafter spans					
12	Douglas Fir-Larch	SS	10-0	15-9	20-9	Note a	Note a	10-0	15-9	20-1	24-6	Note a
	Douglas Fir-Larch	#1	9-8	14-9	18-8	22-9	Note a	9-0	13-2	16-8	20-4	23-7
	Douglas Fir-Larch	#2	9-5	13-9	17-5	21-4	24-8	8-5	12-4	15-7	19-1	22-1
	Douglas Fir-Larch	#3	7-1	10-5	13-2	16-1	18-8	6-4	9-4	11-9	14-5	16-8
	Hem-Fir	SS	9-6	14-10	19-7	25-0	Note a	9-6	14-10	19-7	24-1	Note a
	Hem-Fir	#1	9-3	14-4	18-2	22-2	25-9	8-9	12-10	16-3	19-10	23-0
	Hem-Fir	#2	8-10	13-7	17-2	21-0	24-4	8-4	12-2	15-4	18-9	21-9
	Hem-Fir	#3	7-1	10-5	13-2	16-1	18-8	6-4	9-4	11-9	14-5	16-8
	Southern Pine	SS	9-10	15-6	20-5	Note a	Note a	9-10	15-6	20-5	Note a	Note a
	Southern Pine	#1	9-8	15-2	20-0	24-9	Note a	9-8	14-10	18-8	22-2	Note a
	Southern Pine	#2	9-6	14-5	18-8	22-3	Note a	9-0	12-11	16-8	19-11	23-4
	Southern Pine	#3	7-7	11-2	14-3	16-10	20-0	6-9	10-0	12-9	15-1	17-11
	Spruce-Pine-Fir	SS	9-3	14-7	19-2	24-6	Note a	9-3	14-7	18-8	22-9	Note a
	Spruce-Pine-Fir	#1	9-1	13-9	17-5	21-4	24-8	8-5	12-4	15-7	19-1	22-1
	Spruce-Pine-Fir	#2	9-1	13-9	17-5	21-4	24-8	8-5	12-4	15-7	19-1	22-1
	Spruce-Pine-Fir	#3	7-1	10-5	13-2	16-1	18-8	6-4	9-4	11-9	14-5	16-8
16	Douglas Fir-Larch	SS	9-1	14-4	18-10	23-9	Note a	9-1	13-9	17-5	21-3	24-8
	Douglas Fir-Larch	#1	8-9	12-9	16-2	19-9	22-10	7-10	11-5	14-5	17-8	20-5
	Douglas Fir-Larch	#2	8-2	11-11	15-1	18-5	21-5	7-3	10-8	13-6	16-6	19-2
	Douglas Fir-Larch	#3	6-2	9-0	11-5	13-11	16-2	5-6	8-1	10-3	12-6	14-6
	Hem-Fir	SS	8-7	13-6	17-10	22-9	Note a	8-7	13-6	17-1	20-10	24-2
	Hem-Fir	#1	8-5	12-5	15-9	19-3	22-3	7-7	11-1	14-1	17-2	19-11
	Hem-Fir	#2	8-0	11-9	14-11	18-2	21-1	7-2	10-6	13-4	16-3	18-10
	Hem-Fir	#3	6-2	9-0	11-5	13-11	16-2	5-6	8-1	10-3	12-6	14-6
	Southern Pine	SS	8-11	14-1	18-6	23-8	Note a	8-11	14-1	18-6	23-8	Note a
	Southern Pine	#1	8-9	13-9	18-1	21-5	25-7	8-8	12-10	16-2	19-2	22-10
	Southern Pine	#2	8-7	12-6	16-2	19-3	22-7	7-10	11-2	14-5	17-3	20-2
	Southern Pine	#3	6-7	9-8	12-4	14-7	17-4	5-10	8-8	11-0	13-0	15-6
	Spruce-Pine-Fir	SS	8-5	13-3	17-5	22-1	25-7	8-5	12-9	16-2	19-9	22-10
	Spruce-Pine-Fir	#1	8-2	11-11	15-1	18-5	21-5	7-3	10-8	13-6	16-6	19-2
	Spruce-Pine-Fir	#2	8-2	11-11	15-1	18-5	21-5	7-3	10-8	13-6	16-6	19-2
	Spruce-Pine-Fir	#3	6-2	9-0	11-5	13-11	16-2	5-6	8-1	10-3	12-6	14-6

(continued)

TABLE 2308.10.3(3)—continued
RAFTER SPANS FOR COMMON LUMBER SPECIES
(Ground Snow Load = 30 pounds per square foot, Ceiling Not Attached to Rafters, $L/\Delta = 180$)

RAFTER SPACING (inches)	SPECIES AND GRADE		DEAD LOAD = 10 pounds per square foot					DEAD LOAD = 20 pounds per square foot				
			2 × 4	2 × 6	2 × 8	2 × 10	2 × 12	2 × 4	2 × 6	2 × 8	2 × 10	2 × 12
			(ft. - in.)	(ft. - in.)	(ft. - in.)	(ft. - in.)	(ft. - in.)	(ft. - in.)	(ft. - in.)	(ft. - in.)	(ft. - in.)	(ft. - in.)
			Maximum rafter spans									
19.2	Douglas Fir-Larch	SS	8-7	13-6	17-9	21-8	25-2	8-7	12-6	15-10	19-5	22-6
	Douglas Fir-Larch	#1	7-11	11-8	14-9	18-0	20-11	7-1	10-5	13-2	16-1	18-8
	Douglas Fir-Larch	#2	7-5	10-11	13-9	16-10	19-6	6-8	9-9	12-4	15-1	17-6
	Douglas Fir-Larch	#3	5-7	8-3	10-5	12-9	14-9	5-0	7-4	9-4	11-5	13-2
	Hem-Fir	SS	8-1	12-9	16-9	21-4	24-8	8-1	12-4	15-7	19-1	22-1
	Hem-Fir	#1	7-9	11-4	14-4	17-7	20-4	6-11	10-2	12-10	15-8	18-2
	Hem-Fir	#2	7-4	10-9	13-7	16-7	19-3	6-7	9-7	12-2	14-10	17-3
	Hem-Fir	#3	5-7	8-3	10-5	12-9	14-9	5-0	7-4	9-4	11-5	13-2
	Southern Pine	SS	8-5	13-3	17-5	22-3	Note a	8-5	13-3	17-5	22-0	25-9
	Southern Pine	#1	8-3	13-0	16-6	19-7	23-4	7-11	11-9	14-9	17-6	20-11
	Southern Pine	#2	7-11	11-5	14-9	17-7	20-7	7-1	10-2	13-2	15-9	18-5
	Southern Pine	#3	6-0	8-10	11-3	13-4	15-10	5-4	7-11	10-1	11-11	14-2
	Spruce-Pine-Fir	SS	7-11	12-5	16-5	20-2	23-4	7-11	11-8	14-9	18-0	20-11
	Spruce-Pine-Fir	#1	7-5	10-11	13-9	16-10	19-6	6-8	9-9	12-4	15-1	17-6
	Spruce-Pine-Fir	#2	7-5	10-11	13-9	16-10	19-6	6-8	9-9	12-4	15-1	17-6
	Spruce-Pine-Fir	#3	5-7	8-3	10-5	12-9	14-9	5-0	7-4	9-4	11-5	13-2
24	Douglas Fir-Larch	SS	7-11	12-6	15-10	19-5	22-6	7-8	11-3	14-2	17-4	20-1
	Douglas Fir-Larch	#1	7-1	10-5	13-2	16-1	18-8	6-4	9-4	11-9	14-5	16-8
	Douglas Fir-Larch	#2	6-8	9-9	12-4	15-1	17-6	5-11	8-8	11-0	13-6	15-7
	Douglas Fir-Larch	#3	5-0	7-4	9-4	11-5	13-2	4-6	6-7	8-4	10-2	11-10
	Hem-Fir	SS	7-6	11-10	15-7	19-1	22-1	7-6	11-0	13-11	17-0	19-9
	Hem-Fir	#1	6-11	10-2	12-10	15-8	18-2	6-2	9-1	11-6	14-0	16-3
	Hem-Fir	#2	6-7	9-7	12-2	14-10	17-3	5-10	8-7	10-10	13-3	15-5
	Hem-Fir	#3	5-0	7-4	9-4	11-5	13-2	4-6	6-7	8-4	10-2	11-10
	Southern Pine	SS	7-10	12-3	16-2	20-8	25-1	7-10	12-3	16-2	19-8	23-0
	Southern Pine	#1	7-8	11-9	14-9	17-6	20-11	7-1	10-6	13-2	15-8	18-8
	Southern Pine	#2	7-1	10-2	13-2	15-9	18-5	6-4	9-2	11-9	14-1	16-6
	Southern Pine	#3	5-4	7-11	10-1	11-11	14-2	4-9	7-1	9-0	10-8	12-8
	Spruce-Pine-Fir	SS	7-4	11-7	14-9	18-0	20-11	7-1	10-5	13-2	16-1	18-8
	Spruce-Pine-Fir	#1	6-8	9-9	12-4	15-1	17-6	5-11	8-8	11-0	13-6	15-7
	Spruce-Pine-Fir	#2	6-8	9-9	12-4	15-1	17-6	5-11	8-8	11-0	13-6	15-7
	Spruce-Pine-Fir	#3	5-0	7-4	9-4	11-5	13-2	4-6	6-7	8-4	10-2	11-10

For SI: 1 inch = 25.4 mm, 1 foot = 304.8 mm, 1 pound per square foot = 47.9 N/m^2.

a. Span exceeds 26 feet in length. Check sources for availability of lumber in lengths greater than 20 feet.

TABLE 2308.10.3(4)
RAFTER SPANS FOR COMMON LUMBER SPECIES
(Ground Snow Load = 50 pounds per square foot, Ceiling Not Attached to Rafters, $L/\Delta = 180$)

RAFTER SPACING (inches)	SPECIES AND GRADE		DEAD LOAD = 10 pounds per square foot					DEAD LOAD = 20 pounds per square foot				
			Maximum rafter spans									
			2 × 4	2 × 6	2 × 8	2 × 10	2 × 12	2 × 4	2 × 6	2 × 8	2 × 10	2 × 12
			(ft. - in.)	(ft. - in.)	(ft. - in.)	(ft. - in.)	(ft. - in.)	(ft. - in.)	(ft. - in.)	(ft. - in.)	(ft. - in.)	(ft. - in.)
12	Douglas Fir-Larch	SS	8-5	13-3	17-6	22-4	26-0	8-5	13-3	17-0	20-9	24-10
	Douglas Fir-Larch	#1	8-2	12-0	15-3	18-7	21-7	7-7	11-2	14-1	17-3	20-0
	Douglas Fir-Larch	#2	7-8	11-3	14-3	17-5	20-2	7-1	10-5	13-2	16-1	18-8
	Douglas Fir-Larch	#3	5-10	8-6	10-9	13-2	15-3	5-5	7-10	10-0	12-2	14-1
	Hem-Fir	SS	8-0	12-6	16-6	21-1	25-6	8-0	12-6	16-6	20-4	23-7
	Hem-Fir	#1	7-10	11-9	14-10	18-1	21-0	7-5	10-10	13-9	16-9	19-5
	Hem-Fir	#2	7-5	11-1	14-0	17-2	19-11	7-0	10-3	13-0	15-10	18-5
	Hem-Fir	#3	5-10	8-6	10-9	13-2	15-3	5-5	7-10	10-0	12-2	14-1
	Southern Pine	SS	8-4	13-0	17-2	21-11	Note a	8-4	13-0	17-2	21-11	Note a
	Southern Pine	#1	8-2	12-10	16-10	20-3	24-1	8-2	12-6	15-9	18-9	22-4
	Southern Pine	#2	8-0	11-9	15-3	18-2	21-3	7-7	10-11	14-1	16-10	19-9
	Southern Pine	#3	6-2	9-2	11-8	13-9	16-4	5-9	8-5	10-9	12-9	15-2
	Spruce-Pine-Fir	SS	7-10	12-3	16-2	20-8	24-1	7-10	12-3	15-9	19-3	22-4
	Spruce-Pine-Fir	#1	7-8	11-3	14-3	17-5	20-2	7-1	10-5	13-2	16-1	18-8
	Spruce-Pine-Fir	#2	7-8	11-3	14-3	17-5	20-2	7-1	10-5	13-2	16-1	18-8
	Spruce-Pine-Fir	#3	5-10	8-6	10-9	13-2	15-3	5-5	7-10	10-0	12-2	14-1
16	Douglas Fir-Larch	SS	7-8	12-1	15-10	19-5	22-6	7-8	11-7	14-8	17-11	20-10
	Douglas Fir-Larch	#1	7-1	10-5	13-2	16-1	18-8	6-7	9-8	12-2	14-11	17-3
	Douglas Fir-Larch	#2	6-8	9-9	12-4	15-1	17-6	6-2	9-0	11-5	13-11	16-2
	Douglas Fir-Larch	#3	5-0	7-4	9-4	11-5	13-2	4-8	6-10	8-8	10-6	12-3
	Hem-Fir	SS	7-3	11-5	15-0	19-1	22-1	7-3	11-5	14-5	17-8	20-5
	Hem-Fir	#1	6-11	10-2	12-10	15-8	18-2	6-5	9-5	11-11	14-6	16-10
	Hem-Fir	#2	6-7	9-7	12-2	14-10	17-3	6-1	8-11	11-3	13-9	15-11
	Hem-Fir	#3	5-0	7-4	9-4	11-5	13-2	4-8	6-10	8-8	10-6	12-3
	Southern Pine	SS	7-6	11-10	15-7	19-11	24-3	7-6	11-10	15-7	19-11	23-10
	Southern Pine	#1	7-5	11-7	14-9	17-6	20-11	7-4	10-10	13-8	16-2	19-4
	Southern Pine	#2	7-1	10-2	13-2	15-9	18-5	6-7	9-5	12-2	14-7	17-1
	Southern Pine	#3	5-4	7-11	10-1	11-11	14-2	4-11	7-4	9-4	11-0	13-1
	Spruce-Pine-Fir	SS	7-1	11-2	14-8	18-0	20-11	7-1	10-9	13-8	16-8	19-4
	Spruce-Pine-Fir	#1	6-8	9-9	12-4	15-1	17-6	6-2	9-0	11-5	13-11	16-2
	Spruce-Pine-Fir	#2	6-8	9-9	12-4	15-1	17-6	6-2	9-0	11-5	13-11	16-2
	Spruce-Pine-Fir	#3	5-0	7-4	9-4	11-5	13-2	4-8	6-10	8-8	10-6	12-3

(continued)

TABLE 2308.10.3(4)—continued
RAFTER SPANS FOR COMMON LUMBER SPECIES
(Ground Snow Load = 50 pounds per square foot, Ceiling Not Attached to Rafters, $L/\Delta = 180$)

RAFTER SPACING (inches)	SPECIES AND GRADE		DEAD LOAD = 10 pounds per square foot					DEAD LOAD = 20 pounds per square foot				
			2 × 4	2 × 6	2 × 8	2 × 10	2 × 12	2 × 4	2 × 6	2 × 8	2 × 10	2 × 12
			\-\- Maximum rafter spans \-\-									
			(ft. - in.)	(ft. - in.)	(ft. - in.)	(ft. - in.)	(ft. - in.)	(ft. - in.)	(ft. - in.)	(ft. - in.)	(ft. - in.)	(ft. - in.)
19.2	Douglas Fir-Larch	SS	7-3	11-4	14-6	17-8	20-6	7-3	10-7	13-5	16-5	19-0
	Douglas Fir-Larch	#1	6-6	9-6	12-0	14-8	17-1	6-0	8-10	11-2	13-7	15-9
	Douglas Fir-Larch	#2	6-1	8-11	11-3	13-9	15-11	5-7	8-3	10-5	12-9	14-9
	Douglas Fir-Larch	#3	4-7	6-9	8-6	10-5	12-1	4-3	6-3	7-11	9-7	11-2
	Hem-Fir	SS	6-10	10-9	14-2	17-5	20-2	6-10	10-5	13-2	16-1	18-8
	Hem-Fir	#1	6-4	9-3	11-9	14-4	16-7	5-10	8-7	10-10	13-3	15-5
	Hem-Fir	#2	6-0	8-9	11-1	13-7	15-9	5-7	8-1	10-3	12-7	14-7
	Hem-Fir	#3	4-7	6-9	8-6	10-5	12-1	4-3	6-3	7-11	9-7	11-2
	Southern Pine	SS	7-1	11-2	14-8	18-9	22-10	7-1	11-2	14-8	18-7	21-9
	Southern Pine	#1	7-0	10-8	13-5	16-0	19-1	6-8	9-11	12-5	14-10	17-8
	Southern Pine	#2	6-6	9-4	12-0	14-4	16-10	6-0	8-8	11-2	13-4	15-7
	Southern Pine	#3	4-11	7-3	9-2	10-10	12-11	4-6	6-8	8-6	10-1	12-0
	Spruce-Pine-Fir	SS	6-8	10-6	13-5	16-5	19-1	6-8	9-10	12-5	15-3	17-8
	Spruce-Pine-Fir	#1	6-1	8-11	11-3	13-9	15-11	5-7	8-3	10-5	12-9	14-9
	Spruce-Pine-Fir	#2	6-1	8-11	11-3	13-9	15-11	5-7	8-3	10-5	12-9	14-9
	Spruce-Pine-Fir	#3	4-7	6-9	8-6	10-5	12-1	4-3	6-3	7-11	9-7	11-2
24	Douglas Fir-Larch	SS	6-8	10-3	13-0	15-10	18-4	6-6	9-6	12-0	14-8	17-0
	Douglas Fir-Larch	#1	5-10	8-6	10-9	13-2	15-3	5-5	7-10	10-0	12-2	14-1
	Douglas Fir-Larch	#2	5-5	7-11	10-1	12-4	14-3	5-0	7-4	9-4	11-5	13-2
	Douglas Fir-Larch	#3	4-1	6-0	7-7	9-4	10-9	3-10	5-7	7-1	8-7	10-0
	Hem-Fir	SS	6-4	9-11	12-9	15-7	18-0	6-4	9-4	11-9	14-5	16-8
	Hem-Fir	#1	5-8	8-3	10-6	12-10	14-10	5-3	7-8	9-9	11-10	13-9
	Hem-Fir	#2	5-4	7-10	9-11	12-1	14-1	4-11	7-3	9-2	11-3	13-0
	Hem-Fir	#3	4-1	6-0	7-7	9-4	10-9	3-10	5-7	7-1	8-7	10-0
	Southern Pine	SS	6-7	10-4	13-8	17-5	21-0	6-7	10-4	13-8	16-7	19-5
	Southern Pine	#1	6-5	9-7	12-0	14-4	17-1	6-0	8-10	11-2	13-3	15-9
	Southern Pine	#2	5-10	8-4	10-9	12-10	15-1	5-5	7-9	10-0	11-11	13-11
	Southern Pine	#3	4-4	6-5	8-3	9-9	11-7	4-1	6-0	7-7	9-0	10-8
	Spruce-Pine-Fir	SS	6-2	9-6	12-0	14-8	17-1	6-0	8-10	11-2	13-7	15-9
	Spruce-Pine-Fir	#1	5-5	7-11	10-1	12-4	14-3	5-0	7-4	9-4	11-5	13-2
	Spruce-Pine-Fir	#2	5-5	7-11	10-1	12-4	14-3	5-0	7-4	9-4	11-5	13-2
	Spruce-Pine-Fir	#3	4-1	6-0	7-7	9-4	10-9	3-10	5-7	7-1	8-7	10-0

For SI: 1 inch = 25.4 mm, 1 foot = 304.8 mm, 1 pound per square foot = 47.9 N/m².

a. Span exceeds 26 feet in length. Check sources for availability of lumber in lengths greater than 20 feet.

TABLE 2308.10.3(5)
RAFTER SPANS FOR COMMON LUMBER SPECIES
(Ground Snow Load = 30 pounds per square foot, Ceiling Attached to Rafters, $L/\Delta = 240$)

RAFTER SPACING (inches)	SPECIES AND GRADE		DEAD LOAD = 10 pounds per square foot					DEAD LOAD = 20 pounds per square foot				
			2 × 4	2 × 6	2 × 8	2 × 10	2 × 12	2 × 4	2 × 6	2 × 8	2 × 10	2 × 12
			(ft. - in.)	(ft. - in.)	(ft. - in.)	(ft. - in.)	(ft. - in.)	(ft. - in.)	(ft. - in.)	(ft. - in.)	(ft. - in.)	(ft. - in.)
			Maximum rafter spans									
12	Douglas Fir-Larch	SS	9-1	14-4	18-10	24-1	Note a	9-1	14-4	18-10	24-1	Note a
	Douglas Fir-Larch	#1	8-9	13-9	18-2	22-9	Note a	8-9	13-2	16-8	20-4	23-7
	Douglas Fir-Larch	#2	8-7	13-6	17-5	21-4	24-8	8-5	12-4	15-7	19-1	22-1
	Douglas Fir-Larch	#3	7-1	10-5	13-2	16-1	18-8	6-4	9-4	11-9	14-5	16-8
	Hem-Fir	SS	8-7	13-6	17-10	22-9	Note a	8-7	13-6	17-10	22-9	Note a
	Hem-Fir	#1	8-5	13-3	17-5	22-2	25-9	8-5	12-10	16-3	19-10	23-0
	Hem-Fir	#2	8-0	12-7	16-7	21-0	24-4	8-0	12-2	15-4	18-9	21-9
	Hem-Fir	#3	7-1	10-5	13-2	16-1	18-8	6-4	9-4	11-9	14-5	16-8
	Southern Pine	SS	8-11	14-1	18-6	23-8	Note a	8-11	14-1	18-6	23-8	Note a
	Southern Pine	#1	8-9	13-9	18-2	23-2	Note a	8-9	13-9	18-2	22-2	Note a
	Southern Pine	#2	8-7	13-6	17-10	22-3	Note a	8-7	12-11	16-8	19-11	23-4
	Southern Pine	#3	7-7	11-2	14-3	16-10	20-0	6-9	10-0	12-9	15-1	17-11
	Spruce-Pine-Fir	SS	8-5	13-3	17-5	22-3	Note a	8-5	13-3	17-5	22-3	Note a
	Spruce-Pine-Fir	#1	8-3	12-11	17-0	21-4	24-8	8-3	12-4	15-7	19-1	22-1
	Spruce-Pine-Fir	#2	8-3	12-11	17-0	21-4	24-8	8-3	12-4	15-7	19-1	22-1
	Spruce-Pine-Fir	#3	7-1	10-5	13-2	16-1	18-8	6-4	9-4	11-9	14-5	16-8
16	Douglas Fir-Larch	SS	8-3	13-0	17-2	21-10	Note a	8-3	13-0	17-2	21-3	24-8
	Douglas Fir-Larch	#1	8-0	12-6	16-2	19-9	22-10	7-10	11-5	14-5	17-8	20-5
	Douglas Fir-Larch	#2	7-10	11-11	15-1	18-5	21-5	7-3	10-8	13-6	16-6	19-2
	Douglas Fir-Larch	#3	6-2	9-0	11-5	13-11	16-2	5-6	8-1	10-3	12-6	14-6
	Hem-Fir	SS	7-10	12-3	16-2	20-8	25-1	7-10	12-3	16-2	20-8	24-2
	Hem-Fir	#1	7-8	12-0	15-9	19-3	22-3	7-7	11-1	14-1	17-2	19-11
	Hem-Fir	#2	7-3	11-5	14-11	18-2	21-1	7-2	10-6	13-4	16-3	18-10
	Hem-Fir	#3	6-2	9-0	11-5	13-11	16-2	5-6	8-1	10-3	12-6	14-6
	Southern Pine	SS	8-1	12-9	16-10	21-6	Note a	8-1	12-9	16-10	21-6	Note a
	Southern Pine	#1	8-0	12-6	16-6	21-1	25-7	8-0	12-6	16-2	19-2	22-10
	Southern Pine	#2	7-10	12-3	16-2	19-3	22-7	7-10	11-2	14-5	17-3	20-2
	Southern Pine	#3	6-7	9-8	12-4	14-7	17-4	5-10	8-8	11-0	13-0	15-6
	Spruce-Pine-Fir	SS	7-8	12-0	15-10	20-2	24-7	7-8	12-0	15-10	19-9	22-10
	Spruce-Pine-Fir	#1	7-6	11-9	15-1	18-5	21-5	7-3	10-8	13-6	16-6	19-2
	Spruce-Pine-Fir	#2	7-6	11-9	15-1	18-5	21-5	7-3	10-8	13-6	16-6	19-2
	Spruce-Pine-Fir	#3	6-2	9-0	11-5	13-11	16-2	5-6	8-1	10-3	12-6	14-6

(continued)

TABLE 2308.10.3(5)—continued
RAFTER SPANS FOR COMMON LUMBER SPECIES
(Ground Snow Load = 30 pounds per square foot, Ceiling Attached to Rafters, $L/\Delta = 240$)

RAFTER SPACING (inches)	SPECIES AND GRADE		DEAD LOAD = 10 pounds per square foot					DEAD LOAD = 20 pounds per square foot				
			Maximum rafter spans									
			2 × 4	2 × 6	2 × 8	2 × 10	2 × 12	2 × 4	2 × 6	2 × 8	2 × 10	2 × 12
			(ft. - in.)	(ft. - in.)	(ft. - in.)	(ft. - in.)	(ft. - in.)	(ft. - in.)	(ft. - in.)	(ft. - in.)	(ft. - in.)	(ft. - in.)
19.2	Douglas Fir-Larch	SS	7-9	12-3	16-1	20-7	25-0	7-9	12-3	15-10	19-5	22-6
	Douglas Fir-Larch	#1	7-6	11-8	14-9	18-0	20-11	7-1	10-5	13-2	16-1	18-8
	Douglas Fir-Larch	#2	7-4	10-11	13-9	16-10	19-6	6-8	9-9	12-4	15-1	17-6
	Douglas Fir-Larch	#3	5-7	8-3	10-5	12-9	14-9	5-0	7-4	9-4	11-5	13-2
	Hem-Fir	SS	7-4	11-7	15-3	19-5	23-7	7-4	11-7	15-3	19-1	22-1
	Hem-Fir	#1	7-2	11-4	14-4	17-7	20-4	6-11	10-2	12-10	15-8	18-2
	Hem-Fir	#2	6-10	10-9	13-7	16-7	19-3	6-7	9-7	12-2	14-10	17-3
	Hem-Fir	#3	5-7	8-3	10-5	12-9	14-9	5-0	7-4	9-4	11-5	13-2
	Southern Pine	SS	7-8	12-0	15-10	20-2	24-7	7-8	12-0	15-10	20-2	24-7
	Southern Pine	#1	7-6	11-9	15-6	19-7	23-4	7-6	11-9	14-9	17-6	20-11
	Southern Pine	#2	7-4	11-5	14-9	17-7	20-7	7-1	10-2	13-2	15-9	18-5
	Southern Pine	#3	6-0	8-10	11-3	13-4	15-10	5-4	7-11	10-1	11-11	14-2
	Spruce-Pine-Fir	SS	7-2	11-4	14-11	19-0	23-1	7-2	11-4	14-9	18-0	20-11
	Spruce-Pine-Fir	#1	7-0	10-11	13-9	16-10	19-6	6-8	9-9	12-4	15-1	17-6
	Spruce-Pine-Fir	#2	7-0	10-11	13-9	16-10	19-6	6-8	9-9	12-4	15-1	17-6
	Spruce-Pine-Fir	#3	5-7	8-3	10-5	12-9	14-9	5-0	7-4	9-4	11-5	13-2
24	Douglas Fir-Larch	SS	7-3	11-4	15-0	19-1	22-6	7-3	11-3	14-2	17-4	20-1
	Douglas Fir-Larch	#1	7-0	10-5	13-2	16-1	18-8	6-4	9-4	11-9	14-5	16-8
	Douglas Fir-Larch	#2	6-8	9-9	12-4	15-1	17-6	5-11	8-8	11-0	13-6	15-7
	Douglas Fir-Larch	#3	5-0	7-4	9-4	11-5	13-2	4-6	6-7	8-4	10-2	11-10
	Hem-Fir	SS	6-10	10-9	14-2	18-0	21-11	6-10	10-9	13-11	17-0	19-9
	Hem-Fir	#1	6-8	10-2	12-10	15-8	18-2	6-2	9-1	11-6	14-0	16-3
	Hem-Fir	#2	6-4	9-7	12-2	14-10	17-3	5-10	8-7	10-10	13-3	15-5
	Hem-Fir	#3	5-0	7-4	9-4	11-5	13-2	4-6	6-7	8-4	10-2	11-10
	Southern Pine	SS	7-1	11-2	14-8	18-9	22-10	7-1	11-2	14-8	18-9	22-10
	Southern Pine	#1	7-0	10-11	14-5	17-6	20-11	7-0	10-6	13-2	15-8	18-8
	Southern Pine	#2	6-10	10-2	13-2	15-9	18-5	6-4	9-2	11-9	14-1	16-6
	Southern Pine	#3	5-4	7-11	10-1	11-11	14-2	4-9	7-1	9-0	10-8	12-8
	Spruce-Pine-Fir	SS	6-8	10-6	13-10	17-8	20-11	6-8	10-5	13-2	16-1	18-8
	Spruce-Pine-Fir	#1	6-6	9-9	12-4	15-1	17-6	5-11	8-8	11-0	13-6	15-7
	Spruce-Pine-Fir	#2	6-6	9-9	12-4	15-1	17-6	5-11	8-8	11-0	13-6	15-7
	Spruce-Pine-Fir	#3	5-0	7-4	9-4	11-5	13-2	4-6	6-7	8-4	10-2	11-10

For SI: 1 inch = 25.4 mm, 1 foot = 304.8 mm, 1 pound per square foot = 47.9 N/m².

a. Span exceeds 26 feet in length. Check sources for availability of lumber in lengths greater than 20 feet.

TABLE 2308.10.3(6)
RAFTER SPANS FOR COMMON LUMBER SPECIES
(Ground Snow Load = 50 pounds per square foot, Ceiling Attached to Rafters, $L/\Delta = 240$)

RAFTER SPACING (inches)	SPECIES AND GRADE		DEAD LOAD = 10 pounds per square foot					DEAD LOAD = 20 pounds per square foot				
			Maximum rafter spans									
			2 × 4	2 × 6	2 × 8	2 × 10	2 × 12	2 × 4	2 × 6	2 × 8	2 × 10	2 × 12
			(ft. - in.)	(ft. - in.)	(ft. - in.)	(ft. - in.)	(ft. - in.)	(ft. - in.)	(ft. - in.)	(ft. - in.)	(ft. - in.)	(ft. - in.)
12	Douglas Fir-Larch	SS	7-8	12-1	15-11	20-3	24-8	7-8	12-1	15-11	20-3	24-0
	Douglas Fir-Larch	#1	7-5	11-7	15-3	18-7	21-7	7-5	11-2	14-1	17-3	20-0
	Douglas Fir-Larch	#2	7-3	11-3	14-3	17-5	20-2	7-1	10-5	13-2	16-1	18-8
	Douglas Fir-Larch	#3	5-10	8-6	10-9	13-2	15-3	5-5	7-10	10-0	12-2	14-1
	Hem-Fir	SS	7-3	11-5	15-0	19-2	23-4	7-3	11-5	15-0	19-2	23-4
	Hem-Fir	#1	7-1	11-2	14-8	18-1	21-0	7-1	10-10	13-9	16-9	19-5
	Hem-Fir	#2	6-9	10-8	14-0	17-2	19-11	6-9	10-3	13-0	15-10	18-5
	Hem-Fir	#3	5-10	8-6	10-9	13-2	15-3	5-5	7-10	10-0	12-2	14-1
	Southern Pine	SS	7-6	11-0	15-7	19-11	24-3	7-6	11-10	15-7	19-11	24-3
	Southern Pine	#1	7-5	11-7	15-4	19-7	23-9	7-5	11-7	15-4	18-9	22-4
	Southern Pine	#2	7-3	11-5	15-0	18-2	21-3	7-3	10-11	14-1	16-10	19-9
	Southern Pine	#3	6-2	9-2	11-8	13-9	16-4	5-9	8-5	10-9	12-9	15-2
	Spruce-Pine-Fir	SS	7-1	11-2	14-8	18-9	22-10	7-1	11-2	14-8	18-9	22-4
	Spruce-Pine-Fir	#1	6-11	10-11	14-3	17-5	20-2	6-11	10-5	13-2	16-1	18-8
	Spruce-Pine-Fir	#2	6-11	10-11	14-3	17-5	20-2	6-11	10-5	13-2	16-1	18-8
	Spruce-Pine-Fir	#3	5-10	8-6	10-9	13-2	15-3	5-5	7-10	10-0	12-2	14-1
16	Douglas Fir-Larch	SS	7-0	11-0	14-5	18-5	22-5	7-0	11-0	14-5	17-11	20-10
	Douglas Fir-Larch	#1	6-9	10-5	13-2	16-1	18-8	6-7	9-8	12-2	14-11	17-3
	Douglas Fir-Larch	#2	6-7	9-9	12-4	15-1	17-6	6-2	9-0	11-5	13-11	16-2
	Douglas Fir-Larch	#3	5-0	7-4	9-4	11-5	13-2	4-8	6-10	8-8	10-6	12-3
	Hem-Fir	SS	6-7	10-4	13-8	17-5	21-2	6-7	10-4	13-8	17-5	20-5
	Hem-Fir	#1	6-5	10-2	12-10	15-8	18-2	6-5	9-5	11-11	14-6	16-10
	Hem-Fir	#2	6-2	9-7	12-2	14-10	17-3	6-1	8-11	11-3	13-9	15-11
	Hem-Fir	#3	5-0	7-4	9-4	11-5	13-2	4-8	6-10	8-8	10-6	12-3
	Southern Pine	SS	6-10	10-9	14-2	18-1	22-0	6-10	10-9	14-2	18-1	22-0
	Southern Pine	#1	6-9	10-7	13-11	17-6	20-11	6-9	10-7	13-8	16-2	19-4
	Southern Pine	#2	6-7	10-2	13-2	15-9	18-5	6-7	9-5	12-2	14-7	17-1
	Southern Pine	#3	5-4	7-11	10-1	11-11	14-2	4-11	7-4	9-4	11-0	13-1
	Spruce-Pine-Fir	SS	6-5	10-2	13-4	17-0	20-9	6-5	10-2	13-4	16-8	19-4
	Spruce-Pine-Fir	#1	6-4	9-9	12-4	15-1	17-6	6-2	9-0	11-5	13-11	16-2
	Spruce-Pine-Fir	#2	6-4	9-9	12-4	15-1	17-6	6-2	9-0	11-5	13-11	16-2
	Spruce-Pine-Fir	#3	5-0	7-4	9-4	11-5	13-2	4-8	6-10	8-8	10-6	12-3

(continued)

TABLE 2308.10.3(6)—continued
RAFTER SPANS FOR COMMON LUMBER SPECIES
(Ground Snow Load = 50 pounds per square foot, Ceiling Attached to Rafters, $L/\Delta = 240$)

RAFTER SPACING (inches)	SPECIES AND GRADE		DEAD LOAD = 10 pounds per square foot					DEAD LOAD = 20 pounds per square foot				
			2 × 4	2 × 6	2 × 8	2 × 10	2 × 12	2 × 4	2 × 6	2 × 8	2 × 10	2 × 12
			Maximum rafter spans									
			(ft. - in.)	(ft. - in.)	(ft. - in.)	(ft. - in.)	(ft. - in.)	(ft. - in.)	(ft. - in.)	(ft. - in.)	(ft. - in.)	(ft. - in.)
19.2	Douglas Fir-Larch	SS	6-7	10-4	13-7	17-4	20-6	6-7	10-4	13-5	16-5	19-0
	Douglas Fir-Larch	#1	6-4	9-6	12-0	14-8	17-1	6-0	8-10	11-2	13-7	15-9
	Douglas Fir-Larch	#2	6-1	8-11	11-3	13-9	15-11	5-7	8-3	10-5	12-9	14-9
	Douglas Fir-Larch	#3	4-7	6-9	8-6	10-5	12-1	4-3	6-3	7-11	9-7	11-2
	Hem-Fir	SS	6-2	9-9	12-10	16-5	19-11	6-2	9-9	12-10	16-1	18-8
	Hem-Fir	#1	6-1	9-3	11-9	14-4	16-7	5-10	8-7	10-10	13-3	15-5
	Hem-Fir	#2	5-9	8-9	11-1	13-7	15-9	5-7	8-1	10-3	12-7	14-7
	Hem-Fir	#3	4-7	6-9	8-6	10-5	12-1	4-3	6-3	7-11	9-7	11-2
	Southern Pine	SS	6-5	10-2	13-4	17-0	20-9	6-5	10-2	13-4	17-0	20-9
	Southern Pine	#1	6-4	9-11	13-1	16-0	19-1	6-4	9-11	12-5	14-10	17-8
	Southern Pine	#2	6-2	9-4	12-0	14-4	16-10	6-0	8-8	11-2	13-4	15-7
	Southern Pine	#3	4-11	7-3	9-2	10-10	12-11	4-6	6-8	8-6	10-1	12-0
	Spruce-Pine-Fir	SS	6-1	9-6	12-7	16-0	19-1	6-1	9-6	12-5	15-3	17-8
	Spruce-Pine-Fir	#1	5-11	8-11	11-3	13-9	15-11	5-7	8-3	10-5	12-9	14-9
	Spruce-Pine-Fir	#2	5-11	8-11	11-3	13-9	15-11	5-7	8-3	10-5	12-9	14-9
	Spruce-Pine-Fir	#3	4-7	6-9	8-6	10-5	12-1	4-3	6-3	7-11	9-7	11-2
24	Douglas Fir-Larch	SS	6-1	9-7	12-7	15-10	18-4	6-1	9-6	12-0	14-8	17-0
	Douglas Fir-Larch	#1	5-10	8-6	10-9	13-2	15-3	5-5	7-10	10-0	12-2	14-1
	Douglas Fir-Larch	#2	5-5	7-11	10-1	12-4	14-3	5-0	7-4	9-4	11-5	13-2
	Douglas Fir-Larch	#3	4-1	6-0	7-7	9-4	10-9	3-10	5-7	7-1	8-7	10-0
	Hem-Fir	SS	5-9	9-1	11-11	15-2	18-0	5-9	9-1	11-9	14-5	16-8
	Hem-Fir	#1	5-8	8-3	10-6	12-10	14-10	5-3	7-8	9-9	11-10	13-9
	Hem-Fir	#2	5-4	7-10	9-11	12-1	14-1	4-11	7-3	9-2	11-3	13-0
	Hem-Fir	#3	4-1	6-0	7-7	9-4	10-9	3-10	5-7	7-1	8-7	10-0
	Southern Pine	SS	6-0	9-5	12-5	15-10	19-3	6-0	9-5	12-5	15-10	19-3
	Southern Pine	#1	5-10	9-3	12-0	14-4	17-1	5-10	8-10	11-2	13-3	15-9
	Southern Pine	#2	5-9	8-4	10-9	12-10	15-1	5-5	7-9	10-0	11-11	13-11
	Southern Pine	#3	4-4	6-5	8-3	9-9	11-7	4-1	6-0	7-7	9-0	10-8
	Spruce-Pine-Fir	SS	5-8	8-10	11-8	14-8	17-1	5-8	8-10	11-2	13-7	15-9
	Spruce-Pine-Fir	#1	5-5	7-11	10-1	12-4	14-3	5-0	7-4	9-4	11-5	13-2
	Spruce-Pine-Fir	#2	5-5	7-11	10-1	12-4	14-3	5-0	7-4	9-4	11-5	13-2
	Spruce-Pine-Fir	#3	4-1	6-0	7-7	9-4	10-9	3-10	5-7	7-1	8-7	10-0

For SI: 1 inch = 25.4 mm, 1 foot = 304.8 mm, 1 pound per square foot = 47.9 N/m².

2308.10.4 Ceiling joist and rafter framing. Rafters shall be framed directly opposite each other at the ridge. There shall be a ridge board at least 1-inch (25 mm) nominal thickness at ridges and not less in depth than the cut end of the rafter. At valleys and hips, there shall be a single valley or hip rafter not less than 2-inch (51 mm) nominal thickness and not less in depth than the cut end of the rafter.

2308.10.4.1 Ceiling joist and rafter connections. Ceiling joists and rafters shall be nailed to each other and the assembly shall be nailed to the top wall plate in accordance with Tables 2304.9.1 and 2308.10.1. Ceiling joists shall be continuous or securely joined where they meet over interior partitions and fastened to adjacent rafters in accordance with Tables 2308.10.4.1 and 2304.9.1 to provide a continuous rafter tie across the building where such joists are parallel to the rafters. Ceiling joists shall have a bearing surface of not less than $1^1/_2$ inches (38 mm) on the top plate at each end.

Where ceiling joists are not parallel to rafters, an equivalent rafter tie shall be installed in a manner to provide a continuous tie across the building, at a spacing of not more than 4 feet (1219 mm) o.c. The connections shall be in accordance with Tables 2308.10.4.1 and 2304.9.1, or connections of equivalent capacities shall be provided. Where ceiling joists or rafter ties are not provided at the top of the rafter support walls, the ridge formed by these rafters shall also be supported by a girder conforming to Section 2308.4.

Rafter ties shall be spaced not more than 4 feet (1219 mm) o.c. Rafter tie connections shall be based on the equivalent rafter spacing in Table 2308.10.4.1. Where rafter ties are spaced at 32 inches (813 mm) o.c., the number of 16d common nails shall be two times the number specified for rafters spaced 16 inches (406 mm) o.c., with a minimum of 4-16d common nails where no snow loads are indicated. Where rafter ties are spaced at 48 inches (1219 mm) o.c., the number of 16d common nails shall be two times the number specified for rafters spaced 24 inches (610 mm) o.c., with a minimum of 6-16d common nails where no snow loads are indicated. Rafter/ceiling joist connections and rafter/tie connections shall be of sufficient size and number to prevent splitting from nailing.

2308.10.4.2 Notches and holes. Notching at the ends of rafters or ceiling joists shall not exceed one-fourth the depth. Notches in the top or bottom of the rafter or ceiling joist shall not exceed one-sixth the depth and shall not be located in the middle one-third of the span, except that a notch not exceeding one-third of the depth is permitted in the top of the rafter or ceiling joist not further from the face of the support than the depth of the member.

Holes bored in rafters or ceiling joists shall not be within 2 inches (51 mm) of the top and bottom and their diameter shall not exceed one-third the depth of the member.

2308.10.4.3 Framing around openings. Trimmer and header rafters shall be doubled, or of lumber of equivalent cross section, where the span of the header exceeds 4 feet (1219 mm). The ends of header rafters more than 6 feet (1829 mm) long shall be supported by framing anchors or rafter hangers unless bearing on a beam, partition or wall.

2308.10.5 Purlins. Purlins to support roof loads are permitted to be installed to reduce the span of rafters within allowable limits and shall be supported by struts to bearing walls. The maximum span of 2-inch by 4-inch (51 mm by 102 mm) purlins shall be 4 feet (1219 mm). The maximum span of the 2-inch by 6-inch (51 mm by 152 mm) purlin shall be 6 feet (1829 mm), but in no case shall the purlin be smaller than the supported rafter. Struts shall not be smaller than 2-inch by 4-inch (51 mm by 102 mm) members. The unbraced length of struts shall not exceed 8 feet (2438 mm) and the minimum slope of the struts shall not be less than 45 degrees (0.79 rad) from the horizontal.

2308.10.6 Blocking. Roof rafters and ceiling joists shall be supported laterally to prevent rotation and lateral displacement in accordance with the provisions of Section 2308.8.5.

2308.10.7 Wood trusses.

2308.10.7.1 Design. Wood trusses shall be designed in accordance with the requirements of Chapter 23 and accepted engineering practice. Members are permitted to be joined by nails, glue, bolts, timber connectors, metal connector plates or other approved framing devices.

2308.10.7.2 Bracing. The bracing of wood trusses shall comply with their appropriate engineered design.

2308.10.7.3 Alterations to trusses. Truss members and components shall not be cut, notched, drilled, spliced or otherwise altered in any way without written concurrence and approval of a registered design professional. Alterations resulting in the addition of loads to any member (e.g., HVAC equipment, water heater) shall not be permitted without verification that the truss is capable of supporting such additional loading.

2308.10.8 Roof sheathing. Roof sheathing shall be in accordance with Tables 2304.7(3) and 2304.7(5) for wood structural panels, and Tables 2304.7(1) and 2304.7(2) for lumber and shall comply with Section 2304.7.2.

2308.10.8.1 Joints. Joints in lumber sheathing shall occur over supports unless approved end-matched lumber is used, in which case each piece shall bear on at least two supports.

2308.10.9 Roof planking. Planking shall be designed in accordance with the general provisions of this code.

In lieu of such design, 2-inch (51 mm) tongue-and-groove planking is permitted in accordance with Table 2308.10.9. Joints in such planking are permitted to be randomly spaced, provided the system is applied to not less than three continuous spans, planks are center matched and end matched or splined, each plank bears on at least one support, and joints are separated by at least 24 inches (610 mm) in adjacent pieces.

2308.10.10 Attic ventilation. For attic ventilation, see Section 1202.2.

TABLE 2308.10.4.1
RAFTER TIE CONNECTIONS^g

RAFTER SLOPE	TIE SPACING (inches)	NO SNOW LOAD				GROUND SNOW LOAD (pound per square foot)							
						30 pounds per square foot				50 pounds per square foot			
						Roof span (feet)							
		12	20	28	36	12	20	28	36	12	20	28	36
		Required number of 16d common nails^{a,b} per connection^{c,d,e,f}											
3:12	12	4	6	8	10	4	6	8	11	5	8	12	15
	16	5	7	10	13	5	8	11	14	6	11	15	20
	24	7	11	15	19	7	11	16	21	9	16	23	30
	32	10	14	19	25	10	16	22	28	12	27	30	40
	48	14	21	29	37	14	32	36	42	18	32	46	60
4:12	12	3	4	5	6	3	5	6	8	4	6	9	11
	16	3	5	7	8	4	6	8	11	5	8	12	15
	24	4	7	10	12	5	9	12	16	7	12	17	22
	32	6	9	13	16	8	12	16	22	10	16	24	30
	48	8	14	19	24	10	18	24	32	14	24	34	44
5:12	12	3	3	4	5	3	4	5	7	3	5	7	9
	16	3	4	5	7	3	5	7	9	4	7	9	12
	24	4	6	8	10	4	7	10	13	6	10	14	18
	32	5	8	10	13	6	10	14	18	8	14	18	24
	48	7	11	15	20	8	14	20	26	12	20	28	36
7:12	12	3	3	3	4	3	3	4	5	3	4	5	7
	16	3	3	4	5	3	4	5	6	3	5	7	9
	24	3	4	6	7	3	5	7	9	4	7	10	13
	32	4	6	8	10	4	8	10	12	6	10	14	18
	48	5	8	11	14	6	10	14	18	9	14	20	26
9:12	12	3	3	3	3	3	3	3	4	3	3	4	5
	16	3	3	3	4	3	3	4	5	3	4	5	7
	24	3	3	5	6	3	4	6	7	3	6	8	10
	32	3	4	6	8	4	6	8	10	5	8	10	14
	48	4	6	9	11	5	8	12	14	7	12	16	20
12:12	12	3	3	3	3	3	3	3	3	3	3	3	4
	16	3	3	3	3	3	3	3	4	3	3	4	5
	24	3	3	3	4	3	3	4	6	3	4	6	8
	32	3	3	4	5	3	5	6	8	4	6	8	10
	48	3	4	6	7	4	7	8	12	6	8	12	16

For SI: 1 inch = 25.4 mm, 1 foot = 304.8 mm, 1 pound per square foot = 47.8 N/m^2.

a. 40d box or 16d sinker box nails are permitted to be substituted for 16d common nails.

b. Nailing requirements are permitted to be reduced 25 percent if nails are clinched.

c. Rafter tie heel joint connections are not required where the ridge is supported by a load-bearing wall, header or ridge beam.

d. When intermediate support of the rafter is provided by vertical struts or purlins to a load-bearing wall, the tabulated heel joint connection requirements are permitted to be reduced proportionally to the reduction in span.

e. Equivalent nailing patterns are required for ceiling joist to ceiling joist lap splices.

f. Connected members shall be of sufficient size to prevent splitting due to nailing.

g. For snow loads less than 30 pounds per square foot, the required number of nails is permitted to be reduced by multiplying by the ratio of actual snow load plus 10 divided by 40, but not less than the number required for no snow load.

TABLE 2308.10.9
ALLOWABLE SPANS FOR 2-INCH TONGUE-AND-GROOVE DECKING

SPAN[a] (feet)	LIVE LOAD (pound per square foot)	DEFLECTION LIMIT	BENDING STRESS (f) (pound per square inch)	MODULUS OF ELASTICITY (E) (pound per square inch)
		Roofs		
4	20	1/240 1/360	160	170,000 256,000
	30	1/240 1/360	210	256,000 384,000
	40	1/240 1/360	270	340,000 512,000
4.5	20	1/240 1/360	200	242,000 305,000
	30	1/240 1/360	270	363,000 405,000
	40	1/240 1/360	350	484,000 725,000
5.0	20	1/240 1/360	250	332,000 500,000
	30	1/240 1/360	330	495,000 742,000
	40	1/240 1/360	420	660,000 1,000,000
5.5	20	1/240 1/360	300	442,000 660,000
	30	1/240 1/360	400	662,000 998,000
	40	1/240 1/360	500	884,000 1,330,000
6.0	20	1/240 1/360	360	575,000 862,000
	30	1/240 1/360	480	862,000 1,295,000
	40	1/240 1/360	600	1,150,000 1,730,000
6.5	20	1/240 1/360	420	595,000 892,000
	30	1/240 1/360	560	892,000 1,340,000
	40	1/240 1/360	700	1,190,000 1,730,000
7.0	20	1/240 1/360	490	910,000 1,360,000
	30	1/240 1/360	650	1,370,000 2,000,000
	40	1/240 1/360	810	1,820,000 2,725,000

(continued)

TABLE 2308.10.9–continued
ALLOWABLE SPANS FOR 2-INCH TONGUE-AND-GROOVE DECKING

SPAN[a] (feet)	LIVE LOAD (pound per square foot)	DEFLECTION LIMIT	BENDING STRESS (f) (pound per square inch)	MODULUS OF ELASTICITY (E) (pound per square inch)
Roofs				
7.5	20	1/240 1/360	560	1,125,000 1,685,000
	30	1/240 1/360	750	1,685,000 2,530,000
	40	1/240 1/360	930	2,250,000 3,380,000
8.0	20	1/240 1/360	640	1,360,000 2,040,000
	30	1/240 1/360	850	2,040,000 3,060,000
Floors				
4	40	1/360	840	1,000,000
4.5			950	1,300,000
5.0			1,060	1,600,000

For SI: 1 inch = 25.4 mm, 1 foot = 304.8 mm, 1 pound per square foot = 0.0479 kN/m², 1 pound per square inch = 0.00689 N/mm².

a. Spans are based on simple beam action with 10 pounds per square foot dead load and provisions for a 300-pound concentrated load on a 12-inch width of decking. Random layup is permitted in accordance with the provisions of Section 2308.10.9. Lumber thickness is $1^1/_2$ inches nominal.

2308.11 Additional requirements for conventional construction in Seismic Design Category B or C. Structures of conventional light-frame construction in Seismic Design Category B or C, as determined in Section 1616, shall comply with Sections 2308.11.1 through 2308.11.3, in addition to the provisions of Sections 2308.1 through 2308.10.

2308.11.1 Number of stories. Structures of conventional light-frame construction shall not exceed two stories in height in Seismic Design Category C.

Exception: Detached one- and two-family dwellings are permitted to be three stories in height in Seismic Design Category C.

2308.11.2 Concrete or masonry. Concrete or masonry walls, or masonry veneer shall not extend above the basement.

Exceptions:

1. Masonry veneer is permitted to be used in the first two stories above grade or the first three stories above grade where the lowest story has concrete or masonry walls in Seismic Design Category B, provided that structural use panel wall bracing is used, and the length of bracing provided is 1.5 times the required length as determined in Table 2308.9.3(1).

2. Masonry veneer is permitted to be used in the first story above grade or the first two stories above grade where the lowest story has concrete or masonry walls in Seismic Design Category B or C.

3. Masonry veneer is permitted to be used in the first two stories above grade in Seismic Design Categories B and C provided the following criteria are met:

    3.1. Type of brace per Section 2308.9.3 shall be Method 3 and the allowable shear capacity in accordance with Table 2306.4.1 shall be a minimum of 350 plf (5108 N/m) (ASD).

    3.2. The bracing of the top story shall be located at each end and at least every 25 feet (7620 mm) o.c. but not less than 40 percent of the braced wall line. The bracing of the first story shall be located at each end and at least every 25 feet (7620 mm) o.c. but not less than 35 percent of the braced wall line.

    3.3. Hold-down connectors shall be provided at the ends of braced walls for the second floor to first floor wall assembly with an allowable design of 2,000 pounds (907.0 kg). Hold-down connectors shall be provided at the ends of each wall segment of the braced walls for the first floor to foundation with an allowable design of 3,900 pounds (1768 kg). In all cases, the hold-down connector force shall be transferred to the foundation.

    3.4. Cripple walls shall not be permitted.

2308.11.3 Framing and connection details. Framing and connection details shall conform to Sections 2308.11.3.1 through 2308.11.3.3.

2308.11.3.1 Anchorage. Braced wall lines shall be anchored in accordance with Section 2308.6 at foundations.

2308.11.3.2 Stepped footings. Where the height of a required braced wall panel extending from foundation to floor above varies more than 4 feet (1219 mm), the following construction shall be used:

1. Where the bottom of the footing is stepped and the lowest floor framing rests directly on a sill bolted to the footings, the sill shall be anchored as required in Section 2308.3.3.

2. Where the lowest floor framing rests directly on a sill bolted to a footing not less than 8 feet (2438

mm) in length along a line of bracing, the line shall be considered to be braced. The double plate of the cripple stud wall beyond the segment of footing extending to the lowest framed floor shall be spliced to the sill plate with metal ties, one on each side of the sill and plate. The metal ties shall not be less than 0.058 inch [1.47 mm (16 galvanized gage)] by 1.5 inches (38 mm) wide by 48 inches (1219 mm) with eight 16d common nails on each side of the splice location (see Figure 2308.11.3.2). The metal tie shall have a minimum yield of 33,000 pounds per square inch (psi) (227 Mpa).

3. Where cripple walls occur between the top of the footing and the lowest floor framing, the bracing requirements for a story shall apply.

2308.11.3.3 Openings in horizontal diaphragms. Openings in horizontal diaphragms with a dimension perpendicular to the joist that is greater than 4 feet (1.2 m) shall be constructed in accordance with the following:

1. Blocking shall be provided beyond headers.

2. Metal ties not less than 0.058 inch [1.47 mm (16 galvanized gage)] by 1.5 inches (38 mm) wide with eight 16d common nails on each side of the header-joist intersection shall be provided (see Figure 2308.11.3.3). The metal ties shall have a minimum yield of 33,000 psi (227 Mpa).

2308.12 Additional requirements for conventional construction in Seismic Design Category D or E. Structures of conventional light-frame construction in Seismic Design Category D or E, as determined in Section 1616, shall conform to Sections 2308.12.1 through 2308.12.9, in addition to the requirements for Seismic Design Category B or C in Section 2308.11.

2308.12.1 Number of stories. Structures of conventional light-frame construction shall not exceed one story in height in Seismic Design Category D or E.

Exception: Detached one- and two-family dwellings are permitted to be two stories high in Seismic Design Category D or E.

2308.12.2 Concrete or masonry. Concrete or masonry walls, or masonry veneer shall not extend above the basement.

Exception: Masonry veneer is permitted to be used in the first story above grade in Seismic Design Category D provided the following criteria are met:

1. Type of brace in accordance with Section 2308.9.3 shall be Method 3 and the allowable shear capacity in accordance with Table 2306.4.1 shall be a minimum of 350 plf (5108 N/m) (ASD).

2. The bracing of the first story shall be located at each end and at least every 25 feet (7620 mm) o.c. but not less than 45 percent of the braced wall line.

3. Hold-down connectors shall be provided at the ends of braced walls for the first floor to foundation with an allowable design of 2,100 pounds (1768 kg).

4. Cripple walls shall not be permitted.

2308.12.3 Braced wall line spacing. Spacing between interior and exterior braced wall lines shall not exceed 25 feet (7620 mm).

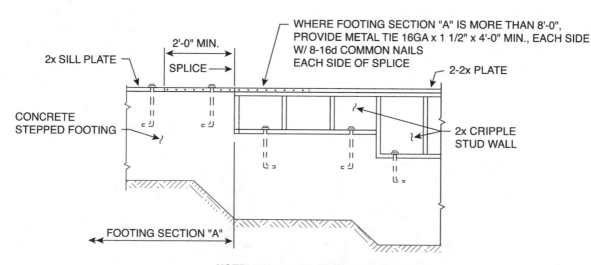

2x SILL PLATE

2'-0" MIN.

SPLICE

WHERE FOOTING SECTION "A" IS MORE THAN 8'-0", PROVIDE METAL TIE 16GA x 1 1/2" x 4'-0" MIN., EACH SIDE W/ 8-16d COMMON NAILS EACH SIDE OF SPLICE

2-2x PLATE

CONCRETE STEPPED FOOTING

2x CRIPPLE STUD WALL

FOOTING SECTION "A"

NOTE: WHERE FOOTING SECTION "A" IS LESS THAN 8'-0" LONG IN A 25'-0" TOTAL LENGTH WALL, PROVIDE BRACING AT CRIPPLE STUD WALL

For SI: 1 inch = 25.4 mm, 1 foot = 304.8 mm.

FIGURE 2308.11.3.2
STEPPED FOOTING CONNECTION DETAILS

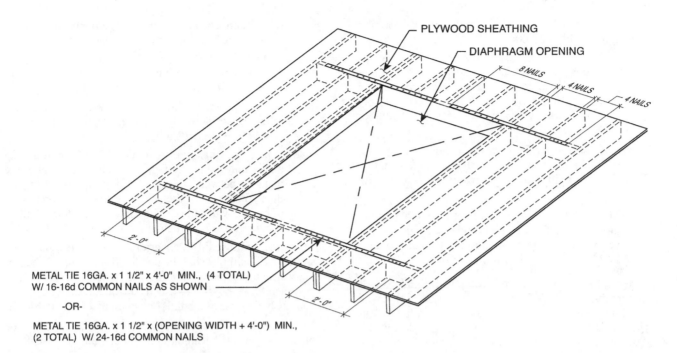

PLYWOOD SHEATHING

DIAPHRAGM OPENING

8 NAILS

4 NAILS

4 NAILS

2'-0"

2'-0"

METAL TIE 16GA. x 1 1/2" x 4'-0" MIN., (4 TOTAL)
W/ 16-16d COMMON NAILS AS SHOWN

-OR-

METAL TIE 16GA. x 1 1/2" x (OPENING WIDTH + 4'-0") MIN.,
(2 TOTAL) W/ 24-16d COMMON NAILS

For SI: 1 inch = 25.4 mm, 1 foot = 304.8 mm.

FIGURE 2308.11.3.3
OPENINGS IN HORIZONTAL DIAPHRAGMS

2308.12.4 Braced wall line sheathing. Braced wall lines shall be braced by one of the types of sheathing prescribed by Table 2308.12.4 as shown in Figure 2308.9.3. The sum of lengths of braced wall panels at each braced wall line shall conform to Table 2308.12.4. Braced wall panels shall be distributed along the length of the braced wall line and start at not more than 8 feet (2438 mm) from each end of the braced wall line. A designed collector shall be provided where the bracing begins more than 8 feet (2438 mm) from each end of a braced wall line. Panel sheathing joints shall occur over studs or blocking. Sheathing shall be fastened to studs and top and bottom plates and at panel edges occurring over blocking. Wall framing to which sheathing used for bracing is applied shall be nominal 2 inch wide (actual $1^1/_2$ inch, 38 mm) or larger members.

Cripple walls having a stud height exceeding 14 inches (356 mm) shall be considered a story for the purpose of this section and shall be braced as required for braced wall lines in accordance with Table 2308.12.4. Where interior braced wall lines occur without a continuous foundation below, the length of parallel exterior cripple wall bracing shall be one and one-half times the lengths required by Table 2308.12.4. Where the cripple wall sheathing type used is Type S-W, and this additional length of bracing cannot be provided, the capacity of Type S-W sheathing shall be increased by reducing the spacing of fasteners along the perimeter of each piece of sheathing to 4 inches (102 mm) o.c.

2308.12.5 Attachment of sheathing. Fastening of braced wall panel sheathing shall not be less than that prescribed in Table 2308.12.4 or 2304.9.1. Wall sheathing shall not be attached to framing members by adhesives.

2308.12.6 Irregular structures. Conventional light-frame construction shall not be used in irregular portions of structures in Seismic Design Category D or E. Such irregular portions of structures shall be designed to resist the forces specified in Chapter 16 to the extent such irregular features affect the performance of the conventional framing system. A portion of a structure shall be considered to be irregular where one or more of the conditions described in Items 1 through 6 below are present.

1. Where exterior braced wall panels are not in one plane vertically from the foundation to the uppermost story in which they are required, the structure shall be considered to be irregular [see Figure 2308.12.6(1)].

   Exception: Floors with cantilevers or setbacks not exceeding four times the nominal depth of the floor joists [see Figure 2308.12.6(2)] are permitted to support braced wall panels provided:

   1. Floor joists are 2 inches by 10 inches (51 mm by 254 mm) or larger and spaced not more than 16 inches (406 mm) o.c.

   2. The ratio of the back span to the cantilever is at least 2:1.

   3. Floor joists at ends of braced wall panels are doubled.

   4. A continuous rim joist is connected to the ends of cantilevered joists. The rim joist is permitted to be spliced using a metal tie not less than 0.058 inch (1.47 mm) (16 galvanized gage) and $1^1/_2$ inches (38 mm) wide fastened with six 16d common nails on each

side. The metal tie shall have a minimum yield of 33,000 psi (227 Mpa).

5. Joists at setbacks or the end of cantilevered joists shall not carry gravity loads from more than a single story having uniform wall and roof loads, nor carry the reactions from headers having a span of 8 feet (2438 mm) or more.

2. Where a section of floor or roof is not laterally supported by braced wall lines on all edges, the structure shall be considered to be irregular [see Figure 2308.12.6(3)].

Exception: Portions of roofs or floors that do not support braced wall panels above are permitted to extend up to 6 feet (1829 mm) beyond a braced wall line [see Figure 2308.12.6(4)].

3. Where the end of a required braced wall panel extends more than 1 foot (305 mm) over an opening in the wall below, the structure shall be considered to be irregular. This requirement is applicable to braced wall panels offset in plane and to braced wall panels offset out of plane as permitted by the exception to Item 1 above in this section [see Figure 2308.12.6(5)].

Exception: Braced wall panels are permitted to extend over an opening not more than 8 feet (2438 mm) in width where the header is a 4-inch by 12-inch (102 mm by 305 mm) or larger member.

4. Where portions of a floor level are vertically offset such that the framing members on either side of the offset cannot be lapped or tied together in an approved manner, the structure shall be considered to be irregular [see Figure 2308.12.6(6)].

Exception: Framing supported directly by foundations need not be lapped or tied directly together.

5. Where braced wall lines are not perpendicular to each other, the structure shall be considered to be irregular [see Figure 2308.12.6(7)].

6. Where openings in floor and roof diaphragms having a maximum dimension greater than 50 percent of the distance between lines of bracing or an area greater than 25 percent of the area between orthogonal pairs of braced wall lines are present, the structure shall be considered to be irregular [see Figure 2308.12.6(8)].

2308.12.7 Exit facilities. Exterior exit balconies, stairs and similar exit facilities shall be positively anchored to the primary structure at not over 8 feet (2438 mm) o.c. or shall be designed for lateral forces. Such attachment shall not be accomplished by use of toenails or nails subject to withdrawal.

2308.12.8 Steel plate washers. Steel plate washers shall be placed between the foundation sill plate and the nut. Such washers shall be a minimum of $^3/_{16}$ inch by 2 inches by 2 inches (4.76 mm by 51 mm by 51mm) in size.

2308.12.9 Anchorage in Seismic Design Category E. Steel bolts with a minimum nominal diameter of $^5/_8$ inch (15.9 mm) shall be used in Seismic Design Category E.

TABLE 2308.12.4
WALL BRACING IN SEISMIC DESIGN CATEGORIES D AND E
(Minimum Length of Wall Bracing per each 25 Linear Feet of Braced Wall Line[a])

STORY LOCATION	SHEATHING TYPE[b]	$0.50 \leq S_{DS} < 0.75$	$0.75 \leq S_{DS} \leq 1.00$	$1.00 < S_{DS}$
Top or only story	G-P[d]	14 feet 8 inches	18 feet 8 inches[c]	25 feet 0 inches[c]
	S-W	8 feet 0 inches	9 feet 4 inches[c]	12 feet 0 inches[c]
Story below top story	G-P[d]	NP	NP	NP
	S-W	13 feet 4 inches[c]	17 feet 4 inches[c]	21 feet 4 inches[c]
Bottom story of three stories	G-P[d]	Conventional construction not permitted; conformance with Section 2301.2.1 or 2301.2.2 is required.		
	S-W			

For SI: 1 inch = 25.4 mm, 1 foot = 304.8 mm.

a. Minimum length of panel bracing of one face of wall for S-W sheathing or both faces of wall for G-P sheathing; h/w ratio shall not exceed 2:1. For S-W panel bracing of the same material on two faces of the wall, the minimum length is permitted to be one-half the tabulated value but the h/w ratio shall not exceed 2:1 and design for uplift is required.

b. G-P = gypsum board, fiberboard, particleboard, lath and plaster, or gypsum sheathing boards; S-W = wood structural panels and diagonal wood sheathing. NP = not permitted.

c. Applies to one- and two-family detached dwellings only.

d. Nailing as specified below shall occur at all panel edges at studs, at top and bottom plates, and, where occurring, at blocking:
 For $^1/_2$-inch gypsum board, 5d (0.113 inch diameter) cooler nails at 7 inches on center;
 For $^5/_8$-inch gypsum board, No. 11 gage (0.120 inch diameter) at 7 inches on center;
 For gypsum sheathing board, $1^3/_4$ inches long by $^7/_{16}$-inch head, diamond point galvanized nails at 4 inches on center;
 For gypsum lath, No. 13 gage (0.092 inch) by $1^1/_8$ inches long, $^{19}/_{64}$-inch head, plasterboard at 5 inches on center;
 For portland cement plaster, No. 11 gage (0.120 inch) by $1^1/_2$ inches long, $^7/_{16}$- inch head at 6 inches on center;
 For fiberboard and particleboard, No. 11 gage (0.120 inch) by $1^1/_2$ inches long, $^7/_{16}$-inch head, galvanized nails at 3 inches on center.

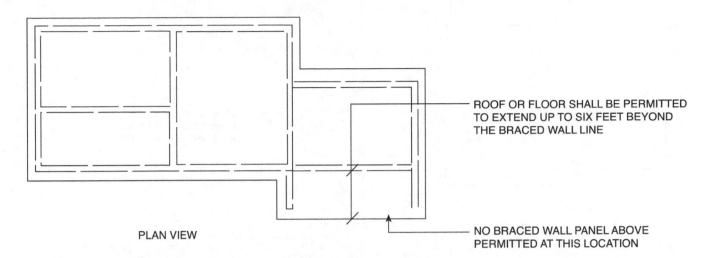

PLAN VIEW

ROOF OR FLOOR SHALL BE PERMITTED TO EXTEND UP TO SIX FEET BEYOND THE BRACED WALL LINE

NO BRACED WALL PANEL ABOVE PERMITTED AT THIS LOCATION

FIGURE 2308.12.6(1)
BRACED WALL PANELS OUT OF PLANE

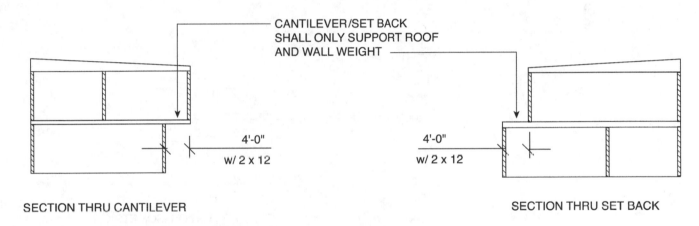

CANTILEVER/SET BACK SHALL ONLY SUPPORT ROOF AND WALL WEIGHT

SECTION THRU CANTILEVER

4'-0"
w/ 2 x 12

4'-0"
w/ 2 x 12

SECTION THRU SET BACK

For SI: 1 foot = 304.8 mm.

FIGURE 2308.12.6(2)
BRACED WALL PANELS SUPPORTED BY CANTILEVER OR SET BACK

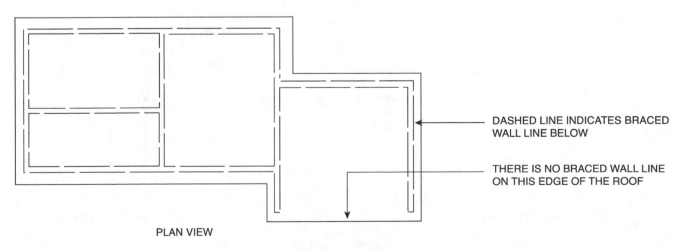

PLAN VIEW

DASHED LINE INDICATES BRACED WALL LINE BELOW

THERE IS NO BRACED WALL LINE ON THIS EDGE OF THE ROOF

FIGURE 2308.12.6(3)
FLOOR OR ROOF NOT SUPPORTED ON ALL EDGES

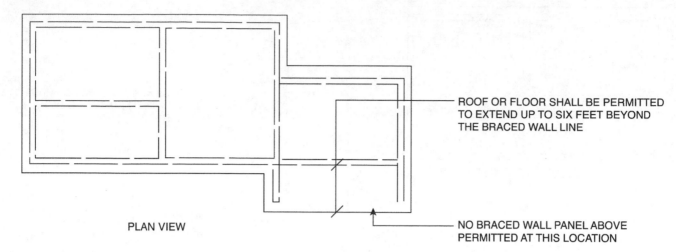

For SI: 1 foot = 304.8 mm.

FIGURE 2308.12.6(4)
ROOF OR FLOOR EXTENSION BEYOND BRACED WALL LINE

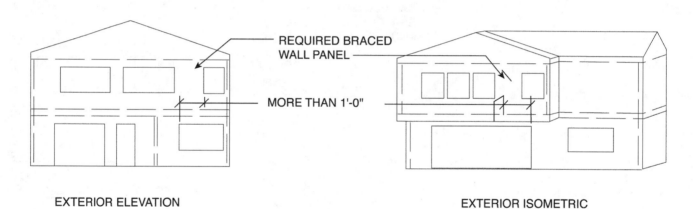

For SI: 1 foot = 304.8 mm.

FIGURE 2308.12.6(5)
BRACED WALL PANEL EXTENSION OVER OPENING

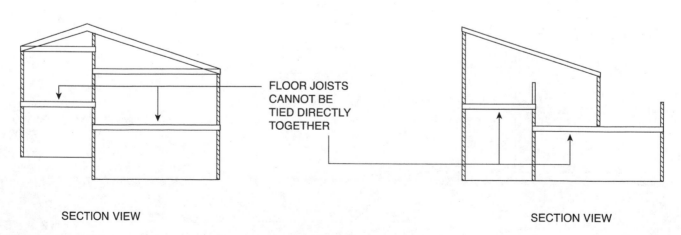

FIGURE 2308.12.6(6)
PORTIONS OF FLOOR LEVEL OFFSET VERTICALLY

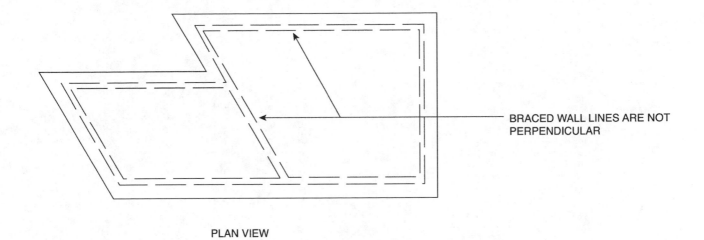

PLAN VIEW

FIGURE 2308.12.6(7)
BRACED WALL LINES NOT PERPENDICULAR

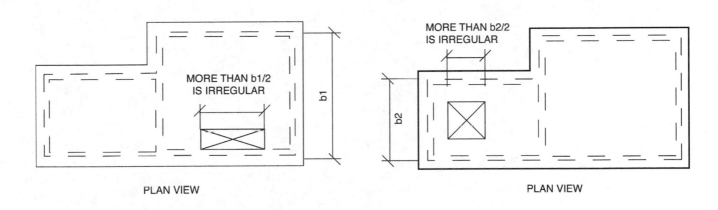

FIGURE 2308.12.6(8)
OPENING LIMITATIONS FOR FLOOR AND ROOF DIAPHRAGMS

CHAPTER 24

GLASS AND GLAZING

SECTION 2401
GENERAL

2401.1 Scope. The provisions of this chapter shall govern the materials, design, construction and quality of glass, light-transmitting ceramic and light-transmitting plastic panels for exterior and interior use in both vertical and sloped applications in buildings and structures.

2401.2 Glazing replacement. The installation of replacement glass shall be as required for new installations.

SECTION 2402
DEFINITIONS

2402.1 Definitions. The following words and terms shall, for the purposes of this chapter and as used elsewhere in this code, have the meanings shown herein.

DALLE GLASS. A decorative composite glazing material made of individual pieces of glass that are embedded in a cast matrix of concrete or epoxy.

DECORATIVE GLASS. A carved, leaded or Dalle glass or glazing material whose purpose is decorative or artistic, not functional; whose coloring, texture or other design qualities or components cannot be removed without destroying the glazing material and whose surface, or assembly into which it is incorporated, is divided into segments.

SECTION 2403
GENERAL REQUIREMENTS FOR GLASS

2403.1 Identification. Each pane shall bear the manufacturer's label designating the type and thickness of the glass or glazing material. The identification shall not be omitted unless approved and an affidavit is furnished by the glazing contractor certifying that each light is glazed in accordance with approved construction documents that comply with the provisions of this chapter. Safety glazing shall be identified in accordance with Section 2406.2.

Each pane of tempered glass, except tempered spandrel glass, shall be permanently identified by the manufacturer. The identification label shall be acid etched, sand blasted, ceramic fired, embossed or shall be of a type that once applied cannot be removed without being destroyed.

Tempered spandrel glass shall be provided with a removable paper marking by the manufacturer.

2403.2 Glass supports. Where one or more sides of any pane of glass are not firmly supported, or are subjected to unusual load conditions, detailed construction documents, detailed shop drawings and analysis or test data assuring safe performance for the specific installation shall be prepared by a registered design professional.

2403.3 Framing. To be considered firmly supported, the framing members for each individual pane of glass shall be designed so the deflection of the edge of the glass perpendicular to the glass pane shall not exceed $^1/_{175}$ of the glass edge length or $^3/_4$ inch (19.1 mm), whichever is less, when subjected to the larger of the positive or negative load where loads are combined as specified in Section 1605.

2403.4 Interior glazed areas. Where interior glazing is installed adjacent to a walking surface, the differential deflection of two adjacent unsupported edges shall not be greater than the thickness of the panels when a force of 50 pounds per linear foot (plf) (730 N/m) is applied horizontally to one panel at any point up to 42 inches (1067 mm) above the walking surface.

2403.5 Louvered windows or jalousies. Float, wired and patterned glass in louvered windows and jalousies shall be no thinner than nominal $^3/_{16}$ inch (4.8 mm) and no longer than 48 inches (1219 mm). Exposed glass edges shall be smooth.

Wired glass with wire exposed on longitudinal edges shall not be used in louvered windows or jalousies.

Where other glass types are used, the design shall be submitted to the building official for approval.

SECTION 2404
WIND, SNOW, SEISMIC
AND DEAD LOADS ON GLASS

2404.1 Vertical glass. Glass sloped 15 degrees (0.26 rad) or less from vertical in windows, curtain and window walls, doors and other exterior applications shall be designed to resist the wind loads in Section 1609 for components and cladding. Glass in glazed curtain walls, glazed storefronts and glazed partitions shall meet the seismic requirements of ASCE 7, Section 9.6.2.10. Glazing firmly supported on all four edges is permitted to be designed by the following provisions. Where the glass is not firmly supported on all four edges, analysis or test data ensuring safe performance for the specific installation shall be prepared by a registered design professional.

The design of vertical glazing shall be based on the following equation:

$$F_{gw} \leq F_{ga} \qquad \text{(Equation 24-1)}$$

where:

F_{gw} is the wind load on the glass computed in accordance with Section 1609 and F_{ga} is the maximum allowable load on the glass computed by the following formula:

$$F_{ga} = c_1 F_{ge} \qquad \text{(Equation 24-2)}$$

where:

F_{ge} = Maximum allowable equivalent load, pounds per square foot (psf) (kN/m²) determined from Figures 2404(1) through 2404(12) for the applicable glass dimensions and thickness.

c_1 = Factor determined from Table 2404.1 based on glass type.

TABLE 2404.1
c_1 FACTORS FOR VERTICAL AND SLOPED GLASS[a]
[For use with Figures 2404(1) through 2404(12)]

GLASS TYPE	FACTOR
Single Glass	
Regular (annealed)	1.0
Heat strengthened	2.0
Fully tempered	4.0
Wired	0.50
Patterned[c]	1.0
Sandblasted[d]	0.50
Laminated—regular plies[e]	0.7/0.90[f]
Laminated—heat-strengthened plies[e]	1.5/1.8[f]
Laminated—fully tempered plies[e]	3.0/3.6[f]
Insulating Glass[b]	
Regular (annealed)	1.8
Heat strengthened	3.6
Fully tempered	7.2
Laminated—regular plies[e]	1.4/1.6[f]
Laminated—heat-strengthened plies[e]	2.7/3.2[f]
Laminated—fully tempered plies[e]	5.4/6.5[f]

a. Either Table 2404.1 or 2404.2 shall be appropriate for sloped glass depending on whether the snow or wind load is dominant (see Section 2404.2). For glass types (vertical or sloped) not included in the tables, refer to ASTM E 1300 for guidance.

b. Values apply for insulating glass with identical panes.

c. The value for patterned glass is based on the thinnest part of the pattern; interpolation between graphs is permitted.

d. The value for sandblasted glass is for moderate levels of sandblasting.

e. Values for laminated glass are based on the total thickness of the glass and apply for glass with two equal glass ply thicknesses.

f. The lower value applies if, for any laminated glass pane, either the ratio of the long to short dimension is greater than 2.0 or the lesser dimension divided by the thickness of the pane is 150 or less; the higher value applies in all other cases.

2404.2 Sloped glass. Glass sloped more than 15 degrees (0.26 rad) from vertical in skylights, sunrooms, sloped roofs and other exterior applications shall be designed to resist the most critical of the following combinations of loads.

$$F_g = W_o - D \qquad \textbf{(Equation 24-3)}$$

$$F_g = W_i + D + 0.5\,S \qquad \textbf{(Equation 24-4)}$$

$$F_g = 0.5\,W_i + D + S \qquad \textbf{(Equation 24-5)}$$

where:

D = Glass dead load (psf)
For glass sloped 30 degrees (0.52 rad) or less from horizontal,

$$D = 13\,t_g \text{ (For SI: } 0.0245\,t_g)$$

For glass sloped more than 30 degrees (0.52 rad) from horizontal,

$$D = 13\,t_g \cos\theta \text{ (For SI: } 0.0245\,t_g \cos\theta).$$

F_g = Total load, psf (kN/m²) on glass.

S = Snow load, psf (kN/m²) as determined in Section 1608.

t_g = Total glass thickness, inches (mm) of glass panes and plies.

W_i = Inward wind force, psf (kN/m²) as calculated in Section 1609.

W_o = Outward wind force, psf (kN/m²) as calculated in Section 1609.

θ = Angle of slope from horizontal.

Exception: Unit skylights shall be designed in accordance with Section 2405.5.

The design of sloped glazing shall be based on the following equation:

$$F_g \leq F_{ga} \qquad \textbf{(Equation 24-6)}$$

where F_g is the maximum load on the glass determined from Equations 24-3 through 24-5, and F_{ga} is the maximum allowable load on the glass.

If F_g is determined by Equation 24-3 or 24-4 above, F_{ga} shall be computed as for vertical glazing in Section 2404.1. If F_g is determined by Equation 24-5 above, F_{ga} shall be computed by the following equation:

$$F_{ga} = c_2 F_{ge} \qquad \textbf{(Equation 24-7)}$$

where:

F_{ge} = Maximum allowable equivalent load (psf) determined from Figures 2404(1) through 2404(12) for the applicable glass dimensions and thickness.

c_2 = Factor determined from Table 2404.2 based on glass type.

TABLE 2404.2
c_2 FACTORS FOR SLOPED GLASS[a]
[For use with Figures 2404(1) through 2404(12)]

GLASS TYPE	FACTOR
Single Glass	
Regular (annealed)	0.6
Heat strengthened	1.6
Fully tempered	3.6
Wired	0.3
Patterned[c]	0.6
Laminated — regular plies[d]	0.3/0.45[e]
Laminated — heat-strengthened plies[d]	0.8/1.2[e]
Laminated — fully tempered plies[d]	1.8/2.7[e]
Insulating Glass[b]	
Regular (annealed)	1.1
Heat strengthened	2.9
Fully tempered	6.5
Laminated — regular plies[d]	0.54/0.81[e]
Laminated — heat-strengthened plies[d]	1.4/2.2[e]
Laminated — fully tempered plies[d]	3.3/4.9[e]

a. Either Table 2404.1 or 2404.2 shall be appropriate for sloped glass depending on whether the snow or wind load is dominant (see Section 2404.2). For glass types (vertical or sloped) not included in the tables, refer to ASTM E 1300 for guidance.

b. Values apply for insulating glass with identical panes.

c. The value for patterned glass is based on the thinnest part of the pattern; interpolation between graphs is permitted.

d. Values for laminated glass are based on the total thickness of the glass and apply for glass with two equal glass ply thicknesses.

e. The lower value applies where, for any laminated glass pane, either the ratio of the long to short dimension is greater than 2.0 or the lesser dimension divided by the thickness of the pane is 150 or less. The higher value applies in all other cases.

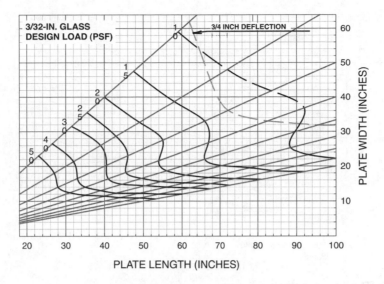

For SI: 1 inch = 25.4 mm, 1 pound per square foot = 0.0479 kPa.

FIGURE 2404(1)[a, b, c, d, e, f]
**MAXIMUM ALLOWABLE LOAD FOR VERTICAL AND SLOPED
RECTANGULAR GLASS SUPPORTED ON ALL EDGES**

NOTES:
a. In each graph, the vertical axis is the lesser dimension; the horizontal axis is the greater dimension.
b. The diagonal number on each graph shows the equivalent design load in psf.
c. The dashed lines indicate glass that has deflection in excess of $^3/_4$ inch.
d. Interpolation between lines is permitted. Extrapolation is not allowed.
e. For laminated glass, the applicable glass thickness is the total glass thickness.
f. For insulating glass panes, the applicable glass thickness is the thickness of one pane.

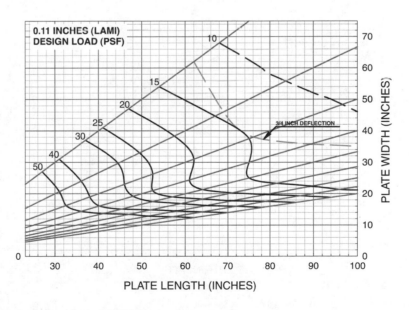

For SI: 1 inch = 25.4 mm, 1 pound per square foot = 0.0479 kPa.

FIGURE 2404(2)[a, b, c, d, e, f]
**MAXIMUM ALLOWABLE LOAD FOR VERTICAL AND SLOPED
RECTANGULAR GLASS SUPPORTED ON ALL EDGES**

NOTES:
a. In each graph, the vertical axis is the lesser dimension; the horizontal axis is the greater dimension.
b. The diagonal number on each graph shows the equivalent design load in psf.
c. The dashed lines indicate glass that has deflection in excess of $^3/_4$ inch.
d. Interpolation between lines is permitted. Extrapolation is not allowed.
e. For laminated glass, the applicable glass thickness is the total glass thickness.
f. For insulating glass panes, the applicable glass thickness is the thickness of one pane.

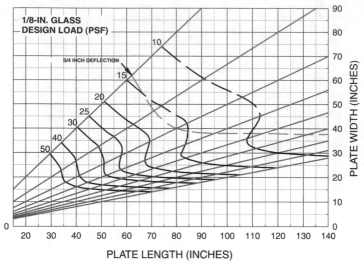

For SI: 1 inch = 25.4 mm, 1 pound per square foot = 0.0479 kPa.

FIGURE 2404(3)[a, b, c, d, e, f]
MAXIMUM ALLOWABLE LOAD FOR VERTICAL AND SLOPED
RECTANGULAR GLASS SUPPORTED ON ALL EDGES

NOTES:
a. In each graph, the vertical axis is the lesser dimension; the horizontal axis is the greater dimension.
b. The diagonal number on each graph shows the equivalent design load in psf.
c. The dashed lines indicate glass that has deflection in excess of $^3/_4$ inch.
d. Interpolation between lines is permitted. Extrapolation is not allowed.
e. For laminated glass, the applicable glass thickness is the total glass thickness.
f. For insulating glass panes, the applicable glass thickness is the thickness of one pane.

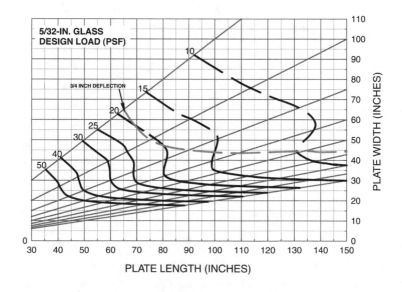

For SI: 1 inch = 25.4 mm, 1 pound per square foot = 0.0479 kPa.

FIGURE 2404(4)[a, b, c, d, e, f]
MAXIMUM ALLOWABLE LOAD FOR VERTICAL AND SLOPED
RECTANGULAR GLASS SUPPORTED ON ALL EDGES

NOTES:
a. In each graph, the vertical axis is the lesser dimension; the horizontal axis is the greater dimension.
b. The diagonal number on each graph shows the equivalent design load in psf.
c. The dashed lines indicate glass that has deflection in excess of $^3/_4$ inch.
d. Interpolation between lines is permitted. Extrapolation is not allowed.
e. For laminated glass, the applicable glass thickness is the total glass thickness.
f. For insulating glass panes, the applicable glass thickness is the thickness of one pane.

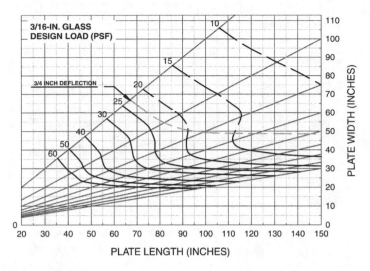

For SI: 1 inch = 25.4 mm, 1 pound per square foot = 0.0479 kPa.

FIGURE 2404(5)[a, b, c, d, e, f]
MAXIMUM ALLOWABLE LOAD FOR VERTICAL AND SLOPED
RECTANGULAR GLASS SUPPORTED ON ALL EDGES

NOTES:
a. In each graph, the vertical axis is the lesser dimension; the horizontal axis is the greater dimension.
b. The diagonal number on each graph shows the equivalent design load in psf.
c. The dashed lines indicate glass that has deflection in excess of $^3/_4$ inch.
d. Interpolation between lines is permitted. Extrapolation is not allowed.
e. For laminated glass, the applicable glass thickness is the total glass thickness.
f. For insulating glass panes, the applicable glass thickness is the thickness of one pane.

For SI: 1 inch = 25.4 mm, 1 pound per square foot = 0.0479 kPa.

FIGURE 2404(6)[a, b, c, d, e, f]
MAXIMUM ALLOWABLE LOAD FOR VERTICAL AND SLOPED
RECTANGULAR GLASS SUPPORTED ON ALL EDGES

NOTES:
a. In each graph, the vertical axis is the lesser dimension; the horizontal axis is the greater dimension.
b. The diagonal number on each graph shows the equivalent design load in psf.
c. The dashed lines indicate glass that has deflection in excess of $^3/_4$ inch.
d. Interpolation between lines is permitted. Extrapolation is not allowed.
e. For laminated glass, the applicable glass thickness is the total glass thickness.
f. For insulating glass panes, the applicable glass thickness is the thickness of one pane.

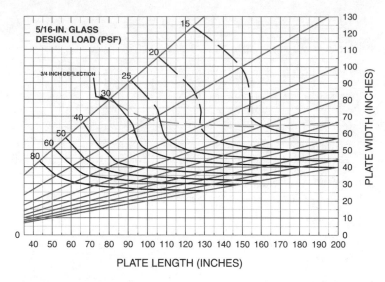

For SI: 1 inch = 25.4 mm, 1 pound per square foot = 0.0479 kPa.

FIGURE 2404(7)[a, b, c, d, e, f]
MAXIMUM ALLOWABLE LOAD FOR VERTICAL AND SLOPED
RECTANGULAR GLASS SUPPORTED ON ALL EDGES

NOTES:
a. In each graph, the vertical axis is the lesser dimension; the horizontal axis is the greater dimension.
b. The diagonal number on each graph shows the equivalent design load in psf.
c. The dashed lines indicate glass that has deflection in excess of $^3/_4$ inch.
d. Interpolation between lines is permitted. Extrapolation is not allowed.
e. For laminated glass, the applicable glass thickness is the total glass thickness.
f. For insulating glass panes, the applicable glass thickness is the thickness of one pane.

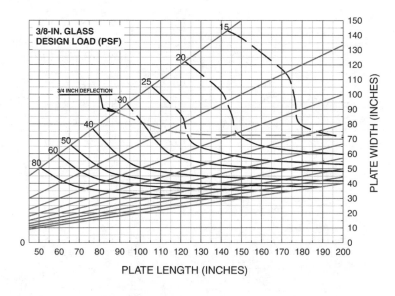

For SI: 1 inch = 25.4 mm, 1 pound per square foot = 0.0479 kPa.

FIGURE 2404(8)[a, b, c, d, e, f]
MAXIMUM ALLOWABLE LOAD FOR VERTICAL AND SLOPED
RECTANGULAR GLASS SUPPORTED ON ALL EDGES

NOTES:
a. In each graph, the vertical axis is the lesser dimension; the horizontal axis is the greater dimension.
b. The diagonal number on each graph shows the equivalent design load in psf.
c. The dashed lines indicate glass that has deflection in excess of $^3/_4$ inch.
d. Interpolation between lines is permitted. Extrapolation is not allowed.
e. For laminated glass, the applicable glass thickness is the total glass thickness.
f. For insulating glass panes, the applicable glass thickness is the thickness of one pane.

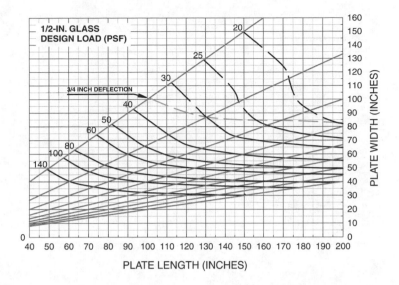

For SI: 1 inch = 25.4 mm, 1 pound per square foot = 0.0479 kPa.

FIGURE 2404(9)[a, b, c, d, e, f]
MAXIMUM ALLOWABLE LOAD FOR VERTICAL AND SLOPED
RECTANGULAR GLASS SUPPORTED ON ALL EDGES

NOTES:
a. In each graph, the vertical axis is the lesser dimension; the horizontal axis is the greater dimension.
b. The diagonal number on each graph shows the equivalent design load in psf.
c. The dashed lines indicate glass that has deflection in excess of $^3/_4$ inch.
d. Interpolation between lines is permitted. Extrapolation is not allowed.
e. For laminated glass, the applicable glass thickness is the total glass thickness.
f. For insulating glass panes, the applicable glass thickness is the thickness of one pane.

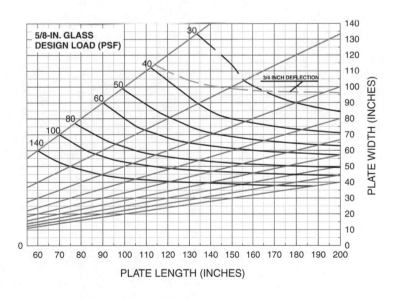

For SI: 1 inch = 25.4 mm, 1 pound per square foot = 0.0479 kPa.

FIGURE 2404(10)[a, b, c, d, e, f]
MAXIMUM ALLOWABLE LOAD FOR VERTICAL AND SLOPED
RECTANGULAR GLASS SUPPORTED ON ALL EDGES

NOTES:
a. In each graph, the vertical axis is the lesser dimension; the horizontal axis is the greater dimension.
b. The diagonal number on each graph shows the equivalent design load in psf.
c. The dashed lines indicate glass that has deflection in excess of $^3/_4$ inch.
d. Interpolation between lines is permitted. Extrapolation is not allowed.
e. For laminated glass, the applicable glass thickness is the total glass thickness.
f. For insulating glass panes, the applicable glass thickness is the thickness of one pane.

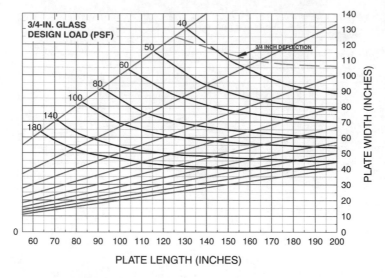

For SI: 1 inch = 25.4 mm, 1 pound per square foot = 0.0479 kPa.

FIGURE 2404(11)[a, b, c, d, e, f]
MAXIMUM ALLOWABLE LOAD FOR VERTICAL AND SLOPED
RECTANGULAR GLASS SUPPORTED ON ALL EDGES

NOTES:
a. In each graph, the vertical axis is the lesser dimension; the horizontal axis is the greater dimension.
b. The diagonal number on each graph shows the equivalent design load in psf.
c. The dashed lines indicate glass that has deflection in excess of $^3/_4$ inch.
d. Interpolation between lines is permitted. Extrapolation is not allowed.
e. For laminated glass, the applicable glass thickness is the total glass thickness.
f. For insulating glass panes, the applicable glass thickness is the thickness of one pane.

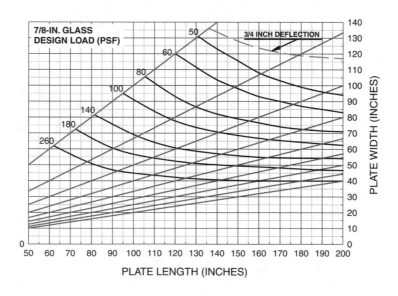

For SI: 1 inch = 25.4 mm, 1 pound per square foot = 0.0479 kPa.

FIGURE 2404(12)[a, b, c, d, e, f]
MAXIMUM ALLOWABLE LOAD FOR VERTICAL AND SLOPED
RECTANGULAR GLASS SUPPORTED ON ALL EDGES

NOTES:
a. In each graph, the vertical axis is the lesser dimension; the horizontal axis is the greater dimension.
b. The diagonal number on each graph shows the equivalent design load in psf.
c. The dashed lines indicate glass that has deflection in excess of $^3/_4$ inch.
d. Interpolation between lines is permitted. Extrapolation is not allowed.
e. For laminated glass, the applicable glass thickness is the total glass thickness.
f. For insulating glass panes, the applicable glass thickness is the thickness of one pane.

SECTION 2405
SLOPED GLAZING AND SKYLIGHTS

2405.1 Scope. This section applies to the installation of glass and other transparent, translucent or opaque glazing material installed at a slope more than 15 degrees (0.26 rad) from the vertical plane, including glazing materials in skylights, roofs and sloped walls.

2405.2 Allowable glazing materials and limitations. Sloped glazing shall be any of the following materials, subject to the listed limitations.

1. For monolithic glazing systems, the glazing material of the single light or layer shall be laminated glass with a minimum 30-mil (0.76 mm) polyvinyl butyral (or equivalent) interlayer, wired glass, light-transmitting plastic materials meeting the requirements of Section 2607, heat-strengthened glass or fully tempered glass.

2. For multiple-layer glazing systems, each light or layer shall consist of any of the glazing materials specified in Item 1 above.

Annealed glass is permitted to be used as specified within Exceptions 2 and 3 of Section 2405.3.

For additional requirements for plastic skylights, see Section 2610. Glass-block construction shall conform to the requirements of Section 2101.2.5.

2405.3 Screening. Where used in monolithic glazing systems, heat-strengthened glass and fully tempered glass shall have screens installed below the glazing material. The screens and their fastenings shall: (1) be capable of supporting twice the weight of the glazing; (2) be firmly and substantially fastened to the framing members and (3) be installed within 4 inches (102 mm) of the glass. The screens shall be constructed of a noncombustible material not thinner than No. 12 B&S gage (0.0808 inch) with mesh not larger than 1 inch by 1 inch (25 mm by 25 mm). In a corrosive atmosphere, structurally equivalent noncorrosive screen materials shall be used. Heat-strengthened glass, fully tempered glass and wired glass, when used in multiple-layer glazing systems as the bottom glass layer over the walking surface, shall be equipped with screening that conforms to the requirements for monolithic glazing systems.

Exception: In monolithic and multiple-layer sloped glazing systems, the following applies:

1. Fully tempered glass installed without protective screens where glazed between intervening floors at a slope of 30 degrees (0.52 rad) or less from the vertical plane shall have the highest point of the glass 10 feet (3048 mm) or less above the walking surface.

2. Screens are not required below any glazing material, including annealed glass, where the walking surface below the glazing material is permanently protected from the risk of falling glass or the area below the glazing material is not a walking surface.

3. Any glazing material, including annealed glass, is permitted to be installed without screens in the sloped glazing systems of commercial or detached noncombustible greenhouses used exclusively for growing plants and not open to the public, provided

that the height of the greenhouse at the ridge does not exceed 30 feet (9144 mm) above grade.

4. Screens shall not be required within individual dwelling units in Groups R-2, R-3 and R-4 as applicable in Section 101.2 where fully tempered glass is used as single glazing or as both panes in an insulating glass unit, and the following conditions are met:

   4.1. Each pane of the glass is 16 square feet (1.5 m²) or less in area.

   4.2. The highest point of the glass is 12 feet (3658 mm) or less above any walking surface or other accessible area.

   4.3. The glass thickness is $^3/_{16}$ inch (4.8 mm) or less.

5. Screens shall not be required for laminated glass with a 15-mil (0.38 mm) polyvinyl butyral (or equivalent) interlayer used within individual dwelling units in Groups R-2, R-3 and R-4 as applicable in Section 101.2 within the following limits:

   5.1. Each pane of glass is 16 square feet (1.5 m²) or less in area.

   5.2. The highest point of the glass is 12 feet (3658 mm) or less above a walking surface or other accessible area.

2405.4 Framing. In Type 1 and 2 construction, sloped glazing and skylight frames shall be constructed of noncombustible materials. In structures where acid fumes deleterious to metal are incidental to the use of the buildings, approved pressure-treated wood or other approved noncorrosive materials are permitted to be used for sash and frames. Framing supporting sloped glazing and skylights shall be designed to resist the tributary roof loads in Chapter 16. Skylights set at an angle of less than 45 degrees (0.79 rad) from the horizontal plane shall be mounted at least 4 inches (102 mm) above the plane of the roof on a curb constructed as required for the frame. Skylights shall not be installed in the plane of the roof where the roof pitch is less than 45 degrees (0.79 rad) from the horizontal.

Exception: Installation of a skylight without a curb shall be permitted on roofs with a minimum slope of 14 degrees (three units vertical in 12 units horizontal) in Group R-3 occupancies as applicable in Section 101.2. All unit skylights installed in a roof with a pitch flatter than 14 degrees (0.25 rad) shall be mounted at least 4 inches (102 mm) above the plane of the roof on a curb constructed as required for the frame unless otherwise specified in the manufacturer's installation instructions.

2405.5 Unit skylights. Unit skylights shall be tested and labeled as complying with *101/I.S.2/NAFS Voluntary Performance Specification for Windows, Skylights and Glass.* The label shall state the name of the manufacturer, the approved labeling agency, the product designation and the performance grade rating as specified in 101/I.S.2/NAFS. If the product manufacturer has chosen to have the performance grade of the skylight rated separately for positive and negative design pressure, then the label shall state both performance grade ratings as specified in 101/I.S.2/NAFS and the skylight shall comply with Section 2405.5.2. If the skylight is not rated separately for positive and negative pressure, then the performance grade rating shown on the label shall be the performance grade rating determined in accordance with 101/I.S.2/NAFS for both positive and negative design pressure, and the skylight shall conform to Section 2405.5.1.

2405.5.1 Unit skylights rated for the same performance grade for both positive and negative design pressure. The design of unit skylights shall be based on the following equation:

$$F_g \leq PG \qquad \text{(Equation 24-8)}$$

where:

F_g is the maximum load on the skylight determined from Equations 24-3 through 24-5 in Section 2404.2.

PG is the performance grade rating of the skylight.

2405.5.2 Unit skylights rated for separate performance grades for positive and negative design pressure. The design of unit skylights rated for performance grade for both positive and negative design pressures shall be based on the following equations:

$$F_{gi} \leq PG_{Pos} \qquad \text{(Equation 24-9)}$$

$$F_{go} \leq PG_{Neg} \qquad \text{(Equation 24-10)}$$

where:

PG_{Pos} is the performance grade rating of the skylight under positive design pressure,

PG_{Neg} is the performance grade rating of the skylight under negative design pressure, and

F_{gi} and F_{go} are determined in accordance with the following:

If $W_O \geq D$, where W_o is the outward wind force, psf (kN/m²) as calculated in Section 1609 and D is the dead weight of the glazing, psf (kN/m²) as determined in Section 2404.2 for glass, or by the weight of the plastic, psf (kN/m²) for plastic glazing.

F_{gi} is the maximum load on the skylight determined from Equations 24-4 and 24-5 in Section 2404.2,

F_{go} is the maximum load on the skylight determined from Equation 24-3.

If $W_O < D$, where W_o is the outward wind force, psf (kN/m²) as calculated in Section 1609 and D is the dead weight of the glazing, psf (kN/m²) as determined in Section 2404.2 for glass, or by the weight of the plastic for plastic glazing.

F_{gi} is the maximum load on the skylight determined from Equations 24-3 through 24-5 in Section 2404.2,

$F_{go} = 0$.

SECTION 2406
SAFETY GLAZING

2406.1 Human impact loads. Individual glazed areas, including glass mirrors, in hazardous locations as defined in Section 2406.3 shall comply with Sections 2406.1.1 through 2406.1.5.

2406.1.1 CPSC 16 CFR 1201. Except as provided in Sections 2406.1.2 through 2406.1.5, all glazing shall pass the test requirements of CPSC 16 CFR 1201, listed in Chapter 35. Glazing shall comply with the CPSC 16 CFR, Part 1201 criteria, for Category I or II as indicated in Table 2406.1.

2406.1.2 Wired glass. In other than Group E, wired glass installed in fire doors, fire windows and view panels in fire-resistant walls shall be permitted to comply with ANSI Z97.1.

2406.1.3 Plastic glazing. Plastic glazing shall meet the weathering requirements of ANSI Z97.1.

2406.1.4 Glass block. Glass-block walls shall comply with Section 2101.2.5.

2406.1.5 Louvered windows and jalousies. Louvered windows and jalousies shall comply with Section 2403.5.

2406.2 Identification of safety glazing. Except as indicated in Section 2406.2.1, each pane of safety glazing installed in hazardous locations shall be identified by a label specifying the labeler, whether the manufacturer or installer, and the safety glazing standard with which it complies, as well as the information specified in Section 2403.1. The label shall be acid etched, sand blasted, ceramic fired or an embossed mark, or shall be of a type that once applied cannot be removed without being destroyed.

Exceptions:

1. For other than tempered glass, labels are not required, provided the building official approves the use of a certificate, affidavit or other evidence confirming compliance with this code.

2. Tempered spandrel glass is permitted to be identified by the manufacturer with a removable paper label.

2406.2.1 Multilight assemblies. Multilight glazed assemblies having individual lights not exceeding 1 square foot (0.09 square meter) in exposed area shall have at least one

TABLE 2406.1
MINIMUM CATEGORY CLASSIFICATION OF GLAZING

EXPOSED SURFACE AREA OF ONE SIDE OF ONE LITE	GLAZING IN STORM OR COMBINATION DOORS (Category class)	GLAZING IN DOORS (Category class)	GLAZED PANELS REGULATED BY ITEM 7 OF SECTION 2406.3 (Category class)	GLAZED PANELS REGULATED BY ITEM 6 OF SECTION 2406.3 (Category class)	DOORS AND ENCLOSURES REGULATED BY ITEM 5 OF SECTION 2406.3 (Category class)	SLIDING GLASS DOORS PATIO TYPE (Category class)
9 square feet or less	I	I	No requirement	I	II	II
More than 9 square feet	II	II	II	II	II	II

For SI: 1 square foot = 0.0929m².

light in the assembly marked as indicated in Section 2406.2. Other lights in the assembly shall be marked "CPSC 16 CFR 1201" or "ANSI Z97.1," as appropriate.

2406.3 Hazardous locations. The following shall be considered specific hazardous locations requiring safety glazing materials:

1. Glazing in swinging doors except jalousies (see Section 2406.3.1).

2. Glazing in fixed and sliding panels of sliding door assemblies and panels in sliding and bifold closet door assemblies.

3. Glazing in storm doors.

4. Glazing in unframed swinging doors.

5. Glazing in doors and enclosures for hot tubs, whirlpools, saunas, steam rooms, bathtubs and showers. Glazing in any portion of a building wall enclosing these compartments where the bottom exposed edge of the glazing is less than 60 inches (1524 mm) above a standing surface.

6. Glazing in an individual fixed or operable panel adjacent to a door where the nearest exposed edge of the glazing is within a 24-inch (610 mm) arc of either vertical edge of the door in a closed position and where the bottom exposed edge of the glazing is less than 60 inches (1524 mm) above the walking surface.

Exceptions:

1. Panels where there is an intervening wall or other permanent barrier between the door and glazing.

2. Where access through the door is to a closet or storage area 3 feet (914 mm) or less in depth. Glazing in this application shall comply with Section 2406.3, Item 7.

3. Glazing in walls perpendicular to the plane of the door in a closed position, other than the wall towards which the door swings when opened, in one- and two-family dwellings or within dwelling units in Group R-2.

7. Glazing in an individual fixed or operable panel, other than in those locations described in preceding Items 5 and 6, which meets all of the following conditions:

7.1. Exposed area of an individual pane greater than 9 square feet (0.84 m²);

7.2. Exposed bottom edge less than 18 inches (457 mm) above the floor;

7.3. Exposed top edge greater than 36 inches (914 mm) above the floor; and

7.4. One or more walking surface(s) within 36 inches (914 mm) horizontally of the plane of the glazing.

Exception: Safety glazing for Item 7 is not required for the following installations:

1. A protective bar 1¹/₂ inches (38 mm) or more in height, capable of withstanding a horizontal load of 50 pounds plf (730 N/m) without contacting the glass, is installed on the accessible sides of the glazing 34 inches to 38 inches (864 mm to 965 mm) above the floor.

2. The outboard pane in insulating glass units or multiple glazing where the bottom exposed edge of the glass is 25 feet (7620 mm) or more above any grade, roof, walking surface or other horizontal or sloped (within 45 degrees of horizontal) (0.78 rad) surface adjacent to the glass exterior.

8. Glazing in guards and railings, including structural baluster panels and nonstructural in-fill panels, regardless of area or height above a walking surface.

9. Glazing in walls and fences enclosing indoor and outdoor swimming pools, hot tubs and spas where all of the following conditions are present:

9.1. The bottom edge of the glazing on the pool or spa side is less than 60 inches (1524 mm) above a walking surface on the pool or spa side of the glazing; and

9.2. The glazing is within 60 inches (1524 mm) horizontally of the water's edge of a swimming pool or spa.

10. Glazing adjacent to stairways, landings and ramps within 36 inches (914 mm) horizontally of a walking surface; when the exposed surface of the glass is less than 60 inches (1524 mm) above the plane of the adjacent walking surface.

11. Glazing adjacent to stairways within 60 inches (1524 mm) horizontally of the bottom tread of a stairway in any direction when the exposed surface of the glass is less than 60 inches (1524 mm) above the nose of the tread.

Exception: Safety glazing for Item 10 or 11 is not required for the following installations where:

1. The side of a stairway, landing or ramp which has a guardrail or handrail, including balusters or in-fill panels, complying with the provisions of Sections 1012 and 1607.7; and

2. The plane of the glass is greater than 18 inches (457 mm) from the railing.

2406.3.1 Exceptions: The following products, materials and uses shall not be considered specific hazardous locations:

1. Openings in doors through which a 3-inch (76 mm) sphere is unable to pass.

2. Decorative glass in Section 2406.3, Item 1, 6 or 7.

3. Glazing materials used as curved glazed panels in revolving doors.

4. Commercial refrigerated cabinet glazed doors.

5. Glass-block panels complying with Section 2101.2.5.

6. Louvered windows and jalousies complying with the requirements of Section 2403.5.

7. Mirrors and other glass panels mounted or hung on a surface that provides a continuous backing support.

2406.4 Fire department access panels. Fire department glass access panels shall be of tempered glass. For insulating glass units, all panes shall be tempered glass.

SECTION 2407
GLASS IN HANDRAILS AND GUARDS

2407.1 Materials. Glass used as structural balustrade panels in railings shall be constructed of either single fully tempered glass, laminated fully tempered glass or laminated heat-strengthened glass. Glazing in railing in-fill panels shall be of an approved safety glazing material that conforms to the provisions of Section 2406.1.1. For all glazing types, the minimum nominal thickness shall be $^1/_4$ inch (6.4 mm). Fully tempered glass and laminated glass shall comply with Category II of CPSC 16 CFR 1201, listed in Chapter 35.

2407.1.1 Loads. The panels and their support system shall be designed to withstand the loads specified in Section 1607.7. A safety factor of four shall be used.

2407.1.2 Support. Each handrail or guard section shall be supported by a minimum of three glass balusters or shall be otherwise supported to remain in place should one baluster panel fail. Glass balusters shall not be installed without an attached handrail or guard.

2407.1.3 Parking garages. Glazing materials shall not be installed in railings in parking garages except for pedestrian areas not exposed to impact from vehicles.

SECTION 2408
GLAZING IN ATHLETIC FACILITIES

2408.1 General. Glazing in athletic facilities and similar uses subject to impact loads, which forms whole or partial wall sections or which is used as a door or part of a door, shall comply with this section.

2408.2 Racquetball and squash courts.

2408.2.1 Testing. Test methods and loads for individual glazed areas in racquetball and squash courts subject to impact loads shall conform to those of CPSC 16 CFR, Part 1201, listed in Chapter 35, with impacts being applied at a height of 59 inches (1499 mm) above the playing surface to an actual or simulated glass wall installation with fixtures, fittings and methods of assembly identical to those used in practice.

Glass walls shall comply with the following conditions:

1. A glass wall in a racquetball or squash court, or similar use subject to impact loads, shall remain intact following a test impact.

2. The deflection of such walls shall not be greater than $1^1/_2$ inches (38 mm) at the point of impact for a drop height of 48 inches (1219 mm).

Glass doors shall comply with the following conditions:

1. Glass doors shall remain intact following a test impact at the prescribed height in the center of the door.

2. The relative deflection between the edge of a glass door and the adjacent wall shall not exceed the thickness of the wall plus $^1/_2$ inch (12.7 mm) for a drop height of 48 inches (1219 mm).

2408.3 Gymnasiums and basketball courts. Glazing in multipurpose gymnasiums, basketball courts and similar athletic facilities subject to human impact loads shall comply with Category II of CPSC 16 CFR 1201, listed in Chapter 35.

SECTION 2409
GLASS IN FLOORS AND SIDEWALKS

2409.1 General. Glass installed in the walking surface of floors, landings, stairwells and similar locations shall comply with Sections 2409.2 through 2409.4.

2409.2 Design load. The design for glass used in floors, landings, stair treads and similar locations shall be determined as indicated in Section 2409.4 based on the load that produces the greater stresses from the following:

1. The uniformly distributed unit load (F_u) from Section 1605;

2. The concentrated load (F_c) from Table 1607.1; or

3. The actual load (F_a) produced by the intended use.

The dead load (D) for glass in psf (kN/m²) shall be taken as the total thickness of the glass plies in inches by 13 (For SI: glass plies in mm by 0.0245). Load reductions allowed by Section 1607.9 are not permitted.

2409.3 Laminated glass. Laminated glass having a minimum of two plies shall be used. The glass shall be capable of supporting the total design load, as indicated in Section 2409.4, with any one ply broken.

2409.4 Design formula. Glass in floors and sidewalks shall be designed to resist the most critical of the following combinations of loads:

$$F_g = 2 F_u + D \qquad \text{(Equation 24-11)}$$

$$F_g = (8 F_c / A) + D \qquad \text{(Equation 24-12)}$$

$$F_g = F_a + D \qquad \text{(Equation 24-13)}$$

where:

A = Area of rectangular glass, ft² (m²).

D = Glass dead load (psf) = $13\, t_g$ (for SI: $0.0245\, t_g$, kN/m²).

t_g = Total glass thickness, inches (mm).

F_a = Actual intended use load, psf (kN/m²).

F_c = Concentrated load, pounds (kN).

F_g = Total load, psf (kN/m²) on glass.

F_u = Uniformly distributed load, psf (kN/m²).

The design of the glazing shall be based on

$$F_g \leq F_{ga} \qquad \text{(Equation 24-14)}$$

where F_g is the maximum load on the glass determined from the load combinations above, and F_{ga} is the maximum allowable load on the glass, computed by the following formula:

$$F_{ga} = 0.67\, c_2 F_{ge} \qquad \text{(Equation 24-15)}$$

where:

F_{ge} = Maximum allowable equivalent load, psf (kN/m^2), determined from Figures 2404(1) through 2404(12) for the applicable glass dimensions and thickness; and

c_2 = Factor determined from Table 2404.2 based on glass type.

The factor, c_2, for laminated glass found in Table 2404.2 shall apply to two-ply laminates only. The value of F_a shall be doubled for dynamic applications.

CHAPTER 25
GYPSUM BOARD AND PLASTER

SECTION 2501
GENERAL

2501.1 Scope.

2501.1.1 General. Provisions of this chapter shall govern the materials, design, construction and quality of gypsum board, lath, gypsum plaster and cement plaster.

2501.1.2 Performance. Lathing, plastering and gypsum board construction shall be done in the manner and with the materials specified in this chapter, and when required for fire protection, shall also comply with the provisions of Chapter 7.

2501.1.3 Other materials. Other approved wall or ceiling coverings shall be permitted to be installed in accordance with the recommendations of the manufacturer and the conditions of approval.

SECTION 2502
DEFINITIONS

2502.1 Definitions. The following words and terms shall, for the purposes of this chapter and as used elsewhere in this code, have the meanings shown herein.

CEMENT PLASTER. A mixture of portland or blended cement, portland cement or blended cement and hydrated lime, masonry cement or plastic cement and aggregate and other approved materials as specified in this code.

EXTERIOR SURFACES. Weather-exposed surfaces.

GYPSUM BOARD. Gypsum wallboard, gypsum sheathing, gypsum base for gypsum veneer plaster, exterior gypsum soffit board, predecorated gypsum board or water-resistant gypsum backing board complying with the standards listed in Tables 2506.2, 2507.2 and Chapter 35.

GYPSUM PLASTER. A mixture of calcined gypsum or calcined gypsum and lime and aggregate and other approved materials as specified in this code.

GYPSUM VENEER PLASTER. Gypsum plaster applied to an approved base in one or more coats normally not exceeding $^1/_4$ inch (6.4 mm) in total thickness.

INTERIOR SURFACES. Surfaces other than weather-exposed surfaces.

WEATHER-EXPOSED SURFACES. Surfaces of walls, ceilings, floors, roofs, soffits and similar surfaces exposed to the weather except the following:

1. Ceilings and roof soffits enclosed by walls, fascia, bulkheads or beams that extend a minimum of 12 inches (305 mm) below such ceiling or roof soffits.
2. Walls or portions of walls beneath an unenclosed roof area, where located a horizontal distance from an open exterior opening equal to at least twice the height of the opening.

3. Ceiling and roof soffits located a minimum horizontal distance of 10 feet (3048 mm) from the outer edges of the ceiling or roof soffits.

WIRE BACKING. Horizontal strands of tautened wire attached to surfaces of vertical supports which, when covered with the building paper, provide a backing for cement plaster.

SECTION 2503
INSPECTION

2503.1 Inspection. Lath and gypsum board shall be inspected in accordance with Section 109.3.5.

SECTION 2504
VERTICAL AND HORIZONTAL ASSEMBLIES

2504.1 Scope. The following requirements shall be met where construction involves gypsum board, lath and plaster in vertical and horizontal assemblies.

2504.1.1 Wood framing. Wood supports for lath or gypsum board, as well as wood stripping or furring, shall not be less than 2 inches (51 mm) nominal thickness in the least dimension.

Exception: The minimum nominal dimension of wood furring strips installed over solid backing shall not be less than 1 inch by 2 inches (25 mm by 51 mm).

2504.1.2 Studless partitions. The minimum thickness of vertically erected studless solid plaster partitions of $^3/_8$-inch (9.5 mm) and $^3/_4$-inch (19.1 mm) rib metal lath or $^1/_2$-inch-thick (12.7 mm) long-length gypsum lath and gypsum board partitions shall be 2 inches (51 mm).

SECTION 2505
SHEAR WALL CONSTRUCTION

2505. 1 Resistance to shear (wood framing). Wood-framed shear walls sheathed with gypsum board, lath and plaster shall be designed and constructed in accordance with Section 2306.4 and are permitted to resist wind and seismic loads. Walls resisting seismic loads shall be subject to the limitations in Section 1617.6.

2505.2 Resistance to shear (steel framing). Cold-formed steel framed shear walls sheathed with gypsum board and constructed in accordance with the materials and provisions of Sections 2211.1, 2211.2, 2211.2.1 and 2211.2.2.3 are permitted to resist wind and seismic loads. Walls resisting seismic loads shall be subject to the limitations in Section 1617.6.

SECTION 2506
GYPSUM BOARD MATERIALS

2506.1 General. Gypsum board materials and accessories shall be identified by the manufacturer's designation to indicate compliance with the appropriate standards referenced in this section and stored to protect such materials from the weather.

2506.2 Standards. Gypsum board materials shall conform to the appropriate standards listed in Table 2506.2 and Chapter 35 and, where required for fire protection, shall conform to the provisions of Chapter 7.

TABLE 2506.2
GYPSUM BOARD MATERIALS AND ACCESSORIES

MATERIAL	STANDARD
Accessories for gypsum board	ASTM C 1047
Gypsum sheathing	ASTM C 79
Gypsum wallboard	ASTM C 36
Joint reinforcing tape and compound	ASTM C 474; C 475
Nails for gypsum boards	ASTM C 514, F 547, F 1667
Steel screws	ASTM C 954; C 1002
Steel studs, nonload bearing	ASTM C 645
Steel studs, load bearing	ASTM C 955
Water-resistant gypsum backing board	ASTM C 630
Exterior soffit board	ASTM C 931
Fiber-reinforced gypsum panels	ASTM C 1278
Gypsum backing board and gypsum shaftliner board	ASTM C 442
Gypsum ceiling board	ASTM C 1395
Standard specification for gypsum board	ASTM C 1396
Predecorated gypsum board	ASTM C 960
Adhesives for fastening gypsum wallboard	ASTM C 557
Testing gypsum and gypsum products	ASTM C 22; C 472; C 473
Glass mat gypsum substrate	ASTM C 1177
Glass mat gypsum backing panel	ASTM C 1178

2506.2.1 Other materials. Metal suspension systems for acoustical and lay-in panel ceilings shall conform with ASTM C 635 listed in Chapter 35 and Section 9.6.2.6 of ASCE 7 for installation in high seismic areas.

SECTION 2507
LATHING AND PLASTERING

2507.1 General. Lathing and plastering materials and accessories shall be marked by the manufacturer's designation to indicate compliance with the appropriate standards referenced in this section and stored in such a manner to protect them from the weather.

2507.2 Standards. Lathing and plastering materials shall conform to the standards listed in Table 2507.2 and Chapter 35 and, where required for fire protection, shall also conform to the provisions of Chapter 7.

TABLE 2507.2
LATH, PLASTERING MATERIALS AND ACCESSORIES

MATERIAL	STANDARD
Accessories for gypsum veneer base	ASTM C 1047
Exterior plaster bonding compounds	ASTM C 932
Gypsum base for veneer plasters	ASTM C 588
Gypsum casting and molding plaster	ASTM C 59
Gypsum Keene's cement	ASTM C 61
Gypsum lath	ASTM C 37
Gypsum plaster	ASTM C 28
Gypsum veneer plaster	ASTM C 587
Interior bonding compounds, gypsum	ASTM C 631
Lime plasters	ASTM C 5; C 206
Masonry cement	ASTM C 91
Metal lath	ASTM C 847
Plaster aggregates Sand	ASTM C 35; C 897
Perlite	ASTM C 35
Vermiculite	ASTM C 35
Plastic cement	ASTM C 1328
Blended cement	ASTM C 595
Portland cement	ASTM C 150
Steel studs and track	ASTM C 645; C 955
Steel screws	ASTM C 1002; C 954
Welded wire lath	ASTM C 933
Woven wire plaster base	ASTM C 1032

SECTION 2508
GYPSUM CONSTRUCTION

2508.1 General. Gypsum board and gypsum plaster construction shall be of the materials listed in Tables 2506.2 and 2507.2. These materials shall be assembled and installed in compliance with the appropriate standards listed in Tables 2508.1 and 2511.1, and Chapter 35.

TABLE 2508.1
INSTALLATION OF GYPSUM CONSTRUCTION

MATERIAL	STANDARD
Gypsum sheathing	ASTM C 1280
Gypsum veneer base	ASTM C 844
Gypsum board	GA-216; ASTM C 840
Interior lathing and furring	ASTM C 841
Steel framing for gypsum boards	ASTM C 754; C 1007

2508.2 Limitations. Gypsum wallboard or gypsum plaster shall not be used in any exterior surface where such gypsum construction will be exposed directly to the weather. Gypsum wallboard shall not be used where there will be direct exposure to water or continuous high humidity conditions. Gypsum sheathing shall be installed on exterior surfaces in accordance with ASTM C 1280.

2508.2.1 Weather protection. Gypsum wallboard, gypsum lath or gypsum plaster shall not be installed until weather protection for the installation is provided.

2508.3 Single-ply application. Edges and ends of gypsum board shall occur on the framing members, except those edges and ends that are perpendicular to the framing members. Edges and ends of gypsum board shall be in moderate contact except in concealed spaces where fire-resistance-rated construction, shear resistance or diaphragm action is not required.

2508.3.1 Floating angles. Fasteners at the top and bottom plates of vertical assemblies, or the edges and ends of horizontal assemblies perpendicular to supports, and at the wall line are permitted to be omitted except on shear resisting elements or fire-resistance-rated assemblies. Fasteners shall be applied in such a manner as not to fracture the face paper with the fastener head.

2508.4 Joint treatment. Gypsum board fire-resistance-rated assemblies shall have joints and fasteners treated.

Exception: Joint and fastener treatment need not be provided where any of the following conditions occur:

1. Where the gypsum board is to receive a decorative finish such as wood paneling, battens, acoustical finishes or any similar application that would be equivalent to joint treatment.

2. On single-layer systems where joints occur over wood framing members.

3. Square edge or tongue-and-groove edge gypsum board (V-edge), gypsum backing board or gypsum sheathing.

4. On multilayer systems where the joints of adjacent layers are offset from one to another.

5. Assemblies tested without joint treatment.

2508.5 Horizontal gypsum board diaphragm ceilings. Gypsum board shall be permitted to be used on wood joists to create a horizontal diaphragm ceiling in accordance with Table 2508.5.

2508.5.1 Diaphragm proportions. The maximum allowable diaphragm proportions shall be $1^1/_2$:1 between shear resisting elements. Rotation or cantilever conditions shall not be permitted.

2508.5.2 Installation. Gypsum board used in a horizontal diaphragm ceiling shall be installed perpendicular to ceiling framing members. End joints of adjacent courses of gypsum board shall not occur on the same joist.

2508.5.3 Blocking of perimeter edges. All perimeter edges shall be blocked using a wood member not less than 2-inch by 6-inch (51 mm by 159 mm) nominal dimension. Blocking material shall be installed flat over the top plate of the wall to provide a nailing surface not less than 2 inches (51 mm) in width for the attachment of the gypsum board.

2508.5.4 Fasteners. Fasteners used for the attachment of gypsum board to a horizontal diaphragm ceiling shall be as defined in Table 2508.5. Fasteners shall be spaced not more than 7 inches (178 mm) on center (o.c.) at all supports, including perimeter blocking, and not more than $^3/_8$ inch (9.5 mm) from the edges and ends of the gypsum board.

2508.5.5 Lateral force restrictions. Gypsum board shall not be used in diaphragm ceilings to resist lateral forces imposed by masonry or concrete construction.

SECTION 2509
GYPSUM BOARD IN SHOWERS AND WATER CLOSETS

2509.1 Wet areas. Showers and public toilet walls shall conform to Sections 1210.2 and 1210.3.

2509.2 Base for tile. When gypsum board is used as a base for tile or wall panels for tubs, shower or water closet compartment walls, water-resistant gypsum backing board shall be used as a substrate. Regular gypsum wallboard is permitted under tile or wall panels in other wall and ceiling areas when installed in accordance with GA-216 or ASTM C 840.

TABLE 2508.5
SHEAR CAPACITY FOR HORIZONTAL WOOD FRAMED GYPSUM BOARD DIAPHRAGM CEILING ASSEMBLIES

MATERIAL	THICKNESS OF MATERIAL (MINIMUM) (inches)	SPACING OF FRAMING MEMBERS (MAXIMUM) (inches)	SHEAR VALUE[a,b] (plf of ceiling)	MIMIMUM FASTENER SIZE
Gypsum board	$^1/_2$	16 o.c.	90	5d cooler or wallboard nail; $1^5/_8$-inch long; 0.086-inch shank; $^{15}/_{64}$-inch head[c]
Gypsum board	$^1/_2$	24 o.c.	70	5d cooler or wallboard nail; $1^5/_8$-inch long; 0.086-inch shank; $^{15}/_{64}$-inch head[c]

For SI: 1 inch = 25.4 mm.

a. Values are not cumulative with other horizontal diaphragm values and are for short-term loading due to wind or seismic loading. Values shall be reduced 25 percent for normal loading.

b. Values shall be reduced 50 percent in Seismic Categories D, E and F.

c. $1^1/_4$-inch, No. 6 Type S or W screws are permitted to be substituted for the listed nails.

2509.3 Limitations. Water-resistant gypsum backing board shall not be used in the following locations:

1. Over a vapor retarder in shower or bathtub compartments.

2. Where there will be direct exposure to water or in areas subject to continuous high humidity.

3. On ceilings where frame spacing exceeds 12 inches (305 mm) o.c. for $^1/_2$-inch-thick (12.7 mm) water-resistant gypsum backing board and more than 16 inches (406 mm) o.c. for $^5/_8$-inch-thick (15.9 mm) water-resistant gypsum backing board.

SECTION 2510
LATHING AND FURRING FOR
CEMENT PLASTER (STUCCO)

2510.1 General. Exterior and interior cement plaster and lathing shall be done with the appropriate materials listed in Table 2507.2 and Chapter 35.

2510.2 Weather protection. Materials shall be stored in such a manner as to protect such materials from the weather.

2510.3 Installation. Installation of these materials shall be in compliance with ASTM C 926 and ASTM C 1063.

2510.4 Corrosion resistance. Metal lath and lath attachments shall be of corrosion-resistant material.

2510.5 Backing. Backing or a lath shall provide sufficient rigidity to permit plaster applications.

2510.5.1 Support of lath. Where lath on vertical surfaces extends between rafters or other similar projecting members, solid backing shall be installed to provide support for lath and attachments.

2510.5.2 Use of gypsum backing board.

2510.5.2.1 Use of gypsum board as a backing board. Gypsum lath or gypsum wallboard shall not be used as a backing for cement plaster.

Exception: Gypsum lath or gypsum wallboard is permitted, with a weather-resistant barrier, as a backing for self-furred metal lath or self-furred wire fabric lath and cement plaster where either of the following conditions occur:

1. On horizontal supports of ceilings or roof soffits.

2. On interior walls.

2510.5.2.2 Use of gypsum sheathing backing. Gypsum sheathing is permitted as a backing for metal or wire fabric lath and cement plaster on walls. A weather-resistant barrier shall be provided in accordance with Section 2510.6.

2510.5.3 Backing not required. Wire backing is not required under expanded metal lath or paperbacked wire fabric lath.

2510.6 Weather-resistant barriers. Weather-resistant barriers shall be installed as required in Section 1404.2 and, where applied over wood-based sheathing, shall include a weather-resistant vapor-permeable barrier with a performance at least equivalent to two layers of Grade D paper.

2510.7 Preparation of masonry and concrete. Surfaces shall be clean, free from efflorescence, sufficiently damp and rough for proper bond. If the surface is insufficiently rough, approved bonding agents or a portland cement dash bond coat mixed in proportions of not more than two parts volume of sand to one part volume of portland cement or plastic cement shall be applied. The dash bond coat shall be left undisturbed and shall be moist cured not less than 24 hours.

SECTION 2511
INTERIOR PLASTER

2511.1 General. Plastering gypsum plaster or cement plaster shall not be less than three coats where applied over metal lath or wire fabric lath and not less than two coats where applied over other bases permitted by this chapter.

Exception: Gypsum veneer plaster and cement plaster specifically designed and approved for one-coat applications.

TABLE 2511.1
INSTALLATION OF PLASTER CONSTRUCTION

MATERIAL	STANDARD
Gypsum plaster	ASTM C 842
Gypsum veneer plaster	ASTM C 843
Interior lathing and furring (gypsum plaster)	ASTM C 841
Lathing and furring (cement plaster)	ASTM C 1063
Portland cement plaster	ASTM C 926
Steel framing	ASTM C 754; C 1007

2511.1.1 Installation. Installation of lathing and plaster materials shall conform with Table 2511.1 and Section 2507.

2511.2 Limitations. Plaster shall not be applied directly to fiber insulation board. Cement plaster shall not be applied directly to gypsum lath or gypsum plaster except as specified in Sections 2510.5.1 and 2510.5.2.

2511.3 Grounds. Where installed, grounds shall ensure the minimum thickness of plaster as set forth in ASTM C 842 and ASTM C 926. Plaster thickness shall be measured from the face of lath and other bases.

2511.4 Interior masonry or concrete. Condition of surfaces shall be as specified in Section 2510.7. Approved specially prepared gypsum plaster designed for application to concrete surfaces or approved acoustical plaster is permitted. The total thickness of base coat plaster applied to concrete ceilings shall be as set forth in ASTM C 842 or ASTM C 926. Should ceiling surfaces require more than the maximum thickness permitted in ASTM C 842 or ASTM C 926, metal lath or wire fabric lath shall be installed on such surfaces before plastering.

2511.5 Wet areas. Showers and public toilet walls shall conform to Sections 1210.2 and 1210.3. When wood frame walls and partitions are covered on the interior with cement plaster or tile of similar material and are subject to water splash, the framing shall be protected with an approved moisture barrier.

SECTION 2512
EXTERIOR PLASTER

2512.1 General. Plastering with cement plaster shall not be less than three coats where applied over metal lath or wire fabric lath and not less than two coats where applied over masonry, concrete or gypsum board backing as specified in Section 2510.5. If the plaster surface is to be completely covered by veneer or other facing material, or is completely concealed by another wall, plaster application need be only two coats, provided the total thickness is as set forth in ASTM C 926.

2512.1.1 On-grade floor slab. On wood framed or steel stud construction with an on-grade concrete floor slab system, exterior plaster shall be applied in such a manner as to cover, but not to extend below, the lath and paper. The application of lath, paper and flashing or drip screeds shall comply with ASTM C 1063.

2512.1.2 Weep screeds. A minimum 0.019-inch (0.48 mm) (No. 26 galvanized sheet gage), corrosion-resistant weep screed with a minimum vertical attachment flange of $3^1/_2$ inches (89 mm) shall be provided at or below the foundation plate line on exterior stud walls in accordance with ASTM C 926. The weep screed shall be placed a minimum of 4 inches (102 mm) above the earth or 2 inches (51 mm) above paved areas and be of a type that will allow trapped water to drain to the exterior of the building. The weather-resistant barrier shall lap the attachment flange. The exterior lath shall cover and terminate on the attachment flange of the weep screed.

2512.2 Plasticity agents. Only approved plasticity agents and approved amounts thereof shall be added to portland cement. When plastic cement or masonry cement is used, no additional lime or plasticizers shall be added. Hydrated lime or the equivalent amount of lime putty used as a plasticizer is permitted to be added to cement plaster or cement and lime plaster in an amount not to exceed that set forth in ASTM C 926.

2512.3 Limitations. Gypsum plaster shall not be used on exterior surfaces.

2512.4 Cement plaster. Plaster coats shall be protected from freezing for a period of not less than 24 hours after set has occurred. Plaster shall be applied when the ambient temperature is higher than 40°F (4°C), unless provisions are made to keep cement plaster work above 40°F (4°C) during application and 48 hours thereafter.

2512.5 Second-coat application. The second coat shall be brought out to proper thickness, rodded and floated sufficiently rough to provide adequate bond for the finish coat. The second coat shall have no variation greater than $^1/_4$ inch (6.4 mm) in any direction under a 5-foot (1524 mm) straight edge.

2512.6 Curing and interval. First and second coats of cement plaster shall be applied and moist cured as set forth in ASTM C 926 and Table 2512.6.

TABLE 2512.6
CEMENT PLASTERS[a]

COAT	MINIMUM PERIOD MOIST CURING	MINIMUM INTERVAL BETWEEN COATS
First	48 hours[a]	48 hours[b]
Second	48 hours	7 days[c]
Finish	—	Note c

a. The first two coats shall be as required for the first coats of exterior plaster, except that the moist-curing time period between the first and second coats shall not be less than 24 hours. Moist curing shall not be required where job and weather conditions are favorable to the retention of moisture in the cement plaster for the required time period.

b. Twenty-four-hour minimum interval between coats of interior cement plaster. For alternate method of application, see Section 2512.8.

c. Finish coat plaster is permitted to be applied to interior portland cement base coats after a 48-hour period.

2512.7 Application to solid backings. Where applied over gypsum backing as specified in Section 2510.5 or directly to unit masonry surfaces, the second coat is permitted to be applied as soon as the first coat has attained sufficient hardness.

2512.8 Alternate method of application. The second coat is permitted to be applied as soon as the first coat has attained sufficiently rigidity to receive the second coat.

2512.8.1 Admixtures. When using this method of application, calcium aluminate cement up to 15 percent of the weight of the portland cement is permitted to be added to the mix.

2512.8.2 Curing. Curing of the first coat is permitted to be omitted and the second coat shall be cured as set forth in ASTM C 926 and Table 2512.6.

2512.9 Finish coats. Cement plaster finish coats shall be applied over base coats that have been in place for the time periods set forth in ASTM C 926. The third or finish coat shall be applied with sufficient material and pressure to bond and to cover the brown coat and shall be of sufficient thickness to conceal the brown coat.

SECTION 2513
EXPOSED AGGREGATE PLASTER

2513.1 General. Exposed natural or integrally colored aggregate is permitted to be partially embedded in a natural or colored bedding coat of cement plaster or gypsum plaster, subject to the provisions of this section.

2513.2 Aggregate. The aggregate shall be applied manually or mechanically and shall consist of marble chips, pebbles or similar durable, moderately hard (three or more on the Mohs hardness scale), nonreactive materials.

2513.3 Bedding coat proportions. The bedding coat for interior or exterior surfaces shall be composed of one-part portland cement, one-part Type S lime and a maximum of three parts of graded white or natural sand by volume. The bedding coat for interior surfaces shall be composed of 100 pounds (45.4 kg) of neat gypsum plaster and a maximum of 200 pounds (90.8 kg) of graded white sand. A factory-prepared bedding coat for interior or exterior use is permitted. The bedding coat for exterior surfaces shall have a minimum compressive strength of 1,000 pounds per square inch (psi) (6895 kPa).

2513.4 Application. The bedding coat is permitted to be applied directly over the first (scratch) coat of plaster, provided the ultimate overall thickness is a minimum of $^7/_8$ inch (22 mm), including lath. Over concrete or masonry surfaces, the overall thickness shall be a minimum of $^1/_2$ inch (12.7 mm).

2513.5 Bases. Exposed aggregate plaster is permitted to be applied over concrete, masonry, cement plaster base coats or gypsum plaster base coats installed in accordance with Section 2511 or 2512.

2513.6 Preparation of masonry and concrete. Masonry and concrete surfaces shall be prepared in accordance with the provisions of Section 2510.7.

2513.7 Curing of base coats. Cement plaster base coats shall be cured in accordance with ASTM C 926. Cement plaster bedding coats shall retain sufficient moisture for hydration (hardening) for 24 hours minimum or, where necessary, shall be kept damp for 24 hours by light water spraying.

SECTION 2601
GENERAL

2601.1 Scope. These provisions shall govern the materials, design, application, construction and installation of foam plastic, foam plastic insulation, plastic veneer, interior plastic finish and trim and light-transmitting plastics. See Chapter 14 for requirements for exterior wall finish and trim.

SECTION 2602
DEFINITIONS

2602.1 General. The following words and terms shall, for the purposes of this chapter and as used elsewhere in this code, have the meanings shown herein.

FOAM PLASTIC INSULATION. A plastic that is intentionally expanded by the use of a foaming agent to produce a reduced-density plastic containing voids consisting of open or closed cells distributed throughout the plastic for thermal insulating or acoustical purposes and that has a density less than 20 pounds per cubic foot (pcf) (320 kg/m³).

LIGHT-DIFFUSING SYSTEM. Construction consisting in whole or in part of lenses, panels, grids or baffles made with light-transmitting plastics positioned below independently mounted electrical light sources, skylights or light-transmitting plastic roof panels. Lenses, panels, grids and baffles that are part of an electrical fixture shall not be considered as a light-diffusing system.

LIGHT-TRANSMITTING PLASTIC ROOF PANELS. Structural plastic panels other than skylights that are fastened to structural members, or panels or sheathing and that are used as light-transmitting media in the plane of the roof.

LIGHT-TRANSMITTING PLASTIC WALL PANELS. Plastic materials that are fastened to structural members, or to structural panels or sheathing, and that are used as light-transmitting media in exterior walls.

PLASTIC, APPROVED. Any thermoplastic, thermosetting or reinforced thermosetting plastic material that conforms to combustibility classifications specified in the section applicable to the application and plastic type.

PLASTIC GLAZING. Plastic materials that are glazed or set in frame or sash and not held by mechanical fasteners that pass through the glazing material.

REINFORCED PLASTIC, GLASS FIBER. Plastic reinforced with glass fiber having not less than 20 percent of glass fibers by weight.

THERMOPLASTIC MATERIAL. A plastic material that is capable of being repeatedly softened by increase of temperature and hardened by decrease of temperature.

THERMOSETTING MATERIAL. A plastic material that is capable of being changed into a substantially nonreformable product when cured.

SECTION 2603
FOAM PLASTIC INSULATION

2603.1 General. The provisions of this section shall govern the requirements and uses of foam plastic insulation in buildings and structures.

2603.2 Labeling and identification. Packages and containers of foam plastic insulation and foam plastic insulation components delivered to the job site shall bear the label of an approved agency showing the manufacturer's name, the product listing, product identification and information sufficient to determine that the end use will comply with the code requirements.

2603.3 Surface-burning characteristics. Unless otherwise indicated in this section, foam plastic insulation and foam plastic cores of manufactured assemblies shall have a flame spread index of not more than 75 and a smoke-developed index of not more than 450 where tested in the maximum thickness intended for use in accordance with ASTM E 84. Loose fill-type foam plastic insulation shall be tested as board stock for the flame spread index and smoke-developed index.

Exceptions:

1. Smoke-developed index for interior trim as provided for in Section 2604.2.
2. In cold storage buildings, ice plants, food plants, food processing rooms and similar areas, foam plastic insulation where tested in a thickness of 4 inches (102 mm) shall be permitted in a thickness up to 10 inches (254 mm) where the building is equipped throughout with an automatic fire sprinkler system in accordance with Section 903.3.1.1. The approved automatic sprinkler system shall be provided in both the room and that part of the building in which the room is located.
3. Foam plastic insulation that is a part of a Class A, B or C roof-covering assembly provided the assembly with the foam plastic insulation satisfactorily passes FM 4450 or UL 1256. The smoke-developed index shall not be limited for roof applications.
4. Foam plastic insulation greater than 4 inches (102 mm) in thickness shall have a maximum flame spread index of 75 and a smoke-developed index of 450 where tested at a minimum thickness of 4 inches (102 mm), provided the end use is approved in accordance with Section 2603.8 using the thickness and density intended for use.
5. Flame spread and smoke-developed indexes for foam plastic interior signs in covered mall buildings provided the signs comply with Section 402.14.

2603.4 Thermal barrier. Except as provided for in Sections 2603.4.1 and 2603.8, foam plastic shall be separated from the interior of a building by an approved thermal barrier of 0.5-inch (12.7 mm) gypsum wallboard or equivalent thermal barrier material that will limit the average temperature rise of the unexposed surface to not more than 250°F (120°C) after 15 minutes of fire exposure, complying with the standard time-temperature curve of ASTM E 119. The thermal barrier shall be installed in such a manner that it will remain in place for 15 minutes based on FM 4880, UL 1040, NFPA 286 or UL 1715. Combustible concealed spaces shall comply with Section 717.

2603.4.1 Thermal barrier not required. The thermal barrier specified in Section 2603.4 is not required under the conditions set forth in Sections 2603.4.1.1 through 2603.4.1.13.

2603.4.1.1 Masonry or concrete construction. In a masonry or concrete wall, floor or roof system where the foam plastic insulation is covered on each face by a minimum of 1 inch (25 mm) thickness of masonry or concrete.

2603.4.1.2 Cooler and freezer walls. Foam plastic installed in a maximum thickness of 10 inches (254 mm) in cooler and freezer walls shall:

1. Have a flame spread index of 25 or less and a smoke-developed index of not more than 450, where tested in a minimum 4 inch (102 mm) thickness.

2. Have flash ignition and self-ignition temperatures of not less than 600°F and 800°F (316°C and 427°C), respectively.

3. Have a covering of not less than 0.032-inch (0.8 mm) aluminum or corrosion-resistant steel having a base metal thickness not less than 0.0160 inch (0.4 mm) at any point.

4. Be protected by an automatic sprinkler system. Where the cooler or freezer is within a building, both the cooler or freezer and that part of the building in which it is located shall be sprinklered.

2603.4.1.3 Walk-in coolers. In nonsprinklered buildings, foam plastic having a thickness that does not exceed 4 inches (102 mm) and a maximum flame spread index of 75 is permitted in walk-in coolers or freezer units where the aggregate floor area does not exceed 400 square feet (37 m²) and the foam plastic is covered by a metal facing not less than 0.032-inch-thick (0.81 mm) aluminum or corrosion-resistant steel having a minimum base metal thickness of 0.016 inch (0.41 mm). A thickness of up to 10 inches (254 mm) is permitted where protected by a thermal barrier.

2603.4.1.4 Exterior walls—one-story buildings. For one-story buildings, foam plastic having a flame spread index of 25 or less, and a smoke-developed index of not more than 450, shall be permitted without thermal barriers in or on exterior walls in a thickness not more than 4 inches (102 mm) where the foam plastic is covered by a thickness of not less than 0.032-inch-thick (0.81 mm) aluminum or corrosion-resistant steel having a base metal thickness of 0.0160 inch (0.41 mm) and the building is equipped throughout with an automatic sprinkler system in accordance with Section 903.3.1.1.

2603.4.1.5 Roofing. Foam plastic insulation under a roof assembly or roof covering that is installed in accordance with the code and the manufacturer's instructions shall be separated from the interior of the building by wood structural panel sheathing not less than 0.47 inch (11.9 mm) in thickness bonded with exterior glue, with edges supported by blocking, tongue-and-groove joints or other approved type of edge support, or an equivalent material. A thermal barrier is not required for foam plastic insulation that is a part of a Class A, B or C roof-covering assembly, provided the assembly with the foam plastic insulation satisfactorily passes FM 4450 or UL 1256.

2603.4.1.6 Attics and crawl spaces. Within an attic or crawl space where entry is made only for service of utilities, foam plastic insulation shall be protected against ignition by 1.5-inch-thick (38 mm) mineral fiber insulation; 0.25-inch-thick (6.4 mm) wood structural panel, particleboard or hardboard; 0.375-inch (9.5 mm) gypsum wallboard, corrosion-resistant steel having a base metal thickness of 0.016 inch (0.4 mm) or other approved material installed in such a manner that the foam plastic insulation is not exposed. The protective covering shall be consistent with the requirements for the type of construction.

2603.4.1.7 Doors not required to have a fire protection rating. Where pivoted or side-hinged doors are permitted without a fire protection rating, foam plastic insulation, having a flame spread index of 75 or less and a smoke-developed index of not more than 450, shall be permitted as a core material where the door facing is of metal having a minimum thickness of 0.032-inch (0.8 mm) aluminum or steel having a base metal thickness of not less than 0.016 inch (0.4 mm) at any point.

2603.4.1.8 Exterior doors in buildings of Group R-2 or R-3. In occupancies classified as Group R-2 or R-3 as applicable in Section 101.2, foam-filled exterior entrance doors to individual dwelling units that do not require a fire-resistance rating shall be faced with wood or other approved materials.

2603.4.1.9 Garage doors. Where garage doors are permitted without a fire-resistance rating and foam plastic is used as a core material, the door facing shall be metal having a minimum thickness of 0.032-inch (0.8 mm) aluminum or 0.010-inch (0.25 mm) steel or the facing shall be minimum 0.125-inch-thick (3.2 mm) wood. Garage doors having facings other than those described above shall be tested in accordance with, and meet the acceptance criteria of DASMA 107.

> **Exception:** Garage doors using foam plastic insulation complying with Section 2603.3 in detached and attached garages associated with one- and two-family dwellings need not be provided with a thermal barrier.

2603.4.1.10 Siding backer board. Foam plastic insulation of not more than 2,000 British thermal units per square feet (Btu/sq. ft.) (22.7 MJ/m²) as determined by NFPA 259 shall be permitted as a siding backer board with a maximum thickness of 0.5 inch (12.7 mm), provided it is separated from the interior of the building by not less than 2 inches (51 mm) of mineral fiber insulation

PLASTIC

or equivalent or where applied as insulation with residing over existing wall construction.

2603.4.1.11 Interior trim. Foam plastic used as interior trim in accordance with Section 2604 shall be permitted without a thermal barrier.

2603.4.1.12 Interior signs. Foam plastic used for interior signs in covered mall buildings in accordance with Section 402.14 shall be permitted without a thermal barrier.

2603.4.1.13 Type V construction. Foam plastic spray applied to a sill plate and header of Type V construction is subject to all of the following:

1. The maximum thickness of the foam plastic shall be $3^1/_4$ inches (82.6 mm).
2. The density of the foam plastic shall be in the range of 1.5 to 2.0 pcf (24 to 32 kg/m³).
3. The foam plastic shall have a flame spread index of 25 or less and an accompanying smoke-developed index of 450 or less when tested in accordance with ASTM E84.

2603.5 Exterior walls of buildings of any height. Exterior walls of buildings of Type I, II, III or IV construction of any height shall comply with Sections 2603.5.1 through 2603.5.7. Exterior walls of cold storage buildings required to be constructed of noncombustible materials, where the building is more than one story in height, shall also comply with the provisions of Sections 2603.5.1 through 2603.5.7. Exterior walls of buildings of Type V construction shall comply with Sections 2603.2, 2603.3 and 2603.4.

2603.5.1 Fire-resistance-rated walls. Where the wall is required to have a fire-resistance rating, data based on tests conducted in accordance with ASTM E 119 shall be provided to substantiate that the fire-resistance rating is maintained.

2603.5.2 Thermal barrier. Any foam plastic insulation shall be separated from the building interior by a thermal barrier meeting the provisions of Section 2603.4, unless special approval is obtained on the basis of Section 2603.8.

Exception: One-story buildings complying with Section 2603.4.1.4.

2603.5.3 Potential heat. The potential heat of foam plastic insulation in any portion of the wall or panel shall not exceed the potential heat expressed in Btu per square feet (mJ/m²) of the foam plastic insulation contained in the wall assembly tested in accordance with Section 2603.5.5. The potential heat of the foam plastic insulation shall be determined by tests conducted in accordance with NFPA 259 and the results shall be expressed in Btu per square feet (mJ/m²).

Exception: One-story buildings complying with Section 2603.4.1.4.

2603.5.4 Flame spread and smoke-developed indexes. Foam plastic insulation, exterior coatings and facings shall be tested separately in the thickness intended for use, but not to exceed 4 inches (102 mm), and shall each have a flame

spread index of 25 or less and a smoke-developed index of 450 or less as determined in accordance with ASTM E 84.

Exception: Prefabricated or factory-manufactured panels having minimum 0.020-inch (0.51 mm) aluminum facings and a total thickness of 0.25 inch (6.4 mm) or less are permitted to be tested as an assembly where the foam plastic core is not exposed in the course of construction.

2603.5.5 Test standard. The wall assembly shall be tested in accordance with and comply with the acceptance criteria of NFPA 285.

Exception: One-story buildings complying with Section 2603.4.1.4.

2603.5.6 Label required. The edge or face of each piece of foam plastic insulation shall bear the label of an approved agency. The label shall contain the manufacturer's or distributor's identification, model number, serial number or definitive information describing the product or materials' performance characteristics and approved agency's identification.

2603.5.7 Ignition. Exterior walls shall not exhibit sustained flaming where tested in accordance with NFPA 268. Where a material is intended to be installed in more than one thickness, tests of the minimum and maximum thickness intended for use shall be performed.

Exception: Assemblies protected on the outside with one of the following:

1. A thermal barrier complying with Section 2603.4.
2. A minimum 1 inch (25 mm) thickness of concrete or masonry.
3. Glass-fiber-reinforced concrete panels of a minimum thickness of 0.375 inch (9.5 mm).
4. Metal-faced panels having minimum 0.019-inch-thick (0.48 mm) aluminum or 0.016-inch-thick (0.41 mm) corrosion-resistant steel outer facings.
5. A minimum 0.875 inch (22.2 mm) thickness of stucco complying with Section 2510.

2603.6 Roofing. Foam plastic insulation meeting the requirements of Sections 2603.2, 2603.3 and 2603.4 shall be permitted as part of a roof-covering assembly, provided the assembly with the foam plastic insulation is a Class A, B or C roofing assembly where tested in accordance with ASTM E 108 or UL 790.

2603.7 Plenums. Foam plastic insulation shall not be used as interior wall or ceiling finish in plenums except as permitted in Section 2604 or when protected by a thermal barrier in accordance with Section 2603.4.

2603.8 Special approval. Foam plastic shall not be required to comply with the requirements of Sections 2603.4 through 2603.7, where specifically approved based on large-scale tests such as, but not limited to, FM 4880, UL 1040, NFPA 286 or UL 1715. Such testing shall be related to the actual end-use configuration and be performed on the finished manufactured foam plastic assembly in the maximum thickness intended for use. Foam plastics that are used as interior finish on the basis of special tests shall also conform to the flame spread requirements of Chapter 8. Assemblies tested shall include seams, joints and other typical details used in the installation of the assembly and shall be tested in the manner intended for use.

SECTION 2604
INTERIOR FINISH AND TRIM

2604.1 General. Plastic materials installed as interior finish or trim shall comply with Chapter 8. Foam plastics shall only be installed as interior finish where approved in accordance with the special provisions of Section 2603.8. Foam plastics that are used as interior finish shall also meet the flame spread index requirements for interior finish in accordance with Chapter 8. Foam plastics installed as interior trim shall comply with Section 2604.2.

[F] 2604.2 Interior trim. Foam plastic used as interior trim shall comply with Sections 2604.2.1 through 2604.2.4.

> **[F] 2604.2.1 Density.** The minimum density of the interior trim shall be 20 pcf (320 kg/m^3).

> **[F] 2604.2.2 Thickness.** The maximum thickness of the interior trim shall be 0.5 inch (12.7 mm) and the maximum width shall be 8 inches (204 mm).

> **[F] 2604.2.3 Area limitation.** The interior trim shall not constitute more than 10 percent of the aggregate wall and ceiling area of any room or space.

> **[F] 2604.2.4 Flame spread.** The flame spread index shall not exceed 75 where tested in accordance with ASTM E 84. The smoke-developed index shall not be limited.

SECTION 2605
PLASTIC VENEER

2605.1 Interior use. Where used within a building, plastic veneer shall comply with the interior finish requirements of Chapter 8.

2605.2 Exterior use. Exterior plastic veneer shall be permitted to be installed on the exterior walls of buildings of any type of construction in accordance with all of the following requirements:

1. Plastic veneer shall comply with Section 2606.4.
2. Plastic veneer shall not be attached to any exterior wall to a height greater than 50 feet (15 240 mm) above grade.
3. Sections of plastic veneer shall not exceed 300 square feet (27.9 m^2) in area and shall be separated by a minimum of 4 feet (1219 mm) vertically.

> **Exception:** The area and separation requirements and the smoke-density limitation are not applicable to plastic veneer applied to buildings constructed of Type VB construction, provided the walls are not required to have a fire-resistance rating.

SECTION 2606
LIGHT-TRANSMITTING PLASTICS

2606.1 General. The provisions of this section and Sections 2607 through 2611 shall govern the quality and methods of application of light-transmitting plastics for use as light-transmitting materials in buildings and structures. Foam plastics shall comply with Section 2603. Light-transmitting plastic materials that meet the other code requirements for walls and roofs shall be permitted to be used in accordance with the other applicable chapters of the code.

2606.2 Approval for use. Sufficient technical data shall be submitted to substantiate the proposed use of any light-transmitting material, as approved by the building official and subject to the requirements of this section.

2606.3 Identification. Each unit or package of light-transmitting plastic shall be identified with a mark or decal satisfactory to the building official, which includes identification as to the material classification.

2606.4 Specifications. Light-transmitting plastics, including thermoplastic, thermosetting or reinforced thermosetting plastic material, shall have a self-ignition temperature of 650°F (343°C) or greater where tested in accordance with ASTM D 1929; a smoke-developed index not greater than 450 where tested in the manner intended for use in accordance with ASTM E 84, or not greater than 75 where tested in the thickness intended for use in accordance with ASTM D 2843 and shall conform to one of the following combustibility classifications:

> **Class CC1:** Plastic materials that have a burning extent of 1 inch (25 mm) or less where tested at a nominal thickness of 0.060 inch (1.5 mm), or in the thickness intended for use, in accordance with ASTM D 635,

> **Class CC2:** Plastic materials that have a burning rate of 2.5 inches per minute (1.06 mm/s) or less where tested at a nominal thickness of 0.060 inch (1.5 mm), or in the thickness intended for use, in accordance with ASTM D 635.

2606.5 Structural requirements. Light-transmitting plastic materials in their assembly shall be of adequate strength and durability to withstand the loads indicated in Chapter 16. Technical data shall be submitted to establish stresses, maximum unsupported spans and such other information for the various thicknesses and forms used as deemed necessary by the building official.

2606.6 Fastening. Fastening shall be adequate to withstand the loads in Chapter 16. Proper allowance shall be made for expansion and contraction of light-transmitting plastic materials in accordance with accepted data on the coefficient of expansion of the material and other material in conjunction with which it is employed.

2606.7 Light-diffusing systems. Unless the building is equipped throughout with an automatic sprinkler system in accordance with Section 903.3.1.1, light-diffusing systems shall not be installed in the following occupancies and locations:

1. Group A with an occupant load of 1,000 or more.
2. Theaters with a stage and proscenium opening and an occupant load of 700 or more.
3. Group I-2.

4. Group I-3.

5. Exit stairways and exit passageways.

2606.7.1 Support. Light-transmitting plastic diffusers shall be supported directly or indirectly from ceiling or roof construction by use of noncombustible hangers. Hangers shall be at least No. 12 steel-wire gage (0.106 inch) galvanized wire or equivalent.

2606.7.2 Installation. Light-transmitting plastic diffusers shall comply with Chapter 8 unless the light-transmitting plastic diffusers will fall from the mountings before igniting, at an ambient temperature of at least 200°F (93°C) below the ignition temperature of the panels. The panels shall remain in place at an ambient room temperature of 175°F (79°C) for a period of not less than 15 minutes.

2606.7.3 Size limitations. Individual panels or units shall not exceed 10 feet (3048 mm) in length nor 30 square feet (2.79 m²) in area.

2606.7.4 Fire suppression system. In buildings that are equipped throughout with an automatic sprinkler system in accordance with Section 903.3.1.1, plastic light-diffusing systems shall be protected both above and below unless the sprinkler system has been specifically approved for installation only above the light-diffusing system. Areas of light-diffusing systems that are protected in accordance with this section shall not be limited.

2606.7.5 Electrical lighting fixtures. Light-transmitting plastic panels and light-diffuser panels that are installed in approved electrical lighting fixtures shall comply with the requirements of Chapter 8 unless the light-transmitting plastic panels conform to the requirements of Section 2606.7.2. The area of approved light-transmitting plastic materials that are used in required exits or corridors shall not exceed 30 percent of the aggregate area of the ceiling in which such panels are installed, unless the building is equipped throughout with an automatic sprinkler system in accordance with Section 903.3.1.1.

2606.8 Partitions. Light-transmitting plastics used in or as partitions shall comply with the requirements of Chapters 6 and 8.

2606.9 Bathroom accessories. Light-transmitting plastics shall be permitted as glazing in shower stalls, shower doors, bathtub enclosures and similar accessory units. Safety glazing shall be provided in accordance with Chapter 24.

2606.10 Awnings, patio covers and similar structures. Awnings constructed of light-transmitting plastics shall be constructed in accordance with provisions specified in Section 3105 and Chapter 32 for projections and appendages. Patio covers constructed of light-transmitting plastics shall comply with Section 2606. Light-transmitting plastics used in canopies at motor fuel-dispensing facilities shall comply with Section 2606 except as modified by Section 406.5.2.

2606.11 Greenhouses. Light-transmitting plastics shall be permitted in lieu of plain glass in greenhouses.

2606.12 Solar collectors. Light-transmitting plastic covers on solar collectors having noncombustible sides and bottoms shall be permitted on buildings not over three stories in height or 9,000 square feet (836.1 m²) in total floor area, provided the light-transmitting plastic cover does not exceed 33.33 percent of the roof area for CC1 materials or 25 percent of the roof area for CC2 materials.

Exception: Light-transmitting plastic covers having a thickness of 0.010 inch (0.3 mm) or less or shall be permitted to be of any plastic material provided the area of the solar collectors does not exceed 33.33 percent of the roof area.

SECTION 2607
LIGHT-TRANSMITTING PLASTIC WALL PANELS

2607.1 General. Light-transmitting plastics shall not be used as wall panels in exterior walls in occupancies in Groups A-1, A-2, H, I-2 and I-3. In other groups, light-transmitting plastics shall be permitted to be used as wall panels in exterior walls, provided that the walls are not required to have a fire-resistance rating and the installation conforms to the requirements of this section. Such panels shall be erected and anchored on a foundation, waterproofed or otherwise protected from moisture absorption and sealed with a coat of mastic or other approved waterproof coating. Light-transmitting plastic wall panels shall also comply with Section 2606.

2607.2 Installation. Exterior wall panels installed as provided for herein shall not alter the type of construction classification of the building.

2607.3 Height limitation. Light-transmitting plastics shall not be installed more than 75 feet (22 860 mm) above grade plane, except as allowed by Section 2607.5.

2607.4 Area limitation and separation. The maximum area of a single wall panel and minimum vertical and horizontal separation requirements for exterior light-transmitting plastic wall panels shall be as provided for in Table 2607.4. The maximum percentage of wall area of any story in light-transmitting plastic wall panels shall not exceed that indicated in Table 2607.4 or the percentage of unprotected openings permitted by Section 704.8, whichever is smaller.

Exceptions:

1. In structures provided with approved flame barriers extending 30 inches (760 mm) beyond the exterior wall in the plane of the floor, a vertical separation is not required at the floor except that provided by the vertical thickness of the flame barrier projection.

2. Veneers of approved weather-resistant light-transmitting plastics used as exterior siding in buildings of Type V construction in compliance with Section 1406.

3. The area of light-transmitting plastic wall panels in exterior walls of greenhouses shall be exempt from the area limitations of Table 2607.4 but shall be limited as required for unprotected openings in accordance with Section 704.8.

TABLE 2607.4
AREA LIMITATION AND SEPARATION REQUIREMENTS FOR
LIGHT-TRANSMITTING PLASTIC WALL PANELS[a]

FIRE SEPARATION DISTANCE (feet)	CLASS OF PLASTIC	MAXIMUM PERCENTAGE AREA OF EXTERIOR WALL IN PLASTIC WALL PANELS	MAXIMUM SINGLE AREA OF PLASTIC WALL PANELS (square feet)	MINIMUM SEPARATION OF PLASTIC WALL PANELS (feet)	
				Vertical	Horizontal
Less than 6	—	Not Permitted	Not Permitted	—	—
6 or more but less than 11	CC1	10	50	8	4
	CC2	Not Permitted	Not Permitted	—	—
11 or more but less than or equal to 30	CC1	25	90	6	4
	CC2	15	70	8	4
Over 30	CC1	50	Not Limited	3[b]	0
	CC2	50	100	6[b]	3

For SI: 1 foot = 304.8 mm, 1 square foot = 0.0929 m^2.
a. For combinations of plastic glazing and plastic wall panel areas permitted, see Section 2607.6.
b. For reductions in vertical separation allowed, see Section 2607.4.

2607.5 Automatic sprinkler system. Where the building is equipped throughout with an automatic sprinkler system in accordance with Section 903.3.1.1, the maximum percentage area of exterior wall in any story in light-transmitting plastic wall panels and the maximum square footage of a single area given in Table 2607.4 shall be increased 100 percent, but the area of light-transmitting plastic wall panels shall not exceed 50 percent of the wall area in any story, or the area permitted by Section 704.8 for unprotected openings, whichever is smaller. These installations shall be exempt from height limitations.

2607.6 Combinations of glazing and wall panels. Combinations of light-transmitting plastic glazing and light-transmitting plastic wall panels shall be subject to the area, height and percentage limitations and the separation requirements applicable to the class of light-transmitting plastic as prescribed for light-transmitting plastic wall panel installations.

SECTION 2608
LIGHT-TRANSMITTING PLASTIC GLAZING

2608.1 Buildings of Type VB construction. Openings in the exterior walls of buildings of Type VB construction, where not required to be protected by Section 704, shall be permitted to be glazed or equipped with light-transmitting plastic. Light-transmitting plastic glazing shall also comply with Section 2606.

2608.2 Buildings of other types of construction. Openings in the exterior walls of buildings of types of construction other than Type VB, where not required to be protected by Section 704, shall be permitted to be glazed or equipped with light-transmitting plastic in accordance with Section 2606 and all of the following:

1. The aggregate area of light-transmitting plastic glazing shall not exceed 25 percent of the area of any wall face of the story in which it is installed. The area of a single pane of glazing installed above the first story above grade plane shall not exceed 16 square feet (1.5 m^2) and the ver-

tical dimension of a single pane shall not exceed 4 feet (1219 mm).

   Exception: Where an automatic sprinkler system is provided throughout in accordance with Section 903.3.1.1, the area of allowable glazing shall be increased to a maximum of 50 percent of the wall face of the story in which it is installed with no limit on the maximum dimension or area of a single pane of glazing.

2. Approved flame barriers extending 30 inches (762 mm) beyond the exterior wall in the plane of the floor, or vertical panels not less than 4 feet (1219 mm) in height, shall be installed between glazed units located in adjacent stories.

   Exception: Buildings equipped throughout with an automatic sprinkler system in accordance with Section 903.3.1.1.

3. Light-transmitting plastics shall not be installed more than 75 feet (22 860 mm) above grade level.

   Exception: Buildings equipped throughout with an automatic sprinkler system in accordance with Section 903.3.1.1.

SECTION 2609
LIGHT-TRANSMITTING PLASTIC ROOF PANELS

2609.1 General. Light-transmitting plastic roof panels shall comply with this section and Section 2606. Light-transmitting plastic roof panels shall not be installed in Groups H, I-2 and I-3. In all other groups, light-transmitting plastic roof panels shall comply with any one of the following conditions:

1. The building is equipped throughout with an automatic sprinkler system in accordance with Section 903.3.1.1.

2. The roof construction is not required to have a fire-resistance rating by Table 601.

3. The roof panels meet the requirements for roof coverings in accordance with Chapter 15.

2609.2 Separation. Individual roof panels shall be separated from each other by a distance of not less than 4 feet (1219 mm) measured in a horizontal plane.

Exceptions:

1. The separation between roof panels is not required in a building equipped throughout with an automatic sprinkler system in accordance with Section 903.3.1.1.

2. The separation between roof panels is not required in low-hazard occupancy buildings complying with the conditions of Section 2609.4, Exception 2 or 3.

2609.3 Location. Where exterior wall openings are required to be protected by Section 704.8, a roof panel shall not be installed within 6 feet (1829 mm) of such exterior wall.

2609.4 Area limitations. Roof panels shall be limited in area and the aggregate area of panels shall be limited by a percentage of the floor area of the room or space sheltered in accordance with Table 2609.4.

Exceptions:

1. The area limitations of Table 2609.4 shall be permitted to be increased by 100 percent in buildings equipped throughout with an automatic sprinkler system in accordance with Section 903.3.1.1.

2. Low-hazard occupancy buildings, such as swimming pool shelters, shall be exempt from the area limitations of Table 2609.4, provided that the buildings do not exceed 5,000 square feet (465 m^2) in area and have a minimum fire separation distance of 10 feet (3048 mm).

3. Greenhouses that are occupied for growing plants on a production or research basis, without public access, shall be exempt from the area limitations of Table 2609.4 provided they have a minimum fire separation distance of 4 feet (1220 mm).

4. Roof coverings over terraces and patios in occupancies in Group R-3 as applicable in Section 101.2 shall be exempt from the area limitations of Table 2609.4 and shall be permitted with light-transmitting plastics.

TABLE 2609.4
AREA LIMITATIONS FOR LIGHT-TRANSMITTING PLASTIC ROOF PANELS

CLASS OF PLASTIC	MAXIMUM AREA OF INDIVIDUAL ROOF PANELS (square feet)	MAXIMUM AGGREGATE AREA OF ROOF PANELS (percent of floor area)
CC1	300	30
CC2	100	25

For SI: 1 square foot = 0.0929 m^2.

SECTION 2610
LIGHT-TRANSMITTING PLASTIC SKYLIGHT GLAZING

2610.1 Light-transmitting plastic glazing of skylight assemblies. Skylight assemblies glazed with light-transmitting plastic shall conform to the provisions of this section and Section 2606. Unit skylights glazed with light-transmitting plastic shall also comply with Section 2405.5.

Exception: Skylights in which the light-transmitting plastic conforms to the required roof-covering class in accordance with Section 1505.

2610.2 Mounting. The light-transmitting plastic shall be mounted above the plane of the roof on a curb constructed in accordance with the requirements for the type of construction classification, but at least 4 inches (102 mm) above the plane of the roof. Edges of light-transmitting plastic skylights or domes shall be protected by metal or other approved noncombustible material, or the light-transmitting plastic dome or skylight shall be shown to be able to resist ignition where exposed at the edge to a flame from a Class B brand as described in ASTM E 108 or UL 790.

Exceptions:

1. Curbs shall not be required for skylights used on roofs having a minimum slope of three units vertical in 12 units horizontal (25-percent slope) in occupancies in Group R-3 as applicable in Section 101.2 and on buildings with a nonclassified roof covering.

2. The metal or noncombustible edge material is not required where nonclassified roof coverings are permitted.

2610.3 Slope. Flat or corrugated light-transmitting plastic skylights shall slope at least four units vertical in 12 units horizontal (4:12). Dome-shaped skylights shall rise above the mounting flange a minimum distance equal to 10 percent of the maximum span of the dome but not less than 3 inches (76 mm).

Exception: Skylights that pass the Class B Burning Brand Test specified in ASTM E 108 or UL 790.

2610.4 Maximum area of skylights. Each skylight shall have a maximum area within the curb of 100 square feet (9.30 m^2).

Exception: The area limitation shall not apply where the building is equipped throughout with an automatic sprinkler system in accordance with Section 903.3.1.1 or the building is equipped with smoke and heat vents in accordance with Section 910.

2610.5 Aggregate area of skylights. The aggregate area of skylights shall not exceed 33$^1/_3$ percent of the floor area of the room or space sheltered by the roof in which such skylights are installed where Class CC1 materials are utilized, and 25 percent where Class CC2 materials are utilized.

Exception: The aggregate area limitations of light-transmitting plastic skylights shall be increased 100 percent beyond the limitations set forth in this section where the building is equipped throughout with an automatic sprinkler system in accordance with Section 903.3.1.1 or the building is equipped with smoke and heat vents in accordance with Section 910.

2610.6 Separation. Skylights shall be separated from each other by a distance of not less than 4 feet (1219 mm) measured in a horizontal plane.

Exceptions:

1. Buildings equipped throughout with an automatic sprinkler system in accordance with Section 903.3.1.1.

2. In Group R-3 as applicable in Section 101.2, multiple skylights located above the same room or space with a combined area not exceeding the limits set forth in Section 2610.4.

2610.7 Location. Where exterior wall openings are required to be protected in accordance with Section 704, a skylight shall not be installed within 6 feet (1829 mm) of such exterior wall.

2610.8 Combinations of roof panels and skylights. Combinations of light-transmitting plastic roof panels and skylights shall be subject to the area and percentage limitations and separation requirements applicable to roof panel installations.

SECTION 2611
LIGHT-TRANSMITTING PLASTIC INTERIOR SIGNS

2611.1 General. Light-transmitting plastic interior wall signs shall be limited as specified in Sections 2611.2 through 2611.4. Light-transmitting plastic interior wall signs in covered mall buildings shall comply with Section 402.14. Light-transmitting plastic interior signs shall also comply with Section 2606.

2611.2 Aggregate area. The sign shall not exceed 20 percent of the wall area.

2611.3 Maximum area. The sign shall not exceed 24 square feet (2.23 m²).

2611.4 Encasement. Edges and backs of the sign shall be fully encased in metal.

CHAPTER 27

ELECTRICAL

SECTION 2701
GENERAL

2701.1 Scope. This chapter governs the electrical components, equipment and systems used in buildings and structures covered by this code. Electrical components, equipment and systems shall be designed and constructed in accordance with the provisions of the ICC *Electrical Code.*

[F] SECTION 2702
EMERGENCY AND STANDBY POWER SYSTEMS

2702.1 Installation. Emergency and standby power systems shall be installed in accordance with the ICC *Electrical Code,* NFPA 110 and NFPA 111.

2702.1.1 Stationary generators. Emergency and standby power generators shall be listed in accordance with UL 2200.

2702.2 Where required. Emergency and standby power systems shall be provided where required by Sections 2702.2.1 through 2702.2.19.

2702.2.1 Group A occupancies. Emergency power shall be provided for voice communication systems in Group A occupancies in accordance with Section 907.2.1.2.

2702.2.2 Smoke control systems. Standby power shall be provided for smoke control systems in accordance with Section 909.11.

2702.2.3 Exit signs. Emergency power shall be provided for exit signs in accordance with Section 1011.5.3.

2702.2.4 Means of egress illumination. Emergency power shall be provided for means of egress illumination in accordance with Section 1006.3.

2702.2.5 Accessible means of egress elevators. Standby power shall be provided for elevators that are part of an accessible means of egress in accordance with Section 1007.4.

2702.2.6 Horizontal sliding doors. Standby power shall be provided for horizontal sliding doors in accordance with Section 1008.1.3.3.

2702.2.7 Semiconductor fabrication facilities. Emergency power shall be provided for semiconductor fabrication facilities in accordance with Section 415.9.10.

2702.2.8 Membrane structures. Standby power shall be provided for auxiliary inflation systems in accordance with Section 3102.8.2. Emergency power shall be provided for exit signs in temporary tents and membrane structures in accordance with the *International Fire Code.*

2702.2.9 Hazardous materials. Emergency or standby power shall be provided in occupancies with hazardous materials in accordance with Section 414.5.4.

2702.2.10 Highly toxic and toxic materials. Emergency power shall be provided for occupancies with highly toxic or toxic materials in accordance with the *International Fire Code.*

2702.2.11 Organic peroxides. Standby power shall be provided for occupancies with silane gas in accordance with the *International Fire Code.*

2702.2.12 Pyrophoric materials. Emergency power shall be provided for occupancies with silane gas in accordance with the *International Fire Code.*

2702.2.13 Covered mall buildings. Standby power shall be provided for voice/alarm communication systems in covered mall buildings in accordance with Section 402.12.

2702.2.14 High-rise buildings. Emergency and standby power shall be provided in high-rise buildings in accordance with Sections 403.10 and 403.11.

2702.2.15 Underground buildings. Emergency and standby power shall be provided in underground buildings in accordance with Sections 405.9 and 405.10.

2702.2.16 Group I-3 occupancies. Emergency power shall be provided for doors in Group I-3 occupancies in accordance with Section 408.4.2.

2702.2.17 Airport traffic control towers. Standby power shall be provided in airport traffic control towers in accordance with Section 412.1.5.

2702.2.18 Elevators. Standby power for elevators shall be provided as set forth in Section 3003.1.

2702.2.19 Smokeproof enclosures. Standby power shall be provided for smokeproof enclosures as required by Section 909.20.

2702.3 Maintenance. Emergency and standby power systems shall be maintained and tested in accordance with the *International Fire Code.*

CHAPTER 28

MECHANICAL SYSTEMS

SECTION 2801
GENERAL

2801.1 Scope. Mechanical appliances, equipment and systems shall be constructed, installed and maintained in accordance with the *International Mechanical Code* and the *International Fuel Gas Code*. Masonry chimneys, fireplaces and barbecues shall comply with the *International Mechanical Code* and Chapter 21 of this code.

CHAPTER 29

PLUMBING SYSTEMS

SECTION 2901
GENERAL

2901.1 Scope. The provisions of this chapter and the *International Plumbing Code* shall govern the erection, installation, alteration, repairs, relocation, replacement, addition to, use or maintenance of plumbing equipment and systems. Plumbing systems and equipment shall be constructed, installed and maintained in accordance with the *International Plumbing Code.* Private sewage disposal systems shall conform to the *International Private Sewage Disposal Code.*

[P] SECTION 2902
MINIMUM PLUMBING FACILITIES

2902.1 Minimum number of fixtures. Plumbing fixtures shall be provided for the type of occupancy and in the minimum number shown in Table 2902.1 Types of occupancies not shown in Table 2902.1 shall be considered individually by the building official. The number of occupants shall be determined by this code. Occupancy classification shall be determined in accordance with Chapter 3.

TABLE 2902.1
MINIMUM NUMBER OF REQUIRED PLUMBING FIXTURES[a]

No.	CLASSIFICATION	USE GROUP	DESCRIPTION	WATER CLOSETS (SEE SECTION 419.2 OF THE *INTERNATIONAL PLUMBING CODE* FOR URINALS) MALE	FEMALE	LAVATORIES MALE	FEMALE	BATHUBS OR SHOWERS	DRINKING FOUNTAINS (SEE SECTION 410.1 OF THE *INTERNATIONAL PLUMBING CODE)*	OTHER
1	Assembly (see Sections 2902.2, 2902.5 and 2902.6)	A-1	Theaters usually with fixed seats and other buildings for the performing arts and motion pictures	1 per 125	1 per 65	1 per 200		—	1 per 500	1 service sink
		A-2	Nightclubs, bars, taverns, dance halls and buildings for similar purposes	1 per 40	1 per 40	1 per 75		—	1 per 500	1 service sink
			Restaurants, banquet halls and food courts	1 per 75	1 per 75	1 per 200		—	1 per 500	1 service sink
		A-3	Auditoriums without permanent seating, art galleries, exhibition halls, museums, lecture halls, libraries, arcades and gymnasiums	1 per 125	1 per 65	1 per 200		—	1 per 500	1 service sink
			Passenger terminals and transportation facilities	1 per 500	1 per 500	1 per 750		—	1 per 1,000	1 service sink
		A-3	Places of worship and other religious services. Churches without assembly halls	1 per 150	1 per 75	1 per 200		—	1 per 1,000	1 service sink
		A-4	Coliseums, arenas, skating rinks, pools and tennis courts for indoor sporting events and activities	1 per 75 for the first 1,500 and 1 per 120 for the remainder exceeding 1,500	1 per 40 for the first 1,500 and 1 per 60 for the remainder exceeding 1,500	1 per 200	1 per 150	—	1 per 1,000	1 service sink
		A-5	Stadiums, amusement parks, bleachers and grandstands for outdoor sporting events and activities	1 per 75 for the first 1,500 and 1 per 120 for the remainder exceeding 1,500	1 per 40 for the first 1,500 and 1 per 60 for the remainder exceeding 1,500	1 per 200	1 per 150	—	1 per 1,000	1 service sink

(continued)

2003 INTERNATIONAL BUILDING CODE®

547

TABLE 2902.1—continued
MINIMUM NUMBER OF REQUIRED PLUMBING FACILITIES[a]

No.	CLASSIFICATION	USE GROUP	DESCRIPTION	WATER CLOSETS (SEE SECTION 419.2 OF THE *INTERNATIONAL PLUMBING CODE* FOR URINALS)		LAVATORIES		BATHUBS OR SHOWERS	DRINKING FOUNTAINS (SEE SECTION 410.1 OF THE *INTERNATIONAL PLUMBING CODE*)	OTHER
				MALE	FEMALE	MALE	FEMALE			
2	Business (see Sections 2902.2, 2902.4, 2902.4.1 and 2902.6)	B	Buildings for the transaction of business, professional services, other services involving merchandise, office buildings, banks, light industrial and similar uses	1 per 25 for the first 50 and 1 per 50 for the remainder exceeding 50		1 per 40 for the first 50 and 1 per 80 for the remainder exceeding 50		—	1 per 100	1 service sink
3	Educational	E	Educational facilities	1 per 50		1 per 50		—	1 per 100	1 service sink
4	Factory and industrial	F-1 and F-2	Structures in which occupants are engaged in work fabricating, assembly or processing of products or materials	1 per 100		1 per 100		See Section 411 of the *International Plumbing Code*	1 per 400	1 service sink
5	Institutional	I-1	Residential care	1 per 10		1 per 10		1 per 8	1 per 100	1 service sink
		I-2	Hospitals, ambulatory nursing home patients[b]	1 per per room[c]		1 per per room[c]		1 per 15	1 per 100	1 service sink
			Employees, other than residential care[b]	1 per 25		1 per 35		—	1 per 100	—
			Visitors, other than residential care	1 per 75		1 per 100		—	1 per 500	—
		I-3	Prisons[b]	1 per cell		1 per cell		1 per 15	1 per 100	1 service sink
		I-3	Reformatories, detention centers and correctional centers[b]	1 per 15		1 per 15		1 per 15	1 per 100	1 service sink
		I-4	Adult day care and child care[b]	1 per 15		1 per 15		1 per 15[d]	1 per 100	1 service sink
6	Mercantile (see Section 2902.2, 2902.5 and 2902.6)	M	Retail stores, service stations, shops, salesrooms, markets and shopping centers	1 per 500		1 per 750		—	1 per 1,000	1 service sink
7	Residential	R-1	Hotels, motels, boarding houses (transient)	1 per guestroom		1 per guestroom		1 per guestroom	—	1 service sink
		R-2	Dormitories, fraternities, sororities and boarding house (not transient)	1 per 10		1 per 10		1 per 8	1 per 100	1 service sink
		R-2	Apartment house	1 per dwelling unit		1 per dwelling unit		1 per dwelling unit	—	1 kitchen sink per dwelling unit; 1 automatic clothes washer connection per 20 dwelling units[e]

(continued)

TABLE 2902.1—continued
MINIMUM NUMBER OF REQUIRED PLUMBING FACILITIES[a]

No.	CLASSIFICATION	USE GROUP	DESCRIPTION	WATER CLOSETS (SEE SECTION 419.2 OF THE *INTERNATIONAL PLUMBING CODE* FOR URINALS)		LAVATORIES		BATHUBS OR SHOWERS	DRINKING FOUNTAINS (SEE SECTION 410.1 OF THE *INTERNATIONAL PLUMBING CODE*)	OTHER
				MALE	FEMALE	MALE	FEMALE			
7	Residential	R-3	One- and two-family dwellings	1 per dwelling unit		1 per dwelling unit		1 per dwelling unit	—	1 kitchen sink per dwelling unit; 1 automatic clothes washer connection per 20 dwelling units[e]
		R-4	Residential care/assisted living facilities	1 per 10		1 per 10		1 per 8	1 per 100	1 service sink
8	Storage (see Sections 2902.2, 2902.4 and 2902.4.1)	S-1 S-2	Structures for the storage of goods, warehouses, storehouses and freight depots, low and moderate hazard	1 per 100		1 per 100		See Section 411 of the *International Plumbing Code*	1 per 1,000	1 service sink

a. The fixtures shown are based on one fixture being the minimum required for the number of persons indicated or any fraction of the number of persons indicated. The number of occupants shall be determined by this code.

b. Toilet facilities for employees shall be separate from facilities for inmates or patients.

c. A single-occupant toilet room with one water closet and one lavatory serving not more than two adjacent patient rooms shall be permitted where such room is provided with direct access from each patient room and with provisions for privacy.

d. For day nurseries, a maximum of one bathtub shall be required.

e. For attached one- and two-family dwellings, one automatic clothes washer connection shall be required per 20 dwelling units.

2902.1.1 Unisex toilet and bath fixtures. Fixtures located within unisex toilet bathing rooms complying with Section 404 of the *International Plumbing Code* are permitted to be included in determining the minimum required number of fixtures for assembly and mercantile occupancies.

2902.2 Separate facilities. Where plumbing fixtures are required, separate facilities shall be provided for each sex.

Exceptions:

1. Separate facilities shall not be required for private facilities.

2. Separate employee facilities shall not be required in occupancies in which 15 or fewer people are employed.

3. Separate facilities shall not be required in structures or tenant spaces with a total occupant load, including both employees and customers, of 15 or less.

4. Separate facilities shall not be required in mercantile occupancies in which the maximum occupant load is 50 or less.

2902.3 Number of occupants of each sex. The required water closets, lavatories and showers or bathtubs shall be distributed equally between the sexes based on the percentage of each sex anticipated in the occupant load. The occupant load shall be composed of 50 percent of each sex, unless statistical data approved by the building official indicate a different distribution of the sexes.

2902.4 Location of employee toilet facilities in occupancies other than assembly or mercantile. Access to toilet facilities in occupancies other than mercantile and assembly shall be from within the employees' working area. Employee facilities shall be either separate facilities or combined employee and public facilities.

Exception: Facilities that are required for employees in storage structures or kiosks, and are located in adjacent structures under the same ownership, lease or control, shall be a maximum travel distance of 500 feet (152 m) from the employees' working area.

2902.4.1 Travel distance. The required toilet facilities in occupancies other than assembly or mercantile shall be located not more than one story above or below the employees' working area and the path or travel to such facilities shall not exceed a distance of 500 feet (152 m).

Exception: The location and maximum travel distances to required employee toilet facilities in factory and industrial occupancies are permitted to exceed that required in Section 2902.4.1, provided the location and maximum travel distance are approved by the building official.

2902.5 Location of employee toilet facilities in mercantile and assembly occupancies. Employees shall be provided with toilet facilities in building and tenant spaces utilized as restaurants, nightclubs, places of public assembly and mercantile occupancies. The employee facilities shall be either separate facilities or combined employee and public facilities. The required toilet facilities shall be located not more than one story above or below the employees' work area and the path of travel to such facilities, in other than covered malls, shall not exceed a distance of 500 feet (152 m). The path of travel to required facilities in covered malls shall not exceed a distance of 300 feet (91 m).

> **Exception:** Employee toilet facilities shall not be required in tenant spaces where the travel distance from the main entrance of the tenant spaces to a central toilet area does not exceed 300 feet (91 m) and such central toilet facilities are located not more than one story above or below the tenant space.

2902.6 Public facilities. Customers, patrons and visitors shall be provided with public toilet facilities in structures and tenant spaces intended for public utilization. Public toilet facilities shall be located not more than one story above or below the space required to be provided with public toilet facilities and the path of travel to such facilities shall not exceed a distance of 500 feet (152 m).

2902.6.1 Covered malls. In covered mall buildings, the path of travel to required toilet facilities shall not exceed a distance of 300 feet (91 m). Facilities shall be installed in each individual store or in a central toilet area located in accordance with this section. The maximum travel distance to the central toilet facilities in covered mall buildings shall be measured from the main entrance of any store or tenant space.

2902.6.2 Pay facilities. Where pay facilities are installed, such facilities shall be in excess of the required minimum facilities. Required facilities shall be free of charge.

2902.6.3 Signage. A legible sign designating the sex shall be provided in a readily visible location near the entrance to each toilet facility. Signs for accessible toilet facilities shall comply with ICC A117.1.

CHAPTER 30
ELEVATORS AND CONVEYING SYSTEMS

SECTION 3001
GENERAL

3001.1 Scope. This chapter governs the design, construction, installation, alteration and repair of elevators and conveying systems and their components.

3001.2 Referenced standards. Except as otherwise provided for in this code, the design, construction, installation, alteration, repair and maintenance of elevators and conveying systems and their components shall conform to ASME A17.1, ASME A90.1, ASME B20.1, ALI ALCTV, and ASCE 24 for construction in flood hazard areas established in Section 1612.3.

3001.3 Accessibility. Passenger elevators required to be accessible by Chapter 11 shall conform to ICC A117.1.

3001.4 Change in use. A change in use of an elevator from freight to passenger, passenger to freight, or from one freight class to another freight class shall comply with Part XII of ASME A17.1.

SECTION 3002
HOISTWAY ENCLOSURES

3002.1 Hoistway enclosure protection. Elevator, dumbwaiter and other hoistway enclosures shall have a fire-resistance rating not less than that specified in Chapter 6 and shall be constructed in accordance with Chapter 7.

3002.1.1 Opening protectives. Openings in hoistway enclosures shall be protected as required in Chapter 7.

3002.1.2 Hardware. Hardware on opening protectives shall be of an approved type installed as tested, except that approved interlocks, mechanical locks and electric contacts, door and gate electric contacts and door-operating mechanisms shall be exempt from the fire test requirements.

3002.2 Number of elevator cars in a hoistway. Where four or more elevator cars serve all or the same portion of a building, the elevators shall be located in at least two separate hoistways. Not more than four elevator cars shall be located in any single hoistway enclosure.

3002.3 Emergency signs. An approved pictorial sign of a standardized design shall be posted adjacent to each elevator call station on all floors instructing occupants to use the exit stairways and not to use the elevators in case of fire. The sign shall read: IN FIRE EMERGENCY, DO NOT USE ELEVATOR. USE EXIT STAIRS. The emergency sign shall not be required for elevators that are part of an accessible means of egress complying with Section 1007.4.

3002.4 Elevator car to accommodate ambulance stretcher. In buildings four stories in height or more, at least one elevator shall be provided for fire department emergency access to all floors. Such elevator car shall be of such a size and arrangement to accommodate a 24-inch by 76-inch (610 mm by 1930 mm) ambulance stretcher in the horizontal, open position and shall be identified by the international symbol for emergency medical services (star of life). The symbol shall not be less than 3 inches (76 mm) high and shall be placed inside on both sides of the hoistway door frame.

3002.5 Emergency doors. Where an elevator is installed in a single blind hoistway or on the outside of a building, there shall be installed in the blind portion of the hoistway or blank face of the building, an emergency door in accordance with ASME A17.1.

3002.6 Prohibited doors. Doors, other than hoistway doors and the elevator car door, shall be prohibited at the point of access to an elevator car unless such doors are readily openable from the car side without a key, tool, special knowledge or effort.

3002.7 Common enclosure with stairway. Elevators shall not be in a common shaft enclosure with a stairway.

[F] SECTION 3003
EMERGENCY OPERATIONS

3003.1 Standby power. In buildings and structures where standby power is required or furnished to operate an elevator, the operation shall be in accordance with Sections 3003.1.1 through 3003.1.4.

3003.1.1 Manual transfer. Standby power shall be manually transferable to all elevators in each bank.

3003.1.2 One elevator. Where only one elevator is installed, the elevator shall automatically transfer to standby power within 60 seconds after failure of normal power.

3003.1.3 Two or more elevators. Where two or more elevators are controlled by a common operating system, all elevators shall automatically transfer to standby power within 60 seconds after failure of normal power where the standby power source is of sufficient capacity to operate all elevators at the same time. Where the standby power source is not of sufficient capacity to operate all elevators at the same time, all elevators shall transfer to standby power in sequence, return to the designated landing and disconnect from the standby power source. After all elevators have been returned to the designated level, at least one elevator shall remain operable from the standby power source.

3003.1.4 Venting. Where standby power is connected to elevators, the machine room ventilation or air conditioning shall be connected to the standby power source.

3003.2 Fire-fighters' emergency operation. Elevators shall be provided with Phase I emergency recall operation and Phase II emergency in-car operation in accordance with ASME A17.1.

SECTION 3004
HOISTWAY VENTING

3004.1 Vents required. Hoistways of elevators and dumbwaiters penetrating more than three stories shall be provided with a means for venting smoke and hot gases to the outer air in case of fire.

Exceptions:

1. In occupancies of other than Groups R-1, R-2, I-1, I-2 and similar occupancies with overnight sleeping quarters, venting of hoistways is not required where the building is equipped throughout with an approved automatic sprinkler system installed in accordance with Section 903.3.1.1 or 903.3.1.2.

2. Sidewalk elevator hoistways are not required to be vented.

3004.2 Location of vents. Vents shall be located below the floor or floors at the top of the hoistway, and shall open either directly to the outer air or through noncombustible ducts to the outer air. Noncombustible ducts shall be permitted to pass through the elevator machine room provided that portions of the ducts located outside the hoistway or machine room are enclosed by construction having not less than the fire protection rating required for the hoistway. Holes in the machine room floors for the passage of ropes, cables or other moving elevator equipment shall be limited so as not to provide greater than 2 inches (51 mm) of clearance on all sides.

3004.3 Area of vents. Except as provided for in Section 3004.3.1, the area of the vents shall not be less than $3^1/_2$ percent of the area of the hoistway nor less than 3 square feet (0.28 m^2) for each elevator car, and not less than $3^1/_2$ percent nor less than 0.5 square foot (0.047 m^2) for each dumbwaiter car in the hoistway, whichever is greater. Of the total required vent area, not less than one-third shall be of the permanently open type unless all vents activate upon detection of smoke from any of the elevator lobby smoke detectors.

3004.3.1 Reduced vent area. Where mechanical ventilation conforming to the *International Mechanical Code* is provided, a reduction in the required vent area is allowed provided that all of the following conditions are met:

1. The occupancy is not in Group R-1, R-2, I-1 or I-2 or of a similar occupancy with overnight sleeping quarters.

2. The vents required by Section 3004.2 do not have outside exposure.

3. The hoistway does not extend to the top of the building.

4. The hoistway and machine room exhaust fan is automatically reactivated by thermostatic means.

5. Equivalent venting of the hoistway is accomplished.

3004.4 Closed vents. Closed portions of the required vent area shall consist of windows or duct openings glazed with annealed glass not more than 0.125 inch (3.2 mm) thick.

3004.5 Plumbing and mechanical systems. Plumbing and mechanical systems shall not be located in an elevator shaft.

Exception: Floor drains, sumps and sump pumps shall be permitted at the base of the shaft provided they are indirectly connected to the plumbing system.

SECTION 3005
CONVEYING SYSTEMS

3005.1 General. Escalators, moving walks, conveyors, personnel hoists and material hoists shall comply with the provisions of this section.

3005.2 Escalators and moving walks. Escalators and moving walks shall be constructed of approved noncombustible and fire-retardant materials. This requirement shall not apply to electrical equipment, wiring, wheels, handrails and the use of $^1/_{28}$-inch (0.9 mm) wood veneers on balustrades backed up with noncombustible materials.

3005.2.1 Enclosure. Escalator floor openings shall be enclosed except where Exception 2 of Section 707.2 is satisfied.

3005.2.2 Escalators. Where provided in below-grade transportation stations, escalators shall have a clear width of 32 inches (815 mm) minimum.

Exception: The clear width is not required in existing facilities undergoing alterations.

3005.3 Conveyors. Conveyors and conveying systems shall comply with ASME B20.1.

3005.3.1 Enclosure. Conveyors and related equipment connecting successive floors or levels shall be enclosed with fire barrier walls and approved opening protectives complying with the requirements of Section 3002 and Chapter 7.

3005.3.2 Conveyor safeties. Power-operated conveyors, belts and other material-moving devices shall be equipped with automatic limit switches which will shut off the power in an emergency and automatically stop all operation of the device.

3005.4 Personnel and material hoists. Personnel and material hoists shall be designed utilizing an approved method that accounts for the conditions imposed during the intended operation of the hoist device. The design shall include, but is not limited to, anticipated loads, structural stability, impact, vibration, stresses and seismic restraint. The design shall account for the construction, installation, operation and inspection of the hoist tower, car, machinery and control equipment, guide members and hoisting mechanism. Additionally, the design of personnel hoists shall include provisions for field testing and maintenance which will demonstrate that the hoist device functions in accordance with the design. Field tests shall be conducted upon the completion of an installation or following a major alteration of a personnel hoist.

SECTION 3006
MACHINE ROOMS

3006.1 Access. An approved means of access shall be provided to elevator machine rooms and overhead machinery spaces.

3006.2 Venting. Elevator machine rooms that contain solid-state equipment for elevator operation shall be provided with an independent ventilation or air-conditioning system to protect against the overheating of the electrical equipment. The system shall be capable of maintaining temperatures within the range established for the elevator equipment.

3006.3 Pressurization. The elevator machine room serving a pressurized elevator hoistway shall be pressurized upon activation of a heat or smoke detector located in the elevator machine room.

3006.4 Machine rooms and machinery spaces. Elevator machine rooms and machinery spaces shall be enclosed with construction having a fire-resistance rating not less than the required rating of the hoistway enclosure served by the machinery. Openings shall be protected with assemblies having a fire-resistance rating not less than that required for the hoistway enclosure doors.

3006.5 Shunt trip. Where elevator hoistways or elevator machine rooms containing elevator control equipment are protected with automatic sprinklers, a means installed in accordance with NFPA 72, Section 3-8.15, Elevator Shutdown, shall be provided to disconnect automatically the main line power supply to the affected elevator prior to the application of water. This means shall not be self-resetting. The activation of sprinklers outside the hoistway or machine room shall not disconnect the main line power supply.

3006.6 Plumbing systems. Plumbing systems shall not be located in elevator equipment rooms.

CHAPTER 31

SPECIAL CONSTRUCTION

SECTION 3101
GENERAL

3101.1 Scope. The provisions of this chapter shall govern special building construction including membrane structures, temporary structures, pedestrian walkways and tunnels, awnings and canopies, marquees, signs, and towers and antennas.

SECTION 3102
MEMBRANE STRUCTURES

3102.1 General. The provisions of this section shall apply to air-supported, air-inflated, membrane-covered cable and membrane-covered frame structures, collectively known as membrane structures, erected for a period of 180 days or longer. Those erected for a shorter period of time shall comply with the *International Fire Code*. Membrane structures covering water storage facilities, water clarifiers, water treatment plants, sewage treatment plants, greenhouses and similar facilities not used for human occupancy, are required to meet only the requirements of Sections 3102.3.1 and 3102.7.

3102.2 Definitions. The following words and terms shall, for the purposes of this section and as used elsewhere in this code, have the meanings shown herein:

AIR-INFLATED STRUCTURE. A building where the shape of the structure is maintained by air pressurization of cells or tubes to form a barrel vault over the usable area. Occupants of such a structure do not occupy the pressurized area used to support the structure.

AIR-SUPPORTED STRUCTURE. A building wherein the shape of the structure is attained by air pressure and occupants of the structure are within the elevated pressure area. Air-supported structures are of two basic types:

Double skin. Similar to a single skin, but with an attached liner that is separated from the outer skin and provides an airspace which serves for insulation, acoustic, aesthetic or similar purposes.

Single skin. Where there is only the single outer skin and the air pressure is directly against that skin.

CABLE-RESTRAINED, AIR-SUPPORTED STRUCTURE. A structure in which the uplift is resisted by cables or webbings which are anchored to either foundations or dead men. Reinforcing cable or webbing is attached by various methods to the membrane or is an integral part of the membrane. This is not a cable-supported structure.

MEMBRANE-COVERED CABLE STRUCTURE. A nonpressurized structure in which a mast and cable system provides support and tension to the membrane weather barrier and the membrane imparts stability to the structure.

MEMBRANE-COVERED FRAME STRUCTURE. A nonpressurized building wherein the structure is composed of a rigid framework to support a tensioned membrane which provides the weather barrier.

NONCOMBUSTIBLE MEMBRANE STRUCTURE. A membrane structure in which the membrane and all component parts of the structure are noncombustible.

3102.3 Type of construction. Noncombustible membrane structures shall be classified as Type IIB construction. Noncombustible frame or cable-supported structures covered by an approved membrane in accordance with Section 3102.3.1 shall be classified as Type IIB construction. Heavy timber frame-supported structures covered by an approved membrane in accordance with Section 3102.3.1 shall be classified as Type IV construction. Other membrane structures shall be classified as Type V construction.

> **Exception:** Plastic less than 30 feet (9144 mm) above any floor used in greenhouses, where occupancy by the general public is not authorized, and for aquaculture pond covers, is not required to be flame resistant.

3102.3.1 Membrane and interior liner material. Membranes and interior liners shall be either noncombustible as set forth in Section 703.4, or flame resistant as determined in accordance with NFPA 701 and the manufacturer's test protocol.

> **Exception:** Plastic less than 20 mil (500 mm) in thickness used in greenhouses, where occupancy by the general public is not authorized, and for aquaculture pond covers, is not required to be flame resistant.

3102.4 Allowable floor areas. The area of a membrane structure shall not exceed the limitations set forth in Table 503, except as provided in Section 506.

3102.5 Maximum height. Membrane structures shall not exceed one story nor shall such structures exceed the height limitations in feet set forth in Table 503.

> **Exception:** Noncombustible membrane structures serving as roofs only.

3102.6 Mixed construction. Membrane structures shall be permitted to be utilized as specified in this section as a portion of buildings of other types of construction. Height and area limits shall be as specified for the type of construction and occupancy of the building.

3102.6.1 Noncombustible membrane. A noncombustible membrane shall be permitted for use as the roof or as a skylight of any building or atrium of a building of any type of construction provided it is at least 20 feet (6096 mm) above any floor, balcony or gallery.

3102.6.1.1 Flame-resistant membrane. A flame-resistant membrane shall be permitted to be used as the roof or as a skylight on buildings of Type IIB, III, IV and V construction provided it is at least 20 feet (6096 mm) above any floor, balcony or gallery.

3102.7 Engineering design. The structure shall be designed and constructed to sustain dead loads; loads due to tension or inflation; live loads including wind, snow or flood and seismic loads and in accordance with Chapter 16.

3102.8 Inflation systems. Air-supported and air-inflated structures shall be provided with primary and auxiliary inflation systems to meet the minimum requirements of Sections 3102.8.1 through 3102.8.3.

3102.8.1 Equipment requirements. This inflation system shall consist of one or more blowers and shall include provisions for automatic control to maintain the required inflation pressures. The system shall be so designed as to prevent overpressurization of the system.

3102.8.1.1 Auxiliary inflation system. In addition to the primary inflation system, in buildings exceeding 1,500 square feet (140 m^2) in area, an auxiliary inflation system shall be provided with sufficient capacity to maintain the inflation of the structure in case of primary system failure. The auxiliary inflation system shall operate automatically when there is a loss of internal pressure and when the primary blower system becomes inoperative.

3102.8.1.2 Blower equipment. Blower equipment shall meet the following requirements:

1. Blowers shall be powered by continuous-rated motors at the maximum power required for any flow condition as required by the structural design.
2. Blowers shall be provided with inlet screens, belt guards and other protective devices as required by the building official to provide protection from injury.
3. Blowers shall be housed within a weather-protecting structure.
4. Blowers shall be equipped with backdraft check dampers to minimize air loss when inoperative.
5. Blower inlets shall be located to provide protection from air contamination. The location of inlets shall be approved.

3102.8.2 Standby power. Wherever an auxiliary inflation system is required, an approved standby power-generating system shall be provided. The system shall be equipped with a suitable means for automatically starting the generator set upon failure of the normal electrical service and for automatic transfer and operation of all of the required electrical functions at full power within 60 seconds of such service failure. Standby power shall be capable of operating independently for a minimum of 4 hours.

3102.8.3 Support provisions. A system capable of supporting the membrane in the event of deflation shall be provided for in air-supported and air-inflated structures having an occupant load of more than 50 or where covering a swimming pool regardless of occupant load. The support system shall be capable of maintaining membrane structures used as a roof for Type I construction not less than 20 feet (6096 mm) above floor or seating areas. The support system shall be capable of maintaining other membranes at least 7 feet (2134 mm) above the floor, seating area or surface of the water.

SECTION 3103
TEMPORARY STRUCTURES

3103.1 General. The provisions of this section shall apply to structures erected for a period of less than 180 days. Tents and other membrane structures erected for a period of less than 180 days shall comply with the *International Fire Code*. Those erected for a longer period of time shall comply with applicable sections of this code.

Exception: Provisions of the *International Fire Code* shall apply to tents and membrane structures erected for a period of less than 180 days.

3103.1.1 Permit required. Temporary structures that cover an area in excess of 120 square feet (11.16 m^2), including connecting areas or spaces with a common means of egress or entrance which are used or intended to be used for the gathering together of 10 or more persons, shall not be erected, operated or maintained for any purpose without obtaining a permit from the building official.

3103.2 Construction documents. A permit application and construction documents shall be submitted for each installation of a temporary structure. The construction documents shall include a site plan indicating the location of the temporary structure and information delineating the means of egress and the occupant load.

3103.3 Location. Temporary structures shall be located in accordance with the requirements of Table 602 based on the fire-resistance rating of the exterior walls for the proposed type of construction.

3103.4 Means of egress. Temporary structures shall conform to the means of egress requirements of Chapter 10 and shall have a maximum exit access travel distance of 100 feet (30 480 mm).

SECTION 3104
PEDESTRIAN WALKWAYS AND TUNNELS

3104.1 General. This section shall apply to connections between buildings such as pedestrian walkways or tunnels, located at, above or below grade level, that are used as a means of travel by persons. The pedestrian walkway shall not contribute to the building area or the number of stories or height of connected buildings.

3104.2 Separate structures. Connected buildings shall be considered to be separate structures.

Exceptions:

1. Buildings on the same lot in accordance with Section 503.1.3.
2. For purposes of calculating the number of Type B units required by Chapter 11, structurally connected buildings and buildings with multiple wings shall be considered one structure.

3104.3 Construction. The pedestrian walkway shall be of noncombustible construction.

Exception: Combustible construction shall be permitted where connected buildings are of combustible construction.

3104.4 Contents. Only materials and decorations approved by the building official shall be located in the pedestrian walkway.

3104.5 Fire barriers between pedestrian walkways and buildings. Walkways shall be separated from the interior of the building by fire barrier walls with a fire-resistance rating of not less than 2 hours. This protection shall extend vertically from a point 10 feet (3048 mm) above the walkway roof surface or the connected building roof line, whichever is lower, down to a point 10 feet (3048 mm) below the walkway and horizontally 10 feet (3048 mm) from each side of the pedestrian walkway. Openings within the l0-foot (3048 mm) horizontal extension of the protected walls beyond the walkway shall be equipped with devices providing a $^3/_4$-hour fire protection rating in accordance with Section 715.

Exception: The walls separating the pedestrian walkway from a connected building are not required to have a fire-resistance rating by this section where any of the following conditions exist:

1. The distance between the connected buildings is more than 10 feet (3048 mm), the pedestrian walkway and connected buildings are equipped throughout with an automatic sprinkler system in accordance with NFPA 13 and the wall is constructed of a tempered, wired or laminated glass wall and doors subject to the following:

    1.1. The glass shall be protected by an automatic sprinkler system in accordance with NFPA 13 and the sprinkler system shall completely wet the entire surface of interior sides of the glass wall when actuated.

    1.2. The glass shall be in a gasketed frame and installed in such a manner that the framing system will deflect without breaking (loading) the glass before the sprinkler operates.

    1.3. Obstructions shall not be installed between the sprinkler heads and the glass.

2. The distance between the connected buildings is more than 10 feet (3048 mm), and both sidewalls of the pedestrian walkway are at least 50 percent open with the open area uniformly distributed to prevent the accumulation of smoke and toxic gases.

3. Buildings are on the same lot, in accordance with Section 503.1.3.

4. Where exterior walls of connected buildings are required by Section 704 to have a fire-resistance rating greater than 2 hours, the walkway shall be equipped throughout with an automatic sprinkler system installed in accordance with NFPA 13.

The previous exceptions shall apply to pedestrian walkways having a maximum height above grade of three stories or 40 feet (12 192 mm), or five stories or 55 feet (16 764 mm) where sprinklered.

3104.6 Public way. Pedestrian walkways over a public way shall also comply with Chapter 32.

3104.7 Egress. Access shall be provided at all times to a pedestrian walkway that serves as a required exit.

3104.8 Width. The unobstructed width of pedestrian walkways shall not be less than 36 inches (914 mm). The total width shall not exceed 30 feet (9144 mm).

3104.9 Exit access travel. The length of exit access travel shall not exceed 200 feet (60 960 mm).

Exceptions:

1. Exit access travel distance on a pedestrian walkway equipped throughout with an automatic sprinkler system in accordance with NFPA 13 shall not exceed 250 feet (76 200 mm).

2. Exit access travel distance on a pedestrian walkway constructed with both sides at least 50 percent open shall not exceed 300 feet (91 440 mm).

3. Exit access travel distance on a pedestrian walkway constructed with both sides at least 50 percent open, and equipped throughout with an automatic sprinkler system in accordance with NFPA 13, shall not exceed 400 feet (122 m).

3104.10 Tunneled walkway. Separation between the tunneled walkway and the building to which it is connected shall not be less than 2-hour fire-resistant construction and openings therein shall be protected in accordance with Table 715.3.

3104.11 Ventilation. Smoke and heat vents shall be provided for enclosed walkways and tunneled walkways as required for Group F-1 occupancies in accordance with Section 910.

SECTION 3105
AWNINGS AND CANOPIES

3105.1 General. Awnings or canopies shall comply with the requirements of this section and other applicable sections of this code.

3105.2 Definition. The following term shall, for the purposes of this section and as used elsewhere in this code, have the meaning shown herein.

RETRACTABLE AWNING. A retractable awning is a cover with a frame that retracts against a building or other structure to which it is entirely supported.

3105.3 Design and construction. Awnings and canopies shall be designed and constructed to withstand wind or other lateral loads and live loads as required by Chapter 16 with due allowance for shape, open construction and similar features that relieve the pressures or loads. Structural members shall be protected to prevent deterioration. Awnings shall have frames of noncombustible material, fire-retardant-treated wood, wood of Type IV size, or 1-hour construction with combustible or noncombustible covers and shall be either fixed, retractable, folding or collapsible.

3105.4 Canopy materials. Canopies shall be constructed of a rigid framework with an approved covering, that is flame resistant in accordance with NFPA 701 or has a flame spread index not greater than 25 when tested in accordance with ASTM E 84.

SECTION 3106
MARQUEES

3106.1 General. Marquees shall comply with this section and other applicable sections of this code.

3106.2 Thickness. The maximum height or thickness of a marquee measured vertically from its lowest to its highest point shall not exceed 3 feet (914 mm) where the marquee projects more than two-thirds of the distance from the property line to the curb line, and shall not exceed 9 feet (2743 mm) where the marquee is less than two-thirds of the distance from the property line to the curb line.

3106.3 Roof construction. Where the roof or any part thereof is a skylight, the skylight shall comply with the requirements of Chapter 24. Every roof and skylight of a marquee shall be sloped to downspouts that shall conduct any drainage from the marquee in such a manner so as not to spill over the sidewalk.

3106.4 Location prohibited. Every marquee shall be so located as not to interfere with the operation of any exterior standpipe, and such that the marquee does not obstruct the clear passage of stairways or exit discharge from the building or the installation or maintenance of street lighting.

3106.5 Construction. A marquee shall be supported entirely from the building and constructed of noncombustible materials. Marquees shall be designed as required in Chapter 16. Structural members shall be protected to prevent deterioration.

SECTION 3107
SIGNS

3107.1 General. Signs shall be designed, constructed and maintained in accordance with this code.

SECTION 3108
RADIO AND TELEVISION TOWERS

3108.1 General. Subject to the provisions of Chapter 16 and the requirements of Chapter 15 governing the fire-resistance ratings of buildings for the support of roof structures, radio and television towers shall be designed and constructed as herein provided.

3108.2 Location and access. Towers shall be located and equipped with step bolts and ladders so as to provide ready access for inspection purposes. Guy wires or other accessories shall not cross or encroach upon any street or other public space, or over above-ground electric utility lines, or encroach upon any privately owned property without written consent of the owner of the encroached-upon property, space or above-ground electric utility lines.

3108.3 Construction. Towers shall be constructed of approved corrosion-resistant noncombustible material. The minimum type of construction of isolated radio towers not more than 100 feet (30 480 mm) in height shall be Type IIB.

3108.4 Loads. Towers shall be designed to resist wind loads in accordance with TIA/EIA-222. Consideration shall be given to conditions involving wind load on ice-covered sections in localities subject to sustained freezing temperatures.

3108.4.1 Dead load. Towers shall be designed for the dead load plus the ice load in regions where ice formation occurs.

3108.4.2 Wind load. Adequate foundations and anchorage shall be provided to resist two times the calculated wind load.

3108.5 Grounding. Towers shall be permanently and effectively grounded.

SECTION 3109
SWIMMING POOL ENCLOSURES AND
SAFETY DEVICES

3109.1 General. Swimming pools shall comply with the requirements of this section and other applicable sections of this code.

3109.2 Definition. The following word and term shall, for the purposes of this section and as used elsewhere in this code, have the meaning shown herein.

SWIMMING POOLS. Any structure intended for swimming, recreational bathing or wading that contains water over 24 inches (610 mm) deep. This includes in-ground, above-ground and on-ground pools; hot tubs; spas and fixed-in-place wading pools.

3109.3 Public swimming pools. Public swimming pools shall be completely enclosed by a fence at least 4 feet (1290 mm) in height or a screen enclosure. Openings in the fence shall not permit the passage of a 4-inch-diameter (102 mm) sphere. The fence or screen enclosure shall be equipped with self-closing and self-latching gates.

3109.4 Residential swimming pools. Residential swimming pools shall comply with Sections 3109.4.1 through 3109.4.3.

> **Exception:** A swimming pool with a power safety cover or a spa with a safety cover complying with ASTM F 1346.

> **3109.4.1 Barrier height and clearances.** The top of the barrier shall be at least 48 inches (1219 mm) above grade measured on the side of the barrier that faces away from the swimming pool. The maximum vertical clearance between grade and the bottom of the barrier shall be 2 inches (51 mm) measured on the side of the barrier that faces away from the swimming pool. Where the top of the pool structure is above grade, the barrier is authorized to be at ground level or mounted on top of the pool structure, the maximum vertical clearance between the top of the pool structure and the bottom of the barrier shall be 4 inches (102 mm).

> **3109.4.1.1 Openings.** Openings in the barrier shall not allow passage of a 4-inch-diameter (102 mm) sphere.

> **3109.4.1.2 Solid barrier surfaces.** Solid barriers which do not have openings shall not contain indentations or protrusions except for normal construction tolerances and tooled masonry joints.

> **3109.4.1.3 Closely spaced horizontal members.** Where the barrier is composed of horizontal and vertical members and the distance between the tops of the horizontal members is less than 45 inches (1143 mm), the horizontal members shall be located on the swimming pool side of the fence. Spacing between vertical mem-

bers shall not exceed 1.75 inches (44 mm) in width. Where there are decorative cutouts within vertical members, spacing within the cutouts shall not exceed 1.75 inches (44 mm) in width.

3109.4.1.4 Widely spaced horizontal members. Where the barrier is composed of horizontal and vertical members and the distance between the tops of the horizontal members is 45 inches (1143 mm) or more, spacing between vertical members shall not exceed 4 inches (102 mm). Where there are decorative cutouts within vertical members, spacing within the cutouts shall not exceed 1.75 inches (44 mm) in width.

3109.4.1.5 Chain link dimensions. Maximum mesh size for chain link fences shall be a 2.25 inch square (57 mm square) unless the fence is provided with slats fastened at the top or the bottom which reduce the openings to no more than 1.75 inches (44 mm).

3109.4.1.6 Diagonal members. Where the barrier is composed of diagonal members, the maximum opening formed by the diagonal members shall be no more than 1.75 inches (44 mm).

3109.4.1.7 Gates. Access gates shall comply with the requirements of Sections 3109.4.1.1 through 3109.4.1.6 and shall be equipped to accommodate a locking device. Pedestrian access gates shall open outward away from the pool and shall be self-closing and have a self-latching device. Gates other than pedestrian access gates shall have a self-latching device. Where the release mechanism of the self-latching device is located less than 54 inches (1372 mm) from the bottom of the gate, the release mechanism shall be located on the pool side of the gate at least 3 inches (76 mm) below the top of the gate, and the gate and barrier shall have no opening greater than 0.5 inch (12.7 mm) within 18 inches (457 mm) of the release mechanism.

3109.4.1.8 Dwelling wall as a barrier. Where a wall of a dwelling serves as part of the barrier, one of the following shall apply:

1. Doors with direct access to the pool through that wall shall be equipped with an alarm which produces an audible warning when the door and its screen are opened. The alarm shall sound continuously for a minimum of 30 seconds immediately after the door is opened and be capable of being heard throughout the house during normal household activities. The alarm shall automatically reset under all conditions. The alarm shall be equipped with a manual means to temporarily deactivate the alarm for a single opening. Such deactivation shall last no more than 15 seconds. The deactivation switch shall be located at least 54 inches (1372 mm) above the threshold of the door.

2. The pool shall be equipped with a power safety cover which complies with ASTM F 1346.

3. Other means of protection, such as self-closing doors with self-latching devices, which are approved by the administrative authority, shall be ac-

cepted so long as the degree of protection afforded is not less than the protection afforded by Section 3109.4.1.8, Item 1 or 2.

3109.4.1.9 Pool structure as barrier. Where an above-ground pool structure is used as a barrier or where the barrier is mounted on top of the pool structure, and the means of access is a ladder or steps, then the ladder or steps either shall be capable of being secured, locked or removed to prevent access, or the ladder or steps shall be surrounded by a barrier which meets the requirements of Sections 3109.4.1.1 through 3109.4.1.8. When the ladder or steps are secured, locked or removed, any opening created shall not allow the passage of a 4-inch-diameter (102 mm) sphere.

3109.4.2 Indoor swimming pools. Walls surrounding indoor swimming pools shall not be required to comply with Section 3109.4.1.8.

3109.4.3 Prohibited locations. Barriers shall be located so as to prohibit permanent structures, equipment or similar objects from being used to climb the barriers.

3109.5 Entrapment avoidance. Where the suction inlet system, such as an automatic cleaning system, is a vacuum cleaner system which has a single suction inlet, or multiple suction inlets which can be isolated by valves, each suction inlet shall protect against user entrapment by an approved antivortex cover, a 12-inch by 12-inch (304 mm by 304 mm) or larger grate, or other approved means.

In addition, all pools and spas shall be equipped with an alternative backup system which shall provide vacuum relief should grate covers be missing. Alternative vacuum relief devices shall include one of the following:

1. Approved vacuum release system.

2. Approved vent piping.

3. Other approved devices or means.

CHAPTER 32

ENCROACHMENTS INTO THE PUBLIC RIGHT-OF-WAY

SECTION 3201
GENERAL

3201.1 Scope. The provisions of this chapter shall govern the encroachment of structures into the public right-of-way.

3201.2 Measurement. The projection of any structure or appendage shall be the distance measured horizontally from the lot line to the outermost point of the projection.

3201.3 Other laws. The provisions of this chapter shall not be construed to permit the violation of other laws or ordinances regulating the use and occupancy of public property.

3201.4 Drainage. Drainage water collected from a roof, awning, canopy or marquee, and condensate from mechanical equipment shall not flow over a public walking surface.

SECTION 3202
ENCROACHMENTS

3202.1 Encroachments below grade. Encroachments below grade shall comply with Sections 3202.1.1 through 3202.1.3.

3202.1.1 Structural support. A part of a building erected below grade that is necessary for structural support of the building or structure shall not project beyond the lot lines, except that the footings of street walls or their supports which are located at least 8 feet (2438 mm) below grade shall not project more than 12 inches (305 mm) beyond the street lot line.

3202.1.2 Vaults and other enclosed spaces. The construction and utilization of vaults and other enclosed space below grade shall be subject to the terms and conditions of the authority or legislative body having jurisdiction.

3202.1.3 Areaways. Areaways shall be protected by grates, guards or other approved means.

3202.2 Encroachments above grade and below 8 feet in height. Encroachments into the public right-of-way above grade and below 8 feet (2438 mm) in height shall be prohibited except as provided for in Sections 3202.2.1 through 3202.2.3. Doors and windows shall not open or project into the public right-of-way.

3202.2.1 Steps. Steps shall not project more than 12 inches (305 mm) and shall be guarded by approved devices not less than 3 feet (914 mm) high, or shall be located between columns or pilasters.

3202.2.2 Architectural features. Columns or pilasters, including bases and moldings shall not project more than 12 inches (305 mm). Belt courses, lintels, sills, architraves, pediments and similar architectural features shall not project more than 4 inches (102 mm).

3202.2.3 Awnings. The vertical clearance from the public right-of-way to the lowest part of any awning, including valances, shall be 7 feet (2134 mm) minimum.

3202.3 Encroachments 8 feet or more above grade. Encroachments 8 feet (2438 mm) or more above grade shall comply with Sections 3202.3.1 through 3202.3.4.

3202.3.1 Awnings, canopies, marquees and signs. Awnings, canopies, marquees and signs shall be constructed so as to support applicable loads as specified in Chapter 16. Awnings, canopies, marquees and signs with less than 15 feet (4572 mm) clearance above the sidewalk shall not extend into or occupy more than two-thirds the width of the sidewalk measured from the building. Stanchions or columns that support awnings, canopies, marquees and signs shall be located not less than 2 feet (610 mm) in from the curb line.

3202.3.2 Windows, balconies, architectural features and mechanical equipment. Where the vertical clearance above grade to projecting windows, balconies, architectural features or mechanical equipment is more than 8 feet (2438 mm), 1 inch (25 mm) of encroachment is permitted for each additional 1 inch (25 mm) of clearance above 8 feet (2438 mm), but the maximum encroachment shall be 4 feet (1219 mm).

3202.3.3 Encroachments 15 feet or more above grade. Encroachments 15 feet (4572 mm) or more above grade shall not be limited.

3202.3.4 Pedestrian walkways. The installation of a pedestrian walkway over a public right-of-way shall be subject to the approval of local authority having jurisdiction. The vertical clearance from the public right-of-way to the lowest part of a pedestrian walkway shall be 15 feet (4572 mm) minimum.

3202.4 Temporary encroachments. Where allowed by the local authority having jurisdiction, vestibules and storm enclosures shall not be erected for a period of time exceeding 7 months in any one year and shall not encroach more than 3 feet (914 mm) nor more than one-fourth of the width of the sidewalk beyond the street lot line. Temporary entrance awnings shall be erected with a minimum clearance of 7 feet (2134 mm) to the lowest portion of the hood or awning where supported on removable steel or other approved noncombustible support.

CHAPTER 33

SAFEGUARDS DURING CONSTRUCTION

SECTION 3301
GENERAL

3301.1 Scope. The provisions of this chapter shall govern safety during construction and the protection of adjacent public and private properties.

3301.2 Storage and placement. Construction equipment and materials shall be stored and placed so as not to endanger the public, the workers or adjoining property for the duration of the construction project.

SECTION 3302
CONSTRUCTION SAFEGUARDS

3302.1 Remodeling and additions. Required exits, existing structural elements, fire protection devices and sanitary safeguards shall be maintained at all times during remodeling, alterations, repairs or additions to any building or structure.

Exceptions:

1. When such required elements or devices are being remodeled, altered or repaired, adequate substitute provisions shall be made.

2. When the existing building is not occupied.

3302.2 Manner of removal. Waste materials shall be removed in a manner which prevents injury or damage to persons, adjoining properties and public rights-of-way.

SECTION 3303
DEMOLITION

3303.1 Construction documents. Construction documents and a schedule for demolition must be submitted when required by the building official. Where such information is required, no work shall be done until such construction documents or schedule, or both, are approved.

3303.2 Pedestrian protection. The work of demolishing any building shall not be commenced until pedestrian protection is in place as required by this chapter.

3303.3 Means of egress. A party wall balcony or horizontal exit shall not be destroyed unless and until a substitute means of egress has been provided and approved.

3303.4 Vacant lot. Where a structure has been demolished or removed, the vacant lot shall be filled and maintained to the existing grade or in accordance with the ordinances of the jurisdiction having authority.

3303.5 Water accumulation. Provision shall be made to prevent the accumulation of water or damage to any foundations on the premises or the adjoining property.

3303.6 Utility connections. Service utility connections shall be discontinued and capped in accordance with the approved rules and the requirements of the authority having jurisdiction.

SECTION 3304
SITE WORK

3304.1 Excavation and fill. Excavation and fill for buildings and structures shall be constructed or protected so as not to endanger life or property. Stumps and roots shall be removed from the soil to a depth of at least 12 inches (305 mm) below the surface of the ground in the area to be occupied by the building. Wood forms which have been used in placing concrete, if within the ground or between foundation sills and the ground, shall be removed before a building is occupied or used for any purpose. Before completion, loose or casual wood shall be removed from direct contact with the ground under the building.

3304.1.1 Slope limits. Slopes for permanent fill shall not be steeper than one unit vertical in two units horizontal (50-percent slope). Cut slopes for permanent excavations shall not be steeper than one unit vertical in two units horizontal (50-percent slope). Deviation from the foregoing limitations for cut slopes shall be permitted only upon the presentation of a soil investigation report acceptable to the building official.

3304.1.2 Surcharge. No fill or other surcharge loads shall be placed adjacent to any building or structure unless such building or structure is capable of withstanding the additional loads caused by the fill or surcharge. Existing footings or foundations which can be affected by any excavation shall be underpinned adequately or otherwise protected against settlement and shall be protected against later movement.

3304.1.3 Footings on adjacent slopes. For footings on adjacent slopes, see Chapter 18.

3304.1.4 Fill supporting foundations. Fill to be used to support the foundations of any building or structure shall comply with Section 1803.5. Special inspections of compacted fill shall be in accordance with Section 1704.7.

SECTION 3305
SANITARY

3305.1 Facilities required. Sanitary facilities shall be provided during construction, remodeling or demolition activities in accordance with the *International Plumbing Code.*

SECTION 3306
PROTECTION OF PEDESTRIANS

3306.1 Protection required. Pedestrians shall be protected during construction, remodeling and demolition activities as required by this chapter and Table 3306.1. Signs shall be provided to direct pedestrian traffic.

TABLE 3306.1
PROTECTION OF PEDESTRIANS

HEIGHT OF CONSTRUCTION	DISTANCE FROM CONSTRUCTION TO LOT LINE	TYPE OF PROTECTION REQUIRED
8 feet or less	Less than 5 feet	Construction railings
	5 feet or more	None
More than 8 feet	Less than 5 feet	Barrier and covered walkway
	5 feet or more, but not more than one-fourth the height of construction	Barrier and covered walkway
	5 feet or more, but between one-fourth and one-half the height of construction	Barrier
	5 feet or more, but exceeding one-half the height of construction	None

For SI: 1 foot = 304.8 mm.

3306.2 Walkways. A walkway shall be provided for pedestrian travel in front of every construction and demolition site unless the authority having jurisdiction authorizes the sidewalk to be fenced or closed. Walkways shall be of sufficient width to accommodate the pedestrian traffic, but in no case shall they be less than 4 feet (1219 mm) in width. Walkways shall be provided with a durable walking surface. Walkways shall be accessible in accordance with Chapter 11 and shall be designed to support all imposed loads and in no case shall the design live load be less than 150 pounds per square foot (psf) (7.2 kN/m²).

3306.3 Directional barricades. Pedestrian traffic shall be protected by a directional barricade where the walkway extends into the street. The directional barricade shall be of sufficient size and construction to direct vehicular traffic away from the pedestrian path.

3306.4 Construction railings. Construction railings shall be at least 42 inches (1067 mm) in height and shall be sufficient to direct pedestrians around construction areas.

3306.5 Barriers. Barriers shall be a minimum of 8 feet (2438 mm) in height and shall be placed on the side of the walkway nearest the construction. Barriers shall extend the entire length of the construction site. Openings in such barriers shall be protected by doors which are normally kept closed.

3306.6 Barrier design. Barriers shall be designed to resist loads required in Chapter 16 unless constructed as follows:

1. Barriers shall be provided with 2-inch by 4-inch (51 mm by 102 mm) top and bottom plates.

2. The barrier material shall be a minimum of $^3/_4$-inch (19.1 mm) boards or $^1/_4$-inch (6.4 mm) wood structural use panels.

3. Wood structural use panels shall be bonded with an adhesive identical to that for exterior wood structural use panels.

4. Wood structural use panels $^1/_4$ inch (6.4 mm) or $^5/_{16}$ inch (23.8 mm) in thickness shall have studs spaced not more than 2 feet (610 mm) on center (o.c.).

5. Wood structural use panels $^3/_8$ inch (9.5 mm) or $^1/_2$ inch (12.7 mm) in thickness shall have studs spaced not more than 4 feet (1219 mm) o.c., provided a 2-inch by 4-inch (51 mm by 102 mm) stiffener is placed horizontally at

midheight where the stud spacing exceeds 2 feet (610 mm) o.c.

6. Wood structural use panels $^5/_8$ inch (15.9 mm) or thicker shall not span over 8 feet (2438 mm).

3306.7 Covered walkways. Covered walkways shall have a minimum clear height of 8 feet (2438 mm) as measured from the floor surface to the canopy overhead. Adequate lighting shall be provided at all times. Covered walkways shall be designed to support all imposed loads. In no case shall the design live load be less than 150 psf (7.2 kN/m²) for the entire structure.

Exception: Roofs and supporting structures of covered walkways for new, light-frame construction not exceeding two stories in height are permitted to be designed for a live load of 75 psf (3.6kN/m²) or the loads imposed on them, whichever is greater. In lieu of such designs, the roof and supporting structure of a covered walkway are permitted to be constructed as follows:

1. Footings shall be continuous 2-inch by 6-inch (51 mm by 152 mm) members.

2. Posts not less than 4 inches by 6 inches (102 mm by 152 mm) shall be provided on both sides of the roof and spaced not more than 12 feet (3658 mm) o.c.

3. Stringers not less than 4 inches by 12 inches (102 mm by 305 mm) shall be placed on edge upon the posts.

4. Joists resting on the stringers shall be at least 2 inches by 8 inches (51 mm by 203 mm) and shall be spaced not more than 2 feet (610 mm) o.c.

5. The deck shall be planks at least 2 inches (51 mm) thick or wood structural panels with an exterior exposure durability classification at least $^{23}/_{32}$ inch (18.3 mm) thick nailed to the joists.

6. Each post shall be knee braced to joists and stringers by 2-inch by 4-inch (51 mm by 102 mm) minimum members 4 feet (1219 mm) long.

7. A 2-inch by 4-inch (51 mm by 102 mm) minimum curb shall be set on edge along the outside edge of the deck.

3306.8 Repair, maintenance and removal. Pedestrian protection required by this chapter shall be maintained in place and kept in good order for the entire length of time pedestrians may be endangered. The owner or the owner's agent, upon the completion of the construction activity, shall immediately remove walkways, debris and other obstructions and leave such public property in as good a condition as it was before such work was commenced.

3306.9 Adjacent to excavations. Every excavation on a site located 5 feet (1524 mm) or less from the street lot line shall be enclosed with a barrier not less than 6 feet (1829 mm) high. Where located more than 5 feet (1524 mm) from the street lot line, a barrier shall be erected when required by the building official. Barriers shall be of adequate strength to resist wind pressure as specified in Chapter 16.

SECTION 3307
PROTECTION OF ADJOINING PROPERTY

3307.1 Protection required. Adjoining public and private property shall be protected from damage during construction, remodeling and demolition work. Protection must be provided for footings, foundations, party walls, chimneys, skylights and roofs. Provisions shall be made to control water runoff and erosion during construction or demolition activities. The person making or causing an excavation to be made shall provide written notice to the owners of adjoining buildings advising them that the excavation is to be made and that the adjoining buildings should be protected. Said notification shall be delivered not less than 10 days prior to the scheduled starting date of the excavation.

SECTION 3308
TEMPORARY USE OF STREETS, ALLEYS AND
PUBLIC PROPERTY

3308.1 Storage and handling of materials. The temporary use of streets or public property for the storage or handling of materials or of equipment required for construction or demolition, and the protection provided to the public shall comply with the provisions of the authority having jurisdiction and this chapter.

3308.1.1 Obstructions. Construction materials and equipment shall not be placed or stored so as to obstruct access to fire hydrants, standpipes, fire or police alarm boxes, catch basins or manholes, nor shall such material or equipment be located within 20 feet (6096 mm) of a street intersection, or placed so as to obstruct normal observations of traffic signals or to hinder the use of public transit loading platforms.

3308.2 Utility fixtures. Building materials, fences, sheds or any obstruction of any kind shall not be placed so as to obstruct free approach to any fire hydrant, fire department connection, utility pole, manhole, fire alarm box or catch basin, or so as to interfere with the passage of water in the gutter. Protection against damage shall be provided to such utility fixtures during the progress of the work, but sight of them shall not be obstructed.

SECTION 3309
FIRE EXTINGUISHERS

[F] 3309.1 Where required. All structures under construction, alteration or demolition shall be provided with not less than one approved portable fire extinguisher in accordance with Section 906 and sized for not less than ordinary hazard as follows:

1. At each stairway on all floor levels where combustible materials have accumulated.
2. In every storage and construction shed.
3. Additional portable fire extinguishers shall be provided where special hazards exist, such as the storage and use of flammable and combustible liquids.

3309.2 Fire hazards. The provisions of this code and the *International Fire Code* shall be strictly observed to safeguard against all fire hazards attendant upon construction operations.

SECTION 3310
EXITS

3310.1 Stairways required. Where a building has been constructed to a height greater than 50 feet (15 240 mm) or four stories, or where an existing building exceeding 50 feet (15 240 mm) in height is altered, at least one temporary lighted stairway shall be provided unless one or more of the permanent stairways are erected as the construction progresses.

3310.2 Maintenance of exits. Required means of egress shall be maintained at all times during construction, demolition, remodeling or alterations and additions to any building.

> **Exception:** Approved temporary means of egress systems and facilities.

[F] SECTION 3311
STANDPIPES

3311.1 Where required. Buildings four stories or more in height shall be provided with not less than one standpipe for use during construction. Such standpipes shall be installed where the progress of construction is not more than 40 feet (12 192 mm) in height above the lowest level of fire department access. Such standpipe shall be provided with fire department hose connections at accessible locations adjacent to usable stairs. Such standpipes shall be extended as construction progresses to within one floor of the highest point of construction having secured decking or flooring.

3311.2 Buildings being demolished. Where a building is being demolished and a standpipe exists within such a building, such standpipe shall be maintained in an operable condition so as to be available for use by the fire department. Such standpipe shall be demolished with the building but shall not be demolished more than one floor below the floor being demolished.

3311.3 Detailed requirements. Standpipes shall be installed in accordance with the provisions of Chapter 9.

> **Exception:** Standpipes shall be either temporary or permanent in nature, and with or without a water supply, provided that such standpipes conform to the requirements of Section 905 as to capacity, outlets and materials.

3311.4 Water supply. Water supply for fire protection, either temporary or permanent, shall be made available as soon as combustible material accumulates.

[F] SECTION 3312
AUTOMATIC SPRINKLER SYSTEM

3312.1 Completion before occupancy. In buildings where an automatic sprinkler system is required by this code, it shall be unlawful to occupy any portion of a building or structure until the automatic sprinkler system installation has been tested and approved, except as provided in Section 110.3.

3312.2 Operation of valves. Operation of sprinkler control valves shall be permitted only by properly authorized personnel and shall be accompanied by notification of duly designated parties. When the sprinkler protection is being regularly turned off and on to facilitate connection of newly completed segments, the sprinkler control valves shall be checked at the end of each work period to ascertain that protection is in service.

CHAPTER 34

EXISTING STRUCTURES

[EB] SECTION 3401
GENERAL

3401.1 Scope. The provisions of this chapter shall control the alteration, repair, addition and change of occupancy of existing structures.

> **Exception:** Existing bleachers, grandstands and folding and telescopic seating shall comply with ICC 300-02.

3401.2 Maintenance. Buildings and structures, and parts thereof, shall be maintained in a safe and sanitary condition. Devices or safeguards which are required by this code shall be maintained in conformance with the code edition under which installed. The owner or the owner's designated agent shall be responsible for the maintenance of buildings and structures. To determine compliance with this subsection, the building official shall have the authority to require a building or structure to be reinspected. The requirements of this chapter shall not provide the basis for removal or abrogation of fire protection and safety systems and devices in existing structures.

3401.3 Compliance with other codes. Alterations, repairs, additions and changes of occupancy to existing structures shall comply with the provisions for alterations, repairs, additions and changes of occupancy in the *International Fire Code, International Fuel Gas Code, International Plumbing Code, International Property Maintenance Code, International Private Sewage Disposal Code, International Mechanical Code, International Residential Code* and ICC *Electrical Code*.

[EB] SECTION 3402
DEFINITIONS

3402.1 Definitions. The following term shall, for the purposes of this chapter and as used elsewhere in the code, have the following meaning:

TECHNICALLY INFEASIBLE. An alteration of a building or a facility that has little likelihood of being accomplished because the existing structural conditions require the removal or alteration of a load-bearing member that is an essential part of the structural frame, or because other existing physical or site constraints prohibit modification or addition of elements, spaces or features which are in full and strict compliance with the minimum requirements for new construction and which are necessary to provide accessibility.

[EB] SECTION 3403
ADDITIONS, ALTERATIONS OR REPAIRS

3403.1 Existing buildings or structures. Additions or alterations to any building or structure shall conform with the requirements of the code for new construction. Additions or alterations shall not be made to an existing building or structure which will cause the existing building or structure to be in violation of any provisions of this code. An existing building plus additions shall comply with the height and area provisions of Chapter 5. Portions of the structure not altered and not affected by the alteration are not required to comply with the code requirements for a new structure.

> **Exception:** For buildings and structures in flood hazard areas established in Section 1612.3, any additions, alterations or repairs that constitute substantial improvement of the existing structure, as defined in Section 1612.2, shall comply with the flood design requirements for new construction and all aspects of the existing structure shall be brought into compliance with the requirements for new construction for flood design.

3403.2 Structural. Additions or alterations to an existing structure shall not increase the force in any structural element by more than 5 percent, unless the increased forces on the element are still in compliance with the code for new structures, nor shall the strength of any structural element be decreased to less than that required by this code for new structures. Where repairs are made to structural elements of an existing building, and uncovered structural elements are found to be unsound or otherwise structurally deficient, such elements shall be made to conform to the requirements for new structures.

3403.2.1 Existing live load. Where an existing structure heretofore is altered or repaired, the minimum design loads for the structure shall be the loads applicable at the time of erection, provided that public safety is not endangered thereby.

3403.2.2 Live load reduction. If the approved live load is less than required by Section 1607, the areas designed for the reduced live load shall be posted in with the approved load. Placards shall be of an approved design.

3403.3 Nonstructural. Nonstructural alterations or repairs to an existing building or structure are permitted to be made of the same materials of which the building or structure is constructed, provided that they do not adversely affect any structural member or the fire-resistance rating of any part of the building or structure.

3403.4 Stairways. An alteration or the replacement of an existing stairway in an existing structure shall not be required to comply with the requirements of a new stairway as outlined in Section 1009 where the existing space and construction will not allow a reduction in pitch or slope.

[EB] SECTION 3404
FIRE ESCAPES

3404.1 Where permitted. Fire escapes shall be permitted only as provided for in Sections 3404.1.1 through 3404.1.4.

3404.1.1 New buildings. Fire escapes shall not constitute any part of the required means of egress in new buildings.

3404.1.2 Existing fire escapes. Existing fire escapes shall be continued to be accepted as a component in the means of egress in existing buildings only.

3404.1.3 New fire escapes. New fire escapes for existing buildings shall be permitted only where exterior stairs cannot be utilized due to lot lines limiting stair size or due to the sidewalks, alleys or roads at grade level. New fire escapes shall not incorporate ladders or access by windows.

3404.1.4 Limitations. Fire escapes shall comply with this section and shall not constitute more than 50 percent of the required number of exits nor more than 50 percent of the required exit capacity.

3404.2 Location. Where located on the front of the building and where projecting beyond the building line, the lowest landing shall not be less than 7 feet (2134 mm) or more than 12 feet (3658 mm) above grade, and shall be equipped with a counterbalanced stairway to the street. In alleyways and thoroughfares less than 30 feet (9144 mm) wide, the clearance under the lowest landing shall not be less than 12 feet (3658 mm).

3404.3 Construction. The fire escape shall be designed to support a live load of 100 pounds per square foot (4788 Pa) and shall be constructed of steel or other approved noncombustible materials. Fire escapes constructed of wood not less than nominal 2 inches (51 mm) thick are permitted on buildings of Type 5 construction. Walkways and railings located over or supported by combustible roofs in buildings of Type 3 and 4 construction are permitted to be of wood not less than nominal 2 inches (51 mm) thick.

3404.4 Dimensions. Stairs shall be at least 22 inches (559 mm) wide with risers not more than, and treads not less than, 8 inches (203 mm) and landings at the foot of stairs not less than 40 inches (1016 mm) wide by 36 inches (914 mm) long, located not more than 8 inches (203 mm) below the door.

3404.5 Opening protectives. Doors and windows along the fire escape shall be protected with 3/4-hour opening protectives.

[EB] SECTION 3405
GLASS REPLACEMENT

3405.1 Conformance. The installation or replacement of glass shall be as required for new installations.

[EB] SECTION 3406
CHANGE OF OCCUPANCY

3406.1 Conformance. No change shall be made in the use or occupancy of any building that would place the building in a different division of the same group of occupancy or in a different group of occupancies, unless such building is made to comply with the requirements of this code for such division or group of occupancy. Subject to the approval of the building official, the use or occupancy of existing buildings shall be permitted to be changed and the building is allowed to be occupied for purposes in other groups without conforming to all the requirements of this code for those groups, provided the new or proposed use is less hazardous, based on life and fire risk, than the existing use.

3406.2 Certificate of occupancy. A certificate of occupancy shall be issued where it has been determined that the requirements for the new occupancy classification have been met.

3406.3 Stairways. Existing stairways in an existing structure shall not be required to comply with the requirements of a new stairway as outlined in Section 1009 where the existing space and construction will not allow a reduction in pitch or slope.

[EB] SECTION 3407
HISTORIC BUILDINGS

3407.1 Historic buildings. The provisions of this code relating to the construction, repair, alteration, addition, restoration and movement of structures, and change of occupancy shall not be mandatory for historic buildings where such buildings are judged by the building official to not constitute a distinct life safety hazard.

3407.2 Flood hazard areas. Within flood hazard areas established in accordance with Section 1612.3, where the work proposed constitutes substantial improvement as defined in Section 1612.2, the building shall be brought into conformance with Section 1612.

> **Exception:** Historic buildings that are:
>
> a. Listed or preliminarily determined to be eligible for listing in the National Register of Historic Places; or
>
> b. Determined by the Secretary of the U.S. Department of Interior as contributing to the historical significance of a registered historic district or a district preliminarily determined to qualify as an historic district; or
>
> c. Designated as historic under a state or local historic preservation program that is approved by the Department of Interior.

[EB] SECTION 3408
MOVED STRUCTURES

3408.1 Conformance. Structures moved into or within the jurisdiction shall comply with the provisions of this code for new structures.

[EB] SECTION 3409
ACCESSIBILITY FOR EXISTING BUILDINGS

3409.1 Scope. The provisions of Sections 3409.1 through 3409.8 apply to maintenance, change of occupancy, additions and alterations to existing buildings, including those identified as historic buildings.

> **Exception:** Type B dwelling or sleeping units required by Section 1107 are not required to be provided in existing buildings and facilities.

3409.2 Maintenance of facilities. A building, facility or element that is constructed or altered to be accessible shall be maintained accessible during occupancy.

3409.3 Change of occupancy. Existing buildings, or portions thereof, that undergo a change of group or occupancy shall have all of the following accessible features:

1. At least one accessible building entrance.

2. At least one accessible route from an accessible building entrance to primary function areas.

3. Signage complying with Section 1110.

4. Accessible parking, where parking is being provided.

5. At least one accessible passenger loading zone, when loading zones are provided.

6. At least one accessible route connecting accessible parking and accessible passenger loading zones to an accessible entrance.

Where it is technically infeasible to comply with the new construction standards for any of these requirements for a change of group or occupancy, the above items shall conform to the requirements to the maximum extent technically feasible. Change of group or occupancy that incorporates any alterations or additions shall comply with this section and Sections 3409.4, 3409.5, 3409.6 and 3409.7.

3409.4 Additions. Provisions for new construction shall apply to additions. An addition that affects the accessibility to, or contains an area of primary function, shall comply with the requirements in Section 3409.6 for accessible routes.

3409.5 Alterations. A building, facility or element that is altered shall comply with the applicable provisions in Chapter 11 and ICC A117.1, unless technically infeasible. Where compliance with this section is technically infeasible, the alteration shall provide access to the maximum extent technically feasible.

Exceptions:

1. The altered element or space is not required to be on an accessible route, unless required by Section 3409.6.

2. Accessible means of egress required by Chapter 10 are not required to be provided in existing buildings and facilities.

3409.5.1 Extent of application. An alteration of an existing element, space or area of a building or facility shall not impose a requirement for greater accessibility than that which would be required for new construction.

Alterations shall not reduce or have the effect of reducing accessibility of a building, portion of a building or facility.

3409.6 Alterations affecting an area containing a primary function. Where an alteration affects the accessibility to, or contains an area of primary function, the route to the primary function area shall be accessible. The accessible route to the primary function area shall include toilet facilities or drinking fountains serving the area of primary function.

Exceptions:

1. The costs of providing the accessible route are not required to exceed 20 percent of the costs of the alterations affecting the area of primary function.

2. This provision does not apply to alterations limited solely to windows, hardware, operating controls, electrical outlets and signs.

3. This provision does not apply to alterations limited solely to mechanical systems, electrical systems, installation or alteration of fire protection systems and abatement of hazardous materials.

4. This provision does not apply to alterations undertaken for the primary purpose of increasing the accessibility of an existing building, facility or element.

3409.7 Scoping for alterations. The provisions of Sections 3409.7.1 through 3409.7.11 shall apply to alterations to existing buildings and facilities.

3409.7.1 Entrances. Accessible entrances shall be provided in accordance with Section 1105.

Exception: Where an alteration includes alterations to an entrance, and the building or facility has an accessible entrance, the altered entrance is not required to be accessible, unless required by Section 3409.6. Signs complying with Section 1110 shall be provided.

3409.7.2 Elevators. Altered elements of existing elevators shall comply with ASME A17.1 and ICC A117.1. Such elements shall also be altered in elevators programmed to respond to the same hall call control as the altered elevator.

3409.7.3 Platform lifts. Platform (wheelchair) lifts complying with ICC A117.1 and installed in accordance with ASME A18.1 shall be permitted as a component of an accessible route.

3409.7.4 Stairs and escalators in existing buildings. In alterations where an escalator or stair is added where none existed previously, an accessible route shall be provided in accordance with Sections 1104.4 and 1104.5.

3409.7.5 Ramps. Where steeper slopes than allowed by Section 1010.2 are necessitated by space limitations, the slope of ramps in or providing access to existing buildings or facilities shall comply with Table 3409.7.5.

TABLE 3409.7.5
RAMPS

SLOPE	MAXIMUM RISE
Steeper than 1:10 but not steeper than 1:8	3 inches
Steeper than 1:12 but not steeper than 1:10	6 inches

For SI: 1 inch = 25.4 mm.

3409.7.6 Performance areas. Where it is technically infeasible to alter performance areas to be on an accessible route, at least one of each type of performance area shall be made accessible.

3409.7.7 Dwelling or sleeping units. Where I-1, I-2, I-3, R-1, R-2 or R-4 dwelling or sleeping units are being altered or added, the requirements of Section 1107 for Accessible or Type A units and Chapter 9 for accessible alarms apply only to the quantity of spaces being altered or added.

3409.7.8 Jury boxes and witness stands. In alterations, accessible wheelchair spaces are not required to be located within the defined area of raised jury boxes or witness stands and shall be permitted to be located outside these spaces where the ramp or lift access restricts or projects into the means of egress.

3409.7.9 Toilet rooms. Where it is technically infeasible to alter existing toilet and bathing facilities to be accessible, an accessible unisex toilet or bathing facility is permitted. The

EXISTING STRUCTURES

unisex facility shall be located on the same floor and in the same area as the existing facilities.

3409.7.10 Dressing, fitting and locker rooms. Where it is technically infeasible to provide accessible dressing, fitting or locker rooms at the same location as similar types of rooms, one accessible room on the same level shall be provided. Where separate-sex facilities are provided, accessible rooms for each sex shall be provided. Separate-sex facilities are not required where only unisex rooms are provided.

3409.7.11 Check-out aisles. Where check-out aisles are altered, at least one of each check-out aisle serving each function shall be made accessible until the number of accessible check-out aisles complies with Section 1109.12.2.

3409.7.12 Thresholds. The maximum height of thresholds at doorways shall be $^3/_4$ inch (19.1 mm). Such thresholds shall have beveled edges on each side.

3409.8 Historic buildings. These provisions shall apply to buildings and facilities designated as historic structures that undergo alterations or a change of occupancy, unless technically infeasible. Where compliance with the requirements for accessible routes, ramps, entrances or toilet facilities would threaten or destroy the historic significance of the building or facility, as determined by the authority having jurisdiction, the alternative requirements of Sections 3409.8.1 through 3409.8.5 for that element shall be permitted.

3409.8.1 Site arrival points. At least one accessible route from a site arrival point to an accessible entrance shall be provided.

3409.8.2 Multilevel buildings and facilities. An accessible route from an accessible entrance to public spaces on the level of the accessible entrance shall be provided.

3409.8.3 Entrances. At least one main entrance shall be accessible.

Exceptions:

1. If a main entrance cannot be made accessible, an accessible nonpublic entrance that is unlocked while the building is occupied shall be provided; or

2. If a main entrance cannot be made accessible, a locked accessible entrance with a notification system or remote monitoring shall be provided.

Signs complying with Section 1110 shall be provided at the primary entrance and the accessible entrance.

3409.8.4 Toilet and bathing facilities. Where toilet rooms are provided, at least one accessible toilet room complying with Section 1109.2.1 shall be provided.

3409.8.5 Ramps. The slope of a ramp run of 24 inches (610 mm) maximum shall not be steeper than one unit vertical in eight units horizontal (12-percent slope).

[EB] SECTION 3410
COMPLIANCE ALTERNATIVES

3410.1 Compliance. The provisions of this section are intended to maintain or increase the current degree of public safety, health and general welfare in existing buildings while permitting repair, alteration, addition and change of occupancy without requiring full compliance with Chapters 2 through 33, or Sections 3401.3, and 3403 through 3407, except where compliance with other provisions of this code is specifically required in this section.

3410.2 Applicability. Structures existing prior to [DATE TO BE INSERTED BY THE JURISDICTION. NOTE: IT IS RECOMMENDED THAT THIS DATE COINCIDE WITH THE EFFECTIVE DATE OF BUILDING CODES WITHIN THE JURISDICTION], in which there is work involving additions, alterations or changes of occupancy shall be made to conform to the requirements of this section or the provisions of Sections 3403 through 3407. The provisions in Sections 3410.2.1 through 3410.2.5 shall apply to existing occupancies that will continue to be, or are proposed to be, in Groups A, B, E, F, M, R, S and U. These provisions shall not apply to buildings with occupancies in Group H or I.

3410.2.1 Change in occupancy. Where an existing building is changed to a new occupancy classification and this section is applicable, the provisions of this section for the new occupancy shall be used to determine compliance with this code.

3410.2.2 Partial change in occupancy. Where a portion of the building is changed to a new occupancy classification, and that portion is separated from the remainder of the building with fire barrier wall assemblies having a fire-resistance rating as required by Table 302.3.2 for the separate occupancies, or with approved compliance alternatives, the portion changed shall be made to conform to the provisions of this section.

Where a portion of the building is changed to a new occupancy classification, and that portion is not separated from the remainder of the building with fire separation assemblies having a fire-resistance rating as required by Table 302.3.2 for the separate occupancies, or with approved compliance alternatives, the provisions of this section which apply to each occupancy shall apply to the entire building. Where there are conflicting provisions, those requirements which secure the greater public safety shall apply to the entire building or structure.

3410.2.3 Additions. Additions to existing buildings shall comply with the requirements of this code for new construction. The combined height and area of the existing building and the new addition shall not exceed the height and area allowed by Chapter 5. Where a fire wall that complies with Section 705 is provided between the addition and the existing building, the addition shall be considered a separate building.

3410.2.4 Alterations and repairs. An existing building or portion thereof, which does not comply with the requirements of this code for new construction, shall not be altered or repaired in such a manner that results in the building being less safe or sanitary than such building is currently. If, in the alteration or repair, the current level of safety or sanitation is to be reduced, the portion altered or repaired shall conform to the requirements of Chapters 2 through 12 and Chapters 14 through 33.

3410.2.5 Accessibility requirements. All portions of the buildings proposed for change of occupancy shall conform to the accessibility provisions of Chapter 11.

3410.3 Acceptance. For repairs, alterations, additions and changes of occupancy to existing buildings that are evaluated in accordance with this section, compliance with this section shall be accepted by the building official.

3410.3.1 Hazards. Where the building official determines that an unsafe condition exists, as provided for in Section 115, such unsafe condition shall be abated in accordance with Section 115.

3410.3.2 Compliance with other codes. Buildings that are evaluated in accordance with this section shall comply with the *International Fire Code* and *International Property Maintenance Code.*

3410.4 Investigation and evaluation. For proposed work covered by this section, the building owner shall cause the existing building to be investigated and evaluated in accordance with the provisions of this section.

3410.4.1 Structural analysis. The owner shall have a structural analysis of the existing building made to determine adequacy of structural systems for the proposed alteration, addition or change of occupancy. The existing building shall be capable of supporting the minimum load requirements of Chapter 16.

3410.4.2 Submittal. The results of the investigation and evaluation as required in Section 3410.4, along with proposed compliance alternatives, shall be submitted to the building official.

3410.4.3 Determination of compliance. The building official shall determine whether the existing building, with the proposed addition, alteration or change of occupancy, complies with the provisions of this section in accordance with the evaluation process in Sections 3410.5 through 3410.9.

3410.5 Evaluation. The evaluation shall be comprised of three categories: fire safety, means of egress and general safety, as defined in Sections 3410.5.1 through 3410.5.3.

3410.5.1 Fire safety. Included within the fire safety category are the structural fire resistance, automatic fire detection, fire alarm and fire suppression system features of the facility.

3410.5.2 Means of egress. Included within the means of egress category are the configuration, characteristics and support features for means of egress in the facility.

3410.5.3 General safety. Included within the general safety category are the fire safety parameters and the means of egress parameters.

3410.6 Evaluation process. The evaluation process specified herein shall be followed in its entirety to evaluate existing buildings. Table 3410.7 shall be utilized for tabulating the results of the evaluation. References to other sections of this code indicate that compliance with those sections is required in order to gain credit in the evaluation herein outlined. In applying this section to a building with mixed occupancies, where the separation between the mixed occupancies does not qualify for any category indicated in Section 3410.6.16, the score for each occupancy shall be determined and the lower score determined for each section of the evaluation process shall apply to the entire building.

Where the separation between the mixed occupancies qualifies for any category indicated in Section 3410.6.16, the score for each occupancy shall apply to each portion of the building based on the occupancy of the space.

3410.6.1 Building height. The value for building height shall be the lesser value determined by the formula in Section 3410.6.1.1. Chapter 5 shall be used to determine the allowable height of the building, including allowable increases due to automatic sprinklers as provided for in Section 504.2. Subtract the actual building height from the allowable and divide by 12 $^1/_2$ feet. Enter the height value and its sign (positive or negative) in Table 3410.7 under Safety Parameter 3410.6.1, Building Height, for fire safety, means of egress and general safety. The maximum score for a building shall be 10.

3410.6.1.1 Height formula. The following formulas shall be used in computing the building height value.

$$\text{Height value, feet} = \frac{(AH) - (EBH)}{12.5} \times CF$$

$$\text{Height value, stories} = (AS - EBS) \times CF$$

(Equation 34-1)

where:

AH = Allowable height in feet from Table 503.

EBH = Existing building height in feet.

AS = Allowable height in stories from Table 503.

EBS = Existing building height in stories.

CF = 1 if $(AH) - (EBH)$ is positive.

CF = Construction-type factor shown in Table 3409.6.6(2) if $(AH) - (EBH)$ is negative.

Note. Where mixed occupancies are separated and individually evaluated as indicated in Section 3410.6, the values AH, AS, EBH and EBS shall be based on the height of the fire area of the occupancy being evaluated.

3410.6.2 Building area. The value for building area shall be determined by the formula in Section 3410.6.2.2. Section 503 and the formula in Section 3410.6.2.1 shall be used to determine the allowable area of the building. This shall include any allowable increases due to open perimeter and automatic sprinklers as provided for in Section 506. Subtract the actual building area from the allowable area and divide by 1,200 square feet (112 m²). Enter the area value and its sign (positive or negative) in Table 3410.7 under Safety Parameter 3410.6.2, Building Area, for fire safety, means of egress and general safety. In determining the area value, the maximum permitted positive value for area is 50 percent of the fire safety score as listed in Table 3410.8, Mandatory Safety Scores.

3410.6.2.1 Allowable area formula. The following formula shall be used in computing allowable area:

$$AA = \frac{(SP + OP + 100) \times (\text{area, Table 503})}{100}$$

(Equation 34-2)

where:

AA = Allowable area.

SP = Percent increase for sprinklers (Section 506.3).

OP = Percent increase for open perimeter (Section 506.2).

3410.6.2.2 Area formula. The following formula shall be used in computing the area value. Determine the area value for each occupancy fire area on a floor-by-floor basis. For each occupancy, choose the minimum area value of the set of values obtained for the particular occupancy.

$$\text{Area value } i = \frac{\text{Allowable area}_i}{1,200 \text{ square feet}} \left[1 - \left(\frac{\text{Actual area}_i}{\text{Allowable area}_i} + \ldots + \frac{\text{Actual area}_n}{\text{Allowable area}_n} \right) \right]$$

(Equation 34-3)

where:

i = Value for an individual separated occupancy on a floor.

n = Number of separated occupancies on a floor.

3410.6.3 Compartmentation. Evaluate the compartments created by fire barrier walls which comply with Sections 3410.6.3.1 and 3410.6.3.2 and which are exclusive of the wall elements considered under Sections 3410.6.4 and 3410.6.5. Conforming compartments shall be figured as the net area and do not include shafts, chases, stairways, walls or columns. Using Table 3410.6.3, determine the appropriate compartmentation value (*CV*) and enter that value into Table 3410.7 under Safety Parameter 3410.6.3, Compartmentation, for fire safety, means of egress and general safety.

3410.6.3.1 Wall construction. A wall used to create separate compartments shall be a fire barrier conforming to

Section 706 with a fire-resistance rating of not less than 2 hours. Where the building is not divided into more than one compartment, the compartment size shall be taken as the total floor area on all floors. Where there is more than one compartment within a story, each compartmented area on such story shall be provided with a horizontal exit conforming to Section 1021. The fire door serving as the horizontal exit between compartments shall be so installed, fitted and gasketed that such fire door will provide a substantial barrier to the passage of smoke.

3410.6.3.2 Floor/ceiling construction. A floor/ceiling assembly used to create compartments shall conform to Section 711 and shall have a fire-resistance rating of not less than 2 hours.

3410.6.4 Tenant and dwelling unit separations. Evaluate the fire-resistance rating of floors and walls separating tenants, including dwelling units, and not evaluated under Sections 3410.6.3 and 3410.6.5. Under the categories and occupancies in Table 3410.6.4, determine the appropriate value and enter that value in Table 3410.7 under Safety Parameter 3410.6.4, Tenant and Dwelling Unit Separation, for fire safety, means of egress and general safety.

TABLE 3410.6.4
SEPARATION VALUES

OCCUPANCY	CATEGORIES				
	a	b	c	d	e
A-1	0	0	0	0	1
A-2	-5	-3	0	1	3
R	-4	-2	0	2	4
A-3, A-4, B, E, F, M, S-1	-4	-3	0	2	4
S-2	-5	-2	0	2	4

3410.6.4.1 Categories. The categories for tenant and dwelling unit separations are:

1. Category a — No fire partitions; incomplete fire partitions; no doors; doors not self-closing or automatic closing.

2. Category b — Fire partitions or floor assembly less than 1-hour fire-resistance rating or not con-

TABLE 3409.6.3
COMPARTMENTATION VALUES

OCCUPANCY	CATEGORIES[a]				
	a Compartment size equal to or greater than 15,000 square feet	b Compartment size of 10,000 square feet	c Compartment size of 7,500 square feet	d Compartment size of 5,000 square feet	e Compartment size of 2,500 square feet
A-1, A-3	0	6	10	14	18
A-2	0	4	10	14	18
A-4, B, E, S-2	0	5	10	15	20
F, M, R, S-1	0	4	10	16	22

For SI: 1 square foot = 0.093 m².

a. For areas between categories, the compartmentation value shall be obtained by linear interpolation.

structed in accordance with Sections 708 or 711, respectively.

3. Category c — Fire partitions with 1 hour or greater fire-resistance rating constructed in accordance with Section 708 and floor assemblies with 1-hour but less than 2-hour fire-resistance rating constructed in accordance with Section 711, or with only one tenant within the fire area.

4. Category d — Fire barriers with 1-hour but less than 2-hour fire-resistance rating constructed in accordance with Section 706 and floor assemblies with 2-hour or greater fire-resistance rating constructed in accordance with Section 711.

5. Category e — Fire barriers and floor assemblies with 2-hour or greater fire-resistance rating and constructed in accordance with Sections 706 and 711, respectively.

3410.6.5 Corridor walls. Evaluate the fire-resistance rating and degree of completeness of walls which create corridors serving the floor, and constructed in accordance with Section 1016. This evaluation shall not include the wall elements considered under Sections 3410.6.3 and 3410.6.4. Under the categories and groups in Table 3410.6.5, determine the appropriate value and enter that value into Table 3410.7 under Safety Parameter 3410.6.5, Corridor Walls, for fire safety, means of egress and general safety.

TABLE 3410.6.5
CORRIDOR WALL VALUES

OCCUPANCY	CATEGORIES			
	a	b	c[a]	d[a]
A-1	-10	-4	0	2
A-2	-30	-12	0	2
A-3, F, M, R, S-1	-7	-3	0	2
A-4, B, E, S-2	-5	-2	0	5

a. Corridors not providing at least one-half the travel distance for all occupants on a floor shall use Category b.

3410.6.5.1 Categories. The categories for corridor walls are:

1. Category a — No fire partitions; incomplete fire partitions; no doors; or doors not self-closing.

2. Category b — Less than 1-hour fire-resistance rating or not constructed in accordance with Section 708.4.

3. Category c — 1-hour to less than 2-hour fire-resistance rating, with doors conforming to Section 715 or without corridors as permitted by Section 1016.

4. Category d — 2-hour or greater fire-resistance rating, with doors conforming to Section 715.

3410.6.6 Vertical openings. Evaluate the fire-resistance rating of vertical exit enclosures, hoistways, escalator openings and other shaft enclosures within the building, and openings between two or more floors. Table 3410.6.6(1) contains the appropriate protection values. Multiply that value by the construction-type factor found in Table 3410.6.6(2). Enter the vertical opening value and its sign

(positive or negative) in Table 3410.7 under Safety Parameter 3410.6.6, Vertical Openings, for fire safety, means of egress and general safety. If the structure is a one-story building, enter a value of 2. Unenclosed vertical openings that conform to the requirements of Section 707 shall not be considered in the evaluation of vertical openings.

3410.6.6.1 Vertical opening formula. The following formula shall be used in computing vertical opening value.

$$VO = PV \times CF \qquad \text{(Equation 34-4)}$$

VO = Vertical opening value.

PV = Protection value [Table 3409.6.6(1)]

CF = Construction type factor [Table 3409.6.6(2)]

TABLE 3410.6.6(1)
VERTICAL OPENING PROTECTION VALUE

PROTECTION	VALUE
None (unprotected opening)	-2 times number floors connected
Less than 1 hour	-1 times number floors connected
1 to less than 2 hours	1
2 hours or more	2

TABLE 3410.6.6(2)
CONSTRUCTION-TYPE FACTOR

	TYPE OF CONSTRUCTION								
	IA	IB	IIA	IIB	IIIA	IIIB	IV	VA	VB
FACTOR	1.2	1.5	2.2	3.5	2.5	3.5	2.3	3.3	7

3410.6.7 HVAC systems. Evaluate the ability of the HVAC system to resist the movement of smoke and fire beyond the point of origin. Under the categories in Section 3409.6.7.1, determine the appropriate value and enter that value into Table 3410.7 under Safety Parameter 3410.6.7, HVAC Systems, for fire safety, means of egress and general safety.

3410.6.7.1 Categories. The categories for HVAC systems are:

1. Category a — Plenums not in accordance with Section 602 of the *International Mechanical Code.* -10 points.

2. Category b — Air movement in egress elements not in accordance with Section 1016.4. -5 points.

3. Category c — Both categories a and b are applicable. -15 points.

4. Category d — Compliance of the HVAC system with Section 1016.4 and Section 602 of the *International Mechanical Code.* 0 points.

5. Category e — Systems serving one story; or a central boiler/chiller system without ductwork connecting two or more stories. 5 points.

3410.6.8 Automatic fire detection. Evaluate the smoke detection capability based on the location and operation of automatic fire detectors in accordance with Section 907 and the *International Mechanical Code.* Under the categories and occupancies in Table 3410.6.8, determine the appropriate value and enter that value into Table 3410.7 under Safety Parameter 3410.6.8, Automatic Fire Detection, for fire safety, means of egress and general safety.

TABLE 3410.6.8
AUTOMATIC FIRE DETECTION VALUES

OCCUPANCY	CATEGORIES				
	a	b	c	d	e
A-1, A-3, F, M, R, S-1	-10	-5	0	2	6
A-2	-25	-5	0	5	9
A-4, B, E, S-2	-4	-2	0	4	8

3410.6.8.1 Categories. The categories for automatic fire detection are:

1. Category a — None.

2. Category b — Existing smoke detectors in HVAC systems and maintained in accordance with the *International Fire Code*.

3. Category c — Smoke detectors in HVAC systems. The detectors are installed in accordance with the requirements for new buildings in the *International Mechanical Code*.

4. Category d — Smoke detectors throughout all floor areas other than individual guestrooms, tenant spaces and dwelling units.

5. Category e — Smoke detectors installed throughout the fire area.

3410.6.9 Fire alarm systems. Evaluate the capability of the fire alarm system in accordance with Section 907. Under the categories and occupancies in Table 3410.6.9, determine the appropriate value and enter that value into Table 3410.7 under Safety Parameter 3410.6.9, Fire Alarm, for fire safety, means of egress and general safety.

TABLE 3410.6.9
FIRE ALARM SYSTEM VALUES

OCCUPANCY	CATEGORIES			
	a	ba	c	d
A-1, A-2, A-3, A-4, B, E, R	-10	-5	0	5
F, M, S	0	5	10	15

a. For buildings equipped throughout with an automatic sprinkler system, add 2 points for activation by a sprinkler water flow device.

3410.6.9.1 Categories. The categories for fire alarm systems are:

1. Category a — None.

2. Category b — Fire alarm system with manual fire alarm boxes in accordance with Section 907.3 and alarm notification appliances in accordance with Section 907.9.

3. Category c — Fire alarm system in accordance with Section 907.

4. Category d — Category c plus a required emergency voice/alarm communications system and a fire command station that conforms to Section 403.8 and contains the emergency voice/alarm

communications system controls, fire department communication system controls and any other controls specified in Section 911 where those systems are provided.

3410.6.10 Smoke control. Evaluate the ability of a natural or mechanical venting, exhaust or pressurization system to control the movement of smoke from a fire. Under the categories and occupancies in Table 3410.6.10, determine the appropriate value and enter that value into Table 3410.7 under Safety Parameter 3410.6.10, Smoke Control, for means of egress and general safety.

TABLE 3410.6.10
SMOKE CONTROL VALUES

OCCUPANCY	CATEGORIES					
	a	b	c	d	e	f
A-1, A-2, A-3	0	1	2	3	6	6
A-4, E	0	0	0	1	3	5
B, M, R	0	2a	3a	3a	3a	4a
F, S	0	2a	2a	3a	3a	3a

a. This value shall be 0 if compliance with Category d or e in Section 3410.6.8.1 has not been obtained.

3410.6.10.1 Categories. The categories for smoke control are:

1. Category a — None.

2. Category b — The building is equipped throughout with an automatic sprinkler system. Openings are provided in exterior walls at the rate of 20 square feet (1.86 m^2) per 50 linear feet (15 240 mm) of exterior wall in each story and distributed around the building perimeter at intervals not exceeding 50 feet (15 240 mm). Such openings shall be readily openable from the inside without a key or separate tool and shall be provided with ready access thereto. In lieu of operable openings, clearly and permanently marked tempered glass panels shall be used.

3. Category c — One enclosed exit stairway, with ready access thereto, from each occupied floor of the building. The stairway has operable exterior windows and the building has openings in accordance with Category b.

4. Category d — One smokeproof enclosure and the building has openings in accordance with Category b.

5. Category e — The building is equipped throughout with an automatic sprinkler system. Each fire area is provided with a mechanical air-handling system designed to accomplish smoke containment. Return and exhaust air shall be moved directly to the outside without recirculation to other fire areas of the building under fire conditions. The system shall exhaust not less than six air changes per hour from the fire area. Supply air by mechanical means to the fire area is not required. Containment of smoke shall be considered as confining smoke to the fire area involved without migration to other

fire areas. Any other tested and approved design which will adequately accomplish smoke containment is permitted.

6. Category f — Each stairway shall be one of the following: a smokeproof enclosure in accordance with Section 1019.1.8; pressurized in accordance with Section 909.20.5; or shall have operable exterior windows.

3410.6.11 Means of egress capacity and number. Evaluate the means of egress capacity and the number of exits available to the building occupants. In applying this section, the means of egress are required to conform to Sections 1003 through 1014 and 1016 through 1023 (except that the minimum width required by this section shall be determined solely by the width for the required capacity in accordance with Table 1005.1). The number of exits credited is the number that are available to each occupant of the area being evaluated. Existing fire escapes shall be accepted as a component in the means of egress when conforming to Section 3404. Under the categories and occupancies in Table 3410.6.11, determine the appropriate value and enter that value into Table 3410.7 under Safety Parameter 3410.6.11, Means of Egress Capacity, for means of egress and general safety.

TABLE 3410.6.11
MEANS OF EGRESS VALUES

OCCUPANCY	CATEGORIES				
	a[a]	b	c	d	e
A-1, A-2, A-3, A-4, E	-10	0	2	8	10
M	-3	0	1	2	4
B, F, S	-1	0	0	0	0
R	-3	0	0	0	0

a. The values indicated are for buildings six stories or less in height. For buildings over six stories in height, add an additional -10 points.

3410.6.11.1 Categories. The categories for means of egress capacity and number of exits are:

1. Category a — Compliance with the minimum required means of egress capacity or number of exits is achieved through the use of a fire escape in accordance with Section 3403.

2. Category b — Capacity of the means of egress complies with Section 1004 and the number of exits complies with the minimum number required by Section 1018.

3. Category c — Capacity of the means of egress is equal to or exceeds 125 percent of the required means of egress capacity, the means of egress complies with the minimum required width dimensions specified in the code and the number of exits complies with the minimum number required by Section 1018.

4. Category d — The number of exits provided exceeds the number of exits required by Section

1018. Exits shall be located a distance apart from each other equal to not less than that specified in Section 1014.2.

5. Category e — The area being evaluated meets both Categories c and d.

3410.6.12 Dead ends. In spaces required to be served by more than one means of egress, evaluate the length of the exit access travel path in which the building occupants are confined to a single path of travel. Under the categories and occupancies in Table 3410.6.12, determine the appropriate value and enter that value into Table 3410.7 under Safety Parameter 3410.6.12, Dead Ends, for means of egress and general safety.

TABLE 3410.6.12
DEAD-END VALUES

OCCUPANCY	CATEGORIES		
	a	b	c
A-1, A-3, A-4, B, E, F, M, R, S	-2	0	2
A-2, E	-2	0	2

a. For dead-end distances between categories, the dead-end value shall be obtained by linear interpolation.

3410.6.12.1 Categories. The categories for dead ends are:

1. Category a — Dead end of 35 feet (10 670 mm) in nonsprinklered buildings or 70 feet (21 340 mm) in sprinklered buildings.

2. Category b — Dead end of 20 feet (6096 mm); or 50 feet (15 240 mm) in Group B in accordance with Section 1016.3 exception 2.

3. Category c — No dead ends; or ratio of length to width (l/w) is less than 2.5:1.

3410.6.13 Maximum exit access travel distance. Evaluate the length of exit access travel to an approved exit. Determine the appropriate points in accordance with the following equation and enter that value into Table 3410.7 under Safety Parameter 3410.6.13, Maximum Exit Access Travel Distance, for means of egress and general safety. The maximum allowable exit access travel distance shall be determined in accordance with Section 1015.1.

$$\text{Points} = 20 \times \frac{\genfrac{}{}{0pt}{}{\text{Maximum allowable}}{\text{travel distance}} - \genfrac{}{}{0pt}{}{\text{Maximum actual}}{\text{travel distance}}}{\text{Max. allowable travel distance}}$$

3410.6.14 Elevator control. Evaluate the passenger elevator equipment and controls that are available to the fire department to reach all occupied floors. Elevator recall controls shall be provided in accordance with the *International Fire Code*. Under the categories and occupancies in Table 3410.6.14, determine the appropriate value and enter that value into Table 3410.7 under Safety Parameter 3410.6.14, Elevator Control, for fire safety, means of egress and general safety. The values shall be zero for a single-story building.

TABLE 3410.6.14
ELEVATOR CONTROL VALUES

ELEVATOR TRAVEL	CATEGORIES			
	a	b	c	d
Less than 25 feet of travel above or below the primary level of elevator access for emergency fire-fighting or rescue personnel	-2	0	0	+2
Travel of 25 feet or more above or below the primary level of elevator access for emergency fire-fighting or rescue personnel	-4	NP	0	+4

For SI: 1 foot = 304.8 mm.

3410.6.14.1 Categories. The categories for elevator controls are:

1. Category a — No elevator.

2. Category b — Any elevator without Phase I and II recall.

3. Category c — All elevators with Phase I and II recall as required by the *International Fire Code*.

4. Category d — All meet Category c; or Category b where permitted to be without recall; and at least one elevator that complies with new construction requirements serves all occupied floors.

3410.6.15 Means of egress emergency lighting. Evaluate the presence of and reliability of means of egress emergency lighting. Under the categories and occupancies in Table 3410.6.15, determine the appropriate value and enter that value into Table 3410.7 under Safety Parameter 3410.6.15, Means of Egress Emergency Lighting, for means of egress and general safety.

TABLE 3410.6.15
MEANS OF EGRESS EMERGENCY LIGHTING VALUES

NUMBER OF EXITS REQUIRED BY SECTION 1010	CATEGORIES		
	a	b	c
Two or more exits	NP	0	4
Minimum of one exit	0	1	1

3410.6.15.1 Categories. The categories for means of egress emergency lighting are:

1. Category a — Means of egress lighting and exit signs not provided with emergency power in accordance with Section 2702.

2. Category b — Means of egress lighting and exit signs provided with emergency power in accordance with Section 2702.

3. Category c — Emergency power provided to means of egress lighting and exit signs which provides protection in the event of power failure to the site or building.

3410.6.16 Mixed occupancies. Where a building has two or more occupancies that are not in the same occupancy classification, the separation between the mixed occupancies shall be evaluated in accordance with this section. Where there is no separation between the mixed occupancies or the

separation between mixed occupancies does not qualify for any of the categories indicated in Section 3410.6.16.1, the building shall be evaluated as indicated in Section 3410.6 and the value for mixed occupancies shall be zero. Under the categories and occupancies in Table 3410.6.16, determine the appropriate value and enter that value into Table 3410.7 under Safety Parameter 3410.6.16, Mixed Occupancies, for fire safety and general safety. For buildings without mixed occupancies, the value shall be zero.

TABLE 3410.6.16
MIXED OCCUPANCY VALUES[a]

OCCUPANCY	CATEGORIES		
	a	b	c
A-1, A-2, R	-10	0	10
A-3, A-4, B, E, F, M, S	-5	0	5

a. For fire-resistance ratings between categories, the value shall be obtained by linear interpolation.

3410.6.16.1 Categories. The categories for mixed occupancies are:

1. Category a — Minimum 1-hour fire barriers between occupancies.

2. Category b — Fire barriers between occupancies in accordance with Section 302.3.2.

3. Category c — Fire barriers between occupancies having a fire-resistance rating of not less than twice that required by Section 302.3.2.

3410.6.17 Automatic sprinklers. Evaluate the ability to suppress a fire based on the installation of an automatic sprinkler system in accordance with Section 903.3.1.1. "Required sprinklers" shall be based on the requirements of this code. Under the categories and occupancies in Table 3410.6.17, determine the appropriate value and enter that value into Table 3410.7 under Safety Parameter 3410.6.17, Automatic Sprinklers, for fire safety, means of egress divided by 2 and general safety.

TABLE 3410.6.17
SPRINKLER SYSTEM VALUES

OCCUPANCY	CATEGORIES					
	a	b	c	d	e	f
A-1, A-3, F, M, R, S-1	-6	-3	0	2	4	6
A-2	-4	-2	0	1	2	4
A-4, B, E, S-2	-12	-6	0	3	6	12

3410.6.17.1 Categories. The categories for automatic sprinkler system protection are:

1. Category a — Sprinklers are required throughout; sprinkler protection is not provided or the sprinkler system design is not adequate for the hazard protected in accordance with Section 903.

2. Category b — Sprinklers are required in a portion of the building; sprinkler protection is not provided or the sprinkler system design is not adequate for the hazard protected in accordance with Section 903.

3. Category c — Sprinklers are not required; none are provided.

4. Category d — Sprinklers are required in a portion of the building; sprinklers are provided in such portion; the system is one which complied with the code at the time of installation and is maintained and supervised in accordance with Section 903.

5. Category e — Sprinklers are required throughout; sprinklers are provided throughout in accordance with Chapter 9.

6. Category f — Sprinklers are not required throughout; sprinklers are provided throughout in accordance with Chapter 9.

3410.6.18 Incidental use. Evaluate the protection of incidental use areas in accordance with Section 302.1.1. Do not include those where this code requires suppression throughout the building including covered mall buildings, high-rise buildings, public garages and unlimited area buildings. Assign the lowest score from Table 3409.6.18 for the building or fire area being evaluated. If there are no specific occupancy areas in the building or fire area being evaluated, the value shall be zero.

TABLE 3410.6.18
INCIDENTAL USE AREA VALUES[a]

PROTECTION REQUIRED BY TABLE 302.1.1	PROTECTION PROVIDED						
	None	1 Hour	AFSS	AFSS with SP	1 Hour and AFSS	2 Hours	2 Hours and AFSS
2 Hours and AFSS	-4	-3	-2	-2	-1	-2	0
2 Hours, or 1 Hour and AFSS	-3	-2	-1	-1	0	0	0
1 Hour and AFSS	-3	-2	-1	-1	0	-1	0
1 Hour	-1	0	-1	0	0	0	0
1 Hour, or AFSS with SP	-1	0	-1	0	0	0	0
AFSS with SP	-1	-1	-1	0	0	-1	0
1 Hour or AFSS	-1	0	0	0	0	0	0

a. AFSS = Automatic fire suppression system; SP = Smoke partitions (See Section 302.1.1.1).
NOTE: For Table 3409.7, see page 596.

3410.7 Building score. After determining the appropriate data from Section 3410.6, enter those data in Table 3410.7 and total the building score.

3410.8 Safety scores. The values in Table 3410.8 are the required mandatory safety scores for the evaluation process listed in Section 3410.6.

TABLE 3410.8
MANDATORY SAFETY SCORES[a]

OCCUPANCY	FIRE SAFETY (MFS)	MEANS OF EGRESS (MME)	GENERAL SAFETY (MGS)
A-1	16	27	27
A-2	19	30	30
A-3	18	29	29
A-4, E	23	34	34
B	24	34	34
F	20	30	30
M	19	36	36
R	17	34	34
S-1	15	25	25
S-2	23	33	33

a. MFS = Mandatory Fire Safety;
   MME = Mandatory Means of Egress;
   MGS = Mandatory General Safety.

3410.9 Evaluation of building safety. The mandatory safety score in Table 3410.8 shall be subtracted from the building score in Table 3410.7 for each category. Where the final score for any category equals zero or more, the building is in compliance with the requirements of this section for that category. Where the final score for any category is less than zero, the building is not in compliance with the requirements of this section.

3410.9.1 Mixed occupancies. For mixed occupancies, the following provisions shall apply:

1. Where the separation between mixed occupancies does not qualify for any category indicated in Section 3410.6.16, the mandatory safety scores for the occupancy with the lowest general safety score in Table 3410.8 shall be utilized (see Section 3410.6.)

2. Where the separation between mixed occupancies qualifies for any category indicated in Section 3410.6.16, the mandatory safety scores for each occupancy shall be placed against the evaluation scores for the appropriate occupancy.

TABLE 3410.7
SUMMARY SHEET — BUILDING CODE

Existing occupancy _____   Proposed occupancy _____

Year building was constructed_____   Number of stories _____ Height in feet _____

Type of construction _____   Area per floor _____

Percentage of open perimeter _____%   Percentage of height reduction _____%

Completely suppressed: Yes _____ No _____   Corridor wall rating _____

Compartmentation: Yes _____ No _____   Required door closers: Yes _____ No _____

Fire-resistance rating of vertical opening enclosures _____

Type of HVAC system_____, serving number of floors _____

Automatic fire detection: Yes _____ No_____,   type and location _____

Fire alarm system: Yes _____ No_____,   type _____

Smoke control: Yes _____ No_____,   type _____

Adequate exit routes: Yes _____ No_____   Dead ends: _____ Yes _____ No _____

Maximum exit access travel distance _____   Elevator controls: Yes _____ No _____

Means of egress emergency lighting: Yes _____No _____   Mixed occupancies: Yes_____ No _____

SAFETY PARAMETERS	FIRE SAFETY (FS)	MEANS OF EGRESS (ME)	GENERAL SAFETY (GS)
3410.6.1 Building Height 3410.6.2 Building Area 3410.6.3 Compartmentation			
3410.6.4 Tenant and Dwelling Unit Separations 3410.6.5 Corridor Walls 3410.6.6 Vertical Openings			
3410.6.7 HVAC Systems 3410.6.8 Automatic Fire Detection 3410.6.9 Fire Alarm System			
3410.6.10 Smoke control 3410.6.11 Means of Egress 3410.6.12 Dead ends	* * * * * * * * * * * *		
3410.6.13 Maximum Exit Access Travel Distance 3410.6.14 Elevator Control 3410.6.15 Means of Egress Emergency Lighting	* * * * * * * *		
3410.6.16 Mixed Occupancies 3410.6.17 Automatic Sprinklers 3410.6.18 Incidental Use		* * * * ÷ 2 =	
Building score — total value			

* * * *No applicable value to be inserted.

TABLE 3410.9
EVALUATION FORMULAS[a]

FORMULA	T.3409.7		T.3409.8		SCORE	PASS	FAIL
FS-MFS ≥ 0	_____	(FS)	− _____	(MFS) =	_____	_____	_____
ME-MME ≥ 0	_____	(ME)	− _____	(MME) =	_____	_____	_____
GS-MGS ≥ 0	_____	(GS)	− _____	(MGS) =	_____	_____	_____

a. FS = Fire Safety   MFS = Mandatory Fire Safety
   ME = Means of Egress   MME = Mandatory Means of Egress
   GS = General Safety   MGS = Mandatory General Safety

CHAPTER 35

REFERENCED STANDARDS

This chapter lists the standards that are referenced in various sections of this document. The standards are listed herein by the promulgating agency of the standard, the standard identification, the effective date and title, and the section or sections of this document that reference the standard. The application of the referenced standards shall be as specified in Section 102.4.

AA

Aluminum Association
900 - 19th Street N.W., Suite 300
Washington, DC 20006

Standard reference number	Title	Referenced in code section number
ADM 1—00	Aluminum Design Manual: Part 1-A Aluminum Structures, Allowable Stress Design; and Part 1-B —Aluminum Structures, Load and Resistance Factor Design of Buildings and Similar Type Structures	1604.3.5, 2002.1
ASM 35—80	Aluminum Sheet Metal Work in Building Construction	2002.1

AAMA

American Architectural Manufacturers Association
1827 Waldon Office Square, Suite 104
Schaumburg, IL 60173

Standard reference number	Title	Referenced in code section number
1402—86	Standard Specifications for Aluminum Siding, Soffit and Fascia	1404.5.1
101/I.S.2—97	Voluntary Specifications for Aluminum, Vinyl (PVC) and Wood Windows and Glass Doors	1714.5.1
101/I.S.2/NAFS—02	Voluntary Performance Specification for Windows, Skylights and Glass Doors	1714.5.1, 2405.5

ACI

American Concrete Institute
P.O. Box 9094
Farmington Hills, MI 48333-9094

Standard reference number	Title	Referenced in code section number
216.1—97	Method for Determining Fire Resistance of Concrete and Masonry Construction Assemblies	Table 721.1(2), 721.1
318—02	Building Code Requirements for Structural Concrete	1604.3.2, Table 1617.6, 1617.6.2.4.3, Table 1704.3, 1704.4.1, Table 1704.4, 1708.3, 1805.4.2.6, 1805.9, 1808.2.23.1.1, 1808.2.23.2, 1808.2.23.2.1, 1808.2.23.2.2, 1809.2.3.2, 1809.2.3.2.2, 1810.1.2.2, 1812.8, 1901.2, 1901.3, 1901.4, 1902, 1903.1, 1903.2, 1903.3, 1903.4, 1903.5.1, 1903.6, 1904.4.2, 1905.1.4, 1905.3, 1905.4, 1905.5, 1905.6.5.5, 1905.8.3, 1905.11.3, 1906.1.5, 1906.3, 1906.4.3, 1907.1, 1907.2, 1907.4.1, 1907.6, 1907.7.2, 1907.7.3, 907.7.4, 1907.7.5, 1907.8, 1907.9, 1907.10, 1907.11, 1907.12, 1907.13, 1908 ,1908.1.1, 1908.1.2, 1908.1.3, 1908.1.4, 1908.1.5, 1908.1.6, 1908.1.7, 1908.1.8, 1908.1.9, 1909.1, 1909.3, 1909.4, 1909.5, 1909.6, 1910, 1910.2.1, 1910.2.2, 1910.2.3, 1910.2.4, 1910.3.1, 1910.4.1, 1910.4.2, 1910.4.3, 1910.4.3.1, 1910.5.2, 1913.1, 2108.3, 2205.3
530—02	Building Code Requirements for Masonry Structures	1405.5, 1405.5.3, 1405.9, 1604.3.4, 1704.5, 1704.5.1, Table 1704.5.1, 1704.5.2, Table 1703.3.1, 1708.1.1, 1708.1.2, 1708.1.3, 1805.5.2, 1812.7, 2101.2.3, 2101.2.4, 2101.2.5, 2103.11.6, 2106.1, 2106.1.1.1, 2106.1.1.2, 2106.1.1.3, 2106.3, 2106.4, 2106.5, 2106.6, 2107.1, 2107.2, 2107.2.1, 2107.2.2, 2107.2.4, 2107.2.5, 2107.2.6, 2108.1, 2108.2, 2108.4, 2109.1, 2109.2.3.1, 2109.2.3.2
530.1—02	Specifications for Masonry Structures	1405.5.1, 1405.9.1, Table1704.5.1, Table 1704.5.3, 1805.5.2, 2103.11.7, 2104.1, 2104.1.1, 2104.3,
TG/T1.1—01	Acceptance Criteria for Moment Frames Based on Testing	1908.1.3

AF&PA

American Forest & Paper Association
1111 19th St, NW Suite 800
Washington, DC 20036

Standard reference number	Title	Referenced in code section number
AF&PA—93	Span Tables for Joists and Rafters	2306.1.1, 2308.8, 2308.10.2, 2308.10.3
AF&PA/ASCE 16—95	Load and Resistance Factor Design (LRFD) for Engineered Wood Construction	2307.1
NDS—01	National Design Specification (NDS) for Wood Construction— with 2001 Supplement	721.6.3.2, 1715.1.1, 1715.1.4, 1805.4.5, 1808.1, 2306.1, 2306.2.1, 2306.3.2, Table 2306.3.1, Table 2306.4.1, 2306.3.4, 2306.3.5, 2306.4.1, Table 2308.9.3(4)
T.R. No. 7—87	Basic Requirements for Permanent Wood Foundation System	1805.4.6, 1807.2, 2304.9.5
WCD No. 4—89	Plank and Beam Framing for Residential Buildings	2306.1.2
WFCM—01	Wood Frame Construction Manual for One- and Two-Family Dwellings	2301.2.3, 2308.1, 2308.2.1

AHA

American Hardwood Association
1210 West N.W. Highway
Palatine, IL 60067

Standard reference number	Title	Referenced in code section number
A135.4—95	Basic Hardboard	1404.3.1, 2303.1.6
A135.5—95	Prefinished Hardboard Paneling	2303.1.6, 2304.6.2
A135.6—98	Hardboard Siding	1404.3.2, 2303.1.6
A194.1—85	Cellulosic Fiber Board	2303.1.5

AISC

American Institute of Steel Construction
One East Wacker Drive, Suite 3100
Chicago, IL 60601-2001

Standard reference number	Title	Referenced in code section number
335—89s1	Specification for Structural Steel Buildings—Allowable Stress Design and Plastic Design, including Supplement No.1, 2001	1604.3.3, Table 1617.6.2, Table 1704.3, 2203.2, 2205.1
341—02	Seismic Provisions for Structural Steel Buildings	1602.1, Table 1617.6.2, 1707.2, 1708.4, 2205.2.1, 2205.2.2, 2205.3, 2205.3.1
HSS (2000)	Load and Resistance Factor Design Specification for Steel Hollow Structural Sections	1604.3.3, Table 1617.6, 2203.2, 2205.1
LRFD (1999)	Load and Resistance Factor Design Specification for Structural Steel Buildings	1604.3.3, Table 1617.6, Table 1704.3, 2203.2, 2205.1, 2205.3

AISI

American Iron and Steel Institute
1140 Connecticut Avenue
Suite 705
Washington, DC 20036

Standard reference number	Title	Referenced in code section number
NASPEC 2001	North American Specification for Design of Cold-Formed Steel Structural Members	1604.3.3, 2209.1
General	Standard for Cold-Formed Steel Framing-General Provisions, 2001	2210.1
Header	Standard for Cold-Formed Steel Framing-Header Design, 2001	2210.2
Truss	Standard for Cold-Formed Steel Framing-Truss Design, 2001	2210.3

AITC

American Institute of Timber Construction
Suite 140
7012 S. Revere Parkway
Englewood, CO 80112

Standard reference number	Title	Referenced in code section number
A 190.1—1992	Structural Glued Laminated Timber	2303.1.3, 2306.1
Technical Note 7—1996	Calculation of Fire Resistance of Glued Laminated Timbers	721.6.3.3
104—84	Typical Construction Details	2306.1
110—01	Standard Appearance Grades for Structural Glued Laminated Timber	2306.1
112—93	Tongue-and-Groove Heavy Timber Roof Decking	2306.1
113—01	Dimensions of Structural Glued Laminated Timber	2306.1
117—01	Standard Specifications for Structural Glued Laminated Timber of Softwood Species—Design Requirements—Standard Specifications for Structural Glued Laminated Timber of Softwood Species—Manufacturing Requirements	2306.1
119—96	Standard Specifications for Structural Glued Laminated Timber of Hardwood Species	2306.1
200—92	Inspection Manual	2306.1
500—91	Determination of Design Values for Structural Glued Laminated Timber	2306.1

ALI

Automotive Lift Institute
P.O. Box 33116
Indialantic, FL 32903-3116

Standard reference number	Title	Referenced in code section number
ALCTV—98	Standard for Automotive Lifts—Safety Requirements for Construction, Testing and Validation	3001.2

ANSI

American National Standards Institute
25 West 43rd Street, Fourth Floor
New York, NY 10036

Standard reference number	Title	Referenced in code section number
A 13.1—96	Scheme for the Identification of Piping Systems	415.9.6.4
A 42.2—71	Portland Cement and Portland Cement Lime Plastering, Exterior (Stucco) and Interior	2109.8.4.6
A 42.3—71	Lathing and Furring for Portland Cement and Portland Cement Lime Plastering, Exterior Stucco and Interior	2109.8.4.6
A108.1A—99	Installation of Ceramic Tile in the Wet-set Method, with Portland Cement Mortar	2103.9
A108.1B—99	Installation of Ceramic Tile, Quarry Tile on a Cured Portland Cement Mortar Setting Bed with Dry-set or Latex-Portland Mortar	2103.9
A108.4—99	Installation of Ceramic Tile with Organic Adhesives or Water Cleanable Tile-setting Epoxy Adhesive	2103.9.7
A108.5—99	Installation of Ceramic Tile with Dry-set Portland Cement Mortar or Latex-Portland Cement Mortar	2103.9.1, 2103.9.2, 2103.9.3
A108.6—99	Installation of Ceramic Tile with Chemical Resistant, Water Cleanable Tile-setting-and-grouting Epoxy	2103.9.4
A108.7—92	Specification for Electrically Conductive Ceramic Tile Installed with Conductive Dry-set Portland Cement Mortar	2103.9.2
A108.8—99	Installation of Ceramic Tile with Chemical Resistant Furan Resin Mortar and Grout	2103.9.5
A108.9—99	Installation of Ceramic Tile with Modified Epoxy Emulsion Mortar/Grout	2103.9.6
A 108.10—99	Installation of Grout in Tilework	2103.9.8
A 118.1—99	Specifications for Dry-set Portland Cement Mortar	2103.9.1
A 118.2—99	Specifications for Conductive Dry-set Portland Cement Mortar	2103.9.2
A 118.3—99	Specifications for Chemical Resistant, Water Cleanable Tile-setting and -grouting Epoxy and Water Cleanable Tile-setting Epoxy Adhesive	2103.9.4
A 118.4—99	Specifications for Latex-portland Cement Mortar	2103.9.3
A 118.5—99	Specifications for Chemical Resistant Furan Mortar and Grouts for Tile Installation	2103.9.5
A 118.6—99	Specifications for Cement Grouts for Tile Installation	2103.9.8
A 118.8—99	Specifications for Modified Epoxy Emulsion Mortar/Grout	2103.9.6
A 136.1—99	Specifications for Organic Adhesives for Installation of Ceramic Tile	2103.9.7
A 137.1—88	Specifications for Ceramic Tile	2103.4
A 208.1—99	Particleboard	2303.1.7, 2303.1.7.1

ANSI—continued

| B 31.3—99 | Process Piping—Including Addendum .... | 415.9.6.1 |
| Z 97.1—84 (R1994) | Safety Glazing Materials Used in Buildings—Safety Performance Specifications and Methods of Test (Reaffirmed 1994) .... | 2406.1.3, 2406.1.2, 2407.1 |

APA

APA - Engineered Wood Association
P.O. Box 11700
Tacoma, WA 98411-0700

Standard reference number	Title	Referenced in code section number
APA PDS	Plywood Design Specification (revised 1998).............................	2306.1, Table 2306.3.1, 2306.3.2, 2306.4.1
APA PDS Supplement 1—90	Design and Fabrication of Plywood Curved Panels (revised 1995)	2306.1
APA PDS Supplement 2—92	Design and Fabrication of Plywood-lumber Beams (revised 1998)	2306.1
APA PDS Supplement 3—90	Design and Fabrication of Plywood Stressed-skin Panels (revised 1996)..	2306.1
APA PDS Supplement 4—90	Design and Fabrication of Plywood Sandwich Panels (revised 1993)	2306.1
APA PDS Supplement 5—95	Design and Fabrication of All-plywood Beams (revised 1995)	2306.1
EWS R540—96	Builders Tips: Proper Storage and Handling of Glulam Beams	2306.1
EWS S475—99	Glued Laminated Beam Design Tables	2306.1
EWS S560—99	Field Notching and Drilling of Glued Laminated Timber Beams	2306.1
EWS T300—99	Glulam Connection Details	2306.1
EWS X440—00	Product Guide-Glulam..	2306.1
EWS X445—97	Glulam in Residential Construction —Southern Edition	2306.1
EWS X450—97	Glulam in Residential Construction —Western Edition	2306.1

ASAE

American Society of Agricultural Engineers
2950 Niles Road
St. Joseph, MI 49085-9659

Standard reference number	Title	Referenced in code section number
EP 484.2 (1998)	Diaphragm Design of Metal-Clad, Wood-Frame Rectangular Buildings	2306.1
EP 486.1 (2000)	Shallow Post Foundation Design..	2306.1
EP 559 (1997)	Design Requirements and Bending Properties for Mechanically Laminated Columns	2306.1

ASCE/SEI

American Society of Civil Engineers
Structural Engineering Institute
1801 Alexander Bell Drive
Reston, VA 20191-4400

Standard reference number	Title	Referenced in code section number
3—91	Structural Design of Composite Slabs...	1604.3.3, 2209.2
5—02	Building Code Requirements for Masonry Structures .... 1405.5, 1405.5.3, 1405.9, 1604.3.4, 1704.5, 1704.5.1, Table 1704.5.1, 1704.5.2, Table 1703.3.1, 1708.1.1, 1708.1.2, 1708.1.3, 1805.5.2, 1812.7, 2101.2.3, 2101.2.4, 2101.2.5, 2103.11.6, 2106.1, 2106.1.1.1, 2106.1.1.2, 2106.1.1.3, 2106.3, 2106.4, 2106.5, 2106.6, 2107.1, 2107.2, 2107.2.1, 2107.2.2, 2107.2.4, 2107.2.5, 2107.2.6, 2108.1, 2108.2, 2108.4, 2109.1, 2109.2.3.1, 2109.2.3.2	
6—02	Specifications for Masonry Structures .... 1405.5.1, 1405.9.1, Table1704.5.1, Table 1704.5.3, 1805.5.2, 2103.11.7, 2104.1, 2104.1.1, 2104.3,	
7—02	Minimum Design Loads for Buildings and Other Structures .... 1605.1, 1605.2.2, 1605.3.1.2, 1605.3.2, 1608.1, 1608.3, 1608.3.4, 1608.3.5, 1608.4, 1608.5, 1608.6, 1608.7, 1608.8, 1608.9, 1609.1.1, 1609.1.4.1, 1609.3, Table 1609.3.1, 1609.7.3, 1612.2, 1614.1, 1616.1, 1616.3, 1616.4.5, 1616.5, Table 1616.5.1.1, Table 1616.5.1.2, 1616.6, 1617.1, 1617.2, 1617.2.1, 1617.2.2.2, 1617.3, 1617.4, 1617.6, 1617.6.1, 1617.6.1.1, 1618.1, 1619, 1620.1, 1620.1.1, 1620.1.2, 1620.1.3, 1620.2.1, 1620.2.7, 1620.3.1, 1621.1, 1621.1.1, 1621.1.2, 1621.1.3, 1622.1, 1622.1.1, 1622.1.2, 1622.1.3, 1623.1, 1623.1.1	

ASME

American Society of Mechanical Engineers
Three Park Avenue
New York, NY 10016-5990

ASTM

ASTM International
100 Barr Harbor Drive
West Conshohocken, PA 19428-2959

ASTM—continued

A 792/A 792M—01a	Specification for Steel Sheet, 55% Aluminum-Zinc Alloy-Coated by the Hot-Dip Process	Table 1507.4.3, 2211.2.2, 2211.2.2.1
A 875/A 875M—01a	Specification for Steel Sheet Zinc-5% Aluminum Alloy-Coated by the Hot Dip Process	2211.2.2, 2211.2.2.1
A 884—99	Specification for Epoxy-Coated Steel Wire and Welded Wire Fabric for Reinforcement	2103.11.7.2
A 898/A 898M—91 (2001)	Specification for Straight Beam Ultrasonic Examination of Rolled Steel Structural Shapes	1708.4
A 899—91 (1999)	Specification for Steel Wire Epoxy-Coated	2103.11.7.2
A 913/A 913M—01	Specification for High-Strength Low-Alloy Steel Shapes of Structural Quality, Produced by Quenching and Self-Tempering Process (QST)	1809.3.1
A 951—00	Specification for Masonry Joint Reinforcement	2103.11.2
A 996/A 996M—00	Specification for Rail-Steel and Axle-Steel Deformed Bars for Concrete Reinforcement	2103.11.1, 2103.11.6
A1008—01a	Specification for Steel, Sheet, Cold-Rolled, Carbon, Structural, High-Strength Low-Alloy and High-Strength Low-Alloy with Improved Formability	2103.11.5
B 42—98	Specification for Seamless Copper Pipe, Standard Sizes	909.13.1
B 43—98	Specification for Seamless Red Brass Pipe, Standard Sizes	909.13.1
B 68—99	Specification for Seamless Copper Tube, Bright Annealed	909.13.1
B 88—99e1	Specification for Seamless Copper Water Tube	909.13.1
B 101—01	Specification for Lead-Coated Copper Sheet and Strip for Building Construction	Table 1507.4.3
B 209—96	Specification for Aluminum and Aluminum-Alloy Steel and Plate	Table 1507.4.3
B 251—97	Specification for General Requirements for Wrought Seamless Copper and Copper-Alloy Tube	909.13.1
B 280—99e1	Specification for Seamless Copper Tube for Air Conditioning and Refrigeration Field Service	909.13.1
B 633—98e01	Specification for Electrodeposited Coatings of Zinc on Iron and Steel	2211.2
C 5—79 (1997)	Specification for Quicklime for Structural Purposes	Table 2507.2
C 22/C 22M—00	Specification for Gypsum	Table 2506.2
C 27—98	Specification for Standard Classification of Fireclay and High-Alumina Refractory Brick	2111.5, 2111.8
C 28/C 28M—00	Specification for Gypsum Plasters	Table 2507.2
C 31/C 31M—98	Practice for Making and Curing Concrete Test Specimens in the Field	Table 1704.4, 1905.6.3.2, 1905.6.4.2
C 33—99ae1	Specification for Concrete Aggregates	Table 1904.2.1
C 33—01a	Specification for Concrete Aggregates	721.3.1.4, 721.4.1.1.3
C 34—96 (2001)	Specification for Structural Clay Load-Bearing Wall Tile	2103.2
C 35—95 (2001)	Specification for Inorganic Aggregates for Use in Gypsum Plaster	Table 2507.2
C 36/C 36M-01	Specification for Gypsum Wallboard	Figure 721.5.1(2), Figure 721.5.1(3), Table 721.5.1(2), Table 2506.2
C 37/C 37M-01	Specification for Gypsum Lath	Table 2507.2
C 39—99ae1	Test Method for Compressive Strength of Cylindrical Concrete Specimens	1905.6.3.2
C 42/C 42M—99	Test Method for Obtaining and Testing Drilled Cores and Sawed Beams of Concrete	1905.6.5.2
C 55—01a	Specification for Concrete Brick	Table 721.3.2, 2103.1, 2105.2.2.1.2
C 56—96 (2001)	Specification for Structural Clay Non-Load-Bearing Tile	2103.2
C 59/C 59M—00	Specification for Gypsum Casting and Molding Plaster	Table 2507.2
C 61/C 61M—00	Specification for Gypsum Keene's Cement	Table 2507.2
C 62—01	Specification for Building Brick (Solid Masonry Units Made from Clay or Shale)	2103.2, 2105.2.2.1.1
C 67—02	Test Methods of Sampling and Testing Brick and Structural Clay Tile	721.4.1.1.1, 1507.3.5, 2104.5, 2105.2.2.1.1, 2109.8.1.1
C 73—99a	Specification for Calcium Silicate Face Brick (Sand-Lime Brick)	Table 721.3.2, 2103.1
C 79—01	Specification for Treated Core and Nontreated Core Gypsum Sheathing Board	Table 2506.2
C 90—01a	Specification for Loadbearing Concrete Masonry Units	Table 721.3.2, 1805.5.2, 2103.1, 2105.2.2.1.2
C 91—01	Specification for Masonry Cement	Table 2103.7(1), Table 2507.2
C 94/C 94M—00	Specification for Ready-Mixed Concrete	109.3.1, 1905.8.2
C 126—99	Specification for Ceramic Glazed Structural Clay Facing Tile, Facing Brick, and Solid Masonry Units	2103.2
C 140—01 ae1	Test Method Sampling and Testing Concrete Masonry Units and Related Units	721.3.1.2, 1507.3.5, 2105.2.2.1.2
C 150—99a	Specification for Portland Cement	1904.1, Table 1904.2.3
C 150—01	Specification for Portland Cement	Table 2103.7(1), Table 2507.2
C 172—99	Practice for Sampling Freshly Mixed Concrete	Table 1704.4, 1905.6.3.1
C 199—84 (2000)	Test Method for Pier Test for Refractory Mortars	2111.5, 2111.8, 2113.12
C 206—84 (1997)	Specification for Finishing Hydrated Lime	Table 2507.2
C 207—91 (1997)	Specification for Hydrated Lime for Masonry Purposes	Table 2103.7(1)
C 208—95 (2001)	Specification for Cellulosic Fiber Insulating Board	2303.1.5
C 212—00	Specification for Structural Clay Facing Tile	2103.2
C 216—01a	Specification for Facing Brick (Solid Masonry Units Made from Clay or Shale)	2103.2, 2105.2.2.1.1
C 270—01a	Specification for Mortar for Unit Masonry	2103.7, Table 2103.7(2)
C 315—00	Specification for Clay Flue Linings	2113.11.1, Table 2113.16(1), Table 2113.16(2)
C 317/C 317M—00	Specification for Gypsum Concrete	1915.1

ASTM—continued

C 330—99	Specification for Lightweight Aggregates for Structural Concrete	721.1.1, 1905.1.4
C 331—01	Specification for Lightweight Aggregates for Concrete Masonry Units	721.3.1.4, 721.4.1.1.3
C 406—00	Specification for Roofing Slate	1507.7.4
C 442/C 442M—01	Specification for Gypsum Backing Board, Gypsum, Coreboard and Gypsum Shaftliner Board	Table 2506.2
C 472— 99	Specification for Standard Test Methods for Physical Testing of Gypsum, Gypsum Plasters and Gypsum Concrete	Table 2506.2
C 473—00	Test Method for Physical Testing of Gypsum Panel Products	Table 2506.2
C 474—01	Test Methods for Joint Treatment Materials for Gypsum Board Construction	Table 2506.2
C 475—01	Specification for Joint Compound and Joint Tape for Finishing Gypsum Wallboard	Table 2506.2
C 476—01	Specification for Grout for Masonry	2103.10, 2105.2.2.1.1, 2105.2.2.1.2
C 503—99e01	Specification for Marble Dimension Stone (Exterior)	2103.3
C 514—01	Specification for Nails for the Application of Gypsum Board	Table 721.1(2), Table 721.1(3), Table 2306.4.5, Table 2506.2
C 516—e01	Specifications for Vermiculite Loose Fill Thermal Insulation	721.3.1.4, 721.4.1.1.3
C 547—00	Specification for Mineral Fiber Pipe Insulation	Table 721.1(2), Table 721.1(3)
C 549—81 (1995)	Specification for Perlite Loose Fill Insulation	721.3.1.4, 721.4.1.1.3
C 557—99	Specification for Adhesives for Fastening Gypsum Wallboard to Wood Framing	Table 2506.2
C 568—99	Specification for Limestone Dimension Stone	2103.3
C 587—97	Specification for Gypsum Veneer Plaster	Table 2507.2
C 588/C 588M—01	Specification for Gypsum Base for Veneer Plasters	Table 2507.2
C 595—00	Specification for Blended Hydraulic Cements	1904.1, Table 1904.2.3
C 595—01	Specification for Blended Hydraulic Cements	Table 2103.7(1), Table 2507.2
C 615—99	Specification for Granite Dimension Stone	2103.3
C 616—99	Specification for Quartz-Based Dimension Stone	2103.3
C 618—99	Specification for Coal Fly Ash and Raw or Calcined Natural Pozzolan for Use as a Mineral Admixture in Concrete	1904.1, Table 1904.2.3
C 629—99	Specification for Slate Dimension Stone	2103.3
C 630/C 630M—01	Specification for Water-Resistant Gypsum Backing Board	Table 2506.2
C 631—00	Specification for Bonding Compounds for Interior Gypsum Plastering	Table 2507.2
C 635—00	Specification for the Manufacturer, Performance, and Testing of Metal Suspension Systems for Acoustical Tile and Lay-In Panel Ceilings	803.9.1.1, 2506.2.1
C 636—96	Practice for Installation of Metal Ceiling Suspension Systems for Acoustical Tile and Lay-In Panels	803.9.1.1
C 645—00	Specification for Nonstructural Steel Framing Members	Table 2506.2, Table 2507.2
C 652—01a	Specification for Hollow Brick (Hollow Masonry Units Made from Clay or Shale)	2103.2, 2105.2.2.1.1
C 685/ C 685M—98a	Specification for Concrete Made by Volumetric Batching and Continuous Mixing	1905.8.2
C 744—99	Specification for Prefaced Concrete and Calcium Silicate Masonry Units	2103.1
C 754—00	Specification for Installation of Steel Framing Members to Receive Screw-Attached Gypsum Panel Products	Table 2508.1, Table 2511.1
C 836—00	Specification for High Solids Content, Cold Liquid-Applied Elastomeric Waterproofing Membrane for Use with Separate Wearing Course	1507.15.2
C 840—01	Specification for Application and Finishing of Gypsum Board	Table 2508.1, 2509.2
C 841—99	Specification for Installation of Interior Lathing and Furring	Table 2508.1, Table 2511.1
C 842—99	Specification for Application of Interior Gypsum Plaster	Table 2511.1, 2511.3, 2511.4
C 843—99	Specification for Application of Gypsum Veneer Plaster	Table 2511.1
C 844—99	Specification for Application of Gypsum Base to Receive Gypsum Veneer Plaster	Table 2508.1
C 845—96	Specification for Expansive Hydraulic Cement	1904.1, Table 1904.2.3
C 847—95 (2000)	Specification for Metal Lath	Table 2507.2
C 887—79a (2001)	Specification for Packaged, Dry, Combined Materials for Surface Bonding Mortar	1807.2.2, 2103.8
C 897—00	Specification for Aggregate for Job-Mixed Portland Cement-Based Plasters	Table 2507.2
C 926—98a	Specification for Application of Portland Cement-Based Plaster	2510.3, Table 2511.1, 2511.3, 2511.4, 2512.1, 2512.1.2 2512.2, 2512.6, 2512.8.2, 2513.7, 2512.9
C 931/C 931M—01	Specification for Exterior Gypsum Soffit Board	Table 2506.2
C 932—98a	Specification for Surface-Applied Bonding Agents for Exterior Plastering	Table 2507.2
C 933—96a (2001)	Specification for Welded Wire Lath	Table 2507.2
C 946—91 (2001)	Specification for Practice for Construction of Dry-Stacked, Surface-Bonded Walls	2103.8, 2109.2.3.2
C 954—00	Specification for Steel Drill Screws for the Application of Gypsum Panel Products or Metal Plaster Bases to Steel Studs from 0.033 inch (0.84 mm) to 0.112 inch (2.84 mm) in Thickness	2211.2.2.2, Table 2506.2, Table 2507.2
C 955—01	Specification for Load-Bearing (Transverse and Axial) Steel Studs, Runners (Tracks), and Bracing or Bridging for Screw Application of Gypsum Panel Products and Metal Plaster Bases	Table 2506.2, Table 2507.2
C 956—97	Specification for Installation of Cast-in-Place Reinforced Gypsum Concrete	1915.1

C 957—93 (1998)	Specification for High-Solids Content, Cold Liquid-Applied Elastomeric Waterproofing Membrane with Integral Wearing Surface	1507.15.2
C 960—01	Specification for Predecorated Gypsum Board	Table 2506.2
C 989—99	Specification for Ground Granulated Blast-Furnace Slag for Use in Concrete and Mortars	1904.1, Table 1904.2.3
C1002—01	Specification for Steel Self-Piercing Tapping Screws for the Application of Gypsum Panel Products or Metal Plaster Bases to Wood Studs or Steel Studs	Table 2506.2, Table 2507.2
C1007—00	Specification for Installation of Load Bearing (Transverse and Axial) Steel Studs and Related Accessories	Table 2508.1, Table 2511.1
C1019—00b	Test Method of Sampling and Testing Grout	2105.2.2.1.1, 2105.2.2.1.2
C1029—96	Specification for Spray-Applied Rigid Cellular Polyurethane Thermal Insulation	1507.14.2
C1032—96	Specification for Woven Wire Plaster Base	Table 2507.2
C1047—99	Specification for Accessories for Gypsum Wallboard and Gypsum Veneer Base	Table 2506.2, Table 2507.2
C1063—99	Specification for Installation of Lathing and Furring to Receive Interior and Exterior Portland Cement-Based Plaster	2510.3, Table 2511.1, 2512.1.1
C1088—01a	Specification for Thin Veneer Brick Units Made from Clay or Shale	2103.2
C1157—00	Performance Specification for Hydraulic Cement	1904.1, Table 1904.2.3
C1167—96	Specification for Clay Roof Tiles	1507.3.4, 1507.3.5
C1177/C1177M—01	Specification for Glass Mat Gypsum Substrate for Use as Sheathing	Table 2506.2
C1178/C1178M—01	Specification for Glass Mat Water-Resistant Gypsum Backing Panel	Table 2506.2
C1186—99	Specification for Flat Nonasbestos Fiber Cement Sheets	1404.10
C1218/ C1218M—99	Test Method for Water-Soluble Chloride in Mortar and Concrete	1904.4.1
C1240—00e1	Specification for Silica Fume for Use as a Mineral Admixture in Hydraulic-Cement Concrete, Mortar, and Grout	1904.1, Table 1904.2.3
C1261—98	Specification for Firebox Brick for Residential Fireplaces	2111.5, 2111.8
C1278/C 1278M—01	Specification for Fiber-Reinforced Gypsum Panels	Table 2506.2
C1280—99	Specification for Application of Gypsum Sheathing	Table 2508.1, 2508.1,Table 2508.2, 2508.2
C1283—00e01	Practice for Installing Clay Flue Liners	2113.12
C1314—02	Test Method for Compressive Strength of Masonry Prisms	2105.2.2.2.2, 2105.3.1, 2105.3.2
C1328—00	Specification for Plastic (Stucco Cement)	Table 2507.2
C1329—00	Specification for Mortar Cement	Table 2103.7(1)
C1395/1395M—01	Specification for Gypsum Ceiling Board	Table 2506.2
D 25—99e01	Specification for Round Timber Piles	1809.1.1
D 41-94 (2000)e01	Specification for Asphalt Primer Used in Roofing, Dampproofing, and Waterproofing	Table 1507.10.2
D 43-94 (2000)	Specification for Coal Tar Primer Used in Roofing, Dampproofing, and Waterproofing	Table 1507.10.2
D 56—01	Test Method for Flash Point By Tag Closed Tester	307.2
D 86—01e01	Test Method for Distillation of Petroleum Products at Atmospheric Pressure	307.2
D 93—00	Test Method for Flash Point By Pensky-Martens Closed Cup Tester	307.2
D 224—89 (1996)	Specification for Smooth-Surfaced Asphalt Roll Roofing (Organic Felt)	1507.2.9.2, 1507.6.4
D 225—01	Specification for Asphalt Shingles (Organic Felt) Surfaced with Mineral Granules	1507.2.5
D 226—97a	Specification for Asphalt-Saturated Organic Felt Used in Roofing and Waterproofing	1404.2, Table 1507.2, 1507.2.3, 1507.3.3, 1507.5.3, 1507.6.3, 1507.7.3, Table 1507.8, 1507.8.3, 1507.9.3, 1507.9.4, Table 1507.10.2
D 227—97a	Specification for Coal-Tar-Saturated Organic Felt Used in Roofing and Waterproofing	Table 1507.10.2
D 249—89 (1996)	Specification for Asphalt Roll Roofing (Organic Felt) Surfaced with Mineral Granules	1507.3.3, 1507.6.4
D 312—00	Specification for Asphalt Used in Roofing	Table 1507.10.2
D 371—89 (1996)	Specification for Asphalt Roll Roofing (Organic Felt) Surfaced with Mineral Granules: Wide-Selvage	1507.6.4
D 422—63	Test Method for Particle-Size Analysis of Soils	1802.3.2
D 450—(2000)e1	Specification for Coal-Tar Pitch Used in Roofing, Dampproofing, and Waterproofing	Table 1507.10.2
D 635—98	Test Method for Rate of Burning and/or Extent and Time of Burning of Self-Supporting Plastics in a Horizontal Position	2606.4
D1143—81 (1994)e01	Test Method for Piles Under Static Axial Compressive Load	1808.2.8.3
D1227-95 (2000)	Specification for Emulsified Asphalt Used as a Protective Coating for Roofing	Table 1507.10.2, 1507.15.2
D1557—00	Test Method for Laboratory Compaction Characteristics of Soil Using Modified Effort (56,000 ft-lb/ft^3 (2,700 kN m/m^3))	1803.5
D1586—99	Specification for Penetration Test and Split-Barrel Sampling of Soils	1615.1.5
D1761—88(2000)	Test Method for Mechanical Fasteners in Wood	1715.1.1, 1715.1.2, 1715.1.3
D1863—93 (2000)	Specification for Mineral Aggregate Used on Built-Up Roofs	Table 1507.10.2
D1929—96 (2000)e01	Test Method for Determining Ignition Properties of Plastics	402.14.4, 406.5.2, 1407.11.2.1, 2606.4
D1970—01	Specification for Self-Adhering Polymer Modified Bituminous Sheet Materials Used as Steep Roof Underlayment for Ice Dam Protection	1507.2.4, 1507.2.9.2
D2166—00	Test Method for Unconfined Compressive Strength of Cohesive Soil	1615.1.5

ASTM—continued

D2178—97a	Specification for Asphalt Glass Felt Used in Roofing and Waterproofing	Table 1507.10.2
D2216—98	Test Method for Laboratory Determination of Water (Moisture) Content of Soil and Rock by Mass	1615.1.5
D2487—00	Practice for Classification of Soils for Engineering Purposes (Unified Soil Classification System)	Table 1610.1, 1802.3.1
D2626—97b	Specification for Asphalt-Saturated and Coated Organic Felt Base Sheet Used in Roofing	1507.3.3, Table 1507.10.2
D2822—91(1997) el	Specification for Asphalt Roof Cement	Table 1507.10.2
D2823—90(1997) el	Specification for Asphalt Roof Coatings	Table 1507.10.2
D2843—99	Test for Density of Smoke from the Burning or Decomposition of Plastics	2606.4
D2850—95(1999)	Test Method for Unconsolidated, Undrained Triaxial Compression Test on Cohesive Soils	1615.1.5
D2898—94 (1999)	Test Methods for Accelerated Weathering of Fire-Retardant-Treated Wood for Fire Testing	1505.1, 2303.2.1, 2303.2.3
D3019-94 (2002)e01	Specification for Lap Cement Used with Asphalt Roll Roofing, Non-Fibered, Asbestos Fibered, and Non-Asbestos Fibered	Table 1507.10.2
D3161—99a	Test Method for a Wind Resistance of Asphalt Shingles (Fan Induced Method)	1507.2.7
D3201—94 (1998) el	Test Method for Hygroscopic Properties of Fire-Retardant Wood and Wood-Base Products	2303.2.4
D3278—96e01	Test Methods for Flash Point of Liquids by Small Scale Closed-Cup Apparatus	307.2
D3462—01e01	Specification for Asphalt Shingles Made from Glass Felt and Surfaced with Mineral Granules	1507.2.5
D3468—99	Specification for Liquid-Applied Neoprene and Chlorosulfonated Polyethylene Used in Roofing and Waterproofing	1507.15.2
D3679—01c	Specification for Rigid Poly (Vinyl Chloride) (PVC) Siding	1404.9, 1405.13
D3689—90 (1995)	Method for Testing Individual Piles Under Static Axial Tensile Load	1808.2.8.5
D3737—01b	Practice for Establishing Allowable Properties for Structural Glued Laminated Timber (Glulam)	2303.1.3
D3746—85 (1996) el	Test Method for Impact Resistance of Bituminous Roofing Systems	1504.7
D3747—e01	Specification for Emulsified Asphalt Adhesive for Adhering Roof Insulation	Table 1507.10.2
D3909—97b	Specification for Asphalt Roll Roofing (Glass Felt) Surfaced with Mineral Granules	1507.6.4, Table 1507.10.2
D4022—94(2000)e1	Specification for Coal Tar Roof Cement, Asbestos Containing	Table 1507.10.2
D4272—99	Test Method for Total Energy Impact of Plastic Films by Dart Drop	1504.7
D4318—00	Test Methods for Liquid Limit, Plastic Limit, and Plasticity Index of Soils	1615.1.5, 1802.3.2
D4434—96	Specification for Poly (Vinyl Chloride) Sheet Roofing	1507.13.2
D4479—00	Specification for Asphalt Roof Coatings - Asbestos-Free	Table 1507.10.2
D4586—00	Specification for Asphalt Roof Cement, Asbestos-Free	Table 1507.10.2
D4601—98	Specification for Asphalt-Coated Glass Fiber Base Sheet Used in Roofing	Table 1507.10.2
D4637—96	Specification for EPDM Sheet Used in Single-Ply Roof Membrane	1507.12.2
D4829—95	Test Method for Expansion Index of Soils	1802.3.2
D4869—88 (1993)e1	Specification for Asphalt-Saturated (Organic Felt) Underlayment Used in Steep Slope Roofing	Table 1507.2, 1507.2.3
D4897—01	Specification for Asphalt-Coated Glass-Fiber Venting Base Sheet Used in Roofing	Table 1507.10.2
D4945—00	Test Method for High-Strain Dynamic Testing of Piles	1808.2.8.3
D4990—97a	Specification for Coal Tar Glass Felt Used in Roofing and Waterproofing	Table 1507.10.2
D5019—96	Specification for Reinforced Non-Vulcanized Polymeric Sheet Used in Roofing Membrane	1507.12.2
D5055—00	Specification for Establishing and Monitoring Structural Capacities of Prefabricated Wood I-joists	2303.1.2
D5456—01ae01	Specification for Evaluation of Structural Composite Lumber Products	2303.1.9
D5516—99a	Test Method of Evaluating the Flexural Properties of Fire-Retardant Treated Softwood Plywood Exposed to the Elevated Temperatures	2303.2.2.1
D5643—94(00)e1	Specification for Coal Tar Roof Cement, Asbestos-Free	Table 1507.10.2
D5664—01	Test Methods for Evaluating the Effects of Fire-Retardant Treatment and Elevated Temperatures on Strength Properties of Fire-Retardant Treated Lumber	2303.2.2.2
D5665—99a	Specification for Thermoplastic Fabrics Used in Cold-Applied Roofing and Waterproofing	Table 1507.10.2
D5726—98	Specification for Thermoplastic Fabrics Used in Hot-Applied Roofing and Waterproofing	Table 1507.10.2
D6083—97a	Specification for Liquid Applied Acrylic Coating Used in Roofing	Table 1507.10.2, 1507.15.2
D6162—00a	Specification for Styrene Butadiene Styrene (SBS) Modified Bituminous Sheet Materials Using a Combination of Polyester and Glass Fiber Reinforcements	1507.11.2
D6163—00 e01	Specification for Styrene Butadiene Styrene (SBS) Modified Bituminous Sheet Materials Using Glass Fiber Reinforcements	1507.11.2
D6164—00	Specification for Styrene Butadiene Styrene (SBS) Modified Bituminous Sheet Metal Materials Using Polyester Reinforcements	1507.11.2
D6222—00e01	Specification for Atactic Polypropylene (APP) Modified Bituminous Sheet Materials Using Polyester Reinforcements	1507.11.2
D6223—00e01	Specification for Atactic Polypropylene (APP) Modified Bituminous Sheet Materials Using a Combination of Polyester and Glass Fiber Reinforcements	1507.11.2
D6298—98	Specification for Fiberglass Reinforced Styrene-Butadiene-Styrene (SBS) Modified Bituminous Sheets with a Factory Applied Metal Surface	1507.11.2
D6305—99e1	Practice for Calculating Bending Strength Design Adjustment Factors for Fire-Retardant-Treated Plywood Roof Sheathing	2303.2.2.1

ASTM—continued

E 84—01	Test Method for Surface Burning Characteristics of Building Materials	402.10, 402.14.4, 406.5.2, 410.3.5.3, 703.4.2, 719.1, 719.4, 802.1, 803.1, 803.5, 803.6.1, 803.6.2, 1407.10, 1407.10.1, 2303.2, 2603.3, 2603.4.1.13, 2603.5.4, 2604.2.4, 2606.4, 3105.3
E 90—99	Test Method for Laboratory Measurement of Airborne Sound Transmission Loss of Building Partitions and Elements	1207.2
E 96—00	Test Method for Water Vapor Transmission of Materials	1203.2
E 108—00	Test Methods for Fire Tests of Roof Coverings	1505.1, 2603.6, 2610.2, 2610.3
E 119—00a	Test Methods for Fire Tests of Building Construction and Materials	410.3.5.2, 703.2, 703.2.1, 703.2.3, 703.3, 704.7, 704.9, 706.7, 711.3.2, 712.3.1, 712.4.1, 712.4.6, 713.1, 713.4, 714.7, 715.2, 716.5.2, 715.5.3.1, 715.6.2, 716.5.2, 716.5.3.1, 716.6.2, Table 721.1(1), 1407.10.2, 2103.2, 2603.3, 2603.4, 2603.4.1.13, 2603.5.4, 2604.2.4, 2606.4
E 136—99e01	Test Method for Behavior of Materials in a Vertical Tube Furnace at 750°C	703.4.1
E 328—86	Methods for Stress Relaxation for Materials and Structures	2103.11.6
E 330—97e01	Test Method for Structural Performance of Exterior Windows, Curtain Walls, and Doors by Uniform Static Air Pressure Difference	1714.5.2
E 331—00	Test Method for Water Penetration of Exterior Windows, Skylights, Doors, and Curtain Walls by Uniform Static Air Pressure Difference	1403.2
E 492—90 (1996)e1	Test Method for Laboratory Measurement of Impact Sound Transmission Through Floor-Ceiling Assemblies Using the Tapping Machine	1207.3
E 605—93 (2000)	Test Method for Thickness and Density of Sprayed Fire-Resistive Material (SFRM) Applied to Structural Members	1704.11.3, 1704.11.3.1, 1704.11.3.2, 1704.11.4
E 681—01	Test Methods for Concentration Limits of Flammability of Chemicals (Vapors and Gases)	307.2
E 736—00	Test Method for Cohesion/Adhesion of Sprayed Fire-Resistive Materials Applied to Structural Members	1704.11.5
E 814—00	Test Method of Fire Tests of Through-Penetration Fire Stops	702.1, 712.3.1.2, 712.4.1.2
E 970—00	Test Method for Critical Radiant Flux of Exposed Attic Floor Insulation Using a Radiant Heat Energy Source	719.3.1
E1300—00	Practice for Determining Load Resistance of Glass in Buildings	Table 2404.1, Table 2404.2
E1592—01	Test Method for Structural Performance of Sheet Metal Roof and Siding Systems by Uniform Static Air Pressure Difference	1504.3.2
E1602—01	Guide for Construction of Solid Fuel-Burning Masonry Heaters	2112.2
E1886—97	Test Method for Performance of Exterior Windows, Curtain Walls, Doors and Storm Shutters Impacted by Missiles and Exposed to Cyclic Pressure Differentials	1609.1.4
E1966—00	Test Method for Fire-Resistant Joint Systems	702.1, 712.3
E1996—01	Specification for Performance of Exterior Windows, Curtain Walls, Doors and Storm Shutters Impacted by Windborne Debris in Hurricanes	1609.1.4
F 547—01	Terminology of Nails for Use with Wood and Wood-Base Materials	Table 2506.2
F1346—91 (1996)	Performance Specification for Safety Covers and Labeling Requirements for All Covers for Swimming Pools, Spas and Hot Tubs	3104.9, 3109.4.1.8
F1667—01a	Specification for Driven Fasteners: Nails, Spikes, and Staples	Table 721.1(2), Table 721.1(3), 1507.2.6, 2303.6, Table 2506.2
G 152—00a	Practice for Operating Open Flame Carbon Arc Light Apparatus for Exposure of Nonmetallic Materials	1504.5
G 154—00a	Practice for Operating Fluorescent Light Apparatus for UV Exposure of Nonmetallic Materials	1504.5
G 155—00a	Practice for Operating Xenon Arc Light Apparatus for Exposure of Non-Metallic Materials	1504.5

AWPA

American Wood-Preservers' Association
P.O. Box 5690
Grandbury, TX 76049

Standard reference number	Title	Referenced in code section number
C1—00	All Timber Products—Preservative Treatment by Pressure Processes	1403.6, 1505.6, 2303.1.8
C2—01	Lumber, Timber, Bridge Ties and Mine Ties—Preservative Treatment by Pressure Processes	1403.6, Table 1507.9.5, 1805.4.5, 1805.7.1, 2303.1.8, 2304.11.2, 2304.11.4, 2304.11.7
C3—99	Piles—Preservative Treatment by Pressure Processes	1403.6, 1805.4.5, 1809.1.2, 2303.1.8
C4—99	Poles—Preservative Treatment by Pressure Processes	1403.6, 1805.7.1, 1808.1.2, 2303.1.8
C9—00	Plywood—Preservative Treatment by Pressure Processes	1403.6, 2303.1.8, 2304.11.2, 2304.11.4, 2304.11.7
C14—99	Wood for Highway Construction, Pressure Treatment by Pressure Process	2303.1.8
C15—00	Wood for Commercial-Residential Construction Preservative Treatment by Pressure Process	1403.6, 2303.1.8
C16—00	Wood Used on Farms, Pressure Treatment by Pressure Process	2303.1.8
C18—99	Standard for Pressure Treated Material in Marine Construction	1403.6
C22—96	Lumber and Plywood for Permanent Wood Foundations—Preservative Treatment by Pressure Processes	1403.6, 1805.4.6, 2303.1.8

AWPA—continued

C23—00	Round Poles and Posts Used in Building Construction—Preservative Treatment by Pressure Processes	1403.6, 2303.1.8
C24—96	Sawn Timber Piles Used to Support Residential and Commercial Structures	1403.6, 1809.1.2, 2303.1.8
C28—99	Standard for Preservative Treatment by Pressure Process of Structural Glued Laminated Members and Laminations before Gluing	1403.6, 2303.1.8
C31—00	Lumber Used Out of Contact with the Ground and Continuously Protected from Liquid Water—Treatment by Pressure Processes	2303.1.8
C33—00	Standard for Preservative Treatment of Structural Composite Lumber by Pressure Processes	2303.1.8
M4—01	Standard for the Care of Preservative-Treated Wood Products	1809.1.2, 2303.1.8
P1/13—01	Standard for Creosote Preservative	1403.6, 2303.1.8
P2—01	Standard for Creosote Solutions	1403.6, 2303.1.8
P3—01	Standard for Creosote-Petroleum Solution	1403.6
P5—01	Standard for Waterborne Preservatives	2303.1.8
P8—01	Standard for Oil-Borne Preservatives	2303.1.8
P9—01	Standard for Solvents and Formulations for Organic Preservative Systems	2303.1.8

AWS

American Welding Society
550 N.W. LeJeune Road
Miami, FL 33126

Standard reference number	Title	Referenced in code section number
D1.1—2000	Structural Welding Code—Steel	Table 1704.3, 1704.3.1, 1708.4
D1.3—1998	Structural Welding Code—Sheet Steel	Table 1704.3
D1.4—1998	Structural Welding Code—Reinforcing Steel	Table 1704.3, 1903.5.2

BHMA

Builders Hardware Manufacturers' Association
355 Lexington Avenue, 17th Floor
New York, NY 10017-6603

Standard reference number	Title	Referenced in code section number
A 156.10—1999	Power Operated Pedestrian Doors	1008.1.3.2
A 156.19—1997	Power Assist and Low Energy Operated Doors	1008.1.3.2

CGSB

Canadian General Standards Board
222 Queens Street
14th Floor, Suite 1402
Ottawa, Ontario, Canada KIA 1G6

Standard reference number	Title	Referenced in code section number
37-GP-52M (1984)	Roofing and Waterproofing Membrane, Sheet Applied, Elastomeric	1504.7, 1507.12.2
CAN/CGSB 37.54—95	Polyvinyl Chloride Roofing and Waterproofing Membrane	1507.13.2
37-GP-56M (1980)	Membrane, Modified, Bituminous, Prefabricated, and Reinforced for Roofing—with December 1985 Amendment	1507.11.2

CPSC

Consumer Product Safety Commission
4330 East West Highway
Bethesada, MD 20814-4408

Standard reference number	Title	Referenced in code section number
16 CFR Part 1201(1977)	Safety Standard for Architectural Glazing Material	2406.1.1, 2406.2.1, 2407.1, 2408.2.1, 2408.3
16 CFR Part 1209 (1979)	Interim Safety Standard for Cellulose Insulation	719.6
16 CFR Part 1404 (1979)	Cellulose Insulation	719.6
16 CFR Part 1500 (1991)	Hazardous Substances and Articles; Administration and Enforcement Regulations	307.2
16 CFR Part 1500.44 (2001)	Method for Determining Extremely Flammable and Flammable Solids	307.2
16 CFR Part 1507 (2001)	Fireworks Devices	307.2
16 CFR Part 1630 (2000)	Standard for the Surface Flammability of Carpets and Rugs	804.5.1

CSSB

Cedar Shake and Shingle Bureau
P.O. Box 1178
Sumas, WA 98295-1178

Standard reference number	Title	Referenced in code section number
CSSB—97	Grading and Packing Rules for Western Red Cedar Shakes and Western Red Shingles of the Cedar Shake and Shingle Bureau . . . . Table 1507.8.4, Table 1507.9.5	

DASMA

Door and Access Systems Manufacturers
Association International
1300 Summer Avenue
Cleveland, OH 44115-2851

Standard reference number	Title	Referenced in code section number
107—97	Room Fire Test Standard for Garage Doors Using Foam Plastic Insulation 2603.4.1.9	

DOC

U.S. Department of Commerce
National Institute of Standards and Technology
100 Bureau Drive Stop 3460
Gaithersburg, MD 20899

Standard reference number	Title	Referenced in code section number
PS-1—95	Construction and Industrial Plywood . . . . 2211.2.2.2, 2303.1.4, 2304.6.2, Table 2304.7(4), 2306.3.2	
PS-2—92	Performance Standard for Wood-based Structural-use Panels . . . . 1809.1.1, 2211.3.1, 2303.1.4, 2304.6.2, Table 2304.7(4), Table 2304.7(5), Table 2306.3.1, 2306.3.2	
PS 20—99	American Softwood Lumber Standard 1809.1.1, 2302.1, 2303.1.1	

DOL

U.S. Department of Labor
c/o Superintendent of Documents
U.S. Government Printing Office
Washington, DC 20402-9325

Standard reference number	Title	Referenced in code section number
29 CFR Part 1910.1000 (1974)	Air Contaminants 902.1	

DOTn

U.S. Department of Transportation
c/o Superintendent of Documents
U.S. Government Printing Office
Washington, DC 20402-9325

Standard reference number	Title	Referenced in code section number
49 CFR Part 172 (1999)	Hazardous Materials Tables, Special Provisions, Hazardous Materials Communications, Emergency Response Information and Training Requirements 307.2	
49 CFR Parts 173-178 (1999)	Specification of Transportation of Explosive and Other Dangerous Articles, UN 0335,UN 0336 Shipping Containers . . 307.2	

FEMA

Federal Emergency Management Agency
Federal Center Plaza
500 C Street S.W.
Washington, DC 20472

Standard reference number	Title	Referenced in code section number
Pub 302 (1997)	NEHRP Recommended Provisions for Seismic Regulations for New Buildings and Other Structures . . . . Figure 1615(7), Figure 1615(8), Figure 1615(9), Figure 1615(10)	
TB 11-01	Crawlspace Construction for Buildings Located in Special Flood Hazard Areas 1807.1.2.1	

FM

Factory Mutual
Standards Laboratories Department
1151 Boston-Providence Turnpike
Norwood, MA 02062

Standard reference number	Title	Referenced in code section number
4450 (1989)	Approval Standard for Class 1 Insulated Steel Deck Roofs— with Supplements through July 1992. . . . . 1504.3.1, 1508.1, 2603.3, 2603.4.1.5	
4470 (1986)	Approval Standard for Class 1 Roof Covers with 1992 Supplements 1504.3.1, 1504.7	
4880 (2001)	Standard for Evaluating Insulated Wall or Wall and Roof/Ceiling Assemblies, Plastic Interior Finish Materials, Plastic Exterior Building Panels, Wall/Ceiling Coating Systems, Interior and Exterior Finish Systems. 2603.4, 2603.8	

GA

Gypsum Association
810 First Street N.E. #510
Washington, DC 20002-4268

Standard reference number	Title	Referenced in code section number
GA 216—00	Application and Finishing of Gypsum Board . . . . Table 2508.1, 2509.2	
GA 600—00	Fire Resistance Design Manual,16th Edition, April, 2000. . . . . Table 721.1(1), Table 721.1(2), Table 721.1(3)	

HPVA

Hardwood Plywood Veneer Association
1825 Michael Faraday Drive
Reston, VA 20190-5350

Standard reference number	Title	Referenced in code section number
HP-1—2000	Standard for Hardwood and Decorative Plywood 2303.3, 2304.6.2	

ICC

International Code Council
5203 Leesburg Pike, Suite 600
Falls Church, VA 22041

Standard reference number	Title	Referenced in code section number
ICC/ANSI A117.1—98	Accessible and Usable Buildings and Facilities 406.2.2, 907.9.1.3,1007.6.5, 1010.1, 1010.6.5, 1010.9, 1011.3, 1101.2, 1102.1, 1103.2.13, 1106.6, 1107.2, 1109.2.2, 1109.3, 1109.4, 1109.8, 1109.15, 3001.3, 3409.5, 3409.7.2, 3409.7.3	
ICC 300—02	ICC Standard on Bleachers, Folding and Telescopic Seating, and Grandstands 1024.1.1	
ICC EC—03	ICC Electrical Code™ 101.4.1, 107.3, 414.5.4, 414.9.2.8.1, 904.3.1, 907.5, 909.11, 909.12.1, 909.16.3, 1205.4.1, 1405.10.4, 2701.1, 2702.1, 3401.3	
IEBC——03	International Existing Building Code™ 101.2	
IECC—03	International Energy Conservation Code® 101.4.7, 1202.3.2, 1301.1.1, 1403.2	

ICC—continued

IFC—03	International Fire Code®	101.4.6, 102.6, 201.3, 307.9, Table 307.7(1), Table 307.7(2), 307.9, 404.2,406.5.1, 406.5.2, 406.6.1, 410.3.6, 411.1, 412.4.1, 413.1, 414.1.1, 414.1.2, 414.1.2.1, 414.2.4, Table 414.2.4, 414.3, 414.5, 414.5.1, Table 414.5.1, 414.5.2, 414.5.4, 414.5.5, 414.6, 415.1, 415.3, 415.3.1, Table 415.3.1, Table 415.3.2, 415.7, 415.7.1, 415.7.1.4, 415.7.2, 415.7.2.3, 415.7.2.5, 415.7.2.7, 415.7.2.8, 415.7.2.9, 415.7.3, 415.7.3.3.3, 415.7.3.5, 415.7.4, 415.8, 415.9.1, 415.9.2.7, 415.9.5.1, 415.9.7.2, 704.8.2, 706.1, 901.2, 901.3, 901.5, 901.6.2, 903.2.6.1, 903.2.11, Table 903.2.13, 903.5, 904.2.1, 905.1, 906.1, 907.2.5, 907.2.12.2, 907.2.14, 907.2.16, 907.19, 909.20, 910.2.3, Table 910.3, 1001.3, 1203.4.2, 1203.5, 2702.2.8, 2702.2.10, 2702.2.11, 2702.2.12, 2702.3, 3102.1, 3103.1, 3309.2, 3401.3, 3410.3.2, 3410.6.8.1, 3410.6.14, 3410.6.14.1
IFGC—03	International Fuel Gas Code®	101.4.2, 201.3, 415.7.3, 2113.11.1.2, 2113.15, 2801.1, 3401.3
IMC—03	International Mechanical Code®	101.4.3, 201.3, 307.9, 406.4.2, 406.6.3, 406.6.5, 409.3, 412.4.6, 414.1.2, 414.1.2.1, 414.1.2.2, 414.3, 415.7.1.4, 415.7.2, 415.7.2.8, 415.7.3, 415.7.4, 415.9.11.1, 416.3, 603.1, 707.2, 716.2.2, 716.5.4, 716.6.1, 716.6.2, 716.6.3, 717.5, 719.1, 903.2.12.1, 904.2.1, 904.11, 908.6, 909.1, 909.10.2, 1014.5, 1016.4.1, 1203.1, 1203.2.1, 1203.4.2, 1203.4.2.1, 1203.5, 1209.3, 2304.5, 3004.3.1, 3410.6.7.1, 3410.6.8
IPC—03	International Plumbing Code®	101.4.4, 201.3, 415.7.4, 717.5, 903.3.5, 1206.3.3, 1503.4, 1807.4.3, 2901.1, 2902.1.1, 3305.1, 3401.3,
IPMC—03	International Property Maintenance Code®	101.4.5, 102.6, 103.3, 3401.3, 3410.3.2.
IPSDC—03	International Private Sewage Disposal Code®	101.4.4, 2901.1, 3401.3
IRC—03	International Residential Code®	101.2, 308.3, 308.5 1706.1.1, 3401.3
IUWIC—03	International Urban—Wildland Interface Code™	Table 1505.1
SBCCI SSTD 10—99	Standard for Hurricane Resistant Residential Construction	1609.1.1, 2308.2.1
SBCCI SSTD 11—97	Test Standard for Determining Wind Resistance of Concrete or Clay Roof Tiles	1715.2.1, 1715.2.2

NAAMM

National Association of Architectural
Metal Manufacturers
8 South Michigan Ave
Chicago, IL 60603

Standard reference number	Title	Referenced in code section number
FP 1001—97	Guide Specifications for Design of Metal Flag Poles	1609.1.1

NCMA

National Concrete Masonry Association
2302 Horse Pen Road
Herndon, VA 22071-3499

Standard reference number	Title	Referenced in code section number
TEK 5-8A (1996)	Details for Concrete Masonry Fire Walls	Table 719.1(2)

NFPA

National Fire Protection Association
1 Batterymarch Park
Quincy, MA 02269-9101

Standard reference number	Title	Referenced in code section number
11—98	Low Expansion Foam	904.7
11A—99	Medium- and High-Expansion Foam Systems	904.7
12—00	Carbon Dioxide Extinguishing Systems	904.8, 904.11
12A—97	Halon 1301 Fire Extinguishing Systems	904.9
13—99	Installation of Sprinkler Systems	704.12, 707.2, 903.3.1.1, 903.3.2, 903.3.5.1.1, 904.11, 907.8, 1621.3.10.1, 3104.5, 3104.9
13D—99	Installation of Sprinkler Systems in One- and Two-Family Dwellings and Manufactured Homes	903.1.2, 903.3.1.3, 903.3.5.1.1
13R—99	Installation of Sprinkler Systems in Residential Occupancies Up to and Including Four Stories in Height	903.1.2, 903.3.1.2, 903.3.5.1.1, 903.3.5.1.2, 903.4
14—00	Installation of Standpipe, Private Hydrants and Hose Systems	905.2, 905.3.4, 905.4.2, 905.8
16—99	Installation Foam-Water Sprinkler and Foam-Water Spray Systems	904.7, 904.11

NFPA—continued

17—98	Dry Chemical Extinguishing Systems	904.6, 904.11
17A—98	Wet Chemical Extinguishing Systems	904.5, 904.11
30—00	Flammable and Combustible Liquids Code	415.3
32—00	Drycleaning Plants	415.7.4
40—97	Storage and Handling of Cellulose Nitrate Motion Picture Film	409.1
61—99	Prevention of Fires and Dust Explosions in Agricultural and Food Product Facilities	415.7.1
72—99	National Fire Alarm Code	505.4, 901.6, 903.4.1, 904.3.5, 907.2, 907.2.1, 907.2.1.1, 907.2.10, 907.2.10.4, 907.2.11.2, 907.2.11.3, 907.2.12.2.3, 907.2.12.3, 907.4, 907.5, 907.9.2, 907.10, 907.14, 907.16, 907.17, 911.1, 3006.5
80—99	Fire Doors and Fire Windows	302.1.1.1, 715.3, 715.4.6.1, 715.4.4, 715.4.7.2, 715.5, 1008.1.3.3
85—01	Boiler and Combustion Systems Hazards Code	415.7.1
	(Note: NFPA 8503 has been incorporated into NFPA 85)	
101—00	Life Safety Code	1024.6.2
110—99	Emergency and Standby Power Systems	2702.1
111—01	Stored Electrical Energy Emergency and Standby Power Systems	2702.1
120—99	Coal Preparation Plants	415.7.1
231C—98	Rack Storage of Materials	507.2
252—99	Methods of Fire Tests of Door Assemblies	715.3.1, 715.3.2, 715.3.3, 715.3.4.1
253—00	Test for Critical Radiant Flux of Floor Covering Systems Using a Radiant Heat Energy Source	406.6.4, 804.2, 804.3
257—00	Fire Test for Window and Glass Block Assemblies	715.3.3, 715.4, 715.4.1, 715.4.2
259—98	Test Method for Potential Heat of Building Materials	2603.4.1.10, 2603.5.3
265—98	Method of Fire Tests for Evaluating Room Fire Growth Contribution of Textile Wall Coverings	803.6.1, 803.6.1.1, 803.6.1.2
268—96	Test Method for Determining Ignitibility of Exterior Wall Assemblies Using a Radiant Heat Energy Source	1406.2.1, 1406.2.1.1, 1406.2.1.2, 2603.5.7
285—98	Method of Test for the Evaluation of Flammability Characteristics of Exterior Non-Load-Bearing Wall Assemblies Containing Combustible Components Using the Intermediate-Scale, Multistory Test Apparatus	1407.10.4, 2603.5.5
286—00	Method of Fire Test for Evaluating Contribution of Wall and Ceiling Interior Finish to Room Fire Growth	402.14.4, 803.2, 803.2.1, 803.5, 2603.4, 2603.8
409—95	Aircraft Hangers	412.2.6, 412.4.5
418—01	Heliports	412.5.6
651—98	Machining and Finishing of Aluminum and the Production and Handling of Aluminum Powders	415.7.1
654—00	Prevention of Fire & Dust Explosions from the Manufacturing, Processing, and Handling of Combustible Particulate Solids	415.7.1
655—93	Prevention of Sulfur Fires and Explosions	415.7.1
664—98	Prevention of Fires Explosions in Wood Processing and Woodworking Facilities	415.7.1
701—99	Methods of Fire Tests for Flame-Propagation of Textiles and Films	802.1, 805.1, 805.2, 3102.3.1, 3105.3,
704—96	System for the Identification of the Hazards of Materials for Emergency Response	414.7.2, 415.2
1124—98	Manufacture, Transportation, and Storage of Fireworks and Pyrotechnic Articles	415.3.1
2001—00	Clean Agent Fire Extinguishing Systems	904.10

NIST

National Institute of Standards and Technology
U.S. Department of Commerce
100 Bureau Dr. – Stop 3460
Gaithersburg, MD 20899-3460

Standard reference number	Title	Referenced in code section number
BMS 071 (1941)	Fire Tests of Wood- and Metal-Framed Partitions	721.7
TRBM-44 (1944)	Fire-Resistance and Sound-Insulation Ratings for Walls, Partitions, and Floors	721.7

PCI

Precast Prestressed Concrete Institute
175 W. Jackson Boulevard, Suite 1859
Chicago, IL 60604-9773

Standard reference number	Title	Referenced in code section number
MNL 124—89	Design for Fire Resistance of Precast Prestressed Concrete	721.2.3.1, Table 721.2.3(4)

MNL 128—01   Recommended Practice for Glass Fiber Reinforced Concrete Panels...1903.8

PTI

Post-Tensioning Institute
1717 W. Northern Avenue, Suite 114
Phoenix, AZ 85021

Standard reference number	Title	Referenced in code section number
PTI 1996	Design and Construction of Post-Tensioned Slabs-on-Ground, 2nd Edition 1805.8.2	

RMA

Rubber Manufacturers Association
1400 K. Street, N.W. #900
Washington, DC 20005

Standard reference number	Title	Referenced in code section number
RP-1—90	Minimum Requirements for Non-reinforced Black EPDM Rubber Sheets 1507.12.2	
RP-2—90	Minimum Requirements for Fabric-reinforced Black EPDM Rubber Sheets 1507.12.2	
RP-3—85	Minimum Requirements for Fabric-reinforced Black Polychloroprene Rubber Sheets.......................1507.12.2	

RMI

Rack Manufacturers Institute
8720 Red Oak Boulevard, Suite 201
Charlotte, NC 28217

Standard reference number	Title	Referenced in code section number
RMI (1997)	Design, Testing and Utilization of Industrial Steel Storage Racks 2208.1	

SJI

Steel Joist Institute
3127 10th Avenue, North
Myrtle Beach, SC 29577-6760

Standard reference number	Title	Referenced in code section number
SJI—1994	Standard Specification for Joist Girders 1604.3.3, 2206	
K-Series Specification—1994	Standard Specification for Open Web Steel Joists, K Series 2206	
SJI—1994	Standard Specification for Longspan Steel Joists, LH Series and Deep Longspan Steel Joists, DLH Series 2206	

SPRI

Single-Ply Roofing Institute
77 Rumford Ave.
Suite 3-B
Waltham, MA 02453

Standard reference number	Title	Referenced in code section number
ES-1—98	Wind Design Standard for Edge Systems Used with Low Slope Roofing Systems 1504.5	
RP-4—88	Wind Design Guide for Ballasted Single-ply Roofing Systems 1504.4	

TIA

Telecommunications Industry Association
2500 Wilson Boulevard
Arlington, VA 22201-3834

Standard reference number	Title	Referenced in code section number
TIA/EIA-222-F—96	Structural Standards for Steel Antenna Towers and Antenna Supporting Structures	1609.1.1, 3108.4

TMS

The Masonry Society
3970 Broadway, Unit 201-D
Boulder, CO 80304-1135

Standard reference number	Title	Referenced in code section number
216—97	Method for Determining Fire Resistance of Concrete and Masonry Construction Assemblies	Table 721.1(2), 721.1
402—02	Building Code for Masonry Structures	1405.5, 1405.5.3, 1405.9, 1604.3.4, 1704.5, 1704.5.1, Table 1704.5.1, 1704.5.2, Table 1703.3.1, 1708.1.1, 1708.1.2, 1708.1.3, 1805.5.2, 1812.7, 2101.2.3, 2101.2.4, 2101.2.5, 2103.11.6, 2106.1, 2106.1.1.1, 2106.1.1.2, 2106.1.1.3, 2106.3, 2106.4, 2106.5, 2106.6, 2107.1, 2107.2, 2107.2.1, 2107.2.2, 2107.2.4, 2107.2.5, 2107.2.6, 2108.1, 2108.2, 2108.4, 2109.1, 2109.2.3.1, 2109.2.3.2
602—02	Specification for Masonry Structures	1405.5.1, 1405.9.1, Table 1704.5.1, Table 1704.5.3, 1805.5.2, 2103.11.7, 2104.1, 2104.1.1, 2104.3

TPI

Truss Plate Institute
583 D'Onofrio Drive, Suite 200
Madison, WI 53719

Standard reference number	Title	Referenced in code section number
TPI 1—2002	National Design Standards for Metal-Plate-Connected Wood Truss Construction	2303.4, 2306.1

UL

Underwriters Laboratories
333 Pfingsten Road
Northbrook, IL 60062-2096

Standard reference number	Title	Referenced in code section number
10A—1998	Tin Clad Fire Doors—with Revisions through July 1998	715.3
10B—1997	Fire Tests of Door Assemblies	715.3.2
10C—1998	Positive Pressure Fire Tests of Door Assemblies—with Revisions through November 2001	715.3.1, 715.3.3
14B—1998	Sliding Hardware for Standard Horizontally Mounted Tin Clad Fire Doors—with Revisions through July 2000	715.3
14C—1996	Swinging Hardware for Standard Tin Clad Fire Doors Mounted Singly and in Pairs	715.3
103—1998	Factory-Built Chimneys for Residential Type and Building Heating Appliances—with Revisions through March 1999	717.2.5
127—1996	Factory-Built Fireplaces—with Revisions through November 1999	717.2.5
268—1996	Smoke Detectors for Fire Protective Signaling Systems—with Revisions through January 1999	407.6, 907.2.6.1
300—1996	Fire Testing of Fire Extinguishing Systems for Protection of Restaurant Cooking Areas —with Revisions through December 1998	904.11
555—99	Fire Dampers—with Revisions through October 2000	716.3
555C—96	Ceiling Dampers	716.3, 716.6.2
555S—99	Smoke Dampers—with Revisions through December 1999	716.3, 716.3.1.1
580—94	Test for Uplift Resistance of Roof Assemblies—with Revisions through February 1998	1504.3.1, 1504.3.2
641—95	Type L Low-Temperature Venting Systems—with Revisions through April 1999	2113.11.1.4
790—97	Tests for Fire Resistance of Roof Covering Materials—with Revisions through July 1998	1505.1, 2603.6, 2610.2, 2610.3
864—96	Control Units for Fire Protective Signaling Systems—with Revisions through March 1999	909.12
1040—96	Fire Test of Insulated Wall Construction—with Revisions through April 2001	1407.10.3, 2603.4, 2603.8
1256—98	Fire Test of Roof Deck Construction—with Revisions through March 2000	1508.1, 2603.3, 2603.4.1.5
1479—94	Fire Tests of Through-Penetration Firestops	712.3.1.2, 712.4.1.2

1715—97	Fire Test of Interior Finish Material 1407.10.2, 1407.10.3, 2603.4, 2603.8
1777—96	Chimney Liners—with Revisions through July 1998 .... 2113.11.1, 2113.19
1784—01	Air Leakage Tests of Door Assemblies 707.14.1, 710.5.2, 715.3.3, 715.3.5.1
1897—98	Uplift Tests for Roof Covering Systems—with Revisions through December 1999 .... 1504.3.1
1975—96	Fire Test of Foamed Plastics Used for Decorative Purposes 402.10, 402.14.5
2079—98	Tests for Fire Resistance of Building Joint Systems... 702.1, 712.3
2200—98	Stationary Engine Generator Assemblies 2702.1.1

ULC

Underwriters Laboratories of Canada
7 Crouse Road
Scarborough, Ontario, Canada M1R3A9

Standard reference number	Title	Referenced in code section number
S102.2—M88	Method of Test for Surface Burning Characteristics of Flooring, Floor Coverings, and Miscellaneous Materials and Assemblies 719.4

USC

United States Code
c/o Superintendent of Documents
U.S. Government Printing Office
Washington, DC 20402-9325

Standard reference number	Title	Referenced in code section number
18 USC Part 1, Ch.40	Importation, Manufacture, Distribution and Storage of Explosive Materials 307.2

WDMA

Window and Door Manufacturers Association
1400 East Touhy Avenue #470
Des Plaines, IL 60018

Standard reference number	Title	Referenced in code section number
AAMA/NWWDA 101/I.S.2—97	Voluntary Specifications for Aluminum, Vinyl (PVC) and Wood Windows and Glass Doors 1714.5.1
AAMA/NWWDA 101/I.S.2/NAFS—02	Voluntary Performance Specification for Window, Skylights and Glass Doors 1714.5.1, 2405.5

WRI

Wire Reinforcement Institute, Inc.
203 Loudon Street, S.W.
2nd Floor, Suite 203C
Leesburg, VA 22075

Standard reference number	Title	Referenced in code section number
WRI/CRSI—96	Design of Slab-on-ground Foundations 1805.8.2

APPENDIX A

EMPLOYEE QUALIFICATIONS

The provisions contained in this appendix are not mandatory unless specifically referenced in the adopting ordinance.

SECTION A101
BUILDING OFFICIAL QUALIFICATIONS

A101.1 Building official. The building official shall have at least 10 years' experience or equivalent as an architect, engineer, inspector, contractor or superintendent of construction, or any combination of these, five years of which shall have been supervisory experience. The building official should be certified as a building official through a recognized certification program. The building official shall be appointed or hired by the applicable governing authority.

A101.2 Chief inspector. The building official can designate supervisors to administer the provisions of the *International Building*, *Mechanical* and *Plumbing Codes*, *International Fuel Gas Code*, and the ICC *Electrical Code*. Each supervisor shall have at least 10 years' experience or equivalent as an architect, engineer, inspector, contractor or superintendent of construction, or any combination of these, five years of which shall have been in a supervisory capacity. They shall be certified through a recognized certification program for the appropriate trade.

A101.3 Inspector and plan examiner. The building official shall appoint or hire such number of officers, inspectors, assistants and other employees as shall be authorized by the jurisdiction. A person shall not be appointed or hired as inspector of construction or plan examiner who has not had at least 5 years' experience as a contractor, engineer, architect, or as a superintendent, foreman or competent mechanic in charge of construction. The inspector or plan examiner shall be certified through a recognized certification program for the appropriate trade.

A101.4 Termination of employment. Employees in the position of building official, chief inspector or inspector shall not be removed from office except for cause after full opportunity has been given to be heard on specific charges before such applicable governing authority.

SECTION A102
REFERENCED STANDARDS

IBC-03	*International Building Code*–A101.2
IMC-03	*International Mechanical Code*–A101.2
IPC-03	*International Plumbing Code*–A101.2
IFGC-03	*International Fuel Gas Code*–A101.2
ICC EC-03	ICC *Electrical Code*–A101.2

APPENDIX B

BOARD OF APPEALS

The provisions contained in this appendix are not mandatory unless specifically referenced in the adopting ordinance.

SECTION B101
GENERAL

B101.1 Application. The application for appeal shall be filed on a form obtained from the building official within 20 days after the notice was served.

B101.2 Membership of board. The board of appeals shall consist of persons appointed by the chief appointing authority as follows:

1. One for five years; one for four years; one for three years; one for two years; and one for one year.

2. Thereafter, each new member shall serve for five years or until a successor has been appointed.

The building official shall be an ex officio member of said board but shall have no vote on any matter before the board.

B101.2.1 Alternate members. The chief appointing authority shall appoint two alternate members who shall be called by the board chairperson to hear appeals during the absence or disqualification of a member. Alternate members shall possess the qualifications required for board membership and shall be appointed for five years, or until a successor has been appointed.

B101.2.2 Qualifications. The board of appeals shall consist of five individuals, one from each of the following professions or disciplines:

1. Registered design professional with architectural experience or a builder or superintendent of building construction with at least ten years' experience, five of which shall have been in responsible charge of work.

2. Registered design professional with structural engineering experience

3. Registered design professional with mechanical and plumbing engineering experience or a mechanical contractor with at least ten years' experience, five of which shall have been in responsible charge of work.

4. Registered design professional with electrical engineering experience or an electrical contractor with at least ten years' experience, five of which shall have been in responsible charge of work.

5. Registered design professional with fire protection engineering experience or a fire protection contractor with at least ten years' experience, five of which shall have been in responsible charge of work.

B101.2.3 Rules and procedures. The board is authorized to establish policies and procedures necessary to carry out its duties.

B101.2.4 Chairperson. The board shall annually select one of its members to serve as chairperson.

B101.2.5 Disqualification of member. A member shall not hear an appeal in which that member has a personal, professional or financial interest.

B101.2.6 Secretary. The chief administrative officer shall designate a qualified clerk to serve as secretary to the board. The secretary shall file a detailed record of all proceedings in the office of the chief administrative officer.

B101.2.7 Compensation of members. Compensation of members shall be determined by law.

B101.3 Notice of meeting. The board shall meet upon notice from the chairperson, within 10 days of the filing of an appeal or at stated periodic meetings.

B101.3.1 Open hearing. All hearings before the board shall be open to the public. The appellant, the appellant's representative, the building official and any person whose interests are affected shall be given an opportunity to be heard.

B101.3.2 Procedure. The board shall adopt and make available to the public through the secretary procedures under which a hearing will be conducted. The procedures shall not require compliance with strict rules of evidence, but shall mandate that only relevant information be received.

B101.3.3 Postponed hearing. When five members are not present to hear an appeal, either the appellant or the appellant's representative shall have the right to request a postponement of the hearing.

B101.4 Board decision. The board shall modify or reverse the decision of the building official by a concurring vote of two-thirds of its members.

B101.4.1 Resolution. The decision of the board shall be by resolution. Certified copies shall be furnished to the appellant and to the building official.

B101.4.2 Administration. The building official shall take immediate action in accordance with the decision of the board.

APPENDIX C

GROUP U - AGRICULTURAL BUILDINGS

The provisions contained in this appendix are not mandatory unless specifically referenced in the adopting ordinance.

SECTION C101
GENERAL

C101.1 Scope. The provisions of this appendix shall apply exclusively to agricultural buildings. Such buildings shall be classified as Group U and shall include the following uses:

1. Livestock shelters or buildings, including shade structures and milking barns.

2. Poultry buildings or shelters.

3. Barns.

4. Storage of equipment and machinery used exclusively in agriculture.

5. Horticultural structures, including detached production greenhouses and crop protection shelters.

6. Sheds.

7. Grain silos.

8. Stables.

SECTION C102
ALLOWABLE HEIGHT AND AREA

C102.1 General. Buildings classified as Group U Agricultural shall not exceed the area or height limits specified in Table C102.1.

C102.2 One-story unlimited area. The area of a one-story Group U agricultural building shall not be limited if the building is surrounded and adjoined by public ways or yards not less than 60 feet (18 288 mm) in width.

C102.3 Two-story unlimited area. The area of a two-story Group U agricultural building shall not be limited if the building is surrounded and adjoined by public ways or yards not less than 60 feet (18 288 mm) in width and is provided with an approved automatic sprinkler system throughout in accordance with Section 903.3.1.1.

SECTION C103
MIXED OCCUPANCIES

C103.1 Mixed occupancies. Mixed occupancies shall be protected in accordance with Chapter 3.

SECTION C104
EXITS

C104.1 Exit facilities. Exits shall be provided in accordance with Chapters 10 and 11.

Exceptions:

1. The maximum travel distance from any point in the building to an approved exit shall not exceed 300 feet (91 440 mm).

2. One exit is required for each 15,000 square feet (1393.5 m^2) of area or fraction thereof.

TABLE C102.1—BASIC ALLOWABLE AREA FOR A GROUP U, ONE STORY IN HEIGHT AND MAXIMUM HEIGHT OF SUCH OCCUPANCY

I		II		III and IV		V	
A	B	A	B	III A and IV	III B	A	B
ALLOWABLE AREA (square feet)[a]							
Unlimited	60,000	27,100	18,000	27,100	18,000	21,100	12,000
MAXIMUM HEIGHT IN STORIES							
Unlimited	12	4	2	4	2	3	2
MAXIMUM HEIGHT IN FEET							
Unlimited	160	65	55	65	55	50	40

For SI: 1 square foot = 0.0929 m^2.

a. See Section C102 for unlimited area under certain conditions.

APPENDIX D

FIRE DISTRICTS

The provisions contained in this appendix are not mandatory unless specifically referenced in the adopting ordinance.

SECTION D101
GENERAL

D101.1 Scope. The fire district shall include such territory or portion as outlined in an ordinance or law entitled "An Ordinance (Resolution) Creating and Establishing a Fire District." Wherever, in such ordinance creating and establishing a fire district, reference is made to the Fire District, it shall be construed to mean the fire district designated and referred to in this appendix.

D101.1.1 Mapping. The fire district complying with the provisions of Section D101.1 shall be shown on a map that shall be available to the public.

D101.2 Establishment of area. For the purpose of this code, the fire district shall include that territory or area as described in Sections D101.2.1 through D101.2.3.

D101.2.1 Adjoining blocks. Two or more adjoining blocks, exclusive of intervening streets, where at least 50 percent of the ground area is built upon and more than 50 percent of the built-on area is devoted to hotels and motels of Group R-1; Group B occupancies; theaters, nightclubs, restaurants of Group A-1 and A-2 occupancies; garages, express and freight depots, warehouses and storage buildings used for the storage of finished products (not located with and forming a part of a manufactured or industrial plant); or Group S occupancy. Where the average height of a building is two and one-half stories or more, a block should be considered if the ground area built upon is at least 40 percent.

D101.2.2 Buffer zone. Where four contiguous blocks or more comprise a fire district, there shall be a buffer zone of 200 feet (60 960 mm) around the perimeter of such district. Streets, rights-of-way and other open spaces not subject to building construction can be included in the 200-foot (60 960 mm) buffer zone.

D101.2.3 Developed blocks. Where blocks adjacent to the fire district have developed to the extent that at least 25 percent of the ground area is built upon and 40 percent or more of the built-on area is devoted to the occupancies specified in Section D101.2.1, they can be considered for inclusion in the fire district, and can form all or a portion of the 200-foot (60 960 mm) buffer zone required in Section D101.2.2.

SECTION D102
BUILDING RESTRICTIONS

D102.1 Types of construction permitted. Within the fire district every building hereafter erected shall be either Type I, II, III or IV, except as permitted in Section D104.

D102.2 Other specific requirements.

D102.2.1 Exterior walls. Exterior walls of buildings located in the fire district shall comply with the requirements in Table 601 except as required in Section D102.2.6.

D102.2.2 Group H prohibited. Group H occupancies shall be prohibited from location within the fire district.

D102.2.3 Construction type. Every building shall be constructed as required based on the type of construction indicated in Chapter 6.

D102.2.4 Roof covering. Roof covering in the fire district shall conform to the requirements of Class A or B roof coverings as defined in Section 1505.

D102.2.5 Structural fire rating. Walls, floors, roofs and their supporting structural members shall be a minimum of 1-hour fire-resistance-rated construction.

Exceptions:

1. Buildings of Type IV construction.
2. Buildings equipped throughout with an automatic sprinkler system in accordance with Section 903.3.1.1.
3. Automobile parking structures.
4. Buildings surrounded on all sides by a permanently open space of not less than 30 feet (9144 mm).
5. Partitions complying with Section 603.1(8).

D102.2.6 Exterior walls. Exterior load-bearing walls of Type II buildings shall have a fire-resistance rating of 2 hours or more where such walls are located within 30 feet (9144 mm) of a common property line or an assumed property line. Exterior nonload-bearing walls of Type II buildings located within 30 feet (9144 mm) of a common property line or an assumed property line shall have fire-resistance ratings as required by Table 601, but not less than 1 hour. Exterior walls located more than 30 feet (9144 mm) from a common property line or an assumed property line shall comply with Table 601.

Exception: In the case of one-story buildings that are 2,000 square feet (186 m²) or less in area, exterior walls located more than 15 feet (4572 mm) from a common property line or an assumed property line need only comply with Table 601.

D102.2.7 Architectural trim. Architectural trim on buildings located in the fire district shall be constructed of approved noncombustible materials or fire-retardant-treated wood.

D102.2.8 Permanent canopies. Permanent canopies are permitted to extend over adjacent open spaces provided:

1. The canopy and its supports shall be of noncombustible material, fire-retardant-treated wood, Type IV construction or of 1-hour fire-resistance-rated construction.

   Exception: Any textile covering for the canopy shall be flame resistant as determined by tests conducted in accordance with NFPA 701 after both accelerated water leaching and accelerating weathering.

2. Any canopy covering, other than textiles, shall have a flame spread index not greater than 25 when tested in accordance with ASTM E 84 in the form intended for use.

3. The canopy shall have at least one long side open.

4. The maximum horizontal width of the canopy shall not exceed 15 feet (4572 mm).

5. The fire resistance of exterior walls shall not be reduced.

D102.2.9 Roof structures. Structures, except aerial supports 12 feet (3658 mm) high or less, flagpoles, water tanks and cooling towers, placed above the roof of any building within the fire district shall be of noncombustible material and shall be supported by construction of noncombustible material.

D102.2.10 Plastic signs. The use of plastics complying with Section 2611 for signs is permitted provided the structure of the sign in which the plastic is mounted or installed is noncombustible.

D102.2.11 Plastic veneer. Exterior plastic veneer is not permitted in the fire district.

SECTION D103
CHANGES TO BUILDINGS

D103.1 Existing buildings within the fire district. An existing building shall not hereafter be increased in height or area unless it is of a type of construction permitted for new buildings within the fire district or is altered to comply with the requirements for such type of construction. Nor shall any existing building be hereafter extended on any side, nor square footage or floors added within the existing building unless such modifications are of a type of construction permitted for new buildings within the fire district.

D103.2 Other alterations. Nothing in Section D103.1 shall prohibit other alterations within the fire district provided there is no change of occupancy that is otherwise prohibited and the fire hazard is not increased by such alteration.

D103.3 Moving buildings. Buildings shall not hereafter be moved into the fire district or to another lot in the fire district unless the building is of a type of construction permitted in the fire district.

SECTION D104
BUILDINGS LOCATED PARTIALLY
IN THE FIRE DISTRICT

D104.1 General. Any building located partially in the fire district shall be of a type of construction required for the fire district, unless the major portion of such building lies outside of the fire district and no part is more than 10 feet (3048 mm) inside the boundaries of the fire district.

SECTION D105
EXCEPTIONS TO RESTRICTIONS
IN FIRE DISTRICT

D105.1 General. The preceding provisions of this appendix shall not apply in the following instances:

1. Temporary buildings used in connection with duly authorized construction.

2. A private garage used exclusively as such, not more than one story in height, nor more than 650 square feet (60 m²) in area, located on the same lot with a dwelling.

3. Fences not over 8 feet (2438 mm) high.

4. Coal tipples, material bins and trestles of Type IV construction.

5. Water tanks and cooling towers conforming to Sections 1509.3 and 1509.4.

6. Greenhouses less than 15 feet (4572 mm) high.

7. Porches on dwellings not over one story in height, and not over 10 feet (3048 mm) wide from the face of the building, provided such porch does not come within 5 feet (1524 mm) of any property line.

8. Sheds open on a long side not over 15 feet (4572 mm) high and 500 square feet (46 m²) in area.

9. One- and two-family dwellings where of a type of construction not permitted in the fire district can be extended 25 percent of the floor area existing at the time of inclusion in the fire district by any type of construction permitted by this code.

10. Wood decks less than 600 square feet (56 m²) where constructed of 2-inch (51 mm) nominal wood, pressure treated for exterior use.

11. Wood veneers on exterior walls conforming to Section 1405.4.

12. Exterior plastic veneer complying with Section 2605.2 where installed on exterior walls required to have a fire-resistance rating not less than 1 hour, provided the exterior plastic veneer does not exhibit sustained flaming as defined in NFPA 268.

SECTION D106
REFERENCED STANDARDS

ASTM E 84-01   Test Method for Surface   D102.2.8
Burning Characteristics
of Building Materials

NFPA 268-96   Test Method for   D105.1
Determining Ignitability of
Exterior Wall Assemblies
Using a Radiant Heat Energy
Source

NFPA 701-99   Methods of   D102.2.8
Fire Tests for Flame-
Propagation of Textiles
and Films

APPENDIX E

SUPPLEMENTARY ACCESSIBILITY REQUIREMENTS

The provisions contained in this appendix are not mandatory unless specifically referenced in the adopting ordinance.

SECTION E101
GENERAL

E101.1 Scope. The provisions of this appendix shall control the supplementary requirements for the design and construction of facilities for accessibility to physically disabled persons.

E101.2 Design. Technical requirements for items herein shall comply with this code and ICC A117.1.

SECTION E102
DEFINITIONS

E102.1 General. The following words and terms shall, for the purposes of this appendix, have the meanings shown herein.

CLOSED-CIRCUIT TELEPHONE. A telephone with a dedicated line such as a house phone, courtesy phone or phone that must be used to gain entrance to a facility.

MAILBOXES. Receptacles for the receipt of documents, packages or other deliverable matter. Mailboxes include, but are not limited to, post office boxes and receptacles provided by commercial mail-receiving agencies, apartment houses and schools.

TRANSIENT LODGING. A building, facility or portion thereof, excluding inpatient medical care facilities and long-term care facilities, that contains one or more dwelling units or sleeping units. Examples of transient lodging include, but are not limited to, resorts, group homes, hotels, motels, dormitories, homeless shelters, halfway houses and social service lodging.

SECTION E103
ACCESSIBLE ROUTE

E103.1 Raised platforms. In banquet rooms or spaces where a head table or speaker's lectern is located on a raised platform, an accessible route shall be provided to the platform.

SECTION E104
SPECIAL OCCUPANCIES

E104.1 General. Transient lodging facilities shall be provided with accessible features in accordance with Sections E104.2 and E104.3. Group I-3 occupancies shall be provided with accessible features in accordance with Sections E104.3 and E104.4.

E104.2 Accessible beds. In rooms or spaces having more than 25 beds, five percent of the beds shall have a clear floor space complying with ICC A117.1.

E104.2.1 Sleeping areas. A clear floor space complying with ICC A117.1 shall be provided on both sides of the accessible bed. The clear floor space shall be positioned for parallel approach to the side of the bed.

Exception: This requirement shall not apply where a single clear floor space complying with ICC A117.1 positioned for parallel approach is provided between two beds.

E104.3 Communication features. Communication features complying with ICC A117.1 shall be provided in accordance with Sections E104.3.1 through E104.3.4.

E104.3.1 Transient lodging. In transient lodging facilities, sleeping units with accessible communication features shall be provided in accordance with Table E104.3.1. Units required to comply with Table E104.3.1 shall be dispersed among the various classes of units.

E104.3.2 Group I-3. In Group I-3 occupancies at least 2 percent, but no fewer than one of the total number of general holding cells and general housing cells equipped with audible emergency alarm systems and permanently installed telephones within the cell, shall comply with Section E104.3.3.

E104.3.3 Dwelling units and sleeping units. Where dwelling units and sleeping units are altered or added, the requirements of Section E104.3 shall apply only to the units being altered or added until the number of units with accessible communication features complies with the minimum number required for new construction.

E104.3.4 Notification devices. Visual notification devices shall be provided to alert room occupants of incoming telephone calls and a door knock or bell. Notification devices shall not be connected to visual alarm signal appliances. Permanently installed telephones shall have volume controls and an electrical outlet complying with ICC A117.1 located within 48 inches (1219 mm) of the telephone to facilitate the use of a TTY.

E104.4 Partitions. Solid partitions or security glazing that separates visitors from detainees in Group I-3 occupancies shall provide a method to facilitate voice communication. Such methods are permitted to include, but are not limited to, grilles, slats, talk-through baffles, intercoms or telephone handset devices. The method of communication shall be accessible to individuals who use wheelchairs and individuals who have difficulty bending or stooping. Hand-operable communication devices, if provided, shall comply with Section E106.3.

TABLE E104.3.1
DWELLING OR SLEEPING UNITS WITH ACCESSIBLE COMMUNICATION FEATURES

TOTAL NUMBER OF DWELLING OR SLEEPING UNITS PROVIDED	MINIMUM REQUIRED NUMBER OF DWELLING OR SLEEPING UNITS WITH ACCESSIBLE COMMUNICATION FEATURES
1	1
2 to 25	2
26 to 50	4
51 to 75	7
76 to 100	9
101 to 150	12
151 to 200	14
201 to 300	17
301 to 400	20
401 to 500	22
501 to 1,000	5% of total
1,001 and over	50 plus 3 for each 100 over 1,000

SECTION E105
OTHER FEATURES AND FACILITIES

E105.1 Water coolers. Where water coolers are provided, at least 50 percent, but not less than one, of such units provided on each floor shall comply with ICC A117.1.

E105.2 Portable toilets and bathing rooms. Where multiple single-user portable toilet or bathing units are clustered at a single location, at least 5 percent, but not less than one toilet unit or bathing unit at each cluster, shall comply with ICC A117.1. Signs containing the International Symbol of Accessibility and complying with ICC A117.1 shall identify accessible portable toilets and bathing units.

> **Exception:** Portable toilet units provided for use exclusively by construction personnel on a construction site.

E105.3 Laundry equipment. Where provided in spaces required to be accessible, washing machines and clothes dryers shall comply with this section.

> **E105.3.1 Washing machines.** Where three or fewer washing machines are provided, at least one shall comply with ICC A117.1. Where more than three washing machines are provided, at least two shall comply with ICC A117.1.

> **E105.3.2 Clothes dryers.** Where three or fewer clothes dryers are provided, at least one shall comply with ICC A117.1. Where more than three clothes dryers are provided, at least two shall comply with ICC A117.1.

E105.4 Depositories, vending machines, change machines and similar equipment. Where provided, at least one of each type of depository, vending machine, change machine and similar equipment shall comply with ICC A117.1.

> **Exception:** Drive-up-only depositories are not required to comply with this section.

E105.5 Mailboxes. Where mailboxes are provided in an interior location, at least 5 percent, but not less than one, of each type shall comply with ICC A117.1. In residential and institutional facilities, where mailboxes are provided for each dwelling unit or sleeping unit, mailboxes complying with ICC A117.1 shall be provided for each unit required to be an Accessible unit.

E105.6 Automatic teller machines and fare machines. Where automatic teller machines or self-service fare vending, collection or adjustment machines are provided, at least one machine of each type at each location where such machines are provided shall be accessible. Where bins are provided for envelopes, wastepaper or other purposes, at least one of each type shall be accessible.

E105.7 Two-way communication systems. Where two-way communication systems are provided to gain admittance to a building or facility or to restricted areas within a building or facility, the system shall comply with ICC A117.1.

SECTION E106
TELEPHONES

E106.1 General. Where coin-operated public pay telephones, coinless public pay telephones, public closed-circuit telephones, courtesy phones or other types of public telephones are provided, accessible public telephones shall be provided in accordance with Sections E106.2 through E106.5 for each type of public telephone provided. For purposes of this section, a bank of telephones shall be considered two or more adjacent telephones.

E106.2 Wheelchair-accessible telephones. Where public telephones are provided, wheelchair-accessible telephones complying with ICC A117.1 shall be provided in accordance with Table E106.2.

E106.3 Volume controls. All public telephones provided shall have volume control complying with ICC A117.1.

E106.4 TTYs. TTYs complying with ICC A117.1 shall be provided in accordance with Sections E106.4.1 through E106.4.9.

TABLE E106.2
WHEELCHAIR-ACCESSIBLE TELEPHONES

NUMBER OF TELEPHONES PROVIDED ON A FLOOR, LEVEL OR EXTERIOR SITE	MINIMUM REQUIRED NUMBER OF WHEELCHAIR-ACCESSIBLE TELEPHONES
1 or more single unit	1 per floor, level and exterior site
1 bank	1 per floor, level and exterior site
2 or more banks	1 per bank

E106.4.1 Bank requirement. Where four or more public pay telephones are provided at a bank of telephones, at least one public TTY shall be provided at that bank.

> **Exception:** TTYs are not required at banks of telephones located within 200 feet (60 960 mm) of, and on the same floor as, a bank containing a public TTY.

E106.4.2 Floor requirement. Where four or more public pay telephones are provided on a floor of a privately owned building, at least one public TTY shall be provided on that floor. Where at least one public pay telephone is provided on a floor of a publicly owned building, at least one public TTY shall be provided on that floor.

E106.4.3 Building requirement. Where four or more public pay telephones are provided in a privately owned building, at least one public TTY shall be provided in the building. Where at least one public pay telephone is provided in a publicly owned building, at least one public TTY shall be provided in the building.

E106.4.4 Site requirement. Where four or more public pay telephones are provided on a site, at least one public TTY shall be provided on the site.

E106.4.5 Rest stops, emergency road stops, and service plazas. Where a public pay telephone is provided at a public rest stop, emergency road stop or service plaza, at least one public TTY shall be provided.

E106.4.6 Hospitals. Where a public pay telephone is provided in or adjacent to a hospital emergency room, hospital recovery room or hospital waiting room, at least one public TTY shall be provided at each such location.

E106.4.7 Transportation facilities. Transportation facilities shall be provided with TTYs in accordance with Sections E109.2.5 and E110.2 in addition to the TTYs required by Sections E106.4.1 through E106.4.4.

E106.4.8 Detention and correctional facilities. In detention and correctional facilities, where a public pay telephone is provided in a secured area used only by detainees or inmates and security personnel, then at least one TTY shall be provided in at least one secured area.

E106.4.9 Signs. Public TTYs shall be identified by the International Symbol of TTY complying with ICC A117.1. Directional signs indicating the location of the nearest public TTY shall be provided at banks of public pay telephones not containing a public TTY. Additionally, where signs provide direction to public pay telephones, they shall also provide direction to public TTYs. Such signs shall comply with ICC A117.1 and shall include the International Symbol of TTY.

E106.5 Shelves for portable TTYs. Where a bank of telephones in the interior of a building consists of three or more public pay telephones, at least one public pay telephone at the bank shall be provided with a shelf and an electrical outlet in accordance with ICC A117.1.

> **Exceptions:**
>
> 1. In secured areas of detention and correctional facilities, if shelves and outlets are prohibited for purposes of security or safety shelves and outlets for TTYs are not required to be provided.
>
> 2. The shelf and electrical outlet shall not be required at a bank of telephones with a TTY.

SECTION E107
SIGNAGE

E107.1 Signs. Required accessible portable toilets and bathing facilities shall be identified by the International Symbol of Accessibility.

E107.2 Designations. Interior and exterior signs identifying permanent rooms and spaces shall be tactile. Where pictograms are provided as designations of interior rooms and spaces, the pictograms shall have tactile text descriptors. Signs required to provide tactile characters and pictograms shall comply with ICC A117.1.

> **Exceptions:**
>
> 1. Exterior signs that are not located at the door to the space they serve are not required to comply.
>
> 2. Building directories, menus, seat and row designations in assembly areas, occupant names, building addresses and company names and logos are not required to comply.
>
> 3. Signs in parking facilities are not required to comply.
>
> 4. Temporary (seven days or less) signs are not required to comply.

E107.3 Directional and informational signs. Signs that provide direction to, or information about, permanent interior spaces of the site and facilities shall contain visual characters complying with ICC A117.1.

> **Exception:** Building directories, personnel names, company or occupant names and logos, menus and temporary (seven days or less) signs are not required to comply with ICC A117.1.

E107.4 Other signs. Signage indicating special accessibility provisions shall be provided as follows:

1. At bus stops and terminals, signage must be provided in accordance with Section E108.4.

2. At fixed facilities and stations, signage must be provided in accordance with Sections E109.2.2 through E109.2.2.3.

3. At airports, terminal information systems must be provided in accordance with Section E110.3.

SECTION E108
BUS STOPS

E108.1 General. Bus stops shall comply with Sections E108.2 through E108.5.

E108.2 Bus boarding and alighting areas. Bus boarding and alighting areas shall comply with Sections E108.2.1 through E108.2.4.

E108.2.1 Surface. Bus boarding and alighting areas shall have a firm, stable surface.

E108.2.2 Dimensions. Bus boarding and alighting areas shall have a clear length of 96 inches (2440 mm) minimum, measured perpendicular to the curb or vehicle roadway edge, and a clear width of 60 inches (1525 mm) minimum, measured parallel to the vehicle roadway.

E108.2.3 Connection. Bus boarding and alighting areas shall be connected to streets, sidewalks or pedestrian paths by an accessible route complying with Section 104.

E108.2.4 Slope. Parallel to the roadway, the slope of the bus boarding and alighting area shall be the same as the roadway, to the maximum extent practicable. For water drainage, a maximum slope of 1:48 perpendicular to the roadway is allowed.

E108.3 Bus shelters. Where provided, new or replaced bus shelters shall provide a minimum clear floor or ground space complying with ICC A117.1, Section 305, entirely within the shelter. Such shelters shall be connected by an accessible route to the boarding area required by Section E108.2.

E108.4 Signs. New bus route identification signs shall have finish and contrast complying with ICC A117.1. Additionally, to the maximum extent practicable, new bus route identification signs shall provide visual characters complying with ICC A117.1.

Exception: Bus schedules, timetables and maps that are posted at the bus stop or bus bay are not required to meet this requirement.

E108.5 Bus stop siting. Bus stop sites shall be chosen such that, to the maximum extent practicable, the areas where lifts or ramps are to be deployed comply with Sections E108.2 and E108.3.

SECTION E109
TRANSPORTATION FACILITIES AND STATIONS

E109.1 General. Fixed transportation facilities and stations shall comply with the applicable provisions of Sections E109.2 and E109.3.

E109.2 New construction. New stations in rapid rail, light rail, commuter rail, intercity rail, high speed rail and other fixed guideway systems shall comply with Sections E109.2.1 through E109.2.8.

E109.2.1 Station entrances. Where different entrances to a station serve different transportation fixed routes or groups of fixed routes, at least one entrance serving each group or route shall comply with Section 1104 and ICC A117.1.

E109.2.2 Signs. Signage in fixed transportation facilities and stations shall comply with Sections E109.2.2.1 through E109.2.2.3.

E109.2.2.1 Tactile signs. Where signs are provided at entrances to stations identifying the station or the entrance, or both, at least one sign at each entrance shall be tactile. A minimum of one tactile sign identifying the specific station shall be provided on each platform or boarding area. Such signs shall be placed in uniform locations at entrances and on platforms or boarding areas within the transit system to the maximum extent practicable. Tactile signs shall comply with ICC A117.1.

Exceptions:

1. Where the station has no defined entrance but signs are provided, the tactile signs shall be placed in a central location.

2. Signs are not required to be tactile where audible signs are remotely transmitted to hand-held receivers, or are user or proximity actuated.

E109.2.2.2 Identification signs. Stations covered by this section shall have identification signs containing visual characters complying with ICC A117.1. Signs shall be clearly visible and within the sightlines of a standing or sitting passenger from within the train on both sides when not obstructed by another train.

E109.2.2.3 Informational signs. Lists of stations, routes and destinations served by the station which are located on boarding areas, platforms or mezzanines shall provide visual characters complying with ICC A117.1 Signs covered by this provision shall, to the maximum extent practicable, be placed in uniform locations within the transit system.

E109.2.3 Fare machines. Self-service fare vending, collection and adjustment machines shall comply with ICC A117.1, Section 707. Where self-service fare vending, collection or adjustment machines are provided for the use of the general public, at least one accessible machine of each type provided shall be provided at each accessible point of entry and exit.

E109.2.4 Rail-to-platform height. Station platforms shall be positioned to coordinate with vehicles in accordance with the applicable provisions of 36 CFR, Part 1192. Low-level platforms shall be 8 inches (250 mm) minimum above top of rail.

Exception: Where vehicles are boarded from sidewalks or street level, low-level platforms shall be permitted to be less than 8 inches (250 mm).

E109.2.5 TTYs. Where a public pay telephone is provided in a transit facility (as defined by the Department of Transportation) at least one public TTY complying with ICC A117.1, Section 704.4, shall be provided in the station. In addition, where one or more public pay telephones serve a particular entrance to a transportation facility, at least one TTY telephone complying with ICC A117.1, Section 704.4, shall be provided to serve that entrance.

E109.2.6 Track crossings. Where a circulation path serving boarding platforms crosses tracks, an accessible route complying with ICC A117.1 shall be provided.

> **Exception:** Openings for wheel flanges shall be permitted to be $2^1/_2$ inches (64 mm) maximum.

E109.2.7 Public address systems. Where public address systems convey audible information to the public, the same or equivalent information shall be provided in a visual format.

E109.2.8 Clocks. Where clocks are provided for use by the general public, the clock face shall be uncluttered so that its elements are clearly visible. Hands, numerals and digits shall contrast with the background either light-on-dark or dark-on-light. Where clocks are mounted overhead, numerals and digits shall comply with ICC A117.1, Section 703.4.

E109.3 Existing facilities: key stations. In rapid rail, light rail, commuter rail, intercity rail, high-speed rail and other fixed guideway systems, altered stations and intercity rail and key stations, as defined under criteria established by the Department of Transportation in Subpart C of 49 CFR Part 37, shall comply with Sections E109.3.1 through E109.3.3.

E109.3.1 Accessible route. At least one accessible route from an accessible entrance to those areas necessary for use of the transportation system shall be provided. The accessible route shall include the features specified in Section E109.2, except that escalators shall comply with Section 3005.2.2. Where technical infeasibility in existing stations requires the accessible route to lead from the public way to a paid area of the transit system, an accessible fare collection machine complying with Section E109.2.3 shall be provided along such accessible route.

E109.3.2 Platform and vehicle floor coordination. Station platforms shall be positioned to coordinate with vehicles in accordance with the applicable provisions of 36 CFR Part 1192. Low-level platforms shall be 8 inches (250 mm) minimum above top of rail.

> **Exception:** Where vehicles are boarded from sidewalks or street level, low-level platforms shall be permitted to be less than 8 inches (250 mm).

E109.3.3 Direct connections. New direct connections to other facilities shall have an accessible route complying with Section 3409.6 from the point of connection to boarding platforms and transportation system elements used by the public. Any elements provided to facilitate future direct connections shall be on an accessible route connecting boarding platforms and transportation system elements used by the public.

SECTION E110
AIRPORTS

E110.1 New construction. New construction of airports shall comply with Sections E110.2 through E110.4.

E110.2 TTYs. Where public pay telephones are provided, at least one TTY shall be provided in compliance with ICC A117.1, Section 704.4. Additionally, if four or more public pay telephones are located in a main terminal outside the security areas, a concourse within the security areas or a baggage claim area in a terminal, at least one public TTY complying with ICC A117.1, Section 704.4, shall also be provided in each such location.

E110.3 Terminal information systems. Where terminal information systems convey audible information to the public, the same or equivalent information shall be provided in a visual format.

E110.4 Clocks. Where clocks are provided for use by the general public, the clock face shall be uncluttered so that its elements are clearly visible. Hands, numerals and digits shall contrast with their background either light-on-dark or dark-on-light. Where clocks are mounted overhead, numerals and digits shall comply with ICC A117.1, Section 703.4.

SECTION E111
QUALIFIED HISTORIC BUILDINGS AND FACILITIES

E111.1 General. Qualified historic buildings and facilities shall comply with Sections E111.2 through E111.5.

E111.2 Qualified historic buildings and facilities. These procedures shall apply to buildings and facilities designated as historic structures that undergo alterations or a change of occupancy.

E111.3 Qualified historic buildings and facilities subject to Section 106 of the National Historic Preservation Act. Where an alteration or change of occupancy is undertaken to a qualified historic building or facility that is subject to Section 106 of the National Historic Preservation Act, the federal agency with jurisdiction over the undertaking shall follow the Section 106 process. Where the State Historic Preservation Officer or Advisory Council on Historic Preservation determines that compliance with the requirements for accessible routes, ramps, entrances or toilet facilities would threaten or destroy the historic significance of the building or facility, the alternative requirements of Section 3409 for that element are permitted.

E111.4 Qualified historic buildings and facilities not subject to Section 106 of the National Historic Preservation Act. Where an alteration or change of occupancy is undertaken to a qualified historic building or facility that is not subject to Section 106 of the National Historic Preservation Act, and the entity undertaking the alterations believes that compliance with the requirements for accessible routes, ramps, entrances or toilet facilities would threaten or destroy the historic significance of the building or facility, the entity shall consult with the State Historic Preservation Officer. Where the State Historic Preservation Officer determines that compliance with the accessibility requirements for accessible routes, ramps, entrances or toilet facilities would threaten or destroy the historical significance of the building or facility, the alternative requirements of Section 3409 for that element are permitted.

E111.4.1 Consultation with interested persons. Interested persons shall be invited to participate in the consultation process, including state or local accessibility officials, individuals with disabilities and organizations representing individuals with disabilities.

E111.4.2 Certified local government historic preservation programs. Where the State Historic Preservation Officer has delegated the consultation responsibility for purposes of this section to a local government historic preservation program that has been certified in accordance with Section 101 of the National Historic Preservation Act of 1966 [(16 U.S.C. 470a(c)] and implementing regulations (36 CFR 61.5), the responsibility shall be permitted to be carried out by the appropriate local government body or official.

E111.5 Displays. In qualified historic buildings and facilities, where alternative requirements of Section 3409 are permitted, displays and written information shall be located where they can be seen by a seated person. Exhibits and signs displayed horizontally shall be 44 inches (1120 mm) maximum above the floor.

SECTION E112
REFERENCED STANDARDS

DOJ 36 CFR Part 1192	Americans with Disabilities Act (ADA) Accessibility Guidelines for Transportation Vehicles (ADAAG). Washington, D.C.: Department of Justice, 1991	E109.2.4, E109.3.2
DOT 49 CFR Part 37	Transportation Services for Individuals with Disabilities (ADA), Washington, D.C.: Department of Transportation, 1999	E109.3, E109.3.2, E109.4
DOJ 28	CFR Part 36, Americans with Disabilities Act (ADA). Washington, D.C.: Department of Justice, 1991	E109.4
ICC/ANSI A117.1-98	Accessible and Usable Buildings and Facilities	E101.2, et al
16 USC Sec. 470	National Historic Preservation Act	E111.2, E111.3, E111.3.2

APPENDIX F

RODENT PROOFING

The provisions contained in this appendix are not mandatory unless specifically referenced in the adopting ordinance.

SECTION F101
GENERAL

F101.1 General. Buildings or structures and the walls enclosing habitable or occupiable rooms and spaces in which persons live, sleep or work, or in which feed, food or foodstuffs are stored, prepared, processed, served or sold, shall be constructed in accordance with the provisions of this section.

F101.2 Foundation wall ventilation openings. Foundation wall ventilator openings shall be covered for their height and width with perforated sheet metal plates no less than 0.070 inch (1.8 mm) thick, expanded sheet metal plates not less than 0.047 inch (1.2 mm) thick, cast iron grills or grating, extruded aluminum load-bearing vents or with hardware cloth of 0.035 inch (0.89 mm) wire or heavier. The openings therein shall not exceed $^1/_4$ inch (6.4 mm).

F101.3 Foundation and exterior wall sealing. Annular spaces around pipes, electric cables, conduits, or other openings in the walls shall be protected against the passage of rodents by closing such openings with cement mortar, concrete masonry or noncorrosive metal.

F101.4 Doors. Doors on which metal protection has been applied shall be hinged so as to be free swinging. When closed, the maximum clearance between any door, door jambs and sills shall not be greater than $^3/_8$ inch (9.5 mm).

F101.5 Windows and other openings. Windows and other openings for the purpose of light or ventilation located in exterior walls within 2 feet (610 mm) above the existing ground level immediately below such opening shall be covered for their entire height and width, including frame, with hardware cloth of at least 0.035 inch (0.89 mm) wire or heavier.

F101.5.1 Rodent-accessible openings. Windows and other openings for the purpose of light and ventilation in the exterior walls not covered in this chapter, accessible to rodents by way of exposed pipes, wires, conduits and other appurtenances, shall be covered with wire cloth of at least 0.035 inch (0.89 mm) wire. In lieu of wire cloth covering, said pipes, wires, conduits and other appurtenances shall be blocked from rodent usage by installing solid sheet metal guards 0.024 inch (0.61 mm) thick or heavier. Guards shall be fitted around pipes, wires, conduits or other appurtenances. In addition, they shall be fastened securely to and shall extend perpendicularly from the exterior wall for a minimum distance of 12 inches (305 mm) beyond and on either side of pipes, wires, conduits or appurtenances.

F101.6 Pier and wood construction.

F101.6.1 Sill less than 12 inches above ground. Buildings not provided with a continuous foundation shall be provided with protection against rodents at grade by providing either an apron in accordance with Section F101.6.1.1 or a floor slab in accordance with Section 101.6.1.2.

F101.6.1.1 Apron. Where an apron is provided, the apron shall not be less than 8 inches (203 mm) above, nor less than 24 inches (610 mm) below, grade. The apron shall not terminate below the lower edge of the siding material. The apron shall be constructed of an approved nondecayable, water-resistant ratproofing material of required strength and shall be installed around the entire perimeter of the building. Where constructed of masonry or concrete materials, the apron shall not be less than 4 inches (102 mm) in thickness.

F101.6.1.2 Grade floors. Where continuous concrete grade floor slabs are provided, open spaces shall not be left between the slab and walls, and openings in the slab shall be protected.

F101.6.2 Sill at or above 12 inches above ground. Buildings not provided with a continuous foundation and which have sills 12 or more inches (305 mm) above the ground level shall be provide with protection against rodents at grade in accordance with any of the following:

1. Section F101.6.1.1 or F101.6.1.2;

2. By installing solid sheet metal collars at least 0.024 inch (0.6 mm) thick at the top of each pier or pile and around each pipe, cable, conduit, wire or other item which provides a continuous pathway from the ground to the floor; or

3. By encasing the pipes, cables, conduits or wires in an enclosure constructed in accordance with Section F101.6.1.1.

APPENDIX G

FLOOD-RESISTANT CONSTRUCTION

The provisions contained in this appendix are not mandatory unless specifically referenced in the adopting ordinance.

SECTION G101
ADMINISTRATION

G101.1 Purpose. The purpose of this appendix is to promote the public health, safety and general welfare and to minimize public and private losses due to flood conditions in specific flood hazard areas through the establishment of comprehensive regulations for management of flood hazard areas designed to:

1. Prevent unnecessary disruption of commerce, access and public service during times of flooding;

2. Manage the alteration of natural flood plains, stream channels and shorelines;

3. Manage filling, grading, dredging and other development which may increase flood damage or erosion potential;

4. Prevent or regulate the construction of flood barriers which will divert floodwaters or which can increase flood hazards; and

5. Contribute to improved construction techniques in the flood plain.

G101.2 Objectives. The objectives of this appendix are to protect human life, minimize the expenditure of public money for flood control projects, minimize the need for rescue and relief efforts associated with flooding, minimize prolonged business interruption, minimize damage to public facilities and utilities, help maintain a stable tax base by providing for the sound use and development of flood-prone areas, contribute to improved construction techniques in the flood plain and ensure that potential owners and occupants are notified that property is within flood hazard areas.

G101.3 Scope. The provisions of this appendix shall apply to all proposed development in a flood hazard area established in Section 1612 of this code.

G101.4 Violations. Any violation of a provision of this appendix, or failure to comply with a permit or variance issued pursuant to this appendix or any requirement of this appendix, shall be handled in accordance with Section 113.

SECTION G102
APPLICABILITY

G102.1 General. This appendix, in conjunction with the *International Building Code*, provides minimum requirements for development located in flood hazard areas, including the subdivision of land, installation of utilities, placement and replacement of manufactured homes, new construction and repair, reconstruction, rehabilitation, or additions to new construction and substantial improvement of existing buildings and structures, including restoration after damage.

G102.2 Establishment of flood hazard areas. Flood hazard areas are established in Section 1612.3 of the *International Building Code*, adopted by the governing body on [INSERT DATE].

SECTION G103
POWERS AND DUTIES

G103.1 Permit applications. The building official shall review all permit applications to determine whether proposed development sites will be reasonably safe from flooding. If a proposed development site is in a flood hazard area, all site development activities, including grading, filling, utility installation and drainage modification, and all new construction and substantial improvements (including the placement of prefabricated buildings and manufactured homes) shall be designed and constructed with methods, practices and materials that minimize flood damage and that are in accordance with this code and ASCE 24.

G103.2 Other permits. It shall be the responsibility of the building official to assure that approval of a proposed development shall not be given until proof that necessary permits have been granted by federal or state agencies having jurisdiction over such development.

G103.3 Determination of design flood elevations. If design flood elevations are not specified, the building official is authorized to require the applicant to:

1. Obtain, review and reasonably utilize data available from a federal, state or other source, or

2. Determine the design flood elevation in accordance with accepted hydrologic and hydraulic engineering techniques. Such analyses shall be performed and sealed by a registered design professional. Studies, analyses and computations shall be submitted in sufficient detail to allow review and approval by the building official. The accuracy of data submitted for such determination shall be the responsibility of the applicant.

G103.4 Activities in riverine flood hazard areas. In riverine situations, until a regulatory floodway is designated, the building official shall not permit any new construction, substantial improvement or other development, including fill, unless the applicant demonstrates that the cumulative effect of the proposed development, when combined with all other existing and anticipated development, will not increase the design flood elevation more than 1 foot (305 mm) at any point within the community.

G103.5 Floodway encroachment. Prior to issuing a permit for any floodway encroachment, including fill, new construction, substantial improvements and other development or land-disturbing activity, the building official shall require submission of a certification, along with supporting technical data, that demonstrates that such development will not cause any increase of the level of the base flood.

G103.5.1 Floodway revisions. A floodway encroachment that increases the level of the base flood is authorized if the applicant has applied for a conditional Flood Insurance Rate Map (FIRM) revision and has received the approval of the Federal Emergency Management Agency (FEMA).

G103.6 Watercourse alteration. Prior to issuing a permit for any alteration or relocation of any watercourse, the building official shall require the applicant to provide notification of the proposal to the appropriate authorities of all affected adjacent government jurisdictions, as well as appropriate state agencies. A copy of the notification shall be maintained in the permit records and submitted to FEMA.

G103.6.1 Engineering analysis. The building official shall require submission of an engineering analysis which demonstrates that the flood-carrying capacity of the altered or relocated portion of the watercourse will not be decreased. Such watercourses shall be maintained in a manner which preserves the channel's flood-carrying capacity.

G103.7 Alterations in coastal areas. Prior to issuing a permit for any alteration of sand dunes and mangrove stands in flood hazard areas subject to high velocity wave action, the building official shall require submission of an engineering analysis which demonstrates that the proposed alteration will not increase the potential for flood damage.

G103.8 Records. The building official shall maintain a permanent record of all permits issued in flood hazard areas, including copies of inspection reports and certifications required in Section 1612.

SECTION G104
PERMITS

G104.1 Required. Any person, owner or authorized agent who intends to conduct any development in a flood hazard area shall first make application to the building official and shall obtain the required permit.

G104.2 Application for permit. The applicant shall file an application in writing on a form furnished by the building official. Such application shall:

1. Identify and describe the development to be covered by the permit.

2. Describe the land on which the proposed development is to be conducted by legal description, street address or similar description that will readily identify and definitely locate the site.

3. Include a site plan showing the delineation of flood hazard areas, floodway boundaries, flood zones, design flood elevations, ground elevations, proposed fill and excavation and drainage patterns and facilities.

4. Indicate the use and occupancy for which the proposed development is intended.

5. Be accompanied by construction documents, grading and filling plans and other information deemed appropriate by the building official.

6. State the valuation of the proposed work.

7. Be signed by the applicant or the applicant's authorized agent.

G104.3 Validity of permit. The issuance of a permit under this appendix shall not be construed to be a permit for, or approval of, any violation of this appendix or any other ordinance of the jurisdiction. The issuance of a permit based on submitted documents and information shall not prevent the building official from requiring the correction of errors. The building official is authorized to prevent occupancy or use of a structure or site which is in violation of this appendix or other ordinances of this jurisdiction.

G104.4 Expiration. A permit shall become invalid if the proposed development is not commenced within 180 days after its issuance, or if the work authorized is suspended or abandoned for a period of 180 days after the work commences. Extensions shall be requested in writing and justifiable cause demonstrated. The building official is authorized to grant, in writing, one or more extensions of time, for periods not more than 180 days each.

G104.5 Suspension or revocation. The building official is authorized to suspend or revoke a permit issued under this appendix wherever the permit is issued in error or on the basis of incorrect, inaccurate or incomplete information, or in violation of any ordinance or code of this jurisdiction.

SECTION G105
VARIANCES

G105.1 General. The board of appeals established pursuant to Section 112 shall hear and decide requests for variances. The board of appeals shall base its determination on technical justifications, and has the right to attach such conditions to variances as it deems necessary to further the purposes and objectives of this appendix and Section 1612.

G105.2 Records. The building official shall maintain a permanent record of all variance actions, including justification for their issuance.

G105.3 Historic structures. A variance is authorized to be issued for the repair or rehabilitation of a historic structure upon a determination that the proposed repair or rehabilitation will not preclude the structure's continued designation as a historic structure, and the variance is the minimum necessary to preserve the historic character and design of the structure.

Exception: Within flood hazard areas, historic structures that are not:

a. Listed or preliminarily determined to be eligible for listing in the National Register of Historic Places; or

b. Determined by the Secretary of the U.S. Department of Interior as contributing to the historical significance of a registered historic district or a district preliminarily determined to qualify as an historic district; or

c. Designated as historic under a state or local historic preservation program that is approved by the Department of Interior.

G105.4 Functionally dependent facilities. A variance is authorized to be issued for the construction or substantial improvement of a functionally dependent facility provided the criteria in Section 1612.1 are met and the variance is the minimum necessary to allow the construction or substantial improvement, and that all due consideration has been given to methods and materials that minimize flood damages during the design flood and create no additional threats to public safety.

G105.5 Restrictions. The board of appeals shall not issue a variance for any proposed development in a floodway if any increase in flood levels would result during the base flood discharge.

G105.6 Considerations. In reviewing applications for variances, the board of appeals shall consider all technical evaluations, all relevant factors, all other portions of this appendix and the following:

1. The danger that materials and debris may be swept onto other lands resulting in further injury or damage;

2. The danger to life and property due to flooding or erosion damage;

3. The susceptibility of the proposed development, including contents, to flood damage and the effect of such damage on current and future owners;

4. The importance of the services provided by the proposed development to the community;

5. The availability of alternate locations for the proposed development that are not subject to flooding or erosion;

6. The compatibility of the proposed development with existing and anticipated development;

7. The relationship of the proposed development to the comprehensive plan and flood plain management program for that area;

8. The safety of access to the property in times of flood for ordinary and emergency vehicles;

9. The expected heights, velocity, duration, rate of rise and debris and sediment transport of the floodwaters and the effects of wave action, if applicable, expected at the site; and

10. The costs of providing governmental services during and after flood conditions including maintenance and repair of public utilities and facilities such as sewer, gas, electrical and water systems, streets and bridges.

G105.7 Conditions for issuance. Variances shall only be issued by the board of appeals upon:

1. A technical showing of good and sufficient cause that the unique characteristics of the size, configuration or topography of the site renders the elevation standards inappropriate;

2. A determination that failure to grant the variance would result in exceptional hardship by rendering the lot undevelopable;

3. A determination that the granting of a variance will not result in increased flood heights, additional threats to public safety, extraordinary public expense, nor create nuisances, cause fraud on or victimization of the public or conflict with existing local laws or ordinances;

4. A determination that the variance is the minimum necessary, considering the flood hazard, to afford relief; and

5. Notification to the applicant in writing over the signature of the building official that the issuance of a variance to construct a structure below the base flood level will result in increased premium rates for flood insurance up to amounts as high as $25 for $100 of insurance coverage, and that such construction below the base flood level increases risks to life and property.

SECTION G201
DEFINITIONS

G201.1 General. The following words and terms shall, for the purposes of this appendix, have the meanings shown herein. Refer to Chapter 2 for general definitions.

G201.2 Definitions.

DEVELOPMENT. Any man-made change to improved or unimproved real estate, including but not limited to, buildings or other structures, temporary or permanent storage of materials, mining, dredging, filling, grading, paving, excavations, operations and other land disturbing activities.

FUNCTIONALLY DEPENDENT FACILITY. A facility which cannot be used for its intended purpose unless it is located or carried out in close proximity to water, such as a docking or port facility necessary for the loading or unloading of cargo or passengers, shipbuilding or ship repair. The term does not include long-term storage, manufacture, sales or service facilities.

MANUFACTURED HOME. A structure that is transportable in one or more sections, built on a permanent chassis, designed for use with or without a permanent foundation when attached to the required utilities, and constructed to the Federal Mobile Home Construction and Safety Standards and rules and regulations promulgated by the U.S. Department of Housing and Urban Development. The term also includes mobile homes, park trailers, travel trailers and similar transportable structures that are placed on a site for 180 consecutive days or longer.

MANUFACTURED HOME PARK OR SUBDIVISION. A parcel (or contiguous parcels) of land divided into two or more manufactured home lots for rent or sale.

RECREATIONAL VEHICLE. A vehicle that is built on a single chassis, 400 square feet (37.16 m²) or less when measured at the largest horizontal projection, designed to be self-propelled or permanently towable by a light-duty truck, and designed primarily not for use as a permanent dwelling but as temporary living quarters for recreational, camping, travel or seasonal use. A recreational vehicle is ready for highway use if it is on its wheels or jacking system, is attached to the site only by quick disconnect-type utilities and security devices and has no permanently attached additions.

VARIANCE. A grant of relief from the requirements of this section which permits construction in a manner otherwise prohibited by this section where specific enforcement would result in unnecessary hardship.

VIOLATION. A development that is not fully compliant with this appendix or Section 1612, as applicable.

SECTION G301
SUBDIVISIONS

G301.1 General. Any subdivision proposal, including proposals for manufactured home parks and subdivisions, or other proposed new development in a flood hazard area shall be reviewed to assure that:

1. All such proposals are consistent with the need to minimize flood damage;

2. All public utilities and facilities, such as sewer, gas, electric and water systems are located and constructed to minimize or eliminate flood damage; and

3. Adequate drainage is provided to reduce exposure to flood hazards.

G301.2 Subdivision requirements. The following requirements shall apply in the case of any proposed subdivision, including proposals for manufactured home parks and subdivisions, any portion of which lies within a flood hazard area:

1. The flood hazard area, including floodways and areas subject to high velocity wave action, as appropriate, shall be delineated on tentative and final subdivision plats;

2. Design flood elevations shall be shown on tentative and final subdivision plats;

3. Residential building lots shall be provided with adequate buildable area outside the floodway; and

4. The design criteria for utilities and facilities set forth in this appendix and appropriate *International Codes* shall be met.

SECTION G401
SITE IMPROVEMENT

G401.1 Development in floodways. Development or land disturbing activity shall not be authorized in the floodway unless it has been demonstrated through hydrologic and hydraulic analyses performed in accordance with standard engineering practice that the proposed encroachment will not result in any increase in the level of the base flood.

G401.2 Flood hazard areas subject to high velocity wave action.

1. Development or land disturbing activity shall only be authorized landward of the reach of mean high tide.

2. The use of fill for structural support of buildings is prohibited.

G401.3 Sewer facilities. All new or replaced sanitary sewer facilities, private sewage treatment plants (including all pumping stations and collector systems) and on-site waste disposal systems shall be designed in accordance with Chapter 8, ASCE 24, to minimize or eliminate infiltration of floodwaters into the facilities and discharge from the facilities into floodwaters, or impairment of the facilities and systems.

G401.4 Water facilities. All new replacement water facilities shall be designed in accordance with the provisions of Chapter 8, ASCE 24, to minimize or eliminate infiltration of floodwaters into the systems.

G401.5 Storm drainage. Storm drainage shall be designed to convey the flow of surface waters to minimize or eliminate damage to persons or property.

G401.6 Streets and sidewalks. Streets and sidewalks shall be designed to minimize potential for increasing or aggravating flood levels.

SECTION G501
MANUFACTURED HOMES

G501.1 Elevation. All new and replacement manufactured homes to be placed or substantially improved in a flood hazard area shall be elevated such that the lowest floor of the manufactured home is elevated to or above the design flood elevation.

G501.2 Foundations. All new and replacement manufactured homes, including substantial improvement of existing manufactured homes, shall be placed on a permanent, reinforced foundation that is designed in accordance with Section 1612.

G501.3 Anchoring. All new and replacement manufactured homes to be placed or substantially improved in a flood hazard area shall be installed using methods and practices which minimize flood damage. Manufactured homes shall be securely anchored to an adequately anchored foundation system to resist flotation, collapse and lateral movement. Methods of anchoring are authorized to include, but are not limited to, use of over-the-top or frame ties to ground anchors. This requirement is in addition to applicable state and local anchoring requirements for resisting wind forces.

SECTION G601
RECREATIONAL VEHICLES

G601.1 Placement prohibited. The placement of recreational vehicles shall not be authorized in flood hazard areas subject to high velocity wave action and in floodways.

G601.2 Temporary placement. Recreational vehicles in flood hazard areas shall be fully licensed and ready for highway use, and shall be placed on a site for less than 180 consecutive days.

G601.3 Permanent placement. Recreational vehicles that are not fully licensed and ready for highway use, or that are to be placed on a site for more than 180 consecutive days, shall meet the requirements of Section G501 for manufactured homes.

SECTION 701
TANKS

G701.1 Underground tanks. Underground tanks in flood hazard areas shall be anchored to prevent flotation, collapse or lateral movement resulting from hydrostatic loads, including the effects of buoyancy, during conditions of the design flood.

Above-ground tanks. Above-ground tanks in flood hazard areas shall be elevated to or above the design flood elevation or shall be anchored or otherwise designed and constructed to prevent flotation, collapse or lateral movement resulting from hydrodynamic and hydrostatic loads, including the effects of buoyancy, during conditions of the design flood.

Tank inlets and vents. In flood hazard areas, tank inlets, fill openings, outlets and vents shall be:

1. At or above the design flood elevation or fitted with covers designed to prevent the inflow of floodwater or outflow of the contents of the tanks during conditions of the design flood.

2. Anchored to prevent lateral movement resulting from hydrodynamic and hydrostatic loads, including the effects of buoyancy, during conditions of the design flood.

SECTION G702
REFERENCED STANDARDS

ASCE 24–98	Flood Resistance Design and Construction	G103.1, G401.3, G401.4
HUD 24 CFR Part 3280 (1994)	Manufactured Home Construction and Safety Standards	G201
IBC-03	*International Building Code*	G102.2

APPENDIX H

SIGNS

The provisions contained in this appendix are not mandatory unless specifically referenced in the adopting ordinance.

SECTION H101
GENERAL

H101.1 General. A sign shall not be erected in a manner that would confuse or obstruct the view of or interfere with exit signs required by Chapter 10 or with official traffic signs, signals or devices. Signs and sign support structures, together with their supports, braces, guys and anchors, shall be kept in repair and in proper state of preservation. The display surfaces of signs shall be kept neatly painted or posted at all times.

H101.2 Signs exempt from permits. The following signs are exempt from the requirements to obtain a permit before erection:

1. Painted nonilluminated signs.
2. Temporary signs announcing the sale or rent of property.
3. Signs erected by transportation authorities.
4. Projecting signs not exceeding 2.5 square feet (0.23 m²).
5. The changing of moveable parts of an approved sign that is designed for such changes, or the repainting or repositioning of display matter shall not be deemed an alteration.

SECTION H102
DEFINITIONS

H102.1 General. Unless otherwise expressly stated, the following words and terms shall, for the purposes of this appendix, have the meanings shown herein. Refer to Chapter 2 of the *International Building Code* for general definitions.

COMBINATION SIGN. A sign incorporating any combination of the features of pole, projecting and roof signs.

DISPLAY SIGN. The area made available by the sign structure for the purpose of displaying the advertising message.

ELECTRIC SIGN. A sign containing electrical wiring, but not including signs illuminated by an exterior light source.

GROUND SIGN. A billboard or similar type of sign which is supported by one or more uprights, poles or braces in or upon the ground other than a combination sign or pole sign, as defined by this code.

POLE SIGN. A sign wholly supported by a sign structure in the ground.

PORTABLE DISPLAY SURFACE. A display surface temporarily fixed to a standardized advertising structure which is regularly moved from structure to structure at periodic intervals.

PROJECTING SIGN. A sign other than a wall sign, which projects from and is supported by a wall of a building or structure.

ROOF SIGN. A sign erected upon or above a roof or parapet of a building or structure.

SIGN. Any letter, figure, character, mark, plane, point, marquee sign, design, poster, pictorial, picture, stroke, stripe, line, trademark, reading matter or illuminated service, which shall be constructed, placed, attached, painted, erected, fastened or manufactured in any manner whatsoever, so that the same shall be used for the attraction of the public to any place, subject, person, firm, corporation, public performance, article, machine or merchandise, whatsoever, which is displayed in any manner outdoors. Every sign shall be classified and conform to the requirements of that classification as set forth in this chapter.

SIGN STRUCTURE. Any structure which supports or is capable of supporting a sign as defined in this code. A sign structure is permitted to be a single pole and is not required to be an integral part of the building.

WALL SIGN. Any sign attached to or erected against the wall of a building or structure, with the exposed face of the sign in a plane parallel to the plane of said wall.

SECTION H103
LOCATION

H103.1 Location restrictions. Signs shall not be erected, constructed or maintained so as to obstruct any fire escape or any window or door or opening used as a means of egress or so as to prevent free passage from one part of a roof to any other part thereof. A sign shall not be attached in any form, shape or manner to a fire escape, nor be placed in such manner as to interfere with any opening required for ventilation.

SECTION H104
IDENTIFICATION

H104.1 Identification. Every outdoor advertising display sign hereafter erected, constructed or maintained, for which a permit is required shall be plainly marked with the name of the person, firm or corporation erecting and maintaining such sign and shall have affixed on the front thereof the permit number issued for said sign or other method of identification approved by the building official.

SECTION H105
DESIGN AND CONSTRUCTION

H105.1 General requirements. Signs shall be designed and constructed to comply with the provisions of this code for use of materials, loads and stresses.

H105.2 Permits, drawings and specifications. Where a permit is required, as provided in Chapter 1, construction documents shall be required. These documents shall show the dimensions, material and required details of construction, including loads, stresses and anchors.

H105.3 Wind load. Signs shall be designed and constructed to withstand wind pressure as provided for in Chapter 16.

H105.4 Seismic load. Signs designed to withstand wind pressures shall be considered capable of withstanding earthquake loads, except as provided for in Chapter 16.

H105.5 Working stresses. In outdoor advertising display signs, the allowable working stresses shall conform to the requirements of Chapter 16. The working stresses of wire rope and its fastenings shall not exceed 25 percent of the ultimate strength of the rope or fasteners.

Exceptions:

1. The allowable working stresses for steel and wood shall be in accordance with the provisions of Chapters 22 and 23.

2. The working strength of chains, cables, guys or steel rods shall not exceed one-fifth of the ultimate strength of such chains, cables, guys or steel.

H105.6 Attachment. Signs attached to masonry, concrete or steel shall be safely and securely fastened by means of metal anchors, bolts or approved expansion screws of sufficient size and anchorage to safely support the loads applied.

SECTION H106
ELECTRICAL

H106.1 Illumination. A sign shall not be illuminated by other than electrical means, and electrical devices and wiring shall be installed in accordance with the requirements of the ICC *Electrical Code.* Any open spark or flame shall not be used for display purposes unless specifically approved.

H106.1.1 Internally illuminated signs. Except as provided for in Sections 402.14 and 2611, where internally illuminated signs have sign facings of wood or approved plastic, the area of such facing section shall not be more than 120 square feet (11.16 m²) and the wiring for electric lighting shall be entirely enclosed in the sign cabinet with a clearance of not less than 2 inches (51 mm) from the facing material. The dimensional limitation of 120 square feet (11.16 m²) shall not apply to sign facing sections made from flame-resistant-coated fabric (ordinarily known as "flexible sign face plastic") that weighs less than 20 ounces per square yard (678 g/m²) and which, when tested in accordance with NFPA 701, meets the requirements of both the small-scale test and the large-scale test, or which, when tested in accordance with an approved test method, exhibits an average burn time for 10 specimens of 2 seconds or less and a burning extent of 15 centimeters or less.

H106.2 Electrical service. Signs that require electrical service shall comply with the ICC *Electrical Code.*

SECTION H107
COMBUSTIBLE MATERIALS

H107.1 Use of combustibles. Wood, approved plastic or plastic veneer panels as provided for in Chapter 26, or other materials of combustible characteristics similar to wood, used for moldings, cappings, nailing blocks, letters and latticing, shall comply with Section H109.1, and shall not be used for other ornamental features of signs, unless approved.

H107.1.1 Plastic materials. Notwithstanding any other provisions of this code, plastic materials which burn at a rate no faster than 2.5 inches per minute (64 mm/s) when tested in accordance with ASTM D 635 shall be deemed approved plastics and can be used as the display surface material and for the letters, decorations and facings on signs and outdoor display structures.

H107.1.2 Electric sign faces. Individual plastic facings of electric signs shall not exceed 200 square feet (18.6 m²) in area.

H107.1.3 Area limitation. If the area of a display surface exceeds 200 square feet (18.6 m²), the area occupied or covered by approved plastics shall be limited to 200 square feet (18.6 m²) plus 50 percent of the difference between 200 square feet (18.6 m²) and the area of display surface. The area of plastic on a display surface shall not in any case exceed 1,100 square feet (102 m²).

H107.1.4 Plastic appurtenances. Letters and decorations mounted on an approved plastic facing or display surface can be made of approved plastics.

SECTION H108
ANIMATED DEVICES

H108.1 Fail-safe device. Signs that contain moving sections or ornaments shall have fail-safe provisions to prevent the section or ornament from releasing and falling or shifting its center of gravity more than 15 inches (381 mm). The fail-safe device shall be in addition to the mechanism and the mechanism's housing which operate the movable section or ornament. The fail-safe device shall be capable of supporting the full dead weight of the section or ornament when the moving mechanism releases.

SECTION H109
GROUND SIGNS

H109.1 Height restrictions. The structural frame of ground signs shall not be erected of combustible materials to a height of more than 35 feet (10668 mm) above the ground. Ground signs constructed entirely of noncombustible material shall not be erected to a height of greater than 100 feet (30 480 mm) above the ground. Greater heights are permitted where approved and located so as not to create a hazard or danger to the public.

H109.2 Required clearance. The bottom coping of every ground sign shall be not less than 3 feet (914 mm) above the ground or street level, which space can be filled with platform decorative trim or light wooden construction.

H109.3 Wood anchors and supports. Where wood anchors or supports are embedded in the soil, the wood shall be pressure treated with an approved preservative.

SECTION H110
ROOF SIGNS

H110.1 General. Roof signs shall be constructed entirely of metal or other approved noncombustible material except as provided for in Sections H106.1.1 and H107.1. Provisions shall be made for electric grounding of metallic parts. Where combustible materials are permitted in letters or other ornamental features, wiring and tubing shall be kept free and insulated therefrom. Roof signs shall be so constructed as to leave a clear space of not less than 6 feet (1829 mm) between the roof level and the lowest part of the sign and shall have at least 5 feet (1524 mm) clearance between the vertical supports thereof. No portion of any roof sign structure shall project beyond an exterior wall.

> **Exception:** Signs on flat roofs with every part of the roof accessible.

H110.2 Bearing plates. The bearing plates of roof signs shall distribute the load directly to or upon masonry walls, steel roof girders, columns or beams. The building shall be designed to avoid overstress of these members.

H110.3 Height of solid signs. A roof sign having a solid surface shall not exceed, at any point, a height of 24 feet (7315 mm) measured from the roof surface.

H110.4 Height of open signs. Open roof signs in which the uniform open area is not less than 40 percent of total gross area shall not exceed a height of 75 feet (22 860 mm) on buildings of Type 1 or Type 2 construction. On buildings of other construction types, the height shall not exceed 40 feet (12 192 mm). Such signs shall be thoroughly secured to the building upon which they are installed, erected or constructed by iron, metal anchors, bolts, supports, chains, stranded cables, steel rods or braces and they shall be maintained in good condition.

H110.5 Height of closed signs. A closed roof sign shall not be erected to a height greater than 50 feet (15 240 mm) above the roof of buildings of Type 1 or Type 2 construction, nor more than 35 feet (10 668 mm) above the roof of buildings of Type 3, 4 or 5 construction.

SECTION H111
WALL SIGNS

H111.1 Materials. Wall signs which have an area exceeding 40 square feet (3.72 m²) shall be constructed of metal or other approved noncombustible material, except for nailing rails and as provided for in Sections H106.1.1 and H107.1.

H111.2 Exterior wall mounting details. Wall signs attached to exterior walls of solid masonry, concrete or stone shall be safely and securely attached by means of metal anchors, bolts or expansion screws of not less than 3/8 inch (9.5 mm) diameter and shall be embedded at least 5 inches (127 mm). Wood blocks shall not be used for anchorage, except in the case of wall signs attached to buildings with walls of wood. A wall sign shall not be supported by anchorages secured to an unbraced parapet wall.

H111.3 Extension. Wall signs shall not extend above the top of the wall, nor beyond the ends of the wall to which the signs are attached unless such signs conform to the requirements for roof signs, projecting signs or ground signs.

SECTION H112
PROJECTING SIGNS

H112.1 General. Projecting signs shall be constructed entirely of metal or other noncombustible material and securely attached to a building or structure by metal supports such as bolts, anchors, supports, chains, guys or steel rods. Staples or nails shall not be used to secure any projecting sign to any building or structure. The dead load of projecting signs not parallel to the building or structure and the load due to wind pressure shall be supported with chains, guys or steel rods having net cross-sectional dimension of not less than 3/8 inch (9.5 mm) diameter. Such supports shall be erected or maintained at an angle of at least 45 percent (0.78 rad) with the horizontal to resist the dead load and at angle of 45 percent (0.78 rad) or more with the face of the sign to resist the specified wind pressure. If such projecting sign exceeds 30 square feet (2.8 m²) in one facial area, there shall be provided at least two such supports on each side not more than 8 feet (2438 mm) apart to resist the wind pressure.

H112.2 Attachment of supports. Supports shall be secured to a bolt or expansion screw that will develop the strength of the supporting chains, guys or steel rods, with a minimum 5/8-inch (15.9 mm) bolt or lag screw, by an expansion shield. Turn buckles shall be placed in chains, guys or steel rods supporting projecting signs.

H112.3 Wall mounting details. Chains, cables, guys or steel rods used to support the live or dead load of projecting signs are permitted to be fastened to solid masonry walls with expansion bolts or by machine screws in iron supports, but such supports shall not be attached to an unbraced parapet wall. Where the supports must be fastened to walls made of wood, the supporting anchor bolts must go through the wall and be plated or fastened on the inside in a secure manner.

H112.4 Height limitation. A projecting sign shall not be erected on the wall of any building so as to project above the roof or cornice wall or above the roof level where there is no cornice wall; except that a sign erected at a right angle to the building, the horizontal width of which sign is perpendicular to such a wall and does not exceed 18 inches (457 mm), is permitted to be erected to a height not exceeding 2 feet (610 mm) above the roof or cornice wall or above the roof level where there is no cornice wall. A sign attached to a corner of a building and parallel to the vertical line of such corner shall be deemed to be erected at a right angle to the building wall.

H112.5 Additional loads. Projecting sign structures which will be used to support an individual on a ladder or other servicing device, whether or not specifically designed for the servicing device, shall be capable of supporting the anticipated additional load, but not less than a 100-pound (445 N) concentrated horizontal load and a 300-pound (1334 N) concentrated vertical load applied at the point of assumed or most eccentric loading. The building component to which the projecting sign is attached shall also be designed to support the additional loads.

SECTION H113
MARQUEE SIGNS

H113.1 Materials. Marquee signs shall be constructed entirely of metal or other approved noncombustible material except as provided for in Sections H106.1.1 and H107.1.

H113.2 Attachment. Marquee signs shall be attached to approved marquees that are constructed in accordance with Section 3106.

H113.3 Dimensions. Marquee signs, whether on the front or side, shall not project beyond the perimeter of the marquee.

H113.4 Height limitation. Marquee signs shall not extend more than 6 feet (1829 mm) above, nor 1 foot (305 mm) below such marquee, but under no circumstances shall the sign or signs have a vertical dimension greater than 8 feet (2438 mm).

SECTION H114
PORTABLE SIGNS

H114.1 General. Portable signs shall conform to requirements for ground, roof, projecting, flat and temporary signs where such signs are used in a similar capacity. The requirements of this section shall not be construed to require portable signs to have connections to surfaces, tie-downs or foundations where provisions are made by temporary means or configuration of the structure to provide stability for the expected duration of the installation.

TABLE 4-A
SIZE, THICKNESS AND TYPE OF GLASS PANELS IN SIGNS

MAXIMUM SIZE OF EXPOSED PANEL		MINIMUM THICKNESS OF GLASS (inches)	TYPE OF GLASS
Any dimension (inches)	Area (square inches)		
30	500	$1/8$	Plain, plate or wired
45	700	$3/16$	Plain, plate or wired
144	3,600	$1/4$	Plain, plate or wired
> 144	> 3,600	$1/4$	Wired glass

For SI: 1 inch = 25.4 mm, 1 square inch = 645 mm².

TABLE 4-B
THICKNESS OF PROJECTION SIGN

PROJECTION (feet)	MAXIMUM THICKNESS (feet)
5	2
4	2.5
3	3
2	3.5
1	4

For SI: 1 foot = 304.8 mm.

SECTION H115
REFERENCED STANDARDS

ASTM D 635-98	Test Method for Rate of Burning and/or Extent and Time of Burning of Self-Supporting Plastics in a Horizontal Position	H107.1.1
ICC EC-03	ICC *Electrical Code*	H106.1, H106.2
NFPA 701-99	Methods of Fire Test for Flame Propagation of Textiles and Films	H106.1.1

APPENDIX I

PATIO COVERS

The provisions contained in this appendix are not mandatory unless specifically referenced in the adopting ordinance.

SECTION I101
GENERAL

I101.1 General. Patio covers shall be permitted to be detached from or attached to dwelling units. Patio covers shall be used only for recreational, outdoor living purposes and not as carports, garages, storage rooms or habitable rooms. Openings shall be permitted to be enclosed with insect screening, approved translucent or transparent plastic not more that 0.125 inch (3.2 mm) in thickness, glass conforming to the provisions of Chapter 24 or any combination of the foregoing.

SECTION I102
DEFINITIONS

I102.1 General. The following word and term shall, for the purposes of this appendix, have the meaning shown herein.

PATIO COVERS. One story structures not exceeding 12 feet (3657 mm) in height. Enclosure walls shall be permitted to be of any configuration, provided the open or glazed area of the longer wall and one additional wall is equal to at least 65 percent of the area below a minimum of 6 feet 8 inches (2032 mm) of each wall, measured from the floor.

SECTION I103
EXTERIOR OPENINGS

I103.1 Light, ventilation and emergency egress. Exterior openings required for light and ventilation shall be permitted to open into a patio structure. However, the patio structure shall be unenclosed if such openings are serving as emergency egress or rescue openings from sleeping rooms. Where such exterior openings serve as an exit from the dwelling unit, the patio structure, unless unenclosed, shall be provided with exits conforming to the provision of Chapter 10.

SECTION I104
STRUCTURAL PROVISIONS

I104.1 Design loads. Patio covers shall be designed and constructed to sustain, within the stress limits of this code, all dead loads plus a minimum vertical live load of 10 pounds per square foot (0.48 kN/m²) except that snow loads shall be used where such snow loads exceed this minimum. Such patio covers shall be designed to resist the minimum wind and seismic loads set forth in this code.

I104.2 Footings. In areas with a frost depth of zero, a patio cover shall be permitted to be supported on a concrete slab on grade without footings, provided the slab conforms to the provisions of Chapter 19 of this code, is not less than $3^{1}/_{2}$ inches (89 mm) thick and further provided that the columns do not support loads in excess of 750 pounds (3.36 kN) per column.

APPENDIX J
GRADING

The provisions contained in this appendix are not mandatory unless specifically referenced in the adopting ordinance.

SECTION J101
GENERAL

J101.1 Scope. The provisions of this chapter apply to grading, excavation and earthwork construction, including fills and embankments. Where conflicts occur between the technical requirements of this chapter and the soils report, the soils report shall govern.

J101.2 Flood hazard areas. The provisions of this chapter shall not apply to grading, excavation and earthwork construction, including fills and embankments, in floodways within flood hazard areas established in Section 1612.3 unless it has been demonstrated through hydrologic and hydraulic analyses performed in accordance with standard engineering practice that the proposed work will not result in any increase in the level of the base flood.

SECTION J102
DEFINITIONS

J102.1 Definitions. For the purposes of this appendix chapter, the terms, phrases and words listed in this section and their derivatives shall have the indicated meanings.

BENCH. A relatively level step excavated into earth material on which fill is to be placed.

COMPACTION. The densification of a fill by mechanical means.

CUT. See Excavation.

DOWN DRAIN. A device for collecting water from a swale or ditch located on or above a slope, and safely delivering it to an approved drainage facility

EROSION. The wearing away of the ground surface as a result of the movement of wind, water or ice.

EXCAVATION. The removal of earth material by artificial means, also referred to as a cut.

FILL. Deposition of earth materials by artificial means.

GRADE. The vertical location of the ground surface.

GRADE, EXISTING. The grade prior to grading.

GRADE, FINISHED. The grade of the site at the conclusion of all grading efforts.

GRADING. An excavation or fill or combination thereof.

KEY. A compacted fill placed in a trench excavated in earth material beneath the toe of a slope.

All slope references in the chapter have been modified to show the horizontal:vertical relationship.

SLOPE. An inclined surface, the inclination of which is expressed as a ratio of horizontal distance to vertical distance.

TERRACE. A relatively level step constructed in the face of a graded slope for drainage and maintenance purposes.

SECTION J103
PERMITS REQUIRED

J103.1 Permits required. Except as exempted in Section J103.2, no grading shall be performed without first having obtained a permit therefor from the building official. A grading permit does not include the construction of retaining walls or other structures.

J103.2 Exemptions. A grading permit shall not be required for the following:

1. Grading in an isolated, self-contained area, provided there is no danger to the public, and that such grading will not adversely affect adjoining properties.

2. Excavation for construction of a structure permitted under this code.

3. Cemetery graves.

4. Refuse disposal sites controlled by other regulations.

5. Excavations for wells, or trenches for utilities.

6. Mining, quarrying, excavating, processing or stockpiling rock, sand, gravel, aggregate or clay controlled by other regulations, provided such operations do not affect the lateral support of, or significantly increase stresses in, soil on adjoining properties.

7. Exploratory excavations performed under the direction of a registered design professional This phrase was added to assure that the "exploratory excavation" is not to begin construction of a building prior to receiving a permit for the sole purpose of preparing a soils report.

Exemption from the permit requirements of this appendix shall not be deemed to grant authorization for any work to be done in any manner in violation of the provisions of this code or any other laws or ordinances of this jurisdiction.

SECTION J104
PERMIT APPLICATION AND SUBMITTALS

J104.1 Submittal requirements. In addition to the provisions of Section 105.3, the applicant shall state the estimated quantities of excavation and fill.

J104.2 Site plan requirements. In addition to the provisions of Section 106, a grading plan shall show the existing grade and finished grade in contour intervals of sufficient clarity to indicate the nature and extent of the work and show in detail that it complies with the requirements of this code. Drafting requirements were deleted here.The plans shall show the existing grade on adjoining properties in sufficient detail to identify how grade changes will conform to the requirements of this code.

J104.3 Soils report. A soils report prepared by registered design professionals shall be provided which shall identify the nature and distribution of existing soils; conclusions and recommendations for grading procedures; soil design criteria for any structures or embankments required to accomplish the proposed grading; and, where necessary, slope stability studies, and recommendations and conclusions regarding site geology.

Exception: A soils report is not required where the building official determines that the nature of the work applied for is such that a report is not necessary.

J104.4 Liquefaction study. For sites with mapped maximum considered earthquake spectral response accelerations at short periods (S_s) greater than 0.5g as determined by Section 1615, a study of the liquefaction potential of the site shall be provided, and the recommendations incorporated in the plans.

Exception: A liquefaction study is not required where the building official determines from established local data that the liquefaction potential is low.

SECTION J105
INSPECTIONS

J105.1 General. Most of this section was deleted or simplified. Inspections shall be governed by Section 109 of this code.

J105.2 Special inspections. The special inspection requirements of Section 1704.7 shall apply to work performed under a grading permit where required by the building official.

SECTION J106
EXCAVATIONS

J106.1 Maximum slope. The slope of cut surfaces shall be no steeper than is safe for the intended use, and shall be no steeper than 2 horizontal to 1 vertical (50 percent) unless the applicant furnishes a soils report justifying a steeper slope.

Exceptions:

1. A cut surface may be at a slope of 1.5 horizontal to 1 vertical (67 percent) provided that all the following are met:

   1.1. It is not intended to support structures or surcharges.

   1.2. It is adequately protected against erosion.

   1.3. It is no more than 8 feet (2438 mm) in height.

   1.4. It is approved by the building official.

2. A cut surface in bedrock shall be permitted to be at a slope of 1 horizontal to 1 vertical (100 percent).

SECTION J107
FILLS

J107.1 General. Unless otherwise recommended in the soils report, fills shall conform to provisions of this section.

J107.2 Surface preparation. The ground surface shall be prepared to receive fill by removing vegetation, topsoil and other unsuitable materials, and scarifying the ground to provide a bond with the fill material.

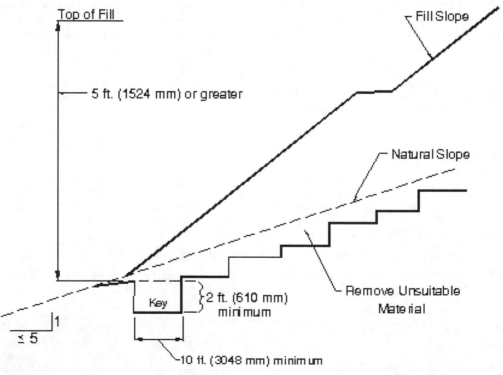

For SI: 1 foot = 304.8 mm

FIGURE J107.3
BENCHING DETAILS

J107.3 Benching. Where existing grade is at a slope steeper than 5 horizontal to 1 vertical (20 percent) and the depth of the fill exceeds 5 feet (1524 mm) benching shall be provided in accordance with Figure J107.3. A key shall be provided which is at least 10 feet (3048 mm) in width and 2 feet (610 mm) in depth.

J107.4 Fill material. Fill material shall not include organic, frozen or other deleterious materials. No rock or similar irreducible material greater than 12 inches (305 mm) in any dimension shall be included in fills.

J107.5 Compaction. All fill material shall be compacted to 90 percent of maximum density as determined by ASTM D1557, Modified Proctor, in lifts not exceeding 12 inches (305 mm) in depth.

J107.6 Maximum slope. The slope of fill surfaces shall be no steeper than is safe for the intended use. Fill slopes steeper than 2 horizontal to 1 vertical (50 percent) shall be justified by soils reports or engineering data.

SECTION J108
SETBACKS

J108.1 General. Cut and fill slopes shall be set back from the property lines in accordance with this section. Setback dimensions shall be measured perpendicular to the property line and shall be as shown in Figure J108.1, unless substantiating data is submitted justifying reduced setbacks.

J108.2 Top of slope. The setback at the top of a cut slope shall not be less than that shown in Figure J108.1, or than is required to accommodate any required interceptor drains, whichever is greater.

J108.3 Slope protection. Where required to protect adjacent properties at the toe of a slope from adverse effects of the grading, additional protection, approved by the building official,

shall be included. Such protection may include but shall not be limited to:

1. Setbacks greater than those required by Figure J108.1.
2. Provisions for retaining walls or similar construction.
3. Erosion protection of the fill slopes.
4. Provision for the control of surface waters.

SECTION J109
DRAINAGE AND TERRACING

J109.1 General. Unless otherwise recommended by a registered design professional, drainage facilities and terracing shall be provided in accordance with the requirements of this section.

> **Exception:** Drainage facilities and terracing need not be provided where the ground slope is not steeper than 3 horizontal to 1 vertical (33 percent).

J109.2 Terraces. Terraces at least 6 feet (1829 mm) in width shall be established at not more than 30-foot (9144 mm) vertical intervals on all cut or fill slopes to control surface drainage and debris. Suitable access shall be provided to allow for cleaning and maintenance.

Where more than two terraces are required, one terrace, located at approximately mid-height, shall be at least 12 feet (3658 mm) in width.

Swales or ditches shall be provided on terraces. They shall have a minimum gradient of 20 horizontal to 1 vertical (5 percent) and shall be paved with concrete not less than 3 inches (76 mm) in thickness, or with other materials suitable to the application. They shall have a minimum depth of 12 inches (305 mm) and a minimum width of 5 feet (1524 mm).

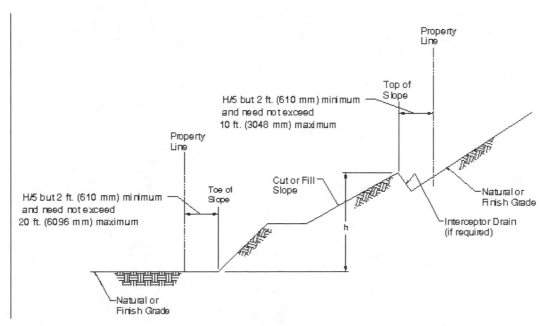

For SI: 1 foot = 304.8 mm

FIGURE J108.1
DRAINAGE DIMENSIONS

A single run of swale or ditch shall not collect runoff from a tributary area exceeding 13,500 square feet (1256 m²) (projected) without discharging into a down drain.

J109.3 Interceptor drains. Interceptor drains shall be installed along the top of cut slopes receiving drainage from a tributary width greater than 40 feet, measured horizontally. They shall have a minimum depth of 1 foot (305 mm) and a minimum width of 3 feet (915 mm). The slope shall be approved by the building official, but shall not be less than 50 horizontal to 1 vertical (2 percent). The drain shall be paved with concrete not less than 3 inches (76 mm) in thickness, or by other materials suitable to the application. Discharge from the drain shall be accomplished in a manner to prevent erosion and shall be approved by the building official.

J109.4 Drainage across property lines. Drainage across property lines shall not exceed that which existed prior to grading. Excess or concentrated drainage shall be contained on site or directed to an approved drainage facility. Erosion of the ground in the area of discharge shall be prevented by installation of nonerosive down drains or other devices.

SECTION J110
EROSION CONTROL

J110.1 General. The faces of cut and fill slopes shall be prepared and maintained to control erosion. This control shall be permitted to consist of effective planting.

> **Exception:** Erosion control measures need not be provided on cut slopes not subject to erosion due to the erosion-resistant character of the materials.

Erosion control for the slopes shall be installed as soon as practicable and prior to calling for final inspection.

J110.2 Other devices. Where necessary, check dams, cribbing, riprap or other devices or methods shall be employed to control erosion and provide safety.

SECTION J111
REFERENCED STANDARDS

ASTM D 1557-00	Test Method for Laboratory Compaction Characteristics of Soil Using Modified Effort [56,000 ft-lb/ft³ (2,700kN-m/m³)].	J107.6

INDEX

O

P

T

EDITORIAL CHANGES – SECOND PRINTING

Page 12, CERAMIC FIBER BLANKET: Section reference now reads . . . 721.1.1

Page 13, CONCRETE CARBONATE AGGREGATE: now reads . . . See Section 721.1.1.

Page 13, CONCRETE , CELLULAR: now reads . . . See Section 721.1.1.

Page 13, CONCRETE, LIGHTWEIGHT AGGREGATE: now reads . . . See Section 721.1.1.

Page 13, CONCRETE, PERLITE: now reads . . . See Section 721.1.1.

Page 13, CONCRETE, SAND-LIGHTWEIGHT: now reads . . . See Section 721.1.1.

Page 13, CONCRETE, SILICEOUS AGGREGATE: now reads . . . See Section 721.1.1.

Page 13, CONCRETE, VERMICULITE: now reads . . . See Section 721.1.1.

Page 15, GLASS FIBERBOARD: now reads . . . See Section 721.1.1.

Page 17, MINERAL BOARD: now reads . . . See Section 721.1.1.

Page 24, 302.2: Section reference in line 3 now reads . . . Section 302.3.2

Page 27, 307.2: Last line of COMBUSTIBLE FIBERS now reads . . . wastepaper, certain synthetic fibers or other like materials.

Page 28, Table 307.7(1): Column 2, row 2, line 2 now reads . . . IIIA

Page 28, Table 307.7(1): Column 4, row 12, line 1 now reads . . . $1^{e,g}$

Page 39, 402.4.1.4: Section reference in 1st sentence now reads . . . Section 1004.

Page 40, 402.7.3: Section reference in last sentence now reads . . . Section 705.

Page 45, 406.1.2: Section reference in last sentence now reads . . . Section 705.

Page 45, 406.2.7: Section reference in last sentence now reads . . . Section 302.3.2.

Page 46, 406.3.4: Last line now reads . . . of Sections 302.3, 402.7.1, 406.3.13, 508.3, 508.4 and 508.7.

Page 49, 408.3.3: Section reference now reads . . . Section 1009.9

Page 49, 408.3.6: Item 2, section reference now reads . . . Section 715.3.

Page 54, 412.1.3: Exception, section reference now reads . . . Section 1019.1.8

Page 64, Table 415.3.2: Column 2, row 3, last line now reads . . . Division 1.6

Page 65, 415.7.3.4.1: Line 8 now reads . . . Section 715.

Page 66, 415.7.3.5.2: Section reference now reads . . . Section 715

Page 66, 415.9.2.2: Exception 2, last line now reads . . . Section 715

Page 75, 505.3: Line 3 now reads . . . Section 1013.3.

Page 75, 505.3: Exceptions 1 and 2 now read . . . Section 1014.1 and 1007, respectively.

Page 77, 507.6: Last line now reads . . . with Table 302.3.2

Page 78, 508.4: Last line now reads . . . Section 302.3.2.

Page 78, 508.7.1: Line 4 now reads . . . in Table 302.3.2 ...

Page 82, 603.1: Item 15 now reads . . . Nailing or furring strips as permitted by Section 803.4.

Page 91, 705.6: Exception 2, line 1 now reads . . . Two-hour fire-resistance-rated walls shall be ...

Page 91, 705.6: Exception 4.3. line 7 now reads . . . by a minimum of 2-inch (51 mm) nominal ledgers ...

Page 92, 706.7: Section references now read . . . Section 715 and 1019.1.1, respectively.

Page 93, 706.8.1: Section reference now reads . . . Section 1019.1.2

Page 93, 707.2: Exception 2.2, line 9 now reads . . . Section 907.10

Page 94, 707.8.1: Exception now reads . . . Section 1019.1.2

Page 95, 708.1: New item added now reads . . . 5. Elevator lobby separation as required by Section 707.14.1.

Page 95, 708.3: Last line now reads . . . wall shall be at least 1 hour.

Page 96, 710.5.2: Last line now reads . . . the ambient temperture test and the elevated temperature exposure test.

Page 102, 715.4: Last line now reads . . . Section 715.4.8.

Page 105, 716.5.3: Last line now reads . . . as permitted by Section 1019.1.2.

Page 106, 716.6.2: Last line now reads . . . with Section 712.4.2, where exhaust ducts are located with the cavity of a wall, and where exhaust ducts do not pass through another dwelling unit or tenant space.

Page 121, Table 720.1(2): Column 3, row 2, line 4 now reads . . .with $2\frac{1}{4}''$ Type S drywall screws, spaced 12" on center, wallboard joints covered with paper tape and joint compound, fastener heads covered with joint compound, ...

Page 160, 803.7: Section reference now reads . . . Section 803.6

Page 160, 803.7: Exception, last line now reads . . . with Sections 803.1 or 803.6.

Page 161, 805.1.2: Exception, last line now reads . . . with Section 803.4.

Page 173, 905.10: Line 2 now reads . . . during construction and demolition operations shall ...

Page 182, 909.5.2: Last 2 lines now read . . . Door openings shall be protected by fire door assemblies complying with Section 715.3.3.

Page 182, 909.5.2: Exception 1, last line now reads . . . in accordance with Section 907.10.

Page 188, 909.20.2.1: Line 5 now reads . . . in accordance with Section 715.3.7.

Page 188, 909.20.3.1: Last line now reads . . . in accordance with Section 715.3.

Page 188, 909.20.3.2: Section references now read . . . Section 715.3.

Page 188, 909.20.4.1: Section references now read . . . Section 715.3.

Page 195, Table 1004.1.2: Column 2, row 6 now reads . . . See Section 1004.7

Page 217, 1021.3: Last line now reads . . . with Section 907.10.

Page 239, 1203.1: Section reference now reads . . . Section 1203.4

Page 240, 1203.4.3: Section reference now reads . . . Section 1206

Page 256, Table 1507.2: Column 2, row 10, line 2 now reads . . . (0.105 inch)

Page 271, 1604.6: Section reference now reads . . . Section 1713

Page 323, 1615.1.4: Equation 16-43, section reference in notation T now reads . . . (see Section 9.5.5.3 of ASCE 7).

Page 331, 1617.2.2.2: Reference to r in notation r_{max} now reads . . . ρ

Page 332, 1617.6.1.1: Item 3, last line now reads . . . factor of 4.

Page 333, 1617.6.1: New subsection 1617.6.1.3 added.

Page 342, 1622.1.3: ASCE 7, Section changed to 9.14.7.9 and section reference in 5th line now reads 9.1.3

Page 350, 1704.5: Last line of Section and Exception 1 now read . . . Table 1604.5 and Section 1617.2).

Page 350, 1704.5.2: Last line now reads . . . Table 1604.5 and Section 1617.2).

Page 352, Table 1704.5.3: Column 5, row 12 now reads . . . Sec. 1.2.2(e), 2.1.4, 3.1.6

Page 371, 1805.9: paragraph 2, line 2 and last line line of exception 2 now reads . . . provisions of ACI 318, Sections 21.10.1 to 21.10.3

Page 377, 1808.2.23.2: Line 6 now reads . . . Provisions of ACI 318, Section 21.10.4; Exception 2, last line now reads . . . Section 21.10.4; Exception 3 now reads . . . Section 21.10.4.4(a) of ACI 318...

Page 399, 1910.4.1: Last line now reads . . . as modified by Section 1908.1.7.

Page 400, 1910.5.2: Last line now reads . . . Section 1908.1.6.

Page 418, 2106.5.1: Section now reads . . . When calculating in-plane shear or diagonal tension stresses by the working stress design method, shear walls that resist seismic forces shall be designed to resist 1.5 times the seismic forces required by Chapter 16. The 1.5 multiplier need not be applied to the overturning moment.

Page 419, 2107.2.6: 2nd paragraph deleted

Page 432, 2113.3: line 5 now reads . . . in Sections 2113.3.1, 2113.3.2 and 2113.4.

Page 459, 2305.2.4.1: line 3 now reads . . . requirements in Section 1620.5 or Section 9.5.2.6.5 of ASCE 7

Page 460, Table 2305.3.3: Note a now reads . . . For design to resist seismic forces, shear wall height-width ratios greater than 2:1, but not exceeding $3\frac{1}{2}$:1, are permitted provided the allowable shear resistance values in Table 2306.4.1 are multiplied by $2w/h$.

Page 510, Table 2308.12.4: Column 5, row 1 now reads . . . $1.00 < S_{DS}$

Page 524, 2406.2: Section reference in line 2 now reads . . . Section 2406.2.1

2003 INTERNATIONAL BUILDING CODE®

EDITORIAL CHANGES – THIRD PRINTING

Page 24, 302.3.2: Exception, line 4 now reads . . . the fire-resistance ratings in Table 302.3.2...

Page 24, Table 302.3.2: Note d now reads . . . See Section 406.1.4.

Page 28, Table 307.7(1): Note n added in title.

Page 29, Table 307.7(1): Note n added in title.

Page 29, Table 307.7(1): Note e now reads . . . e. Maximum allowable quantities shall be increased 100 percent when stored in approved storage cabinets, gas cabinets, exhausted enclosures or safety cans as specified in the *International Fire Code*. Where Note d also applies, the increase for both notes shall be applied accumulatively.

Page 29, Table 307.7(1): Note m now reads . . . m. For gallons of liquids, divide in pounds by 10 in accordance with Section 2703.1.2 of the *International Fire Code*.

Page 29, Table 307.7(1): Note n now reads . . . n. For storage and display quantities in Group M and storage quantities in Group S occupancies complying with Section 414.2.4, see Table 414.2.4.

Page 40, 402.7.3: Exception, last line now reads . . . complying with Section 705.

Page 51, 410.3.1: Exception 1, last line now reads . . . with Section 410.3.4.

Page 56, Section 414.1.2.2 now reads **414.1.2.1 Aerosols** and 414.1.2.2 deleted.

Page 57, Table 414.2.4: Note i now reads . . . i. The permitted quantities shall not be limited in a building equipped throughout with an automatic sprinkler system in accordance with Section 903.3.1.1.

Page 60, LIQUID STORAGE ROOM, last line now reads . . . liquids in a closed condition.

Page 74, Table 503: Group I-3, Type IIB55 now reads . . . 10,000

Page 134, Table 721.2.3(2): Title now reads . . . COVER THICKNESS FOR PRESTRESSED CONCRETE FLOOR OR ROOF SLABS (inches)

Page 178, 907.2.13: Line 4 now reads . . . be activated in accordance with Section 907.6.

Page 197, 1007.1: Exception 3, last line now reads . . . in Section 1024.8.

Page 199, 1008.1.2: Exception 6 now reads . . . Power-operated doors in accordance with Section 1008.1.3.2.

Page 213, 1016.4: now reads . . . **Air movement in corridors.** Exit access corridors shall not serve as supply, return, exhaust, relief or ventilation air ducts.

Page 217, 1021.3: Line 3 now reads . . . smoke detector installed in accordance with Section 907.10.

Page 248, 1405.9.1 has been deleted.

Page 248, 1405.9.1.1: now reads . . . **1405.9.1 Interior adhered masonry veneers.** Interior adhered masonry veneers shall have a maximum weight of 20 psf (0.958 kg/m^2) and shall be installed in accordance with Section 1405.9. Where the interior adhered masonry veneer is supported by wood construction, the supporting members shall be designed to limit deflection to 1/600 of the span of the supporting members.

Page 425, 2110.1.1: Exception 1, line 4 now reads . . . 715 in fire barriers and fire partitions that have a...

Page 428, Table 2111.1: Row 2, column 3 now reads . . . 4-inch minimum thickness for hearth, 2-inch minimum thickness for hearth extension.

Page 441, 2211.2.2: Item 8, last line now reads . . . by Table 2211.2(1).

Page 441, Table 2211.2(2): Title now reads . . . NOMINAL SHEAR VALUES FOR WIND AND SEISMIC FORCES IN POUNDS PER FOOT FOR SHEAR WALLS FRAMED WITH COLD-FORMED STEEL STUDS AND FACED WITH GYPSUM BOARD[a,b]

Page 443, Table 2211.3: Table notes now read . . . a. See Section 2211.3.2, item 2. b. See Section 2211.3.2, item 1.

Page 443, 2211.3.2: Line 3 now reads . . . multiplied by the sum of the widths (ΣL_i) of the Type II shear...

Page 443, 2211.3.3: Line 1 now reads . . . **Anchorage and load path.** Design of Type II shear wall...

Page 444, 2211.3.3.3: Line 3 now reads . . . Section 2211.3.3.1, Type II shear wall bottom plates...

Page 444, 2211.3.3.4: Line 2 now reads . . . each end of each Type II shear wall segment shall be...

Page 468, Table 2306.3.2: Note a, line 2 now reads . . . above for nail size of actual grade, and (3) Multiply value by the following adjustment factor = [1 - (0.5 - SG)], where SG = Specific gravity of the framing lumber.

Page 547, Table 2902.1: Title now reads . . . MINIMUM NUMBER OF REQUIRED PLUMBING FIXTURES[a]